风雨六十年

巢纪平论文选集

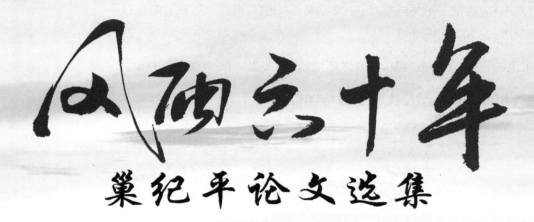

冯立成 李耀锟 陈幸荣 等 编

海洋出版社

2017年·北京

图书在版编目(CIP)数据

风雨六十年:巢纪平论文选集/冯立成等编 .—北京:海洋出版社,2017. 12
ISBN 978-7-5027-9955-7

Ⅰ.①风… Ⅱ.①冯… Ⅲ.①海洋环境预报-文集 Ⅳ.①X321-53

中国版本图书馆 CIP 数据核字(2018)第 043230 号

责任编辑:苏 勤
责任印制:赵麟苏

海洋出版社 **出版发行**

http://www. oceanpress. com. cn
北京市海淀区大慧寺路 8 号 邮编:100081
北京朝阳印刷厂有限责任公司印刷 新华书店北京发行所经销
2017 年 12 月第 1 版 2017 年 12 月第 1 次印刷
开本:889mm×1194mm 1/16 印张:52
字数:1536 千字 定价:298. 00 元
发行部:62132549 邮购部:68038093 总编室:62114335

世纪的气候
证书

宁静致远

在美好的感觉中，幸福是瞬间的，愉悦是片刻的，欢乐也是短暂的，唯有能持久的是平静。一个人没有永远的欢乐，也没有永远的愉悦，更没有永远的幸福，但却可以拥有永远的平静。

静如水的心静是指灵魂深处的安静。高尚的安静是宠辱不惊的欣然，是淡泊名利的超然，是曾经沧海的井然，是狂风暴雨的坦然。人世间奥妙，唯静者能看得透。于是能达到“这个世界很丰富，每当寂静时就会产生灵感，棘手问题往往迎刃而解” 的境界。所谓屈指一算，便知未来，实质源于在静中的分析思考。用于实际便能洞察先机，驰骋闹市而不乱。

内心平静的人深知用潇洒来养心，用谦退来补身，用平和来处世，用涵容来待人。

平静的大敌是躁。唯无求的平柔能克躁。深懂无争者其心静也达上乘，有上乘功力者，能抗七情六欲于体外，更能养颜而驻青春。实在平静是幸福的源泉。

宁静而致远。

编者序

我们作为巢纪平院士的学生，非常高兴能在他 85 岁高寿并从事科学研究 60 年之际为他编辑出版这本论文选集。

巢纪平院士 1932 年出生于江苏无锡，1954 年毕业于南京大学气象系气象专修科，1984 年任博士生导师，1995 年当选中国科学院院士。

巢纪平院士 1954—1984 年在中国科学院地球物理研究所、大气物理研究所和地理研究所工作。历任技术员（1954—1956）、研究实习员（1956—1962）、助理研究员（1962—1964）和副研究员（1964—1978）。于 1978 年提升为研究员。在这期间曾任地理研究所气候研究室主任，大气物理研究所第七研究室（动力气候研究室）主任。1980—1982 年应邀到美国普林斯顿大学地球物理流体力学实验室（GFDL）任高级科学家从事科研工作。

1984 年巢纪平院士转到国家海洋局工作。任研究员、国家海洋环境预报中心主任（1984—1989），国家海洋局科学技术委员会副主任。1989 年转任国家海洋环境预报中心名誉主任。1992—1993 年任夏威夷大学访问教授。

巢纪平院士曾任中国气象学会理事，中国海洋学会常务理事。在国际学术组织中任国际气候委员会（ICCL/IAMAP IUGG）委员（1981—1987），国际海洋和气候变化委员会（CCCL/IOC WMO）委员（1986—1991），热带海洋和全球大气科学指导委员会（SSG/TOGA）成员（1986—1991）。巢纪平院士曾担任过国家科技七五攻关第 76 项——海洋环境数值预报首席科学家，1985—1991 年任中、美热带西太平洋海气相互作用研究计划中方首席科学家。他曾担任过《Advances in Atmospheric Sciences》副主编和《海洋学报》主编。

巢纪平院士发表的著作主要有：《积云动力学》（合作者周晓平，科学出版社出版，1964），此书被美国空军翻译为英文出版《Cumulus dynamics》（United State Airforce，1969，AFCRL-69-0027），《厄尔尼诺和南方涛动动力学》（气象出版社，1993），《热带大气和海洋动力学》（气象出版社，2009），《中国大百科全书 · 大气科学 · 海洋科学 · 水文科学卷》（动力气象部分主编，中国大百科全书出版社，1987），《The Climate of China and Global Climate》（副主编，China Ocean Press，1987），《Storm Surges, Observation & Modelling》（主编，China Ocean Press，1990），《Air-Sea Interaction in Tropical Western Pacific》（主编，China Ocean Press，1993）等。

由于巢纪平院士在科学上的成就，被选入《中国大百科全书 · 大气科学 · 海洋科学 · 水文科学》人物条目（中国大百科全书出版社，1987）和《中国现代科学家传记》（科学出版社，1992)。

由这个论文选集可以看到巢纪平院士在这60年的科学研究生涯中涉及了多个方面，包括大气动力学，海洋动力学，海气相互作用动力学，长期数值天气预报以及近年来开展的生态动力学，在运动的时间尺度上从十几分钟的积云到气候的平衡态。

巢纪平院士在科学上有不少创新性工作。例如中小尺度运动方程的建立，在国际上是同一时期独立提出的，长期数值天气预报距平模式的提出也是首创的；他也是最早研究海气相互作用理论的科学家之一，大气运动多平衡态理论的提出也是相当早的，热带运动半地转适应过程和发展过程的提出具有新意，等。

在这本论文选集出版之际，我们代表他的学生们祝老师身体健康，并在科学研究上继续做出有创见的工作。

该书编辑过程中有大量的文献搜集、核对工作。具体分工如下：史珍博士负责收集核对第一、二部分，亢妍妍博士负责收集核对第三部分，汪雷博士负责收集核对第四部分，李耀锟博士负责收集核对第五部分，冯立成博士负责收集核对第六部分及附录。在此一并致谢！

是为序。

编者

2017年11月15日

自　序

1954年我从南京大学气象系专修科毕业后分配到中国科学院(前)地球物理研究所任技术员，开始科技工作，至今已一个甲子过去了。这60年祖国的科技事业，从百废俱兴到正在营造科技强国梦，是值得大书特书的。但这是那些科技管理学家、科技历史学家的大手笔，他们的职责所在，我，科技界的一介小人物，既无宏伟的视野，更无掌控全局的魄力，是不敢涉足的，只能谈点亲身的感受。

这60年，前30年即从1954年到1984年是在中国科学院度过的，包括1980年到1982年有近两年应美国普林斯顿大学之聘，在NOAA的GFDL任高级科学家(正教授级)工作。从1984年末调国家海洋局国家海洋环境预报中心任职至今。

在阳光普照科技事业的这60年中，有过1956年激动人心的科学大进军，青年人的心沸腾了，正如当时北京晚报报道的"中关村不夜城，一片灯火彻夜达旦"，都在为向科学进军战斗！也有过1979年的科学春天，虽然那个春天太短了，但春天毕竟是美好的！

像天气变化一样，在阳光普照的日子里，有点雾，阴雨，甚至大暴雨也很自然，但小人物如何度过这些岁月，可能会有一些可读的小故事。首先熟悉的朋友会好奇，怎么一改常态出什么选集，这是肺癌切除手术闹的，生活的常态变了，在家休养，真感到无奈和无聊，想想这60年那些事、那些人，历历在目。又看看现在为迎接科技强国的大动作，有的不甚懂，心中有话无人说，也不好意思说。就像小学生看了大教授画的宏伟蓝图，不好意思去问的那种心态。于是想出一招出个论文选集，从一个一般科技人员论文变化的视角，去看这大时代的变化，同时为之写个自序，即使有不懂之处，朋友间也没有什么不好意思交谈的，或用了不敬之词(估计不会多)，敬请谅解。

这种事自然不能靠媒体，就本人性格来讲，一般除介绍一些大科学家的事迹会接见媒体外，这是不得不做的社会责任，除一次外不记得和记者谈过我自己。那次是两位年轻记者，完不成上面交办的采访任务会影响前途，另他们非常诚恳，所以见了。他们的采访发表后，听说还得了奖，我也很高兴。把这篇忠实原意的采访放在此书的附录了。

人在社会上的经历、遭遇，不单主要靠本人的智商和情商，在很大程度上是机遇，通俗地讲就是运气。如果说在气象和海洋社会中，还知道有我这么一个人，那是我比同时代有的人有较好的机遇。若要谈到我个人，或许最感兴趣的是我这个两年制的专修科生怎么会进入最高殿堂中国科学院的，这就是机遇。

我最早的机遇是当时的国家形势，当时抗美援朝急需一批天气预报员，在南京大学、北京大学办了二年制的专修科。我有幸进入这个班，提前毕业了，中国科学院也需要少量科技辅助人员。但能在全班28个学生中进入中国科学院,这主要是系主任朱炳海教授的努力,那时刚院系调整,教授说话还起一定作用,我知道朱炳海教授为我的分配出了很大的力,按现在的用词是出了洪荒之力!

朱炳海教授对我这一生是有知遇之恩的老师,他待我之好无以言表,这不单因为他是江阴人,我祖母也是江阴人,主要还在他重才、护才。我选读气象专修科,自然不是有想去抗美援朝的政治觉悟,实在不爱读气象,我又进过两个大学,如再不读,面临是无书可读,实出无奈,赶快读完走人,所以上课很不用心,在混。有一天上普通气象课,我在看小说,被老师发现了,这在任何时期都是大逆不道的事,被叫站起来训斥一顿后,问:"你知道我在讲什么吗?"我回答:"我知道您在讲气压垂直分布,不过老师您写错了。"

这事闹大了,告到系主任那里,朱先生让我第二天到他办公室。我忐忑不安走进东北大楼他的办公室时,我准备好了,等待我的是处分。朱先生的眼镜估计已有二千度,严肃得令人不寒而栗。朱先生说:"巢纪平,你昨天不好好上课,还说某先生的公式写错了,确有其事?"边说边推过一张纸,"你算一遍。"我推了一遍,朱先生看了就不再提这件事了。化险为夷!这是一个尘封的故事,知道的人不会很多,因留校的都是本科的高材生,我们班只留了陈良栋,不久也离开南大,何况他也不会知道后面发生的事。我从不说朱先生是伯乐,因我只是一匹肯跑的马。其实,伯乐常有,但像朱先生那样正直、大度的人确实不多。可以那么说,当时若不是朱先生是系主任,在气象界里不会有我这个姓巢的。当然,我能进中国科学院,也不单是这一事件。教我们动力气象的是徐尔灏教授,英国绅士派的文质彬彬,他的课讲得很认真,特别讲到尺度分析时,讲得入微的细,而讲到他自己的研究工作散度方程分析时,更是清晰。那时经常有个讨论课,由徐先生亲自主持,讨论到柯利奥来力的折向作用时,徐先生说在北半球从赤道北上的空气是到不了极地的,接着又问,柯利奥来力还有一个向上的分量,为什么空气不向高空走?这对刚学动力气象的学生来讲,的确有点难。好在我学过尺度分析,前一天晚上也注意到这个问题,回答了,因为它太小,小到可以不计。徐先生看了我一眼,说了一句:"你是对的。"幺枕生教授来自浙江大学,带来了浙大严格的学风。他教的气候学不是地理学的描述性气候,实际上是物理气候。南京冬天很冷,寒假前气候学终考,他来了个可带参考书,不受时间限制的开卷考,我考了两个小时不到,听说一般考了五个小时。分数出来我得了80分,是唯一达80分的人。所以说,我进中国科学院,不单是朱先生考虑"同乡"私谊,是有其他理由的。

我要讲的第二个故事,是1962年在中关村原地球物理所大礼堂的那场考试,发起这场考试的是赵九章先生,主持考试的是顾震潮先生,主考官是华罗庚先生,考生是我。考了近三个小时,什么问题都考到了,最后还是被北大周培源先生的代表黄敦先生给考住了,他指出我用的方程无量纲化不规范,不符合兀定律。的确,力学和气象学在对方程无量纲化时,它们的方法论是不一样的。故事未完,考后有一天华先生把我叫到他办公室,态度很亲切,随便聊几句后话入正题,问我愿不愿意跟他学数学,我真受宠若惊。回去我就找了管我的顾震潮先生,原以为他会告诫我天资不够,或者会说这边工作需要人。没想到他来了一句,你要做无产阶级科学家,还是做资产阶级科学家?无言以答,从此华先生那里也不敢去了。近20年后,1980年一个晚上在美国普林斯顿又见到华先生,他在高等研究所访问,是叶铭汉先生带我去的。见到华先生还那么精神很高兴,我问,华先生还认识我吗,他看了我一眼:“巢纪平,我是你主考官怎么能不认识?”异地相逢,谈笑甚欢,后来他告诉我,他母亲也姓巢。再一次见到华先生是1985年胡耀邦同志召开的科技政策小型座谈会上。华先生见了我说:“知道你已离开科学院去了海洋局了,做点实际工作也好。”没想到这是我最后一次见到华先生。很快噩讯传来,华先生在日本讲学,心脏病突发仙逝。我沉默了一天,写了一篇纪念文章在科学报发表了,可参见附录。在这一篇小文中写法大致是这么安排的,第一部分讲了一些不为人知或鲜为人知的人和事。小人物没有什么“军情解码”,写出来给别人饭后消遣看看,但凡这些人有恩于我,或这些事器重或尊重我,会终生难忘的,不说出来心里憋得慌。

从1966年6月“文化大革命”开始,我被关进专政队,专政队也即季羡林先生《牛棚杂忆》书中的“牛棚”。“牛棚”中实际关的不是牛,是一些有这种问题或那种问题的“阶级敌人”。在牛棚中的人除了不时挨批斗或陪批斗外,就是劳动,再有时间写交代、外调,空下来读“红宝书”,那时一本毛主席语录可背出来。但也确有本来很受我敬重的领导干部在“牛棚”中读资本论或马克思主义全集的。关了近两年放出来后,在监督或半监督下工作,去过湖南岳阳的山沟,协助“革命派”观测污染,严冬去过吉林,也是协助“革命派”做松花江雾的土炮消雾试验。虽然这些人过去都是我部下,但是这叫此一时彼一时。近10年的时间没有干过什么正经的事。若有,则是1973年春季南方大涝,国务院上层担心秋季北方会不会大旱?把预测任务下到中国科学院大气物理所。革命派的领导们把这个研究交给我这个半监督对象来做。这种季节预报的信号不能在大气中找,正好我读过1951年左右吕炯先生一篇关于黑潮和亲潮的文章,于是我用前冬黑潮的冷暖来做指标预测来年中国东部的降水分布。预测结果是1973年南方会涝但北方不会旱,预报基本准确。预报做了,文章发表在专集里了,但不能署上我的名字。署不署名字这并不重要,重要的是从此我又多了一个可思考问题的科研领域——海气相互作用。1974年,为了躲避又一次批林批孔运动,无奈用针刺麻醉做了胃次全切手术,在家休养。北大办了一个数值预报培训班,负责该班的陈受钧来找我,要我为该班讲长期数值预报课,此课怎么讲,一共可参考的文章只有两篇。但陈受钧是我一个很好的朋友,他交办的事我会尽量去做的。不久提出“海气相互作用下的距平长期数值预报方法”。提出个理论并不困难,但要在100万次每秒的计算机条件下算出一个例子,这并不容易,感谢几位同事帮我自带咸菜从738厂、天津情报所、燕山石油化工总厂,一路辗转,终于计算出了一个实况例子。

我说这一段历史，是为了讲另一个感恩的故事当前奏，为什么我去了地理研究所。我不会对“文化大革命”做任何评价，但有一点感觉到了，人的灵魂，当然也包括我自己的灵魂，在那种阶级斗争下被扭曲了。往往最不实事求是，整得最狠的人，是过去最接近的人，不这样不足以表明已划清界限！大气物理所当时已无我容身之地。我向院里打报告调出中国科学院，因那时我是副研究员，按当时规定，高级研究员调出必须经院领导批。报告到了主持工作的郁文同志手里，他让秘书通知我，院内调哪个所任我选，但不能出院。于是，我找了地理研究所的气候室副主任丘宝剑、副所长左大康先生、所长黄秉维先生，他们都表示欢迎我去。在此我说句心里话，地理所这些领导同志，比大气所的某些人善良、正直得多。在地理所那段时期，他们怎么善待我的，可参见附录。

知道下面这个故事的主要人物都已仙去，但这确实是一个鲜为人知的有趣故事。“文化大革命”结束后邹竞蒙同志出任气象局局长，章基嘉同志从南京气象学院院长调任副局长，一天章来找我，告，经研究气象局领导建议我去当中国气象科学研究院院长。我说待我考虑，立即去大楼306办公室见叶笃正、陶诗言两位先生，他们出奇的平静，说了一句话：“巢纪平，你还是走了好。”我听了也出奇的平静，但心情有点黯淡，我毕竟在这个所工作20多年了。我立即跟章基嘉打了电话，表示去。但随即给邹、章两位写了封信，鉴于某人年纪比我大，建议由他当院长，我当副院长协助他工作。很快气象局领导让我写一个气科院如何调整学科向现代化发展的规划，我写了也送去了。但调动自此无消息了，后来得知这份规划外泄了，遭到气科院诸多研究员的一致反对。后在邯郸气象学会，邹竞蒙从日内瓦回来，问我怎么还不来报到，我回答，遭到气科院多人反对，邹竞蒙说：“会有此事？”我说：“确有此事，您问章基嘉。”章来了说确有此事。回京后，陶诗言先生把我叫了去，说：“调你是他们局领导定的，现在阻力很大，他们很为难，不如你给气象局写封信，说你不愿意去的。”信写了，调动事结束了，但对他们两位知遇之恩是难忘的，这也许是凡气象局要我帮办的事，我一直尽力去办的一份潜在的情结吧！

这些故事看上去与选集无关，但却是我想讲的，一开始就交代了，出选集，一方面从论文发表的内容和年代，看大时代的变迁，另一方面也是为自己说心里话营造一个平台。这些故事也有一个中心思想，别以为自己多能干，只是机遇好，遇到了肯帮助你的善良的人！多少比我能干的人，或者在人群中消失了，或者眼看倒下来了，他们运气不如我，能悟到这一点，心态就会平和得多，人各有志，不必强求。

当然，科学院的30年，不会就这几个故事。那些与人际关系有关的故事，就让它们永远尘封吧！人活得已经很累，何必在心里再压些令人郁闷的事！忘了吧！

谈后一个30年，我为什么会来到国家海洋局工作的？从根上讲，与我在美国工作有关，倒不是因为在大气所又无容身之地了。我已经建立了一个动力气候研究室，有几位同仁一起工作。事出在1982—1983年那次厄尔尼诺事件的暖水首先出现在赤道太平洋中、西部，美国海洋界的科学家和管理层已经意识到需加强对热带西太平洋的观测，他们鞭长莫及。那时有一个《中美政府间海洋和渔业协定》，于是1982年他们提出在热带西太平洋进行浮标观测的合作意向。紧锣密鼓，他们的科学家先来中国访问，1983年国家海洋局组团回访，组成以海洋局第一海洋研究所所长陈则实为团长的代表团。组团过程看上去是个小插曲，实际上变成主旋律，决定了我后半生的命运。组团自然以国家海洋局的人为主，但也给了中国科学院名额，其中一个点名给我，以代表团科学顾问名义出行。

回国后，中、美两国政府即国家海洋局和NOAA准备开始实质性合作谈判。国家海洋局局长罗钰如同志是位学术型干部，年轻时就读清华大学化学系，“一二·九”运动到延安参加革命，新中国成立后成了海军的一位高级干部。他把我找了去，谈了中、美两国政府间西太平洋海气相互作用研究的重要性，希望我以大局为重调到海洋局从科学层面参加这次谈判。西太平洋的观测对我国无论从国防、经济、科学诸方面的重要性我自然理解，但双边政府要进行西太平洋海气相互作用合作的财政贡献的底线差距不小，我担心谈不下来。所以我回答罗局长，可以借调到海洋局参加谈判，是否正式调入海洋局，看谈判结果。美国主管谈判的是NOAA一位分工外事和气候的司长，博士，是位正直很有事业心的管理者，我们从仪器折价、人员培训多个方面调整经费的差距，谈判成功。调动势在必行，但我以什么身份调进海洋局，罗局长还另有考虑。1982年海洋局把行政处级编制的国家海洋水文环境预报总台，提升到行政级别为正司级的国家海洋环境预报中心，请我出任第一任主任，并考虑建设现代化的海洋环境预报系统。当时除了在大慧寺8号已盖有两座宿舍楼以外，其他办公楼虽有1000万元预算，但尚未动工，不是国家重点工程，连钢筋、水泥指标都没有。技术手段只有一台100万次/秒的计算机。我查了一下海洋局外汇经费，全局只有150万美元的外汇，离购进一台其速度起码可做业务化预报的Cyber2000万次/秒的计算机(这已经是在不监管情况下可向中国进口的最快计算机)经费差额甚大。科技人员的缺少就更不用提了。对一位年近70岁局长的战略思想我很敬重，对我本人的信任和尊重，我也很感动。但巧妇难为无米之炊，我回答罗局长，我去努力下看能否从国家拿到一笔经费，拿到了我就调过来。经前辈科学家和高层管理干部的支持，国家计委同意在已制定的国家“七五”科技攻关项目原定的75项增加成76项，名为海洋环境数值预报，经费五千万元。借这个机会我要特别感谢当时国家计委主持制定“七五”科技攻关项目的秦声涛司长，谢谢他对这个项目重要性的理解和支持，以及对项目大手笔的拨款。就这样，我离开工作30年的中国科学院，跨入另一个科学领域，进入思维方式、工作方式很不相同的政府部门的事业单位。时年已52岁。

中、美西太平洋海气相互作用合作试验于1985年底在广州首航，海洋局严宏谟局长去了，出席的有国家科委主任宋健同志，美国驻华大使洛德先生等。用的是“向阳红14”号。船在夏威夷停靠，美方在那里举行了隆重的欢迎仪式，严局长和我也出席了。浮标标有PRC-US，施放在西太平洋赤道东经165度。这一合作历时5年多，成功完成后于

1991 年在海口结题。机会特别好的是在这一合作试验期间出现了 1986—1987 年的一次中等强度的厄尔尼诺事件。实测资料使我们对厄尔尼诺暴发前后西太平洋海洋物理状态的反应有新的认识。应该说,由于国家海洋局参与并对国际大型海洋试验所做出的贡献,它在国际海洋界受到重视,地位也大大提高。我本人也因此进入国际海洋和气候变化委员会(CCCL/IOC-WMO,1986—1991)及国际热带海洋和全球大气科学指导小组(SSG-TOGA,1986—1991)。遗憾的是,在合作试验完成后,国家海洋局把东经 165 度那个 PRC-US 浮标撤了回来。我本人可以理解,当时国家有关部门和国家海洋局领导层对海洋科学的重要性缺少应有的认识,在财力上也缺少一年两次去西太平洋的经费支持。浮标撤回来了,随之国家海洋局的科学地位在国际海洋界也下降了。不久美国在中-东热带太平洋建成 TAO 观测阵列,后来日本海洋研究开发机构(JAMSTEC)参加进来在热带西太平洋建立了 TRITON 观测阵列,这两个阵列的观测资料至今仍服务于国际厄尔尼诺预测。海洋资料和气象资料一样,除国家自身的国防、社会、经济需要外,它是一种公益事业,服务于全球社会的需要。这是要成为海洋大国应有的基本认识。

借这个机会,中、美西太平洋合作试验得以顺利、胜利完成,我要感谢国家海洋局第一海洋研究所的王宗山、蒲书箴两位航次首席科学家,要感谢时任国家海洋局南海分局的梁松局长,更要感谢严宏谟局长,在当时无专项经费保证的困难情况下,他设法筹集每次顺利出航的经费。

再转向另一个项目,海洋环境数值预报,从 1984 年底破土盖楼,争取到“七五”科技攻关第 76 项的五千万元经费,从科学院、北大、中国科技大学等单位引进各种人才,加上原单位的人才,特别是年轻的管理干部,夜以继日地奋战,在 1987 年中办公大楼落成,CDC 的 Cyber 每秒 2000 万次速度的计算机安装进大楼,遥感卫星天线安装到楼顶,不久,第一张海洋环境预报的实时预报图做出来了,预报产品每天在中央电视台发布。第 76 项海洋环境数值预报在 1991 年正式验收通过。在此我同样要感谢那些日夜奋斗的科研和管理人员。

至于我自己遇到的暴风雨要比预期的大。1989 年“政治风波”的确在劝阻不住的情况下,怕群众控制不住情绪做出越轨的行动,所以跟着去了,送了点饮料很平静地回来了。本来不想谈这类令人不高兴的事,不过“解密”一下让后来人看看,人的灵魂被扭曲后会不实事求是到什么程度。举报信直到国家高层领导人、中组部、中纪委、国务院机关事务管理局,八条罪状,只看第一条,煽动群众,亲自带队,上街游行,支持暴乱,以致院内出现打倒邓小平的口号。是要把我置于死地而后快了。但这一条的真相是什么,1989 年 4 月不知是谁用砖在卡车上写了邓小平三个字,我看到后让保卫干部拍了照,没有想到几个月后,把邓小平三个字用到那个地方,写出这么一条。无所不用其极,叹为观止!随即进驻工作组,一个月后局党组下文,免去国家海洋环境预报中心主任,任为名誉主任。

不当主任挺好，几年主任当下来实在太累。这样可静下来做点研究。但事情并没有完，1991年中国科学院恢复学部委员增选，于是类似的举报信不仅到了地学部，而且至少有50位学部委员收到同样的信。在当时的政治形势下落选是必然的了。我没有把落选看得那么重，但的确感到大慧寺8号的空气被那几位不小的干部弄得有点污浊，飘然而走，到夏威夷大学当访问教授去了。

我的确是要感谢严宏谟同志的，他是一位正派和正直的人。他和我的一席电话令人感动！他说："老巢，我局长已到任了，在我走前一定把你的问题弄清楚。"于是一份盖上国徽章的国家海洋局文件送到中国科学院地学部，主要写了两条，一是经多次审查，巢纪平未发现有经济问题，二是涉及的人事都是司局级干部，巢纪平无权干预，是正常调动。当然，也要感谢当年主持地学部工作的主任，把这封公函在全体院士中宣读了，也要感谢多年和我在科学院共事的院士的支持，1995年我当选院士。

在国家海洋局的故事就这些了！

我是抱着参与发展中国海洋科技事业来到国家海洋局的，是否来错了单位，世上没有后悔的药，不必提，但的确来错了时间，确实来早了。那时即使国家层面也未高度认识到中国同时也是一个海洋大国，重视海洋科技的发展有着不可估量的战略意义！体制的变化就是一个侧面，从国务院直属国家局，到科委代管局，到直属部管的国家局。至于国家海洋局领导层，从严局长后，国家海洋局的领导层更加重视"管理"，但如何做到真正的管理，恕我直言，从来没有说清楚。意见提得很尖锐，但不带任何个人情绪，事实上，我和各任局领导关系都很好，他们待我也很好，只是对事业发展上的不同意见，历来存在，这应是正常现象。

但"壮志"虽不酬，人却未先去，身在曹营心也在曹营。借这一小文，写点个人看法。西太平洋海洋科技的战略地位我们丢了，我是十分谅解地写那段历史的。现在应更战略地看到印度洋这一海域，我们有南极这一海路，在印度洋也做了一些工作。加强从南海到南极的观测布局，这样在南北一线上，就可加大对澳大利亚、新西兰的科技影响力，再向西扩张点，在东南亚就成了领军科技强国，影响可达非洲。如何布局，局内不失有识之士，可不带框框地多听意见。

下面言归正传，对选文作些注释。

在第一部分数值预报，我只放了两篇文章，事实上也是仅有的两篇文章，但却代表一个预报业务化时代的开始。顾震潮先生是位对新生事物十分敏锐和政治性强的科学家。当1952年第一张数值预报图在高等研究所做出来后，他在没有高速计算机的情况下，组织人用图解法做。我记得第一张24小时预报图是气象局的廖洞贤先生做的，我这里给出的是48小时预报图。短期天气预报，我一生就做了这张图。做完这张图，反右派的政治运动开始，我被下放到河北井陉县农村劳动，一年多回来，奉命去兰州搞人工降水和土炮消雹，改行了。第二篇文章是关于海洋环境数值预报业务化实时预报的。当时"一穷二白”的艰苦过程前面已经谈了。这篇文章是应政府间海委会(IOC)秘书长的邀请在1991年巴黎大会上做的Bruun纪念报告。那时能做海洋环境业务化实时预报的国家不多。很有意思,这也是我此生写的海洋环境预报的第一篇文章也是最后一篇文章。

选集的第二部分,安排了中、小尺度系统。从文章发表的时间,可以看到跨了两个不同的政治气候,一是大跃进时期,那是非常辛苦的,不时要进入云层人工降水,下来已经很累,还要念书做研究,后一个时期政治气氛比较宽松些。应该说,这组论文中不乏创新的工作。小尺度运动方程的建立和国外是同步的,甚至更完整些;非线性热对流的发展这是首例,虽然美国Malkus也发表一例,但她算的是对流泡,用的是数值解;双平衡态过去尚未有人提出,当然这是一个很简单的模式。

在这一部分中把冯立成三维f平面西边界流的强化放进去了,它和大尺度海洋环流西边界流的强化有相似之处,但物理过程显然不同,放在这里进一步思考。另外,在海洋方面也放了亢妍妍用有限元法算的沿岸上升流,和冯立成算的非线性地形下上升流的不稳定性。

第三部分是长期数值预报,做这项研究的背景前面已经介绍了,当1981年在美国一次学术会上报告我在GFDL算的一个月预报例子时,引起不少人的兴趣,也有不少质疑,怀疑我做了什么手脚。会上一位非常受人尊敬的老科学家,长期预报创始人之一,对我这个后辈非常厚爱,说:“纪平,没有就直说!”,我说,没有,这个例子是助手算的,我没有插过手。当然这只是一个结果很成功的例子,已经注意到,要加大分辨率,在垂直方向至少加到三层,因有热量垂直输送需要计算。这里也选了三层模式和季节预报结果。Miyakoda写了文章,提到距平模式的优缺点,也用了这个例子。

第四部分是大气和海洋动力学,能称道的工作不多。1957年日本气象学家Gambo来地球物理所访问,做一个系列学术报告,其中有他自己做的正压有限振幅大地形扰动,看了我的斜压有限振幅扰动的文章初稿后,他把要做的地形报告撤回了。我要说一句的是,那时做个工作多困难,把青藏高原和落基山用球函数展开,三个人用手摇计算器整整花了一年时间,在此要再次感谢叶笃正先生,没有他的大力支持并借我两位统计员,这个工作是做不成的,现在这样的球函数展开,用计算机机时不会花一分钟。由刘飞主算的跨赤道多平衡态是创新的,非静力平衡下的经圈环流有新意。陈英仪的那篇波的波作用密度及稳定性也和国际上的水平不相上下。至于运动的多时态特征,是中国气象学家的强项,在此我只是把它整理更清楚而已。李耀锟的海洋经圈环流的两平衡态解是有新意的。

第五部分是热带运动和海气相互作用,我要强调的是,季劲钧是位严谨、数学根基扎实的学者,他的那篇发表于1979年的海气相互作用和赤道辐合带的文章,实际上是最早的热带海气相互作用的文章,1980年我在GFDL报告时,引起广泛兴趣,他们建议我把边界条件改一下,修改后重新发表。我因忙于其他事情抽不出时间去改。β-平面上的热带海气相互作用那个时期做的人很多，季振刚、张人禾的文章也不比他们差。虽然在取海洋物理量时一般用海表温度距平，但考虑到，一是信号弱，也易受降水、蒸发等影响；二是从物理上看，上层海洋变化主要受风应力影响，由于使温跃层的深度发生变化，从而使上层混合层中海温距平发生变化，为此引进上层海洋最大海温距平概念。热带半地转适应和发展过程，相信是我最早提出来的，至于别人是否应用和如何评价，有学术自由！用最大截断模的海气相互作用，这里给出王彰贵、刘琳、高新全的文章，耦合波的不稳定性有新结果。

第六部分涉及用动力学和热力学方法来研究荒漠化、城市化，我相信这是新的思想，不去注释了，读者自己去评吧！

最后想借这个文集平台谈点感想。科技的发展人才是根本。在建国初期从欧美回来一批英才，他们基本上是在抗日苦难中读完大学，抗日战争后出国深造的。建国后他们回来建设新中国。打个不太恰当的比喻，姜太公钓鱼，愿者上钩，鱼饵是两个字——国家！就为这两个字义无反顾地回来了。给他们的待遇是副研究员，到 1956 年调整工资时变成三级研究员，工资 230 元。新中国气象科技、教育事业的发展，建国初期回来的叶笃正、谢义炳、顾震潮……做出了重要的贡献！

现在管理层为营造世界第一流科技强国，百人计划、千人计划、万人计划、长江学者……名目愈来愈多，引进工资漫天开价，把一个科技队伍变成科技市场。本是同门兄弟，转一圈聘回来，三个月走人，工资内外差近 10 倍，如何让留在国内的科研人员心平气和地工作。

引进优秀人才不是仅仅靠价格就能实现的事，转一圈聘回来的也不一定就是有用的人才，国内成长的科研人员也不一定比回来的水平低，还是应当重才、重德，崇尚老一代科学家为祖国无私奉献的精神。一介凡夫之拙见，希望政治强、学历高的制定政策和执行政策者斟酌。有些东西我看不甚懂，小人物不敢也不想多问，留下一个巢纪平之惑吧！

谨以此文缅怀敬爱的赵九章先生！

巢纪平

2017 年 10 月 19 日 于国家海洋环境预报中心

编写说明

巢纪平院士在60多年的科研工作中与同事和学生发表了160余篇科技论文。我们从中精心挑选了72篇编入选集，力求能够反映巢纪平院士不同时期的科研兴趣和成果。论文集按照研究方向分为六大类，每一类中又将同一科学问题不同年份的研究成果编排在一起，方便读者系统地了解巢纪平院士的研究思路及成果。

由于入选文章跨越60年的时间，并且来源于不同的刊物，格式各异。我们在编辑过程中，统一风格的同时又尊重历史，尽量保持原汁原味。例如文中同一人名，在不同英文文章中可能有不一样的拼写法等。对于中英文同时发表的文章，我们选录中文稿。特此说明。

编者

2017.11

目　录

第 1 部分　数值预报

第 2 部分　中、小尺度运动

第 3 部分　长期数值预报

第4部分 大尺度大气和海洋运动

第5部分 热带运动和海气相互作用

第 6 部分 荒漠化、城市化的理论

附 录

第 1 部分　数值预报

准地转两层模式天气数值预报方法的试验*

顾震潮　巢纪平　瞿章

（中国科学院地球物理研究所）

提要：作者用准地转两层模式以图解方法试做了 24 小时及 48 小时亚洲部分的 500 mbar 高度预报。从所做的几个例子来看，有的预报结果较好，但像阻塞高压的一个例子中高压位置南边的负变高没有预报出来，因此预报图上没有阻塞高压的形成。原因主要是在准地转两层模式没有确切地考虑到 500 mbar 的辐散场。作者指出 500 mbar 上天气系统有猛烈发展时 500 mbar 上的辐散场是比较大的，要考虑到后者的作用就必须应用三层模式或不用准地转假定。作者并指出了 Fjϕrtoft 的图解方法也有不小的误差。

1　引言

近年来数值预报有很大的发展[1]，数值预报研究及日常数值预报工作已经证明大范围天气形势的预报，至少已可以和预报员的半经验的预报作比较，有许多例子中还显然比半经验的预报来得好。例如东亚寒潮暴发的一个试报例子[2]，证明预报员所难于报出的高空大范围形势转变过程，却可用简单的两层模式相当满意地在 24 小时以前预报出来。

为了取得经验以发展数值预报方式的工作，我们在这方面也做了些试验研究。这首先是对两层模式来做的。因为在理论上两层模式是最简单的一种斜压模式，在实用上也是比较有价值且能够用图解法及时作出日常预报的。所以对这种两层模式预报能力的了解是很有用的。

过去几年中两层模式虽有不少试报的结构发现（见参考文献[1]及参考文献[3]），但预报的时限一般只有 24 小时，对结果的分析讨论也还不够。在我们这次的试报中，预报的时限是 48 小时（也附带做了一些 24 小时的预报），并对所选个案的预报结果做了较详细的讨论，并由此进一步讨论了两层模式的局限性。

2　48 小时及 24 小时两层模式的试报结果和讨论

方法　试报所根据的两层模式是取平均层 $\bar{p}$（500 mbar）和 p_0（1 000 mbar）。预报的直接对象是 $\bar{p}$ 等压面的绝对高度 $\bar{z}$ 和 $\bar{p}$ 等压面对 p_0 等压面的相对高度（厚度）z_T。在地转风假定下，预报 $\bar{z}$ 和 z_T 的涡度方程可以写成

$$\left.\begin{aligned} \nabla^2 \frac{\partial \bar{z}}{\partial t} &= -\bar{v} \cdot \nabla(\bar{\zeta}+f) - \bar{A}^2 v_T \cdot \nabla \zeta_T, (1.1) \\ \nabla^2 \frac{\partial z_T}{\partial t} &= -\bar{v} \cdot \nabla \zeta - v_T \cdot \nabla(\bar{\zeta}+f) + B\omega(\bar{p}) (1.2) \end{aligned}\right\} \quad (1)$$

* 气象学报，1957，28（1）：41－62。

此处假定 $v=\bar{v}+A(p)v_T$，而对气压平均时 $\bar{A}(p)=A(\bar{p})=0$，并且略去 $\omega\partial\zeta/\partial p, f\partial\omega/\partial p$ 等项。其中 ζ 是涡度，$\omega=\dfrac{dp}{dt}$ 是气压坐标中的垂直速度，B 是一个常数。要使这个预报方程组成为闭合方程组，还要从热力学第一定律引入一个 ω 的关系式，及假定 ω 对高度的分布。

现在我们的目的在于用 Fjφrtoft 图解法[4]来作 48 小时或 24 小时的 $\bar{z}$ 预报，在简单的情况下，我们可根据 Fjφrtoft 取 $H=v_T\cdot\nabla\zeta_T$，而

$$\frac{d_g H}{dt}=0 \tag{2}$$

由此式(1.1)写成 $\dfrac{d_g\bar{\zeta}}{dt}=-H$，而对各个质点

$$\bar{\zeta}=\bar{\zeta}_a+Ht \tag{3}$$

这样我们就可以用 Fjφrtoft 推移 $\bar{\zeta}$ 的方法，由原始的 $\bar{\zeta}_a(x_0,y_0,t_0)$ 求出新分布 $\bar{\zeta}_a(x,y,t_0+\Delta t)$，再加上 Ht 即得未来的 ζ 分布，由此再求 $\bar{\zeta}$ 的局地变化就可以解 Poisson 方程而求出 $\partial\bar{z}/\partial t$。

在实际计算中，H 取的是起始 ζ_T 在 $v_{T,m}=-\dfrac{g}{f}\nabla z_{T,m}xk$ [$z_{T,m}(x,y)$ 是 $(x+d,y)$，$(x-d,y)$，$(x,y-d)$，$(x,y+d)$ 四点上 z_T 的平均值] 场中前后各推移 12 小时所得的 ζ_T 差值（即以 $t=1$ 天为单位）。$H(x_0,y_0)$ 求出后，再把 H 在 $\bar{v}_m$ 场中向前推移，求出预报时限一半时的 $H(x,y)$，再来与 $\bar{\zeta}_a(x,y,t+\Delta t)$ 相加。这是因为 H 由起始时不断作用到预报时限的终了，所以把它在预报时限中间所在的地点上与 $\bar{\zeta}_a(x,y,t_0+\Delta t)$ 上相加是比较合适的。

Fjφgroft[5]曾说明考虑了垂直运动分布，也可能不用式(2)而解出式(1.1)和式(1.2)，但我们暂不预报 $\bar{z}_T$ 场，因此不用详细考虑 $\bar{z}_T$ 的问题。

在实际计算中我们取 $\bar{A}^2=\dfrac{1}{2}$，格宽 $d=500$ km，而解 Poisson 差分方程 $\alpha-\bar{\alpha}=h$ 时，采用的公式是 $a=h+2h_m$，或 $a=h+3h_m$（m 表示某点四面最近四点上的平均值）所用的天气图底图都用双标准（北纬 60°及北纬 30°）的兰勃脱投影。所用的天气图都已经预先重新分析或检查修改了的，特别是 500 mbar 对 1 000 mbar 厚度场，我们用图解减法作为帮助来分析改正了日常分析（如西德 Bad Kissengen 的图）上的一些比较大的错误。

根据实际经验，式(2)的假定还是合适的。这在以后的例子中也可以看出来。

例子 我们这里所取的例子一共有 3 个，其中 2 个是东亚寒潮的例子；1 个是阻塞高压形成的例子。每个例子前后包括 3~5 天，对每个例子都做了 48 小时与 24 小时的预报。

第一例 这是 1955 年 2 月 16—19 日东亚寒潮暴发的例子。廖洞贤[2]对这个例子曾用图解法做过几个 24 小时数值预报。我们做了由 16 日(23)（图 1a）报 18 日(23)及由 17 日(23)报 19 日(23)的两个 48 小时的 500 mbar 形势预报，和 16 日(23)到 17 日(23)，17 日(23)到 18 日(23)的两个 24 小时预报（对于形势发展过程等情况本文从略，请见参考文献[2]）。从 16 日(23)报 17 日(23)的图 1b 至图 1c 中，已表明如果考虑到斜压性，那么，500 mbar 的形势预报是相当令人满意的，这与以前[2]预报的结果相同，虽然在对 $v_T\cdot\nabla\zeta_T$ 的处理上稍有不同（注意 16 日开始的预报用 $\alpha=h+3h_m$，或 17 日开始的改用对短波歪曲较小的 $\alpha=h+2h_m$ 式）。

有意思的是 48 小时的预报。对于 48 小时预报,斜压变高起着更大的作用(见参考文献[2]),考虑了这项之后,由计算所得变高图 1f 和图 1m 与实况图 1g,图 1n 是要接近多了。同样 48 小时的两层模式预报图(图 1h 和图 1o)要比正压模式的预报(图略)接近实况得多。可以看出东亚大陆上经向度是大大增强了,蒙古一带北风很强,寒潮的南下已在所难免。

由于形势迅速转变时,对空间的平均流场 $\bar{z}_m$ 本身仍有比较大的变化,我们所做的 48 小时预报不是把涡度场在 $\bar{v}_m$ 场中一直推移 48 小时,而是用 24 小时的间隔推移了二次,后 24 小时的位移是用 24 小时预报图上的 $\bar{v}_m$ 来做的。这样效果就更好些*。例如由 16 日(23)报 18 日(23),如果只做 48 小时的一次的推移,那么,由于开始时乌拉尔区脊还比较平浅,乌拉尔区的气旋性涡度就可以在 48 小时中沿着 $\bar{v}_m$ 移到乌拉尔以东气旋性涡度比较小的区域,结果在乌拉尔以东产生了一个负变高区,贝加尔湖以东的高压脊就发展不了太强,并分裂为两个较小的脊(图 1j)。

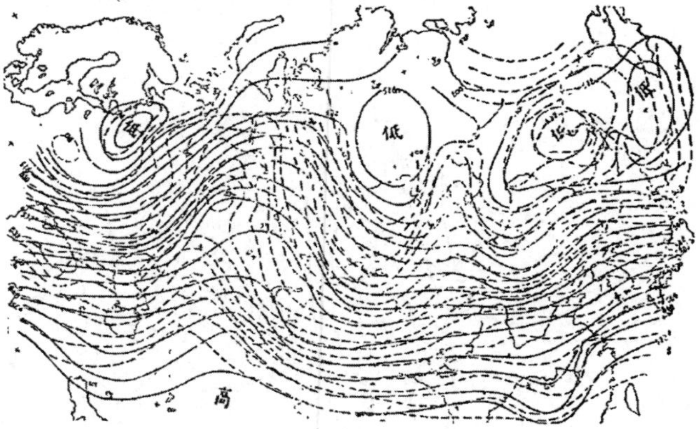

图 1a　1955 年 2 月 16 日(23)500 mbar 图,实线:500 mbar 等高线,
虚线:500/1 000 mbar 相对高度等高线。单位:10 m

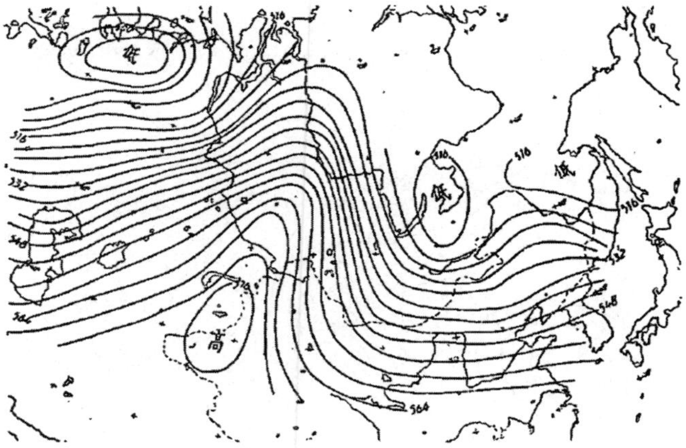

图 1b　1955 年 2 月 17 日(23)500 mbar 预报图(两层模式,24 小时)

由 17 日(23)报 19 日(23)的结果也比较好,48 小时预报中存在的缺点与由 18 日(23)报 19 日(23)的[2]相似。即低压太偏北,脊报得太强,但经向度加大的趋势是正确的。

---

* 如果在 24 小时预报图上重新求出 $\bar{\zeta}$ 场,效果还要好些。

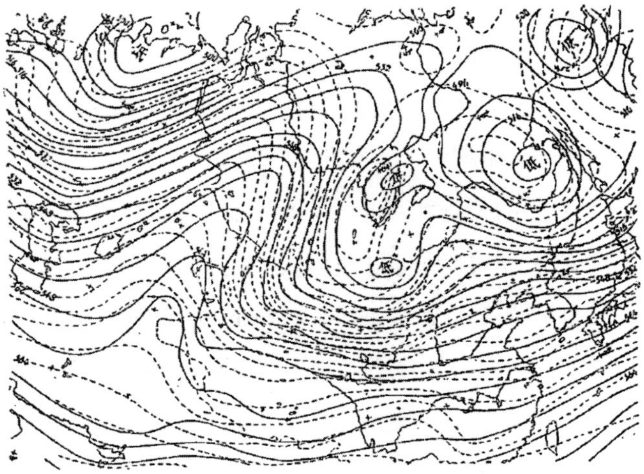

图 1c　1955 年 2 月 17 日(23)500 mbar 图(实况),说明同图 1a

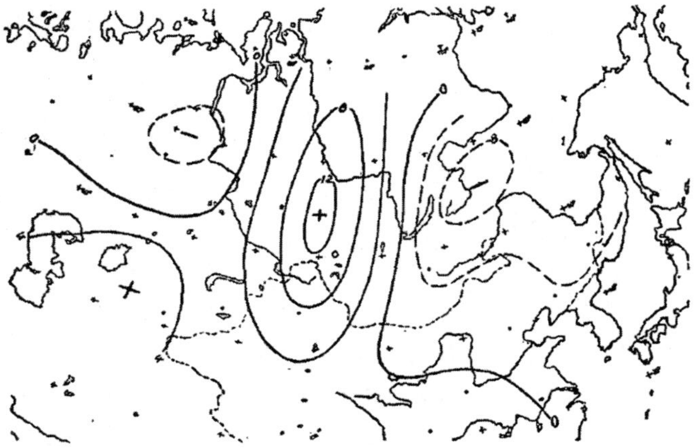

图 1d　1955 年 2 月 16—18 日 500 mbar 48 小时变高(一层模式)

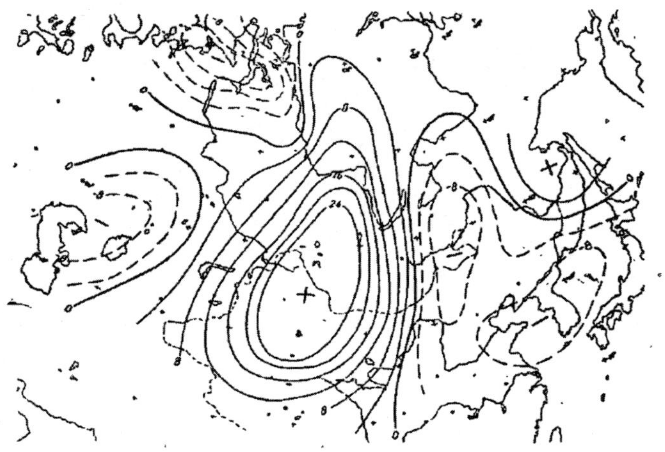

图 1e　1955 年 2 月 16—18 日 500 mbar 48 小时变高(两层模式斜压部分)

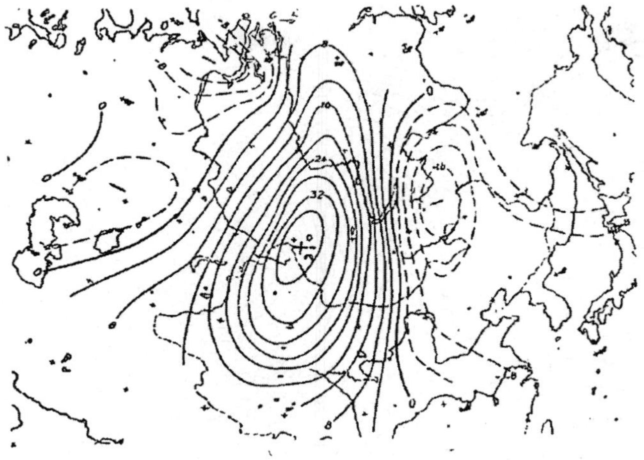

图 1f　1955 年 2 月 16—18 日 500 mbar 48 小时变高预报图(两层模式)

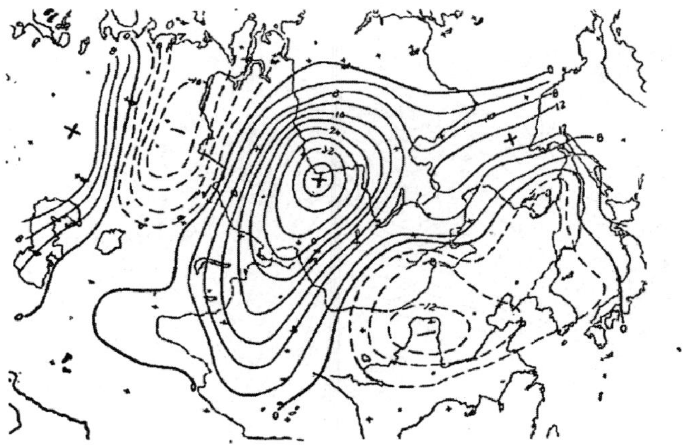

图 1g　1955 年 2 月 16—18 日 500 mbar 48 小时变高实况

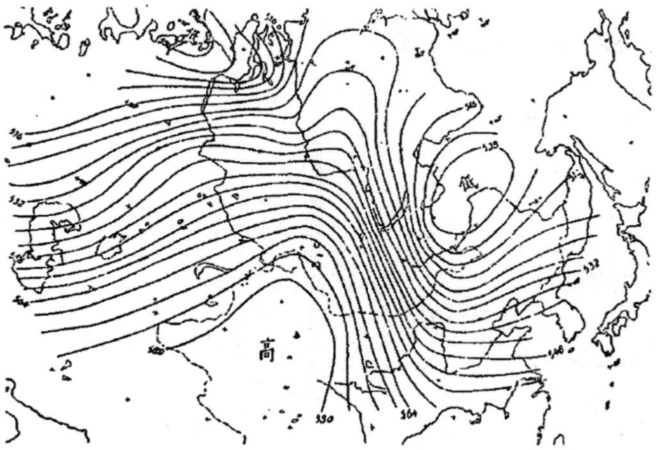

图 1h　1955 年 2 月 18 日(23)500 mbar 预报图(两层模式,48 小时)

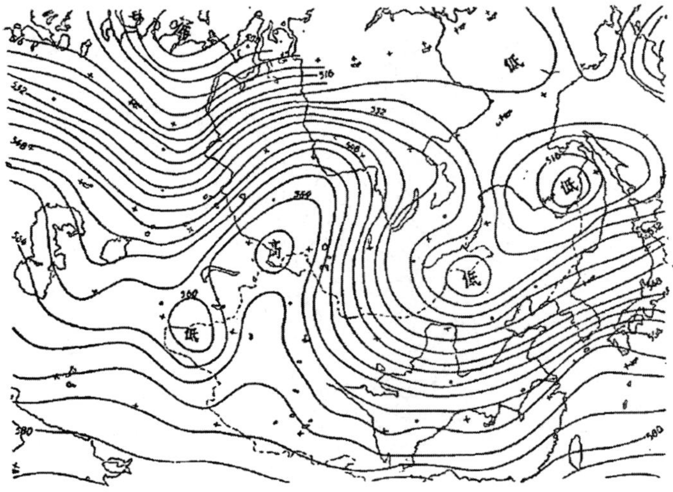

图 1i　1955 年 2 月 18 日(23)500 mbar 图(实况),说明同图 1a

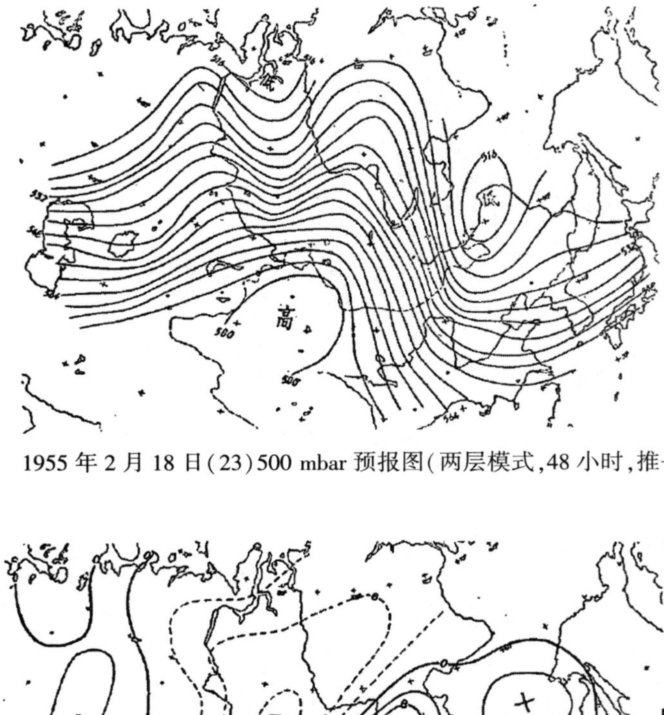

图 1j　1955 年 2 月 18 日(23)500 mbar 预报图(两层模式,48 小时,推一次)

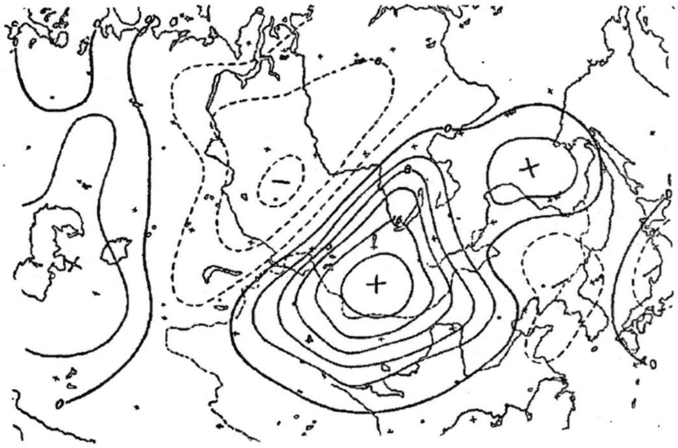

图 1k　1955 年 2 月 17—19 日(23)500 mbar 48 小时变高(一层模式)

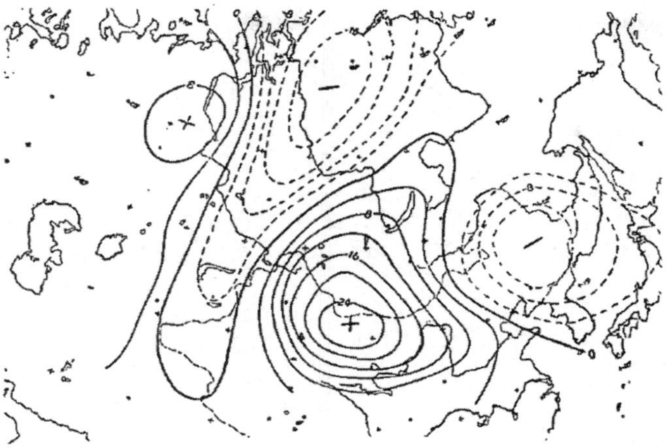

图 1l　1955 年 2 月 17—19 日(23) 500 mbar 48 小时变高(两层模式斜压部分)

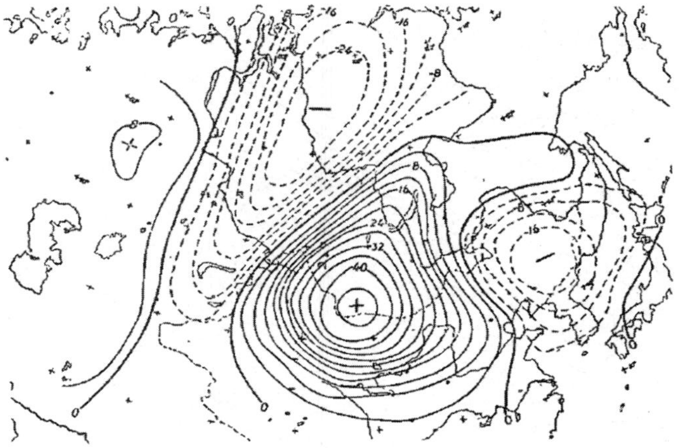

图 1m　1955 年 2 月 17—19 日(23) 500 mbar 48 小时变高预报图(两层模式)

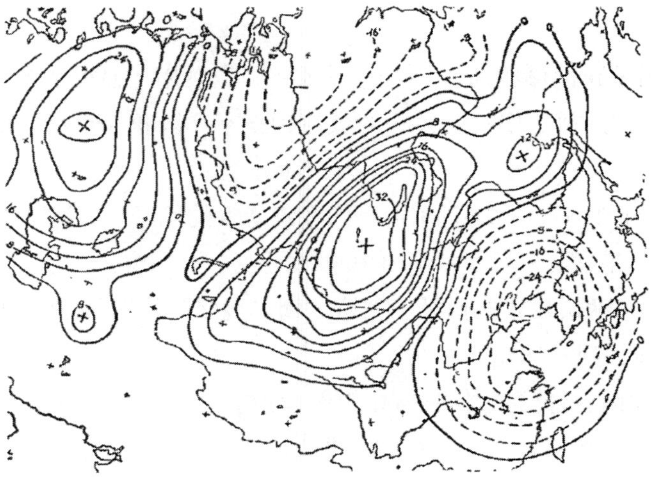

图 1n　1955 年 2 月 17—19 日(23) 500 mbar 48 小时变高实况

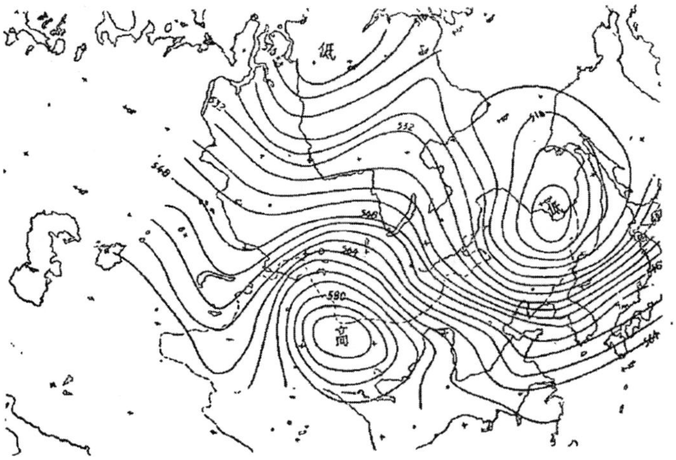

图 1o　1955 年 2 月 19(23)500 mbar 预报图(两层模式,48 小时)

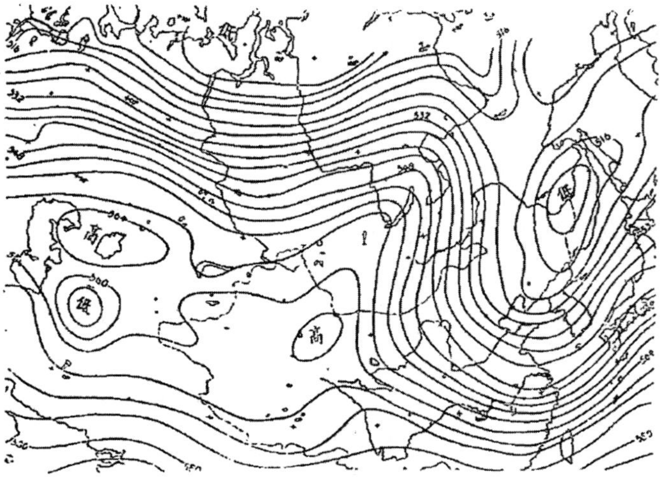

图 1p　1955 年 2 月 19 日(23)500 mbar 图(实况)

总的来看,从这个例子我们可以知道,像这样的寒潮,虽然用半经验的预报技术(无论是 24 小时还是 48 小时)一般不易报好,但是用两层模式的数值预报来作预报,即使是 48 小时的预报效果也是比较好的。

第二例　这是 1956 年 2 月 23—27 日东亚一次寒潮暴发的例子。在这个例子中,贝加尔湖附近的低槽向东移动并且发展起来,但强度没有第一例中那样强。

从预报效果来看,这个两层模式的预报不如第一例来得好。特别是 48 小时预报是不成功的(图 2 中各图),24 小时预报比较好些(图略)。

预报结果比较差的原因,可能是在第一例中上游西伯利亚西部高压脊的发展比较重要,而在第二例中主要只是低槽的发展。大家知道准地转两层模式(或一层模式)中高压脊容易报得强一些而低压槽要弱一些。这样,第二例的预报就不会比第一例的预报来得好。

其他各点与第一例的类似,见图 2 中各图。

要说明的是,在第一例中凡是求涡度及场的加减都是用图解方法,而在第二例中则在差分网格上直接作数值计算求出。第二例中解 Poisson 方程用的公式都是 $\alpha=h+2h_m$。

第三例　无论对东亚的天气形势，还是我国的具体天气过程来说，乌拉尔高空暖高压的形成消退或者说乌拉尔区阻塞形势的形成或崩溃是很重要的。但是在理论上即使到最近对阻塞高压的问题也只有很少的研究[6,7]。因此对于阻塞高压形成等的原因还不太清楚，通过阻塞高压形成的试报，我们一方面可以对它的消长了解得更清楚，另一方面也可以了解我们所取模式的预报能力。

我们选取了一个乌拉尔高压形成的例子。这个高压由 1955 年 12 月 29 日形成一直保持到 1956 年 1 月中旬。它的崩溃还促成了一次东亚寒潮的暴发。

在 1955 年 12 月 27 日(11)500 mbar 图(图 3a)上，可以看到欧洲的经向度还不太大。乌拉尔附近有一个小高压；它是在减弱中的一个系统，不是后来在它东边另外发展出来的高压。但是在北海海面 500 mbar 高压脊的西边，已形成了一个厚度场的高空中心(暖中心，图 3b)。

解 Poisson 方程时用的是 $\alpha=h+2h_m$ 公式。

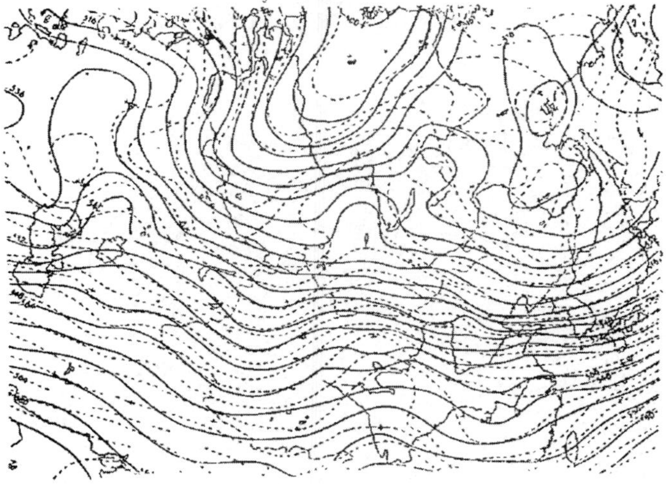

图 2a　1956 年 2 月 23 日(23)500 mbar 图

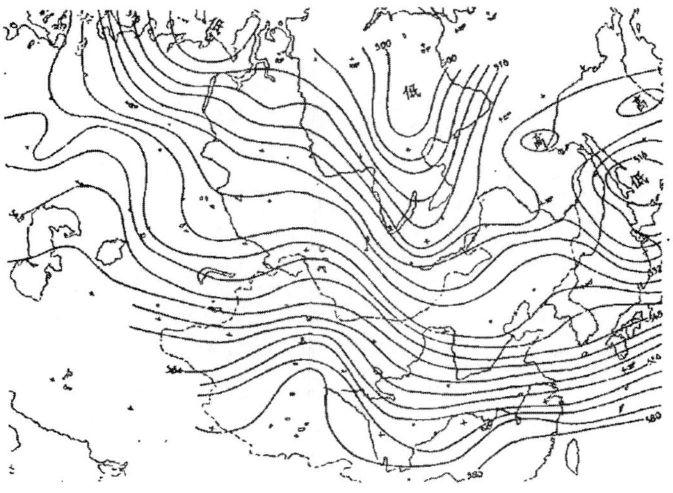

图 2b　1956 年 2 月 25 日(23)500 mbar 预报图

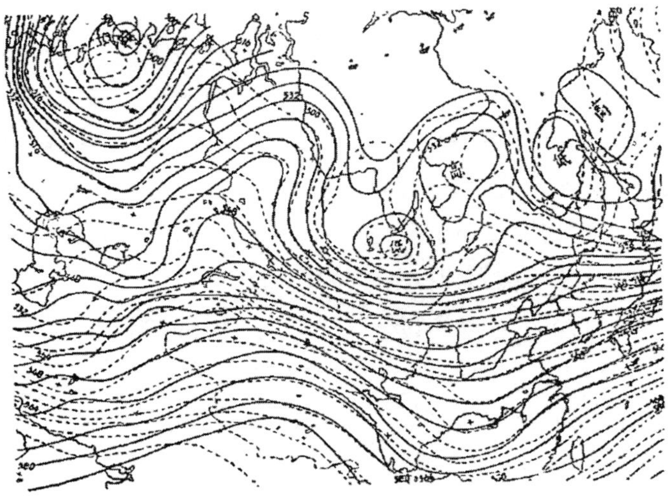

图 2c 1956 年 2 月 25 日(23)500 mbar 图

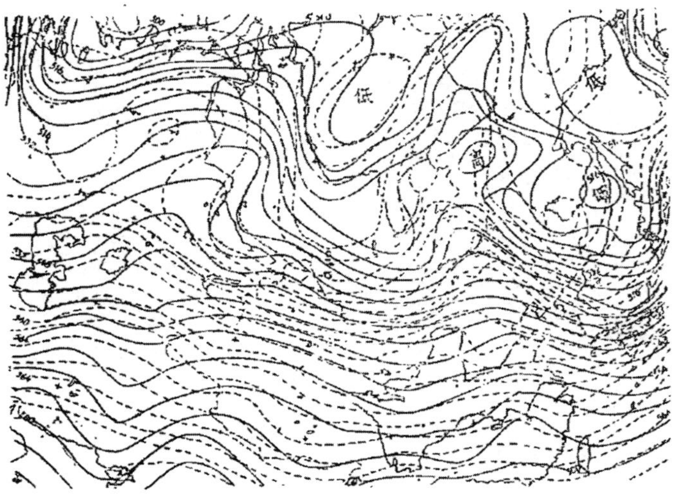

图 2d 1956 年 2 月 24 日(23)500 mbar 图

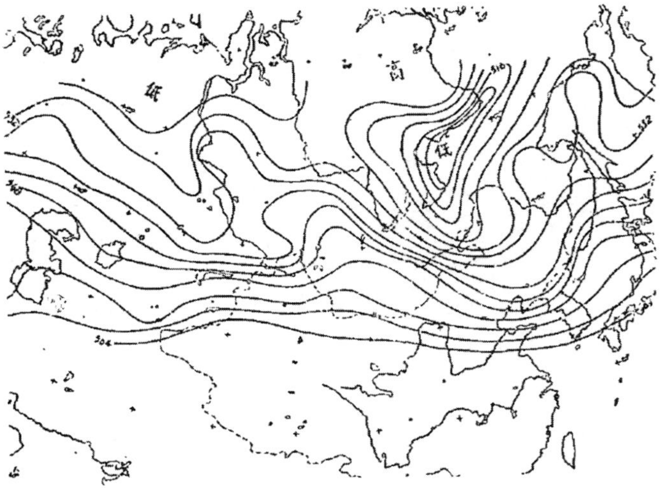

图 2e 1956 年 2 月 26 日(23)500 mbar 预报图

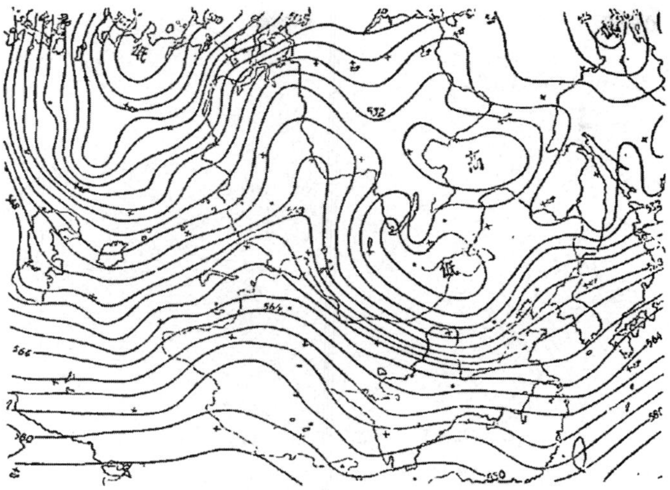

图 2f　1956 年 2 月 26 日(23)500 mbar 图

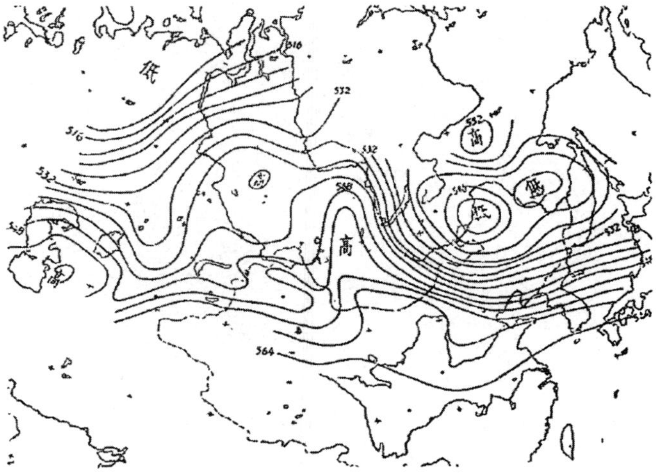

图 2g　1956 年 2 月 27 日(23)500 mbar 预报图

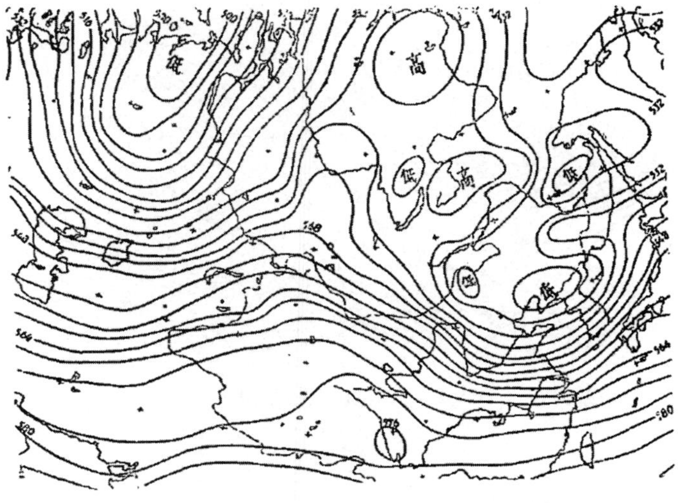

图 2h　1956 年 2 月 27 日(23)500 mbar 图

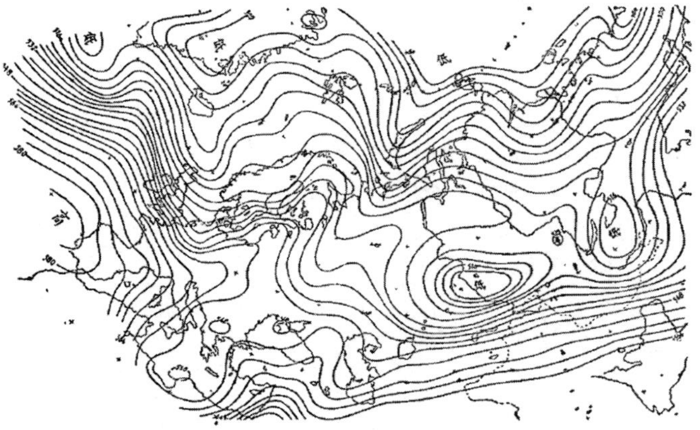

图 3a 1955 年 2 月 27 日(11)500 mbar 高度图

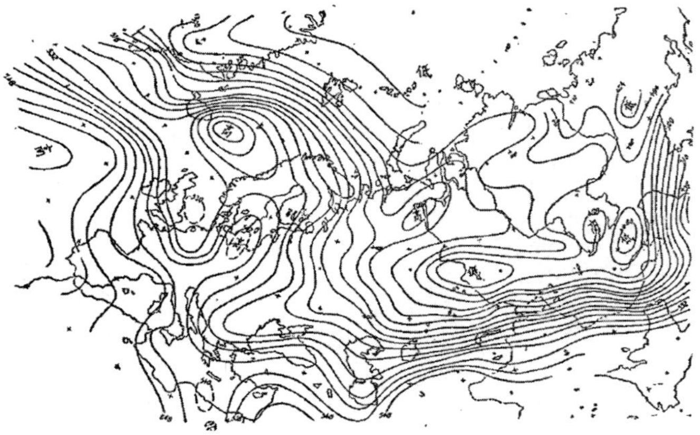

图 3b 1955 年 2 月 27 日(11)500/1 000 mbar 相对高度图

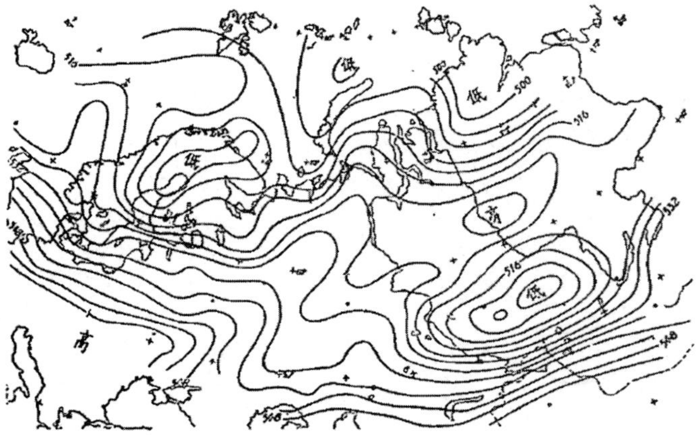

图 3c 1955 年 2 月 28 日(11)500 mbar 预报图(一层模式,24 小时)

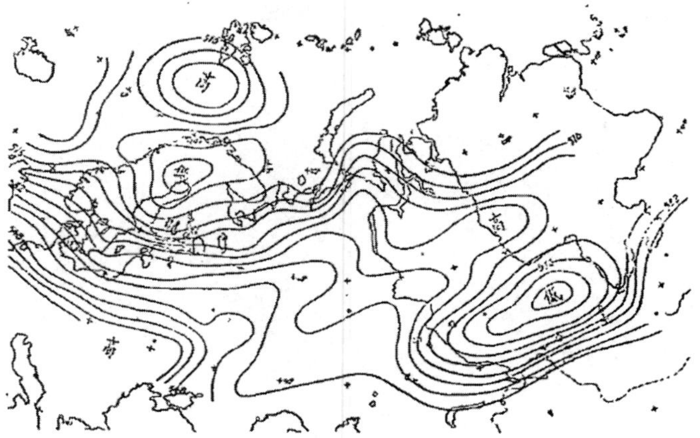

图 3d　1955 年 2 月 28 日(11)500 mbar 预报图(两层模式,24 小时)

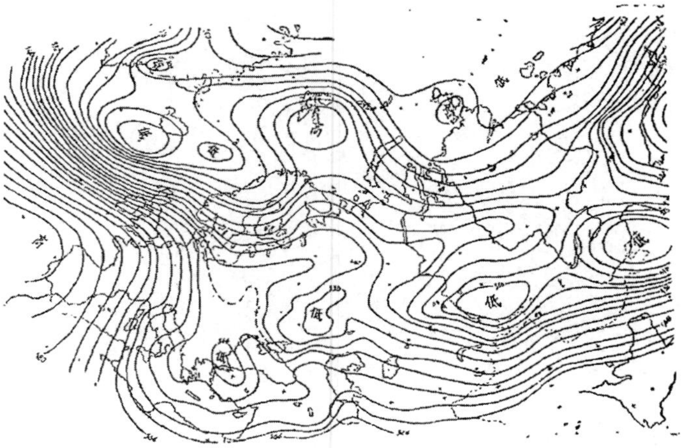

图 3e　1955 年 2 月 28 日(11)500 mbar 高度图(实况)

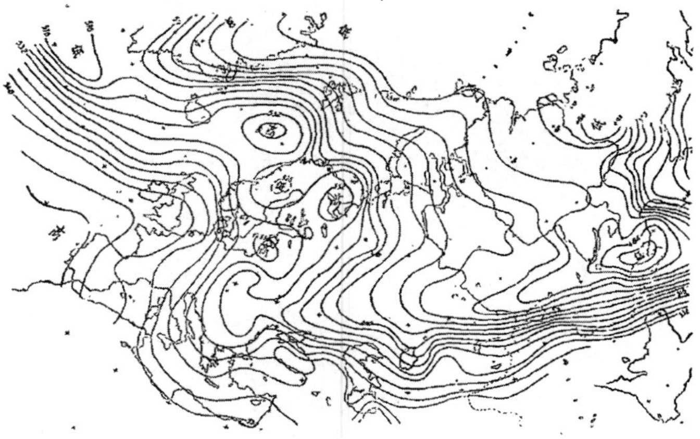

图 3f　1955 年 2 月 28 日(11)500/1 000 mbar 相对高度图

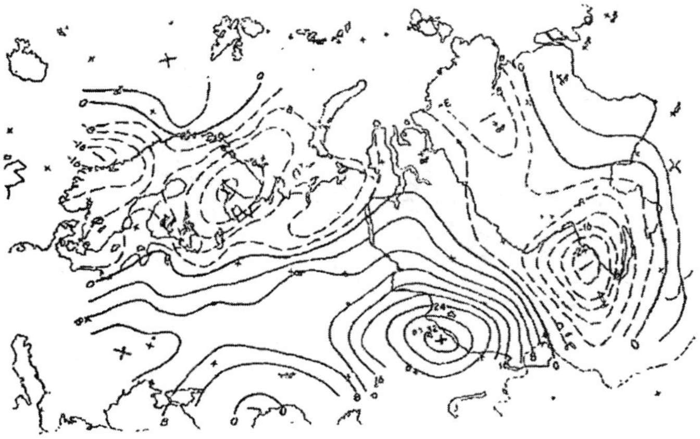

图 3g　1955 年 2 月 27—29 日(11)500 mbar,48 小时变高预报图(两层模式)

图 3h　1955 年 2 月 27—29 日(11)500 mbar,48 小时变高实况

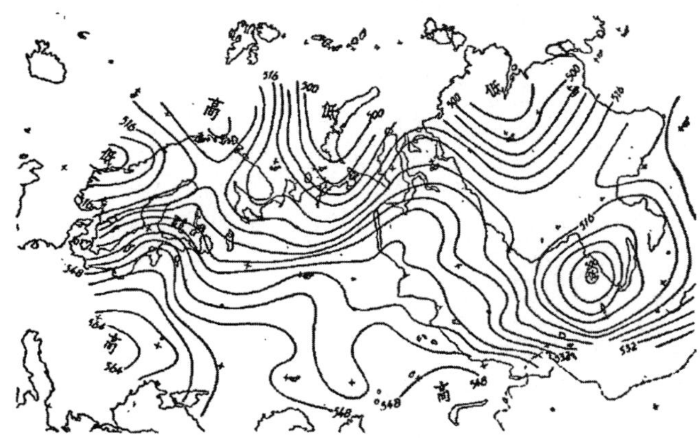

图 3i　1955 年 2 月 29 日预报图(两层模式,48 小时)

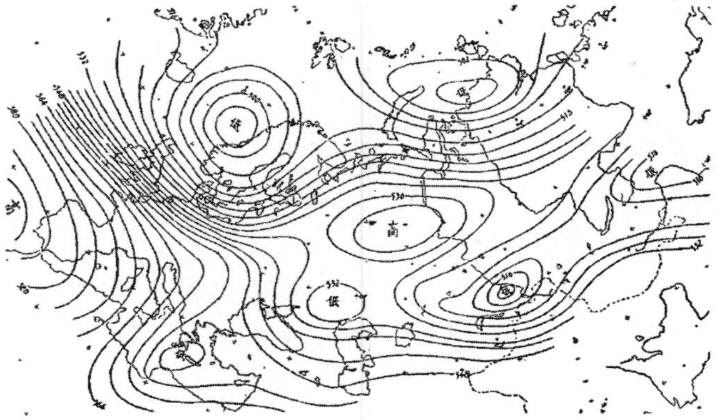

图 3j　1955 年 12 月 29 日 500 mbar 高度图(实况)

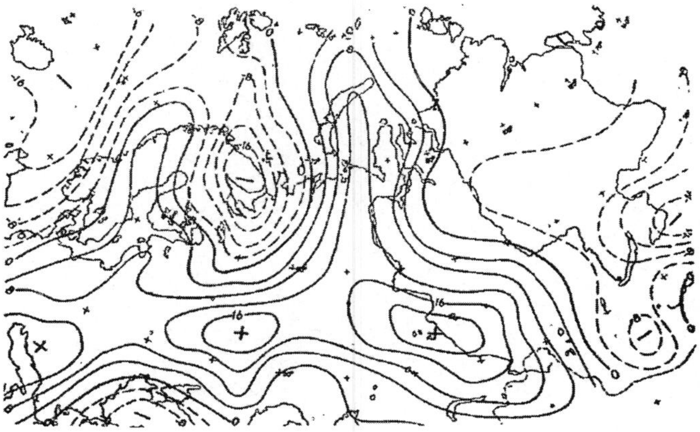

图 3k　1955 年 12 月 28—29 日(11)500 mbar,24 小时变高预报图(两层模式)

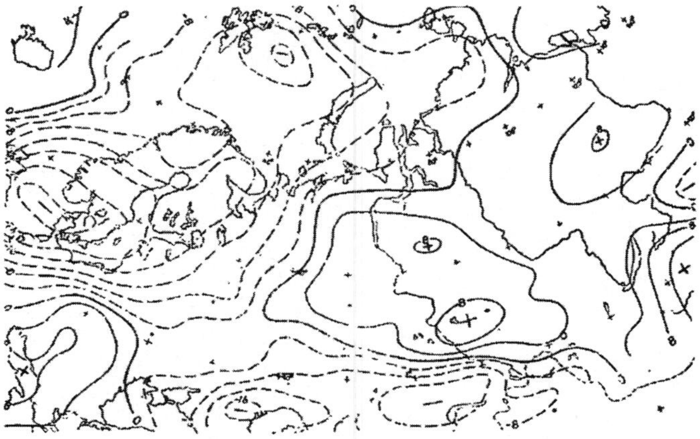

图 3l　1955 年 12 月 28—29 日(11)500 mbar,24 小时变高实况

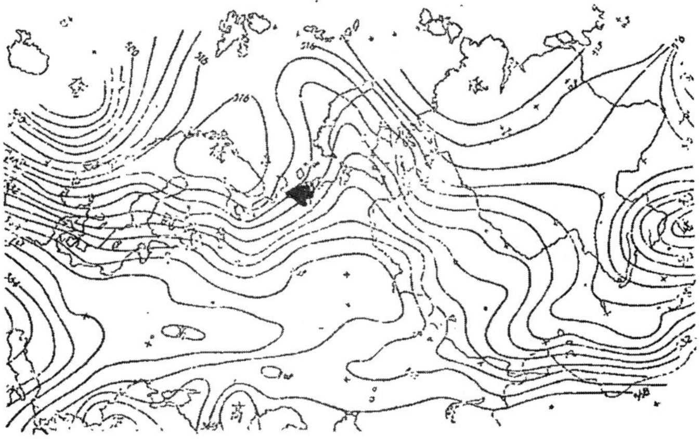

图 3m　1955 年 12 月 29 日(11)500 mbar 预报图(两层模式,24 小时)

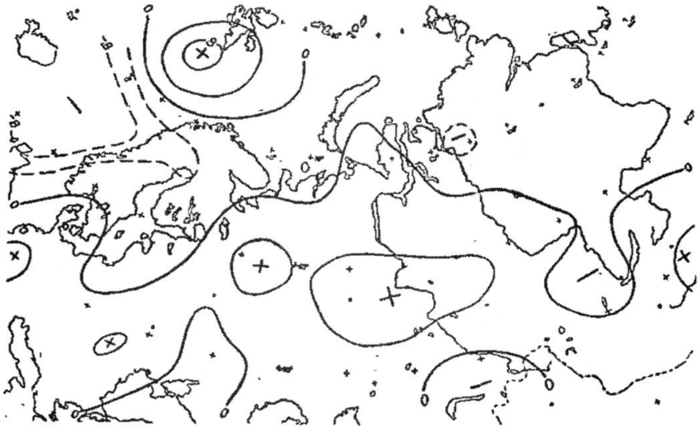

图 3n　1955 年 12 月 27—28 日 500 mbar,24 小时变高图(两层模式斜压部分)

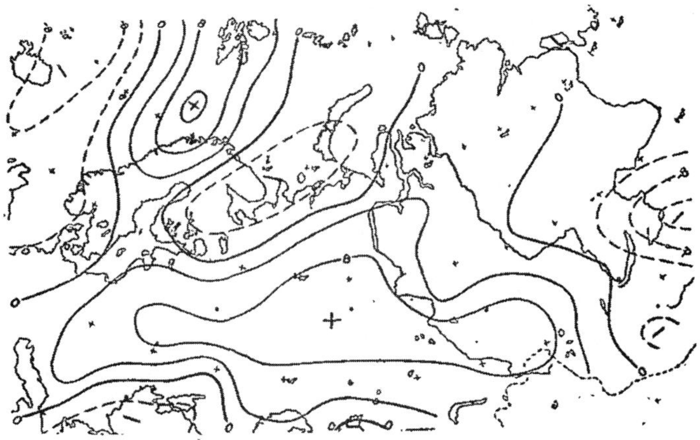

图 3o　1955 年 12 月 28—29 日 500 mbar,24 小时变高图(两层模式斜压部分)

24 小时的两层模式预报说明考虑了斜压性,预报的变化是和实况比较接近的。两层模式预报图中清楚地给出北海东北部新生的高压上正变高虽然很强,总的算得的变高(图 3d)甚至比实况(图 3e)还强,但是在它西南方,即在伏尔加河顿河下游的负变高却没有出现,因此,高压脊虽然加强得很厉害,但却没有形成闭合的高中心。而整个形势便全然不同(图 3f 至图 3h)。

为了了解 48 小时预报的失败是否与时限太长有关,我们又做了 28 日(11)到 29 日(11)的两层模式预报。预报结果(图 3i 至图 3o)说明,虽然 24 小时预报比 48 小时预报来得好些,但是主要的变化即阻塞高压的形成却没有报出来。因此预报仍是不成功的。

因此这个例子的试报结果说明:①至少有些例子中与阻塞高压相联结的正变高不是整个对流层大气辐散[6,8]的结果,而是斜压发展的结果;②单纯的“斜压发展”即 $v_T \cdot \nabla \zeta_T$ 项并不能说明阻塞高压的生成,因为高压脊并没有从南边切断。

以上各例子的预报也是好坏大有不同。就效果而言,形势预报好坏可用网格点上预报高度值和实况高度值之间的相关系数来表示,但是高度图的相关系数只能大致反映高度分布预报结果与实况的相似程度,但不能完全反映预报的要求,也不足以反映预报的本领。因此我们也是用预报变高与变高实况之间的相关系数来表示预报的成功程度。

各例中变高相关系数的数值如下,核对的范围见图 4,每次共对 6×9 点,格宽 500 km(受记录限制,48 小时预报范围向北扩展不够大,北边有时少几点)。

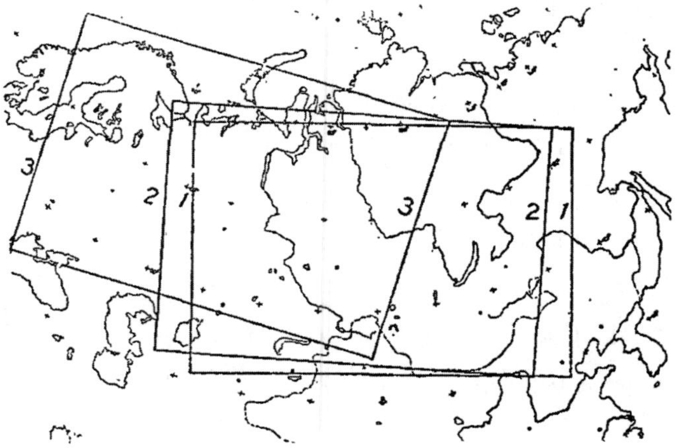

图 4　核对变高的范围

1-指第一例;2-指第二例;3-指第三例

第一例:

2 月16—17 日　$r=0.63$,　　　2 月16—18 日　$r=0.65$,

17—18 日　$r=0.83$,　　　17—19 日　$r=0.84$,

第二例:

2 月23—24 日　$r=0.74$,　　　2 月23—25 日　$r=0.36$,

24—25 日　$r=0.63$,　　　24—26 日　$r=0.40$,

25—26 日　$r=0.36$,　　　25—27 日　$r=0.52$,

第三例:

12 月27—28 日　$r=0.60$,　　　12 月27—29 日　$r=0.39$,

28—29 日 $r=0.32$。

上面这些相关系数比一般报告(参考文献[1])所给的要小些。这可能与图解法有关*。因为 Fjϕrtoft 解 Poisson 方程的图解法中误差大小和符号与解的波长有关,而图上各处分解所得的波长成分各有不同。因此由 Fjϕrtoft 解 Poisson 方程的图解法所求出的结果,其误差在图上是不均匀的。并且它的误差也不小,特别是在波长较短的时候。例如在第二例的第三个 48 小时预报中,由 Fjϕrtoft 方法求出的 48 小时变高[图 5(a)]比用正规的 Relaxation 方法所求出的[图 5(b)]要大了 0.5 倍到 1 倍。由图上可以看出用 Relaxation 方法求出的结果是比用 Fjϕrtoft 方法的要接近变高实况[图 5(c)]得多。前者与实况的相关系数是 0.52,后者是 0.62。但也必须指出的是用计算机的帮助逐步外推也不全都是有很好的结果的(例如 Bushby 和 Hints[10]在用计算机帮助下考虑了涡度垂直输送和辐散,做的两层 24 小时的预报时 500 mbar 变高还得到过 $r=0.18$,对 1 000 mbar 形势还得到过负相关),考虑到这些情况,上面这些例子的结果与一般试报结果是比较接近的。

从以上几个例子来看,对寒潮预报来说,两层模式的数值预报结果比较好。在寒潮暴发初期高空高压脊的发展一般是比较重要的,而这是预报得比较好的。但总的来说有一个共同的偏向,就是高压报得太强而低压报得太弱。特别显著的是负变压的中心总是偏北,这是地转风假定下各层模式预报的通病[1]。

对于阻塞高压的形成来说,预报是不成功的。正如上面所说预报的不成功的原因主要还是它南面的负变压区没报好。由此可见问题还是在低压发展报得不好。我们看 Bolin[8]用一层模式试报阻塞高压的例子(图 6),就可以发现同样的毛病。如果从变高来看(图略)尤其清楚(即使考虑到对流层整层的辐散也只能使北边的正变高加强,但南边的负变高仍然没有报好)。然而要将原因归之于地转风假定本身却不是很完全合适的。Charney 在 1950 年 11 月 25 日的北美低压发展的例子(见参考文献[1])证明,同样在地转风假定下,层次的多少与预报结果就有不同。

我们这个例子的情形也是相似的,根据涡度方程来看,在这种情形下低压生成区的上游可能在 500 mbar 上有比较大的辐合。实际上也是如此,我们用温度个别变化法分别求了 1955 年 12 月 28 日(11)到 29 日(11)的时间中伏尔加河顿河流域 700 mbar,500 mbar,300 mbar 上的垂直运动,发现在这区域中自 700 mbar 到 300 mbar 这段气柱整层平均来说都是水平辐合的,而在伏尔加河中游一带的辐合还很强(平均 2.8×10^{-6} s),符号也一致。从 850 mbar 到 300 mbar 各层上都是辐合的。从测风记录来看在伏尔加河中游也是辐合的(500 mbar 上 4.4×10^{-6} s)。因此辐合的作用是不小的,而低压的发展可能也就是辐合的结果。

从过去的个案分析来看情况是相似的,Bolin 的例子我们没有记录,但从 Charney 三层模式试报的例子中,500 mbar 上辐散是很大的(图 7),发展也更猛烈。在 Eliassen 和 Hubert 的切断低压衰退的例子中,辐散的作用更是明显,低压中心整个对流层有猛烈的辐散且在 500 mbar 附近最大。而低压在 24 小时内就很快减弱了[图 8(a)]。至少从这几个例子来看,在低压发展或较弱时 500 mbar 上的辐散是大的,整个气层中的辐散分布大致会如图 8(b)那样分布。由此看来在涡度方程中辐散项的作用很大,这也就是为什么即使一层模式,考虑了辐散[12],有些预报结果也比较好[13]的原因。

但也得说明,阻塞高压的形成可能是多样的,有一些阻塞高压(如 1950 年 5 月 18 日欧洲西北部

---

* Fjϕrtoft[9]认为他的图解法不比用计算机逐步数值计算外推来得差。然而他所给的数字中有一个例子中图解法的结果($r=0.71$)竟比用计算机逐步外推的($r=0.53$)更好,这是不大可能的。因此这似乎只能反证图解法可有相当的误差($E_r=0.18$!)。

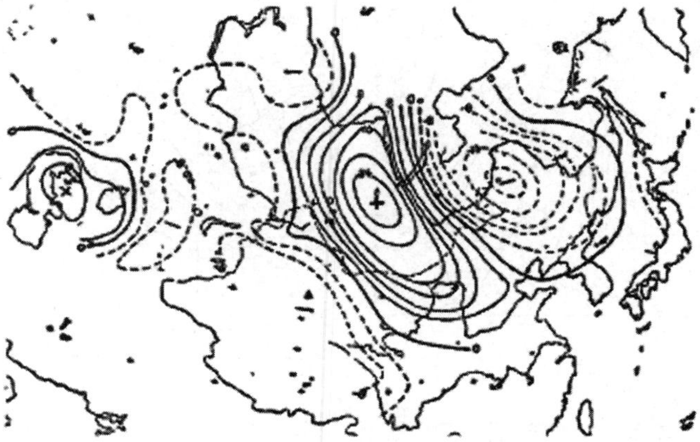

(a)

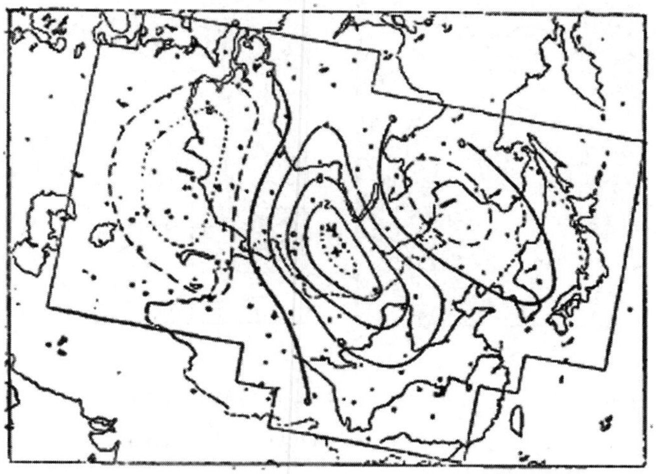

(b)

(c)

图 5　Fjϕrtoft 图解法结果与 Relaxation 法结果比较，1956 年 2 月 25—27 日 500 mbar 变高图

(a)用 Fjϕrtoft 图解法求得值($\alpha=h+2\bar{h}$)；(b)用 Relaxation 法求得值($\Delta\zeta_{48}$值与(a)中相同；(c)实况

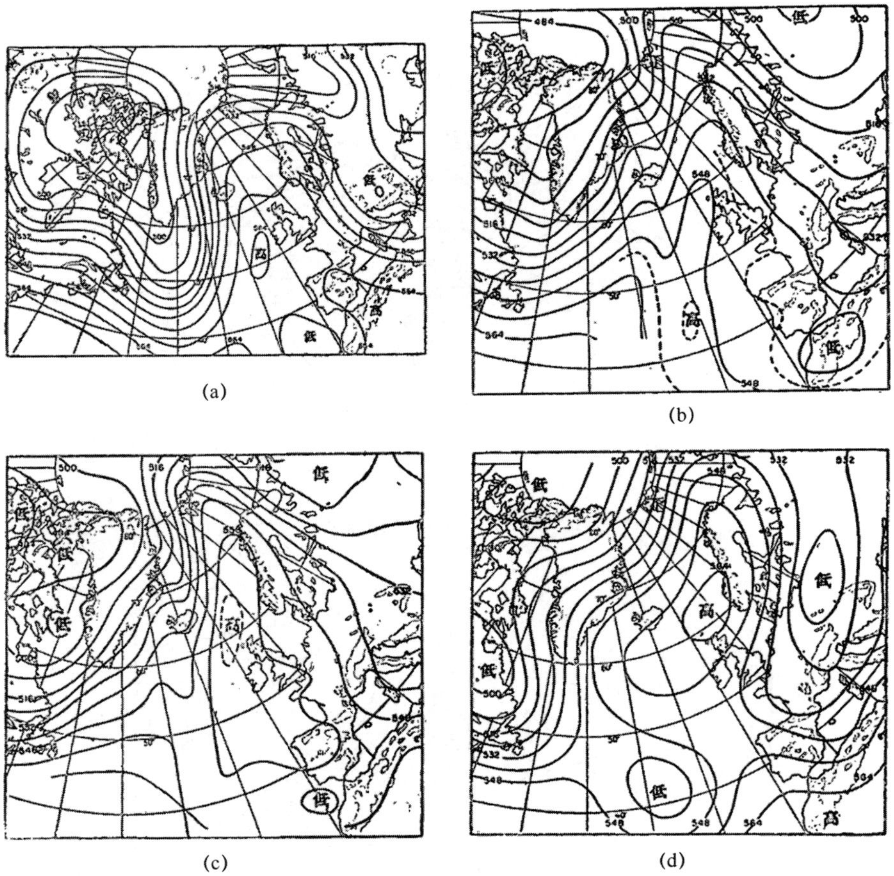

图 6　1948 年 2 月 17 日(11)到 19 日(11)的 500 mbar 预报图

(a)17 日(11)实况;(b)19 日(11)正压模式 500 mbar 预报;

(c)正压预报,对流层有整层辐散;(d)19 日(11)500 mbar 实况(采自 Bolin[8])

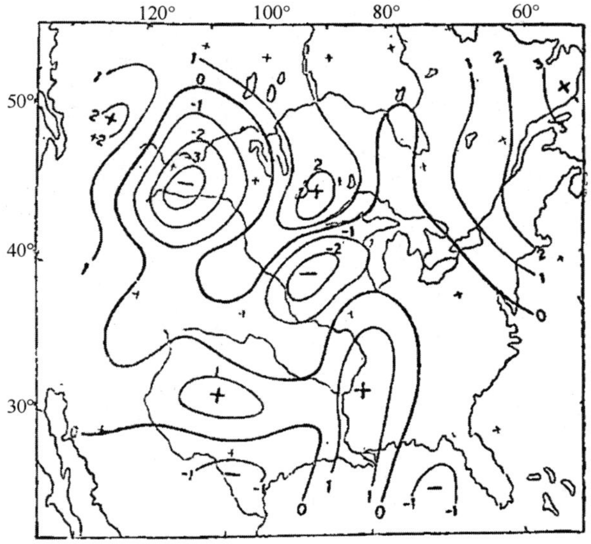

图 7　1950 年 11 月 24 日(23)北美 500 mbar 辐散场(单位 10^{-5}/s)

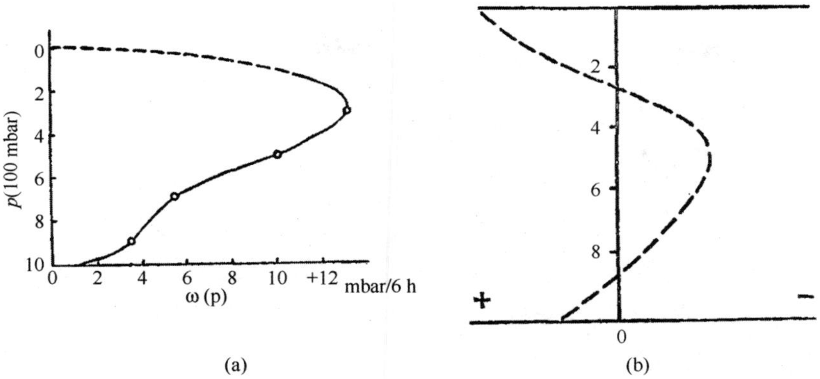

图 8　低压发展前的典型辐散分布

(a) Eliassen-Hubert 例;(b)可能的一般情形

形成的阻塞高压)南边先已有切断低压存在,高压的形成过程就有所不同,不能一概而论。

*　　*　　*　　*

最后,在计算中有几点可以提出来的是:①分别计算涡度平流和$\frac{\partial f}{\partial y}v$ 项的作用,我们发现尽管在量级比较时$\frac{\partial f}{\partial y}v$ 项不小,但在实际数值积分中该项是比较不重要的。一般由此造成的变高都在 40 m/24 h 以下。这是因为空气运动的南北幅度有限,在预报时限中南北位移不大的缘故。②H 的变化一般是不大的,例如从图 3n 和图 3o 变高中心的变高大小的比较,就可以看出在 24 小时中 H 改变所相当的变高改变在 40 m 以下。所以 $\mathrm{d}H/\mathrm{d}t=0$ 的假定还是可以用的。

3　准地转两层模式数值预报的局限性

从以上的讨论,我们可以考察一下两层模式对辐散场分布的表示能力,来了解它的预报效果。

一般准地转两层模式中所取的都是没有整层辐散的大气,即 $\omega(0)=\omega(p_0)=0$,因此像我们这样,把涡度方程对 p 从 $p=p_0$ 到 $p=0$ 积分,所得到的平均高度 $\bar{p}$ 上的涡度方程(1.1)是不会有辐散项的。不但对整层积分处理时平均层上不会有辐散,就是在用垂直方向上取气压差分(像 Charney[16] 那样)来定两层的层次时,只要有 $\omega(0)=\omega(p_0)=0$ 的假定(或 $\omega(0)=\omega(p_0)$),平均层上也仍是没有辐散的。

值得注意的是方程式(1.1)是对“积分大气”的方程,并不能与“平均层”$p=\bar{p}$ 上的涡度方程 $\nabla^2\frac{\partial\bar{z}}{\partial t}=-\bar{v}\cdot\nabla(\zeta+f)+f\mathrm{div}\,\bar{v}$ 比较,而把$-v_T\cdot\nabla\zeta_T$ 看作是辐散项的一个表示。因为$-v_T\cdot\nabla\zeta_T$ 项是把$-v\cdot\nabla\zeta$ 项对高度积分而得的,是斜压性作用的直接结果。也就是这样,所以我们不能简单地取 $p=\bar{p}$ 的涡度方程为 $\nabla^2\frac{\partial\bar{z}}{\partial t}=-\bar{v}\cdot\nabla(\bar{\zeta}+f)+f\ \mathrm{div}\ \bar{v}$。因为“−”在这里代表的是一种运算,而不是简单的某一层次的表示。

从两层模式中的基本假定来看,风的分布是按照 $v(x,y,p,t)=\bar{v}(x,y,t)+A(p)v_T(x,y,t)$ 分布的,在这种情形下 $\mathrm{div}_2 v=\mathrm{div}_2\bar{v}+A(p)\,\mathrm{div}\ v_T$,也必然有无辐散层 p^* 在这层上。

$$A(p^*) = -\operatorname{div}_2\bar{v}(x,y,t)/\operatorname{div}_2 v_T(x,y,t)$$

由于 $A(p)$ 只能按平均情形来定，由地面到平流层下部 $A(p)$ 由负到正作单调增加，因此在适当的 $\operatorname{div}_2\bar{v}$ 及 $\operatorname{div}_2 v_T$ 两个已知函数下也只能有一个 p^* 值，而且时间、地点可以不同。因此这时只能有一个无辐散层，在这层的上下辐散符号相反。但 $\operatorname{div}_2\bar{v}=\overline{\operatorname{div}_2 v}$，由前面所说的上下边界条件得 $\overline{\operatorname{div}_2 v}=0$，所以在这种情形下无辐散层就是平均层 $\bar{p}=p^*$（"积分大气"就是无辐散的，不必要像 Charney（见参考文献[1]）那样要凑出一个无辐散层来）。根据统计所得 $A(p)$ 的分布，两层模式中 $\bar{p}$ 就在 500 mbar 附近，因而在两层模式中 500 mbar 高度上辐散也是很小的。在这种条件下的两层模式中，形式上保留这一项的许多做法（见参考文献[1]）也是不能起多大作用的。

由此也可以推知，在两层模式中只有一个 ω 极大值。它在 500 mbar 高度附近。

然而上界条件 $\omega(0)=0$ 并非必要，因为 $\omega(0)=\omega(p_0)=0$ 的假定虽然可使大气的整层辐散等于零，因而消去重力波，但我们作了地转风近似，重力波已经消掉。不用 $\omega(0)=0$ 的假定也可以求得所要的涡度方程。这种做法也很多，有的直接假定无辐散层的存在，而把涡度方程用到那一层上（如 Petterssen[14]，Estoque[15]），有的就考虑对流层有整层的辐散（Charney[16]，Bolin[8]），但不论怎样都不能改善辐散在垂直方向上的相对分布，也不是对任何波长的波都有显著的改进作用。因此并不能使我们的例子，Bolin[8] 的例子以及 Eliassen-Hubert 的例子等等这些阻塞形势中切断低压的生灭，以及 Charney（见参考文献[1]）和 Petterssen[14] 所提的那类高空低压发展的生灭的预报有基本的改善。

要改进这种预报结果只有两个方法。一是改用三层模式。应该指出三层模式当然比二层模式更好考虑到大气斜压性的实际情况，从系统发展来看，三层模式的具体优点是使涡度方程中的水平辐散可能采取普遍的分布形式，而能反映出发展比较猛烈时的辐散构造，因此它是预报猛烈发展所必要的。问题是在这种情形下由于"平流项"很多，图解法赶不上时间，因而就必须使用高速计算机来做。

另一方法是也可以不用地转风假定，而研究如何用实际风场（包括实际的水平辐散场）来报未来的气流场，这样我们就用实际的辐散场，不用相对高度求 ω 场再求辐散场。上面的那些辐散构造的假定就可以避免了。这不仅对二层问题，就是对一层模式也可以有显著改进。这方面需进行数值积分的试验[可用 Hollmann 的或 Bolin（见参考文献[1]）的方法]，已有一些试验[13] 看来是一次外推 24 小时的，因此还是不够的。

总之，模式的选取一定要根据大量个案中构造和发展的研究才能很好地决定。从气候平均（如 Arnason[17]）来决定还是不够的，因为问题不仅在一般的发展如何预报，从预报的观点来说，更重要的是与危险天气相联系的发展特别强烈的个案中，系统是怎样构造和怎样发展起来的。因此，正像其他预报技术一样，数值预报和理论方法也是需要由个案分析试报及日常数值预报工作的成功和失败例子的分析而不断改进的。

* * * *

本文由曾佑思帮助制图，谢葆良、薛永瞻、梁佩典帮助计算，在此致谢。

参考文献

[1] 顾震潮.大范围温压场的流体力学预报法.气象学报，1955，26：211–230，295–327.

[2] 廖洞贤.简化的两个参数模型的图解数值预报.气象学报，1956，27：153–166.

[3] Thompson，P.D.，and W.L.Gates.A test of numerical prediction methods based on the barotropic and two-parameter baroclinic models，*J.Meteor.*，1956，13：127–141.

[4]　Fjφrtoft, R. On a numerical method of integration the barotropic vorticity equation, *Tellus*, 1952, 4: 179-194.

[5]　Fjφrtoft, R. On the use of space-smoothing in physical weather forecasting, *Tellus*, 1955, 7: 462-480.

[6]　Yeh, T.C. On encrgy dispersion in the atmosphere, *J, Meteor.*, 1949, 6: 1-16.

[7]　ROSSBY, C.G. On the dynamics of certain types of blocking waves, *J. Chin. Geophy.* Soc., 1950, 2: 1-13.

[8]　Bolin, B., An improved barotropic model and some aspects of using the balance equatiion for threedimensional flow, *Tellus*, 1956, 8: 61-75.

[9]　Fjφrtoft, R. "On forecasting with barotropic-model", *Tellus*, 1956, 8: 115.

[10]　Busbby, F. H., and M. K. Hints. The computation of forecasting charts by application of the Sawyer - Bushby two - parameter model, *Q. J. Roy. Met. Soc.*, 1954, 80: 165-173.

[11]　Elassen, A., and W.E. Hubert. Computation of vertical motion and vorrticity budget in a blocking situation, *Tellus*, 1953, 5: 196-206.

[12]　Успии ский, Ъ.Д. Теория локалвных изменений геолотециальных высот изобарических иоверхности, М. иГ., 1954(5).

[13]　Васюков, К.А. Анадиз локальных изменений давления и условий иклогенеза в тросфере, Тр.ЦИП, вы, 1955, 45 (72): 1-58.

[14]　Petterssen, S., A general survey of factors influencing development at sea level, *J. Meteor.*, 1955, 12: 36-42.

[15]　Estoque, M.A. A prediction model for cyclone development integrated by Fjφrtoft's method, *J. Meteor.*, 1956, 13: 195-202.

[16]　Charney, J.G. The use of primitive equations of motion in numerical prediction, *Tellus*, 1955, 7: 22-26.

[17]　Arnason, G., and L. Vuorela, A two-parameter repressentation of the normal temperature distribution of the 1000-500 mb layer, *TELLUS*, 1955, 7: 189-203.

The Real-time Numerical Forecasting System for Marine Environmental Elements in China*

CHAO Jiping, WU Huiding

National Research Center for Marine Environmental Forecast, State Oceanic Administration Beijing, 100081, China

Summary

Like other countries along the ocean coast in the world, China is being challenged by the damages caused by severe oceanic events, such as storm surges, severe waves and sea ice. It is therefore of primary significance in China to investigate marine modelling of various time-scales and to improve the method of short-time scale prediction in recent years. A numerical prediction system including a method for decoding GTS data, objective analysis, atmospheric models combined with boundary layer model, storm surge models and sea wave model, has been established since January 1989.

Now, the predictors can receive reference from our numerical modelling products for doing the routine forecasting. Some numerical forecast products, such as sea ice during the winter season, are being directly issued to users.

The sea wave of the WAM model has been introducted into this system and connected with drive by sea-surface wind given by a five-level atmospheric model for forecasting the sea waves in the area from 15°N to 45°N and from 105°E to 155°E.

A two-dimensional and non-linear storm surge model has been used for modelling surges along the coast of China. Also a moving nested barotropic atmosphere model and as analogy scheme have been adopted to forecast the track and the moving speed of typhoons. The sea surface pressure and stress fields are calculated based on the structure of the idealized typhoon field. The predictions of surge elevations and depth-averaged currents in the areas of the China Sea are obtained by using the storm surge model with these fields. A new baroclinic typhoon prediction model has been developed and will be connected with the storm surge model. The above tide-level atmospheric model with a boundary layer model is also connected with surge model to forecast the water elevations and currents caused by cold-wave and/or extratopical cyclones during the sping and winter seasons.

The detail structure of our models and the results of forecasting will be introduced by a report.

Abstract

The real-time numerical forecasting system for marine environmental elements including ocean wave, sea ice, storm surge, coastal current and sea surface temperature is presented. The introduction deals with the rou-

* Série technique Unesco Commission Océanographioue intergouvernementale, 1992, 39: 27−33.

tine model of waves and storm surges in particluar and show some results of forecasts.

It is indicated that the most difficult one for real-time numerical forecasting is shortage of data both atmospheric and oceanic on the sea. An urgent task, therefore, is to improve the density and accuracy of oceanic data, especially to allocate more buoys and make further advances on the oceanic date llite remote sensing system.

Introduction

With the increasing needs of marine exploration, traffic, engineering, environmental protection and other activities, it has become urgent to set up a modern, objective and quantitative numerical forecasting system for marine environments in China. Therefore, a quasi-routine system has been established initially which started running on the computer system in January 1989 in the National Center for Marine Environmental Forecasts (NCMEF).

This routine numerical forecasting system consists of eight parts, they are: data decoding, objective analysis, atmospheric model, ocean wave model, sea ice model, storm surge model, depth-average coastal current model and sea surface temperature model. The data are received from the Global Telecommunication System (GTS), tidal stations along the coast of China and several buoys. An overall flow chart of this system is shown in Fig.1. It takes 90-mins of CPU time on the computer Cyber 180/840 for running the system for 72 h forecasting. Below, is an introduction to the routine models of waves and storm surges in articular and show some results of forecasts.

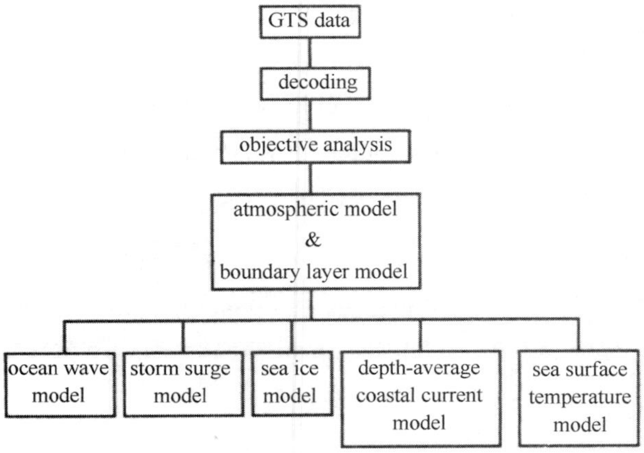

Fig. 1　The flow chart of the numerical forecasting system in NCMEF

Ocean wave forecasting

The third general wave model [WAMDI (1988)] was introduced and used for forecasting in the near-shore shallaw water region of the China Sea. the used wave spectrum equation is

$$\frac{\partial F}{\partial t} + \frac{1}{\cos\phi}\frac{\partial}{\partial\phi}(\dot{\phi}\cos\phi F) + \frac{\partial}{\partial\lambda}(\dot{\lambda}F) + \frac{\partial}{\partial\theta}(\dot{\theta}F) = S \tag{1}$$

where F is the wave spectrum, $\dot{\phi} = \mathrm{d}\phi/\mathrm{d}t$, $\dot{\lambda} = \mathrm{d}\lambda/\mathrm{d}t$ and $\dot{\theta} = \dot{\theta}_{gc} + \dot{\theta}_D$, $\dot{\theta}_{gc}$ and $\dot{\theta}_D$ are the variations of the wave direction caused by the great circle reflection and topographic reflection respectively.The wind forcing, nonlinear interaction, dispersion process and bottom friction effect are involved in the source functions on the right hand side of Eq.(1).

The forecasting area is 15°N−45°N, 105°E−141°E on a mesh of 2°×2°. The time-integral is by a implicit scheme, the interval of time-step is one hour, the resolution of frequency is $\Delta f/f=0.1$ and the resolution of direction is $\Delta\theta=30°$.

The forcing wind fields fare obtained from a shortrange numerical weather prediction model with an atmospheric boundary layer model.The 5-level primitive equation model with the σ-system of coordinates are used, in which large-scale condensation, convective adjustment, orographic effect and boundary layer parameterization are involved.The limited area model with grid interval of 190K.5 km is nested in one way in the hemispheric model with the mesh of 381. 0 km.The finite difference scheme with conservation of total energy and mass is adopted, the lowest layer ($\sigma=0.81-1.00$) of the 5-level model is taken as the boundary layer, which is divided into two sub-layers, surface layer and Ekman-layer.The effects of thermal stratification and baroclinity on wind distribution are taken into account. Based on the boundary layer resistance law, the boundary layer model is connected with the atmospheric model for forecasting the sea surface winds.

The testing forecasts in the past two years show that the wave model connecting with the above model can provide a reference for the routine forecasting of wind waves in the China Sea. According to the atmospheric process, the forecasting results of the wave model are satisfactory for the waves caused by atmospheric coldwave and cyclone moving eastward from the continent into the sea but not always satisfactory for the waves caused by typhoon.The two cases of prediction are given as follows:

(1) The case of atmospheric cold wave

Figure.2 shows the numerical forecasting wave fields caused by the cold-air moving southward on Feb.1, 1989.From the figure, we can see there are two centers of rough sea with significant wave heights of 3−3.5 m and 2.5−3 m.One is in the South China Sea and the other, which is getting intensive, is in the ocean on the east of the Philippines.Meantime a moderate sea with 2−2.5 m high can be also seen in the Japanese Sea. From ship observations at 00 GMT of the 2nd and 3rd February, there was a rough sea with 4−5 m high in South China Sea and a rough sea with 3 m high in Japanese Sea.

(2) The case of cyclone

The wave model gave a fairly good forecast for the sea wave caused by the cyclone moving eastward during the period of 15−18 February 1989.A rough sea with 3 m high, which strengthen with moving north-eastward, was observed in the area to the East China Sea at 00 GMT of 16 February.Figure 3 gives the evolution of the forecast wave fields.It shows that the moving direction of the wave region and the intensity of waves are agreeable with the observation.Figure 4 shows the observation of waves at 00 GME of 18 February.

It trun out that the forecast results and the observations are in reasonable agreement except that the intensity of forecasts is slightly weak.

The forecasting area shown above will be further expanded southward up to the Equator to cover the whole South China Sea and the forecasting range will be extended from 2 − 3 days to 4 − 5 days in the near future.

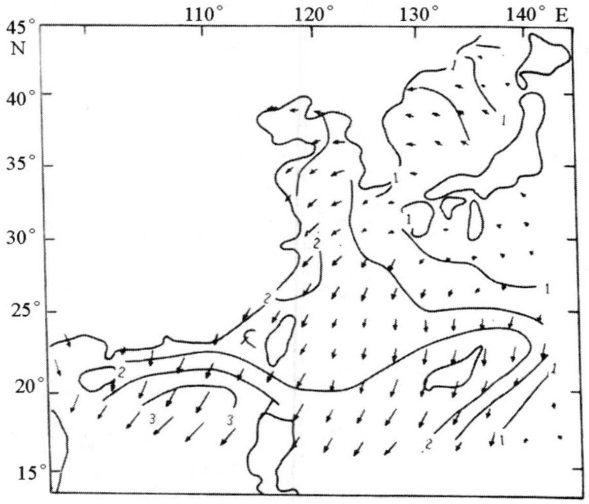

Fig.2　The numerical forecasting wave field for 48-h on 1 Feb.1989

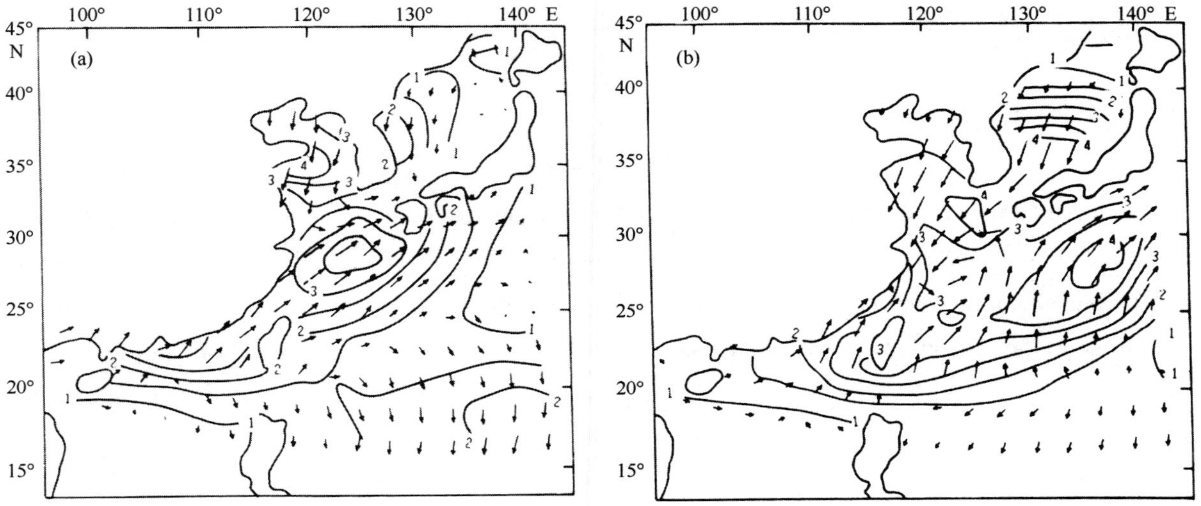

Fig.3　The numerical forecasting wave fields for 36-h on 15(a) and 16(b) Feb.1989

Storm surge forecasting

A two-dimensional nolinear storm surge model has been used for forecasting surges along the coast of China.

The water is assumed to be homogenous and incompressible.The prediction equations for the elevation of sea surface and depth-mean current are

$$\frac{\partial \vec{V}}{\partial t} = -(\vec{V}\cdot\nabla)\vec{V} - f\vec{K}\times\vec{V} - g\nabla\zeta - \frac{1}{\rho}\nabla P_a + \frac{\vec{\tau}_a}{\rho_D} - \frac{\vec{\tau}_b}{\rho_D} \qquad (2)$$

$$\partial\frac{\zeta}{\partial}t = -\nabla\cdot(D\vec{V})$$

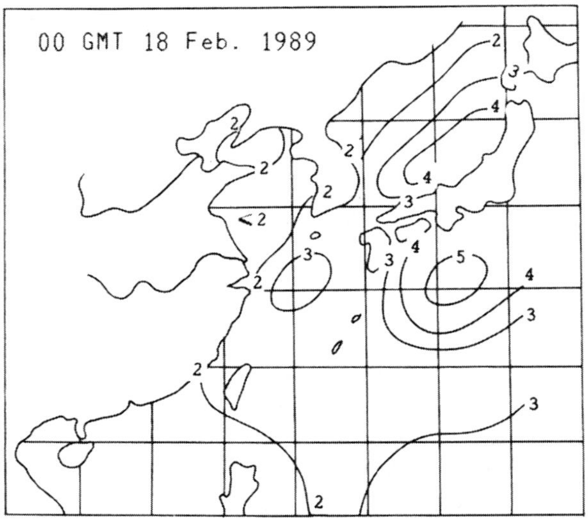

Fig.4 The observed wave fields at 00 GMT on 18 Feb.1989

where

$$\vec{V} = \frac{1}{D}\int_{-h}^{\zeta} \vec{V} dz$$

$$D = \zeta + h$$

ζ is the deviation of the elevation of the sea surface from equilibrium state, h, the undisturbed water depth which is taken from the charted depth, the density of sea water, P the atmospheric pressure on the sea surface, the bottom stress which is supposed to be proportional to the quadratic value of depth-mean current. The wind stress on the sea surface which is expressed as:

$$\vec{\tau}_A = \rho_A C_D \mid \vec{V}_a \mid \vec{V}_a$$

where ρ_s is the density of air on the sea surface and C_D the surface drag coefficient which is taken as 2. 6× 10^{-3}. V_a is the wind vector over the sea.

The computational domain covers the South China Sea, the East China Sea, the Yellow sea and the Bohai sea. The Arakawa's B-type grid with the interval of 0.125° in latitude and in longitude and the split explicit scheme with the time-step of 2 mins and 12 mins are used in the model. According to the idealized typhoon structure, the atmospheric pressure and wind field $V.$ are estimatied by using typhoon parameters including center position, intensity, maximun wind speed and its radius. The typhoon positions are forecasted by using the typhoon prediction system including a moving nested barotropic typhoon model and an analog scheme, other parameters are obtained from the empirical predictions.

During the trial forecasting, serious disaster was caused by the storm surges of Typhoon 8923 along the coast of Zhejiang Province. The peak surge met the highest tide just when Typhoon 8923 landed near the tidal station of Zhejiang Province Haimen on September 15, 1989. The highest tidal level rose to the level of 689 cm and exceeded the warning level by 148 cm. The dashed line in Fig.5 shows the predicted track of Typhoon 8923. The forecast fields of surge elevation and the depth-mean currents are obtained from the storm surge model, as shown in Fig.5. The peak of 122 cm given by the unmerical predication is quite close to the observed peak at the Station Haimen except a phase leg, as shown in Fig.6, for a slower moving speed of the predicted typhoon than that of the observed. But the predicted peak by the empirical-statistical methods is not over 50 cm. The predicted elevation conforms with that observed at the Station Zhenhai (Fig.6).

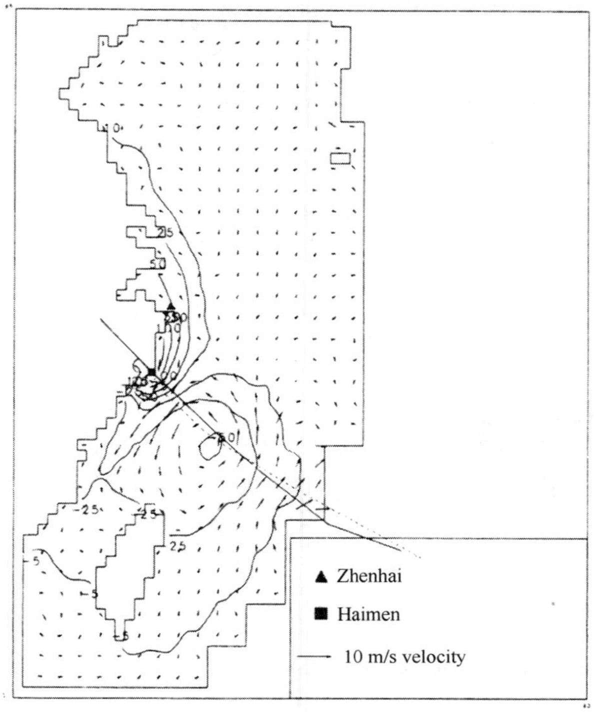

Fig.5　The predicted track of Typhoon 8923, the surge elevation and the depth-mean currents

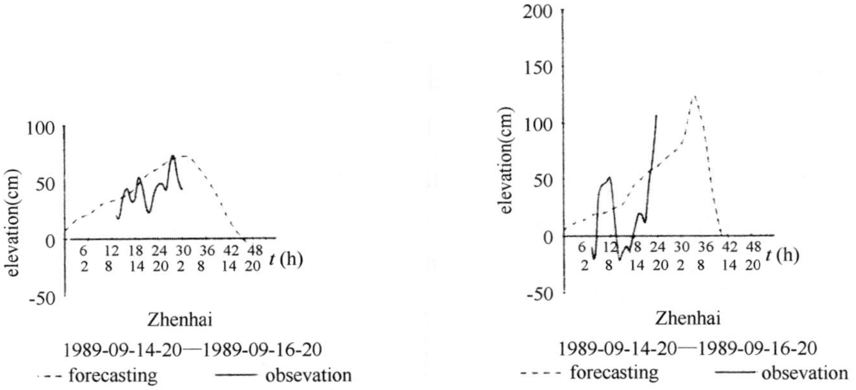

Fig.6　Comparison between the predicted surge elevations of Typhoon 8923 at grid points (dashed lines) and the observed at the tidal stations bear by (solid lines). The observed data were suspended at the station Haimen for the tide gauge was broken down

A hindcasting case is given in order to check the forecasting capability for severe storm surges. Typhoon 8007 hit the coast of Guangdong Province and the sea encroached upon the land of Leizhou Peninsula in 1980. Fig.7 shows the 24 h forecasting fields of surge elevation and depth-mean current in the region of the South China Sea. A sequence of forecasting charts clearly shows the surge development induced by Typhoon

8007.From Fig.7 it is obvious that there are strengthened currents westward along the coast of Guangdong Province which then trun southward.It causes sea surface elevation to rise sharply along the east coast of Leizhou Peninsula during 22 July 1980. Fig.8 shows that the forecasting and simulating elevation at the grid points agree closely with the observation at the tidal stations near by. The observed data were suspended at the station Zhanjiang for the tide gauge was broken down at 09 GMT (01 BLT) 22 July.The star sign notes the visual estimation from the mark of sea water on the wall of coastal buildings.The predicted peak valused and their appearance time are in conformity with the observed information.

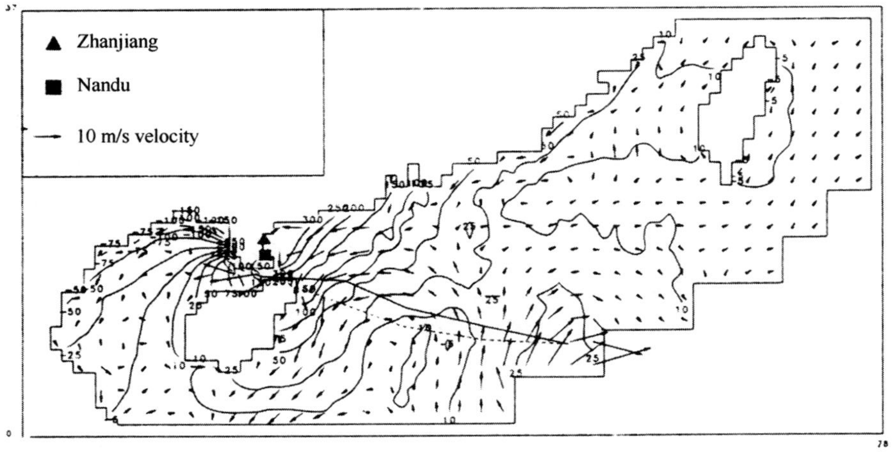

Fig.7 The predicted track of Typhoon 8007, the surge elevation and the depth-mean currents

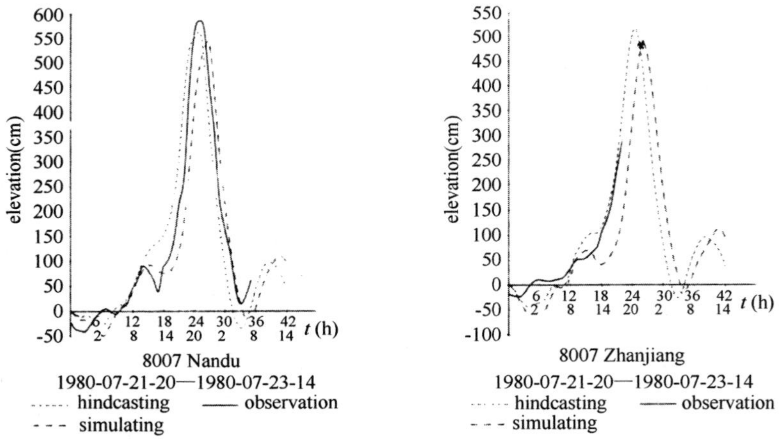

Fig.8 Comparison between forecasting and simulating (dashed lines) elevations of Typhoon 8007 at grid points and observations at the tidal stations near by (solid lines)

A new numerical typhoon forecast system in which a baroclinic typhoon prediction model, the barotropic typhoon motion prediction model, the analog scheme and specialist system are included, has been proposed and will be connected with the storm surge model. A variationa technique is used for improving the forecasts of the typhoon wind-pressure fields.

The case of atmospheric cold-wave: A 5−level numerical weather prediction model with a boundary layer model is connected with the surge model to forecast the water elevations and currents caused by the cold-wave and/or the extratropical cyclone during spring-winter. The cold high moved near the Bohai Sea and caused the water level to rise rapidly in the Laizhou Bay of the Bohai Sea during 15-16 Feb., 1989, as shown in Fig.9. The dashed line is the forecasting elevation at the Station Xiaying. The numerical forecasting value quite agrees with the observation. It is shown clearly from the numerical weather prediction results that the deviation after 14hrs is related to that the forecasting moving speed of the high is slower than the realone. Fig.10 shows the forecasting fields of water elevation and currents caused by the cold-wave.

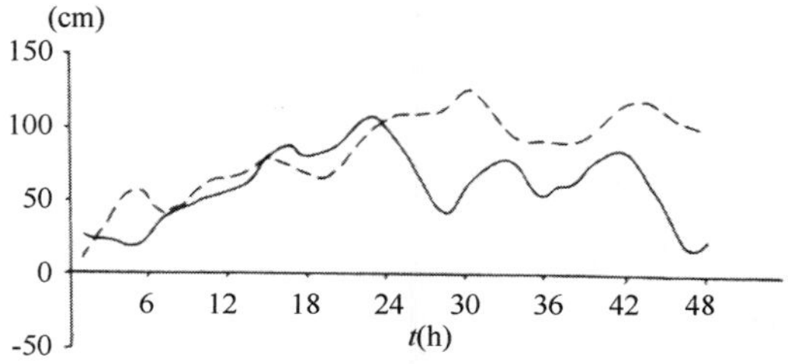

Fig.9　Comparison of the forecasting elevations (dashed line) with the observation (solid line) at the Station Xiaying

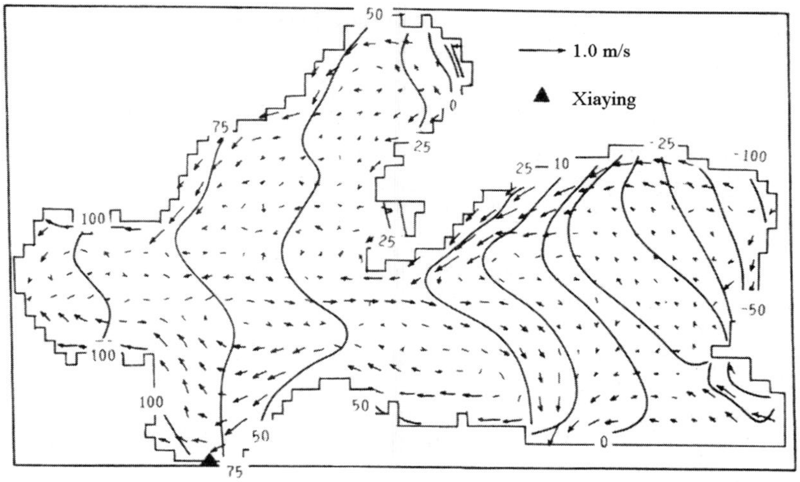

Fig.10　The 36-h forecasting fields of elevations and currents in the Bohai Sea and the North Yellow Sea, at 12 GMT 15 Feb.1989

Verification procedures have been used to check the forecasting system and make objective assessment of the models. Standard deviation of error, weighted root-mean square error by surge peak and absolute correlation between forecasting and observational elevations at tidal stations under the influence of storm surges were estimated and they are, respectively, 30 cm, 37 cm and 65% on the average for the testing forecasts in 1989.

Remark

In brief, because this routine numerical forecasting system was established not long ago, there are still many problems need to be solved, the most difficult one is short of data both atmospheric and oceanic on the sea, it is still hard to improve the reality of the initial field even a good objective analysis was used. Consequently, it is extremely necessary to improve the density and accuracy of oceanic data, especially to allocate more buoys and make further advances on the oceanic statellite remote sensing system so as to get the data for the four-dimensional as simulation.

Apart from providing the numerical forecasting products to the operational forecasting group as a reference to the routine forecasts by conventional methods, now we are going to issue them to users directly in several ways.

Acknowledgements

The authors would like to thank Prof. Yi Zengxin, Mr. Chen Bin, Mr. Du Jianbin, and Mr. Zhang Zhanhai for their help in completing this paper.

References

The WAMDI Group. 1988. The WAM Model—a third generation ocean wave prediction model. *J. Phys. Oceanogr*, 1988, 10.

Wu Huiding and Ji Xiaoyang. 1990. Numerical prediction of typhoon surges. *Physics of Shallow Seas*. Edited by Wang Huatong et al. Beijing: China Ocean Press, 158–166.

Wu Huiding, Yang Shiying. A split explicit scheme for numerical storm surge prediction, Storm Surge: Observations and Modelling. Edited by Chao Jiping et al. Beijing: China Ocean Press (in press)

Discussion

Question: Is the entire coast of China, so charted with respect to bottom topography and details so that you are able to forecast storm surges along the entire coast when a typhoon comes? As you konw it takes a fare amount of information, well in advance, to forecast a storm surge when a typhoon arrives. You have to have the bottom topography in your models, you have to have the details of the coast in your models, etc. Do you have all of this for the entire coast of China now?

Answer: The typhoon wind force is very difficult ot predict for us, so right now we only predict the track and position of typhoons, not the wind force. The wind force and the atmospheric pressure we get that by a theoretical structure model–it is very difficult to do that. But in the next thirty years...no. Right now, we have five layers a model. We want to predict all the typhoon structure but it is not easy to do that. In particular, data is short for the China Sea. So I hope that the developing countries establish some system of oceanic supplies.

Prof. Manule M. Murillo

Are there any more questions for Dr. Chao? No. I would like to thank Dr. Chao for his presentation. Equally I would like to take this opportunity to also thank Drs. Mooers and Sakshaug for their presentations today winthin the field of the Anton Bruun Lectures.

第2部分　中、小尺度运动

层结大气中热对流发展的一个非线性分析*

巢纪平

(中国科学院地球物理研究所)

提要:在本文中应用了由多罗德尼秦(Дородницын)首先提出的方法,解出了描写热对流发展的联立非线性偏微分方程组。并对6种不同的大气不稳定层结的分布,计算了热对流发展的动力过程。计算结果指出,热对流的发展是极其迅速的,在不到20分钟的时间内便能发展成熟。对于发展成熟的热对流,气流在低层是辐合的,在上层是辐散的,并且在高层有补偿的下沉气流出现。文中对下沉气流出现的机制做了分析。最后,应用这一理论计算了积云的发展,得到了积云发展初期的一些有意义的结果。

1 引言

大气中热对流的形成和发展与气团内部积云的生长有密切的关系。众所周知,积云的发展这是小尺度动力气象学和云雾宏观物理学中的一个极为重要的问题。

然而,由于描写热对流发展的微分方程是非线性的,再加上凝结热的释放增加了新的能源,使理论研究对流的发展变得困难。因此,过去的气象学者们不得不只在一些简单的情况下研究它。例如,最早有所谓"气块法",以后又有J. 皮捷克尼斯(Bjerknes)[1]提出的"薄片法"。自然,由于过于简化,这些方法尚不能得到与现实情况相接近的对流发展的图案。以后,H. 斯托海尔(Stommel)[2]提出了对流发展的"动力吸入"(entrainment)模式,使研究对流的发展向前迈进了一步。这个模式后来有很多作者加以发挥,其中最重要的是最近G. J. 哈尔提纳尔(Haltinar)[3]的工作。"动力吸入"模式虽然接触到了对流发展中的一个很重要的方面,即对流气柱和周围空气间质量的交换。但是考虑的办法是简单化了的,并没有把对流气柱和周围的空气作为一个不可分割的统一体来考虑,而是在对流柱和周围空气之间有一个不连续的过渡。由于这样,在"动力吸入"模式中只能得到对流达到常定时的垂直结构,而不能得到发展了的对流的完整的空间图案。在另一方面Л. Н. 古特曼(Гутман)的工作[4]是值得注意的,他解出了描写对流常定结构的非线性偏微分方程组,从而得到了积云发展中期的物理图像。自然,所有这些工作,讨论的都是对流发展生命史中,中期和后期的结构问题,并没有说明局部对流是如何发展起来的。最近,J. S. 马尔库斯(Malkus)等[5]用高速电子计算机探讨了对流单元发展的演变问题,但是由于计算误差的出现,计算只到最初的7分钟便停止了。

这样,直到目前为止,对于热对流特别是属于它的演变过程的研究尚远远不够,有鉴于此,在本文中我们将研究层结大气中,在地面加热的触发下热对流的发展问题。我们推广由多罗德尼秦首先提出的方法,解出了描写对流生长的联立非线性偏微分方程组,得到了对流发展初期的物理图像,并用此理论讨论了气团内部积云发展初期的动力过程。

* 气象学报,1961,31(3):191-204.

2　微分方程和边界条件

在圆柱坐标中,轴对称的自由对流方程式可以写成[6]:

$$\frac{\partial u}{\partial t}+u\frac{\partial u}{\partial r}+w\frac{\partial u}{\partial z}=-\frac{1}{\bar{\rho}}\frac{\partial p}{\partial r}+k\frac{\partial^2 u}{\partial z^2} \tag{1}$$

$$\frac{\partial w}{\partial t}+u\frac{\partial w}{\partial r}+w\frac{\partial w}{\partial z}=-\frac{1}{\bar{\rho}}\frac{\partial p}{\partial z}+k\frac{\partial^2 w}{\partial z^2}+\beta\theta \tag{2}$$

其中,r 是经向坐标,z 是垂直坐标;u 和 w 是分别沿 r 和 z 轴方向的速度分量;p 和 θ 为空气的扰动压力和扰动温度;$\bar{\rho}$ 为空气的平均密度;$\beta=g/\bar{\theta}$,g 为重力加速度,$\bar{\theta}$ 为平均温度;k 为湍流扩散系数。

连续性方程写成不可压缩的形式:

$$\frac{\partial ur}{\partial r}+\frac{\partial wr}{\partial z}=0 \tag{3}$$

层结大气中的热力学第一定律可以近似地写成:

$$\frac{\partial \theta}{\partial t}+u\frac{\partial \theta}{\partial r}+w\frac{\partial \theta}{\partial z}+\alpha(z)w=k\frac{\partial^2 \theta}{\partial z^2} \tag{4}$$

式中,静力稳定度参数定义是:

$$\alpha(z)=\begin{cases}\gamma_a-\gamma(z) & \\ & \text{当相对湿度} \leqslant 100\% \\ \gamma_b(\bar{\theta},\bar{P})-\gamma(z) & \end{cases} \tag{5}$$

此处 γ_a 为干绝热递减率,γ_b 是湿绝热递减率,而 $\gamma(z)$ 为大气的垂直温度梯度。

关于方程组式(1)至式(4)的近似性所带来的物理误差,在笔者最近的一篇文章中有过详细的讨论,并指出用不可压缩条件式(3)来代替一般的连续性方程时可以滤掉对我们研究的对象而言是“噪音”的声波[7]。

我们研究的是热对流的发展,因而问题的下界条件是:

$$z=0,\ \theta=\theta_0 f(r,t),\ w=0 \tag{6}$$

问题的上界条件为:

$$z\to\infty,\ \theta=0,\ w=0 \tag{7}$$

在式(6)中 $f(r,t)$ 是已知函数。

假定在运动开始时大气是静止的,因此初始条件写成:

$$t=0,\ u=w=\theta=p=0 \tag{8}$$

最后,侧向边界条件考虑到轴对称后,自然应为:

$$r=0,\ u=\frac{\partial w}{\partial r}=\frac{\partial \theta}{\partial r}=\frac{\partial p}{\partial r}=0 \tag{9}$$

以及

$$r^2\to\infty,\ w=\theta=p=0 \tag{10}$$

我们将上面的方程和条件写成无因次的形成,引进

$$t=\tau t',\ r=Lr',\ z=Hz',\ u=Vu',\ w=Ww',\ \theta'=\Delta\Theta\theta',\ p=\Delta Pp'$$

此处 τ 是现象的特征时间;L 和 H 为现象的水平和垂直的特征尺度;V 和 W 为特征水平速度和特征垂直速度;$\Delta\Theta$,ΔP 分别为现象的特征温度扰动量和特征气压扰动量。

对于积云这样的小尺度对流现象，一般比值 $H/L \sim 1$，即现象的垂直尺度和水平尺度具有同样的量级。根据连续性方程自然有 $W/V \sim 1$，即现象的垂直特征速度和水平特征速度也具有同一量级。由此，一般所谓的“长波”近似法不能应用，即必须在垂直运动方程中保留加速度项，而不能用静力平衡条件来代替[7]。

我们得到方程的无因次形式为：

$$T^* \frac{\partial u'}{\partial t'} + u' \frac{\partial u'}{\partial r'} + w' \frac{\partial u'}{\partial z'} = -E^* \frac{\partial p'}{\partial r'} + \frac{1}{R^*} \frac{\partial^2 u'}{\partial z'^2} \tag{11}$$

$$T^* \frac{\partial w'}{\partial t'} + u' \frac{\partial w'}{\partial r'} + w' \frac{\partial w'}{\partial z'} = -E^* \frac{\partial p'}{\partial z'} + \frac{1}{R^*} \frac{\partial^2 w'}{\partial z'^2} + \frac{G^*}{r^{*2}} \theta' \tag{12}$$

$$\frac{\partial u' r'}{\partial r'} + \frac{\partial w' r'}{\partial z'} = 0 \tag{13}$$

$$T^* \frac{\partial \theta'}{\partial t'} + u' \frac{\partial \theta'}{\partial r'} + w' \frac{\partial \theta'}{\partial z'} + \frac{F^* G^*}{R^{*2}} \alpha' w' = \frac{1}{R^*} \frac{\partial^2 \theta'}{\partial z'^2} \tag{14}$$

其中各无因次参量群定义成：

$$T^* = \frac{L}{\tau V},\ R^* = \frac{LV}{k},\ G^* = \frac{\beta \Delta \Theta L^3}{k^2},\ E^* = \frac{\Delta P}{\rho V^2},\ F^* = \frac{\alpha V^2}{\beta \Delta \Theta^2}$$

由于考虑的热对流问题，因此将 $-E^* \frac{\partial p'}{\partial z'}$ 与 $\frac{G^*}{R^{*2}} \theta'$ 比较而允许略掉前者。自然，在像锋面积云这样的问题中，可以预料垂直气压梯度力将会有重要的贡献，因而在处理锋面积云时这一项看来是不能省略的。

另外，我们将函数 $\alpha'(z)$ 展成：

$$\alpha'(z') = \alpha'_0 + \alpha'_1 z' + \alpha'_2 z'^2 + \cdots \tag{15}$$

于是，方程(12)和方程(14)变为：

$$T^* \frac{\partial w'}{\partial t'} + u' \frac{\partial w'}{\partial r'} + w' \frac{\partial w'}{\partial z'} = \frac{1}{R^*} \frac{\partial^2 w'}{\partial z'^2} + \frac{G^*}{R^{*2}} \theta' \tag{12'}$$

$$T^* \frac{\partial \theta'}{\partial t'} + u' \frac{\partial \theta'}{\partial r'} + w' \frac{\partial \theta'}{\partial z'} + \frac{F^* G^*}{R^{*2}} (\alpha'_0 + \alpha'_1 z' + \alpha'_2 z' + \cdots) = \frac{1}{R^*} \frac{\partial^2 \theta'}{\partial z'^2} \tag{14'}$$

相应的无因次边界条件和初始条件为：

$$z' = 0,\ w' = 0,\ \theta' = f(r', t') \tag{16}$$

$$z' \to \infty,\ w' = 0,\ \theta' = 0 \tag{17}$$

$$r' = 0,\ u' = \frac{\partial w'}{\partial r'} = \frac{\partial \theta'}{\partial r'} = \frac{\partial p'}{\partial r'} = 0 \tag{18}$$

$$r^2 \to \infty,\ w' = \theta' = p' = 0 \tag{19}$$

$$t' = 0,\ u' = w' = \theta' = p' = 0 \tag{20}$$

方程(11)，方程(12)，方程(13)和方程(14′)构成了对变量 u', w', p', θ' 的闭合方程。在条件式(16)至式(20)下可以定解。

3 问题的解答

现在引进变换：

$$s=(R^*T^*)^{\frac{1}{2}}z'/2\sqrt{t'} \tag{21}$$

并在我们的情况下，根据多罗德尼秦的方法（见参考文献[8]），求下面形式的解：

$$\begin{aligned}\theta'&=\sum_{n=0}^{\infty}\theta_n(r',z')t'^{\frac{n+2}{2}},\ u'=\sum_{n=0}^{\infty}u_n(r',z')t'^{\frac{n+3}{2}}\\ w'&=\sum_{n=0}^{\infty}w_n(r',z')t'^{\frac{n+4}{2}},\ p'=\sum_{n=0}^{\infty}p_n(r',z')t'^{\frac{n+1}{2}}\end{aligned} \tag{22}$$

将加热函数也展开成 t' 的幂级数：

$$f=\sum_{n=0}^{\infty}f_n(r')t'^{\frac{n+2}{2}} \tag{23}$$

考虑式(21)后，将式(22)代入方程(11)，方程(12′)，方程(13)和方程(14′)，我们便得到对于 Θ_n, w_n, u_n 和 p_n 的微分方程为：

$$\begin{aligned}\Lambda_{n+2}(\theta_n)=&\frac{4F^*G^*\alpha'_0}{R^{*2}T^*}w_{n-4}+\frac{8F^*G^*\alpha'_1}{R^*(R^*T^*)^{1/2}}sw_{n-5}+\frac{4}{T^*}\left(u_0\frac{\partial\theta_{n-5}}{\partial r}+\cdots+u_{n-5}\frac{\partial\theta_0}{\partial r}\right)\\&+\frac{2(R^*T^*)^{1/2}}{T^*}\left(w_0\frac{\partial\theta_{n-5}}{\partial s}+\cdots+w_{n-5}\frac{\partial\theta_0}{\partial s}\right)+\frac{16F^*G^*\alpha'_2}{R^*}s^2w_{n-6}+\cdots\end{aligned} \tag{24}$$

$$\begin{aligned}\Lambda_{n+4}(w_n)=&-\frac{4G^*}{R^*T^*}\theta_n+\frac{4}{T^*}\left(u_0\frac{\partial w_{n-5}}{\partial r}+\cdots+u_{n-5}\frac{\partial w_0}{\partial r}\right)\\&+\frac{2(R^*T^*)^{1/2}}{T^*}\left(w_0\frac{\partial w_{n-5}}{\partial s}+\cdots+w_{n-5}\frac{\partial w_0}{\partial s}\right)\end{aligned} \tag{25}$$

$$\frac{\partial(ru_n)}{\partial r}=-\frac{(R^*T^*)^{1/2}}{2}\frac{\partial(rw_n)}{\partial s} \tag{26}$$

$$\begin{aligned}\frac{\partial p_n}{\partial r}=&\frac{T^*}{4E^*}\Lambda_{n+3}(u_n)-\frac{1}{E^*}\left(u_0\frac{\partial u_{n-5}}{\partial r}+\cdots+u_{n-5}\frac{\partial u_0}{\partial r}\right)\\&-\frac{(R^*T^*)^{1/2}}{2E^*}\left(w_0\frac{\partial u_{n-5}}{\partial s}+\cdots+w_{n-5}\frac{\partial u_0}{\partial s}\right)\end{aligned} \tag{27}$$

其中，线性运算子 Λ_m 为：

$$\Lambda_m=\frac{\partial^2}{\partial s^2}+2s\frac{\partial}{\partial s}-2m$$

同样，边界条件为：

$$s=0,\ \theta_n=f_n(r),\ w_n=0 \tag{28}$$

$$s\to\infty,\ \theta_n=w_n=0 \tag{29}$$

$$r=0,\ u_n=\frac{\partial\theta_n}{\partial r}=\frac{\partial w_n}{\partial r}=\frac{\partial p_n}{\partial r}=0 \tag{30}$$

$$r^2\to\infty,\ \theta_n=w_n=p_n=0 \tag{31}$$

我们注意到，当把解写成式(22)的形式时，初始条件已经自然满足。在以上的方程中，为了简便起见，我们略去了各量右上角的撇号。

方程组式(24)至式(27)有一个很重要的特点,当$n<5$时,方程是线性的;当$n\geqslant 5$时,方程的右端虽然出现了非线性项,但这时这些非线性项已经可以用$n<5$的项来表示,也即这些非线性项是已知的。换言之,当$n\geqslant 5$时,方程仍然是线性的,但不是齐次的。这样,对于所有的n方程和条件都是线性的,因此不难求解。

我们注意到方程(24),当$n<4$时具有下面的形式:

$$\Lambda_m(y_n)=\frac{\partial^2 y_n}{\partial s^2}+2s\frac{\partial y_n}{\partial s}-2my_n=0 \tag{32}$$

这个方程的一般解为[9]:

$$y_n=c_1(r)H_m(s)+c_2(r)L_m(s) \tag{33}$$

其中

$$H_m(s)=\frac{1}{i^m}h_m(is),$$

$$L_m(s)=A_m\int_\infty^s\left[\iiint_{(m)}\cdots\int_\infty^s \mathrm{e}^{-s'^2}(\mathrm{d}s')^m\right]\mathrm{d}s'=\frac{A_m}{m!}\int_\infty^s(s-s')^m\mathrm{e}^{-s'^2}\mathrm{d}s'^{①}$$

此处$h_m(s)$是切比雪夫-黑米特(Tehebycheff-Hemite)多项式,$A_m=2m\,A_{m-2}$,$A_0=-\frac{2}{\sqrt{\pi}}$,$A_1=2$,并且

$$L_m(0)=1,\ L_m(\infty)=0 \tag{34}$$

很容易证明:

$$\int_\infty^s L_m(s')\mathrm{d}s'=\frac{A_m}{A_{m+1}}L_{m+1}(s),\ sL_m(s)=\frac{A_m}{A_{m+1}}(m+1)(L_{m+1}(s)-L_{m-1}(s)) \tag{35}$$

由于函数$H_m(s)$,当$s\to\infty$时趋于无限,因此由θ_n的上界条件c_1必须为零。$c_2(r)$则可以用下界条件式(28)决定。这样,我们求得式(24)的解为:

$$\theta_n(r,s)=f_n(r)L_{n+2}(s)(\text{当}\ n<4) \tag{36}$$

当$n\leqslant 9$时,非齐次方程(24)和方程(25)取下面的形式:

$$\Lambda_m(y_n)=\sum_{k=0}^{l}\alpha_k L_k+\sum_{ij}b_{ij}L_iL_j \tag{37}$$

为了解出方程(37),考虑到函数L_m是方程(32)的一个解,不难引进下面的补助公式:

$$a_kL_k=\Lambda_m\left(\frac{a_kL_k}{2(k-m)}\right),$$

及

$$b_{ij}L_iL_j=\Lambda_m\left\{\frac{b_{ij}}{2}\left[\frac{A_iA_j}{A_{i+1}A_{j+1}}L_{i+1}L_{j+1}\right.\right.$$
$$\left.\left.+\sum_{k=2}^{r}\alpha^{k-1}(r-1)(r-2)\cdots(r-k+1)\frac{A_iA_j}{A_{i+k}A_{j+k}}L_{i+k}L_{j+k}\right]\right\}$$

其中,$2r=m-(i+j)$。

应用这两个补助式,我们不难求得方程(24)和方程(25),当$n\leqslant 9$时满足条件式(28),式(29)的解为:

---

① 后面这一等式成立,是考虑到条件$L_m(\infty)=0$,将函数$L_m(s)$在∞处用泰勒公式展开的结果。

$$\theta_n = f_n L_{n+2} + \sum_{k=0}^{l} \frac{a_k}{2(k-m)}(L_k - L_m) + \sum_{ij} \frac{b_{ij}}{2}\left\{\frac{A_i A_j}{A_{i+1}A_{j+1}} \cdot (L_{i+1}L_{j+1} - L_m) + \sum_{k=2}^{r} 2^{k-1}(r-1)(r-2)\cdots(r-k+1) \times \frac{A_i A_j}{A_{i+k}A_{j+k}}(L_{i+k}L_{j+k} - L_m)\right\} \tag{38}$$

$$w_n = \sum_{k=0}^{l} \frac{a_k}{2(k-m)}(L_k - L_m) + \sum_{ij} \frac{b_{ij}}{2}\left\{\frac{A_i A_j}{A_{i+1}A_{j+1}} \cdot (L_{i+1}L_{j+1} - L_m) + \sum_{k=2}^{r} 2^{k-1}(r-1)(r-2)\cdots(r-k+1) \times \frac{A_i A_j}{A_{i+k}A_{j+k}}(L_{i+k}L_{j+k} - L_m)\right\} \tag{39}$$

当 w_n 决定后，u_n 可以从式(26)和 式(30)求出为：

$$u_n = -\frac{(R^* T^*)^{1/2}}{2} \frac{1}{r} \frac{\partial}{\partial s}\int_0^r w_n r \mathrm{d}r \tag{40}$$

同样的，应用条件式(31)积分式(27)后，求得 p_n 为：

$$p_n = \int_o^r \left\{\frac{T^*}{4E^*}(\Lambda_{n+3}(u_n)) - \frac{1}{E^*}\left(u_0 \frac{\partial u_{n-5}}{\partial r} + \cdots + u_{n-5}\frac{\partial u_0}{\partial r}\right) - \frac{(R^* T^*)^{1/2}}{2E^*}\left(w_0 \frac{\partial u_{n-5}}{\partial s} + \cdots + w_{n-5}\frac{\partial u_0}{\partial s}\right)\right\} \mathrm{d}r \tag{41}$$

当 $n>9$ 时，求问题的解答并无原则上的困难，不过如果我们选择的特征时间 $\tau'<1$（例如不大于 0.9）则解答式(22)已经可以用 9 项很好地表示出。因此当 $n>9$ 时方程(24)至方程(27)的解在此未给出。

4 数值计算和结果讨论

我们的目的在于讨论对流发展初期的一些特征，因此将地面加热函数取成①：

$$\theta_{z=0} = 2t'e^{-\tau'^2}$$

各无因次参量群取成：

$$T^* = 0.840,\ R^* = 0.200,\ G^* = 0.667,\ F^* = 0.120$$

或相当于取各特征量及参数为：

$$t = 1.2 \times 10^3\ \mathrm{s},\ \Delta\Theta = 0.5℃,\ V = 1\ \mathrm{m/s},\ L = 10^3\ \mathrm{m},$$

$$\alpha_0 = 10^{-3}℃/\mathrm{m},\ k = 5 \times 10^3\ \mathrm{m^2/s}$$

由于在下面所取的大气层结是不稳定的，在不稳定大气中的湍流扩散系数可以设想要比一般来得大，因此在这里我们将 k 值取大了一个量级。不过，看来这个数值可能偏大了一些。

例 1，我们将层结参数 α 取成 $\alpha(z)=-1-3z+z^2$。这样的层结分布相当于大气在 3 km 以下是不稳定的，最大的不稳定度出现在 1.5 km 附近，在 3 km 左右以上一直到 5 km 左右大气是稳定的，再往上则有一个逆温层（图 1）。

在大气这样的层结分布下，计算了热对流的发展图案。我们注意到在加热开始后 6 分钟时，对流已经发展，这时最大的垂直速度为 7 cm/s，最大的水平速度为 9 cm/s。对流达到的最大高度为 4.5 km。这时的气流是辐合的（图 2）。到 12 分钟时，对流的高度已达到 6.5 km，并且最大的垂直速

① 相当于在 20 分钟时地面最大的加热温度为 2℃。

度为 0.76 m/s，其中心在 1.5 km 高度附近；最大的水平速度出现在地面，为 1.02 m/s。这时，气流整层仍然是辐合的（图 3 至图 5）。当 18 分钟时（图 6 至图 8），最大的垂直速度已发展到 3.56 m/s，地面的水平速度为 5.45 m/s。在图 6 上可以发现一个很醒目的现象，即这时气流在低层是辐合的，在中层是辐散的，而在高层还出现了下沉气流。这是一个很有意思的现象。众所周知，对流云发展旺盛时，顶部幞状云的出现正是说明有下沉气流的存在。

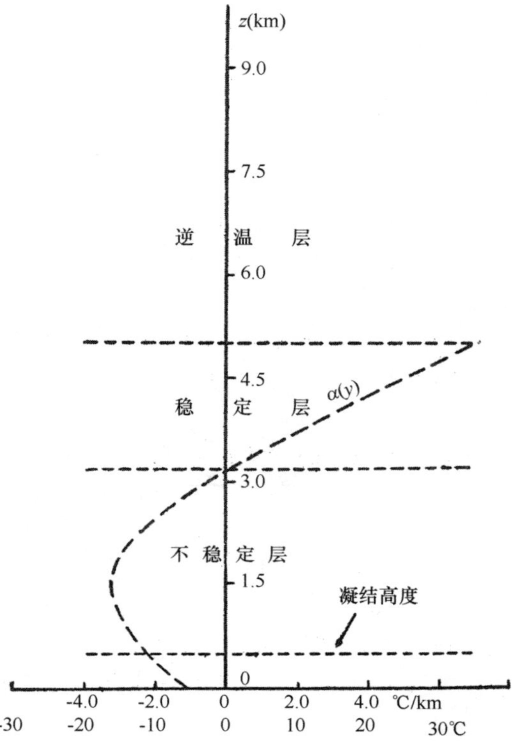

图 1　大气层结分布

$$\alpha(z)=\begin{cases}\gamma_a-\gamma \\ \gamma_b-\gamma\end{cases}，当 f\leqslant 100\% = -1-3z+z^2$$

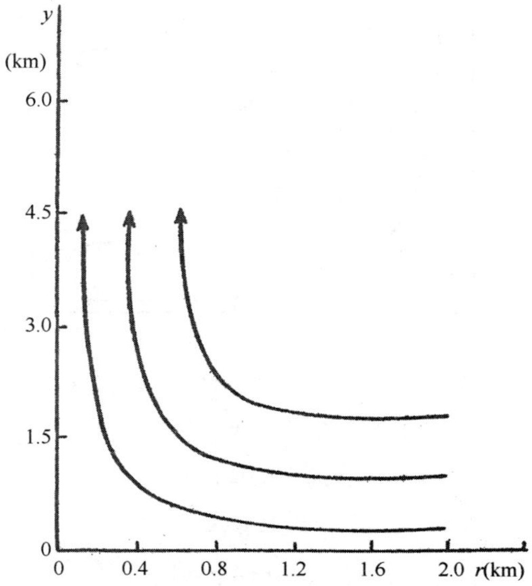

图 2　加热开始后 6 分钟时气流分布

$w_{最大}=0.07$ m/s，$u_{最大}=0.09$ m/s，

$\theta_{最大}=0.19$℃，$M_{最大}=0$ g/m^3

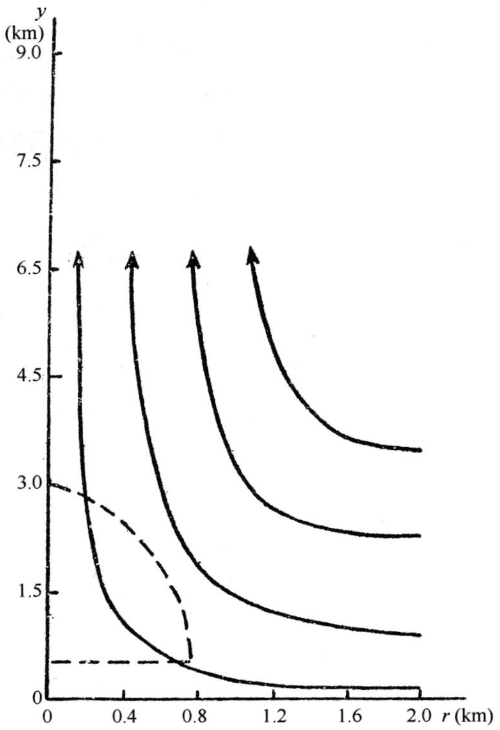

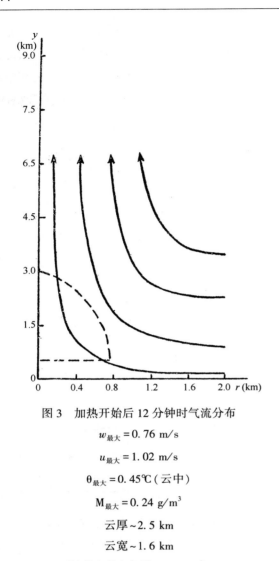

图 3 加热开始后 12 分钟时气流分布

$w_{最大}=0.76\ m/s$

$u_{最大}=1.02\ m/s$

$\theta_{最大}=0.45℃$(云中)

$M_{最大}=0.24\ g/m^3$

云厚~2.5 km

云宽~1.6 km

云侧进入的空气量~$1.5\times10^3\ t/s$

云底进入的空气量~$0.75\times10^3\ t/s$

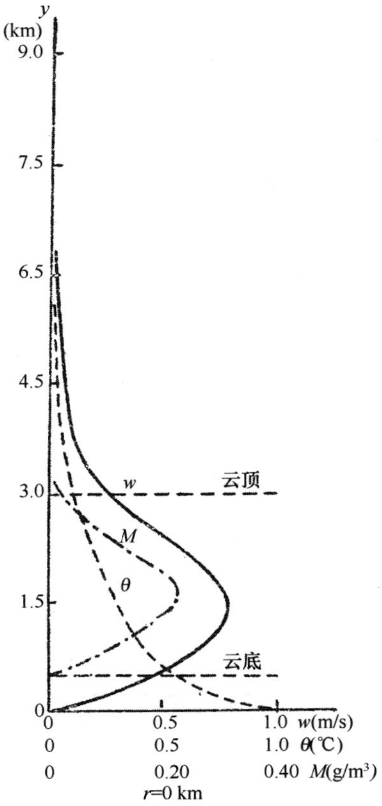

图 4

为什么在顶部会出现下沉气流？一个最直接的解释是由于上层逆温层的存在,阻碍了对流的向上发展,而向两侧辐散,为了补偿辐散气流因而引导了高层下沉气流的发展。然而,事实上下沉气流产生的机制要比这样的解释深刻得多。为此,我们计算了例2(见图9)。这时取大气的层结分布为 $\alpha(z)=-1-3z$,即去掉了逆温层和稳定层,整层大气都是不稳定的。到 18 分钟时对流发展的图案如图 9 所示。我们注意到,除最大的上升速度有微弱的加大外($w_{最大}=3.92\ m/s$),对流气柱的形状仍和例 1 一致。这说明下沉气流的产生与大气原有的逆温层是否存在关系不大。为进一步说明逆温层对下沉气流出现关系不大,我们计算了例3(见图10)。在例 3 中加大了逆温层的强度,将层结参数取成 $\alpha(z)=-1-3z+3z^2$。计算结果见图 10。由图可见,相反的,这时辐散气流和下沉气流都变弱。自然,垂直速度也将变小($w_{最大}=2.70\ m/s$)。在另一方面,如果我们取例 4(见图 11),$\alpha(z)=-3z+z^2$,即使低层的层结变得较为稳定,则对于同样的 18 分钟气流的分布如图 11 所示。我们发现在这种情况下,中层的辐散气流不见了,上层的下沉气流也不见了。同时最大的垂直速度这时也变小到只有 2.44 m/s。如果我们减小及加强大气中层的不稳定度，即取例 5（见图 12），$\alpha=-1-1.5z+z^2$，例 6（见图 13），$\alpha=-1-4.5z+z^2$，则对流发展到 18 分钟时相应的图像可见图 12 和图 13。由图可见，这时下沉气流仍然出现，只不过在例 5 中辐散气流开始的高度要比例 6 来得低。前者从 3 km 附近开始，后者则在 5 km 左右才开始。另外，在例 5 中最大的垂直速度为 $w_{最大}=2.5$ m/s，在例 6 中则为 $w_{最大}=4.62$ m/s。

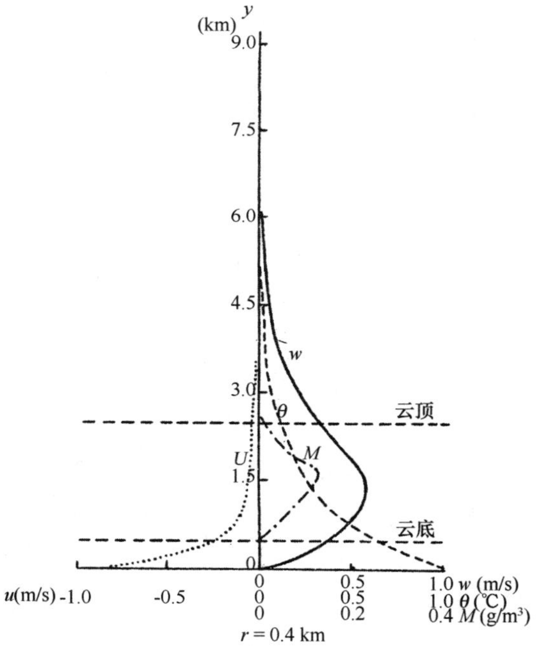

图 5

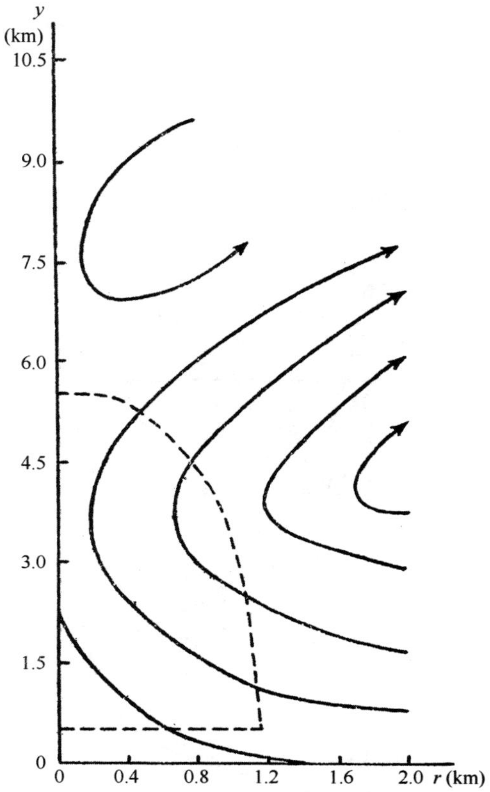

图 6　加热开始后 18 分钟气流分布

$w_{最大}=3.56$ m/s

$u_{最大}=-5.45$ m/s

$\theta_{最大}=1.32$℃（云中）

$M_{最大}=1.22$ g/m^3

云厚～5.0 km

云宽～2.4 km

云侧进入的空气量～25.0×10^3 t/s

云底进入的空气量～7.0×10^3 t/s

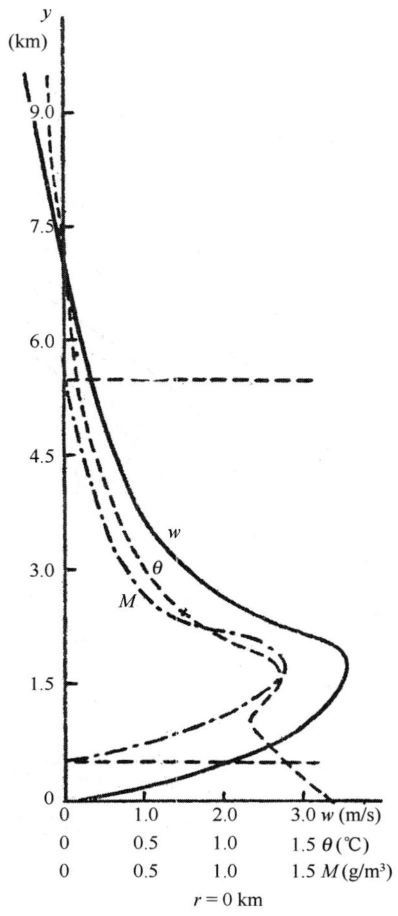
y
(km)
9.0
7.5
6.0
4.5
3.0
1.5
0
w
θ
M
0
1.0
2.0
3.0 w (m/s)
0
0.5
1.0
1.5 θ(℃)
0
0.5
1.0
1.5 M (g/m³)
r = 0 km

图 7

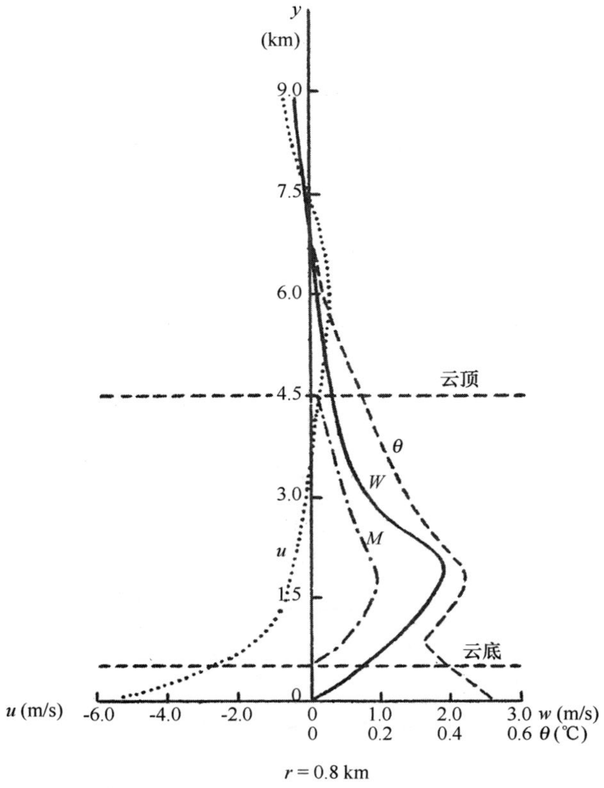
y
(km)
9.0
7.5
6.0
云顶
4.5
θ
W
3.0
M
u
1.5
云底
0
u (m/s)
-6.0
-4.0
-2.0
0
1.0
2.0
3.0 w (m/s)
0
0.2
0.4
0.6 θ(℃)
r = 0.8 km

图 8

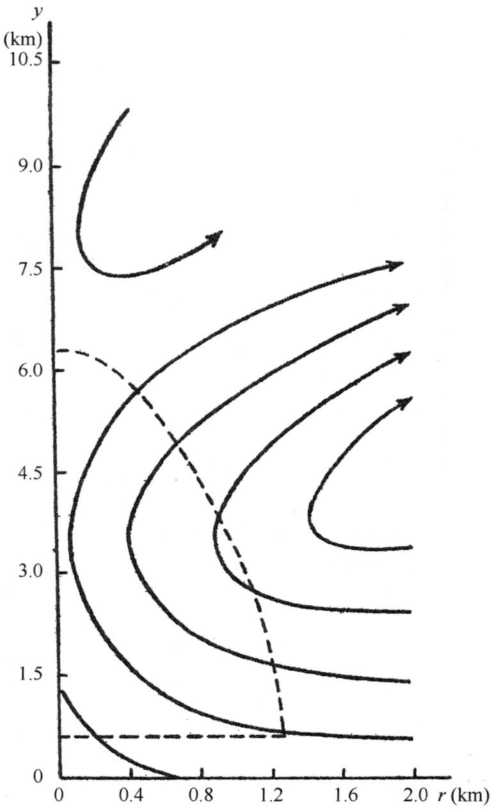

图 9　例 2　18 分钟时气流的分布

[$\alpha(z) = -1-3z$]

$w_{最大} = 3.92$ m/s

$u_{最大} = -5.48$ m/s

$\theta_{最大} = 2.00$℃（云中）

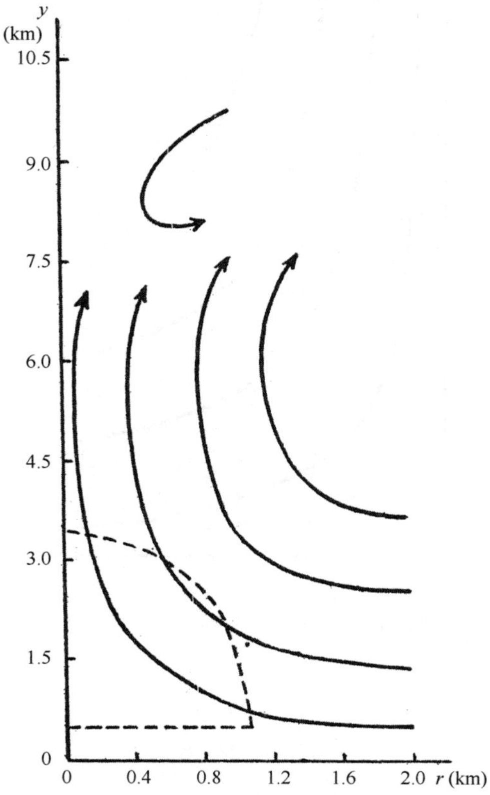

图 10　例 3　18 分钟时气流的分布

[$\alpha(z) = -1-3z+3z^2$]

$w_{最大} = 2.70$ m/s

$u_{最大} = -5.51$ m/s

$\theta_{最大} = 0$℃（云中）

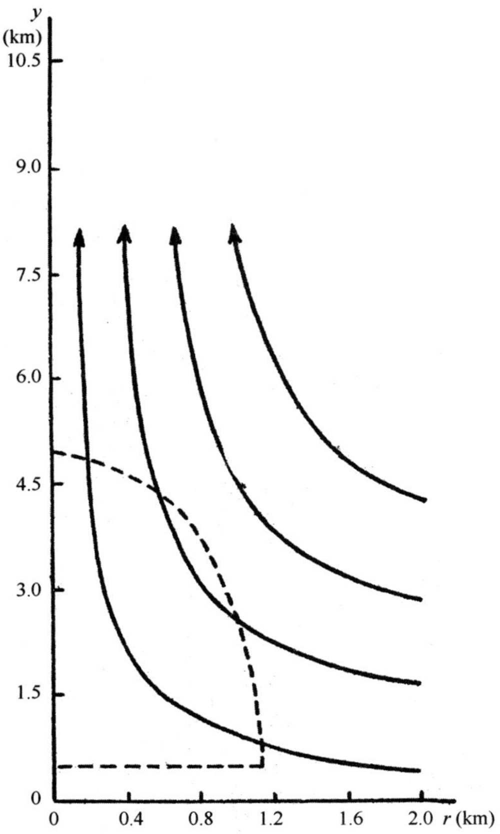

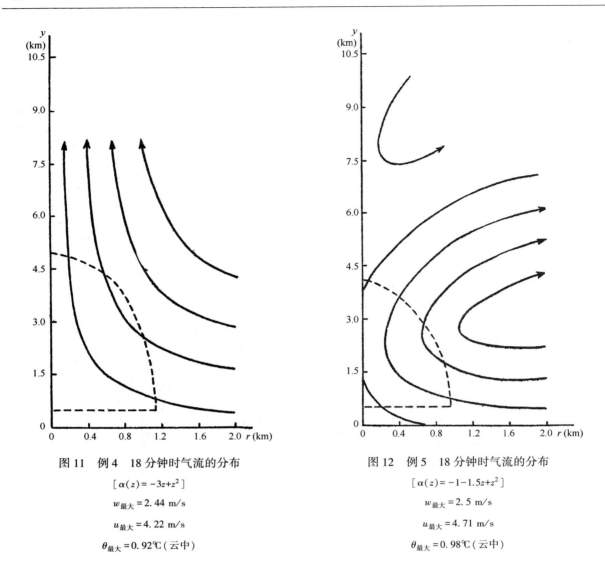

图 11 例 4 18 分钟时气流的分布

[$\alpha(z)=-3z+z^2$]

$w_{最大}=2.44$ m/s

$u_{最大}=4.22$ m/s

$\theta_{最大}=0.92$℃(云中)

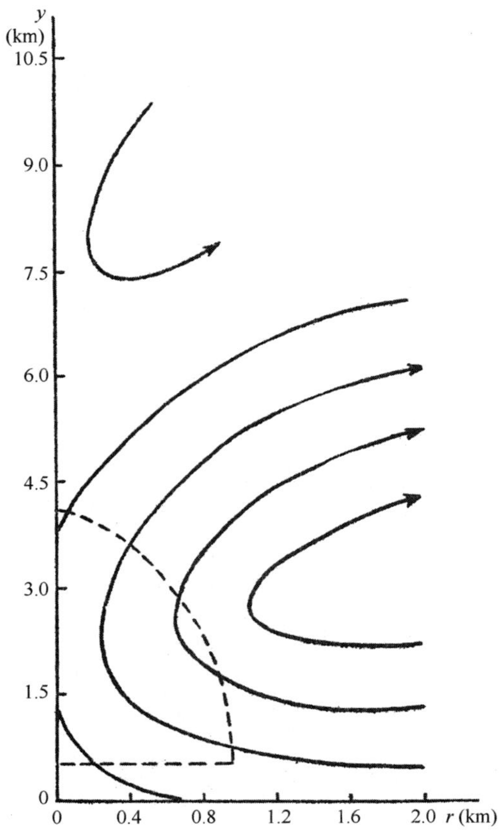

图 12 例 5 18 分钟时气流的分布

[$\alpha(z)=-1-1.5z+z^2$]

$w_{最大}=2.5$ m/s

$u_{最大}=4.71$ m/s

$\theta_{最大}=0.98$℃(云中)

由此可见,在小尺度的热对流中下沉补偿气流的出现,与低层不稳定能量的强弱有很大的关系。这在物理上是很容易理解的,对流从地面开始发展在低层取得较多的不稳定能量后,垂直速度定会有大的增加,由于上升气流这种较大的垂直梯度的存在,周围的空气必然要从四周流进对流气柱中以补偿上升气流。自然,周围较冷的空气流入对流气柱中后,要使得上升气流的速度减小,于是空气的质量在某一高度上会有所"堆积"。为了满足质量的连续性要求在某一高空以上辐散气流自然出现,辐散气流的出现进而又引导了上层补偿性的下沉气流。如果低层的不稳定能量较小,对流气柱中的上升速度的垂直梯度都比较小,中层的辐散气流可以不出现,于是也不会引导出下沉的补偿气流。

我们指出,层结的分布虽然对下沉气流的出现有很大影响,然而对流本身发展的非线性也是一个很重要的因素。我们知道用一个线性模式是算不出下沉补偿气流来的[10]。

以上计算的结果可以综合如下:

(1)对流随时间的发展具有非线性的特征。在加热开始的几分钟后,对流气柱的最大上升速度只有 10 cm/s 的量级,到 10 分钟左右仍然小于 1 m/s,但到 18 分钟时上升速度可以增长到每秒几米的量级。

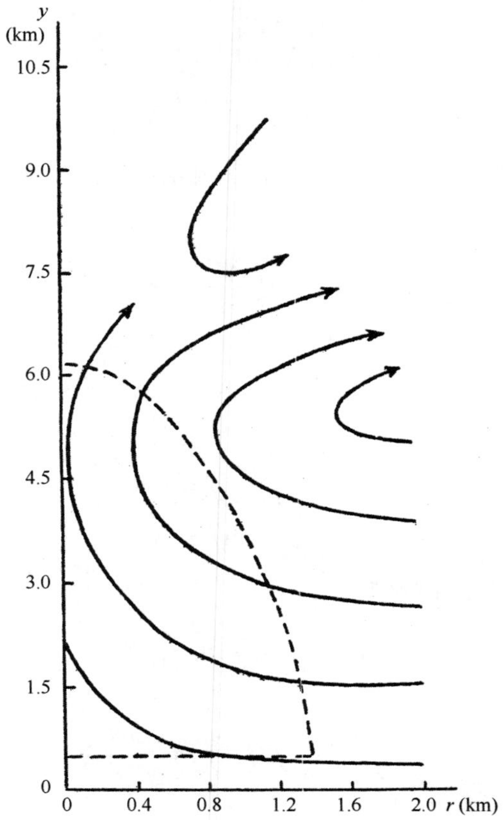

图 13　例 6　18 分钟时气流的分布

[$\alpha(z) = -1-4.5z+z^2$]

$w_{最大} = 4.62$ m/s

$u_{最大} = 6.72$ m/s

$\theta_{最大} = 1.66$℃（云中）

(2) 在热对流发展的初期，气流是整层辐合的，当对流发展到一定时期后，于是在中层出现辐散气流，而在上层则出现补偿性的下沉气流，这时对流发展成熟，进入生命史的中期。

(3) 水平的辐合气流在贴地层最强。

(4) 对流呈一气柱状而不呈“气泡”状。

5　应用到积云的发展

上面所提出的热对流理论，可以作为气团内部积云发展的初步理论。事实上，当对流的速度场和湿绝热梯度知道后，不难求得决定积云中含水量 $M(\mathrm{g/m^3})$ 的方程如下：

$$\frac{\partial M}{\partial t} + u\frac{\partial M}{\partial r} + w\frac{\partial M}{\partial z} + \bar{\rho}\frac{c_p}{L^*}(\gamma_b - \gamma_a)w = k\frac{\partial^2 M}{\partial z^2} \tag{42}$$

$$\gamma_b(\overline{P},\overline{\Theta}) = \gamma_a \frac{\overline{P} + 0.622 \dfrac{L^* E}{R\overline{\Theta}}}{\overline{P} + 0.622 \dfrac{L^*}{c_p} \dfrac{\partial E}{\partial \overline{\Theta}}}$$

假定是已知函数。c_p 为空气的定压比热，L^* 为水汽的凝结潜热，E 是饱和水气压，k 是气体常数。

式(42)的无因次形式为：

$$T^* \frac{\partial M'}{\partial t'} + u' \frac{\partial M'}{\partial r'} + w' \frac{\partial M'}{\partial z'} + Q^* w' = \frac{1}{k^*} \frac{\partial^2 M'}{\partial z'^2} \tag{43}$$

此处 $Q^* = \dfrac{\overline{\rho} c_p(\gamma_b - \gamma_a)L}{\Delta M L^*}$。

在边界条件

$$z \leqslant H_c,\ M' = 0 \tag{44}$$

$$z \to \infty,\ M' = 0 \tag{45}$$

（式中，H_c 为凝结高度）下，式(43)的解为：

$$M' = \sum_{n=0}^{\infty} M_n(r', s') t'^{\frac{n+2}{2}} \tag{46}$$

式中，M_n 求得为：

$$M_n = \sum_{k=0}^{l} \frac{a_k}{2(k-m)}(L_k - L_m) + \sum_{ij} \frac{b_{ij}}{2}\left\{\frac{A_i A_j}{A_{i+1}A_{j+1}}(L_{i+1}L_{j+1} - L_m) + \sum_{k=2}^{r} 2^{k-1}(r-1)(r-2)\cdots(r-k+1)\frac{A_i A_j}{A_{i+k}A_{j+k}}(L_{i+k}L_{j+k} - L_m)\right\} \tag{47}$$

为了计算简单起见，在凝结高度以上参数 $\gamma_b - \gamma_a$ 取为常数，并等于 -4℃/km。凝结高度取在 0.5 km上。这样从式(46)可以决定 M。另外，我们将积云的上界和侧界取在 $M \approx 0.05$ g/m^3 处。积云的界线在图中用虚线表示。

在例 1 中，在加热开始 6 分钟后，云尚未出现（$M \approx 0$），到 12 分钟时积云已经出现，云顶高度在 3 km处。云中最大的含水量为 0.24 g/m^3。到 18 分钟时，云顶已发展到 5 km 高度，云宽为 2.4 km，因此得到的是直展云相当浓积云阶段，在云中最大的含水量为 1.22 g/m^3。含水量最大处出现在云的中部，含水量随高度的分布和垂直速度的分布很接近。在 12 分钟时从云侧进入的空气量为 1.5×10^3 t/s，从云底进入的空气量为 0.75×10^3 t/s。到 18 分钟时，从云侧进入的空气量为 2.5×10^4 t/s，而从云底进入的空气量为 7.0×10^3 t/s。因此，从云侧进入的空气量要比从云底进入的来得多，并且随着积云的发展，从云侧进入的空气愈来愈比从云底进入的多。由此看来，侧向吸入作用是积云发展过程中的一个极重要的方面。

在 18 分钟时，在例 2 中云顶的高度为 6.5 km，例 3 为 3.5 km，例 4 为 4.5 km，例 5 为 4.0 km，例 6 为 6.0 km。由此可见，积云发展的垂直厚度与大气的不稳定度成正比。

6 结语

在这项工作中，虽然初步研究了积云的发展过程，但是由于模式的局限性，对影响积云发展的很多因素，如风速的切变，积云和周围流场相互之间的制约影响，特别是积云发展时，宏观的气流和微观的云滴增长的相互作用都没有加以考虑,因此所得到的结果还是初步的。我们认为,只有考虑了积云发展时宏观和微观的物理过程的相互制约后,才能较好地表现出整个积云发展的生命史,这是我们在下一步工作中需要考虑的。

总之,积云的发展这是一个极为重要的问题,是值得给予多方面注意的。

致谢:作者对赵九章教授、顾震潮和叶笃正两位先生在工作过程中给予的鼓励深表感谢,对参加本文计算工作的同志也表示感谢。

参考文献

[1] Bjerknes, J., Q.J.Roy.*Metes.Soc.*1938, 64: 325-330.

[2] Stommel, H., *J.Metes.*1947, 4: 91-94.

[3] Haltinar, G.J., *Tellus* 1959, 11(1): 4-15.

[4] Гутман.Л.Н., *ДАН СССР*, 1957, 112: 1033-1036.

[5] Malkus, J.S., Witt, G., The Atmosphere and Sea in Motion.1959, 425-429.

[6] 郎道,等.连续介质力学[J].中译本.北京:人民教育出版社,1960.

[7] Ъаев, В, К., *Трубы ЦИП*, 1956, 43(70): 3-18.

[8] Мхитарян, А.Н., *ДАН Армяской СССР*, 1954, 19(2): 33-40.

[9] Haque, M.A., *Q.J.Roy.Meteor.Soc.*1952, 78: 394-406.

论小尺度过程动力学的一些基本问题*

巢纪平

（中国科学院地球物理研究所）

提要：本文对层结大气中小尺度过程动力学的一些基本问题做了研究。

首先用频率法对线性化后的小尺度运动方程做了分析。分析指出，线性化后的运动方程中包含了两种不同类型的波动，即快速传播的声波和相对声波而言的慢波——重力内波。

对于有气象意义的运动而言，声波是一种“噪音”，必须从运动方程中把它滤掉。文中指出，由于静力偏差（非静力平衡）对小尺度运动的发展具有重要的意义，一般用静力平衡来消除声波的方法对小尺度运动并不适用。文中提出了3种不同的消除声波的方法，即①不可压缩条件的应用；②将解按参数 $\varepsilon'(=V^2/gL)$ 的幂次展开；③平衡方程的应用。同时分析了这3种不同滤波方法所带来的误差程度。

作为初值问题，研究了当某一时刻在大气有限区域内的扰动破坏了静力平衡后所激发出的声波的频散性质，指出，从静力平衡的破坏到某种平衡状态的建立，场的这一适应过程完成得极其迅速。

最后，利用所提出的滤波方法，建立了滤掉声波后的二维非线性运动方程组。

1 引言

许多重要的现象，如飑线、对流云、气流过山和龙卷风等，都属于小尺度运动的范畴。近年来，由于观测资料的积累和客观的需要，研究小尺度运动的重要性已日益显著。然而，直到目前为止，对小尺度动力学中某些带有原则性的问题尚缺少系统的研究。

例如，众所周知，原始的欧拉流体力学方程组所包含的往往不是一种性质的运动，在考虑了地球自转和地球自转效应随纬度变化后的动力学方程组中，在未加处理前除包含了低频的行星尺度运动外，还包含了对于行星尺度运动是“噪声”的高频运动——重力波和声波。由于欧拉方程对于波速非常大而振幅很小的这些“噪声” 非常灵敏，因而当在初始记录中存在这类扰动，而在计算过程中计算方法稍为不严格，则计算结果就会大大地受到这类扰动的扰乱，使结果引起相当大的误差。接受L.F.里查逊（Richardson）①的经验教训，近年来大尺度天气数值预报获得较大进展的一个重要方面，即为分清了大尺度运动和作为“噪声” 的扰动的特性，同时将欧拉运动方程做了适当的处理，滤掉了这些“噪声”，这样使上述计算中的困难在一定程度上得到克服。同样性质的问题，可以预料在小尺度运动中也存在。

在本文中，我们将对小尺度运动中“噪声”的性质、滤掉“噪声”的方法以及小尺度系统中流体力学场的适应过程等问题做些探讨；并利用所提出的滤波方法，建立滤掉声波后的非线性方程组，讨论

---

* 气象学报，1962，32(2)：104-118.

① Richardson L.F.Weather Predication by Numerical Process.Cambridge，England，1922.

这些方程组所能够描写的对象。

2　基本运动方程组

在绝热和不考虑湍流扩散的情况下,发生在旋转地球大气中的各种不同尺度的运动,原则上可以用下面形式的欧拉运动方程组来描写

$$\left.\begin{aligned}&\frac{\mathrm{d}u}{\mathrm{d}t}=-\frac{1}{\bar{\rho}}\frac{\partial p'}{\partial x}+fv,\ \frac{\mathrm{d}v}{\mathrm{d}t}=-\frac{1}{\bar{\rho}}\frac{\partial p'}{\partial x}-fu,\\&\frac{\mathrm{d}w}{\mathrm{d}t}=-\frac{1}{\bar{\rho}}\frac{\partial p'}{\partial z}-g\frac{\rho'}{\bar{\rho}},\\&\frac{\partial\rho}{\partial t}=-\left(\frac{\partial\rho u}{\partial x}+\frac{\partial\rho v}{\partial y}+\frac{\partial\rho w}{\partial z}\right),\ \frac{\mathrm{d}p}{\mathrm{d}t}=c^2\frac{\mathrm{d}\rho}{\mathrm{d}t}\end{aligned}\right\}\tag{1}$$

式中符号意义如下:x,y,z 是笛卡尔坐标;t 是时间;u,v,w 为速度分量;ρ,p 为空气的密度和气压,并且 $\rho=\bar{\rho}(z)+\rho'(x,y,z,t)$,$p=\bar{p}(z)+p'(x,y,z,t)$,$\bar{\rho},\bar{p}$ 是大气基本状态的密度和压力,为已知函数,并且满足关系式 $\mathrm{d}\bar{p}/\mathrm{d}z=-\bar{\rho}g$;$g$ 为重力加速度;f 为科里奥利参数;$c^2=kp/\rho$ 为绝热声速的平方,取为常数,$k=c_p/c_v$;c_p,c_v 分别为空气的定压和定容比热;$\frac{\mathrm{d}}{\mathrm{d}t}=\frac{\partial}{\partial t}+u\frac{\partial}{\partial x}+v\frac{\partial}{\partial y}+w\frac{\partial}{\partial z}$。

然而,由于现象的尺度不同,它们的动力学性质就会有很大的不同,因而描写这些现象的方程组也会有所不同。换言之,按照现象的尺度,欧拉原始形式的方程组(1)可以作不同的简化。

如果现象的水平特征尺度为 L,垂直特征尺度为 H,水平特征速度为 V,垂直特征速度为 W,扰动密度的特征值为 $\Delta\rho$,则可以引进下面两个重要的无因次量。

$$\varepsilon=\frac{V}{fL},\ F=\frac{W^2}{gH\cdot\dfrac{\Delta\rho}{\rho}}\tag{2}$$

来表征发生在地球大气中的各种不同尺度运动的动力学特征。这两个数的前者是水平方向的惯性力与科里奥利力之比,称罗斯培(Rossby)数,后者是垂直方向的惯性力与浮力之比,称弗罗德(Froude)数。

对于前述的小尺度系统,观测表明一般有 $L\sim H\sim10^3\sim10^4$ m,$V\sim W\sim10$ m/s,而$\frac{\Delta\rho}{\rho}\sim\frac{\Delta\theta}{\theta}$($\theta$ 为位温)~10^{-3}。由此,对于小尺度运度一般有 $\varepsilon\sim10^{-1}-10^2$,$F\sim10^0$。这表明在研究小尺度运动时可以在水平运动方程中略去因地球自转而产生的科里奥利力,而在垂直运动方程中惯性力必须保留,即不能应用静力平衡近似。由于这样,小尺度现象与大气中其他的天气过程如气旋、反气旋等的动力学性质很不相同。因为在后者的情况下,由于 $\varepsilon\leqslant1$ 及 $F\ll1$,科里奥利力起着重要的作用,而静力平衡条件一般都成立。

当在式(1)中略去了科里奥利力并展开后,便得到描写小尺度过程的原始形式的欧拉运动方程组:

$$
\left.\begin{aligned}
&\bar{\rho}\left(\frac{\partial u'}{\partial t}+u'\frac{\partial u'}{\partial x}+v'\frac{\partial u'}{\partial y}+w'\frac{\partial u'}{\partial z}\right)=-\frac{\partial p'}{\partial x},\\
&\bar{\rho}\left(\frac{\partial v'}{\partial t}+u'\frac{\partial v'}{\partial x}+v'\frac{\partial v'}{\partial y}+w'\frac{\partial v'}{\partial z}\right)=-\frac{\partial p'}{\partial y},\\
&\bar{\rho}\left(\frac{\partial w'}{\partial t}+u'\frac{\partial w'}{\partial x}+v'\frac{\partial w'}{\partial y}+w'\frac{\partial w'}{\partial z}\right)=-\frac{\partial p'}{\partial z}-g\rho',\\
&\frac{\partial \rho'}{\partial t}+\frac{\partial \rho' u'}{\partial x}+\frac{\partial \rho' v'}{\partial y}+\frac{\partial \rho' w'}{\partial z}=-\left(\frac{\partial \bar{\rho} u'}{\partial x}+\frac{\partial \bar{\rho} v'}{\partial y}+\frac{\partial \bar{\rho} w'}{\partial z}\right),\\
&\frac{\partial p'}{\partial t}+u'\frac{\partial p'}{\partial x}+v'\frac{\partial p'}{\partial y}+w'\frac{\partial p'}{\partial z}=-c^2\left(\frac{\partial u'}{\partial x}+\frac{\partial v'}{\partial y}+\frac{\partial w'}{\partial z}\right)\rho'\\
&-c^2\left(\frac{\partial \bar{\rho} u'}{\partial x}+\frac{\partial \bar{\rho} v'}{\partial y}+\frac{\partial \bar{\rho} w'}{\partial z}\right)-\beta\bar{\rho} w'
\end{aligned}\right\} \tag{3}
$$

式中,$\beta=kR(\gamma_a-\gamma)$是描写大气层结的静力稳定度参数,其中 R 为气体常数,$\gamma_a=\frac{\kappa-1}{\kappa}\frac{g}{R}$,$\gamma=-\frac{\partial \bar{T}}{\partial z}$。$T$ 为温度。为了符号统一,速度分量也标了撇号,这表示对于大气的基本状态恒有 $\bar{u}=\bar{v}=\bar{w}=0$。

如果在式(3)中略去了非线性项,则得到下面形式的线性化运动方程组:

$$
\left.\begin{aligned}
&\bar{\rho}\frac{\partial u'}{\partial t}=-\frac{\partial p'}{\partial x},\\
&\bar{\rho}\frac{\partial v'}{\partial t}=-\frac{\partial p'}{\partial y},\\
&\bar{\rho}\frac{\partial w'}{\partial t}=-\frac{\partial p'}{\partial z}-g\rho',\\
&\frac{\partial \rho'}{\partial t}=-\left(\frac{\partial \bar{\rho} u'}{\partial x}+\frac{\partial \bar{\rho} v'}{\partial y}+\frac{\partial \bar{\rho} w'}{\partial z}\right),\\
&\frac{\partial p'}{\partial t}=-c^2\left(\frac{\partial \bar{\rho} u'}{\partial x}+\frac{\partial \bar{\rho} v'}{\partial y}+\frac{\partial \bar{\rho} w'}{\partial z}\right)-\beta\bar{\rho} w'
\end{aligned}\right\} \tag{4}
$$

这一方程组可以变成对于某一单一变量的方程:

$$\mathscr{L}(X)=0, \tag{5}$$

其中

$$\mathscr{L}=\frac{\partial^4}{\partial t^4}-c^2\Delta_3\frac{\partial^2}{\partial t^2}-(\beta+g)\frac{\partial^3}{\partial z\partial t^2}-\beta g\Delta_2. \tag{6}$$

$$X=(\bar{\rho}u',\bar{\rho}v',\bar{\rho}w',p',\rho') \tag{7}$$

而

$$\Delta_3=\frac{\partial^2}{\partial x^2}+\frac{\partial^2}{\partial y^2}+\frac{\partial^2}{\partial z^2},\ \Delta_2=\frac{\partial^2}{\partial x^2}+\frac{\partial^2}{\partial y^2}$$

方程(5)是描写线性化后的可压缩层结大气小尺度过程的基本方程。

3 线性化运动方程的分析——声波和重力波

我们来分析方程(5)所描写的过程的物理性质。求下面形式的简谐波解

$$\chi \sim \exp\left[-\frac{\beta+g}{2c^2}z + i(k_x x + k_y y + k_z z - \omega t)\right] \tag{8}$$

式中，k_x，k_y，k_z 分别为沿 x，y，z 方向的波数，ω 为振动频率。

将式(8)代入式(5)后得到频率的表达式为

$$\omega^2 = \frac{c^2}{2}\left[k_x^2 + k_y^2 + k_z^2 + \left(\frac{\beta+g}{2c^2}\right)^2\right]\left\{1 \pm \sqrt{1 - \frac{4\beta g}{c^4}\frac{k_x^2+k_y^2}{\left[k_x^2+k_y^2+k_z^2+\left(\frac{\beta+g}{2c^2}\right)^2\right]^2}}\right\} \tag{9}$$

当考虑了地球自转后，A.M.奥布霍夫（Обухов）和 A.C.莫宁（Монин）[1]求得过相应的表达式。另外 G.霍尔曼（Hollmann）[2]求得的频率公式也与此类似。

正如奥布霍夫等[1]所指出，式(9)中的正号所表示的是声速的振动频率，负号表示的是重力内波的频率。

对于我们所讨论的小尺度过程一般 $k \sim 10^{-3} \sim 10^{-1}/\text{m}$，而 $\beta \sim \frac{1}{6}g$，因此根号中的量有

$$\frac{4\beta g}{c^4}\frac{k_x^2+k_y^2}{\left[k_x^2+k_y^2+k_z^2+\left(\frac{\beta+g}{2c^2}\right)^2\right]^2} \sim 10^{-1}$$

这样式(9)便可近似地写成

$$\omega_a^2 = c^2\left[k_x^2+k_y^2+k_z^2+\left(\frac{\beta+g}{2c^2}\right)^2\right]（声波） \tag{10}$$

$$\omega_g^2 = \sigma^2\frac{k_x^2+k_y^2}{k_x^2+k_y^2+k_z^2+\left(\frac{\beta+g}{2c^2}\right)^2}（重力波） \tag{11}$$

式中，$\sigma^2 = \frac{\beta g}{c^2} = \frac{g}{\theta}\frac{\partial\theta}{\partial z}$，$\theta$ 为位温。

在一般的静力稳定度值下，这两种频率之比为

$$\frac{\omega_a^2}{\omega_g^2} \sim 10^2 - 10^3 \tag{12}$$

而重力波的周期一般在 10 分钟左右。

由此可见，在可压缩层结大气中存在着两类小尺度的波动：一是快速的（高频的）声波；二是相对声波而言的慢速的（低频的）重力内波。或者更一般地，在大气中存在着两类小尺度过程：一类主要由大气的可压缩性所制约；另一类则主要由大气的层结所制约。

4　在运动方程中滤掉声波的方法

一般来说，大气中有意义的小尺度过程主要决定于大气的层结，与声波有关的过程一般很少具有气象意义。同时，由于声波这样的高频扰动的存在，必然将对讨论有意义的“缓慢”过程带来困难。因此，有必要把声波作为小尺度过程中的“噪音”而在运动方程组中把它滤掉。

然而，如何能在运动方程中把声波滤掉，同时又使得其他的缓慢运动不受到过分的歪曲，这自然是一个具有原则性的重要问题。奥布霍夫和莫宁[1]以及霍尔曼[2]曾指出，应用静力平衡可以滤掉声波。对于大尺度运动，静力平衡是一个极好的近似关系式，这一滤波方法自然可以采用。对于小尺度

运动,正如前面尺度分析已经指出,过程是非静力平衡的,而且静力偏差对小尺度过程的发展还具有重要的作用,因而,用这一方法来滤去小尺度过程中的声波是不能允许的,我们必须寻求另外的滤波方法。

在下面我们提出了 3 种不同精确度的滤波方法。

4.1 不可压缩近似的应用

我们称流体不可压缩,是指在运动方程中凡密度的变化与压力变化有关的部分皆省略,而与温度变化有关的部分则保留,即

$$\frac{T'}{\bar{T}} \approx -\frac{\rho'}{\bar{\rho}}, \tag{13}$$

不可压缩近似也称为自由对流近似[3]。

在不可压缩近似下,式(3)可以改写成

$$\left.\begin{aligned}
&\frac{\partial u'}{\partial t}+u'\frac{\partial u'}{\partial x}+v'\frac{\partial u'}{\partial y}+w'\frac{\partial u'}{\partial z}=-\frac{1}{\bar{\rho}}\frac{\partial p'}{\partial x},\\
&\frac{\partial v'}{\partial t}+u'\frac{\partial v'}{\partial x}+v'\frac{\partial v'}{\partial y}+w'\frac{\partial v'}{\partial z}=-\frac{1}{\bar{\rho}}\frac{\partial p'}{\partial y},\\
&\frac{\partial w'}{\partial t}+u'\frac{\partial w'}{\partial x}+v'\frac{\partial w'}{\partial y}+w'\frac{\partial w'}{\partial z}=-\frac{1}{\bar{\rho}}\frac{\partial p'}{\partial z}+\frac{g}{\bar{T}}T',\\
&\frac{\partial u'}{\partial x}+\frac{\partial v'}{\partial y}+\frac{\partial w'}{\partial z}=0,\\
&\frac{\partial T'}{\partial t}+u'\frac{\partial T'}{\partial x}+v'\frac{\partial T'}{\partial y}+w'\frac{\partial T'}{\partial z}=-(\gamma_a-\gamma)w'
\end{aligned}\right\} \tag{14}$$

线性化方程组为

$$\left.\begin{aligned}
&\frac{\partial u'}{\partial t}=-\frac{1}{\bar{\rho}}\frac{\partial p'}{\partial x},\\
&\frac{\partial v'}{\partial t}=-\frac{1}{\bar{\rho}}\frac{\partial p'}{\partial y},\\
&\frac{\partial w'}{\partial t}=-\frac{1}{\bar{\rho}}\frac{\partial p'}{\partial z}+\frac{g}{\bar{T}}T',\\
&\frac{\partial u'}{\partial x}+\frac{\partial v'}{\partial y}+\frac{\partial w'}{\partial z}=0,\\
&\frac{\partial T'}{\partial t}=-(\gamma_a-\gamma)w'
\end{aligned}\right\} \tag{15}$$

今引进管量 $\psi'=(\psi'_1,\psi'_2,\psi'_3)$ 来分解速度场,即

$$u'=\frac{\partial \psi'_2}{\partial z}-\frac{\partial \psi'_3}{\partial y},v'=\frac{\partial \psi'_3}{\partial x}-\frac{\partial \psi'_1}{\partial z},w'=\frac{\partial \psi'_1}{\partial y}-\frac{\partial \psi'_2}{\partial x} \tag{16}$$

管量自然具有无辐散特征[4],即

$$\frac{\partial \psi'_1}{\partial x}+\frac{\partial \psi'_2}{\partial y}+\frac{\partial \psi'_3}{\partial z}=0 \tag{17}$$

将式(16)代入式(15)并考虑到式(17)后，不难得到

$$\Delta_3 \frac{\partial \psi'_1}{\partial t} = \frac{g}{\overline{T}} \frac{\partial T'}{\partial y} \tag{18}$$

$$\Delta_3 \frac{\partial \psi'_2}{\partial t} = \frac{g}{\overline{T}} \frac{\partial T'}{\partial x} \tag{19}$$

$$\frac{\partial T'}{\partial t} = \frac{\partial \theta}{\partial z}\left(\frac{\partial \psi'_2}{\partial x} - \frac{\partial \psi'_1}{\partial y}\right) \tag{20}$$

方程(17)至方程(20)对变量 $\psi'_1, \psi'_2, \psi'_3, T'$ 是闭合的。由这一方程组也可以求得下面的方程式：

$$\mathscr{L}(\chi) = 0 \tag{21}$$

式中

$$\mathscr{L} = \Delta_3 \frac{\partial^2}{\partial t^2} + \sigma^2 \Delta_2 \tag{22}$$

$$\chi = (\psi'_1, \psi'_2, T') \tag{23}$$

当 ψ'_1, ψ'_2 求得后，ψ'_3 由条件式(17)决定。

取式(21)的解为

$$\chi \sim \exp[i(k_x x + k_y y + k_z z - \omega t)] \tag{24}$$

则得频率公式

$$\omega^2 = \sigma^2 \frac{k_x^2 + k_y^2}{k_x^2 + k_y^2 + k_z^2} \tag{25}$$

这自然是重力内波的振动频率。由此可见，应用这一近似可以滤掉声波。

应用不可压缩近似虽然是一种方便的滤波方法，但比较式(25)与式(11)，我们发现在式(25)的分母中少了一个因子 $\left(\frac{\beta+g}{2c^2}\right)^2$。这表明应用这一方法虽然滤掉了声波，但重力波也受到了歪曲。由于 $\left(\frac{\beta+g}{2c^2}\right)^2 \sim 10^{-4}/\text{m}$，因此，当现象的特征尺度 $L \sim 10$ km 时，丢掉这一因子便会带来较大的误差。由此可见，不可压缩近似这一滤波方法只有对于尺度较小的运动($L \sim 1$ km)才是合适的。

4.2 小参数展开法

引进无因次变量

$$\bar{t} = t/\frac{L}{V}, (\bar{x}, \bar{y}, \bar{z}) = \left(\frac{x}{L}, \frac{y}{L}, \frac{z}{L}\right), (\bar{u}', \bar{v}', \bar{w}') = \left(\frac{u'}{V}, \frac{v'}{V}, \frac{w'}{V}\right),$$

$$\bar{p}' = \frac{p'}{\bar{\rho} V^2}, \bar{\rho}' = \rho'/\frac{\bar{\rho} V^2}{Lg}$$

将上式代入式(4)，便得到无因次的运动方程组①：

① 在最后两个方程中略掉了一密度随高度变化的项。

$$\left.\begin{aligned}&\frac{\partial\bar{u}'}{\partial\bar{t}}=-\frac{\partial\bar{p}'}{\partial\bar{x}},\ \frac{\partial\bar{v}'}{\partial\bar{t}}=-\frac{\partial\bar{p}'}{\partial\bar{y}}\\&\frac{\partial\bar{w}'}{\partial\bar{t}}=-\frac{\partial\bar{p}'}{\partial\bar{z}}-\bar{\rho}'\\&\varepsilon'\frac{\partial\bar{\rho}'}{\partial\bar{t}}=-\left(\frac{\partial\bar{u}'}{\partial\bar{x}}+\frac{\partial\bar{v}'}{\partial\bar{y}}+\frac{\partial\bar{w}'}{\partial\bar{z}}\right)\\&\varepsilon'\mu\frac{\partial\bar{p}'}{\partial\bar{t}}=-\left(\frac{\partial\bar{u}'}{\partial\bar{x}}+\frac{\partial\bar{v}'}{\partial\bar{y}}+\frac{\partial\bar{w}'}{\partial\bar{z}}\right)-\varepsilon'\alpha\mu\bar{w}'\end{aligned}\right\}\tag{26}$$

式中无因次参数 ε',μ,α 定义为

$$\varepsilon'=\frac{V^2}{gL},\ \mu=\frac{L}{L_0},\ \alpha=\frac{L\beta}{V^2}\left(L_0=\frac{c^2}{g}\right)\tag{27}$$

参数 ε'即为通常的弗罗德数,其数值见表 1。

表 1

ε' L(m) / V(m/s)	10	10^2	10^3	10^4
10	1	0.1	0.01	0.001
20	4	0.4	0.04	0.004
30	9	0.9	0.09	0.009

由此可见,当 $L>10^2$ m,一般都有 $\varepsilon'\ll 1$。

现在用无因次势量场 $\bar{\varphi}'$和无因次管量场 $\bar{\psi}'=(\bar{\psi}'_1,\bar{\psi}'_2,\bar{\psi}'_3)$来分解速度场[4],

$$\bar{u}'=\frac{\partial\bar{\psi}'_2}{\partial\bar{z}}-\frac{\partial\bar{\psi}'_3}{\partial\bar{y}}+\frac{\partial\bar{\varphi}'}{\partial\bar{x}},\ \bar{v}'=\frac{\partial\bar{\psi}'_3}{\partial\bar{x}}-\frac{\partial\bar{\psi}'_1}{\partial\bar{z}}+\frac{\partial\bar{\varphi}'}{\partial\bar{y}},\ \bar{w}'=\frac{\partial\bar{\psi}'_1}{\partial\bar{y}}-\frac{\partial\bar{\psi}'_2}{\partial\bar{x}}+\frac{\partial\bar{\varphi}'}{\partial\bar{z}}\tag{28}$$

对于 $\bar{\psi}'$仍有约束条件

$$\frac{\partial\bar{\psi}'_1}{\partial\bar{x}}+\frac{\partial\bar{\psi}'_2}{\partial\bar{y}}+\frac{\partial\bar{\psi}'_3}{\partial\bar{z}}=0\tag{29}$$

将式(28)代入式(26)并考虑到式(29)后,得到

$$\Delta_3\frac{\partial\bar{\psi}'_1}{\partial\bar{t}}=-\frac{\partial\bar{\rho}'}{\partial\bar{y}}\tag{30}$$

$$\Delta_3\frac{\partial\bar{\psi}'_2}{\partial\bar{t}}=-\frac{\partial\bar{\rho}'}{\partial\bar{x}}\tag{31}$$

$$\Delta_3\frac{\partial\bar{\varphi}'}{\partial\bar{t}}=-\Delta_3\bar{p}'-\frac{\partial\bar{\rho}'}{\partial\bar{z}}\tag{32}$$

$$\varepsilon' \frac{\partial \bar{\rho}'}{\partial \bar{t}} = \Delta_3 \bar{\varphi}' \tag{33}$$

$$\varepsilon' \mu \frac{\partial \bar{p}'}{\partial \bar{t}} = - \Delta_3 \bar{\varphi}' - \varepsilon' \mu \alpha \left(\frac{\partial \bar{\psi}'_1}{\partial \bar{y}} - \frac{\partial \bar{\psi}'_2}{\partial \bar{x}} + \frac{\partial \bar{\varphi}'}{\partial \bar{z}} \right) \tag{34}$$

不难看出，式(30)，式(31)分别是绕 x 和 y 轴的涡度方程，而式(32)为辐散方程。

今将方程组式(30)至式(34)的解按小参数 ε'的幂次展开，即令

$$\bar{\psi}'_1 = \bar{\psi}'_{10} + \varepsilon' \bar{\psi}'_{11} + \cdots,\ \bar{\psi}'_2 = \bar{\psi}'_{20} + \varepsilon' \bar{\psi}'_{21} + \cdots,\ \bar{\varphi}'_1 = \bar{\varphi}'_0 + \varepsilon' \bar{\varphi}'_1 + \cdots,$$

$$\bar{p}' = \bar{p}'_0 + \varepsilon \bar{p}'_1 + \cdots,\ \bar{\rho}' = \bar{\rho}'_0 + \varepsilon \bar{\rho}'_1 + \cdots \tag{35}$$

将式(35)代入式(30)至式(34)并按 ε'的幂次归并同类项，最后得到下面的无穷微分方程组：

$$\Delta_3 \bar{\varphi}'_0 = 0 \tag{36}$$

$$\Delta_3 \frac{\partial \bar{\psi}'_{10}}{\partial \bar{t}} = - \frac{\partial \bar{\rho}'_0}{\partial \bar{y}} \tag{37}$$

$$\Delta_3 \frac{\partial \bar{\psi}'_{20}}{\partial \bar{t}} = - \frac{\partial \bar{\rho}'_0}{\partial \bar{x}} \tag{38}$$

$$\Delta_3 \frac{\partial \bar{\varphi}'_0}{\partial \bar{t}} = - \Delta_3 \bar{p}'_0 - \frac{\partial \bar{\rho}'_0}{\partial \bar{z}} \tag{39}$$

$$\frac{\partial \bar{\rho}'_0}{\partial \bar{t}} = - \Delta_3 \bar{\varphi}'_1 \tag{40}$$

$$\mu \frac{\partial \bar{p}'_0}{\partial \bar{t}} = - \Delta_3 \bar{\varphi}'_1 - \alpha\mu \left(\frac{\partial \bar{\psi}'_{10}}{\partial \bar{y}} - \frac{\partial \bar{\psi}'_{20}}{\partial \bar{x}} + \frac{\partial \bar{\varphi}'_0}{\partial \bar{z}} \right) \tag{41}$$

由式(36)，按调和函数的性质，只要在边界上 $\bar{\varphi}'_0 = 0$，则在整个定义域上恒有

$$\bar{\varphi}'_0 = 0 \tag{42}$$

考虑到式(42)后，对于零级近似的其他各量满足下面的方程

$$\mathscr{L}(\chi) = 0 \tag{43}$$

其中

$$\mathscr{L} = \Delta_3 \frac{\partial^2}{\partial \bar{t}^2} + \mu \frac{\partial^3}{\partial \bar{z} \partial \bar{t}^2} + \alpha\mu\Delta_2 \tag{44}$$

$$\chi = (\bar{\psi}'_{10}, \bar{\psi}'_{20}, \bar{p}'_0, \bar{\rho}'_0) \tag{45}$$

回到有因次量后不难求得方程(43)所决定的频率公式为

$$\omega^2 = \sigma^2 \frac{k_x^2 + k_y^2}{k_x^2 + k_y^2 + k_z^2 + \left(\dfrac{g}{2c^2} \right)^2} \tag{46}$$

这显然是重力内波的振动频率。由此可见，用小参数展开法同样消掉了运动方程中的声波。

比较式(46)与式(25)，在目前的情况下多了一附加项 $\left(\dfrac{g}{2c^2} \right)^2$。这自然是压缩性所引起。由于这

一因子其大小正比于$\mu=L/L_0$，因此当$L\ll L_0$时，这一附加项的贡献不大，但当$L\sim L_0$时，就变得重要。如取$c=280\ \mathrm{m/s}$，则$L_0\approx 8\ \mathrm{km}$。

这一分析表明了，压缩性对缓慢过程的作用，只有当尺度接近或超过8 km时才是重要的。R.R.郎(Long)[5]用另外的方法也得到过这一结论。

在另一方面，将频率公式(46)与式(11)相比，我们发现在目前的情况下还少了一个与稳定度有关的因子$(\beta/2c^2)^2$。

4.3 用平衡方程来代替辐散方程

直接从方程组式(30)至式(34)出发，在辐散方程(32)中，令辐散随时间的变化为零，即得

$$\Delta_3\bar{p}' + \frac{\partial\bar{\rho}'}{\partial\bar{z}'} = 0 \tag{47}$$

我们称这一关系式为线性化情况下小尺度过程的平衡方程。

由式(30)，式(31)，式(33)和式(34)以及式(47)不难得到

$$\mathscr{L}(X) = 0 \tag{48}$$

其中

$$\mathscr{L} = \Delta_3\frac{\partial^2}{\partial\bar{t}^2} + (\mu + \mu\alpha\varepsilon')\frac{\partial^3}{\partial\bar{z}\partial\bar{t}^2} + \alpha\mu\Delta_2 \tag{49}$$

$$X = (\bar{\psi}'_1, \bar{\psi}'_2, \bar{\varphi}', \bar{p}, \bar{\rho}') \tag{50}$$

式(48)所决定的频率公式为

$$\omega^2 = \sigma^2\frac{k_x^2 + k_y^2}{k_x^2 + k_y^2 + k_z^2 + \left(\dfrac{\beta + g}{2c^2}\right)^2} \tag{51}$$

显然，这即为式(11)。

根据以上的分析可以得到结论：用小参数展开的方法是较用不可压缩近似精确的方法，而用平衡方程来代替辐散方程则精确度更高。事实上不难了解，3种不同滤波方法的本质在于如何在不产生声波的前提下将辐散(压缩性)处理得更好，使它尽可能对缓慢过程发生作用。

最后我们指出，现在所用的第二种滤去快波的方法与大尺度运动中由И.А.基培尔(Кибель)[6]首先提出的，后经莫宁[7]、A.M.雅格龙(Яглом)[8]等发展的将解按小参数$\varepsilon=V/fL$展开而消去"噪音"的方法是相应的，只不过由于运动的性质不同所用的小参数不同而已。而与莫宁[9]最近在大尺度运动中所提出的将解按马赫数$M=V^2/c^2$展开的方法则更相近。第三种滤波方法与大尺度过程中采用平衡方程来消除快波的方法是相应的，当然两者的平衡方程是不同的。另外，在重力场中与压缩性有关的线性长度$L_0=c^2/g$的物理意义相当于奥布霍夫[10]所提出的在科里奥利力场中的线性长度$L_1=c/f$。由此可见，大气中的小尺度过程与大尺度过程的动力学性质虽然很不相同，但是却有很多等价的性质，这是很有意思的。

5 重力场中声波的频散适应过程

以上的分析明确了某一时刻在大气中的某一有限区域中，受到一个小扰动破坏了静力平衡后，便会激发出快速的声波和"缓慢的"重力内波。由于声波的频散速度要比重力波来得快，当在某一有限

区域中声波频散后，剩下的自然只有“缓慢”变化的重力内波。在只有重力内波的场中，气压和密度间存在着平衡关系式(47)。从静力平衡的破坏到这种平衡状态的建立，场的这一建立过程，我们称之为小尺度过程中流体力学场的适应过程。事实上，正如奥布霍夫等[1]所指出，这一过程也就是声波频散的过程。在本节中我们将进一步来分析这一过程。

不难证明，在方程(5)中略去了带有算子 $\beta g\Delta_2$ 这一项后的方程中，只包含快速声波。我们对速度势 φ 写出这一方程

$$\frac{\partial^2\varphi}{\partial t^2}=c^2\left(\frac{\partial^2\varphi}{\partial x^2}+\frac{\partial^2\varphi}{\partial y^2}+\frac{\partial^2\varphi}{\partial z^2}\right)+(\beta+g)\frac{\partial\varphi}{\partial z} \tag{52}$$

事实上，很容易验证在式(52)中只包含声波的频率

$$\omega^2=c^2\left[k_x^2+k_y^2+k_z^2+\left(\frac{\beta+g}{2c^2}\right)^2\right] \tag{53}$$

此即为式(10)。

为了便于比较，我们只考虑在(x,z)平面中所发生的运动。引进变换

$$\varphi(x,z,t)=\varphi'(x,z,t)\mathrm{e}^{-\left(\frac{\beta+g}{2c^2}\right)z} \tag{54}$$

则式(52)变为

$$\frac{\partial^2\varphi'}{\partial t^2}=c^2\left(\frac{\partial^2\varphi'}{\partial x^2}+\frac{\partial^2\varphi'}{\partial z^2}\right)-l^2\varphi' \tag{55}$$

式中，$l^2=\left(\frac{\beta+g}{2c}\right)^2$。

我们注意到描写在层结大气中在重力作用下的声波传播的方程式(55)，与科里奥利力场中正压情况下地转平衡破坏后的快波的传播方程极相似[10]。不过在后者的情况下，波动的传播是在(x,y)平面上进行的，传播速度为 $c=\sqrt{gH_0}$（H_0 为均质大气的厚度），并且 $l=2\omega_z$ 是折向参数。在目前的情况下，声波的传播是在(x,z)平面上进行的，传播速度为 $c=\sqrt{\kappa RT}$，同时 l 表示的是大气层结和重力场的联合影响。

设在初始时刻 $\varphi'\partial\varphi'/\partial t$ 是坐标的已知函数，即

$$t=0,\ \begin{cases}\varphi'(x,z,0)=f(x,z),\\ \dfrac{\partial}{\partial t}\varphi'(x,z,0)=g(x,z)\end{cases} \tag{56}$$

同时为了便于相互参照，我们将初始条件 f,g 依变量 z 解析开拓到区域$(0,-\infty)$上，则半无穷平面上的方程(55)满足条件式(56)的解，与全平面上同一问题的解是完全一致的。对于方程(55)满足条件式(56)的解可参照奥布霍夫[10]而直接写出，为

$$\begin{aligned}\varphi'(x,z,t)=&\frac{1}{2\pi c}\frac{\partial}{\partial t}\iint_{r\leqslant ct}\frac{f(x+r\cos\theta,z+r\sin\theta)}{\sqrt{c^2t^2-r^2}}\cos\left(\frac{1}{c}\sqrt{c^2t^2-r^2}\right)r\mathrm{d}r\mathrm{d}\theta\\&+\frac{1}{2\pi c}\iint_{r\leqslant ct}\frac{g(x+r\cos\theta,z+r\sin\theta)}{\sqrt{c^2t^2-r^2}}\cos\left(\frac{1}{c}\sqrt{c^2t^2-r^2}\right)r\mathrm{d}r\mathrm{d}\theta\end{aligned} \tag{57}$$

关于这一个解的详细讨论可参见奥布霍夫的文章。在我们现在的情况下，由于扰动传播速度为 $c=\sqrt{\kappa RT}\backsimeq 300$ m/s，因此在 1 分钟内扰动便可影响到约 20 km 外的区域。扰动传播时显然是带有特征频率为 $l=\frac{\beta+g}{2c}$的阻尼振荡。如取 $\gamma_a-\gamma\sim 4\times10^{-3}$℃/m，那么，$l\sim\frac{1}{60}$/s，即周期约 1 分钟。在波所影响

到的范围内，扰动的振幅将随时间成反比而趋向于零。

根据式(57)的渐近表达式不难估计扰动振幅随时间的衰减量级。如果 $\varphi'_{t=0}=f,\left(\frac{\partial\varphi'}{\partial t}\right)_{t=0}=g$ 只有限初始扰动区域——半径为 R 的一个圆中不等于零(这里要求 $R\ll L_0$)，则在初始扰动中心($r=0$)解的渐近表达式为[10]

$$\varphi'(0,0,t)\approx-\frac{\bar{f}}{2}\left[\frac{R}{L_0}\text{sinlt}+\frac{R}{ct}\text{coslt}\right]\frac{R}{ct}+\frac{\bar{g}}{2}\frac{R^2}{c^2t^2}\text{coslt} \tag{58}$$

其中，$(ct)^2\gg R^2$。

不难看出，当 $t\sim L_0/c$ 时(即 t 约为半分钟)，对于 $\bar{f}$ 而言的那部分扰动振幅其大小约为

$$\left|\varphi'\left(0,0,\frac{L_0}{c}\right)\right|\approx\bar{f}\frac{R^2}{L_0^2} \tag{59}$$

因此

$$\frac{\varphi'\left(0,0,\frac{L_0}{c}\right)}{\varphi'(0,0,0)=\bar{f}}\approx\frac{R^2}{L_0^2}\ll 1 \tag{60}$$

如取 $R=1$ km，则 $R^2/L_0^2\sim10^{-2}$，即约在半分钟后初始扰动中心扰动的振幅已经衰减到只有初始值的 1/100。

这一简单的分析表明，在小尺度过程中作为“噪音”的声波，它的频散是极其迅速的。换言之，从静力平衡的破坏到另一种平衡状态的建立，流体力学场的这一适应过程完成得极其迅速。由此也不难理解，声波这一快速过程对一般有气象意义的小尺度过程很少具有重要性，因此一开始就在运动方程中将它滤掉是完全适合的。

6 小尺度过程的非线性方程组

由于小尺度过程一般都具有非线性特色(如对流云、飑线等)，因此有必要建立滤掉“噪音”后的非线性方程组。

在目前的讨论中我们把运动限制在(x,z)平面中。

先建立应用不可压缩近似时的非线性运动方程组。这一方程组即为方程组(14)。或者，对于二维问题有

$$\frac{\partial u'}{\partial t}+u'\frac{\partial u'}{\partial x}+w'\frac{\partial u'}{\partial z}=-\frac{1}{\bar{\rho}}\frac{\partial p'}{\partial x} \tag{61}$$

$$\frac{\partial w'}{\partial t}+u'\frac{\partial w'}{\partial x}+w'\frac{\partial w'}{\partial z}=-\frac{1}{\bar{\rho}}\frac{\partial p'}{\partial z}+\frac{g}{\bar{T}}T' \tag{62}$$

$$\frac{\partial u'}{\partial x}+\frac{\partial w'}{\partial z}=0 \tag{63}$$

$$\frac{\partial T'}{\partial t}+u'\frac{\partial T'}{\partial x}+w'\frac{\partial T'}{\partial z}=-(\gamma_a-\gamma)\omega' \tag{64}$$

由式(63)引进流函数 φ'，并且

$$u'=\frac{\partial\psi'}{\partial z},\ w'=-\frac{\partial\psi'}{\partial x} \tag{65}$$

由式(61)、式(62)消去p'后得绕y轴的涡度方程

$$\Delta_2 \frac{\partial \psi'}{\partial t} = \frac{\partial(\psi', \Delta_2 \psi')}{\partial(x, z)} - \frac{g}{\overline{T}} \frac{\partial T'}{\partial x} \tag{66}$$

在此 $\Delta_2 = \frac{\partial^2}{\partial x^2} + \frac{\partial^2}{\partial z^2}$。

式(64)考虑到式(65)后,得

$$\frac{\partial T'}{\partial t} = \frac{\partial(\psi', T')}{\partial(x, z)} + (\gamma_a - \gamma) \frac{\partial \psi'}{\partial x} \tag{67}$$

式(66)和式(67)两式组成了变量ψ',T'的闭合方程。

自然,这一方程组只能用来研究尺度较小的小尺度过程。必须指出,我们不可以用由不可压缩近似得到的三维运动方程组来研究龙卷风。因为对于龙卷风这样的现象,虽然它的尺度很小,但是它的气压变化很大,因气压变化引起的密度变化必须考虑。Л. Н. 古特曼(Гутман)[11]应用不可压缩近似得到的方程组来研究龙卷风的做法不能认为是合适的。

在小参数展开的情况下,二维无因次方程组为

$$\frac{\partial \overline{u}'}{\partial \overline{t}} + \overline{u}' \frac{\partial \overline{u}'}{\partial \overline{x}} + \overline{w}' \frac{\partial \overline{u}'}{\partial \overline{z}} = - \frac{\partial \overline{p}'}{\partial \overline{x}} \tag{68}$$

$$\frac{\partial \overline{w}'}{\partial \overline{t}} + \overline{u}' \frac{\partial \overline{w}'}{\partial \overline{x}} + \overline{w}' \frac{\partial \overline{w}'}{\partial \overline{z}} = - \frac{\partial \overline{p}'}{\partial \overline{z}} - \overline{\rho}' \tag{69}$$

$$\varepsilon' \left(\frac{\partial \overline{\rho}'}{\partial \overline{t}} + \frac{\partial \overline{\rho}' \overline{u}'}{\partial \overline{x}} + \frac{\partial \overline{\rho}' \overline{w}'}{\partial \overline{z}} \right) = - \left(\frac{\partial \overline{u}'}{\partial \overline{x}} + \frac{\partial \overline{w}'}{\partial \overline{z}} \right) \tag{70}$$

$$\varepsilon' \mu \left(\frac{\partial \overline{p}'}{\partial \overline{t}} + \overline{u}' \frac{\partial \overline{p}'}{\partial \overline{x}} + \overline{w}' \frac{\partial \overline{p}'}{\partial \overline{z}} \right) = - \varepsilon' \left(\frac{\partial \overline{u}'}{\partial \overline{x}} + \frac{\partial \overline{w}'}{\partial \overline{z}} \right) \overline{\rho}' - \left(\frac{\partial \overline{u}'}{\partial \overline{x}} + \frac{\partial \overline{w}'}{\partial \overline{z}} \right) - \varepsilon' \mu \alpha \overline{w}' \tag{71}$$

或将式(68),式(69)改写成涡度方程和辐散方程

$$\frac{\partial \overline{\zeta}'}{\partial \overline{t}} + \overline{u}' \frac{\partial \overline{\zeta}'}{\partial \overline{x}} + \overline{w}' \frac{\partial \overline{\zeta}'}{\partial \overline{z}} + \overline{\zeta}' \overline{D}' = \frac{\partial \overline{\rho}'}{\partial \overline{x}} \tag{72}$$

$$\frac{\partial \overline{D}'}{\partial \overline{t}} + \overline{u}' \frac{\partial \overline{D}'}{\partial \overline{x}} + \overline{w}' \frac{\partial \overline{D}'}{\partial \overline{z}} + \overline{D}'^2 + 2 \frac{\partial(\overline{w}', \overline{u}')}{\partial(\overline{x}, \overline{z}')} = - \Delta_2 \overline{p}' - \frac{\partial \overline{\rho}'}{\partial \overline{z}'} \tag{73}$$

引进流函数$\overline{\psi}'$和速度势$\overline{\varphi}'$,定义成

$$\overline{u}' = \frac{\partial \overline{\psi}'}{\partial \overline{z}} + \frac{\partial \overline{\varphi}'}{\partial \overline{x}}, \quad \overline{w}' = - \frac{\partial \overline{\psi}'}{\partial \overline{x}} + \frac{\partial \overline{\varphi}'}{\partial \overline{z}} \tag{74}$$

因此

$$\overline{\zeta}' = \Delta_2 \overline{\psi}', \quad \overline{D}' = \Delta_2 \overline{\varphi}' \tag{75}$$

将式(74),式(75)代入式(72),式(73),式(70)和式(71)后得

$$\frac{\partial \Delta_2 \overline{\psi}'}{\partial \overline{t}} + \left(\frac{\partial \overline{\psi}'}{\partial \overline{z}} + \frac{\partial \overline{\varphi}'}{\partial \overline{x}} \right) \frac{\partial \Delta_2 \overline{\psi}'}{\partial \overline{x}} + \left(- \frac{\partial \overline{\psi}'}{\partial \overline{x}} + \frac{\partial \overline{\varphi}'}{\partial \overline{z}} \right) \frac{\partial \Delta_2 \overline{\psi}'}{\partial \overline{z}}$$

$$+\Delta_2\bar{\psi}'\cdot\Delta_2\bar{\varphi}'=\frac{\partial\bar{\rho}'}{\partial\bar{x}} \tag{76}$$

$$\frac{\partial\Delta_2\bar{\varphi}'}{\partial\bar{t}}+\left(\frac{\partial\bar{\psi}'}{\partial\bar{z}}+\frac{\partial\bar{\varphi}'}{\partial\bar{x}}\right)\frac{\partial\Delta_2\bar{\varphi}'}{\partial\bar{x}}+\left(-\frac{\partial\bar{\psi}'}{\partial\bar{x}}+\frac{\partial\bar{\varphi}'}{\partial\bar{z}}\right)\frac{\partial\Delta_2\bar{\varphi}'}{\partial\bar{z}}+(\Delta_2\bar{\varphi}')^2+$$

$$+2\frac{\partial\left(-\frac{\partial\bar{\psi}'}{\partial\bar{x}}+\frac{\partial\bar{\varphi}'}{\partial\bar{z}},\frac{\partial\bar{\psi}'}{\partial\bar{z}}+\frac{\partial\bar{\varphi}'}{\partial\bar{x}}\right)}{\partial(\bar{x},\bar{z})}=-\Delta_2\bar{p}'-\frac{\partial\bar{\rho}'}{\partial\bar{z}} \tag{77}$$

$$\varepsilon'\left[\frac{\partial\bar{\rho}'}{\partial\bar{t}}+\left(\frac{\partial\bar{\psi}'}{\partial\bar{z}}+\frac{\partial\bar{\varphi}'}{\partial\bar{x}}\right)\frac{\partial\bar{\rho}'}{\partial\bar{x}}+\left(-\frac{\partial\bar{\psi}'}{\partial\bar{x}}\cdot\frac{\partial\bar{\varphi}'}{\partial\bar{z}}\right)\frac{\partial\bar{\rho}'}{\partial\bar{z}}+\bar{\rho}'\Delta_2\bar{\varphi}'\right]=-\Delta_2\bar{\varphi}' \tag{78}$$

$$\varepsilon'\mu\left[\frac{\partial\bar{p}'}{\partial\bar{t}}+\left(\frac{\partial\bar{\psi}'}{\partial\bar{z}}+\frac{\partial\bar{\varphi}'}{\partial\bar{x}}\right)\frac{\partial\bar{p}'}{\partial\bar{x}}+\left(-\frac{\partial\bar{\psi}'}{\partial\bar{x}}+\frac{\partial\bar{\varphi}'}{\partial\bar{z}}\right)\frac{\partial\bar{p}'}{\partial\bar{z}}\right]$$

$$=-\varepsilon'\Delta_2\bar{\varphi}'\cdot\bar{\rho}'-\Delta_2\bar{\varphi}'-\varepsilon\mu\alpha\left(-\frac{\partial\bar{\psi}'}{\partial\bar{x}}+\frac{\partial\bar{\varphi}'}{\partial\bar{z}}\right) \tag{79}$$

今将解按 ε' 的幂次展开，即令

$$\begin{aligned}&\bar{\psi}'=\bar{\psi}'_0+\varepsilon'\bar{\psi}'_1+\cdots,\ \bar{\varphi}'=\bar{\varphi}'_0+\varepsilon'\bar{\varphi}'_1+\cdots,\\&\bar{\rho}'=\bar{\rho}'_0+\varepsilon'\bar{\rho}'_1+\cdots,\ \bar{p}'=\bar{p}'_0+\varepsilon'\bar{p}'_1+\cdots\end{aligned} \tag{80}$$

将式(80)代入式(76)至式(79)归并 ε' 的同幂次项，得到下面的无穷微分方程组

$$\bar{\varphi}'_0=0 \tag{81}$$

$$\Delta_2\frac{\partial\bar{\psi}'_0}{\partial\bar{t}}=\frac{\partial(\bar{\psi}'_0,\Delta_2\bar{\psi}'_0)}{\partial(\bar{x},\bar{z})}+\frac{\partial\bar{\rho}'_0}{\partial\bar{x}} \tag{82}$$

$$2\left[\left(\frac{\partial^2\bar{\psi}'_0}{\partial\bar{x}\partial\bar{z}}\right)^2-\frac{\partial^2\bar{\psi}'_0}{\partial\bar{x}^2}\frac{\partial^2\bar{\psi}'_0}{\partial\bar{z}^2}\right]=-\Delta_2\bar{p}'_0-\frac{\partial\bar{\rho}'_0}{\partial\bar{z}} \tag{83}$$

$$\frac{\partial\bar{\rho}'_0}{\partial\bar{t}}=\frac{\partial(\bar{\psi}'_0,\bar{\rho}'_0)}{\partial(\bar{x},\bar{z})}-\Delta_2\bar{\varphi}'_1 \tag{84}$$

$$\mu\frac{\partial\bar{p}'_0}{\partial\bar{t}}=\mu\left[\frac{\partial(\bar{\psi}'_0,\bar{p}'_0)}{\partial(\bar{x},\bar{z})}+\alpha\frac{\partial\bar{\psi}'_0}{\partial\bar{x}}\right]-\Delta_2\bar{\varphi}'_1 \tag{85}$$

方程组式(82)至式(85)组成了对变量 $\bar{\psi}'_0,\bar{\varphi}'_1,\bar{p}'_0,\bar{\rho}_0$ 的闭合方程组。这组方程可以化成较简单的形式。令

$$\bar{\pi}'_0=\mu\bar{p}'_0-\bar{\rho}'_0 \tag{86}$$

则由式(84)，式(85)消去 $\bar{\varphi}'_1$ 后，得

$$\frac{\partial\bar{\pi}'_0}{\partial\bar{t}}=\frac{\partial(\bar{\psi}'_0,\bar{\pi}'_0)}{\partial(\bar{x},\bar{z})}+\mu\alpha\frac{\partial\bar{\psi}'_0}{\partial\bar{x}} \tag{87}$$

考虑到式(86)后由式(82)，式(83)得(省去中间的运算)

$$\left(\Delta_2+\mu\frac{\partial}{\partial\bar{z}}\right)\frac{\partial\bar{\psi}'_0}{\partial\bar{t}}=\frac{\partial\left(\bar{\psi}'_0,\Delta_2\bar{\psi}'_0+\mu\dfrac{\partial\bar{\psi}'_0}{\partial\bar{z}}\right)}{\partial(\bar{x},\bar{z})}-\frac{\partial\bar{\pi}'_0}{\partial\bar{x}}\tag{88}$$

由于对于一般的小尺度过程 $\varepsilon'\ll1$，因此这一零级近似的方程组式(87)、式(88)已经可以相当精确地逼近问题的解。不过对于 ε'接近于 1 的运动，例如对龙卷风($V\sim50$ m，$L\sim10^2$ m $\varepsilon'\sim1$)，只考虑零级近似是不够的。

与不可压缩近似所得的方程(66)，式(67)相比，在现在的情况下，在涡度方程中多了一个与压缩性参数 μ 有关的项。因此，对于一些尺度较大的现象，如强大积云的发展，飑线、气流流过一个较大的山脊等，用现在这一方程组来研究看来要更合适一些。

最后指出，根据线性化的分析不难了解，速度场分解式中势量的那一部分，对于我们感兴趣的小尺度过程一般是很小的。因此，除了连续性方程和绝热方程中保留与辐散有关的项外，在其他地方凡与速度势有关的部分皆可省掉。特别是，从式(77)我们得到小尺度过程的非线性平衡方程

$$2\left[\left(\frac{\partial^2\bar{\psi}'}{\partial\bar{x},\partial\bar{z}}\right)^2-\frac{\partial^2\bar{\psi}'}{\partial\bar{x}^2}\cdot\frac{\partial^2\bar{\psi}'}{\partial\bar{z}^2}\right]=-\Delta_2\bar{p}'-\frac{\partial\bar{\rho}'}{\partial\bar{z}}\tag{89}$$

其他各方程则为

$$\Delta_2\frac{\partial\bar{\psi}'}{\partial\bar{t}}=\frac{\partial(\bar{\psi}',\Delta\bar{\psi}')}{\partial(\bar{x},\bar{z})}+\frac{\partial\bar{\rho}'}{\partial\bar{x}}\tag{90}$$

$$\varepsilon'\frac{\partial\bar{\rho}'}{\partial\bar{t}}=\varepsilon'\frac{\partial(\bar{\psi}',\bar{\rho}')}{\partial(\bar{x},\bar{z})}-(\varepsilon'\bar{\rho}'+1)\Delta\bar{\varphi}'\tag{91}$$

$$\varepsilon'\mu\frac{\partial\bar{p}'}{\partial\bar{t}}=\varepsilon'\mu\left[\frac{\partial(\bar{\psi}',\bar{p}')}{\partial(\bar{x},\bar{z})}+\alpha\frac{\partial\bar{\psi}'}{\partial\bar{x}}\right]-(\varepsilon'\bar{\rho}'+1)\Delta\bar{\varphi}'\tag{92}$$

显然，在现在的情况下，在连续性方程和绝热方程中，比前一方法得到的结果多考虑了一部分辐散的影响，当 $\varepsilon'\sim1$ 时，这些附加项就变得重要。另外，方程组式(89)至式(92) 与方程组式(82)至式(85)不同者还在于，后者是对零级近似写出的(但考虑了速度势的一级近似)，而现在则无此限制。

由此来看，即使对于发展猛烈的小系统，如龙卷风，用现在这一方法得到的方程组来研究也是合适的。自然，当研究龙卷风时必须改用圆柱坐标。

同样地，如果引进

$$\bar{\pi}'=\mu\bar{p}'-\bar{\rho}'\tag{93}$$

则由式(91)、式(92)得

$$\frac{\partial\bar{\pi}'}{\partial\bar{t}}=\frac{\partial(\bar{\psi}',\bar{\pi}')}{\partial(\bar{x},\bar{z})}+\mu\alpha\frac{\partial\bar{\psi}'}{\partial\bar{x}}\tag{94}$$

考虑到式(93)后，从式(89)，式(90)得

$$\left(\Delta_2+\mu\frac{\partial}{\partial\bar{z}}\right)\frac{\partial\bar{\psi}'}{\partial t}=\frac{\partial\left(\bar{\psi}',\Delta\bar{\psi}'+\mu\dfrac{\partial\bar{\psi}'}{\partial\bar{z}}\right)}{\partial(\bar{x},\bar{z})}-\frac{\partial\bar{\pi}'}{\partial\bar{x}}\tag{95}$$

至此，本文所要讨论的问题已经完结。

应用不可压缩近似滤掉声波后的非线性方程组，来定量计算热对流的发展，我们在最近曾做过研究[12]。

参考文献

[1] Обухов.А.М.和 Монин.А.С.，*Tellus*.1959，11：159-162.

[2] Hollmann，G.，*Beitr.z.Physik der Atmosphare*，Bd.1959，31：5-30.

[3] 郎道，等.连续介质力学（中译本）.北京：高等教育出版社，1958.

[4] Lamb，H.，*Hydrodynamics*.1932：148.

[5] Long，R. R.，Symposium on the use of Models in Geophysics Fluid Dynmics，Ist Johns Hopking University，1953：135-147.

[6] Кибель，И.А.，*Нзв.АН СССР*，*гер.zеоф.и геог*.1940，5：627-638.

[7] Монин，А.С.，*Нзв.АН СССР*，*сер.геоф*.1952，4：76-85.

[8] Яглом，А.М.，*Изв.АН СССР*，*сер.геоф*.1953，4：346-369.

[9] Монин，А.С.，*Изв.АН СССР*，*сер.геоф*.1961，4：602-612.

[10] Обухов，А.М.，*Изв.АН СССРР*，*еср.геоф*.1949，4：281-306.

[11] Гутман，Л.Н.，*Изв.АН СССР*，*еср.zоеф*.1957，1：79-93.

[12] 巢纪平.气象学报，1961，31：191-204.

论层结和风场对小尺度扰动发展的非线性影响*

巢纪平

（中国科学院地球物理研究所）

提要：本文考虑了小尺度扰动发展时，扰动场和平均场（环境）间的非线性相互作用后，分析了大气层结和盛行风对扰动发展的影响。分析结果指出，在层结是中性或不稳定分布时，扰动总能够得到发展；在稳定层结的情况下，只要某一临界条件满足，扰动也能发展。一般来说，盛行风将抑制扰动的发展；但对某些类型的风速廓线，盛行风对扰动的发展也能起到积极的作用。

1　引言

一个众所周知的观测事实是，在大气层结不稳定（或条件性不稳定）的条件下，对流云可以得到旺盛的发展。然而，在我国南方曾多次观测到，即使在副热带高压控制下，大气层结为稳定时，仍有对流活动，甚至在某些时间，在某些局部地区，还能发展成浓积云。

我们曾注意到，在大范围稳定层结的条件下，在某一地区对流的发展，往往要通过几次反复的生灭过程后，才能发展旺盛而形成浓积云。由于这种对流发展的整个过程所历经的时间，一般都很短，对流最后能得到旺盛发展的这一事实，是很难用在这一段时间过程中，大范围层结条件的改变来解释的。对流发展的这一过程有可能表明，在对流发展尚未最后达到旺盛的酝酿时期时，对流的逐次发展和环境间的相互作用，改变了局部地区大气层结的稳定度，从而为以后时刻对流的发展，创立了局部天气条件。

在另一方面，观测也表明，盛行风强时，对流云一般很难得到旺盛的发展。同时风场对对流发展的影响是极为复杂的。它与风速垂直廓线的形状有很大关系。J. 库特纳（Kuettnar）[1] 和 V. G. 普兰克（Plank）[2] 就会注意到这种关系。

对流发展条件的理论分析虽然开始得很早，但对上述这些有意思的现象，却还没有较好的动力学解释。从气块法出发，所得到的结论是，在稳定层结下，对流不可能发展。由薄片法[3,4] 所求得的对流发展判据，也过于偏高。同时在这些工作中，也没有考虑到风场的影响。风速垂直切变在对流发展过程中的作用，J. S. 马尔库斯（Malkus）[5,6] 有过分析，但她的模式是过分简化过了的，同时着眼点也与本文的目的不同。

我们认为，重要问题在于，对流的发展是一个动力学的问题。因而要分析对流发展的条件，必须联系到对流发展的动力学过程，才能期望得到较符合实际的结果。在参考文献［7］中，我们从描写对流发展的运动方程入手分析了问题的初值稳定性后，得到一个考虑了盛行风影响后的对流发展判据。不过由于运动方程做过线性化的处理，这样就割断了扰动场与环境之间相互制约的过程。在本文中，我们将分析对流发展时扰动场与平均场之间的非线性相互作用，从而探讨环境的层结和盛行风对对

* 气象学报，1962，32（2）：164－176.

流发展的影响。同样的问题在参考文献[8]中已做过渐近分析。在以下的分析中,我们近似地用扰动场来表示对流活动。

2 平均场的变化方程和对流发展的判据

如果不考虑湍流对动量和热量的耗散作用,则在(x,z)平面中的小尺度运动,可以用下面的方程组来描写:

$$\frac{\partial u}{\partial t}+u\frac{\partial u}{\partial x}+w\frac{\partial u}{\partial z}=-\frac{1}{\rho}\frac{\partial p}{\partial x} \tag{1}$$

$$\frac{\partial w}{\partial t}+u\frac{\partial w}{\partial x}+w\frac{\partial w}{\partial z}=-\frac{1}{\rho}\frac{\partial p}{\partial z}-g \tag{2}$$

$$\frac{\partial u}{\partial x}+\frac{\partial w}{\partial z}=0 \tag{3}$$

$$\frac{\partial T}{\partial t}+u\frac{\partial T}{\partial x}+w\frac{\partial T}{\partial z}+\gamma_a w=0 \tag{4}$$

$$p=\rho RT \tag{5}$$

式中,u,w 是沿水平坐标 x 和垂直坐标 z 方向的速度分量;ρ,p,T 分别为大气的密度、压力和温度;g 是重力加速度;R 为气体常数;γ_a 为干绝热温度递减率,如空气已达到饱和状态则可以将 γ_a 改写成 γ_b,γ_b 为湿绝热温度递减率。

将各量分成平均的和扰动的两个部分:

$$\begin{gathered}u=\bar{u}(z,t)+u'(x,z,t),\ w=w'(x,z,t),\\ T=\bar{T}(z,t)+T'(x,z,t),\ p=p'(z,t)+p'(x,z,t),\\ \rho=\bar{\rho}(z,t)+\rho'(x,z,t)\end{gathered} \tag{6}$$

其中平均量的定义为:

$$\bar{A}=\frac{1}{2L}\int_0^{2L}A\mathrm{d}x$$

L 为扰动的水平宽度,$2L$ 则为扰动的水平波长。

取式(1)和式(4)两式对 x 的平均,考虑到式(3)后便得到平均状态的变化方程为:

$$\frac{\partial\bar{u}}{\partial t}=-\frac{\partial\overline{u'w'}}{\partial z} \tag{7}$$

$$\frac{\partial\bar{T}}{\partial t}=-\frac{\partial\overline{w'T'}}{\partial z} \tag{8}$$

或者由于 $\bar{\gamma}=-\frac{\partial\bar{T}}{\partial z}$($\bar{\gamma}$ 为平均温度的垂直梯度),由式(8)对 z 微分后得到大气层结的变化方程如下:

$$\frac{\partial\bar{\gamma}}{\partial t}=\frac{\partial^2\overline{w'T'}}{\partial z^2} \tag{9}$$

将式(7)至式(9)对 t 微分后,我们得到平均场对时间的二阶导数:

$$\frac{\partial^2\bar{u}}{\partial t^2}=-\frac{\partial}{\partial z}\left(\frac{\partial\overline{u'w'}}{\partial t}\right) \tag{10}$$

$$\frac{\partial^2 \overline{T}}{\partial t^2} = -\frac{\partial}{\partial z}\left(\frac{\partial\,\overline{w'T'}}{\partial t}\right) \tag{11}$$

$$\frac{\partial^2 \overline{\gamma}}{\partial t^2} = \frac{\partial^2}{\partial z^2}\left(\frac{\partial\,\overline{w'T'}}{\partial t}\right) \tag{12}$$

仿此可以求得平均场对时间更高级的导数。当这些导数求得后，下一时刻的平均场就可以决定。然而要将各高阶导数都求出来，是很困难的。因此我们只局限在对一阶和二阶导数的分析上。这样，下面的分析结果，只适用于对流发展初期。

在垂直气柱中，平均动能和平均内能的变化，由下面各式决定：

$$E_t = \int_0^H \overline{uu_t}\,\mathrm{d}z,\ E_{tt} = \int_0^H (\overline{uu_{tt}} + \overline{u_t^2})\,\mathrm{d}z \tag{13}$$

$$I_t = \frac{C_v}{T_m}\int_0^H \overline{TT_t}\,\mathrm{d}z,\ I_{tt} = \frac{C_v}{T_m}\int_0^H (\overline{TT_{tt}} + \overline{T_t^2})\,\mathrm{d}z \tag{14}$$

式中，H 是对流柱的垂直厚度；C_v 是空气的定容比热；T_m 为平均温度的平均值，取作常量。下角 t 表示对时间的微商。

由于大气的平均状态和扰动状态组成了一个统一的整体，因此在不考虑湍流耗损的作用下，平均状态总能量的减少，则表示扰动总能量要增加；反之，则表示扰动总能量要减小。根据能量转换这一原则，我们自然可以这样来定义对流发展的判据：

$$(E_t + I_t) + (E_{tt} + I_{tt})\Delta t \lessgtr 0 \quad 对流\ \begin{matrix}发展\\衰减\end{matrix} \tag{15}$$

3　扰动场的变化方程

将式(6)代入式(1)至式(5)，便近似地得到扰动运动的变化方程：

$$\frac{\partial u'}{\partial t} + u'\frac{\partial u'}{\partial x} + w'\frac{\partial u'}{\partial z} + \overline{u}\frac{\partial u'}{\partial x} + \frac{\partial \overline{u}}{\partial z}w' = -\frac{1}{\overline{\rho}}\frac{\partial p'}{\partial x} \tag{16}$$

$$\frac{\partial w'}{\partial t} + u'\frac{\partial w'}{\partial x} + w'\frac{\partial w'}{\partial z} + \overline{u}\frac{\partial w'}{\partial x} = -\frac{1}{\overline{\rho}}\frac{\partial p'}{\partial z} + BT' \tag{17}$$

$$\frac{\partial u'}{\partial x} + \frac{\partial w'}{\partial z} = 0 \tag{18}$$

$$\frac{\partial T'}{\partial t} + u'\frac{\partial T'}{\partial x} + w'\frac{\partial T'}{\partial z} + (\gamma_a - \overline{\gamma})w' + \overline{u}\frac{\partial T'}{\partial x} = 0 \tag{19}$$

式中，$\beta = g/\overline{T}$，取作常量。

由式(18)引进流函数 ψ'，并且

$$u' = -\frac{\partial \psi'}{\partial z},\ w' = -\frac{\partial \psi'}{\partial x} \tag{20}$$

从式(16)，式(17)两式消去 p'后得到绕 y 轴的涡度方程：

$$\Delta_2\frac{\partial \psi'}{\partial t} = \left(\frac{\partial \psi'}{\partial z}\frac{\partial \Delta_2\psi'}{\partial x} - \frac{\partial \psi'}{\partial x}\frac{\partial \Delta_2\psi'}{\partial z}\right) - \overline{u}\frac{\partial \Delta_2\psi'}{\partial x} + \frac{\partial^2 \overline{u}}{\partial z^2}\frac{\partial \psi'}{\partial x} + \beta\frac{\partial T'}{\partial x} \tag{21}$$

将式(20)代入式(19)得

$$\frac{\partial T'}{\partial t}=\left(\frac{\partial\psi'}{\partial z}\frac{\partial T'}{\partial x}-\frac{\partial\psi'}{\partial x}\frac{\partial T'}{\partial z}\right)-(\gamma_a-\bar{\gamma})\frac{\partial\psi'}{\partial x}-\bar{u}\frac{\partial T'}{\partial x}\tag{22}$$

在式(21)中 $\Delta_2=\frac{\partial^2}{\partial x^2}+\frac{\partial^2}{\partial z^2}$。

在式(21),式(22)两式中,如果方程的右端由初始场决定,则由此可求得 ψ',T' 对时间的一次导数。这样描写平均场的式(10)至式(12)的右端也为已知。

4 层结分布对对流发展的影响

在本节中,我们首先分析大气层结对对流发展的影响,以及对流的发展反过来对层结的影响。假定在开始时刻 $\bar{u}=0$,并取平均温度的分布为:

$$\bar{T}=\bar{T}_0-\bar{\gamma}z\tag{23}$$

式中,$\bar{\gamma}$ 在初始时刻与高度无关。

将初始时刻的对流活动用下面的扰动场来近似地表示:

$$\psi'=a\sin lz\sin kx,\ T'=b\sin lz\sin kx\tag{24}$$

式中,$l=\frac{\pi}{H}$,$k=\frac{\pi}{L}$。我们指出,对于式(24)所表示的扰动有 $\overline{w'T'}=0$。

将式(24)代入式(21),式(22)的右端,得到

$$\Delta_2\frac{\partial\psi'}{\partial t}=\beta bk\sin lz\cos kx\tag{25}$$

$$\frac{\partial T'}{\partial t}=-(\gamma_a-\bar{\gamma})ak\sin lz\cos kx\tag{26}$$

设式(25)的解为:

$$\frac{\partial\psi'}{\partial t}\beta bkF(z)\cos kx\tag{27}$$

式中 $F(z)$ 根据式(25)由下式决定:

$$\frac{\mathrm{d}^2F}{\mathrm{d}z^2}-k^2F=\sin lz\tag{28}$$

条件为:

$$F(0)=F(H)=0\tag{29}$$

由式(28),式(29)两式定出 F 后代回式(27),我们求得:

$$\frac{\partial\psi'}{\partial t}=-\frac{\beta bk}{l^2+k^2}\sin lz\cos kx\tag{30}$$

注意到对于式(24)所表示的扰动,由式(7)至式(9)三式恒有:

$$\bar{u}_t=\bar{T}_t=\bar{\gamma}_t=0\tag{31}$$

在另一方面将式(30)和式(26)代入式(10)至式(12),则得:

$$\bar{u}_{tt}=0\tag{32}$$

$$\bar{T}_{tt}=-\frac{\pi}{2H}\left(\frac{\beta b^2\mu^2}{1+\mu^2}-a^2\mu^2\frac{\pi^2}{H^2}(\gamma_a-\bar{\gamma})\right)\sin 2\pi\eta\tag{33}$$

$$\bar{\gamma}_{tt}=\frac{\pi^2}{H^2}\left(\frac{\beta b^2\mu^2}{1+\mu^2}-a^2\mu^2\frac{\pi^2}{H^2}(\gamma_a-\bar{\gamma})\right)\cos 2\pi\eta \tag{34}$$

式中，$\mu=k/l,\eta=z/H$。

当式(33)决定后，考虑到式(14)由式(15)求得对流是否发展的判据为：

$$-\frac{C_v\bar{\gamma}\Delta tH}{4T_m}\left(\frac{\beta b^2\mu^2}{1+\mu^2}-a^2\mu^2\frac{\pi^2}{H^2}(\gamma_a-\bar{\gamma})\right)\lessgtr 0\quad \text{对流}\begin{matrix}\text{发展}\\ \text{衰减}\end{matrix} \tag{35}$$

由此不难看出，由式(24)所决定的对流能够发展的条件为：

$$\frac{\beta b^2}{1+\mu^2}>a^2\frac{\pi^2}{H^2}(\gamma_a-\bar{\gamma}) \tag{36}$$

由此可见，除了在中性层结($\gamma_a-\bar{\gamma}=0$)和不稳定层结($\gamma_a-\bar{\gamma}<0$)条件下，对流总可以得到发展外，当大气层结是稳定时，只要条件式(36)满足，对流也能得到发展。

考虑到 $b=T'_{\max},\frac{a\pi}{L}=w'_{\max}$，则式(36)可以换成下式：

$$\gamma_a-\bar{\gamma}<\beta\frac{\mu^2}{1+\mu^2}\frac{T'^2_{\max}}{w'^2_{\max}} \tag{37}$$

式(37)指出，对于一定的稳定度，温度越高的扰动，越容易得到发展；垂直气流越强的扰动，越不容易得到发展。

一般来说，温度的测量要比垂直速度的测量来得容易，因而式(37)还可以用来估算在稳定层结下，对流发展对可能到达的最大垂直速度。这个最大垂直速度显然为：

$$w^2_{\max}=\beta\frac{\mu^2}{1+\mu^2}\cdot\frac{T'^2_{\max}}{\gamma_a-\bar{\gamma}} \tag{38}$$

如取 $\mu=1,T'_{\max}=0.5℃,\gamma_a-\bar{\gamma}=3\times10^{-3}℃/\mathrm{m}$，则 $w\approx1.2\ \mathrm{m/s}$。

显然，在现在的情况下，扰动发展的能量来自大气的平均内能，然而大气的平均内能，只有当$\overline{w'T'}>0$时才能释放出来，而在我们所给的初始扰动中$\overline{w'T'}\equiv0$，因此要平均内能释放，必然要有 $\int_0^H\frac{\partial\overline{w'T'}}{\partial t}\mathrm{d}z>0$ 方可。由式(24)，式(26)，式(30)不难算出：

$$\int_0^H\frac{\partial\overline{w'T'}}{\partial t}\mathrm{d}z=\frac{\mu^2}{4}\left[\frac{\beta b^2}{1+\mu^2}-a^2\frac{\pi^2}{H^2}(\gamma_a-\bar{\gamma})\right] \tag{39}$$

由此可见，式(36)即为 $\int_0^H\frac{\partial\overline{w'T'}}{\partial t}\mathrm{d}z>0$ 的条件。

对于发展和不发展的对流所引起的局地层结的改变，可由式(34)算出。数值例子见图 1a 和图 1b。图 1a 是不发展的情况，这时参数取 $\mu=1,H=1\ \mathrm{km},\gamma_a-\bar{\gamma}=3\times10^{-3}℃/\mathrm{m},b=1℃,w_{\max}=3\ \mathrm{m/s}$。图 1b 是发展的情况，取 $H=4\ \mathrm{km},b=3.6℃,w_{\max}=0.8\ \mathrm{m/s}$，其余同前。由图可见，在不发展的情况下，中层的稳定度要减小，而在低层和高层稳定度要增大；在发展情况下则相反，中层稳定度要增大，低层和高层的稳定度要减小。变化的最大值，在 100 s 后可以到达 $1.3\times10^{-3}℃/\mathrm{m}$，即约改变了原来层结的 40%。

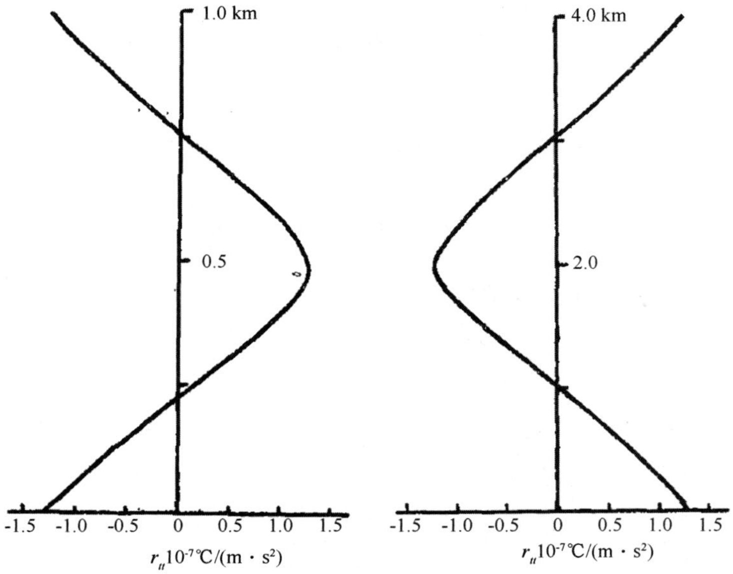

图 1

这一简单的分析,说明了扰动场与平均场之间的相互作用,在对流的发展过程中起着重要的影响。我们在研究对流发展的文章中[9],虽然由于考虑了扰动方程中的非线性项后,得到了对流发展过程中的一些有意义的结果,但是,由于没有考虑扰动场与平均场之间的相互作用,因而并没有算出大气平均层结如此显著的改变。同时在那篇文章中,计算结果也表明,如果在开始时刻大气层结是稳定的,则对流很难得到显著的发展。同样地 Л. Н. 古特曼(Гутман)[10]从常定非线性扰动方程出发,也得到在稳定层结大气中对流云不可能发展这一结论。由此可见,只有考虑了这种相互作用后,才能解释层结稳定的大气中热对流能够发展这一观测事实。

5 盛行风对对流发展的影响

设在起始时刻大气中存在着盛行风 $\bar{u}(z)$,$\bar{u}$ 随高度的分布是已知的。为了分析简便起见,初始时刻的平均温度分布和扰动的形式仍取式(23)和式(24)。

这时式(21)和式(22)化为:

$$\Delta_2^2 \frac{\partial \psi'}{\partial t} = ak\left[(k^2 + l^2)\bar{u} + \bar{u}_{zz} + \frac{\beta b}{a}\right]\sin lz \cos kx \tag{40}$$

$$\frac{\partial T'}{\partial t} = -ak\left[(\gamma_a - \bar{\gamma}) + \frac{b}{a}\bar{u}\right]\sin lz \cos kx \tag{41}$$

式中,$\bar{u}_{zz} = \dfrac{\partial^2 \bar{u}}{\partial z^2}$。

假设式(40)的解为下面的形式:

$$\frac{\partial \psi'}{\partial t} = akF(z)\cos kx \tag{42}$$

其中 $F(z)$ 满足方程:

$$\frac{\mathrm{d}^2F}{\mathrm{d}z^2}-k^2F=\left[(k^2+l^2)\bar{u}+\bar{u}_{zz}+\frac{\beta b}{a}\right]\sin lz \tag{43}$$

式(43)的边界条件仍同式(29)。

方程(43)满足条件式(29)的解为：

$$\bar{F}(\eta)=\int_0^1 G(\eta,\xi)f(\xi)\,\mathrm{d}\xi \tag{44}$$

式中，$f(\xi)$为式(43)右端的非齐次部分，而影响函数为：

$$G(\eta,\xi)=\begin{cases}\dfrac{\sinh n\eta\ \sinh n(\xi-1)}{n\sinh n} & (\eta\leqslant\xi)\\ \dfrac{\sinh n\xi\ \sinh n(\eta-1)}{n\ \sinh n} & (\eta\geqslant\xi)\end{cases} \tag{45}$$

其中，$n=kH$。

下面我们分别以几种具有一定代表性的盛行风速廓线来讨论风场对小尺度对流发展的影响。

(1)$\bar{u}$=常数(均匀气流)

这时我们求得：

$$\frac{\partial\psi'}{\partial t}=-ak\left(\bar{u}+\frac{\beta b}{a(k^2+l^2)}\right)\sin lz\ \cos kx \tag{46}$$

由计算不难证明，在这种情况下，平均场的变化与初始时刻无盛行风时完全相同。平均动能保持不变，平均内能的变化规律仍同上节。所以，对流发展的判据仍同式(35)。这表明均匀气流对对流的发展并无影响。

(2)$\bar{u}=\bar{u}_0+Az$

其中，$\bar{u}_0$ 为地面风速；A 为风速垂直切变，取为常数。

这时涡度方程的解为：

$$\frac{\partial\psi'}{\partial t}=-ak\left[\left((\bar{u}_0+Az)+\frac{\beta b}{a(k^2+l^2)}\right)\sin lz+\frac{2l}{k^2+l^2}A\cos lz+\frac{2l}{k^2+l^2}\frac{A}{\sinh kH}(\sinh kz-\sinh k(H-z))\right]\cos kx \tag{47}$$

由直接计算得知，平均场的一阶导数恒为零，二阶导数则为：

$$\bar{u}_{tt}=\left(\frac{\pi}{H}\right)^3\mu^2a^2A\left\{-\frac{1}{2}\sin 2\pi\eta+\frac{1}{\sinh\mu\pi}[\sinh\mu\pi(1-\eta)-\sinh\mu\pi\eta]\cdot\sin\pi\eta\right\} \tag{48}$$

$$\begin{aligned}\bar{T}_{tt}=&-\frac{1}{2}\left\{\left[\frac{\beta b^2\mu^2}{1+\mu^2}-a^2\mu^2\frac{\pi^2}{H^2}(\gamma_a-\bar{\gamma})\right]\frac{\pi}{H}\sin 2\pi\eta\right.\\&+\frac{2ab\mu^2}{1+\mu^2}\left(\frac{\pi}{H}\right)^2A\left[\frac{\cos\pi\eta(\sinh\mu\pi\eta-\sinh\mu\pi(1-\eta))}{\sinh\pi\mu}\right.\\&\left.\left.+\frac{\mu\sin\pi\eta(\cosh\mu\pi\eta+\cosh\mu\pi(1-\eta))}{\sinh\mu\pi}-\cos 2\pi\eta\right]\right\}\end{aligned} \tag{49}$$

$$\begin{aligned}\bar{\gamma}_{tt}=&\left\{\left[\frac{\beta b^2\mu^2}{1+\mu^2}-a^2\mu^2\frac{\pi^2}{H^2}(\gamma_a-\bar{\gamma})\right]\left(\frac{\pi^2}{H^2}\right)\cos 2\pi\eta\right.\\&+\frac{ab\mu^2}{1+\mu^2}\left(\frac{\pi}{H}\right)^3\cdot A\cdot\left[2\sin 2\pi\eta+2\mu\frac{\cos\pi\eta(\cosh\mu\pi\eta+\cosh\mu\pi(1-\eta))}{\sinh\pi\mu}\right.\end{aligned}$$

$$+(\mu^2-1)\frac{\sin\pi\eta(\sinh\mu\pi\eta-\sinh\mu\pi(1-\eta))}{\sinh\mu\pi}\Big]\Big\} \tag{50}$$

不难求得平均动能的变化为：

$$E_u=A^2\frac{a^2\pi^2}{4H}\frac{\mu^2}{(1+\mu^2)^2}\left[(\mu^2+1)(\mu^2-3)+\frac{16}{\pi}\mu\coth\frac{\mu\pi}{2}\right] \tag{51}$$

由于上式中括号内的量对所有的μ值恒取正值，因此无论风速垂直切变是正值或是负值平均动能总要增加，即在这种情况下，扰动不可能从平均动能中取得发展的能量。

很有意思，虽然式(49)与式(33)很不一样，但平均内能的变化却完全一样，即仍为：

$$I_u=-\frac{C_v\bar{\gamma}H}{4T_m}\left(\frac{\beta b^2\mu^2}{1+\mu^2}-a^2\mu^2\frac{\pi^2}{H^2}(\gamma_a-\bar{\gamma})\right) \tag{52}$$

这表明在小尺度运动中，当风速的廓线随高度呈线性变化时，扰动不可能从盛行气流中取得大气的平均内能。这是一个很有兴趣的结果。

由于风速垂直一次切变唯一的作用是使大气的平均动能增加，因此，风速垂直一次切变是不利于小尺度对流发展的因子。

联合式(51)与式(52)，我们得到对流发展的特别式为：

$$A^2\frac{a^2\pi^2}{H}\frac{\mu^2}{(1+\mu^2)^2}\left((\mu^2+1)(\mu^2-3)+\frac{16}{\pi}\coth\frac{\mu\pi}{2}\right)$$
$$-\frac{C_v\beta\bar{\gamma}b^2H}{T_m}\left(\frac{\mu^2}{1+\mu^2}-\mu^2\frac{a^2\pi^2(\gamma_a-\bar{\gamma})}{\beta b^2H^2}\right)\lessgtr 0\quad \text{对流}\ \begin{matrix}\text{发展}\\ \text{衰减}\end{matrix} \tag{53}$$

由此可见，在现在的情况下，满足条件式(36)或式(37)，还只是给定了扰动发展的必要条件，要使条件成为充分，还必须对风速垂直切变的大小给予下面的限制：

$$|A|\lessgtr A_c\quad \text{对流}\ \begin{matrix}\text{发展}\\ \text{衰减}\end{matrix} \tag{54}$$

其中临界风速垂直切变A_c为：

$$A_c=\frac{|T'_{\max}|}{|w'_{\max}|}(1+\mu^2)\cdot\sqrt{\frac{C_v\beta\bar{\gamma}}{T_m}}\cdot\sqrt{\frac{\dfrac{\mu^2}{1+\mu^2}-\dfrac{(\gamma_a-\bar{\gamma})}{\beta}\cdot\dfrac{w'^2_{\max}}{T'^2_{\max}}}{(\mu^2+1)(\mu^2-3)+\dfrac{16}{\pi}\mu\coth\dfrac{\mu\pi}{2}}} \tag{55}$$

不同参数值下，A_c的大小见图2。

关于由于对流的发展和衰减所引起大气平均气流和层结的改变，在图3中给出了数值例子。

(3) $\bar{u}=4\bar{u}_m\eta(1-\eta)$

我们指出在这一风速廓线中，风速的垂直二次切变为$\bar{u}_{zz}=-8u_m/H^2$，恒小于零。

将式(43)的$\bar{F}$代入式(42)，得到在这一例子中的涡度方程的解为：

$$\frac{\partial\psi'}{\partial t}=-ak\left\{\left(\frac{\beta bH^2}{a(\mu^2+1)\pi^2}+\bar{u}-\frac{2(\mu^2-1)H^2}{(\mu^2+1)^2\pi^2}\bar{u}_{zz}\right)\sin\pi\eta\right.$$
$$+\frac{2H}{(\mu^2+1)\pi}\bar{u}_z(z)\cos\pi\eta+\frac{2H}{(\mu^2+1)\pi}\frac{1}{\sinh\mu\pi}$$
$$\left.\times[\bar{u}_z(H)\sinh\mu\pi\eta-\bar{u}_z(0)\sinh\mu\pi(1-\eta)]\right\}\cos kx \tag{56}$$

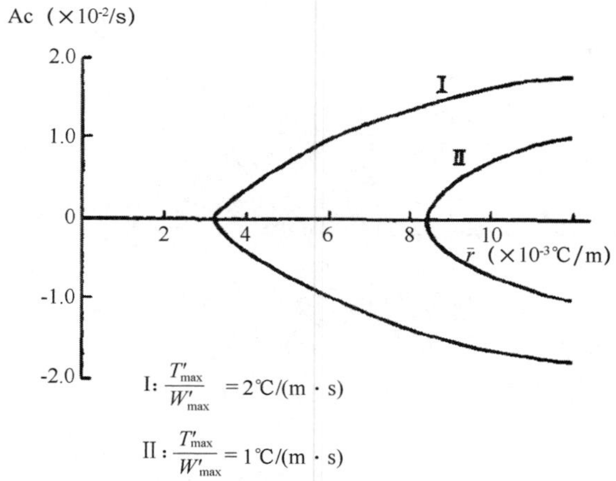

图 2

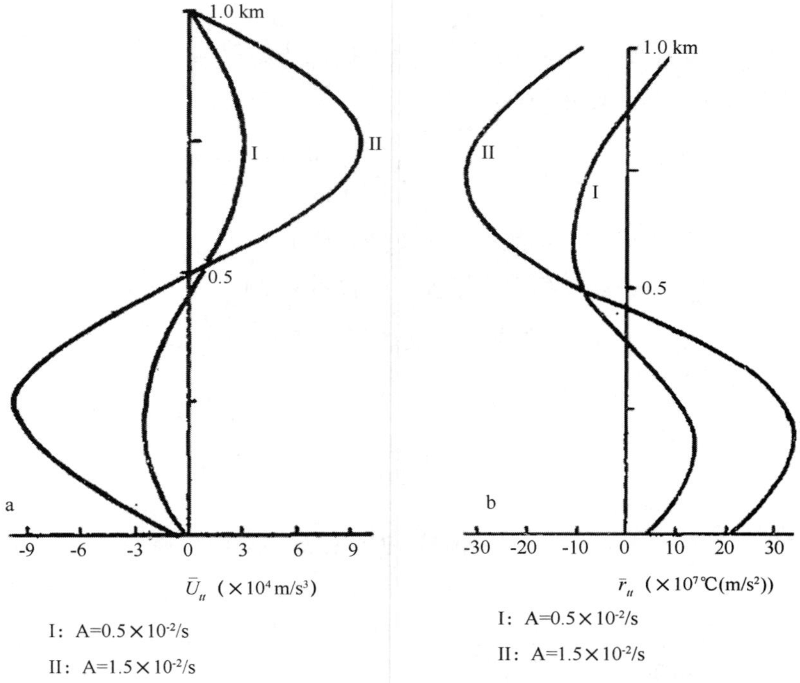

图 3

由式(10)至式(12)求得平均场的变化为：

$$\bar{u}_{tt} = \frac{a^2k^2\bar{u}_m}{H^2}\left\{-2\pi(1-2\eta)\sin 2\pi\eta + \frac{4(\mu^2-3)}{\mu^2+1}\sin^2\pi\eta + \frac{4\pi\sin\pi\eta}{\sinh\mu\pi}\left[\sinh\mu\pi\eta + \sinh\mu\pi(1-\eta)\right]\right\} \tag{57}$$

$$\bar{T}_{tt} = -\frac{1}{2}\left\{\left[\frac{\beta b^2\mu^2}{\mu^2+1} - a^2\mu^2\frac{\pi^2}{H^2}(\gamma_a - \bar{\gamma}) - \frac{8ab\bar{u}_m(\mu^2-3)}{H^2(\mu^2+1)^2}\right]\frac{\pi}{H}\sin 2\pi\eta\right.$$

$$- \bar{u}_m\frac{8ab\mu^2}{\mu^2+1}\frac{\pi^2}{H^3}\cdot\left[\frac{\cos\pi\eta(\sinh\pi\mu\eta + \sinh\pi\mu(1-\eta))}{\sinh\mu\pi}\right.$$

$$+\mu\frac{\sin\pi\eta(\cosh\pi\mu\eta-\cosh\pi\mu(1-\eta))}{\sinh\mu\pi}+(1-2\eta)\cos2\pi\eta\Bigg]\Bigg\} \tag{58}$$

$$\bar{\gamma}_{tt}=\left\{\left[\frac{\beta b^2\mu^2}{\mu^2+1}-a^2\mu^2\frac{\pi^2}{H^2}(\gamma_a-\bar{\gamma})-\frac{8ab\bar{u}_m(\mu^2-3)}{H^2(\mu^2+1)^2}\right]\frac{\pi^2}{H^2}\cos2\pi\eta\right.$$
$$-\frac{4ab\mu^2}{\mu^2+1}\frac{\pi^3}{H^3}\bar{u}_m\left[-\frac{2}{\pi}\cos2\pi\eta-2(1-2\eta)\sin2\pi\eta+2\mu\frac{\cos\pi\eta(\cosh\mu\pi\eta-\cosh\mu\pi(1-\eta))}{\sinh\mu\pi}\right.$$
$$\left.\left.+(\mu^2-1)\frac{\sin\mu\eta(\sinh\mu\pi\eta+\sinh\mu\pi(1-\eta))}{\sinh\mu\pi}\right]\right\} \tag{59}$$

不难求得平均功能的二阶导数为：

$$E_{tt}=\frac{a^2(k^2+l^2)\bar{u}_m^2}{4H}\left\{\frac{16}{3}\left(1-\frac{6}{\pi^2}\right)\frac{\mu^2}{\mu^2+1}-\frac{64\mu^2}{3(\mu^2+1)^2}\left(1+\frac{3}{\pi^2}\right)\right.$$
$$\left.-\frac{256}{\pi^2}\frac{\mu^2(3\mu^2-1)}{(\mu^2+1)^4}+\frac{256\mu^3}{\pi(\mu^2+1)^3}\tanh\frac{\mu\pi}{2}\right\}. \tag{60}$$

由直接计算得知，式(60)对所有的μ值恒取正值。因此，这表示扰动不可能从平均气流中取得功能。

大气平均内能的二阶导数这时为：

$$I_{tt}=-\frac{\bar{\gamma}C_vH}{4T_m}\left\{\left[\frac{\beta b^2\mu^2}{\mu^2+1}-a^2\mu^2\frac{\pi^2}{H^2}(\gamma_a-\bar{\gamma})\right]-\frac{8ab\bar{u}_m}{H^2(\mu^2+1)}\left[\frac{5\mu^2-3}{\mu^2+1}+2\mu^2\operatorname{csch}\mu\pi\right]\right\} \tag{61}$$

由于在大气中的小尺度对流一般μ值接近于1，而$ab\sim w'_{\max}T'_{\max}$一般也大于零，因此式(61)中与风场有关的那一项总要使大气中的平均内能增加。联合对动能的分析，可以得出结论：具有二次垂直切变为负值的盛行气流，将抑制对流的继续发展。当取$H=L=1$ km，$b=1$℃，$w_{\max}=3$ m/s，$\bar{\gamma}=7\times10^{-3}$℃/m，$\bar{u}_m=10$ m/s时$\bar{u}_{tt}$，$\bar{\gamma}_{tt}$的值见图4。

6 结果讨论

以上的分析表明，盛行风的存在一般不利于对流的发展。这是与观测事实相符合的。但是，在某些情况下，例如H. 第逊(Dessens)[11]指出，强烈发展的对流云一般在高空伴随有强烈的盛行风。他并认为对流在高层将从盛行风中取得维持它发展的功能。因此我们要问，是否存在着对对流发展有利的那样的风速廓线？下面来分析这一问题。

事实上，由以上的分析不难看出，涡度方程式(21)中与温度有关的那一项，对平均动能的释放并不起作用，但去掉了这一项后，在(x,z)平面中的小尺度运动的涡度方程与在(x,y)平面中的涡度方程具有相同的构造。因而，在(x,z)平面的小尺度运动中，平均动能的释放规律应该与(x,y)平面中运动平均动能的释放规律完全相同，而后者的规律性郭晓岚有过详细的讨论[12]，并指出：当风速廓线中存在着拐点(即二次切变为零的点)时，平均运动将输送动能给扰动；反之，如在风速廓线中不存在拐点时，平均动能则要增加，即动能的输送方向倒转。因此，只要盛行风的垂直廓线中存在着拐点，对流从盛行风中取得平均动能是可能的。不过在这种情况下，我们还要进一步分析，由于这一盛行风的存在，会不会阻碍平均内能的释放。事实上平均动能和平均内能相反的输送过程是存在的。为了说明这一点，我们举下面的例子。

取$\bar{u}=\bar{u}_m\sin^2lz$，在这一风速廓线中存在着两个拐点$\left(\frac{H}{4},\frac{3H}{4}\right)$。平均温度场的分布和初始扰动的形

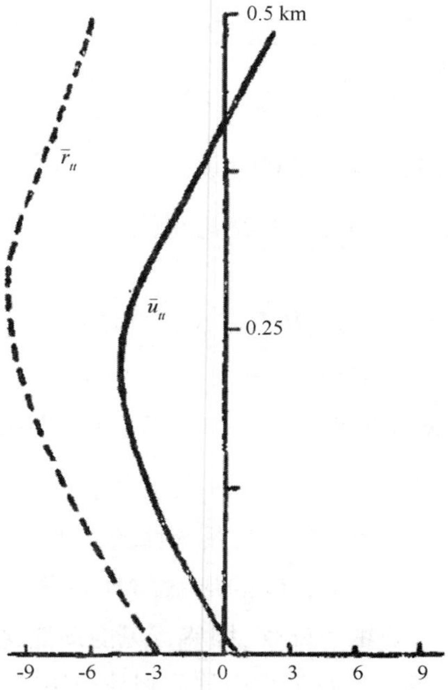

图 4

成仍同前。

这时涡度方程的解为：

$$\frac{\partial\psi'}{\partial t}=-ak\left\{\left[\frac{\beta b}{a(k^2+l^2)}+\bar{u}_m\frac{3k^2-l^2}{4(k^2+l^2)}\right]\sin lz-\frac{k^2-3l^2}{4(k^2+9l^2)}\bar{u}_m\sin 3lz\right\}\cos kx \tag{62}$$

平均场的变化为：

$$\bar{u}_{tt}=-\frac{k^2a^2\pi^2\bar{u}_m(\mu^2-3)}{(\mu^2+9)H^2}\sin\pi\eta\sin 3\pi\eta \tag{63}$$

$$\begin{aligned}\bar{T}_{tt}=-\frac{1}{2}\Bigg\{&\left[\frac{\beta b^2\mu^2}{\mu^2+1}-a^2\mu^2\frac{\pi^2}{H^2}(\gamma_a-\bar{\gamma})\right]\frac{\pi}{H}\sin 2\pi\eta\\&+\frac{ab}{2}\mu^2\left(\frac{\pi}{H}\right)^3\bar{u}_m\left[\left(1-2\frac{\mu^2-3}{\mu^2+9}\right)\sin 4\pi\eta-\left(\frac{\mu^2+5}{2(\mu^2+1)}+\frac{\mu^2-3}{\mu^2+9}\right)\sin 2\pi\eta\right]\Bigg\}\end{aligned} \tag{64}$$

$$\begin{aligned}\bar{\gamma}_{tt}=\Bigg\{&\left[\frac{\beta b^2\mu^2}{\mu^2+1}-a^2\mu^2\frac{\pi^2}{H^2}(\gamma_a-\bar{\gamma})\right]\frac{\pi^2}{H^2}\cos 2\pi\eta\\&+\frac{ab}{2}\mu^2\left(\frac{\pi}{H}\right)^4\bar{u}_m\left[2\left(1-2\frac{\mu^2-3}{\mu^2+9}\right)\cos 4\pi\eta-\left(\frac{\mu^2+5}{2(\mu^2+1)}+\frac{\mu^2-3}{\mu^2+9}\right)\cos 2\pi\eta\right]\Bigg\}\end{aligned} \tag{65}$$

平均动能的变化不难求得为：

$$E_{tt}=\frac{a^2(k^2+l^2)u_m^2}{4H}\left\{\frac{\pi^2}{2}\frac{(\mu^2-3)\mu^2}{(\mu^2+1)(\mu^2+9)}\right\} \tag{66}$$

由此可见，当$\mu^2<3$，即$\frac{H}{L}<\sqrt{3}$时，平均动能要减小，即对于$\frac{H}{L}<\sqrt{3}$的对流可以从盛行风中取得平均动能。由于一般来说$\frac{H}{L}\approx 1$，因此这一条件通常都可以满足。

平均内能的二阶层数为：

$$I_{tt}=-\frac{\overline{\gamma}C_v H}{4T_m}\left\{\left[\frac{\beta b^2\mu^2}{\mu^2+1}-a^2\mu^2\frac{\pi^2}{H^2}(\gamma_a-\overline{\gamma})\right]-\frac{ab\pi^2\mu^2}{2H^2}u_m\frac{\mu^4+30}{(\mu^2+1)(\mu^2+9)}\right\} \tag{67}$$

显然，由于这一盛行风的存在将减少平均内能的释放。

因此，对于$\frac{H}{L}<\sqrt{3}$的扰动，盛行风是否对扰动的发展有利要看条件

$$\overline{u}_m>\frac{\overline{\gamma}C_v HT'_{\max}(\mu^4+30)\mu}{T_m\pi w'_{\max}(\mu^2+1)(3-\mu^2)}=\overline{u}_m^* \tag{68}$$

是否成立。当此条件成立时，盛行风将促进对流的发展，反之将抑制对流的发展。如取$\overline{\gamma}=7\times10^{-3}$℃/m，$H=1$ km，$\mu=1$，$b=T'_{\max}=1$℃，$w'_{\max}=3$ m/s，则$\overline{u}_m^*\approx15$ m/s，如取$H=5$ km，其余参数相同，则$\overline{u}_m^*\approx75$ m/s。一般来说，在气团内部这一条件不易满足，只有在系统性条件下，才有可能出现这样大的盛行风。我们注意到在第逊所根据的资料中，其风速廓线的形状与现在这个例子很接近，同时最大风速约为80 m/s。因此，看来第逊的假说有一定的可能性，即盛行风的动能将是强烈对流发展的一个能源。

在以上所分析的几个例子中，盛行风对平均内能的释放皆不能起到促进的作用，究竟要怎么样的风速廓线才能对平均内能的释放起到积极的作用呢？我们不难算出当风速廓线能用一个对z的二次多项式来迫近，并且二次切变为常数时，在平均内能变化的公式中与风速有关的一项为：

$$I_{tt}=-\frac{\overline{\gamma}C_v Hab}{4T_m}\overline{u}_{zz}\left(\frac{5\mu^2-3}{(\mu^2+1)^2}+\frac{2\mu^2}{\mu^2+1}\operatorname{csch}\mu\pi\right) \tag{69}$$

由此可见，平均内能的释放与风速二次切变有关，当$\overline{u}_{zz}>0$时平均内能要减小，即释放出来；反之，当$\overline{u}_{zz}<0$时则平均内能要增加。

自然，本文所用的这种近似方法具有一定的局限性。

首先，我们只计算了泰勒展开式中的二次项，如果再考虑高次项的影响，结论将会做某些修改。不过高次项的影响可以做下面的定性估计。设初始时刻的平均风速为$\overline{u}_0$，以后时刻为$\overline{u}$，则平均功能的改变为[12]：

$$\Delta E=\frac{1}{2}\int_0^H(\overline{u}^2-\overline{u}_0^2)\mathrm{d}z=\int_H^0\overline{u}_0\Delta\overline{u}\mathrm{d}z+\frac{1}{2}\int_H^0(\Delta\overline{u})^2\mathrm{d}z \tag{70}$$

如果$\Delta\overline{u}$等于$\frac{1}{2}t^2\overline{u}_{tt}$，那么式中右端第一项对应于$E$对时间的泰勒展开式中的二次项$\frac{1}{2}t^2E_{tt}$（即我们所计算的项），而右端的第二项则对应于展开式中的$\frac{1}{4!}t^4\frac{\partial^4E}{\partial t^4}$那一项。因此只要$\Delta\overline{u}$不恒等于零，式(70)右端第二项恒取正值，即总要使平均动能增加。同样地，也可以将平均内能写成与式(70)相类似的形式：

$$\Delta I=\frac{C_v}{2T_m}\int_0^H(\overline{T}^2-\overline{T}_0^2)\mathrm{d}z=\frac{C_v}{T_m}\int_H^0\overline{T}_0\Delta\overline{T}\mathrm{d}z+\frac{C_v}{2T_m}\int_H^0(\Delta T)^2\mathrm{d}z \tag{71}$$

因此，平均内能对时间的泰勒展开式中的四次项也总要使平均内能增加。由此可见，在以上的分析中，考虑了高次项后要加强对流衰减的趋势和减弱对流发展的趋势。

其次，分析是针对某一特定扰动进行的，对于不同形式的扰动，定量的结果会有一定的差异。不

过上述能量转换的机制看来是成立的。

最后，我们指出，在以上所求得的对流发展的判据中，包含了扰动场中量 T'，w'，而这两个量只有当对流已经发展后才能测定，事先并不知道，这就使得我们应用这些判据时受到一定的限制。因此上面求得的条件，只能作为对流已经发生后看它是否能进一步发展的判据。

参考文献

[1]　Kuettnar，J. *Tellus*，1959，11：267–294.

[2]　Plank，V.G.，Cumulus dynamics，Conference on cumulus convection portsmonth，New Hampshire，1959：109–118.

[3]　Bjerkues，J.，*Q.J.Roy.Meteor.Soc*，1938，64：328–330.

[4]　Шишкин，Н.С.，*Трубы ГГО*，1958，82.

[5]　Malkus，J.S.，*Trnas.Amer.Geopohys Un*，，1949，30：19–25.

[6]　Malkus，J.S.，*Q.J.Roy.Meteor.Soc*，1952，78：530–542.

[7]　巢纪平.对流发展局部气象条件的初步理论分析.气象学报，1962，32：87–90.

[8]　巢纪平.小尺度对流的发展和环境间相互作用的一个近似分析.气象学报，1962，32：11–18.

[9]　巢纪平.层结大气中热对流发展的一个非线性分析.气象学报，1961，31：191–204.

[10]　Гутман，Л.Н.，ДАН СССР т，1957，112：1033–1036.

[11]　Dessens，H.，Physics of Precipitation cloud physics conference，Woods Hole，Mass.，1959：333–336.

[12]　郭晓岚，*Tellus*.1953，5：475–493.

对流云动力学研究的进展*

巢纪平

（中国科学院地球物理研究所）

在我国，对流云的活动是极频繁的。研究对流云的发生、发展条件和结构，对国民经济有着重要的意义。一方面，很多灾害性天气如雷暴、冰雹和暴雨等，都与对流云的猛烈发展有关。研究对流云发展和活动的规律，将直接有助于提高这种小范围灾害性天气预报的准确率。另一方面，对流云是人工降水作业的对象。虽然目前人工降水的催化方法，一般只是从影响云滴增长的微观过程入手，但由于云滴增长的微观过程受到云的宏观条件的制约，因此，研究对流云发展的宏观条件和特征，同研究云滴增长的微观过程一样，对进一步提高人工降水和人工消雹的效果，都是重要的环节。由于对流云的研究涉及很多方面，本文只对动力学方面的研究成果作些介绍。

目前观测事实已经为动力学的研究提出了不少问题。从预报和人工降水的角度来看，我们认为有两个问题是更为重要的。这两个问题是：① 在什么局地条件下，对流云能够发生、发展？② 对流云的生命史具有哪些特征以及哪些因子制约了它的强度和结构？

实际上在这方面的理论研究开始得很早，例如很早就有用绝热气流法以及到20世纪30年代提出的用薄层法来研究对流云的发展条件。然而，实践表明，用这种简单的理论来预计对流云的发生、发展，常常并不是有效的。事实上，由于控制对流云发生、发展的条件是复杂的，如环境的层结、湿度、盛行风以及下垫面的属性等，在不同情况下都会产生重要的影响。因此一个研究对流云发生、发展的理论，绝不可能是简单的单因子理论。

不过，在过去，要进行这种多因子影响的定量研究，事实上是不可能的。因为当时还缺乏一种有效的计算工具。只有当电子计算机问世后，才为这种研究提供了可能性。同时近年来大气环流数值试验和数值天气预报的成功，也鼓舞了人们尝试去应用数值试验这种新方法来研究对流云的理论问题。

像数值研究天气变化必须首先建立天气方程组一样，要研究对流云的发生、演变和结构，也必须首先建立对流云动力学方程组。一组能描写对流云生命史中各种过程的方程，是极其复杂的。因为这里不仅有宏观气流的发展问题，同时还有云滴增长的微观过程、雷电形成的物理过程等问题。即使不包括雷电过程，这样一组方程至少也有15个变量[1]。从方程的性质上看，这里面有描写宏观气象要素演变的流体力学方程，也有描写云滴增长的属于统计力学的非线性随机碰并方程。在目前即使应用高速电子计算机，要直接求出这组方程的解也是困难的。不过，作为研究的第一步，可以把宏观过程和微观过程分开来处理。这样，问题就大大简化了。对于研究宏观过程来说，现在只需建立一组描写对流云发展的宏观方程组。

宏观对流云动力学方程组，实质上是一组考虑了水汽相变后的重力场中的可压缩流体力学方程组。由于对流云本质上是一种自由对流现象，而它的生命期一般不超过2小时，比起一个太阳日来要

* 科学通报，14(7)，8-12(1963)；10.1360/csb1963-8-7-8.

短很多，因此与一般大尺度运动方程组不同，其中用阿基米德浮力来代换重力，也不考虑因地球自转而引起的柯利奥来力。还有一个重要的不同点是：由于对流云的垂直尺度一般与水平尺度一样，也有几千米到十几千米，因此不能像大尺度运动那样，可以用压力的静力平衡关系来代替垂直运动方程。

另一方面，由于空气介质的可压缩性，在这样一组原始形式的方程组中必然包含有声波。比起对流运动来，声波是一种高频的振荡，因此，当求这样一组方程的数值解时，需要相当小的时间和空间步长，否则计算误差容易发展。像处理大尺度运动时要从原始方程中滤掉重力惯性波这种“气象噪音”一样，也必须设法从现在的原始方程组中滤掉成为“噪音” 的声波。如何能尽量不歪曲对流运动的本质，又能方便地滤去声波，这是一个具有原则意义的问题。对于线性方程，这个问题已分别有人解决了[2,3,4]。因此对流云宏观动力学方程组的形式也基本上确定了。

于是研究一开始提出的两个问题，就归到给定一定的初始、边界和环境条件，求对流云动力学方程组的积分。自然，由于这仍然是一组非常复杂的非线性偏微分方程，在求解时，根据不同的情况还得进行不同的简化，即拟订各种所谓“模式”。

在第一个问题方面，梅逊等[5]把对流云看成是与外界有混合过程的浮升气块。应用这一模式他们分析了对流云发生、发展的条件*。如果不考虑云内外空气有混合以及摩擦阻力，那么这个模式就变成经典的绝热气块模式。在这种情况下，计算表明，气块将做强烈的振荡运动。这种运动的性质和强度都与对流云的实际发展情况不符。这说明为什么过去简单的绝热气块法常常过分夸大了对流云的发展强度。由此也说明了，由云内外空气混合所造成的“阻力”，对对流云的发展的确是重要的。

当考虑了混合过程并给了一定的数值后，计算表明，制约对流云发生、发展的重要因素，是环境的层结、湿度、初始温度偏差。并且对流云的发生、发展“敏锐” 地依赖于这些参数的数值。例如，当其他参数相同，相对湿度为 70%时，甚至层结取 9℃/km，云也不能发生。但当相对湿度增大到 80%时，即使层结取 8℃/km，也可能有云形成。又如，在一定条件下，层结从 7℃/km 增加到 8℃/km 时，云的强度很少有改变，但当层结从 8℃/km 增加得 9℃/km 时，云厚将从 0. 9 km 增大到 4. 5 km，垂直速度的最大值从 2. 5 m/s 增大到 12 m/s，含水量从 0. 25 g/kg 增大到 1. 9 g/kg。这种“敏锐性” 更突出地表现在对初始温度偏差 ΔT_0 的数值上。如果当其他参数相同，$\Delta T_0 = 0.5$℃时，云尚不能发生，但当 $\Delta T_0 = 0.6$℃时，云就发展起来了。浅井富雄[7]的一维模式试验结果也表明，对流云能否发生与参数值有很大关系。

从以上的数值试验结果可以得出一条重要的结论：对于一定的环境条件(层结、湿度等)，要激发起对流云，初始扰动的强度必须达到某一临界值。进一步利用更接近实际的模式给出对流云发生、发展的临界条件，这不仅对对流云预报是重要的，而且还具有更积极的意义。因为参照理论分析的结果，在比较合适的情况下，人们可以根据已经出现的环境条件，用人工的办法制造某种初始扰动来激发对流云的发展，为局部降水创造必要的条件。事实上，这种人工造云的试验也有人在做了。虽然这里面还存在不少需要解决的问题，但不失为是人工控制局部天气的一个探索方向。

关于第二个问题。观测表明，对流云往往是由很多个云泡迭积而成。因此研究者首先注意研究云泡的生命史。从马尔古斯开始的很多数值试验都表明[8-10]：初始温度扰动只要经过 3~5 分钟便能发展成一个成熟的云泡，成熟了的云泡呈蘑菇状，流场的结构宛如涡环。上升的云泡是暖的，四周的干冷空气从泡的底部楔入泡中，使温度梯度在泡的前沿部分增大(图 1)。计算也表明，环境空气不稳

---

* 陈瑞荣最近也做过类似的研究。

定度的增大,虽然有利于云泡的发展,但却使云泡的尺度变小。由于目前缺乏云泡生命史特征的详细观测,这些图案的可靠程度尚有待验证。但这些计算结果与一些对流泡的实验结果(见参考文献[11])是相近的。

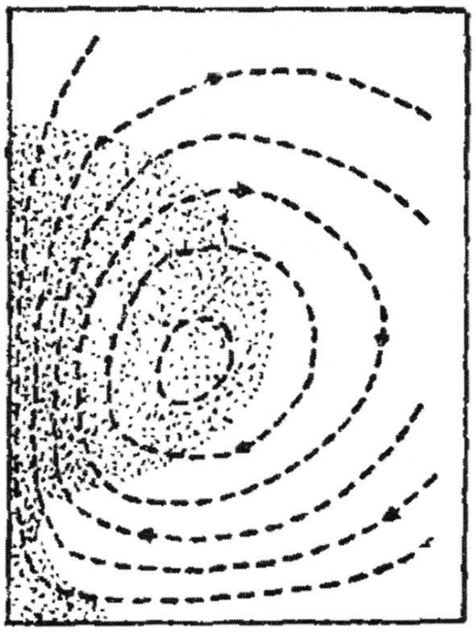

图 1 云泡结构阴影区是暖空气区

在以上的数值计算中,四周一般都采用固定边界,大气除下界地面外,是不存在固定边界的。周晓平[12]采用边界随时间外推的办法,进一步做了云泡发展的数值研究。云泡生命史的特征与上面的结果有所不同(图 2)。在发展初期,在云泡主环流的下面还有一个小的、方向与主环流相反的逆环流,以后这个逆环流逐渐消失,而在主环流的顶部,又逐渐发展出一个下沉环流。云顶则出现下沉气流,这在已有观测中指出过。云底逆环流的出现,虽然尚未被观测所注意,但仔细看一下商德尔[13]的对流实验,便可以发现这个逆环流是存在的。

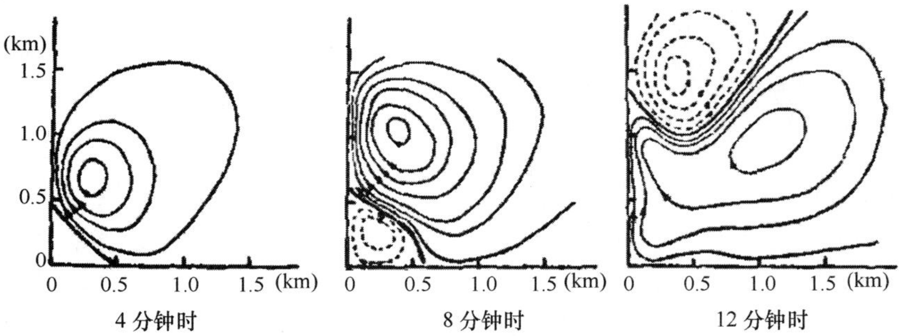

图 2 云泡的演变

在上述的数值试验中,尚未考虑水汽的作用。最近,周晓平等[14]进一步计入了水汽的影响。

另一方面,也有很多关于尺度较大的对流云塔的研究。哈尔挺纳[15]遵循斯图墨耳[16]和霍顿等[17]的工作,应用一个考虑了侧向混合过程后的一维定常模式,数值地研究了对流云塔发展到中期时的垂直结构。计算所得的垂直速度分布与观测较接近,即呈抛物线状,最大值发生在云的中上部。含水量的最大值发生在云顶,与观测结果还比较符合。计算并指出,影响对流云强度最重要的参数是环境的相对湿度和大气本身的平均温度。这两个参数对云厚、最大速度和最大含水量值的影响如图 3

所示。环境的层结对云厚、垂直速度的大小有显著的影响，但对含水量大小的影响却不大。这些结果是合理的，说明了对流云为什么经常在“温高湿重”的天气里会有猛烈的发展。

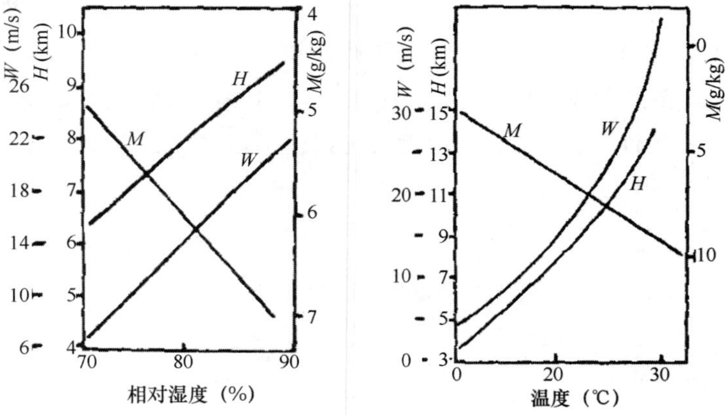

H—含水量；*M*—云厚；*W*—垂直速度

图 3　对流云的强度与相对湿度和温度的关系

在构成一维模式的连续性方程时，哈尔挺纳假定云形是随高度不变的，即把对流云塔看成是圆柱体，但对空气的侧向混合率却未加限制。斯快尔司等[18]则相反，他们假定侧向混合率正比于当时的垂直速度，而对云形不加限制。在这种情况下计算指出，当云底的质量流大到一定时，对流云的顶部可以出现钻状结构，即形成积雨云。

在对流云定常态的研究方面，古特曼[19]曾给出过一个更完善的理论。他求出了二维非钱性对流云动力学方程组（不考虑湍流扩散项）在齐次边界条件下的闭合解，并给出一个积雨云形成的例子（图 4）。与斯快尔司等的见解不同，他认为钻状结构的出现与环境逆温层的存在有直接关系。很有意思，根据古特曼的理论分析，二维非线性对流云动力学方程组，在齐次边界条件下只有当层结有不稳定层存在时才有非零解，否则只有零解。在不稳定层结下存在非零解，这表明发展成熟了的对流云可以不靠云根而只靠凝结潜热便能维持。由此也可以看出，凝结潜热的确是对流云发展的一个重要能源。

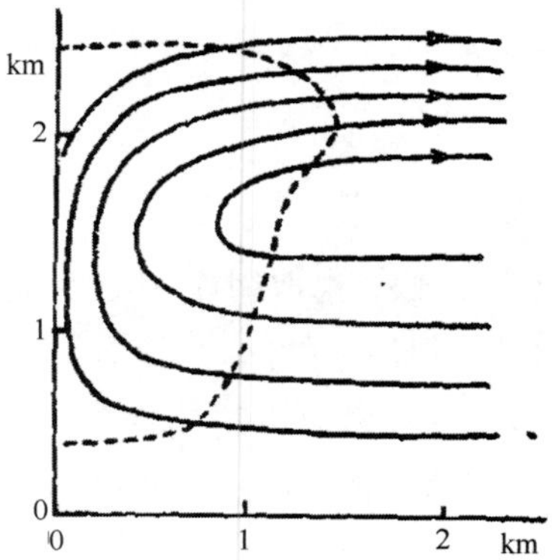

（虚线是云界）

图 4　对流云的结构

为了进一步了解对流云塔的发展过程，巢纪平[20]研究了在地面加热作用下非线性方程组的非定常解。图5给出了一个计算例子。在这个例子中,地面加热强度随时间线性增加, 到18分钟对加热约2℃; 层结在3 km以下是不稳定的, 以上是稳定的, 并在5 km起有逆温层。计算结果表明, 对流云发展到一定时间后流场将发生突变, 从整层是辐合气流变为在低层是幅合气流而在上层是辐散气流, 并在顶部还出现下沉气流。在这个例子中, 我们看到, 云顶不但能发展到稳定层中, 而且还可以穿入逆温层。这表明,只要有足够的能量支持, 对流云是可以穿透对流层顶而进入到平流层的。另外一个例子也表明, 在一定条件下, 对流云的厚度又可以比不稳定层的厚度小很多。由此看来,精确地预报积云的发展强度是一个很复杂的问题。

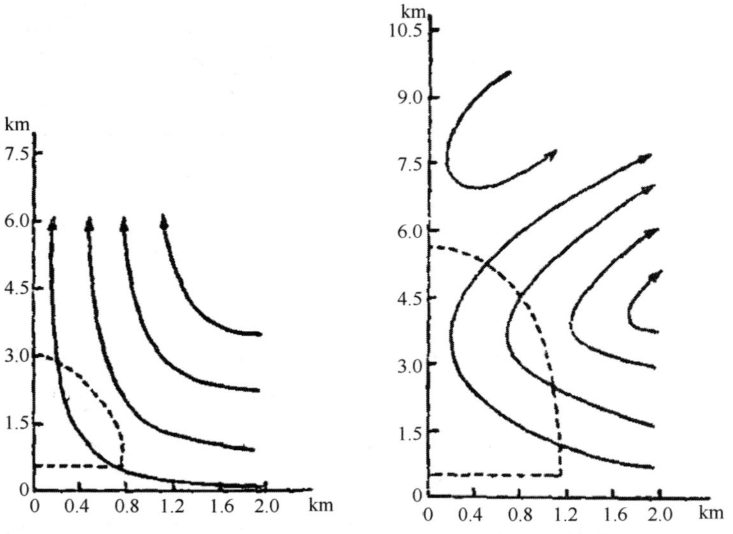

图5 对流云的演变

胡广兴的计算并指出[21], 在一定条件下云顶的下沉气流可以继续向下延伸, 甚至最后可以抑制下部上升气流的发展。

在以上的工作中, 只研究了对流云生命史中早期或中期的一些特征, 并未给出对流云的整个生命史。在这方面, 浅井富雄[7]应用一维的非定常模式给出了一个例子。在他的例子中, 在运动开始后20~30分钟, 对流云发展到最盛期, 这时云厚约5 km, 最大垂直速度约20 m/s, 以后云的发展停止,到50分钟后变成微弱的重力振荡而消失。

* * *

近几年来对流云动力学的研究进展很快, 也取得了一定的成绩, 但离解决问题还有很大差距。在进一步的研究中, 有两方面的问题是值得注意的。

(1) 拟订更合理的模式,针对我国各地经常出现的气象条件, 通过数值试验进一步了解对流云发生、发展的临界条件和生命史的特征。在拟订模式时, 应考虑如何逐步加入与微观过程的相互影响, 以及与环境之间的相互影响。事实上, 初步的分析已经表明,这种相互制约是重要的[22,23]。通过模式试验,最后应该建立一个宏观与微观结合的、能反映我国对流性降水发展特征的理论。

(2) 猛烈发展的对流云是经常与中小系统的活动联系在一起的。目前的理论研究尚局限在对个体云泡和云塔的分析方面, 如何把对流云的活动和中小系统的活动结合起来, 是值得考虑的。

此外, 如观测表明, 冰雹经常发生在系统风强并有垂直切变的情况下, 看来盛行风的存在并不总是不利于云的发展的，究竟在什么条件下，盛行风的存在能促进对流云的发展呢？又如，观测肯定地表明，雷雨云系统的移动与风场有系统的偏离，其原因何在？再如，对流云塔经常是由一些云泡继续发展迭积而成，一簇云泡如何构成了一个对流云塔等等。这些具体问题还有待澄清。

参考文献

[1] 顾震潮.气象学报,1962,32:267-284.

[2] 巢纪平.气象学报,1962,32:104-118.

[3] Ogura, Y.& Charney, J. G., Proc, International Symposium on Numerical Weather Predication in Tokyo, Tokyo, *Meteor. Soc. Japan*, 1960, 431-452.

[4] Qgura, Q.& Phillps, N.A., *J. Atomos. Sci.*, 1962, 19: 173-179.

[5] Mason, B.J.& Emig, R., *Q. J. Roy, Meteor, Soc.*, 1961, 87: 212-22.

[6] 陈瑞荣.气象学报,1962,32:285-300.

[7] Asai, T., Proc. International Symposium on Numerical Weather Predication in Tokyo, Tokyo, Meteor. Soc. Japan, 1960: 469-476.

[8] Malkus, J.S.& Witt, G., The Atmosphere and Sea in Motion, New York, Rockefeller Institute Press, 1959: 452-439.

[9] Lilly, D.K., *Tellus*, 1962, 14: 148-172.

[10] Ogura, Y., *J. Atmos. Sci.*, 1962, 19: 492-502.

[11] Scorer, R.S., *J. Fluid Mech.*, 1957, 2: 583-594.

[12] 周晓平.ДАН СССР, Серня Геофиз., 1962: 548-557.

[13] Saunder, P.M., *Tellus*, 1962, 14: 177-194.

[14] 周晓平,等.积云发展理论(未刊稿)1963.

[15] Haltiner, G.J., *Tellus*, 1959, 11: 4-15.

[16] Stommel, H., *J. Meteor*, 1947, 4: 91-94.

[17] Houghton, H.G.& Gramer, H.E. *J. Meteor*, 1951, 8: 95-102.

[18] Squires, P.& Turner, T.S., *Tellus*, 1992, 4: 422-474.

[19] Гутман, Д.Н., ДАН СССР, 1957, 112: 1033-1036.

[20] 巢纪平.气象学报,1961,31:191-204.

[21] 胡广兴.气象学报,1962,32:154-163.

[22] 巢纪平.气象学报,1962,32:164-176.

[23] 巢纪平,胡广兴.论雨滴对气流拖带作用的动力影响(未刊稿).1962.

论云中水滴对气流拖带的动力效果*

巢纪平　胡广兴**

（1. 中国科学院地球物理研究所）

提要　本文进一步研究了对流云中，降水质点对气流的拖带作用。把降水质点和空气作为混合介质看待，根据泥砂输运理论[1]，重新推导了问题的基本方程组。在这一基础上，分析了问题的常定和非常定解。得到两点主要的结果：①按云中实际可能出现的含水量数量，通过拖带作用后，可以产生与观测结果同量极的下沉气流强度；②云中降水泡（降水集中区），在降落过程中，水量将向泡的前沿区集中，由于水量的这种聚集，降水到达地面时有可能出现阵性，即降水强度一开始就很大。

1　引言

对流云在降水阶段，云中会出现每秒几米的下沉气流。Byers[2]等曾指出，这种下沉气流可能是通过水滴对空气的拖带作用而造成的。

研究这个问题很有意义。因为如果一定强度的降水量，就能造成相当大的下沉气流的话，则除了解释云消散的一种自然机制外，人们就有可能在云中播散某些悬浮质后，人工地造成下沉气流，而促使云消散。

已经有很多人在理论分析中注意到这种作用。如 Squires[3]、Haltinar[4]、Mason 和 Emig[5]以及最近的陈瑞荣[6]等，在研究不同提法的问题中，都曾加入了拖带作用的影响。不过，值得指出，在这些工作中，拖带作用是用直观的方式在方程中引进的，并未经过严格的推演。

我们注意到，这一问题与泥砂输运理论有一定的相似之处。在泥砂输运问题中，水流挟带泥砂运行，而泥砂通过拖带作用反过来又将影响水流的运动。由这种相似性，本文将根据 Баренблатт[1]的泥砂输运理论，重新建立本问题的基本方程组。并在此基础上进一步分析水滴对气流的拖带作用。

通过分析，本文得到两点主要的结果：①通过拖带作用，可以产生与观测结果同量级的下沉气流。因而，降水质点对云中空气的拖带是致使云崩溃的一种重要作用；②考虑了拖带作用后，降水将以阵性的形式出现。

2　基本方程组的推导

以下基本方程的推导，按参考文献[1]进行。

取直角坐标 x_1, x_2, x_3。x_3 轴指向上。采用以下符号：ρ_1 为空气的密度，ρ_2 为降水质点的密度；τ 为质点的相对体积；v_i 为空气沿 x_i 轴的速度分量；$i=1,2,3$；w_i 为降水质点沿 x_i 轴的速度分量；f_i 为单位体积混合物内，空气质点与降水质点之间相互作用力沿 x_i 轴的分量；g_i 为沿 x_i 轴的重力加速度分量。

---

* 气象学报，1963，33(4)：449-458.

** 上海气象台工作。

空气的运动方程可写成(按张量形式)

$$\frac{\partial}{\partial t}\rho_1(1-\tau)v_i+\frac{\partial}{\partial x_a}\rho_1(1-\tau)v_iv_a=-\frac{\partial sp}{\partial x_i}-\rho_i(1-\tau)g_i-f_i \tag{1}$$

式中,sp 表示作用在空气质点上的压力分量。

如果把降水质点群作为连续介质看待,则有运动方程

$$\frac{\partial}{\partial t}\rho_2\tau w_i+\frac{\partial}{\partial x_a}\rho_2\tau w_iw_a=-\frac{\partial(1-s)p}{\partial x_i}-\rho_2\tau g_i+f_i \tag{2}$$

式中,$(1-s)p$ 表示作用在降水质点上的压力分量。

将式(1)、式(2)相加,得到不均质流体运动量的方程

$$\frac{\partial DV_i}{\partial t}+\frac{\partial\pi_{ia}}{\partial x_a}=-\frac{\partial p}{\partial x_i}-Dg_i \tag{3}$$

式中

$$D=\rho_1(1-\tau)+\rho_2\tau \tag{4}$$

是不均质流体的密度,而 V_i,π_{ia}为

$$V_i=\frac{\rho_1(1-\tau)v_i+\rho_2\tau w_i}{D} \tag{5}$$

$$\pi_{ia}=\rho_1(1-\tau)v_iv_a+\rho_2\tau w_iw_a \tag{6}$$

空气的质量守恒方程为

$$\frac{\partial}{\partial t}\rho_1(1-\tau)+\frac{\partial}{\partial x_a}\rho_1(1-\tau)v_a=0 \tag{7}$$

假定降水质点在输送过程中,不与云滴相碰,也没有蒸发等效应发生,即降水质点的质量是守恒的。这时有质量守恒方程

$$\frac{\partial}{\partial t}\rho_2\tau+\frac{\partial}{\partial x_a}\rho_2\tau w_a=0 \tag{8}$$

降水质点在降落过程中与云滴有重力碰并的情形,不难由此而引申。我们将另文讨论。

将式(7)、式(8)相加,得到不均质流体的质量平衡方程

$$\frac{\partial D}{\partial t}+\frac{\partial DV_a}{\partial x_a}=0 \tag{9}$$

将式(7)、式(8)分别除以 ρ_1 和 ρ_2,再相加,就得到不均质流体的连续性方程

$$\frac{\partial}{\partial x_a}[(1-\tau)v_a+\tau w_a]=0 \tag{10}$$

方程组式(3)、式(9)和式(10)可以简化。在我们所要讨论的问题中,空气及降水质点的加速度均较重力加速度小很多,因此可以假定:降水质点和空气的水平速度分量相同,而垂直方向的速度分量则相差某一数值 a,即

$$w_i=v_i-a\delta_{i3}(\text{当 } i\neq 3 \text{ 时},\delta_{i3}=0,i=3 \text{ 时 } \delta_{i3}=1) \tag{11}$$

根据这一假定,则有

$$V_i=v_i-\frac{1}{D}\rho_2\tau a\delta_{i3}=w_i+\frac{1}{D}\rho_1(1-\tau)a\delta_{i3} \tag{12}$$

$$\pi_{ia}=DV_iV_a+\frac{1}{D}\rho_1\rho_2\tau(1-\tau)a^2\delta_{i3}\delta_{a3} \tag{13}$$

$$(1-\tau)v_a+\tau w_a=V_a+\frac{1}{D}(\rho_2-\rho_1)a\tau(1-\tau)\delta_{a3} \tag{14}$$

将式(12)、式(13)和式(14)代入式(3)及式(10),得到

$$\frac{\partial DV_i}{\partial t}+\frac{\partial DV_iV_a}{\partial x_a}=-Dg_i-\frac{\partial p}{\partial x_i}-\rho_1\rho_2\frac{\partial}{\partial x_3}\left[\frac{a^2\tau(1-\tau)}{D}\right]\delta_{i3} \tag{15}$$

$$\frac{\partial V_a}{\partial x_a}=-(\rho_2-\rho_1)\frac{\partial}{\partial x_3}\left[\frac{a\tau(1-\tau)}{D}\right] \tag{16}$$

将式(16)代入式(9),得到

$$\frac{\partial D}{\partial t}+V_a\frac{\partial D}{\partial x_a}=(\rho_2-\rho_1)D\frac{\partial}{\partial x_3}\left[\frac{a\tau(1-\tau)}{D}\right] \tag{17}$$

由于 a 可以事先从理论上或者用实验确定,所以方程组式(15)、式(16)和式(17)对变量 V_i,P,D 封闭。事实上,不难看出,a 是降水质点等速降落的速度,即为末速度。

一般在云雾降水问题中,下面条件

$$\tau\ll 1,\ \sigma\tau\ll 1,\ \sigma=\frac{\rho_2-\rho_1}{\rho_1}$$

成立。因此,方程组还可以进一步简化成

$$\frac{\partial V_i}{\partial t}+V_a\frac{\partial V_i}{\partial x_a}=-(1+\tau\sigma)g_i-\frac{1}{\rho_1}\frac{\partial p}{\partial x_i} \tag{18}$$

$$\frac{\partial \tau}{\partial t}+V_a\frac{\partial \tau}{\partial x_a}=a\frac{\partial \tau}{\partial x_3} \tag{19}$$

$$\frac{\partial V_a}{\partial x_a}=-a\sigma\frac{\partial \tau}{\partial x_3} \tag{20}$$

由于空气的密度 ρ_1 要比降水质点的密度 ρ_2 小很多,因此 $\sigma\approx\frac{\rho_2}{\rho_1}$,而 $M=\sigma\tau$,是单位质量的空气中,所含有的降水质点的质量。如果把云滴的质量也考虑在内,则显然 M 就是含水量。由此,方程(18)可以写成

$$\frac{\partial V_i}{\partial t}+V_a\frac{\partial V_i}{\partial x_a}=-(1+M)g_i-\frac{1}{\rho_i}\frac{\partial p}{\partial x_i} \tag{21}$$

将方程(19)乘以$\frac{\rho_2}{\rho_1}$后,得到

$$\frac{\partial M}{\partial t}+V_a\frac{\partial M}{\partial x_a}=a\frac{\partial M}{\partial x_a} \tag{22}$$

方程(20)右端由于很小,可以略去,得

$$\frac{\partial V_a}{\partial x_a}=0 \tag{23}$$

方程组式(21)、式(22)和式(23)是本问题的基本方程式。

与在引言中所指出的一些工作不同, 现在引进了新的质量输运方程(22)。由于式(22)中包含了质点的末速度 a, 因此, 即使在方程(21)中的 M 相同(即含水量一样), 而组成 M 的质点的大小不同,则问题的结果显然也不会相同。值得指出, M 本质上是一个可观测的宏观量, 因此, 如果只在式(21) 中引进 M,而不考虑式(22),则问题仍然是在宏观动力学范畴中的问题。只有引进了式(22)后,

由于其中包含了与降水质点大小有关的量 a 后，宏观和微观的相互制约过程，才在问题中得到初步的体现。

在另一方面，从形式上看，方程(21)与一般通用的形式是相同的(见参考文献[6])，但在现在的方程中，V 是空气和降水质点所组成的混合介质的速度，不单是空气的速度。不过，当降水质点的含量 M，不是出乎寻常的大时，这一差别是不重要的。这是因为，由于这时

$$D=\rho_1\left[(1-\tau)+\frac{\rho_2}{\rho_1}\tau\right]=\rho_1[(1-\tau)+M]\approx\rho_1,$$

而由式(12)，得

$$V_i=v_i-Ma\delta_{i3} \tag{24}$$

一般来说，$|Ma|\ll|v_i|$，所以 $V_i\approx v_i$。

或者，只要开始时假定，降水质点的加速度较小，而略去式(2)左边的加速度项，如果再假定，降水质点的含量也很小，则有 $s\approx1$，而可以略去右方的气压梯度力，因此，式(2)可以简化成

$$f_i=\rho_2\tau g_i \tag{25}$$

将此式代入式(1)，经过简化，便得到一般所通用的方程。

下面对方程组式(21)至式(23)的解作分析。

3　常定问题

假定在式(21)中，气压梯度力和重力相平衡，即假定不均质的加速度完全由降水质点的拖带所引起。将式(23)代入式(21)和式(22)两式中，得到

$$\frac{\partial V_a V_i}{\partial x_a}=-Mg_i \tag{26}$$

$$\frac{\partial V_a M}{\partial x_a}=a\frac{\partial M}{\partial x_3} \tag{27}$$

取一维问题，由此两式得

$$\frac{\partial V_3^2}{\partial x_3}=-Mg \tag{28}$$

$$\frac{\partial V_3 M}{\partial x_3}=a\frac{\partial M}{\partial x_3} \tag{29}$$

取边界条件为

$$x_3=H(\text{云顶}),\ M=M_0,\ V_3=-M_0a \tag{30}$$

由式(24)，第二个条件相当假定在云顶空气的速度为零。

式(29)对 x_3 积分，应用边界条件式(30)后，得到

$$(V_3-a)M=-a(1+M_0)M_0 \tag{31}$$

将式(31)代入式(28)，得到

$$2V_3(V_3-a)\frac{\mathrm{d}V_3}{\mathrm{d}x_3}=ag(1+M_0)M_0 \tag{32}$$

对式(32)积分，应用式(30)后，得到

$$\frac{2}{3}(V_3^3+M_0^3a^3)-a(V_3^2-M_0^2a^2)=ag(x_3-H)(1+M_0)M_0 \tag{33}$$

为了便于作数值计算,引进

$$\eta^* = \frac{x_3}{H},\ w^* = \frac{V_3}{a},\ \mu_0^2 = gH(1+M_0)\frac{M_0}{a^2}$$

这样,式(33)可以变成无因次形式,为

$$\frac{2}{3}(w^{*3} + M_0^3) - (w^{*2} - M_0^2) = -\mu_0^2(1-\eta^*) \tag{34}$$

当 w^* 由此式决定后,由式(31)可以决定 M_0。

对下面6个例子做了计算①:(1)$M_0=1$ g/kg,R(水滴半径)= 250 μm;(2)$M_0=1$ g/kg,$R=500$ μm;(3)$M_0=1$ g/kg,$R=1\ 000$ μm;(4)$M_0=2$ g/kg,$R=250$ μm;(5)$M_0=2$ g/kg,$R=500$ μm;(6)$M_0=2$ g/kg,$R=1\ 000$ μm。在各例中,H(云厚)皆取为5 km。计算结果见表1。

表1

v_3(m/s) \ 离云顶距离(km) \ 例	1	2	3	4	5
1	-2.38	-3.17	-3.7	-4.14	-4.53
2	-2.70	-3.60	-4.32	-4.68	-5.86
3	-2.80	-3.85	-4.55	-5.11	-5.67
4	-3.17	-4.14	-4.84	-5.46	-6.03
5	-3.64	-4.86	-5.72	-6.48	-7.02
6	-3.85	-5.11	-6.09	-6.86	-7.56

这一计算结果表明,降水的拖带作用是重要的,它可以产生与观测事实相同量级的下沉气流。因此,降水发展后的拖带作用,看来是使云崩溃的一个因子。

4 一维非常定问题

一维非常定问题的基本方程为

$$\frac{\partial V_3}{\partial t} + V_3\frac{\partial V_3}{\partial x_3} = -Mg \tag{35}$$

$$\frac{\partial M}{\partial t} + V_3\frac{\partial M}{\partial x_3} = a\frac{\partial M}{\partial x_3} \tag{36}$$

今将 V_3,M 改作自变量,而将 t 和 x_3 作应变量,为此,将式(35)、式(36)两式重新用雅可比形式写出

$$\frac{\partial(V_3,x_3)}{\partial(t,x_3)} + V_3\frac{\partial(V_3,t)}{\partial(x_3,t)} = -Mg \tag{37}$$

$$\frac{\partial(M,x_3)}{\partial(t,x_3)} + V_3\frac{\partial(M,t)}{\partial(x_3,t)} = a\frac{\partial(M,t)}{\partial(x_3,t)} \tag{38}$$

① 降水质点取水滴,末速度和半径的关系,按参考文献[7]。

将以上两式分别乘以 $\partial(t,x_3)/\partial(V_3,M)$，得

$$\frac{\partial(V_3,x_3)}{\partial(V_3,M)} - V_3\frac{\partial(V_3,t)}{\partial(V_3,M)} = -Mg\frac{\partial(t,x_3)}{\partial(V_3,M)} \tag{39}$$

$$\frac{\partial(M,x_3)}{\partial(V_3,M)} - V_3\frac{\partial(M,t)}{\partial(V_3,M)} = -a\frac{\partial(M,t)}{\partial(V_3,M)} \tag{40}$$

展开后得

$$\frac{\partial t}{\partial V_3} = -\frac{1}{Mg} \tag{41}$$

$$\frac{\partial x_3}{\partial V_3} = (V_3 - a)\frac{\partial t}{\partial V_3} \tag{42}$$

对式(41)、式(42)积分，得到问题的普遍解为

$$t + \frac{V_3}{gM} = \overline{H}_1(M) \tag{43}$$

$$x_3 + at + \frac{V_3^2}{2gM} = \overline{H}_2(M) \tag{44}$$

只要给出不同的初始条件，定出普遍函数 $\overline{H}_1$，$\overline{H}_2$ 后，就可以求得不同的特解。

引进无因次量

$$\eta^* = \frac{x_3}{H},\ \tau^* = \frac{a}{H}t,\ M^* = \frac{M}{M_0},\ w^* = \frac{V_3}{M_0 a},\ \mu_1^2 = \frac{a^2}{gH}$$

则式(43)、式(44)取下面的无因次形式

$$\tau^* + \mu_1^2\frac{W^*}{M^*} = \overline{H}_1(M^*) \tag{45}$$

$$\eta^* + \tau^* + \frac{M_0\mu_1^2}{2}\frac{W^{*2}}{M^*} = \overline{H}_2(M^*) \tag{46}$$

今求下面的特解。取初始时刻云中降水量的分布为

$$\tau^* = 0:\ M^* = \eta^*(1-\eta^*),\ M^* = 0,\ 当\begin{cases}\eta^* \geqslant 1\\ \eta^* \leqslant 0\end{cases} \tag{47}$$

初始时刻空气的速度为零，由式(24)得到初始时刻混合介质的速度为

$$\tau^* = 0:\ W^* = -\eta^*(1-\eta^*),\ M^* = 0,\ 当\begin{cases}\eta^* \geqslant 1\\ \eta^* \leqslant 0\end{cases} \tag{48}$$

由式(47)得

$$\tau^* = 0,\ \eta^* = \frac{1}{2} \pm \sqrt{\frac{1}{4} - M^*} \tag{49}$$

将这组条件代入式(45)、式(46)，求得

$$\overline{H}_1 = -\mu_1^2,$$

$$\overline{H}_2 = \frac{1}{2} + \frac{M_0}{2}\mu_1^2 M^* \pm \sqrt{\frac{1}{4} - M^*}$$

由此得到问题的特解为

$$\tau^* = -\mu_1^2 - \mu_1^2 \frac{W^*}{M^*} \tag{50}$$

$$\eta^* = \frac{1}{2} - \tau + \frac{M_0}{2}\mu_1^2\left(M^* - \frac{W^{*2}}{M^*}\right) \pm \sqrt{\frac{1}{4} - M^*} \tag{51}$$

或者，由式(50)、式(51)消去 M^* 后，得到

$$\eta^* = \frac{1}{2} - \tau^* + \frac{M_0}{2}\left[(\tau^* + \mu_1^2) - \frac{\mu_1^4}{\mu_1^2 + \tau^*}\right]W^* \pm \sqrt{\frac{1}{4} + \frac{\mu_1^2}{\mu_1^2 + \tau^*}W^*} \tag{52}$$

当 W^* 由式(52)决定后，M^* 由下式

$$M^* = -\frac{\mu_1^2}{\mu_1^2 + \tau^*}W^* \tag{53}$$

决定。注意到，由于 M^* 必须为正，因此，由式(53)，W^* 必须为负。

作为例子，取 $\mu_1^2 = 10^{-2}$，$M_0 = 8$ g/kg(相当初始时刻降水区的中心强度为 $M = 2$ g/kg)。计算结果见图 1。如取 $H = 500$ m，$a = 7$ m/s，则当 $\tau^* = 1$，或相当 $t = 70$ s 时，空气的最大下沉速度 $v_{3\max} = -1.3$ m/s；当 $t = 140$ s 时，$v_{3\max} = -2.6$ m/s。由此可见，下沉气流的强度随时间增大。

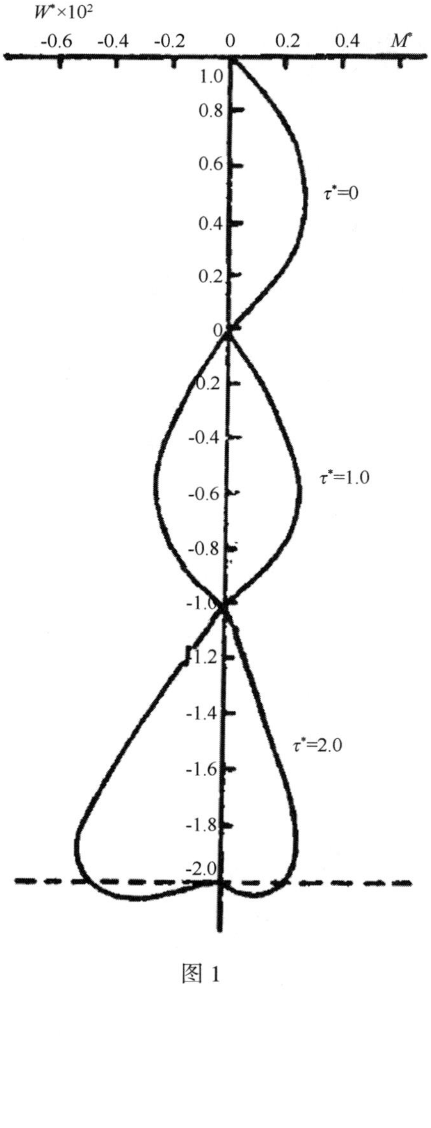

图 1

不难直接由式(52)求得最大下沉速度为

$$W_{\max}^{*}=-\frac{1}{4}\frac{\mu_1^2+\tau^*}{\mu_1^2}\approx-\frac{1}{4}\frac{\tau^*}{\mu_1^2}\tag{54}$$

回到有因次量后,得

$$V_{3\max}\approx-\frac{1}{4}gM_0t\tag{55}$$

即最大下沉速度与降水质点含量成正比,并随时间呈线性增长。

计算表明,降水中心的下沉速度,要比降水质点的末速度大。例如,在 70 s 到 140 s 这一段时间间隔中,平均下沉速度约为 9.3 m/s。不难由式(52)直接算出最大降水中心的下沉速度为

$$\left(\frac{\mathrm{d}x_3}{\mathrm{d}t}\right)_{\max}\approx-a+\frac{V_{3\max}}{2}\approx-a-\frac{1}{8}gM_0t\tag{56}$$

由此可见,降水中心的降落速度,在降落过程中,也是加速的。另外,计算指出,降水区上界和下界的下降速度,正好等于水滴的末速度。

这是一个很有意思的现象,由于降水区中心的下降速度较其下界的下降速度来得大,于是经过一段时间后,水量就要向降水区的前沿部分聚集,以致最后像气体动力学中骇波的形成那样,形成一不连续的分布(如图 1 中,$\tau^*=2$ 时的粗横断线所示)。

显然,如果降水到达地面前,水量已经聚集在前沿部分,那么可以预料,降水到达地面时将以阵性形式出现。即降水突然开始,其强度一开始就很大。如果真是这样,那么这一个简单的模式,就为阵性降水的形成,提供了一个探索的方向。

根据气体动力学理论(见参考文献[8]),求得不连续面形成的时间和高度分别为

$$\tau_0^*=-\mu_1^2+\mu_1\sqrt{\frac{2}{M_0}\left(1+\frac{M_0\mu_1^2}{2}\right)}\approx\mu_1\sqrt{\frac{2}{M_0}}\tag{57}$$

$$\eta_0^*=-\tau_0^*\tag{58}$$

应用到上面的例子,用 $\mu_1^2=10^{-2}$, $M_0=8$ g/kg 代入后,取 $\tau_0^*\approx1.58$, $\eta\approx-1.58$。回到有因次量后,得 $t\approx110$ s, $x_3\approx-770$ m。

Салвман[9] 对雷阵雨的雷达观测结果表明,最大回波强度区的下降速度要比回波区上、下界的下降速度来得快,亦即最大回波强度区,在下降过程中逐渐偏向回波区的前沿。这一个观测结果,定性地支持了上面的理论分析。

5　二维非常定问题

这时方程组为

$$\frac{\partial V_1}{\partial t}+V_1\frac{\partial V_1}{\partial x_1}+V_3\frac{\partial V_1}{\partial x_3}=-\frac{1}{\rho_1}\frac{\partial p'}{\partial x_1}\tag{59}$$

$$\frac{\partial V_3}{\partial t}+V_1\frac{\partial V_3}{\partial x_1}+V_3\frac{\partial V_3}{\partial x_3}=-Mg-\frac{1}{\rho_1}\frac{\partial p}{\partial x_3}\tag{60}$$

$$\frac{\partial V_1}{\partial x_1}+\frac{\partial V_3}{\partial x_3}=0\tag{61}$$

$$\frac{\partial M}{\partial t}+V_1\frac{\partial M}{\partial x_1}+(V_3-a)\frac{\partial M}{\partial x_3}=0\tag{62}$$

在式(59)和式(60)两式中,p'是扰动气压。压力的平均部分假定满足静力关系,与重力已相抵消。

由式(61)引进流函数 ψ,定义成

$$V_1 = -\frac{\partial\psi}{\partial x_3},\ V_3 = \frac{\partial\psi}{\partial x_1} \tag{63}$$

由式(59)、式(60)削去 p'后,得

$$\Delta_2 \frac{\partial\psi}{\partial t} = \frac{\partial(\Delta_2\psi,\psi)}{\partial(x_1,x_3)} - \frac{\partial M}{\partial x_1}g \tag{64}$$

将式(62)写成

$$\frac{\partial M}{\partial t} = \frac{\partial(M,\psi)}{\partial(x_1,x_3)} + a\frac{\partial M}{\partial x_3} \tag{65}$$

式中,$\Delta_2 = \frac{\partial^2}{\partial x_1^2} + \frac{\partial^2}{\partial x_3^2}$。

要求出方程(64)和方程(65)的闭合解,是困难的。下面用数值解法。方程(64)对于$\frac{\partial\psi}{\partial t}$是一个 Poisson 方程,当 ψ,M 的初始值给定后,可以方便地借张弛法求出初始时刻的 $\partial\psi/\partial t$ 值。同样,由式(65),可以求得初始时刻的 $\partial M/\partial t$ 值。将所得的一次导数乘以时间增量 Δt 后,便得到下一时刻的 ψ 和 M 场。仿此,可进行再一次的迭代。数值积分的差分格式采用中央差,空间网格取 400 m,时间步长取 50 s。计算迭代了二次,即算到 100 s。

初始时刻的 M 值分布如图 2 实线所示,最大值为 2 g/kg。初始时刻空气的速度等于零,因此,$V_1 = 0$,$V_3 = -Ma$。水滴的末度速仍取 7 m/s。

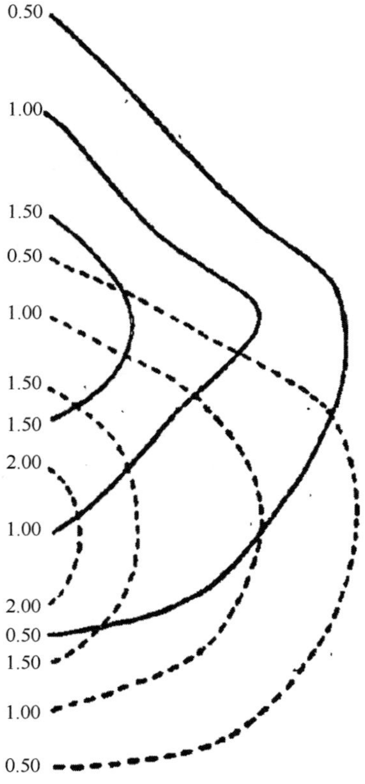

图 2

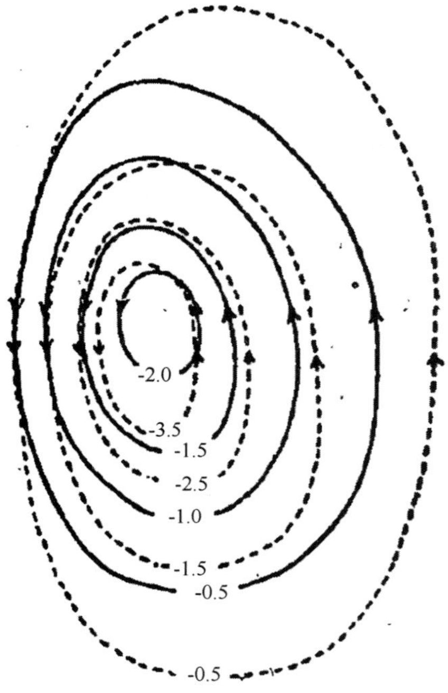

图 3

计算所得 100 s 后的 M 场由图 2 虚线表示。这时降水中心下降了约 800 m。平均下降速度约 8 m/s。由水滴拖带所造成的流场如图 3 所示。图中实线是 50 s 时的情况，虚线是 100 s 时的情况。由图可见，在中轴附近是下沉气流，两旁则为补偿性的上升气流。50 s 和 100 s 时气流的强度，分别见图 4 和图 5。图中实线是等垂直速度线，虚线是等水平速度线。在 50 s 时，下沉气流最大速度约为 0. 36 m/s，上升气流最大速度约为 0. 12 m/s，水平速度最大约为 0. 11 m/s。到 100 s 时，强度比 50 s 时增加了约一倍。

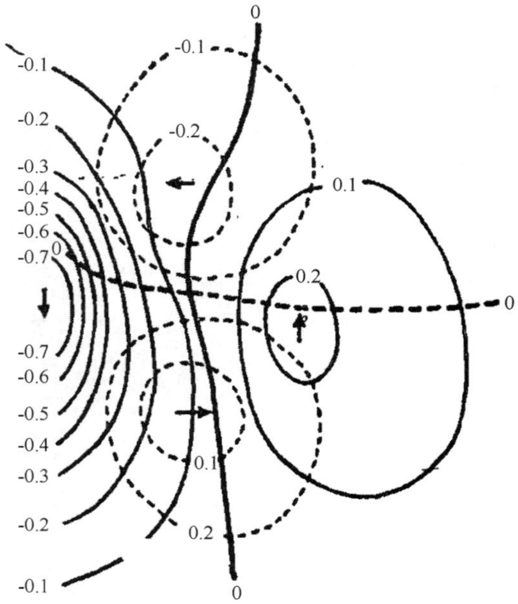

图 4

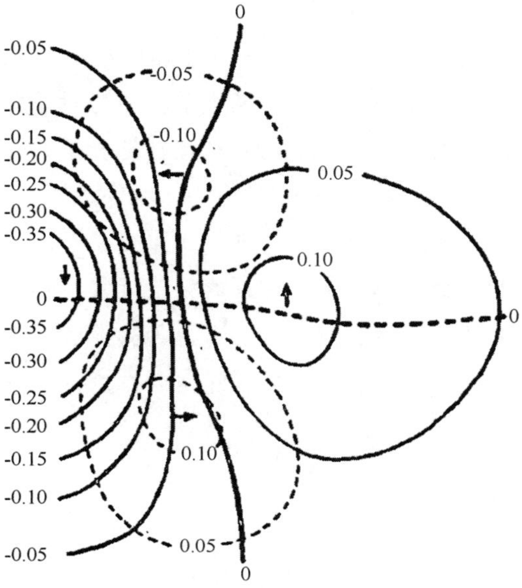

图 5

6　结语

以上初步的分析表明，考虑宏观过程和微观过程的相互制约，对对流云及其降水的发展有重要的作用。进一步研究这种相互制约规律，看来是今后积云动力学及降水理论研究中，值得注意的一个方面。

参考文献

[1]　Ьаренблатт, Г.И., *Прп.М.М.*, 1953, 17:261-274.

[2]　Byers H.R., & R.R.Braham. The Thunderstorm.Report of the Thunderstorm Project, Washington, 1949:24.

[3]　Squires, P., *Tellus*, 1958, 10:381-385.

[4]　Haltinar, *Tellus*, 1960, 12:393-398.

[5]　Mason, B.J.& Emig.R.*Q.J.Roy.Meteor.Soc.*, 1961, 87:212-222.

[6]　陈瑞荣.气象学报, 1963, 33:257-270.

[7]　Mason, B.J.Cloud Physics, Oxford, 1957:421.

[8]　郎道学.连续介质力学(第二册), 中译本.人民教育出版社, 1960:474-482.

[9]　Салвман, E.M., *Tp ГГО*, 1957, 72:46-65.

风速垂直切变对于对流的发展和结构的影响*

巢纪平[1] 陈历舒[2]

(1. 中国科学院地球物理研究所;2. 中国科学技术大学)

提要:本文分两个部分。在第一部分中，分析了具有速度垂直切变的盛行风，对于对流发展强度的影响。指出，风速垂直切变对于对流的发展是否有利，决定于对流流场的结构。对于一种特定的流场，通过扰动增长率的计算，定量地讨论了风速垂直切变和不稳定层结对扰动发展的相对重要性。

在第二部分中，进一步求得了考虑风速垂直切变影响后的非线性对流运动方程组在边界热源作用的解，对解的数值计算表明，由于风速垂直切变的作用，对流流场的结构与静止大气中的情况不同，不再是轴对称的，而主要由两个顺时针流动的涡环组成，这种流场结构，与一般移动的雷阵雨云中的流场结构很相似。

1 引言

强大对流云的发展，一般是与天气系统的活动分不开的。由于天气系统的活动，在对流云发展的背景中，一般存在有盛行的风场。盛行风对对流云的发展有多方面的影响。例如，观测表明[1-3]，对流云经常在高空急流的下方有猛烈的发展，有时云顶甚至能穿透对流层顶进入平流层中。观测又指出，在盛行风场中发展起来的对流云，它们的流场结构是不对称的，在云的前方(相对盛行风的方向)是上升气流，后方是下沉气流，中间形成一强的切变区[4-6]，如图 1 所示。

图 1 阵雨云中的气流模型[5]

根据这些观测事实，可以提出两个值得探讨的理论问题:第一，对流云除从凝结潜热中获得发展的能量外，能不能从有垂直切变的盛行风场中也获得发展的能量；第二，对流云中的非对称的流场结构，是由什么原因所造成的，是否与风的垂直切变有关。

就我们所知，到目前为止，这两个问题尚没有很好的理论解释。目前，在第一个问题的理论分

* 气象学报，1964，34(1):94-102.

析方面,Taylor[7]、Eliassen 等[8]以及最近郭晓岚[9]都曾分析过速度随高度呈线性增加的层结流体中小扰动的稳定性。他们的结果表明，如果流体的层结是静力稳定的,那么小扰动是动力稳定的，换言之，气流的垂直切变并不能促进小扰动振幅的增长。由于他们用的是小扰动方法，这一结果对有限振幅的扰动不能是定论。笔者曾分析过线性风速廓线对有限振幅扰动的影响[10]，得到的结论与上面相同。不过，值得指出，当时的理论分析，是针对特定形式的简谐形扰动的，显然，大气中的对流云不是一个简谐形扰动。对于其他类型的风速廓线,Drazin[11]与笔者[10]用不同的方法分析后得到了相同的结果,即如果风速廊线中存在有反折点,那么层结流体中的扰动可以从盛行风场中取得发展的能量。虽然这表明一定形式的风速廓线是有利于对流发展的,不过这种形式的风速廊线太特殊,对于大气中的实际情况来说,一般很难恰巧满足这样的要求。

对于第二个问题,除了一些定性的讨论外[4,6],迄今尚未有人在理论上做过定量的探讨。

在本文中,我们准备进一步研究上面提出的两个问题。

2　动力学方程和能量积分

研究 x-z 平面中的二维问题。使 x 轴与盛行 $\widetilde{U}(z)$ 的方向一致,z 轴与之正交,并指向上。这样,问题的动力学方程组为

$$\frac{\partial u}{\partial t}+u\frac{\partial u}{\partial x}+w\frac{\partial u}{\partial z}=-\frac{1}{\bar{\rho}}\frac{\partial p}{\partial x}-\widetilde{U}\frac{\partial u}{\partial x}-\frac{\mathrm{d}\widetilde{U}}{\mathrm{d}z}w+k\frac{\partial^2\widetilde{U}}{\partial z^2}+k\Delta_2 u \tag{1}$$

$$\frac{\partial w}{\partial t}+u\frac{\partial w}{\partial x}+w\frac{\partial w}{\partial z}=-\frac{1}{\rho}\frac{\partial p}{\partial z}-\widetilde{U}\frac{\partial w}{\partial x}+\beta\theta+k\Delta_2 w \tag{2}$$

$$\frac{\partial u}{\partial x}+\frac{\partial w}{\partial z}=0 \tag{3}$$

$$\frac{\partial \theta}{\partial t}+u\frac{\partial \theta}{\partial x}+w\frac{\partial \theta}{\partial z}=-\widetilde{U}\frac{\partial \theta}{\partial x}-\alpha w+k\Delta_2\theta \tag{4}$$

式中,u,w 分别为扰动速度的水平和垂直分量;p 和 θ 为空气的扰动压力和扰动位温;$\bar{\rho}$ 为空气的平均密度;$\beta=g/\bar{\theta}$,g 为重力加速度,$\bar{\theta}$ 为平均位温;$\alpha=\gamma_a-\gamma$,为静力稳定度;γ_a,当水汽未饱和时是干绝热递减率,当饱和时是湿绝热递减率,γ 为平均温度的垂直梯度;k 是湍流交换系数;$\Delta_2=\frac{\partial^2}{\partial x^2}+\frac{\partial^2}{\partial z^2}$。显然,运动总的水平速度应为 $u_{总}=\widetilde{U}(z)+u$。

在下面的研究中,取盛行风的廓线是高度的线性函数,即

$$\widetilde{U}(z)=\widetilde{U}'z \tag{5}$$

我们来给出能量积分。由于湍流耗散项的作用已为人们所熟知,在以下的能量积分中略去这一项的作用。将式(1)乘以 u 后加上乘以 w 后的式(2),取所得式子的面积分,应用式(3),并设在积分边解上法向速度为零,于是得到动能积分

$$\frac{\partial}{\partial t}\iint\frac{1}{2}(u^2+w^2)\,\mathrm{d}s=-\widetilde{U}'\iint uw\,\mathrm{d}s+\beta\iint w\theta\,\mathrm{d}s \tag{6}$$

同样地,由式(4)得到内能积分

$$\frac{\partial}{\partial t}\iint \frac{1}{2}\theta^2\mathrm{d}s = -\alpha\iint w\theta\mathrm{d}s \tag{7}$$

方程组式(1)至式(4)和式(6)、式(7)是本文的基本方程组。在下面的第一部分中,将应用式(6)、式(7)两式来讨论风速垂直切变对对流发展强度的影响;在第二部分中,将直接求出方程组式(1)至式(4)在一个边值问题下的解,在这一基础上讨论风速垂直切变对对流流场结构的影响。

第一部分

3 考虑了风速切变后对流的发展条件

引进流函数 ψ,定义成

$$u = \frac{\partial\psi}{\partial z},\quad w = -\frac{\partial\psi}{\partial x} \tag{8}$$

设在所讨论的时间间隔中,对流的形状不变,而振幅可以变化,因此可以假定

$$\psi = A(t)\phi(x,z),\quad \theta = B(t)\Theta(x,z) \tag{9}$$

在这里用 ϕ 和 Θ 来表示流场和温度场的结构,用 $A(t)$ 和 $B(t)$ 来表示振幅。

将式(8)、式(9)两式代入式(6)、式(7)两式后,得到

$$\frac{\mathrm{d}A}{\mathrm{d}t} = -\widetilde{U}'k_1A + \beta k_2B \tag{10}$$

$$\frac{\mathrm{d}B}{\mathrm{d}t} = -\alpha l_1A \tag{11}$$

式中

$$k_1 = \frac{\iint W\cdot V\mathrm{d}s}{\iint(W^2+V^2)\mathrm{d}s},\ k_2 = \frac{\iint W\Theta\mathrm{d}s}{\iint(W^2+V^2)\mathrm{d}s},\ l_2 = \frac{\iint W\Theta\mathrm{d}s}{\iint\Theta^2\mathrm{d}s} \tag{12}$$

而

$$V = \frac{\partial\phi}{\partial z},\quad W = -\frac{\partial\phi}{\partial x} \tag{13}$$

由式(10)、式(11)两式消去 B 后,得到描写流场振幅变化的方程

$$\frac{\mathrm{d}^2A}{\mathrm{d}t^2} + \widetilde{U}'k_1\frac{\mathrm{d}A}{\mathrm{d}t} + k_2l_1\alpha\beta A = 0 \tag{14}$$

这一方程的解为

$$A = c_1\mathrm{e}^{\sigma_1 g} + c_2\mathrm{e}^{\sigma_2 g} \tag{15}$$

式中增长率 $\sigma_{1,2}$ 为

$$\sigma_{1,2} = -\frac{\widetilde{U}'k_1}{2} \pm \sqrt{\left(\frac{\widetilde{U}'k_1}{2}\right) - \alpha\beta k_2l_1} \tag{16}$$

根据特征根式(16)的性质,对流的发展条件如下。

(1)对流的振幅成指数增长的情形为:

① $\alpha<0$(静力不稳定),或者

② α>0(静力稳定)，$\widetilde{U}'k_1<0$，并且

$$R_i \leqslant \frac{1}{4}\frac{k_1^2}{k_2 l_1} \tag{17}$$

(2)对流的振幅成指数振荡增长的情形为：

$$\alpha > 0, \widetilde{U}'k_1 < 0,\ 而\ R_i > \frac{1}{4}\frac{k_1^2}{k_2 l_1} \tag{18}$$

(3)对流的振幅成指数衰减的情形为：

$$\alpha > 0, \widetilde{U}'k_1 > 0,\ 而\ R_i \leqslant \frac{1}{4}\frac{k_1^2}{k_2 l_1} \tag{19}$$

(4)对流的振幅成指数振荡衰减的情形为：

$$\alpha > 0, \widetilde{U}'k_1 > 0,\ 而\ R_i > \frac{1}{4}\frac{k_1^2}{k_2 l_1} \tag{20}$$

在以上各式中，$R_i = \frac{\alpha\beta}{\widetilde{U}'^2}$，为李查德逊数。

由此可见，除了不稳定层结是有利于对流发展的一个因子外，对于一定结构的对流环流，风速垂直切变也可以是一个有利的因子。

4　发展对流的理想模型和增长率

上节分析表明，只要满足条件 $\widetilde{U}'k_1<0$，风速垂直切变也有利于对流的发展。大气中经常出现的情况是 $\widetilde{U}'>0$，即风速随高度增加。在这种情况下，最能发展的对流，它的扰动流场的结构必须满足 $k_1<0$ 的条件。亦即要成立条件

$$\iint W \cdot V \mathrm{d}s < 0 \tag{21}$$

不难看出，这时要求扰动流场中垂直速度与水平速度间有净的负相关。图 2 中给出了满足这种条件的一种理想的对流流场的结构。这种由两个同向涡环组成的后倾流场，与实际观测到的雷、阵雨云中的流场(见图 1) 是相似的。当然，图 1 所示的是包括盛行风在内的总的流场，不过即使去掉平均风的部分后，看来也会具有发展所要求的结构。

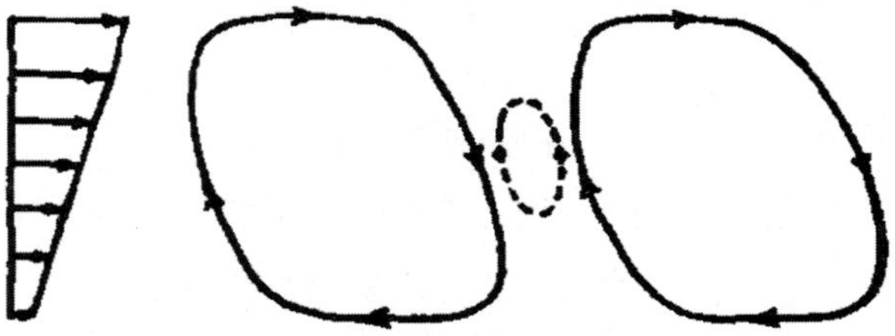

图 2　对流流场的理想模型

为了进一步分析风速切变作用与静力稳定度作用的相对重要性，我们来计算发展扰动的增长率。由于在式(16)中包含了依赖于对流结构的参数 k_1 和 $k_2 l_1$，因此计算只能对某一特定结构的对流进行。

容易看出，$|k_1|\leqslant 1, k_2 l_1 \geqslant 0$，作为例子，我们假定有这样一个对流环流，它所决定的 $k_1=-\frac{1}{2}, k_2 l_1=\frac{1}{2}$。对这一特定的对流，它的增长率与风速切变和静力稳定度的依赖关系见图 3。

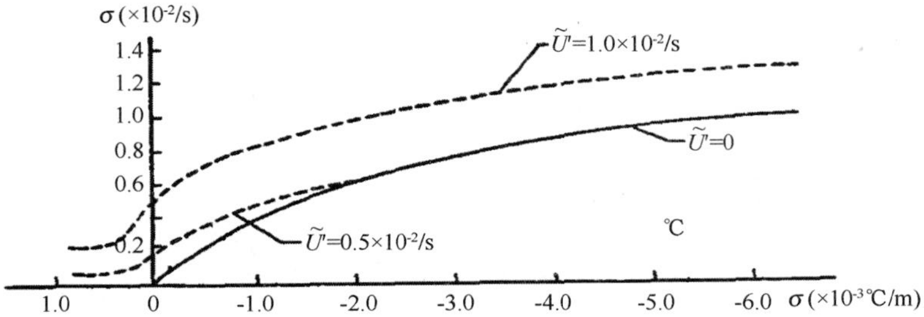

图 3 一个特定扰动的增长率

由图可见，当风速切变值大到 10^{-2}/s 左右时，它对于对流发展的影响是极显著的。例如，当静力稳定度取 $\alpha=-2\times10^{-3}$℃/m，在没有风速切变时，增长率为 0.58×10^{-2}/s，当考虑了风速切变后，为 0.90×10^{-2}/s，加大了 55%。当 $\alpha=0$ 时，单纯由 $\widetilde{U}'=10^{-2}$/s 大小的风速切变所引起的增长率到 0.5×10^{-2}/s，这相当于单纯由 $\alpha=-2\times10^{-3}$℃/s 这样大的不稳定度所产生的增长率。

虽然，对于不同的对流，风速切变的相对重要性是不一样的，但上面的估计可以给出一个初步的量的了解。一般来说，大气中水汽是向上减小的，而风速垂直切变在对流层高层高空急流的下方最大。因此，相对来说，在中、低层，湿绝热不稳定对于对流的发展更重要；在高层，风速垂直切变的作用更重要。由此来看，一般能穿透对流层顶而进入平流层的强大对流云，需要从高空急流中取得能量。

第二部分

5 无因次方程组和边界条件

在上节中给出了一种能够从风速垂直切变中获取能量的对流的理想模型，但是在大气中，这种类型的对流流场如何能够形成，这是一个需要进一步研究的问题。

我们知道，在静止的背景条件下，如果激发对流的初始扰动或者边界源是对称形式，那么对流流场也是对称的[12,13]。当背景条件中存在着具有垂直切变的盛行风后，对流流场具有怎么样的结构，这个问题虽然尚未有过研究，但不难事先作个估计。

如果基本流场的速度随高度是增加的，则在基本流场中就具有沿 y 轴方向的正的涡度。在这种情况下，通过扰动流场和基本流场之间的作用，基本流场中的正涡度要传输给扰动，使扰动流场获得正的涡度。于是本来是对称形式的对流流场中，原来具有正涡度的环流要加强，具有负涡度的环流要减弱。如果本来对流中具有负涡度的环流不强，那么在适当的条件下，就有可能使它反向，从而破坏本来流场的对称性，使其变成具有两个正涡度的对流流场。

为了进一步定量地表明上述的可能性，下面直接研究在地面对称热源激发下的对流运动。

取问题的定解条件为

$$z=0,\ w=0,\ \theta=\Delta\Theta_0 f(x,t) \tag{22}$$

$$z \to \infty,\ w = \theta = p = \frac{\partial p}{\partial z} = 0 \tag{23}$$

$$x \to -\infty,\ u = 0 \tag{24}$$

$$t = 0,\ u = w = \theta = p = 0 \tag{25}$$

引进无因次量

$$t' = t \Big/ \frac{L_0^2}{k}, (x', z') = \frac{1}{L_0}(x, z), (u', w') = \frac{L_0}{k}(u, w)$$

$$\theta' = \theta / \Delta\theta_0, p' = p \Big/ \bar{\rho}\left(\frac{k}{L_0}\right)^2 \tag{26}$$

式中,外参数 $\Delta\theta_0$ 为地面热源的强度;L_0 为它的尺度。

考虑到式(26)后,基本方程组式(1)至式(4)的无因次形式为(略去右上角的撇号)

$$\frac{\partial u}{\partial t} + u\frac{\partial u}{\partial x} + w\frac{\partial u}{\partial z} = -\left(\frac{GF}{R_i}\right)^{1/2} z\frac{\partial u}{\partial x} - \left(\frac{GF}{R_i}\right)^{1/2} w - \frac{\partial p}{\partial x} + \Delta_2 u \tag{27}$$

$$\frac{\partial w}{\partial t} + u\frac{\partial w}{\partial x} + w\frac{\partial w}{\partial z} = -\left(\frac{GF}{R_i}\right)^{1/2} z\frac{\partial w}{\partial x} + G\theta - \frac{\partial p}{\partial z} + \Delta_2 w \tag{28}$$

$$\frac{\partial u}{\partial x} + \frac{\partial w}{\partial z} = 0 \tag{29}$$

$$\frac{\partial \theta}{\partial t} + u\frac{\partial \theta}{\partial x} + w\frac{\partial \theta}{\partial z} = Fw - \left(\frac{GF}{R_i}\right)^{1/2} z\frac{\partial \theta}{\partial x} + \Delta_2 \theta \tag{30}$$

式中,无因次特征数的定义为

$$G = \frac{\beta\Delta\theta_0 L_0}{k^2},\ F = -\frac{\alpha L_0}{\Delta\Theta_0},\ R_i = -\frac{\alpha\beta}{\widetilde{U}'^2} \tag{31}$$

为了求解时的方便,在以下的计算中,将不直接用式(28),而改用由式(28)至式(30)导出的平衡方程[14]

$$\frac{\partial^2 p}{\partial z^2} = -\frac{\partial^2 p}{\partial x^2} + G\frac{\partial \theta}{\partial z} - 2\left(\frac{GF}{R_i}\right)^{1/2}\frac{\partial w}{\partial x} + 2\left(\frac{\partial u}{\partial x}\frac{\partial w}{\partial z} - \frac{\partial w}{\partial x}\frac{\partial u}{\partial z}\right) \tag{32}$$

相应的无因次边界条件和初始条件为

$$z = 0,\ w = 0,\ \theta = f(x, t) \tag{33}$$

其余在形式上仍同式(23)至式(25)。

6　问题的解法

今将 f 展成关于 t 的幂级数,即

$$f(x, t) = \sum_{n=0}^{\infty} f_n(x) t^{\frac{m+2}{2}} \tag{34}$$

引进变换

$$S = z / 2\sqrt{t} \tag{35}$$

并求问题为如下形式的解

$$\theta = \sum_{n=0}^{\infty} \Theta_n(x, s) t^{\frac{n+2}{2}},\ p = \sum_{n=0}^{\infty} \Pi_n(x, s) t^{\frac{n+3}{2}},$$

$$u=\sum_{n=0}^{\infty}U_n(x,s)t^{\frac{n+3}{2}},\ w=\sum_{n=0}^{\infty}W_n(x,s)t^{\frac{n+4}{2}} \tag{36}$$

式中,Θ_n,Π_n,U_n 和 W_n 由下面的常微分方程组决定。这些方程为

$$\begin{aligned}\Lambda_{n+2}(\Theta_n)=&-4\frac{\partial^2\Theta_{n-2}}{\partial x^2}+8\left(\frac{GF}{R_i}\right)^{1/2}S\frac{\partial\Theta_{n-3}}{\partial x}-4FW_{n-4}\\&+2\left(W_{n-5}\frac{\partial\Theta_0}{\partial s}+\cdots+W_0\frac{\partial\Theta_{n-5}}{\partial s}\right)+4\left(U_{n-5}\frac{\partial\Theta_0}{\partial x}+\cdots+U_0\frac{\partial\Theta_{n-5}}{\partial x}\right)\end{aligned} \tag{37}$$

$$\begin{aligned}\frac{\partial^2\Pi_n}{\partial s^2}=&2G\frac{\partial\Theta_m}{\partial s}-4\frac{\partial^2\Pi_{n-2}}{\partial x^2}-8\left(\frac{GF}{R_i}\right)^{1/2}\frac{\partial W_{n-3}}{\partial x}\\&+4\left(\frac{\partial U_{n-5}}{\partial x}\cdot\frac{\partial W_0}{\partial s}+\cdots+\frac{\partial U_0}{\partial x}\cdot\frac{\partial W_{n-5}}{\partial s}\right)-4\left(\frac{\partial W_{n-5}}{\partial x}\cdot\frac{\partial U_0}{\partial s}+\cdots+\frac{\partial W_0}{\partial x}\cdot\frac{\partial U_{n-5}}{\partial s}\right)\end{aligned} \tag{38}$$

$$\begin{aligned}\Lambda_{n+4}(W_n)=&-4G\Theta_n+2\frac{\partial\Pi_n}{\partial s}-4\frac{\partial^2W_{n-2}}{\partial x^2}+8\left(\frac{GF}{R_i}\right)^{1/2}S\frac{\partial W_{n-3}}{\partial x}\\&+4\left(U_{n-5}\frac{\partial W_0}{\partial x}+\cdots+U_0\frac{\partial W_{n-5}}{\partial x}\right)+2\left(W_{n-5}\frac{\partial W_0}{\partial s}+\cdots+W_0\frac{\partial W_{n-5}}{\partial s}\right)\end{aligned} \tag{39}$$

$$\frac{\partial U_n}{\partial x}=-\frac{1}{2}\frac{\partial W_n}{\partial s} \tag{40}$$

其中,运算子 Λ_m 为

$$\Lambda_m=\frac{\partial^2}{\partial s^2}+2s\frac{\partial}{\partial s}-2m \tag{41}$$

这组常微分方程组的边界条件为

$$s=0,\ \Theta_n=f_n(x),\ W_n=0 \tag{42}$$

$$s\to\infty,\ \Theta_n=W_n=\Pi_n=\frac{\partial\Pi_n}{\partial s}=0 \tag{43}$$

$$x\to-\infty,\ U_n=0 \tag{44}$$

容易看出,当 $n=0$ 时,方程(37)是线性齐次的,它满足边界条件式(42)、式(43)的解可以方便地求出。当 Θ_0 求得后,由式(38),应用条件式(43)直接积分两次后求得 Π_0。将 Θ_0,Π_0 代入式(39)的右方,解此线性非齐次方程,可以求出 W_0。最后,当 W_0 求得后,在条件式(44)下积分式(40)便求得 U_0。当 $n\geqslant1$ 时,可以用同样的步骤求解。

当 $n\leqslant8$ 时,方程的解可以按 Мхитарян[15]、Зейтунян[16]所给出的方法方便地求得。当 $n>8$ 时,可以像 Баев[17]那样用格林函数求解。由于解法是现成的,在此不再给出,可参见提到的有关文献。

7 数值计算结果

作为一例,取加热函数的形式为

$$f(x)=te^{-x^2} \tag{45}$$

计算时取 $G=400$,$F=0.3$,$R_i=30$。级数共算了 8 项。

计算结果列在图 3 至图 5 中。图 3 是无因次时间 为 0.6 没有盛行风作用下的流场,图 4 是考虑了风速垂直切变后的扰动流场,图 5 是加上基本风场后的流场。基本风场的强度取 $\widetilde{U}'=1.0\times10^{-3}$/s,

特征尺度取 $L_0=1$ km。另外,湍流交换系数取 500 m^2/s。

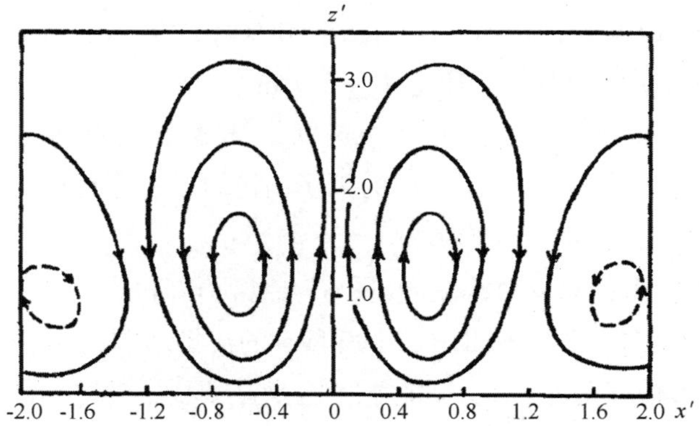

图4　没有风速切变时的对流流场

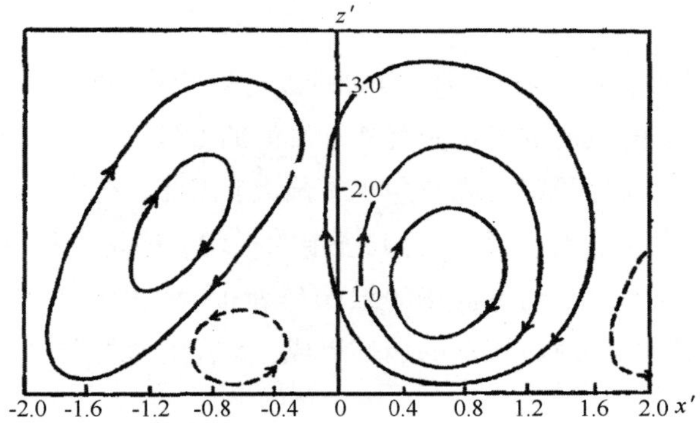

图5　考虑了风速切变后的扰动流场

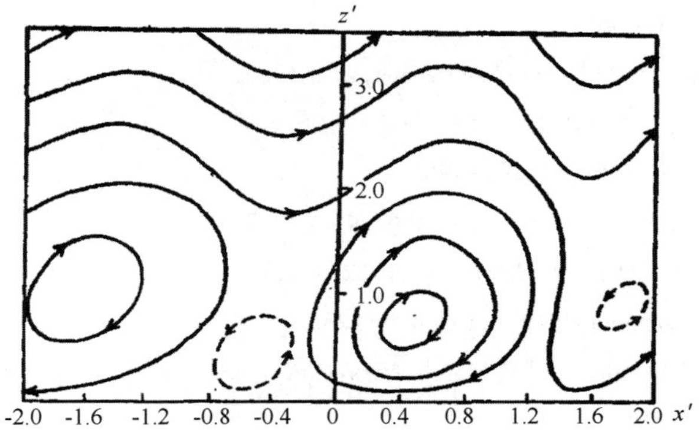

图6　加上盛行风场后的对流流场

比较图4和图5,一个醒目的现象是,当考虑了风速垂直切变作用后,扰动流场的结构发生了根本的改变。这时流场不再具有对称的性质,而是主要由两个同向的顺时针流动的涡环组成。所得的这个扰动流场它的负相关性,虽然不如图2那么理想,但由于在右边涡环上升区的负相关部分中,水平速度和垂直速度都很大,因此,这个对流流场的总效果仍然是负相关的。

这是很有意思的,风速垂直切变可以产生一个具有负相关的流场,这个流场形成后,根据第四节的分析,自然又能进一步从风速垂直切变中获得使它发展的能量。因此,风速垂直切变是一个有利于对流发展的因子。事实上,这个直接的计算也确实表明,考虑了风速的垂直切变后,对流的强度是加大了。例如,当 $L_0=1$ km,$\Delta\theta_0=3$℃,$k=500$ m^2/s,$\alpha=-0.9\times10^{-3}$℃/m,$\widetilde{U}'=1.0\times10^{-3}$/s 时,在没有风场作用的情况下,当 $t=20$ min 时,最大的上升速度为 6.7 m/s,最大的下沉速度为−5.0 m/s;当考虑了风切变后,最大的上升速度为 13.5 m/s,最大的下沉速度为−8.5 m/s,上升气流的强度增加了一倍左右。

风速垂直切变既然是一个有利于对流发展的因子,我们在一定程度上就可以了解,为什么强大的冰雹云会经常发生在高空急流的天气形势下[2]。

在另一方面,将计算所得的总的对流流场(图 6)与实际雷阵雨云的流场相比(图 1),可以看出,主要的结构是相似的。由此可见,一般雷雨云中所具有的流场结构,与盛行风的存在有直接的关系,而 Browning 等[6]提出的水滴蒸发等影响,可能不一定是必要的。当然,由于水滴对气流的拖带等作用,可以加强云中后部下沉环流的强度[18]。

8 总结

本文研究了盛行风速的垂直切变对于对流运动的影响,得到了两点主要的结果:

(1) 在合适的条件下,正的风速垂直切变有利于对流的发展。由于在大气中水汽的含量在中、低空较大,而风速垂直切变在对流层顶附近高空急流中最大,因此,在中、低空湿绝热不稳定能量的释放,是对流云发展的主要能源。在高空,则盛行风场中的平均动能是一种主要的能源。

(2) 由于风速垂直切变的作用,对流流场的结构与静止大气中一般的热对流流场很不一样,不再是轴对称的,而是主要由两个顺时针流动的涡环组成,这样的流场与一般移动的雷、阵雨云中的流场结构是很接近的。

参考文献

[1] Malkus,J.S.*Tellus*,1954,6:351−366.
[2] Dessens,H.,*Geophys.Monogr.*,5:333−338,Amer.geophys.Un.1960.
[3] 中山章.研究时报,1962,14:135−142,453−458.
[4] Newtwon,C.W.,et al.*J.Meteor.*,1959,16:483−496.
[5] Probert-Jones,J.R.& Harper,W.G.,*Meteor.Mag.*,1962,91:273−284.
[6] Browning,K.A.& Ludlum,f.H.,*Q.J.Roy.Meteor.Soc.*,1962,88:117−135.
[7] Taylor,G.I.,*Proc.Roy.Soc.(London) A*,1931,132:499−523.
[8] Eliassen,A.et al,Institute of Weather and Research,Norwegian Acad.Sci.Letters.publ.,No.1,1953:1−30.
[9] 郭晓岚.*Phys.Fluids*,1963,6:195−211.
[10] 巢纪平.气象学报,1962,32:164−176.
[11] Drazin,G.G.,*J.Fluid Mech.*,1958,4:214−224.
[12] 周晓平.*ДАН СССР*,*Серня.Геофпз.*1962:548−557.
[13] 巢纪平.气象学报,1961,31:191−205.
[14] 巢纪平.气象学报,1962,32:104−118.
[15] Мхитарян А.Н.,*ДАН Армянской ССР*,1954,19(2):33−40.
[16] Зейтунян Х.Н.,*ДАН СССР*,1960,133:1319−1322.
[17] Ъаев В.К.,*Труды Цип*,1956,43(70):3−8.
[18] 巢纪平,胡广兴.气象学报,1963,33:449−458.

二层模式中小地形对于气压跳跃形成的初步研究*

巢纪平[1]　章光锟[1]　袁孝明[2]

（1. 中国科学院地球物理研究所；2. 中国科学技术大学）

提要　本文应用一个两层密度不同的流体所组成的模式，讨论了小地形对于气流的影响。由一维常定问题的解指出，对于气流一定的上游条件，山脉存在一临界高度；山区气流的被扰状态，决定于上游条件和山的高度。当上游状态中的弗罗德（Froude）数小于 1，而山脉的高度等于临界高度时，背风面将出现常定的“气压跳跃”。也由一维非常定的解，研究了“气压跳跃”的形成过程。

1　引言

小地形对于气流的流动有明显的影响，这可以间接地从山区云系的分布看出。在山的背风面常常可以观测到有好几排荚状云，在荚状云下面紧靠山处，有时还可以观测到有强烈发展的积云。由于这种积云看上去好像绕水平轴在转动，所以它称为旋转云（rotor cloud）。目前观测已经肯定，荚状云由气流的背风波动所造成，而旋转云则由转子气流造成。

对于背风波的性质和形成条件已经有了不少研究（见参考文献[1]）。对于转子气流，由于其中上升运动的剧烈性，可以估计到这可能不是一个线性理论可以解决的问题。目前转子气流的形成机制虽然也有了好几种解释（见参考文献[2]），但问题远不如背风波研究得那么清楚。

有很多人都把转子气流的出现，看成是类似于“水跃”的一种“气压跳跃”现象的伴随物[2-5]。在这方面，Long[4] 曾经对二层密度不同的层结流体中阻碍物背后“气压跳跃”发生的条件，进行过详细的实验研究。实验表明，“气压跳跃”的出现，与流体的弗罗德数、阻碍物的高度等有密切的关系。

在本文中，我们将应用流体密度不同的二层模式对背风面“气压跳跃”的形成和发展过程做进一步的讨论。

2　两层模式的基本方程

注意到在转子气流出现时山区上空一般都存在一逆温层[2]，为了数学上处理的方便，我们把逆温层看成是两个密度各不相同的气团的分界面。设逆温层的高度为 $\eta(x)$；上层空气的密度为 ρ_1，下层空气的密度为 ρ_2；流体上界的高度为 $\overline{H}$，$\overline{H}$ 是均一不变的；山脉高度为 $\delta(x)$。我们再把水平坐标轴 x 放在海平面上，垂直坐标轴 z 与之正交，指向上方为正（图 1）。

对于下层流体，有运动方程

$$\frac{\partial u}{\partial t}+u\frac{\partial u}{\partial x}=-\frac{1}{\rho_2}\frac{\partial p}{\partial x} \tag{1}$$

* 气象学报，1964，34（2）：233-241.

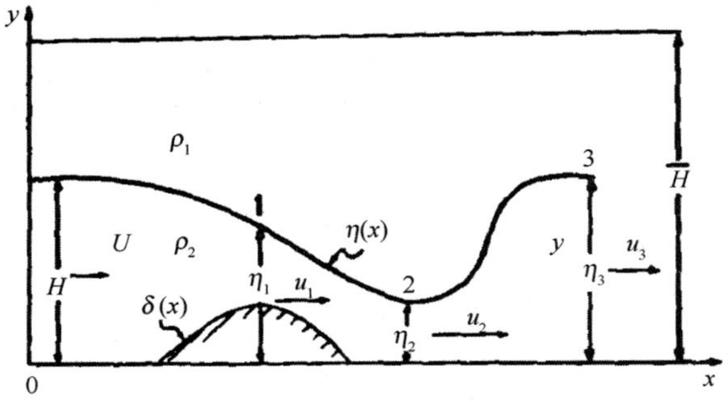

图 1

和连续方程：

$$\frac{\partial u}{\partial x}+\frac{\partial w}{\partial z}=0 \tag{2}$$

式中，u 和 w 分别为水平速度和垂直速度。

设流体是静力平衡的，于是下层流体中任一高度为 z 处的压力为：

$$p=\overline{p}+\rho_{1_g}(\overline{H}-\eta)+\rho_{2_g}(\eta-z) \tag{3}$$

式中，$\overline{p}$ 为 $\overline{H}$ 高度上的气压，是一个常量。

将式(3)代入式(1)后，得到

$$\frac{\partial u}{\partial t}+u\frac{\partial u}{\partial x}=-\frac{\rho_2-\rho_1}{\rho_2}g\frac{\partial \eta}{\partial x}\equiv -g^*\frac{\partial \eta}{\partial x} \tag{4}$$

从 δ 到 η 对式(2)积分，得到

$$(\eta-\delta)\frac{\partial u}{\partial x}=w_\delta-w_\eta \tag{5}$$

式中

$$w_\delta=u\frac{\partial \delta}{\partial x},\ w_\eta=u\frac{\partial \eta}{\partial x}+\frac{\partial \eta}{\partial t} \tag{6}$$

在求式(5)时，我们假定了 u 随高度是均一的。

将式(6)代入式(5)，得到连续性方程为：

$$\frac{\partial \eta}{\partial t}+\frac{\partial u(\eta-\delta)}{\partial x}=0 \tag{7}$$

由于山高 δ 和 g^* 都是已知的，因此式(4)、式(7)两式组成了对 u 和 η 的闭合方程组。

3 常定问题

(1)问题的解

关于常定问题，Long[4]，Schweitzer[3]都做过讨论。在本节中我们用不同的方法做进一步的讨论。我们先给出问题的显式解，在这基础上引进临界山高，再进一步讨论背风面“气压跳跃”的强度和上游条件的关系。

对于常定问题，式(4)，式(7)两式变成：

$$u\frac{\partial u}{\partial x} = -g^*\frac{\partial \eta}{\partial x} \tag{8}$$

$$\frac{\partial(\eta-\delta)u}{\partial x} = 0 \tag{9}$$

问题的边界条件取成

$$x=0,\ \delta=0,\ \eta=H,\ u=U \tag{10}$$

式(9)对 x 积分,应用条件式(10)后,得到

$$(\eta-\delta)u = HU \tag{11}$$

同样,由式(8)得到

$$\frac{1}{2}u^2 + g^*\eta = \frac{1}{2}U^2 + g^*H \tag{12}$$

将式(11)代入式(12)消去 u 后,得到

$$g^*(\eta-\delta)^3 + \left(g^*\delta - \frac{1}{2}U^2 - g^*H\right)(\eta-\delta)^2 + \frac{1}{2}U^2H^2 = 0 \tag{13}$$

引进无因次量:

$$\bar{\eta} = \frac{\eta}{H},\ \bar{\delta} = \frac{\delta}{H}$$

于是式(13)为

$$\bar{\xi}^3 + \left(\bar{\delta} - 1 - \frac{1}{2}F^2\right)\bar{\xi}^2 + \frac{1}{2}F^2 = 0 \tag{14}$$

式中

$$\bar{\xi} = \bar{\eta} - \bar{\delta},\ F = \frac{U}{\sqrt{g^*H}} \tag{15}$$

其中,F 为弗罗德数,是表征运动特征的一个重要的无因次量。

今将式(14)化为标准型,为此引进

$$\phi = \bar{\xi} - \frac{1}{3}\left(1 + \frac{1}{2}F^2 - \bar{\delta}\right) \tag{16}$$

代入式(14)后,得到

$$\phi^3 + P\phi + Q = 0 \tag{17}$$

其中

$$P = -\frac{1}{3}\left(1 + \frac{1}{2}F^2 - \bar{\delta}\right)^2,\ Q = -\frac{2}{27}\left(1 + \frac{1}{2}F^2 - \bar{\delta}\right)^3 + \frac{1}{2}F^2 \tag{18}$$

方程(17)是一个三次的代数方程,其根容易用标准方法求得,即

$$\phi = \sqrt[3]{-\frac{Q}{2} + \sqrt{\frac{Q^2}{4} + \frac{P^3}{27}}} + \sqrt[3]{-\frac{Q}{2} - \sqrt{\frac{Q^2}{4} + \frac{P^3}{27}}} \tag{19}$$

此即为问题的解。

(2)临界高度

解式(19)的性质由判别式

$$\Delta = \frac{Q^2}{4} + \frac{P^3}{27} \gtreqless 0 \tag{20}$$

决定或者将 P,Q 值代入后，得

$$\Delta = \frac{1}{2}F^2\left[\frac{1}{8}F^2 - \frac{1}{27}\left(1 + \frac{1}{2}F^2 - \bar{\delta}\right)^3\right] \gtreqless 0 \tag{21}$$

为了使这一判别式的物理意义更清楚，我们将它改写成

$$\bar{\delta} \gtreqless 1 + \frac{1}{2}F^2 - \frac{3}{2}\sqrt[3]{F^2} = \bar{\delta}_c \tag{22}$$

Long[4]用不同的方法也给出过这个临界高度。

由式(22)表明，在现在的问题中存在一个临界山高 $\bar{\delta}_c$，这临界山高是由上游条件，即由 F 所决定的。当上游条件给定后，解的性质将随实际山的高度而异。换言之，山区气流的被扰状态，除决定于上游条件外，也决定于山脉的高度。

临界山高与 F 的关系见图 2。由图 2 可见，当 F 增大时，$\bar{\delta}_c$ 变小；当 F 增大到 1 时，$\bar{\delta}_c$ 为零；当 F 再继续增大时，$\bar{\delta}_c$ 也相应增加。

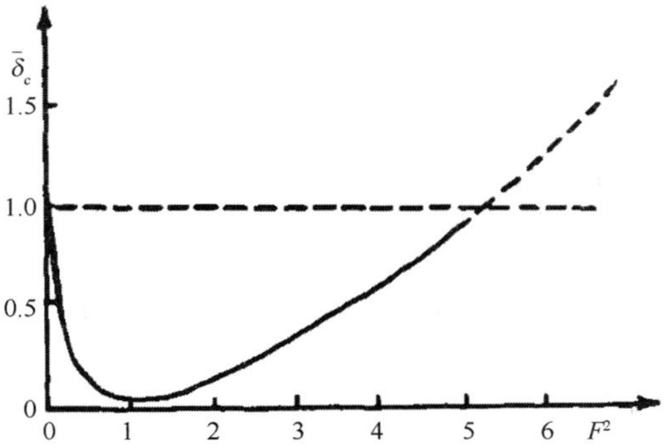

图 2

(3)解的性质

上面指出，当 F 给定后，解的性质随式(22)而异。计算表明，当 $\bar{\delta}<\bar{\delta}_c$ 时，给定一个 $\bar{\delta}$ 值后，存在两个不同的 $\bar{\eta}>\bar{\delta}$ 的值；当 $\bar{\delta}=\bar{\delta}_c$ 时，有两个不同的 $\bar{\eta}$ 值，其中一个 $\bar{\eta}>\bar{\delta}$，另一个 $\bar{\eta}<\bar{\delta}$；当 $\bar{\delta}>\bar{\delta}_c$ 时，只有一个 $\bar{\eta}<\bar{\delta}$的根。由于 $\bar{\eta}<\bar{\delta}$ 表示逆温层的高度比山还低，这显然是没有意义的解。这表明当 $\bar{\delta}>\bar{\delta}_c$ 时，问题不存在常定解。

当 $F=\sqrt{5.0}$（相应的 $\bar{\delta}_c=0.935$）和 $F=\sqrt{0.1}$（相应的 $\bar{\delta}_c=0.355$）时，$\bar{\eta}$ 和 $\bar{\delta}$ 的依赖关系分别见图 3a和图 3b。

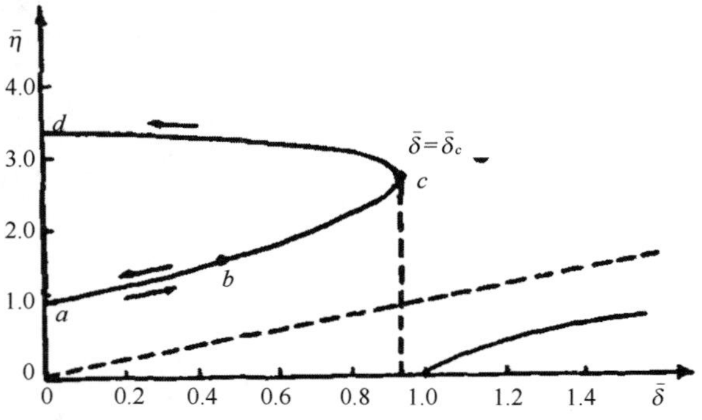

图 3a

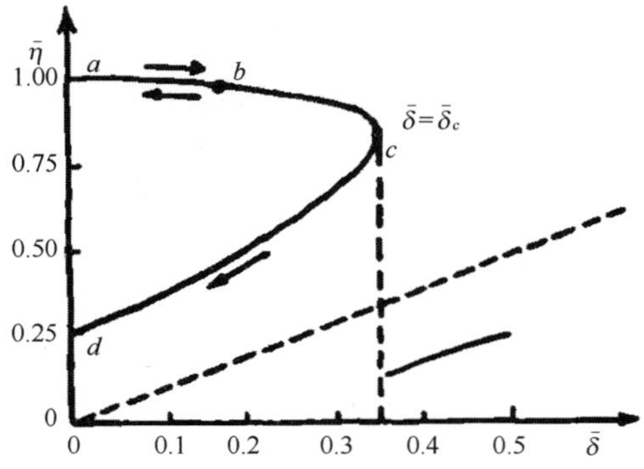

图 3b

由图 3a 和图 3b,我们可以对逆温层高度在山区的变化做一分析。取山脉是对称的,其最大高度 $\bar{\delta}_{max}<\bar{\delta}_c$。当 $F>1$,由图 3a 可见逆温层高度 $\bar{\eta}$ 在山前由 $\bar{\eta}=1$ 起,随着山的增高逆温层高度也相应增高。如果到山顶时,$\bar{\eta}$ 已增到图上的 b 点,过山顶后,$\bar{\delta}$ 将减小。考虑到解的连续性,$\bar{\eta}$ 值将沿 ba 而减小,到山脚($\bar{\delta}=0$),$\bar{\eta}$ 又回到 a 点($\bar{\eta}=1$)。可见,在 $F>1$,$\bar{\delta}<\bar{\delta}_c$ 的情况下,受扰逆温层在山的两侧取对称分布(图 4a)。如果山的最大高度恰好等于临界高度,即 $\bar{\delta}=\bar{\delta}_c$,那么,到山顶时,$\bar{\eta}$ 的值由 1 增到 c 点的最大值,当过山后,$\bar{\eta}$ 的值将沿 cd 而继续增大,到山脚 $\bar{\eta}$ 达到最大值。这样,受扰逆温层在山的两侧呈不对称分布(图 4b)。当 $F<1$ 时,也可以按图 3b 作类似的讨论。这时,当 $\bar{\delta}<\bar{\delta}_c$ 时,逆温层在山的两侧也是对称的,不过与 $F>1$ 时不同,$\bar{\eta}$ 在山顶取最小值(图 4c);当 $\bar{\delta}=\bar{\delta}_c$ 时,逆温层的高度在过山后将继续下降,一直到山脚达到最小值(图 4d)。

这一计算结果的可靠性,不难从 Long[4,5] 的实验结果中得到定性的验证。

(4)“气压跳跃”的形成条件和强度

结合式(8),式(9)两式,可以得到

$$\left(1-\frac{g^*(\eta-\delta)}{u^2}\right)\frac{\mathrm{d}\eta}{\mathrm{d}x}=\frac{\mathrm{d}\delta}{\mathrm{d}x}. \tag{23}$$

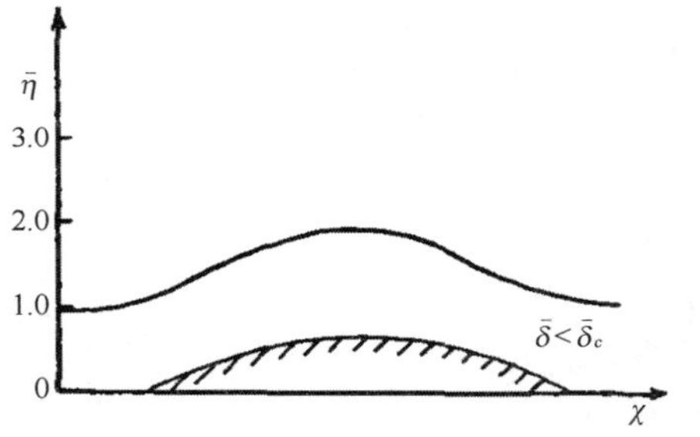

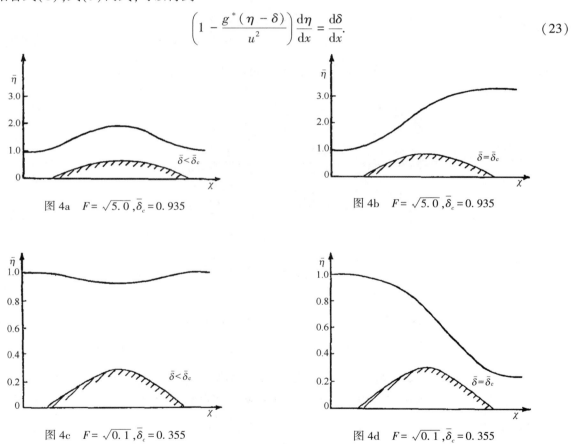

图 4a　$F=\sqrt{5.0}$,$\bar{\delta}_c=0.935$

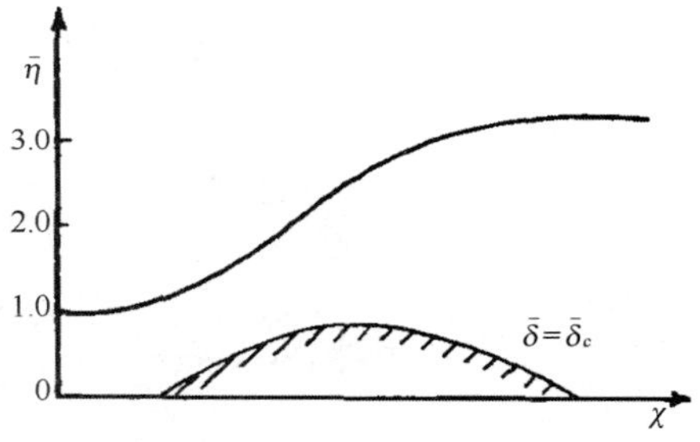

图 4b　$F=\sqrt{5.0}$,$\bar{\delta}_c=0.935$

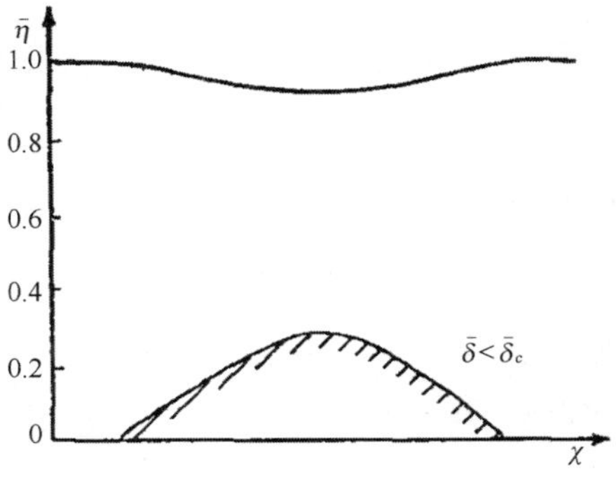

图 4c　$F=\sqrt{0.1}$,$\bar{\delta}_c=0.355$

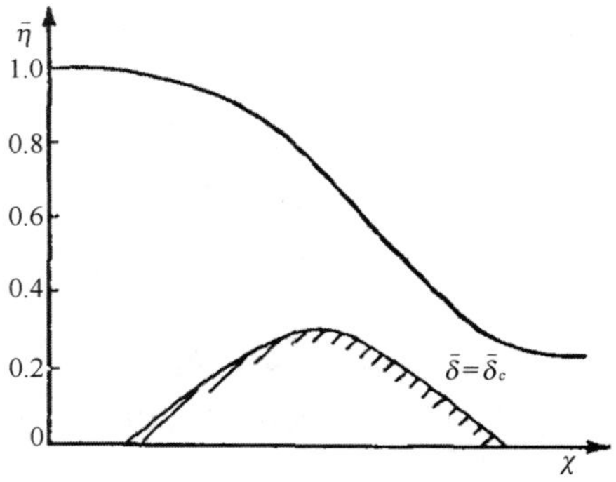

图 4d　$F=\sqrt{0.1}$,$\bar{\delta}_c=0.355$

由此式容易看出,当 $u>\sqrt{g^*(\eta-\delta)}$ 时,η 将随 δ 的增高而增高;当 $u<\sqrt{g^*(\eta-\delta)}$ 时,则 η 将随 δ 的增高而减小。同时也可以看出,只有在山顶 $\left(\frac{\mathrm{d}\delta}{\mathrm{d}x}=0\right)$,气流的速度 u 才能达到它的临界值 $\sqrt{g^*(\eta-\delta)}$。显然,如果山的高度 δ,达到某一临界高度 δ_c 后,使得在山顶的气流有 $u=\sqrt{g^*(\eta-\delta)}$,那么本来在山的向

风面逆温层高度是增高的，过山后仍然将继续增高，或者本来是降低的，过山后仍然将继续降低。

因此，如果山的高度 $\delta=\delta_c$，那么在上游本是次临界的气流，在过山时流速要不断增大，到山顶达到临界值；过山顶后由于高度继续降低，于是速度仍将继续增大，以致过山顶后形成超临界气流。显然，这就是图 4d 所表明的情况。

如果在图 1 所示中的 2 处，气流是超临界的，那么由于气流继续向前流动时，受摩擦等作用的影响，流速必将减小，最后又恢复到次临界流。这样在超临界流和次临界流之间的区域中必然将形成"气压跳跃"。如果"气压跳跃"的强度较大，那么在"气压跳跃"中会出现转子气流。

根据上面的分析，我们就可以得到结论，常定状态的"气压跳跃"和转子气流出现的条件有二：一是上游的弗罗德数 F 必须小于 1；二是山的高度必须达到临界高度 δ_c。

Kuettner[2] 认为，这种与"水跃"相似的理论虽然是转子气流的一个很好的解释，但是也有不足之处。他指出，在这种情况下"气压跳跃"可能到达的高度 η（图 1 中 $\bar{\eta}_3$）要低于山顶逆温层的高度（图 1 中 $\bar{\eta}_1$），而观测却表明，在一般情况下，旋转云顶的高度总要比山顶"焚风墙"云（Fōhn wall）的高度来得高。为了使得旋转云的高度增大到超过"焚风墙"云的高度，于是他提出需要在原来的理论中考虑到"气压跳跃"区域中由于湍流的发展所造成的该处逆温强度的减弱效应。

我们指出，Kuettner 的意见并不完全正确。因为即使从上面这种简单的"气压跳跃"理论出发，也并不总是 $\bar{\eta}_3<\bar{\eta}_1$ 的。事实上，根据"水跃"理论[6]，容易求得跳跃前后的高度比为

$$\left(\frac{\bar{\eta}_3}{\bar{\eta}_2}\right)^2+\left(\frac{\bar{\eta}_3}{\bar{\eta}_2}\right)-2F_2^2=0 \tag{24}$$

式中 $F_2^2=\dfrac{u_2^2}{g^*\eta_2}$。由此得到

$$\bar{\eta}_3=-\frac{1}{2}\bar{\eta}_2+\sqrt{\frac{1}{4}+2F_2^2}\cdot\bar{\eta}_2 \tag{25}$$

或者改写成

$$\bar{\eta}_3=-\frac{1}{2}\bar{\eta}_2+\sqrt{\frac{1}{4}\bar{\eta}_2^2+\frac{2F_0^2}{\bar{\eta}_2}} \tag{26}$$

由式(26)，在气压跳跃能够发生的参数区域中（$F_0<1,\bar{\delta}=\bar{\delta}_c(F_0)$），$\bar{\eta}_2$ 的值可以由问题的解式(19)算出。因此 $\bar{\eta}_3$ 的值可以按上游条件算出(图 5 中实线)。同时 $\bar{\eta}_1$ 的值也可以算出来(图 5 中虚线)。由图 5 可以看出，只有在 $F_0<0.4$ 时，才有 $\bar{\eta}_3<\bar{\eta}_1$；而当 $F_0>0.4$ 后，$\bar{\eta}_3>\bar{\eta}_1$。因此来看，即使上面这种简单的"气压跳跃"理论也可以在一定程度上解释转子气流的观测事实。Kuettner 之所以得到 $\bar{\eta}_3<\bar{\eta}_1$ 的结论，是因为他没有直接分析问题的解，而认为 $F_2\gg1$，因而由式(25)得到的近似式

$$\bar{\eta}_3\approx\sqrt{2}F_2\bar{\eta}_2$$

出发来作计算。事实上，在"气压跳跃"能够出现的情况下，F_2 并不总是很大的。

4 非常定问题

由于在 $\delta>\delta_c$ 的情况下问题不存在常定解，同时由于常定解所要求的"气压跳跃"形成的条件也太

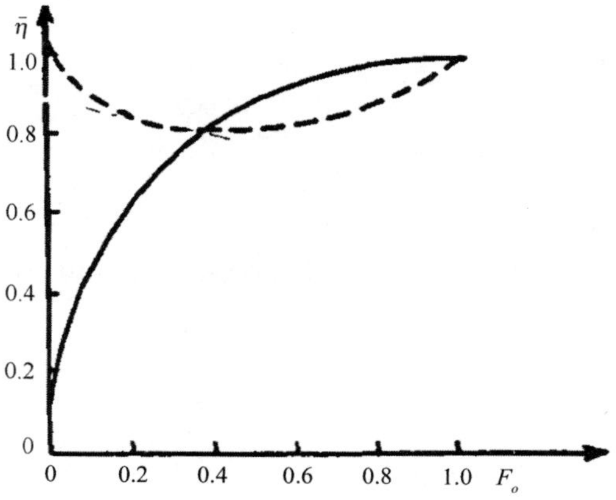

图 5

强，在实际情况下很难经常成立，因此我们在本节中进一步研究非常定解。

为了方便，用 $\eta' = \eta - H$ 来代替式(4)、式(7)两式中的 η。这样我们就有下列方程：

$$\frac{\partial u}{\partial t} + u\frac{\partial u}{\partial x} = -g^{*}\frac{\partial \eta'}{\partial x} \tag{27}$$

$$\frac{\partial \eta}{\partial t} + \frac{\partial [u(H + \eta - \delta)]}{\partial x} = 0 \tag{28}$$

引进无因次量：

$$\bar{x} = \frac{x}{H},\ \bar{t} = \frac{t}{\sqrt{\frac{H}{g^{*}}}},\ \bar{\delta} = \frac{\delta}{H},\ \bar{\eta} = \frac{\eta}{H},\ \bar{u} = \frac{u}{\sqrt{g^{*}H}}$$

这样就得到无因次方程：

$$\frac{\partial \bar{u}}{\partial \bar{t}} + \bar{u}\frac{\partial \bar{u}}{\partial \bar{x}} = -\frac{\partial \bar{\eta}}{\partial \bar{x}} \tag{29}$$

$$\frac{\partial \bar{\eta}}{\partial \bar{t}} + \frac{\partial}{\partial \bar{x}}[\bar{u}(1 - \bar{\delta} + \bar{\eta})] = 0 \tag{30}$$

由式(29)和式(30)两式容易得到：

$$\left(\frac{\partial}{\partial \bar{t}} + (\bar{u} + \bar{c})\frac{\partial}{\partial \bar{x}}\right)(\bar{u} + 2\bar{c}) = -\frac{\partial \bar{\delta}}{\partial \bar{x}} \tag{31}$$

$$\left(\frac{\partial}{\partial \bar{t}} + (\bar{u} - \bar{c})\frac{\partial}{\partial \bar{x}}\right)(\bar{u} - 2\bar{c}) = -\frac{\partial \bar{\delta}}{\partial \bar{x}} \tag{32}$$

式中

$$\bar{c}^{2} = 1 - \bar{\delta} + \bar{\eta} \tag{33}$$

是逆温层上波动的传播速度。

方程(31)和方程(32)存在两条特征线

$$\left.\begin{aligned}\left(\frac{d\bar{x}}{d\bar{t}}\right)_{\mathrm{I}} &= \bar{u} + \bar{c},\\ \left(\frac{d\bar{x}}{d\bar{t}}\right)_{\mathrm{II}} &= \bar{u} - \bar{c}.\end{aligned}\right\} \tag{34}$$

方程(31),方程(32)可以用沿特征线作差分的办法来求解。相应的差分式为:

沿特征线Ⅰ有:

$$\bar{u} + 2\bar{c} = \bar{u} + 2\bar{c}_1 - \frac{\Delta\bar{\delta}}{\Delta\bar{x}}\Delta t_{\mathrm{I}} \tag{35}$$

沿特征线Ⅱ有:

$$\bar{u} - 2\bar{c} = \bar{u}_2 - 2\bar{c}_2 - \frac{\Delta\bar{\delta}}{\Delta\bar{x}}\Delta t_{\mathrm{II}} \tag{36}$$

差分格式如图6所示。这样任意一点的 $\bar{u}$ 和 $\bar{c}$ 都可以用在它前一时刻的量来表示,

$$\bar{u} = \frac{1}{2}(\bar{u}_1 + \bar{u}_2) + (\bar{c}_1 - \bar{c}_2) - \frac{1}{2}(\Delta t_{\mathrm{I}} + \Delta t_{\mathrm{II}})\frac{\Delta\bar{\delta}}{\Delta\bar{x}} \tag{37}$$

$$\bar{c} = \frac{1}{4}(\bar{u}_1 - \bar{u}_2) + \frac{1}{2}(\bar{c}_1 + \bar{c}_2) - \frac{1}{4}(\Delta t_{\mathrm{I}} - \Delta t_{\mathrm{II}})\frac{\Delta\bar{\delta}}{\Delta\bar{x}} \tag{38}$$

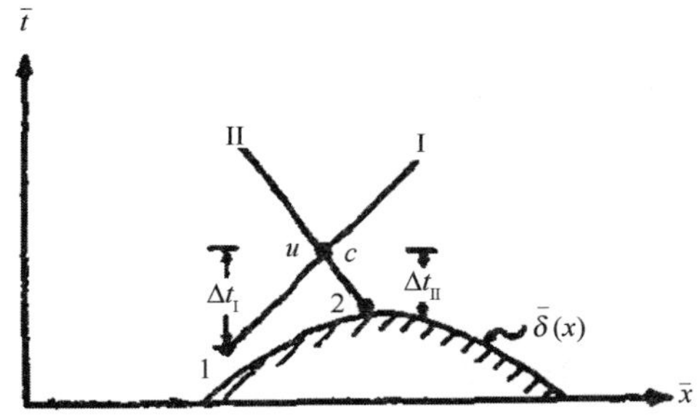

图6

在以下的各例计算中,取 $\bar{t}=0$ 时,$\bar{\eta}=0$,而 $\bar{u}$ 则给以不同的值,但在 $\bar{x}$ 方向是均匀的。

图7a是 $\bar{u}_{i=0}=0$ 的情况,图7b是 $\bar{u}_{i=0}=0.5$ 的情况,图7c是 $\bar{u}_{i=0}=1.0$ 的情况,图7d是 $\bar{u}_{i=0}=1.5$ 的情况。

比较这些例子,我们可以得到下面几点结果:

(1)当上游的风速强时(图7c,图7d),过山后可以产生较强的下滑气流,随着下滑气流的加强,在背风面的某个地区,逆温层将上跳,最后形成不连续的分布。

(2)上游的风速越大,背风面不连续跳跃的形成时间越短。如在 $\bar{u}_{i=0}=0.5$ 时为 $\bar{t}=0.45$,在 $\bar{u}_{i=0}=1.0$ 时为 $\bar{t}=0.35$,$\bar{u}_{i=0}=1.5$ 时为 $\bar{t}=0.30$。

(3)上游的风速越大,不连续跳跃的位置离山越远。

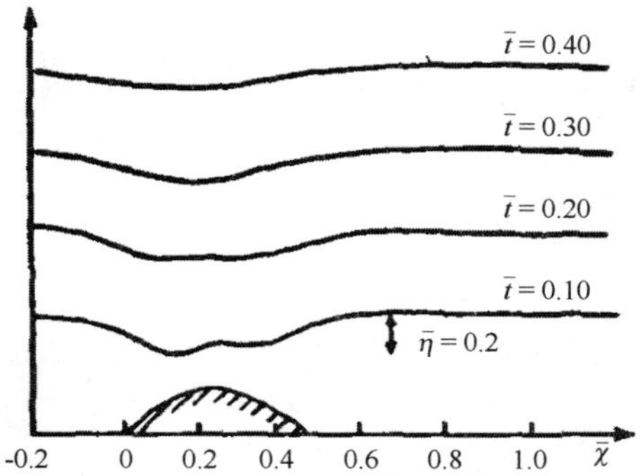

图 7a

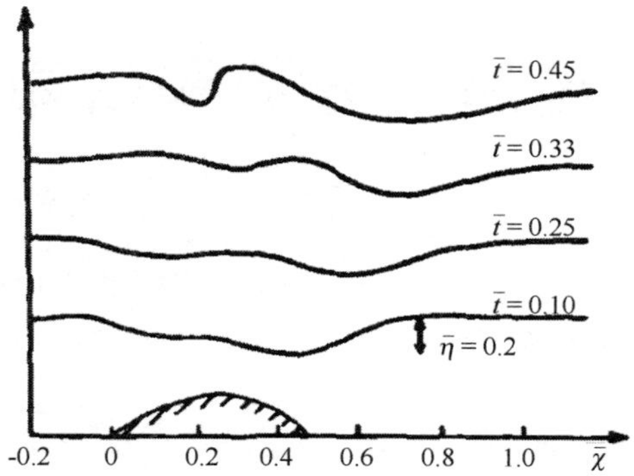

图 7b

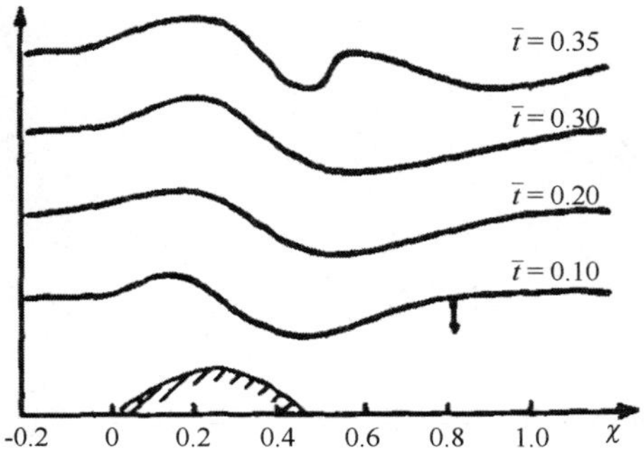

图 7c

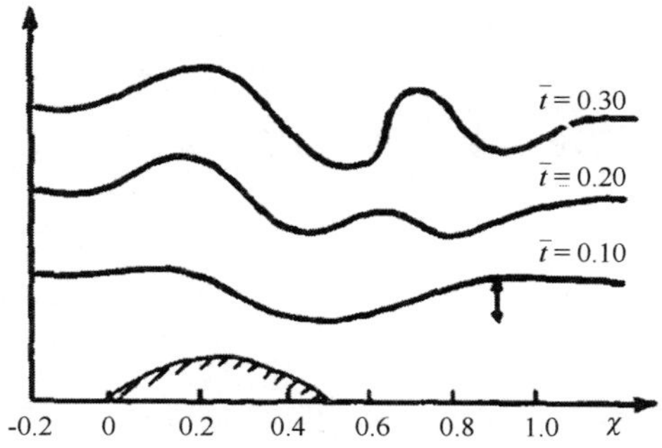

图 7d

从以上几个计算例子来看，只要山有一定的高度，上游风速有一定的强度，背风面一般总是可以产生“跳跃”现象的。由于我们求的是数值解，因此“跳跃”产生的时间、地点和山高、气流强度、逆温层高度和强度间的定量关系在此尚不能给出。但作为例子可以看出非线性理论和线性理论的结果是很不一样的，由线性理论是算不出这种不连续性的跳跃来的。

我们认为拟订更合适的模式，进一步分析地形对气流影响的非线性效应，是一个值得研究的题目。

参考文献

[1]　叶笃正，气象学报，1956，27：243-262.

[2]　Kuettner，J.，*Aero.Revue*，1958，33：208-214.

[3]　Schweitzer，H.，*Arch.Meteor.Geophy.biokl.*，1953，5：350-371.

[4]　Long，R.R.，*Tellus.*1954，6：97-115.

[5]　Bell，F.K.，*Austral.J.Phys.*，1956，9：373-386.

[6]　郎道，П.Д.，栗弗席兹，E.M.连续介质力学第二册.北京：人民教育出版社，1960：514-516.

沿岸上升流和沿岸急流的一个半解析理论*

巢纪平[1,2,3] 陈显尧[2,3]

（1. 国家海洋局第一海洋研究所，青岛 266061；
2. 中国科学院海洋研究所，青岛 266071；3. 国家海洋局海洋环境科学和数值模拟重点实验室，青岛 26601）

摘要：在考虑了陆架地形后，在垂直海岸的 $x-z$ 剖面上，对 Boussinesq 流体的非线性海洋运动方程求得了总动量守恒、温度守恒和位势涡度守恒的普适形式，进而得到流函数所满足的椭圆形二阶偏微分方程，在给定流体沿地形运动的条件下，算出问题的解。计算结果表明，沿岸可以出现上升流也可以出现下沉流，它依赖于海洋的大尺度背景条件。计算所得的上升流、沿岸急流、温度的锋区结构与一些观测事实接近。

关键词：动量、温度、位势涡度守恒；上升、下沉流和急流；非线性解析理论

1 引言

在陆架海中的流体受陆架坡度的影响极易在海岸附近产生上升（下沉）流，伴随上升（下沉）流的是底层的向岸（离岸）流和表层的离岸（向岸）流，在旋转地球上，由于科里奥利力的作用，由向岸或离岸流，又会导出沿岸方向的急流，并由此有与这样的流动相适应的温度场。如此相互制约的陆架环流是如何形成的，一直是陆架海洋动力学研究的中心问题之一。陆架环流直接关系到近海海洋的生态平衡，影响渔场的发展以及海洋环境污染净化等问题，因此对陆架环流研究具有非常重要的意义。

关于沿岸上升流及沿岸急流形成和发展的理论，在线性理论的范畴内，Hsueh 等[1]研究过连续层结介质中上升流形成的定常解，Allen[2]研究过层结介质中的非定常解。考虑到陆架坡度可以很陡，上升流可以很强，小振幅的线性化假设并不总是合适的，因此人们一般改用对非线性方程的数值积分，如最近 Allen 等[3]应用 Blumberg-Mellor[4]模式对 Oregon 陆架所做的工作，以及较早 Hamilton 和 Rattery[5,6]对西北非洲沿岸上升流所做的工作。数值模拟虽然是研究非线性问题的一种有力的方法，但往往对问题的物理过程看不太清楚。我们认为研究像沿岸上升流这类非线性问题，采用半解析半数值的方法更为适宜，这样可使影响上升流形成和发展的物理因子和物理过程较为清楚，而同时又能得到非线性问题的解。

中国的海洋学家很早就对沿岸上升流进行过观测研究，如胡敦欣等指出[7]，根据观测资料，渤海、黄海、东海和南海均有沿岸上升流被观测到，其中又以浙江沿岸的上升流最为显著，同时他们又讨论了沿岸上升流产生的原因，提出海洋本身的环境条件如黑潮的作用要大于风吹的作用。以后颜廷壮[8]也指出，浙江沿岸上升流是台湾暖流和地形共同作用的结果，风的影响一般不是主要的。潘

* 地球物理学报，2003，46（1）：26-30.

基金项目：国家重点基础研究规划项目（G1999043809）、国家自然科学基金项目（49736019）与国家海洋局青年海洋科学基金项目（2000206）。

玉球等[9]也认为，浙江沿岸上升流锋，是由黑潮次表层水向岸水平辐合，并在斜坡上抬升引起的。刘先炳和苏纪兰[10]的数值试验则进一步指出，浙江沿岸上升流按其成因可分成两个区域，即近岸区和远岸区。在近岸区风应力起主要作用，在远岸区台湾暖流的诱生作用很重要。

根据上升流形成的观测研究和问题本身的特点，本文将在考虑陆架地形的情况下，求 Boussinesq 流体的海洋非线性运动方程的解，研究陆架海外侧大尺度背景条件(如黑潮的强度)对沿岸上升流及沿岸急流的影响。

2　基本方程及边界条件

设海岸呈南北走向，y 轴与之平行，向北为正，x 轴与海岸垂直，离岸为正，z 轴垂直向上，海洋的平均深度为 D，相对 D 的陆架地形为 $h(x)$。设海水是 Boussinesq 流体，即其密度只是温度的函数，而发生在其中的运动是无辐散的。这样，对于发生在(x,z)平面上的定常运动，如在边界内部的流体中不考虑湍流过程，其非线性运动方程组为

$$u\frac{\partial u}{\partial x}+w\frac{\partial u}{\partial z}-fv=-\frac{1}{\rho_0}\frac{\partial p}{\partial x} \tag{1}$$

$$u\frac{\partial v}{\partial x}+w\frac{\partial v}{\partial z}+fu=0 \tag{2}$$

$$\frac{\partial u}{\partial x}+\frac{\partial w}{\partial z}=0 \tag{3}$$

$$u\frac{\partial w}{\partial x}+w\frac{\partial w}{\partial z}=-\frac{1}{\rho_0}\frac{\partial p}{\partial z}-(1-\alpha T)g \tag{4}$$

$$u\frac{\partial T}{\partial x}+w\frac{\partial T}{\partial z}=0 \tag{5}$$

式中，u, v, w 分别是沿 x, y, z 轴的速度分量；p 和 T 是流体的压强和温度；ρ_0 是流体的平衡密度；α 是流体的热膨胀系数。f 是科里奥利参数，为 $2\Omega\sin\phi$，Ω 是地球自转角速度，ϕ 是纬度，在这个理论中 f 取为常数，g 为重力加速度并没有采用静力平衡假设，这是因为考虑到在海岸附近允许出现惯性边界层，同时也允许陆架地形可以有急剧的变化。

问题的运动学边界条件取为

$$z=h(x),\quad w=u\frac{\partial h_c}{\partial x} \tag{6}$$

$$z=D,\quad w=0 \tag{7}$$

$$x=0,\quad u=0 \tag{8}$$

$$x\to\infty,\quad w\to 0 \tag{9}$$

其中，h_c 为海底地形。对于近海海洋环流中的某些问题，只用运动学边界条件约束是不够的，来自大气的强迫是需要考虑的，例如对温度的分布，观测表明它除受海流的水平和垂直方向的平流影响外，还与海表面和大气之间的热量交换有关，例如感热通量的影响。

3　经圈动量守恒方程和温度守恒积分

由方程(3)引进流函数 φ，定义为

$$u=-\frac{\partial\varphi}{\partial z},\quad W=\frac{\partial\varphi}{\partial x}\tag{10}$$

方程(2)写为总动量守恒形式

$$J(\varphi,v+fx)=0\tag{11}$$

其中，$J(A,B)=\frac{\partial A}{\partial x}\frac{\partial B}{\partial z}-\frac{\partial A}{\partial z}\frac{\partial B}{\partial x}$。方程(5)改为

$$J(\varphi,T)=0\tag{12}$$

方程(11)和方程(12)可写成

$$v+fx=F_1(\varphi)\tag{13}$$

$$T=F_2(\varphi)\tag{14}$$

式中，F_1,F_2 均为 φ 的普适函数。下面确定普适函数 F_1,F_2。

普适函数 F 是 φ 的函数簇，可以取不同的关于 φ 的函数形式，其中最简单的形式是 φ 的线性函数，即 $F_1(\varphi)\propto\varphi,F_2(\varphi)\propto\varphi$，为了确定比例系数的量纲和数量级，可采取下面的估计方法。

设当 x 充分大时有一横向速度 u_∞，由方程(11)有 $\varphi_\infty=-u_\infty z$ 或 $z=-\varphi_\infty/u_\infty$。同时设 $v_\infty=v_0+\Lambda z$，$T_\infty=T_0+\Gamma z$。v_0,T_0 分别为平均速度与温度。对式(13)关于 z 微分，可得

$$\frac{\partial v_\infty}{\partial z}=F'_1(\varphi_\infty)\frac{\partial\varphi_\infty}{\partial z}=-u_\infty F'_1(\varphi_\infty)\tag{15}$$

由此有

$$F'_1(\varphi_\infty)=-\frac{\Lambda}{u_\infty}$$

积分，并用到区域内(因为是普适函数)有

$$F_1(\varphi)=-\frac{\Lambda}{u_\infty}\varphi\tag{16}$$

代入式(13)，有

$$v+fx=-\frac{\Lambda}{u_\infty}\varphi\tag{17}$$

类似地，对于温度有

$$T=T_0-\frac{\Gamma}{u_\infty}\varphi\tag{18}$$

由此得到经圈速度、温度与流函数的关系。自然，这只是在给定条件下的一种关系，而这种关系只在流体内部成立，在边界上失效。作为物理上理解的参考，参数 u_∞、Γ、Λ 可看成是无穷远处大尺度海洋背景场中的环境参数值。

4 位势涡度守恒方程

将方程(1)写成

$$J\left(\varphi,-\frac{\partial\varphi}{\partial z}\right)-fv=-\frac{1}{\rho_0}\frac{\partial p}{\partial x}\tag{19}$$

方程(4)写成

$$J\left(\varphi,\frac{\partial\varphi}{\partial x}\right)=-\frac{1}{\rho_0}\frac{\partial p}{\partial z}-(1-\alpha T)g\tag{20}$$

式(19)对 z 微分，式(20)对 x 微分，后一式减前一式可得

$$J\left(\varphi,\ \frac{\partial^2\varphi}{\partial x^2}+\frac{\partial^2\varphi}{\partial z^2}\right)+f\frac{\partial v}{\partial z}-g\alpha\frac{\partial T}{\partial x}=0 \tag{21}$$

利用普适关系式(17)和式(18)，简单推导可得

$$\frac{\partial^2\varphi}{\partial x^2}+\frac{\partial^2\varphi}{\partial z^2}+\frac{g\alpha\Gamma}{u_\infty}z+\frac{f\Lambda}{u_\infty}x=F_3(\varphi) \tag{22}$$

采用与前述方法类似的过程可确定普适函数 F_3，即为

$$F_3(\varphi)=-\frac{g\alpha\Gamma}{u_\infty^2}\varphi \tag{23}$$

5　流函数与垂直运动的控制方程

将方程(23)式代入式(22)可得

$$\frac{\partial^2\varphi}{\partial x^2}+\frac{\partial^2\varphi}{\partial z^2}+\frac{g\alpha\Gamma}{u_\infty^2}\varphi=-\frac{g\alpha\Gamma}{u_\infty}z-\frac{f\Lambda}{u_\infty}x \tag{24}$$

此方程即为本问题的一个控制方程。

方程(24)对 φ 是线性的，在确定这个线性方程的过程中，没有对原来非线性原始方程作任何小扰动线性化的假设，所以现在运动可以不是小振幅,而是有限振幅的，因此本文提出的方法是陆架上升流的一个有限振幅理论。同时，也可以给出普适函数的非线性形式，以背景场的温度垂直分布为参考,如果大尺度环境的温度分布具有温跃层结构，例如下面的垂直温度分布形式

$$T_\infty=T_0+\Gamma z-\Pi^2 z \tag{25}$$

Π 为背景场温度分布参数，采用相同的过程，可以确定区域内温度分布与流函数的关系

$$T_\infty=T_0-\frac{\Gamma}{u_\infty}\varphi_\infty-\frac{\Pi}{u_\infty^2}\varphi_\infty^2$$

类似地，方程(22)中的普适函数 F_3 为

$$F_3(\varphi)=-g\alpha\left(\frac{\Gamma}{u_\infty^2}\varphi+\frac{2\,\Pi}{u_\infty^3}\varphi^2\right) \tag{26}$$

控制方程为

$$\frac{\partial^2\varphi}{\partial x^2}+\frac{\partial^2\varphi}{\partial z^2}+\left(\frac{g\alpha\Gamma}{u_\infty^2}+\frac{2g\alpha\Pi}{u_\infty^2}z\right)\varphi+\frac{2g\alpha\Pi}{u_\infty^2}\varphi^2=-\frac{g\alpha\Gamma}{u_\infty}z-\frac{f\Lambda}{u_\infty}x \tag{27}$$

可见，如果大尺度海洋背景温度场的垂直分布具有温跃层结构时，由此得到的普适函数及控制方程是非线性的。需指出的是方程式(24)与式(27)都只能在边界附近成立，即由它求算得出的解只是局部解,这是由方程式(17)所表示的经圈总动量守恒性所决定的，由于 fx 的出现，当 x 很大时,式(17)将很难成立，这意味着在离岸很远的地区将会有别的物理过程参与进来。下面讨论问题的边界条件。

考虑到流体必须沿着陆架地形运动，而不能穿越，因此有

$$z=h(x),\quad \varphi=\varphi_0(\text{常数}) \tag{28}$$

当流体流到海岸时也不能穿越，需沿着岸壁运动，同时考虑到流体运动的连续性，故有

$$x=0,\quad \varphi=\varphi_0 \tag{29}$$

如果海面不是一个自由面，并考虑到连续性，给出“刚盖”条件为

$$z=D,\quad \varphi=\varphi_0 \tag{30}$$

另外，当 x 充分大时

$$\frac{\partial \varphi}{\partial x} \to 0, \text{或} \quad \varphi = \varphi_0 \tag{31}$$

对垂直速度可以利用下面的控制方程计算，对式(24)、式(27)关于 x 微分，给出

$$\frac{\partial^2 w}{\partial x^2} + \frac{\partial^2 w}{\partial z^2} + \frac{g\alpha\Gamma}{u_\infty^2}w = -\frac{f\Lambda}{u_\infty} \tag{32}$$

$$\frac{\partial^2 w}{\partial x^2} + \frac{\partial^2 w}{\partial z^2} + \left(\frac{g\alpha\Gamma}{u_\infty^2} + \frac{2g\alpha\Pi}{u_\infty^2}z\right)w + \frac{4g\alpha\Pi}{u_\infty^3}\varphi w = -\frac{f\Lambda}{u_\infty} \tag{33}$$

边界条件为式(6)至式(9)，给出数值计算结果。以非线性普适函数为例，主要参数取值分别为 $f=0.9\times10^{-4}$ s，$g=9.8$ m/s^2，$\alpha=1.8\times10^{-4}$/℃，海底地形为 $h_c(x)=h_0\exp(-\beta x^2)$，其中，$h_0=60$ m，$\beta=3.4\times10^{-10}$/m^2。根据实际情况，假设陆架外海洋在 100 m 水深的情况下，温度垂直梯度与南北向流速的垂直梯度分别为，$\Gamma=0.004$℃/m，$\Lambda=0.003$/s，令 $T_0=19$℃，$v_0=20$ cm/s。图 1 给出了当 $u_\infty=0.6$ cm/s 时(x, z)平面上海洋的定常运动，整个流场表现为上升流，上升流的最大速度在海底陆坡处，为 0.01 cm/s。可以注意到，在上面两个方程中，如将 u_∞ 改取 $-u_\infty$，则 w 的符合也反过来，即变成下沉流，但结构不变。

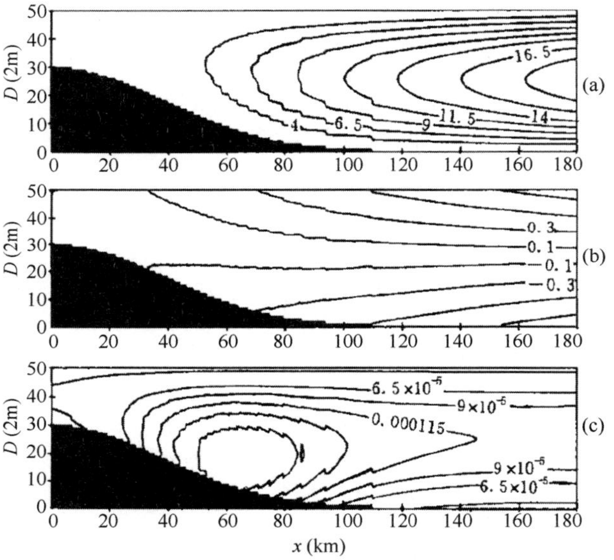

图 1 上升流 (a)流线；(b)水平流速 u；(c)垂直流速 w

6 温度和沿岸流的计算

需要指出，不能由计算出来的流函数通过式(17)或式(18)直接计算温度场和沿岸流场，这是因为式(17)或式(18)所表示的 T、v 和 φ 的关系只在区域内部成立，在边界上失效。而图 1 中计算所得的 φ 场，除受方程式(24)或式(27)制约外，还受到边界条件式(28)至式(31)的约束，因此对于 T、v 的计算除了仍要用到式(18)、式(17)外，尚要确立适当的具有物理意义的边界条件来约束，这样才能问题适定。

对温度场的计算，为了给出合适的海底边界条件，首先将式(19)代入方程(25)并对 x 微分，从

而给出关于 $T^\circ=\dfrac{\partial T}{\partial x}$的微分方程

$$\frac{\partial^2 T^\circ}{\partial x^2}+\frac{\partial^2 T^\circ}{\partial z^2}+\frac{g\alpha\Gamma}{u_\infty^2}T^\circ=\frac{f\Lambda\Gamma}{u_\infty^2} \tag{34}$$

此时利用式(10)、式(18)可以给出海底边界条件

$$z=h(x)\ 时,\quad T^\circ=-\frac{\Gamma}{u_\infty}w \tag{35}$$

考虑到感热通量的影响，给定上表面边界条件为

$$\rho_a C_{P\nu T}D(T^\circ-T^\circ_{\text{air}})=\frac{\mathrm{d}Q(x)}{\mathrm{d}x} \tag{36}$$

其中，T°_{air}为大尺度空气温度场，计算时可取为定值 $T^\circ_{\text{air}}=20℃$，$Q(x)$为感热通量，计算时取 $Q(x)=Q_0\exp\left(-\dfrac{x-x_0}{L}\right)$，其中 $Q_0=3.0\ \text{W/m}^2$，$x_0=100\ \text{km}$，$L=180\ \text{km}$，$\rho_a=1.293\ \text{kg/m}^3$ 为空气密度，空气比热 $C_P=1.008\ \text{kJ/(kg·℃)}$，比例参数 $v_T=1\ 000\ \text{m}^2/\text{s}$。

侧边界条件为

$$x=0\ 或\ x\to\infty\ 时,\quad T^\circ=0 \tag{37}$$

计算得到 $T^\circ=\dfrac{\partial T}{\partial x}$后，从 $x=L$ 积分至 $x=0$ 可以确定温度场 T。

图 2 给出了在海表面有热输入的情况下整个温度场的分布，与实际观测对比[10]（见图 1）可以看出本文所提出的动力学分析过程能够反映出实际海洋中上升流区域温度场在海底地形与表面热输入共同作用下所反映出的分布形态。

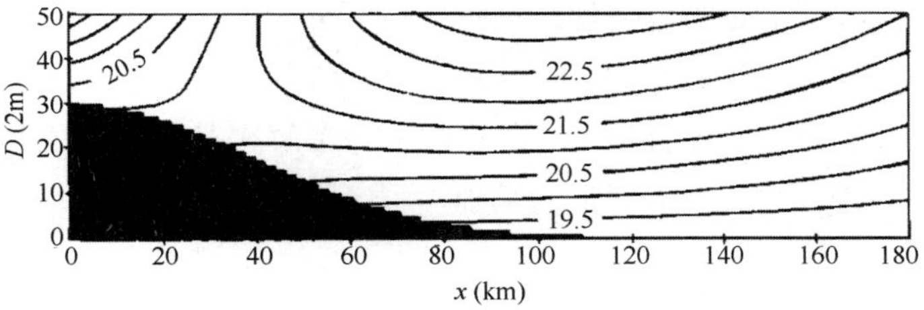

图 2　海面有热量输入时的温度分布

类似地，我们可以同样给出计算沿岸流的过程，其中 $v^\circ=\dfrac{\partial v}{\partial x}$所满足的微分方程为

$$\frac{\partial^2 v^\circ}{\partial x^2}+\frac{\partial^2 v^\circ}{\partial z^2}+\frac{g\alpha\Gamma}{u_\infty^2}v^\circ=\frac{f}{u_\infty^2}(\Lambda^2-g\alpha\Gamma), \tag{38}$$

如考虑到在海底固体边界附近会有摩擦作用，因此不妨设

$$z=h(x)\ 时,\quad v^\circ=0 \tag{39}$$

考虑到上表面风应力的作用，可以给出上边界条件

$$\rho A_z=\frac{\partial v^\circ}{\partial z}=\frac{\mathrm{d}\tau(x)}{\mathrm{d}x} \tag{40}$$

其中，$\tau(x)$为表面风应力，计算中取 $\tau(x)=\tau_0\exp\left(-\dfrac{x-x_0}{L}\right)$，其中 $\tau_0=0.1\ \text{N/m}^2$，$\rho=1\ 025\ \text{kg/m}^3$ 为海水

的密度，$A_z=0.001\ m^2/s$ 为垂直湍流黏性系数。对理想流体运动方程（38）取式（40）这样的边界条件时，意味着在海表附近有一非常薄的湍流边界层，由于假设它非常薄，因此作为近似可以不考虑这一层的过渡作用，边界影响直接进入理想流体内部。

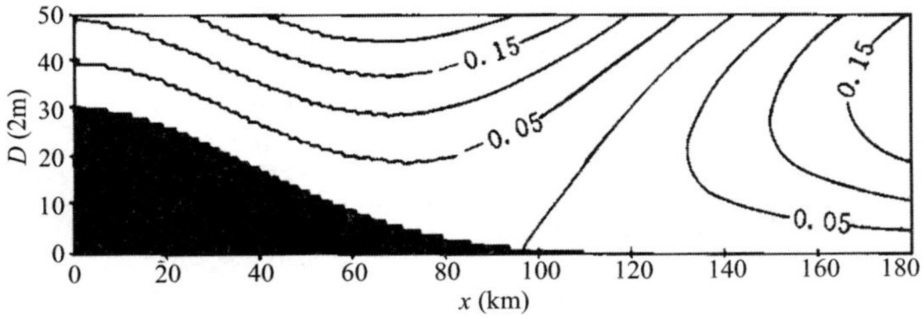

图 3　海面风应力作用时的沿岸急流

其中侧边界条件为

$$x=0 \text{ 或 } x\to\infty \text{ 时}, \quad v^{\circ}=0 \tag{41}$$

其他主要参数同上，计算结果如图 3 所示。注意到大气风应力是向北的，在远离海岸时应吹引出向北的洋流，而在海岸附近流体内部的动力过程变得十分重要，洋流是向南的，最大强度达 0.47 m/s，其结构呈急流状。

7　结语

本文利用在流体内部经向总动量守恒积分、温度守恒积分和位涡度守恒积分，将一个高度非线性的地球流体力学方程组简化成线性或非线性的一个椭圆形二阶微分方程，在一定边界条件下，可以求得用特殊函数表示的解析解，也可以方便地进行数值积分。文中对最后的微分方程采用了数值求解，因此可看成为半解析理论。计算表明，所得到的上升流和温度分布能够定性地反映出实际上升流区域的分布形态，且在物理上也是合理的，因此这是一个解决某些非线性地球流体力学问题可使用的方法，其结果可作为进一步的数值模拟的基础。

参考文献

[1] Hsueh Y, Kenney R N.Steady coastal upwelling in a continuously stratified ocean.*J.Phys.Oceanogr.*, 1972,2:27–33.

[2] Allen J S.Upwelling and coastal jet in a continuously st ratified ocean.*J.Phys.Oceanogr.*,1973,3:245–257.

[3] Allen J S.Upwelling circulation on the Oregon continental shelf.Part 1.Response to idealized forcing.*J.Phys.Oceanogr.*, 1995,25:1843–1866.

[4] Blumberg A E,Mellor G L.A description of a three-dimensional coastal circulation model.Three-dimensional coastal and estuarine.In:*Sci.Ser.*,4,Heaps N,ed.American Geophysical Union.,1987:1–16.

[5] Hamilton P,Rattery M.A numerical model of the depth-dependent wind driven upwelling.*J.Phys.Oceanogr.*,1978,8:430–457.

[6] SU Ji lan,Pan Yuqiu.On the shelf circulation north of Taiwan.*Acta Oceanologica Sinica*,1988,6:1–20.

[7] HU Dunxin.A study of coastal upwelling off SE China.*Kexue Tongbao*,1980,25:159–163.

[8] 颜廷壮.浙江和琼东沿岸上升流的成因分析.海洋学报,1992,14:12–18.

[9] 潘玉球,苏纪兰.浙江沿岸上升流锋区特征及其成因的初步探讨.海洋湖沼通报,1982,3:1–7.

[10] 刘先炳,苏纪兰.浙江沿岸上升流和沿岸锋面的数值研究.海洋学报,1991,13:305–314.

A Numerical Study of the Effect of the Marginal Sea on Coastal Upwelling in a Non-linear Inertial Model*

Chao JiPing[2], Kang YanYan[1,3], Li JianPing[2]

(1. LASG, Institute of Atmospheric Physics, Chinese Academy of Sciences, Beijing 100029, China; 2. National Marine Environmental Forecasting Center, Beijing 100081, China; 3. College of Earth Sciences, University of Chinese Academy of Sciences, Beijing 100049, China)

Abstrct: Inertia theory and the finite element method are used to investigate the effect of marginal seas on coastal upwelling. In contrast to much previous research on wind-driven upwelling, this paper does not consider localized wind effects, but focuses instead on temperature stratification, the slope of the continental shelf, and the background flow field. Finite element method, which is both faster and more robust than finite difference method in solving problems with complex boundary conditions, was developed to solve the partial differential equations that govern coastal upwelling. Our results demonstrate that the environment of the marginal sea plays an important role in coastal upwelling. First, the background flow at the outer boundary is the main driving force of upwelling. As the background flow strengthens, the overall velocity of cross-shelf flow increases and the horizontal scale of the upwelling front widens, and this is accompanied by the movement of the upwelling front further offshore. Second, temperature stratification determines the direction of cross-shelf flows, with strong stratification favoring a narrow and intense upwelling zone. Third, the slope of the continental shelf plays an important role in controlling the intensity of upwelling and the height that upwelling may reach: the steeper the slope, the lower height of the upwelling. An additional phenomenon that should be noted is upwelling separation, which occurs even without a local wind force in the nonlinear model.

Key words: coastal upwelling; inertia theory; Finite Element Method (FEM)

Upwelling raises cold, nutrient-rich waters to the surface, which encourages seaweed growth and supports the bloom of phytoplankton. It plays a vital role in fishery production and determines the location of fishing grounds. Furthermore, the inner shelf region, because of the upwelling separation, acts as a barrier to cross-shelf transport, which enhances the retention of fish eggs and larvae within the coastal environment (Roy et al., 1998). Upwelling also plays an important role in the global carbon cycle (Anderson et al.,

* SCIENCE CHIN: Earth Sciences, 2014, 57(11): 2587–2596.

2009). Therefore, the intensity and location of upwelling is important to marine ecosystems, which has motivated decades of study on the dynamics of the upwelling process.

Previous research has mostly focused on wind-driven upwelling. However, there is also much research indicating that the local wind stress is not the main force driving the upwelling along some coasts. Zhao et al. (2003) reported that upwelling off the Yangtze River estuary is caused by interaction between the Taiwan Warm Current and the topography. Hu et al. (1980) suggested that upwelling off the Zhejiang coast was mainly a result of the rise of the Kuroshio Current along the bottom shoreward. Previous results also showed that coastal upwelling is highly correlated with the seasonal and annual variability of the Taiwan Warm Current (Yan et al., 1992). Oke et al. (2000) indicated that coastal upwelling off eastern Australia is mainly caused by alongshelf topographic variations. Besides the wind and background current, the tide has been demonstrated to be a dominant driver of costal upwelling off the Yangtze River estuary and Zhejiang coast in China via barotropic and baroclinic processes (Huang et al., 1996; Lü et al., 2007). The present paper therefore discusses the effects of factors such as the background current, topography and stratification without a local wind force.

Present theory predicts that a steep continental shelf favors narrow and intense upwelling, while a gentle slope favors broad and weak upwelling. Allen et al. (1995) concluded that, over a wide shelf, onshore flow is confined to the bottom boundary layer and the upwelling front is weak; while over a steep shelf, the onshore flow occurs primarily through the interior, and the upwelling front is stronger, and near the coast. Bakun (1996) found that upwelling separation may not occur in the absence of a steep slope. Estrade et al. (2008) demonstrated that, in the case of an alongshore wind, the cross-shore width of upwelling scaled with D/S, where D is the thickness of Ekman layer, S is the bottom topographic slope. Peffley et al. (1976) and Rodrigues et al. (2001) studied the effects of the bottom topography and the coastline on wind-driven upwelling off the Oregon coast and Brazilian coast. The relative importance of topography and coastline depends on the specific sea; where the coastline is straight, the topography contributes more to the driving of upwelling. Gan et al. (2009a) showed that the alongshore variation of bottom topography caused the upwelling center in Regional Ocean Model System (ROMS), that the upwelling intensified over a widened shelf between Shantou and Shanwei, and there were upwelling centers in Shantou. The previous studies show that bottom topography plays an important role in coastal upwelling.

Stratification also greatly affects coastal upwelling. Allen et al. (1995) found that, with reduced temperature stratification, the coastal upwelling and along-shore jet stream weakened, and the front was farther offshore. In the stratified case, a major part of the upwelling cell was within 3 km of the coast, while in a homogeneous experiment, the upwelling cell was located 8 km offshore. Austin et al. (2002) reported similar results showing that cross-shore transport reduced as stratification weakened. Lentz (2001) reported that there was a marked difference in cross-shelf transport between summer, when waters were more stratified, and fall, when waters were weakly stratified. Kirincich (2005) simulated the separation of upwelling as stratification weakens. Gan et al. (2009b) pointed out that cross-shore circulation strengthens at a higher level of stratification by a plume. Additionally, recent research has suggested that upwelling has intensified off northwest Africa and in the northern South China Sea (NSCS) as the planet has warmed (Liu et al., 2009;

McGregor et al., 2007).

Most of the previous work is based on numerical models, whereas little has been done on the basic theory of upwelling. The numerical simulations always depend heavily on parameterization of the turbulent mixing and momentum, and therefore it is difficult to explain the physical mechanism. In this paper, we examine non-wind-driven upwelling with nonlinear inertia theory. The inertia theory was first used for the Gulf Stream by Charney (1955), and Pedlosky (1978) then applied the inertia theory to coastal upwelling in predicting velocity and density fields. The nonlinear inertia model used in the present paper is derived from the inertia model of Chao et al. (2003), who used the model to obtain the coastal jet and temperature front. In contrast to Chao's theory, this study considers a non-linear temperature stratification and different bottom topography, and uses the finite element method (FEM) to solve the governing equation. Three experiments are conducted to investigate the effects of stratification, background circulation and topography.

1　Model equations

A left-handed "west coast" coordinate system is used, with positive x being the offshore direction, positive y the northward direction, and positive z the upward direction. The $x-z$ dimensional equations eliminating along-shelf variability are

$$u\frac{\partial u}{\partial x}+w\frac{\partial u}{\partial z}-fv=-\frac{1}{\rho_0}\frac{\partial p}{\partial x} \tag{1}$$

$$u\frac{\partial v}{\partial x}+w\frac{\partial v}{\partial z}+fu=0 \tag{2}$$

$$u\frac{\partial w}{\partial x}+w\frac{\partial w}{\partial z}=-\frac{1}{\rho_0}\frac{\partial p}{\partial z}-(1-\alpha T)g \tag{3}$$

$$u\frac{\partial T}{\partial y}+w\frac{\partial T}{\partial z}=0 \tag{4}$$

$$\frac{\partial u}{\partial x}+\frac{\partial w}{\partial z}=0 \tag{5}$$

where α is the coefficient of the thermal expansion of the seawater, and the other symbols are those commonly used in the literature.

A stream function is introduced according to continuity eq. (5):

$$u=-\frac{\partial \psi}{\partial z},\quad w=\frac{\partial \psi}{\partial x} \tag{6}$$

Equations (2) and (4) then take the form

$$J(\psi,v+fx)=0 \tag{7}$$

$$J(\psi,T)=0 \tag{8}$$

and it follows that

$$v+fx=F_1(\psi) \tag{9}$$

$$T=F_2(\psi) \tag{10}$$

where $F_i(\psi)$ is an arbitrary function of ψ to be determined. To evaluate $F_i(\psi)$, we introduce cross-shore velocity u_∞ for the outer boundary condition ($x=\infty$). Eq. (6) then gives z as a function of ψ_∞, $z=-\psi_\infty/u_\infty$. In addition, we assume $v_{\infty} = v_0 + \Lambda z$ and $T_{\infty} = T_0 + \Gamma_1 z + \Gamma_2 z^2 + \Gamma_3 z^3$ for the outer boundary condition. Differentiationg eq.(9), we have

$$\frac{\partial V_{\infty}}{\partial z} = \frac{\partial F_1(\psi_{\infty})}{\partial \psi_{\infty}} \frac{\partial \psi_{\infty}}{\partial z} = \Lambda$$

so that

$$\frac{\mathrm{d}F_1}{\mathrm{d}\psi_{\infty}} = -\frac{\Lambda}{\mu_{\infty}} \tag{11}$$

In the same way,

$$\frac{\mathrm{d}F_2}{\mathrm{d}\psi_{\infty}} = -\frac{\Gamma_1 + 2\Gamma_2 z + 3\Gamma_3 z^2}{\mu_{\infty}} = -\frac{\Gamma_1}{\mu_{\infty}} + 2\frac{\Gamma_2}{\mu_{\infty}^2}\psi_{\infty} - 3\frac{\Gamma_3}{\mu_{\infty}^3}\psi_{\infty}^2 \tag{12}$$

Eliminating p from eqs.(1) and (3), we have

$$u\frac{\partial}{\partial x}\left(\frac{\partial u}{\partial z} - \frac{\partial w}{\partial x}\right) + w\frac{\partial}{\partial z}\left(\frac{\partial u}{\partial z} - \frac{\partial w}{\partial x}\right) - f\frac{\partial v}{\partial z} + g\alpha\frac{\partial T}{\partial x} = 0 \tag{13}$$

Using eqs. (11) and (12), eq.(13) can be written as

$$\vec{U} \cdot \left(\frac{\partial^2\psi}{\partial x^2} + \frac{\partial^2\psi}{\partial z^2} - f\frac{\mathrm{d}F_1}{\mathrm{d}\psi} - g\alpha\frac{\mathrm{d}F_2}{\mathrm{d}\psi}\right) = 0 \tag{14}$$

or

$$\frac{\partial^2\psi}{\partial x^2} + \frac{\partial^2\psi}{\partial z^2} - f\frac{\mathrm{d}F_1}{\mathrm{d}\psi} - g\alpha\frac{\mathrm{d}F_2}{\mathrm{d}\psi} = F_3(\psi) \tag{15}$$

Using eqs. (11) and (12), eq. (15) can be written as

$$\frac{\partial^2\psi}{\partial x^2} + \frac{\partial^2\psi}{\partial z^2} + \frac{f\Lambda}{\mu_{\infty}}x - g\alpha\left(\frac{\Gamma_1}{\mu_{\infty}} + 2\frac{\Gamma_2}{\mu_{\infty}^2}\psi_{\infty} - 3\frac{\Gamma_3}{\mu_{\infty}^3}\psi_{\infty}^2\right)z = F_3(\psi) \tag{16}$$

Along the outer boundary,

$$F_3(\psi_{\infty}) = \frac{f\Lambda}{\mu_{\infty}}x_{\infty} + g\alpha\frac{\psi_{\infty}}{\mu_{\infty}}\left(\frac{\Gamma_1}{\mu_{\infty}} + \frac{2\Gamma_2\psi_{\infty}}{\mu_{\infty}^2} - \frac{3\Gamma_3\psi_{\infty}^2}{\mu_{\infty}^3}\right) \tag{17}$$

where x_{∞} is the width of the inertia domain, or the offshore distance of the outer boundary. F_1, F_2, F_3 are determined along the outer boundary and are consequently applicable to every point in the interior region. We then have the governing equation:

$$\frac{\partial^2\psi}{\partial x^2} + \frac{\partial^2\psi}{\partial z^2} - g\alpha\left(-\frac{\Gamma_1}{\mu_{\infty}^2} - 2\frac{\Gamma_2}{\mu_{\infty}^3}\psi - 3\frac{\Gamma_3}{\mu_{\infty}^4}\psi^2 + 2\frac{\Gamma_2}{\mu_{\infty}^2}z\right.$$
$$\left. - 3\frac{\Gamma_3}{\mu_{\infty}^3}\psi z\right)\psi = -\frac{f\Lambda}{\mu_{\infty}}(x - x_{\infty}) - g\alpha\frac{\Gamma_1}{\mu_{\infty}}z \tag{18}$$

The nonlinear partial differential equations have been converted into a second-order elliptic equation of a single variable. In the equation, $f = 0.9\times10^{-4}\ \mathrm{s}^{-1}$, $g = 9.8\ \mathrm{m\cdot s^{-2}}$, $\alpha = 1.8\times10^{-4}\,℃^{-1}$, the vertical along-shore velocity gradient $\Lambda = 0.003\ \mathrm{s}^{-1}$, and the bottom topography is taken to be $h_c(x) = h_0\exp(-\beta(x/100)^2)$, where $h_0 = 195$ m, $\beta = 3.4\times10^{-6}\ \mathrm{m}^{-2}$. The stratification parameters are given according to the measured temperature in the NSCS.

The boundary conditions for $\psi(x, z)$ are

$$
\begin{aligned}
&\psi(x,0)=0,\\
&\psi(x,h_c(x))=0,\\
&\psi(0,z)=0,\\
&\frac{\partial\psi}{\partial x}(x_\infty,z)=0
\end{aligned} \tag{19}
$$

2 Numerical method

To handle the computational complexity of the terrain, the FEM is used here to solve the nonlinear elliptical eq. (18).In contrast to the finite difference method widely used in previous papers, the FEM is fast and robust in solving a problem with complex boundary conditions.

2.1 Calekin method (weighted residual method)

The Calekin method is also known as a "weighted residual method". The idea is to satisfy the differential equation in an average sense by converting it into an integral equation.The differential equation is multiplied by a weighting function and then averaged over the domain.

The governing eq. (18) can be simplified as

$$
\frac{\partial^2\psi}{\partial x^2}+\frac{\partial^2\psi}{\partial z^2}-a(\psi)\psi=F(x,z) \tag{20}
$$

with the boundary conditions

$$
\Gamma_1:\psi=\psi_0,\psi_0 \text{ is a constant},
$$

$$
\Gamma_2:\frac{\partial\psi}{\partial n}=G,G \text{ is a constant}
$$

where

$$
a(\psi)=g\alpha\left(-\frac{\Gamma_1}{\mu_\infty^2}+2\frac{\Gamma_2}{\mu_\infty^3}\psi-3\frac{\Gamma_3}{\mu_\infty^4}\psi^2+2\frac{\Gamma_2}{\mu_\infty^2}z-3\frac{\Gamma_3}{\mu_\infty^3}\psi z\right)
$$

and $f(x,z)=-\frac{f\Lambda}{\mu_\infty}(x-\xi)-g\alpha\frac{\Gamma_1}{\mu_\infty}z.$

To solve eq.(20), we set

$$
R=\frac{\partial^2\psi}{\partial x^2}+\frac{\partial^2\psi}{\partial z^2}-a(\psi)\psi-f(x,z) \tag{21}
$$

where R is called the residual. Hence, solving eq. (20) is equivalent to calculating the weak form of $R(\psi)=0$,

$$
\langle R,N\rangle=\int_s N^T\left(\frac{\partial^2\psi}{\partial x^2}+\frac{\partial^2\psi}{\partial z^2}-a(\psi)\psi-f(x,z)\right)\mathrm{d}x\mathrm{d}z \tag{22}
$$

where $\psi=(N\psi)=N_1\psi_1+N_2\psi_2+\cdots+N_n\psi_n$, N is a shape function and n is the total number of nodes.

After introducing the boundary conditions into eq. (22), we have

$$
\rho(\psi)=\langle R,N\rangle=\int_s N^T\left(\frac{\partial^2(N\psi)}{\partial x^2}+\frac{\partial^2(N\psi)}{\partial z^2}-a((N\psi))(N\psi)-f(x,z)-g\right)\cdot\mathrm{d}x\mathrm{d}z \tag{23}
$$

Quation (23) can be rearranged and solved using the Newton-Raphson method. We start the process with an initial guess $\psi^{(n)}$, and set $\psi^{(n+1)}$ to satisfy $\rho(\psi^{(n+1)})=0$, then

$$\frac{\partial \rho(\psi^n)}{\partial \psi}(\psi^{n+1} - \psi^n) = \rho(\psi^{(n+1)}) - \rho(\psi^n) = -\rho(\psi^n) \tag{24}$$

$$W = \psi^{n+1} - \psi^n = \frac{-\rho(\psi^n)}{\dfrac{\partial \rho(\psi^n)}{\partial \psi}} \tag{25}$$

The process is repeated until a sufficiently accurate value is reached.

2.2 Element type: Eight-node quadrangular element

Usually, a quadrangular element provides higher accuarcy than a triangular element. To increase the computational accuracy, an eight-node quadrangular element with secondorder Lagrangian interpolation is used in the simulation. The Lagrangian interpolation can be expressed as

$$\psi^e(\varepsilon,\eta) = a_1 + a_2\varepsilon + a_3\eta + a_4\varepsilon^2 + a_5\varepsilon\eta + a_6\eta^2 + a_7\varepsilon^2\eta + a_8\varepsilon\eta^2 \tag{26}$$

To check the accuracy of our FEM code, we solved the nonlinear Ginzburg-Landau equation (eq. (27)) using our FEM code and compared the results with those obtained by Alberty et al. (1999) using a four-node quadrilateral element (Figure 1):

$$\varepsilon\Delta\psi = \psi^3 - \psi,\ \psi = 0 \quad \text{on } \Gamma, where\ \varepsilon = 1/100 \tag{27}$$

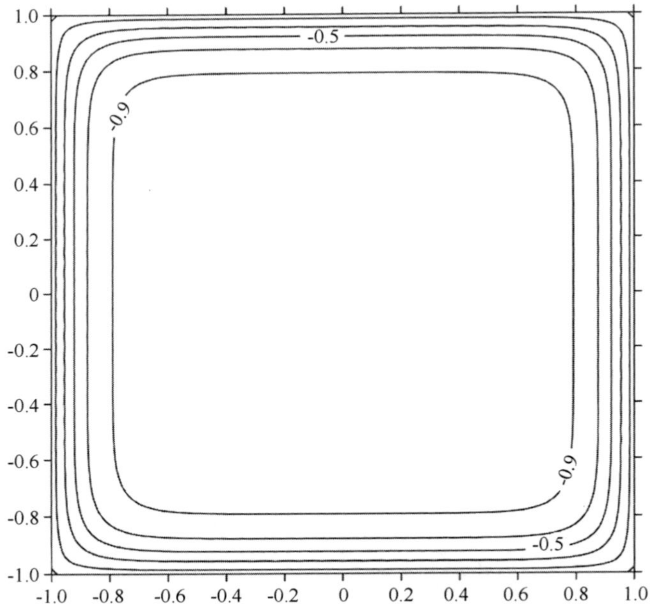

Fig. 1 Solution for the Ginzburg-Landau equation

3 Observation

We consider the case of the NSCS to determine the parameters associated with stratification. Figure 2 shows the temperature profile from 0 to 200 m below the surface of the NSCS in July. Because of the limited data resolution, the bottom topography cannot be seen clearly, especially in the Taiwan Strait where the depth is less than 40 m and there is only one layer. However, we are still able to identify the low-temperature region near the shore, and this suggests the existence of upwelling as indicated by the arrow.

Taking 19.5°N as an example, the longitude-depth crosssection of temperature between January and December is shown in Figure 3. It is clear that the highest sea surface temperature occurred in July, and the lowest in January, so that the stratification is strongest in July and weakest in January. To examine the effect of stratification on upwelling, we used the observed stratification from July, September, November and January (stratifications 1, 2, 3, and 4, respectively, in Figure 4) in the following simulations. Data used in this study are from the CORA National Marine Information Center (http://www.cora.net.cn/).

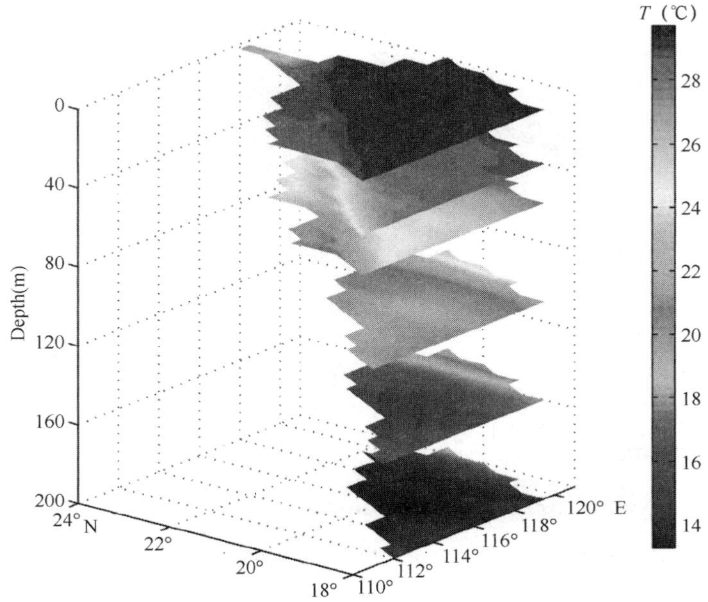

Fig. 2 Temperature slice (from 0 to 200 m below the sea surface) of the NSCS in July(见书后彩插)

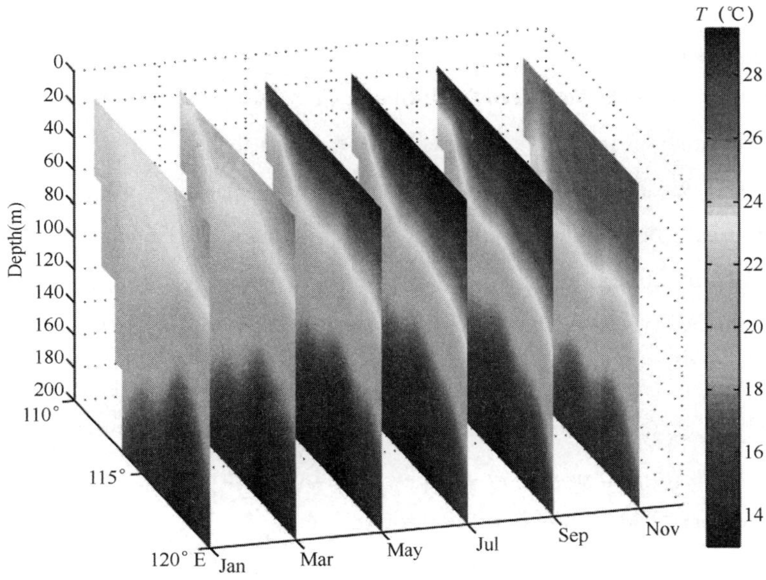

Fig. 3 Longitude-depth cross-section of temperature from January to December at 19.5°N(见书后彩插)

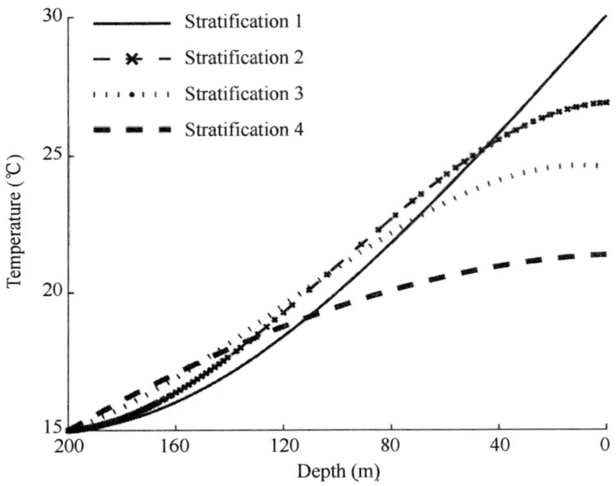

Fig. 4 Four stratification scenarios used in simulations

4 Simulation results

4.1 Effects of stratification

Our results are for a depth of 200 m, offshore distance of 130 km, and $u_\infty = -0.06\ \mathrm{m \cdot s^{-1}}$. Figure 5 shows that coastal upwelling occurred in stratification cases 1–4. With the lower sea surface temperatures, stratification is reduced, upwelling weakens, and the width of the upwelling increases. Hence, strong temperature stratification favors strong and narrow upwelling. However, in the case of stratification 4, during winter, there is downwelling. In all four temperature stratification scenarios, the inflow and outflow were confined to the bottom and surface boundary layers, respectively. Another important phenomenon, termed upwelling separation, is a zone of increased vertical velocity that developed 12 km from the coast, which was not reported by Chao et al. (2003). To confirm that this upwelling separation is not caused by the FEM code, we solved Chao's linear equation using the FEM code. We found that the results are consistent with Chao's linear model (Figure 6). We therefore conclude that upwelling separation is caused by the nonlinear field accelerations associated with the nonlinear temperature stratification. The inner shelf is the region inshore of the upwelling front (Austin, 2002). In Figure 5(a)–(d) correspond to stratification scenarios 1–4, respectively, and it is seen that the width of the inner shelf does not vary with the stratification.

4.2 Effects of u_∞

The effects of u_∞ are examined using the same terrain and stratification as used in experiment 1, and by increasing u_∞ sfrom −0.06 to $-0.3\ \mathrm{m \cdot s^{-1}}$ (Figure 7). The flow pattern does not change, and upwelling still occurs in July, September, and November, and downwelling in January. However, the intensity and horizontal scale of the upwelling front increases, and the boundary layer thickens. The second distinctive feature is that the offshore distance of the upwelling front increases from 12 km when $u_\infty = -0.06\ \mathrm{m \cdot s^{-1}}$ to 35 km when $u_\infty = -0.3\ \mathrm{m \cdot s^{-1}}$.

A further experiment considered the effect of gradually increasing u_∞ from −0.06 to $-0.3\ \mathrm{m \cdot s^{-1}}$ under

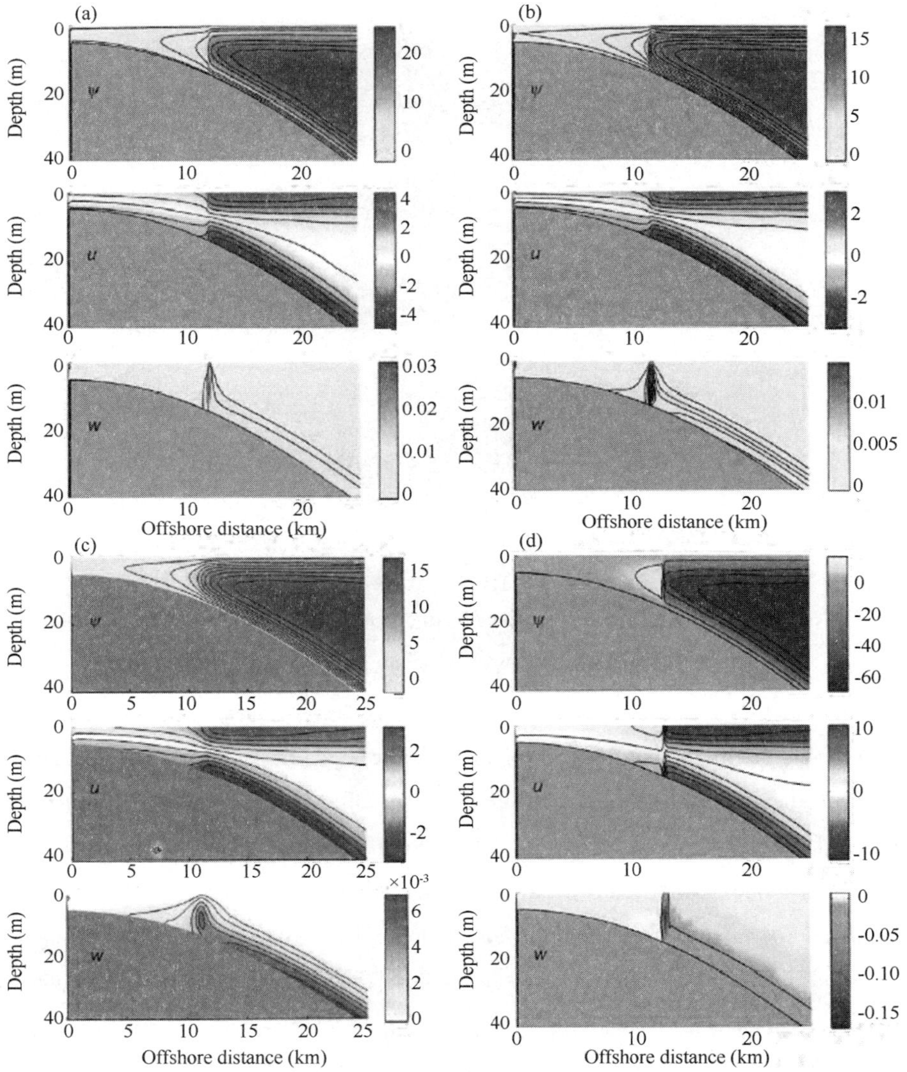

Fig. 5 Streamfunction, cross-shelf velocity and vertical velocity fields for $u_{\infty} = -0.06$ m s^{-1}

(a)-(d) correspond to stratifications 1-4 respectively. The units of u and w are m s^{-1}(见书后彩插)

the conditions of stratification scenario 2. We see that the offshore distance and the width of the inner shelf increase from 12 to about 60 km (Figure 8). Because fish tend to avoid spawning in the vicinity of upwelling centers (Roy et al., 1998), the widened inner shelf would notably affect the nearshore ecosystem.

4.3 Effect of topography

To examine the role played by topography, a steep, narrow continental shelf is used in the following simulation. Compared with experiment 1, the water depth remains at 200 m and $u_{\infty} = -0.06$ m · s^{-1}, but the offshore distance decreases to 13 km. The results (Figure 9) show that the flow pattern does not change under stratification scenarios 1 and 4, although the intensity is enhanced. However, in the cases of stratification scenarios 2 and 3, the flow pattern changes dramatically, in that the upwelling cannot reach the sea surface and the return flow is in the interior. Xu et al. (1983) and Pan et al. (1982) pointed out that the upwelling

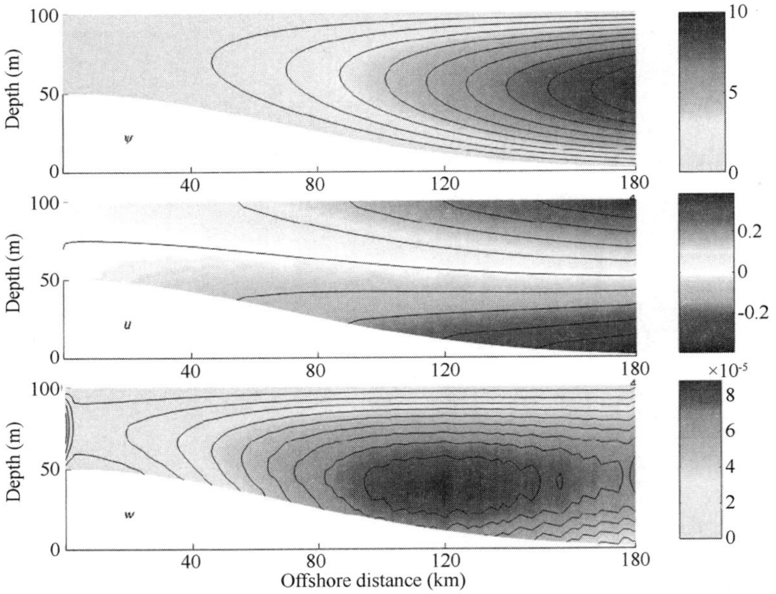

Fig. 6 Streamfunction, cross-shelf velocity and vertical velocity fields been obtained with the FEM. The units of u and w are m · s^{-1}(见书后彩插)

cell off the Zhejiang coast mainly exists 5 m below the sea surface, and it reaches the sea surface only in certain months. However, they did not discuss what factors may affect the height that the upwelling can reach. Additionally, Su et al. (2009) and Dong et al. (2004) found two counterrotating cells in the vertical direction using an analytical model. MacCready et al. (1993) indicated that the location of the return flow depends on the relative importance of stratification and topography for wind-driven upwelling (Austin et al., 2002; MacCready et al., 1993). The location of the offshore flow in our study also depends on the topography, even though it is not driven by local wind.

To further examine the effect of the continental slope on the location of the offshore flow, we increase the continental shelf slope by shortening the offshore distance to 42, 39, 26, and then 13 km, with water depth remaining the same, $u_\infty = -0.06$ m · s^{-1}, and using stratification scenario 2. It is fairly clear that the height that the upwelling can reach is reduced over the deeper continental shelf, and that the upwelling cannot reach the ocean surface when the offshore distance is less than 42 km (Table 1). That is, a wider shelf favors a surface offshore flow, while a steeper continental shelf favors an interior offshore flow. Gan et al. (2009a) has also pointed out that upwelling strengthened over a widened shelf between Shantou and Shanwei.

Table 1 Height that upwelling can reach lowers over the deeper continental shelf

Offshore distance (km)	42	39	26	13
The height that upwelling can reach (m)	0	10	50	90

5 Discussion and conclusion

The simple nonlinear model presented in this paper successfully reproduced the structure of coastal up-

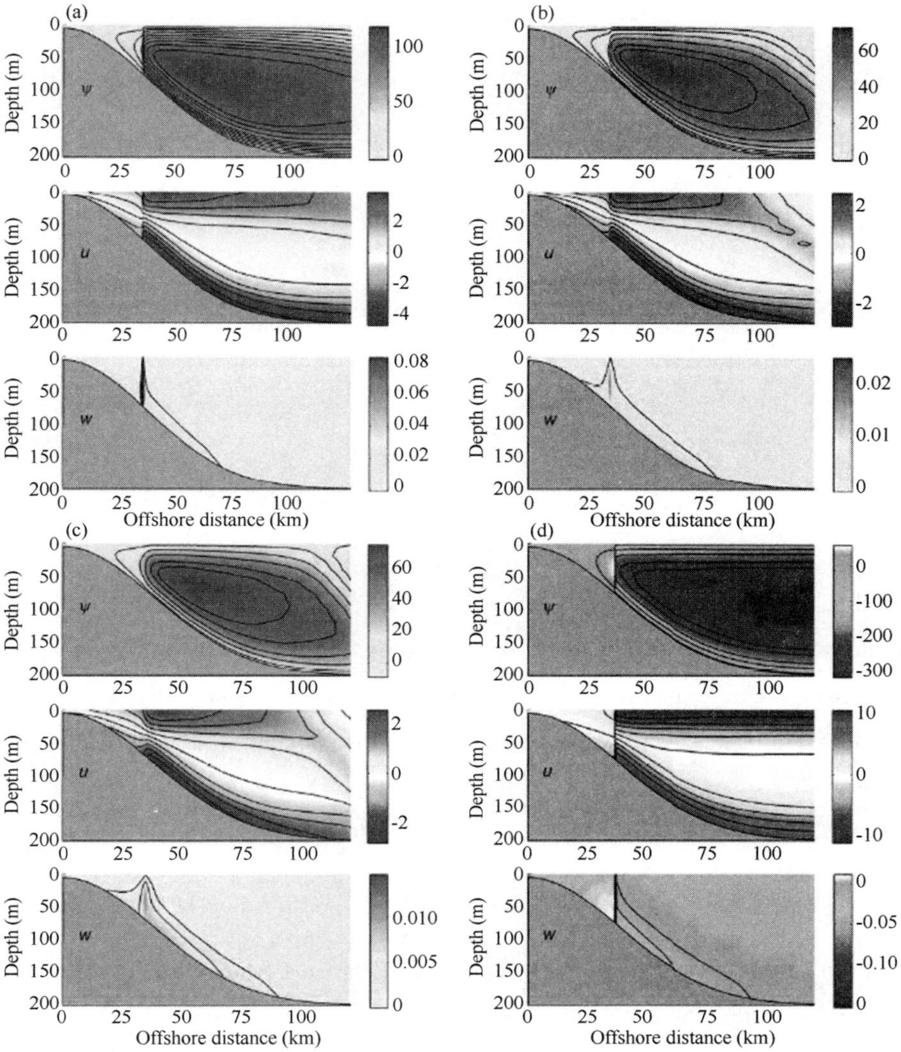

Fig. 7　Streamfunction, cross-shelf velocity and vertical velocity fields for $u_\infty = -0.3\ \mathrm{m\cdot s^{-1}}$

(a)-(d) correspond to stratifications 1-4 respectively. The units of u and w are $\mathrm{m\cdot s^{-1}}$(见书后彩插)

welling that is not driven by wind effects. Compared with Chao's linear model (Chao et al., 2003), we successfully simulated boundaryconcentrated upwelling and the phenomenon of upwelling separation.

Strong stratification tends to generate strong and narrow upwelling, but with reduced stratification, upwelling weakens and the horizontal scale increases. Consequently, during summer, the coastal upwelling front is strong and narrow, while in winter, when temperature stratification is weakest, downwelling may occur. In addition, the width of the inner shelf does not vary with the stratification. The well-known mechanism of upwelling separation is caused by the overlap of the surface and bottom Ekman layers, and reduces crossshore transport. Kirincich (2005) and Estrade et al. (2008) showed that cross-shelf transport decreased, and upwelling separated, during periods of reduced stratification. Additionally, Estrade et al. (2008) found that the separation point was located near the isobath $h \approx 0.4D$, where D is the thickness of the Ekman layer. Nevertheless, the assumption made in this paper is of no-viscosity (i.e., no Ekman layer), and instead we used the inertia boundary layer theory, which was first proposed by Charney (1955).

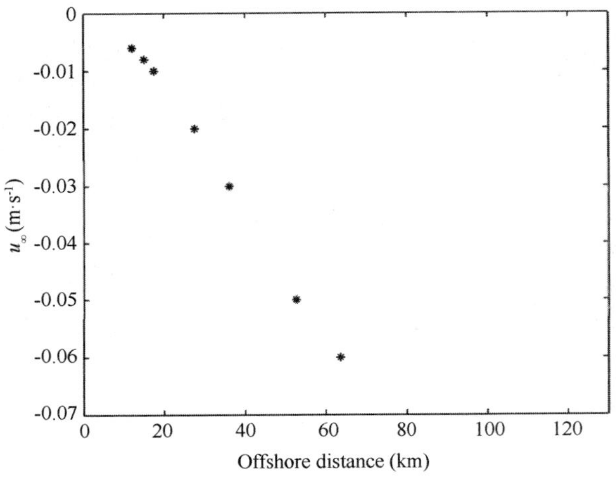

Fig. 8 Offshore distance of separation point as a function of u_∞

Charney defined the inertia boundary layer as the region in which the nonlinear field accelerations are comparable in magnitude to the Coriolis and pressure forces per unit mass. The crossshore velocity in the inertia boundary layer is much higher than that in the internal ocean. Hydrodynamic jump occurs during the transition from fast to slow fluid movement. In technical terms, the phenomenon occurs when the movement of water changes from being supercritical to being subcritical, which is observed when a fluid with a high critical mass meets a fluid with lower critical mass. The jump is determined by the Froude number (defined as $\sqrt{\frac{u^2}{gH}}$, where u is the cross-shore velocity, H is the water depth, and g is acceleration due to gravity). Therefore, the width of the inner shelf in inertia theory is related to the cross-shore velocity and bottom topography.

With increasing u_∞, the intensity and horizontal scale of the upwelling front increase, the inertia boundary thickens, and the separating point moves further offshore, leading to widening of the inner shelf. According to previous research on wind-driven upwelling, this widening of the inner shelf is caused by the thickening of the boundary layer. Under inertial theory, the boundary layer is related to u_∞ (i.e., increasing u_∞ results in a thicker boundary layer), which leads to an increase in the width of the inner shelf.

The response of upwelling to the topography depends on stratification. Over a steep continental shelf, the flow cannot reach the sea surface, and the height that the upwelling can attain decreases as the continental shelf steepens. However, during winter, there is always downwelling. The inertial model does not allow us to conclude any turbulent mixing coefficient or linearization hypothesis, and it thus cannot describe the mixing process. This would be better addressed if we consider a two-layer model in which the upper layer is the Ekman layer and the lower layer is the inertial layer. Another deficiency of this study is its inability to examine the effect of alongshore topography. Our future work will include a three-dimensional finite element model and twolayer model.

The study was supported by the National Basic Research Program of China (Grant No. 2010CB950400) and the program in National Marine Environmental Forecasting Center. The CORA data used in this study have been developed and kindly provided by

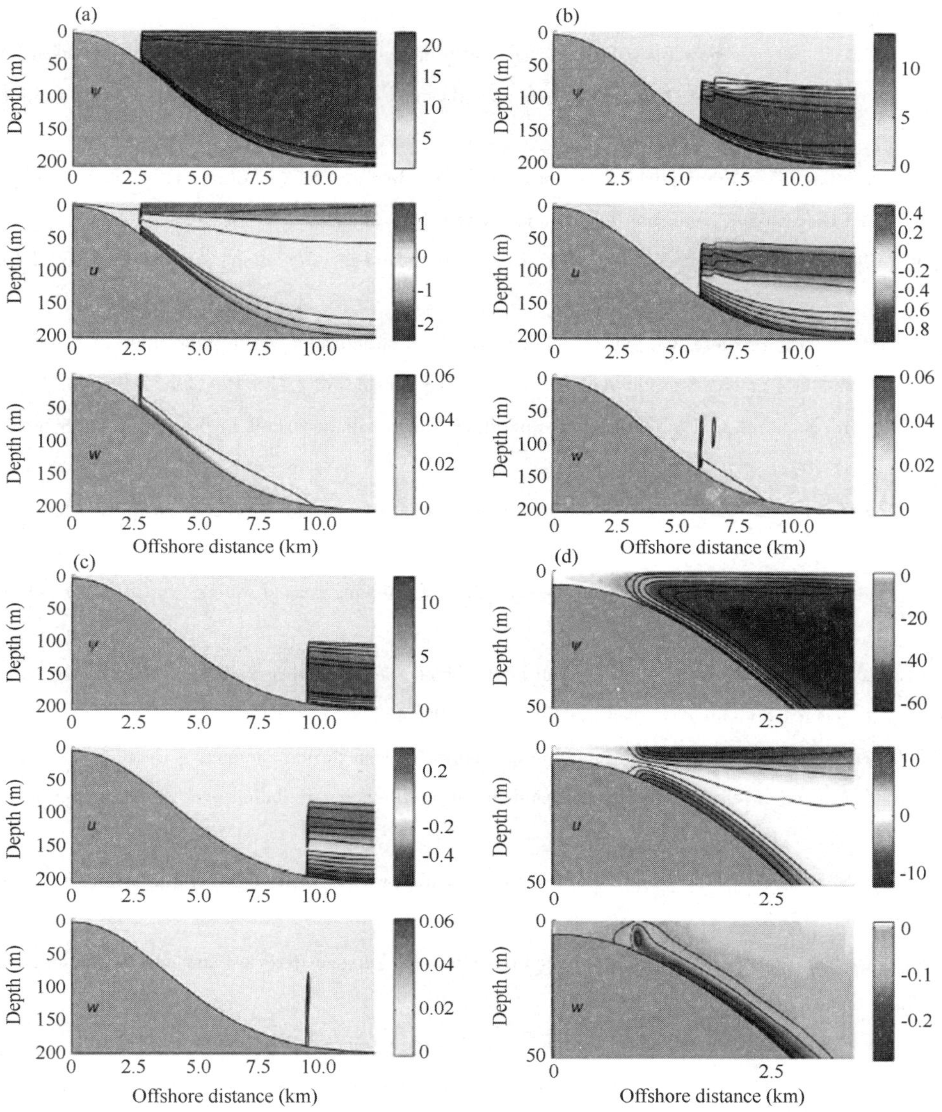

Fig. 9　Streamfunction, cross-shelf velocity and vertical velocity fields over the deeper continental shelf (depth of 200 m, offshore distance of 13 km) for $u_\infty = -0.06\ \mathrm{m \cdot s^{-1}}$ (a)-(d) correspond to stratifications 1-4 respectively. The units of u and w are $\mathrm{m \cdot s^{-1}}$(见书后彩插)

National Marine Data. We greatly appreciate assistance and valuable suggestions provided Dr. Li Yaokun. We also thank two anonymous reviewers for their helpful advice and comments.

References

Alberty J, Carstensen C, Funken S A. 1999. Remarks around 50 lines of Matlab: Short finite element implementation. Numer Algorithms, 20:117-137.

Allen J, Newberger P, Federiuk J. 1995. Upwlling circulation on the oregon continental shelf. Part I: Response to idealized forcing. J Phys Oceanogr, 25: 1843-1866.

Anderson R, Ali S, Bradtmiller L, et al. 2009. Wind-driven upwelling in the Southern Ocean and the deglacial rise in atmospheric CO_2. Science,323: 1443.

Austin J A, Lentz S J. 2002. The inner shelf response to wind-driven upwelling and downwelling. J Phys Oceanogr, 32: 2171-2193.

Bakun A. 1996. Patterns in the Ocean: Ocean Processes and Marine Population Dynamics. California: California University Press, 323.

Chao J P, Chen X Y. 2003. A semi-analytical theory of coastal upwelling and jet (in Chinese). Chin J Geophys, 46: 26-30.

Charney J G. 1955. The Gulf Stream as an inertial boundary layer. Proc Natl Acad Sci USA, 41: 731.

Dong C, Ou H W, Chen D, et al. 2004. Tidally induced cross-frontal mean circulation: Analytical study. J Phys Oceanogr, 34: 293-305.

Estrade P, Marchesiello P, De Verdiere A C, et al. 2008. Cross-shelf structure of coastal upwelling: A two dimensional extension of Ekman's theory and a mechanism for inner shelf upwelling shut down. J Mar Res,66: 589-616.

Gan J P, Cheung A, Guo X, et al. 2009a. Intensified upwelling over a widened shelf in the northeastern South China Sea. J Geophys Res, 114:C09109.

Gan J P, Li L, Wang D X, et al. 2009b. Interaction of a river plume with coastal upwelling in the northeastern South China Sea. Cont Shelf Res,29: 728-740.

Hu D X, Lü L H, Xiong Q C, et al. 1980. Study on the upwelling of Zhejiang coastal waters (in Chinese). Chin Sci Bull, 25: 131-133.

Huang Z K, Yu G Y, Luo Y Y, et al. 1996. Numerical modeling of tide-induced upwelling in coastal areas of the East China Sea (in Chinese).J Ocean Univ Qingdao (Nat Sci Ed), 26: 405-412.

Kirincich A R. 2005. Wind-driven inner-shelf circulation off central Oregon during summer. J Geophys Res, 110: C10S03.

Lentz S J. 2001. The influence of stratification on the wind-driven cross-shelf circulation over the North Carolina Shelf. J Phys Oceanogr,31: 2749-2760.

Liu Y, Peng Z C, Wei G J, et al. 2009. Variation of summer coastal upwelling at northern South China Sea during the last 100 years. Geochimica,38: 317-322.

Lü X G, Qiao F L, Xia C S, et al. 2007. Tidally induced upwelling off Yangtze River estuary and in Zhejiang coastal waters in summer. Sci China Ser D-Earth Sci, 50: 462-473.

MacCready P, Rhines P B. 1993. Slippery bottom boundary layers on a slope. J Phys Oceanogr, 23: 5-22.

McGregor H, Dima M, Fischer H, et al. 2007. Rapid 20th-century increase in coastal upwelling off northwest Africa. Science, 315: 637.

Oke P R, Middleton J H. 2000. Topographically induced upwelling off Eastern Australia. J Phys Oceanogr, 30: 512-531.

Pan Y Q, Cao X Z, Xu J P. 1982. A preliminary investigation of the cause and characteristics of the upwelling front zone off Zhejiang (in Chinese).Trans Oceanol Limnol, 3: 1-7.

Pedlosky J. 1978. An inertial model of steady coastal upwelling. J Phys Oceanogr, 8: 171-177.

Peffley M B, O'Brien J J. 1976. A three-dimensional simulation of coastal upwelling off Oregon. J Phys Oceanogr, 6: 164-180.

Roy C. 1998. An upwelling-induced retention area off Senegal: A mechanism to link pwelling and retention processes. S Afr J Marine Sci, 19: 89-98.

Rodrigues R R, Lorenzzetti J A. 2001. A numerical study of the effects of bottom topography and coastline geometry on the Southeast Brazilian coastal upwelling. Cont Shelf Res, 21: 371-394.

Su J, Pohlmann T. 2009. Wind and topography influence on an upwelling system at the eastern Hainan coast. J Phys Oceanogr, 114: C06017.

Xu J P, Cao X Z, Pan Y Q. 1983. Evidence for the coastal upwelling off Zhejiang (in Chinese). Trans Oceanol Limnol, 4: 17-25.

Yan T Z. 1992. Mechanism analysis of the coastal upwelling off Zhejiang and Qiongdong (in Chinese). Acta Oceanol Sin, 14: 12-18.

Zhao B, Li H, Yang Y. 2003. Numerical simulation of upwelling in the Changjiang river mouth area (in Chinese). Studia Marina Sin, 45:64-76.

地转气流中的重力惯性内波*

巢纪平[1]，吴钦岳[2]

（1. 中国科学院地球物理研究所；2. 中国科学技术大学）

提要：本文用一个一维线性的二层模式，分析了地转气流中重力惯性内波的动力学性质。讨论了重力惯性内波的不稳定性、相速度和群速度后，指出：

(1)当地转风速大于重力内波波速时，某些尺度的重力惯性内波可以产生不稳定性，这种不稳定性可以用来解释某些中系统的发展。

(2)某些波长的重力惯性内波，它的群速度可以大于相速度；因此在某一地区有中系统发展时，它的能量可以很快地传到下游，而在某下游产生新的系统。

1 引言

一般地将需要考虑地球自转偏向力的作用，而可以忽略地转参数随纬度变化的运动称为中尺度运动。这类运动的特征尺度为 $10^1 \sim 10^2$ km。由于一些具有危险天气（阵性降水、雷暴、阵风等）的天气系统，如雷暴、飑线等都是属于中尺度运动，因此研究雷暴、飑线这一类中尺度系统有着重要的实际意义。中尺度雷暴系统的研究，这几年来正受到我国气象工作者的注意，并且在观测、分析等方面都已做了不少的工作。

目前的雷达观测和天气分析，已提出了一些关于雷暴系统的很有意思的现象。如经常观测到：当某处有中尺度系统（雷暴、阵风）在发展时，往往在它们下游也常有相应的系统在发展，即所谓“共鸣型雷暴”就是一例。我们知道，对于大尺度运动来讲，由于群速度可以大于相速度，亦即能量可以先位相速而到达下游，因此产生了所谓的“上游效应”。对于中尺度运动来说，有没有这种类似的机制呢？这是一个值得探讨的问题。又如有些观测表明，许多雷暴系统经常在夜间层结相当稳定的条件下发展起来，这些雷暴发展的能量是从哪里来的呢？这也是值得探讨的问题。

原则上看，层结大气中的中尺度运动是在重力和地球偏向力的联合作用下发展起来的。因而，上述现象可能是重力惯性内波在大气的具体条件下的一些反映。为此在本文中，我们将通过对地转风场中的重力惯性内波特征的分析来讨论这些问题。

2 基本方程

作为研究的第一步，我们考虑理想流体的正压运动。为了考虑实际大气的层结性，我们取二层不可压缩流体。二层流体的分界面，在大气中的典型例子便是逆温层。令 ρ_1 和 ρ_2 分别表示上下层流体的密度，H 和 h 分别表示扰动前和扰动后分界面的高度，u 和 v 分别表示下层流体中的运动速度分量。设运动满足静力平衡的条件。如果上层流体有足够的厚度，则下层流体的运动将不会使上层流体的

---

* 气象学报，1964，34(4)：523-530.

运动受到大的影响。这时取作近似，可以假定上层流体的运动完全不受下层流体运动的影响。做了这样的一些假定后，容易得到下层流体的运动方程为[1]：

$$\frac{\partial u}{\partial t}+u\frac{\partial u}{\partial x}+v\frac{\partial u}{\partial y}=-g^{*}\frac{\partial h}{\partial x}+fv \tag{1}$$

$$\frac{\partial v}{\partial t}+u\frac{\partial v}{\partial x}+v\frac{\partial v}{\partial y}=-g^{*}\frac{\partial h}{\partial y}-fu \tag{2}$$

$$\frac{\partial h}{\partial t}+\frac{\partial(uh)}{\partial x}+\frac{\partial(vh)}{\partial y}=0 \tag{3}$$

式中，x,y 为笛卡儿坐标轴；t 为时间；$g^{*}=\frac{\rho_2-\rho_1}{\rho_1}g$，$g$ 为重力加速度；f 为地球的自转参数。

设：扰动运动是迭加在地转运动

$$u_g=-\frac{g^{*}}{f}\frac{\partial H}{\partial y} \tag{4}$$

上的，并设：

$$u=u_g+u',\ v=v',\ h=H+h' \tag{5}$$

其中，u',v'分别为扰动的速度分量；h'为分界面的扰动高度。将式(5)代入式(1)至式(3)后，在扰动不依赖 y 和略去二级微量的假定下，得到：

$$\frac{\partial u'}{\partial t}+u_g\frac{\partial u'}{\partial x}=-g^{*}\frac{\partial h'}{\partial x}+fv' \tag{6}$$

$$\frac{\partial v'}{\partial t}+u_g\frac{\partial v'}{\partial x}=-fu' \tag{7}$$

$$\frac{\partial h'}{\partial t}+u_g\frac{\partial h'}{\partial x}+H\frac{\partial u'}{\partial x}-\frac{fu_g v'}{g^{*}}=0 \tag{8}$$

或者可以得到对于一个应变量的单一方程为：

$$\mathscr{L}(u',v',h')=0 \tag{9}$$

式中算子 $\mathscr{L}$ 为：

$$\mathscr{L}=\frac{\mathrm{D}^3}{\mathrm{D}t^3}-C_0^2\frac{\partial^3}{\partial x^2\mathrm{D}t}+f^2\frac{\mathrm{D}}{\mathrm{D}t}-f^2u_g\frac{\partial}{\partial x} \tag{10}$$

其中，$\frac{\mathrm{D}}{\mathrm{D}t}=\frac{\partial}{\partial t}+u_g\frac{\partial}{\partial x}$，$C_0=\sqrt{g^{*}H}$ 为重力内波波速。方程(9)是我们所要研究的基本方程。Hinkelmann[2]曾给出了类似的方程，不过它讨论的是重力惯性外波，而我们现在讨论的是流体内部分界面上的重力惯性内波。

3　相速度的公式

设扰动是沿气流的方向传播的，因此可以设(以 v'为例)：

$$v'(x,t)=v'_0\mathrm{e}^{ik(x-ct)} \tag{11}$$

式中，k 为 x 方向的波数；c 为相速度 。将式(11)代入式(9)后，得到频率公式为：

$$k^2(u_g-c)^3-(f^2+c_0^2k^2)(u_g-c)+f^2u_g=0 \tag{12}$$

为了讨论方便，今将式(12)无量纲化，取 $\bar{c}=c/c_0$，于是式(12)可以改写为：

$$(F-\bar{c})^3-(1+2R^2)(F-\bar{c})+2R^2F=0 \tag{13}$$

式中，$R=f/\sqrt{2}kc_0$，$F=u_g/c_0$ 为 Froude 数。这是控制地转风场中重力惯性内波动力学性质的两个基本参数。由于像飑线等中尺度系统是一种浅层的现象，它的垂直厚度 H 很小，c_0 一般不会太大；因此，$F>1$的情况也是可以出现的。一般来说，我们称 $u_g>c_0$ 的地转气流是超临界流，而 $u_g<c_0$ 的地转气流为次临界流。在不同的 c_0 值下，R 和波长 L 的转换见表 1。

表 1

R \ L(km) \ c_0	4(m/s)	6(m/s)	8(m/s)	10(m/s)
0.2	80	120	200	200
0.4	160	240	320	400
0.6	240	360	480	600
0.8	320	480	640	800
1.0	400	600	800	1 000

表 1 中取 $\varphi=38°$，即 $f=0.88\times10^{-4}$。

三次代数方程(13)的 3 个根分别为：

$$\left.\begin{aligned}\bar{c}_1&=F-\frac{1}{2}(a+b)-\mathrm{i}\frac{\sqrt{3}}{2}(a-b),\\ \bar{c}_2&=F-\frac{1}{2}(a+b)+\mathrm{i}\frac{\sqrt{3}}{2}(a-b),\\ \bar{c}_3&=F+(a+b)\end{aligned}\right\} \tag{14}$$

其中，

$$\left.\begin{aligned}a&=\sqrt[3]{R^2F+\sqrt{R^4F^2-\left(\frac{1+2R^2}{3}\right)^3}},\\ a&=\sqrt[3]{R^2F-\sqrt{R^4F^2-\left(\frac{1+2R^2}{3}\right)^3}}\end{aligned}\right\} \tag{15}$$

如果 $F=0$，即没有平均气流；那么式(14)变为：

$$\left.\begin{aligned}\bar{c}_1&=\sqrt{1+2R^2},\\ \bar{c}_2&=-\sqrt{1+2R^2},\\ \bar{c}_3&=0\end{aligned}\right\} \tag{16}$$

或者回到有因次后为：

$$\bar{c}_{1,2}=\pm\sqrt{g^*H+f^2/k^2} \tag{17}$$

此即为静止大气中重力惯性内波的相速度[2,3]。

如果取 $R=0$，即不考虑地球的自转，则式(14)可以改写为：

$$\left.\begin{aligned}\bar{c}_1 &= F + 1,\\ \bar{c}_2 &= F - 1,\\ \bar{c}_3 &= F\end{aligned}\right\} \tag{18}$$

或者有因次的形式为：

$$\left.\begin{aligned}\bar{c}_1 &= u_g + c_0,\\ \bar{c}_2 &= u_g - c_0,\\ \bar{c}_3 &= u_g\end{aligned}\right\} \tag{19}$$

此公式即为一般在盛行气流中重力内波的波速公式。

4　不稳定性的讨论

首先，当考虑了盛行地转气流后，重力惯性内波所具有的一个带有原则意义的性质是：它可以是不稳定的。由公式(14)容易知道，当 $\bar{c}_1$ 和 $\bar{c}_2$ 为共轭复根，它就出现不稳定。其条件是：

$$(R^2F)^2 - \left(\frac{1 + 2R^2}{3}\right) > 0 \tag{20}$$

或者为

$$F^2 > \frac{1}{R^4}\left(\frac{1 + 2R^2}{3}\right)^3 \tag{21}$$

容易证明，式(21)右端大于或等于 1。因此，重力惯性内波出现不稳定的必要条件是：

$$F \geqslant 1, \text{或}\ u_g \geqslant c_0 \tag{22}$$

即：只有当地转风的速度超过作为临界速度的重力内波波速时，才有产生不稳定性的可能。

重力惯性内波出现不稳定的临界参数曲线见图 1。图 1 表明：当 $F=1$ 时，只有 $R=1$ 或者 $k=f/\sqrt{2}c_0$ 的波有可能不稳定；而当 $F>1$ 时，在 $k=f/\sqrt{2}c_0$ 附近有一不稳定波数区；而当 F 值越大，这个不稳定区就越宽，亦即能够出现不稳定的波就越多。如果 $c_0=10$ m/s，$f=10^{-4}$/s；则 $F=1.5$ 时，最短的不稳定波长为 400 km。

现在，我们来求不稳定波的发展速度，令：

$$\bar{c}_j = \bar{c}_{jr} + i\bar{c}_{ji},\quad (j = 1,2) \tag{23}$$

由式(14)求得：

$$\left.\begin{aligned}\bar{c}_{1i} &= -\frac{\sqrt{3}}{2}(a - b),\\ \bar{c}_{2i} &= \frac{\sqrt{3}}{2}(a - b)\end{aligned}\right\} \tag{24}$$

由于 $a \geqslant b$，因此波速为 $\bar{c}_2$ 的这一簇波在不稳定时，它的振幅是随时间而增长的；而 $\bar{c}_1$ 这一簇波则在不稳定参数区中，振幅将逐渐衰减。不稳定波振幅的增长速度见图 2。当 $F=1.5$ 时，最不稳定波的增长速度约为 $c_i=0.7c_0$。由于这时最不稳定波所对应的 R 约为 1.5，相应的波数约为 $k=f/1.5\sqrt{2}c_0$。因此振幅 A 的增加约为 $|A_t/A_0| \approx \exp(kc_it) = \exp(0.7t/1.5\sqrt{2})$，所以当取 $f=10^{-4}$/s 时，约经过 19 个小时，

就能够使小扰动的振幅增大 10 倍。

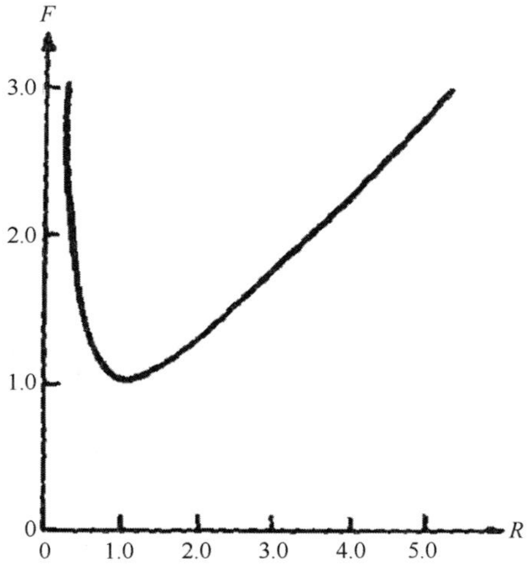

图 1 不稳定区

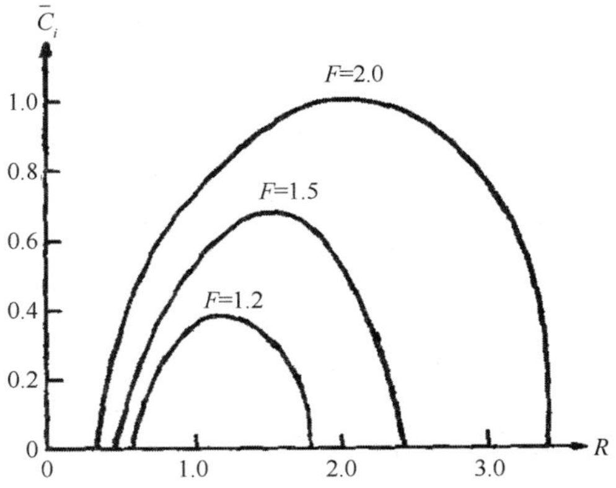

图 2 增长率

这种超临界流的不稳定性,最早海洋学家 Stommel[4] 在研究湾流的弯曲时,曾简单地提到过。在气象学中,这种不稳定性就笔者所知,还尚未被人们所注意。由于这种不稳定性所要求的条件、扰动的空间尺度和增长速度都是在大气中能够实现的,因此,笔者认为:这种由平均位能供给能量的重力惯性内波的不稳定性,可能与大气中某些中尺度扰动的发展有关。这是值得在今后的实际工作中加以注意的。

5 相速度的讨论

由三次代数方程根的判别式知道,当条件式(20)不满足时,方程具有 3 个不同的实根;并且当 $F^2R^4=\left(\frac{1+2R^2}{3}\right)^3$ 时,3 个实根中有 2 个为重根。在这种情况下,改用下面关于根的三角函数表达式讨论问题会更方便些。表达式为:

$$\left.\begin{aligned}\bar{c}_1 &= F + 2A\cos\frac{\varphi}{3},\\ \bar{c}_2 &= F - 2A\cos\left(\frac{\pi}{3}-\frac{\varphi}{3}\right),\\ \bar{c}_3 &= F - 2A\cos\left(\frac{\pi}{3}+\frac{\varphi}{3}\right)\end{aligned}\right\} \tag{25}$$

式中,

$$A=\left(\frac{1+2R^2}{3}\right)^{\frac{1}{2}},\ \varphi=\cos^{-1}B,\ B=FR^2\bigg/\left(\frac{1+2R^2}{3}\right)^{\frac{3}{2}}$$

当条件式(20)成立时,式(14)中的虚部为增长率,实部为相速度。即为:

$$\left.\begin{aligned}\bar{c}_{1r} &= F - \frac{1}{2}(a+b),\\ \bar{c}_{2r} &= F - \frac{1}{2}(a+b),\\ \bar{c}_{3r} &= F + (a+b)\end{aligned}\right\}\tag{26}$$

亦即 $\bar{c}_{1r}$ 和 $\bar{c}_{2r}$ 为重根。

当重力惯性内波不出现不稳定时，相速度的性质比较简单。由式(25)，当 $R\to0$ 时，便得到式(18)。当 $R\to\infty$ 时，容易看出 $\bar{c}_1\to\infty$，$\bar{c}_2\to-\infty$，$\bar{c}_1\to0$。当 $F=0$，$F=0.5$ 时，相速度随 R 的变化分别见图 3 和图 4 的实线。

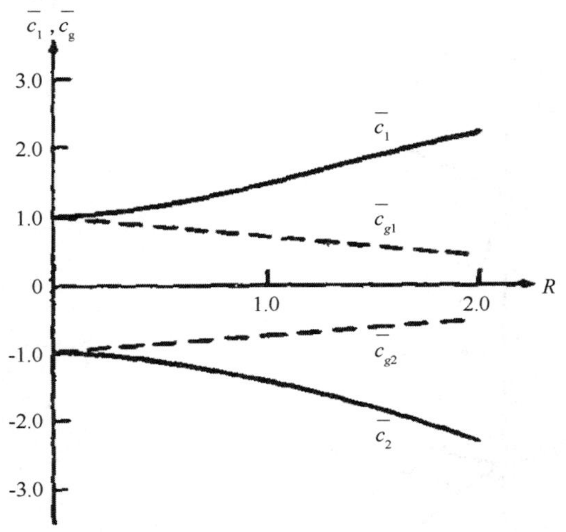

图 3　$F=0$ 时的相速、群速

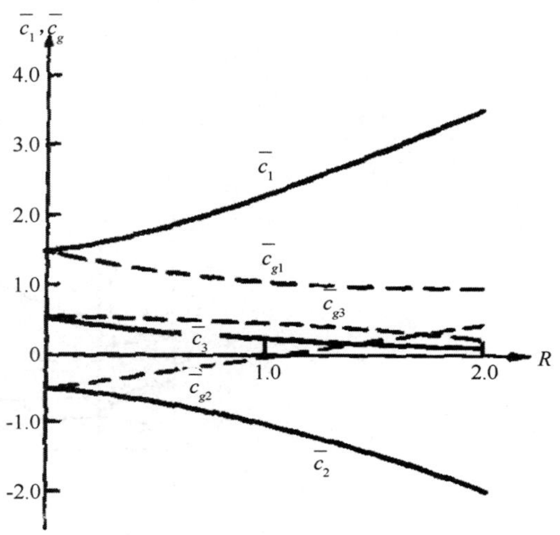

图 4　$F=0.5$ 时的相速、群速

当重力惯性内波出现不稳定时，相速度性质比较复杂。这时由稳定区趋向不稳定区的边界时，即当 $R\to R_-^*$（R_-^* 为不稳定区域左边界线，下标"-"表示左极限），由于

$$B=\frac{FR^2}{\left(\dfrac{1+2R^2}{3}\right)^{3/2}}=1,\ A=\left(\frac{1+2R^2}{3}\right)^{\frac{1}{2}}=3\sqrt{FR^2},$$

$$\varphi=0^\circ$$

所以由式(25)得到由稳定区趋向不稳定区左边界的相速度为：

$$\text{当 } R\to R_-^*,\quad \left.\begin{aligned}\bar{c}_{1r} &= F + 2\sqrt[3]{FR^2},\\ \bar{c}_{2r} &= F - \sqrt[3]{FR^2},\\ \bar{c}_{3r} &= F - \sqrt[3]{FR^2}\end{aligned}\right\}\tag{27}$$

即这时的 $\bar{c}_2$，$\bar{c}_3$ 变为一对重根。另一方面，在不稳定区中，当 $R\to R_+^*$ 时，由于 $a=b=\sqrt[3]{R^2F}$，故由式(26)得到：

$$
当 R \to R_+^*, \quad \left.\begin{aligned} \bar{c}_{1r} &= F - \sqrt[3]{FR^2}, \\ \bar{c}_{2r} &= F - \sqrt[3]{FR^2}, \\ \bar{c}_{3r} &= F + 2\sqrt[3]{FR^2} \end{aligned}\right\} \tag{28}
$$

比较式(27),式(28)两式后,可以看出:通过不稳定区的左边界时,$\bar{c}_1$ 和 $\bar{c}_3$ 两个根互相置换。同样的分析表明:当通过不稳定区的右边界时,$\bar{c}_1$ 和 $\bar{c}_3$ 又再一次互相置换。

当 $F=1.5$ 时,相速度随 R 的改变见图 5 中的实线。根据计算结果,我们很容易看出:当考虑了盛行地转风后,重力惯性内波有 3 个相速度。参考了式(18)后,容易知道:$\bar{c}_1$ 和 $\bar{c}_2$ 分别是两组快波,其中 $\bar{c}_1$ 是相对气流的前进波,$\bar{c}_2$ 是后退波;而 $\bar{c}_3$ 是一组慢波,当不考虑地球自转的影响时,慢波的相速就是风速。

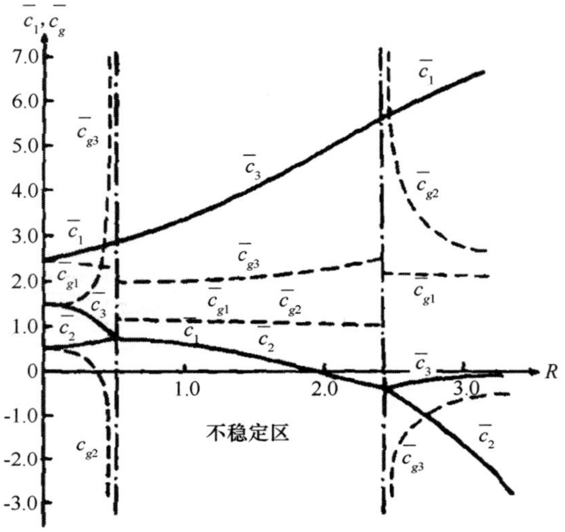

图 5　$F=1.5$ 时的相速、群速

考虑了盛行气流之后,正如图 5 所示,相速度具有一个很有意思的特性是:当通过不稳定区的边界时,它的变化是不连续的。如 $\bar{c}_3$ 当波长大到使波变为不稳定时,相速度将突然变大;以后当波长再一次增大到进入稳定区时,相速度又突然变小。同样地,$\bar{c}_1$ 在不稳定区交界处也有类似的情况。

6　群速度的讨论

由于重力慢性内波是一种频散波,因此群速度不等于相速度。在无因次形式下的群速度公式为:

$$
\bar{c}_g = \bar{c} - R\frac{\partial \bar{c}}{\partial R} \tag{29}
$$

在稳定区域中的群速度,可将式(25)代入式(29)则得到:

$$\left.\begin{aligned}\bar{c}_{g1} &= DF + \bar{c}_1(1 - D) - E\sin\frac{\varphi}{3} \\ \bar{c}_{g2} &= DF + \bar{c}_2(1 - D) - E\sin\left(\frac{\pi}{3} - \frac{\varphi}{3}\right) \\ \bar{c}_{g3} &= DF + \bar{c}_3(1 - D) + E\sin\left(\frac{\pi}{3} + \frac{\varphi}{3}\right)\end{aligned}\right\} \tag{30}$$

式中,

$$D = \frac{2R^2}{3A^2}, \quad E = \frac{4R^2F(1 - R^2)}{9A^2\sqrt{1 - B^2}}$$

在不稳定区域中的群速度,可由式(26)代入式(29)后得到:

$$\left.\begin{aligned}\bar{c}_{g1} &= \bar{c}_1 + \frac{1}{2}G(d + e) \\ \bar{c}_{g2} &= \bar{c}_2 + \frac{1}{2}G(d + e) \\ \bar{c}_{g3} &= \bar{c}_3 - G(d + e)\end{aligned}\right\} \tag{31}$$

式中,

$$G = \frac{2}{3}R^2, \quad d = \frac{1}{a^2}\left[F + \frac{F^2R^2 - \left(\frac{1 + 2R^2}{3}\right)^2}{\sqrt{F^2R^4 - \left(\frac{1 + 2R^2}{3}\right)^3}}\right]$$

$$e = \frac{1}{b^2}\left[F - \frac{F^2R^2 - \left(\frac{1 + 2R^2}{3}\right)^2}{\sqrt{F^2R^4 - \left(\frac{1 + 2R^2}{3}\right)^3}}\right]$$

由式(30)可知,当 $R \to 0$,由于 $A \to \frac{\sqrt{3}}{3}$,$D = B = E \to 0$,因此容易得到:

$$\text{当 } R \to 0, \quad \left.\begin{aligned}\bar{c}_{g1} &= \bar{c}_1 = F + 1 \\ \bar{c}_{g2} &= \bar{c}_2 = F - 1 \\ \bar{c}_{g3} &= \bar{c}_3 = F\end{aligned}\right\} \tag{32}$$

当 $R \to \infty$ 时,由于 $D \to 1, B \to 0, E \to -F, \varphi \to \frac{\pi}{2}, A \to \sqrt{\frac{2}{3}}R$,而 $\bar{c}_1 \to +\infty$,$\bar{c}_2 \to -\infty$,$\bar{c}_3 \to 0$;因此求得 $\bar{c}_{g1} \to \frac{3}{2}F$,$\bar{c}_{g2} \to \frac{3}{2}F$,$\bar{c}_{g3} \to 0$。此外,如不稳定区的左边界和右边界的值分别为 R^* 和 R^{**},则由于 $R \to R_-^*$ 或 R_+^{**} 时,$B \to 1$,$\varphi \to 0$,而 $E \to \infty$ 或 $-\infty$,因此得到当 $R \to R_-^*$ 时,$\bar{c}_{g2} \to -\infty$,$\bar{c}_{g3} \to -\infty$;当 $R \to R_+^{**}$ 时,$\bar{c}_{g2} \to +\infty$,$\bar{c}_{g3} \to -\infty$,而 $\bar{c}_{g1}$ 则可以证明为有限。在不稳定区域中,式(31)表明:$\bar{c}_{g1} = \bar{c}_{g2}$,而且 $\bar{c}_{g3}$ 在不稳定区域的边界上取有限值。但与相速度所不同的是:通过不稳定区的边界时,两边的 $\bar{c}_{g1}$ 和 $\bar{c}_{g2}$ 值出现跳跃。群速度的计算结果,分别见图 3 至图 5 中的虚线。

根据计算,首先我们注意到在考虑了盛行地转风后,在不稳定区域中对于相速为 $\bar{c}_1$ 和 $\bar{c}_2$ 的快波,它的群速度也可以大于相速度 。这表明,对于中尺度系统也可以像大尺度系统那样有“上游效应”[5];即当一个中尺度系统(雷暴系统,大风区)在发展时,在它的下游可以激起新的系统产生。

重力惯性内波的群速度可以大于相速度,这一结果带有原则意义。这表明了当中尺度系统在发展时,在它的前方可能不断地有新的系统产生,这是与一些观测结果相符合的。如所谓“共鸣型雷暴”可能就是由于这种机制所产生的。

7 结论

本文讨论了地转气流中的重力惯性内波的性质,得到下面两点主要结果:

(1)当地转风速大于重力内波波速时,某些尺度的重力惯性内波可以产生不稳定性,这种不稳定性可以用来解释某些中系统的发展。

(2)某些波长的重力惯性内波,它的群速度可以大于相速度。这表明了一些发展的中尺度系统的能量可以很快地被传到下游,而在下游产生新的系统。

参考文献

[1] Teppcr, M. *J. Met.*, 195512, :287-297.

[2] Hinkelmann. K. *Tellus*, 1951, 3:285-296.

[3] Rossby, C. G., *J. Met.*, 1945, 2:187-204.

[4] Stommel, H. *J. Mar. Res.*, 1953, 12:184-195.

[5] 叶笃正. *J. Met.*, 1949, 6:1-16.

风生边界急流稳定性的渐近理论*

巢纪平[1, 2]　冯立成[3,4]　王东晓[5]

(1. 国家海洋环境预报中心，北京　100081；2. 国家海洋局第一海洋研究所，青岛　266061；3. 中国科学院大气物理研究所大气科学和地球流体力学数值模拟国家重点实验室，北京　100029；4. 中国科学院研究生院，北京　100049；5. 中国科学院南海海洋研究所热带海洋环境动力学重点实验室，广州　510301)

提要：观测表明，当冬季盛吹北风时，在中国南海西边界附近将形成一支向南的急流，在一定条件下这支急流可弯曲成波动甚至形成涡旋。本文应用等值浅水模式，采用截断模方法，分析了急流的稳定性，并给出急流上不稳定波出现的条件。分析表明只有当向南的风生急流很强很窄时，由变性的 Kelvin 波和风应力强迫出的地形 Rossby 波在长波波段耦合而出现不稳定，不稳定波在波长约 200 km 时向北传播的相速度约为 0.2 m/s，波振幅增长到 e 倍所需的时间约 1.5 天。分析进一步表明，夏季向北的风生流在海洋的西边界附近是稳定的．这些结果在一定程度上解释了观测结果。

关键词：中国南海；西边界急流；边界波的稳定性；耦合波；等值浅水模式

1　引言

在陆架海中由海底地形可产生沿岸急流及与之相伴随的沿岸上升流[1~5]，而当风沿陆架海的海岸线方向盛吹时，也可产生类似的现象。由地形或风产生的沿岸急流和沿岸上升流在南海尤为明显。

南海环流的一个显著特征就是存在环流的西边界强化。研究表明冬季南海西边界流是一支强大、自北(海南岛以东)向南(卡里马达海峡)的贯通流[6, 7]。关于西边界流的强度目前尚缺乏充分的观测依据。刘秦玉等[8]在将南海考虑为一个封闭海盆的情况下，发现南海海盆尺度环流的结构可以从 Sverdrup 流函数分布得到反映，并且由海盆内区 Sverdrup 流函数分布可以得到，南海西边界流的强度在冬季为 5~6Sv，夏季为 3~4Sv。根据对从风场诊断出的流函数场分布的分析，王卫强等[9]发现南海西边界流强度的年变化与季风相关联。由此可见南海西边界流的季节和年际变化都十分明显。

南海中尺度涡旋发育很丰富。环流的多涡结构的生消与迁移与南海环流季节演变的关系密切。苏京志等[10]通过分析卫星高度计数据，揭示了南海涡动的变化特征，南海涡旋涡动动能的季节变化定量地表述了越南外海涡旋的时空演变规律。Wang 等[11]针对卫星海面高度异常资料对 1993—2000 年的涡旋进行了统计，发现南海涡旋的生成和运动可以划分为 4 个具有不同特点的海区，其中之一为(110°—116°E，8°—16°N)的越南外海。上述研究结果初步表明了南海西边界流与越南外海局地涡旋之间可能存在动力联系。

* 地球物理学报，2006，49(3)：642-649.

基金项目：国家自然科学基金项目(40231012 和 40325015)资助。

对南海环流特征的这些观测，提出了两个很值得在动力学上研究的问题：一是吕宋海峡在什么情况下可以将太平洋的水注入南海；二是南海西边界强化的向南急流与涡旋发生之间的关系。关于第一个问题我们将在另文讨论，而本文采用对等值浅水模式的截断模方法，研究南海西边界流急流不稳定的若干条件，试图用这些解析结果初步表明南海西边界流与越南外海局地涡旋之间可能存在一定的动力联系。

2 风生地转流

不可能对实际的南海海盆进行理论分析，只能引进简化的理论框架。设海洋西边界是南北走向(即沿 y 轴)，海岸的坐标为 $x=0$,海洋的未扰深度为 H_0。当研究由风应力 $\tau(x,y)$ 产生的气候态时，可近似地用均质流体模型进行，由此有

$$g\Delta\hat{H}=\frac{\tau}{H_0} \tag{1}$$

由海洋深度的不均匀可产生风吹地转流，南北方向的地转流为

$$fV_g=g\frac{\partial\hat{H}}{\partial x} \tag{2}$$

如果风应力只是南北方向的,式(1)为

$$g\frac{\partial\hat{H}}{\partial y}=\frac{\tau_y}{H_0} \tag{3}$$

设风应力及风生地转流在南北方向是均匀的，则由式(2)、式(3)两式给出

$$\beta V_g=\frac{1}{H_0}\frac{\partial\tau_y}{\partial x} \tag{4}$$

式中,g 为重力加速度,f 是科里奥利参数，$\beta=\mathrm{d}f/\mathrm{d}y$ 为科里奥利参数随纬度的变化率，V_g 为地转流流速。

3 等值浅水运动

在分析海洋运动的距平状态时，可将海洋分出一活动层，其深度为 D，偏离风吹气候洋流的运动只发生在这一层中，即它们是等值浅水运动，或者可称应用的是 $1\frac{1}{2}$层模式。

等值浅水运动方程为

$$\frac{\partial u}{\partial t}+u\frac{\partial u}{\partial x}+v\frac{\partial u}{\partial y}-fv=-g'\frac{\partial h}{\partial x} \tag{5}$$

$$\frac{\partial v}{\partial t}+u\frac{\partial v}{\partial x}+v\frac{\partial v}{\partial y}+fu=-g'\frac{\partial h}{\partial y} \tag{6}$$

$$\frac{\partial h}{\partial t}+\frac{\partial(hu)}{\partial x}+\frac{\partial(hv)}{\partial y}=0 \tag{7}$$

令

$$u=u',\ v=V_g+v',\ h=\hat{H}+h' \tag{8}$$

式中,u,v 分别是沿 x,y 轴的速度分量;u',v'分别为沿 x,y 轴的速度分量的扰动量；h'为扰动深度。

线性化方程为

$$\varepsilon \frac{\partial u'}{\partial t} + \delta\left(V_g \frac{\partial u'}{\partial y}\right) - fv' = - g' \frac{\partial h'}{\partial x} \tag{9}$$

$$\frac{\partial v'}{\partial t} + V_g \frac{\partial v'}{\partial y} + u' \frac{\mathrm{d}V_g}{\mathrm{d}x} + fu' = - g' \frac{\partial h'}{\partial y} \tag{10}$$

$$\frac{\partial h'}{\partial t} + V_g \frac{\partial h'}{\partial y} + H_0\left(\frac{\partial u'}{\partial x} + \frac{\partial v'}{\partial y}\right) + u' \frac{\mathrm{d}\hat{H}}{\mathrm{d}x} = 0 \tag{11}$$

在式(11)中未加微分号的 $\hat{H}$ 已用 H_0 近似。考虑到式(2)，式(11)式可写成

$$\frac{\partial h'}{\partial t} + V_g \frac{\partial h'}{\partial y} + H_0\left(\frac{\partial u'}{\partial x} + \frac{\partial v'}{\partial y}\right) + \frac{f}{g'}V_g u' = 0 \tag{12}$$

在式(9)中 ε，δ 为标识符。当 $\varepsilon=0$，$\delta=0$ 时称半地转近似；当 $\varepsilon=0$，而 $\delta=1$ 时称准半地转近似。本文分析准半地转近似的情况，使基本流的作用能起到多方面的影响。

3.1　无量纲形式

引进重力波波速 $C=\sqrt{g'H_0}$，$g'=g\dfrac{\Delta\rho}{\rho}$为约化重力，$\Delta\rho$ 为上、下层海水的密度差，特征尺度 $l=C/f$，$(x,y)\sim l$，特征时间 $T\sim 1/f$，$t\sim T$，$(u,v)\sim C$，$h'\sim H_0$，为方便，暂时取 $V_g\sim C$。事实上风吹流的量级应由风应力的强度来决定，这将在下面作量级调整。上述方程的无量纲形式为(仍按原符号)

$$\varepsilon \frac{\partial u'}{\partial t} + \delta V_g \frac{\partial u'}{\partial y} - v' = - \frac{\partial h'}{\partial x} \tag{13}$$

$$\frac{\partial v'}{\partial t} + V_g \frac{\partial v'}{\partial y} + u' \frac{\mathrm{d}V_g}{\mathrm{d}x} + u' = - \frac{\partial h'}{\partial y} \tag{14}$$

$$\frac{\partial h'}{\partial t} + V_g \frac{\partial h'}{\partial y} + \frac{\partial u'}{\partial x} + \frac{\partial v'}{\partial y} + u'V_g = 0 \tag{15}$$

在一般情况下地转流的旋度要比行星涡度小很多，因此式(14)可简化成

$$\frac{\partial v'}{\partial t} + V_g \frac{\partial v'}{\partial y} + u' = - \frac{\partial h'}{\partial y} \tag{16}$$

3.2　风生流对基本运动的影响

在定量分析运动方程的本征值问题前，先对风生流对基本运动的影响作一个简单的分析。假定大尺度的风生流在 x 方向是均匀的，或变化是缓慢的，对后者可用 Wentzel - Kramers - Brillouin solutions(WKB 方法)分析。

3.2.1　Kelvin 波

按常用定义，令区域中无垂直于边界的流动，即 $u'=\partial u'/\partial x=0$，由式(15)、式(16)两式给出

$$\left[\left(\frac{\partial}{\partial t} + V_g \frac{\partial}{\partial y}\right)^2 - \frac{\partial^2}{\partial y^2}\right]h' = 0 \tag{17}$$

当取行波解即 $\exp[(\mathrm{i}(ky-\sigma t)]$时，式(17)的色散关系为

$$\sigma = k(\pm 1 + V_g) \tag{18}$$

注意到，当无基本流时，$\sigma=-k$ 的运动其振幅满足地转关系

$$v' = \partial h'/\partial x \tag{19}$$

且当 $x\to\infty$ 时振幅有限。可见这是沿 y 轴负方向，即向南传播的 Kelvin 波，风生流将改变波的传播速度，如风生流是向南的，将使波的传播速度变快。

3.2.2 地形 Rossby 波

引进管量和势量场，定义为

$$u' = -\frac{\partial \Psi}{\partial y} + \frac{\partial \varphi}{\partial x}, \quad v' = \frac{\partial \Psi}{\partial x} + \frac{\partial \varphi}{\partial y} \tag{20}$$

式中，Ψ 为流函数；φ 为势函数。暂时去掉式(13)的标识符，对式(13)和式(16)取交叉微分后相减，并联立式(15)得到

$$\left(\frac{\partial}{\partial t} + V_g \frac{\partial}{\partial y}\right)(\Delta^2 \Psi - h') + \left(\frac{\partial \Psi}{\partial y} - \frac{\partial \varphi}{\partial x}\right) V_g = 0 \tag{21}$$

如果运动是准地转的，有 $h' \approx \Psi$，准地转运动是准无辐散的，因此式(21)与势量有关的项 $\partial\varphi/\partial x \approx 0$。于是式(21)简化成

$$\left(\frac{\partial}{\partial t} + V_g \frac{\partial}{\partial y}\right)(\Delta^2 - 1)\Psi + V_g \frac{\partial \Psi}{\partial y} = 0 \tag{22}$$

这是考虑了风生流后的位势涡度方程。由于 $V_g \sim \partial \hat{H}/\partial x$，因此式(22)的基本运动为地形 Rossby 波[12]。

由此可见，Kelvin 波和地形 Rossby 波是这个动力系统中最基本的运动形式。

4 准半地转近似控制方程

注意到在准半地转近似下，高频的重力惯性波将被过滤。这时的基本方程为

$$V_g \frac{\partial u'}{\partial y} - v' = -\frac{\partial h'}{\partial x} \tag{23}$$

及式(15)和式(16)。

4.1 代入背景流

取风应力的分布为 $\tau_y = \tau_y^0 x e^{-\alpha x^2}$。当 $\tau_y^0 < 0$ 时为北风应力，这样的风应力分布表示其最强处在海洋中部。注意到风应力是由大气运动决定的大尺度过程，x 的特征尺度应取 $\sqrt{gH_0/f}$，而不应取上面的 C/f，代入式(4)得到无量纲的风生流为

$$V_g = \frac{\tau_y^0 f}{\beta g H_0^2} \sqrt{\frac{\rho}{\Delta\rho}} (1 - 2\alpha x^2) e^{-\alpha x^2} \tag{24}$$

可见当北风应力时，沿岸附近是一支向南的急流，离开海岸一段距离后是向北流，转折的经度依赖于 α 的取值。将式(24)写成

$$V_g = \Lambda (1 - 2\alpha x^2) e^{-\alpha x^2} \tag{25}$$

式中，$\Lambda = \tau_y^0 f / \beta g H_0^2 \sqrt{\Delta\rho/\rho}$。如取 $\tau_y^0 \sim -9\ m^2/s^2$，$H_0 \sim 10^3$ m，$f \sim 2\times10^{-5}$/s(低纬)，$\beta \sim 2\times10^{-11}$/(s·m)，$g \sim 9.8\ m/s^2$，$\Delta\rho/\rho \sim 10^{-2}$，得 $\Lambda \approx -10$，因此 Λ 的合理取值范围为 $10^0 \sim 10^1$，其符号由风应力的方向决定。

代入背景流后有

$$\Lambda(1 - 2\alpha x^2) e^{-\alpha x^2} \frac{\partial u'}{\partial y} = v' - \frac{\partial h'}{\partial x} \tag{26}$$

$$\frac{\partial v'}{\partial t} + \Lambda(1 - 2\alpha x^2) e^{-\alpha x^2} \frac{\partial v'}{\partial y} + u' = -\frac{\partial h'}{\partial y} \tag{27}$$

$$\frac{\partial h'}{\partial t}+\Lambda(1-2\alpha x^2)\mathrm{e}^{-\alpha x^2}\frac{\partial h'}{\partial y}+\frac{\partial v'}{\partial y}+\frac{\partial u'}{\partial x}+\Lambda(1-2\alpha x^2)\mathrm{e}^{-\alpha x^2}u'=0 \tag{28}$$

这是要分析的基本方程。

4.2　Weber 函数展开方程

不影响问题的基本性质，现将定义域(0, +∞)扩展到(−∞，+∞)，之后，只要将有关的 x 置换成 $-x$ 即可将结果直接用来讨论海洋的边界在东侧的情况。

分析中用到的 Weber 函数 $D_n(x)$ 满足方程

$$\frac{\mathrm{d}^2D_n}{\mathrm{d}x^2}+\left(n+\frac{1}{2}-\frac{1}{4}x^2\right)D_n=0 \tag{29}$$

并有

$$\frac{\mathrm{d}D_n}{\mathrm{d}x}=\frac{1}{2}(nD_{n-1}-D_{n+1}) \tag{30}$$

$$D_{n+1}-xD_n+nD_{n-1}=0 \tag{31}$$

$$D_{2n}(0)=(-1)^n(2n-1)!!,\quad D_{2n+1}(0)=0 \tag{32}$$

$$\int_{-\infty}^{\infty}D_n(x)D_m(x)\mathrm{d}x=n!\ \sqrt{2\pi}\delta_{nm} \tag{33}$$

以及积分

$$I=\int_{-\infty}^{\infty}\mathrm{e}^{-\alpha x^2}\mathrm{d}x=\sqrt{\frac{\pi}{a}} \tag{34}$$

于是对形如下面的积分容易积出，如

$$I=\int_{-\infty}^{\infty}x^2\mathrm{e}^{-\alpha x^2}\mathrm{d}x=-\frac{\partial I}{\partial\alpha}=\frac{1}{2}\pi^{1/2}\alpha^{-5/2} \tag{35}$$

类似地可算出 $\int_{-\infty}^{\infty}x^{2n}\mathrm{e}^{-\alpha x^2}\mathrm{d}x$。

对 v',h'用 Weber 函数展成

$$(v',h')=\sum_{n=0}^{\infty}[v'(y),h'_n(y)]D_n(x) \tag{36}$$

考虑到当 $x=0$ 时，$u'=0$，因此 u'用 Weber 函数展开为

$$u'=\sum_{n=0}^{\infty}u'_{2n+1}(y)D_{2n+1}(x) \tag{37}$$

方程(26)给出

$$\sum_{n=0}^{\infty}v'_nD_n=\frac{1}{2}\sum_{n=0}^{\infty}[nh'_nD_{n-1}-h'_nD_{n+1}]+\frac{\partial}{\partial y}\sum_{n=0}^{\infty}\Lambda(1-2\alpha x^2)\mathrm{e}^{-\alpha x^2}u'_{2n+1}D_{2n+1} \tag{38}$$

或者写成

$$\sum_{n=0}^{\infty}\left\langle v'_n-\frac{1}{2}[(n+1)h'_{n+1}-h'_{n-1}]\right\rangle D_n=\frac{\partial}{\partial y}\sum_{n=0}^{\infty}\Lambda(1-2\alpha x^2)\mathrm{e}^{-\alpha x^2}u'_{2n+1}D_{2n+1} \tag{39}$$

应用 Weber 函数展开后，方程(27)、方程(28)为

$$\sum_{n=0}^{\infty}\left\langle\frac{\partial v'_n}{\partial t}+\Lambda(1-2\alpha x^2)\mathrm{e}^{-\alpha x^2}\frac{\partial v'_n}{\partial y}\right\rangle D_n=-\sum_{n=0}^{\infty}u'_{2n+1}D_{2n+1}-\sum_{n=0}^{\infty}\frac{\partial h'_n}{\partial y}D_n \tag{40}$$

$$\sum_{n=0}^{\infty}\left\langle \frac{\partial h'_n}{\partial t}+\Lambda(1-2\alpha x^2)e^{-\alpha x^2}\frac{\partial h'_n}{\partial y}+\frac{\partial v'_n}{\partial y}\right\rangle D_n$$

$$=-\sum_{n=0}^{\infty}\left\langle \frac{1}{2}[(2n+1)D_{2n}-D_{2n+2}]+\Lambda(1-2\alpha x^2)e^{-\alpha x^2}D_{2n+1}\right\rangle u'_{2n+1} \tag{41}$$

4.3 最大简化截断模方程

为了解运动的宏观性质，现分析最大简化截断模方程的解。对 v'，h'取 $n=0$，1，即一个偶对称模和一个奇对称模，对 u'取 $n=1$。由式(39)至式(41)给出

$$v'_0=\frac{1}{2}h'_1 \tag{42}$$

$$v'_1=-\frac{1}{2}h'_0+A_{u,1}\frac{\partial u'_1}{\partial y} \tag{43}$$

$$\left(\frac{\partial}{\partial t}+A_{v,0}\frac{\partial}{\partial y}\right)v'_0=-\frac{\partial h'_0}{\partial y} \tag{44}$$

$$\left(\frac{\partial}{\partial t}+A_{v,1}\frac{\partial}{\partial y}\right)v'_1=-u'_1-\frac{\partial h'_1}{\partial y} \tag{45}$$

$$\left(\frac{\partial}{\partial t}+A_{h,0}\frac{\partial}{\partial y}\right)h'_0=-\frac{\partial v'_0}{\partial y}-\frac{1}{2}u'_1 \tag{46}$$

$$\left(\frac{\partial}{\partial t}+A_{h,1}\frac{\partial}{\partial y}\right)h'_1=-A_{u,1}u'_1-\frac{\partial v'_1}{\partial y} \tag{47}$$

式中

$$A_{v,0}=A_{h,0}=A_0=\frac{\Lambda}{\sqrt{2\pi}}\int_{-\infty}^{\infty}(1-2\alpha x^2)e^{-\alpha x^2}D_0^2(x)dx \tag{48a}$$

$$A_{v,1}=A_{h,1}=A_{u,1}=A_1=\frac{\Lambda}{\sqrt{2\pi}}\int_{-\infty}^{\infty}(1-2\alpha x^2)e^{-\alpha x^2}D_1^2(x)dx \tag{48b}$$

5 本征值分析

引进变量

$$q'_0=h'_0+v'_0 \tag{49}$$

方程(44)、方程(46)相加得到

$$\left(\frac{\partial}{\partial t}+(A_0+1)\frac{\partial}{\partial y}\right)q'_0=-\frac{1}{2}u'_1 \tag{50}$$

由方程(42)、方程(43)给出

$$q'_0=\frac{1}{2}h'_1-2v'_1+2A_{u1}\frac{\partial u'_1}{\partial y} \tag{51}$$

代入方程(50)，得到

$$\left[\frac{\partial}{\partial t}+(1+A_0)\frac{\partial}{\partial y}\right]\times\left[h'_1-4v'_1+4A_1\frac{\partial u'_1}{\partial y}\right]=-u'_1 \tag{52}$$

方程(52)连同方程(45)、方程(47)构成对变量 v'_1, h'_1, u'_1 的本征值问题。

取行波解，即变量~ $\exp[\mathrm{i}(kx-\sigma t)]$，$k$ 为波数，σ 为频率，得到方程(45)、方程(47)、方程(52)的系数行列式为

$$\begin{vmatrix} 4(A_1^2k^2 - A_1k\sigma - 1)(k + A_0k - \sigma) + \sigma - A_1k & (1 + 4A_1k^2)(k + A_0k - \sigma) - k \\ A_1^2k - k - A_1\sigma & \sigma \end{vmatrix}$$

由此得到色散关系的代数方程为

$$A\sigma^3 + B\sigma^2 + C\sigma + D = 0 \tag{53}$$

其中 A,B,C,D 为关于 α,Λ 的系数，其表达式见附录1。或者写成标准形式，即

$$\sigma^3 + P\sigma + Q = 0 \tag{54}$$

其中 P,Q 的表达式见附录2。

6　波动性质的参数域

由上面得到的色散关系式可见波动能否发展取决于两个参数，即风生流的特征宽度或转向参数 α，及特征强度 Λ。这里给出了波动稳定性的参数域(图1)。

当风生流十分宽时可取 $\alpha=0.1$，波动不稳定可发生在 Λ 取较小绝对值(~ −3)或 k 取得较大波数(~ 1)时(图1(a))。当 $\alpha=0.3$(图1(b))时，Λ 要取较大绝对值(~ −16)才能使得波动不稳定。$\alpha=0.5$(图1(c))，$\alpha=0.7$(图1(d))时，欲使波动出现不稳定，向南的风吹流特征速度 Λ 要小于−5，并且当 $\alpha>0.7$ 后，随 α 的增大，风生流特征速度 Λ 绝对值也要相应增大，波动才能出现不稳定。这些结果表明只有当向南的风生流较强，并且是长波时才能出现不稳定，这在一定程度上与观测事实相符。

由此可见，上面各例的计算结果没有偏离上面给出的可能估值范围。但计算发现一个很有意思的结果是，当 Λ 取正值，即南风应力时，当海洋边界置于 $x=0$ 处(海洋边界在西边)，方程(54)不出现共轭复根，即波都是稳定的。

然而如将 $x\to -x$，$\tau_y\to -\tau_y$，问题的本征值不变，即由南风应力产生的向北风生流，在东边界附近也可以是不稳定的。

7　波的色散性质

图2给出了基本波动和在几种参数取值下的波动稳定性。

由图2a可见当风应力为零时，只存在向南传播的 Kelvin 波。当风应力存在时，图2(b)和图2(c)为波动稳定的情况，可见当风生流十分宽时(图2(b))，风应力强迫出一支频率很高并向北传播的波动和一支频率很低先向北传播后转向南传播的波动，后者由前面的分析可以确认基本上是“地形 Rossby 波”。当风生流较窄时(图2(c))，本质上为 Kelvin 波的那支波仍向南传播，但频率变低，而由风应力的作用，强迫出一支频率很高并向南传播的波动和一支频率逐渐增高的向北传播的波动，后者是从“地形 Rossby 波”演变而来。由图2(b)和图2(c)可见，风生流弱而宽时，“地形 Rossby 波”基本上是向北传播的，但频率很低；当风生流增强变窄时，“地形 Rossby 波”的频率明显地随波数增大而增大，这与对位势涡度方程(22)的定性分析基本一致；难以预计的是另一支高频强迫波，它的传播方向很不稳定，当风生急流弱而宽时是向北传播的，当风生急流变强变窄时，它转向南传播，但两

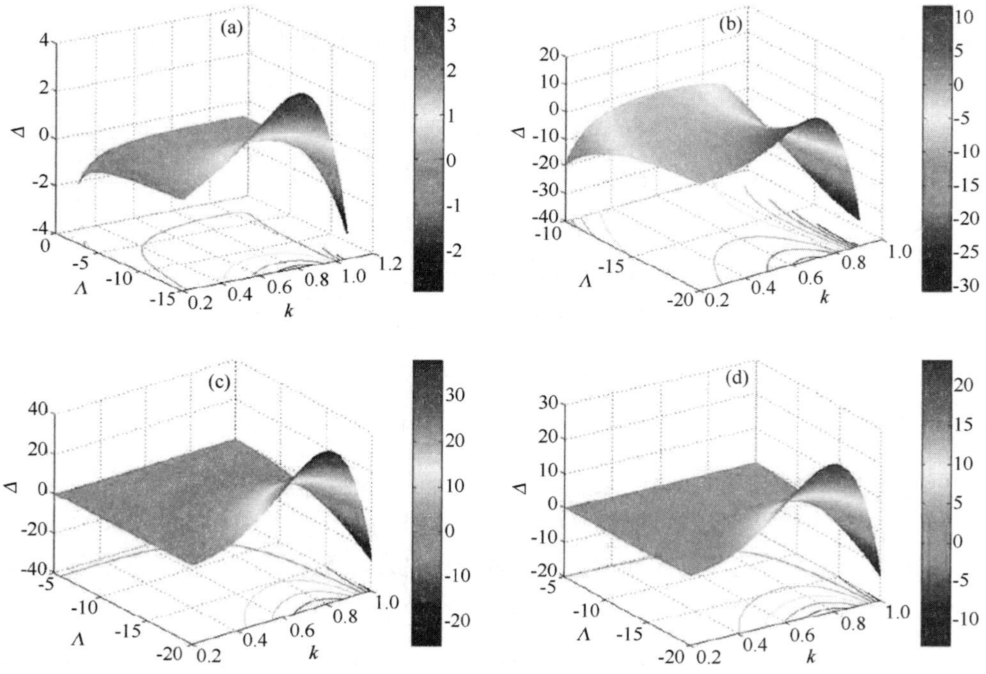

图 1　固定参数 α 时关于 Λ 和 k 的波动稳定性参数域

(a) $\alpha=0.1$；(b) $\alpha=0.3$；(c) $\alpha=0.5$；(d) $\alpha=0.7$。图中 k 为波数，Λ 为风吹流特征强度，$\Delta=(p/2)^2+(Q/3)^3$ 为方程(54)根性质的判别式，$\Delta>0$ 的区域出现共轭复根，这时有一个根是指数增长的，即波动不稳定，$\Delta\leqslant 0$ 时波动稳定(见书后彩插)

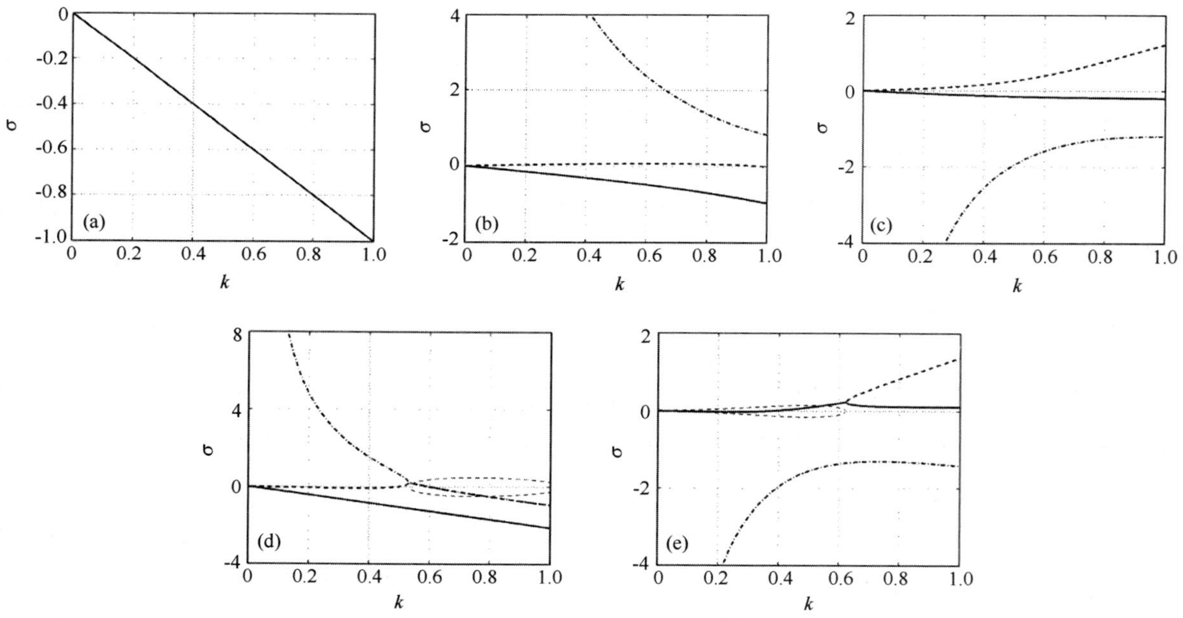

图 2　波动色散关系

(a)为风应力为零时的基本波动；(b,c)为波动稳定的情况，(b)中 $\alpha=0.1, \Lambda=-2.0$，(c)中 $\alpha=0.5, \Lambda=-5.0$；(d,e)为波动不稳定的情况，(d)中 $\alpha=0.1, \Lambda=-4.0$，(e)中 $\alpha=0.5$，$\Lambda=-6.0$，长虚线为不稳定增长率。k 为无量纲波数，σ 为无量纲频率

者也有共同点，都随着波数增大传播速度变小。图 2(d) 和图 2(e) 为波动不稳定的情况，当风生流十分宽时，风应力强迫出的两支波动耦合成不稳定波(图 2(d))，在不稳定波段，从向北传播随波数增大变为向南传播，但这支耦合不稳定波传播速度很慢，不稳定增长率也很小。而风生流较窄并强度较大时，由变性 Kelvin 波和变性"地形 Rossby 波"耦合成向北传播的不稳定波(图 2(e))，当 $k=0.5$ 时(波长约 200 km)，位相速约 0.2 m/s，增长到 e 倍的时间约 1.5 天。从这两个例子的比较可知，后面这个例子即风生流较强、较窄更符合观测事实。

8　风应力结构的影响

以上讨论是基于风应力为奇对称结构。为了研究风应力结构对波动的影响，取风应力结构为偶对称的 $\tau_y=\tau_y^0\dfrac{1}{2}x^2\mathrm{e}^{-\alpha x^2}$，这时特征风吹流速 $V_w=\Lambda x(1-\alpha x^2)\mathrm{e}^{-\alpha x^2}$。代入式(15)，式(16)，式(23)，解得

$$\sigma^2\left(\sigma^2-k^2\frac{-2A^3+3A^2+6A-5}{2A+5}\right)=0 \tag{55}$$

其中

$$A=\frac{\Lambda}{4\sqrt{2}}\left(2\left(\alpha+\frac{1}{2}\right)^{-\frac{3}{2}}-3\alpha\left(\alpha+\frac{1}{2}\right)^{-\frac{5}{2}}\right) \tag{56}$$

公式(55)表明，三支波中的一支是 $\sigma=0$ 的定常运动，其他两支均为非频散波动.当参数处在

$$\Delta=-\frac{2A^3+3A^2+6A-5}{2A+5}>0$$

时，波动稳定，分别为一对反向传播的行波；当参数处在 $\Delta<0$ 时，波动不稳定，其中一支为阻尼驻波，另一支为发展驻波。

9　结论

在海岸呈南北走向的半无界海洋中，在盛行纬向不均匀的经向风驱动下产生的急流型洋流，在一定条件下可以是不稳定的。不稳定意味着波动可以向涡旋发展。

本文应用等值浅水模式，在最大简化截断模的情况下，分析了动力系统的本征值问题。指出，不稳定波出现的条件依赖于风生急流的强度和宽度。当风生急流的宽度窄时，只有强的急流才有可能不稳定。

分析进一步指出，由北风应力产生的向南风生急流，在西边界附近才能出现不稳定，由南风应力产生的向北风生急流是稳定的。然而，由南风应力产生的向北风生急流，在海洋的东边界附近也可以出现不稳定。

如把这一简单理论用于南海，则表明在冬季北风驱动下，在西边界附近，即越南附近，出现不稳定波是可能的，进而可以预计，在非线性情况下，不稳定波的振幅进一步加大而发展成涡旋是可能的。

本文的结果是在线性等值浅水模式中应用最大简化截断模得到的，因此只能认为是带有一定启示的结果，进一步的研究需在这些启示下用数值模拟进行。

附录 1

$$A = 4A_1 k,$$

$$B = -8A_1^2 k^2 - 4A_0 A_1 k^2 - 4A_1 k^2 - A_1 + 5,$$

$$C = 8A_0 A_1^2 k^3 - 4A_0 k + A_1^2 k - 4A_1 k^3 - 5k + A_0 A_1 k + 8A_1^2 k^3 - A_1 k + 4A_1^3 k^3,$$

$$D = -4A_A^3 k^4 + 4A_1 k^4 - 4A_0 A_1^3 k^4 - A_0 A_1^2 k^2 + A_0 k^2 + 4A_0 A_1 k^4,$$

$$A_0 = \frac{\Lambda}{\sqrt{2}}\left(\left(\alpha + \frac{1}{2}\right)^{-\frac{1}{2}} - \alpha\left(\alpha + \frac{1}{2}\right)^{-\frac{3}{2}}\right), A_1 = \frac{\Lambda}{2\sqrt{2}}\left(\left(\alpha + \frac{1}{2}\right)^{-\frac{3}{2}} - 3\alpha\left(\alpha + \frac{1}{2}\right)^{-\frac{5}{2}}\right).$$

附录 2

$$P = b - \frac{a^2}{3},\ Q = \frac{2a^3}{27} - \frac{ab}{3} + c;\ a = \frac{B}{A},\ b = \frac{C}{A},\ c = \frac{D}{A}$$

参考文献

[1] Hsueh Y, Kenney R N.Steady coastal upwelling in a continuously stratified ocean. *J. Phys. Oceanogr.*, 1972, 2:27-33.

[2] Allen J S.Upwelling and coastal jet in a continuously stratified ocean. *J. Phys. Oceanogr.*, 1973, 3:245-257.

[3] Hu D X.A study of coastal upwelling off SE China.KeXue Tongbao, 1980, 25:159-163.

[4] 刘先炳，苏纪兰.浙江沿岸上升流和沿岸锋面的数值研究.海洋学报，1991,13:305-314.

[5] 巢纪平，陈显尧.沿岸上升流和沿岸急流的一个半解析理论.地球物理学报，2003，46(1):26-30.

[6] Wyrtki K.Physical Oceanography of the Southeast Asia Waters. *NAGA Report*, 1961, 2:1-195.

[7] Fang G H, Fang W D, Fang Y, et al.A survey of studies on the South China Sea upper ocean circulation. *Acta Oceanogr. Taiwanica*, 1998, 37:1-16.

[8] 刘秦玉，杨海军，刘征宇.南海 Sverdrup 环流的季节变化特征.自然科学进展，2000，10(11):1035-1039.

[9] 王卫强，王东晓，施 平.南海大尺度动力场年循环和年际变化.热带海洋学报，2001，20(1):61-68.

[10] 苏京志，卢 筠，侯一筠,等.南海表层流场的卫星跟踪浮标观测结果分析. 海洋与湖沼，2002，33(2):121- 127.

[11] Wang G H, Su J L, Chu P C.Mesoscale eddies in the South China Sea observed with altimeter data.*Geophys.Res. Lett.*, 2003, 30:doi:10.1029 2003GL018532.

[12] Pedlosky J.*Geophysical Fluid Dynamics.*New York, Heidelberg,Berlin:Springer-Verlag, 1979.

非线性陆架坡度作用下的陆架地形波*

冯立成[1] 巢纪平[1,2]

(1. 国家海洋环境预报中心, 北京 100081; 2. 国家海洋局第一海洋研究所, 青岛 266061)

提要:应用浅水模式研究了非线性陆架坡度作用下陆架地形波的性质, 指出在地形强迫下陆架海域存在两支陆架地形-Rossby 波和两支惯性重力波。其中向北传播的陆架地形-Rossby 波和惯性重力波在长波波段耦合为不稳定波动。随着陆架坡度的增大向北传播的陆架地形-Rossby 波仍在长波波段和惯性重力波耦合为不稳定波动, 但频率减小。将其应用于南海, 则表明在陆架地形的强迫下西边界附近会出现不稳定波动, 在非线性情况下, 不稳定振幅进一步增大发展成涡旋是可能的。实际观测中南海西边界存在强的边界流, 且南海中尺度涡旋发育丰富。环流的多涡结构的生消和迁移与南海环流变化关系密切。因此本文的研究给出了南海涡旋生成的一种可能机制。

关键词 非线性陆架坡度;陆架地形-Rossby 波;波动不稳定

陆架海域是海洋重要的组成部分, 陆架海中存在多种形式的运动,如陆架波、沿岸上升流、沿岸急流和陆架中尺度涡旋等。陆架海域的运动虽然是在 f 平面上进行的, 但由于陆架坡度可以起到类似于 β 的作用, 因而在 f 平面上的运动也具有某些 β 平面上运动的特色, 形成陆架波动[1]。Wang 等[2]及 Hong 和 Wang[3]还研究了非线性陆架坡与斜压层结的相互作用。Robinson[1]研究陆架波时, 指出陆架坡度可以激发出类似于 Rossby 波的慢波, 自此陆架波动成为陆架动力学研究热点之一。回顾以往的大部分工作都直接应用无辐散近似(如 Buchwald 和 Adams[4],Adams 和 Buchwald[5], Allen[6]),有的还加上对沿岸流的地转平衡, 即长波近似[7], 由于这些近似的引入,因此从未给出陆架波系中应有的成员。巢纪平和徐小标[8,9]利用浅水模式研究了陆架波动系统成员, 指出陆架区域存在着惯性重力波、Kelvin 波、Rossby 波和有条件的 Rossby 重力混合波, 并提出了一种 El Niño 信号传播的机制。简单起见, 上述研究中陆架地形取线性变化, 实际地形多呈非线性变化, 因而有些波动特征没有揭示出来。本文将在上述工作基础上, 进一步讨论非线性陆架坡度下的波动。

1 基本运动方程

设海洋的深度为 h_s+h_c,h_s 为海面扰动高度,$h_c(x)$为陆架的高度(海底至平均海平面). 海岸线沿 y 方向,x 轴垂直于海岸, 海盆的特征宽度为 L。不考虑风应力的作用, 在 f 平面上线性化的浅水运动方程为

$$\frac{\partial v}{\partial t}+fu=-g'\frac{\partial h}{\partial y} \tag{1}$$

* 中国科学:地球科学,2013,43(3):499-502.

国家自然科学基金(批准号: 40906014, 40976015)和国家海洋局青年海洋科学基金(编号: 2010218)资助。

$$\frac{\partial u}{\partial t} - fv = - g' \frac{\partial h}{\partial x} \tag{2}$$

$$\frac{\partial h}{\partial t} + H\left(\frac{\partial u}{\partial x} + \frac{\partial v}{\partial y}\right) + \frac{\mathrm{d}h_c}{\mathrm{d}x}u = 0 \tag{3}$$

式中,g'为视重力,未加微分号的 h 已用 H 代替。引进重力波波速:

$$C = \sqrt{g'H} \tag{4}$$

取特征量

$$t \propto 1/f,\ (x,y) \propto C/f,\ (u,v) \propto C,\ h \propto H,\ h_c \propto H \tag{5}$$

于是有无量纲方程

$$\frac{\partial v}{\partial t} + u = - \frac{\partial h}{\partial y} \tag{6}$$

$$\frac{\partial u}{\partial t} - v = - \frac{\partial h}{\partial x} \tag{7}$$

$$\frac{\partial h}{\partial t} + \left(\frac{\partial u}{\partial x} + \frac{\partial v}{\partial y}\right) + \frac{\mathrm{d}h_c}{\mathrm{d}x}u = 0 \tag{8}$$

由式(6),式(7)得到

$$\frac{\partial^2 v}{\partial t^2} + v = \frac{\partial h}{\partial x} - \frac{\partial^2 h}{\partial t \partial y} \tag{9}$$

$$\frac{\partial^2 u}{\partial t^2} + u = - \frac{\partial h}{\partial y} - \frac{\partial^2 h}{\partial t \partial x} \tag{10}$$

式(8)乘以$\left(\frac{\partial^2}{\partial t^2}+1\right)$后,再利用式(9)和式(10),最后有

$$\left[\frac{\partial^2}{\partial t^2} + 1 - \left(\frac{\mathrm{d}h_c}{\mathrm{d}x}\right)\frac{\partial}{\partial x} - \left(\frac{\partial^2}{\partial y^2} + \frac{\partial^2}{\partial x^2}\right)\right]\frac{\partial h}{\partial t} - \left(\frac{\mathrm{d}h_c}{\mathrm{d}x}\right)\frac{\partial h}{\partial y} = 0 \tag{11}$$

考虑到实际的陆架坡度在近岸会很陡,取无量纲形式为

$$h_c = 1 - e^{-\alpha x} \tag{12}$$

2 色散关系

设解的形式为

$$h = \hat{h}(x)\mathrm{e}^{i(ky+\sigma t)} \tag{13}$$

并作坐标变换,令

$$\xi = \mathrm{e}^{-\alpha x} \tag{14}$$

式(11)变为

$$\xi^2 \frac{\mathrm{d}^2 \hat{h}}{\mathrm{d}\xi^2} - \xi(\xi - 1)\frac{\mathrm{d}\hat{h}}{\mathrm{d}\xi} + \left(\frac{\sigma^2 - k^2 - 1}{\alpha^2} + \frac{k}{\alpha\sigma}\xi\right)\hat{h} = 0 \tag{15}$$

此方程的边界条件,在岸边 $u=0$,由式(10)给出

$$\xi = 1,\quad \frac{\mathrm{d}\hat{h}}{\mathrm{d}\xi} - \frac{k}{\alpha\sigma}\hat{h} = 0 \tag{16}$$

在离岸远处

$$\xi \to 0, \quad \hat{h} \to 0 \tag{17}$$

方程(15)在近岸区域可求得解析解。这时 $\xi \to 1$,方程(15)近似地有

$$\xi^2 \frac{\mathrm{d}^2 \hat{h}}{\mathrm{d}\xi^2} + \left(\frac{\sigma^2 - k^2 - 1}{\alpha^2} + \frac{k}{\alpha\sigma}\xi \right) \hat{h} = 0 \tag{18}$$

作变换(见参见文献[10]):

$$a = \frac{\sigma^2 - k^2 - 1}{\alpha^2}, \quad b = \frac{k}{\alpha\sigma},$$

$$l = \frac{1}{2}\left(1 - \sqrt{1 - \frac{4}{\alpha^2}(\sigma^2 - k^2 - 1)}\right), \quad \eta = 2\sqrt{b\xi} \tag{19}$$

$$\hat{h} = \xi^l \phi(\eta) \tag{20}$$

由式(19),为了保证 l 是实数,要求

$$\sigma^2 < \frac{\alpha^2}{4} + k^2 + 1 = \sigma_c^2 \tag{21}$$

即上述变化只对低频成立,因此在这个意义上可称为低频近似。这时有方程

$$\eta \frac{\mathrm{d}^2\phi}{\mathrm{d}\eta^2} + (4l - 1)\frac{\mathrm{d}\phi}{\mathrm{d}\eta} + \eta\phi = 0 \tag{22}$$

方程(22)有一特解:

$$\phi = \eta^{1-2l} J_p(\eta) \tag{23}$$

为 p 阶 Bessel 函数,式中 $p = 1-2l$。

注意到,$x \to \infty$, $\xi \to 0$ 时 $\eta \to 0$,因此这个函数 $J_p \to 0$,它自动满足无穷远处的条件。在 $x=0$ 处,条件式(16)在新变量下写成:

$$\frac{\mathrm{d}\phi}{\mathrm{d}\eta} + \left(\frac{2l}{\eta} - \frac{\eta}{2}\right)\phi = 0 \tag{24}$$

由于 $x=0, \xi=1$ 时,$\eta = 2\sqrt{b}$,故有

$$\frac{\mathrm{d}\phi}{\mathrm{d}\eta} + \left(\frac{1}{\sqrt{b}} - \sqrt{b}\right)\phi = 0 \tag{25}$$

取解为

$$\phi = \eta^{1-2l} J_p(\eta) \tag{26}$$

将式(26)代入式(25),给出

$$\frac{\mathrm{d}J_p}{\mathrm{d}\eta} + \left(\frac{(1-2l)}{\eta} + \left(\frac{l}{\sqrt{b}} - \sqrt{b}\right)\right) J_p(\eta) = 0 \tag{27}$$

注意到,对任意阶的 Bessel 函数 C_v 有公式

$$\eta \frac{\mathrm{d}C_v}{\mathrm{d}\eta} + vC_v(\eta) = \eta C_{v-1}$$

因此

$$\frac{\mathrm{d}J_p}{\mathrm{d}\eta} = J_{p-1}(\eta) - \frac{p}{\eta} J_p(\eta)$$

将此式代入式(27),并当 $x=0, \eta=2\sqrt{b}$ 时,给出

$$J_{p-1}(\eta)+\left(\frac{(1-2l)}{2\sqrt{b}}+\left(\frac{1}{\sqrt{b}}-\sqrt{b}-\frac{p}{2\sqrt{b}}\right)\right)J_p(\eta)=0 \tag{28}$$

重新写出式中参数,分别为

$$b=\frac{k}{\alpha\sigma},l=\frac{1}{2}\left(1-\sqrt{1-\frac{4}{\alpha^2}(k^2-\sigma^2-1)}\right),$$

$$p=1-2l=\sqrt{1-\frac{4}{\alpha^2}(\sigma^2-k^2-1)} \tag{29}$$

因此式(28)是当海底地形呈指数分布时在近岸区的色散公式。

当陆架坡度陡时(α 较大),可求出式(28)的近似色散关系。

即取近似

$$p=\sqrt{1-\frac{4}{\alpha^2}(\sigma^2-k^2-1)}\approx 1 \tag{30}$$

式(28)的近似形式为

$$J_0(\eta)+\left(\frac{(1-2l)}{2\sqrt{b}}+\left(\frac{1}{\sqrt{b}}-\sqrt{b}-\frac{1}{2\sqrt{b}}\right)\right)J_1(\eta)=0 \tag{31}$$

应用 Bessel 函数展开式的最低阶项,给出

$$1+\left(\frac{1-p}{2}-b\right)=0 \tag{32}$$

将各符号值代入后,得到

$$\sigma^4+(2\alpha^2-k^2-1)\sigma^2-3k\alpha\sigma+k^2=0 \tag{33}$$

这是一个关于 σ 的一元四次方程,其解法见参考文献[11]。

3 波动色散性质

由上面得到的色散关系式可见波动的稳定性与陆架坡度 α 相关。图 1(a)给出了南海(SCS)沿 12°N 断面的地形(资料来自 Etopo5),图 1(b)给出了不同 α 下的理想化地形图。

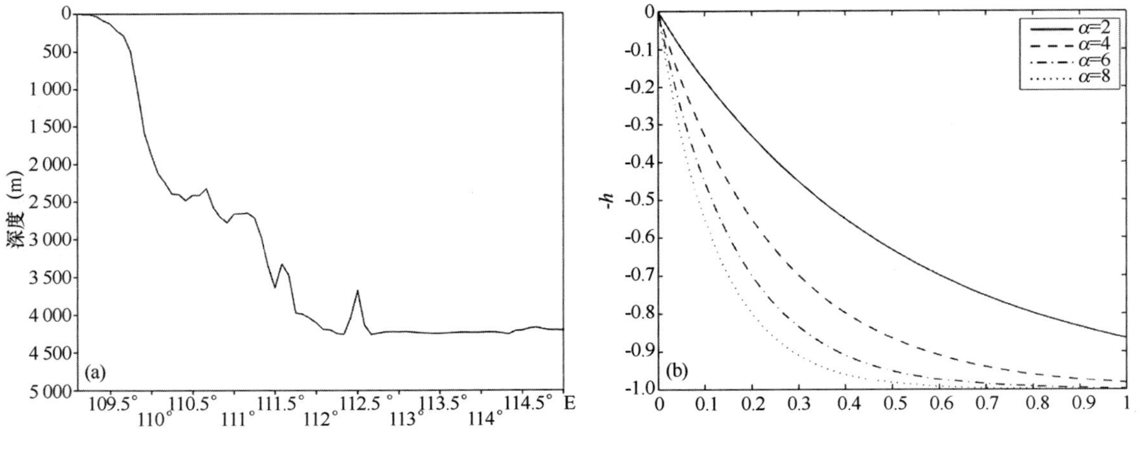

图 1 中国南海沿 12°N 断面地形(a)及理想化地形(b)

图 2 给出了不同 α 取值下的式(33)求得的波动。

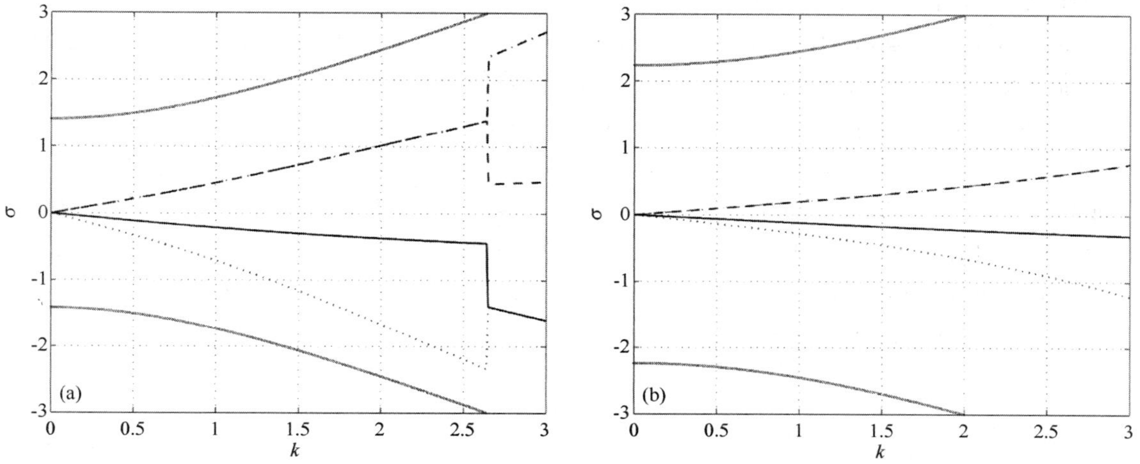

图 2　波动色散关系

(a)和(b)分别为 $\alpha=2$ 及 $\alpha=4$ 时的色散关系；横坐标 k 为无量纲波数，纵坐标 σ 为无量纲频率；图中空心圆线为 σ_c

由图 2 可见，此系统中存在一支向南传播的陆架地形-Rossby 波(实线)，一支向南传播的惯性重力波(点线)，一支向北传播的陆架地形-Rossby 波(虚线)和一支向北传播的惯性重力波(点划线)，且均满足式(21)的低频条件。其中向北传播的陆架地形-Rossby 波和惯性重力波在长波波段耦合为不稳定波动，向南传播的陆架地形-Rossby 波和惯性重力波在短波波段耦合为不稳定波动。随着陆架坡度 α 的增大(图 2(b))向北传播的陆架地形-Rossby 波仍在长波波段和惯性重力波耦合为不稳定波动，但频率减小。向南传播的陆架地形-Rossby 波和惯性重力波则在长波波段保持稳定，且频率减小。

4　讨论和结论

本文应用线性化的浅水模式通过一系列的变化得到了色散关系方程，研究了非线性陆架坡度下陆架地形波的性质，指出在地形强迫下陆架海域存在两支陆架地形-Rossby 波和两支惯性重力波。其中向北传播的陆架地形-Rossby 波和惯性重力波在长波波段耦合为不稳定波动。

实际观测表明中国南海有两个重要的特征：一个是西边界存在强的边界流；另一个是多涡结构，两者之间存在密切联系。南海存在非线性陆架坡度[图 1(a)]，因此西边界有可能出现不稳定波动，并进一步发展成涡旋. 本文的研究从陆架地形波和惯性重力波耦合的角度给出了南海涡旋生成的一种可能机制。需要指出的是上述结果是在简化模型下得到的，因而还需要用原始方程模式进行进一步的验证。

参考文献

[1]　Robinson A R. Continental shelf waves and the response of sea level to weather system. J Geophys Res, 1964, 69: 367-368.

[2]　Wang D X, Hong B, Gan J P, et al. Numerical investigation on propulsion of the counter-wind current in the northern South China Sea in winter. Deep-Sea Res Part I-Oceanogr Res Pap, 2010, 57: 1206-1221.

[3] Hong B, Wang D X. Sensitivity study of the seasonal mean circulation in the northern South China Sea. Adv Atmos Sci, 2008, 25: 824-840.doi: 10.1007/s00376-008-0824-8.
[4] Buchwald V T, Adams J K. The propagation of shelf waves. Proc R Soc A-Math Phys Eng Sci, 1968, 305: 225-250.
[5] Adams J K, Buchwald V T. The generation of shelf waves. J Fluid Mech, 1969, 35: 815-826.
[6] Allen J S. On forced, long continental shelf waves on an f-plane. J Phys Oceanogr, 1976, 6: 426-431.
[7] Gill A E. Schumann E H. The generation of long shelf waves by wind. J Phys Oceanogr, 1974, 4: 83-90.
[8] 巢纪平, 徐小标. 论陆架波系动力学及其在厄尔尼诺信号传播中的作用. 气象学报, 2001, 59: 515-523.
[9] 巢纪平, 徐小标. 一类陆架地形波系在 El Niño 信号传播中的作用. 自然科学进展, 2002, 12: 398-402.
[10] Murphy G M. Ordinary Differential Equations and Their Solution. Princeton: Van Nostrand, 1960, 401.
[11] 叶其孝, 沈永欢. 实用数学手册. 第 2 版. 北京: 科学出版社, 2006, 17-18.

风驱动下 f-平面准地转三维海洋环流的形成及总质量守恒的应用*

巢纪平[1,2]　冯立成[3,4]

(1. 国家海洋环境预报中心,北京 100081;2. 国家海洋局第一海洋研究所,青岛 266061;3. 中国科学院大气物理研究所大气科学和地球流体力学数值模拟国家重点实验室,北京 100029;4. 中国科学院研究生院,北京 100049)

摘要:将理想化的南海海盆在垂直方向上划分为 Ekman 层、惯性层和摩擦层。Ekman 层中的运动由大气风应力驱动,其底部的扰动压力将作为其下惯性层中运动的上边界条件。惯性层中的运动是由 f-平面三维非线性方程在准地转近似下位势涡度守恒控制,由此得到控制惯性层中运动关于扰动压力的三维椭圆型方程。在惯性层以下考虑到深层的海盆水平尺度很小,由此引进带有底部摩擦的线性控制方程,方程的边界条件为惯性层和摩擦层交界面上的扰动压力连续,沿海盆边界假定海水与相邻的固壁间无热量交换,由此设在海盆边界上扰动温度为零。在此基础上分别利用惯性层和摩擦层中的椭圆型控制方程计算了相应层次上冬、夏季的扰动压力和准地转流。结果表明冬季各层上以气旋式环流为主,且随深度的增加流速减小;夏季各层上以反气旋式环流为主,流速也随深度增加而减小。这在一定程度上与观测事实相符。

关键词:南海;准地转环流;斜压惯性运动;海水总质量守恒

1　引言

南海是西太平洋最大的边缘海(图 1),其东为吕宋岛,西部和北部为亚洲大陆,南为印度尼西亚,形成一个半封闭海盆。在其东半部为深水海域,位于 4°—22°N 及 110°—120°E 的水体平均深度约 2 000 m,西北海域为宽幅 200~300 km 的大陆架,西南海域则为较宽但水深较浅的巽他陆架与泰国湾。南海通过窄而浅的水道同东海、苏禄海、爪哇海及印度洋相连。与西太平洋的连接则为宽而深的吕宋海峡,其南北距离约为 400 km,水深在海峡中央可达 2 000 m 以上。

基于早期的水文观测、海表记录和船载浮标资料,Wyrtki[1] 对南海环流进行了最早的系统性海洋学研究,指出南海中上层环流主要受季节性变化的季风系统及通过吕宋海峡的黑潮的影响。随后许多学者的工作证实了这一点。如 Shaw[2] 的研究表明季节性反转的季风在上层环流的形成中起着重要的作用。Chu 等[3] 在 Wyrtki[1],Hellerman 等[4] 研究结果的基础上进一步指出在 4—8 月,弱的西南季风(~0. 1 N/m^2)使得越南外海产生向北的沿岸急流,南海中部形成反气旋式环流;11 月到翌年 3 月,强的东北季风(~0.3 N/m^2)驱动越南外海向南的沿岸急流,并在南海中部形成气旋式环流。通过

* 地球物理学报,2007,50(5):1319-1329.

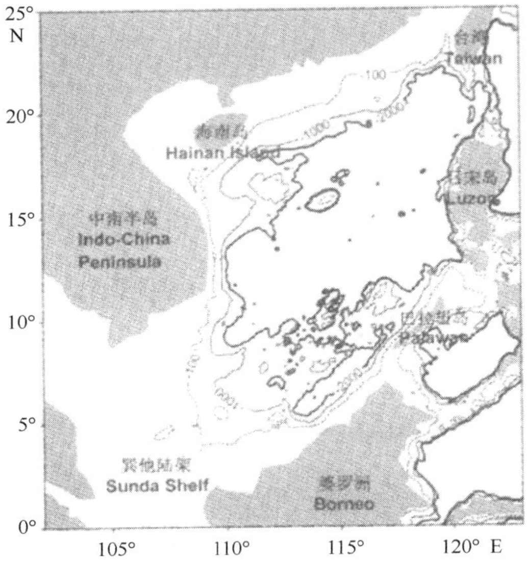

图 1 南海地形及等深线(单位:m)

吕宋海峡的质量输送及黑潮以何种形式影响南海环流尚无定论,不同的作者给出了不同的结论[1,5~8]。

由于对流场的直接观测资料很少[9],以往的研究多用历史水文资料[1,10,11]或海洋数值模式进行模拟[2,3,7,12],近年来卫星高度计资料也被一些学者应用到南海环流的研究中[13,14]。但从动力学的角度对南海环流进行的研究工作则较为欠缺,并且由于资料的缺乏和模式的局限性几乎没有对南海深层环流做过分析研究。巢纪平等[15]通过分析动力系统的本征值问题,给出了南海西边界急流形成的初步解释。刘秦玉等[16]和 Liu 等[17]从 β-平面近似下斜压 Rossby 波调整的角度研究了季风驱动下的海洋环流,提出了南海海盆尺度环流满足 Sverdrup 平衡,风驱动是冬、夏季海洋环流的主要形成机制。本文尝试不考虑斜压 Rossby 波的调整过程,只注重风驱动的结果,采用 f-平面准地转动力学来研究季节以上尺度南海环流。

本文试图用一个理想化的 f-平面准地转模型对南海三维环流的形成进行理论尝试。将理想化的南海海盆在垂直方向上划分为 Ekman 层、惯性层和摩擦层,Ekman 层是一浅层,其中运动受大气风应力驱动,其下是由理想流体非线性方程控制的斜压准地转运动,运动遵从若干守恒律,再下面由于海盆的水平尺度愈来愈小,海盆边界的侧向摩擦变得重要而形成一个摩擦层。在这个理论模式中,考虑了海水在南边与大洋的交换以及在吕宋海峡与黑潮水的交换,并假定进、出南海海水的总质量是守恒的。在这三个不同物理过程耦合的分层模式的基础上,半解析、半数值地计算了冬、夏各层上的扰动压力及准地转流。

2 模式结构

考虑当冬季盛吹北风时,有一部分海水在南边流入印度洋,但又有一部分海水从吕宋海峡流出。一是为了满足连续性要求;二是由运动造成的动压力分布在吕宋海峡形成内外压力差,驱使黑潮水流入南海。夏季在南风应力的作用下,海水运动方向与冬季相反,从南边流入南海,并从吕宋海峡流出。

作为理论研究不可能对完全真实的南海海盆进行实地研究,这里取南海的主体部分(4°~22°N,

110°～120°E）作为研究区域。设这一区域的北边界和西边界均为固体边界，南边界是完全开放的，东边界的东北部（吕宋海峡处）有一开口，其余部分则为固体边界。半封闭海的区域为：$x(0,L)$，$y(0,-Y)$，L 表示 x 轴方向长度，Y 为 y 轴方向长度。

在垂直方向上将海洋分成三层，表层是浅的 Ekman 层，厚度为 D，受风应力驱动，由 Ekman 抽引在底部造成垂直运动，进而底部压力场或高度场的不均匀驱动中层运动。中层（惯性层）是理想流体，厚度为 H_g，运动是非线性的，应用物理量的惯性守恒律来建立控制方程。在惯性层以下由于海盆尺度较小，海盆边界由地形产生的摩擦作用很强，这时惯性解失效，因此设为摩擦层，厚度为 H_f，由摩擦控制下的运动来构成控制方程。设 Ekman 层和惯性层交界处 $z=0$，则惯性层和摩擦层的交界面为 $z=-H_g$，在交界面上压力场或高度场连续。

3　Ekman 层的运动方程

对于定常、线性、黏性的流体，其运动方程为

$$-fv=-\frac{\partial p'}{\partial x}+v\frac{\partial^2 u}{\partial z^2} \tag{1}$$

$$fu=-\frac{\partial p'}{\partial y}+v\frac{\partial^2 v}{\partial z^2} \tag{2}$$

$$0=-\frac{\partial p'}{\partial z}+g\alpha\vartheta \tag{3}$$

$$\frac{\partial u}{\partial x}+\frac{\partial u}{\partial y}+\frac{\partial w}{\partial z}=0 \tag{4}$$

$$\Gamma w=-\frac{1}{\tau}\vartheta \tag{5}$$

式中，v 为湍流黏滞系数；$p'=p/\rho_0$ 为静压力的偏差；α 是流体的热膨胀系数；τ 为牛顿冷却的张弛时间；$\Gamma=\frac{\mathrm{d}T}{\mathrm{d}z}$为基本温度的垂直递减率（在此取成常数）；$\vartheta$ 为扰动温度。

方程（1）和方程（2）对 x、y 交叉微分并由后者减去前者得到

$$f\left(\frac{\partial u}{\partial x}+\frac{\partial v}{\partial y}\right)=v\frac{\partial^2}{\partial z^2}\left(\frac{\partial v}{\partial x}-\frac{\partial u}{\partial y}\right) \tag{6}$$

应用连续性方程（4）并假定：在海表 $z=D$，$w=0$，$v\partial u/\partial z=\tau^z$，$v\partial v/\partial z=\tau^y$；在 Ekman 层底 $z=0$，$w=w_e$，$v\partial u/\partial z=v\partial v/\partial z=0$，并且如果只考虑经向风应力的作用，得到

$$z=0,\quad w_e=\frac{1}{f}\left(\frac{\partial\tau^y}{\partial x}\right) \tag{7}$$

此即 Ekman 抽引。将式（5）置换 w_e，给出

$$z=0,\quad \vartheta=-\frac{\Gamma\tau}{f}\left(\frac{\partial\tau^y}{\partial x}\right) \tag{8}$$

式（7），式（8）表明，气旋性风应力旋度在 Ekman 层底（温跃层顶）产生上升流或负温度距平。

进而，应用式（3）得

$$Z=0,\quad \frac{\partial p'}{\partial z}=-\frac{g\alpha\Gamma\tau}{f}\left(\frac{\partial\tau^y}{\partial x}\right) \tag{9}$$

如设在这一高度上运动已具有准地转性，即

$$v = \frac{1}{f}\frac{\partial p'}{\partial x} \tag{10}$$

则

$$\frac{\partial v}{\partial z} = -\left(\frac{g\alpha\Gamma\tau}{f^2}\right)\frac{\partial^2\tau^y}{\partial x^2} \tag{11}$$

进一步设在海表 $z=D, v=0$，在 Ekman 层底 $z=0, v=v_w$，从而得到

$$|y| > \delta, \quad v_w = \left(\frac{g\alpha D\Gamma\tau}{f^2}\right)\frac{\partial^2\tau^y}{\partial x^2} \equiv \Lambda\frac{\partial^2\tau^y}{\partial x^2} \tag{12}$$

关于风应力在研究区域中可设为 $\tau^y=\tau_0^y e^{-\alpha x}$。

注意到 $y=0$ 是区域的北边界，由向南（北）流必然引起边界附近的涌升流（下沉流），但涌升流（下沉流）主要发生在边界附近的侧向边界层中。设边界层的宽度为 δ，则涌升流（下沉流）可从下面的公式中计算出。由连续性方程，并考虑前面假定的 Ekman 层上、下层垂直速度，可得

$$\frac{\partial v}{\partial y} = -\frac{\partial w}{\partial z} \approx \frac{w_e}{D} \tag{13}$$

设涌升流（下沉流）的分布为

$$|y| < \delta, \quad w_e = w_e^0 e^{y/\delta} e^{-\alpha x} \tag{14}$$

由此得侧向边界层中的离岸流（向岸流）为

$$|y| < \delta, \quad v_b = -\frac{\delta}{D} w_e^0 (1 - e^{y/\delta}) e^{-\alpha x} \tag{15}$$

注意到固壁上无质量流出（流入）$y=0, v_b=0$。由于离岸流（向岸流）实际上由风生流引起，因此它在 x 方向的分布与风生流相同。

取重力波波速为 $C=\sqrt{g\alpha\Gamma H^2}$，$v_b \sim C, x, y \sim l = C/f$（Rossby 变形半径），$w_e^0 \sim \frac{D}{\delta}C, \tau^y \sim \tau_0^y, v_w \sim (D/H)(\tau\tau^0/H)$，则式（12）无量纲化后得

$$v_w = a\alpha^2 e^{-\alpha x} \tag{16}$$

式（15）无量纲化得

$$v_b = -b(1 - e^{y/\delta}) e^{-\alpha x} \tag{17}$$

设在 $|y|=\delta$ 处，风生流与离岸流（向岸流）衔接，则有

$$v_w = v_b, \quad a\alpha^2 = -b(1 - e^{-1}) \tag{18}$$

其中，a, b 为可调因子，确保 v_b, v_w 的量级在合理范围内。

4 惯性层中的运动方程

对于斜压、非线性、定常的理想流体，其运动方程为

$$u\frac{\partial u}{\partial x} + v\frac{\partial u}{\partial y} - fv = -\frac{\partial p'}{\partial x} \tag{19}$$

$$u\frac{\partial v}{\partial x} + v\frac{\partial v}{\partial y} + fu = -\frac{\partial p'}{\partial y} \tag{20}$$

$$0 = -\frac{\partial p'}{\partial z} + g\alpha\vartheta \tag{21}$$

$$\frac{\partial u}{\partial x}+\frac{\partial v}{\partial y}+\frac{\partial w}{\partial z}=0 \tag{22}$$

$$u\frac{\partial \vartheta}{\partial x}+v\frac{\partial \vartheta}{\partial y}+\Gamma w=0 \tag{23}$$

方程(19)、方程(20)对 x、y 交叉微分,并由后者减去前者得

$$u\frac{\partial \Omega}{\partial x}+v\frac{\partial \Omega}{\partial y}+(\Omega+f)\left(\frac{\partial u}{\partial x}+\frac{\partial v}{\partial y}\right)=0 \tag{24}$$

式中

$$\Omega=\frac{\partial v}{\partial x}-\frac{\partial u}{\partial y} \tag{25}$$

为相对涡度。如果 $u,v\sim10^{-1}$ m/s,$x,y\sim10^{5}$ m,估计出 $\Omega\sim10^{-6}$/s,而 $f\sim10^{-4}$/g,因此 $\Omega\ll f$,式(24)在考虑了式(22)后为

$$u\frac{\partial \Omega}{\partial x}+v\frac{\partial \Omega}{\partial y}-f\frac{\partial w}{\partial z}=0 \tag{26}$$

对于水平尺度大于 100 km 的运动可引进地转近似

$$u=-\frac{1}{f}\frac{\partial p'}{\partial y},\quad v=\frac{1}{f}\frac{\partial p'}{\partial x} \tag{27}$$

由此

$$\Omega=\frac{1}{f}\nabla^{2}p' \tag{28}$$

式(26)改写成

$$J\left(p',\frac{1}{f}\nabla^{2}p'\right)=f^{2}\frac{\partial w}{\partial z} \tag{29}$$

考虑式(21),式(23)改写成

$$J\left(p',\frac{\partial p'}{\partial z}\right)=-fg\alpha\Gamma w \tag{30}$$

再对 z 微分得到

$$J\left(p',\frac{\partial^{2}p'}{\partial z^{2}}\right)=-fg\alpha\Gamma\frac{\partial w}{\partial z} \tag{31}$$

与式(29)联合给出

$$J\left(p',\nabla^{2}p'+\frac{f}{g\alpha\Gamma}\frac{\partial^{2}p'}{\partial z^{2}}\right)=0 \tag{32}$$

由此得到

$$\nabla^{2}p'+\frac{f^{2}}{g\alpha\Gamma}\frac{\partial^{2}p'}{\partial z^{2}}=F(p') \tag{33}$$

F 为 p' 的普适函数,设为 rp'

取 $z\sim H,x,y\sim l=C/f$,式中 C 的定义同上,考虑到下层流体有其自身的动力学约束,Ekman 层的影响通过上边界条件起作用,因此取 $u,v\sim C,w\sim\frac{H}{l}C,p'\sim C^{2},\vartheta\sim\Gamma H$。因此式(27)的无量纲形式为

$$u=-\frac{\partial p'}{\partial y},\quad v=\frac{\partial p'}{\partial x} \tag{34}$$

式(33)的无量纲形式为

$$\nabla^2 p' + \frac{\partial^2 p'}{\partial z^2} = rp' \tag{35}$$

5 摩擦层中的运动方程

$$-fv = -\frac{\partial p'}{\partial x} - Ku \tag{36}$$

$$fu = -\frac{\partial p'}{\partial y} - Kv \tag{37}$$

$$\frac{\partial u}{\partial x} + \frac{\partial v}{\partial y} + \frac{\partial w}{\partial z} = 0 \tag{38}$$

$$0 = -\frac{\partial p'}{\partial z} + g\alpha\vartheta \tag{39}$$

$$\Gamma w = -\frac{1}{\tau}\vartheta \tag{40}$$

式中,K 为摩擦系数。

式(36)、式(37)对 x、y 交叉微分,并由后者减去前者得

$$f\left(\frac{\partial u}{\partial x} + \frac{\partial v}{\partial y}\right) = -K\left(\frac{\partial v}{\partial x} - \frac{\partial u}{\partial y}\right) \tag{41}$$

考虑连续性方程(38),右端涡度仍采用准地转近似,式(41)改写为

$$\frac{f^2}{K}\frac{\partial w}{\partial z} = \frac{\partial^2 p'}{\partial x^2} + \frac{\partial^2 p'}{\partial y^2} \tag{42}$$

联合式(39)、式(40)得到

$$\frac{\partial w}{\partial z} = -\frac{1}{g\alpha\tau\Gamma}\frac{\partial^2 p'}{\partial z^2} \tag{43}$$

将式(43)代入式(42),得到控制方程

$$\left(\frac{f^2}{g\alpha\tau\Gamma K}\right)\frac{\partial^2 p'}{\partial z^2} + \frac{\partial^2 p'}{\partial x^2} + \frac{\partial^2 p'}{\partial y^2} = 0 \tag{44}$$

取 $z \sim H, x, y \sim l = C/f$,式中 C 的定义同上,式(44)无量纲化后得

$$\left(\frac{1}{\tau K}\right)\frac{\partial^2 p'}{\partial z^2} + \frac{\partial^2 p'}{\partial x^2} + \frac{\partial^2 p'}{\partial y^2} = 0 \tag{45}$$

6 质量守恒

对于半封闭海南端的坐标为 $y \to -Y$,在那里海洋是开放的,但只有经向流流入或流出半封闭海,而经向速度的分布设为

$$v_{-Y} = v_0 c_s \frac{(z + kH)^m}{H^m} e^{-\beta x} \tag{46a}$$

其中,c_s 为可调参数,式(46a)的无量纲形式为

$$v_{-Y} = v_0 c_s (z + k)^m e^{-\beta x} \tag{46b}$$

式中，v_0 是无量纲的，即已经除去了 C。再利用式(34)可得南边界的扰动压力分布

$$p'|_{y=-Y} = v_0 c_a (z+k)^m \frac{1-e^{-\beta x}}{\beta} + p'_{b0} \tag{47}$$

从海洋南边开口处即 $y=-Y$ 处，如不考虑 Ekman 层(很浅)的作用，则单位时间内流出的水量为

$$M_{y=-Y} = \int_0^{L/l}\int_0^{-k} v_{y=-Y}(x,z)\,dxdz \tag{48}$$

考虑式(46b)后得

$$M_{y=-Y} = -v_0 c_a \frac{k^{m+1}}{m+1}\frac{1-e^{-\beta\frac{L}{l}})}{\beta} \tag{49}$$

这部分从半封闭海南端流出大洋的质量，在保持海洋总质量守恒的条件下，需要从右上角开口处(吕宋海峡)流入相同的质量。

设吕宋海峡的流场形式为

$$u_{\frac{L}{l}} = -u_0 c_e \frac{(z+kH)^m}{H^m}\frac{y}{4l}\left(1+\frac{y}{4l}\right) \tag{50a}$$

其中，c_e 为可调参数，式(50a)的无量纲形式为

$$u_{\frac{L}{l}} = -u_0 c_e (z+k)^n \frac{y}{4}\left(1+\frac{y}{4}\right) \tag{50b}$$

这样从海洋东边开口处(宽为 y_0)，不考虑 Ekman 层的作用，单位时间内流入的水量为

$$N_{z=\frac{L}{l}} = \int_0^{-y_0}\int_0^{-k} u(y,z)\,dydz = u_0 c_e \frac{k^{n+1}}{n+1}\frac{y_0^2(6+y_0)}{16\times 3} \tag{51}$$

由总水量守恒有

$$M_{y=-y} + N_{z=\frac{L}{l}} = 0 \tag{52}$$

即

$$v_0 c_s \frac{k^{m+1}}{m+1}\frac{(1-e^{\beta\frac{L}{l}})}{\beta} = u_0 c_e \frac{k^{n+1}}{n+1}\frac{y_0^2(6+y_0)}{16\times 3} \tag{53}$$

7　边界条件

7.1　惯性层中边界条件

设在海洋西边界无海水进入，即

$$x=0,\quad u_{z=0}=0 \rightarrow p'_{z=0}=p'_0 \rightarrow \text{cons} \tag{54}$$

在海洋北边界也没有海水流入，即

$$y=0,\quad v_{y=0}=0 \rightarrow p'_{y=0}=p'_0 \rightarrow \text{cons} \tag{55}$$

令扰动压力等于 p'_0 是为了使得 $x=0,y=0$ 处的扰动压力连续。海洋东边界的扰动压力，在吕宋海峡即 $y\in(-y_0,0)$ 处，由式(50b)利用地转关系求得

$$x=\frac{L}{l},\ p'_l = u_0 c_e (z+k)^n \frac{y^2(6+y)}{16\times 3} \tag{56}$$

在固体边界部分即 $y\in(-Y,-y_0)$ 处为

$$x=\frac{L}{l},\ p'_e = u_0 c_e (z+k)^n \frac{y_0^2(6+y_0)}{16\times 3} \tag{57}$$

海洋南边界的扰动压力由式(47)决定。

上边界风生流部分的扰动压力可由式(16)根据地转关系得到

$$p'_w = a\alpha(1 - e^{-\alpha x}) \tag{58}$$

离岸流部分的扰动压力由式(17)根据地转关系得到

$$p'_b = -b(1 - e^{y/\delta})\frac{1 - e^{-\alpha x}}{\alpha} \tag{59}$$

下边界与摩擦层相连接。

7.2 摩擦层中边界条件

摩擦层上边界和惯性层下边界的扰动压力连续。即 $z=-H_g, p'_g=p'_f$。

摩擦层中下边界及侧边界均为固壁,可近似认为是绝热的,即沿固壁 $\partial\vartheta/\partial n=0$。但由于本文采用了实际地形资料,边界条件取为 $\partial\vartheta/\partial n=0$ 在计算上难以实现,因此取沿海盆扰动温度 $\vartheta=0$ 来近似。

8 计算实例

8.1 冬季扰动压力及流场

首先考虑冬季的情况,对于研究区域 4°—22°N,110°—120°E,其东西方向长 L=1 000 km,南北方向宽 Y=2 000 km,惯性层与摩擦层交界面高度设为 z=−1 000 m。假定惯性层中完全为水体,摩擦层则利用相应区域真实的海底地形边界。其余参数取值见表 1。计算中水平方向取 Δx=10 km,Δy=10 km,垂直方向 Δz=200 m,即将水平方向划分为 100×200 个网格,垂直方向划分为 26 层。利用第 7 节给出的边界条件及控制方程(35)和方程(45)求得区域各网格点上的 p',再利用地转关系得到相应的 u,v 场。

表 1 试验参数设置

季节	u_0	v_0	y_0	α	β	δ	k
冬	-0. 3	-0.1	-4.0	0.795	0.046	0.5	1.0
夏	0.15	0.08	-4.0	0.624	0.142	0.5	1.0
	a	b	m	n	c_e	c_s	r
冬	-1.0	1.0	2	2	4.0	1.0	1.0
夏	06	-0.4	2	2	4.0	1.0	-1.0

图 2 为惯性层中的流场及扰动压力图,可见表层除了北边界侧向边界层中由于离岸流的经向变化而由地转近似引起的纬向流外,其余均为向南的经向流,流速自西向东逐渐减小,流向与等 p'线相平行。在−200 m 层上,区域内出现 p'的闭合低值中心,区域南部出现向东流,东部出现向北流,这样整个区域形成一个闭合气旋式环流。再往下这一环流更加明显,并且 p'场和流场随深度增加开始变得不规则起来,凸显出地形的作用。下部海水较深处对应于较大的 p'绝对值,较浅处对应于较小的 p'绝对值,在−800 m 层上,在一些原来岛屿所在位置的周围甚至形成了小的反气旋式环流。总的来说,从上往下 p'的绝对值减小,流速减小,地形作用加强。

图 3 为摩擦层中的流场及扰动压力图,图中灰色阴影区域为海底。在−1 000 m 层上,由于实际地形的加入,导致一些岛屿附近扰动压力梯度加大,因而根据地转关系计算出的流场也出现一些虚假的

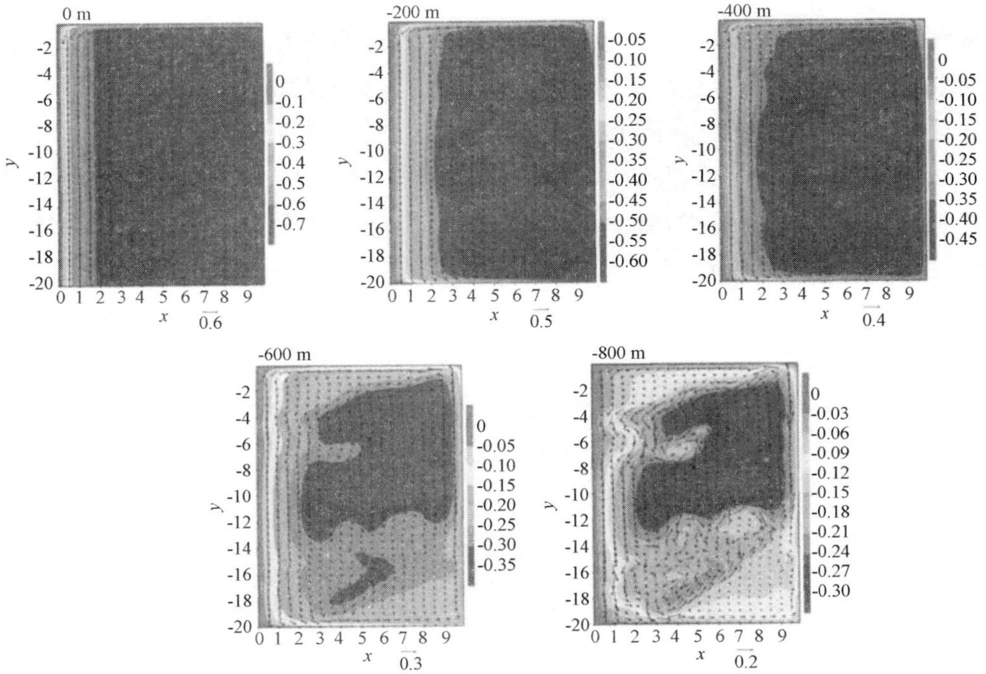

图 2　冬季惯性层中无量纲的准地转流及 p'场（见书后彩插）

惯性层和摩擦层交界面高度为-1 000 m，0 m 为惯性层顶

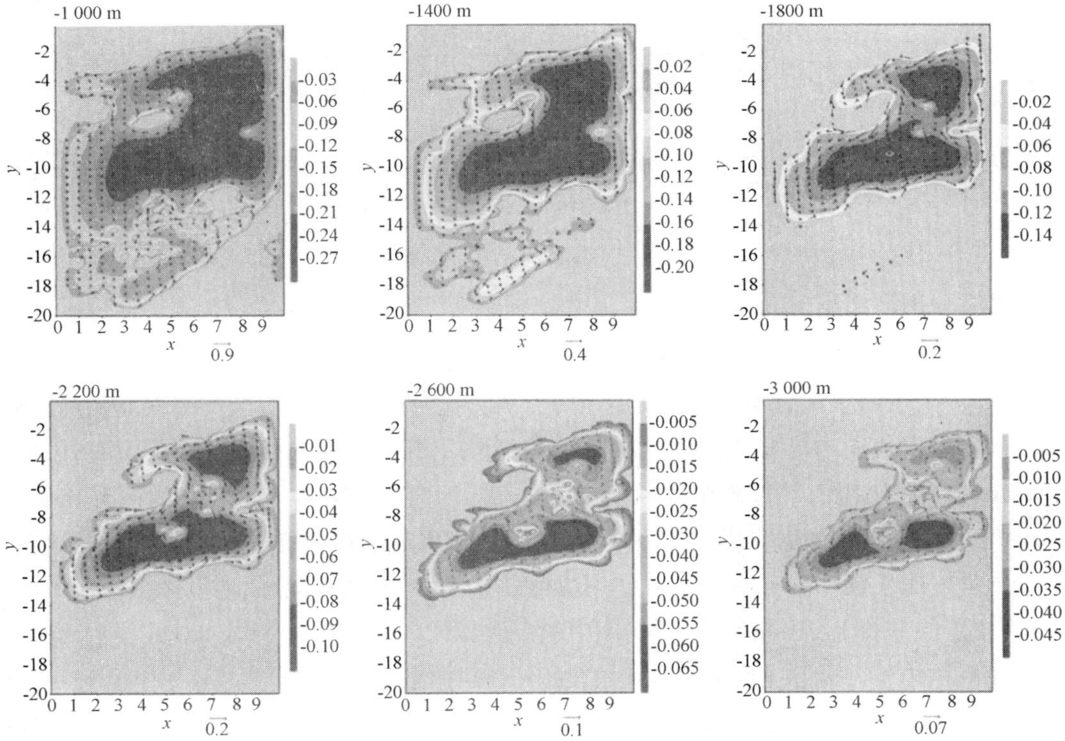

图 3　冬季摩擦层中无量纲的准地转流及 p'场（见书后彩插）

大值,但总体上流速小于-800 m 层,这是与实际相符合的。摩擦层中的环流仍然以气旋式环流为主,在四周为海水所包围的陆地周围形成一些小的反气旋式环流,在下层(-2200 m 以下)海水被分割为几个围绕深水区的小的气旋式环流。整个摩擦层从上往下 p'绝对值减小,流速减小。

由于没有观测资料可供参考,从哪个深度运动开始从惯性层进入摩擦层只能给一个估计,上面计算取 $z=-1\ 000$ m 是基于在这个深度南海海域整体上为水体,不规则的岛屿一般在这个深度以下。改变惯性层和摩擦层交界面的高度,当取:$z=-2\ 000$ m 时(图略),流场及扰动压力的变化较为缓和,惯性层和摩擦层中从上到下扰动压力及流速仍为递减,物理量场的形势无实质性变化。

在这个例子中,由吕宋海峡进入南海的海水速度分布见图 4,可见海水大部分在上层进入,并且流速从上往下,从中间向两边递减。

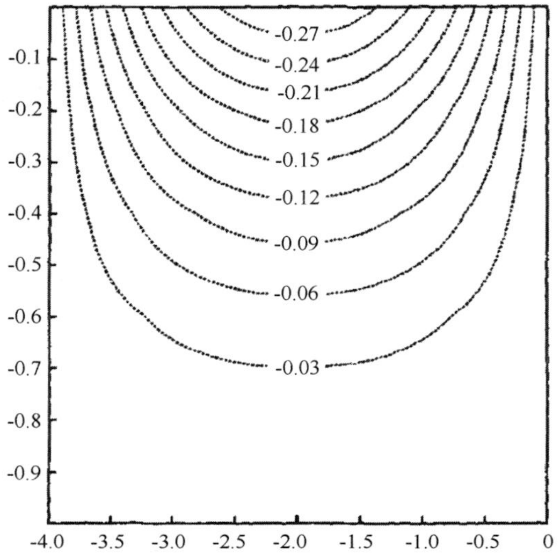

图 4　冬季吕宋海峡流速剖面

横轴表示经向方向,纵轴为垂向,均为无量纲形式。

等值线为无量纲流速,负值表示向西流

8.2　夏季扰动压力及流场

夏季受南风应力影响,参数取值见表 1。图 5 为惯性层中的流场及扰动压力图,在表层除了北边界侧向边界层中由于向岸流的经向变化而由地转近似引起的向东流外,其余均为向北的经向流,流速自西向东逐渐减小,流向与等 p'线相平行。在-200 m 层上开始出现 p'的闭合高值中心,区域南部出现向西流,东部出现向南流,这样整个区域形成一个闭合反气旋式环流。再往下这一环流更加明显,并且 p'场和流场随深度增加而变得更加不规则,地形的作用逐渐加大。下部海水较深处对应于较大的 p'值,较浅处对应于较小的 p'值。总体上,从上往下 p'值减小,流速减小,地形作用逐渐加强。

图 6 为摩擦层中的流场及扰动压力图,图中灰色阴影区域为海底。在-1 000 m 层上,由于实际地形的加入,导致一些岛屿附近扰动压力梯度加大,因而根据地转关系计算出的流场也出现一些虚假的大值。摩擦层中的环流仍然以反气旋式环流为主,在四周为海水所包围的陆地周围形成一些小的气旋式环流,在下层(-2 200 m 以下)环流被分割为几个围绕深水区的小的反气旋式环流。整个摩擦层从上往下 p'值减小,流速减小。

在这个例子中从吕宋海峡流出的海水的流速分布见图 7,可见流速从上到下,从中间向两边递减。

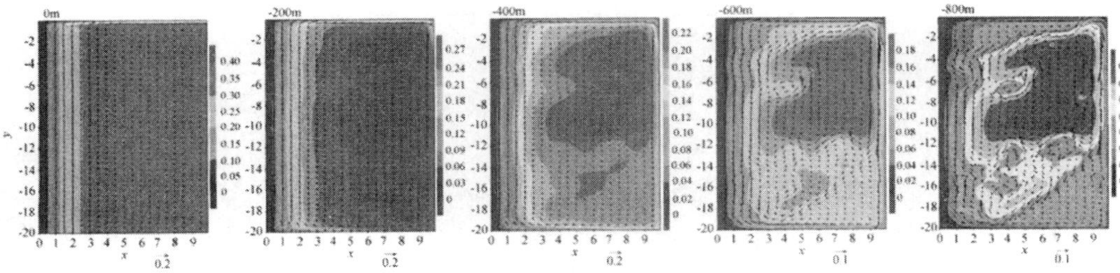

图 5　夏季惯性层中无量纲的准地转流及 p'场（见书后彩插）

惯性层和摩擦层交界面高度为-1 000 m

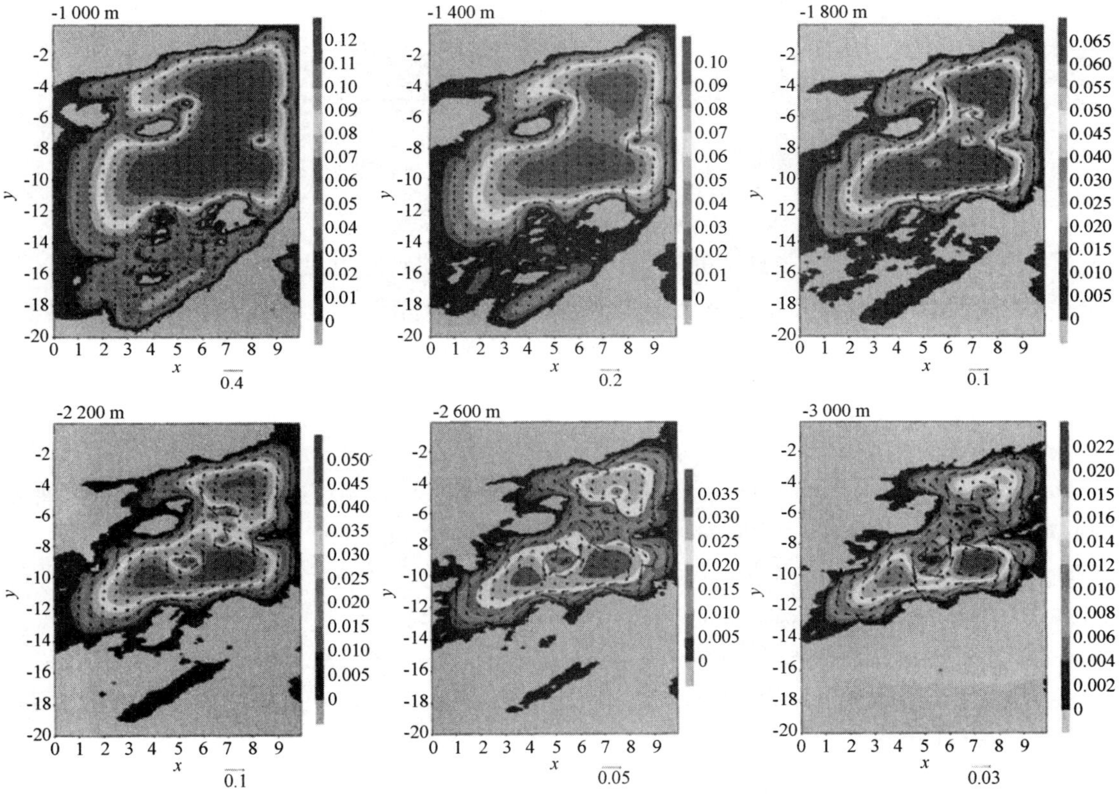

图 6　夏季摩擦层中无量纲的准地转流及 p'场（见书后彩插）

9　结论

本文试图用一个理想化的 f-平面准地转模型对南海三维环流的形成进行理论尝试。在将理想的南海区域垂直方向分为 Ekman 层、惯性层和摩擦层的基础上计算了各层上冬、夏季的准地转流及扰动压力。结果表明冬季各层上以气旋式环流为主，且随深度的增加流速减小；夏季各层上以反气旋式环流为主，且随深度增加流速减小，这在一定程度上与上层已有的观测事实相一致。深层因无实际观测资料可比较，流场是否和本文的计算结果一致，上、下层的流场形势是否发生逆转，还需要进一步的探讨。但这一动力学方法的计算结果可以为这一海域建立区域模式提供参考。

需要指出的是由于理论分析必然是对实际情况的理想化，这个理想化将会引进哪些重要的误差，

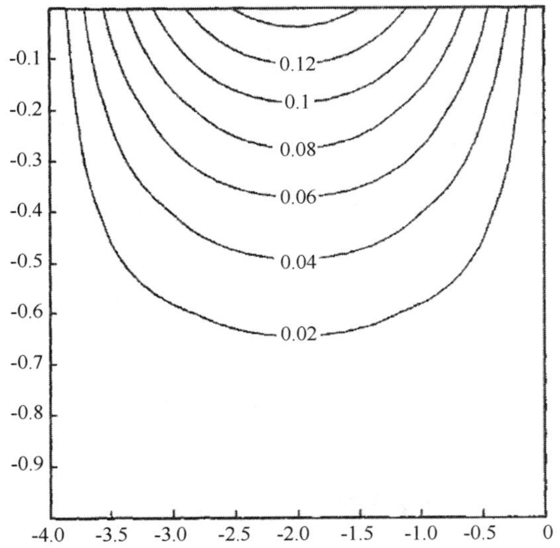

图 7 说明同图 4,但为夏季,正值表示向东流

因无观测可比尚不清楚。但一些具体的缺陷,如北边界上层的侧向边界层中由于采用地转近似导致出现偏离实际的地转流,及惯性层和摩擦层交界处较大的虚假流速以及摩擦层近似采用扰动温度为零而可能引起的误差等,这些都需要在以后的工作中加以改进。

参考文献

[1] Wyrtki K.Physical oceanography of the Southeast Asian waters.Nnga Report,Vo1.2.Scientific Results of Marine Investigations of the South China Sea and the Gulf of Thailand 1959-1961.Scripps Institution of Oceanography,University of California at San Diego,1961:1-195.

[2] Shaw P T,Clmo S Y.surface circulation in the South China Sea. *Deep-Sea Res.*,1994,41:1663-1683.

[3] Chu P C,Rong Li.South China Sea isopycnal-surface circulation.*J.Phys.Oceanogr.*,2000,30:2419-2438.

[4] Hellerman S, Rosenstein M. Normal monthly wind stress over the wored ocea with error estimates. *J. Phys. Oceanogr.*, 1983,13:1093-1104.

[5] Huang Q Z,Wang WZ,Li Y S,et al.Current characteristics of the South China Sea.In:Di Z,Yuan-Bo L,Cheng-Kui Z, eds.Oceanology of China Seas.Khwer.1994:39-46.

[6] Qu T D.Upper layer circulation in the South China Sea.*J.Phys.Oceanogr.*,2000,30:1450-1460.

[7] Metzger E J,Hurlburt H E.Coupled dynamics of the South China Sea,the Sutu Sea,and the Pacific Ocean.*J.Geophys.Res.*, 1996,101:12331-12352.

[8] Metzger E J,Hurlburt H E.the nondeterministic nature of Kuroshio penetration and eddy shedding in the South China Sea.*J.Phys.Oceanogr.*,2001,31:1712-1732.

[9] Xue H J,Chai F,Pettigmw N,et al.Kuroshio intrusion and the circulation in the South China Sea.*J.Geophys.Res.*,2004, 109,C02017,doi:10.1029/2002JC001724.

[10] Shaw P T.the intrusion of water masses into the sea southwest of Taiwan.*J.Geophys.Res.*,1989,94:18213-18226.

[11] Shaw P T.The seasonal variation of the intrusion of the Philippine Sea water into the South China Sea.*J.Geophys.Res.*, 1991,96:821-827.

[12] 翟丽,方国洪,王凯.南海风生正压环流动力机制的数值研究.海洋与湖沼,2004,35(4):289-298.

[13] Shaw P T, Chao S Y, Fu L. Sea surface height variations in the South China Sea from satellite altimetry. *Oceanol. Acta*, 1999, 22: 1-17.

[14] 李立，许金电，靖春生，等. 南海海面高度、动力地形和环流的周年变化——TOPEX/Poseidon卫星测高应用研究. 中国科学（D辑），2002，32(12)：978-986.

[15] 巢纪平，冯立成，王东晓. 风生边界急流稳定性的渐进理论. 地球物理学报，2006，49(3)：642-649.

[16] 刘秦玉，杨海军，刘征宇. 南海Sverdrup环流的季节变化特征. 自然科学进展，2000，10(11)：1035-1039.

[17] Liu Z Y, Yang H J, Liu Q Y. Regional dynamics of seasonal variability in the South China Sea. *J. Phys. Oceanogr.*, 2001, 31: 272-284.

f-平面中尺度海盆热力驱动边界流的形成*

冯立成[1,2] 巢纪平[3,4]

(1. 中国科学院大气物理研究所大气科学和地球流体力学数值模拟国家重点实验室，北京 100029；2. 中国科学院研究生院，北京 100049；3. 国家海洋环境预报中心，北京 100081；4. 国家海洋局第一海洋研究所，青岛 266061)

摘要：采用一个 f-平面准地转但未作线性化假定的惯性模型，考虑了西侧固壁附近摩擦层的作用，在热量守恒条件下，研究了理想化的长方体海盆区域内的扰动温度、边界急流及上升(下沉)流。设研究区域上表面有净的热量输入，相应的西侧边界有等量的热量耗散，其余边界与外界无热量交换，从而整个海域海水热量守恒。结果表明，在西侧边界扰动温度密集出现温度锋；扰动压力及流场存在上下层翻转现象，下层西侧为向北的沿岸急流，扰动压力极大值中心位于西部，上层东侧为向南的急流，扰动压力极大值中心位于东部。西侧较窄的范围内出现较强的垂向流，中部区域也有较大的垂向运动。文中还研究了不同形式的上表面热力强迫的影响，结果表明对于不同形式的上边界热力强迫，均可在海盆西侧出现扰动温度密集，边界急流，亦有上下层流场的翻转现象，但垂向流的分布则有很大不同。

关键词：边界急流；垂向流；温度锋；热力驱动；总热量守恒

关于沿岸上升流和沿岸急流形成和发展的理论已有较多研究。如在线性理论的范畴内，Hsueh 等[1]研究过连续层结介质中上升流形成的定常解，Allen[2]研究过层结介质中的非定常解，巢纪平等[3]研究了南海西边界急流的稳定性。在非线性理论的范畴内，巢纪平等[4]采用半解析半数值的方法研究了沿岸上升流，此外沿岸流的数值模拟工作也有所发展。

在另一方面，近年来还出现了一些针对理想化海盆环流及边界流形成的热力学机制的研究。如王东晓等[5]通过对线性两层海洋模式进行正交模求解，得到了热带矩形海盆在热力强迫下的海洋动力场水平结构。Spall[6]利用数值模式和解析模型研究了理想化边缘海的温盐环流。在他的研究工作中，边缘海内部上表面施以周期性的冷却强迫，其下的海水冷却增密，并主要以中尺度涡旋的形式向侧边界传输。边缘海与外海之间由一窄的海峡相连，因此边缘海内部的热量和质量损失可以通过与外海之间的交换得到补偿。虽然冷却加在海盆中部，但海水质量下沉却集中于特征宽度为 $O(L_d)$ (L_d 为内部变形半径)的侧边界层并依赖于次中尺度的动量和密度混合过程。这一结论与 Spall 和 Pickart[7]，Marotzke 和 Scott[8]所指出的垂直浮力通量和垂直质量通量不需发生在同样的位置也不需由相同的物理过程来携带的观点相一致。Spall[6]认为之所以产生这种现象，是因为只有在侧向涡动黏滞系数较大的边界层，海盆内部由垂直运动拉伸所产生的相对涡度才能被消耗掉。

Pedlosky[9]用理想化的线性黏性模式对此做了进一步的研究。他将实际海盆理想化为侧边界均

* 中国科学 D 辑：地球科学，2007，37(10)：1417-1424.

为固壁的长方体区域，上表面有加热(冷却)强迫，东侧面的温度取为定值，另外三个侧面及底面是绝热的，且海水与外界无质量交换。海盆中的海水层结稳定，浮力频率取为常数。结果表明在海盆内部有加热(冷却)的情况下，上升(下沉)运动集中于窄的黏性侧边界层中，而不是直接发生在加热(冷却)区。这进一步证实了 Spall[6] 的结论。

本文试图用一个 f-平面准地转但未作线性化假定的无黏性的惯性模型，对理想化的长方体海盆区域进行研究。海盆上表面有净的热量输入，相应的西侧边界有等量的热量耗散，其余边界与外界无热量交换，从而整个海盆的海水满足热量守恒。据此给出各边界上扰动温度的梯度，并导出惯性区域关于扰动温度的控制方程，采用数值方法计算海盆区域的扰动温度，沿岸急流及沿岸上升(下沉)流。

1　模式结构及边界条件

本文以理想化的长方体海盆区域为研究对象，设海盆 x 方向长度为 L_x，y 方向长度为 L_y，垂直方向的深度为 H。不考虑上混合层和下摩擦层中的运动，只考虑惯性层中的海水运动，海水为理想流体，其运动是非线性的。假定在海盆的上表面有净的热量输入，在海盆西侧边界有等量的热量耗散，其余边界与外界无热量交换，整个海盆海水的总热量守恒。由此得出海盆区域需满足的热力学边界条件

$$x = 0,\quad \frac{\partial T}{\partial x} = h_x(y,z) \tag{1}$$

$$x = L_x,\quad \frac{\partial T}{\partial x} = 0 \tag{2}$$

$$y = 0,\quad L_y, \frac{\partial T}{\partial y} = 0 \tag{3}$$

$$z = 0,\quad \frac{\partial T}{\partial z} = 0 \tag{4}$$

$$z = H,\quad \frac{\partial T}{\partial z} = h_z(x,y) \tag{5}$$

上边界和西侧边界的热通量相等，
即

$$\iint_{x\,y} \lambda_z \frac{\partial T}{\partial z}\bigg|_{z=H} \mathrm{d}x\mathrm{d}y = \iint_{y\,z} \lambda_x \frac{\partial T}{\partial x}\bigg|_{x=0} \mathrm{d}y\mathrm{d}z \tag{6}$$

其中，λ_x，λ_z 分别为水平、垂直方向的“有效混合热交换系数”(包含分子热传导、热平(对)流和湍流热交换等效应在内的一个参数化系数)。

2　理想流体运动方程

对于斜压、非线性、定常的理想流体，其运动方程为

$$u\frac{\partial u}{\partial x} + v\frac{\partial u}{\partial y} - fv = -\frac{\partial p'}{\partial x} \tag{7}$$

$$u\frac{\partial v}{\partial x} + v\frac{\partial v}{\partial y} + fu = -\frac{\partial p'}{\partial y} \tag{8}$$

$$0=-\frac{\partial p'}{\partial z}+g\alpha\vartheta \tag{9}$$

$$\frac{\partial u}{\partial x}+\frac{\partial v}{\partial y}+\frac{\partial w}{\partial z}=0 \tag{10}$$

$$u\frac{\partial \vartheta}{\partial x}+v\frac{\partial \vartheta}{\partial y}+\Gamma w=0 \tag{11}$$

式中，u, v, w 分别为沿 x, y, z 轴的速度分量；f 为科里奥利参数；g 为重力加速度；$p'=p/\rho_0$ 为静压力的偏差；α 是流体的热膨胀系数；$\Gamma=\frac{\mathrm{d}\,\overline{T}}{\mathrm{d}z}$ 为平均温度（$\overline{T}$）的垂直递减率（T 表示实际温度；Γ 在此取成常数），ϑ 为扰动温度。

方程(7)，方程(8)对 x 和 y 交叉微分，并由第二式减去第一式得

$$u\frac{\partial \Omega}{\partial x}+v\frac{\partial \Omega}{\partial y}+(\Omega+f)\left(\frac{\partial u}{\partial x}+\frac{\partial v}{\partial y}\right)=0 \tag{12}$$

式中

$$\Omega=\frac{\partial v}{\partial x}-\frac{\partial u}{\partial y} \tag{13}$$

为相对涡度。如果 u, $v\sim10^{-1}$m/s，x,$y\sim10^{5}$ m，估计出 $\Omega\sim10^{-6}$/s，而 $f\sim10^{-4}$/s，因此 $\Omega\ll f$，式(12)在考虑了式(10)后为

$$u\frac{\partial \Omega}{\partial x}+v\frac{\partial \Omega}{\partial y}-f\frac{\partial w}{\partial z}=0 \tag{14}$$

2.1 准地转运动

对于水平尺度大于 100 km 的运动可引进地转近似

$$u=-\frac{1}{f}\frac{\partial p'}{\partial y},\quad v=\frac{1}{f}\frac{\partial p'}{\partial x} \tag{15}$$

由此

$$\Omega=\frac{1}{f}\nabla^2 p' \tag{16}$$

式(14)改写成

$$J\left(p',\frac{1}{f}\nabla^2 p'\right)=f^2\frac{\partial w}{\partial z} \tag{17}$$

考虑式(9)，式(10)式改写成

$$J\left(p',\frac{\partial p'}{\partial z}\right)=-fg\alpha\Gamma w \tag{18}$$

再对 z 微分得到

$$J\left(p',\frac{\partial^2 p'}{\partial z^2}\right)=-fg\alpha\Gamma\frac{\partial w}{\partial z} \tag{19}$$

与(17)式联合给出

$$J\left(p',\nabla^2 p'+\frac{f^2}{g\alpha\Gamma}\frac{\partial^2 p'}{\partial z^2}\right)=0 \tag{20}$$

由此得到

$$\nabla^2 p' + \frac{f^2}{g\alpha\Gamma}\frac{\partial^2 p'}{\partial z^2} = F(p') \tag{21}$$

F 为 p' 的普适函数，其形式待定。

2.2　无量纲化

取特征尺度 $z \sim H$，x，$y \sim l = C/f$，重力波波速 $C = \sqrt{g\alpha\Gamma H^2}$。特征速度 $u, v \sim C, w \sim \frac{H}{l}C$，并取 $p' \sim C^2, \vartheta \sim \Gamma H$。

由此式(9)的无量纲形式为

$$\frac{\partial p'}{\partial z} = \vartheta \tag{22}$$

式(15)的无量纲形式为

$$u = -\frac{\partial p'}{\partial y},\quad v = -\frac{\partial p'}{\partial x} \tag{23}$$

式(21)的无量纲形式为

$$\nabla^2 p' + \frac{\partial p'}{\partial z^2} = F(p') \tag{24}$$

将式(22)带入式(24)得

$$\nabla^2 \vartheta + \frac{\partial^2 \vartheta}{\partial z^2} = F(\vartheta) \tag{25}$$

这是惯性层中关于扰动温度 ϑ 的控制方程，考虑到区域外为未扰动海洋，因此可以在众多的普适函数中取其中最简单的形式 $F(\vartheta) = 0$。

3　西侧黏性边界层中的运动方程

海盆西侧固壁附近存在一薄的边界层，边界层中黏性系数很大，海水能量在此耗散。方程为

$$-fv = -\frac{\partial p'}{\partial x} + v\frac{\partial^2 u}{\partial x^2} \tag{26}$$

$$fu = -\frac{\partial p'}{\partial y} + v\frac{\partial^2 v}{\partial x^2} \tag{27}$$

$$\frac{\partial p'}{\partial z} = g\alpha\vartheta \tag{28}$$

$$\frac{\partial u}{\partial x} + \frac{\partial v}{\partial y} + \frac{\partial w}{\partial z} = 0 \tag{29}$$

$$\Gamma w = k\frac{\partial^2 \vartheta}{\partial x^2} \tag{30}$$

其中，v 为湍流黏性系数，k 为热量扩散系数。

式(26)，式(27)两式分别对 x、y 交叉微分，并由第二式减去第一式得

$$f\left(\frac{\partial u}{\partial x} + \frac{\partial v}{\partial y}\right) = v\frac{\partial^2}{\partial x^2}\left(\frac{\partial v}{\partial x} - \frac{\partial u}{\partial y}\right) \tag{31}$$

考虑连续性方程(29)，右端涡度采用准地转近似，上式改写为

$$-\frac{f^2}{v}\frac{\partial w}{\partial z}=\frac{\partial^2}{\partial x^2}\left(\frac{\partial^2 p'}{\partial x^2}+\frac{\partial^2 p'}{\partial y^2}\right) \tag{32}$$

联合式(28)，式(30)两式得到

$$\frac{\partial w}{\partial z}=\frac{k}{g\alpha\Gamma}\frac{\partial^2}{\partial x^2}\left(\frac{\partial^2 p'}{\partial z^2}\right) \tag{33}$$

将式(33)带入式(32)，得到控制方程

$$\left(\frac{f^2}{g\alpha\Gamma\sigma}\right)\frac{\partial^2 p'}{\partial z^2}+\frac{\partial^2 p'}{\partial x^2}+\frac{\partial^2 p'}{\partial y^2}=0 \tag{34}$$

其中 $\sigma=\nu/k$ 为 Prandtl 数。

再将式(28)带入式(34)得

$$\left(\frac{f^2}{g\alpha\Gamma\sigma}\right)\frac{\partial^2 \vartheta}{\partial z^2}+\frac{\partial^2 \vartheta}{\partial x^2}+\frac{\partial^2 \vartheta}{\partial y^2}=0 \tag{35}$$

取 $x\sim\delta$，其余参数特征尺度如前，将上式无量纲化得到

$$\frac{1}{\sigma}\frac{\partial^2 \vartheta}{\partial z^2}+\left(\frac{1}{\delta}\right)^2\frac{\partial^2 \vartheta}{\partial x^2}+\frac{\partial^2 \vartheta}{\partial y^2}=0 \tag{36}$$

将式(36)改写成

$$\left(\frac{1}{\sigma}\left(\frac{\delta}{l}\right)^2\right)\frac{\partial^2 \vartheta}{\partial z^2}+\frac{\partial^2 \vartheta}{\partial x^2}+\left(\frac{\delta}{l}\right)^2\frac{\partial^2 \vartheta}{\partial y^2}=0 \tag{37}$$

考虑到 $\delta\ll l$，因此第三项是一小项，可略去，同时要求 $\frac{1}{\sigma}\left(\frac{\delta}{l}\right)^2\sim(10^0)$。式(37)近似写成

$$A^2\frac{\partial^2 \vartheta}{\partial z^2}+\frac{\partial^2 \vartheta}{\partial x^2}=0 \tag{38}$$

其中，$A^2=\frac{1}{\sigma}\left(\frac{\delta}{l}\right)^2$。

关于 x 积分上式得

$$\left.\frac{\partial \vartheta}{\partial x}\right|_{x=0}=\left.\frac{\partial \vartheta}{\partial x}\right|_{x=\delta}+A^2\int_0^\delta\frac{\partial^2 \vartheta}{\partial z^2}\mathrm{d}x \tag{39}$$

由于 δ 很小，上式近似地写成

$$\left.\frac{\partial \vartheta}{\partial x}\right|_{x=0}=\left.\frac{\partial \vartheta}{\partial x}\right|_{x=\delta}+A^2\delta\left.\frac{\partial^2 \vartheta}{\partial z^2}\right|_{x=\delta} \tag{40}$$

进一步可设边界层内的扰动温度随 x 的分布为

$$\vartheta=\vartheta_g\mathrm{e}^{(\delta-x)} \tag{41}$$

ϑ_g 为边界层和惯性层连接处($x=\delta$)的扰动温度。由此得到边界上的扰动温度为

$$x=0,\quad \vartheta=\vartheta_g\mathrm{e}^{\delta} \tag{42}$$

而

$$\left.\frac{\partial \vartheta}{\partial x}\right|_{x=\delta}=-\vartheta_g \tag{43}$$

由此式(40)改写为

$$\left.\frac{\partial \vartheta}{\partial x}\right|_{x=0}=-\vartheta_g+A^2\delta\left.\frac{\partial^2 \vartheta_g}{\partial z^2}\right|_{x=\delta} \tag{44}$$

这是加入了摩擦层作用后的西边界热通量密度，根据式(6)，上式的面积积分和上边界热通量平衡，即

$$\lambda_x \iint_{y,z} \left[-\vartheta_g + A^2\delta \frac{\partial^2 \vartheta_g}{\partial z^2} \bigg|_{x=\delta} \right] \mathrm{d}y\mathrm{d}z = \lambda_z \iint_{x,y} \frac{\partial \vartheta}{\partial z} \mathrm{d}x\mathrm{d}y \tag{45}$$

为使得这一等式成立，有参数 λ_x,δ,A 可调整。

4　计算实例

4.1　实际海域的理想化

以中国近海为例，设海岸呈南北走向，y 轴与之平行，向北为正，x 轴与海岸垂直，离岸为正，z 轴垂直向上。暂不考虑陆架地形的作用，沿海岸取一长方体区域，其西边界为海岸，东边界及南、北边界均为液体边界。在这一区域，假定有向下的感热通量。虽然这一作用在海表的感热通量，要通过混合层才能到达惯性层顶，但如假定混合层很薄，可近似地把海表的热通量直接作用于惯性层顶。要在惯性层上分出一混合层，这在数学处理上并无困难，但多少带来一些处理上的繁琐，因此这里取惯性层顶的热通量即为海表热通量，当然也不妨认为现在所用的热通量只是海表热通量的一部分。考虑到在西侧固体边界附近黏滞系数很大，因此引进了侧向湍流边界层的订正，海水运动通过侧向湍流边界层后造成净的热量损失，这一损失的热量假定等于上混合层传给惯性层的热量。其余边界与外界无热量交换，因此整个区域满足热量守恒。计算中长方体区域的几何参数及差分网格的配置如表 1 所示。

表 1　长方体区域的几何参数及差分网格的配置

	L_x	L_y	H	Δx	Δy	Δz
有量纲值(km)	300	300	1	3	3	50 m
无量纲值	3	3	1	0.03	0.03	0.05

4.2　扰动温度的边界条件

我们这里根据控制方程式(25)来求海盆中扰动温度 ϑ，由于背景(或气候)温度在 100 km 尺度上比较均匀，因此可认为 $\frac{\partial \vartheta}{\partial n} \approx \frac{\partial T}{\partial n}$，所以上面式(1)至式(6)对 ϑ 仍然成立。设研究区域上边界热通量密度的分布为

$$z = H,\quad q_z = -\lambda_z \frac{\partial T}{\partial z} = h_z \sin\frac{\pi x}{L_x} \sin\frac{\pi y}{L_y} \tag{46}$$

即

$$z = H,\quad \frac{\partial \vartheta}{\partial z} = -\frac{h_z \sin\frac{\pi x}{L_x} \sin\frac{\pi y}{L_y}}{\lambda_z} \tag{47}$$

其中，h_z 为垂向(混合层-惯性层交界面)特征热通量密度，其量级取为 10^1。

西边界的热通量密度由式(44)给出，可见其决定于 ϑ_g。由于初始时刻并不知道 ϑ_g 的分布形式，因此在这里预先假定一种 $\frac{\partial \vartheta}{\partial x}$ 的分布形式进行求解，待求出 ϑ_g 后，再改用式(44)作为西边界条件，并

采用逐级近似法调整其中参数以达到热量守恒。

设西边界的热通量密度随 y 轴、z 轴向北向下减小，初始猜值的分布为

$$x = 0, \quad q_x = -\lambda_x \frac{\partial T}{\partial x} = h_x e^{\frac{z}{a}} e^{-\frac{y}{b}} \tag{48}$$

即

$$x = 0, \quad \frac{\partial \vartheta}{\partial x} = -\frac{h_x e^{\frac{z}{a}} e^{-\frac{y}{b}}}{\lambda_x} \tag{49}$$

其中，h_x 为侧向(流体-固壁交界面)特征热通量密度，其量级取为 10^4。

式(47)中的垂向"有效混合热交换系数" λ_z 及式(49) 中的水平"有效混合热交换系数" λ_x 为计算简便取作常数，这里分别取 $\lambda_z = 10^2$ W/(m·℃)，$\lambda_x = 10^6$W/(m·℃)。

用 ϑ 代替 T 并利用表 1 中的无量纲参数，对式(47)，式(48)及式(2)至式(4)进行无量纲化，得到关于扰动温度的边界条件

$$x = 0, \quad \frac{\partial \vartheta}{\partial x} = -h_x^0 \mathrm{e}^{\frac{z}{a}} \mathrm{e}^{-\frac{y}{b}} \tag{50}$$

$$x = 3, \quad \frac{\partial \vartheta}{\partial x} = 0 \tag{51}$$

$$y = 0,3, \quad \frac{\partial \vartheta}{\partial y} = 0 \tag{52}$$

$$z = 0, \quad \frac{\partial \vartheta}{\partial z} = 0 \tag{53}$$

$$z = 1, \quad \frac{\partial \vartheta}{\partial z} = -h_z^0 \sin\frac{\pi x}{3} \sin\frac{\pi y}{3} \tag{54}$$

根据热量守恒式(6)并考虑式(50)和式(54)得

$$\iint_{x\,y} h_z^0 \sin\frac{\pi x}{3} \sin\frac{\pi y}{3} \mathrm{d}x\mathrm{d}y = \iint_{y\,z} h_z^0 \mathrm{e}^{\frac{z}{a}} \mathrm{e}^{\frac{y}{b}} \mathrm{d}x\mathrm{d}y \tag{55}$$

给定 $a = 2.0$，$b = 3.0$，$h_z^0 = -1.0$，$h_x^0 = -1.484$。

4.3 扰动压力及流场

扰动压力 p' 的分布可由式(24)求得，假定东西、南北边界上 $p' = 0$，这样由地转关系求得的南北边界上 $v = 0$，东西边界上 $u = 0$。由于前面已求得扰动温度 ϑ，上下边界可由静力方程式(9)给定。

计算出 p' 场后由地转关系即可得到 u，v 场分布。热力学方程(11)对 z 求导得

$$\frac{\partial w}{\partial z} = -\left(u \frac{\partial^2 \vartheta}{\partial x \partial z} + v \frac{\partial^2 \vartheta}{\partial y \partial z}\right) \tag{56}$$

考虑到上边界 $z = 1$，$w = 0$ 得

$$w = \int_z^1 \left(u \frac{\partial^2 \vartheta}{\partial x \partial z} + v \frac{\partial^2 \vartheta}{\partial y \partial z}\right) \mathrm{d}z \tag{57}$$

4.4 计算结果分析

用数值方法对椭圆型控制方程(25)求解得到扰动温度 ϑ 的分布(图 1)。可见从下向上负的扰动温度减弱，正的扰动温度增强，扰动温度分布呈现出在西边界密集的特征，尤其在西边界南部最为明显，形成温度锋(2℃/15 km)。上层扰动温度分布(图 1(d))受上表面给定扰动温度影响较大，呈

现中间温度高，四周温度低的特征。

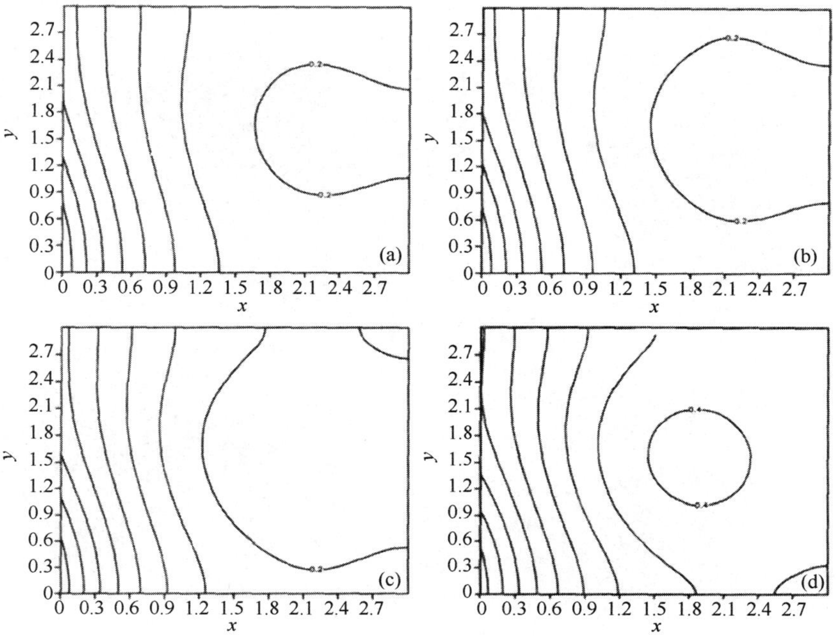

图 1　不同层次上的无量纲扰动温度

(a) $z=0.2$; (b) $z=0.4$; (c) $z=0.6$; (d) $z=0.8$

由控制方程式(24)及相应的边界条件求出 p' 场及流场分布(图 2)。下层 p' 在西部出现正值中心，相应的流场为反气旋式，西侧为向北的沿岸急流可达 0.7 m/s，同时东部有弱的向北流(图 2(a))。向上 p' 的正值中心逐渐东移，流速减小(图 2(b)，图 2(c))。上层 p' 在中部形成正值中心，东侧出现向南的急流(-0.5 m/s)，西南角则形成一个较小的负值中心，伴随气旋式环流(图 2(d))。从下向上 p' 及流速先减小后增大，物理场呈现上下层翻转现象，东西边界流的翻转最为明显。

由式(57)可得到垂直速度 w 的分布(图 3)，由图可见整个区域可大致以东北—西南向对角线划分为两部分，东南部为上升流区，西北部为下沉流区。在西南角上有很强的下沉运动(-0.04 cm/s)，对应于图 1 中西南角密集的等扰动温度线。在我们这里垂向速度分布并不对应上边界的强迫热量分布，这点与 Spall 和 Pickat[7] 及 Marotzke 和 Scott[8] 的结论是一致的。需要指出的是在惯性模型中既可有边界上升、下沉流，也可在内区出现同量级的上升下沉流(0.02 cm/s)，这区别于 Pedlosky[9] 的结论。

4.5　不同类型上边界条件的影响

为了研究不同类型的上表面扰动温度分布对扰动温度场、水平流场及垂直速度的影响，我们取上表面扰动温度随 y 轴向北减小

$$z = 1, \quad \frac{\partial \vartheta}{\partial z} = -h_z^1 e^{-\frac{y}{3}} \tag{58}$$

其他边界条件不变，给定

$$a = 2.0,\ b = 3.0,\ h_z^1 = -1.0,\ h_x^1 = -2.313$$

这时同样在西侧出现扰动温度密集现象(图 4)，p' 及流场具有上下层翻转的特征(图 5)，但垂

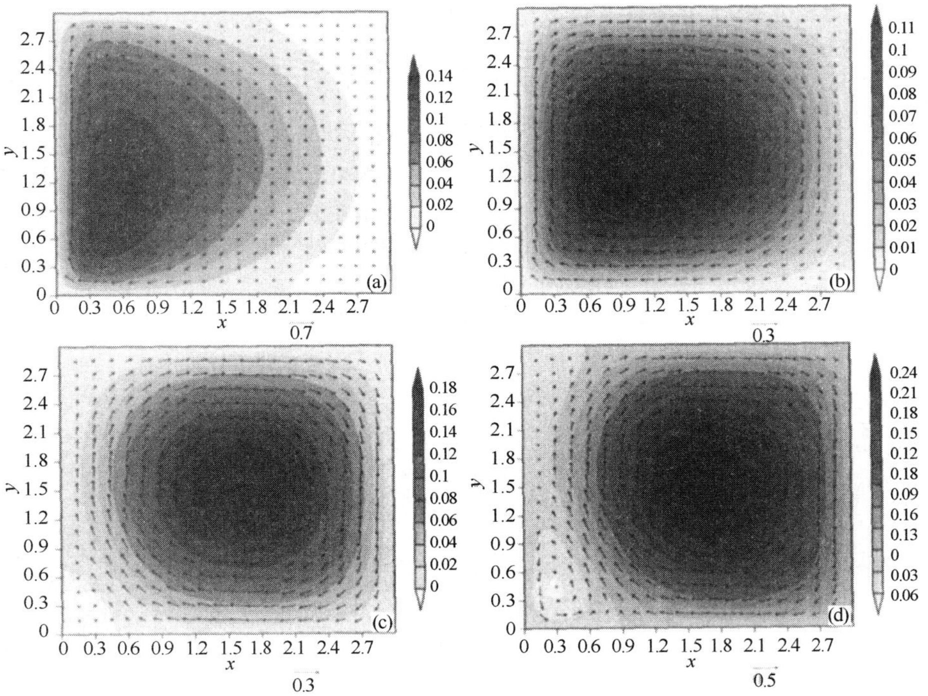

图 2 不同层次上的无量纲扰动压力及水平流场分布

(a) $z=0.2$; (b) $z=0.4$; (c) $z=0.6$; (d) $z=0.8$

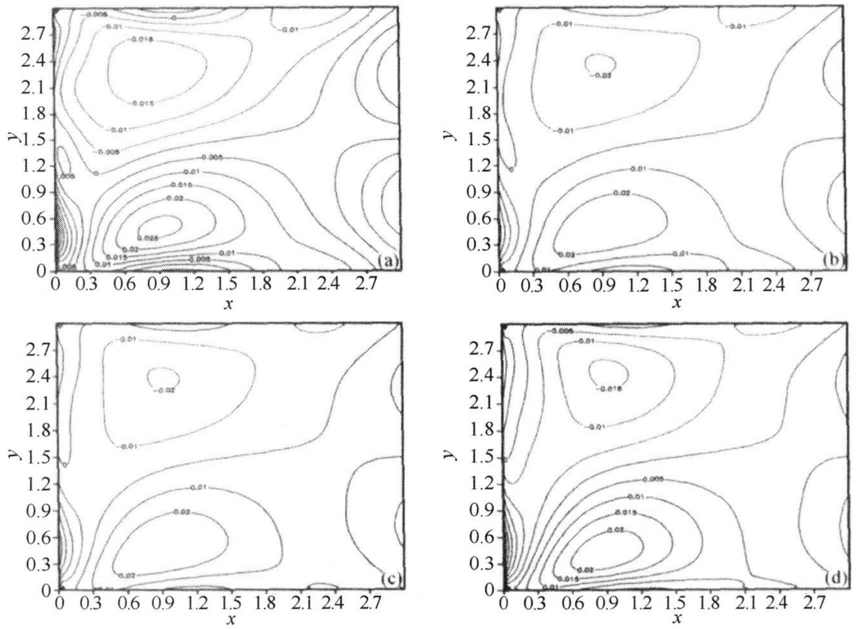

图 3 不同层次上的无量纲垂直速度分布

(a) $z=0.2$; (b) $z=0.4$; (c) $z=0.6$; (d) $z=0.8$

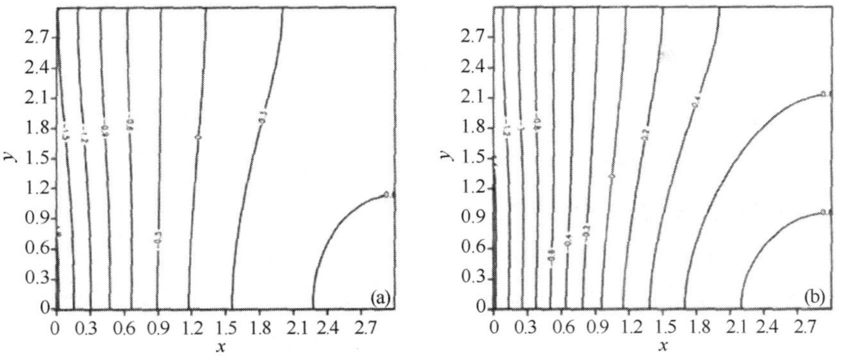

图 4

说明同图 1，但对(a) $z=0.4$；(b) $z=0.8$

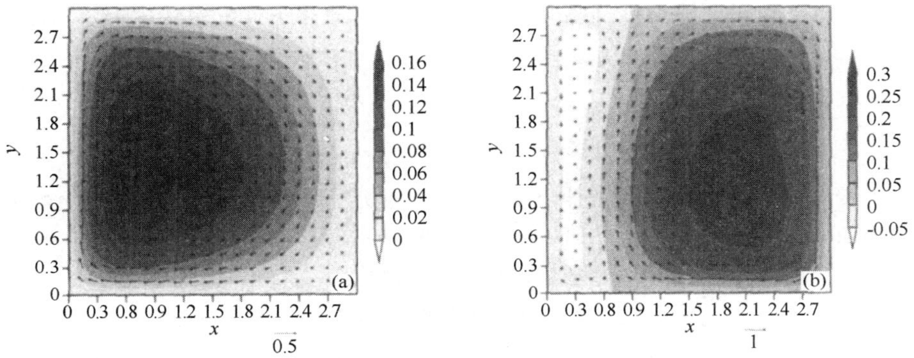

图 5

说明同图 2，但对(a) $z=0.4$；(b) $z=0.8$

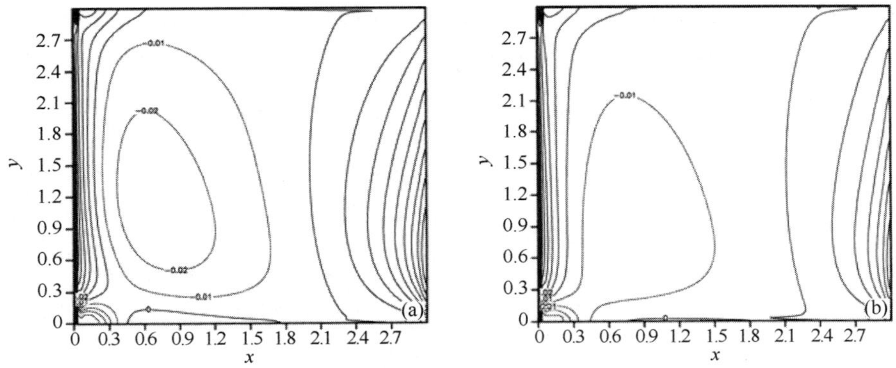

图 6

说明同图 3，但对(a) $z=0.4$；(b) $z=0.8$

直速度场则表现出很大的不同(图 6)，中部区域为较小的下沉运动，东西边界则为上升运动，且西侧边界上升运动出现在较窄的范围(~30 km)。可见不同的上边界加热对沿岸温度锋和沿岸急流的出现无大的影响，但可影响垂向流分布。

5 结论

本文采用一个f-平面准地转但未作线性化假定的惯性模型，考虑了西侧固壁附近摩擦层的作用，在热量守恒条件下，研究了理想化的长方体海盆区域内的扰动温度、边界急流及上升(下沉)流。设研究区域上表面有净的热量输入，相应的西侧边界有等量的热量耗散，其余边界与外界无热量交换，从而整个海域海水热量守恒。结果表明，在西侧边界扰动温度密集出现温度锋(~2℃/15 km)；扰动压力及流场存在上下层翻转现象，下层西侧为向北的沿岸急流，扰动压力极大值中心位于西部，上层东侧为向南的急流，扰动压力极大值中心位于东部。西侧较窄的范围内出现较强的垂向流，中部区域也有较大的垂向运动。文中还研究了不同形式的上表面热力强迫的影响，结果表明对于不同形式的上边界热力强迫，均可在海盆西侧出现扰动温度密集，边界急流，亦有上下层流场的翻转现象，但垂向流的分布则有很大不同。

我们认为这是海岸附近形成温度锋、边界急流和垂向流的一个新的理论，表明即使在没有风的驱动下，由海洋内部热量的再分布也可以在边界附近出现物理量的密集现象。

参考文献

[1] Hsueh Y, Kenney R N. Steady coastal upwelling in a continuously stratified ocean. J Phys Oceanogr, 1972, 2: 27-33.

[2] Allen J S. Upwelling and coastal jet in a continuously stratified ocean. J Phys Oceanogr, 1973, 3: 245-257

[3] 巢纪平，冯立成，王东晓. 风生西边界急流稳定性的渐近理论. 地球物理学报，2006，49(3)：642-649.

[4] 巢纪平，陈显尧. 沿岸上升流和沿岸急流的一个半解析理论.地球物理学报，2003，46(1)：26-30.

[5] 王东晓，陈举，杜岩，等. 热带海盆对热力强迫的线性响应. 地球物理学报，2002，45(增刊)：75-83.

[6] Spall M A. On the thermohaline circulation in semi-enclosed marginal seas. J Mar Res, 2003, 61: 1-25.

[7] Spall M A, Pickart R S. Where does dense water sink? A subpolar gyre example. J Phys Oceanogr, 2001, 31: 810-826.

[8] Marotzke J, Scott J R. Convective mixing and the thermohaline circulation. J Phys Oceanogr, 1999, 29: 2962-2970.

[9] Pedlosky J. Thermohaline driven circulation in small oceanic basins. J Phys Oceanogr, 2003, 33: 2333-2340.

第 3 部分　长期数值预报

一种长期数值天气预报方法的物理基础*

长期数值天气预报研究小组**

摘要:本文对一种新的长期数值天气预报方法的物理基础做了探讨。在考虑了各种热源(汇)参数化形式后,建立了海-气联合系统中的闭合运动方程组。对这组方程的频率分析指出,存在两类在性质上完全相异的运动形式,其中一类是振荡周期为几天的短期波,它本质上是 Rossby 行波;另一类则是时间尺度为月的长期波。在非绝热作用下,这两类波可以是发展的(不稳定)。为了避免短期波以及误差所造成的虚拟过程的强烈发展,而把作为预报对象的、振幅较小的长期波的演变给掩盖掉,作者认为,在方法论上可以把短期波看成是一种"干扰"长期波的"噪音"而滤掉它。文中给出了一种滤掉短期波的方法。最后,分析了海洋的表层温度和大气运动之间的相互适应过程,指出,对于长周期运动,大气运动主要向初始的海温场适应。这就进一步为滤波后的长期预报方法提供了理论根据。

20 多年来,短期天气的数值预报已有相当发展,随着短期天气数值预报方法的获得成功,自然会提出制作长期天气数值预报的可能性问题。虽然,早年 Блинова[1] 和近几年来 Adem[2] 在这方面作了尝试,但并不能认为一个有效的长期天气预报方法已经令人满意的建立起来了。

注意到大气这部热机,其运动状态在时间上是多频的,可以从短到几秒钟的湍流直至长到以月、季和年为特征的长期过程。大气中所出现的这些运动形式虽然是复杂的,但每一种现象的发生和发展都有制约它的主要的物理过程,即有它不同于其他运动形式的矛盾的特殊性。目前已普遍认为,在研究长期天气过程时需要考虑大气和海洋(广泛地为地球表面)耦合系统中的非绝热过程[3,4]。但考虑了加热过程,并不意味着已经可以数值地作长期预报了。本文将讨论这方面存在的困难,并提出解决的途径,从而建立一种有效的长期天气的数值预报方法。

1　地-气交界面上的热量平衡

物理气候学的研究表明[5],在地球表面上的热量平衡关系为:

$$\mathscr{R} = LE + \mathscr{P} + \mathscr{A}, \tag{1}$$

其中,$\mathscr{R}$ 为辐射热通量(即辐射平衡);$\mathscr{P}$ 为地球表面与其上面的大气之间的湍流热通量;$\mathscr{A}$ 为地球表面与其下层之间的热通量;LE(L 为蒸发潜热,E 为蒸发速度)表示蒸发消耗的热通量。

辐射热通量等于下垫面所吸收到的太阳总辐射与有效辐射之差,即

$$\mathscr{R} = (S_0 + s_0)(1 - \tilde{a}) - I \tag{2}$$

式中,S_0 为直接辐射总量;s_0 为散射辐射总量;I 为有效辐射;$\tilde{a}$ 为地球反照率。一般可以经验地分别

---

* 中国科学,1977,20(2):162-172.

** 先后参加过本工作的有:中国科学院地理研究所巢纪平、季劲钧,大气物理研究所何家骅、刘克武,中央气象局气象科学研究所吕越华、周琴芳,北京大学数值预报进修班学员裴巨才、潘华盛、宣德旺、洪明慧、仲雅琴、魏禧。

用下式来计算总辐射和有效辐射：

$$(S_0+s_0)=(S_0+s_0)^*(1-c_s n)$$
$$I=I^*(1-c_i n)+\sigma(T_{s_n}^4-T_0^4)$$

其中，带“ * ”号者表示碧空条件下的辐射值；n 为云量；c_s, c_i 分别为表征云量对总辐射和有效辐射影响的经验常数；σ 为司蒂芬-玻茨曼常数；T_{s_n}, T_0 分别为地球表面及其上面的空气的温度。辐射平衡总的表达式为：

$$\mathscr{R}=(S_0+s_0)^*(1-\tilde{a})(1-c_s n)-I^*(1-c_i n)-\sigma(T_{s_0}^4-T_0^4) \tag{3}$$

我们感兴趣的是异常情况，即对气候平均值的距平情况。设云量的气候平均值为 $\tilde{n}$，距平值为 n'，$n=\tilde{n}+n'$，由此异常的辐射通量可以近似地写成：

$$\mathscr{R}'\approx-[\widetilde{(S'_0+s_0)}^*(1-\tilde{a})c_s-\tilde{I}^*c_i]n'-4\sigma\tilde{T}_0^3(T'_{s_0}-T'_0) \tag{4}$$

式中，“~”号表示气候平均值，这里假定 $\tilde{T}_s\approx\tilde{T}$，“′”表示距平值，下标“0”表示地面值。

如果所考虑的云量主要由于边界层中大范围摩擦辐合所造成，在这种情况下可以认为云量将与摩擦层顶的垂直速度 W_b 成正比，即

$$n=\left(\frac{W_b}{W^*}\right) \tag{5}$$

式中，W^* 为一经验系数，由行星边界层理论可以得到：

$$W_b=l_b\zeta_{0g} \tag{6}$$

其中，l_b 为行星边界层的厚度，ζ_{0g} 为地面的地转涡度。由此

$$n=\left(\frac{l_b}{W^*}\right)\zeta_{0g} \tag{7}$$

如果 $\zeta_{0g}=\xi_{0g}+\zeta'_{0g}$，则云量的距平值近似地为：

$$n'=\left(\frac{\tilde{l}_b}{W^*}\right)\zeta'_{0g} \tag{8}$$

将式(8)代入式(4)，得到：

$$\mathscr{R}'\approx-C\left(\frac{\tilde{l}_b}{W^*}\right)\zeta'_{0g}-4\sigma\tilde{T}^3(T'_{s_0}-T'_0) \tag{9}$$

式中

$$C=\widetilde{(S_0+s_0)}^*(1-\tilde{a})c_s,-\tilde{I}^*c_i$$

这样，通过对云量的参数化，把下垫面接受到的异常辐射量和大气的异常运动联系起来了。

湍流热通量为：

$$\mathscr{P}=-\rho c_p K_T\frac{\partial T}{\partial z}\approx\rho c_p c_D|V|(T_{s_0}-T_0). \tag{10}$$

其距平值为：

$$\mathscr{P}=-\rho c_p K_T\frac{\partial T'}{\partial z}\approx\rho c_p c_D|V|(T'_{s_0}-T'_0) \tag{11}$$

式中，ρ 为空气密度；c_p 为空气的定压比热；K_T 为湍流导热系数；c_D 为曳力系数；$|V|$ 为地面风速的绝对值。

地表面与其下层之间的热通量为：

$$\mathscr{A} = \rho_s c_{ps} K_s \left(\frac{\partial T_s}{\partial z}\right)_0 \tag{12}$$

其距平值为：

$$\mathscr{A}' = \rho_s c_{ps} K_s \left(\frac{\partial T'_s}{\partial z}\right)_0 \tag{13}$$

式中，ρ_s，c_{ps}，K_s 分别为下垫面物质（土壤或水）的密度、比热和导热系数。

蒸发热通量为：

$$LE = - L\rho K_q \left(\frac{\partial q}{\partial z}\right)_0 \tag{14}$$

其距平值为：

$$LE' = - L\rho K_q \left(\frac{\partial q'}{\partial z}\right)_0 \tag{15}$$

式中，q 为比湿；K_q 为对水汽的湍流交换系数，近似地取 $K_q = K_T$。

一般来说，蒸发只有在海面上才是重要的，在海面上蒸发热通量可以取另一种方便的参数化形式。设在海面上空气的比湿已达到饱和值，即

$$q = q_s = \frac{R_e}{R_w} \frac{e_s(T_0)}{p}$$

式中，R_e，R_w 分别为空气和水汽的气体常数；p 为气压；e_s 为饱和水汽压。取上式的对数并微分，近似地得到：

$$\left(\frac{\partial q_s}{\partial z}\right)_0 \approx - \left(\gamma \frac{\partial \ln e_s}{\partial T}\right)_0 q_{s_n}$$

其中，$\gamma = -\partial T / \partial z$。对距平部分近似地有：

$$\left(\frac{\partial q'_s}{\partial z}\right)_0 \approx - \widetilde{\left(\gamma \frac{\partial \ln e_s}{\partial T}\right)_0} q_{s0}$$

由于水汽刚从海面蒸发出来，可以假定

$$q'_{s0} \approx \frac{\partial \widetilde{q}_s}{\partial \widetilde{T}} \cdot T'_{s0} \tag{16}$$

因此蒸发热通量的距平值为：

$$LE' \approx \left(L\rho K_T \widetilde{\gamma} \frac{\partial \ln \widetilde{e}_s}{\partial \widetilde{T}} \cdot \frac{\partial \widetilde{q}_S}{\partial \widetilde{T}}\right)_0 T'_{s0} \tag{17}$$

即异常的蒸发热通量与海表温度的异常值成正比。

把以上各有关公式汇总起来，就得到地-气交界面上在异常情况下的热量平衡方程为：

$$z = 0,\ - \rho c_p K_T \left(\frac{\partial T'}{\partial z}\right)_0 + \rho_s c_{ps} K_s \left(\frac{\partial T'_s}{\partial z}\right)_0 + \left(L\rho K_T \widetilde{\gamma} \frac{\partial \ln \widetilde{e}_s}{\partial \widetilde{T}} \frac{\partial \widetilde{q}_s}{\partial \widetilde{T}}\right)_0 T'_{s0}$$

$$= -\frac{Cl_b}{W^*}\zeta'_{0g} - 4\sigma\tilde{T}^3(T'_{s_0} - T'_0) \tag{18}$$

2 大气中的热源(汇)

由热力学第一定律,在(x,y,p,t)坐标系中,大气温度的变化方程为:

$$\rho c_p\left(\frac{\mathrm{d}T}{\mathrm{d}t} - \sigma_p\omega\right) = \varepsilon \equiv \varepsilon_T + \varepsilon_R + \varepsilon_Q \tag{19}$$

式中,σ_p 为静力稳定度,而

$$\frac{\mathrm{d}}{\mathrm{d}t} \equiv \frac{\partial}{\partial t} + u\frac{\partial}{\partial x} + v\frac{\partial}{\partial y}$$

其中,u,v,ω 分别为空气质点沿 x,y,p 方向的速度。

湍流热流量 ε_T 的形式比较简单,为:

$$\frac{1}{\rho c_p}\varepsilon_T = \frac{\partial}{\partial p}\left(K\frac{\partial T}{\partial p}\right) \tag{20}$$

式中,$K=\rho^2g^2K_T$,K_T 同前。

辐射热流量 ε_R,根据郭晓岚的格式[6]为:

$$\frac{1}{\rho c_p}\varepsilon_R = \frac{\partial}{\partial p}\left(K_R\frac{\partial T}{\partial p}\right) - \frac{1}{\tau_R}(T - T_e) + \frac{k'}{\rho c_p}s \tag{21}$$

式中,s 为太阳的短波辐射;k'为空气对短波辐射的吸收系数;T_e 为环境温度,而

$$K_R \equiv \frac{8\gamma\sigma\tilde{T}^3}{c_pk_s}\rho g^2,\quad \tau_R \equiv \frac{\rho c_p}{8(1-\gamma)\sigma\tilde{T}^3k_w} \tag{22}$$

其中,k_s,k_w 分别为吸收物质在强吸收区和弱吸收区中的平均吸收系数;γ 表示强吸收区内的黑体辐射强度占总的黑体辐射强度的百分数。这表明,辐射能的传递在强吸收区中相当于一类扩散过程,而在弱吸收区中则相当于牛顿冷却形式。

对于水汽相变热流量 ε_Q,可以采用下面简单的参数化方法,

$$\frac{1}{\rho c_p}\varepsilon_Q = -\frac{L}{c_p}\frac{\mathrm{d}q_s}{\mathrm{d}t} \approx -\frac{L}{c_p}\frac{\partial q_s}{\partial z}W_b \tag{23}$$

在这里已近似地用 W_b 代替 W,而 W_b 由式(6)决定,因此上式可近似地写成:

$$\frac{1}{\rho c_p}\varepsilon_Q = \tilde{T}^*\zeta_{0g} \tag{24}$$

式中

$$\tilde{T}^*(p) = \frac{L}{c_p}\tilde{\gamma}\frac{\mathrm{d}\ln\tilde{e}_s}{\mathrm{d}\tilde{T}}l_b\tilde{q}_s(p)$$

考虑了以上 3 种热流量后,热力学第一定律的最后形式为:

$$\frac{\mathrm{d}T}{\mathrm{d}t} - \sigma_p\omega = \frac{\partial}{\partial p}(K + K_R)\frac{\partial T}{\partial p} - \frac{1}{\tau_R}(T - T_e) + \tilde{T}^*\zeta_{0g} + \frac{k'}{\rho c_p}s \tag{25}$$

3　海-气联合系统中的非绝热波

为简单起见，设地球表面均为海洋，同时大气中的气候状态简单地取成：$\widetilde{T}(y,p)$，$\widetilde{u}(p)$，由静力学关系，温度和重力位势 ϕ 之间存在关系式：

$$\widetilde{T}=-\frac{p}{R_e}\frac{\partial\widetilde{\phi}}{\partial p} \tag{26}$$

热成风关系式为：

$$\frac{\partial\widetilde{T}}{\partial y}=f\frac{p}{R_e}\widetilde{u}_p\left(\widetilde{u}_p=\frac{\mathrm{d}\widetilde{u}}{\mathrm{d}p},\text{为常数}\right) \tag{27}$$

由式(25)，得到温度距平 T' 的变化方程为：

$$\frac{\mathrm{D}T'}{\mathrm{D}t}-\sigma_p\omega'+\frac{\partial\widetilde{T}}{\partial y}v'=\frac{\partial}{\partial p}(K+K_R)\frac{\partial T'}{\partial p}-\frac{1}{\tau_R}T'+\widetilde{T}^*\zeta'_{0g} \tag{28}$$

在这里已用了线性化假定，故

$$\frac{\mathrm{D}}{\mathrm{D}t}\equiv\frac{\partial}{\partial t}+\widetilde{u}(p)\frac{\partial}{\partial x}$$

另外，已取大气的环境温度 $T_e=\widetilde{T}$，并略去了异常的太阳短波辐射的影响。

考虑到在静力平衡下有：

$$T'=-\frac{p}{R_e}\frac{\partial\phi'}{\partial p} \tag{29}$$

以及准地转近似：

$$u'=-\frac{1}{f}\frac{\partial\phi'}{\partial y},\quad v'=\frac{1}{f}\frac{\partial\phi'}{\partial x} \tag{30}$$

$$\zeta'=\frac{1}{f}\left(\frac{\partial^2\phi'}{\partial x^2}+\frac{\partial^2\phi'}{\partial y^2}\right)\equiv\frac{1}{f}\Delta\varphi' \tag{31}$$

于是方程(28)可以写成：

$$\left(\frac{D}{Dt}-\frac{\partial}{\partial p}(K+K_R)\frac{\partial}{\partial p}\right)\frac{\partial\phi'}{\partial p}-\widetilde{u}_p\frac{\partial\phi'}{\partial x}+\frac{1}{\tau_R}\frac{\partial\phi'}{\partial y}=-\sigma_p\omega'-\frac{R_e\widetilde{T}^*}{pf}(\Delta\phi')_0 \tag{32}$$

在另一方面，线性化的运动方程(异常状态)为：

$$\frac{\mathrm{D}u'}{\mathrm{D}t}-fv'=-\frac{\partial\phi'}{\partial x}+g\frac{\partial\tau_x}{\partial p} \tag{33a}$$

$$\frac{\mathrm{D}v'}{\mathrm{D}t}+fu'=-\frac{\partial\phi'}{\partial y}+g\frac{\partial\tau_y}{\partial p} \tag{33b}$$

式中，τ_x，τ_y 分别为沿 x，y 方向的摩擦应力。连续性方程为：

$$\frac{\partial u'}{\partial x}+\frac{\partial v'}{\partial y}+\frac{\partial\omega'}{\partial p}=0 \tag{34}$$

由此得到

$$\frac{\mathrm{D}\Delta\phi'}{\mathrm{D}t}+\beta\frac{\partial\phi'}{\partial x}=f^2\frac{\partial\omega'}{\partial p}+gf\frac{\partial}{\partial p}\left(\frac{\partial\tau_y}{\partial x}-\frac{\partial\tau_x}{\partial y}\right) \tag{35}$$

式中，$\beta\equiv \mathrm{d}f/\mathrm{d}y$。

应用式(32)消去式(35)式中的 ω'后，得到非绝热地转涡度方程为：

$$\frac{\mathrm{D}}{\mathrm{D}t}\left(\Delta\phi'+\frac{f^2}{\sigma_p}\frac{\partial^2\phi'}{\partial p^2}\right)+\beta\frac{\partial\phi'}{\partial x}=-\frac{f^2}{\sigma_p\tau_R}\frac{\partial^2\phi'}{\partial p^2}+\frac{f^2}{\sigma_p}\frac{\partial^2}{\partial p^2}(K+K_R)\frac{\partial^2\phi'}{\partial p^2}$$
$$-\left(\frac{f}{\sigma_p}\frac{R_e}{p}\frac{\partial\widetilde{T}^*}{\partial p}\right)(\Delta\phi)_0+gf\frac{\partial}{\partial p}\left(\frac{\partial\tau_y}{\partial x}-\frac{\partial\tau_x}{\partial y}\right) \tag{36}$$

方程(36)的一个边界条件可以取

$$p=p_0(\text{海平面}),\quad \omega'=0 \tag{37}$$

由于在海平面上，方程(25)牛顿辐射冷却项中的环境温度 T_e 应取成海表温度 T_s，而 $T_s=\widetilde{T}_s+T'_s$，因此如设 $\widetilde{T}=\widetilde{T}_s$，则考虑到式(37)后，有

$$p=p_0,\quad \left(\frac{\mathrm{D}_0}{\mathrm{D}t}-\frac{\partial}{\partial p}(K+K_R)\frac{\partial}{\partial p}\right)\frac{\partial\phi'}{\partial p}-\widetilde{u}_p\frac{\partial\phi'}{\partial x}=-\frac{1}{\tau_R}\left(\frac{\partial\phi'}{\partial p}+\frac{R_e}{p_0}T'_s\right) \tag{38}$$

方程(36)的另外的边界条件可以取

$$p=0,\quad \phi'=\frac{\partial\phi'}{\partial p}=\frac{\partial^2\phi'}{\partial p^2}=\frac{\partial^3\phi'}{\partial p^3}=0 \tag{39}$$

应用边界条件式(38)和式(39)，将方程(36)从 $p=p_0$ 到 $p=0$ 积分，并设式中各系数均与高度无关，得到：

$$\int_{p_0}^{0}\left(\frac{\mathrm{D}\Delta\phi'}{\mathrm{D}t}+\beta\frac{\partial\phi'}{\partial x}\right)\mathrm{d}p=-\frac{f^2R_e}{\sigma_p p_0\tau_R}(T'_s)_{p=p_0}$$
$$+p_0\frac{fR_e}{\sigma_p\widetilde{p}}\frac{\partial\widetilde{T}^*}{\partial p}(\Delta\phi')_{p=p_0}-gf\left(\frac{\partial\tau_y}{\partial x}-\frac{\partial\tau_z}{\partial y}\right)_{p=p_0} \tag{40}$$

式中最后一项可写成：

$$\left(\frac{\partial\tau_y}{\partial x}-\frac{\partial\tau_x}{\partial y}\right)_{p=p_0}=-k\frac{p_0}{g}\zeta'_{0g} \tag{41}$$

其中，k 为常数因子。

在另一方面，方程(40)的左方可以看成是平均层(500 mbar)上的运动，设地面的涡度 ζ'_0 与平均层上的值 ζ^* 成正比，即取 $\zeta'_0=b\zeta^*$，这样就得到：

$$\frac{\mathrm{D}\Delta\phi'^*}{\mathrm{D}t}+\beta\frac{\partial\phi'^*}{\partial x}+F\Delta\phi'^*=\frac{f^2R_e}{\sigma_p p_0^2\tau_R}(T'_s)_{p=p_0} \tag{42}$$

式中

$$F=b\left[k-\frac{fR_e}{\sigma_p\widetilde{p}}\frac{\partial\widetilde{T}^*}{\partial\widetilde{p}}\right]_0$$

由此可见，平均层上的异常运动，除受初始场的影响外，还为下垫面(现为海洋)的温度距平所控制。

如果不考虑海流的作用，海温服从热传导方程，即

$$\frac{\partial T'_s}{\partial t}=K_s\frac{\partial^2 T'_s}{\partial z^2} \tag{43}$$

定解条件为:

$$z=-D,\quad T'_s=0 \tag{44}$$

另外,由海洋到大气过渡时,两个介质的温度保持连续,即

$$z=0,\quad T'_{s_0}=T'_0 \tag{45}$$

热量平衡方程考虑到式(11)和式(45)后为:

$$z=0,\quad \rho_s c_{ps}K_s\left(\frac{\partial T'_s}{\partial z}\right)_0+\left(\rho LK_T\bar{\gamma}\frac{\mathrm{dln}\tilde{e}_s}{\mathrm{d}\tilde{T}}\frac{\partial\tilde{q}_s}{\partial\tilde{T}}\right)_0 T'_{s_n}=-\frac{Cl_b}{W^*f}(\Delta\phi'^*)_0 \tag{46}$$

由于在写出此式时已用了条件式(45),所以可以把式(46)作为方程(43)的另一个定解条件。

在这个海气相互作用的模式中所包含的物理过程有:海洋给大气运动以蒸发潜热,大气中的云量调节着海表的辐射平衡,而云量又受大气运动所制约。这个模式与 Монин 等[3] 所设计的接近,但大气运动的形式两者是不同的,所以结果是很不一样的。

现在对以上各式进行无量纲化,引进

$$t=\tau\bar{t},\ z=D\bar{z},\ x=l\bar{x},\ y=l\bar{y},\ T_s=\delta T_s\bar{T}_s,$$

$$\phi'^*=\frac{f^2R_e l\delta T_s}{\sigma_p p_0^2\tau_R\beta}\bar{\phi}'^*$$

式中,τ 为特征时间;l 为水平特征尺度;D 为海洋的深度;δT_s 为海水温度距平的特征值。方程(42)和方程 (43)的无量纲形式分别为(略去"–"号):

$$\varepsilon\frac{\partial\Delta\phi'^*}{\partial t}+\tilde{U}\frac{\partial\Delta\phi'^*}{\partial x}+\tilde{F}\Delta\phi'^*+\frac{\partial\phi'^*}{\partial x}=(T'_s)_{z=0} \tag{47}$$

$$\mu\frac{\partial T'_s}{\partial t}=\frac{\partial^2 T'_s}{\partial z^2} \tag{48}$$

定解条件式(44)和式(46)的无量纲形式分别为:

$$z=-1,\quad T'_{s_0}=0 \tag{49}$$

$$z=0,\left(\frac{\partial T'_s}{\partial z}\right)_0+\lambda_Q T'_{s_0}=-\lambda_s(\Delta\phi'^*)_0 \tag{50}$$

其中无量纲数分别为:

$$\varepsilon=\frac{1}{\tau\beta l},\ \mu=\frac{D^2}{\tau K_s},\ \tilde{U}=\frac{U}{\beta l^2},\ \tilde{F}=\frac{F}{\beta l},\ \lambda_Q=\frac{D}{D_Q},\ \lambda_s=\frac{D}{D_s}$$

而

$$D_Q=\frac{\rho_s c_{ps}K_s}{\rho K_T L\tilde{\gamma}\frac{\mathrm{dln}\tilde{e}_s}{\mathrm{d}\tilde{T}}\frac{\partial\tilde{q}_s}{\partial\tilde{T}}},\quad D_s=\frac{\rho_s c_{ps}K_s}{\frac{Cl_b}{W^*}\cdot\frac{bfR_e}{\sigma_p p_0^2\tau_R\beta l}}$$

为简单起见,设运动与 y 无关,并设

$$\phi'^*=\phi'^*_0\mathrm{e}^{-i(mx-\sigma t)},\quad T'_s=T'_{s0s}(z)e^{-i(mx-\sigma t)} \tag{51}$$

由方程(47)得到:

$$m\mid m(\varepsilon\sigma-\widetilde{U}m)+1-\mathrm{i}\widetilde{F}m\mid \phi'^{*}_{0}=iT'_{s0}(0) \tag{52}$$

由方程(48)得到:

$$\frac{\mathrm{d}^2T_{s0}}{\mathrm{d}z^2}-i\mu\sigma T'_{s0}=0 \tag{53}$$

其解为:

$$T'_{s0}=a'\mathrm{e}^{-\sqrt{\mathrm{i}\sigma\mu z}}+b'\mathrm{e}^{\sqrt{\mathrm{i}\sigma\mu z}}$$

应用条件式(49)后,得到

$$z=0,\quad\begin{cases}T'_{s0}(0)=a'(1-\mathrm{e}^{2\sqrt{\mathrm{i}\sigma\mu}}),\\ \dfrac{\partial T'_{s0}}{\partial z}=-a'\sqrt{\mathrm{i}\sigma\mu}(1+\mathrm{e}^{2\sqrt{\mathrm{i}\sigma\mu}})\end{cases} \tag{54}$$

将式(52)、式(54)两式代入式(50),得到频率方程为:

$$\sqrt{\mathrm{i}\sigma\mu}\,\mathrm{ctnh}\sqrt{\mathrm{i}\sigma\mu}+\lambda_Q=\frac{\mathrm{i}\lambda_s m}{m(\varepsilon\sigma-\widetilde{U}m)+1-\mathrm{i}\widetilde{F}m} \tag{55}$$

由于海洋的混合层约 100 m 深,K_s 一般为 $10^0\sim10^1\ \mathrm{cm^2/s}$,因此 $\tau_s=D^2\ K_s\sim10^7-10^8$ s,如果所研究的运动其特征时间为月的量级,则这时 $\mu\gg1$,而 $\mathrm{ctnh}\sqrt{\mathrm{i}\sigma\mu}\approx1$,于是式(55)简化成

$$\sqrt{\mathrm{i}\sigma\mu}+\lambda_Q=\frac{\mathrm{i}\lambda_s m}{m(\varepsilon\sigma-\widetilde{U}m)+1-\mathrm{i}\widetilde{F}m} \tag{56}$$

将 λ_Q 移至右端,平方后就得到一个对于 σ 的三次方代数方程,这表明在这样一个海气联合系统中存在三组波动。在数值计算中略去 $O(\delta^2)$ 的小项,其中 $\delta=\varepsilon m/\mu\ll1$,并取 $K_s=10\ \mathrm{cm^2/s}$,$\tau_\beta=1/\beta l=1.22\times10^5$ s,$D_Q=0.667\times10^4$ cm,$D_S=1.15\times10^4$ cm。

当 $\widetilde{U}=\widetilde{F}=0$ 时,三组波中有二组波的振荡频率见图 1a。这两组波的频纺十分接近。当波长为 6 000 km时,其值约为 $1.3\times10^{-5}/\mathrm{s}^5$,相当于周期 5~6 天,可见这是一类短期波。

为了阐明这类短期波的物理本质,现不考虑海洋的影响,即取 $D\to0$,于是 $\mu\to0$,$\lambda_Q=\lambda_i\to0$,由于式(56)中 $\sigma\mu$ 的值为有限(即 σ 可以很大),所以要有

$$m(\varepsilon\sigma-\widetilde{U}m)+1-\mathrm{i}\widetilde{F}m=0$$

取 $\widetilde{U}=\widetilde{F}=0$,并回到有量纲量,得到

$$\sigma_R=-\frac{\beta}{m} \tag{57}$$

这是正压大气中 Rossby 行波的振荡频率。它的值见图 1a 中的虚线,可见与另外两组波的频率非常接近。由此表明,这类短期波本质上是在海洋加热影响下的非绝热 Rossby 波。

图 1b 给出这两组短期波的增长率(σ 的虚部),可以看到其中一组是阻尼波,另一组是发展波。当波长为 6 000 km 时,发展波的增长率约为 $1.5\times10^{-6}/\mathrm{s}$,即 7~8 天后振幅增大 e 倍。

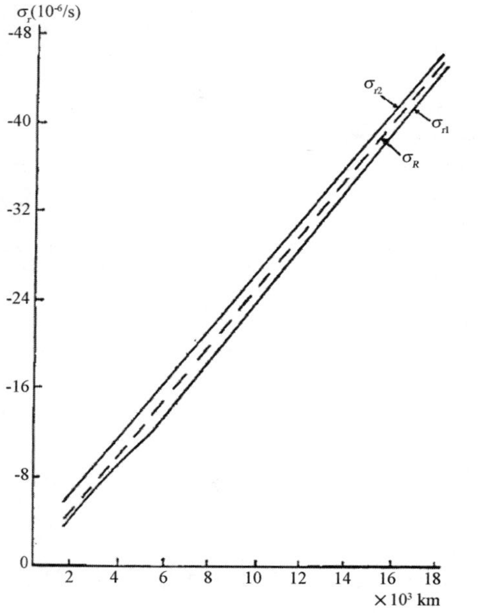

图 1a　短期波和正压 Rossby 波的振荡频率

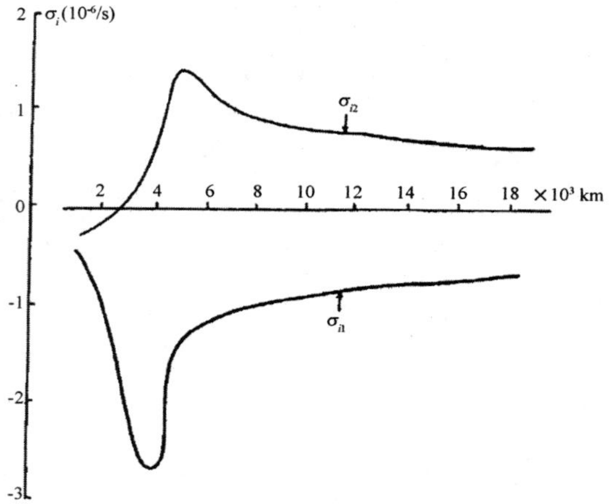

图 1b　短期波的增长率

代数方程(56)中的另一组波,其振荡频率见图 2a(实线),当波长为 6 000 km 时,约 0.5×10^{-6}/s,即周期约 3 个月。当波长大于 3 000~4 000 km 时,这是一类后退波,其移速为 1 000~2 000 km/月。这类移动缓慢的波,可以认为它代表了海洋上空大气中一类与海洋活动相适应的半永生活动中心。

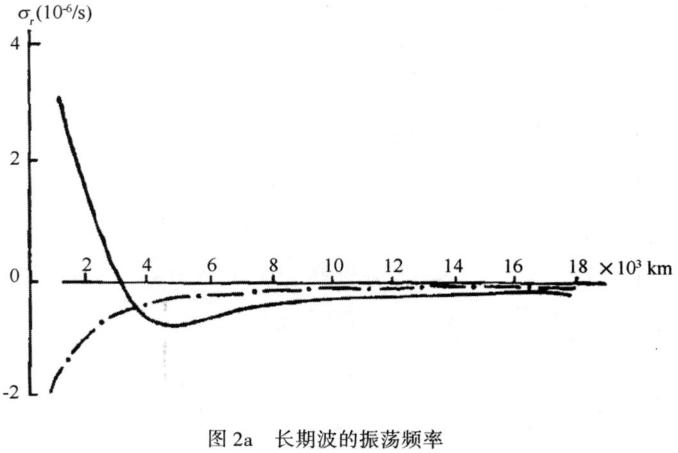

图 2a　长期波的振荡频率

图 2b 给出这类长期波的增长率(实线)。这类波在短波部分是阻尼的,当波长大于 5 000 km 后是发展的,增长率约为 0.5×10^{-6}/s,即相当于 1 个月左右振幅增长 e 倍。

计算表明,盛行风和摩擦的作用对长期波的增长率影响不大,只有对短波长,部分摩擦作用将使增长率减小。

4　短期波的过滤

上节的计算表明,在一个海气联合系统中,的确存在一类缓慢变化的运动,自然可以认为这类缓

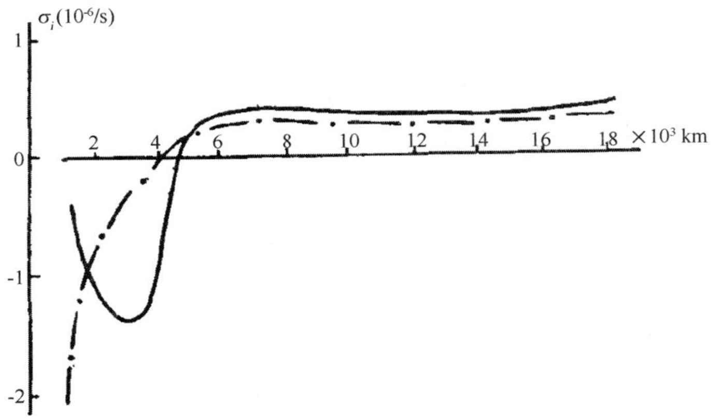

图 2b 长期波的增长率

慢变化的运动，代表了大气中一类在非绝热影响下的长期天气过程。长期天气数值预报的任务就是要把这类过程定量地预报出来。同时也看到，在海气联合系统中也存在着中短期预报对象的“快变”运动，而且由于加热影响，使得这类运动获得了新的发展能源。这种现象在大气中是经常可以观测到，如新生气旋或高空槽出海后，或者台风移至暖海上空时，由于从海洋上获得了感热和潜热而很快地发展起来。

在这种长、短期过程在一个模式中共存的情况下，以为只要把预报时间延长，就可以做出长期天气的预报了，这种想法未免过于简单。事实上，从上面的计算可以看出，由于短期波的增长率几乎比长期波大了一个量级。因此如果不做适当的处理，只是把预报时间拉长，那么小振幅的长期过程必然要被大振幅的短期过程和误差造成的虚拟过程所掩盖掉。如目前可预测性的研究所表明的那样，预报时效不超过两周。故至少在目前，对长期天气过程在物理上了解得尚不十分清楚，以及计算方法尚有待改进的情况下，需要另行建立长期天气过程的预报方法。

这种情况和短期数值发展初期所遇到的困难有类似之处。当年 Richardson 对数值预报尝试的失败，其中主要一个原因就是在他所用的预报方程中，既包含有作为预报对象的天气过程，又包含了一类与天气变化无密切关系的“噪音”，如声波和重力波。由于这类噪声的存在，当计算方案不合适时，它们将不稳定地发展起来而掩盖了有意义的天气过程。以后 Charney 等采用了一种滤波技术（如准地转近似），把这些噪音过滤后，预报才获得成功。作为一种新的尝试，我们认为可以把以 Rossby 行波为主导的短期天气过程，看作是一种“干扰”长期天气过程的噪音而过滤掉，使预报方程中单纯包含长期天气过程。

滤掉 Rossby 波是容易做到的，只要在方程（47）中把与 ε 有关的时间变化项省掉就可以了。事实上，在式（56）中令 $\varepsilon=0$，并令 $\sigma=\sigma_r-\mathrm{i}\sigma_i$，则当 $\widetilde{U}=\widetilde{F}=0$ 时，得到

$$\sigma_r=-\frac{2\lambda_s\lambda_Q m}{\mu} \tag{58}$$

$$\sigma_s=-\frac{1}{\mu}(\lambda_s^2 m-\lambda_Q^2) \tag{59}$$

容易看出，这是长期波的振荡频率和增长率，其值分别见图 2a、图 2b 中的虚线。可见，当波长大于 4 000 km后，过滤后的结果与原来的十分接近。

这种取 $\varepsilon=0$ 的过滤方法，可以称为准定常方法，因为它相当于把描写大气运动的方程(47)简化为一种大气运动与海温场之间的平衡关系式，即

$$\widetilde{U}\frac{\partial\Delta\phi'^{*}}{\partial x}+\widetilde{F}\Delta\phi'^{*}+\frac{\partial\phi'^{*}}{\partial x}=(T'_s)_{z=0} \tag{60}$$

这个方程虽然在形式上是定常的，但并不意味着高度距平没有时间变化，这只是表明大气运动已完全适应了海温场。因为海温场是有时间变化的，所以高度场也将随着海温场的变化而演变，同时通过边界条件式(50)，大气运动对海温的变化仍有所馈影响。

5　大气运动对海温场的适应

现在进一步从理论上来讨论准定常近似成立的物理依据。

令式(47)中 $\widetilde{U}=\widetilde{F}=0$，并对 $v=\dfrac{\partial\phi^{*}}{\partial x}$ 重新写出方程，在一维情况下为：

$$\varepsilon\frac{\partial^2 v}{\partial t\partial x}+v=T'_{s0} \tag{61}$$

下标"0"在此表示海表温度。重新写出式(48)及相应的边界条件

$$\mu\frac{\partial T'_s}{\partial t}=\frac{\partial^2 T'_s}{\partial z^2} \tag{62}$$

$$z=0,\quad \left(\frac{\partial T'_s}{\partial z}\right)_0+\lambda_s T'_{s0}=-\lambda_s\frac{\partial v}{\partial x} \tag{63}$$

$$z\to-\infty,\quad T'_s=0 \tag{64}$$

问题的初始条件为：

$$t=0,\quad \frac{\partial v}{\partial x}=\zeta'^0,\quad T_s=T'^0_s \tag{65}$$

采取下面的近似解法。注意当 $\varepsilon=0$ 时，由式(64)得到风场和海温场之间的平衡关系为：

$$v=T'_{s0} \tag{66}$$

如果所研究的运动，其时间尺度 $\tau>\tau_\beta=1/\beta l$，即大于 Rossby 波的特征时间，这时有 $\varepsilon<1$，所以方程(61)可以近似地写成：

$$v=T'_{s0}-\varepsilon\frac{\partial^2 T'_{s0}}{\partial t\partial x} \tag{67}$$

将此式代入式(63)，得到：

$$\left(\frac{\partial T'_s}{\partial z}\right)_0+\lambda_Q T'_{s0}=-\lambda_s\left(\frac{\partial T'_{s0}}{\partial x}-\varepsilon\frac{\partial^3 T'_{s0}}{\partial t\partial x^2}\right) \tag{68}$$

在这样的近似下，问题的解为：

$$\begin{aligned}T'_{s0}=&\frac{\sqrt{\mu}}{\lambda_s}\int_{-\infty}^{x}T'^0_s(x')\,\mathrm{e}^{-\frac{\lambda_Q}{\lambda_s}(x-x')}\chi\left(\frac{\sqrt{\mu}(x-x')}{\lambda_s},t\right)\mathrm{d}x'\\&+\frac{\sqrt{\mu}}{\lambda_s}\int_{\chi}^{\infty}\int_0^t T'^0_{s0}(x')\,\mathrm{e}^{-\frac{\lambda_Q}{\lambda_s}(x'-x)}J_0\left(2\sqrt{\frac{(x'-x)(t-t')}{\varepsilon}}\right)\cdot\frac{1}{2t'}\\&\times\chi\left(\frac{\sqrt{\mu}(x'-x)}{\lambda_s},t'\right)\cdot\left(\frac{\mu(x'-x)^2}{2\lambda_s^2t'}-1\right)\mathrm{d}t'\mathrm{d}x'\end{aligned}$$

$$
- \varepsilon\int_{-\infty}^{x} \zeta'^{0}(x') \frac{\partial}{\partial\alpha}\left(\mathrm{e}^{-\frac{\lambda_Q}{\lambda_s}a\alpha}\phi\left(\frac{\sqrt{\mu}\,\alpha}{\lambda_s},t\right)\right)\Bigg|_{\alpha=x-x'} \mathrm{d}x'
$$

$$
+ \varepsilon\int_{x}^{\infty}\int_{0}^{t}\zeta_0^{t}(x') \frac{\partial}{\partial\beta}\left(J_0\left(2\sqrt{\frac{\beta(t-t')}{\varepsilon}}\right)\cdot\left(\frac{\mu\beta^2}{2\lambda_s^2 t}-3\right)\cdot\frac{1}{2t'^2}\right.
$$

$$
\left.\times\sqrt{\frac{\mu\beta^2}{\lambda_s^2}}\chi\left(\frac{\sqrt{\mu}\beta}{\lambda_s},t'\right)\right)\Bigg|_{\beta=x'-x} \mathrm{d}t'\mathrm{d}x' \tag{69}
$$

式中，J_0 为零阶贝塞尔函数，而

$$
\chi(a,b)=\frac{1}{\sqrt{ab}}\mathrm{e}^{-\frac{a^2}{4b}},\phi(a,b)=\frac{a}{2\sqrt{\pi b^2}}\mathrm{e}^{\frac{a^2}{4b}}
$$

当 $\varepsilon=0$ 时，式(69)简化成：

$$
T'_{s0}=\frac{\sqrt{\mu}}{\lambda_s}\int_{-\infty}^{x} T'^{0}_{s0}(x')\,\mathrm{e}^{-\frac{\lambda_Q}{\lambda_s}(x-x')}\chi\left(\frac{\sqrt{\mu}(x-x')}{\lambda_s},t\right)\mathrm{d}x' \tag{70}
$$

这相当于准定常近似下的解。当时间适当长后，解将随时间呈 $O(t^{-\frac{1}{2}})$ 的渐近性态而衰减。

当 $\varepsilon\neq0$ 时，式(69)的其他三个积分，其变化最缓慢的部分的渐近性态分别为 $O(t^{-\frac{3}{4}})$，$O(t^{-\frac{3}{2}})$ 和 $O(t^{-\frac{5}{4}})$，亦即最后两项的衰减，要比前两项快得多。因此，当时间适当长后，解的主要部分为前面两项。这表明对 $\tau>\tau_\beta$ 的过程来讲，海温场的演变在很大程度上决定于初始海温场的强度和分布，大气运动只是通过云量对辐射的调节作用而对海温场的演变发生作用。

当海温场已知后，速度场可以近似地由式(67)决定。显然速度场也决定于初始的海温场，也即当时间充分长后，大气运动也要向初始的海温场进行调整。因此，在一般的情况下，当场达到完全适应后，就将建立起式(60)那样的准定常关系。这样在理论上证明了准定常近似对于长期过程是一种可用的近似。然而具体的数值计算表明，式(69)中第二个积分，如初始海温扰动的尺度大于3 000 km时，则到一个月时其值仍有第一个积分的20%~30%。这表明，单纯由准定常近似算得的结果式(70)是很粗糙的，也即准定常近似虽然是一种过滤掉短期波的方法，但却不是精确度很高的方法。当然，设计另外精度较高的滤波方法在原则上是不困难的，但却不会如现在的方法那样简单。

应用本文提出的原理和方法，将在另一个报告中给出长期天气数值预报的模式和预报试验结果。

参考文献

[1] Блинова, Е.И., *ДАН.СССР.ТОМ.*, 1943, 39: 1343.

[2] Adem.J.*Mon. Wea. Rew.*, 1964, 92: 91-103.

[3] Монин, А, С.*Прогноз погобы как заә ача фпзики*, Издателвство《Hayka》, Москва.

[4] Sawyer, J.S.*WMO Tech Note*, 1965, 66: 227-248.

[5] Budyko, M.I.*The Heat Balance of the Earth's Surface*, PB 1956, 3: 692.

[6] Kuo, H.L.(郭晓岚)*Pure, & Aappl. Geophy.*, 1973, 109: 1870-1876.

长期数值天气预报的滤波方法*

长期数值天气预报研究小组**

摘要：在《一种长期数值天气预报方法的物理基础》一文[1]的理论指导下，本文具体地建立了 500 mbar 高度距平场和地表温度距平场的月预报方法，并给出了预报试验结果。这一尝试性的研究表明，这是长期数值天气预报一条有希望的途径。

在参考文献[1]中曾讨论了一种长期数值天气预报方法的物理基础。在理论上对发生在海气联合系统中的过程作分析后指出，当过程的特征时间大于 Rossby 行波的周期后，大气运动将向海表温度场适应，场的适应时间不到一个月。在此基础上，文中设想了一种长期数值天气预报方法，在这种方法中可以把 Rossby 行波作为干扰长期天气过程的"噪音"而过滤掉，所以称这种长期数值天气预报方法为滤波方法。

本文将建立实现这种滤波方法的具体方案，并用于预报月平均高空环流形势和地球表面(以下简称地表)温度。首先推导出 500 mbar 月平均环流形势对地表温度场的适应关系，应用这一适应方程，根据已知的地表温度场，计算出同月 500 mbar 高度距平。并与实况场作比较，通过两者之间相关系数的计算，定量地检验这种适应过程存在的客观性。这样同时也就检验了理论上建立适应方程的准确程度。

然后，在考虑了有关的地-气相互作用后，建立月平均地表温度场的预报方法，并给了预报试验结果，应用预报的地表温度距平场，通过适应方程的计算，就得到 500 mbar 高度距平场的预报。

1　非绝热涡度方程

在(x,y,p,t)坐标系中，有涡度方程

$$\frac{\mathrm{D}}{\mathrm{D}t}(\zeta+f)=f\frac{\partial\omega}{\partial p}+\left(\frac{\partial F_y}{\partial x}-\frac{\partial F_x}{\partial y}\right) \tag{1}$$

另一方面，根据参考文献[1]所引进的辐射、感热和凝结热流量的参数化形式，热力学第一定律为：

$$\frac{\mathrm{D}T}{\mathrm{D}t}-\sigma_p\omega=\frac{\partial}{\partial p}(k+k_R)\frac{\partial T}{\partial p}-\frac{1}{\tau_R}(T-T_e)+\widetilde{T}^*\zeta_{0g}+\frac{k'}{\rho c_p}s \tag{2}$$

式中，F_x,F_y 为包括摩擦应力在内的作用在长期天气过程中的力；ζ_{0g}表示摩擦层顶的地转涡度；T_e 为参考温度；$\sigma_p=(\gamma_a-\gamma)/\rho g$ 为静力稳定度；$k=\rho^2g^2k_T$，k_T 为对感热的湍流导热系数；k'为空气对短波辐射的吸收系数；s 为太阳的短波辐射，而

* 中国科学，1979，9(1)：75-84.

** 参加研究小组的有：巢纪平、季劲钧、朱志辉(中国科学院地理研究所)，吕越华(中央气象局气象科学研究所)，梁幼林、何家骅、李维亮(中国科学院大气物理研究所)，钟强、吴士杰(中国科学院兰州高原大气物理研究所)，潘华盛(黑龙江省气象局)。

$$k_R=\frac{8r\rho\widetilde{T}^3}{c_p k_s}\rho g^2,\quad \tau_R=\frac{\rho c_p}{8(1-r)\rho\widetilde{T}^3 k_w}$$

如考虑长波辐射影响后引进的两个参数,其意义见参考文献[2],以及

$$\widetilde{T}^*(p)=\frac{L}{c_p}\gamma\frac{\mathrm{d}\ln\bar{e}_s}{\mathrm{d}\bar{T}}l_b\widetilde{q}_s(p)$$

为对凝结热流量作参数化后所引进的一个量纲为温度的量。其他有关符号的意义均见参考文献[1]。

现将各物理量分解成

$$A=\bar{A}+A'$$

其中,“-”号表示多年月平均值;“′”号表示对多年月平均值的偏差(距平)。假定多年月平均过程分别满足方程

$$\left(\frac{\partial}{\partial t}+\bar{u}\frac{\partial}{\partial x}+\bar{v}\frac{\partial}{\partial y}\right)(\bar{\zeta}+f)=f\frac{\partial\bar{\omega}}{\partial p}+\left(\frac{\partial F_y}{\partial x}-\frac{\partial F_x}{\partial y}\right)\tag{3}$$

$$\left(\frac{\partial}{\partial t}+\bar{u}\frac{\partial}{\partial x}+\bar{v}\frac{\partial}{\partial y}\right)\bar{T}-\sigma_p\bar{\omega}=\frac{\partial}{\partial p}(k+k_R)\frac{\partial\bar{T}}{\partial p}+\widetilde{T}^*\bar{\zeta}_{0g}+\frac{k'}{\rho c_p}s\tag{4}$$

这里已假定取参考温度 T_e 为多年月平均气温 $\bar{T}$。

引进地转关系

$$u'=-\frac{1}{f}\frac{\partial\phi'}{\partial y},\ v'=\frac{1}{f}\frac{\partial\phi'}{\partial x},\ \zeta'=-\frac{\partial v'}{\partial x}-\frac{\partial u'}{\partial y}=\frac{1}{f}\Delta\phi'\tag{5}$$

以及在静力平衡下

$$T'=-\frac{p}{R}\frac{\partial\phi'}{\partial p}\tag{6}$$

由式(1)、式(2)考虑了式(3)、式(4)式后,分别描写距平状态的方程

$$\frac{\mathrm{d}\Delta\phi'}{\mathrm{d}t}+\widetilde{\beta}_y\frac{\partial\phi'}{\partial x}-\widetilde{\beta}_x\frac{\partial\phi'}{\partial y}=f^2\frac{\partial\omega'}{\partial p}\tag{7}$$

$$\left(\frac{\mathrm{d}}{\mathrm{d}t}-\frac{\partial}{\partial p}(k+k_R)\frac{\partial}{\partial p}\right)\frac{\partial\phi'}{\partial p}-\left(\frac{\partial\bar{u}}{\partial p}\frac{\partial\phi'}{\partial x}+\frac{\partial\bar{v}}{\partial p}\frac{\partial\phi'}{\partial y}\right)+\frac{1}{\tau_R}\frac{\partial\phi'}{\partial p}=-\widetilde{\sigma}_p\omega'-\frac{R\widetilde{T}^*}{pf}(\Delta\phi')\tag{8}$$

式中 $\widetilde{\beta}_x=\partial(\zeta+f)/\partial x$,$\widetilde{\beta}_y=\partial(\zeta+f)/\partial y$,$\widetilde{\sigma}p=\frac{k}{p}\sigma_p$,以及算子

$$\frac{\mathrm{d}}{\mathrm{d}t}\equiv\frac{\partial}{\partial t}+(\bar{u}+u')\frac{\partial}{\partial x}+(\bar{v}+v')\frac{\partial}{\partial y}$$

由式(7)、式(8)两式消去 ω'后,得到非绝热涡度方程

$$\frac{\mathrm{d}}{\mathrm{d}t}\Delta\phi'+\beta_y\frac{\partial\phi'}{\partial x}-\beta_x\frac{\partial\phi'}{\partial y}=-\frac{f^2}{\tau_R\widetilde{\sigma}_p}\frac{\partial^2\phi'}{\partial p^2}+\frac{f^2}{\widetilde{\sigma}_p}\frac{\partial^2}{\partial p^2}(k+k_R)\frac{\partial^2\phi'}{\partial p^2}$$

$$-\frac{f^2}{\widetilde{\sigma}_p}\frac{\partial}{\partial p}\left[\frac{\mathrm{d}}{\mathrm{d}t}\left(\frac{\partial\phi'}{\partial p}\right)-\left(\frac{\partial\bar{u}}{\partial p}\frac{\partial\phi'}{\partial x}+\frac{\partial\bar{v}}{\partial p}\frac{\partial\phi'}{\partial y}\right)\right]-fR\frac{\partial}{\partial p}\left(\frac{\widetilde{T}^*}{p\widetilde{\sigma}_p}\right)(\Delta\phi')\tag{9}$$

2　适应方程及数值试验

方程(9)的下边界条件取

$$p = p_0(\text{海平面}),\quad \omega' = 0 \tag{10}$$

另一方面,如将方程(2)用到海平面上,考虑到 $\omega=0$,同时在海平面参考温度(即环境温度)T_e 可以取成是地表温度 T_s,由于 $T_s=\overline{T}_s+T'_s$,如在海平面上取 $\overline{T}=\overline{T}_s$,则这时相应的距平状态方程为:

$$p = p_0,\quad \left(\frac{\mathrm{d}}{\mathrm{d}t} - \frac{\partial}{\partial p}(k + k_R)\frac{\partial}{\partial p}\right)\frac{\partial\phi'}{\partial p} - \left(\frac{\partial\overline{u}}{\partial p}\frac{\partial\phi'}{\partial x} + \frac{\partial\overline{v}}{\partial p}\frac{\partial\phi'}{\partial y}\right) = -\frac{1}{\tau_R}\left(\frac{\partial\phi'}{\partial p} - \frac{R}{p_0}T'_s\right) \tag{11}$$

这里考虑到凝结热通量只发生在凝结高度以上,所以与凝结热通量有关的一项消失。另外的边界条件可以取成

$$p = 0,\quad \phi' = \frac{\partial\phi'}{\partial p} = \frac{\partial^2\phi'}{\partial p^2} = \frac{\partial^3\phi'}{\partial p^3} = 0 \tag{12}$$

将方程(9)用于 500 mbar 层,变量的垂直微分用差分代替,考虑到条件式(11)和式(12)后,并设 $(\Delta\phi')_0=b(\Delta\phi')_{p=500\text{ mbar}}$,最后得到该层上的非绝热涡度方程为

$$\frac{\mathrm{d}}{\mathrm{d}t}\Delta\phi' + \beta_y\frac{\partial\phi'}{\partial x} - \beta_x\frac{\partial\phi'}{\partial y} - K\Delta\phi' = FT'_s \tag{13}$$

式中,$F = f^2R/\widetilde{\sigma}_p p_0^2\tau_R > 0, K = -bfR\dfrac{\partial}{\partial p}\left(\dfrac{\widetilde{T}^*}{p\widetilde{\sigma}_p}\right) > 0$。

在参考文献[1]中指出,将线性化后的方程(13),联合海水温度的变化方程,再考虑到海-气交界面上的热量平衡方程,在这样一个海气联合系统中将产生两类不同时间尺度的运动,一类本质上为非绝热 Rossby 波的短期波,另一类是长期波。为了滤去作为干扰长期波的短期波,一种简单的方法可以对大气运动方程采用准定常近似,即略去方程中的局地时间变化项。在物理意义上,这意味着当 Rossby 行波频散后,高度距平场和海表温度距平场之间将建立起一种适应关系。在参考文献[1]中还进一步指出,这种适应关系的建立为时不到一个月。在现在的情况下,类似适应后建立的关系式为:

$$(\overline{u} + u')\frac{\partial\Delta\psi'}{\partial x} + (\overline{v} + v')\frac{\partial\Delta\psi'}{\partial y} + \beta_y\frac{\partial\psi'}{\partial x} - \beta_x\frac{\partial\psi'}{\partial y} - K\Delta\psi' = \frac{F}{f}T'_s \tag{14}$$

式中,$\psi'=\phi'/f$ 为地转流函数。我们称这个平衡关系式为适应方程。

为了检验这种适应关系在实际大气中的现实性,下面将应用实际的 T'_s 场,由适应方程算出 ϕ' 场,然后与同月的实况 ϕ' 场作比较,以确定这种适应关系的可靠程度。

当 T'_s 为已知时,方程(14)是对变量 ψ' 的非齐次非线性偏微分方程,现在采用以下两种数值解法。

方案 1　一般叠代法。微分方程(14)在边界条件 $\phi'=0$ 下的解,可以取成是非定常微分方程

$$\frac{1}{D^2}\frac{\partial\phi'}{\partial t} + \Lambda\phi' = \frac{F}{f}T'_s \tag{15}$$

当 $t\to\infty$ 时的极限解。式中 D 为具有长度量纲的参数,算子

$$\Lambda = \frac{\partial(\overline{\psi} + \psi')}{\partial x}\frac{\partial\Delta}{\partial y} - \frac{\partial(\overline{\psi} + \psi')}{\partial y}\frac{\partial\Delta}{\partial x} + \frac{\partial f}{\partial y}\frac{\partial}{\partial x} - \frac{\partial f}{\partial x}\frac{\partial}{\partial y} - K\Delta$$

这里考虑到$\bar{\zeta} \ll f$,故略去$\bar{\zeta}$。由于算子Λ是正定的,即其内积

$$(\Lambda\psi',\psi') = \iint_{Q}(\Lambda\psi')\psi'\mathrm{d}\Omega > 0$$

因此差分方程

$$\frac{1}{D^2}\frac{\psi'^{n+1} - \psi'^{n}}{\tau} + \Lambda^n\psi'^n = \frac{F}{f}T'_s \tag{16}$$

将收敛到一定常解[3]。其中τ是时间步长。叠代用超松驰法进行。

方案2　局地一维法[3]。考虑到$\psi=\bar{\psi}+\psi'$后,将方程(14)改造成

$$\Lambda\psi = S \tag{17}$$

式中算子

$$\Lambda = \frac{\partial(f+\zeta')}{\partial y}\frac{\partial}{\partial x} - \frac{\partial(f+\zeta')}{\partial x}\frac{\partial}{\partial y} - K\left(\frac{\partial^2}{\partial x^2} + \frac{\partial^2}{\partial y^2}\right)$$

源函数

$$S = \frac{F}{f}T'_s + \frac{\partial(\bar{\psi},f)}{\partial(x,y)} - k\Delta\bar{\psi}$$

以及

$$\zeta = \Delta(\psi - \bar{\psi}) \tag{18}$$

现在问题变成由式(17)和式(18)联立解出ψ和ζ'。类似地,方程(17)可以由非定常方程

$$\frac{\partial\psi}{\partial t} + \Lambda\psi = S \tag{19}$$

在边界条件

$$(x,y) \in \Gamma,\quad \psi = \bar{\psi} = \text{常数},\quad \zeta' = 0 \tag{20}$$

下,求$t\to\infty$时相应差分方程的定常解。将算子Λ分解成两个分别为正定的算子Λ_1和Λ_2,即

$$\Lambda = \Lambda_1 + \Lambda_2$$

式中

$$\Lambda_1 = \frac{\partial(f+\zeta')}{\partial y}\frac{\partial}{\partial x} - K\frac{\partial^2}{\partial x^2} + \frac{1}{2}\frac{\partial^2(f+\zeta')}{\partial x\partial y},$$

$$\Lambda_2 = -\frac{\partial(f+\zeta')}{\partial x}\frac{\partial}{\partial y} - K\frac{\partial^2}{\partial y^2} - \frac{1}{2}\frac{\partial^2(f+\zeta')}{\partial x\partial y}$$

引用n次迭代后的残差

$$\varepsilon^n = \Lambda\psi^n - S \tag{21}$$

以及分步计算

$$\left(E + \frac{1}{2}\tau_{n-1}\Lambda_1\right)Y^{n+\frac{1}{2}} = \varepsilon^n \tag{22}$$

$$\left(E + \frac{1}{2}\tau_{n-1}\Lambda_2\right)Y^{n+1} = Y^{n+\frac{1}{2}} \tag{23}$$

$$Z^{n+1} = \Lambda Y^{n+1} \tag{24}$$

式中,E为单位矩阵;迭代参数τ_n由变分法极值原理,取

$$\tau_n = (Z^{n+1} \cdot \varepsilon^n)/(Z^{n+1} \cdot Z^{n+1}) \tag{25}$$

当 Y^{n+1}求得后,解 ψ^{n+1}为:

$$\psi^{n+1} = \psi^n - \tau_n Y^{n+1} \tag{26}$$

算子 Λ_1 中的 ζ'为:

$$\zeta'^n = \Delta(\psi^n - \overline{\psi}) \tag{27}$$

算子 Λ_2 中的 ζ'则取成 $\zeta'^{n+\frac{1}{2}}$。

应用上面这两种计算方案,分别对 1965 年 12 个月的实况地表温度距平,算出同月与之相适应的 500 mbar 高度距平场。计算只限于北半球,网格距离取 $\delta x=\delta y=540$ km,全场共 1 089 个点。应用极射赤平面投影。大陆上的土壤表层温度距平因缺资料,用海平面气温距平代替①。K 值在计算中一般取 2×10^{-6}/s,F 一般取 $1.3\times10^{-4}/(\mathrm{s}^3\cdot℃)$。

表 1 给出用适应方程计算出的 500 mbar 高度距平场与同月实况之间的相关系数 γ。全年平均为 0.49。在表 1 的最后一行中,同时给出相邻两个月实况高度距平场之间的相关系数 γ'。

表 1

月份	1	2	3	4	5	6	7	8	9	10	11	12
γ	0.22	0.41	0.79	0.55	0.56	0.41	0.63	0.41	0.45	0.74	0.41	0.34
月份	12(64)-1	1-2	2-3	3-4	4-5	5-6	6-7	7-8	8-9	9-10	10-11	11-12(65)
γ	-0.12	0.07	-0.10	0.17	-0.01	0.18	0.21	0.49	0.24	-0.33	0.02	0.36

图 1a 为计算所得的 10 月的 500 mbar 高度距平场(每 40 位势米一根等值线),与图 1b 相应月份的实况相比,可以看到两者总的形势都较接近。

图 1a　1965 年 10 月 500 mbar 高度距平(计算)

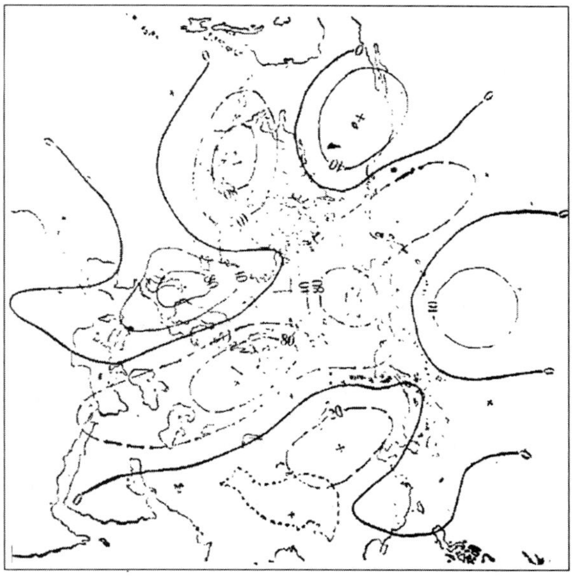

图 1b　1965 年 10 月 500 mbar 高度距平(实况)

① 我们曾用中国境内地温距平和气温距平做过比较,就大范围趋势来看,两者是接近的。

上面给出的计算例子是用一般叠代法算得的，用局地一维法计算的结果，与用叠代法计算结果基本上是一致的。

通过上面的计算和对比，有理由认为高空环流形势在一定程度上将向地表温度场适应，同时也检验了所建立的适应方程(14) 具有一定的客观性。

另外，由表1可以看到，高度场月距平逐月之间的持续性一般都很差，因此采用惯性预报的外推法来预报次月的高度距平场，其准确率一般将很低。但如果能较准确地把地表温度距平场预报好，则根据高度场向地表温度场的适应，就有可能数值地预报好次月的高度场。

3 地表温度距平的预报方法

设在地球陆面表层中的温度变化服从热传导方程，而在海洋表层中，还受洋流平流过程的影响，于是有方程

$$\frac{\partial T_s}{\partial t} + \delta \frac{\partial(\psi_s, T_s)}{\partial(x, y)} = K_s \frac{\partial^2 T_s}{\partial z^2} \tag{28}$$

式中对海洋 $\delta=1$，对陆地 $\delta=0$，ψ_s 为洋流的流函数。

目前，如设多年月平均地球表层温度的变化(气候变化)满足方程

$$\frac{\partial \bar{T}_s}{\partial t} + \delta \frac{\partial(\bar{\psi}_s, \bar{T}_s)}{\partial(x, y)} = K_s \frac{\partial^2 \bar{T}_s}{\partial z^2} \tag{29}$$

则对多年月平均值的距平部分有

$$\frac{\partial^2 T'_s}{\partial z^2} - \frac{1}{K_s}\frac{\partial T'_s}{\partial t} = \frac{\delta}{K_s}\left[\frac{\partial(\bar{\psi}_s, T'_s)}{\partial(x, y)} + \frac{\partial(\bar{\psi}_s, \bar{T}'_s + T'_s)}{\partial(x, y)}\right] \equiv H_1 \tag{30}$$

式中，$\bar{\psi}_s$ 可以由洋流的月平均气候值算出，距平部分 ψ'_s，假定由大气流的吹动而引起，由 Ekman 风吹洋流理论，有

$$u'_s = \frac{\sqrt{2}}{2}\frac{0.0126}{\sqrt{\sin\phi_0}}(u'_0 + v'_0) \tag{31}$$

$$v'_s = \frac{\sqrt{2}}{2}\frac{0.0126}{\sqrt{\sin\phi_0}}(v'_0 - u'_0)$$

式中，ϕ_0 为纬度，而

$$u'_s = -\frac{\partial \psi'_s}{\partial y}, \quad v'_s = \frac{\partial \psi'_s}{\partial x}$$

另外，在地-气交界面上有热量平衡方程，其参数化形式为[1]：

$$z = 0, \quad \rho_s c_{ps} K_s \frac{\partial T'_s}{\partial z} - \rho c_p K_T \frac{\partial T'}{\partial z} + \delta\left(L\rho K_T \gamma \frac{\partial \ln \bar{e}_s}{\partial T}\cdot\frac{\partial \bar{q}_s}{\partial T}\right) T'_s = -\frac{S_0}{W_0} l_b \zeta'_{0g} \tag{32}$$

式中，左方第一、第二两项，分别为地表向其下活动层及向其上大气层中输送的异常感热通量；第三项为地面的异常蒸发热通量，在陆面上略去蒸发的作用；右端为地表受到云量调节后的异常辐射平衡，而云量的距平值设与垂直运动成正比，并通过摩擦层顶的涡度 ζ'_{0g} 来决定垂直运动的大小。各个符号的意义均参见参考文献[1]。

另一个条件为：

$$z = -\infty, \quad T'_s = 0 \tag{33}$$

为了算出式(32)中的 $\partial T'/\partial z$,将式(8)用到近地层,这时垂直速度和凝结热流量均消失,考虑到式(6),应用准定常近似,并回到 z 坐标后,得到

$$\frac{\partial^2 T'}{\partial z^2} - \frac{1}{K\tau_R}T' = \frac{1}{K}\left[\frac{\partial(\overline{\psi} + \psi', T')}{\partial(x,y)} + \frac{\partial(\psi', \overline{T})}{\partial(x,y)}\right] \equiv H_2 \tag{34}$$

将此式用到某一指定的局地点(标以下标 ij),则在水平方向用中央差分来代替微分算子时,方程右端的平流项只与这一点周围的量有关。因此,对于所指定的点,可以把式(34)看成是对温度 T'的局地边界层方程。在条件

$$z = 0, \quad T' = T'_s \tag{35}$$

$$z \to \infty, \quad T' = 0 \tag{36}$$

下,这个对局地点而言的二阶非齐次常微分方程的解容易算得,算出

$$\left.\frac{\partial T'_{ij}}{\partial z}\right|_{z=0} = -\frac{1}{\sqrt{K\tau_R}}T'_{sij}(0) - \int_0^{\infty} e^{-\frac{z'}{\sqrt{K\tau_R}}} H_{2ij}(z')\,dz'$$

$$\approx -\frac{1}{\sqrt{K\tau_R}}T'_{sij}(0) - \sqrt{K\tau_R}\widetilde{H}_{2ij} \tag{37}$$

式中,$T'_s(0)$为地表温度距平值;$\widetilde{H}_2$ 为 H_2 在边界层中对高度的平均值,在现在的模式中近似地取500 mbar等压面上的值。

将式(37)代入式(32),得到

$$z = 0, \quad \frac{\partial T'_{sij}}{\partial z} + \left(\frac{\delta}{D_Q} + \frac{1}{D_S}\right)T'_{sij} = -\frac{\rho c_p K_T}{\rho_s c_{ps} K_s}\sqrt{K\tau_R}\widetilde{H}_{2ij} - \frac{b s_0 l_b}{W_0 \rho_s c_{ps} K_s}\zeta'_{ij} \tag{38}$$

式中,已设 $\zeta'_{0g} = b\zeta'$,ζ'为 500 mbar 上的涡度距平,而

$$D_Q = \frac{\rho_s c_{ps} K_s}{L\rho K_T \gamma \dfrac{\partial \ln \overline{e}_s}{\partial T}\dfrac{\partial \overline{q}_s}{\partial T}}, \quad D_S = \frac{\rho_s c_{ps} K_s}{\rho c_p K_T}\sqrt{K\tau_R}$$

对地球表层温度距平变化方程(30)中的时间变化项用差分写出,得到

$$\frac{\partial^2 T''^{t+\delta t}_{sij}}{\partial z^2} - \frac{1}{K_s \delta t}T''^{t+\delta t}_{sij} = -\frac{1}{K_s \delta t}T''^{t}_{sij} + H_{1ij} \tag{39}$$

条件式(38)左端以及条件式(33)可以取成是 $t+\delta t$ 时刻的值,它们将作为方程(39)的两个定解条件。方程(39)满足这两个条件,并在 $z=0$ 处的解为:

$$T''^{t+\delta t}_{sij}(0) = -D\sqrt{K_s\delta t}\left\{\left(\frac{\rho c_p K_T}{\rho_s c_{ps} K_s}\sqrt{K\tau_R}\widetilde{H}_{2ij} + \frac{b s_o l_b}{W_0 \rho_s c_{ps} K_s}\zeta'_{ij}\right) + \int_{-\infty}^{0} e^{\frac{z'}{\sqrt{K_s\delta t}}}\left(-\frac{1}{K_s\delta t}T''^{t}_{sij} + H_{1ij}\right)dz'\right\} \tag{40}$$

由于地表温度变化的活动层,即使在月的时间量级上也仍然很薄,因此上式积分核中的 T''^{t}_{sij}和 H_{1ij}均可以近似地用表层的值代替。于是最后得到

$$T''^{t+\delta t}_{sij}(0) = DT''^{t}_{sij}(0) - DK_s\delta t\left[\frac{\rho c_p K_T\sqrt{K\tau_R}}{\rho_s c_{ps} K_s\sqrt{K_s\delta t}}\widetilde{H}_{2ij} + \widetilde{H}_{1ij}\right] - D\sqrt{K_s\delta t}\,\frac{b s_0 l_b}{W_0 \rho_s c_{ps} K_s} - \zeta'_{ij} \tag{41}$$

考虑到式(30),式(31),并设$(u'_0, v'_0) = b(u', v')$,u',v'为 500 mbar 层上的距平地转风,式(41)右端的 ψ'_s 均可以表示成 500 mbar 上的 ϕ'值,或 ψ'值,$\widetilde{H}_{1ij}$是 H_{1ij}在海洋表层的平均值。

当方程(41)右端的 ψ', T'_s 为取成 $t+\delta t$ 时刻的值,则这个方程与适应方程(14)组成对变量 ψ', T'_s 的闭合方程组。当然,为了使适应方程(14)成立,要求 δt 取得大于 Rossby 行波的特征周期,例如取 10 天。另一种方法,式(41)右端取成 t 时刻的值,并取 $\delta t=1$ 个月,这样由这个方程应用 t 时刻的 ψ', T'_s,以及其他有关的气候资料,便可以直接算出下一个月的地表温度,然后由适应方程(14)算出下一个月的 500 mbar 高度距平。再下一个月的预报(即预报两个月)可以用预报的 ψ', T_s 再类似地算一次而得到。

Adem[4] 曾用上述后一种显式方法,预报过一个月的月平均海表温度,但所考虑的物理过程以及所做的参数化与我们现在的方法是不一样的。同时在他的预报方法中没有给出预报高空环流形势的途径。

4 预报试验结果

下面给出预报例子。所用参数:K_s 在海洋为湍流导温系数,取 10 cm^2/s,在陆面为一般的导温系数,取 0.005 cm^2/s;K_T 为空气的湍流导温系数,取 10 m^2/s;体积热容量 $\rho_s c_{ps}$,海洋和大陆分别取 1cal/(cm^3 · ℃)和 0.5 cal/(cm^3 · ℃);空气的 ρc_p 取 3×10^{-4} cal/(cm^3 · ℃);$K\approx K_T$。牛顿辐射冷却的特征时间 τ_R 取 5×10^5 s 左右。l_b 为行星边界层的厚度,取 500 m,D_Q 为表征海面蒸发大小的一个综合参数,取 50 m;表征感热垂直输送的参数 D_s,在海洋约取 500 m,在陆地约为 10 cm,经验参数 b/H'_0 取10 s/cm。

由以上参数,算得

$$D \equiv \frac{1}{1+\frac{\sqrt{K_s\delta_t}}{D_s}+\delta\frac{\sqrt{K_s\delta_t}}{D_Q}} \tag{42}$$

在海洋约为 0.5, 在大陆约为 0.1。D 在物理上是表征相邻两个月地表温度持续性的一个量度。由此可见, 在月的时间尺度上, 陆面温度的热惯性一般只有海洋的 20%左右。

图 2a 和图 2b, 分别是 1965 年 8 月的地表温度距平和 500 mbar 高度距平(初始场)。图 3a 和图 3b 是预报的 9 月地表温度距平和 500 mbar 高度距平。图 4a 和图 4b 是 9 月相应的实况。

图 2a 1965 年 8 月地表温度距平实况

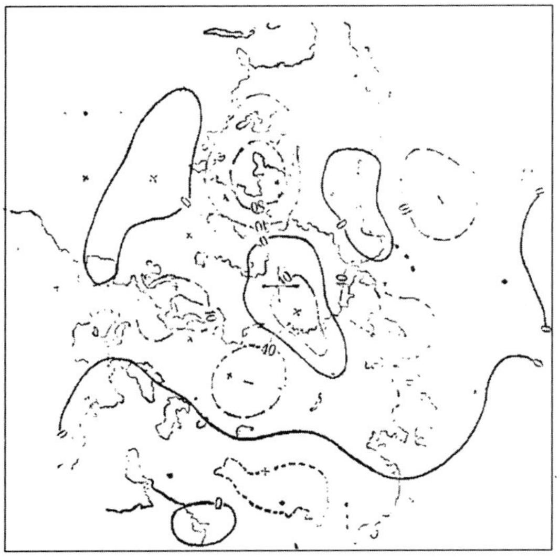

图 2b 1965 年 8 月 500 mbar 高度距平实况

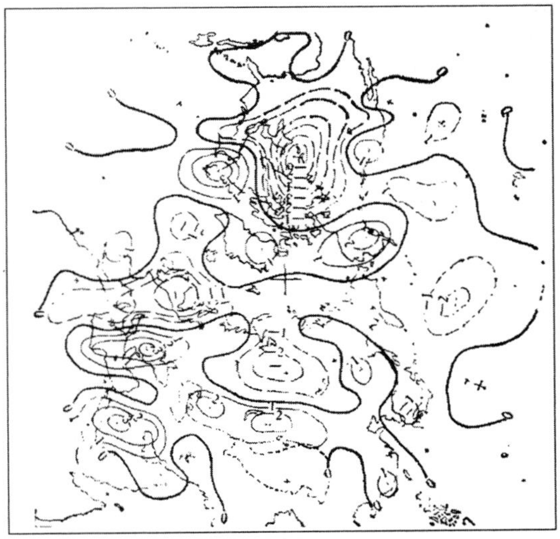

图 3a　1965 年 9 月地表温度距平预报

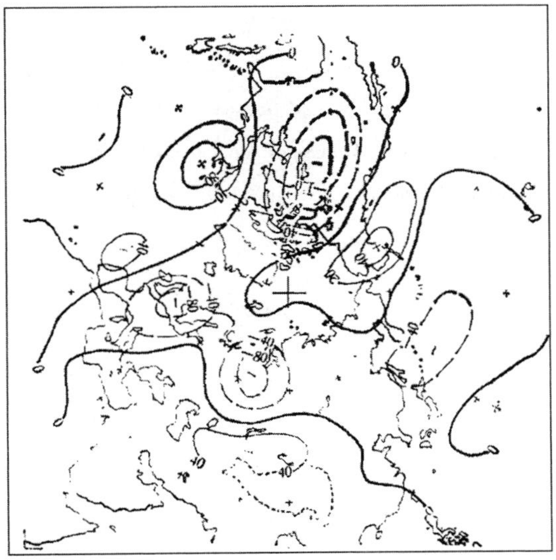

图 3b　1965 年 9 月 500 mbar 高度距平预报

图 4a　1965 年 9 月地表温度距平实况

图 4b　1965 年 9 月 500 mbar 高度距平实况

在地表温度距平的预报图上，与实况相比除白令海峡的负距平中心强度报得太弱，以及极地没有报好外，其他一些主要的距平中心位置和强度都与实况较接近。在高度场的预报方面，主要距平中心位置都与实况接近，但欧洲的一个小范围正距平没有报出来，极地有一个正距平中心也没有报好。从 1965 年 8 月到 9 月，高度距平场实况间的持续性相关系数为 0. 24(见表 1)，而现在预报的高度距平与实况之间的相关系数为 0. 36，高于惯性预报水平。

从这一预报例子看(我们还试验过另外几个例子)，说明这一尝试性的长期天气数值预报方法，是一个值得进一步提高的预报方法。

参考文献

[1] 长期数值天气预报研究小组.中国科学,1977,2:162-172.
[2] Kuo,H.L.*Pure. & Appl.Gcophy.*,1973,109:1870-1876.
[3] Marchuk,G.I.,*Numerical Method on Weather Prediction*,1973.
[4] Adem,J.*Mon.Weu.Rer.*,1964,92:91-103.

A Theory and Method of Long-range Numerical Weather Forecasts*

Chao Jih-Ping, Guo Yu-Fu, Xing Ru-Nan

(*Institute of Atmospheric Physics, Academia Sinica, Beijing*(*Manuscript received* 24 *September* 1981))

Abstract: By parameterizing various types of heat sources (sinks) a linear model of coupled ocean-atmospheric system has been established. By means of frequency analysis it is shown that in this model there are two different types of waves, one corresponding essentially to the travelling Rossby wave in non-adiabatic atmosphere which has a period of several days, and the other is a slowly varying wave which are driven by anomaly heating in ocean. Evidently the existence of this short time-scale weather process presents difficulties for the long-range numerical weather forecasting, because the evolution of the long-range weather process of smaller amplitude will be distorted by the short-range weather process of large amplitude. In order to overcome these difficulties the travelling Rossby wave are filtered as a "noise" from the model of the long-range weather forecast. A simple method of filtering is given. Then a practical model for predicting the monthly mean anomalous fields of 500 mb geopotential height and earth's surface temperature is given and the experimental results are reported. This tentative study shows that the filtering method mentioned in this paper may be a promising way for making long-range weather forecast.

1. Introduction

The success of short-range numerical weather prediction has aroused the interest of meteorologists in the possibility of long-range numerical weather prediction in past two decades. One of goals of recent rapid development of numerical simulation of global circulation model (GCM) is to find out the successful way of long-range weather forecasting. There are several ways for making long-range numerical weather forecast. Usually we may design a non-adiabatic GCM and integrate it numerically as the short-range numerical forecast. On the other hand, we may take another entirely different way. As well known, Namias (1948) used 30-day mean map and predicted the change of the two successive monthly mean maps. It may also be possible to do it along this line by means of numerical method instead of Namias's synoptic method. Naturally, some difficulties have to be overcome. In this aspect, Adem (1964) suggests a model for predicting monthly mean temperature field, but his model did not forecast the monthly mean air-flow. We follow the second line mentioned above, a theory and method have been developed (GLRNWF① 1977, 1979). In this paper, this model will

* Journal of the meteorological society of Japan, 1982, 60(1): 262-291.

① Group of Long-Range Numerical Weather Forecasting.

be further discussed and the predicted examples of one month mean anomalous state of atmospheric circulation and the temperature of earth's surface will be given.

2. Basic equations

In x, y, p, t coordinates, the vorticity equation and the first law of thermodynamics become respectively

$$\frac{\mathrm{d}}{\mathrm{d}t}(\zeta + f) = f\frac{\partial \omega}{\partial p} \tag{1}$$

$$\rho C_p\left(\frac{\mathrm{d}T}{\mathrm{d}t} - \sigma_T \omega\right) = \varepsilon = \varepsilon_T + \varepsilon_R + \varepsilon_Q \tag{2}$$

where f is Coriolis parameter; ρ is the density of air and C_p is the specific heat under constant pressure; σ_T is the static stability; the operator

$$\frac{\mathrm{d}}{\mathrm{d}t} = \frac{\partial}{dt} + u\frac{\partial}{\partial x} + v\frac{\partial}{\partial y}$$

in which u, v, ω are the velocity components respectively.

In Eq. (2), the turbulent heat exchange ε_T is

$$\frac{1}{\rho C_p}\varepsilon_T = \frac{\partial}{\partial p}\left(K_p \frac{\partial T}{\partial p}\right) \tag{3}$$

where $K_p = \rho^2 g^2 K_T$, K_T is the coefficient heat conductivity, g is gravity. The radiational heat exchange ε_R, according to the scheme of Kuo (1968) may be given by

$$\frac{1}{\rho C_p}\varepsilon_R = \frac{\partial}{\partial p}\left(K_R \frac{\partial T}{\partial p}\right) - \frac{1}{\tau_R}(T - T_e) + \frac{k'}{\rho C_p}s \tag{4}$$

where s is the solar radiation, k' is the absorption coefficient of the short wave radiation, T_e is the temperature of the environment, and

$$K_R = \frac{8r\sigma \bar{T}^3}{C_p k_s}\rho g^2, \tau_R = \frac{\rho C_p}{8(1 = r)\sigma \bar{T}^3 k_w}$$

where k_s and k_ω are the mean absorption coefficients of the absorptive matter in the region of strong absorption and weak absorption respectively r is the proportion (in %) of the black body radiation in the region of strong absorption to the total black body radiation. For the heat exchange of condensation ε_Q, it may be simply parameterized as follows:

$$\frac{1}{\sigma C_p}\varepsilon_R = -\frac{L}{C_p}\frac{\mathrm{d}q_s}{\mathrm{d}t} \approx -\frac{L}{C_p}\frac{\partial q_s}{\partial z}W_b \quad z > z_b \tag{5}$$

where q_s is saturated specific humidity, L is the latent heat and Z_b is the level of condensation. According to the theory of the boundary layer we may obtain

$$W_b = l_b \zeta_{0g} \tag{6}$$

where l_b is the thickness of the boundary layer, ζ_{0g} is the geostrophic vorticity on the earthg's surface, substituting (6) into (5), we get

$$\frac{1}{\rho C_p}\varepsilon_Q = \bar{T}_\zeta \zeta_{0g} \tag{5'}$$

where

$$\bar{T}_\zeta = \frac{L}{C_p} v \frac{d\ln \bar{e}_s}{dT} l_b qs, \quad v = -\frac{\partial T}{\partial z}$$

Substituting Eqs. (3), (4) and (5′) into Eq. (2), then the first law of the thermodynamics becomes

$$\frac{dT}{dt} - \sigma_T \omega = \frac{\partial}{\partial p} K \frac{dT}{\partial p} - \frac{1}{\tau_R}(T - T_e) + \bar{T}_\zeta \zeta_{0g} + \frac{k'}{\rho C_p} \tag{7}$$

where $K = K_p + K_R$.

It is assumed that the climatological monthlymean process satisfies the following equation

$$\left(\frac{\partial}{\partial t} + \bar{u}\frac{\partial}{\partial x} + \bar{v}\frac{\partial}{\partial y}\right)(\zeta + f) = f\frac{\partial \bar{\omega}}{\partial p} \tag{8}$$

$$\left(\frac{\partial}{\partial t} + \bar{u}\frac{\partial}{\partial x} + \bar{v}\frac{\partial}{\partial y}\right)\bar{T} - \sigma_T \bar{\omega} = \frac{\partial}{\partial p} K \frac{\partial \bar{T}}{\partial p} + \bar{T}_\zeta \bar{\zeta}_{0g} + \frac{k'}{\rho C_p} s \tag{9}$$

here we have assumed the reference temperature T_e to be equal to the climatological monthly mean air temperature $\bar{T}$.

By introducing the geostrophic approximation

$$u' = -\frac{1}{f}\frac{\partial \phi'}{\partial y}, \quad v' = \frac{\partial \phi'}{\partial x},$$

$$\zeta' = \frac{\partial u'}{\partial x} - \frac{\partial v'}{\partial y} = \frac{1}{f}\Delta\phi' \tag{10}$$

and considering the hydrostatic relation

$$T' = -\frac{p}{K}\frac{\partial \phi'}{\partial p} \tag{11}$$

and Eqs. (8) and (9), the equations of anomalous state may be obtained from Eqs. (2) and (7) as follows

$$\frac{d\Delta\phi'}{dt} + \bar{\beta}_y \frac{\partial \phi'}{\partial x} - \bar{\beta}_x \frac{\partial \phi'}{\partial y} = f^2 \frac{\partial \omega'}{\partial p} \tag{12}$$

$$\left(\frac{d}{dt} - \frac{\partial}{\partial p} K \frac{\partial}{\partial p}\right)\frac{\partial \phi'}{\partial p} - \left(\frac{\partial \bar{u}}{\partial p}\frac{\partial \phi'}{\partial x} + \frac{\partial \bar{v}}{\partial p}\frac{\partial \phi'}{\partial y}\right) + \frac{1}{\tau_R}\frac{\partial \phi'}{\partial p} = -\sigma_p \omega' - \frac{k\bar{T}_s}{pf}(\Delta\phi_0') \tag{13}$$

where R is the gas constant of the air, and

$$\bar{\beta}_x = \frac{\partial(\bar{\zeta} + f)}{\partial x}, \quad \bar{\beta}_y = \frac{\partial(\bar{\zeta} + f)}{\partial y}, \quad \sigma_p = \frac{R}{p}\sigma_T$$

and the operator d/dt now becomes

$$\frac{d}{dt} = \frac{\partial}{\partial t} + (\bar{u} + u')\frac{\partial}{\partial x} + (\bar{v} + v')\frac{\partial}{\partial y}$$

Eliminating ω' from Eqs. (12) and (13), we obtain the non-adiabatic vorticity equation as follows

$$\frac{d\Delta\phi'}{dt} + \bar{\beta}_y \frac{\partial \phi'}{\partial x} - \bar{\beta}_x \frac{\partial \phi'}{\partial y} = -\frac{f^2}{\tau\kappa\sigma_p}\frac{\partial^2 \phi'}{\partial p^2} + \frac{f^2}{\sigma_p}\frac{\partial^2}{\partial p^2} K \frac{\partial^2 \phi'}{\partial p^2} - \frac{f^2}{\sigma_p}\frac{\partial}{\partial p}$$

$$\left[\frac{d}{dt}\left(\frac{\partial \phi'}{\partial p}\right) - \left(\frac{\partial \bar{u}}{\partial p}\frac{\partial \phi'}{\partial x} + \frac{\partial \bar{v}}{\partial p}\frac{\partial \phi'}{\partial y}\right)\right] + G_1 \Delta\phi'_0 \tag{14}$$

where

$$G_1 = -fR\frac{\partial}{\partial p}\left(\frac{\overline{T}_s}{p\sigma_p}\right)$$

On the other hand, we suppose that the temperature variation in the soil obeys the heat conductivity equation, and in the ocean it is also affected by the advection of ocean current, then we get the following equation

$$\frac{\partial T_s}{\partial t} + \delta\frac{\partial(\psi_s, T_s)}{\partial(x,y)} = K_s\frac{\partial^2 T_s}{\partial z^2} \tag{15}$$

where $\delta=1$ for ocean and $\delta=0$ for land, ϕ_s is the stream function of ocean current.

Similarly, if the variation of the climatological monthly mean earth's surface temperature satisfies the equation

$$\frac{\partial \overline{T}_s}{\partial t} + \delta\frac{\partial(\overline{\psi}_s, \overline{T}_s)}{\partial(x,y)} = K_s\frac{\partial^2 \overline{T}_s}{\partial z^2} \tag{16}$$

then the equation for the anomalous part from the climatological monthly mean may be written in the form

$$\frac{\partial^2 T'_s}{\partial z^2} - \frac{1}{K_s}\frac{\partial T'_s}{\partial t} = \frac{\delta}{K_s}\left[\frac{\partial(\overline{\psi}_s, \overline{Ts})}{\partial(x,y)} + \frac{\partial(\psi'_s, T'_s + T'_s)}{\partial(x,y)}\right] \tag{17}$$

where $\overline{\psi}_s$ can be calculated from the climatologlcal monthly average of oceanic current. Assuming that the anomaly part ψ'_s be yielded by the air flow, then, the use of Ekman winddriven current theory leads to

$$u'_s = \frac{\sqrt{2}}{2}\frac{0.0126}{\sqrt{\sin\phi_0}}(u'_0 + v'_0), \quad v'_s = \frac{\sqrt{2}}{2}\frac{0.0126}{\sqrt{\sin\phi_0}}(v'_0 - u'_0) \tag{18}$$

where ϕ_0 is the latitude and

$$u'_s = -\frac{\partial\psi'_s}{\partial y}, v'_s = \frac{\partial\psi'_s}{\partial x} \tag{19}$$

3. Heat balance on the earth's surface

The heat balance on earth's surface (land or ocean) may be expressed by (Budyko 1956)

$$R = LE + P + A \tag{20}$$

where R is the radiational flux of heat, or the radiational balance; P is the turbulent heat flux between the underlying surface and the atmosphere; A is the heat flux between the underlying surface and the lower layer; LE is the expenditure of heat by evaporation (L, the latent heat of evaporation and E, the rate of evaporation).

The radiation balance is the difference of total solar radiation received by the earth's surface and the effective outgoing radiation. This may be expressed as follows:

$$R = (S_0 + s_0)(1 - a) - I \tag{21}$$

where S_0 and s_0 are total direct and diffused radiations respectively; I is the effective radia a is the earth's albedo. In general, these radiations may be calculated as follows:

$$(S_0 + s_0) = \widetilde{(S_0 + s_0)}(1 - c_s n), I = \tilde{I}(1 - c_i n)$$

where the symbol "~" denotes the value in cloudless condition, and symbol "0" denotes the value on the earth's surface; c_s, c_i are the coefficients showing respectively the effects of cloud on the total radiation and on the effective radiation; n is the amount of cloudiness. Thus, the expression of the radiational flux of heat becomes

$$R = \widetilde{(S_0 + s_0)}(1 - a)(1 - c_s n) - \tilde{I}(1 - c_i n) \tag{22}$$

What we are interested in is the anomaly rather than the climatological mean itself. Therefore we express n by $n+n'$, where the symbols "$-$" and "$'$" denote the climatological monthly mean and the anomalous value respectively. Then the anomalous flux of radiation becomes approximately

$$R' \approx -[\widetilde{(S_0 + s_0)}(1 - \bar{a})c_s - \tilde{I}c_i]n' \tag{23}$$

Now we suppose that the cloud is mainly produced by the large-scale vertical motion induced by frictional convergence in the boundary layer, and further that the cloudiness n is proportional to W_b which is the vertical velocity on the top of boundary layer. We get

$$n = \left(\frac{W_b}{W^*}\right) \tag{24}$$

where W^* is an empirical parameter. From Eq.(6) we get

$$n = \left(\frac{l_b}{W^*}\right)\zeta_{0g} \tag{25}$$

By assuming $\zeta_{0g} = \bar{\zeta}_{0g} + \zeta'_{0g}$, we may give the anomaly cloudiness

$$n' = \left(\frac{l_b}{W^*}\right)\zeta'_{0g} \tag{26}$$

Substituting Eq. (26) into Eq. (23), we get

$$R' = -G_4\zeta'_{0g} \tag{27}$$

where

$$G_2 = [\widetilde{(S_0 + s_0)}(1 - \bar{a})c_s - \tilde{I}c_i]\frac{l_b}{W^*}$$

The turbulent heat flux is

$$P = -\rho C_p K_T \frac{\partial T}{\partial z} \tag{28}$$

Its anomaly value may be given by

$$P' = -\rho C_p K_T \frac{\partial T'}{\partial z} \tag{29}$$

The heat flux between the underlying surface and the lower layer is

$$A = \rho_s C_{ps} K_s \left(\frac{\partial T_s}{\partial z}\right)_0 \tag{30}$$

with its anomaly value

$$A' = \rho_s C_{ps} K_s \left(\frac{\partial T'_s}{\partial z}\right)_0 \tag{31}$$

where ρ_s, C_{ps}, K_s are the density, the specific heat and the coefficient of heat conductivity of the soil or the water respectively.

The heat flux of evaporation is

$$LE = - L\rho K_Q \left(\frac{\partial q}{\partial z} \right)_0 \tag{32}$$

Its anomaly value may be written in the form

$$LE' = - L\rho K_T \left(\frac{\partial q'}{\partial z} \right)_0 \tag{33}$$

where q is the specific humidity, K_q is the coefficient of turbulent exchange, and $K_q \sim K_T$ approximately.

In general, the evaporation is important only on the sea surface. In this case the process of evaporation may be parameterized by another convenient form. Assuming the specific humidity of the air above the sea surface reaches its saturated value, we get

$$q = q_s = \frac{R}{R_w} \frac{e_s(T_0)}{P}$$

where R_w is the gas constant of the water vapour, e_s is the saturation vapour pressure. Differentiating the logarithm of the above equation, we may obtain the following approximate expression:

$$\left(\frac{\partial q_s}{\partial z} \right)_0 \approx - \left(v \frac{\partial \ln e_s}{\partial T} \right)_0 q_{s0}$$

Its anomaly value becomes

$$\left(\frac{\partial q'_s}{\partial z} \right)_0 \approx - \left(v \frac{\partial \ln e_s}{\partial T} \right)_0 q'_{s0}$$

Supposing that the water vapour in the air immediately above the sea surface be just saturated at the sea surface and this water vapour evaporated acquires the temperature of the sea surface, then we may assume reasonably

$$q'_{s0} \approx \frac{\partial q_s}{\partial T} T'_{s0} \tag{34}$$

The anomalous part of the heat flux of evaporation becomes

$$LE' \approx G_3 T'_{s0} \tag{35}$$

where

$$G_3 = \left(L\rho K_T v \frac{\partial \ln e_s}{\partial T} \frac{\partial q_s}{\partial T} \right)$$

nd its states that the anomalous flux of evaporation is proportional to the anomalous SST.

According to the results mentioned above, the heat balance of the anomalous state of the earth's surface may be written as

$$z = 0$$

$$- \rho C_p K_T \left(\frac{\partial T'}{\partial z} \right)_0 + \rho_r C_{ps} K_s \left(\frac{\partial T'_s}{\partial z} \right)_0 + G_3 T'_{s0} = - G_2 \zeta'_{0g} \tag{36}$$

4. Waves in the ocean–atmospheric system

Let us next discuss the characters of this model.

One of the boundary conditions for Eq. (14) is that at $p = p_0$ (sea level)

$$\omega' = 0 \tag{37}$$

By applying the condition (37) to Eq. (13) at sea level, replacing the reference temperature T_e in Eq. (7) by the earth's surface temperature $T_e=\overline{T}_s+T'_s$, and taking $\overline{T}=\overline{T}_s$ at sea level, and the condensation is disappeared, then we get at $p=P_0$,

$$\left(\frac{\mathrm{d}}{\mathrm{d}t}-\frac{\partial}{\partial p}K\frac{\partial}{\partial p}\right)\frac{\partial\phi'}{\partial p}-\left(\frac{\partial\overline{u}}{\partial p}\frac{\partial\phi'}{\partial x}+\frac{\partial\overline{v}}{\partial y}\frac{\partial\phi'}{\partial y}\right)=-\left(\frac{\partial\phi'}{\partial p}-\frac{R}{p}T'_s\right) \tag{38}$$

This is one of the boundary conditions for solving Eq. (14).

The upper boundary condition that we take motionless, *i.e.*,

at $p=0$,

$$\phi'=\frac{\partial\phi'}{\partial p}=\frac{\partial^2\phi'}{\partial p^2}=\frac{\partial^3\phi'}{\partial p^3}=0 \tag{39}$$

For simplicity, only one layer geopotential surface *i.e.*, 500mb is considered. Using the boundary conditions (38) and (39), we integrate Eq. (14) from $p=p_0$ to $p=0$, and taking the mean value over height on the right hand, then we get from Eq. (14)

$$\frac{\partial\Delta\phi'}{\partial t}+(\overline{u}+u')\frac{\partial\Delta\phi'}{\partial x}+(\overline{v}+v')\frac{\partial\Delta\phi'}{\partial y}+\overline{\beta}_y\frac{\partial\phi'}{\partial x}-\overline{\beta}_x\frac{\partial\phi'}{\partial y}-b\overline{G}_1\Delta\phi'=FT'_s \tag{40}$$

where $\overline{G}_1$ is the average value of G_1 over whole layers of condensation, and

$$F=\frac{f^2R}{\rho_p p_0^2\tau_R}$$

here the assumption $(\Delta\phi')_0=b(\Delta\phi')$ has been used in the term of condensation, ϕ' is the geopotential height at 500 mb level.

Furthermore, it is assumed that the earth is covered totally by an ocean and the climatological states of atmosphere is taken as $\overline{u}$.Linearizing Eq. (40) and one-dimensional problem is considered, we have

$$\frac{\partial^2 v'}{\partial t\partial x}+\overline{u}\frac{\partial^2 u'}{\partial x^2}+\beta v'-b\overline{G}_1\frac{\partial v'}{\partial x}=\frac{F}{f}T'_s \tag{41}$$

Omitting the terms of advection, Eq. (17) becomes

$$\frac{\partial T'_s}{\partial t}=K_s\frac{\partial^2 T'_s}{\partial z^2} \tag{42}$$

The sensible heat flux in heat balance equation at the earth's surface is also ignored, and then we get from (36)

$$z=0,$$

$$\rho_s C_{ps}K_s\left(\frac{\partial T'_s}{\partial z}\right)_0+G_3T_{s0}=-G_2 f\frac{\partial v'}{\partial x} \tag{43}$$

At the bottom of mixed layer in ocean, we mayuse the condition

$$Z=-D,\quad \frac{\partial T'_s}{\partial z}=0 \tag{44}$$

Integrating Eq. (42) with Z from $-D$ to 0, we obtain

$$\frac{\partial T'_{s0}}{\partial t}=-\frac{K_s}{D}\left[\frac{G_3}{\rho_s C_{ps}K_s}T_{s0}+\frac{G_2 f}{\rho_s C_{ps}K_s}\frac{\partial v'}{\partial x}\right] \tag{45}$$

where an assumption was made:

$$\int_{-0}^{0}\frac{\partial T'_s}{\partial t}\mathrm{d}z \approx D\frac{\partial T'_{s0}}{\mathrm{d}t}$$

There are two time scales in this oceanatmosphere coupled system that are

$$t_1 = (\bar{u}\beta)^{-1/2},\quad t_2 = K_s/D^2 \tag{46}$$

t_1 is the characteristic time of Rossby wave, whereas t_2 is the characteristic time of mixing process of heat in the ocean. The ratio of these two time scales is smaller than the order of one, *i.e.*,

$$\lambda = (\bar{u}\beta)^{-1/2}\left|\frac{D^2}{K_s} < 0(1)\right. \tag{47}$$

because in the common cases we may take $\bar{u}=10^3$ cm/s, $\beta=10^{-13}/(\text{cm}\cdot\text{s})$, $K_S=10$ cm^2/s and $D=3\times 10^3$ cm.

Introducing the non-dimensional variables

$$t^* = t(\bar{u}\beta)^{-1/2},\ x^* = x/(\bar{u}/\beta)^{-1/2},\ v^* = v'/\bar{u},\ T_s^* = T'_s/(\delta T_s) \tag{48}$$

we transform the equations to the non-dimensional form as

$$\frac{\partial^2 v^*}{\partial t^*\partial x^*} + \frac{\partial^2 v^*}{\partial x^{*2}} + v^* C^*\frac{\partial v^*}{\partial x} = F^* F_{s0}^* \tag{49}$$

$$\frac{\partial T_{s0}{}^*}{\partial t^*} = -\lambda\left[\frac{D}{D_Q}T_s{}^* + \frac{D}{D_s}\frac{\partial v^*}{\partial x}\right] \tag{50}$$

where

$$C^* = bG_1/(\bar{u}\beta)^{1/2},\ F^* = f^{-1}F(\delta T_s)/\bar{u}\beta),D_q = \rho_s C_{p_s}K_s/G_3,$$

$$D_s = \rho_s C_{p_s}K_s/G_2(\bar{u}\beta)^{1/2}(\delta T_s)^{-1}f$$

Assuming the solutions of Eqs. (49) and (50) to be of the form

$$V^*(x^*,t^*) = V\mathrm{e}^{i(k^*x^*-\sigma^*t^*)}$$

$$T_{s0}{}^*(x^*,t^*) = T_{s0}\mathrm{e}^{i(k^*x^*-\sigma^*t^*)} \tag{51}$$

we obtain the frequencies of the system as follows

$$\sigma^* = \frac{1}{2}\left\{\left[\left(k^* - \frac{1}{k^*}\right) + \mathrm{i}\left(C^* - \lambda\frac{D}{D_Q}\right)\right] \pm \left[\left(\left(k^* - \frac{1}{k^*}\right) + \mathrm{i}\left(C^* - \lambda\frac{D}{D_Q}\right)\right)^2 - 4\lambda\left(\frac{D}{D_Q}C^* - F^*\frac{D}{D_S} - \mathrm{i}\frac{D}{D_Q}\left(k^* - \frac{1}{k^*}\right)\right)\right]^{1/2}\right\} \tag{52}$$

Considering $\lambda<0(1)$, we get the approximate formula

$$\sigma^* \approx \frac{1}{2}\left\{\left[\left(k^* - \frac{1}{k^*}\right) + \mathrm{i}\left(C^* - \lambda\frac{D}{D_Q}\right)\right] \times \left[1 - 2\lambda\frac{\frac{D}{D_Q}C^* - \frac{D}{D_S}F^* - \frac{D}{D_Q}\left(k^* - \frac{1}{k^*}\right)}{\left(\left(k^* - \frac{1}{k^*}\right) + \mathrm{i}\left(C^* - \lambda\frac{D}{D_Q}\right)\right)^2}\right]\right\} \tag{53}$$

Therefore, there are two roots as follows

$$\sigma_1^* \approx \left[\left(k^* - \frac{1}{k^*}\right) + \mathrm{i}\left(C^* - \lambda\frac{D}{D_Q}\right)\right] - \lambda\frac{\frac{D}{D_Q}C^* - \frac{D}{D_S}F^* - i\frac{D}{D_Q}\left(k^* - \frac{1}{k^*}\right)}{\left(k^* - \frac{1}{k^*}\right) + \mathrm{i}\left(C^* - \lambda\frac{D}{D_Q}\right)} \tag{54}$$

$$\sigma_2^* \approx \lambda \frac{\frac{D}{D_Q}C^* - \frac{D}{D_S}F^* - \mathrm{i}\frac{D}{D_Q}\left(k^* - \frac{1}{k^*}\right)}{\left(k^* - \frac{1}{k^*}\right) + \mathrm{i}\left(C^* - \lambda\frac{D}{D_Q}\right)} \tag{55}$$

Obviously, the order of the ratio of the two frequencies is that

$$\frac{\sigma_2^*}{\sigma_1^*} \sim O(\lambda) \tag{56}$$

It shows that the frequency σ_1^* is fast one whereas σ_2^* is slow one.

For illustrating the character of the fast one, let us omit the heating effect of ocean, *i.e.*, putting $\lambda = 0$, then we obtain

$$\sigma_1^* = \left(k^* - \frac{1}{k^*}\right) + iC^* \tag{57}$$

This is clearly the frequency of Rossby wave with the growth rate which comes from the condensation. However, under this situation another solution disappears, *i.e.*, $\sigma_2^* = 0$.

From this that we may understand that the slow one is produced by the heating of ocean.

On the other hand, if we omit the term of local time variation in equation (41), we get

$$\sigma_1^* = 0 \tag{58}$$

$$\sigma_2^* = \lambda \frac{\frac{D}{D_Q}C^* - \frac{D}{D_S}F^* - \mathrm{i}\frac{D}{D_Q}\left(k^* - \frac{1}{k^*}\right)}{\left(k^* - \frac{1}{k^*}\right) + \mathrm{i}C^*} \tag{59}$$

The latter formula agree well with (55), except it loses a small term with the order of λ in denominator, and also the fast Rossby wave is filtered out.

5. Adaptation equation

The calculation of the above section indicates that in an ocean–atmospheric system there are two basic types of dynamical processes corresponding to two different time-scales. Obviously, the evolution of the long-range process of smaller amplitude will be distorted by the short-range process with large amplitude, because the growth rate of the shorter time-scale waves is about one order of magnitude larger than that of the long time-scale waves. This is a difficulty for making the long-range numerical forecast. As well known, a similar problem also occurs in short-range numerical forecast, the solution of the model of primitive equation contains both meteorologically significant motions and meteorologically noises (gravity waves and sound waves). The existence of these noise waves gives on one hand rapid oscillations in the hydrodynamical field and thus cause large fictitious tendency which will completely distort the slower change connected with the meteorologically significant motion. On the other hand, these noise-waves may grow very rapidly if they are not controlled in some way in the model. One way to overcome this difficulty is to filter out these noises in the model. The other is to control their growth in the model. In making the long-range numerical weather forecast and overcoming the same difficulty, we may apply the similar treatments as done in short-range numerical weather forecast. We now employ the first method and treat the travelling Rossby wave as a "noise-wave" and filter it from the model of long-range numerical weather forecasting.

As mentioned above, a simple method of filtering is that, we may omit the term of local time variation in vorticity equation for atmosphere. It is equivalent to replace the equation of motion Eq. (41) by the relation between the field of atmospheric flow and that of heating anomaly, for two dimensional case that is

$$\bar{u}\frac{\partial \Delta\phi'}{\partial x} - b\bar{G}_1\Delta\phi + \beta\frac{\partial\phi'}{\partial x} = FT'_{s0} \tag{60}$$

Although this equation is stationary in form, the geopotential field still varies with time because the equation describing the field of the underlying temperature anomaly is non-stationary.

In the non-linear case we have a similar relation

$$(\bar{u}+u')\frac{\partial \Delta\phi'}{\partial x} + (\bar{v}+v')\frac{\partial \Delta\phi'}{\partial y} + \bar{\beta}_y\frac{\partial\phi'}{\partial x} - \bar{\beta}_x\frac{\partial\phi'}{\partial y} - b\bar{G}_1\Delta\phi' = FT'_{s0} \tag{61}$$

This equilibrium relation may be called adaptation equation. Physically it means that after the travelling Rossby wave are dispersed, an adjustment relationship between the height anomalous field and the earth's surface temperature field would be established.

In order to test this adjustment relationship, we may use the observed T'_S field to compute the ϕ' field from Eq . (61), and then compared it with the observed ϕ' field of the same month to examine the accuracy of this adaptation equation. The over-relaxation method may be used for getting the solution of this equation.

The corresponding 12 monthly mean anomaly field of 500 mb height are calculated from the observed 12 monthly anomaly field on the earth's surface temperature in 1965. The calculation covers only the northern hemisphere, the grid distance being 540 km and the polar stereographic projection being used. Since data of soil surface temperature on land are indequate now, we use the air temperature at sea level instead. Table 1 gives the correlation coefficients between the relation coefficients between the anomaly fields calculated anomaly field of 500 mb height and of two successive months.

the observed ones of the corresponding months. Figs. 1a and 1b are respectively the calculated The mean correlation coefficient for whole year and observed anomaly fields of 500 mb height is 0.49. The last line in Table 1 gives the cor- for October 1965, Comparing these two figures, it can be seen that their general patterns are quite similar. The calculations and comparisons made above naturally lead to the conclusion that the atmospheric motion tends to adapt itself to the earth's surface temperature to a certain degree for the process of the time-scale of one month.

Table 1 Correlation coefficients

Month	Jan.	Feb.	Mar.	Apr.	May	June	July	Aug.	Sep.	Oct.	Nov.	Dec.
r	0. 22	0.41	0.79	0.55	0.56	0.41	0.63	0.41	0.45	0.74	0.34	0.34
Month	12(64)-1	1-2	2-3	3-4	4-5	5-6	6-7	7-8	8-9	9-10	10-11	11-12(1965)
r	-0.12	0.07	-0.10	0.17	-0.01	0.18	0.21	0.49	0.24	-0.33	0.02	0.36

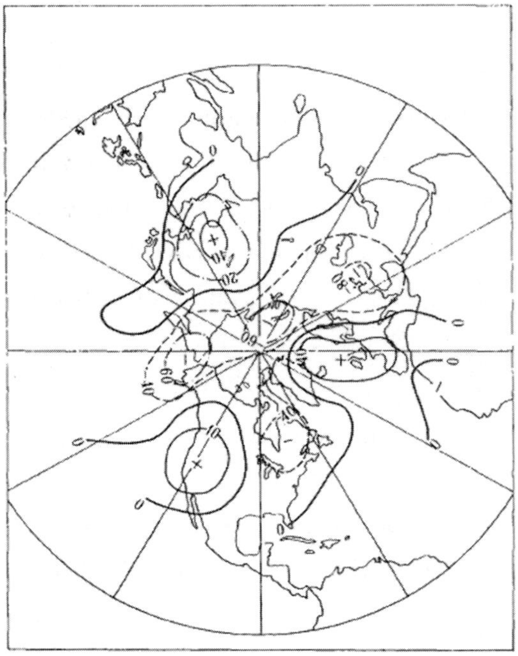

Fig.1a　Anomalous field of 500 mb height for October 1965 (calculated)

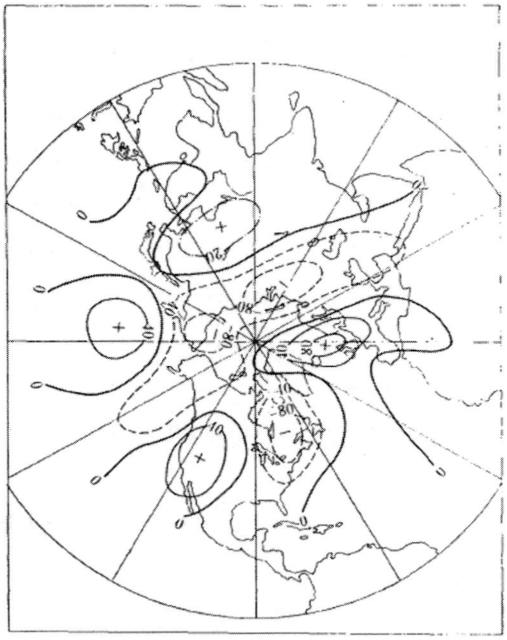

Fig.1b　Anomalous field of 500 mb height for October 1965 (observed)

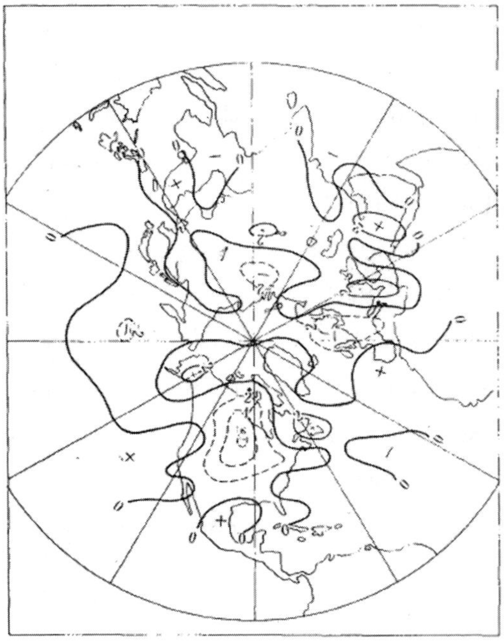

Fig.2a　Anomalous field of earth's surface temperature for September 1965 (predicted)

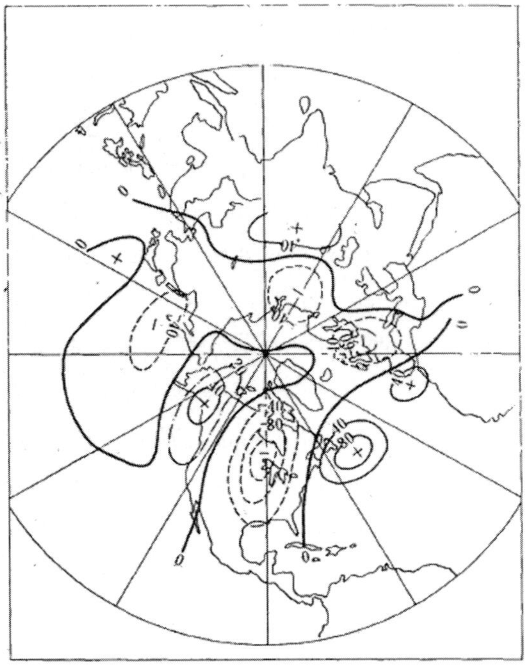

Fig.2b　Anomalous field of 500 mb height for September 1965 (predicted)

6. Forecasting experiments

It may be seen from Table 1 that the persistence between successive monthly mean anomaly fields of 500 mb height is rather poor, so the accuracy of the extrapolation of the persistent prediction for forecasting the anomaly field of 500 mb height of next month would generally be very low, whereas if the prediction of the anomaly field of the earth's surface temperature could be made accurately, it would be possible to predict the height field of the following month numerically by adjusting the height field to the earth's surface temperature field.

From this idea mentioned above, a predicted method of monthly mean state will be developed. At first, we may predicted the earth's surface temperature field by Eq. (17) with boundary conditions (36) and

$$z = -\infty, \quad T_s' = 0 \tag{62}$$

and monthly mean anomaly state of initial month. The predicted method see other paper (GLRNWF 1979). Once the temperature of earth's surface has been obtained, the 500 mb anomalous geopotential height field is calculated by the adaptation equation (61).

In the following, we will give two forecasting examples. This is forecast for one month.

a. Example of September 1965. Figs. 2a and 2b are the predicted anomaly fields of the earth's surface temperature and 500 mb height in September 1965. Figs. 3a and 3b are the corresponding observed maps respectively in samemonth. It may be seen from these figures that the location and intensity of the main centers of anomaly field in the prediction chart of the earth´s surface temperature are close with the observed except that the intensity of the minus anomaly center near Berin sea is too weak and the result of the prediction near artic region is not satisfactory. The location of the main centers of height anomaly are all close with the observed except that a small positive anomaly center in Europe has not been predicted and the predicted positive anomaly center near the pole has not been observed. The correlation coefficient of the persistence of height anomaly from August to September in 1965 is 0.24 (seeTable 1), but the correlation coefficient between the predicted and the observed is 0.36, which is higher than the level of the inertial forecast.

b. Example of February 1978. Figs. 4a and 4b indicate the predicted anomaly fields of earth's surface temperature and 500 mb height in February 1978 respectively. Figs. 5a and 5b denote the corresponding observed maps of the same month. The main discrepancy between prediction and observations is the location of positive anomaly center of temperature and also 500 mb height over America. The calculated center is further northwest than the observed. However, this discrepancy can be improved by three layers model (for atmosphere). The new result of 500mb anomaly height is given in Fig.6.

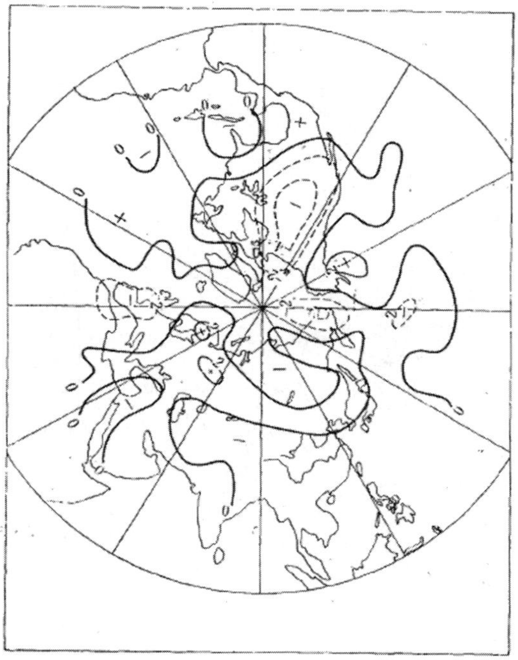

Fig.3a　Anomalous field of earth's surface temperature for September 1965 (observed)

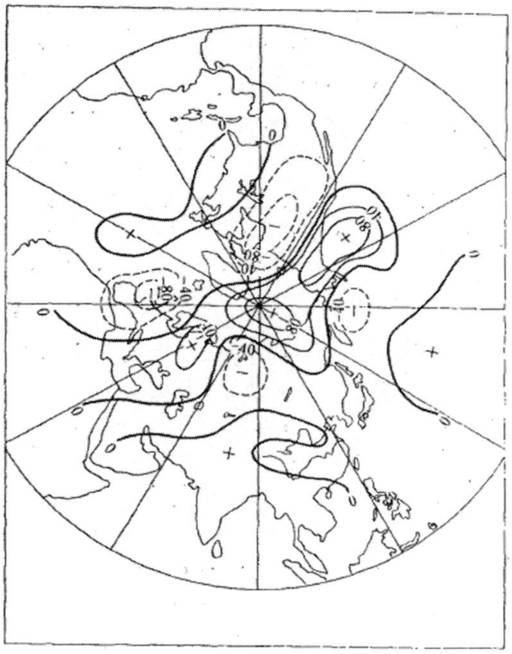

Fig.3b　Anomalous field of 500 mb height for September 1965 (observed)

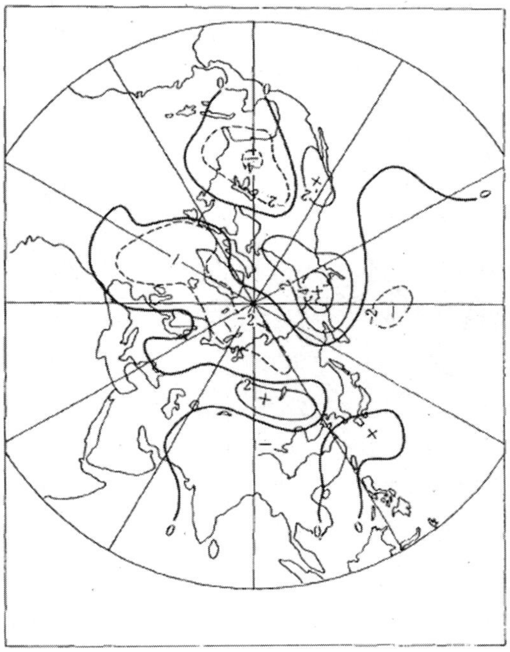

Fig.4a　Anomalous field of eath's surface temperature for February 1978 (predicted)

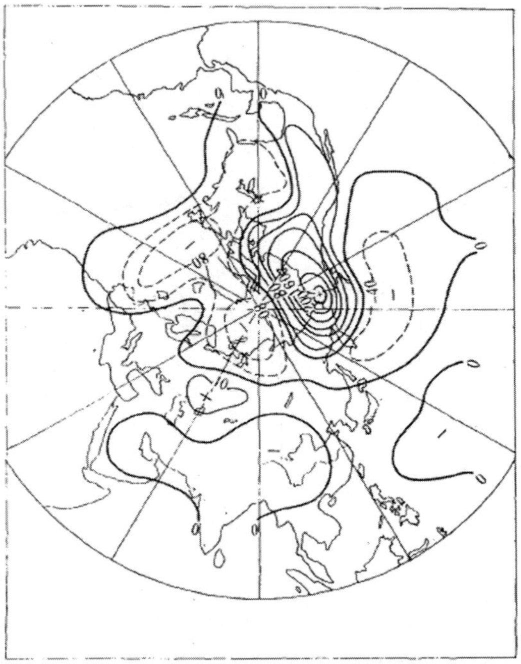

Fig.4b　Anomalous field of 500 mb height for February 1978 (predicted)

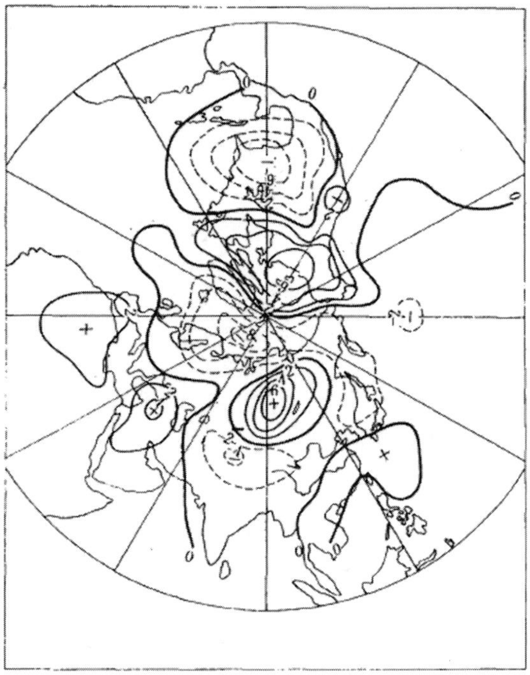

Fig.5a Anomalous field of 500 mb height for February 1978 (observed)

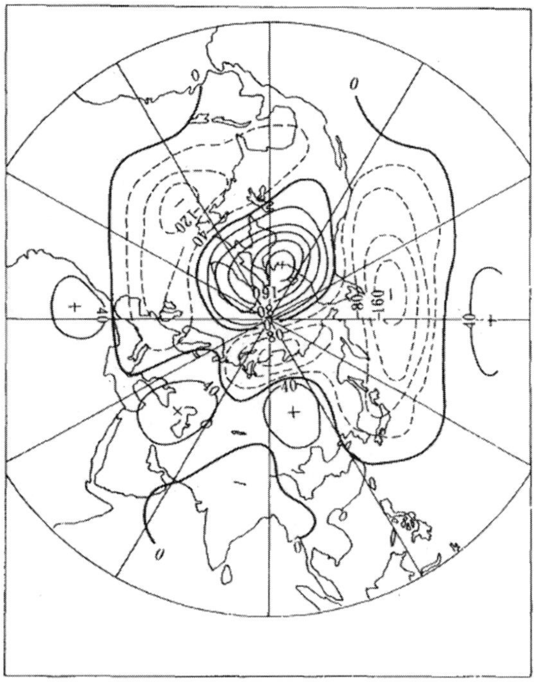

Fig.5b Anomalous field of 500 mb height for Feb. 1978 (observed)

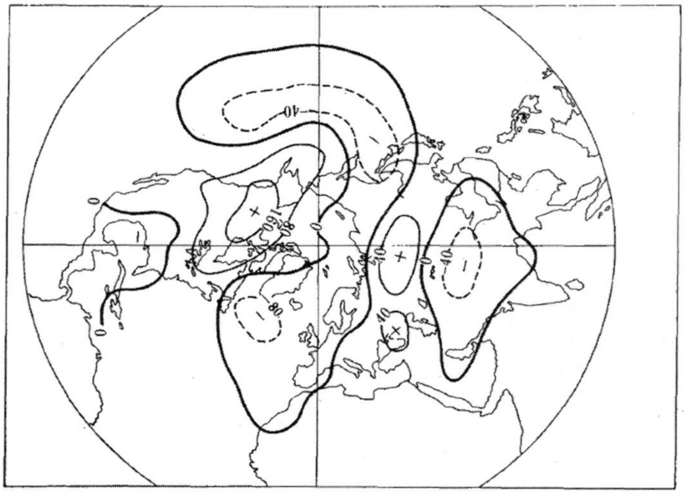

Fig.6 Anomalous field of 500 mb height for February 1978 by three layers model (predicted)

From these predicted examples it may be seen that although the predicted skill is not so high, but this tentative investigation on long-range numerical weather forecasts is a promising way worthy of further study.

References

Adem, J.1964.On the physical basis for the numerical prediction of monthly and seasonal temperatures in the troposphere-ocean-continent system.*Mon. Wea. Rev.*, 92: 91-104.

Budyko, M.I.1956.The heat balance of earth's surface.PB 131692.

Group of Long-Range Numerical Weather Forecasting. 1977. On the physical basis of a model of long-range numerical weather forecasting. *Scientia Sinica*, 20: 377−390.

1979-A filtering method for long−range numerical weather forecasting. *Scinentia Sinica*, 22: 661−674.

Kuo, H − L. 1968-On a simpilfied radiative − condunctive heat transfer equation, Scientific Report, No. 14. The planetery circulation, project, the University of Chicago.

Namias, J. 1948. Evolution of monthly mean circulation and weather patterns. Trans. Amer. Geophys. Univon, 29: 777−788.

大尺度海气相互作用和长期天气预报*

巢纪平

（中国科学院地理研究所）

在 20 世纪 50 年代初，我国气象工作者就注意到西北太平洋特别是黑潮海域，海表温度的冷暖与我国东部地区汛期的旱涝有密切的关系[1]。60 年代开始，大尺度海气相互作用的研究在国外普遍受到重视[2-4]。新中国成立以后在党的领导下，通过广大台站预报员和专业科研人员的努力，我国长期天气预报有了较大的进展，但是，预报准确率和预报时效仍不能满足各方面日益增长的需要。因此，需要多方面探索提高长期天气预报水平的途径和方法，从海气相互作用来研究长期预报无疑也是值得重视的。近年来，一些气象台已在试用海表温度作长期预报，并取得了一定的预报效果。

1 海洋在长期天气过程中的作用

海洋在长期天气演变中的重要性，至少可以从以下两个方面来分析。

1.1 从能量平衡的观点分析

运转大气这部热机的根本能量来自太阳辐射. 太阳给地球的能量其功率约 1.8×10^{14} kW，除被直接反射外，地球实际上收到的能量其功率约 1.0×10^{14} kW，这部分能量以位能的形式储存在大气中。大气这部热机把位能转换成动能的效率仅为 2%，即其转换率 $\partial E/\partial t$（E 为总能量）约 2.0×10^{12} kW。大气的总质量约为 5.3×10^{21} g，因此单位质量动能的制造率为

$$\frac{1}{M}\frac{\partial E}{\partial t} \backsimeq 4\ \mathrm{cm^2/s^3}$$

整个大气把位能转换成动能的时间量级称天气系统的能量生成时间。由于大气的总能量平均约为 10^{21} J，所以能量的生成时间量级为

$$\tau = \left(\frac{1}{E}\frac{\partial E}{\partial t}\right)^{-1} = 10^{21}\ \mathrm{J}/2\times10^{12}\ \mathrm{kW} \backsimeq 5\times10^{5}\ \mathrm{s}$$

即约 1 周。

在另一方面，由于湍流的发展，大气的动能最后要耗散成热能，按 Brunt 的估计，平均耗散率约为 $\varepsilon \backsimeq 5\ \mathrm{cm^2/s^3}$，这与动能的制造率接近平衡。耗散作用的大小可以用有效黏性系数表征成

$$\Delta \backsimeq \varepsilon^{1/3} l^{4/3}$$

式中 l 为天气系统的特征尺度，平均取 3 000 km，由此算得黏性耗散的张弛时间，即动能的耗散时间应为

$$\tau = l^2/\Delta \backsimeq \varepsilon^{-1/3} l^{2/3} \backsimeq 3\times10^{5}\ \mathrm{s}$$

这与动能的生成时间同量级。

在一个天气系统中，如果没有外界能量的继续补充，那么由位能转换成的动能，在 1 周左右的时

* 大气科学，1977，1(3)：223-233.

间就要被湍流耗散光。如果天气系统要继续维持和发展，外界就要不断补充能量。根据以上能量平衡的考虑，Монин[5]把

$$t - t_0 > \tau$$

的过程，称为长期天气过程。显然，这是一类非绝热过程。

这样，我们可以从分析地球-大气系统中能量的收支情况来看海洋的作用。先分析热量的收支情况。到达地球大气上界的短波太阳辐射能量约 1.95 K/(cm^2·min)(太阳常数)，如把这个量以 100 单位计，则从整个地球-大气系统的年平均看，这部分辐射进入大气层后，直接被水、水汽、尘埃和臭氧吸收了 16%，被云吸收了 3%，被空气分子散射掉 6%，被云反射掉 20%，被地面直接反射掉 4%，剩下的 5 1%被地球表面所吸收。被地球表面吸收的这部分太阳辐射将转换成其他形式的能量供给大气，其中以长波辐射形式向上发射了 21%，以潜热形式提供了 23%，以感热形式提供了 7%。从这个意义上看，如 Malkue[2] 所说，运转大气热机的主要"燃料" 来自地球表面。

由于海洋在几何面积上占整个地球表面的 71%，其热容量要比大气大 1 200 倍之多，这样大的热惯性将使海洋能储存更多的热量，并在适当的条件下，以适当的方式向大气提供所储存的部分热量，以推动大气运动。

同时由于地球-大气系统中相当部分热量是以潜热的形式表现的，所以需要再分析水分收支情况。据估计，地球上各种形式的总水量约 1 384×10^6 km^3，其中 97.4%是海水，0.5% 是地下水，0.1%是河水及湖水，2.0%是冻结物，还有 0.0009%是大气中的水汽。也就是说，整个大气所含的水汽量约 1.2×10^{19} g，或相当于 24 mm 厚的水层。但是年平均降水量为 3.96×10^{20} g，其中 2.97×10^{20} g 降在洋面上，0.99×10^{20} g 降在陆面上，总降水量相当于 780 mm 厚的水层。因此，大气中的水分平均每年要更替 780/24≌32 次，或平均每隔 11 天就要更替一次。对全球来讲，为了达到水分平衡，年蒸发量应与年降水量同量级。据估计，来自海洋的蒸发量占总蒸发量的 84%，即约相当 3.34×10^{20} g，蒸发量大于降水量。来自陆面的蒸发量仅为 0.66×10^{20} g，小于降水量. 如果蒸发潜热以 2.4×10^3 J/g 计，则消耗在蒸发上的功率为 3×10^{13} kW，即 15 倍于大气中动能的制造率。因此，从水分收支看，主要的潜热源地也在海洋。

1.2　从非绝热加热的时间尺度看

如果把长期天气过程看成是大气这部热机的"输出"，海洋的加热看成是热机的"输入"，为了使输入能激发出有意义的输出，则要求输入信号的频率(或周期) 与输出信号接近(至少对线性系统如此)。由前述，长期天气过程的特征时间至少应为 5×10^5 s，现在来分析海洋加热的时间尺度。

海洋中储存的一部分能量将表现在海水温度的增高上，如海表温度的异常值为 T'_s，由此向大气发射的异常长波辐射量为

$$E' = \frac{\partial E}{\partial T} \cdot T'_s \backsimeq 4\sigma \widetilde{T}_s^3 \cdot T'_s$$

式中，σ 为史蒂芬-玻茨曼常数。设空气中的吸收物质为水汽，吸收系数为 k(取 6×10^{-6}/cm)，则在单位时间单位体积空气中所吸收的辐射量为 kE'，这部分能量使空气温度增加 $T' \backsimeq T'_s$ 所需要的时间为

$$\tau_r = \rho C_p / 4k\sigma \widetilde{T}_s^3$$

其中，ρ、C_p 分别为空气的密度和比热；$\widetilde{T}_s$ 为平均海表温度。由此算得 $\tau_r \backsimeq 5\times10^5$ s，达到长期天气过程的时间尺度。

同样可以对蒸发过程的特征时间作出估计。海水蒸发后要通过形成云的过程而释放潜热，但个别云块的加热效应不足影响大尺度环流的变化，潜热对大尺度环流的影响是云团的统计效应，所以蒸发的作用需要统计地或参数化地加以考虑。一个简单的参数化方法是设海面上空气的温度为 T，海表温度为 T_s，单位时间单位面积上蒸发潜热可以参数化地表示成

$$LE' = \rho LC_D \mid V \mid (q_s(T_s) - q_s(T)) \backsimeq \rho LC_D \mid V \mid \frac{\partial q_s}{\partial T}\Delta T_s$$

式中，$\Delta T_s = T_S - T$，q_s 为饱和比湿；L 为蒸发潜热；$|V|$ 为风速值；C_D 为曳力系数，其值取 10^{-3}。设这部分蒸发热被垂直气流或湍流带到对流层中下部，如平均取 5 km 高度，则单位空气体积中因这部分水汽相变而获得的潜热为 $\rho LC_D |V| \frac{\partial q_s}{\partial T}\Delta T_s / D (D = 5\ \text{km})$，于是使空气温度改变 ΔT_s 的时间为 $\tau_e = C_p D / LC_D |V| \frac{\partial q_s}{\partial T}$，取 $|V| = 10$ m/s，$\frac{\partial q_s}{\partial T} \backsimeq 3\times10^{-4}$/℃，则得 $\tau_e \backsimeq 10^6$ s，这接近上述大气中水分的更潜时间 11 天。

可见，无论是以长波辐射形式，还是蒸发的形式，海洋加热对大气的长期变化都是有意义的。

另一方面，海洋发生的过程在一定程度上又受大气运动的制约。例如，由于风对海水的搅拌，将使表层海水中的湍流加强，从而影响到表层海洋中热量的分布，以致又影响到加给大气的热量. 如取海洋表面活动层的厚度为 D_s，则通过混合，将活动层底部的热量带到海表所需的时间量级为

$$\tau_s = D_s^2 / \kappa_s$$

式中，κ_s 为对热量的湍流交换系数，如取 $\kappa_s = 10\ \text{cm}^2/\text{s}$，$D_s$ 取 30 m，则有 $\tau_s \backsimeq 9\times10^5$ s，这与长期天气过程的特征时间同级，表明长期天气变化反过来也可以使海洋过程发生有意义的响应。

以上的分析表明，在研究长期气天过程时，不仅需要考虑海洋的作用，而且还要进一步考虑海气之间的相互作用。

2 若干统计事实及在长期预报中的应用

2.1 北大西洋地区的海–气相互作用

Bjerknes[6] 在分析了多年的资料后指出，当冰岛低压加强时，从冰岛西南到 50°N 间的海表温度要变冷；或者，当 50°—60° N 之间西风带强度增强时，相应地海表温度要降低。当然，这种同时期的统计相关，只表明大气环流与海温之间存在相互依存的关系，不能说明两者之间的因果联系。然而，Namias[7] 曾注意到，从 1958 年开始到 1960 年，北欧上空出现了持续性的阻塞活动，使斯堪的那维亚半岛出现了严重的干旱气候。他认为这是由于北大西洋持续性的东西向海温距平梯度，加强了热成风的南风分量，使得气旋路径比常年偏北的结果。Ratcliffe 和 Murray[8] 指出，湾流区的海温与欧洲气压形势之间存在时滞相关，当海表温度比平均为冷时，次月在西北欧将有利于阻塞形势的出现，反之，则有利于出现纬向环流。英国气象局已把海气之间的这种时滞相关用于一个月的长期预报业务。

陈烈庭[9] 以 1972 年前冬北半球出现大范围的环流异常为例，分析了大西洋海温对环流的影响。他指出这与该区冷水的异常发展有关。进而，他用湾流区固定船舶站 E 从 1651—1972 年 1 月的海表温度与翌年 1 月北半球 500 mbar 高度场做了相关计算，相关系数的分布见图 1。这 22 年资料的相关分析除进一步证实 1972 年初异常环流的出现与大西洋持续存在的冷水有关外，同时表明海洋对大气环流的影响可以向下游传播到很远的地区。

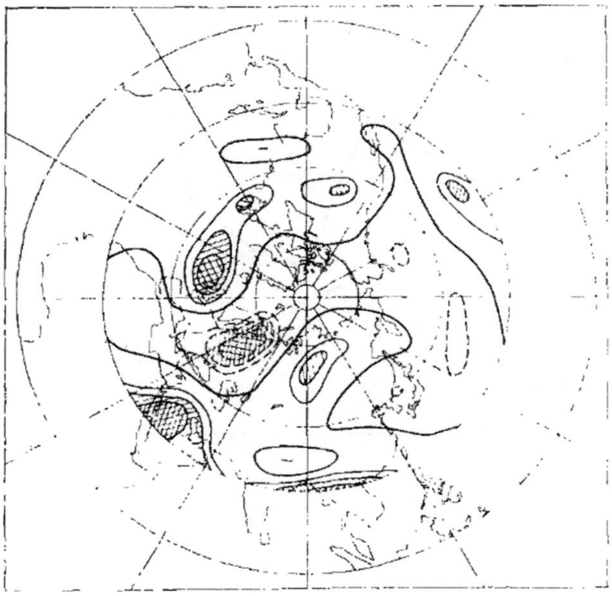

图 1 墨西哥湾流区固定船舶站 D 和 E 10—12 月的海温距平均累积量与翌年 1 月北半球,500 mbar 高度场的相关分布。粗线线相关零线;实线和其中的单斜线区, 双斜线区分别是相关系数等于 0.36 和大于 0.42, 0.54 的区域, 它相当于 0. 10,0. 05,0. 01 的信度。虚线和其中的各种区是负相关情况

2. 2 北太平洋地区的海–气相互作用

Namias 分析了 20 世纪 60 年代北太平洋地区海气相互作用的气候背景[10], 并进一步分析过一些个例。例如他指出, 1962—1963 年冬季太平洋地区的天气异常是海气相互作用所造成[11]。在 1962 年的夏秋两季, 在太平洋中部发展出一片巨大的暖水区, 它引导阿留申低压向南伸展到异于常年的纬度, 使得冬季在这一片暖水区上出现了比常年低 15 mbar 的地面气压, 相应地又使 700 mbar 高度场比常年低达 400 呎(英尺,1 呎=0. 304 8 米)。太平洋中部这个长波槽的异常加强,通过能量频散作用, 在阿拉斯加和美洲西海岸一线发展为一个异常强的长波脊, 在其下游美国东海岸附近又发展出另一个长波槽, 使得冷空气沿美洲东部的长波槽后南下, 造成美国东部大部分地区的严冬天气。类似地, 像 1968—1969 年冬, 在美国西海岸加利福尼亚地区发生的暴雨也是前期海气相互作用的结果[12]。另外, 从多年统计看(如图 2a 所示), 沿 50°N 在 160°W 到 130°W 之间, 夏季的海表温度与秋季海平面气压场之间存在着负相关的关系(相关系数为-0. 67), 这表明在这一地区夏季的暖水将使秋季的海平面气压降低。但又如图 2b 所示, 在海洋对大气的作用过程中, 它本身的温度同时要发生调整, 到了秋季海温距平和同期的气压距平之间已调整到正相关的关系(相关系数达 0. 92)。从这一统计事实可以看到, 海气之间的确存在着相互作用和调整的过程。

归佩兰①分析了 1949—1962 年 14 年中北太平洋月平均海表温度的大尺度特征, 发现可以归纳为 4 个型, 型的持续时间平均约 10 个月, 最长的达 23 个月。相应于不同的海温型, 同时期 700 mbar 月平均高度距平也相对稳定地具有不同形式。一般就大尺度特征来看, 在海表温度正(负) 距平上空, 高度距平基本上也是正(负) 的。由于分析用的是月平均图, 这一事实说明, 就平均来讲, 或统计地

① 归佩兰,北太平洋海表温度特征及相应的大气环流型, 中国科学院大气物理研究所集刊第 6 号, “海气相互作用与旱涝长期天气预报”(以下简称《文集》)。

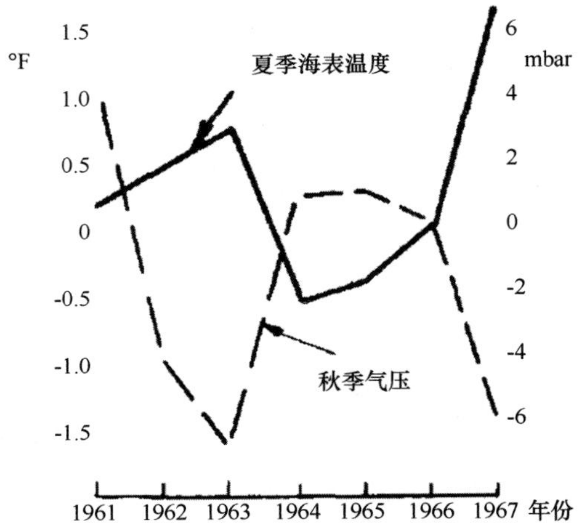

图 2a 沿 50°N,160°~120°W 之间,夏季海表温度与秋季海平面气压的相关

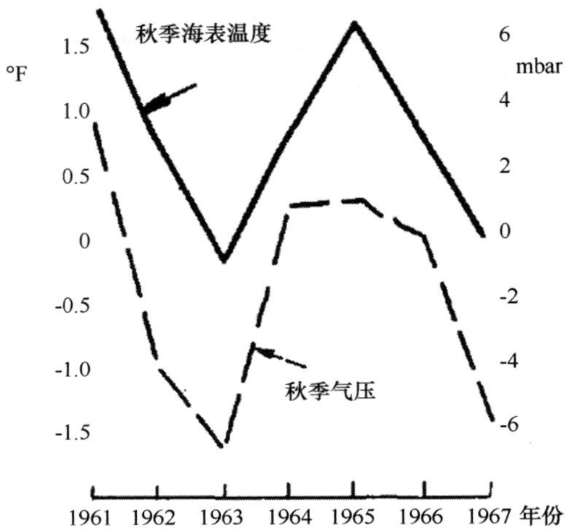

图 2b 沿 50°N,160°~130°W 之间,秋季海表温度与秋季海平面气压的相关

海温场与大气运动之间的大尺度相互适应时间不超过一个月。当然，就个例来讲，相互适应的关系要比统计情况复杂得多。

我国的地理位置相对西风带来讲，位于太平洋上游，太平洋的热状况能不能对我国的天气产生影响，这是我国长期预报工作者很关心的问题。林学椿在这方面做了研究[13]，根据统计分析，他指出当前冬太平洋中部出现大范围海温正(负)距平时，其后期东亚地区的环流指数要加大(减小)，同时海温热状况对大气环流的影响除不断向下游传播外，也将在后期某一段时期内向上游传播一段距离。陈烈庭①进一步指出黑潮热状况除影响本区上空的环流外，其影响也可波及欧洲地区。

长江流域 6 月的梅雨天气对我国工农业生产有重要的影响。大气物理研究所长期预报组②取长江中下游的武汉、九江、芜湖、南昌、安庆、屯溪 6 个站 6 月的月平均雨量与前期太平洋 1949—1962 年的海温距平做了逐月的相关计算。图 3 是与前期 1 月的关系数图。由图可见,在黑潮海域出现了一

① 陈烈庭，1972 年世界性天气异常的环流特征及其与海表水温的联系，《文集》。

② 大气物理研究所长期预报组，冬季太平洋海水温度异常对我国汛期降水的影响，《文集》。

条狭长的正相关带,最大系数达 0.77。图 4 给出沿黑潮海域从 20°N、120°E 向东北到 50°N,165°E 沿线各点的相关系数的逐月变化,可以看到黑潮的影响从秋末开始一直持续到初春,并从低纬开始逐渐向北伸展。

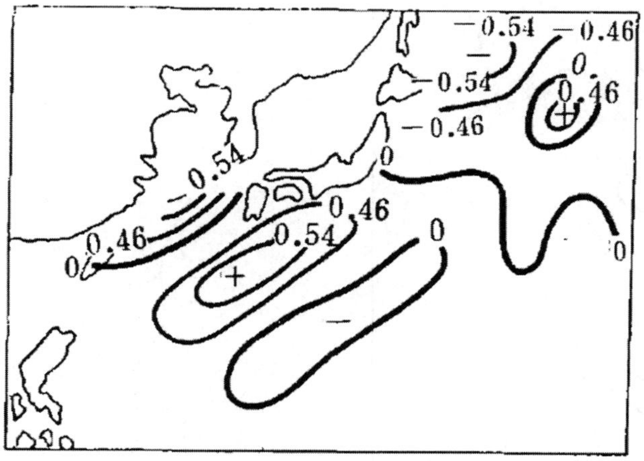

图 3　长江中下游 6 月雨量与前期 1 月黑潮海表层温度的相关

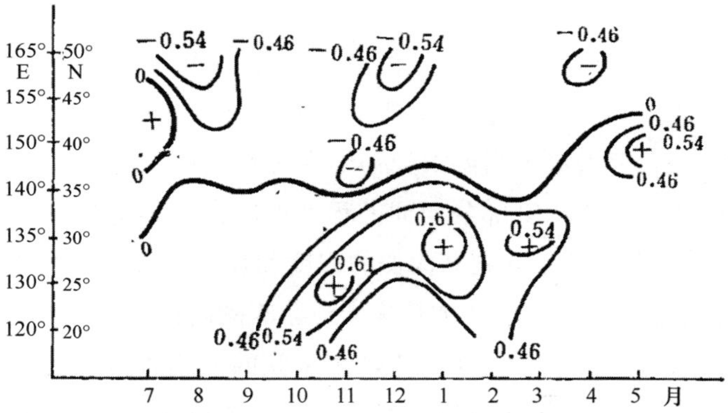

图 4　黑潮冬季海温与长江流域 6 月降水相关系数的逐月变化

为了进一步了解黑潮的热状况可能对我国哪些地区的降水量会有影响，他们又取 1949—1971 年冬季黑潮区的月平均海表温度与后期全国 65 个站(分布在全国各省、自治区）的月平均雨量做了逐月的相关计算。结果表明，冬季黑潮的热状况从 5 月开始对长江流域及江南地区有影响(图 5a)，到 6 月，相关场的形势突变(图 5b)，除在长江中下游维持一条高值相关带外，在 40°N 一带的河北平原也出现另一条高值相关带。7 月，河北平原的高值相关带仍维持外，长江中下游的相关带其值已降到信度 0. 10 以下(图略)。这种雨带和前期黑潮海温之间的相关场季节演变的特征，基本上与东亚从春到夏的季节转换过程的分析一致[14,15]。由于雨带在一定程度上是大气环流在各个方面相互制约的结果，因此可以认为，东亚大气环流的季节性突变和前期太平洋海表温度的异常有相当的联系，至少是制约东亚大气环流演变的一个重要因子。

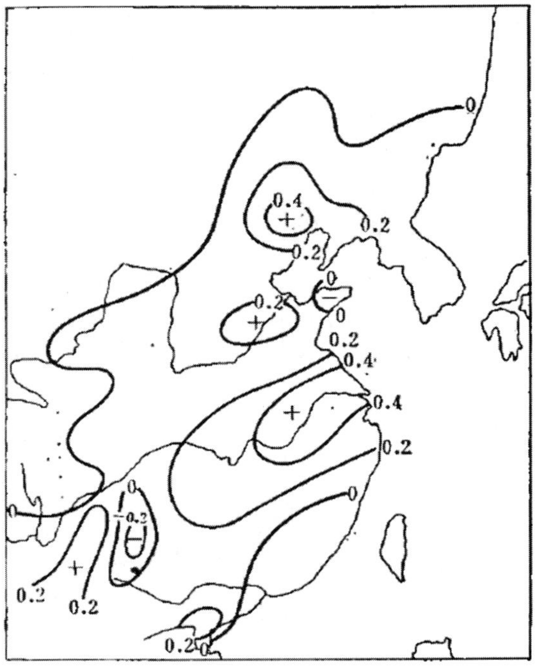

图 5a　海温与 5 月降水的相关场

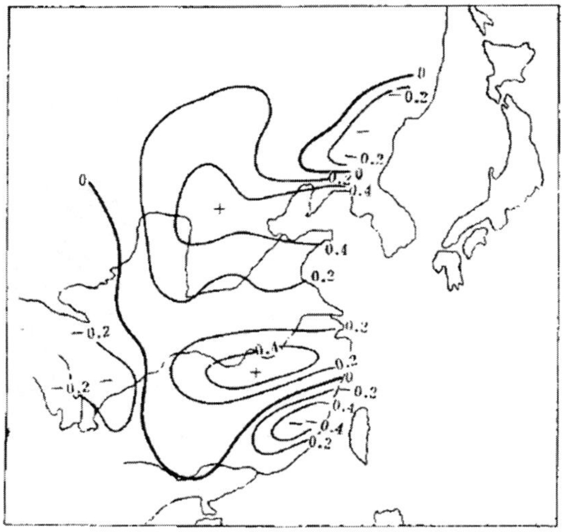

图 5b　海温与 6 月降水的相关场

黑潮海域前冬的热状况为什么能对我国东部汛期的降水发生影响？潘怡航[16]在计算了黑潮海域海气之间感热和潜热的交换后指出，当冬末到初春在黑潮南端，海洋向大气提供的感热和潜热量较正常值为多时，5 月副热带高压边缘部分西伸的程度也较正常年份为强. 而天气学的分析表明，5 月、6 月两湖盆地和长江中下游的雨量与副高西伸的程度和强弱有密切的关系。长江流域规划办公室预报科[17]和章淹[18]等的工作也表明，海洋加热是通过影响副高活动再影响梅雨天气的。

前冬黑潮区的海表温度距平与长江中下游汛期雨量间的上述时滞相关，可以直接应用于雨量的长期预报。图 6 是 1951—1972 年长江中下游 6 月降水距平和前冬黑潮海温距平的逐年变化曲线，可见这两条曲线的趋势是一致的，特别像 1954 年的特大洪涝，在前期黑潮的高海温上已有反映。应用这种关系试做了 1973—1976 年的预报，效果较好。

前期黑潮海域的热状况对华北平原汛期降水的影响①②，对华南秋旱的影响③也有了研究，有的工作④还指出鄂霍次克海高压的强弱也与太平洋的热状况有关。此外，西太平洋和青藏高原的热状况对江南地区汛期降水的共同影响也有分析[19]。

2.3　热带地区的海–气相互作用

热带是推动大气环流能量的主要源地，该地区的海–气相互作用无疑具有重要的意义。Bjerknes[20-22]指出，赤道地区海水温度的异常可以影响到中纬度的大气环流。当冬季赤道海表温度高时（如 1957—1958 年），副热带高压从亚洲到美洲呈东西向分布，轴线也稍向南移，东太平洋中纬

---

① 李鸿洲，海气相互作用对河北平原汛期降水的影响，《文集》。

② 中央气象台长期天气预报组，我国北方东部地区雨季划分和河北平原雨季降水强度预报；华北地区夏季降水趋势及其环流特征分析，中央气象台长期天气预报技术经验总结，1976 年。

③ 陈增强，黑潮海温与华南秋旱，中央气象局气象科学研究所，内部稿，1976 年。

④ 黑龙江省气象台，北太平洋海表温度与黑龙江省的夏季降水，1975 年。

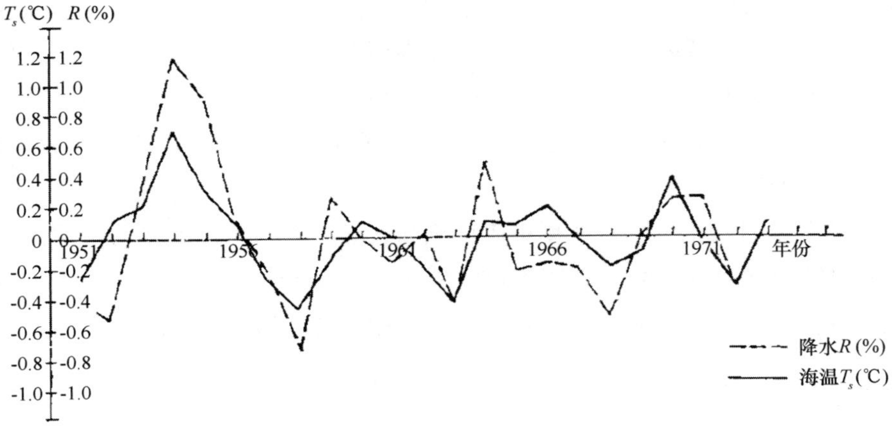

图 6　长江中下游 6 月降水距平和冬季黑潮区海温距平逐年变化曲线

度西风带比常年要强。这表明赤道海洋向大气提供较多的热量,加强了哈特莱环流的上升分支，把低纬的角动量大量输送到中纬，从而加强了西风的强度并使阿拉斯加低压发展。反之，当赤道海温低时(如 1955—1956 年)，阿留申低压在堪察加和阿拉斯加分成两个，使定常行星波的波长变短，西风减弱。Bjerknes 还指出，在 60 年代赤道地区的海水温度有准两年振荡周期。

中国科学院地理研究所长期预报组进一步用多年资料分析了热带海洋对副热常高压长期演变的影响[23]。对这两组时间序列做谱分析后发现，除存在 2~3 个的振荡周期外,长周期的振荡周期主要是三年半。看来 Bjerkens 指出的准两年周期只在 60 年代明显。值得注意的是，像北大西洋湾流区的海表温度、冰岛低压强度等也存在这一周期。图 7 是 1950(1951)—1974(1975) 年热带海表温度距平 ΔT_s(所取范围为 5°N 至 10°S,80°W 至 180°)和副热带高压面积指数 ΔM (以副高单体内≥588 位势什米内的网格总数表示)，这两条曲线已经过 6 个月滑动平均处理，另外在图中还给出 20°—50°N, 120°—160°E 地区 300 mbar 的环流指数 ΔH^* 。如图 7 所示,ΔM 和 ΔT_s 之间具有明显的滞后相关，当海温开始增暖约 1~2 个季度后，副高也开始增强，同时西风环流指数也相应加大。应用这种时滞相关,可以对副高的强弱以及有关地区的汛期雨量做长期预报。统计结果表明，当前一年秋、冬热带海温高时，翌年 8 月淮河流域雨量偏少，而黄河流域的河套地区雨量将偏多。

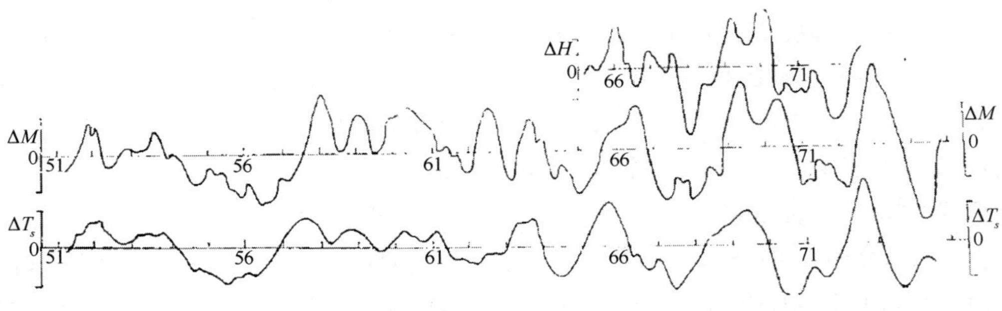

图 7　1950(1951)—1974(1975) 年热带海表温度 ΔT_s、副热带高压面积指数 ΔM 以及环流指数 ΔH 的逐年演变曲线

另一方面，在 Bjerkens 的上述研究中，还指出赤道区的海温分布能引起东西向环流的变化。由于大气对海洋的风吹流(漂流) 作用，东风将使漂流向赤道南北两侧辐散，这样底层的海水将上翻，

使表层海水变冷。所以当东风加强时，将使赤道太平洋东部地区变冷，出现东西向温度差异，这样在东西两端海洋对大气提供的热量将有所不同，于是在西部地区产生上升气流，在东部地区产生下沉气流，这样的东西向环流又将使海平面的东风进一步加强。反之，当东风减弱时，东西向海温差变小，沿赤道的环流变弱，相应地东风就更弱。作为对早年沃克提出的"南方涛动"的一种解释，Bjerkens 称这种东西向环流为沃克环流。

徐群[24]对冬春沃克环流的强弱对初夏东亚环流和长江中下游入梅迟早的影响做了研究，指出：冬春秘鲁沿海海温剧升(降)——4 月南方涛动弱(强)——沃克环流东支退缩(西伸)，6 月，东亚中纬度槽、西太平洋副高和印缅槽都随之偏东(西)，长江中下游入梅推迟(提早)。

3 海-气相互作用的理论分析

Гаърилин 和 Монин[25,5]设计了一个考虑了海-气之间有热量交换、云量对洋面所接受到的太阳辐射量有调节作用，同时云量又受大气运动控制的模式，频率分析表明，在这样一个海-气相互作用模式中存在着以月为特征时间的长周期振荡。

Петухоь 和 Фейгельсон[26,27]进一步分析了云对海-气之间热量交换的调节作用，指出由于云型不同造成云对辐射的吸收率和反射率不同，从而对热量交换过程发生影响。虽然他们的模式纯粹是热力学的，没有考虑大气运动及其反馈过程，但在温度场和湿度场上也出现了以月的周期振荡。

在 Pedlosky 所设计的模式中[28]，海气之间只有感热交换，大气运动对海流有反馈影响，但没有考虑对辐射的调节作用。计算表明，通过海-气之间有限振幅扰动的相互作用，在海洋环流回旋整个海洋一圈的时间尺度上，将产生一个正反馈使海表温度加强。

注意到在上面的理论工作中，或者没有考虑大气的运动，或者只考虑了气旋波的作用，即略去了由于 $\beta = df/dy$（f 为柯里奥利参数）效应而产生的罗斯贝行星波。行星波是大气中一类重要的大尺度运动，在研究大尺度海-气相互作用过程时，不考虑它的作用看来未必合适。

长期数值预极研究小组[29]①设计了一个海-气相互作用模式，在热力学的相互作用方面，在海-气交界面上海洋除接受到太阳辐射外，并向大气提供蒸发潜热，同时通过湍流向表层以下海水输送热量；在动力学的相互作用方面，由大气中行星波的涡度控制云量，通过云量调节海面所接受到的太阳辐射量。对这个模式中所包含运动的频率分析指出，在大气中将出现两类性质完全不同的运动。一类是两组非绝热行星波，其振荡频率如图 8 所示，图中虚线是正压情况下的绝热行星波频率，可见海气相互作用对这类运动的频率影响不大。但如图 9 所示，这两组非绝热行星波中的一组是发展的，另一组则是阻尼的，对波长约 6 000 km 的波，发展波的增长率约 1 周增长到 e 倍。值得注意的是另一类运动，其振荡频率见图 10。当波长为 6 000 km 时，约 0.5×10^{-6}/s，即相当周期约 3 个月。这类长周期波，当波长大于 5 000 km 后是发展的(图 11)，增长率约一个月增长 e 倍。

同时上面的计算还表明，至少对特征时间大于行星波周期的长期过程，存在着大气运动向海温适应的阶段，适应时间不到一个月。这说明在大气运动和海洋加热这一对相互制约的矛盾对立体中，对于长期过程来说，海洋加热场在一个阶段中可以是矛盾的主要方面，大气的长周期运动将受海洋加热的制约。

① 参加者有中国科学院地理研究所、大气物理研究所、中央气象局气象科学研究所和北京大学地球物理系有关人员。

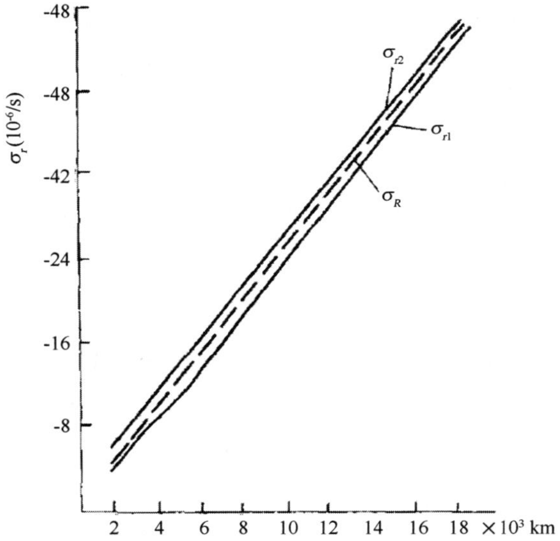

图 8　短期波的振荡频率

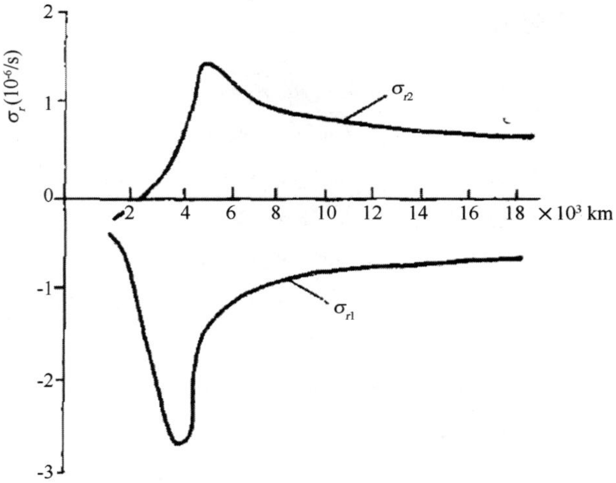

图 9　短期波的增长率

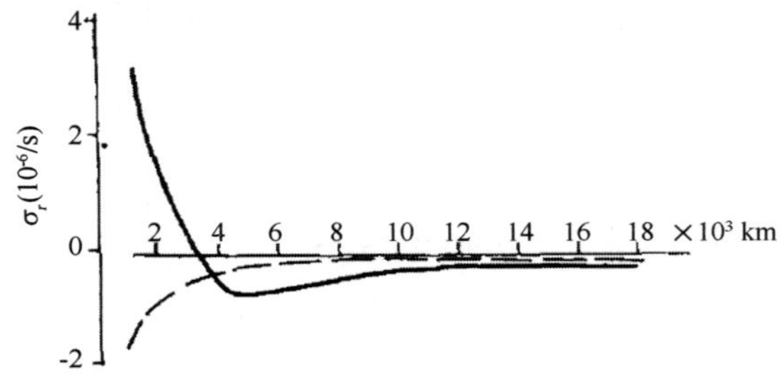

图 10　长期波的振荡频率

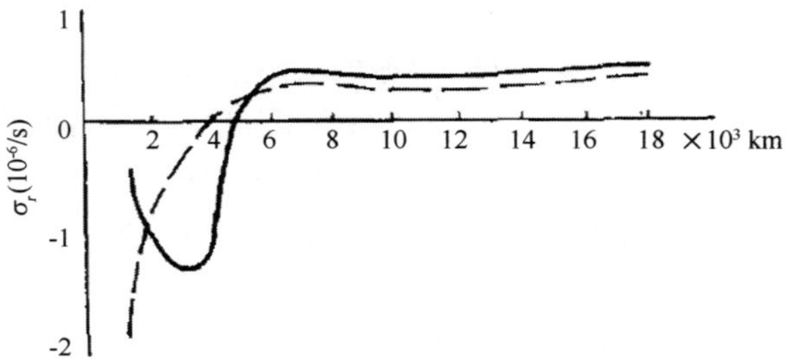

图 11　长期波的增长率

在短期天气的数值预报获得成功后，长期天气的数值预报该如何进行？这是当前动力气象和数值预报工作者正在探索的问题。目前已普遍接受这样的观点，在长期数值预报中需要考虑海洋的加热作用。Adem[10] 曾提出了一种考虑海气有热量交换的热力学模式，用以预报月平均大气和海表的温度。由于模式中没有考虑大气中的动力过程，所以不能预报大气环流的长期演变。通过上面的理论分析，可以看到，在一个海气相互作用的模式中，同时并存着长期和短期过程，而在加热作用下，短期过程的增长率要比长期过程大得多。因此，如果长期数值预报也采用一天天延长预报时效的做法，那么由于短期天气过程和计算误差所造成的虚拟过程的发展，势必把作为预报对象的小振幅的长期过程给掩盖掉。如像目前某些可预测性研究表明的，预报时效不超过两周。众所周知，在短期数值预报的发展史中，Richardson 曾遇到过同样的困难。后来 Charney 用一种滤波技术（准地转近似）把运动方程中所包含的声波和重力波过滤掉，使模式中单纯包含有意义的短期天气过程，使短期数值预报获得了成功。长期数值预报研究小组提出，在制作长期数值预报的方法上，至少在目前对一些过程尚未弄清和计算方法尚不完善的情况下，也可以类似地把行星波为主导的短期天气过程，看成是一种干扰长期天气过程的“噪声”而滤掉它，使模式中单纯只存在长期过程，同时给出了一种过滤行星波的方法。图 10 和图 11 中的虚线分别是滤波后的频率和增长率，可见滤去行星波并没有使长期波受到大的歪曲。初步的数值试验表明，用这样的方案来预报一个月后的月平均 500 mbar 的环流和海表（以及地表）温度，预报结果尚令人满意。当然，这只是一种可行的长期数值预报方案，也可以探索其他的途径。

参考资料

[1] 吕炯.地理学报,1951,18:69-88.

[2] J.S.Malkus.Large-Sale interaction,The Sea.Editior M.N.Hill,Vol.1,1962.

[3] G.M.Hidy.*Bull.Amer.Meteor.Soc.*,1972,53(11).

[4] 朝仓一正.海ち空,1971,47(2-3).

[5] A.C.Монин.作为物理问题的数值预报,1967.

[6] J.Bjerknes.Atlantic air-sea interaction,*Adv.in Geophysics*,Vol.10,Ed.By H.E.Landsberg & J.Vn Meighem,1964.

[7] J.Namias,*Tellus*,1964,16.

[8] R.A.S.Ratcliffe & R.Murray.*Quart.J.Roy.Meteor.Soc.*,1970,96(408).

[9] 陈烈庭.科学通报,1974.

[10] J.Namias.*Mon,Wea.Rev.*,1969,97(3).

[11] J.Namias.Proc.Inter.Cont.Cloud Phys.Toronto,1968:735-743.

[12] J.Namias.*J.Physical Oceanography*,1971,1(2).

[13] 林学椿.北太平洋海表温度异常及其对东亚大气环流的影响. 1975 年长江流域长期水文气象预报讨论会技术经验交流文集(以下简称技术经验交流文集), 127-138.

[14] Yeh,T.C.(叶笃正),等.The Atmosphere and the sea in motion,1959:249-267.

[15] 陶诗言,等. 气象学报, 1957, 28(3): 234-247.

[16] 潘怡航. 冬春海洋加热场与两湖盆地汛期降水的初步分析.技术经验交流文集, 1975:154-163.

[17] 长江流域规划办公室水文处预报科. 西太平洋海表热状况对初夏副高活动的影响及其与长江、洞庭、鄱阳两湖地区汛期降水的关系.技术经验交流文集, 1975: 118-126.

[18] 章淹,等. 初夏西太平洋副热带高压活动与梅雨和海温的关系的初步探讨.技术经验交流文集, 1975:164-177.

[19] 湖南省气象台. 冬半年青藏高原和西太平洋表面热状况与湖南汛期降水的初步分析.技术经验交流文集,1975:139-141.

[20] J.Bjerknes.*Tellus*,1966,18(4).

[21] J.Bjerknes.*Mon,Wea.Rev.*,1969,97(3).

[22] J.Bjerknes.*J.Peysical Oceanograph*,1972,2(3).

[23] 中国科学院地理究研所长期预报组. 热带海洋对副热带高压长期变化的影响.1976,即将发表.

[24] 徐群. 冬春南方涛动对初夏东亚环流和长江中下游入梅迟早的影响.技术经验交流文集,1975:142-153.

[25] Ъ.Л.Гаьрнлин,A.C.Монин.*Дан.Акаб.СССР*,1967,Том 176,No.4.

[26] B.K.Пегухоь,E.M.Фейгельсон,*ИЗВ.АН* СССР.*Физика Атмосферы и Океана*,1973,9(4).

[27] B.K.Пегухоь.*ИЗВ АН СССР.Физика Атмоферы и Океана*,1974,10(3).

[28] J.Pedolsky.*J.Atmos.Sci.*,1975,32(8):1501-1514.

[29] 长期数值天气预报究研小组. 长期数值天气预报的物理基础. 中国科学, 1977, 2.

[30] J.Adem.*Mon.Wea.Rev.*,1964,92(3).

An Anomaly Model and its Application to Long-range Forecasts*

J.P.Chao[1**], R.Caverly[2]

(1. Geophysical Fluid dynamics Laboratory; 2. NOAA Princeton University Princeton, New Jersey 08540)

1. The Anomaly Model and Basic Equations

The idea of an anomaly model is quite simple. Since the climatology is already know, why introduce errors in the forecast trying to predict it? The anomaly model system is by definition free of this climatological model bias which gives it a distinct advantage over the total field model. The climatological component can be removed from the total field equations by dividing all the system variables into their climatological and anomalous components. Formally, this results in the separation of the total field equation into the inter-dependent climatological and anomalous system. This procedrue is the same used by Reynolds to remove the turbulent component from the time mean flow. Unlike Reynolds, we are only interested in the time evolution of the anomaly rather than the mean or climatological flow which is supposedly well known. Since the climatological components are assumed time independent, this system of equations can be ignored leaving only the time dependent anomalous system.

By way of a simple illustration, we have chosen an anomaly model that consists of two prognostic equations. One is the anomalous nonadiabatic geostrophic potential vorticity equation

$$\frac{\mathrm{d}}{\mathrm{d}t}\Delta\phi' + \bar{\beta}_y \frac{\partial\phi'}{\partial x} - \bar{\beta}_x \frac{\partial\phi'}{\partial y} = \frac{f^2}{\bar{\sigma}}\left\{\frac{\partial^2}{\partial p^2}(K_T + K_R)\frac{\partial^2\phi'}{\partial p^2} - \frac{1}{\tau_R}\frac{\partial^2\phi'}{\partial p^2} - \frac{\partial}{\partial p}\left[\frac{\mathrm{d}}{\mathrm{d}t}\left(\frac{\partial\phi'}{\partial p}\right) - \left(\frac{\partial\bar{u}}{\partial p}\frac{\partial\phi'}{\partial x} + \frac{\partial\bar{v}}{\partial p}\frac{\partial\phi'}{\partial y}\right)\right]\right\} - \eta(p)\Delta\phi'_0 \tag{1}$$

where

$$\frac{\mathrm{d}}{\mathrm{d}t} = \frac{\partial}{\partial t} + (\bar{u} + u')\frac{\partial}{\partial x} + (\bar{v} + v')\frac{\partial}{\partial y} \tag{2}$$

and the bar and prime quantities refer to the climatological and anomaly components respectively. All the heating processes are parameterized. On the right hand side of equation (1), the last term comes from the heat exchange of condensation and the term with the parameters τ_R and K_R comes from the radiational heat exchange according to the Kuo scheme. The other prognostic equation in our system is the thermal equation for the under lying ocean and land

$$\frac{\partial T'_s}{\partial t} = K_s\frac{\partial^2 T'_s}{\partial z^2} - \delta\left[\frac{\partial(\psi_s, T'_s)}{\partial(x,y)} + \frac{\partial(\bar{\psi}_s, \bar{T}_s + T'_s)}{\partial(x,y)}\right] \tag{3}$$

* Natiomal Weather Service Proc. of the 6thAnn.Climate Diagn.Workshop1982, 316-319.

** On leave from the Institute of Atmospheric Physics, Academia Sinica, Peking, The People's Republic of China.

where $\delta=1$ for ocean, $\delta=0$ fro land. $\overline{\psi}_s$, the climatological ocean stream function, can be calculated from the climatological monthly mean ocean current, and the anomalous part, ψ'_s, can be calculated using the Ekman wind-driven theory. At the surface the following heat balance is used as a boundary condition

$$\rho_s C_{ps} K_s \left(\frac{\partial T'_s}{\partial z}\right)_0 - \rho C_p K_T \left(\frac{\partial T'}{\partial z}\right)_0 + G_1 T'_{s0} = - G_2 \Delta\phi'_0 \tag{4}$$

where G_1 and G_2 parameterizes the evaporation and the effect of cloudiness upon the radiational balance, respectively.

2. Two Time-Scale Waves and Adaptation Equations

In the model discussed here only the 500 mb geopotential surface is considered. Also it is assumed that the earth is covered totally by an ocean where only the mixed layer will be taken into account. There are two time scales in this simple atmosphere-ocean coupled system, namely,

$$t_1 = (\overline{u}\beta)^{-\frac{1}{2}}, \quad t_2 = (K_s/D^2)$$

where K_S is the coefficient of turbulent exchange of heat in the mixed layer, D is the depth of the mixed layer and $\overline{u}$ is the climatological mean wind. Obviously, t_1 is the characteristic time scale for Rossby waves, whereas, t_2 is the characteristic time of mixing processes of heat in the ocean. If $\overline{u}=10^3$ cm/s, $\beta=10^{-13}$ cm/s, $K_S=10$ cm²/sec and $D=3\times10^3$ cm, the ration of these two time scales is smaller than the order of one, i.e.,

$$(\overline{u}\beta)^{-\frac{1}{2}}/(D^2/K_s) < O(1)$$

Thus, if we are only interested in the monthly mean anomalous state, a simple method of long-range weather forecasting may be designed. Assuming the dynamical effects of the transient rossby waves will be smoothed out to some degree in the time scale of one month, we can treat the transient Rossby waves as the "noise waves" and filter them out from the anomaly model soluting by omitting the term of local time variation in the anomalous vorticity equation; this is equivalent to replacing the vorticity equation by the balanced relation between the field of atmospheric flow and the surface heating sources. In our nonlinear one layer anomaly model, this relation is

$$(\overline{u} + u')\frac{\partial\Delta\phi'}{\partial x} + (\overline{v} + v')\frac{\partial\Delta\phi'}{\partial y} + \overline{\beta}_y \frac{\partial\phi'}{\partial x} - \overline{\beta}_x \frac{\partial\phi'}{\partial x} - K^* \Delta\phi' = FT'_s \tag{5}$$

whrere K^* and F are constants coming from the parameterizations of condensation and radiation, respectively. Although, this is a balance relation, we can prove that the 500 mb geopotential height anomaly ϕ' will adapt itself to the underlying surface for the long-range process. Therefore, this balance relation may be called an adaptation equation.

3. An Example of a One-Month Anomaly Forecast

Two steps are needed in order to make a hemispheric forecast of the 500 mb geopotential height anomaly with our simple one-layer anomaly model. First, the temperature anomaly at the earth's surface is predicted by the thermal equation for the underlying ocean and land (eq.3). A one-month time step is used along with the previous monthly mean surface temperature and 500 mb geopotential height anomalies as initial conditions.

Once the temperautre of the earth's surface has been obtained, the 500 mb anomalous geopotential height field is calculated by the adaptation equation (Eq.5).

To show only one example, we have chosen the January 1977 blocking event. The observed and predicted 500 geopotential monthly anomaly, which compares favorably, are shown in Fig.1. For this case the correlation coefficients between the observed and predicted 500 mb geopotential height anomalies is 0.72. Some readers may be interested to know the computer time needed for these calculations. Using GFDL's ASC (Advanced Scientific Computer), the anomaly model only took 15 seconds, whereas, the GFDL's finite difference GCM took 60 hours to obtain a similar result. In all fairness, however, the January 1977 case is our best. We are not as successful with other cases but considerable improvement in the forecasts can be achieved using a 3-layer model. Presently, we are working towards the development of a hight resolution global anomaly GCM which utilizes all the sophisticated parameterization schemes in the GFDL GCM's.

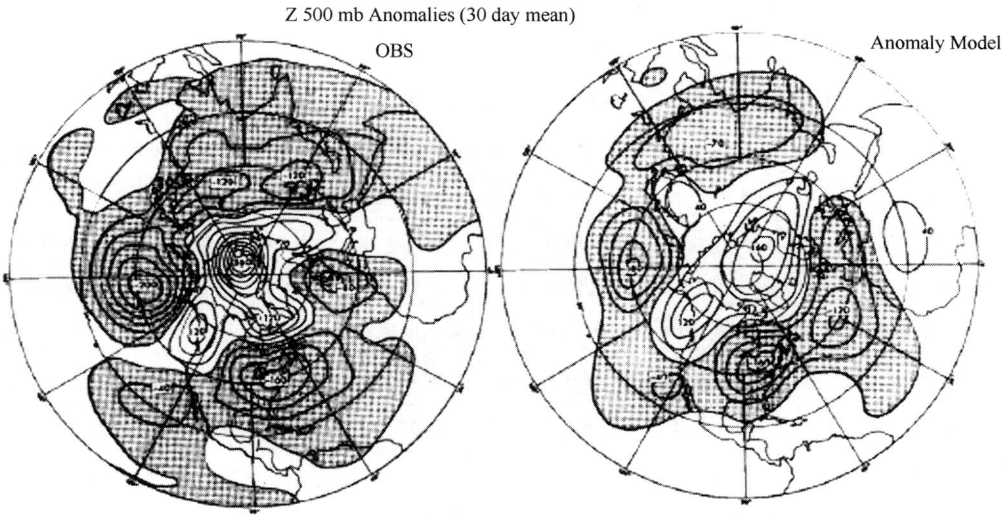

Fig.1　The January 1977 mean 500 mb geopotential height anomalies for the observations (left) and anomaly model prediction (right)

Essay on Dynamical Long-Range Forecasts of Atmospheric Circulation*

K. Miyakoda[1], Chao Jin-Ping**

(1. Geophysical Fluid Dynamics Laboratory/NOAA Princeton University, Princeton, New Jersey 08540;
2. Geophysical Fluid Dynamics Program Princeton University, Princeton, New Jersey 08540)

Abstract: The feasibility of monthly and seasonal forecasts is considered. The gross features of departures of meteorological variables from climatology (anomalies) are the targets of forecasts, and the anomalies can be divided into two modes, i.e., free modes and forced modes. The free modes are the anomalies that are predicted under the specification of climatological external forcings for the surface temperature, that are free from the anomalous forcings, whereas the forced modes are the anomalies that correspond to the anomalous components of external forcings. The GCM (general circulation model) is, in some cases, capable of predicting the free mode at least one month ahead (particularly the most extraordinary blocking event in January, 1977), and is, in other cases, marginal. However, the capability could be increased further by improving the GCM. In addition, recent studies have revealed that there are growing evidences for the feasibility of prediction of forced modes over the United States through the teleconnection process from the sea surface temperature anomalies over the equatorial Pacific.

Yet the GCM approach is expensive and may be limited in improving mathematical accuracy, to a satisfactory extent. As a remedy, the possibility of anomaly models are being investigated.

1 Introduction

What are the long-range forecasts (LRF) of atmospheric circulation? Let us define it in this paper as the "Forecasts of time-mean state of atmospheric circulation or gross weather, the time range of which is beyond the limit of deterministic prediction (say about 2 weeks)." The feasibility of the LRF is, of course, a crucial question. The nature of the LRF is very likely probabilistic in the sense that the ensemble mean as well as the standard deviation are only meaningful quantities in the forecasts.

As is the case in prediction problems in general, two approaches can be considered for the LRF, i.e., the empirical and the dynamical predictions. With the empirical methods, there has been a great deal of studies (see review by Nicholls, 1981; Namias, 1968, 1978). The methods are based on statistical theories

* Journal of the Meteorological Society of Japan, 1982, 60(1): 292-308.

** On leave from the Institute of Atmospheric Physics, Beijing, China.

such as the regression, the trend, and the analogue methods. They are relatively easy to implement, and there have been many ongoing researches and operations. According to the survey of the WMO (World Meteorological Organization) in 1979, national meteorological services in at least 32 countries are interested in the LRF, and are currently issuing monthly (25 countries) or seasonal (15 countries) outlooks, using the empirical technique. The only problem is that these empirical forecasts are marginal in skill.

The second approach for the LRF, is the dynamical method, which is based on the numerical simulation with general circulation models (GCM) one way or another. Mathematically speaking, it is a unique advantage for this approach to handle highly non-linear relations, which can not be yielded by the statistical methods. The implementation is more costly, and yet the demand for the investigation along this line is strong.

2 Atmospheric variability

For many years in the past, searches have been made for the existence of natural periodicities in the atmospheric parameters. If there is any, it would be useful for understanding the mechanism of variation, and it could be a help for the LRF.

In the 1940's it was not easy to detect the periodicity longer than 5 days except the obvious ones. At present, one can mention a number of frequency spectral peaks, that is, the 42 month broad-band periodicity in the surface pressure or oceanic temperature in the equator, QBO (quasi-biennial oscillation), the semi-annual oscillation, the 50-day periodicity in the tropics, the 16 day planetary waves in middle latitudes, and equatorial Kelvin waves. Most of these oscillations are not simple normal modes (except the 16 day wave and Kelvin wave), but the consequences of non-linear processes.

The study of mid-latitude planetary waves has its own long history. The approach is closely connected with the dynamic instability theories and the general circulation studies. As a result, the basic characteristics have been gradually made clear, such as the three-dimensional structure of mid-latitude westerlies, the lateral and vertical propagation of ultra-long waves, and the generation and maintenance of standing and transient planetary waves (for example, van Loon et al., 1973; Hirota and Sato, 1969; Hayashi and Golder, 1977; Lau, 1978).

However, for the purpose of the LRF, it is important to investigate the overall behavior of the hemispheric circulation rather than to decompose the circulation fields spectrally and to study the detail of dynamics for each component.

Investigation on the hemispheric teleconnection pattern of the surface pressure and of geopotential height is another effort of searching for existence of regular standing waves (Lorenz, 1951; Kutzbach, 1970). In fact, forecasters in the United States have long utilized certain patterns of strong teleconnection for the LRF following the classic work of Walker and Bliss (1932). These studies have established the empirical facts on the large-scale atmospheric oscillation. Van Loon and Rogers (1978) confirmed the seasaw oscillation in the Atlantic. Wallace and Gutzler (1981) identified the finite numbers of action centers on the hemisphere, and suggested that there are 5 categories of teleconnection patterns. Yet the multiple equilibria were not proved to exist in the monthly mean state. If they existed, (Charney and DeVore, 1979) the LRF could become substantially simpler. Two kinds of teleconnection patterns are fairly well defined in reality, and the blocking action patterns have been classified accordingly. Can one resynthesize the real atmospheric variability with some of the known periodic processes? The answer appears to be no.

In the global analysis, the component of the Southern Oscillation explains 18% of the normalized variances of monthly mean surface pressure (Kidson, 1975). Brier (1968) found the QBO to explain only about 2% of the monthly variance in the sea-level zonal wind index in the Northern Hemisphere (Trenberth, 1978). That is to say, the dynamic model that will be used for forecast is desirable to test the capability of simulating these periodicities. But even if it is achieved, it is not practically sufficient only with these regular fluctuations alone to attain a reasonable depiction of the observed monthly mean pattern. Thus, the overall behavior of the real atmosphere remains complex.

Dividing the observed surface pressure data in the monthly mean and its deviation, Madden (1976) estimated the signal-to-noise ratio, where the noise is the natural variability of monthly means and the signal is the interannual variability in his definition. The results revealed that the ratio is not large in the middle latitude (40°-60°N); however, it is large outside these latitudes. An interpretation of this result is that the variability in the middle latitudes is possibly unpredictable. A question is whether this interpretation is correct. Even if the interpretation is not correct, this fact may hint that there is considerable challenge lying ahead for the LRF.

Wishful thinking is that an advanced knowledge of GCM's may handle the prediction of the atmospheric internal variability (free modes), including simulation of blockings, some of quasiperiodic processes, and some of the teleconnection patterns. But even if the predicted internal variability turns out to be below the noise level, another hope is that the response to anomalous external forcings can be detected, particularly utilizing equator-midlatitude teleconnection, for example (Bjerknes, 1966). It is important to note that empirical evidences of feasibility is being accumulated (Namias, 1976; Davis, 1978; Barnett, 1981). It is a recent view that the predictability may depend on season as well as location.

An extensive study has been carried out on the phenomena of El Nino, Southern Oscillation, and teleconnection processes, and a systematic behavior is being unraveled (for example, Wyrtki, 1975; Rasmusson and Carpenter, 1981; Horel and Wallace, 1981). These findings have raised a new surge of excitement and enthusiasm on the LRF in the research community. It is at least certain that the dynamic approach lends itself to study causal relationship. We are now on the verge of intense investigation and assessment of feasibility for the LRF.

3 An example of the GCM forecasts of a blocking case

An example of one-month integration with a GCM will be displayed by a case of blocking event for January, 1977. This case includes a long-sustained stationary ridge over the Aleution area, resulting in the most widespread record cold in the eastern part of North America. A number of GCM's was run for one month, starting with the initial condition on 00 GMT, January 1 or 2.

Figure 1 illustrates the results of a most successful simulation (Miyakoda et al., 1982). It shows the geopotential height pattern at 500 mb level, averaged over 30 days for Day 0-30. The GCM employed was the N48L9-E4 finite difference model, where N is the horizontal resolution, representing the grid number between a pole to the equator, L9 is the 9 vertical levels, and the E4 is a physics, which is advanced compared with the A2-physics. Among a number of GCM's applied to this case, the performance of the N48L9-E4 was the second best, and that of the R30L9-A2 was one of the poorest. The R30L9-A2 denotes the spectral truncation in the rhomboidal version at the zonal wavenumber 30, L9 is again the 9 vertical levels, and the A2 physics.

Figure 2 is the Hovmoller diagram, which is the time-longitude chart of the geopotential height in the zonal belt between 40° and 50°N latitude, delineating the time sequence of circulation for one-month. The figure shows the comparison of three Hovmoller diagrams, i.e., the observation, the good simulation with the N48L9-E4 model, and the poor simulation with the R30L9-A2 model. In the successful case, the stationary ridges are well maintained throughout January as the reality did, whereas in the other case, the erroneously developed trough over the North Pacific pushed the ridge downstream, resulting in a zonal flow pattern around the Northern Hemisphere, and the historical coldness was missed by a large margin. It is our view that a model qualified for the use of the LRF should be capable of simulating blockings, because they are linked to almost all the extreme anomalies in climate on record.

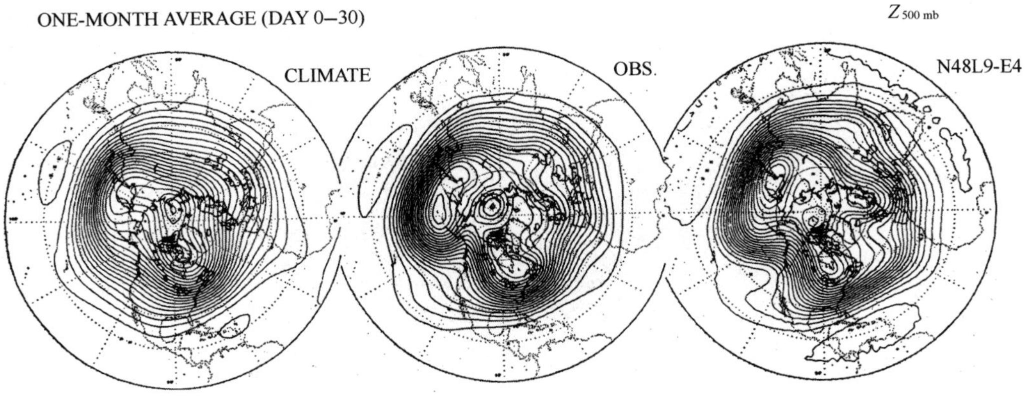

Fig.1　Comparison of the monthly forecast and observation. The maps are the monthly mean 500 mb geopotential height from Day 0 to Day 30 in January 1977: the prognostics by the N48L9-E model (right), the observation (middle), and the climatology for January (left). The contour interval is 30 meters.

4　Time-mean predictability

Figure 3 is th e skill score with the N48L9-E model for January, 1977. The score is represented by the correlation coefficients of the anomaly of the geopotential height at 500 mb. There are a number of scores for the same simulation, only differing in time averaging length. As was demonstrated by several investigators such as Smagorinsky (1969) and Gilchrist (1977), the predictability is raised as the averaging length is increased. Leith (1973), Jones (1975) and Madden (1976) discussed that the standard deviation of a time series in the first order Markov process is reduced by increasing the averaging length. It is interesting to note that not only the standard deviation or the rms error decreases, but also the score is also improved in terms of correlation coefficients, and probably the ratio of rms error and persistence. This is a favorable and important aspect for the LRF.

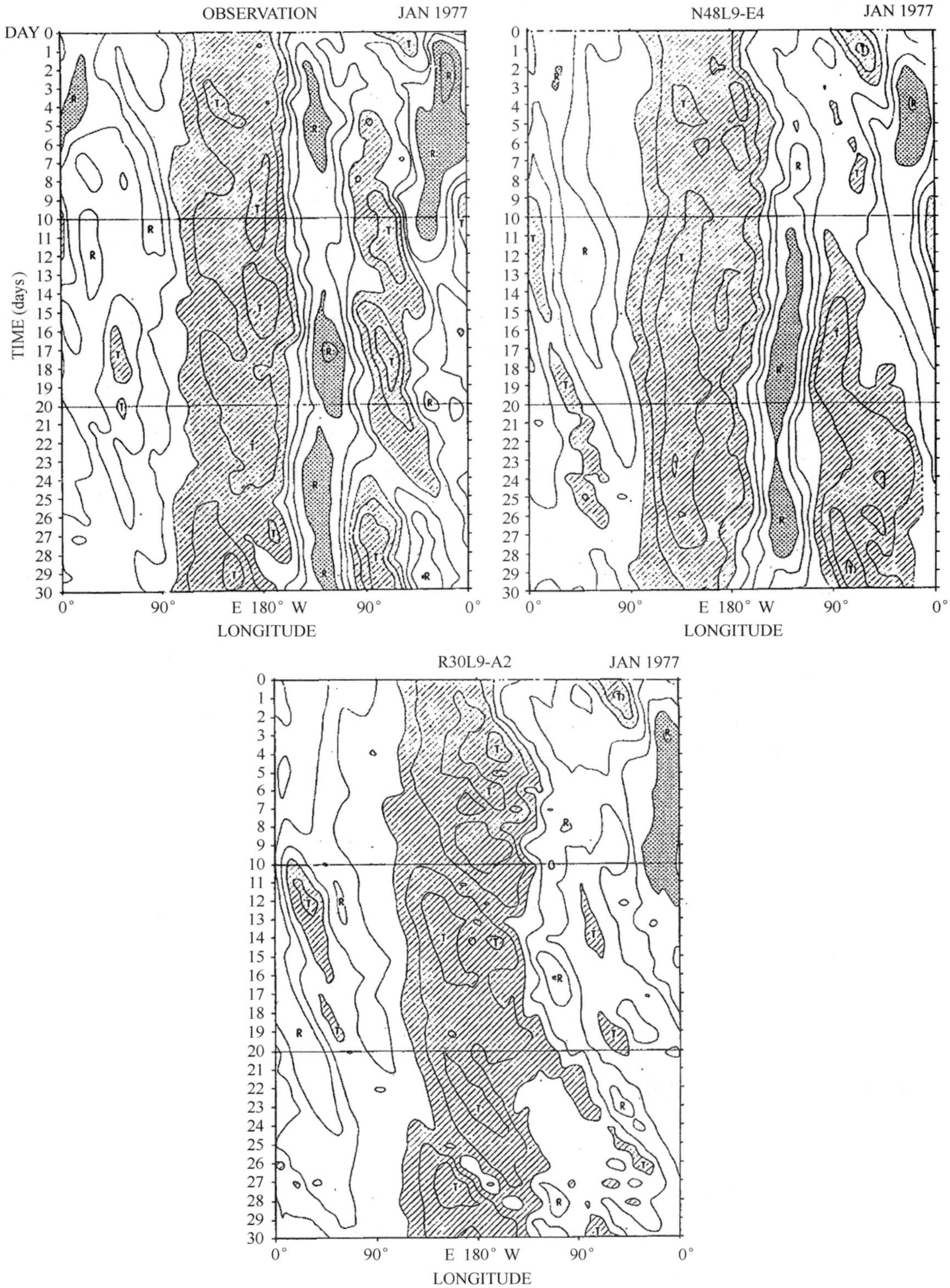

Fig.2 Comparison of the best and the poorest forecasts in Hovmöller (time-longitude) diagram of the 500 mb geopotential height between 40° and 50° latitude from Day 0 to Day 30 in January 1977: the prediction by the N48L9-E model (middle), the prediction by the R30L9-A model (right) and the observations (left). The contour interval is 60 meters. The troughs and the ridges are marked by *T* and *R*, respectively, and the extreme value regions are shaded or stippled.

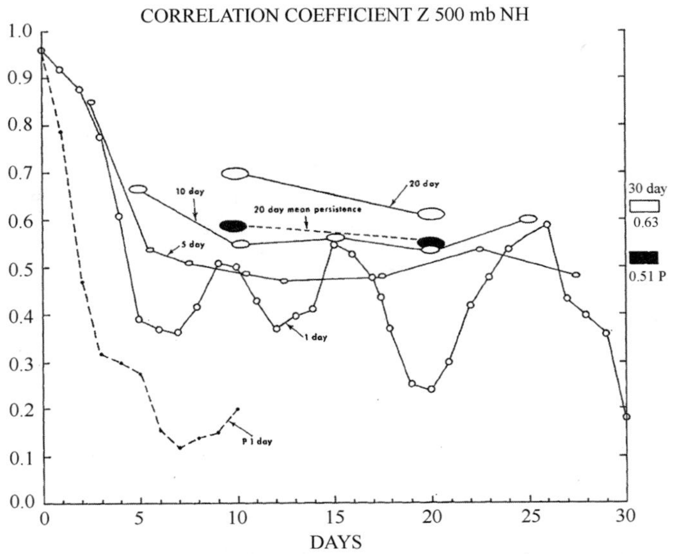

Fig.3　The time-mean predictability is shown by the correlation coefficients for the temporarily averaged anomalies of geopotential height at 500 mb between the prediction by the N48L9−E model and the observation in the case of January 1977. The coefficients for 1, 5, 10 and 20 day mean height fields are indicated by white circles or ovals , which are connected by solid lines. The 20 day mean persistence and the 1 day mean persistences are indicated by black circles or ovals, which are connected by dashed lines. The coefficient for the 30 day mean persistence are shown at the right outside of the diagram.

Let us next proceed to show an example of various realizations of forecasts by changing only the initial conditions. Figure 4 is the correlation and rms error curves for the ensemble mean of Z500 mb anomaly which were all obtained by the R30L9−E4 model. The initial conditions are: January 1, 00 GMT, 1977 with the GFDL analysis and the NMC (National Meteorological Center, Washington, DC), and January 2, 00 GMT, with the GFDL analysis. All these initial conditions are supposed to be good estimates of the truth. The figure indicates that scores do not deviate appreciably from each other until about Day 15, and then they start to diverge after about Day 20, implying that for onemonth integration a single realization is marginally adequate as a representing a probabilistic value. It is interesting and important to note, however, that the divergence of curves is not too large in the 30 day range. In other words, the GCM could calculate successfully some signals even below the so-called noise level of monthly mean variability.

The same proc edure can be applied to the 20 day mean scores. The resulting curves are located a little better than those of the 10 day mean. The degree of divergence of three realizations is slightly less than the 10 day mean score.

Shukla (1981) recently discussed the predictability of time-mean circulation, using the control and perturbation run approach for three different year Januaries. He concluded that the effect of initial condition is retained up to one month or beyond, and that predictability for the planetary scale waves is more than a month. Thus, it is reasonable to regard monthly forecasts as the initial value problem.

In order to reach more general conclusions about monthly predictability, more samples are required. At GFDL, six cases experiment of the GCM one-month prediction is underway; three cases have been calculated.

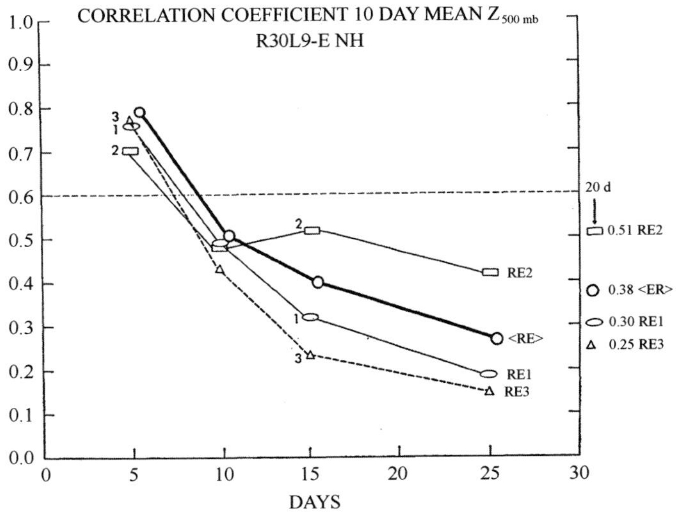

Fig.4 The skill of probabilistic forecasts is shown by the correlation coefficients of 10 day mean 500 mb geopotential height between the prediction by the R30L9 – E model and the observation. The curves marked by RE1, RE2 and RE3 are for three realizations corresponding to three different initial conditions (thin solid and dashed lines), and a curve marked by <RE> is for the ensemble mean of the three realization (thick solid lines). The correlation coefficients for the last 20 day mean fields are shown at the right outside of the diagram.

At this moment, it is too early to discuss these results.

Spar et al. (1976; 1978; 1979) who pioneered the one-month GCM forecast experiment concluded that "the model simulation exhibits no skill in reproducing monthly mean sea level pressure field, but the model does show some small but consistent skill relative to climatology in its simulation of the fields of 500 mb height and 850 mb temperature."

Gilchrist (1977) described the long-range forecast experiment over a period of 30–40 days, and mentioned as follows. "The experiments appear to suggest that the (5–layer) GCM used may, at least on occasions, exhibit some skill in simulating long wave behavior; and that an assessment of whether the use of actual rather than climatological SST improves the forecast is inconclusive."

Almost all investigators , who have studied the monthly forecasts using reasonable GCM's, have mentioned that the predictability limit of the time-means (30 day mean, but even 20 day and 10 day mean) appears to be a month or longer, though they have been cautious. It is also true, that the quality of real GCM forecasts have been currently marginal. Can the quality be improved in the near future? The answer is yes, and the key is further advancement of the GCM and its accuracy in long-term integrations.

In the paper of Miyakoda et al. (1982), ensemble mean curves for various models are shown, indicating that three realizations for each model tend to be clustered together and that curves of different models are distributed systematically in the decreasing order of the skills. So far as one-month prediction is concerned, the quality of the GCM is extremely important. A poor model will not provide a good probabilistic forecast, no matter how many realizations are produced.

It is projected that the GCM prediction could be improved further by increasing the spatial resolution, refining the initial condition, incorporating better SGS (subgrid-scale processes), and perhaps properly including the anomalous external forcings (see Cubasch, 1981).

5 Probabilistic forecasts

Recognizing the inherent error in the initial condition, the forecast process is formulated to determine probabilistic distribution of phase points in the same context as classical statistical mechanics. Namely, the forecasts, starting with an initial cloud, produces the evolution of the projected cloud. Although standard deviation is not large initially, it will soon grow to a large value. The problem is how to obtain the ensemble of phase points. Epstein (1969) proposed the stochastic-dynamic method, in which uncertainty (or error) in the initial time is assumed, the equation of the first moment of errors for prognostic variables and of the second moment of errors are derived; and the system of equation is closed by ignoring the third moments. This approach was further developed by Epstein and Fleming (1971). Pitcher (1977) applied the model to the real case, though the model was simple. It is the authors' opinion, however, that this method has been limited. First of all, the stochastic-dynamic formulation can be possible, only if the dynamic equation is expressed by algebraic equations. The algebraic equations can be obtained in practice only for spectral "interaction" model and not for the spectral "transform" model. Secondly, the system of equations consist of quadratically nonlinear terms, and the number of derived equations is horrendous, so that it is prohibitively costly to solve the equations. Thus, although this theory is supposed to provide the most reasonable and legitimate approach to the probabilistic forecast, we do not think that it is worth all-out efforts of persuance.

Leith (1974) suggested the Monte Carlo method as the substitute of the stochastic-dynamic method. This is the so-to-speak brute-force method, in which a finite number of equally likely forecasts are generated, based on a single analysis, by perturbing the initial condition with a random number, and then statistics are applied to the multiple realizations in the forecasts.

Spar et al. (1978), who made three forecasts, all starting with the same initial data but randomly perturbed in different ways, concluded that random error in the initial state do not appear to represent the major source of forecast error, but major error in the monthly mean prognostic maps are either unknown systematic large-scale errors in the initial analysis or defects in the model itself. Hollingsworth and Savijarvi (1980) also mentioned in their ten day forecast experiments that the individual forecasts within the ensemble of perturbed forecasts deviated quite slowly from each other; all forecasts had the same failure in over-developing a trough over certain areas, for example. A problem here is the random perturbation; the uncertainty in reality is neither random nor that small. Figure 4 is also an example of an ensemble mean forecast with brute force method, but not Monte Carlo. The uncertainty derived from different analysis in the initial data is much larger than the usual specification of random perturbation.

A question is raised: how many samples are needed to establish a statistically valid result? Leith (1974) suggested that 8 samples are acceptable. Of course, even a factor of 8 requires a large amount of computing power compared with a single realization forecast. Is there any other way to make the calculation more economical? If not, the speed-up of the model's prediction is very vital in pursuing this approach. The efficiency of the GCM calculation ought to be radically improved.

6 External forcings

The definition of "external" may need an explanation. What is meant here is: the anomalous effects that are slowly varying in time scale of equal to or more than one-month. For example, the predictive anomalies of sea surface temperature (SST) are determined in the framework of the atmosphere-ocean model, but they are initially specified, and thence subject to slow evolution. Thus the SST anomalies are classified as the external anomaly forcing. On the other hand, the cloud coverage are, irrespective of the initial condition, quickly adjusted to the model (about 5 days—Gordon—personal communication). Thus the cloudiness is classified as the internal forcing in the same way as the precipitation. With this in mind, the external (anomalous) forcings related to the LRF are: (a) ocean, (b) soil moisture, and (c) snow and ice cover. It appears that the studies on these effects may hold great promise for the GCM integration beyond one-month, because the targets of the LRF are the determination of anomalous components, and these media hold good memories for these components. A similar opinion was expressed 15 years ago by Sawyer (1964). It is noted that for the time-scale of climate variation, there are other effects, that is the anomalous forcings in the vegetation, ozone, carbon dioxide, permafrost, continental ice-sheet, solar variability, volcanic activity, the aerosol, mountain glaciers, and the so-called anthropogenic effects. These factors will not be considered here, because of the likelihood of small effects on the seasonal timerange. In this respect, the problem of seasonal weather variation belongs to a disciplinary area different from that of climate variation.

6.1 Oceanic forcings

There were at least two ardent advocators of the ocean-atmosphere interaction in the past decades, e.g. Namias (1959; 1963) and Bjerknes (1959; 1966). Thanks to the perception and enthusiasm of these pioneers, the study of the SST impact has been appreciably advanced compared with other external forcing effects.

Perhaps the issue may be divided into three categories, (a) tropics *in situ*, (b) equator-midlatitude teleconnection, and (c) midlatitude *in situ* and downstream effect.

(a) Tropics *in situ*

Since the tropical atmosphere is conditionally unstable, any small trigger can generate a sizeable scale of cumulus convection. The trigger can be the SST anomaly, and that is what's happening along the ITCZ (Intertropical Convergence Zone). Shukla (1975), based upon a GCM study, indicated that the Indian rainfall decreased in the monsoon season when SST over the western Arabian Sea becomes colder.

(b) Equator-midlatitude teleconnection

Bjerknes (1966; 1969), based on his remarkable insight into observational data, reached a hypothesis that there is a teleconnection process from the equatorial ocean to the mid-latitude atmospheric circulation. Certainly enough, Rowntree (1972), Julian and Chervin (1978), and more recently, Keshavamurti (1981), confirmed Bjerknes' postulate, using the GCM. It is concluded that the SST warm anomaly over the equatorial eastern Pacific (El Nino region) induces anomalous convective rain, releasing latent heat, which in turn increases momentum transport in the Hadley circulation, and leads to strengthening of the subtropical jet, accompanied by shifting and deepening of the Aleutian low.

It was in the last several years, however, that the real impact of this process on the LRF over the middleand high-latitudes has been realized. The recent strategy of short-term climate variability study is to focus on the global teleconnection from the Southern Oscillation. The study of Southern Oscillation phenomenon has its own long history of research from Walker and Bliss (1932) through Troup (1965), and Bjerknes (1969) and Krueger and Gray (1965) to Wyrtki (1973), Krueger and Winston (1975) and Rasmusson and Carpenter (1981) for the Northern Hemisphere and from Pittock (1973) through Streten (1975) and Trenberth (1975) to Nicholls (1977) for the Southern Hemisphere.

From the forecasting standpoint, Quinn and Burt (1972) suggested the use of the Southern Oscillation Index (SOI) (monthly mean pressure differences between Easter Island and Darwin) for the prediction on occurrence or non-occurrence of the heavy rainfall over the central and western Equatorial Pacific in more than 1−2 months lead time. Harnack (1979) and Henricksen (1979) may be the first who used the SOI as one of the predictors for winter temperature in the eastern United States. Barnett (1981) found, based on linear prediction technique, that SST can be used to predict air temperatures over North America one and more seasons in advance, particularly in the winter time, and that the SST and sea level pressure over the equatorial and tropical Pacific Ocean are superior as predictors to their mid-latitude counterparts. A recent study has emphasized further this point, describing that the most crucial SST source region to the United States appears to be the 140°−180°W longitude equatorial zone. Warm episodes in equatorial Pacific SST tend to be accompanied by below-normal heights in the North Pacific and the southeastern United States and above-normal heights over western Canada (Horel and Wallace, 1981). Chen (1981) showed that lagged cross-correlations between SOI and the 700 mb, heights are significantly high in winter with SOI leading the height by one to two seasons. An interesting feature is the tendency for more pronounced North American negative anomalies of height and North Atlantic positive anomalies associated with the high SOI values. The impact of the SST over west equatorial Pacific to the climatic variability over China was reported by Fu (1979a, b). In connection with the equatorial ocean, statistical forecasts of a wind anomaly associated with the Pacific Hadley circulation using SST data, have significant and appreciable skill (Barnett and Hasselmann, 1979). A similar study was conducted by Fu and Li (1979), who found that the behavior of the North Pacific High is predictable, using the equatorial SST information of prior to one or two seasons. An observational study of Pan and Oort (1981)① has revealed a high correlation of variations between the SST anomalies over the middle Equatorial Pacific and the 200 mb westerlies in situ as well as over the North Pacific. A similar but weak effect was observed for the sector of the Atlantic Ocean by Rowntree (1976).

It is thus not surprising that the chain of processes may be extended to the Asian Monsoon phenomena, and eventually to the rainy season events over China and Japan. Rowntree (1979) noted that global mean temperature variations on a scale of a few years are driven to a substantial extent by variations in equatorial Pacific ocean temperature.

---

① Pan, Y-H, and A. H. Oort, 1981: Observed sea surface temperature in the Equatorial Pacific Ocean and global climate anomaly. A seminar presented at GFDL.

(c) Midlatitude *in situ* and remote effect

Namias (1963; 1975) pointed out the close relationship between the ocean and atmosphere in the North Pacific in the form of both the simultaneous and "downstream" response. Ratcliff and Murray (1970) discussed the lag association between the North Atlantic SST and the European surface pressure. Adem (1964; 1965) claimed that the SST anomaly provides useful information on the monthly forecasts. The groups of long-range forecasting in China discussed the lag association between the Kuroshio SST and the rainfall in the eastern part of China.

However, the GCM approaches have so far not been successful in detecting the clear-cut relation between the atmospheric circulation and the SST anomalies (Miyakoda—unpublished; Spar, 1973; Kutzbach et al., 1977; Houghton et al., 1974; Chervin et al., 1976; Shukla and Bangaru, 1980).

Only the experiment of exaggerated anomalies of SST produced significant realistic local response in the models' atmospheric temperature. The ocean primarily serves as the moderator to the atmosphere in terms of thermal effect, so it is not surprising that the ocean effect outside the tropics is not strong.

Using a simple model, Webster (1981) studied the mechanisms of the local and remote atmospheric response to SST anomalies at various latitudes, and found that the magnitude of the total diabatic heating diminishes significantly as the SST anomalies are placed progressively poleward. At high latitude, the response is small due to the creation of an indirect zonal circulation in the vicinity of the anomaly which is related to the strength of the local basic flow.

Thus the reason for the failure in detecting a strong signal in the model study is partly due to the weak signal-to-noise ratio, and partly due to the poor simulation capability of the applied GCM's. Chervin et al. (1976) and Gilman (1978) stressed the necessity of adequate statistical design for the sensitivity experiments, particularly on the midlatitude SST anomaly effect. Egger (1977) computed the atmospheric response to a pool of the warm water in Newfoundland in the linearized equation, and argued the benefit of the linearity assumption for the relatively small anomalies.

Difficulty in detection of the causative factor was also once experimented in an observational study such as Davis (1976), who used 28 years of records of monthly anomalies of the North Pacific SST and sea level pressure, and concluded that the only evidence of predictable forcing influence was the SST field's response to anomalous sea level pressure, and not vice-versa.

It is only in recent studies that Namias (1978) and Davis (1978) have started to indicate that the North Pacific SST anomalies in summer are significantly correlated with next season's surface pressure particularly in the area of Aleutian Islands. Harnack and Landsberg (1978) also suggested that prior SST in the North Pacific may be effective predictors of subsequent atmospheric changes over the North American continent.

Marchuk (1975a, b) proposed to estimate the source of SST for the teleconnection process, and stressed that energy active regions of the world ocean are responsible for development of major weather anomalies.

6.2 Land forcings

Compared with ocean forcings, the knowledge on this subject is meager. The heat storage in the land surface is much less than that in the ocean, and therefore, it may be natural to consider that the land forcings are less important than the ocean forcings. However, in certain ways, the land can give a significant impact on the atmospheric circulation. These are realized through the snow-cover over the ground and the moisture content in the soil.

(a) Snow- and ice-cover

The primary effect of land snow/ ice-cover is an albedo which affects not only the heat balance but also the surface melting snow. On the other hand, the effect of sea-ice is the insulation; the sea-ice modifies the thermal conduction between air and sea.

To give an example of the former effect, we refer to Chen's study on the comparison of the summer circulation between the years of heavy and little snow cover in winter over the Tibetan plateau. The study revealed that typical summer circulation comes about one month later in the years of heavy snow cover than in the years of little snow cover. Yeh and Fu (1980) further investigated the summer circulation for the years of heavy and little snow cover in March over Eurasia, and found that the 200 mb summer westerlies at 40°N over Eurasia are much stronger in the years of heavy snow cover than in the years of little snowcover.

(b) Soil moisture

The role of the soil moisture is to control the intensity of evapotranspiration, consequently affecting the surface temperature, and to influence the rainfall. The other role is to affect the surface albedo, and to modify the soil heat conduction.

It is now increasingly aware that the effect of soil moisture is important not only on the climate numerical experiment (Manabe and Holloway, 1975, Charney et al., 1977; Mintz, 1981), but also on the short-range rainfall forecasts (Walker and Rowntree, 1977; Miyakoda et al., 1981; Yeh and Chen, 1980).

Yeh and Fu (1980) made an observational study on the influence of soil moisture on the atmospheric circulation by comparing the case of extreme flood summer (very wet soil) with the case of an extreme drought summer (very dry soil) in eastern China. They found that the monthly mean soil temperature difference between the flood and drought July reaches as high as 8℃ in large areas, and that the monthly mean air temperature between the two cases reached as high as 4℃. This widespread temperature difference led to drastic differences in the atmospheric circulation. They also noted that the influence of the soil moisture could last more than two months, whereas the snow cover effect could last more than three months. But overall, it has not been easy in other parts of the world except Africa to demonstrate the quantitative impact of the soil moisture in the empirical LRF either due to relatively weak signal or due to our ignorance.

Almost all current GCM's have been following the ground hydrology parameterization of Manabe (1969). The vegetation canopy is supposed to play a role in the evapotranspiration process, but most of the models have not yet incorporated this effect except NCAR (National Center for Atmospheric Research) model. It is certain, however, that the accuracy of soil moisture in the initial data is crucial for the LRF. This is because the temporal variation of soil moisture (internally predicted) is slow, and the initially specified condition of this parameter dominates at least the subsequent two months.

An attempt is underway to estimate the large scale distribution of near-surface soil moisture by remote sensing from satellites. Past experience indicated that this task is not easy at all (NASA, 1980). The problem of soil moisture has been a common concern for many decades by hydrologists, agronomists, geographers, meteorologists, climatologists and paleo-climatologists, and they concur that the problem has not been solved. It is known that the hydrological parameters in soil vary over several orders of magnitude within relatively short distance, say 10 meters. Therefore, the distribution of soil moisture at 200 km grid networks has been regarded as a mathematical outcome or at best a product of parameterized hydrological process in GCM. The GCM's treat the soil moisture over the scale of entire continents, whereas the hydrological models handle the soil moisture only over the scale of catchment. In other words, the fundamental difference between the two schools is the disparity of the grid size. In future, in order to increase the accuracy of hydrological parameters in the LRF, this gap has to be narrowed. Indeed, the estimate of initial soil moisture pattern (or the evapotranspiration pattern) is a brand new venture.

7 Anomaly model approach

In the present and the next sections, we will focus on anomaly models. This approach is entirely different from the GCM approach we have dealt with in the preceeding sections. The LRF with GCM will be practically limited, because the system becomes inevitably huge, cumbersome and expensive for increasing accuracy. Obviously any GCM still involves bias, which is by no means small. In order to remove this bias, it may take decades of hard and tedious work. In addition, the precision of the computer might pose the intrinsic limit for further improvement of accuracy in long-range integration of the GCM.

7.1 The principle of the anomaly model

In the situation of impending limitation, an alternative that can be thought of is the anomaly model. This model is one of the families of twoseparate equation system. An idea is to divide variables into two components, *i.e.*, the basic component and the deviation. As is known, there are various ways of division depending upon the purposes, that is, zonal averages and the perturbation (eddies); ensemble means and the deviation (turbulence); time means and the departure (fluctuation), and the climatological normals and the anomalies.

Concerning the ensemble average system, we have already discussed in Section 5, in connection with the stochastic-dynamic approaches (Epstein, 1969; Friedman-Keller equation in turbulence theory). The zonal mean model was proposed by Thompson (1957), who first mentioned the inherited limit of predictability with the conventional numerical weather prediction and pointed out the merit of the zonal average approach. The system of zonal mean equations was used extensively by Blinova (1957) and Saltzman (1978), the reason being presumably that this system is well connected with the most of the theoretical studies. This category, however, branches into two schools. One school was to use the equation for zonal mean variables and the linearized perturbation equation (Saltzman, 1968; Derome and Wiin-Nielsen, 1971; Verneker and Chang, 1978). Another school was to set up the equations of first and second moments, terminating the series by closure assumption (Gambo and Arakawa, 1958; Kurihara, 1970). Blinova mentioned, based on the practical experience, that the chief difficulty of the first school lies in determination of the zonal mean flow. In this respect, the second school is capable of predicting zonal mean quantity theoretically. Indeed these approaches have contributed profoundly to understanding of dynamics, but practical application has not been attempted.

In this respect, the anomaly model approach appears somewhat different. The variables are divided into the climatological normal and the anomaly, and yet only the equation for the anomaly components is used. Namely, omitting the equation for the normal, the observed normals are utilized. The uniqueness of this model lies in this point. This anomaly model was proposed by the Long-Range Numerical Weather Forecasting Group in Peking (1977; 1979), and it has been extensively applied to real data. Opsteegh and Van den Dool (1980) also used this type of model for the study of LRF, though the equations are linear.

We will, however, discuss mostly the anomaly model of Peking. The equation for the anomaly includes the Reynolds terms, but in the present model, these terms are ignored, and besides the current model uses the geostrophic approximations. The system consists of two major equations. One is the potential vorticity equation for the atmosphere, and the other is the thermal equation for the underlying ocean and land. The two equations are connected at the earth's surface, through the interfacial conditions, which are the heat balance relation, and the mass, heat and stress continuities.

Then the most crucial assumption is introduced, *i.e.*, the stationarity of the atmospheric vorticity equation, but the thermal equations for the ocean and land are kept time-dependent. The reason for this arrangement is to simplify the system by filtering out the solution of transient Rossby waves in the atmosphere. This treatment could also be derived by the concept of two time systems, *i.e.*, slow and fast. The monthly mean standing components are considered to be subject to the slow process of adaptation (adjustment) to the ocean and land thermal forcings.

Monin (1972) described a similar idea in terms of the A-AL system, that is, the atmosphere and the active layer of the underlying surface. He stated that "the simplification of the A-AL system, filtering out the short-term synoptic processes from their solutions, can be derived by neglecting the partial derivatives with respect to time in all equations except those of the heat content of the active layer of the ocean; and numerical experiments with such simplified equations could help to clarify the feasibility of this approach to long-range weather forecasts."

7.2 Results of the anomaly model

Figure 5 is an example of one-month forecast for the spectacular month of January 1977, based on the coupled ocean-atmosphere anomaly model. The initial conditions for the anomaly model are: SST and the land surface temperature and 500mb anomaly geopotential height for onemonth mean from December 1 to 31, 1976. The climatological atmospheric wind vectors for January were applied. The atmospheric model has the horizontal resolution of 5° and the onelayer in the vertical.

The figure shows the predicted anomaly of geopotential height at 500mb for one-month average from 1 to 31 January, 1977. The solution corresponds to "forced mode" in the sense that it is the adapted solution to the SST anomaly and land surface temperature anomaly.

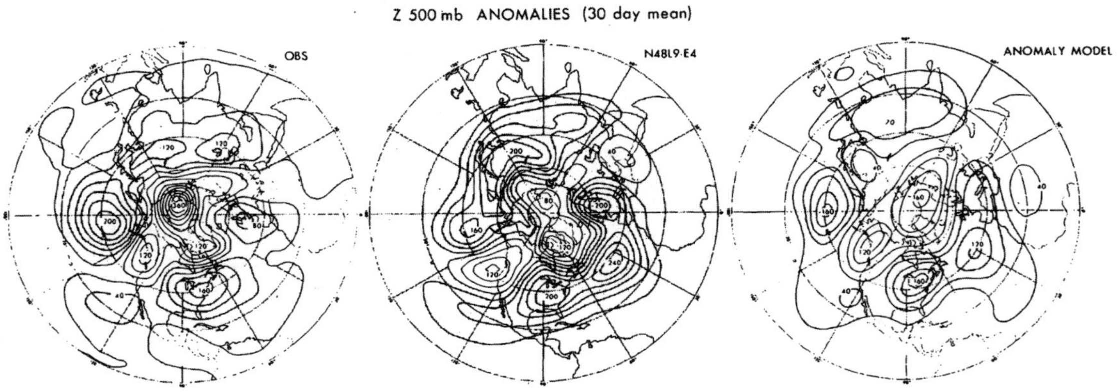

Fig.5 Comparison of the monthly mean predicted anomalies of 500 mb geopotential height: the observation (left), the prognostic of free mode by the N48L9-E GCM (middle), and the prognostic of forced mode by the one-layer anomaly model (right). The contour interval is 30 meters, and the negative regions are shaded

Table 1 Correlation coefficients of the anomalies of Z' (geopotential height) and T' (temperature) between the prediction and observation over the Northern Hemisphere

Cases				Z'		T'	
				Correl.	Persist.	Correl.	Persist.
Aug.	→	Sept.	1965	0.36	0.24		
Dec. 1976	→	Jan.	1977	0.72	0.56		
Jan.	→	Feb.	1978	0.22	0.30	0.29	0.58
Jan.	→	Feb.	1977	0.05	0.34	0.18	0.35
Jan.	→	Feb.	1976	-0.16	0.07	0.07	-0.01
April	→	May	1978	0.05	0.19	0.25	0.32
April	→	May	1977	0.21	0.07	0.33	0.28
April	→	May	1976	0.30	0.25	0.20	0.26
July	→	Aug.	1978	0.16	0.23	0.28	0.40
July	→	Aug.	1977	0.17	0.42	0.26	0.46
July	→	Aug.	1976	0.29	0.47	0.15	0.27
Oct.	→	Nov.	1978	-0.06	0.32	-0.01	0.30
Oct.	→	Nov.	1977	-0.29	-0.07	-0.01	-0.05
Oct.	→	Nov.	1976	0.16	0.21	0.32	0.50

The same figure includes the counterpart of the GCM, *i.e.*, N48L9-E4, which we described earlier. Therefore the GCM solution corresponds to "free mode", *i.e.*, the consequence of the anomaly in the initial condition and dynamical interaction. An interesting and an important point is that the GCM and the anomaly forcing model, in essence, are quite different, and yet the solutions in both models are similar to each other in this case. The total prediction should be the free plus the forced modes.

Some readers may be interested to know the computer time consumed for these calculations. Using the ASC (Advanced Scientific Computer) of Texas Instruments at Princeton, the GCM took 60 hours to obtain the result in this figure, whereas the anomaly model took 15 seconds, though the anomaly model has the hemispheric domain. Note that the same anomaly model needed 45 min. with the computer in Peking. The three-layer anomaly model is also working, and in fact this gives better results than a onelayer anomaly model. The computer time with the three-layer model is 1 min by the Princeton computer for the same calculation above, whereas 2 hours were needed by the Peking computer.

At the Institute of Atmospheric Physics at Peking, 14 cases of one-month forecasts have been performed, using atmospheric one-layer models. The skill score of these results is shown in the Table. In addition, two-month forecast, based on three-layer model, has been made. These preliminary studies appear to indicate that the forced mode of the anomaly model has some skill.

7.3 Anomaly model in the future

The performance of the current anomaly model is far from satisfactory. In order to improve the anomaly model, what can be done? A number of processes should be refined to the level of quality close to the GCM. Above all, the primitive equation system is required instead of the geostrophic approximation, because the important signal from equatorial SST has to be included. Opsteegh and Van den Dool (1980) and Hoskins and Karoly (1981) used the primitive equation for their anomaly model, and discussed the teleconnection effect of Bjerknes. Chen and Xin in Peking (personal communication) have already succeeded in constructing the primitive equation anomaly model.

A question may arise as to whether the stationarity for the atmospheric equations is absolutely essential. If the stationarity is abandoned, the problem becomes the initial value problem, and consequently not only forced mode but also the free mode come in to the solutions.

An investigation is underway at Princeton on this problem with the anomaly model. It is not yet clear that the anomaly model approach is worthy of an all-out endeavor. A serious ambush might be waiting for us. In the non-stationary system, the prediction is no longer deterministic but probabilistic. More accuracy may be required to obtain the proper solution of transient components. In that case, should the sophistication of the anomaly model be reconsidered (in the area of space resolution, SGS processes and numerical algorithm)?

The original equations for the anomaly components include Reynolds terms and anomaly external forcing terms. These terms are the source of potential problems. Do these terms generate the ambush?

Even for the seasonal forecasts, should the free mode be considered, and therefore, the nonstationarity be retained? Our view is that the free modes sometimes are predictable even beyond one month, and for this reason, it is better to keep both, but the system becomes expensive.

In summary, as the overall strategy for the LRF, the GCM study has to be continued and developed further. The basic research must go along this line. The GCM approach is indispensable for understanding of the atmospheric and oceanic phenomena and processes. To what extent the accuracy can be increased only by the pure GCM is an important and interesting question.

8 Conclusions

Summarizing this article, the conclusions are listed below.

(1) It is essential to investigate and establish the range of time-mean predictability. So far as monthly forecasts are concerned, free mode components are dominant over forced mode components. Time-mean free mode anomaly components appear to be predictable for onemonth, but more samples are needed.

(2) Search for the oceanic external forcing has been successful. On the other hand, the knowledge on the land external forcing is extremely meager. Predictive capability of these forcings is crucial for the seasonal forecasts.

(3) For the proper development of the LRF study, the GCM approach is important and indispensable. Yet there may be a limitation on the pure GCM forecasts for seasonal range. An anomaly model may provide a remedy as the accurate and economical forecasting method. It may be wise to pursue both approaches for the study of the LRF. The computational burden for the anomaly model may be substantially small. However, an investigation on the potential of the anomaly model has not been completed.

(4) For the free mode components in particular, the probabilistic forecasts are required instead of the deterministic forecasts. It is desirable to have reasonal and economical methods for dynamical and statistical treatment in terms of attaining the adequate mean and standard deviation of the forecasts and specifying the initial data.

The light at the end of the tunnel remains dim. But a number of hopeful clues have been collected.

Acknowledgements

The authors gratefully acknowledge the cooperation of Dr. T. Gordon, Messrs. R. Ca.verly and W. Stern, and Ms. Xin. Thanks also go to Drs. N. G. Lau, E. Rasmussen, and J. Winston for providing information and reading the manuscript.

References

Adem, J. 1964. On the physical basis for the numerical prediction of monthly and seasonal temperatures in the troposphere-ocean-continent system. *Mon. Wea. Rev.*, 92:91-103.

——. 1965. Experimental aiming at monthly and seasonal numerical weather prediction. *Mon. Wea. Rev.*, 93:495-503.

Barnett, T. P. and K. Hasselmann. 1979. Techniques of linear prediction with application to oceanic and atmospheric fields in the tropical Pacific. *Rev.Geophys. Space Phys.*, 17:949-968.

Barnett, T. P. 1981. Statistical prediction of North American air temperatures from Pacific predictors. *Mon. Wea. Rev.*, 109:1021-1041.

Bjerknes, J. 1959. The recent warming of the North Atlantic. *The Atmosphere and Sea in Motion*, B. Bolin, Ed., The Rockefeller Inst. Press:65-73.

——. 1966. A possible response of the atmospheric Hadley circulation to equatorial anomalies of ocean temperature. *Tellus*, 18:820-829.

——. 1969. Atmospheric teleconnections from the equatorial Pacific. *Man.Wea. Rev.*, 97:163-172.

Blinova, E. N. 1957. Long-range forecasting. Part II of hydrodynamical methods of short- and long-range weather forecasting in the USSR. *Tellus*, 9: 453–463.

Brier, G. W. 1968. Long range prediction of the zonal westerlies and some. problems in data analysis. *Rev. of Geophys.*, 6: 525–551.

Charney, J. G., W. J. Quirk, S. H. Chow and J. Kornfield. 1977. A comparative study of the effects of albedo change on drought in semi-arid regions. *J. Atmos. Sci.*, 4: 1366–1385.

Charney, J. G. and J. G. DeVore. 1979. Multiple flow equilibria in the atmosphere and blocking. *J. Atmos. Sci.*, 36: 1205–1216.

Chen, W. Y. 1981. Fluctuations in Northern Hemisphere 700 mb height field associated with the Southern Oscillation. (To be submitted to *Mon. Wea. Rev.*)

Chervin, R. M., W. M. Washington, and S. H. Schneider. 1976. Testing the statistical significance of the response of the NCAR General Circuation Model to North Pacific Ocean surface temperature anomalies. *J. Atmos. Sci.*, 33: 413–423.

Cubasch, U. 1981. Preliminary assessment of long range integrations performed with the ECMWF global model. Technical Memorandum No. 28, European Centre for Medium Range Weather Forecasts: 21.

Davis, R. E. 1976. Predictability of sea-surface temperature and sea-level pressure anomalies over the North Pacific Ocean. *J. Phys. Oceanogr.*, 6: 249–266.

——. 1978. Predictability of sea level pressure anomalies over the North Pacific. *J. Phys. Oceanogr.*, 8: 233–246.

Derome, J. and A. Wiin-Nielsen. 1971. Response of a middle-latitude model atmosphere to forcing by topography and stationary heat sources. *Mon. Wea. Rev.*, 99: 564–576.

Egger, J. 1977. On the linear theory of the atmospheric response to sea surface temperature anomalies. *J. Atmos. Sci.*, 34: 603–614.

Epstein, E. 1969. Stochastic dynamic prediction. *Tellus*, 21: 739–759.

Epstein, E. S. and R. J. Fleming. 1971. Depicting stochastic dynamic forecasts. *J. Atmos. Sci.*, 28: 500–511.

Fleming, R. J. 1971. On stochastic dynamic prediction: II. Predictability and utility. *Mon. Wea. Rev.*, 99: 927–938.

Fu, C-B. 1979a. The atmospheric vertical circulation during anomalous periods of sea surface temperature over equatorial Pacific region. *Scientia Atmospherica Simica*, 3: 50–57.

——. 1979b. On response of atmospheric temperature field to the variation of SST in equatorial Pacific region. *Geography*, 12: 158–168.

Fu, C-B, and K.–L. Li. 1979. On long–term variation of SST in Pacific Ocean and its effects on subtropical high. *Geography*, 12: 146–157.

Gambo, K. and A. Arakawa. 1958. Prognostic equations for predicting the mean zonal current. Tech. Rept. No. 1, Numerical Weather Prediction Group, Tokyo, Japan.

Gilchrist, A. 1977. An experiment on extended range prediction using a general circulation model and including the influence of sea surface temperature anomalies. *Beitr.z. Phys. d. Atmos.*, 50: 25–40.

Gilman, D. L. 1978. General circulation models, sea-surface temperatures, and short-term climate prediction. Proceedings of the English Technical Exchange Conference. November 28–Decem ber 1, 1978, Air Force Academy, Colorado Springs, Colo: 1–6.

Group of Long-Range Numerical Weather Forecasting. 1977. On the physical basis of a model of long-range numerical weather forecasting. *Scientia Sinica*, 20: 377–390.

——. 1979. A filtering method for long-range numerical weather forecasting. *Scientia Sinica*, 22: 661–674.

Harnack, R. P. and H. E. Landsberg. 1978. Winter season temperature outlooks by objective methods. *J. Geophys. Res.*, 83: 3601–3616.

Harnack, R. P. 1979. A further assessment of winter temperature predictions using objective methods. *Mon. Wea. Rev.*, 107:

250–267.

Hayashi, Y. and D. G. Golder. 1977. Spectral analysis of mid-latitude disturbances appearing in a GFDL general circulation model. *J. Atmos. Sci.*, 34:237–262.

Henricksen, G. C. 1979. An attempt to project winter temperature departure for the eastern United States. *National Weather Digest*.4:27–30.

Hirota, I. and Y. Sato. 1969. Periodic variation of the winter statospheric circulation and intermittent vertical propagation of planetary waves. *J. Meteor. Soc. Japan*,47:390–402.

Hollingsworth, A. and H. Savijarvi. 1980. An experiment in Monte Carlo forecasting. The collection of papers presented at the WMO Symposium on Probabilistic and Statistical Methods in Weather Forecasting, Nice, France, 8–12 September, 1980. WMO:45–47.

Horel, J.D. and J. M. Wallace. 1981. Planetary scale atmospheric phenomena associated with the interannual variability of sea-surface temperature 813–829.

in the Equatorial Pacific. *Mon. Wea. Rev.*:109,

Hoskins, B.J. and D. J. Karoly. 1981. The steady linear response of a spherical atmosphere to thermal and orographic forcing. *J. Atmos. Sci.*:1179–1196.

Houghton, D. D., J. E. Kutzback, M. McClintock, and D. Suchman. 1974. Response of a general circulation model to a sea surface temperature perturbation. *J. Atmos. Sci.*,31:857–868.

Jones, R. H. 1975. Estimating the variance of time averages. *J. Appl. Meteor.*, 14:159–163.

Julian, P. R. and R. M. Chervin. 1978. A study of the southern oscillation and Walker Circulation phenomenon. *Mon. Wea. Rev.*, 106:1433–1451.

Keshavamurty, R. N. 1981. Response of the atmosphere to sea surface temperature anomalies over the equatorial Pacific and the teleconnections of the Southern Oscillation. (Submitted to *J.Atmos. Sci.*)

Kidson, J. W. 1975. Tropical eigenvector analysis and the Sotuhern Oscillation. *Mon. Wea. Rev.*, 103:187–216.

Krueger, A. F. and T. I. Gray, Jr. 1969. Long-term variations in equatorial circulation and rainfall. *Mon. Wea. Rev.*, 97: 700–711.

Krueger, A. F. and J. S. Winston. 1975. Large-scale circalation anomalies over the tropics during 1971–72. *Mon. Wea. Rev.*, 103:465–473.

Kurihara, Y. 1970. A statistical-dynamical model of the general circulation of the atmosphere. *J. Atmos. Sci.*, 27:847–870.

Kutzbach, J. E. 1970. Large–scale features of monthly mean Northern Hemisphere anomaly ma ps of sea-level pressure. *Mon. Wea. Rev.*, 98:708–716.

Kutzbach, J. E., R. M. Chervin and D. D. Houghton. 1977. Response of the NCAR general circulation model to prescribed changes in ocean surface temperature, Part I: Mid-latitude changes. *J.Atmos. Sci.*, 34:1200–1213.

Lau, N. G. 1978. On the three-dimensional structure of the observed transient eddy statistics of the Northern Hemisphere wintertime circulation. *J. Atmos. Sci.*, 35:1900–1923.

Leith, C. E. 1973. The standard error of timeaverage estimates of climatic means. *J. Appl. Meteor.*, 12:1066–1069.

——. 1974. Theoretical skill of Monte Carlo forecasts. *Mon. Wea. Rev.*, 102:409–418.

Lorenz, E. M. 1951. Seasonal and irregular variations of the Northern Hemisphere sea-level pressure profile. *J. Meteor.*, 8:52–59.

Madden, R. A. 1976. Estimations of the natural variability of time-averaged sea-level pressure. *Mon. Wea. Rev.*, 104: 942–952.

Manabe, S. 1969. Climate and the ocean circulation: I. The atmospheric circulation and the hydrology of the earth's surface. *Mon. Wea. Rev.*,97:739–774.

Manabe, S. and J. L. Holloway, Jr. 1975. The seasonal variation of the hydrologic cycle as simulated by a global model of the atmosphere. *J.Geophys. Res.*, 80:617–649.

Marchuk, G. I. 1975a. Formulation of the theory of perturbations for complicated models. Part I: The estimation of the climate change. *Geofisica International*, 15:103–156.

——. 1975b. Formulation of the theory of perturbations for complicated models. Part II. Weather Prediction. *Geofisica International*, 15:169–183.

Mintz, Y. 1981. The influence of soil moisture on rainfall and circulation: A review of simulation experiments.

Miyakoda, K. and R. F. Strickler. 1981. Cumulative results of extended forecast experiment. III. Precipitation. Mon. Wea. Rev., 109:830–842.

Miyakoda, K., T. Gordon, R. Caverly, W. Stern, J. Sirutis and W. Bourke. 1982. Simulation of a blocking event in January 1977. (To be submitted to JAS).

Monin, A. S. 1972. Weather forecasting as a problem in physics. MIT Press, Cambridge, Mass. and London, England:199.

Namias, J. 1959. Recent seasonal interactions between North Pacific waters and the overlying atmospheric circulation. *J. Geophys. Res.*, 64:631–646.

——. 1963. Large-scale air-sea interactions over the North Pacific from summer 1962 through the subsequent winter. *J. Geophys. Res.*, 68:6171–6186.

——. 1968. Long–range weather forecasting history, current status and outlook. *Bull. Amer. Meteor. Soc.*, 49:438–470.

——. 1975. *Short period climatic variations*, *Collected Works of J.Namias*, 1934 through 1974 (2 vols) Univ. California, San Diego:905.

——. 1976. Negative ocean-air feedback system over the North Pacific in the transition from warm to cold seasons. *Mon. Wea Rev.*, 104:1107–1121.

——. 1978. Multiple causes of the North American abnormal winter 1976–77. *Mon. Wea. Rev.*, 106:279–297.

NASA. 1980. Plan of research for integrated soil moisture studies. NASA Goddard Space Flight Center, Greenbelt, Md.

Nicholls, N. 1977. Tropical-extratropical interactions in the Australian region. *Mon. Wea. Rev.*, 105:826–832.

——. 1981. Long-range weather forecastingvalue, status and prospects. *Rev. Geophys. Space Phys.*

Opsteegh, J. D. and H. M. Van den Dool. 1980. Seasonal differences in the stationary response of a linearized primitive equation model: Prospects for long range forecasting. *J. Atmos. Sci.*, 37:2169–2185.

Pitcher, E. J. 1977. Applications of stochastic dynamic prediction to real data. *J. Atmos. Sci.*, 34:3–24.

Pittock, A. B. 1973. Global meridional interactions in stratosphere and troposphere. *Quart. J. Roy. Meteor. Soc.*, 99:424–437.

Quinn, W. N., and W. Burt. 1972. Use of the Southern Oscillation in weather prediction. *J. Appl. Meteor.*, 11:616–628.

Rasmusson, E. M. and T. H. Carpenter. 1981. Variation in tropical sea surface temperature and surface wind fields associated with Southern Oscillation/El Niño (submitted to *Mon. Wea. Rev.*)

Ratcliffe, R. A. S. and R. Murray. 1970. New lag associations between North Atlantic sea temperature and European pressure applied to longrange weather forecasting. *Quart. J. Roy. Meteor. Soc.*, 102:607–625.

Rowntree, P. R. 1972. The influence of tropical east Pacific Ocean temperatures on the atmos-pher. *Quart. J. Roy. Meteor. Soc.*, 98:290–321.

——. 1976. Response of the atmosphere to a tropical Atlantic Ocean temperature anomaly, *Quart. J. Roy. Meteor. Soc.*, 102: 607–625.

——. 1979. The effects of changes in ocean temperature on the atmosphere. *Dynamics of Atmospheres and Oceans*, 3:373–390.

Saltzman, B. 1968. Surface boundary effects on the general circulation and macroclimate: A review of the theory of the quasistationary perturbations in the atmosphere. *Meteor. Monograph*, 8:4–19.

——. 1978. A survey of statistical-dynamical models of the terrestrial climate. *Advances in Geophysics*, 20, Academic Press, New York.

Sawyer, J. S. 1965. Notes on the, possible physical cause of long-term weather anomalies. *WMO Tech. Note*, 66:227-248.

Shukla, J. 1975. Effect of Arabian sea-surface temperature anomaly on Indian summer monsoon: a numerical experiment with the GFDL model. *J. Atmos. Sci.*, 32:503-511.

Shukla, J. and B. Bangaru. 1980. Effect of a Pacific sea-surface temperature anomaly on the circulation over North America. A numerical experiment with the GLAS model. *GARP Publication Series*, 22:501-518.

Shukla, J. 1981. Predictability of time averages. Part I. Dynamical predictability of monthly means. *J. Atmos. Sci.*

Smagorinsky, J. 1969. Problems and promises of deterministic extended range forecasting. Bull. *Amer. Meteor. Sci.*, 50:286-311.

Spar, J. 1973. Some effects of surface anomalies in a global general circulation model. *Mon. Wea. Rev.*, 101:91-100.

Spar, J., R. Atlas, and E. Kuo. 1976. Monthly mean forecast experiments with the GISS model. *Mon. Wea. Rev.*, 104:1215-1241.

Spar, J., J. J. Notario, and W. J. Quirk. 1978. An initial state perturbation experiment with the GISS model. *Mon. Wea. Rev.*, 106:89-100.

Spar, J. and R. Lutz. 1979. Simulations of the monthly mean atmosphere for February 1976 with the GISS model. *Mon. Wea. Rev.*, 107:181-192.

Streten, N. S. 1975. Satellite derived inferences to some characteristics of the South Pacific atmospheric circulation associated with the Nina event of 1972-73. *Mon. Wea. Rev.*, 103:989-995.

Thompson, P. D. 1957. Uncertainty of initial state as a factor in the predictability of large scale atmospheric flow patterns. Tellus, 9:275-295.

Trenberth, K. E. 1978. Fluctuations in short term climate. The notes of a summer colloquium at NCAR. "The general circulation: theory, modeling, and observations." 339-357.

——. 1975. A quasi-bienniel standing wave in the Southern Hemisphere and interactions with sea-surface temperature. Quart. *J. Roy. Meteor. Soc.*, 101:55-74.

Troup, A. J. 1965. The Southern Oscillation. *Quart. J. Roy. Meteor. Soc.*, 91:490-506.

Van Loon, H., R. L. Jenne and K. Labitzke. 1973. Zonal harmonic standing waves. *J. Geophys. Res.*, 78:4413-4471.

Van Loon, H. and J. C. Rogers. 1978. The see saw in winter temperature between Greenland and Northern Europe. Part I: General description *Mon. Wea. Rev.*, 106:296-310.

Vernekar, A. D, and H. D. Chang. 1978. A statistical-dynamical model for stationary perturbations in the atmosphere. *J. Atmos. Sci.*, 35:433-444.

Walker, G. T. and E. W. Bliss. 1932. World Weather V. *Mem. Roy. Meteor. Soc.*, 4:53-84.

Walker, J., and P. R. Rowntree. 1977. The effect of soil moisture on circulation and rainfall in atropical model. *Quart. J. Roy. Meteor. Soc.*, 103:29-46.

Wallace, J. M. and D. S. Gutzler. 1981. Teleconnections in the geopotential height field during the Northern Hemisphere winter. *Mon.. Wea. Rev.*, 109:784-812.

Webster, P. J. 1981. Mechanisms determining the atmospheric response to sea surface temperatur eanomalies. *J. Atmos. Sci.*, 38:554-571.

Wyrtki, K. 1975. El Niño, the dynamic response of the equatorial Pacific Ocean to atmo spheric forcing. *J. Phys. Oceanogr.*, 5:572-584.

Yeh, T. C. and X. S. Chen. 1980. A numerical experiment of the influences of the heavy winter snow cover over Qinghai-Tibetan plateau and severe drought over eastern China on the general circulation of the atmosphere (tobe publi shed).

Yeh, T. C, and C. B. Fu. 1980. The time-lag feedback processes of large-scale rainfall and drought on the atmospheric circulation and climate (to be published).

长期数值预报的三层滤波模式*

邢如楠，郭裕福，巢纪平

（中国科学院大气物理研究所）

摘要：在滤波理论的基础上，本文将大气考虑成斜压的，并用三层模式即 850 mbar、500 mbar 和 200 mbar 高空场，做出了地面温度和高度环流的月距平预报。初步试验结果表明，三层滤波模式比一层模式的预报结果有相当的改进。

1　引言

用动力学方法做长期天气预报，由于物理过程和预报对象均不同，理应与短期数值预报方法有所不同。但目前一般是用大气环流数值模式像短期天气预报一样一天天积分，虽然最近也有成功的例子（Miyakoda 等），但花费的计算机时间十分大。另一方面可用完全不同的方法，如 Adem（1964）[1]提出的一个月平均地面温度预报模式，预报对象是对气候值的月平均距平。但是他设计的模式不可能预报月平均环流距平，因此预报时效只能一个月。近年来，我国已在理论和实验上均提出了自己的方法，这在参考文献[2,3]中已有详细的介绍。在这个理论中认为天气尺度的 Rossby 波可以作为干扰长期天气过程的"噪音"而滤去，并证明大气运动是向着地表温度场适应的，所以这个模式称为滤波模式。以这一理论为基础，在一个简单的一层模式中进一步做出了地面温度距平和 500 mbar高度月距平环流预报，效果尚好。在这一简单模式中，由于垂直方向只有一层，在做地面温度预报用到近地面层的资料时，不得不引进一些近似，这样影响了某些例子的预报效果。为了进一步改进预报效果，我们用滤波方法建立了一个三层长期数值预报模式。并在模式中更详细地考虑了地面反照率对辐射的影响，及下垫面的热力性质等物理因子，这样使效果有了相当的改进。

2　地表温度距平预报方程

参考文献[3]给出了多年月平均距平温度变化方程为：

$$\frac{\partial^2 T_s'}{\partial z^2}-\frac{1}{K_s}\frac{\partial T_s'}{\partial t}=\frac{\delta}{K_s}\left[\frac{\partial(\overline{\Psi}_s,T_s')}{\partial(x,y)}+\frac{\partial(\overline{\Psi}_s',\overline{T}_s+T_s')}{\partial(x,y)}\right]\equiv H_1 \tag{1}$$

式中，对海洋 $\delta=1$，对陆地 $\delta=0$，$\overline{\Psi}_s$ 为平均洋流流函数，距平部分 $\overline{\Psi}_s$ 可根据 Ekman 风吹流理论，用近地面风场算出。K_s 为海洋中垂直湍流交换系数，在陆地为热传导系数。

在 z 坐标中，取垂直分层如图 1。

我们将热力学方程用到近地面层，

$$\left(\frac{\partial^2 T}{\partial z^2}\right)_5-\frac{1}{K\tau_B}T'_5=\frac{1}{K}\left[\frac{\partial(\overline{\Psi}+\Psi',T')}{\partial(x,y)}+\frac{\partial(\Psi',\overline{T})}{\partial(x,y)}\right]_5\equiv H_2 \tag{2}$$

---

* 中国科学院 B 辑，1982(2)：186-192.

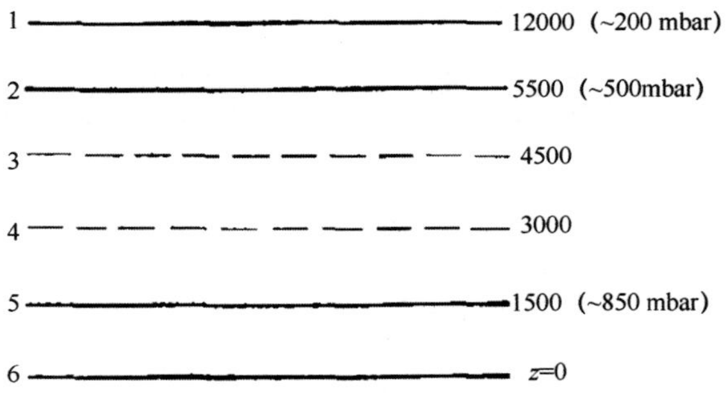

图 1 模式的垂直结构

式(2)的有限差分形式为：

$$\frac{1}{\Delta z^2}(T'_3 - T'_5) - \frac{1}{\Delta z}\left(\frac{\partial T'}{\partial z}\right)_6 - \frac{1}{K\tau_R}T'_5 = H_2 \tag{3}$$

假定在近地面层温度距平和地表温度距平相同，则式(3)中 T'_5 用 T'_s 代入，我们得到，

$$\left(\frac{\partial T}{\partial z}\right)_{z=0} = -\Delta z H_2 + \frac{1}{\Delta z}T'_3 - \left(\frac{1}{\Delta z} + \frac{\Delta z}{K\tau_R}\right)T'_s \tag{4}$$

在地-气交界面上有热量平衡方程[3]，

$$z = 0,\ \rho_s C_{ps} K_s \frac{\partial T'_s}{\partial z} - \rho C_p K_T \frac{\partial T'}{\partial z} + \delta\left(L_\rho K_T r \frac{\partial \ln \bar{e}_s}{\partial T}\frac{\partial \bar{q}_s}{\partial T}\right)T'_s = -\frac{S_0}{W_0}l_b \cdot \zeta'_{0g} \tag{5}$$

式中右端表示由云量调节的异常辐射平衡，云量的参数化形式是假定它与通过摩擦层顶的涡度 ζ'_{0g} 大小来决定的垂直运动成正比。

将式(4)代入方程(5)，得到

$$z = 0,\ \frac{\partial T'_s}{\partial z} + \left(\frac{\delta}{D_Q} + \frac{1}{D_s}\right)T_s = -\frac{S_0 l_b \cdot b}{\rho_s C_{ps} K_s W_0}\zeta'_5 - \frac{\rho C_p K_T}{\rho_s C_{ps} K_s}\left(-\frac{T'_3}{\Delta z} + \Delta z H_2\right) \tag{6}$$

其中，$\zeta'_{0g} \sim b\zeta'_5$，$\zeta'_5$，是近地面层上的涡度距平，

$$D_Q = \frac{\rho_s C_{p_s} K_s}{L K_T \gamma \dfrac{\partial \ln \bar{e}_s}{\partial T}\dfrac{\partial q_s}{\partial T}},\quad D_s = \frac{\rho_s C_{p_s} K_s K \tau_R \Delta z}{\rho C_p K_T (K\tau_R + \Delta z^2)} \tag{7}$$

方程(1) 中地表温度距平对时间的变化项写成差分形式，我们有

$$\frac{\partial^2 T''^{t+\Delta t}_s}{\partial z^2} - \frac{1}{K_s \delta t}T''^{t+\Delta t}_s = -\frac{1}{K_s \delta t}T''^{t}_s + H_1 \tag{8}$$

方程(8) 的两个定解条件，一个是方程(6)，另一个条件是

$$z = -\infty,\quad T'_s = 0 \tag{9}$$

方程(8)满足这两个定解条件，在 $z=0$ 处的解为：

$$T''^{t+\Delta t}_s(0) = -D\sqrt{K_s \delta t}\left\{\frac{\rho C_p K_T}{\rho_s C_{p_s} K_s}\left(H_2 \Delta z - \frac{T'_3}{\Delta z}\right) + \frac{S_0 l_b b}{\rho_s C_{p_s} K_s W_0}\zeta'_5 + \int_{-\infty}^{0} e^{\frac{z}{\sqrt{k_s \sigma t}}}\left(-\frac{1}{K_s \delta t}T''^{t}_s + H_1\right)dz'\right\} \tag{10}$$

积分核中的 T'_s 和 H_1 近似地用表层值代替，最后得到地表温度的预报方程为：

$$T''^{t+\Delta t}_s(0) = D\left\{T''^{t}_s(0) - \sqrt{K_s \delta t}\frac{\rho C_p K_T}{\rho_s C_{ps} K_s}\left(\Delta z H_2 - \frac{T'_s}{\Delta z}\right) - K_s \delta t \widetilde{H}_1 - \sqrt{K_s \delta t}\frac{S_0 l_b b}{\rho_s C_{ps} K_s W_0}\zeta'_5\right\} \tag{11}$$

其中

$$D = \frac{1}{1 + \frac{\sqrt{K_s \delta t}}{D_s} + \delta \frac{\sqrt{K_s \delta t}}{D_e}} \tag{12}$$

$\widetilde{H}_1$ 是 H_1 在海洋表层的值。

在给出了 t 时刻的 T'_s 及高空 1 层、2 层、5 层上的高度距平值后，用静力公式可算出中间层上的温度距平 T'_2，再用拉格朗日插值公式，从 T'_2 和 T'_5 插出 T'_3，加上气候平均资料，即可预报出 $t+\delta t$ 时刻的地面温度距平值。

3　三层适应方程

在 p 坐标系中，垂直方向取三层，即 250 mbar，500 mbar 和 750 mbar（以下分别用下标（1，3，5）表示层次）。考虑了辐射、感热、凝结等非绝热过程后的温度方程为：

$$\frac{\mathrm{D}T}{\mathrm{D}t} - \sigma_p \omega = \frac{\partial}{\partial p}(k + k_R)\frac{\partial T}{\partial p} - \frac{1}{\tau_R}(T - T_e) + \widetilde{T}^* \zeta_{0g} + \frac{k'}{\rho C_p} S \tag{13}$$

其中，ζ'_{0g}为摩擦层顶的地转涡度；T_e 为参考温度；$k=\rho^2 g^2 k_T$，k_T 是对感热的湍流交换系数；k_R 是辐射扩散系数；$\sigma_p=(\gamma_d-\gamma)/\rho g$ 是静力稳定度；τ_R 为牛顿辐射冷却特征时间，各符号意义详见参考文献[2]。

温度方程（13）用到海平面和最下层，同时将牛顿辐射冷却项中的参考温度，取成是地表温度 $T_s=\overline{T}_s+T'_s$，并取 $\overline{T}=\overline{T}_s$。另外考虑到在海平面上 $\omega=0$ 及在第 5 层上的垂直运动 ω_5 近似地用边界层的垂直运动，

$$\omega_b = l_b \zeta_{0g} \tag{14}$$

乘上一个系数 a 即得

$$\omega_5 = - a\rho g l_b \zeta_{0g} \tag{15}$$

其中，l_b 是行星边界层的厚度；ζ_{0g}为地面的地转涡度，我们得到：

$$p = p_0,\quad \left[\frac{\mathrm{D}}{\mathrm{D}t} - \frac{\partial}{\partial p}(k + k_R)\frac{\partial}{\partial p}\right]\frac{\partial \phi'}{\partial p} + u'\frac{\partial}{\partial x}\left(\frac{\partial \overline{\phi}}{\partial p}\right) + v'\frac{\partial}{\partial y}\left(\frac{\partial \overline{\phi}}{\partial p}\right) + \frac{1}{\tau_R}\frac{\partial \phi'}{\partial p} = -\frac{R}{\tau_R p_0}T'_s \tag{16}$$

$$p = 750\ \text{mbar},\quad \left\{\left[\frac{\mathrm{D}}{\mathrm{D}t} - \frac{\partial}{\partial p}(k + k_R)\frac{\partial}{\partial p}\right]\frac{\partial \phi'}{\partial p} + u'\frac{\partial}{\partial x}\left(\frac{\partial \overline{\phi}}{\partial p}\right) + v'\frac{\partial}{\partial y}\left(\frac{\partial \overline{\phi}}{\partial p}\right) + \frac{1}{\tau_R}\frac{\partial \phi'}{\partial p}\right\}_5$$

$$= - b\left(\frac{\widetilde{T'}R}{p} - \widetilde{\sigma}_p l_b \rho g a\right)\zeta'_5 \tag{17}$$

式中，p_0 为海平面气压，$\zeta'_{0g}=b\cdot\zeta'_5$，$b$ 为一参数。

根据参考文献[2] 的理论分析，在长期天气过程中，Rossby 波可作为干扰长期波的短期波而滤去。滤波方法可简单地采用准定常近似，即微分算子变成：

$$\frac{D}{Dt} = (\overline{u} + u')\frac{\partial}{\partial x} + (\overline{v} + v')\frac{\partial}{\partial y} \tag{18}$$

我们将滤波后的非绝热涡度方程写到 1 层，3 层，5 层上，同时用到边界条件

$$p = 0,\quad \phi' = \frac{\partial \phi'}{\partial p} = \frac{\partial^2 \phi'}{\partial p^2} = \frac{\partial^3 \phi'}{\partial p^3} = 0 \tag{19}$$

以及式(16),式(17),得到三层适应方程如下,

$$(\bar{u}_1 + u'_1)\frac{\partial \Delta \Psi'_1}{\partial x} + (\bar{v}_1 + v'_1)\frac{\partial \Delta \Psi'_1}{\partial y} + \beta_y \frac{\partial \Psi'_1}{\partial x} - \beta_x \frac{\partial \Psi'_1}{\partial y}$$

$$- K_1 \Delta \Psi'_1 = \frac{fR(k + k_R)}{\Delta p^3 p_0} T'_s - YY \tag{20}$$

$$(\bar{u}_3 + u'_3)\frac{\partial \Delta \Psi'_3}{\partial x} + (\bar{v}_3 + v'_3)\frac{\partial \Delta \Psi'_3}{\partial y} + \beta_y \frac{\partial \Psi'_3}{\partial x} - \beta_x \frac{\partial \Psi'_3}{\partial y} - K_3 \Delta \Psi'_3$$

$$= FT'_s + E(\Delta \Psi')_5 + \frac{f^2}{\Delta p^2 \widetilde{\sigma}_{p1}}\left[(\bar{u}_1 + u'_1)\frac{\partial \Psi'_3}{\partial x} + (\bar{v}_1 + v'_1)\frac{\partial \Psi'_3}{\partial y} + u'_1 \frac{\partial \Psi'_3}{\partial x} + v'_1 \frac{\partial \Psi'_3}{\partial y}\right.$$

$$\left. + \frac{1}{\tau_R}\Psi'_3 - \frac{4(k + k_R)}{\Delta p}(\Psi'_1 - 2\Psi'_3 + \Psi'_5)\right] \tag{21}$$

$$(\bar{u}_5 + u'_5)\frac{\partial \Delta \Psi'_5}{\partial x} + (\bar{v}_5 + v'_5)\frac{\partial \Delta \Psi'_5}{\partial y} + \beta_y \frac{\partial \Psi'_5}{\partial x} - \beta_x \frac{\partial \Psi'_5}{\partial y} - K_5 \Delta \Psi'_5$$

$$- \left[F + \frac{fR(k + k_R)}{\Delta p^3 p_0}\right] T'_s + YY \tag{22}$$

其中,$\Psi' = \varphi'/f$ 为地转流函数,f 取 45°纬度上的值。

$$F = \frac{fR}{\Delta p \widetilde{\sigma}_p \tau_R p_0},$$

$$K_i = -bfR\frac{\partial}{\partial p}\left(\frac{\widetilde{T}^*}{p\widetilde{\sigma}_p}\right), \quad i = 1,3,5,$$

$$E = \left(\frac{R\widetilde{T}^* f^2}{p_5 \Delta p \widetilde{\sigma}_{p_5}} - \frac{f^2}{\Delta p} l_b \rho g a\right) b,$$

$$YY = \frac{f^2}{\Delta p^2 \widetilde{\sigma}_{p_3}}\left[(\bar{u}_3 + u'_3)\frac{\partial(\psi'_5 - \psi'_1)}{\partial x} + (\bar{v}_3 + v'_3)\frac{\partial(\psi'_5 - \psi'_1)}{\partial y}\right.$$

$$\left. + u'_3 \frac{\partial(\psi_5 - \psi_1)}{\partial x} + v'_3 \frac{\partial(\psi_5 - \psi_1)}{\partial y} + \frac{1}{\tau_R}(\psi'_5 - \psi'_1) + \frac{2}{\Delta p^2}(\psi'_5 - \psi'_1)\right]$$

方程(20),方程(21)和方程(22)可用超松弛法联合迭代求解。

4 试验结果

我们用三层滤波模式做了5个个例试验。这里给出1978年2月地面温度及高空环流的月距平预报,并与一层滤波模式的预报结果进行了比较。预报范围为北半球极地赤射投影,网格距是540 km。初始场是1978年1月的月平均地面温度距平,200 mbar,500 mbar和850 mbar的高度距平及气候平均值。数值计算中用到的参数与参考文献[3]相同。在地面温度的计算上,考虑了地表反照率和下垫面热力性质的不同,资料取自参考文献[4,5]。

图2(a)是三层模式预报的地面温度距平。与观测实况(图2(b))比较,我们看到,几个主要的正、负距平区是报出来了,其中心位置与实况接近,除了新地岛西北部的负中心位置以外。但是预报

的强度普遍偏弱。

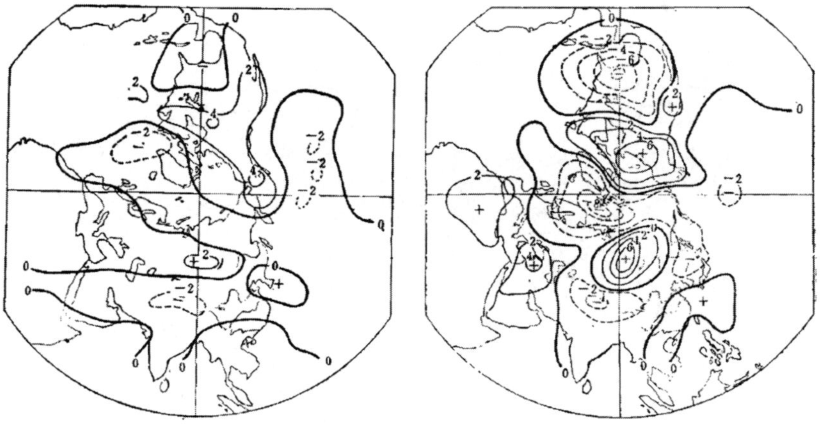

(a)　　(b)

图 2　1978 年 2 月地面温度距平(单位:℃)

(a)三层模式预报;(b)观测实况

用三层和一层模式预报的 500 mbar 高度距平及 1978 年 2 月 500 mbar 观测实况见图 3。从图中可以看到，三层模式预报的高度距平，除了大西洋西部实况是负距平报成了正距平反相了以外，其他地区的预报与实况是接近的。特别是北美的正距平中心位置预报是相当成功的。但一层模式预报的这个正中心与实况相比，位置报得太偏西北。由此可见，三层模式比一层模式的预报结果大有改进。

预报的 200 mbar 高度距平(图 4(a))与实况(图 4(b)) 也较接近。为了节省篇幅，在此未给出 850 mbar 的预报图。

表 1 是 1978 年 2 月个例预报的三层高度距平场与实况之间相同符号的百分数。

表 1　1978 年 2 月预报高度距平场与实况间相同符号的百分数

高度	200 mbar	500 mbar	850 mbar
r	0. 72	0. 71	0. 72

表 2 是 5 个个例预报的地面温度距平场与实况间相同符号的百分数。

表 2　预报的地面温度距平与实况之间相同符号的百分数

日期	1978 年 2 月	1978 年 3 月	1976 年 2 月	1978 年 7 月	1978 年 8 月
r	0. 73	0. 63	0. 63	0. 52	0. 64

在所计算的 5 个个例中，其中有 4 个个例的地面温度预报与一层模式的预报结果进行了比较，对一层模式预报结果有明显改进的有 3 个个例。因此，以上的初步试验结果表明，三层模式可以改进一层模式的预报结果。

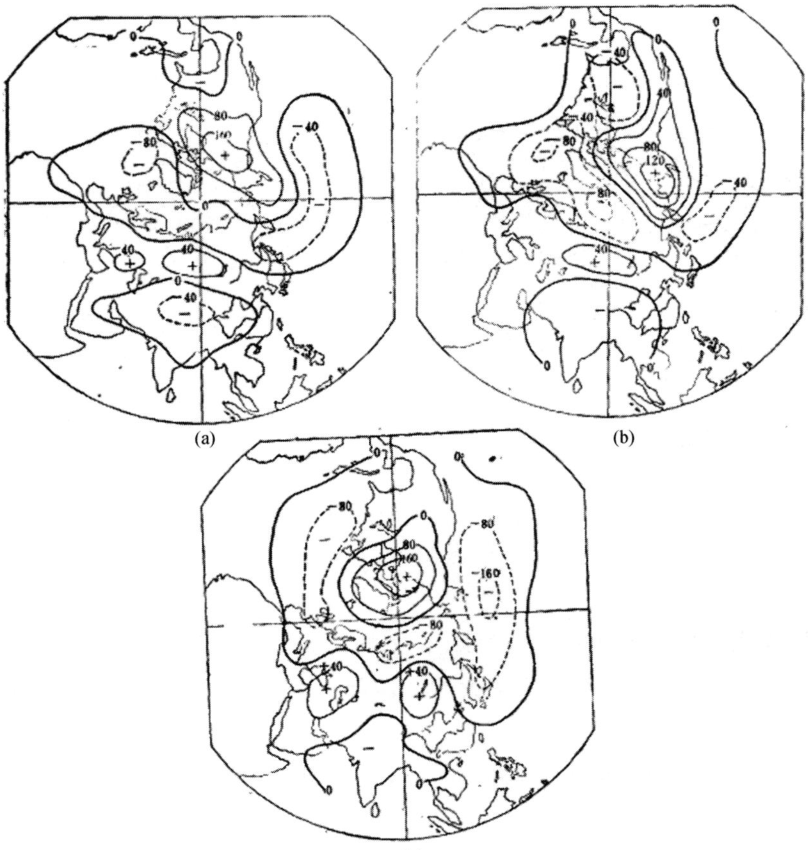

图 3 1978 年 2 月 500 mbar 高度距平场(单位:位势什米)

(a)三层模式预报;(b)一层模式预报;(c)观测实况

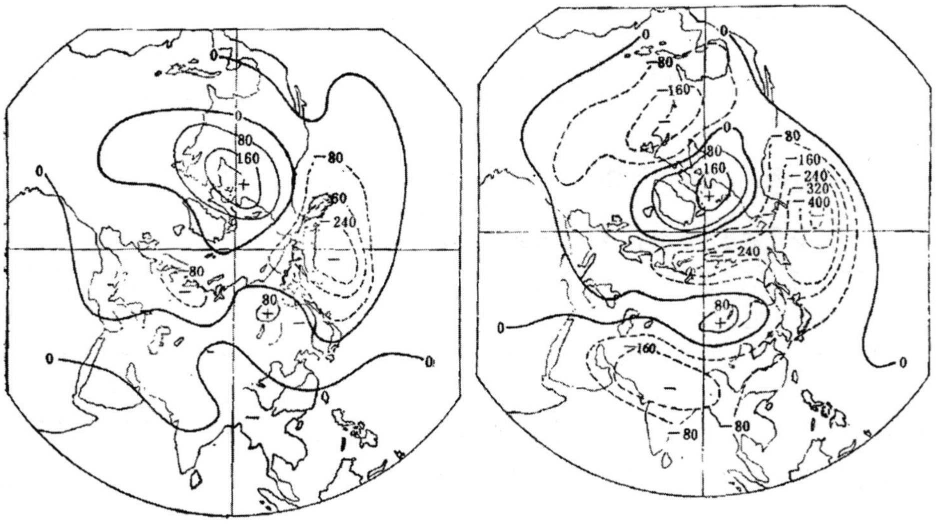

图 4 1978 年 2 月 200 mbar 高度距平场(单位:位势什米)

(a)三层模式预报;(b)观测实况

参考文献

[1]　Adem, J. *Mon. Wea. Rev.*, 1964, 92:91.

[2]　长期数值天气预报研究小组,中国科学. 1977, 2:162-172.

[3]　长期数值天气预报研究小组,中国科学. 1979, 1:75-84.

[4]　Posey, J. & Clapp, P.F. *Global Distribution of Normab Albedo*, Geofisic International, 1964.

[5]　Johnson, J.C. *Physical Metenology*, New York, 1954.

用三层滤波模式做季节预报的试验*

邢如楠　巢纪平

（中国科学院大气物理研究所）

1　引言

在世界上许多国家广泛采用经验的方法制作2周以上的长期预报，而用动力学方法做长期预报还处在研究阶段，这是一个许多气象学家和研究小组都十分感兴趣的课题。一些人用包括复杂物理过程的一般大气环流模式（GCM）做延伸预报，欧洲中心做了广泛的中期预报研究，Spar等[1,2]用GISS九层模式做了一个月的月平均环流预报。最近Miyakoda等（1981）用GFDL的9层有限差分大气环流模式做了1977年1月的月平均环流预报，成功地模拟了冬季的阻塞形势。但是使用复杂的大气环流模式，用短期预报延伸的方法做季节预报仍是困难的，因为为了长时间的数值积分和提高精度，GCM不可避免地变得繁杂和耗费巨大的计算时间，并要求更大更快的计算机配置。

国内自1977年以来，巢纪平等[3,4]提出了滤波的一层模式用于制作长期预报。最近又建立了斜压的三层滤波模式[5]，并做了5个一个月的地面温度和高度距平场的预报。试验结果表明，三层模式对原一层模式的预报结果有相当的改进。这一试验结果鼓舞了我们在此基础上进一步制作二个月以上的预报，以检验三层模式的预报能力。因为三层模式不但报出了地面温度距平，而且可以同时预报出高空的高度月距平场，这为连续做二个月以上的预报提供了所需要的资料。本文给出用三层滤波模式做两个月和三个月的预报得到的初步试验结果。

2　模式

本文中用到的三层滤波模式与参考文献[5]中的相同。这是一个包括了实际海陆分布的大气-海洋耦合的动力学模式。模式中海洋仅包括上层混合层并且把天气尺度的Rossty波作为干扰长期天气过程的“噪音”而滤去了。计算范围为北半球。

模式中包括的主要物理过程有：① 辐射、云对辐射的调节作用和地面反照率对辐射的影响。② 蒸发和凝结过程。③ 大气和海洋中的垂直热量交换，土壤中的热传导和风与洋流对热量的水平输送。计算公式详见参考文献[5]。

计算的初值是月平均地面温度距平，20 mbar、500 mbar和850 mbar上的月平均高度距平以及气候平均值。若取时间步长为一个月，用地面温度距平可以计算出下一个月的地面温度距平场，然后用3个适应方程联合迭代求解，可以得到下个月的三层高度月距平场。若以预报出来的地面温度和三层高度距平为初值，像前面的程序一样，即可以做出第二个月的预报。如此还可以做出第三个月的预报。应当注意在做不同月份的预报时，所需要的气候平均值也需要进行相应的改变。

* 科学通报，1982，12：738-740.

3　试验结果

用三层滤波模式做了 3 个二个月预报个例和 1 个三个月预报试验，预报取得了较好的效果。表 1 给出了预报的和观测的温度距平场间的符号相同的百分数。

表 1　预报的地面温度距平场与观测值之间的符号相同的百分数

日期	1978 年 3 月	1978 年 8 月	1976 年 3 月	1978 年 4 月
r_2	0.73	0.60	0.55	
r_3				0.74

注：r_2 为对二个月预报的，r_3 为对三个月预报的。

下面给出 1978 年 3 月（二个月预报）和 1978 年 4 月（三个月预报）的预报结果。

图 1、图 3 分别是 1978 年 1 月、3 月和 4 月观测的地面温度距平场。我们看到，从 1978 年 1 月到 3 月，北半球地面温度距平分布在北美、格陵兰、美国西海岸和亚洲大陆都有明显的变化，距平符号 3 月与 1 月刚好相反。这样一些变化在二个月的预报图（图 2）上都能看到。预报图与观测实况是相当接近的。

到 1978 年 4 月，乌拉尔山北部地区和欧洲大陆都变成负距平，格陵兰由负距平变为正距平区，1978 年 4 月三个月的预报图与实况（图 4）基本上也是接近的，虽然在某些地区的预报强度与实况有偏差，如新地岛以南的负中心预报偏弱，而美洲大陆的正距平预报又偏强。

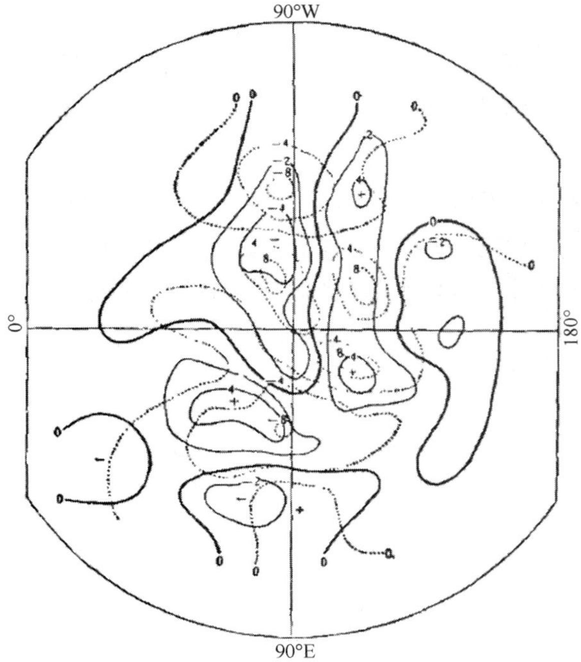

图 1　1978 年 1 月（虚线）和 3 月（实线）观测的地面温度距平场

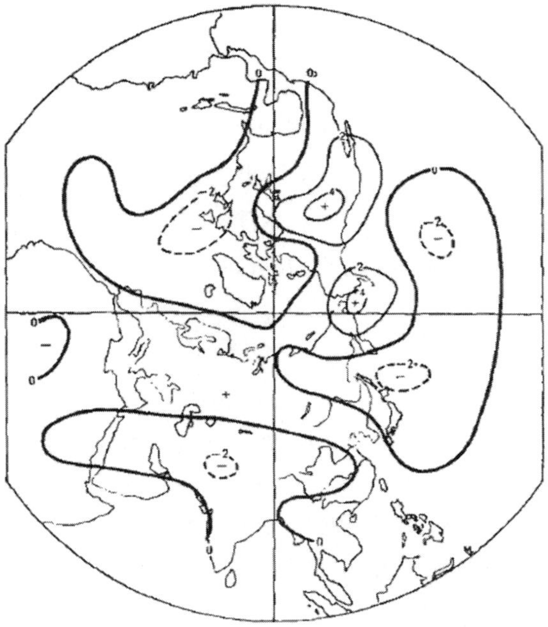

图 2　1978 年 3 月二个月预报的地面温度距平场

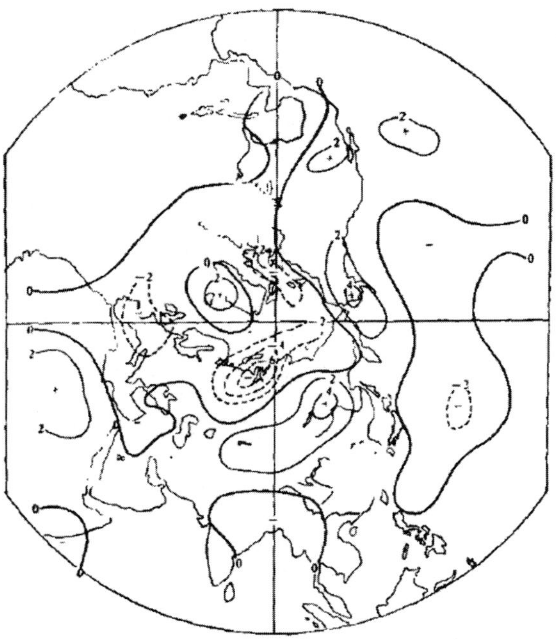

图 3　1978 年 4 月观测的地面温度距平场

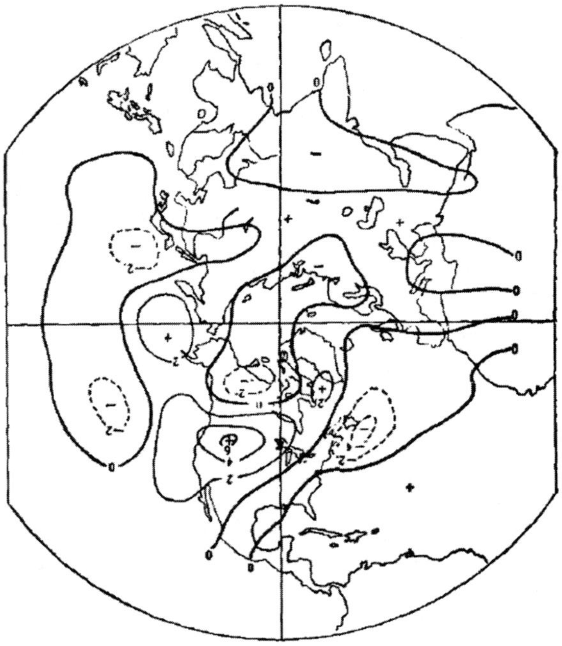

图 4　1978 年 4 月三个月预报的地面温度距平场

限于篇幅，高度场的预报图这里省略了。

此外对 1978 年 8 月 2 个月的预报，成功地报出了这年夏季我国东北地区温度持续偏高的趋势和我国广大地区地面温度为正距平，与实况十分接近。

在试验中还对表 1 给出的个例分别做了一个月的预报。例如，对 1978 年 3 月个例是以 1978 年 2 月观测资料为初始场计算的，并与二个月和三个月的预报结果进行了比较，结果表明二个月和三个月预报的距平符号相同的百分数与一个月预报是相近的。

因此，上述的试验结果表明用动力学方法做二个月以上的长期预报是可行的。三层滤波模式具有做季节预报的能力，并且由于这个模式花费的计算时间远远小于 GCM，在 TQ-6(约 90 万次/s) 上做二个月的预报只要 30 分钟，做三个月的预报为 45 分钟。因此这为业务上使用争取了时效。

参考文献

[1] Spar, J. et al. *Mon. Wea. Rev.*, 1976, 104: 1215-1247.

[2] Spar, J. et al., *Mon. Wea. Rev.*, 1979, 107: 181-192.

[3] 巢纪平，等. 中国科学，1977, 2: 162-172.

[4] 巢纪平，等. 中国科学，1979, 2: 75-84.

[5] 邢如楠，等. 中国科学，1982, 2: 113-119.

海-气耦合距平滤波模式的月、季数值预报*

巢纪平　王晓晞　陈英仪　王立治

（国家海洋环境预报研究中心）

提要　本文用海-气耦合三层距平滤波模式(AFM)做了 1976—1977 年以及 1982—1983 年两个 El Niño 事件年冬季 8 个月预报个例实验。其结果表明，该模式成功地预报了大尺度月地表温度距平场，预报与实况的相关系数基本上都超过了惯性预报。与距平大气环流模式(AGCM)相比，两者的结果相差不大，但 AFM 可节省近 100 倍的计算时间。与此同时，我们还做了季节预报试验，即提前三个月作月预报，完成了 1977 年 2 月、3 月和 1983 年 2 月、3 月 4 个试验例子。结果表明，用该模式作大尺度环流的季节异常预报的潜力是存在的。最后，结合本文的结果，我们把提出距平滤波模式以来近 10 年的工作做了小结，对模式的预报能力做了评价。

1　引言

巢纪平等[1-3]首先提出可避免预报不容易报好的气候场，而只预报距平场，同时把瞬变 Rossby 波作为“高频噪音”滤掉，从而发展了一个海洋、陆地和大气耦合的长期预报的距平滤波模式(简称 AFM)。并且，曾在大气取一层的情况下，用该模式做了月地表温度距平场和 50 hPa 上距平位势高度场的大量预报试验例子[4,5]。结果表明，该模式具有一定的预报潜力，计算时间也较短。在没有“巨型”计算机的情况下，有相当高的实用价值。

然而，由于大气只考虑一层，在计算地表温度距平时，大气中各种动力学量就不得不用 50 hPa 上的值，从而带来了一定的误差，而且，这样也限制了大气斜压性的作用以及高层和低层间的相互调整。改进的途径之一是增加模式的层次，使模式成为对大气的三层距平滤波模式。

2　模式

与以前所用的模式稍有不同，在涡度距平方程中，考虑到应有雷诺应力项出现，故采用混合长度理论把这项表示为水平的湍流交换，因此有：

$$\frac{\partial}{\partial t}(\Delta\phi') + \frac{1}{f}J(\overline{\phi} + \phi', \Delta\phi') + J\left(\phi', f + \frac{1}{f}\Delta\overline{\phi}\right) = f^2\frac{\partial\omega'}{\partial P} + K'\nabla^4\phi' \tag{1}$$

相应的热力学第一定律为：

$$\frac{\partial}{\partial t}\left(\frac{\partial\phi'}{\partial P}\right) + \frac{1}{f}\left[J\left(\overline{\phi} + \phi', \frac{\partial\phi}{\partial P}\right) - J\left(\phi', \frac{R}{P}\overline{T}\right) - \frac{\partial}{\partial P}(k + k_r)\frac{\partial^2\phi'}{\partial P^2}\right]$$

* 气象学报，1986，44(4)：417-425.

$$+\frac{1}{\tau_r}\frac{\partial\phi'}{\partial P}=-\widetilde{\sigma}_P\omega'-\frac{R\varepsilon_0}{P\rho C_P} \tag{2}$$

上式最后一项表示凝结热交换，其参数化形式[6]也作了改进，表示为：

$$\frac{\varepsilon_0}{\rho C_P}=-\frac{L}{C_P}H_bD_b\gamma\left(\frac{\partial\ln\bar{e}_s}{\partial T}\right)_0\left(\frac{\partial q_s}{\partial T}\right)T'_s \tag{3}$$

其中

$$D_b=(\mathrm{div}\ V)_b$$

从方程(1)和方程(2)中消去 ω'，并取 Rayleigh 摩擦，即

$$K'\nabla^4\phi'\approx-K'a\nabla^2\phi'=-K\nabla^2\phi' \tag{4}$$

则可以得到非绝热的涡度距平方程：

$$\frac{\partial}{\partial t}\left(\Delta+\frac{f^2\partial^2}{\widetilde{\sigma}_P\partial P^2}\right)\phi'+\frac{1}{f}J(\bar{\phi}+\phi',\Delta\phi')+J\left(\phi',f+\frac{1}{f}\Delta\bar{\phi}\right)$$

$$+K\Delta\phi'-\delta_1K^*T'_s=\frac{f^2}{\widetilde{\sigma}_P}\frac{\partial G}{\partial P} \tag{5}$$

其中

$$G=-\frac{1}{f}\left[J\left(\bar{\phi}+\phi',\frac{\partial\phi'}{\partial P}\right)-J\left(\phi',\frac{R}{P}\bar{T}\right)\right]+\frac{\partial}{\partial P}(k+k_r)\frac{\partial^2\phi'}{\partial P^2}-\frac{1}{\tau_r}\frac{\partial\phi'}{\partial P} \tag{6}$$

$$K^*=fRH_bD_b\frac{L}{C_P}\gamma\left(\frac{\partial\ln\bar{e}_s}{\partial T}\right)_0\left(\frac{\partial q_s}{\partial T}\right)_0\frac{\partial}{\partial P}\left(\frac{1}{P\widetilde{\sigma}_P}\right) \tag{7}$$

$$\delta_1=\begin{cases}1 & \text{当}\quad D_b<0\\ 0 & \quad\ \ D_b>0\end{cases} \tag{8}$$

考虑到上下边界 $\omega'=0$,且无凝结,则有边界条件:

$$P=P_s,\quad \frac{\partial}{\partial t}\left(\frac{\partial\phi'}{\partial P}\right)-G=0 \tag{9}$$

$$P\to0,\quad \frac{\partial}{\partial t}\left(\frac{\partial\phi'}{\partial P}\right)-G=0 \tag{10}$$

$$P=P_s,\quad \frac{\partial\phi'}{\partial P}=-\frac{R}{P_s}T'_s \tag{11}$$

$$P\to0,\quad \frac{\partial\phi'}{\partial P}=0 \tag{12}$$

对于地球表面温度预报方程,我们仍取参考文献[6]或参考文献[2]中的形式,即:

$$z=0,\quad \frac{\partial T'_{si,j}}{\partial t}+\left(\frac{\delta}{D_Q}+\frac{1}{D_s}\right)\frac{K_s}{h}T'_{si,j}+K_sH_{1i,j}$$

$$+\frac{K\tau_r}{D}\frac{K_R}{h}H'_{2i,j}-\frac{S_0l_b}{W_0\rho_sC_{p_s}K_s}(\zeta'_{0g})_{i,j} \tag{13}$$

其中

$$D_Q=p_sC_{p_s}K_s\Big/\left(L\rho K_T\gamma\frac{\partial\ln\bar{e}_s}{\partial T}\right)_0\left(\frac{\partial q_s}{\partial T}\right)_0 \tag{14}$$

$$D_s=\rho_sC_{p_s}K_s\sqrt{K_T\tau_r}/\rho C_PK_T \tag{15}$$

3　计算方法

方程(13)可以写成差分形式：

$$T'^{t+\delta t}_{si,j} = \left[1 - \frac{K_s \delta t}{h}\left(\frac{\delta}{D_Q} + \frac{1}{D_s}\right)\right] T'^{t}_{si,j} + \delta t \left[K_s H_{1i,j} + \frac{K\tau_t}{D_s h} H_{2i,j} + \frac{s_0 l_b b}{W_0 \rho_s C_{p_s} K_s} \Delta\phi'_{1i,j}\right]^t \tag{16}$$

以上一个月的地表温度距平及大气位势高度场距平为初值，便可得到预报月的地表温度距平场。

求出预报月的 T'_s 后，代入方程(5)，可得到预报月的位势高度距平场。对大气取三层，即 300 hPa，500 hPa 以及 700 hPa。滤瞬变 Rossby 波的方法是去掉方程(5) 的时间偏导数项，变成一个适应方程，用超松弛法求解。

4　月预报实验结果

根据上节的计算方案，δt 取一个月，网格距在水平方向取 540 km。我们计算了 1976—1977 年及 1982—1983 年冬季各 4 个月的逐月预报例子，预报与实况场的相关系数以及预报月与上个月实况场间的相关系数(惯性相关) 列于表 1。

表 1　8 个月预报例子的相关系数比较

编号	预报月份	T'_s		$\phi'_{700\ hPa}$		$\phi'_{500\ hPa}$		$\phi'_{300\ hPa}$	
		预报	惯性	预报	惯性	预报	惯性	预报	惯性
1	1976. 11—12	0. 42	0. 39	0. 28	-0. 15	0. 31	0. 10	0. 27	0. 17
2	1976. 12—1977. 01	0. 41	0. 19	0. 48	0. 38	0. 58	0. 01	0. 64	0. 28
3	1977. 01—02	0. 56	0. 25	0. 61	0. 47	0. 70	0. 53	0. 66	0. 42
4	1977. 02—03	0. 59	0. 31	9. 27	-0.14	0.32	-0.12	0.38	0.15
5	1982. 11—12	0.60	0.47	0.26	0.20	0.34	0.42	0.16	0.15
6	1982. 12—1983. 01	0.48	0.56	0.27	0.35	0.23	0.12	0.33	0.38
7	1983. 01—02	0.44	0.51	0.17	0.61	0.40	0.58	0.21	0.36
8	1983. 02—03	0.62	0.56	0.51	0.44	0.47	0.22	0.53	0.46
平均		0.52	0.41	0.35	0.27	0.42	0.21	0.40	0.25

由表 1 可知，地表温度距平场的预报结果，除了 1983 年 1 月和 1983 年 2 月比惯性顶报稍差外，其他例子都超过了惯性预报。高度场有 5 个例子比惯性预报好，2 个例子接近惯性预报的水平，还有 1 个例子，即 1983 年 1 月报 1983 年 2 月的结果很差。值得注意的是，预报的好坏并不完全取决于环流形势的持续性。如在例 3 和例 8 中，环流的持续性很强(惯性相关很高)，预报结果却都高于惯性相关，而在例 7 中，环流的惯性相关也很高，但预报结果却比惯性预报差。另一方面，在例 1 和例 4 中，高度场的月惯性相关是负的，但预报结果却接近中等水平(相关系数在 0. 3 左右)。

就平均来说，无论是地表温度还是各层的高度场预报水平都高于惯性预报。而且，也比一层模式的结果好。图 1 和图 2 分别给出 1977 年 3 月地表温度场距平以及 500 hPa 的位势高度距平场的结果。1983 年 3 月 700 hPa 的高度场预报放在图 3 中。上述各图中图 a 为预报结果;图 b 为实况。位势高度距平均的单位为 10 m。

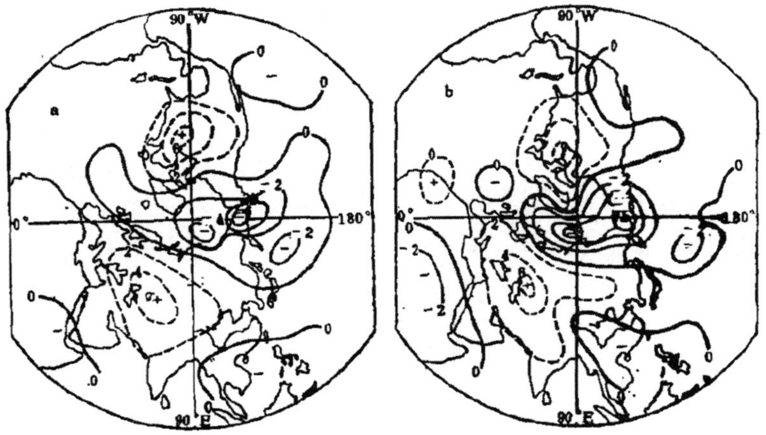

图 1　1977 年 3 月地表温度距平场

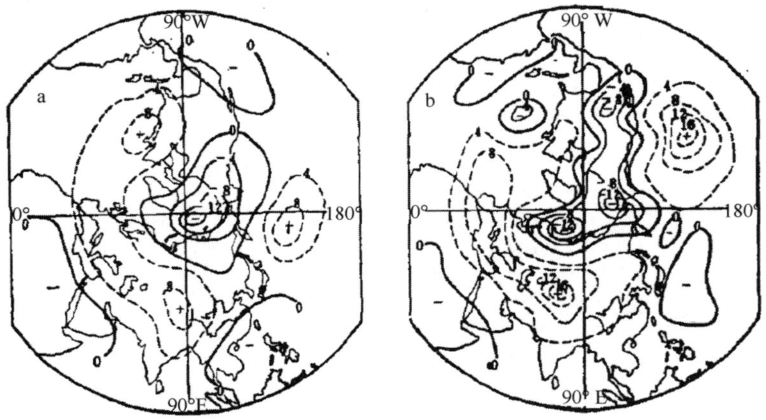

图 2　1977 年 3 月 500 hPa 位势高度距平场

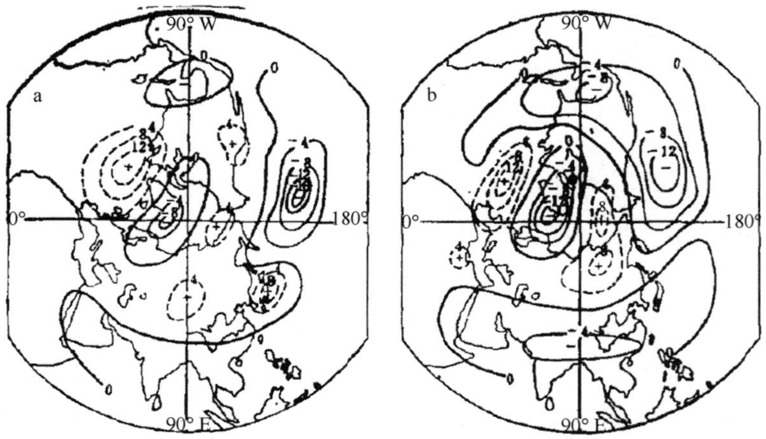

图 3　1983 年 3 月 700 hPa 位势高度距平场

5　与大气环流模式的比较

我们仍采用 AFM 的程序预报地表温度距平，保留方程(5) 的时间偏导数项(即保留瞬变 Rossby 波)，我们称这种方法为距平大气环流模式，简称 AGCM。积分时间步长取 1 小时，积分一个月后取平均值。我们用这种方法完成了 6 个试验例子，并把 AGCM 与 AFM 两种方法预报的结果列于表 2 中。

表 2　AGCM 和 AFM 预报相关系数的比较

预报月份	ϕ'					
	700 hPa		500 hPa		300 hPa	
	AGCM	AFM	AGCM	AFM	AGCM	AFM
1976. 12—1977. 01	0.56	0.48	0.54	0.58	0.60	0.64
1977. 01—02	0.68	0.61	0.67	0.70	0.65	0.66
1977. 02—03	0.29	0.27	0.35	0.32	0.38	0.38
1982. 12—1983. 01	0.24	0.27	0.29	0.23	0.39	0.33
1983. 01—02	0.20	0.17	0.35	0.40	0.28	0.21
1983. 02—03	0.57	0.51	0.50	0.47	0.51	0.53
平均	0.42	0.39	0.45	0.45	0.47	0.46

由表 2 可见，AGCM 和 AFM 的预报能力是很接近的。在低层(700 hPa)，AGCM 比 AFM 要略好一些，但所用计算机时大不相同，用 ND-560 机，AGCM 需 180 分钟，而 AFM 只需 2 分钟。

用 AGCM 做的 1983 年 3 月 700 hPa 位势高度的预报表示在图 4 中。

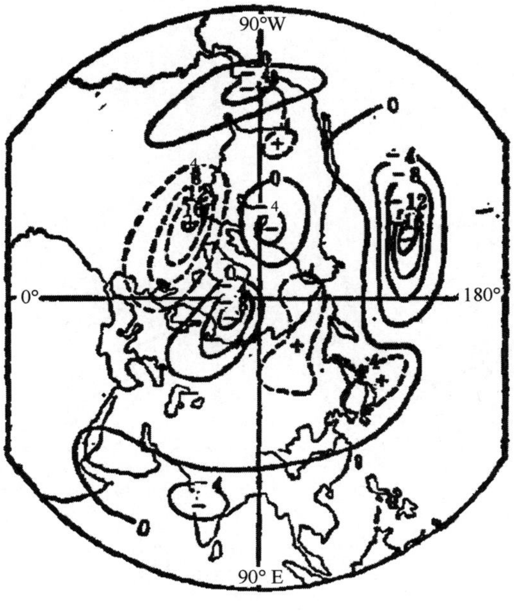

图 4　用 AGCM 预报的 1983 年 3 月 700 hPa 位势高度距平

6　季节预报的潜力

初值取三个月前的地表温度距平和大气位势高度距平值，我们做了 1977 年 2 月、3 月和 1983 年 2 月、3 月 4 个预报例子。δt 可取一个月算三步或取三个月算一步，结果表明后者不比前者差。由于用前一方法调整参数较难，所以我们采用后者。季预报结果及惯性相关的相关系数列于表 3。在这 4 个例子中，有 3 个的相关系数不仅高于惯性预报，而且达到了中等水平(大于 0.3)。只有 1983 年 2 月的位势高度预报失败。平均而言，比惯性相关高。尽管我们的例子不多，预报的水平也尚未达到稳定的程度，但看来用该模式做大尺度环流的季节异常预报的潜力是存在的。图 5 和图 6 分别为 1977 年 3 月地表温度距平和 500 hPa 位势高度距平的季节预报结果。

表 3　季预报试验的相关系数

预报月份	T'_s		ϕ'					
			700 hPa		500 hPa		300 hPa	
	预报	惯性	预报	惯性	预报	惯性	预报	惯性
1976. 11—1977. 02	0.46	0.39	0.58	−0.17	0.74	−0.13	0.76	−0.03
1976. 12—1977. 03	0.36	0.02	0.16	−0.10	0.29	−0.05	0.34	0.07
1982. 11—1983. 02	0.16	0.16	−0.03	0.18	−0.09	0.06	−0.09	−0.27
1982. 12—1983. 03	0.45	0.44	0.36	0.31	0.40	0.39	0.48	0.41
平均	0.36	0.25	0.28	0.06	0.34	0.07	0.38	0.05

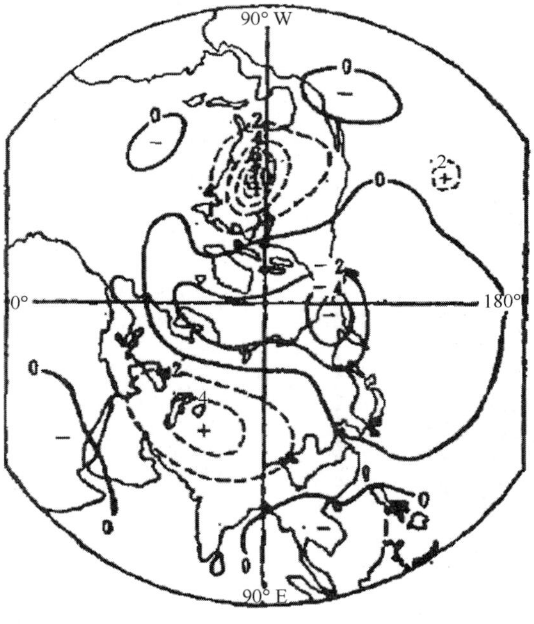

图 5　1977 年 3 月地表温度的季预报结果

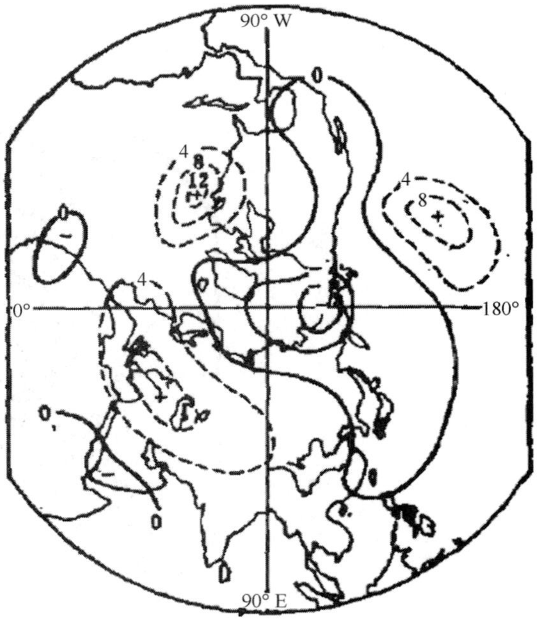

图 6　1977 年 3 月 500 hPa 位势高度的季预报结果

试验结果还表明，位势高度场的预报对初值的依赖极小。换言之，位势高度场距平主要取决于外源(在此为地表温度距平)的形式，或者说是大气环流场对加热场的适应。但是，地表温度场预报的好坏，却与初始场有关。这在物理上很容易理解，因为地表温度(特别是海温)具有较大的热惯性，而大气运动自身的记忆却很短。当时间尺度足够长后，初始场的作用就不重要了，主要依赖大气内部各种过程的调整对外源的适应。

7　对模式和结果的讨论

自 1977 年[1]提出制作长期数值预报距平滤波模式的概念和物理基础及 1979 年[2]发表了第一个月预报试验例子以来，作为研究长期数值预报的一种途径，已在国内外同行中引起了不同的反响和评论[7-10]。事物总是一分为二的，任何一个长期预报方法也不例外。结合本文的结果，我们对该模式做个小结。

距平模式的优点是显然的，它避开了至今连数值模拟都尚未逼真的气候场本身的预报，而把实际观测到的月平均气候场作为已知量输入到距平模式中去，这样我们就可以不管气候场如何形成的问题。当然，气候场本身对距平场的发展起一定作用。在作长期预报方面，AGCM 比一般的 GCM 具有优越性，这一点似乎尚未有人提出过怀疑，尽管人们似乎更偏爱用 GCM 来做长期预报。

事实上，Navarro 和 Miyakoda[12]曾做过一个比较试验，他们先给定一个参考性的大气环流距平场，当他们用谱的 GCM 去"预报"这个参考场时，纬向波数要取到 30 个，"预报场"才能逼近参考场，但纬向波数取到 15 个波时，距平意义下的大气环流形势几乎与参考场完全反了过来。然而，他们发现如果用 AGCM 去"预报"这个参考场时，即使只用 15 个波，所得的结果比 GCM 用 30 个波所得的结果好。

不过，要建立一个严格的 AGCM 存在一定的困难，就是如何把距平场从总场中分离出来。如果对大气运动方程组取时间平均(如时间平均取一个月)，这样由于方程的非线性就必然出现雷诺应力项。这一项如何处理则是湍流经典理论的困难。在我们过去的距平模式中干脆去掉了这一项。当然，也可以采取通常的作法，将雷诺应力项用参数化方法变成对距平分量的湍流耗散项。本文就是这样做的，这种方法的结果是众所周知的，雷诺应力项在此只起耗散和平滑距平场发展的作用。然而，Navarro 和 Miyakoda 的试验表明，雷诺应力项的存在并不对预报出的距平场的形势存在明显的影响，而只是使距平中心的强度有所减弱。如果在所有场合下雷诺应力项的作用只是如此，这自然是一件幸运的事。否则，就会像经典的湍流理论那样，为闭合方程组而陷入死胡同。

争议较多的是对 Rossby 波的过滤问题。由于 Rossby 波在大气和海洋中不仅存在，而且实践也表明了它在短期预报中的重要作用。因而，即使在长期数值预报中滤掉 Rossby 波，也确实容易引起争论。Egger[9]用一个简单化的模式从理论上讨论过这一问题。他认为如果滤去扰动的快变部分(相应地可以看成是 Rossby 波)，则由于去掉了扰动快变部分和低频部分之间的相互作用，将对预报结果造成重大误差。然而，Egger 用于理论分析的方程并不是真正的距平模式，而在距平模式中，由于还包括了气候场部分，因此，正如丑纪范[11]所指出的那样，Egger 过高估计了滤掉 Rossby 波对于预报造成的相对误差。本文作者认为丑纪范已把问题说清楚了，故不再对 Egger 的文章作进一步的评价。然而，有争议的问题仍然存在。近年来，有很多人认为超长波在中、长期天气过程中扮演主要角色。而在 AFM 中连超长波也去掉了，保留的仅是准定常的强迫 Rossby 驻波。这里称 Rossby 驻波是准定常的，是因为地球表面的加热场是缓变的，大气环流的气候场也是随时间缓变的，只是月与月间不同。因此，一个值得讨论的问题是滤掉了超长波而只保留准定常的 Rossby 驻波，将会给月和季这样时间尺度的预报带来多大误差？另一方面，滤掉 Rossby 波所带来的误差还可以通过数值试验来估计。在上述 Navarro 和 Miyakoda 的文章中就作过估计。他们认为 AFM 的预报结果与 AGCM 或 GCM 的结果相比，对于大尺度环流的形势场并无明显影响，只是预报的中心强度要弱。而在本文中连强度也差不多。如果 AFM 对所有的预报例子强度都弱的话，那只要调整一下参数问题就可以解决了。不过这要经过大量预报例子的实验，选出一些在统计上比较稳定的参数，使预报水平稳定。这种试验从一层模式开始我们就着手进行了。同时由于 AFM 将大量节省计算时间，在没有巨型计算机的情况下是重要的。即使有巨型计算机，用 Navarro 和 Miyakoda 的话来说，AFM 仍可以作为一种“快照”(Quick look)，先把预报场的正负距平区确定下来。

除了上述原则性的问题外，AFM 仍有一些技术性的问题有待于改进。根据我们所做的大量例子来看，极区的预报结果一般都不好，看来这要对极冰和反照率之间的反馈过程有所考虑。一个简单的方法是像参考文献[13]那样做。另外，就统计结果看，美洲部分一般比亚洲部分要好，看来这与没有考虑青藏高原的影响有关，因此，考虑地形影响是势在必行的。再者，辐射和云量间的相互作用不仅对气候的形成是重要的，对长期预报也是重要的，应寻求一个切合实际的有效参数化方法把它加在模式中。

最后一个带有理论性的问题是热带与中纬度大气环流之间的相互作用，特别是在 El Niño 年。分析[14]和理论[15]均表明，当赤道“ENSO” 出现时，中纬度大气将要有重大调整，多年统计事实的研究也表明了这一点[16]。然而，本文所用的是一个准地转模式，自然不适应于热带地区的预报。但本文所取的两个年份都是“ENSO”年。除了 1983 年 1 月报 2 月的例子没有报好外，其他例子均接近和超过惯性预报。由此提出一个问题，即使在“ENSO” 年中，中纬度大气环流对赤道地区海温的响应比起对局地温度场的响应来讲，究竟要占多大比例。当然，我们不能排除在“ENSO” 年中纬度的局地温度(特别是海温) 本身与赤道海温间就有一种关系。但究竟要多长的时间尺度赤道海温的巨变才会影响到中纬度是值得讨论的。这就需要对模式加以改进，使它能把热带包括进去，这对预报总是有好处的。

参考文献

[1] Group of long-range unmerical weather forecasting.On the physical basis of a model of long-rang numerical weather forecasting, *Scientia Sinica*, 1977, 20:377-390.

[2] Group of long-range unmerical weather forecasting. A filtering method for long-Range numerical weather forecasting. *Scientia Sinica*, 1979, 22:661-674.

[3] CHAO, Jiping, GUO Yufu & XIN Runan. A Theory and method of long-range numerical weather forecasts, *J. Meteor. Soc. Japan*, 1982, 60:282-291.

[4] Chao Jiping. Filtered anomaly model and its application to monthly and seasonal climate forecasts, WMO/ICSU Study Conf. Phys., Basis for climate prediction, Leningrad. 1982.

[5] Guo Yufu & Chao Jiping. Simplified dynamical anomaly model for long-range numerical forecasts. *Advances in Atmospheric Sciences*, 1984, 1:30-39.

[6] Chao Jiping, Wang Xiaoxi, Chen Yingyi & Wang Lizhi. An stmosphere-ocean/lancl coupled anomaly model for monthly and seasonal forecasts, Proc. International Climate Symposium, Beijing, 1984(to be published).

[7] Chao Jiping & R.Caverly, An anomaly model and its application to long-range forecasts, Proc. sixth annual climate diagnostic workshop, Lamont-Ponerty Geological Observatory Columbia University, Palisades, New York, 1982:316-319.

[8] Miyakda, K., & J.P.Chao. Essay on dynamical long-range forecasts of atmospheric circulation, *J. Meteor. Soc. Japan*, 1982, 60:292-306.

[9]　Egger, J., Simplified dynamics approach, Proc, WMO – CAS/JSC expert study meeting on long-range forecasting, Princeton, 1982: 207–218.

[10]　Barnett, T.P., & R.C.J.Somervill, Advances in short term climate prediction, 18 th general assembly international union of geodesy and geophysics, Hamburg, August 15–27 1983, Contribution in Meteorology, U.S. National Report, 1979–1982: 1098–1102.

[11]　丑纪范.长期数值天气预报(即将由气象出版社出版),1985.

[12]　Navarro, A. & K.Miyakoda.Anomaly model using a barotropic vorticity equation, Proc.sevcnth annual climatc diagnostics workshop, NCAR, Boulder, Colorado, 1982: 456–460.

[13]　Chen Yingyi & Chao Jiping.A two-dimensional energy balance climate model including radiation and ice caps–albedo fcedback.*Adrances in Atmoapheric Sciences*, 1984, 1: 234–245.

[14]　Horel, J.D. & J.M.Wallace.Planetary-scale atmospheric phenomena associated with the southcrn oscillation.*Mon. Wea. Rec.*, 1981, 109: 813–829.

[15]　Hoskins, B.J., & D.Karaly.The steady linear response of a spherical atmosphere to thermal and orographic forcing.*J. Atmos.Sci.*, 1981, 38: 1179–1196.

[16]　Pan Yihong & A.H.Oort.Global climate variations connected with sea surface temperature anomalies in the Eastern Equatorial Pacific Ocean for the 1958–73 period.*Mon. Wea.Rev.*, 1983, 111: 1244–1258.

第 4 部分　大尺度大气和海洋运动

斜压西风带中大地形有限扰动的动力学*

巢纪平

（中国科学院地球物理研究所）

摘要：本文指出，对于无黏性的、绝热的斜压大气的大尺度常定有限扰动，可以用一个三维的 Helmholtz 方程来描写。在地形存在的边界条件下，获得了该方程的解答。

用西藏高原（包括亚洲山系）和洛矶山的实际地形做了数值计算，计算结果表明，地形扰动可以解释西风带平均槽脊的位置，但扰动的强度与实况有一些出入。计算结果亦表明，大地形对西风带的扰动随高度很快阻尼，因此看来大地形对西风带的扰动在对流层低层更为重要。

1　引言

大地形对西风气流的动力扰动是近年来气象学家们所讨论的中心问题之一。我国的气象学者对此亦开展了研究[1,2]。Charney 和 Eliassen[3] 考虑了地形所引起的动量幅数，用正压涡度方程处理了气流越过山脉的问题。后来 Bolin[4] 把他们的工作推广到二维空间，并且把大地形对大气环流的重要性提高到显要的地位。在另一方面，苏联学者 Мусаелян[5] 第一个解出了斜压大气中大地形扰动的球体空间问题。在他以前大气的斜压性和地球的曲率作用，在大地形扰动的研究中是没有人考虑过的。然而在他的理论中，所取的基本气流是随高度不变的，即没有考虑西风的垂直切变作用，所以实质上 Мусаглян 只考虑了大气的稳定度（层结）作用。然而，西风垂直切变的作用是斜压大气最重要的特色之一，是必须考虑的。最近我国的朱抱真[6] 用二层模式计算了西藏高原和洛矶山对西风气流的影响，得出的数值解答能够在一定程度上解释西风带平均槽脊的形成。与此同时，日本气象学家村上多喜雄[7] 亦发表了与朱抱真相类似的工作。所有上面这些工作，处理非线性运动方程的方法是用的小扰动法，即在地形引起的扰动振幅是微小的假定下，使方程线性化。我们认为对于行星尺度的扰动问题，小扰动假定是一个限制。在地形扰动的问题中，第一个不用小扰动假定的是 Steward[8]，他定量地计算了正压流体绕过一个无限长的圆柱体问题。然而他所得到的解答不唯一。最近岸保勘三郎[9] 也去掉了小扰动的限制，讨论了大地形对正压流体的常定有限扰动。但是他的数值计算的结果，却在西藏高原上空出现了一个槽，在山脉上空出现槽这在物理上是不易理解的。

本文的目的，企图在更一般的条件下研究大地形对西风气流的动力影响，进而解释西风带平均槽脊的形成。在现在所设计的模式中，大气是斜压的，扰动的振幅是有限的，同时也将计入地球的曲率作用，亦即我们将解决大地形对西风气流影响的球体空间问题。

* 气象学报，1957，28（4）：303－313.

2 控制常定运动的微分方程

在大气是无黏性的和绝热的情况下,大尺度的大气运动可以用下面形式的涡度方程和热力学方程来描写

$$\frac{\partial \zeta}{\partial t}+\frac{v_\theta}{R}\frac{\partial}{\partial \theta}(\zeta+2\Omega\cos\theta)+\frac{v_\lambda}{R\sin\theta}\frac{\partial \zeta}{\partial \lambda}=f\frac{\partial \omega}{\partial p} \tag{1}$$

$$\frac{\partial^2 \phi}{\partial t\partial p}+\frac{v_\theta}{R}\frac{\partial}{\partial \theta}\left(\frac{\partial \phi}{\partial p}\right)+\frac{v_\lambda}{R\sin\theta}\frac{\partial}{\partial \lambda}\left(\frac{\partial \phi}{\partial p}\right)+\sigma\omega=0 \tag{2}$$

这里采用的是球面坐标系统。式中,p 是气压;$\theta=\frac{\pi}{2}-\varphi$,其中 φ 是地方纬度;θ 向南增加;λ 是地方经度,向东增加;v_θ 和 v_λ 是沿 θ 和 λ 方向的速度分量;$\omega=\frac{\mathrm{d}p}{\mathrm{d}t}$ 是垂直速度;R 是地球半径;$f=2\Omega\cos\theta$ 是科里奥利参变数,其中,Ω 是地球转动角速度;ϕ 是重力位势;$\sigma=-\frac{1}{\rho}\frac{\partial \ln\Theta}{\partial p}$ 是静力稳定度,其中 Θ 是位温;ρ 是空气的密度;ζ 是空气质点的相对涡度,而且

$$\zeta=\frac{1}{R\sin\theta}\left[\frac{\partial}{\partial \theta}(v_\lambda\sin\theta)-\frac{\partial v_\theta}{\partial \lambda}\right]=\frac{1}{f}\nabla^2\phi \tag{3}$$

$$v_\theta=R\dot{\theta}=-\frac{1}{fR\sin\theta}\frac{\partial \phi}{\partial \lambda},\quad v_\lambda=R\sin\theta\dot{\lambda}=\frac{1}{fR}\frac{\partial \phi}{\partial \theta} \tag{4}$$

(3)式中

$$\nabla^2\equiv\frac{1}{R^2\sin\theta}\left[\frac{\partial}{\partial \theta}\left(\sin\theta\frac{\partial}{\partial \theta}\right)+\frac{1}{\sin\theta}\frac{\partial^2}{\partial \lambda^2}\right]$$

是球面坐标系统中的拉普拉斯算子。式(4)中 $\dot{\theta}$ 和 $\dot{\lambda}$ 是沿 θ 和 λ 方向的空气质点的相对角速度。式(3)、式(4)两式中的 f 皆取平均值。

在式(1)中略去了涡度的垂直输送项 $\omega\frac{\partial \zeta}{\partial p}$ 和旋转项 $k\cdot\nabla\omega\times\frac{\partial v}{\partial p}$,并在右端考虑到 $\zeta\ll f$ 而略去了 $\zeta\frac{\partial \omega}{\partial p}$。式(1)右端原来的水平辐散现已通过连续性方程用垂直辐散来代换,并把 f 视作常数。在式(2)中把 σ 取作常量。

引进新的自变量 $\eta=\cos\theta$,假定运动是常定的(即在式(1)和式(2)二式中包含对时间微分的项为零),则式(1)和式(2)可写成

$$J\left(\frac{\nabla^2\phi}{f}+2\Omega\cos\theta,\phi\right)=R^2f^2\frac{\partial \omega}{\partial p} \tag{5}$$

$$\left(\frac{\partial \phi}{\partial p},\phi\right)=-R^2f\sigma\omega \tag{6}$$

式中,$J(A,B)=\frac{\partial A}{\partial \eta}\frac{\partial B}{\partial \lambda}-\frac{\partial A}{\partial \lambda}\frac{\partial B}{\partial \eta}$ 是雅可比算子。

从式(5)和式(6)中消去 ω 得

$$J\left(\frac{f}{\sigma}\frac{\partial^2\phi}{\partial p^2}+\frac{\nabla^2\phi}{f}+2\Omega\cos\theta,\phi\right)=0 \tag{7}$$

在二维的常定运动中，可以得到量$\frac{\nabla^2\phi}{f}+2\Omega\cos\theta$ 和等 ϕ 线是重合的，即等绝对涡度线就是等高线，在三维的常定运动中，由于考虑了辐散的作用，等绝对过度线已经不再与等高线相重合，然而量$\frac{f}{\sigma}\frac{\partial^2\phi}{\partial p^2}+\frac{\nabla^2\phi}{f}+2\Omega\cos\theta$ 与等高线是重合的。因此，式(7)的一般积分为

$$\frac{f}{\sigma}\frac{\partial^2\phi}{\partial p^2}+\frac{\nabla^2\phi}{f}+2\Omega\cos\theta=F(\phi) \tag{8}$$

令

$$\phi(\theta,\lambda,p)=\bar{\phi}(\theta,p)+\phi'(\theta,\lambda,p) \tag{9}$$

因为

$$\bar{v}_\lambda(\theta,p)=R\sin\theta\,\bar{\dot{\lambda}}(p) \tag{10}$$

所以

$$\bar{\phi}(\theta,p)=-fR^2\,\bar{\dot{\lambda}}(p)\cos\theta \tag{11}$$

设$\bar{\dot{\lambda}}(p)$是气压的线性函数，取成

$$\bar{\dot{\lambda}}(p)=\bar{\dot{\lambda}}_0+\bar{\dot{\lambda}}_p(p_0-p) \tag{12}$$

式中，$\bar{\dot{\lambda}}_0$ 表示地面气流；p_0 是地面气压；$\bar{\dot{\lambda}}_p=-\frac{\mathrm{d}\bar{\dot{\lambda}}}{\mathrm{d}p}>0$ 为气流的垂直切变（取作常量）。在上面各式中，符号“-”表示基本流场，“′”表示叠加在基本流场上的扰动流场。将式(9)、式(11)和式(12)三式代入式(8)，并设在二极附近扰动流场恒等于零，由此得

$$F(\bar{\phi})=2(\bar{\dot{\lambda}}+\Omega)\cos\theta=-\frac{2(\bar{\dot{\lambda}}+\Omega)}{fR^2\bar{\dot{\lambda}}}\bar{\phi} \tag{13}$$

这样，我们就定出了 $F(\phi)$的表达式为

$$F(\phi)=-\frac{2(\bar{\dot{\lambda}}+\Omega)}{fR^2\bar{\dot{\lambda}}}\phi \tag{14}$$

因为 $\bar{\dot{\lambda}}\ll\Omega$，式(14)可以简化成

$$F(\phi)=-\frac{2\Omega}{fR^2\bar{\dot{\lambda}}}\phi \tag{15}$$

将式(15)代入式(8)得

$$\frac{f}{\sigma}\frac{\partial^2\phi}{\partial p^2}+\frac{\nabla^2\phi}{f}+2\Omega\cos\theta=-\frac{2\Omega}{fR^2\bar{\dot{\lambda}}}\phi \tag{16}$$

再将式(9)代入式(16)即得

$$\frac{f^2}{\sigma}\frac{\partial^2\phi'}{\partial p^2} + \nabla^2\phi + \frac{2\Omega}{R^2\dot{\bar{\lambda}}}\phi' = 0 \tag{17}$$

令 $P=\frac{\sqrt{\sigma}}{f}p$,式(17)可以写成最后的形式

$$\frac{\partial^2\phi'}{\partial P^2} + \frac{1}{R^2\sin\theta}\left[\frac{\partial}{\partial\theta}\sin\theta\frac{\partial\phi'}{\partial\theta} + \frac{\partial^2\phi'}{\partial\lambda^2}\right] + \frac{2\Omega}{R^2\dot{\bar{\lambda}}}\phi' = 0 \tag{18}$$①

由此可见,三维空间大气的大尺度常定有限扰动可以用一个经典的 Helmholtz 方程来描写。当运动是正压时,式中第一项为零,由此三维的波动方程就变成了二维的波动方程。式(18)是一般的,不仅可以解地形扰动问题,亦可解其他适当的边值问题。

3 地形扰动的边界条件

下面我们将寻求地形扰动问题的边界条件的表达式。热力学方程式(6)当 $p=p_0(\theta,\lambda)$②时可以近似地写成

$$J\left(\frac{\partial\phi}{\partial p}\bigg|_{p=p_0},\phi\right) = fR^2\sigma\rho_0 gW_0 \tag{19}$$

式中,W_0 是由于地形所引起的垂直运动,为

$$W_0 = \frac{v_{\theta_0}}{R}\frac{\partial h}{\partial\theta} + \frac{v_{\lambda_0}}{R\sin\theta}\frac{\partial h}{\partial\lambda} = \frac{1}{R^2 f}J(h,\phi_0) \tag{20}$$

式中,$h=h(\theta,\lambda)$为地形的方程。

因此,如略去了地面密度的水平变化,式(19)可以写成

$$J\left(\frac{\partial\phi}{\partial p}\bigg|_{p=p_0} - \sigma\rho_0 gh,\phi_0\right) = 0 \tag{21}$$

式(21)的一般积分为

$$\frac{\partial\phi}{\partial p}\bigg|_{p=p_0} - \sigma\rho_0 gh = G(\phi_0) \tag{22}$$

同样将式(9)、式(11)和式(12)代入式(22),并设在二极附近 $\phi'\equiv 0$,同时也没有地形存在,即 $h=0$,这样可以求得

$$G(\bar{\phi}_0) = fR^2\dot{\bar{\lambda}}_p\cos\theta = -\frac{\dot{\bar{\lambda}}_p}{\dot{\bar{\lambda}}_0}\bar{\phi}_0 \tag{23}$$

由此求得 $G(\phi)$的表达式为

$$G(\phi_0) = -\frac{\dot{\bar{\lambda}}_p}{\dot{\bar{\lambda}}_0}\phi_0 \tag{24}$$

---

① 在平面坐标中,可以求得相应的方程为

$$\frac{\partial^2\phi'}{\partial x^2} + \frac{\partial^2\phi'}{\partial y^2} + \frac{f^2}{\sigma}\frac{\partial^2\phi'}{\partial p^2} + \frac{\beta}{U_\infty}\phi' = 0,\left(\beta = \frac{df}{dy}\right)$$

② 值得指出,由于地形的存在,地面不可能是一个等压面,并且 p_0 的变化很大,但在工作[3,6,7,9]中都把 p_0 取成是常数,在我们这里就没有这个限制。

将式(24)代入式(22)得

$$\frac{\partial \phi}{\partial p}\Big|_{p=p_0} + \frac{\overline{\dot{\lambda}}_p}{\overline{\dot{\lambda}}_0}\phi_0 = \sigma\rho_0 g h(\theta,\lambda) \tag{25}$$

将式(9)代入式(25),并以 $P=\frac{\sqrt{\sigma}}{f}p$ 为垂直坐标,则得方程(18)的下界条件为

当 $P=P_0(\theta,\lambda)$ 时,

$$\frac{\partial \phi'}{\partial P} + c\phi' = H(\theta,\lambda) \tag{26}$$

式中,$c=f\overline{\dot{\lambda}}_p/\sqrt{\sigma}\overline{\dot{\lambda}}$;$H(\theta,\lambda)=f\rho_0 g\sqrt{\sigma}h(\theta,\lambda)$。

方程(18)的上界条件,我们取成在大气的上界,地形对大气运动的影响消失,即当 $P=0$ 时,

$$\phi' = 0 \tag{27}$$

4　问题的解答

我们把地形按球函数展开

$$h(\theta,\lambda) = \sum_{n=1}^{\infty}\sum_{m=1}^{n}(h_n^m\cos m\lambda + h'^m_n\sin m\lambda)p_n^m(\cos\theta) \tag{28}$$

式中,$p_n^m(\cos\theta)$ 为联属勒让德多项式。

同样地,把 $H(\lambda,\theta)$ 亦按球函数展开

$$H(\theta,\lambda) = \sum_{n=1}^{\infty}\sum_{m=1}^{n}(H_n^m\cos m\lambda + H'^m_n\sin m\lambda)p_n^m(\cos\theta) \tag{29}$$

式中,$H_n^m = f\rho_0 g\sqrt{\sigma}h_n^m$,$H'^m_n = f\rho_0 g\sqrt{\sigma}h'^m_n$。

将要寻求的解答 $\phi'(\theta,\lambda,p)$ 亦按球函数展开:

$$\phi'(\theta,\lambda,p) = \sum_{n=1}^{\infty}\sum_{m=1}^{n}(\phi_n^m(p)\cos m\lambda + \phi'^m_n(p)\sin m\lambda)p_n^m(\cos\theta) \tag{30}$$

将式(30)代入式(18),合并 $\cos m\lambda$ 和 $\sin m\lambda$ 的同类项,考虑到联属勒让德多项式所要求满足的方程,省去中间的运算,直接写出展开式(30)中系数所应满足的方程为:

$$\frac{d^2\phi_n^m}{dP^2} = \left[\frac{n(n+1)}{R^2} - \frac{2\Omega}{R^2\overline{\dot{\lambda}}}\right]\phi_n^m \text{①} \tag{31a}$$

$$\frac{d^2\phi'^m_n}{dP^2} = \left[\frac{n(n+1)}{R^2} - \frac{2\Omega}{R^2\overline{\dot{\lambda}}}\right]\phi'^m_n \tag{31b}$$

将式(29)和式(30)代入式(26),最后得到方程式(31)的下界条件为

$$\frac{d\phi_n^m}{dP} + c\phi_n^m = H_n^m \tag{32a}$$

① 非常有兴趣的是:如果在此假定 $\overline{\dot{\lambda}}$ 是常数,则当 $\frac{n(n+1)}{R^2}-\frac{2\Omega}{R^2\overline{\dot{\lambda}}}>0$ 时,我们最后可以求得一个与 Мусаелян 的解答相当一致的解答。

$$\frac{\mathrm{d}\phi'^m_n}{\mathrm{d}P} + c\phi'^m_n = H'^m_n \tag{32b}$$

由式(27)得到方程式(31)的上界条件为

$$\phi^m_n = 0 \tag{33a}$$

$$\phi'^m_n = 0 \tag{33b}$$

将式(31a)写成

$$\frac{\mathrm{d}^2\phi^m_n}{\mathrm{d}P^2} = \left(k^2 - \frac{l}{\bar{\dot{\lambda}}}\right)\phi^m_n \tag{34}$$

其中，$k^2 = \frac{n(n+1)}{R^2}, l = \frac{2\Omega}{R^2}$。

作自变量和函数的变换：

$$\xi = \frac{2k\sqrt{\sigma}}{f\bar{\dot{\lambda}}_p}\left[\bar{\dot{\lambda}}_0 + \bar{\dot{\lambda}}_p\frac{f}{\sqrt{\sigma}}(P_0 - P)\right] \tag{35}$$

$$\phi^m_n = \mathrm{e}^{-\frac{\xi}{2}}\psi(\xi) \tag{36}$$

则式(34)变为

$$\frac{\mathrm{d}^2\psi}{\mathrm{d}\xi^2} - \frac{\mathrm{d}\psi}{\mathrm{d}\xi} + \frac{r}{\xi}\psi = 0 \tag{37}$$

式中 $r = l/2k\bar{\dot{\lambda}}_p\frac{f}{\sqrt{\sigma}}$。式(37)为汇合超越几何方程(confluent hypergeometric equation)，它的两个特解为(见参考文献[10])

$$\psi_1 = \frac{\sin\pi a}{\pi}\left\{a\xi M(a+1,2,\xi)\cdot\left[\ln\xi + \frac{\Gamma'(a)}{\Gamma(a)} - 2\frac{\Gamma'(1)}{\Gamma(1)} + 1\right] + \sum_{n=1}^{\infty}B_n\frac{a(a+1)\cdots(a+n-1)}{(n-1)!\ n!}\xi^n\right\} \tag{38}$$

$$\psi_2 = \xi M(a+1,2,\xi) \tag{39}$$

其中

$$a = -r,\quad B_n = \sum_{v=0}^{n-1}\left(\frac{1}{a+v} - \frac{2}{1+v}\right) + \frac{1}{n},$$

$$M(a,b,\xi) = 1 + \frac{a}{1\cdot b}\xi + \frac{a(a+1)}{2!\ b(b+1)}\xi^2 + \cdots,$$

$\Gamma(a)$为 Gamma 函数。

因此方程式(37)的一般解为

$$\psi = A\psi_1 + B\psi_2 \tag{40}$$

或式(34)的一般解为

$$\phi^m_n = A\mathrm{e}^{-\frac{\xi}{2}}\psi_1 + B\mathrm{e}^{-\frac{\xi}{2}}\psi_2 \tag{41}$$

其中积分常数 A,B 由式(32a)和式(33a)决定。

将式(32a)和式(33a)用新的自变量写出，则

$$当\qquad \xi = \frac{2k\sqrt{\sigma}\,\overline{\dot{\lambda}}_0}{f\overline{\dot{\lambda}}_p} = \xi_0 \text{ 时},(p = p_0)$$

$$\frac{\mathrm{d}\phi_n^m}{\mathrm{d}\xi} - \frac{c}{2R}\phi_n^m = -\frac{H_n^m}{2R} \tag{42}$$

$$当\qquad \xi = \frac{2k\sqrt{\sigma}}{f\overline{\dot{\lambda}}_p}\left(\overline{\dot{\lambda}}_0 + \overline{\dot{\lambda}}_p \frac{f}{\sqrt{\sigma}}P_0\right) = \xi_1 \text{ 时},(p = 0)$$

$$\phi_n^m = 0 \tag{43}$$

省去中间的运算,直接写出式(34)的解答为

$$\phi_n^m = \frac{\mathrm{e}^{-\frac{1}{2}(\xi-\xi_0)}[\psi_1(\xi_1)\psi_2(\xi) - \psi_2(\xi_1)\psi_1(\xi)]}{2k\Delta}\cdot H_n^m \tag{44a}$$

同理得式(32b)的一般解答为

$$\phi'^m_n = \frac{\mathrm{e}^{-\frac{1}{2}(\xi-\xi_0)}[\psi_1(\xi_1)\psi_2(\xi) - \psi_2(\xi_1)\psi_1(\xi)]}{2k\Delta}\cdot H'^m_n \tag{44b}$$

式中

$$\Delta = \begin{vmatrix} \dfrac{\left[\dfrac{\mathrm{d}\psi_1}{\mathrm{d}\xi} - \left(\dfrac{1}{2} + \dfrac{c}{2k}\right)\psi_1\right]\Big|_{\xi=\xi_0}}{\psi_1(\xi_1)} & \dfrac{\left[\dfrac{\mathrm{d}\psi_2}{\mathrm{d}\xi} - \left(\dfrac{1}{2} + \dfrac{c}{2k}\right)\psi_2\right]\Big|_{\xi=\xi_0}}{\psi_2(\xi_1)} \end{vmatrix}$$

将式(44a)和式(44b)代入式(30),即得问题的解答为

$$\phi'(\theta,\lambda,p) = f\rho_0 g\sqrt{\sigma}\sum_{n=1}^{\infty}\sum_{m=1}^{n}\frac{\mathrm{e}^{-\frac{1}{2}(\xi-\xi_0)}[\psi_1(\xi_1)\psi_2(\xi) - \psi_2(\xi_1)\psi_1(\xi)]}{2k\Delta} \times [h_n^m\cos m\lambda + h'^m_n\sin m\lambda]p_n^m(\cos\theta) \tag{45}$$

式(45)指出,扰动的强度决定于 $\overline{\dot{\lambda}}_0$ 和 $\dfrac{\sqrt{\sigma}}{\overline{\dot{\lambda}}_p}$ 两个物理参数,亦即决定于地面的基本气流和 Richardson 数。

Bolin 认为地形对西风气流的影响在对流层上层更为重要。但是式(45)表明扰动强度随高度的变化是较复杂的,数值计算结果指出,扰动强度随高度递减很快(图 1)。因此,与 Bolin 相反,我们认为大地形对西风带的影响在对流层低层更为重要。对于大地形的常定扰动,Mycaелян 亦得到了相同的结论。

这是一个很有趣的现象,大地形和小地形在垂直方向上的影响是截然不同的。小地形在垂直方向上一般可以产生波系[11]。Mycaелян 指出,决定这两种地形的不同影响的主要因子是地形的尺度和地球的自转。当地形的尺度大时,地球的自转起着重要的作用,在这种情况下可以得到扰动向上呈递减的解答;当地形的尺度小时,地球自转比较不重要,这时扰动在垂直方向上可以产生波系,亦即扰动的强度可以向上增加。

5　西藏高原和洛矶山对西风气流的扰动

我们分别计算了西藏高原(包括亚洲山系)和洛矶山对西风气流的扰动。我们把西藏高原和洛矶

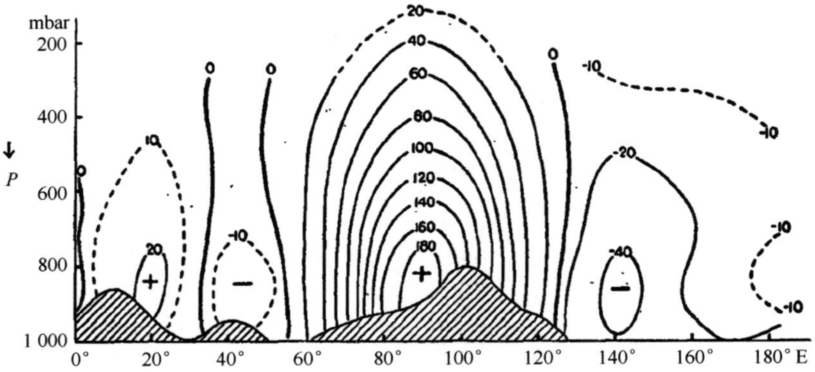

图 1　西藏高原引起的扰动量随高度的分布

（位置:45°N;单位:位势米）

山的地形分别用球函数来逼近。展开式的系数由下面大家所熟知的公式决定

$$h_n^m = \frac{(2n+1)(n-m)!}{2\pi(n+m)!}\int_0^{2\pi}\int_0^{\pi} h(\theta,\lambda)\cos m\lambda p_n^m(\cos\theta)\sin\theta \mathrm{d}\theta \mathrm{d}\lambda,$$

$$h'^m_n = \frac{(2n+1)(n-m)!}{2\pi(n+m)!}\int_0^{2\pi}\int_0^{\pi} h(\theta,\lambda)\sin m\lambda p_n^m(\cos\theta)\sin\theta \mathrm{d}\theta \mathrm{d}\lambda,$$

其中,$h(\theta,\lambda)$是西藏高原和洛矶山的实际高度。

我们把 3 个物理参数的数值取得和冬季的情况相当。为了应用 Charney[10]已经算好的函数ψ_1和ψ_2的表,我们将物理参数做了一些适当的调整,取成:$\sqrt{\sigma}=0.128$ m/(mbar · s),$\bar{\dot{\lambda}}_0=0.63\times10^{-6}$/s,$\bar{\dot{\lambda}}_p=0.68\times10^{-8}$ mbar · s。一般来说,这样取的静力稳定度比正常的情况小了一些,而气流的垂直切变稍大了一些。为了减少繁重的计算量,我们取了一个与赤道成对称的解答,此时,m 加 n 之和为奇数的项都为零。

计算结果表明(图 1),扰动的位相向上变化很小,亦即槽脊线的轴是准垂直的。因此这里只给出了 700 mbar 西藏高原和洛矶山的分别扰动图(图 2 和图 3),以及两者的共同扰动(图 4)。

图 2 和图 3 表明,地形的扰动在山的迎风坡和背风坡产生槽,而在山的上空产生脊。图 2 和图 3 亦表明,山下游的槽线其位置与实况相当接近,美洲洛矶山下游的槽稍向东偏了一些。但是槽的强度与实况比较却相差很多。另外,我们注意到实况的槽,它的强度是北深南浅的,而现在计算出来的槽是北浅南深,这与实况是相反的。这个现象,亚洲海岸的大槽表现得更为明显。

相当有趣的是,西藏高原所引起的欧洲的槽,无论在强度上或走向上,它都与实况相当一致。

由数值计算西藏高原的扰动的绝对值要比洛矶山大得多。一般来说,前者要比后者约大 4 倍之多。由于这样,因此西藏高原和洛矶山的共同扰动①(图 4)在亚洲部分变化很少,而在美洲部分却有了改变,使美洲东海岸的槽更向东移了一些。另外,在高纬度我们算出了 4 个槽,比实况多了一个美洲西海岸的槽。我们认为如果在理论中计入了阻尼因子(摩擦作用),可以减少一些所指出的偏差,使结果更合理些。

上面已经指出,虽然我们取的是斜压槽式,然而槽脊线的位相是准垂直的。Smagorinsky[12]在热源

---

① 由于方程(18)和边界条件式(26)都是线性的,因此可以先分别计算西藏高原和洛矶山的影响,然后再把它们的结果叠加起来,看它们的共同影响。

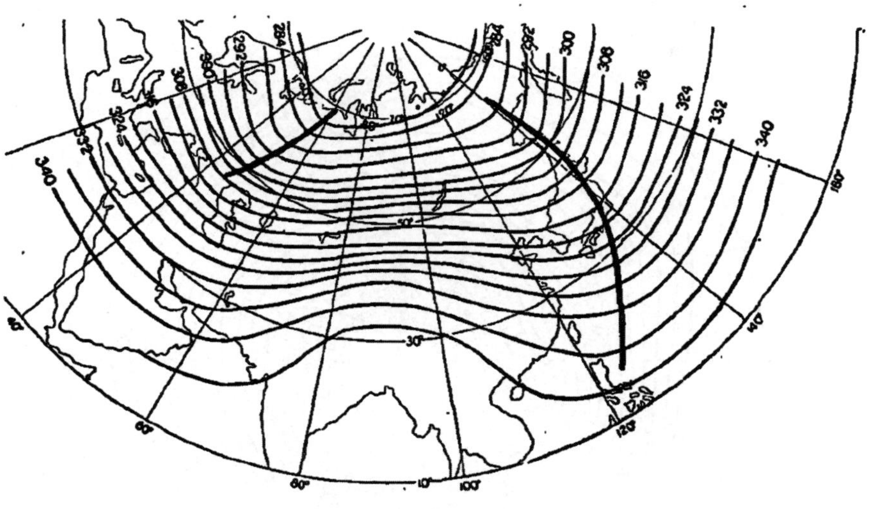

图 2　只考虑了西藏高原的扰动所计算出的 700 mbar 等压面形势图

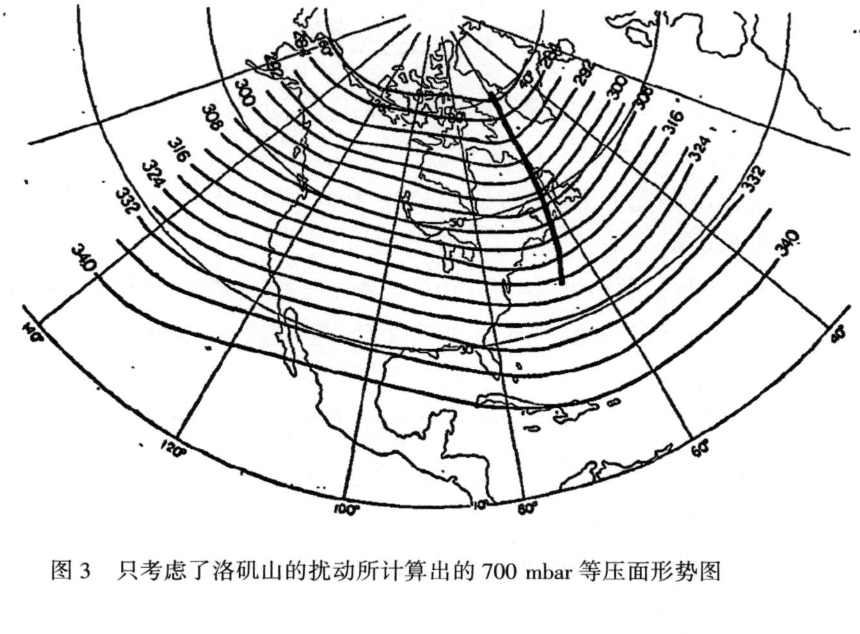

图 3　只考虑了洛矶山的扰动所计算出的 700 mbar 等压面形势图

扰动的研究中,当没有考虑摩擦时,也得到了槽是垂直的结果。很容易证明,摩擦作用对常定情况下槽脊线随高度的倾斜是起着重要的作用的。因此如何进一步在我们的模式中加入摩擦,这是一个很有意义的问题。

6　总结

根据以上各节讨论,本文指出以下几点。

(1)对于无黏性的、绝热的斜压大气的常定运动,与正压大气的常定运动一样,可以用一个 Helmholtz 方程来描写,然而在正压的情况下方程是二维的,考虑了大气的斜压性后,方程就变成三维了。

(2)大地形对西风带的影响,在斜压大气中决定于地面的基本气流和 Richardson 数 $R_i = \sigma / \bar{\dot{\lambda}}_p^2$。扰

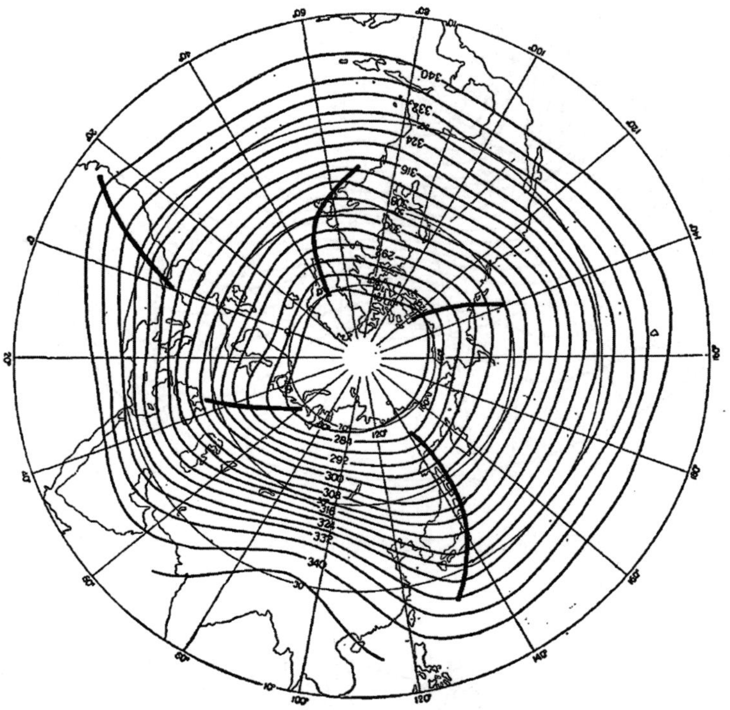

图 4 考虑了西藏高原和洛矶山共同扰动后所计算出的 700 mbar 等压面形势图

动的强度随高度的变化是复杂的，但向上递减很快。因此与 Bolin 相反，我们认为大地形对西风带的影响在对流层低层更为重要。

(3) 大地形对西风带扰动的结果，使得在山的上游和下游各产生一个槽，而在山的上空产生一个脊。槽、脊的位置一般来说与实况相当接近，但强度有一定的出入。因此我们认为在决定槽脊的强度上热源扰动起着重要的作用。

(4) 西藏高原(包括亚洲山系)对西风带的动力扰动的绝对值比洛矶山要大。

(5) 由于没有计入摩擦作用，槽的数目比实况多了一个，同时槽脊线的位相向上都是成垂直的。因此在讨论常定问题时，摩擦作用是相当重要的。

致谢：在本文工作过程中，叶笃正、顾震潮二位教授给予了热忱的指导与鼓励，文信和薛永瞻二位同志协助了数值计算，在此向他们表示诚挚的感谢。

参考文献

[1] 叶笃正．西藏高原对大气环流的季前变化．气象学报，1952，22：33-47.

[2] 顾震潮．西藏高原对东亚环流的动力影响和它的重要性．中国科学，1951，2：283-303.

[3] Charney, J. G. and A. Eliassen. A numerical method for predicting the pertur-bations of the middle latitute westerlies. *Tellus*, 1949, 1: 38-54.

[4] Bolin, B. On the influence of the earth's orography on the general character of westerlies. *Tellus*, 1950, 2: 184-195.

[5] Мусаелян Ш. А. ПространсТвонная аадача обтекання неровностей аомной ловерхностн сучетом сферичиостн вемли. ДАН СССР, 1955, 103: 815-818.

[6] 朱抱真．大尺度热源、热压和地形对西风带的常定扰动(二)．气象学报，1957，28：198-224.

[7] Murakami. The topographical effect upon stationary upper flow putterns. *Paper in Meteo. and Geophy.*,1956,7:69-89.

[8] Steward, H. J. A theory of the effect of obstacles on the waves in westerlies, *J. Meteo.*, 1948, 5:236-238.

[9] Gambo, K. The topographical effect upon the jet stream in the westerlies. *J. Meteo. Soc. Japan*, 1956,34:3.

[10] Charney, J. G. The dynamics of long waves in a baroclinic westerly current. *J. Meteo.*, 1947, 4:125-162.

[11] 叶笃正. 小地形对气流的影响(综合报告). 气象学报,1956, 27:243-262.

[12] Smagorinsky, J. The dynamical influence of large-scale heat sources and sinks on the quasi-stationary mean motion of atmosphere. *Q.J.Roy. Meteo. Soc.*, 1953, 79:342-366.

正压大气中的螺旋行星波*

巢纪平[1]　叶笃正[2]

（1. 中国科学院地理研究所；2. 中国科学院大气物理研究所）

提要：本文对天气图上常见的螺旋状的行星波（Rossby 波）的形式、发展以及其他方面的动力学性质做了理论分析，并讨论了它在维持大气环流中所起的作用。

1 引言

大气环流的基本状态是极地的纬向气流，并在纬向气流上叠加了种种不均匀的波动。Rossby[1]，最早研究了其中的行星波。行星波理论已成为近代天气预报、数值预报和大气环流研究的基础。30多年来，在动力气象学中行星波理论虽已有了重大的进展，例如关于它的能量频散过程[2]、在正压和斜压大气中的不稳定性[3,4]等，但一般研究它的都是对称的正弦波，即波的等位相线，或相当于天气图上观测到的槽、脊线，是沿经线方向的（即南北走向）。然而，在实际大气中的行星波波型一般在高纬度其槽线呈西北—东南走向，而在中、低纬度则自东北向西南倾斜，亦即波的等位相线具有螺旋状的结构。

正弦和螺旋行星波，不仅在几何形状上有所不同，主要的还在于它们的动力学性质是很不一样的。因此在影响天气的类型以及构成大气环流的很多现象中所起的作用也将会有不少差别。例如，为了维持大气环流的平均状态，需要有一种过程把低纬东风带从地球得到的角动量，在高空大量地向中、高纬输送，同时极地东风带从地球得到的少量的角动量也要输送到西风带去，总的来说，角动量的这种南北输送是依靠大型水平涡旋来完成的，这种大型水平涡旋就是观测到的行星波。Starr 曾指出（见参考文献[5]），为了能把低纬的角动量输送到较高的纬度，槽脊线需要从东北向西南倾斜，使在一个波长范围内，沿纬向风速 u 与经向风速 v 之间有净的正相关，即$\overline{uv}>0$，而纯粹对称的正弦波其 $\overline{uv}=0$，不能起到南北输送角动量的作用。叶笃正进而指出[5]，若从大气中的涡度平衡出发，要使中纬度西风得到加强，则不仅要槽线倾斜，还要槽线的倾斜度向北减小，为使极地东风带得到的少量角动量向南输送到西风带，则槽线需呈微弱的西北—东南向，因此槽线需呈上述的螺旋状。可见，正弦和螺旋行星波在维持大气环流中所起的作用是很不一样的。

一般来说，在一个基本流场中，一个扰动的位相速有两部分：一是基本流速的本身；二是由动力学制约引起的附加项。如基本场是绕一极点旋转的，而其角速 Ω 又是随 r（以极点为中心）增加或减少的函数，则其上的乱动将是螺旋状。这种现象大至银河系的旋臂[6,7]，小至台风中的雨带无不如此。本文将用一简单的正压模式，对大气中螺旋状行星波的形成、发展和能量的频散等性质在理论上给以

* 大气科学，1977，1(2)：81-88.

初步的分析。

2 基本方程和螺旋波的表达

取原点在极地的极坐标系(r,θ)，r 指向低纬为正，θ 逆时针为正，相应的空气运动速度为(u,v)。在这个坐标系中，旋转地球大气的运动方程和连续性方程分别为

$$\frac{\partial u}{\partial t}+u\frac{\partial u}{\partial r}+\left(\frac{u}{r}\right)\frac{\partial u}{\partial \theta}-\frac{v^2}{r}-fv=-\frac{\partial}{\partial r}\left(\frac{p}{\rho}\right) \tag{1}$$

$$\frac{\partial v}{\partial t}+u\frac{\partial v}{\partial r}+\left(\frac{v}{r}\right)\frac{\partial v}{\partial \theta}+\frac{uv}{r}+fu=-\frac{\partial}{r\partial \theta}\left(\frac{p}{\rho}\right) \tag{2}$$

$$\frac{\partial ru}{\partial r}+\frac{\partial v}{\partial \theta}=0 \tag{3}$$

式中，p、ρ 分别为空气的压力和密度，f 为柯里奥利参数。

设大气运动的基本状态为梯度风平衡，即

$$\Omega^2 r+f\Omega r=\frac{\partial}{\partial r}\left(\frac{P}{\rho}\right) \tag{4}$$

其中 Ω 为基本气流的旋转角速度。在这基本环流上叠加了一小扰动，即

$$v=\Omega(r)\cdot r+v',\quad u=u',\quad p=P(r)+p' \tag{5}$$

线性化方程组为(略去“,”号)

$$\frac{\partial u}{\partial t}+\Omega\frac{\partial u}{\partial \theta}-(2\Omega+f)v=-\frac{\partial}{\partial r}\left(\frac{p}{\rho}\right) \tag{6}$$

$$\frac{\partial v}{\partial t}+\Omega\frac{\partial v}{\partial \theta}+\left(\frac{K^2}{2\Omega}+f\right)u=-\frac{\partial}{r\partial \theta}\left(\frac{p}{\rho}\right) \tag{7}$$

$$\frac{\partial ru}{\partial r}+\frac{\partial v}{\partial \theta}=0 \tag{8}$$

式中

$$K^2=(2\Omega)^2\left(1+\frac{r}{2\Omega}\frac{\mathrm{d}\Omega}{\mathrm{d}r}\right)$$

而

$$\frac{K^2}{2\Omega}\equiv\left(2\Omega+r\frac{\mathrm{d}\Omega}{\mathrm{d}r}\right)=\frac{1}{r}\frac{\mathrm{d}\Omega r^2}{\mathrm{d}r}=\frac{1}{r}\frac{\mathrm{d}M}{\mathrm{d}r}$$

其中 M 为基本气流的相对角动量。因此

$$\frac{K^2}{2\Omega}+f\equiv\zeta_a$$

为基本气流的绝对涡度。

由方程组式(6)至式(8)得到涡度方程

$$\frac{\partial r\zeta}{\partial r}+\Omega\frac{\partial r\zeta}{\partial \theta}+\frac{\mathrm{d}\zeta_a}{\mathrm{d}r}(ru)=0 \tag{9}$$

式中

$$\zeta\equiv\frac{1}{r}\left(\frac{\partial rv}{\mathrm{d}r}-\frac{\partial u}{\partial \theta}\right) \tag{10}$$

为扰动运动的相对涡度。引进流函数 ψ,定义成

$$v = \frac{\partial \psi}{\partial r}, u = -\frac{1}{r}\frac{\partial \phi}{\partial \theta} \tag{11}$$

由此

$$\zeta = \frac{1}{r}\frac{\partial}{\partial r}\left(r\frac{\partial \psi}{\partial r}\right) + \frac{\partial^2 \psi}{r^2 \partial \theta^2} \tag{12}$$

于是方程(9)为

$$\left(\frac{\partial}{\partial t} + \Omega\frac{\partial}{\partial \theta}\right)\left[\frac{\partial}{\partial r}\left(r\frac{\partial \psi}{\partial r}\right) + \frac{\partial^2 \psi}{r\partial \theta^2}\right] - \frac{\mathrm{d}\zeta_a}{\mathrm{d}r}\frac{\partial \psi}{\partial \theta} = 0 \tag{13}$$

或者,取对数坐标

$$\xi = \ln(r/r_0) \tag{14}$$

上面的方程为

$$\left(\frac{\partial}{\partial t} + \Omega\frac{\partial}{\partial \theta}\right)\left(\frac{\partial^2 \psi}{\partial \xi^2} + \frac{\partial^2 \psi}{\partial \theta^2}\right) - \frac{\mathrm{d}\zeta_a}{\mathrm{d}\xi}\frac{\partial \psi}{\partial \theta} = 0 \tag{15}$$

这是我们要研究的基本方程。

取方程(15)的简谐波解

$$\psi \sim e^{i\varphi} \tag{16}$$

式中位相函数

$$\varphi = \omega t + k\xi - m\theta \tag{17}$$

并且

$$\omega = \frac{\partial \varphi}{\partial t}, \quad k = \frac{\partial \varphi}{\partial \xi}. \tag{18}$$

显然,由 φ=常数,得到

$$\theta = \frac{1}{m}(\omega t + k\xi) + 常数 \tag{19}$$

这表明,于对任何指定时刻 t,当频率 ω 固定时,有

$$\theta = \frac{1}{m}k\ln r + C(t) \tag{20}$$

由此可见,等位相线(例如槽、脊线)是一簇对数螺旋曲线,并且当 $k>0$ 时,θ 将随 ξ 的增加而增加,此为导式(leading)对数螺旋,亦即槽、脊线自西北伸向东南;当 $k<0$ 时,θ 将随 ξ 的增加而减小,此为曳式(trailing)对数螺旋,亦即槽脊线自东北伸向西南。因此形为式(16)的解表征了一类螺旋波。

引进特征量 Ω^*,ψ^*,f^*,相应的无量纲量为

$$\tilde{t} = \Omega^* t, \quad \tilde{\psi} = \psi/\psi^*, \quad \tilde{\zeta}_a = \zeta_a/f^*, \quad \tilde{\Omega} = \Omega/\Omega^* \tag{21}$$

基本方程(15)的无量纲形式为(略去“~”号)

$$\left(\frac{\partial}{\partial t} + \Omega\frac{\partial}{\partial \theta}\right)\left(\frac{\partial^2 \psi}{\partial \xi^2} + \frac{\partial^2 \psi}{\partial \theta^2}\right) - \frac{f^*}{\Omega^*}\frac{\mathrm{d}\zeta_a}{\mathrm{d}\xi}\frac{\partial \psi}{\partial \theta} = 0 \tag{22}$$

现设波振幅是时间和经向距离的缓变函数,则形为式(16)的解可以写成

$$\psi = \Psi(T,R)e^{i\varphi} \tag{23}$$

式中

$$T = \varepsilon t, \quad R = \varepsilon \xi \tag{24}$$

$\varepsilon = \Omega^* / f^* \ll 1$，为一小参数。假设 ω, k 也是 R, T 的缓变函数，则式(22)可写成

$$\left(i(\omega - m\Omega) + \varepsilon \frac{\partial}{\partial T}\right)\left[-(k^2 + m^2)\Psi + \varepsilon i\left(2k\frac{\partial \Psi}{\partial R} + \frac{\partial k}{\partial R}\Psi\right) + \varepsilon^2 \frac{\partial^2 \Psi}{\partial R^2}\right] + \mathrm{i}m\frac{\mathrm{d}\zeta_a}{\mathrm{d}R}\Psi = 0 \tag{25}$$

这是当解取形式(23)时，所要研究的基本方程。

在以下的分析中，并将解按 ε 的幂次展开为

$$\Psi = \Psi_0 + \varepsilon \Psi_1 + \cdots \tag{26}$$

3　频散关系和波振幅方程

将式(26)代入方程(25)得到零级近似方程为

$$\mathrm{i}\left[-(\omega - m\Omega)(k^2 + m^2) + m\frac{\mathrm{d}\zeta_a}{\mathrm{d}R}\right]\Psi_0 = 0 \tag{27}$$

由于 $\Psi_0 \neq 0$，所以有频散关系

$$\omega = m\Omega + \frac{m}{k^2 + m^2}\frac{\mathrm{d}\zeta_a}{\mathrm{d}R} \tag{28}$$

或者令

$$\Omega_p = \frac{\omega}{m} \tag{29}$$

Ω_p 为沿 θ 方向波的相速度，由此有

$$\Omega_p = \Omega + \frac{1}{k^2 + m^2}\frac{\mathrm{d}\zeta_a}{\mathrm{d}R} \tag{30}$$

这是螺旋行星波的位相速公式，也就是在所取的基本气流中的 Rossby 波速公式。

沿经向的群速度为

$$C_g = -\frac{\mathrm{d}\omega}{\mathrm{d}k} \tag{31}$$

正值指向低纬。由式(30)得到

$$C_g = \frac{2mk}{(k^2 + m^2)^2}\frac{\mathrm{d}\zeta_a}{\mathrm{d}R} \tag{32}$$

一级近似方程为

$$\mathrm{i}\left[-(\omega - m\Omega)(k^2 + m^2) + m\frac{\mathrm{d}\zeta_a}{\mathrm{d}R}\right]\Psi_1 - \left[\frac{\partial}{\partial T}((k^2 + m^2)\Psi_0) + (\omega - m\Omega)\left(2k\frac{\partial \Psi_0}{\partial R} + \frac{\partial k}{\partial R}\Psi_0\right)\right] = 0 \tag{33}$$

考虑到式(28)和式(32)，得到

$$\frac{\partial \Psi_0}{\partial T} + C_g\frac{\partial \Psi_0}{\partial R} = -\frac{1}{2}C_g\frac{\partial \ln k}{\partial R}\Psi_0 - \frac{\partial \ln(k^2 + m^2)}{\partial T}\Psi_0 \tag{34}$$

或者乘以 Ψ_0 后，有

$$\frac{\partial \Psi_0^2}{\partial T} + C_g\frac{\partial \Psi_0^2}{\partial R} = -C_g\frac{\partial \ln k}{\partial R}\Psi_0^2 - \frac{\partial \ln(k^2 + m^2)}{\partial T}\Psi_0^2 \tag{35}$$

由于波动的能量正比于波的振幅平方，即 $E \propto \Psi_0^2$，所以这是波能量变化方程。

将方程(35)改写成

$$\frac{\partial \ln(k\Psi_0^2)}{\partial T}+C_g\frac{\partial \ln(k\Psi_0^2)}{\partial R}=-\frac{\partial \ln}{\partial T}\left(\frac{(k^2+m^2)^2}{k}\right) \tag{36}$$

也可以写成

$$\frac{\partial(k\Psi_0^2)}{\partial T}+C_g\frac{\partial(k\Psi_0^2)}{\partial R}=-\left(\frac{k}{k^2+m^2}\right)^2\frac{\partial}{\partial T}\left(\frac{(k^2+m^2)^2}{k}\right)\Psi_0^2 \tag{37}$$

量 $k\Psi_0^2$ 可以解释成单位波长中的能量。可见当波列以群速度沿经向方向传播时,其单位波长中能量所以有变化,是由于经向波数和纬向波数随时间有改变而引起。

在大气中,$|k|$平均在2附近,m 在6附近,如果 k 和 m 的时间变化很小,那么方程(37)右端可以略去,在这种情况下简化成

$$\frac{\partial(k\Psi_0^2)}{\partial T}+C_g\frac{\partial(k\Psi_0^2)}{\partial R}=0 \tag{38}$$

这表明波列以群速度传播时,单位波长中的能量近似地守恒。

在定常情况下,由式(38)进而有

$$k\Psi_0^2=\text{常数} \tag{39}$$

即单位波长中的能量将不随纬度而改变。

4 结果讨论

根据上面的理论分析,可以作出如下几点讨论。

(1)位相速和波形

在绕极传播的波列中,等位相线(例如槽、脊线)将以式(30)的相速度移动,这样等位相线的形状将随时间变化,并趋向位相速随纬度的分布,因此,由位相速随纬度的分布可以间接地推测等位相线的形状。

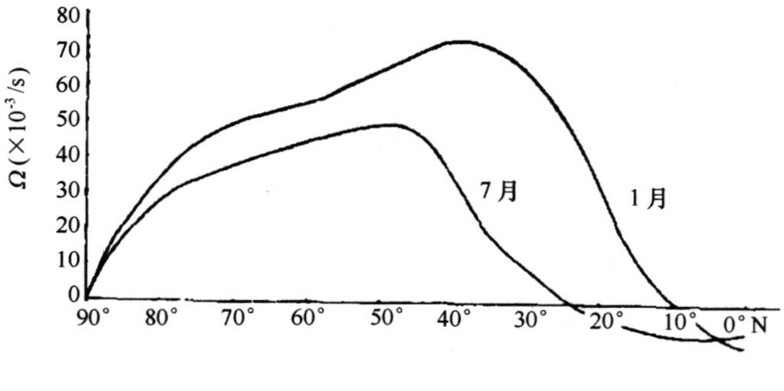

图1 500 mbar Ω 随纬度分布

图1是500 mbar 1月和7月 Ω 随纬度的分布,再考虑 f 随纬度的变化后,可以算出绝对涡度梯度 $d\zeta_a/d\xi$ 的分布,计算表明,绝对涡度梯度在各个纬度皆取负值(图2)。这样由式(30)可见,局地的位相速皆小于同纬度气流的角速度。图3a和图3b分别为$|k|$平均取1、m 取3(图3a)和$|k|$平均取2、m 取6(图3b)情况下1月的位相速。由此推测出在高纬度是导波,在较低纬度是曳波。1月西风角速度极值在40°N附近,在$|k|=1$、$m=3$ 时,导波和曳波的分界线北移到75°N左右,并在50°N以南,波将自东向西传播。在$|k|=2$、$m=6$ 时,在60°N以北是导波,以南是曳波,在38°N以南,波将自东向西传播。

图 4 是 7 月的情况，这时西风角速度在 47°N 左右最大，当 $|k|=1, m=3$ 时，在 75°N 以北是导波，即槽、脊线由西北指向东南，以南是曳波即槽、脊线由东北指向西南；当 $|k|=2, m=6$ 时，导波和曳波的分界纬度在 65°N 附近。

除负位相速（自东向西传播的后退波）较偏北外，计算所得的波形和天气图上观测到的槽、脊线的走向是接近的。

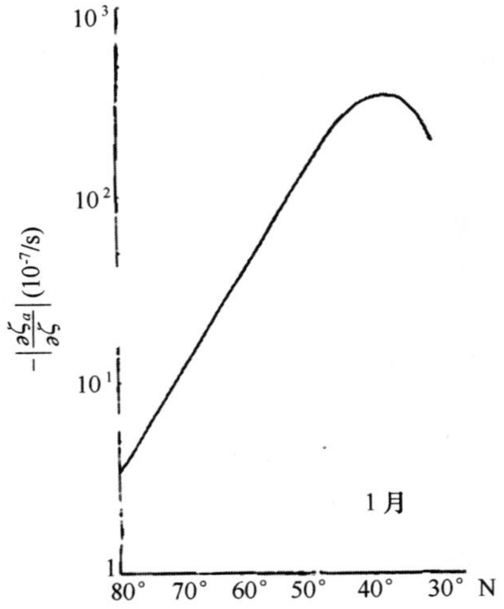

图 2a　500 mbar 1 月 $-\left|\frac{\partial \zeta_a}{\partial \xi}\right|$

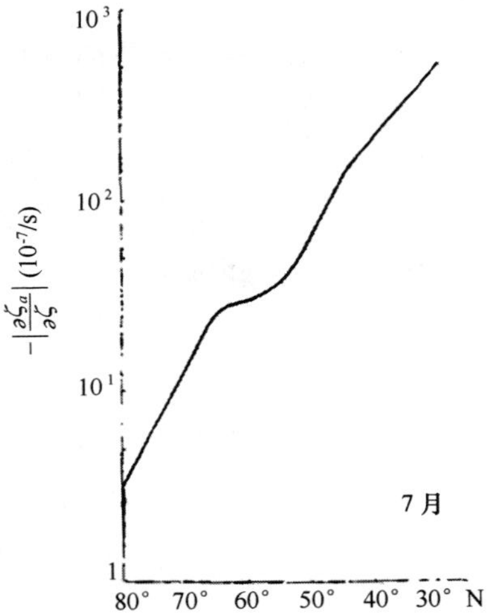

图 2b　500 mbar 7 月 $-\left|\frac{\partial \zeta_a}{\partial \xi}\right|$

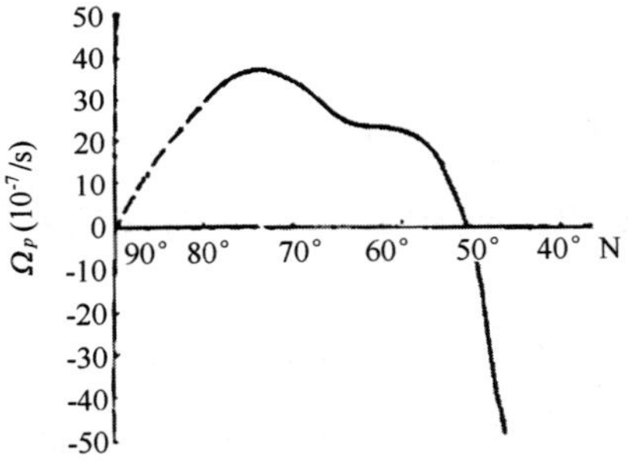

图 3a　500 mbar 1 月 Ω_p($m=3, |k|=1$)

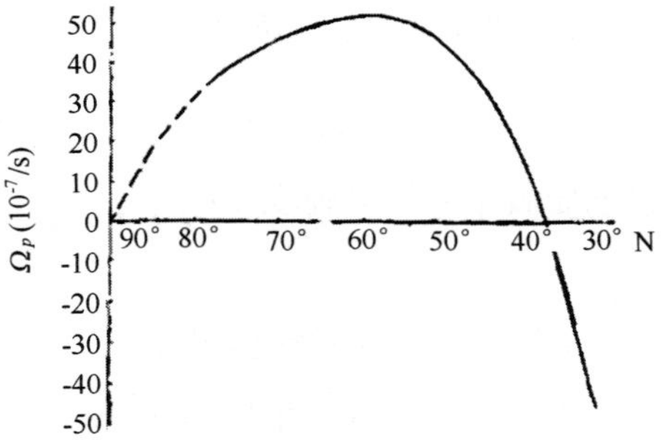

图 3b　500 mbar 1 月 Ω_p($m=6, |k|=2$)

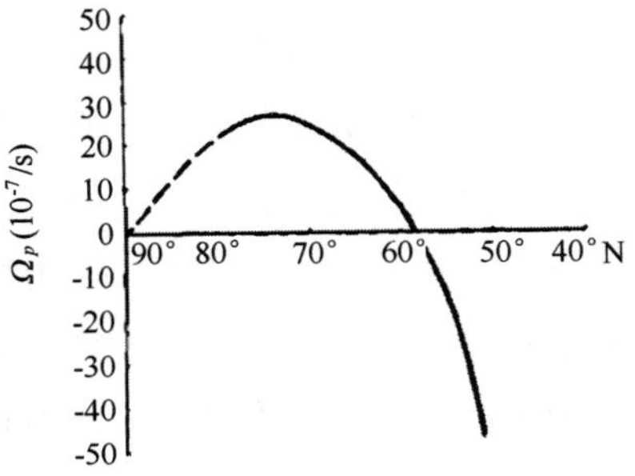

图 4a　500 mbar 7 月 Ω_p($m=3, |k|=1$)

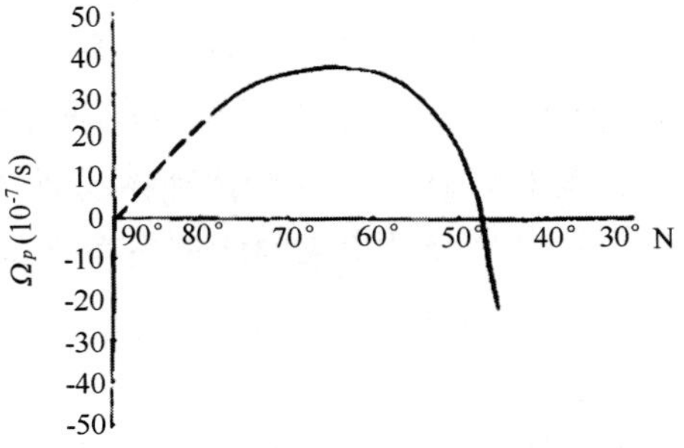

图 4b　500 mbar 7 月 Ω_p($m=6, |k|=2$)

(2)群速度和波能量的径向传播

由上面的观测资料,计算出在各个纬度都有 $d\zeta_a/d\xi<0$,由式(32)可见,对导式波,能量将从低纬传向高纬,对曳式波,能量将从高纬传向低纬。由上面的计算,在中、高纬是导波,在中、低纬是曳波,而中纬度是斜压性最强的地区,由于斜压不稳定,背景场中储藏的部分内能将通过斜压扰动的发展而被释放出来,这部分释放出来的能量将通过这类螺旋行星波传输到高纬和低纬去。

(3)螺旋波的维持和发展

在第1点中讨论了在较差转动气流中螺旋波的形成,现进一步讨论维持它的能源。

将方程(38)改写成

$$\frac{\partial(k\Psi_0^2)}{\partial T}+\frac{\partial C_g(k\Psi_0^2)}{\partial R}=k\frac{\partial C_g}{\partial R}\Psi_0^2 \tag{40}$$

设在区域的边界 $R=R_1$ 和 $R=R_2$ 上,$\Psi_0=0$,积分上式得

$$\frac{\partial}{\partial T}\int_{R_1}^{R_2}(k\Phi_0^2)\,dR=\int_{R_1}^{R_2}\frac{\partial C_g}{\partial R}(k\Psi_0^2)\,dR \tag{41}$$

如果在同一类波形的区域中,波数$|k|$的经向变化不大,则群速度的变化主要决定于 $d\zeta_a/dR$ 的分布。由图2,由于 $d\zeta_a/dR<0$,并愈向低纬其绝对值愈大,所以曳波的群速度为正(指向低纬)同时是辐散的,即 $\partial C_g/\partial R<0$,因此如果在整个区域中都为曳波,则由方程(41)得

$$\frac{\partial}{\partial T}\int_{R_1}^{R_2}|k|\ \Psi_0^2 dR>0$$

这表明在整个区域中单位波长中的波能量要增加,也即这时曳波可以得到维持和发展。由于群速度辐散是绝对涡度梯度在经向方向的不均匀性造成,所以波发展的能量主要来自基本气流。

反之,对导波,当 $d\zeta_a/dR<0$ 时,$C_g<0$,但后者的绝对值也是辐散的,因此有

$$\frac{\partial}{\partial T}\int_{R_2}^{R_2}|k|\ \Psi_0^2 dR<0$$

这表明单纯的导波不能维持。实际上,在天气图上,至少在长时间的平均图上没有观测到有大范围的导式波动。

由于整体的曳式波动在区域中的总能量要不断增加,因此要出现一个准平衡状态的波形,必然要在区域中导式和曳式部分兼有,这就是大气中常见的在极区高纬度为导式而在中高纬和低纬为曳式的波形。

(4)大气环流维持的物理图案

根据以上的讨论,可以绘出维持大气环流的一个简单的物理图案。当旋转的地球表面上接受到不均匀分布的太阳辐射后,在辐射平衡条件下形成了纬向的东、西风带。由于斜压不稳定,中纬度西风带中蕴藏的内能将通过扰动的发展而释放出来,由于基本气流角速度的分布和科里奥利参数随纬度的变化(球面性作用),这些扰动将逐渐演化成导式和曳式的螺旋行星波。通过这类行星波的波动过程,将把中纬度释放出来的能量向高纬和低纬输送。由于能量的不断释放和向外输送,中纬度的西风强度将解弱。但在另一方面,通过极区为导式和以南为曳式的扰动气流,又要将低纬东风带和极地东风带从地球得到的角动量带到中纬去,并使西风加强。这几种过程相互制约的结果,就出现了长时期所观测到的准平衡态的大气环流图式。

(5)推论

从观测看,旋转银河系中有螺旋状的星系臂,旋转地球大气中有螺旋状的行星波,旋转的台风中

也有螺旋状的雨带。这三者的空间和时间尺度相差非常大,其中运动的动力学也有巨大的区别。但三者有一个共同特点,即其基本流场都是围绕一点旋转的,而其角速度 Ω 又是 r 的函数(较差转动),这就使得它们都有螺旋状的扰动。由此推理,其他旋转的系统(如中纬度的气旋和雷暴以及龙卷等)中,也将存在螺旋状的扰动,如果它们的位相速随 r 增加或减小的话,随 r 增加将成导式,随 r 减小则呈曳式。

参考文献

[1] C.G.Rossby. *J.Marine. Res.* 1939,2:38-55.

[2] T.C.Ych. (叶笃正). *J. Met*. 1949, 6:1-16.

[3] H.L.Kuo. (郭晓岚). *J. Met*. 1949, 6:105-122.

[4] J.G.Charney. *J. Met*. 1947, 4, 5, pp. 135-162.

[5] 叶笃正,朱抱真. 大气环流的若干基本问题. 科学出版社,1958.

[6] C.C.Lin, H.H.Shu. *Astrophys. J.* 1964, 140:646-655.

[7] 解伯民,巢纪平. 旋涡星系密度波的演化. 1976.

二维能量平衡模式中极冰-反照率的反馈对气候的影响*

巢纪平　陈英仪

(中国科学院大气物理研究所)

摘要:本文应用一个二维能量平衡气候模式,研究了 Budyko,Sellers 提出的极冰-温度-反照率的反馈对气候的影响问题。如将该模式简化,则可变成 Sellesr 型的一维能量平衡模式。二维模式的计算表明,欲使极冰界线从现在的 72°N 南移到出现冰河气候的 50°N,太阳常数要比现在值减小 15%左右,而不像一维模式研究所指出的,只要太阳常数减小 2%左右。根据计算表明,气候不会因太阳辐射能量稍一变小而急剧地趋于"恶化"。

极冰的存在将加大地表反照率,使低层空气接受的太阳辐射量减小,从而降低地表的温度。但极冰的生消又取决于温度的高低(例如极冰的冰界温度一般为-10℃),因此通过反照率的变化,极冰与温度之间存在一个正反馈过程,这个过程对气候的形成和变迁都有重要的作用。

这方面的定量理论,开始于 Budyko[1] 和 Sellers[2]。他们设计了一个纬圈平均的、垂直积分的地气系统能量平衡模式,其中包括了辐射能量平衡和热量沿纬向的涡旋输送过程,而更主要的是冰面和陆面的反照率不同。他们除算出了在现在的太阳常数下模式的温度纬向分布与现在气候的实况一致外,同时发现,只要太阳常数稍一减小(如比现在值小 2%),冰界将从现在的 72°N 南移到 50°N 左右,即出现冰河期的寒冷气候。

这一理论提出后,立即引起大家的注意。虽然近几年来相继发表了不少研究报告[3-13],并进一步讨论了解的多值性和稳定性,但太阳辐射能量稍一减小气候将趋于恶化这一基本结论不变。

气候是否会因太阳辐射微小的变化而变得如此的坏(寒冷),这不仅是一个具有理论意义的科学问题,而且也是一个对未来人类活动具有实际意义的重大问题,因此需要进一步仔细研究。注意到,Budyko,Sellers 的模式不仅是沿纬圈平均了的,也是在垂直方向积分了的,这虽然是描写地球大气气候状态的一个初步近似,但这样一来发生在实际大气中的很多物理过程就被平滑掉了。例如辐射能的垂直传输和热量的垂直交换过程就不再能反映,而这些过程对于调节地球表面的温度场显然是重要的。在本文中将设计一个纬圈平均了的但具有垂直结构的二维能量平衡模式,重新研究 Budyko 和 Sellers 提出的问题。

1 理论模式

在辐射能量和湍流热量输送相平衡下,有下面的基本方程

* 中国科学,1979,12:1199-1207.

$$\frac{K}{a}\frac{\partial}{\partial x}(1-x^2)\frac{\partial T}{\partial x}+\frac{\partial}{\partial z}\left(k_t\frac{\partial T}{\partial z}\right)+\sum_j\alpha'_j\rho_c(A_j+B_j-2E_j)+\alpha''\rho_cQ=0 \tag{1}$$

$$\frac{\partial A_j}{\partial z}=\alpha'_j\rho_c(A_j-E_j) \tag{2}$$

$$\frac{\partial B_j}{\partial z}=\alpha'_j\rho_c(E_j-E_j) \tag{3}$$

$$\frac{\partial Q}{\partial z}=\alpha''\rho_cQ \tag{4}$$

式中，T 为空气温度；α'_j 和 α''分别为对波长为 λ_j 的长波辐射吸收系数以及对太阳辐射的平均吸收系数；A_j，B_j 分别为在波长 $\Delta\lambda_j$ 区间内向下和向上的长波辐射通量；Q 为太阳辐射通量；E_j 为在 $\Delta\lambda_j$ 内的黑体辐射能量；ρ_c 是吸收介质的密度；K，k_t 分别为水平和垂直湍流交换系数；a 为地球半径；$\sum$是对整个吸收谱范围内的求和符号；z 为垂直坐标；$x=\sin\theta$，θ 为纬度。

由于方程(1)要求对整个吸收谱求积分，这不仅计算量繁重，而且也不便在理论上做解析处理。郭晓岚[14]曾提出一个计算长波辐射通量传递的简化方案。将体积吸收系数按 $\alpha_j\rho_c\gtreqless\mathrm{d}/\mathrm{d}z$，把介质的吸收谱分成强(用 s 表示)、弱(用 w 表示)两个吸收区，在这两个区域中分别定义平均吸收系数，为

$$\alpha_{s,w}=\frac{\int_{s,w}\alpha_\lambda E_\lambda\mathrm{d}\lambda}{E_{s,w}}=\frac{\int_{s,w}\alpha_\lambda G_\lambda\mathrm{d}\lambda}{G_{s,w}} \tag{5}$$

式中

$$E_{s,w}=\int_{s,w}E_\lambda\mathrm{d}\lambda,\quad G_{s,w}=\int_{s,w}G_\lambda\mathrm{d}\lambda,\quad G_{s,w}=A_{s,w}+B_{s,w}$$

同时引进

$$E_s=rE,\quad E_w=(1-r)E,\quad E=\sigma T^4 \tag{6}$$

r 表示在强吸收区中物质的辐射能量占总辐射能量的部分，σ 为 Stefan-Boltzmann 常数。

由方程(2)，方程(3)得到

$$\left(\frac{\partial^2}{\partial z^2}-\alpha_{s,w}^2\rho_c^2\right)G_{s,w}=-2\alpha_{s,w}^2\rho_c^2E_{s,w} \tag{7}$$

对方程(1)进行适当的运算并考虑到 $\alpha_s^2\rho_c^2\gg\mathrm{d}^2/\mathrm{d}z^2$ 而略去后者，$\alpha_w^2\rho_c^2\ll\mathrm{d}^2/\mathrm{d}z^2$ 而略去前者，于是得到[14]

$$\frac{\partial^2}{\partial z^2}\left[\frac{K}{a^2}\frac{\partial}{\partial x}(1-x^2)\frac{\partial T}{\partial x}+\frac{\partial}{\partial z}\left(k_t\frac{\partial T}{\partial z}\right)+\alpha''\rho_cQ\right]=2\left[(1-r)\alpha_w\rho_c-\frac{r}{\alpha_s\rho_c}\frac{\partial^2}{\partial z^2}\right]\frac{\partial^2E}{\partial z^2}$$

积分两次得到

$$\frac{K}{a^2}\frac{\partial}{\partial x}(1-x^2)\frac{\partial T}{\partial x}+\frac{\partial}{\partial z}\left(k_t\frac{\partial T}{\partial z}\right)+\alpha''\rho_cQ=2\left[(1-r)\alpha_w\rho_cE-\frac{r}{\alpha_s\rho_c}\frac{\partial^2E}{\partial z^2}\right]+C_0+C_1z \tag{8}$$

考虑到 $z\to\infty$ 时，各物理量均为有限，因此 $C_1=0$

近似地取

$$\frac{\partial T}{\partial z}\approx\frac{1}{4\sigma\overline{T}^3}\frac{\partial E}{\partial z}$$

并引进光学厚度

$$\xi = \frac{\alpha''}{\alpha_s \xi_0}\int_z^{\infty} \alpha_s \rho_c \mathrm{d}z, \quad \xi_0 = \frac{\alpha''}{\alpha_s}\int_0^{\infty} \alpha_s \rho_c \mathrm{d}z \tag{9}$$

于是式(8)为：

$$D\frac{\partial}{\partial x}(1 - x^2)\frac{\partial E}{\partial x} + \frac{\partial}{\partial \xi}(k_t + k_r)\frac{\partial E}{\partial \xi} - N^2 E = -\widetilde{S}\xi_0 Q + C \tag{10}$$

而式(4)为：

$$\frac{\partial Q}{\partial \xi} = -\xi_0 Q \tag{11}$$

在式(10)中，

$$D = \frac{\xi_0^2 K}{(\alpha''\rho_c)^2 a^2}, \quad k_r = \frac{8r\sigma \overline{T}^3}{\alpha_s \rho_c}, \quad \widetilde{S} = \frac{4\xi_0 \sigma \overline{T}^3}{\alpha''\rho_c},$$

$$N^2 = \frac{8(1 - r)\alpha_w \rho_c \xi_0^2 \sigma \overline{T}^3}{(\alpha''\rho_c)^2}$$

与 k_t 相比较，k_r 可以称为辐射交换系数，而 N^2 则为牛顿辐射冷却系数。另外，由式(11)

$$Q = Q_0(x)\mathrm{e}^{-\xi_0 \xi} \tag{12}$$

式中，Q_0 为大气上界净的太阳辐射通量。

现在来给出问题的边界条件。

1)在大气上界 $\xi=0$，假定

(1)没有向下的长波辐射通量，即

$$A_s = A_w = 0 \tag{13}$$

(2)没有湍流过程，即成立辐射平衡关系，于是由式(1)及式(13)得到

$$\alpha_s \rho_c (B_{s_0} - 2E_{s_0}) + \alpha_w \rho_c (B_{w_0} - 2E_{w_0}) + \alpha''\rho_c Q_0 = 0 \tag{14}$$

由式(7)，根据简化约定，再考虑到式(13)后，有

$$B_{s_0} = 2E_{s_0} = 2rE_0 \tag{15}$$

由式(14)立即得到

$$B_{w_0} = 2(1 - r)E_0 - \frac{\alpha''}{\alpha_w}Q_0 \tag{16}$$

(3)进入大气层净的太阳辐射通量与外逸到太空的长波辐射通量，在全球应达到平衡，即

$$\int_0^1 (B_{s_0} + B_{w_0})\mathrm{d}x = \int_0^1 Q_0(x)\mathrm{d}x \tag{17}$$

按定义

$$\int_0^1 Q_0(x)\mathrm{d}x = \overline{Q}_0$$

如设 $Q_0(x) = \overline{Q}_0 S(x)$，则有

$$\int_0^1 S(x)\mathrm{d}x = 1$$

因此在考虑了式(15)，式(16)后，式(17)为：

$$\xi = 0, \quad \int_0^1 E_0 \mathrm{d}x = \frac{1}{2}\left(1 + \frac{\alpha''}{\alpha_w}\right)\overline{Q}_0 \tag{18}$$

这是我们所需要的一个上界条件。

(4)为一等温层,即

$$\xi = 0, \quad \frac{\partial E}{\partial \xi} = 0 \tag{19}$$

2)在地球表面$\xi = 1$,有热量平衡条件

$$A + (1 - \Gamma) Q = B - k_t \frac{\partial T}{\partial z} \tag{20}$$

式中,Γ为地表的反照率。或改写成

$$A - B = \frac{k_t \alpha'' \rho_c}{4\xi_0 \sigma \bar{T}^3}\left(\frac{\partial E}{\partial \xi}\right) - (1 - \Gamma)\bar{Q}_0 \mathrm{e}^{\bar{\xi}_0} S(x) \tag{21}$$

在强吸收区中,由式(2),式(3)得到

$$A_s - B_s = -\frac{\alpha''}{\alpha_s \xi_0}\frac{\partial G_s}{\partial \xi} \tag{22}$$

按简化约定,由式(7)近似地得到$G_s = 2E_s$,所以上式为:

$$A_s - B_s = -\frac{2r\alpha''}{\alpha_s \xi_0}\frac{\partial E}{\partial \xi} \tag{23}$$

另一方面,在弱吸收区中,由式(7)近似地有

$$\frac{\partial^2 G_w}{\partial \xi^2} = -2(1 - r)\frac{\xi_0^2 \alpha_w^2}{\alpha''^2}E \tag{24}$$

而由式(2)和式(3)得到

$$\frac{\partial G_w}{\partial z} = \alpha_w \rho_c (A_w - B_w) \tag{25}$$

考虑到式(24)和式(16)后容易得到

$$\xi = 1, \quad A_w - B_w = 2(1 - r)\frac{\xi_0 \alpha_w}{\alpha''}\int_0^1 E \mathrm{d}\xi - 2(1 - r)E_0 + \frac{\alpha''}{\alpha_w}Q_0 \tag{26}$$

与式(23)相加,最后得到

$$\xi = 1, \quad A - B = -\frac{2r\alpha''}{\xi_0 \alpha_s}\frac{\partial E}{\partial \xi} + 2(1 - r)\frac{\xi_0 \alpha_w}{\alpha''}\int_0^1 E \mathrm{d}\xi - 2(1 - r)E_0 + \frac{\alpha''}{\alpha_w}Q_0 \tag{27}$$

代入式(21),得到所需的另一个边界条件,为

$$\xi = 1, (k_t + k_r)\frac{\partial E}{\partial \xi} - N^2 \int_0^1 E \mathrm{d}\xi = -\frac{\alpha'' N^2}{\alpha_w \xi_0}E_0 + \widetilde{S}\left[\frac{\alpha''}{\alpha_w} + (1 - \Gamma)\mathrm{e}^{-\xi_0}\right]Q_0 \tag{28}$$

另外,关于侧向条件取成

$$x = 0,1; \quad (1 - x^2)^{\frac{1}{2}}\frac{\partial E}{\partial x} = 0 \tag{29}$$

现在来定方程(10)中的积分常数C。取全球积分平均,应用条件式(19),条件式(28)和式(29)后,得到

$$C = \widetilde{S}Q_0\left[r\left(1 + \frac{\alpha''}{\alpha_w}\right) - \mathrm{e}^{-\xi_0}\int_0^1 \Gamma S(x)\mathrm{d}x\right] \tag{30}$$

最后理论模式归结为:

$$D\frac{\partial}{\partial x}(1-x^2)\frac{\partial E}{\partial x}+\frac{\partial}{\partial \xi}(k_t+k_r)\frac{\partial E}{\partial \xi}-N^2E=-\widetilde{S}\overline{Q}_0\xi_0 e^{-\xi_0\xi}S(x)+\widetilde{S}\overline{Q}_0\left[r\left(1+\frac{\alpha''}{\alpha_w}\right)-e^{-\xi_0}\int_0^1 \Gamma S(x)\mathrm{d}x\right] \tag{31}$$

$$\xi=0,\quad \int_0^1 E_0(x,\mathrm{d}x=\frac{1}{2}\left(1+\frac{\alpha''}{\alpha_w}\right)\overline{Q}_0, \tag{32}$$

$$\xi=1,\quad (k_t+k_r)\frac{\partial E}{\partial \xi}-N^2\int_0^1 E\mathrm{d}\xi=-\frac{\alpha''N^2}{\alpha_w\xi_0}E_0+\widetilde{S}\overline{Q}_0\left[\frac{\alpha''}{\alpha_w}+(1-\Gamma)e^{-\xi_0}\right]S(x) \tag{33}$$

以及侧向条件式(29)。至此,问题的方程组完全闭合。

2 问题的解

按 Budyko[1],取反照率为:

$$\Gamma(x,x_s)=\begin{cases}\Gamma_1=0.62, & 当\ x>x_s\\ \Gamma_0=0.32, & 当\ x<x_s\end{cases} \tag{34}$$

式中,$x_s=\sin\theta_s$,θ_s 为冰界纬度,冰界由温度 $T=-10$℃确定。

令

$$H=\int_0^1 \Gamma(x,x_s)S(x)\mathrm{d}x \tag{35}$$

$$h(x,x_s)=\Gamma(x,x_s)S(x) \tag{36}$$

现将解按勒让德多项式展开成

$$E(x,\xi)=\sum_n E^{(n)}P_n(x),\quad n=0,2,4\cdots \tag{37}$$

若取对赤道对称的解,则侧向条件式(29)将自动满足。同时将太阳辐射以及 H,h 也展成:

$$S(x)=\sum_n S^{(n)}P_n(x),\quad H=\sum_n H^{(n)}P_n(x),$$

$$h(x,x_s)=\sum_n h^{(n)}P_n(x)$$

参照 North[11],取

$$S(x)=S^{(0)}+S^{(2)}P_2(x),\quad S^{(0)}=1,\quad S^{(2)}=-0.482, \tag{38}$$

还可算得展开系数为:

$$H^{(0)}=h^{(0)}=\Gamma_0+(\Gamma_1-\Gamma_0)\left[(1-x_s)-\frac{S^{(2)}}{2}(x_s^3-x_s)\right] \tag{39}$$

$$H^{(2)}=H^{(4)}=\cdots=0 \tag{40}$$

$$h^{(2)}=-\frac{9}{4}S^{(2)}(\Gamma_1-\Gamma_0)x_s^5+\frac{5}{2}(S^{(2)}-1)(\Gamma_1-\Gamma_0)x_s^3$$
$$+\frac{5}{2}\left(1-\frac{S^{(2)}}{2}\right)(\Gamma_1-\Gamma_0)x_s+\Gamma_1 S^{(2)} \tag{41}$$

当 $n=0$ 时,方程和相应的边界条件为:

$$\frac{\mathrm{d}^2E^{(0)}}{\mathrm{d}\xi^2}-q_0^2E^{(0)}=-\widetilde{S}^*\overline{Q}_0\xi_0 e^{-\xi_0\xi}+\widetilde{S}^*\overline{Q}_0 r\left(1+\frac{\alpha''}{\alpha_w}\right)-\widetilde{S}^*\overline{Q}_0 e^{-\varepsilon_0}H^{(0)} \tag{42}$$

$$\xi = 0, \quad E^{(0)} = \frac{1}{2}\left(1 + \frac{\alpha''}{\alpha_w}\right)\overline{Q}_0 \tag{43}$$

$$\xi = 1, \quad \frac{\mathrm{d}E^{(0)}}{\mathrm{d}\xi} - q_0^2\int_0^1 E^{(0)}\mathrm{d}\xi = \widetilde{S}^*\overline{Q}_0\left[r\left(1 + \frac{\alpha''}{\alpha_w}\right) + (\mathrm{e}^{-\xi_0} - 1)\right] - \widetilde{S}^*\overline{Q}_0\mathrm{e}^{-\xi_0}h^{(0)} \tag{44}$$

式中

$$q_0^2 = \frac{N^2}{k_t + k_r}, \quad \widetilde{S}^* = \frac{\widetilde{S}}{k_t + k_r}$$

解为：

$$E^{(0)}(\xi) = a^{(0)}\mathrm{e}^{-q_0\xi} + b^{(0)}\mathrm{e}^{-q_0\xi} + \widetilde{S}^*\overline{Q}_0\left\{\frac{\xi_0}{q_0^2 - \xi_0^2}\mathrm{e}^{-\xi_0\xi} + \frac{1}{q_0^2}\left[\mathrm{e}^{-\xi_0}H^{(0)} - r\left(1 + \frac{\alpha''}{\alpha_w}\right)\right]\right\} \tag{45}$$

式中

$$b^{(0)} = \frac{1}{4}\left(1 + \frac{\alpha''}{\alpha_w}\right)\overline{Q}_0 + \frac{\widetilde{S}^*\overline{Q}_0}{2q_0}\left[r\left(1 + \frac{\alpha''}{\alpha_w}\right) - \mathrm{e}^{-\xi_0}H^{(0)} - \frac{\xi_0}{q_0 + \xi_0}\right]$$

$$a^{(0)} = \frac{1}{2}\left(1 + \frac{\alpha''}{\alpha_w}\right)\overline{Q}_0 + \frac{\widetilde{S}^*\overline{Q}_0}{q_0}\left[r\left(1 + \frac{\alpha''}{\alpha_w}\right) - \mathrm{e}^{-\xi_0}H^{(0)} - \frac{q_0\xi_0}{q_0^2 - \xi_0^2}\right] - b^{(0)}$$

当 $n \neq 0$ 时，方程和相应的边界条件为：

$$\frac{\mathrm{d}^2E^{(0)}}{\mathrm{d}\xi^2} - q_n^2E^{(n)} = -\widetilde{S}^*\overline{Q}_0\xi_0\mathrm{e}^{-\xi_0\xi}S^{(n)} \tag{46}$$

$$\xi = 0, \quad E^{(n)} = 0 \tag{47}$$

$$\xi = 1, \quad \frac{\mathrm{d}E^{(n)}}{\mathrm{d}\xi} - q_0^2\int_0^1 E^{(n)}\mathrm{d}\xi = \widetilde{S}^*\overline{Q}_0\left[\left(\frac{\alpha''}{\alpha_w} + \mathrm{e}^{-\xi_0}\right)S^{(n)} - \mathrm{e}^{-\xi_0}h^{(n)}\right] \tag{48}$$

式中

$$q_n^2 = \frac{N^2 + D_n(n + 1)}{k_t + k_r}.$$

解为：

$$E^{(n)}(\xi) = a^{(n)}\mathrm{e}^{-q_n\xi} + b^{(n)}\mathrm{e}^{q_n\xi} + \frac{\widetilde{S}^*\overline{Q}_0\xi_0}{q_n^2 - \xi_0^2}\mathrm{e}^{-\xi_0\xi}S^{(n)} \tag{49}$$

式中

$$b^{(n)} = \frac{\widetilde{S}^*\overline{Q}_0}{2}\left[\frac{q_n^2 - q_0^2}{q_n}\mathrm{ch}hq_n + \frac{q_0^2}{q_n}\right]^{-1}$$

$$\times\left\{\left[-\frac{\xi_0}{q_n^2 - \xi_0^2}\left(\frac{q_n^2 - q_0^2}{q_n}\mathrm{e}^{-q_n} + \frac{q_0^2}{q_n}\right) + \frac{\alpha''}{\alpha_w} + \frac{q_0^2}{q_n^2 - \xi_0^2} + \mathrm{e}^{-\xi_0}\frac{q_n^2 - q_0^2}{q_n^2 - \xi_0^2}\right]S^{(n)} - \mathrm{e}^{-\xi_0}h^{(n)}\right\},$$

$$a^{(n)} = -\widetilde{S}^*\overline{Q}_0\left[\frac{\xi_0}{q_n^2 - \xi_0^2}S^{(n)}\right] - b^{(n)}$$

3　冰界的确定及海平面的温度分布

当式(37)的系数由上节求得后，在此解式中令 $\xi = 1$（地面），$x = x_s$（冰界），并考虑到 $E = \sigma T^4$ 后，得

到决定 x_s 的方程为：

$$\sigma T_s^4 = \sum_n E^{(n)}(1, x_s) P_n(x_s) \tag{50}$$

在 $E^{(n)}$ 的表达式中，包含了 $\overline{Q}_0 = S_0/4$，S_0 为太阳常数，因此由式(50)可以形式地把冰界方程写成：

$$S_0 = F(x_s) \tag{51}$$

另外，在现在气候情况下，即 $S_0 = 1.92\ \mathrm{K/(cm^2 \cdot min)}$，$x_s = 0.95(\theta_s = 72°\mathrm{N})$ 时，地面温度分布满足：

$$\sigma T^4 = \sum_n E^{(n)}(1, x_s) P_n(x) \tag{52}$$

数值计算中所用参数值取 $\overline{T} = 283°\mathrm{K}$，$r = 0.5$，$\rho_c = 6 \times 10^{-6}\ \mathrm{g/cm^3}$，$\alpha'' = 0.25\ \mathrm{cm^2/g}$，$\alpha_w = 1.25\ \mathrm{cm^2/g}$，$\alpha_s = 100\ \mathrm{cm^2/g}$，$K = 3 \times 10^6\ \mathrm{K/(cm \cdot s \cdot °K)}$，$k_t = 50.53\ \mathrm{cat/(cm \cdot s \cdot °K)}$。根据地球大气中吸收辐射介质的分布和上述有关参数值，算得 $\xi_0 = 0.4$。

根据公式(51)和所给参数，算得冰界纬度与太阳辐射的关系见图 1 中的实线。由图 1 可见，按上述参数值，则冰界在 72°N 与现在的气候状态一致。同时看到，如冰界要南移到 50°N，即出现寒冷的冰河气候，则太阳常数比现在值减小 15%左右。在现在的解中也存在分支点，但位置要比 Budyko[3] 算得的 50°N，North[10] 算得的 37°N 左右都偏南，约在 15°N 左右。因此现在这个计算结果表明，气候不会因太阳辐射通量稍一减小而急剧地趋于“恶化”。

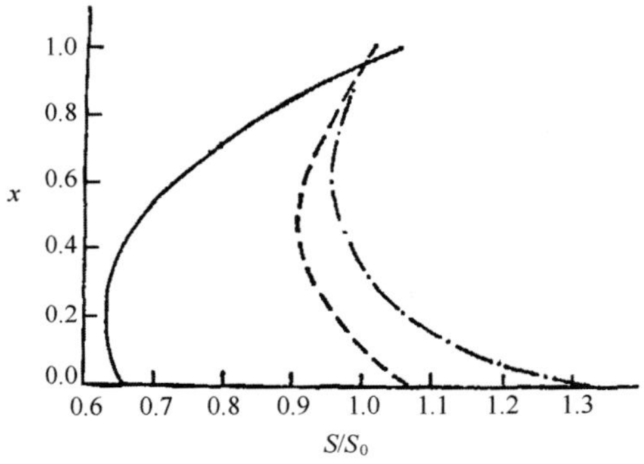

图 1 冰界纬度对太阳常数的依赖性

($S_0 = 1.92\ \mathrm{cat/(cm^2 \cdot min)}$，实线是二维模式的结果，虚线是未简化的一维模式的结果，点划线是简化的一维模式的结果)

图 2 中的实线是二维模式在上述参数下计算所得的海平面温度随纬度的分布。可见与图中以叉号表示的观测值接近。

对不同的垂直交换系数 k_t 也做了计算，表明当 k_t 加大时，要出现相同纬度的冰界，所需太阳辐射通量值增大。反之，当 k_t 值减小时，所需太阳辐射通量也相应减少。

4 与一维模式的比较

如果对第一节得出的二维模式及边界条件(式(31)，式(18)，式(33)和式(29))等方程组求 ξ 从 0 到 1 的积分，可以化成一维模式，结果为：

$$D^* \frac{\mathrm{d}}{\mathrm{d}x}(1 - x^2)\frac{\mathrm{d}\overline{E}}{\mathrm{d}x} = 8(1 - r)\sigma \overline{T}^3 E_0 - 4\sigma \overline{T}^3 \left[(1 - \Gamma)\mathrm{e}^{-\xi_0} + \frac{\alpha''}{\alpha_w}\right] Q_0 + 4\sigma \overline{T}^3 \overline{Q}_0(\mathrm{e}^{-\xi_0} - 1) S(x)$$

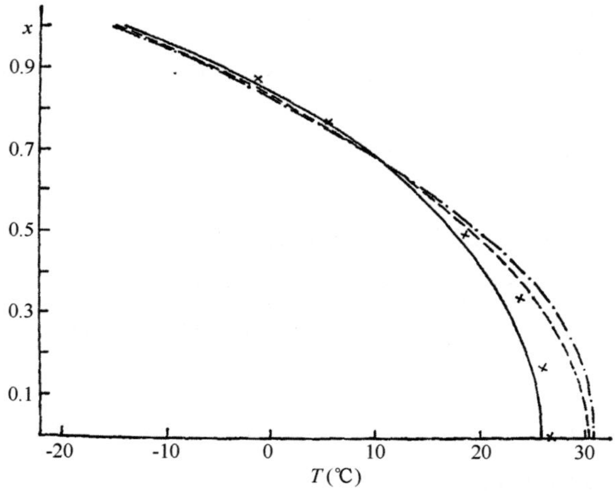

图 2　现在气候情况下,海平面温度随纬度的分布

(×表示观测值,其余说明同图 1)

$$+4\sigma\overline{T}^3\overline{Q}_0\left[r\left(1+\frac{\alpha''}{\alpha_w}\right)-\int_0^1 e^{-\xi_0}\Gamma S(x)\mathrm{d}x\right] \tag{53}$$

其中, $\overline{E}=\int_0^1 E\mathrm{d}\xi$ 为辐射能量在整个高度上的平均值。$D^*=K/a^2$。

式(53)还可做进一步的简化,若把大气看成薄到光学厚度趋于零,即 $\xi_0\to 0$,则在这一薄层中对太阳辐射的吸收可以忽略,这可以用 $a''\to 0$ 来表示这一效果。同时上界的辐射能量 $E_0\to\overline{E}$。如果再假定反射率的平均值 $\int_0^1\Gamma S(x)\mathrm{d}x$ 和参数 r 均等于 0.5①,则式(53)可简化成:

$$D^{**}\frac{d}{\mathrm{d}x}(1-x^2)\frac{\mathrm{d}\overline{E}}{\mathrm{d}x}=\overline{E}-(1-\Gamma)Q_0 \tag{54}$$

其中,$D^{**}=\dfrac{D^*}{4\sigma\overline{T}^3}$。这正是 North[10],所用的 Sellers 型的一维模式。

同 North,用谱方法解这一维方程,得到式(54)的解为:

$$\overline{E}=\overline{Q}_0\sum_n\frac{(2n+1)\int_0^1(1-\Gamma)S(x)P_n(x)\mathrm{d}x}{D^{**}n(n+1)+1}P_n(x). \tag{55}$$

当取 $D^{**}=0.3$,并为与 North[10] 计算作比较,取

$$\overline{E}=A+BT, \tag{56}$$

其中,$A=201.4\mathrm{W/m^2}$,$B=1.45\mathrm{W/(m^2\cdot ℃)}$,则得到冰界纬度与太阳常数的依赖关系如图 1 中的点划线所示,其结果与 North 的结果是一致的。

近似地在式(53)中取 $E_0\approx\overline{E}$,则用同样方法求得解为:

$$\overline{E}=\overline{Q}_0\sum_n\frac{(2n+1)\int_0^1 M(x)P_n(x)\mathrm{d}x}{D^{**}n(n+1)+1}P_n(x) \tag{57}$$

---

① 事实上,考虑到 $\int_0^1\Gamma S(x)\mathrm{d}x\approx\overline{\Gamma}\int_0^1 S(x)\mathrm{d}x=\overline{\Gamma}=\dfrac{0.62+0.32}{2}\approx 0.5$。

其中

$$M(x)=\left(\Gamma e^{-\xi_0}-1-\frac{\alpha''}{\alpha_w}\right)S(x)+\frac{1}{2}\left(1+\frac{\alpha''}{\alpha_w}\right)-\int_0^1 e^{-\xi_0}\Gamma S(x)\,dx$$

当选 $D^{**}=0.4$,$a''/a_w=0.1$,其余参数与上面选取相同时,得到冰界纬度与太阳常数的关系如图 1 的虚线所示。

由图 1 的结果可以看出,当模式不同时,计算结果是不同的。欲要出现冰河期气候,即冰界线从现在的 72°N 南移到 50°N,则用最简化的一维模式(即 North 的 Sellers 型),太阳常数只要减少 2%左右;用未简化的垂直平均一维模式[即式(53)],需要太阳常数减少 5%左右;而用二维模式太阳常数要减少 15%。

当冰界为 72°N 时由 3 种模式计算出的海平面温度随纬度的分布都表示在图 2 中。由图也可见,二维模式算得的温度分布与观测值最接近,而模式做的简化越多,与实际情况越偏离。这也表明二维模式算出的冰界与太阳辐射的依赖关系也应当比简化的一维模式更反映实际情况。这也说明能量的垂直交换过程是重要的。

5 结论

应用本文设计的二维能量平衡模式,冰界对太阳常数的依赖性,要比一维能量平衡模式稳定得多。当太阳常数比现在值减小 15%时,冰界才能从现在气候的 72°N 南移到冰河气候的 50°N 左右,而要出现冰河气候,一维能量平衡模式只需太阳常数减小 2%左右。

由此可见,Budyko、Sellers 提出的极冰-温度-反照率反馈过程对气候影响的理论虽然重要,但定量的结果的可信性是值得怀疑的。从现在的计算看,气候不会像他们估计的那样容易因太阳常数的减少而急剧恶化。当然,这个问题尚需要用更现实的模式做进一步仔细的研究。

参考文献

[1] Budyko, M. I. *Tellus*, 1969,21:611.

[2] Sellers, W. D. *J. Appl. Meteor.*, 1969,8:392.

[3] Budyko, M. I., *EOS Trams AGU*, 1972,10:868.

[4] Schneider, S. H. & T. Gal-Chen. *J. Geophy. Res.*, 1973,78:6182.

[5] Faegre, A. *J. Appl. Meleor.*, 1972,11:4.

[6] Chylek, P. & Coakley. J. A. *J. Atmos. Sci.*, 1975,32:675.

[7] Held, I. M. & Suarez. M. J. *Tellus*, 1974,26:613.

[8] Su, C. H. & Hsien, D. Y. *J. Atmos. Sci.*, 1976,33:2273.

[9] Frederiksen, J. S. *J. Atmos. Sci.*, 1976, 33:2267.

[10] North, G. R. *J. Atmos. Sci.*, 1975, 32:1301.

[11] North, G. R. *J. Atmos. Sci.*, 1975, 32:2033.

[12] Ghil, M. *J. Atmos. Sci.*, 1976,33:3.

[13] Drazin, P. G. & Griffel, D. H., *ibid.*, 1977,34:1696.

[14] 郭晓岚(H-L, Kuo). *Pure & Appl. Geophy.*, 1973,109:1870.

旋转正压大气中的椭圆余弦(Cnoidal)波*

巢纪平[1]　黄瑞新[2]

(1. 中国科学院大气物理研究所;2. 中国科技大学研究生院)

摘要:本文在考虑了 β 效应后,由位势涡度守恒方程导出了极坐标中非线性扰动所满足的 *KdV* 方程。文中还讨论了旋转正压大气中的椭圆余弦波的动力学特点,并算得了类似于切断低压形式的波动。

大气环流的基本状态是绕极的旋转气流,并在其上叠加了若干个波动。Rossby[1] 最早研究了这类波动,并称之为行星波;叶笃正[2] 详细地分析过行星波的色散性。关于行星波的动力学虽已有大量的研究(参见郭晓岚的综合评论[3]),但一般都是在方程线性化范畴内研究的,非线性的行星波虽也有所讨论(如 Pedlosky[4]),但着重是分析波的不稳定性。

一般来说,由于介质的色散性和非线性效应,在一定条件下可以形成一类由 *KdV* 方程所描写的大振幅不变形的长波。当波长为无限时,即形成所谓孤立波,当波是周期性时则形成椭圆余弦波[5]。注意到在地球大气中,由于 $\beta(\equiv \mathrm{d}f/a\mathrm{d}\varphi$, a 为地球半径, φ 为纬度)效应将使得介质变成色散的,如再加上非线性的作用,原则上可以形成这类不变形的长波。对于大气中的孤立波,Long[6],Benny[7] 以及最近 Redekopp[8] 都有过研究,但是这些研究都是对带状的区域进行的,因此这些结果尚不能直接用到球形的地球上去。在天气图上(特别在月平均天气图上)有时可以观测到全球沿纬圈有两三个波数的所谓超长波。这种超长波的移速很慢,甚至有时以阻塞高压或切断低压的形式出现。对于超长波虽也有不少研究(见参考文献[9]),但当已经形成阻塞或切断形式时,这表明波振幅已大到超过线性理论所允许的范围,同时这种带阻塞或切断形式的超长波可以维持相当长的时间,而且波形基本上不变。根据波的这些现象,有理由认为这可能是更接近于一类球面大气中的椭圆余弦波。

1　基本方程

采用平面极坐标(r,θ),r 指向低纬为正,θ 逆时针为正,气流速度分别为(u,v),则正压位势涡度守恒方程为:

$$\frac{\partial \zeta}{\partial t} + u\frac{\partial \zeta}{\partial r} + \frac{v}{r}\frac{\partial \zeta}{\partial \theta} + \frac{\mathrm{d}f}{\mathrm{d}r}u = 0 \tag{1}$$

其中

$$\zeta = \frac{1}{r}\frac{\partial}{\partial r}\left(r\frac{\partial \psi}{\partial r}\right) + \frac{1}{r^2}\frac{\partial^2 \psi}{\partial \theta^2} \tag{2}$$

* 中国科学,1980,10(7):696-705.

是涡度,ψ 为流函数,并满足无辐散条件

$$u=-\frac{1}{r}\frac{\partial\psi}{\partial\theta},\quad v=\frac{\partial\psi}{\partial r}\tag{3}$$

由此,式(1)可以写成:

$$\left[\frac{\partial}{\partial t}+\left(\frac{1}{r}\frac{\partial\psi}{\partial r}\frac{\partial}{\partial\theta}-\frac{1}{r}\frac{\partial\psi}{\partial\theta}\frac{\partial}{\partial r}\right)\right]\left[\frac{1}{r}\frac{\partial}{\partial r}\left(r\frac{\partial\psi}{\partial r}\right)+\frac{1}{r^2}\frac{\partial^2\psi}{\partial\theta^2}\right]+\frac{\beta}{r}\frac{\partial\psi}{\partial\theta}=0\tag{4}$$

其中

$$\beta=-\frac{\mathrm{d}f}{\mathrm{d}r}=\frac{2\omega}{a}\cos\varphi\tag{5}$$

(ω 为地球自转角速度)。

设基本场为一以角速度 $\Omega=\Omega(r)$ 绕极地旋转的气流,我们将流函数写成:

$$\psi=\int_{r_1}^{r}[\Omega(r)-c_0]r\mathrm{d}r+\varepsilon\phi(r,\theta,t),\tag{6}$$

式中,c_0 是一待定常数,它相当于扰动的角位相速,$\varepsilon\phi$ 是流函数的扰动部分,$\varepsilon\ll1$ 为小参数。

由此方程(4)可化为:

$$\left[\frac{\partial}{\partial t}+(\Omega-c_0)\frac{\partial}{\partial\theta}+\varepsilon\left(\frac{1}{r}\frac{\partial\phi}{\partial r}\frac{\partial}{\partial\theta}-\frac{1}{r}\frac{\partial\phi}{\partial\theta}\frac{\partial}{\partial r}\right)\right]$$
$$\times\left[\frac{1}{r}\frac{\partial}{\partial r}\left(r\frac{\partial\phi}{\partial r}\right)+\frac{1}{r^2}\frac{\partial^2\phi}{\partial r^2}\right]+\frac{1}{r}\left[\beta-\frac{\mathrm{d}}{\mathrm{d}r}\left(\frac{1}{r}\frac{\mathrm{d}r^2\Omega}{\mathrm{d}r}\right)\right]\frac{\partial\phi}{\partial\theta}=0\tag{7}$$

为了讨论非线性长波,我们可以采用长波近似中的坐标延伸法,令

$$\Theta=\varepsilon^{1/2}\theta,\quad \tau=\varepsilon^{3/2}t\tag{8}$$

而

$$\phi(r,\theta,t)=\phi^{(1)}(r,\Theta,\tau)+\varepsilon\phi^{(2)}(r,\Theta,\tau)+\cdots\tag{9}$$

代入式(7)中,比较 ε 的同次幂项后可得 $\varepsilon^{1/2}$ 的方程为:

$$\mathscr{L}\phi^{(1)}=0\tag{10}$$

其中定义算子

$$\mathscr{L}=\frac{1}{r}\frac{\partial}{\partial\Theta}\left\{(\Omega-c_0)\left[\frac{\partial}{\partial r}\left(r\frac{\partial}{\partial r}\right)\right]+\left[\beta-\frac{\mathrm{d}}{\mathrm{d}r}\left(\frac{1}{r}\frac{\mathrm{d}r^2\Omega}{\mathrm{d}r}\right)\right]\right\}\tag{11}$$

设在边界上扰动为零,即有边界条件

$$r=r_1,r_2\text{ 时},\quad \phi^{(1)}=\phi^{(2)}=\cdots=0\tag{12}$$

由式(11)及式(12),可以假定 $\phi^{(1)}$ 是可以分离变量的,设

$$\phi^{(1)}=F(\Theta,\tau)G(r)\tag{13}$$

这样得到 $G(r)$ 应满足的方程和边界条件为:

$$(\Omega-c_0)\frac{\mathrm{d}}{\mathrm{d}r}\left(r\frac{\mathrm{d}G}{\mathrm{d}r}\right)+\left[\beta-\frac{\mathrm{d}}{\mathrm{d}r}\left(\frac{1}{r}\frac{\mathrm{d}r^2\Omega}{\mathrm{d}r}\right)\right]G=0\tag{14}$$

$$G(r_1)=G(r_2)=0\tag{15}$$

另一方面,幂次为 $\varepsilon^{3/2}$ 的方程为:

$$\mathscr{L}\phi^{(2)}+\frac{\Omega-c_0}{r^2}\frac{\partial^3\phi^{(1)}}{\partial\Theta^3}+\left[\frac{\partial}{\partial\tau}+\frac{1}{r}\frac{\partial\phi^{(1)}}{\partial r}\frac{\partial}{\partial\Theta}-\frac{1}{r}\frac{\partial\phi^{(1)}}{\partial\Theta}\frac{\partial}{\partial r}\right]\times\left[\frac{1}{r}\frac{\partial}{\partial r}\left(r\frac{\partial\phi^{(1)}}{\partial r}\right)\right]=0\tag{16}$$

利用式(13),式(14)加以整理后,两边乘以$\frac{r}{\Omega-c_0}G$,对 r 从 r_1 到 r_2 积分,并利用边界条件,可得

$$F_r - RFF_{\Theta} - SF_{\Theta\Theta\Theta} = 0 \tag{17}$$

这是在极坐标中考虑 β 效应后的 *KdV* 方程,其中系数分别为:

$$R = I_2/I_1 \tag{18}$$

$$S = I_3/I_1 \tag{19}$$

$$I_1 = \int_{r_1}^{r_2} \frac{\beta - \frac{\mathrm{d}}{\mathrm{d}r}\left(\frac{1}{r}\frac{\mathrm{d}r^2\Omega}{\mathrm{d}r}\right)}{(\Omega - c_0)^2} G^2 \mathrm{d}r \tag{20}$$

$$I_2 = \int_{r_1}^{r_2} \frac{G^3}{\Omega - c_0} \frac{\mathrm{d}}{\mathrm{d}r}\left[\frac{\beta - \frac{\mathrm{d}}{\mathrm{d}r}\left(\frac{1}{r}\frac{\mathrm{d}r^2\Omega}{\mathrm{d}r}\right)}{r(\Omega - c_0)}\right] \mathrm{d}r \tag{21}$$

$$I_3 = \int_{r_1}^{r_2} \frac{G}{r} \mathrm{d}r \tag{22}$$

2　特征值问题

现在的问题归结为求方程(14)在边界条件式(15)下的特征值和特征函数,并使 *KdV* 方程具有非零的系数。为简单起见,我们讨论 $\Omega(r)=\Omega_0$ 为常数的情况,在大气中这相当于西风急流的北侧。这时式(14)化为:

$$\frac{\mathrm{d}}{\mathrm{d}r}\left(r\frac{\mathrm{d}G}{\mathrm{d}r}\right) + \frac{\beta}{\Omega_0 - c_0}G = 0 \tag{23}$$

显然若要求在齐次边界条件下方程有非零解,必须有 $\Omega_0 > c_0$,即基本气流须是超临界的。引入变数

$$\rho = \sqrt{\frac{4\beta r}{\Omega_0 - c_0}} \tag{24}$$

则式(23)化成 Bessel 方程,

$$\rho^2 \frac{\mathrm{d}^2 G}{\mathrm{d}\rho^2} + \rho \frac{\mathrm{d}G}{\mathrm{d}\rho} + \rho^2 G = 0 \tag{25}$$

其通解为:

$$G(\rho) = c_1 J_0(\rho) + c_2 Y_0(\rho) \tag{26}$$

由边界条件式(15)得到联系特征值 c_0 的关系:

$$J_0(\rho_1) Y_0(\rho_2) - Y_0(\rho_1) J_0(\rho_2) = 0 \tag{27}$$

当特征值 c_0 确定后,利用关系式

$$c_1/c_2 = -\frac{Y_0(\rho_1)}{J_0(\rho_1)} = -\frac{Y_0(\rho_2)}{J_0(\rho_2)} \tag{28}$$

及归一化条件

$$\| G \| = \max_{\rho_1 \leqslant \rho \leqslant \rho_2} | G(\rho) | = 1 \tag{29}$$

可以将这两个常数 c_1, c_2 定出。

实际上,在通常情况下,如下面将证明的 ρ_1, ρ_2 一般都较大,因而可以利用 Bessel 函数的渐近展开。如只取 J_0, Y_0 渐近展开的主项,则式(27)化为:

$$\frac{2}{\pi}\sqrt{\frac{1}{\rho_1\rho_2}}\sin(\rho_2-\rho_1)=0 \tag{30}$$

故得

$$\rho_2=\rho_1+n\pi,\quad n=1,2,\cdots \tag{31}$$

由式(24),式(31)得

$$\rho_2=\frac{n\pi}{1-\sqrt{r_1/r_2}} \tag{32}$$

如讨论区间为 $0.3\leqslant r_1/r_2\leqslant 1$,并取 $n=1$,则相应得 $\rho_2\geqslant 6.95$,$\rho_1\geqslant 3.8$。因此采用渐近展开是允许的,而所得的特征值 c_0 的渐近表达式为:

$$c_0=\Omega_0-\frac{4r_2}{\pi}(1-\sqrt{r_1/r_2})^2\beta \tag{33}$$

(在上述条件下,上式与式(27)的精确解的误差小于 5%)。特别当带宽 $\Delta r=r_2-r_1\ll r_2$ 时,又有近似式

$$c_0\approx\Omega_0-\frac{(\Delta r)^2}{r_2\pi^2}\beta \tag{34}$$

利用归一化条件式(29),同样取 J_0,Y_0 渐近展开的主项,可得到 c_1,c_2 的关系为:

$$c_1/c_2=-\operatorname{tg}(\rho_2-\pi/4) \tag{35}$$

而式(29)化成

$$\sqrt{\frac{2}{\pi\rho^*}}\cdot\left|c_1\cdot\frac{\sin(\rho_2-\rho^*)}{\sin(\rho_2-\pi/4)}\right|=1 \tag{36}$$

其中,ρ^* 为 $|G(\rho)|$ 取极大值的点,近似为 $\rho^*=\rho_2-\pi/2$,故得

$$c_1=\pm\sqrt{\frac{\pi\rho^*}{2}}\sin(\rho_2-\pi/4)$$

$$c_2=\mp\sqrt{\frac{\pi\rho^*}{2}}\cos(\rho_2-\pi/4) \tag{37}$$

此时相应的特征函数为:

$$G(\rho)=\pm\sqrt{\rho^*/\rho}\sin(\rho_2-\rho) \tag{38}$$

在以下讨论中我们取式中的负号。

KdV 方程(17)中的系数相应为:

$$R=-\frac{16\beta^2}{(\Omega_0-c_0)^2}\int_{\rho_1}^{\rho_2}\frac{G^3}{\rho^3}\mathrm{d}\rho\Big/\int_{\rho_1}^{\rho_2}\rho G^2\mathrm{d}\rho \tag{39}$$

$$S=4(\Omega_0-c_0)^2\int_{\rho_1}^{\rho_2}\left(\frac{G^2}{\rho}\right)\mathrm{d}\rho\Big/\int_{\rho_1}^{\rho_2}\rho G^2\mathrm{d}\rho \tag{40}$$

把式(38)代入式(39)、式(40),并应用中值定理,近似地得到

$$R=\frac{128}{3\pi r_2^2}(1+\sqrt{r_1/r_2})^{-4} \tag{41}$$

$$S=\frac{64r_2}{\pi^4}\frac{(1-\sqrt{r_1/r_2})^4}{(1+\sqrt{r_1/r_2})^2}\beta \tag{42}$$

以及

$$R/S = \frac{2\pi^3}{3r_2^3} \cdot \frac{1}{(1+\sqrt{r_1/r_2})^2(1-\sqrt{r_1/r_2})^4\beta} \tag{43}$$

由于 R,S 均不为零,故 KdV 方程有解。

3 KdV 方程的椭圆余弦波解

方程(17)在恢复到 θ,t 坐标时为:

$$F_t - \varepsilon RFF_\theta - SF_{\theta\theta\theta} = 0 \tag{44}$$

如取

$$M = \varepsilon F \tag{45}$$

则流函数的扰动部分为:

$$\varepsilon\phi = MG + O(\varepsilon^2) \tag{46}$$

这时 M 满足方程

$$M_t - RMM_\theta - SM_{\theta\theta\theta} = 0 \tag{47}$$

设方程(47)具有行波解,

$$M = h(\theta - U_0 t) = h(\theta') \tag{48}$$

代入式(47),得

$$U_0 h' + Rhh' + Sh''' = 0 \tag{49}$$

积分两次,得

$$h'^2 = -\frac{R}{3S}\left(h^3 + \frac{3U_0}{R}h^2 + g_1 h + g_2\right) \tag{50}$$

其中,g_1,g_2 是积分常数。

由于要求式(50)的解是有界的,故其右端的三次多项式的 3 个根应当全是实数。同时,对周期为有限的波动,这 3 个根应是分立的。考虑到 h 与扰动流函数的振幅成比例,故其中相邻的两个根应一个取正值,另一个取负值,以代表波峰和波谷。因此这 3 个根可以取成 $\alpha,-\sigma,\alpha-\gamma$。其中 $\alpha,\sigma>0$,$\gamma>\alpha+\sigma>0$。式(50)可写成为:

$$h'^2 = -\frac{R}{3S}(h-\alpha)(h+\sigma)(h-\alpha+\gamma) \tag{51}$$

积分后即得椭圆余弦波解

$$h = (\alpha+\sigma)cn^2\sqrt{\frac{R\gamma}{12S}}\theta' - \sigma \tag{52}$$

其中椭圆函数的模为:

$$k = \sqrt{\frac{\alpha+\sigma}{\gamma}} \tag{53}$$

波长为:

$$\lambda = 4\cdot\sqrt{\frac{3S}{R\gamma}}K(k) \tag{54}$$

$$K(k) = \int_0^{\pi/2}\frac{\mathrm{d}t}{\sqrt{1-k^2\sin^2 t}} \tag{55}$$

另外,由三次多项式的根关系,可得

$$-\frac{3U_0}{R}=2\alpha-\sigma-\gamma \tag{56}$$

由于 θ 相当于经度,故波长 λ 应当是 2π 的某个分数

$$\lambda=\frac{2\pi}{m} \tag{57}$$

其中 m 为全球波数。代入式(54)得

$$\frac{12S}{R\gamma}=\left(\frac{\pi}{mK}\right)^2 \tag{58}$$

故椭圆余弦波解式(52)最后可表达为:

$$h=(\alpha+\sigma)cn^2\frac{mK\theta'}{\pi}-\sigma \tag{59}$$

假定扰动流函数对 θ 的全球积分平均为零,

$$\frac{1}{2\pi}\int_0^{2\pi}h\mathrm{d}\theta'=0 \tag{60}$$

将式(59)代入后,可得

$$\frac{\alpha+\sigma}{K}\int_0^K cn^2u\mathrm{d}u-\sigma=0 \tag{61}$$

利用第二种完全椭圆积分,

$$E=\int_0^K \mathrm{dn}^2u\mathrm{d}u \tag{62}$$

可得

$$\int_0^K cn^2u\mathrm{d}u=\frac{E-K(1-k^2)}{k^2} \tag{63}$$

代入式(61),得到

$$\alpha+\sigma=\frac{k^2}{E/K+k^2-1}\sigma \tag{64}$$

另外,设波峰和波谷的绝对值之和为 H,即

$$\alpha+\sigma=H \tag{65}$$

则当波数 m 给定时,由方程

$$kK(k)=\sqrt{\frac{RH}{12S}}\cdot\frac{\pi}{m} \tag{66}$$

可以求出 k,而 α,σ,γ 可由式(53),式(64),式(65)定出;U_0 可由下式定出:

$$U_0=\frac{R}{3}\cdot\frac{3(E/K)+k^2-2}{k^2}H \tag{67}$$

现在分别讨论几种情况:

1)无限小振幅的波

此时,α,σ 很小,$|h-\alpha|\ll\gamma$,所以式(51)化成

$$h'^2=-\frac{R}{3S}(h-\alpha)(h+\sigma)\gamma \tag{68}$$

根关系式(56)化为：

$$\frac{3U_0}{R}=\gamma \tag{69}$$

由式(68)积分得

$$h=\frac{\alpha-\sigma}{2}+\frac{\alpha+\sigma}{2}\cos\sqrt{\frac{R\gamma}{3S}}\theta' \tag{70}$$

设全球波数为 m,则

$$\sqrt{\frac{R\gamma}{3S}}=m \tag{71}$$

由式(69),式(71)得小振幅波的角位相速为：

$$U_0=Sm^2 \tag{72}$$

利用式(42),并设带宽 $\Delta r\ll r_2$ 时,有

$$U_0\approx\frac{(\Delta r)^4}{r_2^3\pi^4}m^2\beta \tag{73}$$

由于椭圆余弦波是叠加在角相速度为 C_0 的扰动上,故实际上的角速度应为(C_0+U_0),利用式(34),式(73),可得

$$C_0+U_0\approx\Omega_0-\left(\frac{\Delta r}{r_2}\right)^2\frac{r_2}{\pi^2}\left[1-\left(\frac{\Delta r}{r_2}\right)^2\frac{m^2}{\pi^2}\right]\beta \tag{74}$$

注意到在 β-平面近似下,并考虑到扰动宽度 b 为有限时的 Rossby 波公式为：

$$C=U-\frac{1}{4\pi^2}\cdot\frac{1}{1/L^2+1/b^2}\beta \tag{75}$$

对于我们讨论的情况,式(31)中若取 $n=1$ 是相当于南北方向只有半个波,故有对应关系

$$\Delta r=b/2,\quad L=\frac{2\pi r_2}{m}$$

因我们只讨论长波,可以认为波长 L 远比扰动宽度 b 为大,故 $b^2/L^2\ll 1$,因而

$$(1/L^2+1/b^2)^{-1}\approx b^2(1-b^2/L^2) \tag{76}$$

将式(75)两边除 r_2,换成角速度方程,并利用式(76),则不难证明式(74),式(75)是完全相同的。

由此可见,在振幅无限小时,椭圆余弦波退化为余弦波,其波速与振幅无关,而与波长的平方成反比(见式(73))。而且在带宽很窄时,椭圆余弦波的行波波速与经典的 Rossby 波公式是一致的。

2)有限振幅波

此时波参数应由式(53),式(64)至式(67)等计算。

(1)若 $H\to 0$,则由式(66)知,$k\to 0$,故椭圆余弦函数退化为余弦函数,$K(k)\to\pi/2$,$3E/K+k^2-2\to 1$,而由式(66)、式(67)得 $U_0\to Sm^2$,这正是前述的式(72)。

(2)若 H 很大,而 m 又很小(如 $m=1,2$),则由式(66)$k\to 1$,$\frac{3E/K+k^2-2}{k^2}\to\left(\frac{3}{K}-1\right)$,故得

$$U_0\approx\frac{R}{3}\left(\frac{3}{K}-1\right)H \tag{77}$$

由此可见,在振幅很大时,椭圆余弦波的速度与振幅成比例。

(3)在一般情况下,由式(32),式(42),式(66),式(67)得到波的角速度公式为：

$$C_0 + U_0 = \Omega_0 - \frac{4r_2(1-\sqrt{r_1/r_2})^2}{\pi^2}\left[1 + \frac{4(2-k^2-3E/K)K^2}{\pi^2}\cdot\frac{16(1-\sqrt{r_1/r_2})^2 m^2}{(1+\sqrt{r_1/r_2})^2\pi^2}\right]\beta \quad (78)$$

$$\equiv \Omega_0 - \Omega_C$$

如 $k \leqslant 0.5$ 时，则 $\frac{4(2-k^2-3E/K)K^2}{\pi^2} \approx -1$，故

$$C_0 + U_0 \approx \Omega_0 - \frac{4r_2(1-\sqrt{r_1/r_2})^2}{\pi^2}\left[1 - \frac{16(1-\sqrt{r_1/r_2})^2 m^2}{(1+\sqrt{r_1/r_2})^2\pi^2}\right]\beta \quad (79)$$

如再假定 $\Delta r/r_2 \ll 1$，则可得到式(74)。

4 计算结果

4.1 关于波速的讨论

取 $\bar{h} = \frac{H}{(r_2^2 - r_1^2)\omega}$，$\bar{h} \to 0$ 表示线性情况下的小振幅波。图 1 给出了 KdV 方程所定出的 U_0 值[由式(66)，式(67)算出]。由图 1 可知，当波振幅(正比于 $\bar{h}$)不同时，U_0 值不同，这正是波的非线性所造成。注意当带宽(正比于 $\Delta r/r_2$)较窄时，大振幅波的 U_0 值为负。

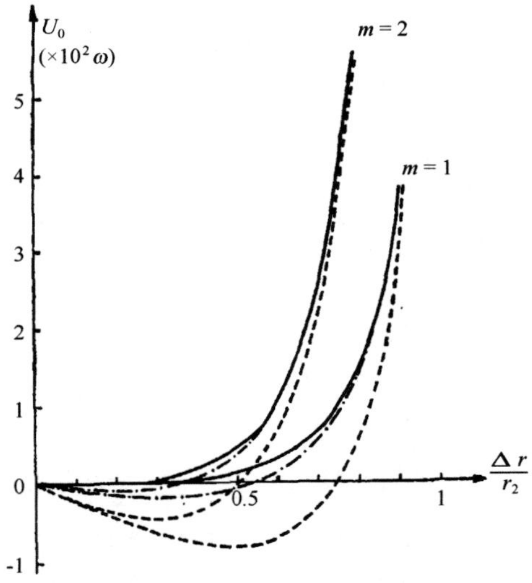

图 1 由 KdV 方程决定的行波速度 U_0

(……$\bar{h}=0.04$，—·—$\bar{h}=0.01333$，——$\bar{h}\to 0$)

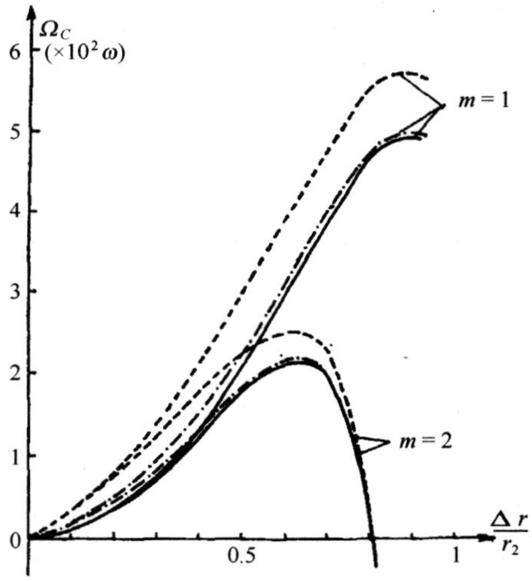

图 2 椭圆余弦波的 Ω_c 值

(说明同图 1)

考虑到 C_0 相当于坐标系的移动角速度，故 C_0+U_0 为波相对于静止坐标系的移动角速度。由式(78)注意到，若 $\Omega_0=\Omega_c$，即基本流场的角速度 Ω_0 与临界值 Ω_c 相等时，波将变成驻波。当 $\bar{h}$ 取不同值及 $m=1,2$ 时，Ω_c 值见图 2。考虑到地球自转角速度 $\omega=7.29\times10^{-5}/s$，若西风急流轴位于中纬度，则 r_2 约为 4×10^6 m。这时若西风风速取 10~25 m/s 的量级，则 Ω_0/ω 为$(3\sim8)\times10^{-2}$的量级。由图 2 可见，这时对 $m=1$ 的长波有可能出现天气图上观测到的驻波，如西风速度小一些，则 $m=2$ 出现驻波形势也是可能的。

另外，注意到由于 $\Omega_c=\Omega_0-(C_0+U_0)$，因此在一般情况下，波相对于盛行气流都是后退的，而在 $C_0<\Omega_0<\Omega_c$ 的时候，则表现为相对于地面后退（西行）波。这种波速很慢或后退的波正是大气中超长波的一种常见的现象。

4.2　关于波的图形

在图 3(a) 中给出 $m=1,\bar{h}=0.04,k=0.991$，而 $\Omega_0=\Omega_c$（驻波）时的流线图，这时 Ω_0/ω 约为 4.8×10^{-2}。如前述，这在大气中是能够出现的西风气流速度。在这种情况下，流型很像天气图上驻定的切断低压。在图 3(b) 中则给出以 $\dfrac{\Omega_0-\Omega_c}{\omega}=2.5\times10^{-2}$ 向东移动的两个切断图形。相应地，如波振幅减小到 0.01333，而基本流场的 Ω_0 不变，则如图 3(c) 所示，这时出现一个以 $\dfrac{\Omega_0-\Omega_c}{\omega}=0.8\times10^{-2}$ 缓慢东移的波，这很像一类偏心极涡的天气形势。在图 3(d) 中，由于波长变短（$m=2$），故波的顺行速度加大到 $\dfrac{\Omega_0-\Omega_c}{\omega}=2.78\times10^{-2}$，这时波表现为对称图形。

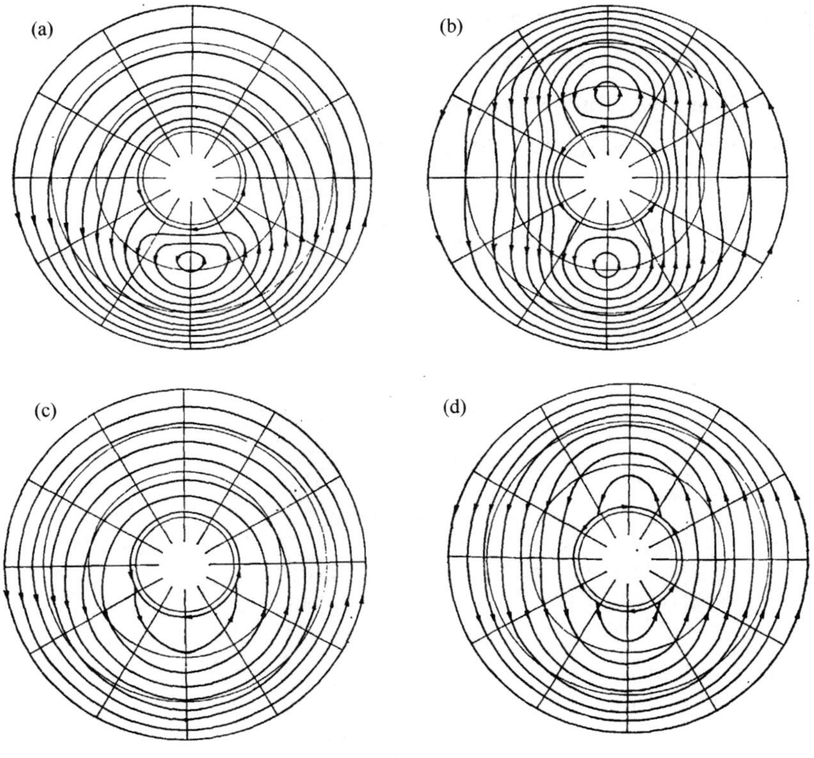

图 3　椭圆余弦波流线图（$r_1/r_2=0.3,\varphi_2=50°\mathrm{N},\Omega_0=0.04797\omega$）

(a) $h=0.04,m=1,k=0.991,\Omega_0-\Omega_c=0$（驻波）；(b) $\bar{h}=0.04,m=2,k=0.828,\Omega_0-\Omega_c=0.02496\omega$；(c) $\bar{h}=0.01333,m=1,k=0.882,\Omega_0-\Omega_c=0.00828\omega$；(d) $\bar{h}=0.01333,m=2,k=0.567,\Omega_0-\Omega_c=0.02775\omega$

我们注意到，由于基本气流取的是相当于西风急流北侧的情况，这时基本流场为气旋性涡度，故在大振幅情况下可以出现类似于切断低压的流型。当然，这里只提出了非线性在形成闭合流型中的重要性,实际上大气中切断低压形成的机制,自然要比现在讨论的情况复杂得多。

5 结语

本文研究了极坐标中在β-平面近似下的非线性*KdV*方程的椭圆余弦波解。解的图形很像大气中具有切断低压形势的超长波。当然这只是说明非线性和色散性在形成切断形势中的重要性,现实大气中切断低压形成的物理机制自然要比现在讨论的情况复杂得多。但研究非线性波动力学无疑有助于进一步认识行星大气中的波动行为。

参考文献

[1] Rossby, C. G. *J. Marine Res.*, 1939,2:38-55.

[2] Yeh, T. C. (叶笃正) *J. Met.*, 1949,6:1-16.

[3] Kuo, H. L. (郭晓岚) *Adv. in Applied Mech.*, 1973,13:247-330.

[4] Pedlosky, J. *J. Atmos. Sci.*, 1972, 29:53-63.

[5] Whitham, G. B. *Linear and Nonlinear Waves*, 1974.

[6] Long. R. R. *J. Aimos. Sci.*, 1966,21:197-200.

[7] Benney, D. J. *J. Math. & Phys.*, 1966,45:52-63.

[8] Redekopp, L. G. *J. Flmd Mech.*, 1977,82:725-745.

[9] Phillips, N. A. *Rev. Geophys.*, 1963,1:123-176.

螺旋 Rossby 波的波作用守恒和稳定性*

陈英仪　巢纪平

（中国科学院大气物理研究所，北京）

摘要：本文应用 WKB 方法导出了在一般较接近实际大气的模式中，螺旋 Rossby 波的广义波作用量所满足的方程。当背景场不随时间变化时，波作用量具有守恒性。从这守恒性质出发讨论了螺旋 Rossby 波的稳定性，指出扰动发展的必要条件是广义绝对涡度梯度至少在区域中的某一条空间曲线上为零。在某些简化条件下，与 Charney 和 Stern[1]，郭晓岚[2]以及 Eady[3]的结果一致。若广义绝对涡度梯度处处不为零，则扰动发展须满足其他的条件，它与基本气流的分布及 Rossby 波等位相线的倾斜状况有关。

1　引言

在高空天气图上观测到的行星波，其槽、脊线（或称位相线）无论在水平或垂直方向一般都呈倾斜的结构。这种倾斜结构对大气环流的维持和发展是重要的，如水平方向的倾斜将造成角动量的南北输送，而维持大气环流的基本状态。但由于数学处理上的困难，过去在研究行星波的稳定性和其他方面时，一般取扰动的位相线沿经线方向垂直的简谐波，而提出一个本征值问题来解决。然而，简谐波和槽、脊线呈弯曲的螺旋波，不仅几何形式不同，而某些动力学性质也有差异。可以说简谐波只是螺旋波的一个特殊情况。

巢纪平和叶笃正[4]首先研究了螺旋行星波的某些运动学和动力学性质，但在这一初步研究中，曾作了某些简化假定，因此一些结论还待修正。继参考文献[4]螺旋行星波概念提出后，刘式适、杨大升[5]对螺旋行星波的传播性质等做过探讨。

在本文中，我们仍采用 WKB 方法，进一步讨论斜压大气中螺旋行星波的动力学，而把正压大气看作是斜压大气的一个特例。

2　基本方程

为方便，本文采用 β 平面近似，若用完整的球面坐标其结果并无原则差别。在 β-平面近似下，大气运动的准地转位涡方程可写成：

$$\frac{\partial q}{\partial t}+\frac{\partial \psi}{\partial x}\left(\frac{\partial q}{\partial y}+\beta\right)-\frac{\partial \psi}{\partial y}\frac{\partial q}{\partial x}=0 \tag{1}$$

其中

* 中国科学 B 辑，1983，13(7)：663-672.

$$q = \nabla_h^2 \psi + \frac{f^2}{\rho_0} \frac{\partial}{\partial z}\left(\frac{\rho_0}{N^2} \frac{\partial \varphi}{\partial z}\right)$$

为相对位涡。ρ_0 为未扰动的大气密度，它仅是高度的函数，式中

$$\psi = P/f_0\rho_0, \quad \beta = \frac{\mathrm{d}f}{\mathrm{d}y}$$

以及

$$N = \left(\frac{g}{\theta_0} \frac{\partial \theta_0}{\partial z}\right)^{1/2}$$

为 Brant-Väisälä 频率，线性化的扰动位涡方程为：

$$\left(\frac{\partial}{\partial t} + U \frac{\partial}{\partial x}\right) q' + \frac{\partial \psi'}{\partial x}\left[\beta - \frac{\partial^2 U}{\partial y^2} - \frac{f^2}{\rho_0} \frac{\partial}{\partial z}\left(\frac{\rho_0}{N^2} \frac{\partial U}{\partial z}\right)\right] = 0 \tag{2}$$

基本气流 U 是(y,z,t)的函数，定义成

$$U = -\frac{\partial \psi_0}{\partial y}$$

而扰动位势涡度为：

$$q' = \nabla_h^2 \psi' + \frac{f^2}{\rho_0} \frac{\partial}{\partial z}\left(\frac{\rho_0}{N^2} \frac{\partial \psi'}{\partial z}\right)$$

引进特征量 L, Ω, H_0 相应的无量纲量为：

$$\tilde{t} = \Omega t, \quad (\tilde{x}, \tilde{y}) = (x, y)/L, \quad \tilde{z} = z/H_0,$$

$$\tilde{U} = U/L\Omega, \quad \tilde{\psi} = \psi'/\psi^*$$

上式的 ψ^* 的量级正比于 ΩL^2，得到无量纲方程如下（已略去“～”号）：

$$\left(\frac{\partial}{\partial t} + U \frac{\partial}{\partial x}\right)\left\{\nabla_h^2 \psi + \frac{1}{\rho_0} \frac{\partial}{\partial z}\left(\rho_0 \alpha^2 \frac{\partial \psi}{\partial z}\right)\right\} + \beta_A \frac{\partial \psi}{\partial x} = 0, \tag{3}$$

其中

$$\frac{1}{\alpha^2} = \frac{N^2}{f^2} \frac{H_0^2}{L^2} = \left(\frac{l_c}{L}\right)^2, \quad l_c = \frac{NH_0}{f} \tag{4}$$

l_c 可称为 Rossby 变形半径。而

$$\beta_A = 2\cos\phi - \frac{\partial^2 U}{\partial y^2} - \frac{1}{\rho_0} \frac{\partial}{\partial z}\left(\rho_0 \alpha^2 \frac{\partial U}{\partial z}\right) \tag{5}$$

可称为广义位势涡度梯度。

为方便起见，令

$$\psi = \Psi(x, y, z, t) \mathrm{e}^{-rz} \tag{6}$$

这里

$$r = \frac{1}{2} \frac{\partial}{\partial z} \ln(\rho_0 \alpha^2) \tag{7}$$

并设 r=常数，则式(3)可写成：

$$\left(\frac{\partial}{\partial t} + U \frac{\partial}{\partial x}\right)\left(\nabla_h^2 \Psi + \alpha^2 \frac{\partial^2 \Psi}{\partial z^2} - \alpha^2 r^2 \Psi\right) + \beta_A \frac{\partial \Psi}{\partial x} = 0 \tag{8}$$

采用 WKB 方法，设波振幅是空间和时间的缓变函数

$$\Psi = A(T, X, Y, Z)\mathrm{e}^{i\varphi} \tag{9}$$

其中位相

$$\varphi = mx + ky + nz - \omega t \tag{10}$$

而

$$\omega = -\frac{\partial\varphi}{\partial t},\quad m = \frac{\partial\varphi}{\partial x},\quad k = \frac{\partial\varphi}{\partial y},\quad n = \frac{\partial\varphi}{\partial z} \tag{11}$$

分别为波的频率、纬向、径向和垂直方向的波数，而缓变坐标为：

$$T = \varepsilon t,\quad X = \varepsilon x,\quad Y = \varepsilon y,\quad Z = \varepsilon z \tag{12}$$

其中，ε 是小于 1 的小参数。

由式(10)可知，对任一时刻 t，在 z = 常数的平面上，等位相线由 $mx+ky$ = 常数所确定。m 是纬向波数，它是正数，当 $k>0$ 时，意味着等位相线呈西北—东南走向（称导波），当 $k<0$ 时，等位相线呈东北—西南走向（称曳波）。同样，在每一时刻 t，在 x = 常数的子午平面上，等位相线由 $ky+nz$ = 常数所确定。当 k,n 同号时，等位相线由下向上向低纬倾斜；当 k,n 异号时，等位相线由下向上向高纬倾斜。同理，每一时刻 t，在 y = 常数的纬圈平面上，等位相线由 $mx+nz$ = 常数所确定，当 $n>0$ 时，等位相线向上、向西倾斜，当 $n<0$ 时，等位相线向上、向东倾斜。

当解取式(9)的形式时，方程(8)可写成

$$\begin{aligned}&\left[-i(\omega - Um) + \varepsilon\frac{\partial}{\partial T} + \varepsilon U\frac{\partial}{\partial x}\right] \times \left\{\varepsilon^2\left(\frac{\partial^2 A}{\partial X^2} + \frac{\partial^2 A}{\partial y^2} + \alpha^2\frac{\partial^2 A}{\partial z^2}\right)\right.\\ &+ \varepsilon i\left(2m\frac{\partial A}{\partial X} + A\frac{\partial k}{\partial X} + 2k\frac{\partial A}{\partial y} + A\frac{\partial k}{\partial y} + 2n\alpha^2\frac{\partial A}{\partial z} + \alpha^2 A\frac{\partial n}{\partial z}\right)\\ &\left.- (m^2 + k^2 + n^2\alpha^2 + \alpha^2 r^2)A\right\} + \beta_A\left(\varepsilon\frac{\partial A}{\partial X} + imA\right) = 0\end{aligned} \tag{13}$$

在以下的分析中，将解按 ε 的幂次展开为：

$$A = A_0 + \varepsilon A_1 + \varepsilon^2 A_2 + \cdots \tag{14}$$

3　相速度及群速度

将式(14)代入式(13)，得到零级近似方程为：

$$[(\omega - Um)(m^2 + k^2 + \alpha^2 n^2 + \alpha^2 r^2) + \beta_A m]A_0 = 0$$

由于 $A_0 \neq 0$，故有频散关系

$$\omega = Um - \frac{\beta_A m}{K^2 + \alpha^2 r^2} \equiv F(m, k, n, Y, Z, T) \tag{15}$$

其中

$$K^2 = m^2 + k^2 + \alpha^2 n^2 \tag{16}$$

为波数的平方和。在此已假设 U 是坐标的缓变函数。由式(15)可得群速度的各分量为：

$$\left.\begin{aligned} C_{gx} &= \frac{\partial \omega}{\partial m} = U - \frac{\beta_A}{(K^2 + \alpha^2 r^2)} + \frac{2\beta_A m^2}{(K^2 + \alpha^2 r^2)^2}, \\ C_{gy} &= \frac{\partial \omega}{\partial k} = \frac{2\beta_A m \cdot k}{(K^2 + \alpha^2 r^2)^2}, \\ C_{gz} &= \frac{\partial \varphi}{\partial n} = \frac{2\alpha^2 mn\beta_A}{(K^2 + \alpha^2 r^2)^2}. \end{aligned}\right\} \tag{17}$$

由式(11)和式(12),可以得到下列的偏微商关系式:

$$\left.\begin{aligned} &\frac{\partial \omega}{\partial X} = -\frac{\partial m}{\partial T}, \quad \frac{\partial \omega}{\partial Y} = -\frac{\partial k}{\partial T}, \quad \frac{\partial \omega}{\partial Z} = -\frac{\partial n}{\partial T}, \\ &\frac{\partial m}{\partial Y} = \frac{\partial k}{\partial X}, \quad \frac{\partial m}{\partial Z} = \frac{\partial n}{\partial X}, \quad \frac{\partial k}{\partial Z} = \frac{\partial n}{\partial Y}. \end{aligned}\right\} \tag{18}$$

另外,在非均匀流体介质中,由 ω, m, k, n 的运动学关系[6]及式(15),有

$$\left.\begin{aligned} \frac{D_g \omega}{DT} &= +\left(\frac{\partial F}{\partial T}\right)_{X,Y,Z,m,k,n} = -m\frac{\partial U}{\partial T} + \frac{m}{(K^2 + \alpha^2 r^2)}\frac{\partial \beta_A}{\partial T}, \\ \frac{D_g m}{DT} &= -\left(\frac{\partial F}{\partial X}\right)_{T,Y,Z,m,k,n} = 0, \\ \frac{D_g k}{DT} &= -\left(\frac{\partial F}{\partial Y}\right)_{T,X,Z,m,k,n} = -m\frac{\partial U}{\partial Y} + \frac{m}{(K^2 + \alpha^2 r^2)}\frac{\partial \beta_A}{\partial Y}, \\ \frac{D_g n}{DT} &= -\left(\frac{\partial F}{\partial Z}\right)_{T,X,Y,m,k,n} = -m\frac{\partial U}{\partial Z} + m\frac{\partial}{\partial Z}\left[\frac{\beta_A}{K^2 + \alpha^2 r^2}\right] \end{aligned}\right\} \tag{19}$$

式中

$$\frac{D_g}{DT} = \frac{\partial}{\partial T} + C_{gx}\frac{\partial}{\partial X} + C_{gy}\frac{\partial}{\partial Y} + C_{gz}\frac{\partial}{\partial Z}$$

能量是以群速度传播的,关于沿纬圈的能量频散性质,叶笃正[7]已有过详细的研究。这里我们分析一下沿经圈(即南北方向)的能量频散特性,特别是研究南北半球能量传播的条件。首先,若有能量从北半球穿过赤道向南半球输送,则 $C_{gy}<0$。而在赤道附近,绝对涡度的梯度一般是大于零的,可见只有当 $k<0$(即槽线为东北—西南向)才能使能量穿过赤道。同理,若能量从南半球向北半球输送,在南半球则需要 $k>0$,即波的槽线呈导式形式,这与 Hoskins 等[8]的数值试验和理论是一致的。他们指出,北半球赤道附近若有东北—西南倾斜的气旋产生时,接着在南半球就出现一个反气旋中心,随着时间的推移,在南半球将出现多个涡旋中心。在另一方面,若有能量从低空向高空输送,在 $\beta_A>0$ 的区域,要求 $n>0$。即等位相线应向上,向西倾斜。反之,在 $\beta_A<0$ 的地区,等位相线应向上,向东倾斜。

4 广义波作用守恒

式(13)的一级近似为:

$$\left(\frac{\partial}{\partial T} + U\frac{\partial}{\partial X}\right)[(K^2 + \alpha^2 r^2)A_0] - (\omega - Um)\left(2m\frac{\partial A_0}{\partial X} + A_0\frac{\partial m}{\partial X}\right.$$
$$\left. + 2k\frac{\partial A_0}{\partial Y} + A_0\frac{\partial k}{\partial Y} + 2n\alpha^2\frac{\partial A_0}{\partial Z} + \alpha^2 A_0\frac{\partial n}{\partial Z}\right) - \beta_A\frac{\partial A_0}{\partial X} = 0 \tag{20}$$

上式乘以 $2A_0(K^2+\alpha^2r^2)/\beta_A m$ 后，变成

$$\frac{(K^2+\alpha^2r^2)^2}{\beta_A}\left(\frac{\partial}{\partial T}+U\frac{\partial}{\partial X}\right)A_0^2+\frac{2A_0^2(K^2+\alpha^2r^2)}{\beta_A}\left(\frac{\partial}{\partial T}+U\frac{\partial}{\partial X}\right)(K^2+\alpha^2r^2)$$
$$-(K^2+\alpha^2r^2)\frac{\partial A_0^2}{\partial X}+2m\left(m\frac{\partial A_0^2}{\partial X}+A_0^2\frac{\partial m}{\partial X}+k\frac{\partial A_0^2}{\partial Y}+A_0^2\frac{\partial k}{\partial Y}\right)$$
$$+2m\left(n\alpha^2\frac{\partial A_0^2}{\partial Z}+\alpha^2A_0^2\frac{\partial n}{\partial Z}\right)=0 \tag{21}$$

令 $B=(K^2+\alpha^2r^2)^2A_0^2/\beta_A$，再考虑到群速度的表达式(17)，则式(21)可写成：

$$\frac{\partial B}{\partial T}+\frac{\partial C_{gx}B}{\partial X}+\frac{\partial C_{gy}B}{\partial Y}+\frac{\partial C_{gz}B}{\partial Z}+\frac{B}{\beta_A}\frac{\partial\beta_A}{\partial T}-2mnA_0^2\frac{\partial\alpha^2}{\partial Z}=0 \tag{22}$$

上式乘以 $1/\alpha^2$ 后，变成

$$\frac{\partial\mathscr{E}}{\partial T}+\frac{\partial C_{gx}\mathscr{E}}{\partial X}+\frac{\partial C_{gy}\mathscr{E}}{\partial Y}+\frac{\partial C_{gz}\mathscr{E}}{\partial Z}+\frac{\mathscr{E}}{\beta_A}\frac{\partial\beta_A}{\partial T}=0 \tag{23}$$

其中

$$\mathscr{E}=\frac{(K^2+\alpha^2r^2)^2A_0^2}{\alpha^2\beta_A}=\frac{m^2\beta_AA_0^2}{\alpha^2(\omega-Um)^2} \tag{24}$$

式中后面这个表达式已用了色散关系式(15)。

若基本场是定常的，即 $\partial\beta_A/\partial T=0$，我们得到波作用守恒方程：

$$\frac{\partial\mathscr{E}}{\partial T}+\nabla_3\cdot(C_g\mathscr{E})=0 \tag{25}$$

通常波作用量的一般定义为：

$$E_0=A^2/(\omega-Um)$$

因此，我们可以称 $\mathscr{E}$ 为广义波作用。

Koroly 和 Hoskins[10] 按 Whitham[11] 和 Andrews 及 McIntyre[12] 的一般原理直接写出了波作用守恒，但除他们的工作中未考虑基本场随时间的变化外，所得的波作用量的形式与我们的也稍有差异。

5　扰动的稳定性

设在所考虑的区域 V 的边界上，扰动为零。由式(23)取积分，得

$$\iiint_V\frac{1}{\beta_A}\frac{\partial\beta_A\mathscr{E}}{\partial T}\mathrm{d}V=\iiint_V\frac{\beta_A}{(K^2+\alpha^2r^2)^2(C-U)^2}\frac{\partial}{\partial T}\left[\frac{(K^2+\alpha^2r^2)^2A_0^2}{\alpha^2}\right]\mathrm{d}V=0 \tag{26}$$

式中已应用了 $\mathscr{E}$ 的定义和色散关系式(15)。$C=\omega/m$ 为相速度。

当背景场为定常时，对方程(25)取体积分，可得

$$\iiint_V\frac{\partial\mathscr{E}}{\partial T}\mathrm{d}V=\frac{\partial}{\partial T}\iiint_V\frac{\beta_AA_0^2}{\alpha^2(C-U)^2}\mathrm{d}V=0 \tag{27}$$

不失一般性，假定初始扰动为零，则上式变成

$$\iiint_V\frac{\beta_AA_0^2}{\alpha^2(C-U)^2}\mathrm{d}V=0 \tag{28}$$

根据上式，可以对扰动不稳定性存在的条件进行讨论。

首先,如扰动是非中性的,即扰动的振幅或波长将随时间变化。此时无论从式(26)(背景场非定常)或从式(28)(背景场定常),都要求广义位势涡度梯度至少在区域中的某一条空间曲线 l 上为零,即

$$l \in V, \quad \beta_A = 2\cos\phi - \frac{\partial^2 U}{\partial y^2} - \frac{1}{\rho_0}\frac{\partial}{\partial z}\left(\rho_0 \alpha^2 \frac{\partial U}{\partial z}\right) = 0 \tag{29}$$

这与 Charney 和 Stern[1] 以及 Pedlosky[13] 用不同的方法得到的扰动不稳定的必要条件是一致的。

特别是对正压情况,式(29)退化成(有量纲量)

$$\beta - \frac{\mathrm{d}^2 U}{\mathrm{d}y^2} = 0 \tag{30}$$

这是郭晓岚[2] 的结果。

很有意思,由于 Eady 波[3] 无 β 效应,用的是 Boussinesq 近似,且基本气流只是高度的线性函数,所以条件式(29)自动满足。由于式(29)只是扰动发展的必要条件,因此对于螺旋 Rossby 波在什么样的背景条件下才能发展尚需做进一步研究。

令

$$E = (K^2 + \alpha^2 r^2) A_0^2 \tag{31}$$

它正比于扰动动能。这时式(25)可写成:

$$\frac{\partial E}{\partial T} + \nabla_3 \cdot (C_g E) = - A_0^2 \alpha^2 \beta_A \frac{D_g}{DT}\left(\frac{K^2 + \alpha^2 r^2}{\alpha^2 \beta_A}\right) \tag{32}$$

考虑到式(15)至式(19),上式可写成:

$$\frac{\partial E}{\partial T} + \nabla_3 \cdot (C_g E) = A_0^2 \left\{2mk \frac{\partial U}{\partial Y} + 2mn\alpha^2 \frac{\partial U}{\partial Z} + \frac{2mn\beta_A}{(K^2 + \alpha^2 r^2)} \frac{\partial \alpha^2}{\partial Z}\right\} \tag{33}$$

将 β_A 的表达式代入,并稍加整理,得到

$$\frac{\partial E}{\partial T} + \nabla_3 \cdot (C_g E) = A_0^2 \left\{2mk \frac{\partial U}{\partial Y} + 2mn\alpha^2 \frac{\partial U}{\partial Z}(1 - g_1) - 2mng_2\left(2\cos\phi - \frac{\partial^2 U}{\partial y^2} - \alpha^2 \frac{\partial^2 U}{\partial z^2}\right)\right\} \tag{34}$$

式中

$$g_1 = g_2\left(\sigma_z + \alpha^2 \frac{\partial}{\partial z}(1/\alpha^2)\right) \tag{35}$$

$$g_2 = \alpha^2 \frac{\partial}{\partial z}\left(\frac{1}{\alpha^2}\right) \cdot \frac{\alpha^2}{(K^2 + \alpha^2 r^2)} \tag{36}$$

而

$$\sigma_z = - \frac{1}{\rho_0} \frac{\partial \rho_0}{\partial z} \tag{37}$$

对于正压大气,式(34)可简化成:

$$\frac{\partial E}{\partial T} + \nabla_2 \cdot (C_g E) = A_0^2 2mk \frac{\partial U}{\partial Y} \tag{38}$$

此式最早由伍荣生推出①。

在边界扰动为零的条件下,对式(34)取体积分,得到

① "正压大气中波动的发展",1978 年江苏省气象学会论文汇编。

$$\iiint_V \frac{\partial E}{\partial T}\mathrm{d}V = 2\pi a\int_{Z_1}^{Z_2}\int_{Y_1}^{Y_2}\left\{2mk\frac{\partial U}{\partial Y} + 2mn\alpha^2(1-g_1)\frac{\partial U}{\partial Z} - 2mng_2\left(2\cos\phi - \frac{\partial^2 U}{\partial y^2} - \alpha^2\frac{\partial^2 U}{\partial z^2}\right)\right\}A_0^2\mathrm{d}Y\mathrm{d}Z \tag{39}$$

下面分别讨论几种简化情况。

(1) 只有 $\partial U/\partial y\neq 0$ 的纯正压大气

这时式(39)化简成

$$\iint_{S_1}\frac{\partial E}{\partial T}\mathrm{d}S = 2\pi a\int_{Y_1}^{Y_2}2mk\frac{\partial U}{\partial Y}A_0^2\mathrm{d}Y \tag{40}$$

如伍荣生指出，在西风急流以南，$\partial U/\partial Y>0$，故导式的螺旋 Rossby 波($k>0$)能发展，而在急流以北，$\partial U/\partial Y<0$，曳式的螺旋 Rossby 波($k<0$)能发展①。这也是卢佩生、曾庆存的结果[14]。

(2) 静力稳定度为常数，但$\dfrac{\partial U}{\partial y}=0$ 的斜压大气$\left(即\dfrac{\partial U}{\partial z}\neq 0\right)$

这时我们有

$$\iint_{S_2}\frac{\partial E}{\partial T}\mathrm{d}S = 2\pi a\int_{Z_1}^{Z_2}2mn\alpha^2\frac{\partial U}{\partial Z}A_0^2\mathrm{d}Z \tag{41}$$

可见，在高空急流的下方，由于 $\partial U/\partial Z>0$，因此需 $n>0$，即扰动的槽脊线向上向西倾斜。而在高空急流的上方，由于 $\partial U/\partial Z<0$，因此只有对槽脊线向上向东倾斜的扰动(即 $n<0$)，才能发展。当然，在此只讨论风的垂直切变作用，在现实大气的平流层中，扰动的发展尚需考虑其他因素。

(3) 静力稳定度为常数，但 $\partial U/\partial y\neq 0$ 的斜压大气

这时我们有

$$\iiint_V \frac{\partial E}{\partial T}\mathrm{d}V = 2\pi a\int_{Z_1}^{Z_2}\int_{Y_1}^{Y_2}\left(2mk\frac{\partial U}{\partial Y} + 2mn\alpha^2\frac{\partial U}{\partial z}\right)A_0^2\mathrm{d}Y\mathrm{d}Z \tag{42}$$

如果(Y_0, Z_0)为西风急流的轴线，则联合上述两种简单情况，发展性扰动的结构将如图 1 所示。也就是说，若扰动发展，则等位相线在急流以南须呈西北—东南走向，在急流以北呈东北—西南走向，在急流下方须向上向西倾斜，在急流上方应向上向东倾斜。

(4) $\beta=0$，且 $\partial U/\partial z=$常数，$\partial U/\partial y=0$，但静力稳定度随高度有变化的斜压大气

如取 Boussinesq 近似，则有

$$\iint_{S_2}\frac{\partial E}{\partial T}\mathrm{d}S = 2\pi a\int_{Z_1}^{Z_2}2mn\alpha^2\frac{\partial U}{\partial z}(1-g)\mathrm{d}Z \tag{43}$$

在这种情况下，不管静力稳定度随高度的变化如何，由于

$$g = 4\left(\frac{1}{N^2}\frac{\partial N^2}{\partial z}\right)^2\Big/\left[4n^2 + \left(\frac{1}{N^2}\frac{\partial N^2}{\partial z}\right)^2\right] < \left(\frac{1}{n}\frac{1}{N^2}\frac{\partial N^2}{\partial z}\right)^2 \sim O\left(\left(\frac{1}{n}\frac{\Delta N^2}{N^2}\right)^2\right)$$

式中，ΔN^2 是 N^2 的垂直变化量级，在一般情况下，$\Delta N^2/N^2<1$，因此只要 $n^2\geqslant 1$，总有

$$0 < g < O(1)$$

由此可见，在这种情况下，扰动发展的条件与(2)相同，只不过如要达到相同的扰动能量变化，需更强的风速垂直切变而已。

如考虑基本场的密度随高度变化，即在 g_1 中保留 σ_z，但考虑到在一般情况下有

---

① m 总取正值。

$$\frac{1}{N^2}\frac{\partial N^2}{\partial z} < \sigma_Z,$$

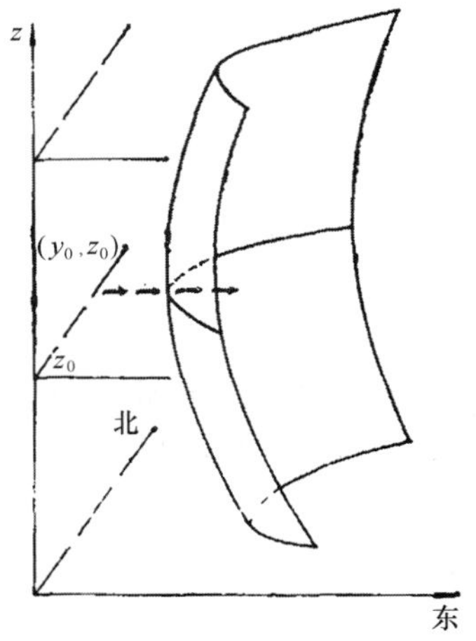

图 1　西风在(Y_0, Z_0)有极大值时，扰动发展所需的等位相线结构
（箭头表示急流位置）

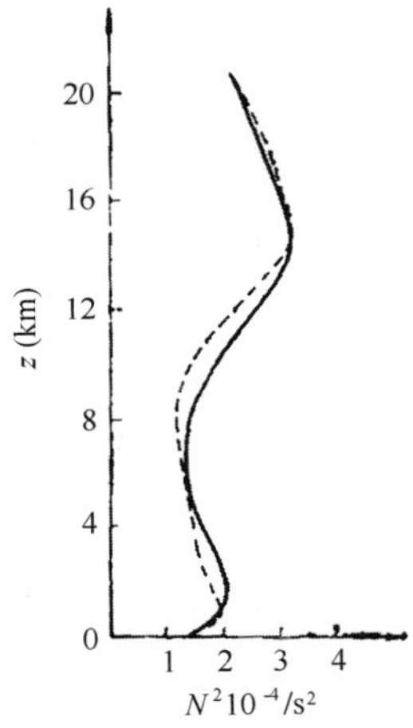

图 2　1 月（实线）、7 月（虚线）N^2 随高度的分布

所以 $g_1 = g_2\sigma_Z$，此时与上式相对应的 g 值应为：

$$g = \frac{1}{N^2}\frac{\partial N^2}{\partial z}\cdot\frac{4\sigma_z}{(4n^2+\sigma_z^2)} < \sigma_z\frac{1}{N^2}\frac{\partial N^2}{\partial z}/n^2$$

可见，这时扰动是否发展，要视静力稳定度的垂直分布而定。图 2 给出了 1 月和 7 月沿纬圈平均的 N^2 随高度的分布曲线（资料取自参考文献[9]）。由曲线可见，在行星边界层（即 1 km 左右）以上的对流层中下部，以及在 14 km 以上的平流层中，$\partial N^2/\partial z<0$，所以式(43)的$(1-g)>0$，扰动发展的条件，当 $\partial u^2/\partial z>0$ 时，仍相似于情况 b。但在对流层顶附近，即 8～14 km 之间，由于 $\partial N^2/\partial z>0$，这时扰动在 $\partial U/\partial z>0$，$n>0$ 的条件下是否发展，要视 g 值的大小而定。事实上，由于 $\sigma_z \sim O(1)$，$O\left(\frac{\Delta N^2}{N^2}\right)<1$，而 $n^2\geqslant 1$，所以 $g<1$。换言之，只要风速随高度线性增加，在这样的条件下，无论静力稳定度的垂直变化如何，向上向西倾斜的扰动总是发展的。

(5) β 的作用

现在来讨论在静力稳定度随高度有变化的情况下，β 对扰动发展所起的作用。先不考虑 $\partial^2U/\partial y^2$ 和 $\partial^2U/\partial z^2$ 的影响，在这种情况下，式(39)变为：

$$\iiint_V \frac{\partial E}{\partial T}\mathrm{d}V = 2\pi a\int_{Z_1}^{Z_2}\int_{Y_1}^{Y_2}\left\{\left[2mk\frac{\partial U}{\partial Y}+2mn\alpha^2(1-g_1)\frac{\partial U}{\partial Z}\right]-2mng_2 2\cos\phi\right\}A_0^2\mathrm{d}Y\mathrm{d}Z \qquad (44)$$

可见，这时 β（即 $2\cos\phi$）的作用视 g_2 的符号，也即由静力稳定度随高度的变化而定。在 $g_2<0$ 的区域，即当静力稳定度随高度减小时，β 的作用有利于向上向西倾斜扰动的发展。而在静力稳定度随高度增加的区域，β 对向上向西倾斜的扰动将起稳定的作用。根据这一分析，由于 β 的作用总是和静力稳定度的垂直变化联系在一起的，所以在静力稳定度为常数的模式中，β 基本上对扰动的发展条件无贡献，这就是为什么在 Eady 的模式中，虽然没有 β 效应，仍可得到与观测接近的发展的扰动。因为在斜压大气中扰动发展的条件主要决定于风速垂直切变，即基本场位能的释放。

(6)急流的作用

这时式(39)成立。设急流在水平和垂直方向上,风廓线的二次切变不处处为零,或存在有限的反折点(即存在 $\partial^2 U/\partial y^2$ 或 $\partial^2 U/\partial z^2$ 为零的点),而在急流轴上 $\partial U/\partial y=0$,$\partial^2 U/\partial y^2<0$;$\partial U/\partial z=0$,$\partial^2 U/\partial z^2<0$,这时把式(39)写成:

$$\iiint_V \frac{\partial E}{\partial T}\mathrm{d}V = 2\pi a\int_{Z_1}^{Z_2}\int_{Y_1}^{Y2}\left\{2mk\frac{\partial U}{\partial Y} + 2mn\alpha^2(1-g_1)\frac{\partial U}{\partial Z} - 2mng_2\beta^*\right\}A_0^2\mathrm{d}Y\mathrm{d}Z \tag{45}$$

在急流轴上及其邻近区域,有

$$\beta^* = 2\cos\phi - \frac{\partial^2 U}{\partial y^2} - \alpha^2\frac{\partial^2 U}{\partial z^2} \geqslant 2\cos\phi \equiv \beta \tag{46}$$

因此,可以认为至少在急流轴附近的区域中,急流的作用相当于增大了 β 效应。而 β 的作用在上面已加以分析过。至于在远离急流轴的区域中,其作用要视风廓线的形式而定。

6　结论

本文应用 WKB 方法导出了在一般较接近实际大气的模式中(包括背景场空间的不均匀和时间的变化),螺旋 Rossby 波的广义波作用量所满足的方程,当背景场不随时间变化时,波作用量具有守恒性。

应用所得到的波作用量方程,分别讨论了螺旋 Rossby 波不稳定的必要条件。首先得到了 Charney 和 Stern 以及郭晓岚的必要条件,即在一般正压和斜压两者具存的大气中,广义位势涡度梯度至少在区域中的某一条空间曲线上为零,而在纯正压大气中,至少在某一条平面曲线上绝对涡度为零。这些条件对螺旋和一般型 Rossby 波均成立。其次指出,若不满足上述条件,螺旋 Rossby 波仍然可以发展。在正压情况下,急流以南导波发展,急流以北曳波发展。在纯斜压情况下且静力稳定度不随高度改变时,在高空急流的下方,由于 $\partial U/\partial Z>0$,只有向上向西倾斜的扰动能发展,而在高空急流上方,由于 $\partial U/\partial Z<0$,只有槽脊线向上向东倾斜的扰动能发展。进而讨论了静力稳定度随高度变化的一般情况下,它对扰动发展虽起一定作用,但不会改变问题的本质。最后讨论了 β 和急流的作用,指出 β 的作用总是与静力稳定度 N^2 随高度的变化联系在一起的。若 N^2 随高度减小,β 的作用有利于向上向西扰动的发展。在 N^2 为常数的模式中,β 基本上对扰动的发展无贡献。由于扰动发展的条件主要决定于风速垂直切变,即基本场位能的释放,这就是为什么在 Eady 模式中,虽然没有 β 效应,仍可得到与观测接近的发展的扰动。

参考文献

[1]　Charney, J. G. & Stern, M. *J.Atmos.Soi.*, 1962,19:159-172.

[2]　Kuo, H. L. *J. Meteor.*, 1949,6:105-122.

[3]　Eady. E. T. *Tellus*, 1949,1:33-52.

[4]　巢纪平,叶笃正. 大气科学,1977,2:81-88.

[5]　刘式适,杨大升. 气象学报,1979,37:14-27.

[6]　Bretherton. F. P. *Mathematical Problems in the Geophysical Sciences* (Ed. Reld, W. H.), Amer.61-102.

[7]　Yeh. T. C.(叶笃正) *J. Met.*, 1949,6:1-16.

[8]　Hoskins, B. J., Simmons, A. J. & Andrews, D. G., *Quart. J. Roy. Meteo. Soc.*, 1977,193:553-568.

[9]　Oort, A. H. & Rasmusson, E. M. *Atmospheric Ciroulation Statistics.*

[10] Koroly, D. J. & Hoskins, B. J. *J. Meteo. Soc. of Japan*, 1982,60:109-123.

[11] Whitham, G. B. *J. Fluid Mech.*, 1970,44:373-395.

[12] Andrews, D. G. & McIntyre, M. E. *J. Fluid. Mech.*, 1978b,89:647-664.

[13] Pedlosky, J. *Geophysical Fluid Dynamics*, Springer-Verlag New York Inc., 440.

[14] 卢佩生,曾庆存. 大气科学,1981,5:1-8.

论大气运动的多时态特征*

——适应、发展和准定常演变

叶笃正[1]　巢纪平[2]

(1. 中国科学院大气物理研究所,北京 100080;2. 国家海洋环境预报研究中心,北京 100081)

摘要:对中国气象学家在地转适应和运动多时态特征方面的研究做了概要性总结。其中主要介绍在风、压场的地转适应以及适应过程完成后,尚可区别出以 Rossby 波频散为特征的发展过程,和以平流过程为特征的准定常演变过程。文中指出,运动的多时态特征是由于动力系统中存在多种物理过程造成,因此在小尺度、中尺度以及热带大尺度运动中都存在适应、发展和准定常演变过程。文中进而指出,这种多时态特征也存在于海气耦合的气候系统中,以及更复杂的气候系统中,也即带有一事实上的普遍性。

关键词:大气动力学;地转适应;大气运动

1　问题的提出

在重力场和地球旋转作用下的大气运动,其最基本的状态是准静力平衡和准地转平衡,即

$$p_z \sim -\rho g \tag{1}$$

$$p_x \sim p_y \sim -\rho f V \tag{2}$$

式中,p 和 ρ 分别为气压和空气的密度,g 为重力,$f=2\Omega\sin\varphi$ 为 Coriolis 参数(Ω 为地球自转角速度,φ 为纬度),V 为速度,下标 x、y、z 分别代表对该项取微商。

在物理学上式(1)是容易理解的,由于地球重力场的作用,流体要向靠近地球固体边界的一层中集中,由于质量的垂直分布造成了压力在垂直方向的变化,其变化通常由静力方程所决定。这意味着对于大尺度运动,基本上是准水平的,因为不可能在相当大的范围内空气克服重力场而产生强的垂直加速度。

在物理学上式(2)只表明压力场和风场之间的相互依存关系。但古典的或一般的看法是,气压场是第一位的,由气压场决定了风场。对于这种观点可以给出一简单的论证。注意到,由式(1),压力随高度的变化其量级相当于两个高度上的压力差值,如大气的厚度为 D,则有

$$p_z \sim \frac{P}{D} \tag{3}$$

对于理想气体,由状态方程估计出

$$D \sim \frac{P}{\rho g} = \frac{R\overline{T}}{g} = H \tag{4}$$

* 大气科学,1998,22(4):385-398.

式中，$\overline{T}$为大气的辐射平衡温度。注意到，大气运动的根本能量来自太阳辐射，而地球与太阳的相对位置在宇宙中是固定了的，因此当组成大气的吸收辐射介质及地球表层吸收辐射的物质被确定之后，由辐射能量造成的平衡温度 T 以及它的经向分布随之也是确定了的。由于气体常数 R 和重力加速度 g 都是确定了的量，因此 H 是一个确定了的量，称为等温大气的厚度，它与均质流体的厚度同量级，约为 10 km。

另一方面，当大气运动的尺度达到地球的旋转作用不能被忽视时，由 g、H 和 f 可以组成一个具有长度量纲的量 L_0，即为

$$L_0 = \frac{\sqrt{gH}}{f} \tag{5}$$

它也是旋转地球大气中一个确定了的固有尺度，同样不决定于运动。L_0 通常称 Rossby 变形半径，约为 3 000 km。由式(2)，考虑了状态方程并略去密度变化后可得

$$\frac{\overline{T}_y}{\overline{T}} \sim \frac{P_y}{P} \sim \frac{\rho f V}{P} \sim \frac{fV}{gH} \tag{6}$$

由此得

$$V \sim \frac{gH}{f}\frac{\overline{T}_y}{\overline{T}} \sim \frac{\delta\,\overline{T}}{\overline{T}}\frac{gH}{fa} = \frac{\delta\,\overline{T}}{\overline{T}}\frac{L_0}{a}C \tag{7}$$

式中，δT 为赤道到极地的经圈温度差，a 为地球半径，$C=\sqrt{gH}$ 称重力波波速。由此可见，速度被经圈温度差所决定，而由式(6)，气压场被温度场所决定，因而，风场是被压力场决定的。考虑到 $\delta\,\overline{T}/\overline{T} \sim 10^{-1}$，$L_0/a \sim 0.5$，$C \sim 300$ m/s，所以 $V \sim 15$ m/s，这就是通常地转风的速度量级。

上述由气压场决定地转风场的经典理论，一直沿袭到 20 世纪 30 年代末时，才有 Rossby 对此提出了挑战[1,2]。他指出，当风应力作用于海洋并引导出一支洋流后，质量场将随之调整产生横切于洋流的压力梯度并使洋流得以维持，而调整的最终状态，是地转平衡的。这一理论与上面经典的看法不同，它表明流场可以是第一位的，而压力场向变化后的流场调整，并把压力场和流场之间的这种调整称为地转适应。继后，Oboukhov[3] 也支持 Rossby 的压力场向流场适应的观点。

1957 年，叶笃正[4]对在地转平衡中谁是主导的方面通过计算后给出了一个辩证的观点，他指出，对于大尺度扰动气压场是主导的，风场地转地向气压场适应；而对于小尺度扰动，风场是主导的，气压场地转地向风场适应。1963 年，曾庆存[5]进一步指出，所谓尺度的大小有一准则，即存在一个临界尺度 L_0，当尺度 $L>L_0$ 时，风场向气压场适应，当尺度 $L<L_0$ 时，气压场向风场适应，而 L_0 即为式(5)表明的 Rossby 变形半径。继后，中国气象学家在这方面做了一系列研究，如陈秋士[6]研究了斜压大气热成风的地转适应，曾庆存[7]提出大气运动在时间上的可区分性，即可以分成快的适应过程和慢的演变过程。叶笃正和李麦村[8]则进一步指出，即使在演变过程中尚可分成较快的发展阶段和准定常的缓变阶段。这样，对大气中的大尺度运动，至少存在 3 个可以区别的时态。这自然是对大气运动认识的深化。

叶笃正和李麦村的研究还表明[9]，即使对于中、小尺度运动也存在风场和气压场之间的适应以及适应后的准平衡演变。最近，巢纪平、林永辉的研究表明[10]，即使在热带，虽然 Coriolis 力很小，但气压场相对来讲也比较均匀，即气压梯度较弱，因此也可以存在纬圈方向或经圈方向的半地转适应过程，以及适应后的演变过程。这表明，大气运动的多时态特征，不仅对不同尺度的运动存在，在不同的

纬度上也同样存在。可以认为这是大气运动一种普遍性的规律。

本文将对中国气象学家在这方面的研究成果做一个概要性的总结。

2　运动的多时态特征

大尺度的大气运动具有静力平衡的特征，在可压缩大气中，静力平衡可以通过声波的频散而达到[9,15]，声波的频散是极其迅速的，其过程对大尺度运动无大的影响，因此我们将不讨论这一过程，并认为大气运动已处在静力平衡状态下。

关于旋转大气中运动的多时态特征，今以正压运动方程来分析。事实上，对于斜压大气（和斜压海洋）如果方程是线性的，则在垂直方向可以用本征模展开，对于任一个本征模（本征值在物理上即为等值厚度）其水平结构方程和正压运动方程并无差别（见参考文献[11]）。

正压运动的控制方程为

$$\frac{\partial u}{\partial t} + u\frac{\partial u}{\partial x} + v\frac{\partial u}{\partial y} - fv = -\frac{\partial \varphi}{\partial x} \tag{8}$$

$$\frac{\partial v}{\partial t} + u\frac{\partial v}{\partial x} + v\frac{\partial v}{\partial y} + fu = -\frac{\partial \varphi}{\partial y} \tag{9}$$

$$\frac{\partial \varphi}{\partial t} + u\frac{\partial \varphi}{\partial x} + v\frac{\partial \varphi}{\partial y} + C^2\left(\frac{\partial u}{\partial x} + \frac{\partial v}{\partial y}\right) = 0 \tag{10}$$

式中，φ 为自由面高度上的重力位势。

引进特征量

$$(u,v) = V(u',v'),\quad \varphi = f_0LV\varphi',\quad f = f_0f'(x,y) = L(x',y'),\quad t = Tt' \tag{11}$$

于是得到无量纲方程（略去撇号）

$$\varepsilon\frac{\partial u}{\partial t} + Ro\left(u\frac{\partial u}{\partial x} + v\frac{\partial u}{\partial y}\right) - fv = -\frac{\partial \varphi}{\partial x} \tag{12}$$

$$\varepsilon\frac{\partial v}{\partial t} + Ro\left(u\frac{\partial v}{\partial x} + v\frac{\partial v}{\partial y}\right) + fu = -\frac{\partial \varphi}{\partial y} \tag{13}$$

$$\left(\frac{L}{L_0}\right)^2\left[\varepsilon\frac{\partial \varphi}{\partial t} + Ro\left(u\frac{\partial \varphi}{\partial x} + v\frac{\partial \varphi}{\partial y}\right)\right] + \left(\frac{\partial u}{\partial x} + \frac{\partial v}{\partial y}\right) = 0 \tag{14}$$

式中

$$\varepsilon = \frac{1}{f_0T},\qquad Ro = \frac{V}{f_0L} \tag{15}$$

Ro 称 Rossby 数，L 是运动在准地转平衡附近的特征尺度。这里应注意到对于运动的时间尺度并没有确定。事实上，运动的特征时间并不是唯一的，可以根据运动本身的特征，区分出多种特征时间，对波动运动来说即多种特征周期。

由参数 f_0、L、C、V 以及地球球面性的 $\beta[=(1/a)(\mathrm{d}f/\mathrm{d}\varphi)]$ 效应，可以组成 4 个不同的特征时间或周期，它们分别为

$$T_1 = \frac{L}{C},\quad T_2 = \frac{1}{f_0},\quad T_3 = \frac{1}{\beta L},\quad T_4 = \frac{L}{V} \tag{16}$$

其物理意义为：T_1 为重力波传播的特征时间；T_2 为惯性振荡的特征时间；T_3 为 Rossby 波的频散时间；T_4 为平流特征时间。在这 4 个特征周期中，有

$$\frac{T_1}{T_2}=\frac{L}{L_0} \tag{17}$$

注意到,如果运动的水平特征尺度接近 Rossby 变形半径 L_0,则 $T_1 \approx T_2$,即这两个时间是不可区别的,事实上这时它们表征了重力惯性波的特征周期。

在另一方面,T_3 和 T_4 给出

$$\frac{T_3}{T_4}=\frac{V}{\beta L^2}=\left(\frac{L_C}{L}\right)^2 \tag{18}$$

式中

$$L_C=\sqrt{\frac{V}{\beta}} \tag{19}$$

在一个基本流为 V 的背景中,当 Rossby 波被激发出来后,L_C 即为 Rossby 波的特征波长。注意到如取 $V \approx 10$ m/s,则 $L_C \approx 10^3$ km,而 $L_0=3\times10^3$ km,所以 $L_0>L_C$。因此如取 $L\approx L_0$,则 $T_3<T_4$,即 Rossby 波的频散时间要短于流体质点的平流时间。

此外,尚有

$$\frac{T_1}{T_4}=\frac{V}{C}\ll 1 \tag{20}$$

而

$$\frac{T_1}{T_3}=\frac{V}{C}\left(\frac{L}{L_C}\right)^2 \tag{21}$$

如取 $L\approx L_0$,则 $(L/L_C)^2\approx 10$,而 $V/C\approx 3\times10^{-2}$,因此有 $T_1<T_3$。于是当取 $L\approx L_0$ 时,有

$$T_1 \approx T_2 < T_3 < T_4 \tag{22}$$

这表明,大尺度大气运动的动力学特征至少存在 3 个特征时间[8,12],即地转适应时间、Rossby 波的频散时间和平流特征时间。如果存在缓变的强迫源,并当强迫源的特征时间 $T_5>T_4$ 时,则大气运动在完成上述诸过程后,将在强迫作用下运动。于是对于一个非定常的初值问题,大气运动将经历以下特征时间为表征的诸物理过程,即

$$T_1(\approx T_2) < T_3 < T_4 < T_5 \tag{23}$$

这就是大气运动的多时态特征。很清楚,这种多时态特征对应于不同的物理过程。

在无外强迫的大气系统中最慢的特征时间为 T_4,下面来分析在 T_4 前(包括 T_4)所发生的动力学过程。注意到当 T 趋近于 T_4 时,有

$$\varepsilon = Ro \tag{24}$$

以及

$$\frac{T_1}{T_4}=\frac{V}{C}=Fr^{1/2} \tag{25}$$

Fr 为 Froude 数,其大小为 $O(10^{-3})$,所以有

$$Ro = Fr^{1/2}(L_0/L) \tag{26}$$

这时方程式(12)至式(14)为

$$Fr^{1/2}\left(\frac{L_0}{L}\right)\left(\frac{\partial u}{\partial t}+u\frac{\partial u}{\partial x}+v\frac{\partial u}{\partial y}\right)-fv=-\frac{\partial \varphi}{\partial x} \tag{27}$$

$$Fr^{1/2}\left(\frac{L_0}{L}\right)\left(\frac{\partial v}{\partial t}+u\frac{\partial v}{\partial x}+v\frac{\partial v}{\partial y}\right)+fu=-\frac{\partial \varphi}{\partial y} \tag{28}$$

$$Fr^{1/2}\left(\frac{L}{L_0}\right)\left(\frac{\partial \varphi}{\partial t}+u\frac{\partial \varphi}{\partial x}+v\frac{\partial \varphi}{\partial y}\right)+\left(\frac{\partial u}{\partial x}+\frac{\partial v}{\partial y}\right)=0 \tag{29}$$

3　地转适应过程

现在来分析地转适应过程。在初始时刻附近，当 $L=L_0$ 时，引进

$$\tau=\frac{t}{\delta} \tag{30}$$

其中，δ 为时间边界层的厚度，时间边界层可看成是在初值附近物理场变化剧烈的一个时间区间，这样式(27)至式(29)可写成

$$\frac{Fr^{1/2}}{\delta}\frac{\partial u}{\partial \tau}-fv+\frac{\partial \varphi}{\partial x}=-Fr^{1/2}\left(u\frac{\partial u}{\partial x}+v\frac{\partial u}{\partial y}\right) \tag{31}$$

$$\frac{Fr^{1/2}}{\delta}\frac{\partial v}{\partial \tau}-fu+\frac{\partial \varphi}{\partial y}=-Fr^{1/2}\left(u\frac{\partial v}{\partial x}+v\frac{\partial v}{\partial y}\right) \tag{32}$$

$$\frac{Fr^{1/2}}{\delta}\frac{\partial \varphi}{\partial \tau}+\frac{\partial u}{\partial x}+\frac{\partial v}{\partial y}=-Fr^{1/2}\left(u\frac{\partial \varphi}{\partial x}+v\frac{\partial \varphi}{\partial y}\right) \tag{33}$$

如取边界层的厚度为 $\delta=Fr^{1/2}$（这样的取法在物理上是容易理解的，即在这一层中主要反映以传播速度为 C 的重力惯性波，而准地转平衡是通过重力惯性波的频散而建立的），则有

$$\frac{\partial u}{\partial \tau}-fv+\frac{\partial \varphi}{\partial x}=-Fr^{1/2}\left(u\frac{\partial u}{\partial x}+v\frac{\partial u}{\partial y}\right) \tag{34}$$

$$\frac{\partial v}{\partial \tau}-fu+\frac{\partial \varphi}{\partial x}=-Fr^{1/2}\left(u\frac{\partial v}{\partial x}+v\frac{\partial v}{\partial y}\right) \tag{35}$$

$$\frac{\partial \varphi}{\partial \tau}+\frac{\partial u}{\partial x}+\frac{\partial v}{\partial y}=-Fr^{1/2}\left(u\frac{\partial \varphi}{\partial x}+v\frac{\partial \varphi}{\partial y}\right) \tag{36}$$

由于 $Fr^{1/2}\ll 1$，上式可略去右端非线性项，而得到

$$\frac{\partial u}{\partial \tau}-fv+\frac{\partial \varphi}{\partial x}=0 \tag{37}$$

$$\frac{\partial v}{\partial \tau}-fu+\frac{\partial \varphi}{\partial y}=0 \tag{38}$$

$$\frac{\partial \varphi}{\partial \tau}+\frac{\partial u}{\partial x}+\frac{\partial v}{\partial y}=0 \tag{39}$$

由于这组方程只适用于时间边界层内，因此它所描写的是地转适应过程。

引进势量 φ 和管量 ψ（流函数），令

$$u=\frac{\partial \varphi}{\partial x}-\frac{\partial \psi}{\partial y},\quad v=\frac{\partial \varphi}{\partial y}+\frac{\partial \psi}{\partial x} \tag{40}$$

有涡度 ζ 和辐散 D 为

$$\zeta=\frac{\partial v}{\partial x}-\frac{\partial u}{\partial y}=\nabla^2\psi,\quad D=\frac{\partial u}{\partial x}+\frac{\partial v}{\partial y}=\nabla^2\varphi \tag{41}$$

当 f 取常数时，由式(37)至式(39)给出

$$\frac{\partial \psi}{\partial \tau} + f\varphi = 0 \tag{42}$$

$$\frac{\partial \varphi}{\partial \tau} - f\psi + \varphi = 0 \tag{43}$$

$$\frac{\partial \varphi}{\partial \tau} + \nabla^2 \varphi = 0 \tag{44}$$

由此得到两个重要的方程,其一为

$$\frac{\partial^2 \varphi}{\partial t^2} - \nabla^2 \varphi + f^2 \varphi = 0 \tag{45}$$

这表明表征辐散的势量将以重力惯性波的形式频散。由它的初值问题的解表明,当时间充分大时,$\partial\varphi/\partial\tau \to 0$ 于是由式(43)给出:

$$f\psi = \varphi \tag{46}$$

即运动是地转平衡的。

另一个重要方程为

$$\frac{\partial}{\partial \tau}(\nabla^2 \psi - f\varphi) = 0 \tag{47}$$

积分后有

$$\nabla^2 \psi - f\varphi = \nabla^2 \psi |_{\tau=0} - f\varphi |_{\tau=0} \tag{48}$$

这表明在适应过程中,位势涡度是时间不变式。当时间充分大,地转平衡达到后,考虑到式(46),上式变为

$$\nabla^2 \psi - f^2 \psi = \nabla^2 \psi |_{\tau=0} - f\varphi |_{\tau=0} \tag{49}$$

因此,适应后的场并不需要解式(45)而得到,可以直接解式(49)而得适应后的流函数(即风场),再由式(46)求适应后的压力场。

Oboukhov[3]分析过一个地转适应的例子,设 $t=0$ 时,$\varphi_0=0$,而速度场由流函数表示成 ψ_0,

$$\psi_0(x,y) = A\left[2 + \left(\frac{R}{L_0}\right)^2 - \left(\frac{r}{R}\right)^2\right] e^{-(r^2/2R^2)} \tag{50}$$

式中,$r^2=x^2+y^2$,R 为扰动的特征尺度。由方程(49) 的解,算出适应后的流函数为

$$\psi(x,y) = A\left[2 - \left(\frac{r}{R}\right)^2\right] e^{-(r^2/2R^2)} \tag{51}$$

而适应后的气压场,由式(46)给出为

$$\varphi = Af\left[\tau - \left(\frac{r}{R}\right)^2\right] e^{-(r^2/2R^2)} \tag{52}$$

Oboukhov 取 $R=500$ km,在这种情况下,$(R/L_0)^2 \ll 1$,因此适应后的流场相对于初始流场变化很小,而适应后的气压场与初始流场相比,变化很大(因初始压力场为零),由此 Oboukhov 得到结论,在流场与气压场的地转平衡关系中,是气压场向流场适应。这一结果支持了 Rossby 的观点。

然而,这样的结论是有条件的,事实上注意到初始函数与适应后的流函数之比为

$$\frac{\psi_0}{\psi} = \left[1 + \frac{\left(\frac{R}{L_0}\right)^2}{2 - \left(\frac{r}{R}\right)^2}\right] \tag{53}$$

可以看到这个比值依赖于初始扰动的尺度，当 R 取 500 km 时，$(R/L_0)^2 \ll 1$，因此 $\Psi_0/\Psi \approx 1$，即适应后的流场变化不大，但如取$(R/L_0)^2 \gg 1$，则适应后的流场可以与初始流场相差很远。注意到这种情况，叶笃正[4]在做了数值计算后指出，对于小尺度扰动，气压场将向风场适应，而对于大尺度扰动则反过来，风场向气压场适应。这样就用一个尺度把上面提到的两种相反的观点辩证地统一起来了，也把地转平衡中气压场与风场之间的因果关系说清楚了。

但是，叶笃正的上述理论，关于扰动尺度的概念是定性的，大小是一个相对的概念，在这方面，曾庆存[5]进一步在理论上证明尺度的大小是相对于 Rossby 变形半径 L_0 而言的，当扰动尺度 $L>L_0$ 时称大尺度扰动，这时风场向气压场适应，当扰动尺度 $L<L_0$ 时称小尺度扰动，这时主要是气压场向风场适应。地转适应中的尺度关系，事实上可以不通过对位势涡度方程的求解而可用更简捷的方法得到。注意到，由方程式(27)至式(29)可直接引进风场和气压场的时间边界层厚度，它们分别为[12]

$$\delta_V = Fr^{1/2}\frac{L_0}{L},\quad \delta_\varphi = Fr^{1/2}\frac{L}{L_0} \tag{54}$$

可见，当 $L>L_0$ 时，$\delta_\varphi>\delta_V$，这意味着当风场已经结束其快变过程而进入缓变过程的区域时，气压场尚未完成其快变过程而仍停留在时间边界层内，因此如取同一时段来比较，风场的变化远比气压场的变化迅速和剧烈，从这个意义上说，是风场向气压场适应的。反之，当 $L<L_0$ 时，$\delta_\varphi<\delta_V$，亦即气压场变化迅速而剧烈，风场变化缓慢和缓和，这表明气压场是向风场适应的。

为什么小尺度的风场可以维持，并且气压场向风场适应，而大尺度的气压场可以维持，风场向气压场适应。对此，叶笃正、朱抱真[13]曾给出一个物理解释。设有一支经圈尺度不大的西风单独存在，由于 Coriolis 力将产生北风，这样质量将由北向南跨过西风输运而建立南高北低的气压梯度，这样将很快达到地转平衡而使西风得到维持。对另一种情况，如果西风的经圈尺度很大，这时在北风向南调动质量的过程中，由于尺度大时间长，无气压平衡的北风将产生东风以削弱西风，从而将进一步削弱北风，减慢向南调动质量的速度，于是在尚未建立足以平衡西风的南北气压梯度时，西风已被大大削弱而难以维持了。如果单独存在的是经向尺度很小的南高北低的气压场，这样质量很快向北输送，在没有建立起平衡这一气压场的西风前，气压场就南北均一化了。相反，如果南高北低的气压场的经向尺度很大，由质量从南向北输送要使气压场均一化的时间很长，这时由南风将引导出西风，并由西风来支持南高北低的气压场，使气压场得以维持并达到风压场之间的地转平衡。

4　发展过程

当 $L=L_0$ 时，对方程式(27)至式(29)进行如下的展开，即

$$在流\begin{bmatrix} u \\ v \\ \varphi \end{bmatrix} = \sum_{n=0} + (Fr^{1/2})^n \begin{bmatrix} u^{(n)} \\ v^{(n)} \\ \varphi^{(n)} \end{bmatrix} \tag{55}$$

$$f = 1 + Fr^{1/2}\beta y \tag{56}$$

其中 $\beta = \beta(L^2/\bar{u})$，$\beta = \mathrm{d}f/\mathrm{d}y$。由此，零级和一级近似方程分别为

$$(Fr^{1/2})^0: \quad \begin{cases} v^{(0)} = \dfrac{\partial \varphi^{(0)}}{\partial x}, \\ u^{(0)} = -\dfrac{\partial \varphi^{(0)}}{\partial y}, \\ \dfrac{\partial u^{(0)}}{\partial x} + \dfrac{\partial v^{(0)}}{\partial y} = 0 \qquad \text{平衡} \end{cases} \tag{57}$$

$$(Fr^{1/2})^1: \quad \begin{cases} \dfrac{\partial u^{(0)}}{\partial t} + u^{(0)}\dfrac{\partial u^{(0)}}{\partial x} + v^{(0)}\dfrac{\partial u^{(0)}}{\partial y} - \beta' y v^{(0)} - v^{(1)} = -\dfrac{\partial \varphi^{(1)}}{\partial x}, \\ \dfrac{\partial v^{(0)}}{\partial t} + u^{(0)}\dfrac{\partial v^{(0)}}{\partial x} + v^{(0)}\dfrac{\partial v^{(0)}}{\partial y} + \beta' y u^{(0)} + u^{(1)} = -\dfrac{\partial \varphi^{(1)}}{\partial y}, \\ \dfrac{\partial \varphi^{(0)}}{\partial t} + u^{(0)}\dfrac{\partial \varphi^{(0)}}{\partial x} + v^{(0)}\dfrac{\partial \varphi^{(0)}}{\partial y} + \dfrac{\partial u^{(1)}}{\partial x} + \dfrac{\partial v^{(1)}}{\partial y} = 0 \end{cases} \tag{58}$$

注意到方程(57)的前两式即为地转风关系，而第三式表明地转风是无辐散的。方程(58)即为曾庆存指出的大气运动可分为适应过程和演变过程中的演变过程[7]。在这里演变过程是准地转的，这一组方程在数值天气预报中有过广泛的应用。

然而，叶笃正和李麦村[8]进一步指出，即使在演变过程中仍然可以区分出相对快变的发展阶段和非常缓慢的准定常演变阶段，事实上后者即为平流过程，这容易由下面的方法将这两种过程区分出来。

注意到由方程(57)和方程(58)可以给出

$$\frac{\partial \Omega_p}{\partial t} + V^{(0)} \cdot \nabla \Omega_p + \beta v^{(0)} = 0 \tag{59}$$

式中

$$\Omega_p = \nabla^2 \varphi^{(0)} - \varphi^{(0)} \tag{60}$$

回到有量纲方程，并引进地转流函数 $\psi = \varphi^{(0)}/f$，得到

$$\left(\frac{\partial}{\partial t} + V \cdot \nabla\right)\Omega_p + \beta v = 0 \tag{61}$$

式中

$$\Omega_p = \nabla^2 \psi - \frac{1}{L_0^2}\psi \tag{62}$$

$$u = -\frac{\partial \psi}{\partial y}, \qquad v = \frac{\partial \psi}{\partial x} \tag{63}$$

方程(61)即为数值天气预报中常用的正压位势涡度方程。

引进

$$(x, y) = L(x', y'), \quad t = \frac{L}{V}y', \quad \psi = \Psi \psi', \quad (u, v) = V(u', v') = \frac{\Psi}{L}(u', v') \tag{64}$$

代入式(61)，得到无量纲方程(略去撇号)

$$\left(\frac{L_C}{L}\right)^2 \frac{\partial \Omega_p}{\partial t} + v = -\left(\frac{L_C}{L}\right)^2 (V \cdot \Omega_p) \tag{65}$$

引进第二时间边界层 δ_2，并令

$$\tau_2 = \frac{t}{\delta_2} \tag{66}$$

于是式(65)给出

$$\left(\frac{L_C}{L}\right)^2 \frac{1}{\delta_2}\frac{\partial \Omega}{\partial \tau} + v = -\left(\frac{L_C}{L}\right)^2 (V \cdot \nabla \Omega_p) \tag{67}$$

可以取第二时间边界层的厚度 $\delta_2 = (L_C/L)^2$,注意到这一时间边界层中包含了 L_C,因此在这一边界层中主要反映了 Rossby 波的频散。而在这时间边界层内的运动方程为

$$\frac{\partial \Omega_p}{\partial \tau_2} + v = -\left(\frac{L_C}{L}\right)^2 (V \cdot \nabla \Omega_p) \tag{68}$$

令

$$\psi = \sum_{n=0} \left(\frac{L_C}{L}\right)^{2n} \psi^{(n)} \tag{69}$$

得到展开式的零级和一级方程分别为

$$\left(\frac{L_C}{L}\right)^0 : \frac{\partial \Omega_p^{(0)}}{\partial \tau_2} + v^{(0)} = 0 \tag{70}$$

$$\left(\frac{L_C}{L}\right)^0 : \frac{\partial \Omega_p^{(1)}}{\partial \tau_2} + v^{(1)} = - V^{(0)} \cdot \nabla \Omega_p^{(0)} \tag{71}$$

此时各级近似的自由项均描述了 Rossby 波动过程,这是边界层内的主要特征,也即叶笃正、李麦村所称为的发展过程。

在发展过程中所描写的是 Rossby 波的能量频散,对此叶笃正[14]曾做过详细的研究,并提出了上游效应,即如果在上游有一气压槽(脊)发展时,由于 Rossby 波的能量频散作用,在一个 Rossby 波波长的距离下将会有一个气压脊(槽) 新生。这一理论无论在理论上或者天气学上都有重要的实用意义。

5　准定常的演变过程

在第二时间边界层外,即进入内部区后,将式(65)按式(69)展开,得到

$$v^{(0)} = 0 \tag{72}$$

$$v^{(1)} = \frac{\partial \Omega_p^{(0)}}{\partial t} + V^{(0)} \nabla \Omega_p^{(0)} \tag{73}$$

如果 $v^{(1)}$ 很小,则有

$$\frac{\partial \Omega_p^{(0)}}{\partial t} + u^{(0)} \frac{\partial}{\partial x} \Omega_p^{(0)} = 0 \tag{74}$$

这是一类沿 x 方向的平流过程,过程的特征时间由 $T_4 = L/V$ 决定,这是在诸特征时间中最慢的一种过程,可以称为准定常的演变过程。或者,在时间边界层外的内部区域中,直接应用式(59)或式(61)的位势涡度平流方程来研究准定常的演变过程。

6　运动多时态特征的普遍性

由以上的分析可以看到,在一个动力系统中运动呈现出不同的、可以区分的多时态特征,事实上这反映在动力系统中包括诸多的物理过程。由于这些物理过程的特征时间不同,因此将由快到慢逐

次地表现出来而形成运动的不同时态。下面再举 4 例来说明。

6.1 小尺度运动

小尺度运动中的积云发展是具有多时态特征的，在一定有利的天气条件下，当在某一局部地区大气受到扰动后，先是由于大气的可压缩性而激发出声波[15]，由于声波的传播速度约为 300 m/s，所以声波的频散过程是极其迅速的。进而，如果大气大尺度背景是稳定层结，则可激发出重力内波，重力内波的振荡频率为$[(g/\theta_0)(d\theta_0/dz)]^{1/2}$（式中 θ_0为静止状态下的位温），其特征周期约为 10 min。再次，当由于上升运动使水汽相变而释放出潜热后，运动由于强烈的发展使得非线性过程变得重要，于是积云将在一个时段（例如为 1 小时左右）作准定常的演变[16]。

6.2 中尺度运动

如将运动方程（8）和方程（9）改写成

$$\frac{\partial u}{\partial t} - (f + \zeta)v = -\frac{\partial P}{\partial x} \tag{75}$$

$$\frac{\partial v}{\partial t} - (f + \zeta)u = -\frac{\partial P}{\partial y} \tag{76}$$

式中

$$P = \varphi + \left(\frac{u^2 + v^2}{2}\right) \tag{77}$$

为空气单位质量的能量。与大尺度运动不同，对中尺度运动相对涡度 ζ 可以与 f 同量级，因此代替地转平衡的平衡状态为[9]

$$(f + \zeta)v = \frac{\partial P}{\partial x} \tag{78}$$

$$(f + \zeta)u = -\frac{\partial P}{\partial y} \tag{79}$$

这表明在中尺度运动中，风基本上沿着等能量线吹。

将方程（75）、方程（76）和方程（10）改写成

$$\frac{\partial u}{\partial t} + \frac{\partial P}{\partial x} = N_u \tag{80}$$

$$\frac{\partial u}{\partial t} + \frac{\partial P}{\partial y} = N_v \tag{81}$$

$$\frac{\partial P}{\partial t} + C^2\left(\frac{\partial u}{\partial x} + \frac{\partial v}{\partial y}\right) = N_P \tag{82}$$

式中

$$\begin{cases} N_u = (f + \zeta)v, \\ N_v = -(f + \zeta)u, \\ N_P = -\left[u\dfrac{\partial(P + \varphi)}{\partial x} + v\dfrac{\partial(P + \varphi)}{\partial y}\right] - \varphi\left(\dfrac{\partial u}{\partial x} + \dfrac{\partial v}{\partial y}\right) \end{cases} \tag{83}$$

其中 $\varphi = \bar{\varphi} + \varphi'$，$\bar{\varphi} = C^2 =$ 常数。由方程式（80）至式（82）立即得到

$$\frac{\partial^2 P}{\partial t^2} + C^2\left(\frac{\partial^2 P}{\partial x^2} + \frac{\partial^2 P}{\partial y^2}\right) = \text{非线性项} \tag{84}$$

可以看到，这个式子左端描述的是重力波的频散，与大尺度的适应过程类似，当重力波频散后，运动将达到由式(78)和式(79)所表示的平衡状态即适应后的状态。但由于在现在的情况下，方程是非线性的，要求出解析解是困难的。

当适应过程完成后，运动也可以用位势涡度守衡来表示，在斜压情况下，叶笃正、李麦村给出为[8]

$$\frac{\mathrm{d}}{\mathrm{d}t}\left[\ln\Omega + \frac{L^2}{L_0^2}\frac{\partial}{\partial\xi}\left(\xi^2\frac{\partial\varphi}{\partial\xi}\right)\right] = 0 \tag{85}$$

(无量纲形式)，式中 $\Omega=f+\zeta$，$L'_0=C/\overline{\Omega}$。类似于前面的方法，也可以分出第二时间边界，在这一时间中表现为位势涡度的急剧变化，因此可以认为这是运动的发展阶段。当运动渡过时间边界层而进入内部区域后，非线性过程变得重要，运动将呈现为准定常的演变。

6.3　热带运动

以上对中、高纬度的大尺度运动的多时态特征做了讨论，在热带地区，由于 Coriolis 力很小，一般来说，运动的地转性难以成立，但对于行星尺度的运动，如大气中的 Walker 环流，海洋中的南北赤道洋流，其纬圈速度很大，而沿经圈方向的压力梯度不如中纬度那么强，因此在纬圈方向的地转平衡是可以成立的，即[10]

$$\frac{1}{2}yu = -\frac{\partial\varphi}{\partial y} \tag{86}$$

(在这里已无量纲化)，另一种情况是在海洋边界附近，经圈流速很大，因此也可以在经圈方向成立地转平衡，即

$$\frac{1}{2}yv = \frac{\partial\varphi}{\partial x} \tag{87}$$

这表明，在热带地区的大气和海洋运动中，虽然同时在两个方向成立地转平衡是困难的，但可以在一个方向运动呈地转平衡状态。因此可以称为半地转平衡状态。

巢纪平和林永辉[10]研究了半地转适应过程。在这一适应过程中，经圈流将以重力惯性波的形式频散，同时运动还存在纬圈的或经圈的半位势涡度不变式。对于前者，当重力惯性波频散后，有半地转位势涡度的时间不变式

$$\left(\frac{\partial^2}{\partial y^2} - \frac{1}{4}y^2\right)v = \frac{\partial}{\partial x}\left[\left(\frac{\partial u}{\partial x} - \frac{\partial u}{\partial y}\right) - \frac{1}{2}y\varphi\right]_{t=0} \tag{88}$$

对于后者则为

$$\left(\frac{\partial^2}{\partial x^2} - \frac{1}{4}y^2\right)u = \frac{\partial}{\partial y}\left[\frac{1}{2}y\varphi - \left(\frac{\partial v}{\partial x} - \frac{\partial u}{\partial y}\right)\right]_{t=0} \tag{89}$$

他们的研究表明，在纬圈半地转适应中，当$(L_0/L_2)^2 \ll 1$时(L_0为赤道 Rossby 变形半径，L_2为初始扰动的经圈特征尺度)，纬圈流向压力场适应，反之，当$(L_0/L_2)^2 \gg 1$时，压力场向纬圈流适应。经圈半地转适应基本上也遵循这样的尺度准则。

热带运动也存在发展过程，但与中、高纬度的运动不同，在发展过程中除了有 Rossby 波的频散外，尚有 Kelvin 波的作用。如果过程是非线性的，自然也存在以平流为主导的准定常演变过程。

6.4　有强迫源的运动

巢纪平和许有丰[17]在有时间缓变的热源作用下，研究了斜压涡度方程的初值问题，在初值附近，大气响应场并未立即随着热源的形式运动，而是呈振动的形式，10 天左右场的变化才逐渐平缓下来并

向热源的变化适应。显然,在初值附近场的急剧变化,是Rossby波首先被激发出来后的表现。由于上述二作者所用的大气模式是线性的,因此当发展过程结束后,大气逐渐向外源调整,并随着外源的变化而变化,如果考虑非线性项,则可以预见会出现以非线性为特征的平流过程。

这是一个有启发性的例子,如果热源不是给定的,而是受着运动的反馈,例如大气的加热来自海洋,这就形成一个海洋和大气的耦合系统。如果在这一耦合系统中重力惯性波已被过滤,则大气和海洋的Rossby波的特征时间分别为$T_s=(2\beta C_s)^{-1/2}$,$T_a=(2\beta C_a)^{-1/2}$,C_a和C_s分别为大气和海洋的重力波波速。由于C_s和C_a的量级分别为1 m/s和100 m/s,于是有

$$\frac{T_a}{T_s}=\left(\frac{C_s}{C_a}\right)^{1/2}\ll 1, \tag{90}$$

亦即大气运动的特征时间要比海洋快得多。因此大气中的Rossby波很快就频散掉,耦合系统中的运动以海洋的缓变运动为主导。这表明在这一海气耦合系统中,运动也具有多时态特征。当年,巢纪平等[8]的长期数值天气预报模式,正是按照这种思想设计的,即是一个海(包括地面)气耦合系统,其中包括有大气中的Rossby波,但相对海洋过程来讲,它是快变过程,因此可以把大气中以Rossby波为特征的快过程过滤掉,使耦合系统中只包括以海洋(及陆地)过程为主的慢过程,并认为月、季尺度的短期气候变化主要受慢过程制约,并由此而做了月、季数值预报。

事实上,多时态特征也会在更复杂的气候系统中存在。只不过表现为快慢过程的气候系统并不会像上面所讨论的那些物理过程那么简单。

参考文献

[1] Rossby,C.G.1937. On the mutual adjustment of pressure and velocity dist ribution in certain simple current systems, Ⅰ, *J.Marine Res.*,1:15~28.

[2] Rossby,C. G. 1938. On the mutual adjustment of pressure and velocity dist ribution in certain simple current systems, Ⅱ, *J. Marine Res.* ,2:239~263.

[3] Oboukhov. A. M. 1949. The problem of geost rophic adaptation, *Izuestiya of Academy of Science USSR*, Ser. Geography and Geophysics,13:281~289.

[4] Yeh,T. C. (叶笃正) 1957. On the formation of quasi-geost rophic motion in the atmosphere, *J. Met. Soc. Japan*, The 75th Anniversary Volume,130~137.

[5] 曾庆存. 1963. 初始扰动结构对适应过程的影响及观测风场的应用. 气象学报,33:37-50.

[6] 陈秋士. 1963. 在简单斜压大气中热成风的形成和破坏. 气象学报,33:153-161.

[7] 曾庆存. 1963. 大气中的适应过程和发展过程(一),(二). 气象学报,35:163-174,281-189.

[8] Yeh,T. C. (叶笃正) and Li Mai-tsun (李麦村). 1982. On the characteri stics of scales of the atmospheric motions, *J. Met. Soc. Japan*, Ser. II,60:16-23.

[9] 叶笃正,李麦村. 1965. 大气运动中的适应问题. 北京:科学出版社.

[10] Chao Jiping (巢纪平) and Lin Yonghui (林永辉). 1996. The foundation and movement of tropical semi-geostrophic adapt at ion, *Acta Meteorologica Sinica*, 10:129-141.

[11] Moore, D. W. and Philander, S. G. H. 1977. Modelling of the tropical ocean circulation, in: *The Sea* (Goldberg, E.D. et al.eds.), Vol.6, Wiley (Interscience), New York, 319-362.

[12] 伍荣生,巢纪平. 1978. 旋转大气中运动的多时态特征和时间边界层. 大气科学,2:267-275.

[13] 叶笃正,朱抱真. 1957. 大气环流若干基本问题. 北京:科学出版社.

[14] Yeh,T. C. 1949. On euergy di spersion in the atmosphere, *J. Met.* ,6:1-16.

[15]　巢纪平. 1962. 论小尺度过程动力学的一些基本问题. 气象学报,32:104-118.

[16]　巢纪平. 1961. 层结大气中热对流发展的一个非线性分析. 气象学报,31:191-204.

[17]　巢纪平,许有丰. 1961. 二层线性模式长期过程的一些计算. 动力气象论文集,北京:科学出版社,90-95.

[18]　长期数值预报小组(巢纪平等). 1977. 一种长期数值预报方法的物理基础. 中国科学,2:162-172.

非静力平衡下平均 Hadley 环流的惯性理论*

巢纪平[1]　刘　飞[2,3]

(1. 国家海洋环境预报中心,北京 100081;2. 中国科学院大气物理研究所东亚区域气候-环境重点实验室,北京 100029; 3. 中国科学院研究生院,北京 100039)

摘要:将热带大气分成两层,在地面附近是行星边界层,在这一层中运动是线性的,当边界层顶辐射平衡温度给定后,可得到边界层顶的垂直运动; 在边界层上是无黏性的理想大气,其中的运动是非线性的,求出了角动量、位温和位势涡度的守恒律,从而建立了在边界层垂直运动驱动下 Hadley 环流的惯性理论,给出了这个理论的若干解例。

关键词:Hadley 环流;惯性理论;角动量、位温、位势涡度守恒

1　引言

平均经圈环流在各纬圈之间热量、动量、涡度和水汽的输送和平衡中起重要作用,资料分析表明,从赤道到极地有 3 个环流圈,在赤道附近空气上升在副热带某一纬度下沉的垂直环流称 Hadley 环流,这是一个热力环流,在极区也有一个热力环流,在两个热力环流之间的中纬度是一个尺度较小的间接环流,称 Ferrel 环流。三圈环流的形成一直是气象学中研究的重要问题之一。在 20 世纪 50 年代,叶笃正、朱抱真在“大气环流的若干基本问题”[1]中就指出,在旋转地球上形成三圈环流的基本因素是加热,大型涡旋输送和摩擦耗散,虽然如此,要在理论上计算出三圈环流仍然是一个难题。不过如果只着眼于研究 Hadley 环流的形成机理问题就要简单得多。Charney[2]把 Hadley 环流的形成作为一个行星动力学问题来研究,问题的提法是: 一个转动的薄壳,当上、下边界加热时,流体的行为如何? 结果表明向极地和向赤道的流动都发生在黏性边界层中,这个问题称为 Charney 问题。

在这以后,Schneider 和 Lindzen[3],郭晓岚[4],Fang 和 Tung[5]发展了黏性、层结大气中的 Charney 问题,除边界加热外,也考虑了流体内部的牛顿辐射冷却型的加热过程的作用。Schneider 和 Lindzen[6],Schneider[7]在模式中进一步引进了积云参数化的加热过程。Held 和 Hou[8]的解析分析指出,在无黏性的非线性系统中存在两个区:一个是绝对角动量守恒区,在这个区中伴随着经圈环流,当加热对赤道对称时,这个区位在邻近赤道的地区;另一个是靠近极地一侧的辐射温度平衡区,在这个区中不存在经圈环流。Lindzen 和 Hou[9]以及 Hou 和 Lindzen[10]又进一步讨论了加热区所在的纬度对 Hadley 环流的影响。Hack 等[11]和 Fang 等[12]分析了 ITCZ 中对流的加热在 Hadley 环流发展中的作用。进而,Satoh[13]更广泛地分析了在辐射-对流平衡下 Hadley 环流的发展。Plumb 等[14]扩展了 Held 和 Hou 的分析,指出如果加热中心不在赤道而在副热带,则角动量守恒的 Hadley 环流区和辐射温度

* 中国科学 D 辑:地球科学,2006,36(3):297-304.

平衡(即无经圈环流)区的位置依赖于辐射温度的强度和构造,存在一个过渡的阈值。通过这一系列研究,对 Hadley 环流的形成和发展机理及规律有了更深入的认识,应该说问题已基本上清楚了。

注意到,除 Charney,Schneider 和 Lindzen 及郭晓岚用的是黏性线性模式的解析分析外,所有计算 Hadley 环流的工作都是用线性、非线性的黏性模式及数值积分进行的。Held 和 Hou 以及 Plumb 和 Hou 的理论分析虽然用的是非线性无黏性模式,但只对 Hadley 环流存在区的一些准则和条件做了研究,并没有应用非线性无黏性模式具体地计算出 Hadley 环流,这是因为在他们的模式中,加热函数是在流体内部用牛顿冷却型来表示的,在这种情况下要求出 Hadley 环流的非线性解析解是困难的。然而由 Ertel 定理我们注意到,对没有摩擦和热源的理想流体中,位势涡度是守恒的,而 Hadley 环流在本质上是位势涡度的某种表现形式,因此如果像 Charney 那样把加热放在边界上,且不考虑黏性,在原则上是可以给出 Hadley 环流的非线性解析解或惯性解的。在另一方面也注意到,ITCZ 的纬向尺度是非常窄的,其中的对流加热如不用参数化形式表示,则在 ITCA 中的运动静力平衡不是都能成立的,运动可以是非静力平衡的。

在本文中,我们将进一步研究 Hadley 环流的发展问题,将大气分成两层,近地面是行星边界层,其中的运动是线性、黏性的,当辐射平衡温度的纬向分布给定后,由边界层动力学可求出边界层顶的垂直运动,以此作为驱动边界层以上大气的运动。边界层以上的大气运动是非静力、非线性的,我们求出了运动的角动量、位温和位势涡度守恒,从而构成一个未经小扰动假设的惯性模式,解析地求得 Hadley 环流解。

2　基本方程和守恒律

在赤道 β-平面近似下的无量纲方程为

$$v\frac{\partial u}{\partial y}+w\frac{\partial u}{\partial z}-yv=0 \tag{1}$$

$$v\frac{\partial v}{\partial y}+w\frac{\partial v}{\partial z}+yu=-\frac{\partial \varphi}{\partial y} \tag{2}$$

$$\delta^2\left(v\frac{\partial w}{\partial y}+w\frac{\partial w}{\partial z}\right)=-\frac{\partial \varphi}{\partial z}+\theta \tag{3}$$

$$v\frac{\partial \theta}{\partial y}+w\frac{\partial \theta}{\partial z}+\Gamma_m^{\cdot}w=0 \tag{4}$$

$$\frac{\partial v}{\partial y}+\frac{\partial w}{\partial z}=0 \tag{5}$$

式中,θ 是扰动位温;$\Gamma_m=\overline{\partial\Theta/\partial z}$是总位温垂直递减率;$\varphi$ 是重力位势,其他符号同常用。

在此已引进特征量:风速 U,特征尺度 $L\sim\sqrt{U/\beta}$;$y=L\tilde{y}$,$z=H\tilde{z}$,$(\theta,\theta_s)=\theta_0(\tilde{\theta},\tilde{\theta}_s)$ $(u,v)=U(\tilde{u},\tilde{v})$,$w=(HU/L)\tilde{w}$,$\varphi=U^2\tilde{\varphi}$,$\theta_0\sim(U^2/gH)\overline{\Theta}$。式中

$$\delta^2=\left(\frac{H}{L}\right)^2,\quad \Gamma_m^{\cdot}=\Gamma_m\left(\frac{gH}{U^2}\frac{H}{\overline{\Theta}}\right) \tag{6}$$

引进流函数,定义成

$$v=-\frac{\partial \psi}{\partial z},\quad w=\frac{\partial \psi}{\partial y} \tag{7}$$

下面给出守恒律：

角动量守恒

式(1)给出

$$u - \frac{1}{2}y^2 = F_1(\psi) \tag{8}$$

位温守恒

式(4)给出

$$\theta + \Gamma_m^{\cdot} z = F_2(\psi) \tag{9}$$

位势涡度守恒

对式(2)，式(3)两式交叉微分，给出

$$v\frac{\partial}{\partial y}\left(\delta^2\frac{\partial w}{\partial y} - \frac{\partial v}{\partial z}\right) + w\frac{\partial}{\partial z}\left(\delta^2\frac{\partial w}{\partial y} - \frac{\partial v}{\partial z}\right) - y\frac{\partial u}{\partial z} - \frac{\partial \theta}{\partial y} = 0 \tag{10}$$

考虑到

$$\frac{\partial u}{\partial z} = \frac{\mathrm{d}F_1}{\mathrm{d}\psi}\frac{\partial \psi}{\partial z}, \qquad \frac{\partial \theta}{\partial y} = \frac{\mathrm{d}F_2}{\mathrm{d}\psi}\frac{\partial \psi}{\partial y} \tag{11}$$

由此给出

$$\delta^2\frac{\partial^2\psi}{\partial y^2} + \frac{\partial^2\psi}{\partial z^2} + \frac{1}{2}y^2\frac{\mathrm{d}F_1}{\mathrm{d}\psi} - z\frac{\mathrm{d}F_2}{\mathrm{d}\psi} = F_3(\psi) \tag{12}$$

以上 $F_{i=1,2,3}(\psi)$ 是普适函数。

3　赤道外运动的特征

假定：当 $|y|$ 充分大时，即离开赤道区进入到副热带区 $y \geqslant y_d$，那里的运动的基本特征是（回到有量纲量）：运动是线性、地转、静力平衡的，因而是热成风平衡的；辐射平衡温度的偏差值记成 $\theta_E(y,z)$，其无量纲形成为

$$\theta_E = \bar{\theta}_0(y)(1-z) \tag{13}$$

即辐射平衡温度的最大偏差值在地面，模式大气顶部为零。梯度为

$$\frac{\partial \theta_E}{\partial y} = \overline{\frac{\partial \bar{\theta}_0}{\partial y}}(1-z) \tag{14}$$

式中，“-”号表示该地区的平均值。热成风平衡为

$$f\frac{\partial u}{\partial z} = -\frac{g}{\Theta}\frac{\partial \theta_E}{\partial y} \tag{15}$$

无量纲形式为

$$\frac{\partial u}{\partial z} = M(1-z) \tag{16}$$

式中无量纲数

$$M = -\frac{gH}{\Theta f U}\overline{\frac{\partial \bar{\theta}_0}{\partial y}} \tag{17}$$

积分式(16)，给出

$$u = u_0 + M\left(z - \frac{1}{2}z^2\right) \tag{18}$$

关于经圈风，假定向赤道的北风都在边界层中，因此在自由大气中是弱的南风 $v_d>0$。

4　普适函数的确定

$$y \to y_d,\quad v_d = -\frac{\partial\psi}{\partial z} \tag{19}$$

积分

$$y \to y_d,\quad z = -\frac{\psi}{v_d} \tag{20}$$

由此普适函数为

$$F_1(\psi) = \left(u_0 - \frac{1}{2}y_d^2\right) - \frac{M}{v_d}\left(\psi + \frac{1}{2v_d}\psi^2\right) \tag{21}$$

$$\frac{\mathrm{d}F_1}{\mathrm{d}\psi} = -\frac{M}{v_d}\left(1 + \frac{\psi}{v_d}\right) \tag{22}$$

$$F_2(\psi) = \bar{\theta}_0(y_d) + (\bar{\theta}_0 - \dot{\Gamma}_m)/v_d \times \psi \tag{23}$$

$$\frac{\mathrm{d}F_2}{\mathrm{d}\psi} = \frac{\bar{\theta}_0 - \dot{\Gamma}_m}{v_d} \tag{24}$$

考虑到 δ^2 是个小量，可以在求普适函数时不予考虑，于是式(11)给出

$$F_3(\psi) = -\frac{1}{2}y_d^2\frac{M}{v_d} + \left(\frac{\theta_0 - \dot{\Gamma}_m}{v_d^2} - \frac{1}{2}y_d^2\frac{M}{v_d^2}\right)\psi \tag{25}$$

由此角动量守恒和位温守恒分别为

$$u = -\frac{1}{2}y^2 = u_0 - \frac{1}{2}y_d^2 - M\left[\frac{\psi}{v_d} + \frac{1}{2}\left(\frac{\psi}{v_d}\right)^2\right] \tag{26}$$

$$\theta = \bar{\theta}_0(y) + \frac{\bar{\theta}_0 - \dot{\Gamma}_m}{v_d}\psi - \dot{\Gamma}_m z \tag{27}$$

位势涡度方程为

$$\delta^2\frac{\partial^2\psi}{\partial y^2} + \frac{\partial^2\psi}{\partial z^2} + \left[\frac{1}{2}(y_d^2 - y^2)\frac{M}{v_d^2} - \frac{\bar{\theta}_0 - \dot{\Gamma}_m}{v_d^2}\right]\psi = -\frac{1}{2}(y_d^2 - y^2)\frac{M}{v_d} + \frac{\bar{\theta}_0 - \dot{\Gamma}_m}{v_d}z \tag{28}$$

5　边界层运动

设在地面以上有一个边界层，边界层中的运动方程为

$$-fv_b = \mu\frac{\partial^2 u_b}{\partial z^2} \tag{29}$$

$$f\frac{\partial u_b}{\partial z} = -\frac{g}{\Theta}\frac{\partial\theta_b}{\partial y} \tag{30}$$

$$\frac{\partial v_b}{\partial y} + \frac{\partial w_b}{\partial z} = 0 \tag{31}$$

由式(31)引入流函数 ψ_b

$$v_b = -\frac{\partial \psi_b}{\partial z}, \quad w_b = \frac{\partial \psi_b}{\partial y} \tag{32}$$

代入式(30)

$$f\frac{\partial \psi_b}{\partial z} = \mu \frac{\partial^2 u_b}{\partial z^2} \tag{33}$$

取条件

$$z \to \infty, \quad \frac{\partial u_b}{\partial z} \to 0, \psi_b \to 0 \tag{34}$$

由此有

$$\psi_b = \frac{\mu}{f}\frac{\partial u_b}{\partial z} \tag{35}$$

由此给出

$$w_b = -\frac{\mu g}{f^2 \Theta}\frac{\partial^2 \theta_b}{\partial y^2} \tag{36}$$

无量纲形式为

$$w_b = -\widetilde{W}\frac{\partial^2 \theta_b}{\partial y^2} \tag{37}$$

式中,$\widetilde{W} = (\mu U)/(f^2 H^2 L)$;$f$ 取低纬度的平均值。

6 定解条件

由式(20)给出

$$y \to y_d, \quad \psi = -v_d z \tag{38}$$

大气层顶必须是条流线,否则空气要流出去,或流进来,于是

$$y_d > y \geqslant 0, \quad z = 1 \quad 均为 \quad \psi = -v_d \tag{39}$$

在赤道,为了与式(39)的连续性,也为了南、北半球的对称性,应有

$$y = 0, \quad \psi = -v_d \tag{40}$$

在边界层顶式(36)可写成

$$z = Z_b, \quad \frac{\partial \psi}{\partial y} = -\widetilde{W}\frac{\partial^2 \theta_b}{\partial y^2} \tag{41}$$

积分

$$\psi(y) - \psi\Big|_{y=0} = -\widetilde{W}\left[\frac{\partial \theta_b}{\partial y} - \frac{\partial \theta_b}{\partial y}\Big|_{y=0}\right] \tag{42}$$

考虑到与式(40)的连接,及对赤道的对称性,有

$$z = Z_b, \quad \psi(y) = -v_d - \widetilde{W}\frac{\partial \theta_b}{\partial y} \tag{43}$$

边界层顶可以不是一条流线,因有垂直运动。考虑到

$$y = y_d, \quad z = Z_b, \quad \psi(y) = -v_d Z_b \tag{44}$$

式(43)在 $y=y_d$ 需与式(44)连接,应有

$$v_d(1-Z_b)=-\widetilde{W}\frac{\partial\theta_b}{\partial y}\bigg|_{y=y_d} \tag{45}$$

若取 $\theta_b=\theta^{\cdot}\theta_E(y)$ 则可以由 v_d 来确定 $y=y_d$ 处 $\theta^{\cdot}$ 的大小。

7　辐射温度分布及解例

取辐射温度为 $\theta_E=\exp(-\alpha y^2)$,式中 y 是无量纲值。

参数的参考值取成:$U\sim10$ m/s,$f\sim5\times10^{-5}$/s (低纬平均值),$\beta\sim2\times10^{-11}$/(m · s),$H\sim10^4$ m,$g\approx9.7$ m/s^2,$\Theta\approx300$°K,$m\approx10$ m^2/s,边界层厚度 $H_b\approx10^3$ m。由此导出:$L=\sqrt{U/\beta}\approx700$ km (Rossby 变形半径),$\theta_0=\frac{U^2}{gU}\Theta\approx0.3$℃,及 $\Gamma_m\sim(1\sim3)$℃/10^3m(静力稳定度),$\overline{\frac{\partial\theta_0}{\partial y}}\sim-\frac{1℃}{10^6\ \text{m}}\approx-10^{-6}$℃/m(斜压性),$y_d$ 有量纲时约 2 000 km(副热带边缘),现在的无量纲值约 3,v_d 的无量纲值约 0.6,$\alpha\approx0.2$。这些参数均可适当调整其大小。

计算表明当取上述参数值时,对方程(28)的计算表明,解强烈地依赖于副热带的辐射平衡温度 $\bar{\theta}_0$ 偏差及斜压性$\overline{\partial\bar{\theta}_0/\partial y}$。

例 1　无量纲 $\bar{\theta}_0=29.5,35$。这时取 $\Gamma_m^{\cdot}=32.3$,$\overline{\partial\theta_0/\partial y}=-1.0\text{e}-6$℃/m。计算结果见图 1,图 2。次序从上到下分别为流函数(无量纲),垂直运动(m/s),纬向风(m/s),扰动温度(℃),横坐标用 Rossby 变形半径表示,垂直坐标是无量纲的。由图 1 可以看到,这时出现一个深厚的经圈环流圈,从赤道到约 1.5 个 Rossby 变形半径是上升运动,在邻近赤道很窄的区域内,上升速度可达 90 cm/s,这表明由非静力平衡在赤道形成一强的"动力边界层",由于在这个例子中静力稳定度取得较小($\Gamma_m=1.0$ ℃/km),而小的静力稳定度可以看成积云塔的潜热释放在宏观尺度上的表现,也可以认为动力边界层中积云塔更易发展,在这一层外,上升速度也可到 10 cm/s。从 1.5 个 Rossby 变形半径以北,是下沉运动区,其范围要大于上升区,强度与上升运动相当(动力边界层除外)。由图 1 和图 2 可以看到,随着$\overline{\theta_0}$增大,垂直运动的下沉区向赤道上空入侵,除动力边界层外,强的上升运动转向低空。纬向运动 u 的内部解,可直接从式(26)算出(取下边界纬向基本流 $u_0=0$),从图 1 可以看出在近赤道附近纬向速度有较大的改变。随着$\overline{\theta_0}$增大,图 1 对流层中部的东风急流消失,东风从赤道向极地方向减强,形成一个宽的东风带,强度上下比较均匀。在区域内部的扰动温度可直接从式(27)算出,在赤道附近的低空可达 10℃ 的量级,其分布向上递减,而且随着$\overline{\theta_0}$的增大,扰动温度的分布,由等值线看出从区域中部低变成区域中部高,温度垂直切变变大。在另一方面,我们也注意到经圈环流的变化和纬向风的变化之间的协调性。在图 1 中,由于从 0.5 个 Rossby 变形半径开始,出现强的南风,由于柯利奥来力的作用不利于东风的发展,从图 1 可看到对流层顶附近东风减弱。这种现象在图 2 中表现得更为明显,在对流层中基本上已被南风覆盖(按模式结构假定向赤道的气候意义下的北风出现在行星边界层中),导致东风急流消失,而变成向副热带减弱的东风带,并在副热带边缘也转变为西风。

例 2　取无量纲温度 $\bar{\theta}_0=33$,而分别取斜压性$\overline{\partial\bar{\theta}_0/\partial y}=0$ ℃/m,-2.0×10^{-6}℃/m,如图 3,图 4,图的

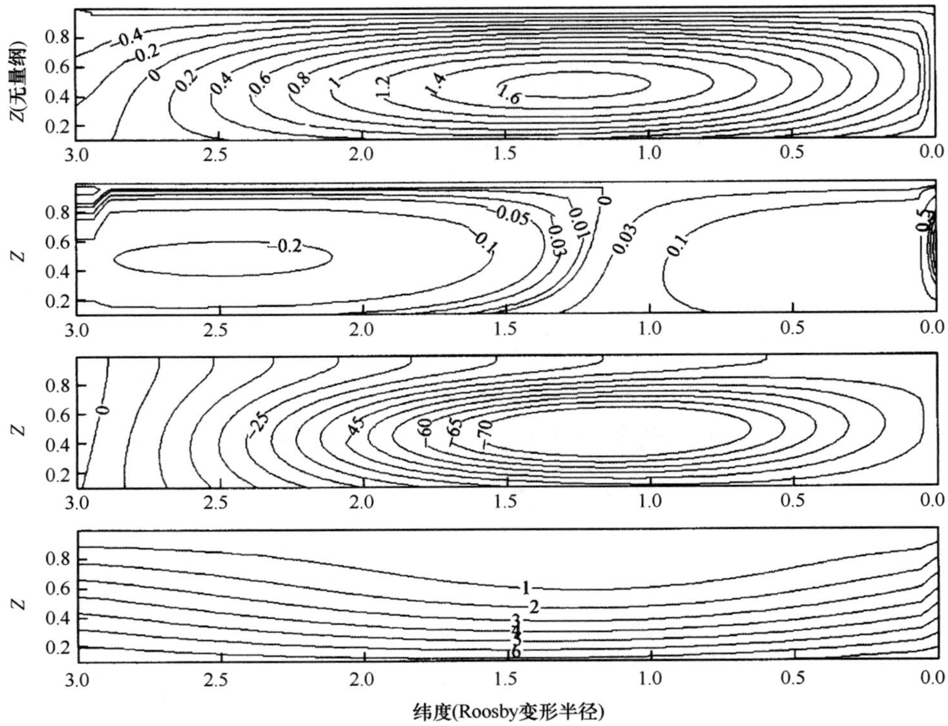

图 1 $\bar{\theta}_0 = 8.85$℃，从上到下分别为流函数(无量纲)、垂直运动(m/s)、纬向风(m/s)、扰动温度(℃)

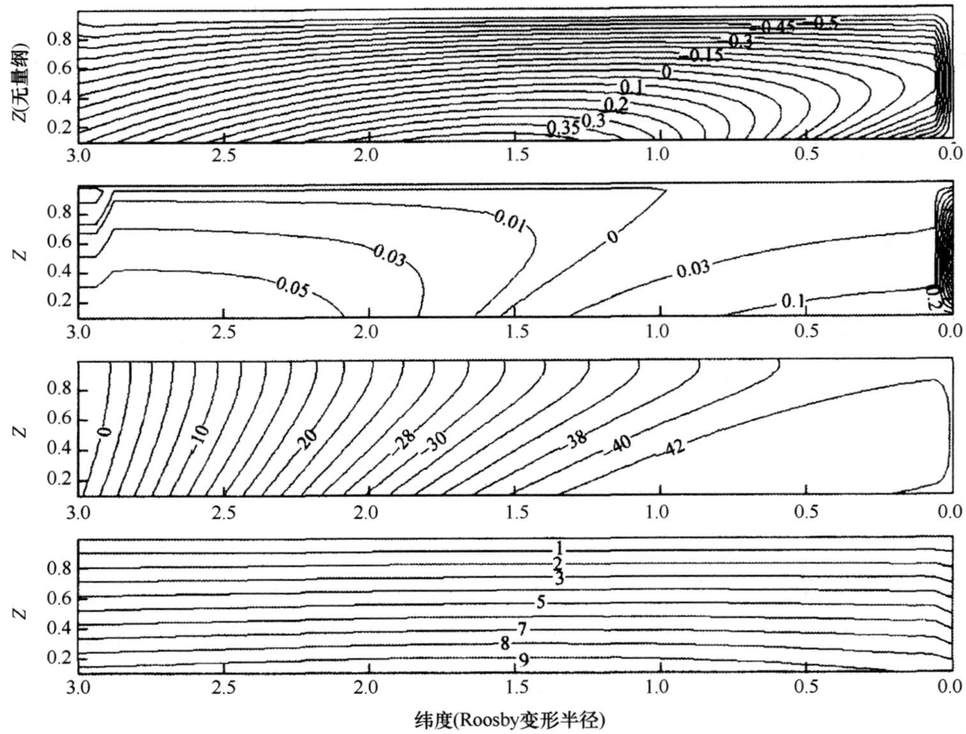

图 2 $\bar{\theta}_0 = 10.5$℃，说明同图 1

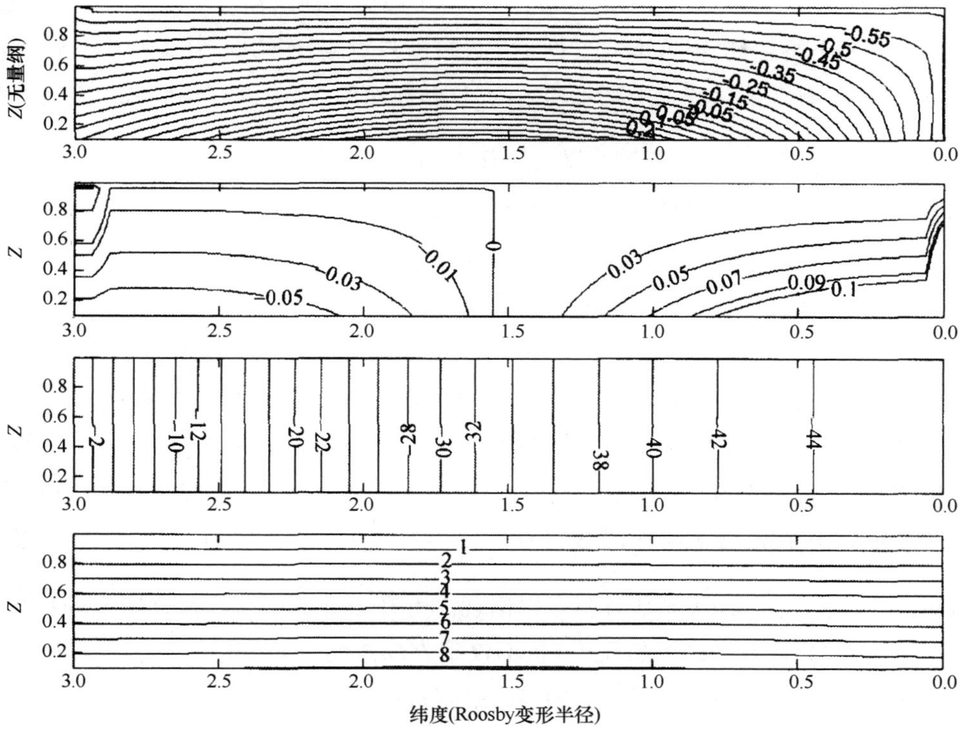

图 3　$\overline{\partial \bar{\theta}_0/\partial y}=0℃/m$，说明同图 1

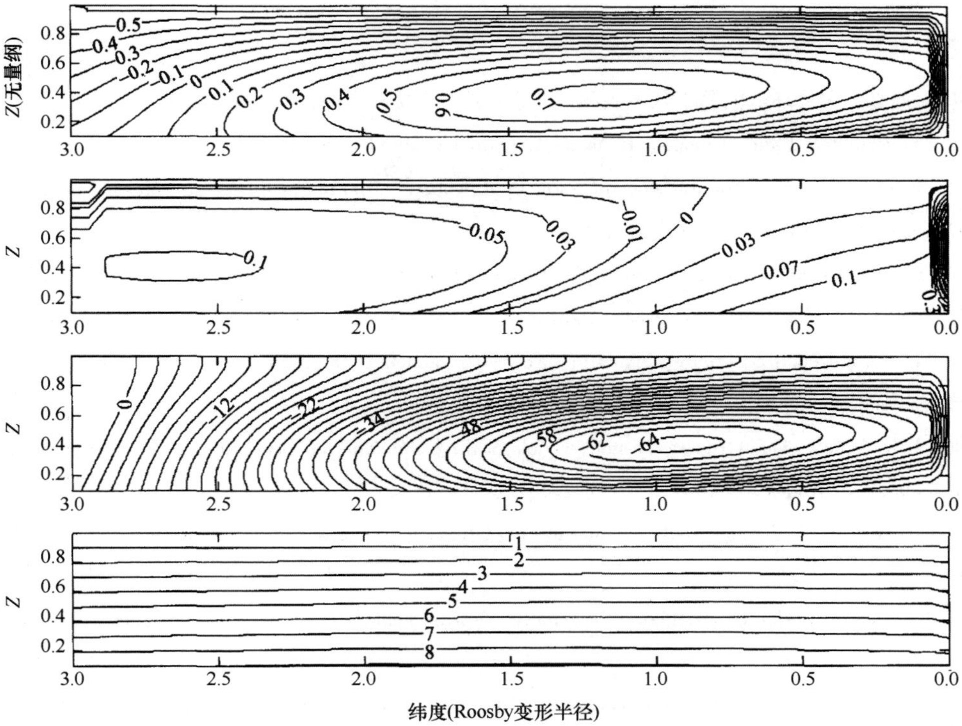

图 4　$\overline{\partial \bar{\theta}_0/\partial y}=-2.0\times10^{-6}℃/m$，说明同图 1

说明同例 1。图 3 显示了一个很有意思的现象，虽然由于边界层顶由辐射平衡温度梯度造成的垂直运动仍可激发出上升和下沉运动，但这本质上是一类大尺度热对流，而纬向风(东风)从赤道向副热带减弱，它们在区域中的所有纬度风速上下都是均匀的，这是显然的，因为没有热成风效应，同时扰动温度在区域内部只有垂直梯度而无经向梯度。随着斜压性$\overline{\partial \bar{\theta}_0/\partial y}$的出现，如图 4 所示，垂直速度的分布开始向赤道倾斜，纬向速度出现垂直梯度并出现东风急流。

8 结语

本文在非静力平衡条件下，对控制平均 Hadley 环流的非线性方程给出了行星边界层上的惯性解，驱动 Hadley 环流的是行星边界层顶由辐射平衡温度梯度造成的垂直运动。当“副热带”的辐射温度和它的纬向梯度给定时，可算出邻近赤道是上升运动区和向极一侧是下沉运动的垂直环流圈。在区域内部绝对角动量守恒下算出了纬向风场，但纬向风的结构对副热带的辐射温度偏差值和斜压性$\overline{\partial \bar{\theta}_0/\partial y}$很敏感，当只有层结性而无斜压性时可以出现由上升和下沉运动的垂直环流，但纬向东风在对流层内部是上下均匀的，只有考虑了斜压性后，不仅垂直环流圈会向赤道倾斜，纬向风也出现垂直梯度并可形成东风急流。由总温度守恒算出的扰动温度场的结构比较稳定。

这是对平均 Hadley 环流理论研究的一个新解，但要指出，没有黏性耗散的惯性解，解对参数域很敏感。另外需指出，当副热带的经向风不是常数时，控制惯性运动的将是二阶非线性的椭圆形方程，在那种情况下解的性质将变得复杂，例如有可能出现多平衡态现象，是值得做进一步研究的。

参考文献

[1] 叶笃正，朱抱真．大气环流的若干基本问题．北京：科学出版社，1957：44-54.

[2] Charney J G. Planetary Fluid Dynamics. Dynamic Meteorology. Morel P, Reidel D, eds. 1973:97-351.

[3] Schneider E K, Lindzen R S. The influence of stable stratification on the thermally driven tropical boundary layer. J Atmos Sci, 1976, 33:1301-1307.

[4] Kuo H L. Planetary boundary layer flow of a stable atmosphere over the globe. J Atmos Sci, 1973, 30:53-65.

[5] Fang M, Tung K K. Solution to the Charney problem of viscous symmetric circulation. J Atmos Sci, 1994, 51:1261-1272.

[6] Schneider E K, Lindzen R S. Axially symmetric steady-state models of the basic state for instability and climate studies. Part I. Linearized calculation. J Atmos Sci, 1977, 34:263-279.

[7] Schneider E K. Axially symmetric steady-state models of the basic state for instability and climate studies. Part Ⅱ. Nonlinear calculations. J Atmos Sci, 1997, 34:280-296.

[8] Held I M, Hou A Y. Nonlinear axially symmetric circulations in a nearly inviscid atmosphere. J Atmos Sci, 1980, 37:515-533.

[9] Lindzen R S, Hou A Y. Hadley circulation for zonally averaged heating centered off the equator. J Atmos Sci, 1988, 45:2416-2427.

[10] Hou A Y, Lindzen R S. The influence of concentrated heating on Hadley circulation. J Atmos Sci, 1992, 49: 1233-1241.

[11] Hack J J, Schubert W H, Stevens D E, et al. Response of the Hadley circulation to convection forcing in the ITCZ. J Atmos Sci, 1989, 46: 2957-2973.

[12] Fang M, Ka Kit Tung. A simple model of nonlinear Hadley circulation with an ITCZ: Analytic and numerical solutions.

J Atmos Sci,1996,53: 1241-1261.

[13] Satoh M. Hadley circulation in radiative-convective equilibrium in an axially symmetric atmosphere. J Atmos Sci,1994, 51: 1947-1968.

[14] Plumb R A,Hou A Y. The response of a zonally symmetric atmosphere to subtropical thermal forcing: Threshold behavior. J Atmos Sci,1992,49:1790-1799.

跨赤道惯性急流的多平衡态*

巢纪平[1,2] 刘飞[3,4]

(1. 国家海洋环境预报中心, 北京 100081; 2. 国家海洋局第一海洋研究所, 青岛 266061; 3. 中国科学院大气物理研究所东亚区域气候-环境重点实验室, 北京 100029; 4. 中国科学院研究生院, 北京 100039)

摘要:发展了 Anderson 和 Moore 跨赤道惯性急流理论,应用非线性等值浅水模式,根据与急流相联系的有旋大尺度气流(大气)或洋流(海洋)特征定出新的普适函数,从而求出跨赤道惯性急流区中沿流线的位势涡度和能量守恒式,由于最后的控制方程是非线性的,因此可以有条件地存在跨赤道的多平衡态的惯性急流解。通过急流区外不同的大尺度气流或洋流的特征分析,讨论了跨赤道惯性急流区中可能存在平衡态的条件,并给出多平衡态的解例。

关键词:多平衡态;跨赤道惯性急流;位势涡度;能量守恒

由于南、北半球海陆分布的不对称性,无论是大气环流或海洋环流对赤道都不对称,它们存在半永久性或季节性的跨赤道气流或洋流。在 20 世纪 60 年代末,Findlater[1] 发现在非洲东岸的低空存在一支从南到北的跨赤道急流(称东非急流,也称 Somali 急流),在赤道的风速可达到 25 m/s,Findlater[2] 描述了这支急流的季节进展。陈隆勋等[3] 指出,除非洲东岸的跨赤道急流外,在东半球还有另外 3 支急流,其中与澳大利亚东南风相联系在南海跨越赤道的急流(称南海急流),强度也不弱,夏季经向风速度可达 6~7 m/s,图 1 是多年 7 月平均 850 hPa 上的流线[4]。

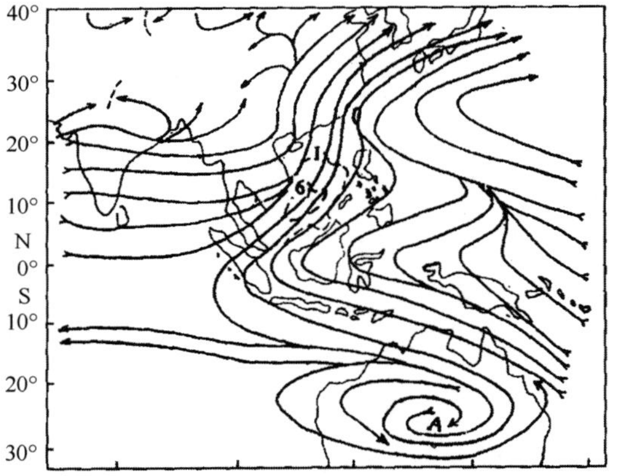

图 1 南海急流多年平均的低空流场[4]

在海洋中同样存在跨赤道的强洋流,其中以与非洲东岸大气中跨赤道相对应的 Somali 流尤为著

* 中国科学D辑:地球科学,2007,37(2):2540-260.

名，其强度可与湾流（Gulf Stream）相比[5]，图 2 是印度洋海表动力高度和表层流[6,7]。

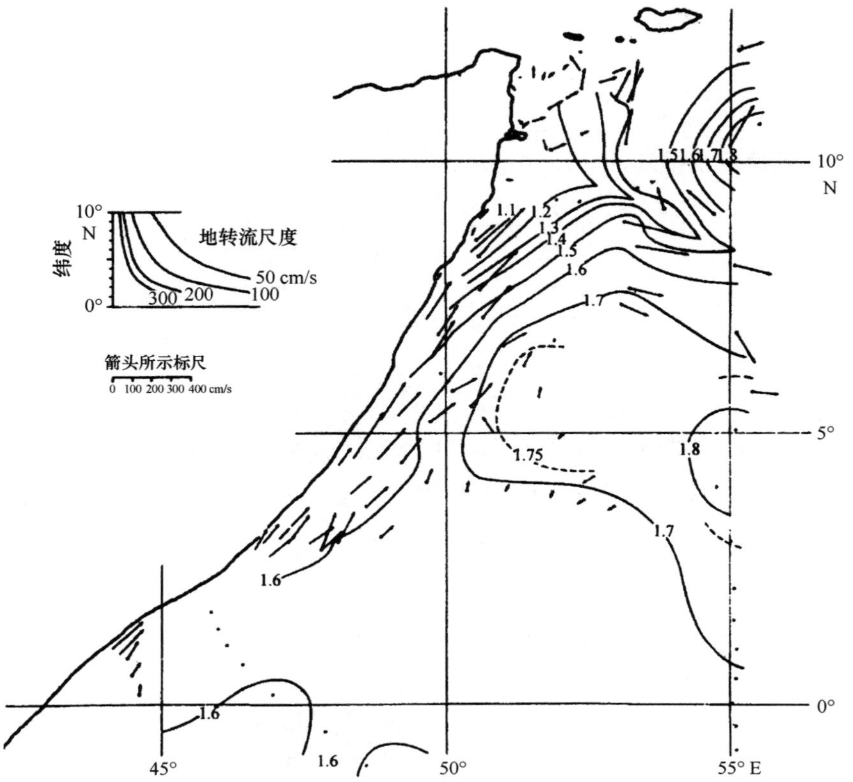

图 2　1954 年 8—9 月印度洋海表动力高度和实测表层流[7]

研究跨赤道流的形成和发展具有重要的意义，在气象学中，非洲东岸低空急流的发展与印度季风的暴发有关，而南海低空急流的发展与东亚季风的暴发有关。在海洋学中，跨赤道急流将通过动量、能量两个半球间的交换而改变水团的物理属性。另一方面，在旋转地球上当流体跨过赤道时，科里奥利参数由大变小再变大，它将使流体运动的性质发生变化，因此跨赤道流的形成也是一个地球物理流体动力学中值得研究的问题。Lighthill[8] 最早提出了印度洋洋流对西南季风的动力响应理论，指出在物理本质上急流的形成是 Rossby 长波的能量在经圈边界附近集中的结果。Anderson[5] 应用线性模式把低空急流看成是一类西边界流，由南半球的高压产生的向西气流遇到经向边界后在边界附近造成能量集中。Krishnamurtt 等[9] 通过一层模式对 Somali 急流数值模拟后指出，它形成的原因：一是非洲的 Madagascar 山的影响；二是地球转动的 beta(β) 效应；三是来自东边的侧向强迫。Bannon[10] 通过数值试验指出，东非急流的形成惯性、科里奥利力、底部摩擦和地形共同起作用。Anderson 和 Moore[11] 最早提出，Somali 急流的形成就像 Charney[12] 的湾流惯性理论那样，同样是一种惯性流。

Anderson 和 Moore[11] 的跨赤道急流惯性理论给出了一些清晰的物理结果，这是对跨赤道急流理论的一个重要发展。但非线性方程的惯性解依赖于决定动量、能量、涡度守恒律成立时的普适函数，一般而言，为了保持运动的连续性，普适函数应根据惯性流区外大尺度流场的特征来决定，而在他们的理论中，普适函数的给定有一定的任意性，同时由此而得到的控制方程是线性的，这就有可能丢掉惯性解中一些重要的动力学结果，例如解的多平衡态。

有鉴于此，本文将在 Anderson 等的惯性理论的框架下，重新根据急流区外大尺度大气和海洋环流特征,构造新的普适函数,使最后的控制方程仍带有一定的非线性性,并在这个基础上讨论多平衡态存在的条件,及多平衡态的解例。

1 模式

赤道β-平面上等值浅水模式的非线性定常方程为

$$u\frac{\partial u}{\partial x}+v\frac{\partial u}{\partial y}-\beta yv=-g'\frac{\partial h}{\partial x} \tag{1}$$

$$u\frac{\partial v}{\partial x}+v\frac{\partial v}{\partial y}+\beta yu=-g'\frac{\partial h}{\partial y} \tag{2}$$

$$\frac{\partial hu}{\partial x}+\frac{\partial hv}{\partial y}=0 \tag{3}$$

式中,h 对海洋可视为温跃层的扰动厚度,对大气可视为低层大气如 850 hPa 的扰动高度;x,y 分别为沿纬圈和经圈的坐标;$g'=g\Delta\rho/\rho$ 为视重力,$\Delta\rho$ 为上下层流体的密度差。引进重力波波速,为

$$C=\sqrt{g'D} \tag{4}$$

D 为未扰厚度。

引进特征量。令 $L_R=(C/2\beta)^{1/2}$ 为赤道 Rossby 变形半径,取特征量为 $(x,y)=L_R(x',y')$,$(u,v)=C(u',v')$,$h=Dh'$,得到无量纲方程为

$$u\frac{\partial u}{\partial x}+v\frac{\partial u}{\partial y}-\frac{1}{2}yv=-\frac{\partial h}{\partial x} \tag{5}$$

$$u\frac{\partial v}{\partial x}+v\frac{\partial v}{\partial y}+\frac{1}{2}yu=-\frac{\partial h}{\partial y} \tag{6}$$

$$\frac{\partial hu}{\partial x}+\frac{\partial hv}{\partial y}=0 \tag{7}$$

由式(7)引进流函数,定义为

$$hu=-\frac{\partial\psi}{\partial y},\quad hv=\frac{\partial\psi}{\partial x} \tag{8}$$

由式(5)和式(6)交叉微分再相减得到:

$$J\left[\psi,\frac{\frac{\partial v}{\partial x}-\frac{\partial u}{\partial y}+\frac{1}{2}y}{h}\right]=0 \tag{9}$$

其中算子 Jacobi $J[A,B]=\frac{\partial A\partial B}{\partial x\,\partial y}-\frac{\partial A\partial B}{\partial y\,\partial x}$,则

$$\frac{\frac{\partial v}{\partial x}-\frac{\partial u}{\partial y}+\frac{1}{2}y}{h}=P(\psi) \tag{10}$$

这是位势涡度沿流线的不变式。在另一方面有

$$J\left[\psi,\frac{1}{2}u^2+\frac{1}{2}v^2+h\right]=0 \tag{11}$$

由此得到:

$$\frac{1}{2}(u^2+v^2)+h=G(\psi) \tag{12}$$

这是能量沿流线的不变式。P, G 是 ψ 的普适函数。

图 3 给出这个研究的模式示意图。它将分成跨赤道的惯性急流区 A，及基本东西向的大尺度来流区 B。下面分别讨论这两个区中的运动特征。

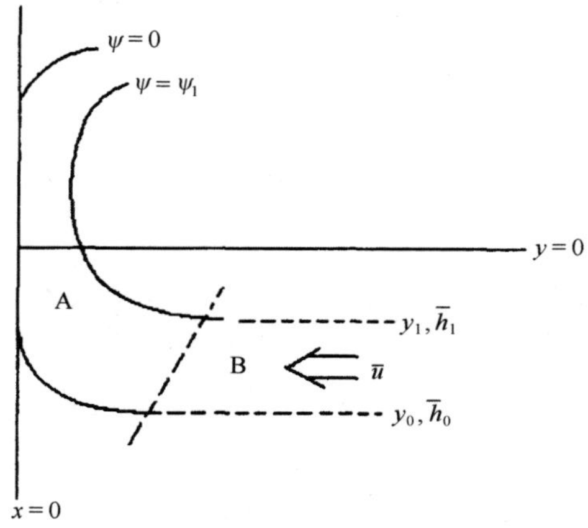

图 3　模式示意图

（1）惯性急流区 A

在图 3 跨赤道急流区 A 中，假定 $\partial v/\partial x \gg \partial u/\partial y$ 及 $v^2 \gg u^2$，于是式(10)和式(12)分别简化为

$$\frac{1}{h}\left(\frac{\partial v}{\partial x}+\frac{1}{2}y\right)=P(\psi) \tag{13}$$

$$\frac{v^2}{2}+h=G(\psi) \tag{14}$$

（2）普适函数的确定区 B

设在赤道以南，在 $x\to\infty$ 的区域中有一支地转向西流，考虑到无论对大气或海洋，这一向西流来自大尺度的反气旋环流，因此该地转向西流速度 $\bar{u}$ 有经向的梯度，设该地转流的南界为 y_0，北界为 y_1，其中 $\bar{u}=\bar{u}_0+ky$（注意因在南半球，对于反气旋的北支，k 应取负值）。由式(6)有

$$\frac{1}{2}y\bar{u}=-\frac{\partial\bar{h}}{\partial y} \tag{15}$$

积分得到：

$$\bar{h}=\bar{h}_0-\frac{1}{4}\bar{u}_0(y^2-y_0^2)-\frac{1}{6}k(y^3-y_0^3) \tag{16}$$

将式(8)代入式(16)并积分可以得到：

$$\bar{\psi}=\frac{1}{30}k^2y^5+\frac{5}{48}k\bar{u}_0y^4+\frac{1}{12}\bar{u}_0^2y^3-\frac{1}{2}aky^2-a\bar{u}_0y+b \tag{17}$$

由 $y=y_0, \psi=0$，则式(17)中，$a=\frac{1}{6}ky_0^3+\frac{1}{4}\bar{u}_0y_0^2+\bar{h}_0$，$b=-\frac{1}{30}k^2y_0^5-\frac{5}{48}k\bar{u}_0y_0^4-\frac{1}{12}\bar{u}_0^2y_0^3+\frac{1}{2}aky_0^2+a\bar{u}_0y_0$。如来流的纬度已定，例如为 $y=y_1$，可得到 $\overline{\psi_1}$。将式(17)改写为

$$\frac{1}{30}k^2y^5+\frac{5}{48}k\bar{u}_0y^4+\frac{1}{12}\bar{u}_0^2y^3-\frac{1}{2}aky^2-a\bar{u}_0y+b-\bar{\psi}=0 \tag{18}$$

对于5次方程不能求出显式的根式解，下面将用数值方法给出近似解 $y^{(j)} = F^{(j)}$，其中 $1 \leqslant j \leqslant 5$，即存在有物理意义的5个实根，也可能有物理意义的实根不到5个。

在区域B中存在条件：

$$x \to \infty, \quad v, \quad \partial v / \partial x \to 0 \tag{19}$$

于是由式(13)和式(14)得出：

$$P(\bar{\psi}) = \frac{1}{2}\frac{y}{\bar{h}} = \frac{1}{2}\frac{F^{(j)}}{\bar{h}_0 - \frac{1}{4}\bar{u}_0(F^{(j)2} - y_0^2) - \frac{1}{6}k(F^{(j)3} - y_0^3)} \tag{20}$$

$$G(\bar{\psi}) = \frac{\bar{u}^2}{2} + \bar{h}_0 - \frac{1}{4}\bar{u}_0(F^{(j)2} - y_0^2) - \frac{1}{6}k(F^{(j)3} - y_0^3) \tag{21}$$

2 控制方程

由于 P 和 G 对 ψ 是普适的，因此式(13)和式(14)变为

$$\frac{\partial v}{\partial x} + \frac{y}{2} = \frac{1}{2}h\frac{F^{(i)}(y_0, \bar{h}_0, \bar{u}, \psi)}{\left(\bar{h}_0 + \frac{1}{4}\bar{u}y_0^2\right) - \frac{1}{4}\bar{u}F^{(i)2}(y_0, \bar{h}_0, \bar{u}, \psi)} \tag{22}$$

$$\frac{1}{2}v^2 + h = \left(\bar{h}_0 + \frac{1}{4}\bar{u}y_0^2\right) - \frac{1}{4}\bar{u}F^{(i)2}(y_0, \bar{h}_0, \bar{u}, \psi) \tag{23}$$

由于 $F^{(i)}$ 中，$1 \leqslant i \leqslant 5$，因此可能有不止一个平衡态，还可能出现2个或3个平衡态，这是本文研究的重点。

将式(8)代入式(22)得到：

$$\frac{\partial v}{\partial \psi}vh + \frac{y}{2} = h\Phi(\psi) \tag{24}$$

式(23)缩写为

$$\frac{1}{2}v^2 + h = \Omega(\psi) \tag{25}$$

由此两式消去 h，得到：

$$\frac{\partial v^2}{\partial \psi} + y\left[\Omega(\psi) - \frac{v^2}{2}\right]^{-1} = 2\Phi(\psi) \tag{26}$$

式中，$y>y_0$，是参变量。

对常微分方程式(26)积分，需给出定解条件，取

(1) 设在急流区中沿 $x=0$ 的 y 轴 $\psi=0$（这一假定在物理上可认为在那里存在固体边界）；

(2) 在分开急流和大尺度背景场的流线 $\psi=\psi_1$ 上 $v=0$；

(3) 为决定 ψ_1 的形态，可认为流体通过任一个截面的流量是守恒的，流入的平均流量为 $-\overline{uh}$，此流量应与急流中的平均流量相等，即有条件：

$$\frac{\int_0^{\psi_1}(vh)\,\mathrm{d}\psi}{\psi_1} = -\overline{uh} \tag{27}$$

由此有

$$\frac{\int_0^{\psi_1} v\left(\Omega(\psi) - \frac{v^2}{2}\right)\mathrm{d}\psi}{\psi_1} = -\overline{uh} \tag{28}$$

将 $\bar{h}$ 的表达式(16)代入式(28)，由此由式(28)和式(26)可以通过打靶方法来决定 $\psi=0$ 上的 v 值和 $\psi_1=0$。

3　多平衡态的分析

在不受外界影响的条件下，系统的性质长时间内不发生任何变化的状态称作为平衡态。本文中，对于急流区外大尺度的气流或者洋流，在惯性区域内形成定常急流，达到一种稳定的状态，且有可能产生几种急流，每一种急流称作一个平衡态。

注意到如地转向西流均匀($k=0$)，能得到如 Chao 等[13]早期得到的一个平衡态，取 $\bar{u}=0.7$，南界 $y_0=-5.0$，北界 $y_1=-4.5$，$\bar{h}_0=3.0$，得到流函数分布图(图 4)。从图中可以看到均匀的向西流在西边界附近迅速向北拐，形成惯性急流区。并且在北纬 6 个 Rossby 变形半径时得到 $x=0$，$h=0$，流线将脱离边界进入流体内部，再往北惯性解失效。对大气如取 $C\approx5$ m/s，则 $L_R\approx500$ km，因此 6 个 Rossby 变形半径相当于急流已到过副热带纬度。需要指出，不管如何改变参数，对无切变或无旋的均匀大尺度来流只存在一个平衡态。

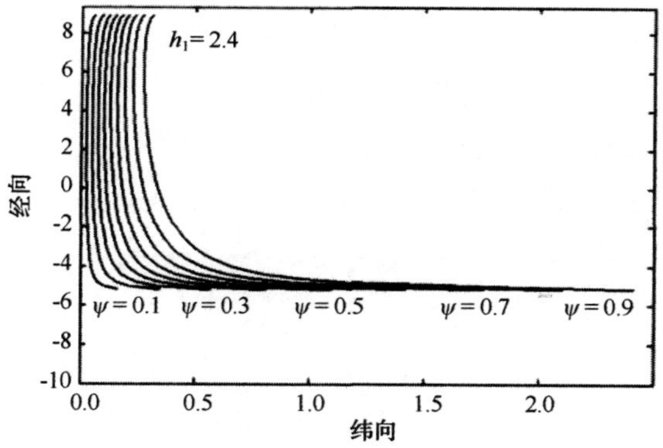

图 4　流函数分布

h_1 为赤道外侧的 h，横坐标和纵坐标单位都为 Rossby 变形半径

然而，如地转向西流有经向梯度，即大尺度来流是切变流或有旋流，这时将存在不止一个平衡态。取南界 $y_0=-4.0$，北界 $y_1=-3.0$，其中 $\bar{u}=\bar{u}_0+ky$。在本次计算中设 $\bar{u}_0=-1.0$，$k=-0.2$，南界 $\bar{h}_0=4.0$，得到北界处流函数 $\bar{\psi}_1=1.112$。在这一组参数下，一元五次方程有两个实根满足式(26)，而且得到了合理的解，所以有两个平衡态。

对于地转向西流存在纬向速度的经向梯度时，在惯性急流区内存在两个平衡态，从图 5 中可以看到对于同一个外区的流函数值，在惯性区中存在着两条流线与其对应，即一条地转向西流中的流线在惯性急流区域内分解成两条流线。两个态都清楚地描述了向西流在西边界附近形成惯性急流区。

经向速度分布如图 6 所示，由于描述的是惯性急流区域内的经向速度分布，由流线的连续性，考虑每一条流线上的速度分布情况，因而把流函数作为横坐标。从图 6(a)和图 6(b)中都可以看到，沿

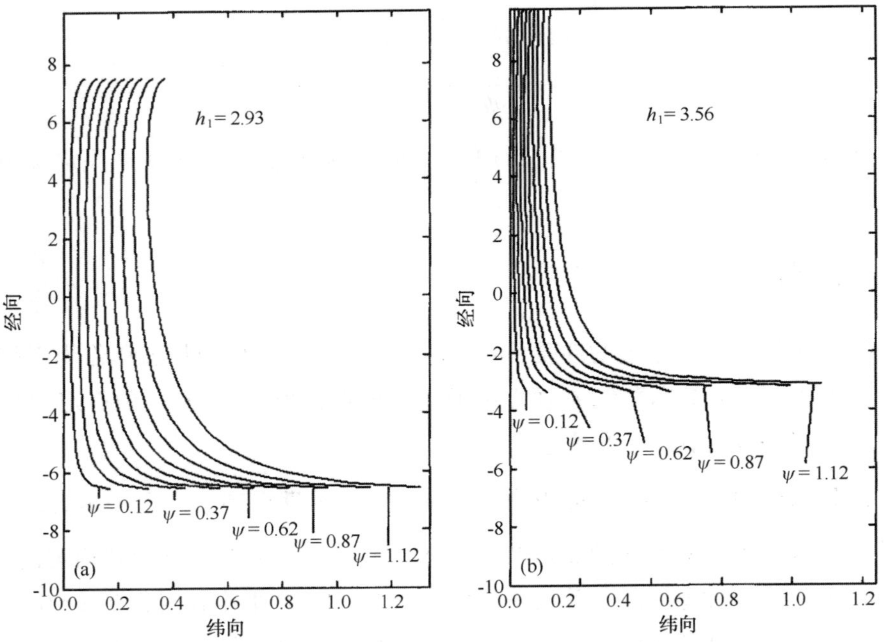

图 5 惯性区域内流函数两个平衡态的分布

(a) 平衡态一;(b) 平衡态二,下同。h_1 为赤道处外侧的 h,横坐标和纵坐标单位同图 4。(a)和(b)中两个平衡态分别在北纬约 7 个 Rossby 变形半径和北纬约 10 个 Rossby 变形半径时得到 $x=0,h=0$,惯性解失效

着流线的经向速度从边界到远离边界处逐渐变小,在边界附近有最大速度,而且越往北经向速度越大(对于大气来说,该速度可达到 12 m/s),经向速度在纬向的梯度也随着纬度向北而增大。这是由于越往北惯性急流区越窄,而由于式(28)约束的流量守恒,所以经向速度变大,而且其纬向梯度也变大。

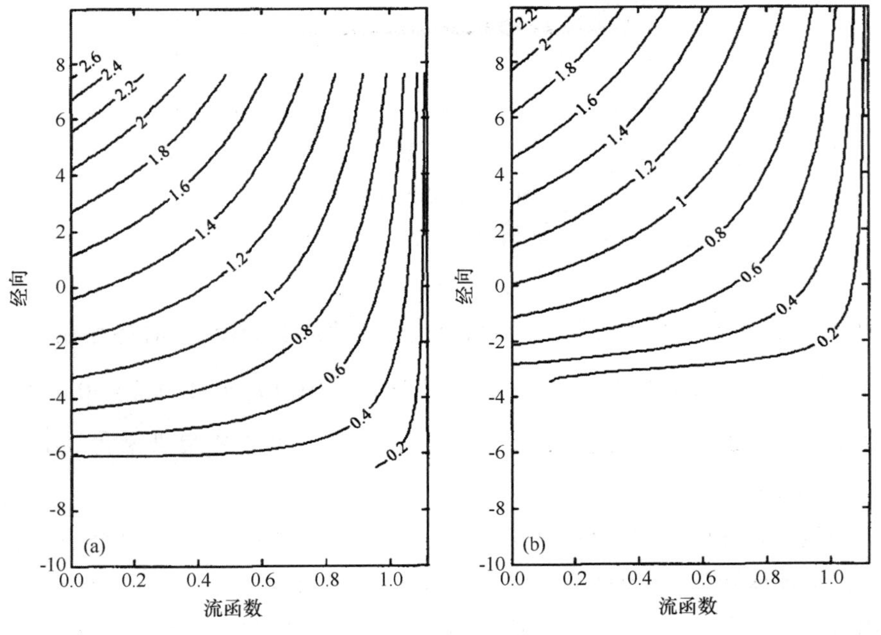

图 6 惯性区域内经向速度两个平衡态的分布

纵坐标单位为 Rossby 变形半径,下同。图中曲线为速度等值线

扰动厚度 h 的两个平衡态分布如图 7 所示。从图 7 中可看到两个态的扰动厚度 h 分布大体一致，随着纬度向北增加，扰动厚度 h 呈递减的趋势，而且在边界上惯性区域最北部，图 7(a)显示第 1 个态在北纬 7.5 个 Rossby 变形半径时 h 减为 0，急流将脱离赤道进入流体内部，惯性解失效；而图 7(b)显示第 2 个态则到北纬 10.2 个 Rossby 变形半径时 h 才减为 0。而且从图 7 中可以看到，两个态在赤道上都呈现出 h 不变的特征。

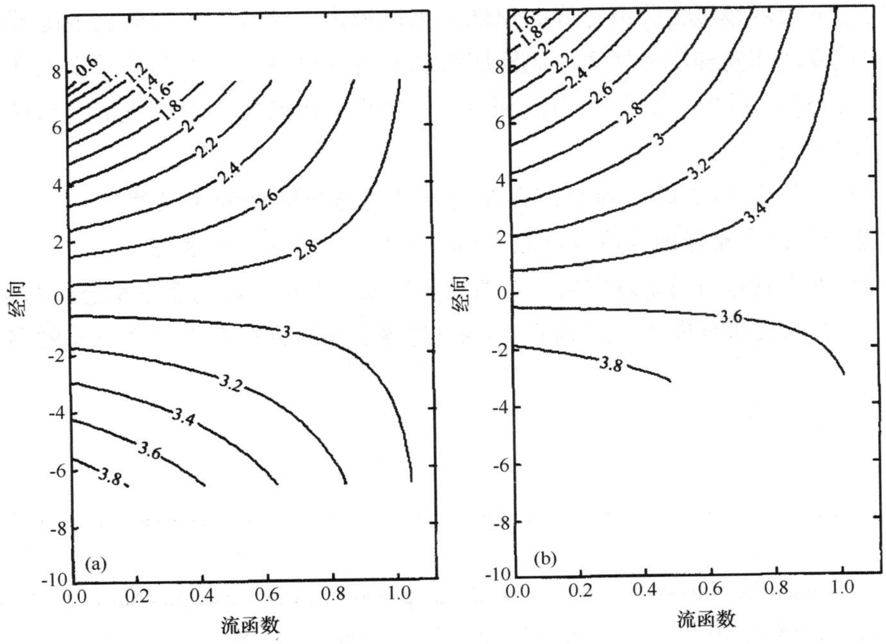

图 7　惯性区域内扰动厚度 h 的两个平衡态的分布

(a)和(b)中曲线为扰动厚度 h 等值线，h 纬向梯度均从负增加到正

4　多平衡态解在气象应用中的意义

当南半球的向西流不均匀时，上节得到两个不同的跨赤道急流平衡态，其中一个态的急流跨过赤道后可达到副热带纬度，另一个态的急流可以达到更北的纬度。当然急流到达的纬度是相对值，可以调整参数值特别是来流的流速值而改变，但相对而言，两个态的急流所渗入到北半球的纬度相差 30%，差别很大。对于东亚地区，如果南海跨赤道急流的发展和南海季风的暴发密切相关，则这两个平衡态所对应的气候特别是雨带的位置非常不同，这无疑给天气或气候预报带来一定的不确定性。然而并不是所有的大尺度天气形势下都会出现多态的环流形势。上面的分析表明，理论上只有当南半球大尺度东风具有切变时才能出现两个跨赤道急流平衡态，若东风的经向分布相对均匀则只有一个较为确定的跨赤道急流态。由此给天气分析一个启示，在基本的大尺度环流背景下次生尺度的天气形势是否可以有不同方向的发展，需要仔细分析。不同天气形势的出现意味着存在多个平衡态势，当然这在实际分析工作中十分困难，因为决定天气的发展形势受多方面物理因素的制约，不会像理论模式那么简单。

5 结论

本文对跨赤道流的多平衡态进行了分析,对控制跨赤道急流的惯性区域的非线性方程给出了多个平衡态的解。通过与惯性急流相联系的大尺度洋流的不同特征得到普适函数,对于均匀地转向西流,只存在一个态,当地转向西流存在经向梯度的时候得到两个平衡态。向西流在西边界时逐渐变为向北的急流,跨过赤道后慢慢会脱离边界。惯性急流区域内各流线上的经向速度随着纬度向北逐渐变大,而且经向速度的纬向梯度也增大。温跃层扰动厚度 h 的纬向梯度随着纬度向北从负值开始一直增加,到赤道,达到 0,到北半球变为正值,而且 h 在边界上逐渐变小为 0,此时急流脱离边界进入流体内部,惯性解将失效。

由上述分析得到多平衡态跨赤道急流存在的条件是急流区外的大尺度环流必须具有切变或涡度。可以预料适当地改变来流的构造有可能得到跨赤道急流的多个平衡态。

在另一方面,本文只是对南半球向西流在西边界形成的惯性急流的多平衡态进行了分析,对于南半球向东流在东边界,北半球向西(东)流在西(东)边界上是否同样存在惯性急流的多平衡态尚有待研究。

参考文献

[1] Findlater J. A major low-lever air current near the Indian Ocean during the northern summer. Q J R Meteorol Soc,1969,95: 362-380.
[2] Findlater J. Mean monthly airflow at low levels over the western Indian Ocean. Geophys Memo,1971,16:1-53.
[3] 陈隆勋,李麦村,李维亮,等. 夏季的季风环流. 大气科学,1979,3: 78-90.
[4] 陈隆勋. 东亚季风系统的结构及其中期变动. 海洋学报,1984,5(6): 744-758.
[5] Anderson D L. The low-level Jet as a western boundary current. Mon Weather Rev,1979,104: 907-921.
[6] Swallow J C,Bruce J G. Current measurements off the Somali coast during the south-west monsoon of 1964. Deep-Sea Res,1966,13: 861-888.
[7] Gill A E. Atmosphere-Ocean Dynamics. San Diego: Academic Press,1982:662.
[8] Lighthill M J. Dynamic response of the Indian Ocean to the onset of the S<W Monsoon. Philos Trans R Soc A-Math Phys Eng Sci, 1969,265: 45-92.
[9] Krishnamurtt T N,Molinari J,Pan H L. Numerical simulation of the Somali jet. J Atmos Sci,1976,33: 2350-2362.
[10] Bannon P R. On the dynamics of the East African jet, Ⅰ: Simulation of mean conditions for July. J Atmos Sci,1979,36: 2139-2152.
[11] Anderson D L, Moore D W. Cross-equatorial inertial jets with special relevance to very remote forcing of Somali current. Deep-Sea Res,1979,26: 1-22.
[12] Charney J G. The Gulf Stream as an inertial boundary layer. Proc Natl Acad Sci U S A,1955,41: 731-740.
[13] Chao J P,Liu J. Inertial theory for oceanic cross-equatorial jet. Acta Meteorol Sin,1996,10: 61-72.

南极绕极流及经圈翻转流的双平衡态理论*

巢纪平[1,2]　李耀锟[3,4]

(1. 国家海洋环境预报中心,北京 100081;2. 国家海洋局第一海洋研究所,青岛 266061;3. 中国科学院大气物理研究所大气科学和地球流体力学数值模拟国家重点实验室,北京 100029;4. 中国科学院研究生院,北京 100049)

摘要:利用非线性惯性理论研究南极绕极流及其经圈环流。模型分为上下两层,上层为 Ekman 层,主要由海表风应力驱动;下层为温跃层,其运动由理想流体的非线性方程控制。通过确定普适函数的形式求得温跃层中惯性模型的解。计算结果表明,在上层条件不变的情况下,温跃层存在两个平衡态解。平衡态一的流函数分布较为平滑,经圈环流深度较浅,纬圈流强度较小;平衡态二的流函数会出现不连续的状况,经圈环流可达 2 000 m 深,纬圈流的强度要比平衡态一大。两个平衡态中纬圈流在温跃层中均存在一个大值区。理论结果特别是平衡态二的结果与资料较为接近。

关键词:南极绕极流;经圈翻转流;惯性理论;双平衡态

南极绕极流(Antarctic Circumpolar Current,ACC)环绕整个南极大陆,连接了太平洋、印度洋、大西洋三大海盆,是南半球海洋环流最主要的特征[1,2]。ACC 及其经圈翻转流(Meridional Overturning Circulation,MOC)对大洋热量、盐分等的输送起着重要作用,由此对全球气候产生重要影响[1,3]。

ACC 主要由海表风应力驱动,但是风应力及温盐强迫的相对重要性及相互作用仍有待研究[4,5]。随着技术的进步和人们对南大洋重要性认识的增加,科学界开展了一系列针对南大洋的观测实验,如 WOCE(World Ocean Circulation Experiment),ISOS(International Southern Ocean Studies),SOOS(The Southern Ocean Observing System)等,对南大洋的观测和认识取得了长足的进步。但人们仍没能对 ACC 系统形成完整的理论认识[6]。因此,采取不同的思路和方法研究将会加深对 ACC 的认识,为进一步理解 ACC 提供帮助。

本文试图利用惯性理论来研究 ACC 及其经圈翻转流。惯性理论在保留了方程中的非线性项的基础上,得到若干物理量的守恒关系,并据此引入普适函数以封闭理论,从而在一定边界条件下求得问题的解。惯性理论的关键在于普适函数的确定,不同的方法会得到不同的普适函数,因此,合理的取法将会使问题的解变得简洁而明确,通常的取法是利用边界外部的信息来确定普适函数的形式。

惯性理论很早就被海洋和大气科学家们所采用。如 Fofonoff 和 Montgomery[7]在 1955 年研究赤道潜流时就曾提出过赤道潜流形成的绝对涡度经圈输送守恒的理论;Charney[8]在同年研究湾流的文章也使用了惯性理论,巢纪平[9]在 1957 年研究青藏高原大地形对西风气流的影响时也使用过类似的理

* 中国科学:地球科学,2011,41(11):1697-1705.

论方法；Pedlosky[10]在1987年发展了Fofonoff等的绝对涡度守恒理论并称之为惯性理论；Johnson和Moore[11]在1997年给出了次表层回流的一个惯性理论；巢纪平等[12]发展了Anderson和Moore[13]的跨赤道惯性急流理论，分析了急流中多平衡态存在的条件。但是目前尚未有利用惯性理论来研究ACC的工作，因此，将惯性理论应用到ACC的研究中会有利于加深对ACC的研究，有助于惯性理论的发展。

1 模型的建立

在经圈平面内，将海洋分为上下两层，上层是Ekman层，下层是温跃层，D，H分别为两层层底的位置。坐标原点取在海表处，y轴向北为正，z轴向上为正（图1）。ACC海域经向范围从0到L。

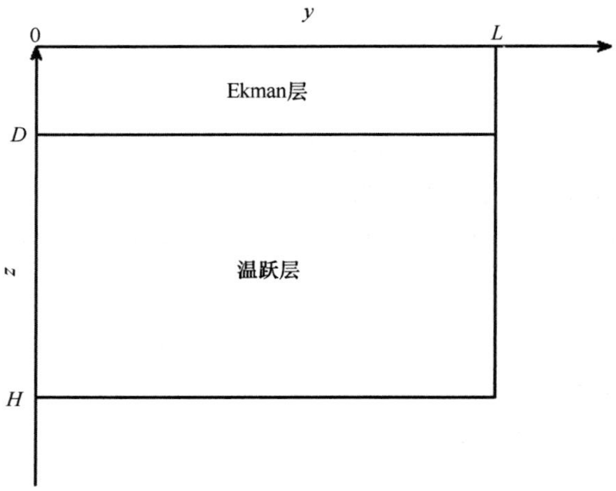

图1 模型框架

1.1 Ekman层

Ekman层中科氏力、气压梯度力和垂直湍流黏性力相平衡：

$$-fv_1 = \kappa \frac{\partial^2 u_1}{\partial z^2} \tag{1}$$

$$fu_1 = \frac{1}{-\rho_0}\frac{\partial p_1}{\partial y} + \kappa \frac{\partial^2 v_1}{\partial z^2} \tag{2}$$

$$\frac{\partial v_1}{\partial y} + \frac{\partial w_1}{\partial z} = 0 \tag{3}$$

水平流速(u_1, v_1)满足海表边界条件：

$$\left(\kappa \frac{\partial u_1}{\partial z}\right)_{z=0} = \frac{1}{\rho_0}\tau^x, \quad \left(\kappa \frac{\partial v_1}{\partial z}\right)_{z=0} = 0 \tag{4}$$

其中κ是运动垂直湍流交换系数，f为科氏参数，ρ_0为参考态密度，τ^x为风应力的纬向分量，其他符号同常用符号。

下边界条件取在无穷深处，并假定在无穷深处$u_1, v_1 = 0$。由于Ekman层中的混合作用十分剧烈，令式（2）中等号右端第一项为常数，即y方向的水平压力梯度力项不随深度变化。在上述条件下，计算得到方程（1），方程（2）的解：

$$u_1 = me^{\frac{z}{h_E}}\sin\left(\frac{\pi}{4} + \frac{z}{h_E}\right) + u_g \tag{5}$$

$$v_1 = me^{\frac{z}{h_E}}\cos\left(\frac{\pi}{4} + \frac{z}{h_E}\right) \tag{6}$$

其中

$$h_E = \sqrt{-\frac{2\kappa}{f}},\quad m = \frac{\tau^x}{\rho_0\sqrt{-f\kappa}},\quad u_g = -\frac{1}{\rho_0}\frac{\partial p_1}{\partial y}$$

由于是在南半球,科氏参数 f 为负数,因此 h_E, m 均为实数。

根据方程(3)引入流函数

$$v_1 = \frac{-\partial\psi_1}{\partial z},\quad w_1 = \frac{\partial\psi_1}{\partial y} \tag{7}$$

式(6)对 z 积分,可求解出 ψ_1:

$$\psi_1 = -\frac{h_E}{2}(u_1 - u_g + v_1) + c \tag{8}$$

其中 c 为积分常数。海表处应为一条流线,令 $\psi_1=0$,于是

$$\psi_1 = -\frac{1}{2}(u_1 - u_g + v_1)h_E + \frac{\sqrt{2}}{2}mh_E \tag{9}$$

1.2　温跃层

利用惯性理论来研究温跃层,温跃层的控制方程组如下:

$$v\frac{\partial u}{\partial y} + w\frac{\partial u}{\partial z} - fv = 0 \tag{10}$$

$$v\frac{\partial v}{\partial y} + w\frac{\partial v}{\partial z} + fu = -\frac{1}{\rho_0}\frac{\partial p}{\partial y} \tag{11}$$

$$v\frac{\partial w}{\partial y} + w\frac{\partial w}{\partial z} = -\frac{1}{\rho_0}\frac{\partial p}{\partial z} + g\alpha\theta \tag{12}$$

$$v\frac{\partial\theta}{\partial y} + w\frac{\partial\theta}{\partial z} + \Gamma w = 0 \tag{13}$$

$$\frac{\partial v}{\partial y} + \frac{\partial w}{\partial z} = 0 \tag{14}$$

其中,θ 为扰动位温;$\Gamma=\mathrm{d}\bar{\theta}/\mathrm{d}z$ 为参考态位温垂直变化率;α 为海水的热膨胀系数;p 为扰动压力。

由式(14)引入流函数,则方程(10)可以写为

$$-\frac{\partial\psi}{\partial z}\left(\frac{\partial u}{\partial y} - f\right) + \frac{\partial\psi}{\partial y}\frac{\partial u}{\partial z} = 0 \tag{15}$$

令

$$\chi = \int\left(\frac{\partial u}{\partial y} - f\right)\mathrm{d}y = u - f_0 y - \frac{1}{2}\beta y^2$$

注意,上式已利用 β 平面近似,即,$f=f_0+\beta y$。式中 f_0 为坐标原点处的科氏参数值。

式(15)可以写为

$$\frac{\partial\psi}{\partial y}\frac{\partial\chi}{\partial z} - \frac{\partial\psi}{\partial z}\frac{\partial\chi}{\partial y} = 0 \tag{16}$$

式(16)的一般积分为

$$u - f_0 y - \frac{1}{2}\beta y^2 = F_1(\psi) \tag{17}$$

式中,F_1 为 ψ 的普适函数,下文中 F_2,F_3 均为 ψ 的普适函数,此处一并指出。

同理,由方程(13)可以推导得到

$$\theta + \bar{\theta}(z) = F_2(\psi) \tag{18}$$

式(11),式(12)交叉求导消去 p,得到涡度方程如下:

$$v\frac{\partial \xi}{\partial y} + w\frac{\partial \xi}{\partial z} - f\frac{\partial u}{\partial z} - g\alpha\frac{\partial \theta}{\partial y} = 0 \tag{19}$$

其中,$\xi = \frac{\partial w}{\partial y} - \frac{\partial v}{\partial z}$为相对涡度。

利用式(17)和式(18),对式(19)变形:

$$\bar{v}\cdot\nabla\left[\frac{\partial^2\psi}{\partial y^2} + \frac{\partial^2\psi}{\partial z^2} + \int f\mathrm{d}y\frac{\mathrm{d}F_1}{\mathrm{d}\psi} - g\alpha z\frac{\mathrm{d}F_2}{\mathrm{d}\psi}\right] = 0 \tag{20}$$

式中 $\bar{v} = (v, w)$ 是经圈平面内的速度矢量。由式(20)可得

$$\frac{\partial^2\psi}{\partial y^2} + \frac{\partial^2\psi}{\partial z^2} + \left(f_0 y + \frac{1}{2}\beta y^2\right)\frac{\mathrm{d}F_1}{\mathrm{d}\psi} - g\alpha z\frac{\mathrm{d}F_2}{\mathrm{d}\psi} = F_3(\psi) \tag{21}$$

1.3 普适函数的确定

根据 Ekman 抽吸,Ekman 层底的垂直速度为

$$w_d = -\frac{\partial}{\partial y}\left(\frac{\tau^x}{\rho_0 f}\right) \tag{22}$$

令温跃层内的垂直速度分布为

$$w = w_d\lambda(z) \tag{23}$$

利用 NCEP Global Ocean Data Assimilation System (GODAS)提供的 1980—2009 年 30 年平均的垂直速度资料,给出三次多项式形式的 $\lambda(z)$,如下:

$$\lambda(z) = a_0 + a_1 z + a_2 z^2 + a_3 z^3 \tag{24}$$

将式(23)代入连续方程(14)中,可得

$$v = \frac{\tau^x}{\rho_0 f}\lambda'(z) \tag{25}$$

又根据流函数定义,有

$$\psi = -\frac{\tau^x}{\rho_0 f}\lambda(z) \tag{26}$$

令 $y = y_s > L$ 在 ACC 边界外部的南半球副热带海洋,此时,上式化为

$$\psi_s = -\left[\frac{\tau^x}{\rho_0 f}\right]_s \lambda(z_s) \tag{27}$$

式中下标 s 表示在 y_s 处的值。

将式(24)代入式(27)中,利用三次方程的求根公式可以求得式(27)的三个根,三个根较为复杂,此处没有给出具体形式。可以将三个根统一写为

$$z_s^{(i)} = \mathscr{F}_i(\psi_s) \tag{28}$$

其中 $i=1,2,3$ 表示方程的三个根。

令在 y_s 处的副热带海洋扰动温度为零,则守恒关系式(18)可以写为

$$\Gamma z_s = F_2(\psi_s) \tag{29}$$

将式(28)代入到式(29)中,即可得到 F_2 的形式:

$$F_2^{(i)}(\psi_s) = \Gamma \mathscr{F}_i(\psi_s) \tag{30}$$

令在 y_s 处的副热带海洋满足热成风平衡:

$$\left[\frac{\partial u}{\partial z}\right]_s = -\left[\frac{g\alpha}{f}\frac{\partial \theta}{\partial y}\right]_s \tag{31}$$

利用守恒关系式(17),式(18),上式可以化简为

$$v_s \frac{\mathrm{d}F_1^{(i)}}{\mathrm{d}\psi_s} = \frac{g\alpha}{f_s} w_s \frac{\mathrm{d}F_2^{(i)}}{\mathrm{d}\psi_s} \tag{32}$$

整理化简,有

$$\frac{\mathrm{d}F_1^{(i)}}{\mathrm{d}\psi_s} = -\left[\frac{g\alpha}{f}\frac{\partial}{\partial y}\left(\frac{\tau^x}{\rho_{of}}\right)^{-1}\right]_s \frac{1}{\lambda'(z_s)}\Gamma \psi_s \mathscr{F}_i(\psi_s) \tag{33}$$

由于在 y_s 处满足热成风平衡,守恒关系式(21)可以化简为

$$\left[\frac{\partial^2 \psi}{\partial y^2} + \frac{\partial^2 \psi}{\partial z^2}\right]_s = F_3(\psi_s) \tag{34}$$

将式(27)代入,整理化简可得

$$F_3^{(i)}(\psi_s) = \frac{\left[\frac{\partial^2}{\partial y^2}\left(\frac{\tau^x}{\rho_a f}\right)\right]_s}{\left[\frac{\tau^x}{\rho_a f}\right]_s}\psi_s - \left[\frac{\tau^x}{\rho_a f}\right]_s \lambda''(z_s^{(i)}) \tag{35}$$

式(30),式(33)及式(35)三式即为三个普适函数的形式,注意到,由于式(28)是三个解,因此算得的三个普适函数各有三个。

1.4 控制方程

由于式(30),式(33)及式(35)三式的普适性,将其运用到 ACC 区域时,只需要把三式中的 ψ_s 改为 ψ 即可,函数的形式并不发生变化。将三式代入式(21),可以得到

$$\frac{\partial^2 \psi}{\partial y^2} + \frac{\partial^2 \psi}{\partial z^2} - \left(f_0 y + \frac{1}{2}\beta y^2\right)\left[\frac{g\alpha}{f}\frac{\partial}{\partial y}\left(\frac{\tau^x}{\rho_0 f}\right)^{-1}\right]_s \times \frac{1}{\lambda'(z_s^{(i)})}\Gamma \psi \mathscr{F}_i(\psi) - g\alpha z \Gamma \mathscr{F}_i(\psi)$$
$$= \frac{\left[\frac{\partial^2}{\partial y^2}\left(\frac{\tau^x}{\rho_a f}\right)\right]_s}{\left[\frac{\tau^x}{\rho_a f}\right]_s}\psi - \left[\frac{\tau^x}{\rho_a f}\right]_s \lambda''(z_s^{(i)}) \tag{36}$$

边界条件取为

$$y = L,\quad \psi = 0;\quad y = 0, \psi = 0;\quad z = H,\quad \psi = 0;\quad z = D,\quad \psi = \frac{\tau^x}{\rho_0 f}$$

式(17)求拉普拉斯后与式(36)联立,消去 $\Delta\psi$,即可得到纬圈流 u 满足的方程:

$$\frac{\partial^2 u}{\partial y^2} + \frac{\partial^2 u}{\partial z^2} + \left(f_0 y + \frac{1}{2}\beta y^2\right)\frac{\mathrm{d}F_1^{(i)}}{\mathrm{d}\psi}\frac{\mathrm{d}F_1^{(i)}}{\mathrm{d}\psi} - g\alpha z\frac{\mathrm{d}F_1^{(i)}}{\mathrm{d}\psi}\frac{\mathrm{d}F_2^{(i)}}{\mathrm{d}\psi} - \beta$$

$$-L\frac{\mathrm{d}^2F_1^{(i)}}{\mathrm{d}\psi^2}\left[\left(\frac{\partial\psi}{\partial y}\right)^2+\left(\frac{\partial\psi}{\partial z}\right)^2\right]-\frac{\mathrm{d}F_1^{(i)}}{\mathrm{d}\psi}\left[\frac{\left[\frac{\partial^2}{\partial y^2}\left(\frac{\tau^x}{\rho_a f}\right)\right]_s}{\left[\frac{\tau^x}{\rho_a f}\right]_s}\psi-\left[\frac{\tau^x}{\rho_a f}\right]_s\lambda''(z_s^{(i)})\right]=0 \tag{37}$$

边界条件取为

$$y=L,u=0;\quad y=0,u=0;\quad z=H,u=0;\quad z=D,u=u_p(y)$$

其中，$u_p(y)$为 Ekman 底的 u 分布。

类似，式(18)求拉普拉斯后与式(36)联立，消去 $\Delta\psi$，即可得到扰动位温 θ 满足的方程，如下：

$$\frac{\partial^2\theta}{\partial y^2}+\frac{\partial^2\theta}{\partial z^2}+\left(f_0y+\frac{1}{2}\beta y^2\right)\frac{\mathrm{d}F_1^{(i)}}{\mathrm{d}\psi}\frac{\mathrm{d}F_2^{(i)}}{\mathrm{d}\psi}-g\alpha z\frac{\mathrm{d}F_2^{(i)}}{\mathrm{d}\psi}\frac{\mathrm{d}F_2^{(i)}}{\mathrm{d}\psi}$$

$$-\frac{\mathrm{d}^2F_2^{(i)}}{\mathrm{d}\psi^2}\left[\left(\frac{\partial\psi}{\partial y}\right)^2+\left(\frac{\partial\psi}{\partial z}\right)^2\right]-\frac{\mathrm{d}F_2^{(i)}}{\mathrm{d}\psi}\left[\frac{\left[\frac{\partial^2}{\partial y^2}\left(\frac{\tau^x}{\rho_a f}\right)\right]_s}{\left[\frac{\tau^x}{\rho_a f}\right]_s}\psi-\left[\frac{\tau^x}{\rho_a f}\right]_s\lambda''(z_s^{(i)})\right]=0 \tag{38}$$

边界条件为

$$y=L,\theta=0;\quad y=0,\theta=0;\quad z=H,\theta=0;\quad z=D,\theta=\theta_p(y)$$

其中，$\theta_p(y)$为 Ekman 层底的 θ 分布。

注意，式(36)，式(37)及式(38)各自均为三个方程，它们均源于式(28)的三个解。

2 计算结果分析

根据方程式(36)至式(38)计算流函数及经圈速度，纬圈流和扰动温度的分布。计算时参数的取值分别为：$\alpha=1.6\times10^{-4}/\mathrm{K}$，$\beta=5.9\times10^{-12}/(\mathrm{s\cdot m})$，$f_0=-1.4\times10^{-4}/\mathrm{s}$，$\Gamma=4\times10^{-3}\ \mathrm{K/m}$，$\kappa=3.1\times10^{-2}\ \mathrm{m^2/s}$，$g=9.8\ \mathrm{m/s^2}$，$D=-100\ \mathrm{m}$，$H=-3000\ \mathrm{m}$，$L=4.9\times10^6\ \mathrm{m}$，$a_0=2.3860$，$a_1=0.7367\times10^{-2}/\mathrm{m}$，$a_2=0.5177\times10^{-5}/\mathrm{m^2}$，$a_3=0.9953\times10^{-9}/\mathrm{m^3}$。

τ^x 取自 The International Comprehensive Ocean-Atmosphere Data Set (ICOADS) 提供的1971—2000年平均的2°×2°的风应力资料。计算中用多项式拟合风应力，给出风应力的解析表达式，拟合效果见图2。图2中红线为多项式拟合得到的风应力，蓝线为资料给出的风应力，由图可见，拟合效果很好。事实上，作者就是根据风应力的分布来选取 ACC 海域的范围的，$y=0$(75°S)处风应力接近零，$y=L$(约30°S)处风应力也为零，即 ACC 海域的范围是从南极大陆向北一直到西风应力为零的区域。这个区域包括了靠近南极大陆的东风漂流区，经向范围大致在0~0.2L之间(75°-65°S)，东风漂流虽然强度较弱，但是主要也是由海表风应力驱动，且也围绕整个南极大陆。因此一并考虑在计算中。

取 $y_s=1.2L$(约23°S)处为确定普适函数所用的副热带海洋的位置，并对不同的 y_s 进行了验证，如当 $y_s=1.1L$ 或 $y_s=1.3L$ 时，计算结果并没有显著的差异(结果略)。这一点也可以从图2看出，即 y_s 所取位置海洋的物理性质是一致的(均在副热带海洋处，海表风应力均为东风应力)。当然，y_s 也不能不加限制随意选取，必须要满足在副热带海洋的条件。

根据给定的参数，利用差分迭代的方法计算(36)，(37)和(38)三式。注意到三式中每个式子代表了三个解，其中两个解收敛，一个解发散，因此流函数、纬圈流和扰动温度均存在两个平衡态。下面分别对每个平衡态进行分析和讨论。

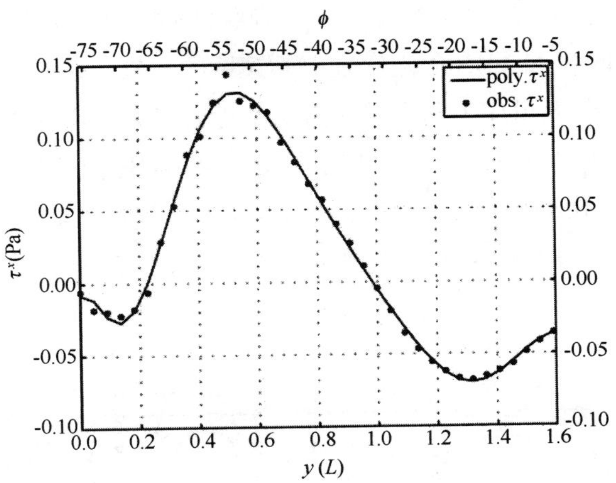

图 2　风应力的分布

为了更好地解释两个平衡态解的意义，下图给出了式(28)的解的图像，如图 3 所示。从图中可以看出，当 $\psi<-0.4598$ 及 $\psi>0.4764$ 时，方程只有一个解，对应于图中的 A，C 两点当 $-0.4598<\psi<0.4764$ 时，方程有三个解，对应于图中的 A，B，C 三点。与 λ 的形式相比较，可知 A 点所在的一支解(图中绿线，下文称为解 Ⅰ)受到了 Ekman 层强烈的影响，B，C 两点所在的解(蓝线和红线，称为解 Ⅱ，解 Ⅲ)则基本体现了温跃层内垂直速度自身的变化。而在计算中，解 Ⅰ 对应的流函数不收敛，解 Ⅱ，解 Ⅲ 对应的流函数收敛，且解 Ⅱ 的收敛速度要慢于解 Ⅲ。因此，可以认为计算得到的两个平衡态是由于解 Ⅱ 和解 Ⅲ 在温跃层内分布不同而造成的，它们是海洋内部的物理过程的反映。在温跃层顶，两个平衡态主要体现了风应力的强烈影响，两个海洋内部物理过程相对于风应力的影响可以忽略，此时两个平衡态几乎是类似的。随着深度的增加，风应力的影响逐步减弱，海洋内部的物理过程的重要性开始体现，两个平衡态的差异逐步变大。由于计算时给定了静止的下边界条件，因此随着深度的继续增加，两个平衡态又会逐步趋同。

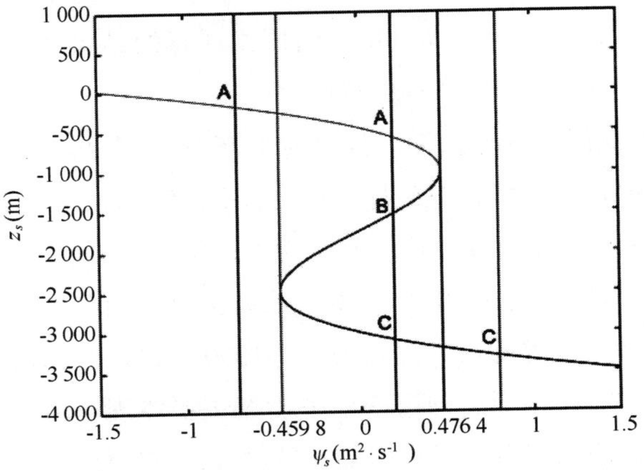

图 3　式(28)解的分布

图 4 是流函数的两个平衡态的分布图。图中横坐标分别是计算区域的经向宽度和相应的纬度值，纵坐标是深度。从图中可以看到，两个平衡态在基本型方面是一致的，都表现为 75°-65°S 之间的反环流，65°-30°S 之间的正环流。当流函数值大于 0.4764(见图 3)时，两个平衡态的大小和分布是

非常类似的；当流函数值小于 0.4764 时，平衡态二流函数（图 4（b））的范围会比平衡态一流函数的范围要大，深度要深，并且在 65°S 和 33°S 附近存在着不连续的现象。这两个纬度对应着纬向风应力为零的区域。即，在这种情况下，西风应力和东风应力的交界处对着海洋深处流函数的不连续面。这种不连续现象的产生也可以由图 3 来解释。由图 3，当 $\psi>0.4764$ 时，方程只有一个解（解Ⅲ，图 3 中 C 点），此时解是一致的；当 $\psi<0.4764$ 时，方程有三个解，其中解Ⅱ和解Ⅲ收敛，解Ⅲ（图 3 中 C 点）是连续的；解Ⅱ（图 3 中 B 点）与 $\psi>0.4764$ 时的解Ⅲ（图 3 中 C 点）并不连续，而这就决定了流函数在计算中存在着不连续的情况。

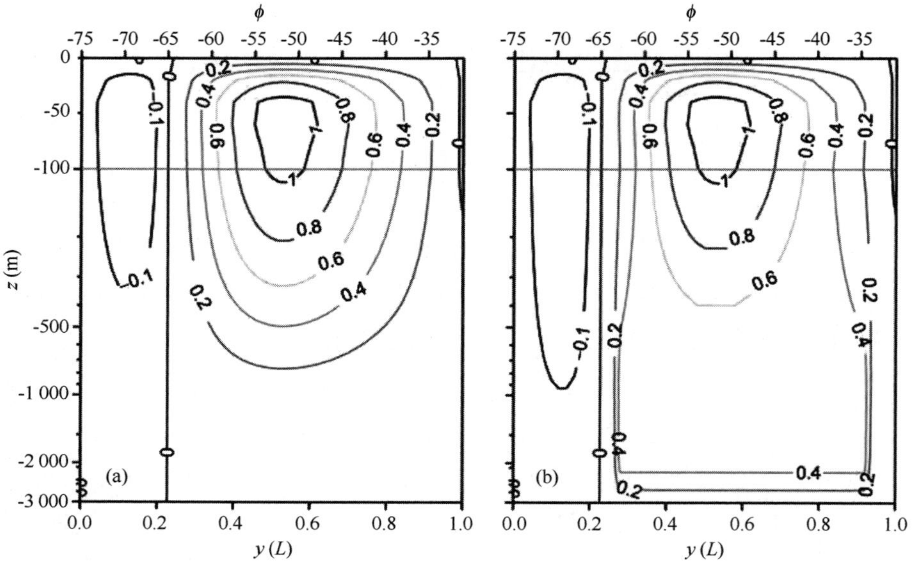

图 4 流函数的两个平衡态的分布（见书后彩插）

（a）平衡态一；（b）平衡态二

令 ACC 所在纬圈的平均长度为 20000 km，根据图 4 计算经圈流输送最大为 22 Sv，中心大约在 50°S 附近。这与 Manabe 等[14]及 Döös 和 Webb[15]用模式计算的结果非常接近。第二个平衡态在范围上似乎与参考文献[14，15]的结果更为一致一些。

图 5 给出了两个平衡态流函数的分岔图。计算时平衡态一、平衡态二流函数均取每个纬度上的最大值。图中横坐标是流函数的最大值，纵坐标为深度，由于 Ekman 层的流函数是相同的，因此图中没有画出 Ekman 层。图中蓝色线条与绿色线条分别为平衡态一、平衡态二流函数的最大值。由图可知，从温跃层层顶开始，随着深度的增加，两个平衡态的流函数开始分岔，深度继续增加，两个平衡态又会逐步靠近，当到达温跃层层底时流函数值均为零。1000～2000 m 深度处是两个平衡态流函数差异最大的区域。这也与前面的分析一致，即第一个平衡态主要体现了风应力的作用随着深度的增加而减小的过程，而第二个平衡态则强调了海洋内部物理过程的重要性。

图 6 给出了经圈流 v 的分布，横纵坐标说明如图 4 所示。v 在混合层内的数值较大，温跃层中数值较小，这是与资料一致的。平衡态二（图 6（b））的零线范围较大，分布较为杂乱，与资料有一定的相似性。受海表风应力的影响，在 ACC 区域，经圈流在表层为从南向北流，底层为从北向南流，南极大陆附近的东风漂流区则相反。因此，在 65°S 处会因为辐散而引起上升运动，这可以从图 7 中看出。图 7 是垂直速度 w 的分布图。两个平衡态的 w 分布有较大的不同，平衡态一在 75°–70°S 区域内为下

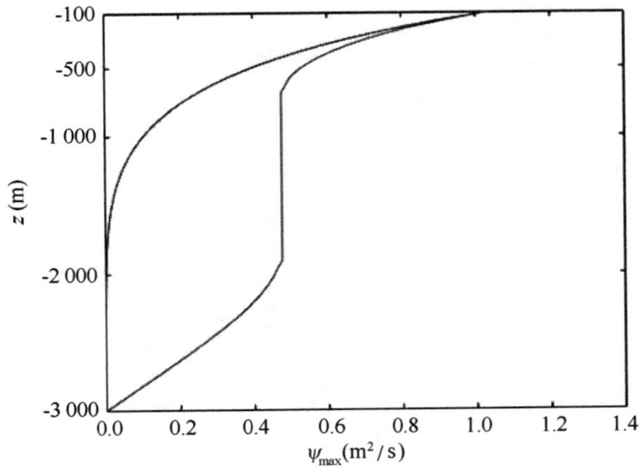

图 5　流函数分岔(见书后彩插)

沉运动,70°−50°S 区域为上升运动,50°−30°S 的区域内为下沉运动,上升下沉运动的深度大约在 1000 m 深度,平衡态二中上升下沉运动的范围变窄,上升运动局限在 65°S 附近的区域内,下沉运动则局限在 35°S 附近处,运动的深度加深到 3000 m 左右。与资料相比,平衡态二似乎更好一些。垂直速度与经圈速度一起,构成了大洋经圈平面内的闭合环流。

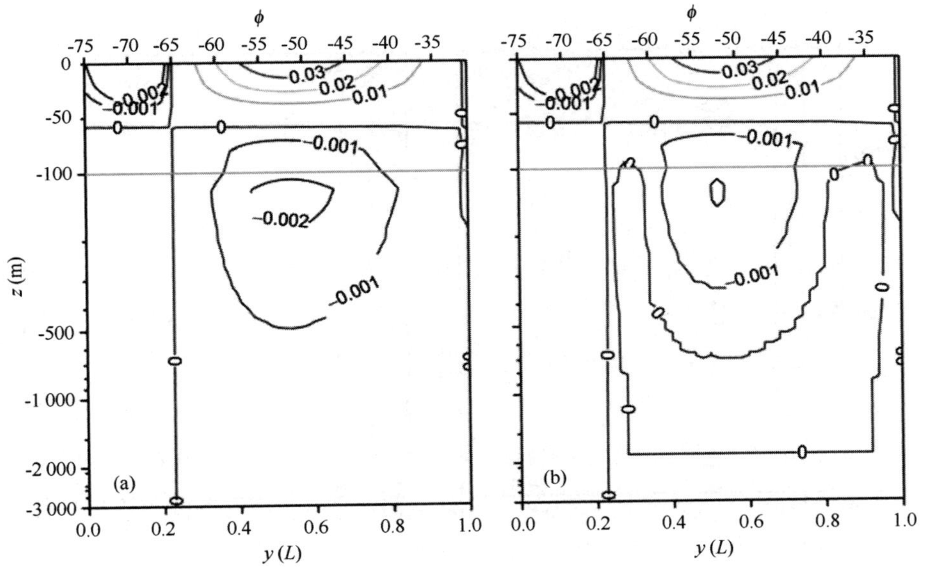

图 6　经圈流的两个平衡态的分布(见书后彩插)

(a) 平衡态一; (b) 平衡态二

纬圈流 u,即 ACC 的分布如图 8 所示。平衡态一和平衡态二的 ACC 在分布形态上较为类似,但是平衡态二的纬圈流强度要更大一些;与资料相比,两个平衡态较为合理地表征了纬圈流,并且位置、强度均比较合适。但是 u 在温跃层中存在一个极大值区,该区域与上层的大值区断裂,这在资料中没有反映。65°−55°S 范围内的纬圈流输送约为 103 Sv(大致为德雷克海峡的范围,比实际海峡范围要宽),而 60°−50°S 范围的输送约为 170 Sv,与参考文献[4]里给出的 135 Sv 在量级上是相当的。

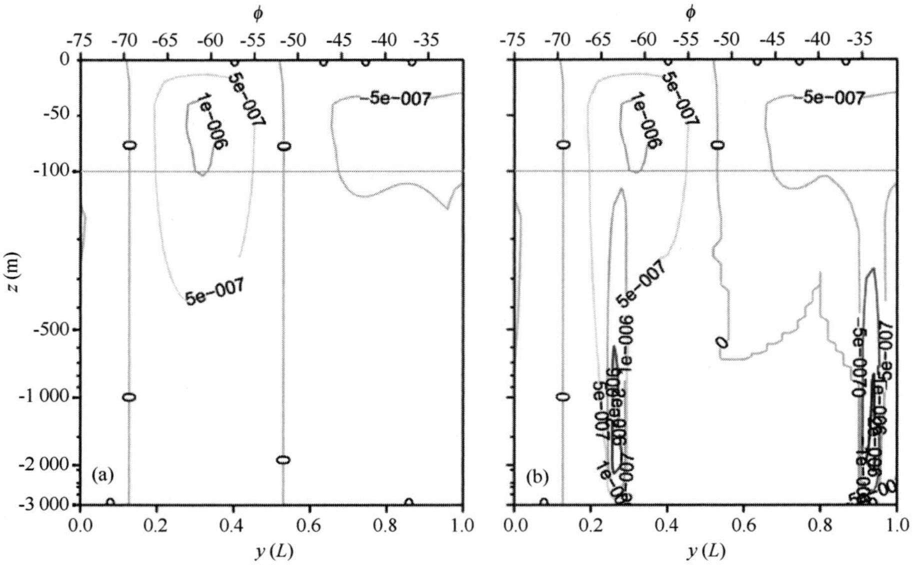

图 7　垂直速度的两个平衡态的分布(见书后彩插)

(a) 平衡态一；(b) 平衡态二

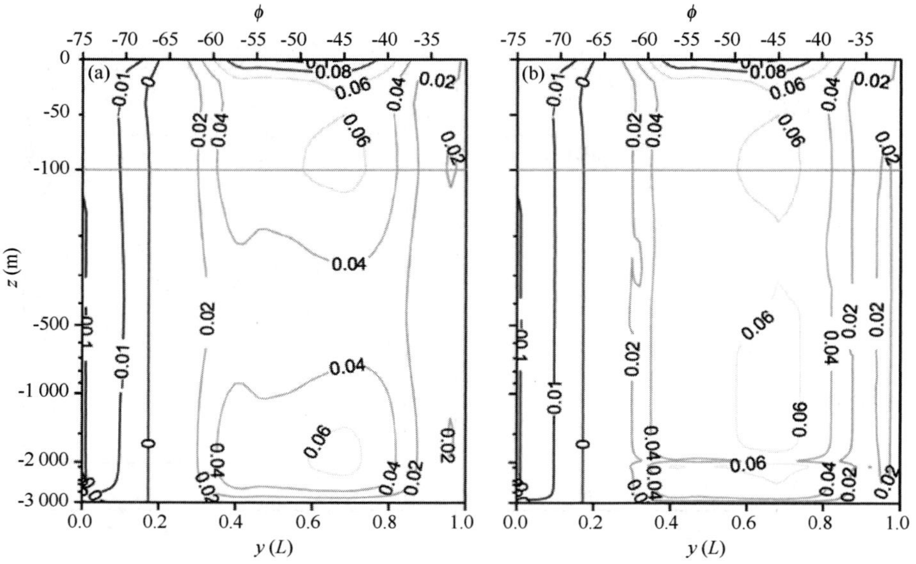

图 8　纬圈流的两个平衡态的分布(见书后彩插)

(a) 平衡态一；(b) 平衡态二

扰动温度的分布如图 9 所示。由于在 Ekman 层中没有涉及扰动温度,因此只计算了温跃层扰动温度的分布,温跃层顶的扰动温度分布取为与风应力一致的扰动型,即,扰动温度在靠近南极大陆的东风漂流区为负,而在 ACC 海域为正。计算结果表明,扰动温度随着深度变化很小,有相当的正压的分布结构。两个平衡态的扰动温度分布差异不大,这说明两个平衡态解可以在温度垂直廓线分布基本相同的情况下产生。由此可推断,与热力过程相比,海洋内部的动力过程扮演着更加重要的角色。

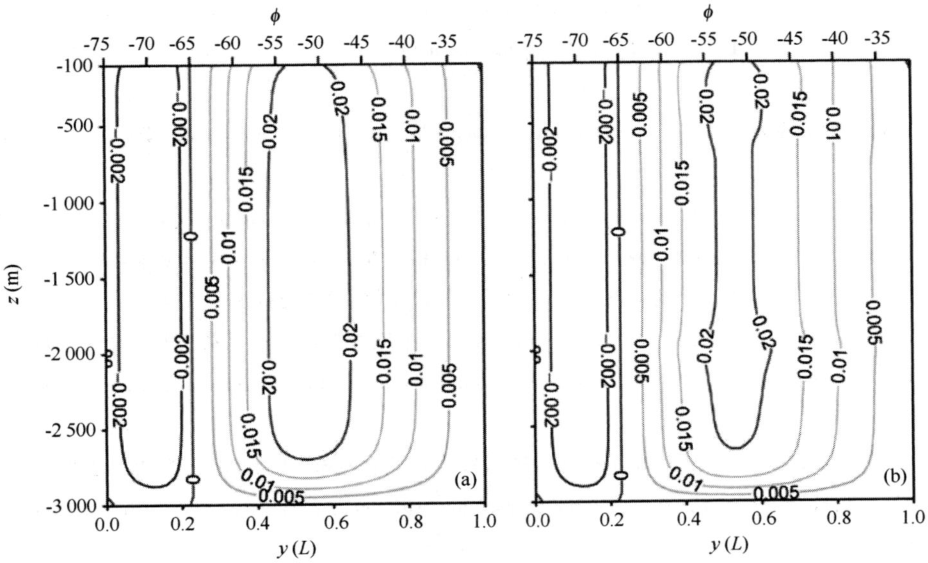

图 9　扰动温度的两个平衡态的分布(见书后彩插)

(a) 平衡态一；(b) 平衡态二

3　总结和讨论

本文利用惯性理论初步分析讨论了南大洋海洋环流。结果表明惯性理论能够较好地反映南大洋海洋环流结构,如计算能够得到 ACC 及其经圈翻转流、靠近南极大陆的东风漂流区,主要特征与资料较为一致。除了与资料相吻合,惯性理论还得到了南大洋环流经圈翻转流的两个平衡态解。第一个平衡态解较为连续,深度较浅;第二个平衡态在纬向风应力为零的区域附近存在着不连续的现象,环流的深度要比第一个平衡态更深一些。而这两个平衡态对应的上部混合层的情形是完全一致的。ACC 两个平衡态间的差异要比经圈翻转流小一些,第二个平衡态的强度稍强。这说明即使上层海洋不发生变化,下层的海洋环流也能够存在不同的形态,并且这些形态都满足给定的约束条件。这两个平衡态的存在性以及二者相互转化还需要进一步研究。

南大洋海域独特的地理环境造就了其对全球气候的重要影响,而对 ACC 及其经圈翻转流的认识还远未达到完善的阶段,对于一些重要的物理因子的作用还存在争议,通过不同的手段研究 ACC,有助于完善整个南大洋环流的知识体系。

致谢:本文使用的海洋速度、温度等变量取自 NOAA 的 GODAS 资料,风应力资料取自 NOAA 的 ICOADS 资料。网址均为 http://www.esrl.noaa.gov/psd/. 刘飞和孙诚博士对本文的计算给予了帮助,特此致谢。同时,感谢审稿专家及阅过初稿并支持工作的同志。

参考文献

[1] Rintoul S C, Hughes C, Olbers D. The Antarctic Circumpolar Current system. In: Siedler G, Church J, Gould J, eds. Ocean Circulation and Climate. London: Academic Press, 2001: 271-302.

[2] Fyfe J C, Saenko O A. Human-induced change in the Antarctic Circumpolar Current. J Climate, 2005, 18: 3068-3073.

[3] Dijkstra H A. Dynamical Oceanography. Verlag Berlin Heidelberg: Springer, 2008: 327-350.

[4] Nowlin W D, Klink J M. The physics of the Antarctic Circumpolar Current. Rev Geophys, 1986, 24: 469-491.

[5] Ivchenko V O, Richards K J, Stevens D P. The dynamics of the Antarctic Circumpolar Current. J Phys Oceanography, 1996, 26: 753-774.

[6] Gallego B, Cessi P, McWilliams J C. The Antarctic Circumpolar Current in equilibrium. J Phys Oceanography, 2006, 34: 1571-1587.

[7] Fofonoff N P, Montgomery R B. The equatorial undercurrent in the light of the vorticity equation. Tellus, 1955, 7: 518-521.

[8] Charney J G. The Gulf Stream as an inertial boundary layer. Proc Natl Acad Sci USA, 1955, 41: 731-740.

[9] 巢纪平．斜压西风带中大地形中有限扰动的动力学．气象学报, 1957, 28: 303-314.

[10] Pedlosky J. An inertial theory of the equatorial undercurrent. J Phys Oceanogr, 1987, 17: 1978-1985.

[11] Johnson G C, Moore D W. The Pacific subsurface countercurrents and an inertial model. J Phys Oceanogr, 1997, 27: 2448-2459.

[12] 巢纪平, 刘飞．跨赤道惯性急流的多平衡态．中国科学 D 辑：地球科学, 2007, 37: 254-260.

[13] Anderson D L T, Moore D W. Cross-equatorial inertial jets with special relevance to very remote forcing of the Somali Current. Deep-Sea Res Part Ⅰ-Oceanogr Res Pap, 1979, 26: 1-22.

[14] Manabe S, Bryan K, Spelman M J. Transient response of a global ocean-atmosphere model to a doubling of atmospheric carbon dioxide. J Phys Oceanogr, 1990, 20: 722-749.

[15] Döös K, Webb D J. The deacon cell and the other meridional cells of the Southern Ocean. J Phys Oceanogr, 1994, 24: 429-442.

第 5 部分　热带运动和海气相互作用

热带海气耦合系统中的长周期振荡及大气中的赤道辐合带*

季劲钧　巢纪平

（中国科学院地理研究所）

摘要：在考虑到大气中各种非绝热加热和对海洋的反馈过程后，建立了一个纬向平均的海洋、大气耦合系统模型。对系统的频率分析表明：存在着一类周期在月以上的长周期振荡。周期的长短依赖于海洋的混合层深度、所在纬度和各种物理过程（如对流凝结加热、辐射冷却、海面蒸发、海水上翻和云量对辐射平衡的调节等）的强弱。

在上述的基础上，讨论了赤道辐合带的形成和变化。计算所得的辐合带宽度和纬度与观测事实是比较一致的。

近年来，许多观测事实表明，低纬度海表温度和热带信风带、副热带高压以及与此相联系的低纬度大气经向运动都存在着月以上的长周期振荡。在这一海洋、大气的耦合系统中，大气的经向运动对动量、热量和水汽的输送，起着重要的作用。因此，讨论这一过程的物理原因、特点，不仅在理论上，而且对建立长期预报模式，都有一定的意义。下面，我们就对这一海气耦合系统中的非绝热加热的振荡过程，做一个简单的分析。先讨论低纬度经向运动长周期振荡的一般特点，然后建立一个赤道辐合带的简单模式。

1　系统方程组的建立

在(x,y,z,t)坐标系中，纬向平均大气运动方程组经线性化后为

$$\frac{\partial u}{\partial t} - f'v = 0 \tag{1}$$

$$\frac{\partial v}{\partial t} + fu = -\frac{1}{\bar{\rho}}\frac{\partial p}{\partial y} \tag{2}$$

$$\frac{\partial v}{\partial y} + \frac{\partial w}{\partial z} = 0 \tag{3}$$

$$\beta_1 T - \frac{1}{\bar{\rho}}\frac{\partial p}{\partial z} = 0 \tag{4}$$

u, v, w, p, T是扰动量，$\beta_1 = g/\bar{T}$，$f' = f - \frac{\partial U_0}{\partial y}$，$U_0$是基本纬向气流，其他符号如通常所用。热力学第一定

* 气象学报，1979，37(3)：32-43.

律是

$$\frac{\partial T}{\partial t}+\sigma_1 w-\frac{f}{\beta_1}\left(\frac{\partial U_0}{\partial z}\right)v=Q \tag{5}$$

Q 是单位质量的加热率。根据长期数值预报研究小组(1977)(下简称研究小组),考虑到牛顿辐射冷却和对流凝结潜热释放的参数化,式(5)可以写成

$$\left(\frac{\partial}{\partial t}+\frac{1}{\tau_R}\right)T+\sigma_1 w=-T^*\left(\frac{\partial u}{\partial y}\right)_b \tag{6}$$

下标“b”表示摩擦层顶的值(下同),σ_1 是稳定度参数($\gamma_a-\gamma$),τ_R 是辐射冷却系数,式中

$$T^*=\frac{L}{C_p}\left(\gamma\frac{\partial \ln e_s}{\partial \bar{T}}\right)l_b\cdot\bar{q}_s(z) \tag{7}$$

这是潜热释放强度的一个度量。l_b 是行星边界层厚度。考虑到强对流区只占正涡度区的一部分,故需对潜热释放强度乘一个修正系数 η,其数值由观测事实决定,为 1/6~1/10。这样

$$\widetilde{T}^*=T^*\cdot\eta$$

在考虑了垂直湍流热交换和海水上下翻腾后,海水温度变化方程可以写成

$$\frac{\partial T_s}{\partial t}+w^*\cdot\frac{\partial \bar{T}_s}{\partial z}=K_s\cdot\frac{\partial^2 T_s}{\partial z^2} \tag{8}$$

混合层中,平均垂直运动主要是由于风应力的埃克曼吸引所引起,因此有

$$w_s^*=-\frac{\rho C_D\mid V\mid}{\rho_s\cdot f}\cdot\left(\frac{\partial u}{\partial y}\right)_0 \tag{9}$$

下标“0”是指海表面的值(下同)。将式(8)从混合层底部积分到海表面,则得到海表温度变化方程是

$$\frac{\partial T_s}{\partial t}=-\frac{\bar{T}_{s,0}-\bar{T}_{s,-D}}{D}w_s^*+\frac{K_s}{D}\left(\frac{\partial T}{\partial z}\right)_0 \quad z=0 \tag{10}$$

式中,D 是混合层深度;$\bar{T}_{s,0}-\bar{T}_{s,-D}$是海表温度与斜温层下温度之差。

在海气交界面上,海水中向下的湍流热交换,向上的蒸发耗热和云对辐射平衡的调节而使之保持平衡,根据研究小组(1977)为

$$\rho_s C_{ps}\cdot K_s\frac{\partial T_s}{\partial z}+L\rho K_q\cdot\gamma_{b,c}\frac{\partial \ln e_s}{\partial \bar{T}}\cdot\frac{\partial \bar{q}_s}{\partial \bar{T}}T'_s=\frac{S_0}{W_0}l_b\cdot\left(\frac{\partial u}{\partial y}\right)_b \tag{11}$$

S_0,W_0 是与短波辐射和垂直运动有关的常数;K_q 是水汽湍流交换系数;$\gamma_{b,c}$是边界层中的温度递减率。

边界条件在海气交界面上取

$$w=0,\quad T=T_s,\quad z=0 \tag{12}$$

在大气上界(对流层顶)

$$\frac{\partial u}{\partial z}=0,\quad z=H \tag{13}$$

H 是大气标高。

现在我们讨论在海气耦合系统中的长周期振荡。把海水混合层中的湍流热交换,作为基本物理过程,并用 D^2/K_s 作为特征时间,同时取如下特征量

$$(y,z,z_s,T_s)=(l,H,D,\Delta T_s)$$

则变量的特征值为

$$(u,v,w,p,T)=\left(\frac{\beta_1 H\Delta T_s}{fl},\frac{\beta_1 H\Delta T_s K_s}{lff'D^2},\frac{\beta_1 H^2\Delta T_s K_s}{l^2ff'D^2},\bar{\rho}\beta_1 H\Delta T_s,\Delta T_s\right)$$

取下列特征值和参数

$$D=10^4\ \mathrm{cm},H=10^6\ \mathrm{cm},l=10^8\ \mathrm{cm},\Delta T_s=2℃,$$

$$f'=f=0.5\times10^{-4}/\mathrm{s}^{-1}(\varphi=20°),K_s=10\ \mathrm{cm^2/s}$$

$$\beta_1=3.4\ \mathrm{cm\cdot s^2/℃^{-1}},\gamma=0.55\times10^{-4}℃/\mathrm{cm},\tau_R=5\times10^5\ \mathrm{s}$$

$$C_D=2\times10^{-3},\Delta\bar{T}_s=\bar{T}_{\rho,0}-\bar{T}_{s,-D}=10℃,|V|=5\times10^2\ \mathrm{cm/s}$$

将这些数值代入方程式(1)至式(4),式(6),式(10)就得到无量纲方程组

$$\frac{\partial u}{\partial t}-v=0 \tag{1'}$$

$$\frac{1}{F^2}\frac{\partial v}{\partial t}+u=-\frac{\partial p}{\partial y} \tag{2'}$$

$$\frac{\partial v}{\partial y}+\frac{\partial w}{\partial z}=0 \tag{3'}$$

$$T=\frac{\partial p}{\partial z} \tag{4'}$$

$$\left(\frac{\partial}{\partial t}+G\right)T-M\cdot v+\frac{K}{F^2}w=-\frac{K}{F}C_v\left(\frac{\partial u}{\partial y}\right)_b \tag{6'}$$

$$\frac{\partial T_s}{\partial t}=-N\left(\frac{\partial u}{\partial y}\right)_0+\left(\frac{\partial T_s}{\partial z}\right)_0 \tag{10'}$$

和边界条件

$$\frac{\partial T_s}{\partial z}+\lambda_Q T_s=\lambda_s\left(\frac{\partial u}{\partial y}\right)_0\qquad z=0 \tag{11'}$$

$$w=0\qquad T=T_s\qquad z=0 \tag{12'}$$

$$\frac{\partial u}{\partial z}=0\qquad z=1 \tag{13'}$$

以及

$$\frac{\partial u}{\partial z}=-\frac{\partial T_s}{\partial y}\qquad z=0 \tag{14}$$

方程组中的无量纲参数分别为

$$G=\frac{D^2}{K_s\cdot\tau_R},\quad F=\frac{fD^2}{K_s},\quad K=\frac{H^2}{l^2}\cdot\sigma_1\beta_1\left(\frac{D^2}{K_s}\right)^2$$

$$C_v=\frac{\widetilde{T}^*}{\sigma_1 H},\quad N=\frac{\Delta\bar{T}_s\rho C_D|V|\beta_1 H\cdot D}{f^2l^2\rho_s\cdot K_s},\quad M=\frac{H}{lf}\cdot\frac{\partial U_0}{\partial z}$$

$$\lambda_Q=\frac{D\rho K_q\cdot\gamma_{b,c}\dfrac{\partial\ln e_s}{\partial\bar{T}}\cdot\dfrac{\partial\bar{q}_s}{\partial\bar{T}}}{\rho_s C_{ps}K_s},\quad \lambda_s=\frac{S_0D\cdot\beta_1H\cdot l_b}{W_0\rho_s C_{ps}\cdot K_s\cdot l^2\cdot f}$$

因 $F\sim5\times10^2$，所以 $F^{-2}\ll1$，由式(2)′得

$$u=-\frac{\partial p}{\partial y} \tag{15}$$

这是地转风关系。采取这一近似方程组，就滤去了重力惯性波。由上述方程组消去 v,w,p,T 后就得到了对于 u 的无量纲方程。由于重力惯性波已不存在，于是方程降为对时间的一阶方程

$$\left(\frac{\partial}{\partial t}+G\right)\frac{\partial^2 u}{\partial z^2}+\frac{K}{F^2}\cdot\frac{\partial^3 u}{\partial t\partial y^2}+M\frac{\partial^3 u}{\partial t\partial y\partial z}=\frac{K}{F}\tilde{\alpha}\left(\frac{\partial^2 u}{\partial y^2}\right)_b \tag{16}$$

其中

$$\tilde{\alpha}=\gamma\frac{\partial\ln e_s}{\partial\bar{T}}\tilde{T}^*$$

2 频率方程

把大气分为二层，设凝结加热只发生在下层，上层无凝结。将地面、750 mbar、500 mbar、250 mbar 和大气上界的变量分别标以 0，1，2，3，4。把方程(16)写在层 1 和层 3 上，在热带可以略去较小的斜压项

$$\text{层 1：}\quad\left(\frac{\partial}{\partial t}+G\right)\frac{\partial^2 u_1}{\partial z^2}+\frac{K}{F^2}\frac{\partial^3 u_1}{\partial t\partial y^2}=\frac{K}{F}\tilde{\alpha}\left(\frac{\partial^2 u}{\partial y^2}\right)_b \tag{17}$$

$$\text{层 3：}\quad\left(\frac{\partial}{\partial t}+G\right)\frac{\partial^2 u_3}{\partial z^2}+\frac{K}{F^2}\cdot\frac{\partial^3 u_3}{\partial t\partial y^2}=0 \tag{18}$$

设

$$u=U(y,z)e^{\sigma t},\quad T_s=\theta_s(y)e^{\sigma t} \tag{19}$$

把式(19)代入式(10)′，式(11)′和式(14)，就得到关系式

$$\left(\frac{\partial U}{\partial z}\right)_0=-\left(\frac{N+\lambda_s}{\sigma+\lambda_Q}\right)\left(\frac{\partial^2 U}{\partial y^2}\right)_0\quad z=0$$

把式(19)代入式(17)，式(18)的垂直差分方程中，并利用上式，就有

$$2\left(\frac{\partial U}{\partial z}\right)_2-\frac{N+\lambda_s}{\sigma+\lambda_Q}\left(\frac{\partial^2 U}{\partial y^2}\right)_0+\frac{K\sigma}{F^2(\sigma+G)}\left(\frac{\partial^2 U}{\partial y^2}\right)_1=\frac{K\tilde{\alpha}}{F(\sigma+G)}\left(\frac{\partial^2 U}{\partial y^2}\right)_b \tag{20}$$

$$-2\left(\frac{\partial U}{\partial z}\right)_2+\frac{K\sigma}{F^2(\sigma+G)}\left(\frac{\partial^2 U}{\partial y^2}\right)_3=0 \tag{21}$$

为简化起见，假设

$$\left(\frac{\partial^2 U}{\partial y^2}\right)_0=\left(\frac{\partial^2 U}{\partial y^2}\right)_b=b\cdot\left(\frac{\partial^2 U}{\partial y^2}\right)_1$$

$$\left(\frac{\partial U}{\partial y}\right)_2=2(U_3-U_1)$$

式中，b 表示垂直风速切变的系数。应用这些假定，则式(20)，式(21)可合并为

$$\frac{\partial^2(U_3-U_1)}{\partial y^2}-\left[\left(\frac{2}{S}\right)^2+\left(\frac{2}{R}\right)^2\right](U_3-U_1)=0 \tag{22}$$

其中

$$R^2 = \frac{K\sigma}{F^2(\sigma + G)}$$

$$S^2 = R^2 + \frac{2b(\lambda_s + N)}{\lambda_Q + \sigma} - b\frac{K\widetilde{\alpha}}{F(\sigma + G)}$$

式(22)可改写为

$$\frac{\partial^2(U_3 - U_1)}{\partial y^2} + \hat{\gamma}^2(U_3 - U_1) = 0 \qquad (22)'$$

其中

$$\hat{\gamma}^2 = -4\left(\frac{1}{R^2} + \frac{1}{S^2}\right)$$

如果 $\hat{\gamma}^2$ 是一正实数 $N>0$,则方程(22)′就有对 y 的周期解。因 $\hat{\gamma}$ 是各过程参数的函数,因而也是系统振荡频率 σ 的函数。随着参数值和频率的变化,$\hat{\gamma}$ 可取实数或复数。只要这些参数值和频率满足一定的条件,就可以使 $\hat{\gamma}$ 落在正实轴上。下面我们先假定方程(22)′有周期解,然后便可得到当 $\hat{\gamma}^2=N>0$ 时的参数值和频率所应满足的条件。

设式(22)′有周期解

$$U_3 - U_1 = e^{\pm i\hat{\gamma}y} \qquad (23)$$

相应的连接条件是

$$(U_3 - U_1)\mid_{y=y_1} = (U_3 - U_1)\mid_{y=y_2}$$
$$\left.\frac{\partial(U_3 - U_1)}{\partial y}\right|_{y=y_1} = \left.\frac{\partial(U_3 - U_1)}{\partial y}\right|_{y=y_2} \qquad (24)$$

由式(23)、式(24),就得到本征值

$$\hat{\gamma} = \frac{2n\pi}{y_2 - y_1} = \frac{2n\pi}{\Delta y} \qquad (25)$$

和本征函数

$$U_3 - U_1 = \mathrm{e}^{\pm i\frac{2n\pi}{\Delta y}y} \qquad (26)$$

式(25)又可写作

$$-4\left(\frac{1}{S^2} + \frac{1}{R^2}\right) = \frac{4\pi^2 n^2}{(\Delta y)^2} > 0 \qquad (25)'$$

上式右端是正实数,将 S^2,R^2 的表达式代入并展开,就得到

$$a_0\sigma^3 + a_1\sigma^2 + a_2\sigma + a_3 = 0 \qquad (27)$$

其中系数分别为

$$a_0 = 2 + \alpha_0^2$$
$$a_1 = 2(G + \lambda_Q) + 2S_n - A_c + \alpha_0^2\lambda_Q + 2\alpha_0^2 S_n - \alpha_0^2 \cdot A_c$$
$$a_2 = 2\lambda_Q G + 4G \cdot S_n - (G + \lambda_Q)A_c + 2\alpha_0^2 GS_n - \alpha_0^2 A_c \cdot \lambda_Q$$
$$a_3 = 2G^2 \cdot S_n - G\lambda_Q \cdot A_c$$

而

$$\alpha_0^2 = \frac{K}{F^2}\left(\frac{n\pi}{\Delta y}\right)^2, \quad A_c = bF\widetilde{a}, S_n = \frac{bF^2}{K}(\lambda_s + N)$$

式(27)就是使 $\hat{\gamma}^2$ 取正实数(同时满足式(24)的连接条件)所应满足的条件。可见,只要频率 σ 是这个方程的根(不论是实根或复数根),上面的讨论就自然成立了。式(26)也正是式(22)′满足条件式(24)的唯一的解。

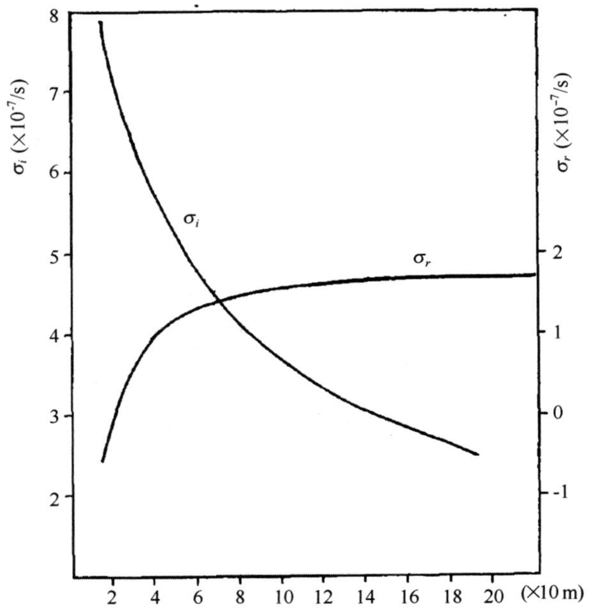

图 1 振荡频率随混合层深度的变化

(单位:10^{-7}/s)

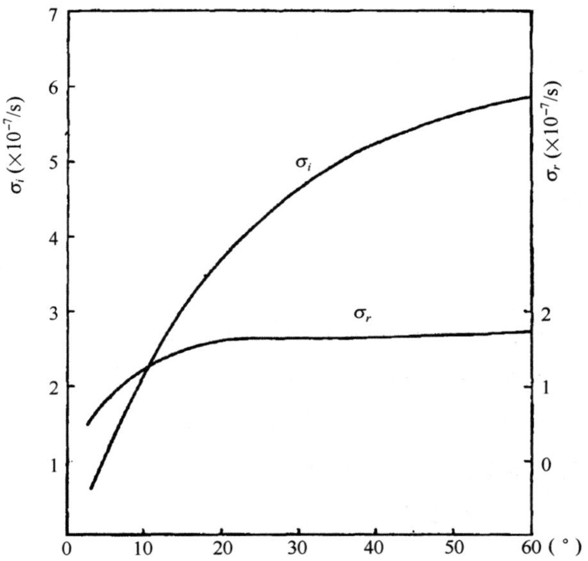

图 2 振荡频率随纬度的变化

(单位:10^{-7}/s)

方程(27)的系数都是实数,它可以有 3 个实根或 1 个实根和 1 对共轭复根,这要看判别式

$$\Delta = \left(\frac{q}{2}\right)^2 + \left(\frac{p}{3}\right)^3$$

如 $\Delta>0$,则有一对共轭复根,否则都是实根。这里

$$q = \frac{2a_1^3}{27a_0^3} - \frac{a_1 a_2}{3a_0^2} + \frac{a_3}{a_0}$$

$$p = \frac{-a_1^2}{3a_0^2} + \frac{a_2}{a_0}$$

在下面实际的计算中,我们选取的参数变化范围都满足 $\Delta>0$ 这一条件(见图 1 至图 3)。这样就得到了系统的振荡频率。

3 计算结果和讨论

将上述各有关参数代入式(27),求得方程的根为(这里取 $\Delta y=3$)

$$\sigma_1 = -4.64;\quad \sigma_{2,3} = 1.59 \pm 3.69\mathrm{i}$$

σ_1 是一个衰减的运动,没有现实意义。$\sigma_{2,3}$ 代表一个发展的振荡解,振荡周期约为 6.5 个月,振幅增长 e 倍的时间约为 2.4 个月。这表明,在低纬度,在海洋与大气热力、动力的相互作用下,大气经向运动存在着一类长周期振荡,它叠加在平均经圈环流上,随着时间的推移,将加强或减弱经圈环流,影响着低纬度与中高纬度地区间的热量和动量的输送,并把低纬度海表热状况,大气环流的变化与中高纬度

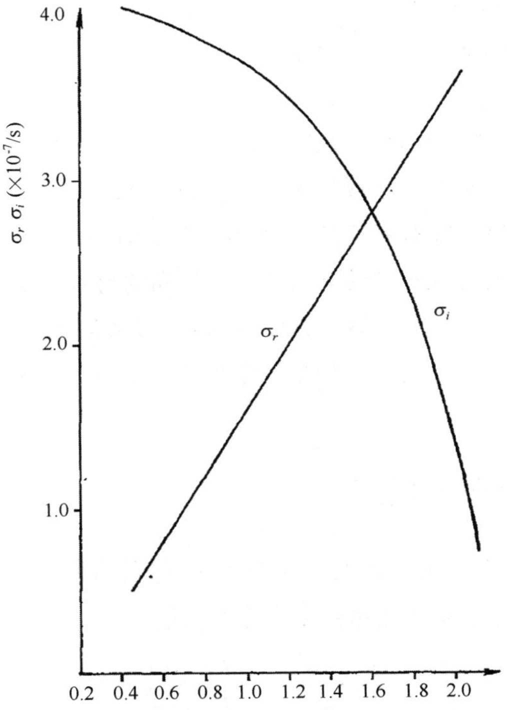

图 3　振荡频率随对流强度的变化

（单位：10^{-7}/s）

环流的变化联系起来。

由方程组式(1)′至式(6)′和边界条件可以看到，u、p、T 是同位相的，u、v 的位相差可由式(1)′确定。

$$A_v = A_u + A_\sigma$$

$\Delta A_{v,u} = A_v - A_u = A_\sigma = \dfrac{\sigma_i(\text{虚部})}{\sigma_r(\text{实部})}$，在上面 6 个多月的振荡中，$\Delta A$ 为 2～3 个月。同样可以估计海面温度 T_s 的位相比 u 早 1.5 个月左右。这就是说，海表温度、纬向风系的变化，要先于经向运动的变化，从而又影响副热带和中高纬度环流的变化。这就为我们用赤道地区海表温度的变化去预测副热带、中高纬度的环流提供了依据。

频率方程(27)的解将依赖于系统中各种物理过程和参数的变化。下面我们来讨论频率、混合层深度和纬度的关系。

图 1 是振荡频率(σ_r，σ_i)随混合层深度变化的曲线。显然深度越深，振荡周期越长，振幅的增长率也随深度成正比，但当深度大于 80～100 m 时，增长率趋向于常数。对应着深度为 20～30 m 的强烈机械混合层，$\sigma_i \sim 7\times10^{-6}$/s，即周期为 3 个月左右。这时 K_s 很大，运动将得到充分发展。深度到达 200 m时($\varphi=20°$)，系统的振荡周期为 10～11 个月。在图 2 上可以看到频率与纬度 φ 的关系，纬度越高周期越短，增长率当 $\varphi>10°$时基本不变。

各种物理过程对系统振荡的作用，由数值计算可以看到，对于 $D=100$ m，$\varphi=20°$时有

① $\tilde{\alpha}=0$，　　$\sigma_{2,3} = -0.23 \pm 3.93i$

② $N=0$，　　$\sigma_{2,3} = 1.55 \pm 2.96i$

③ $\lambda_s = 0$, $\sigma_{2,3} = 1.43 \pm 1.13i$

④ $N = \lambda_s = 0$, $\sigma_2 = 186$, $\sigma_3 = -1.91$;

⑤ $G = 0$, $\sigma_2 = -1.09$, $\sigma_3 = 4.25$。

这说明,如果没有凝结加热供给大气以足够的能量,那么,振荡将是衰减的。相反,如果不考虑辐射冷却,运动将一直发展下去,而不会形成振荡。当不计海水上翻或云对海面辐射的调节作用时,振荡的增长率明显减小。而如果大气对海洋的反馈都不考虑时($N=\lambda_s=0$),则在海洋不断加热的情况下,就会形成大气中强烈的发展运动。计算还表明,在相同的海水上翻和云对海面辐射状况调节的情况下,凝结加热越强,振荡增长率越大。另一方面,在同一凝结强度时,N 和 λ_s 越大,也就是大气对海洋的反馈作用越大(同时这也是大气、海洋耦合程度的一个度量)则周期越短,但振荡增长率改变不大。

4 一个赤道辐合带的简单模式

假定在某一纬度 φ 附近,在赤道一侧有一定宽度 y_0 的对流凝结加热区,而极地一侧没有凝结加热,与前面一样,我们简单地假定凝结加热只发生在对流层下半部。这样,在对流区

$$\left(\frac{\partial}{\partial t} + G\right)\frac{\partial^2 u}{\partial z^2} + \frac{K}{F^2}\cdot\frac{\partial^3 u}{\partial t \partial y^2} = \frac{K}{F}\tilde{\alpha}\left(\frac{\partial^2 u}{\partial y^2}\right)_b \qquad y_0 \geqslant y \geqslant 0 \tag{28}$$

在非对流区

$$\left(\frac{\partial}{\partial t} + G\right)\frac{\partial^2 u}{\partial z^2} + \frac{K}{F^2}\frac{\partial^3 u}{\partial t \partial y^2} = 0 \qquad \infty > y \geqslant y_0 \tag{29}$$

对流区与非对流区有连续性条件

$$u\Big|_{Y_{OE}} = u\Big|_{Y_{OP}}, \qquad \frac{\partial u}{\partial y}\Big|_{Y_{OE}} = \frac{\partial u}{\partial y}\Big|_{Y_{OP}} \tag{30}$$

下标"E"指赤道一侧,"P"指极地一侧。把方程分别写在层1、层3上,整理后就得到

$$\frac{\partial^2 \Delta U}{\partial y^2} - \left[\left(\frac{2}{S_E}\right)^2 + \left(\frac{2}{R}\right)^2\right]\Delta U = 0 \qquad y_0 \geqslant y \geqslant 0 \tag{31}$$

$$\frac{\partial^2 \Delta U}{\partial y^2} - \left[\left(\frac{2}{S_p}\right)^2 + \left(\frac{2}{R}\right)^2\right]\Delta U = 0 \qquad \infty > y \geqslant y_0 \tag{32}$$

其中

$$\Delta U = U_3 - U_1$$

$$S_p^2 = R^2 + \frac{2b(\lambda_s + N)}{\lambda_Q + \sigma}$$

$$S_E^2 = S^2$$

边界条件是

$$U_1 = U_3 = \Delta U = 0 \qquad y = 0, \quad y = \infty \tag{33}$$

方程的解是

$$\Delta U = U_3 - U_1 = 2b_1 \sinh(\gamma_E y) \qquad y_0 \geqslant y \geqslant 0 \tag{34}$$

$$\Delta U = U_3 - U_1 = b_2 e^{-\gamma_p y} \qquad \infty > y \geqslant y_0 \tag{35}$$

式中 b_1, b_2 是待定常数,且

$$\gamma_E^2 = \frac{4}{S_E^2} + \frac{4}{R^2}$$

$$\gamma_P^2 = \frac{4}{S_P^2} + \frac{4}{R^2}$$

利用连续性条件式(30)，就得到方程组式(28)和式(29)的本征值方程

$$\gamma_E + \gamma_P \tan h(\gamma_E y) = 0 \tag{36}$$

由此式就可以确定系统的本征振荡频率。

图 4 是赤道辐合带振荡频率与混合层深度的关系。由图可见，深度越大，振荡周期越长，而增长率稍有增加；深度继续加大，增长率迅速下降，这种相互关系不是直线的。其间有一个对赤道辐合带长周期振荡适当的深度，约有 100 m。与此相对应的振荡周期是 6 个月。频率随纬度的变化从图 5 中看得很清楚，增长率在纬度 10°附近有一个峰值。这和北半球的 ITGZ 徘徊于 5°–10°N 之间的观测事实很相近。

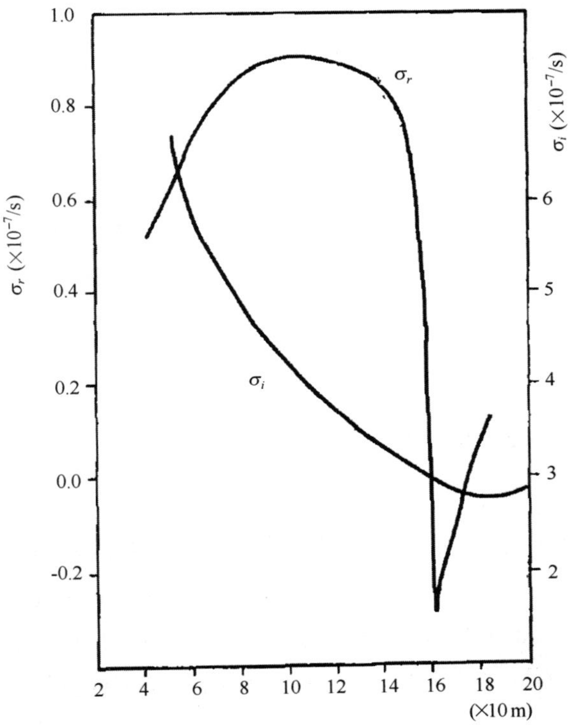

图 4　ITCZ 中频率随混合层深度的变化

($\varphi=10°$, $\eta=0.1$, $y_o=1.5\times10^6$ m)

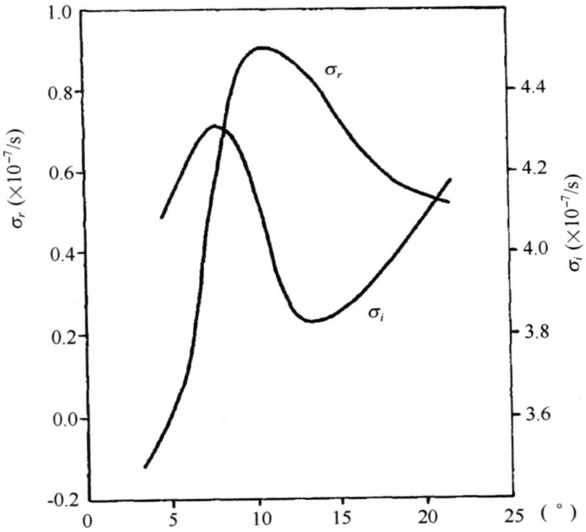

图 5　ITCZ 中频率随纬度的变化

($D=100$ m, $\eta=0.1$, $y_o=1.5\times10^6$ m)

将上面求得的本征振荡频率代入方程的解式(34)，式(35)，并利用连续性条件消去一个常数(如 b_2)，我们就得到了本征函数的解析表达式。在对流区

$$y_0 \geqslant y \geqslant 0$$

$$\frac{U_1}{b_1} = -\frac{8}{\gamma_E^2 S_E^2}\sinh(\gamma_E y) + \frac{8y\sinh\gamma_E y_0}{y_0}\left[\left(\frac{1}{\gamma_E S_E}\right)^2 - \left(\frac{1}{\gamma_p S_p}\right)^2\right] \tag{37}$$

$$\frac{U_3}{b_1} = 2\sinh(\gamma_E y)\left(1 - \frac{4}{\gamma_E^2 S_E^2}\right) + \frac{8y\sinh\gamma_E y_0}{y_0}\left[\left(\frac{1}{\gamma_E S_E}\right)^2 - \left(\frac{1}{\gamma_p S_p}\right)^2\right] \tag{38}$$

在非对流区

$$\infty > y \geqslant y_0$$

$$\frac{U_1}{b_1} = \frac{8\sinh(\gamma_E y)}{S_p^2 \gamma_p^2} e^{-\gamma_p(y-y_0)} \tag{39}$$

$$\frac{U_3}{b_1} = 2\sinh(\gamma_E y) e^{-\gamma_p(y-y_0)} \left(1 - \frac{4}{S_p^2 \gamma_p^2}\right) \tag{40}$$

把 U_1, U_3 代入无量纲方程，就得到 V, W, T_s 的函数形式。图 6，图 7 是本征函数的分布。正如前面已经讨论过的，T_s, u 的位相早于 v, w，函数图像不是同一时刻的。

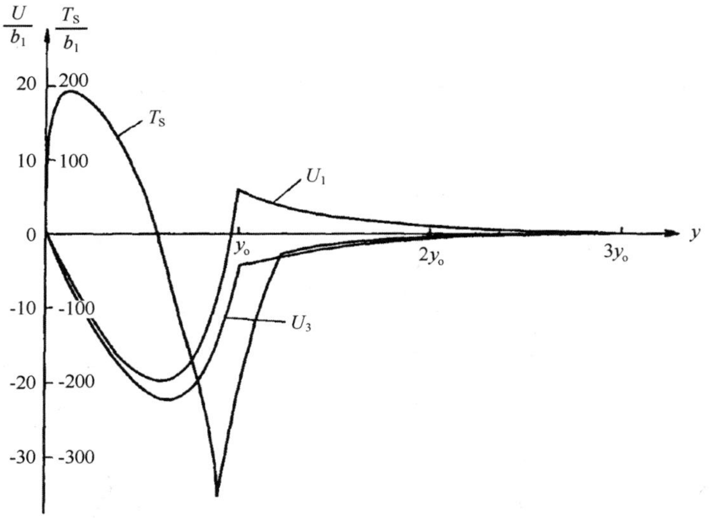

图 6　ITCZ 中，$\frac{U_1}{b_1}, \frac{U_3}{b_1}, \frac{T_s}{b_1}$ 的分布

$(D = 100\ \text{m}, \varphi = 10°, \sigma = (0.8+4.0\text{i}) \times 10^{-7}/\text{s})$

从图 7 中我们看到，在对流层上层，都盛行向极地的气流；在下层，对流区是流向极地，非对流区流向赤道。对流区中强烈的上升区，集中在分界线附近，宽度约为 500 km，这和赤道辐合带的实际宽度是比较一致的。非对流区是下沉气流。这样在子午面上就构成了一个完整的直接环流圈(图 8)。同时在对流区中部，有微弱的下沉气流，在赤道一侧又有一个上升区，显然这是与强上升区对应的补偿气流。

5　结语

分析表明，在上述海洋、大气耦合系统中，存在着一类长周期振荡，其频率与各种物理过程的强度有关。在一般的参数范围内，振荡周期在半年左右。海水混合层越深，周期也越长。同时我们也看到，海水温度变化先于经向运动，从而又影响中高纬环流。所以，为用海温去预测环流提供了依据。

用类似的模式讨论赤道辐合带的特点，结果说明，在对流凝结区中，强烈的上升区集中在一个很窄的带状区域里，与最大振荡增长率相对应的周期约为半年。强烈上升区的纬度约 10°，这些都与观测事实比较一致。

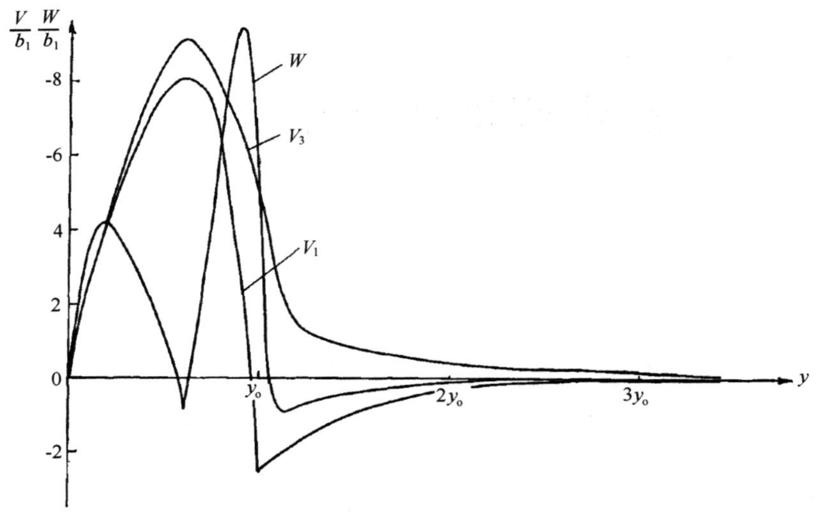

图 7　ITCZ 中，$\frac{V_1}{b_1}$，$\frac{V_3}{b_1}$，$\frac{W}{b_1}$的分布

（D,φ,σ 的含义同图 6）

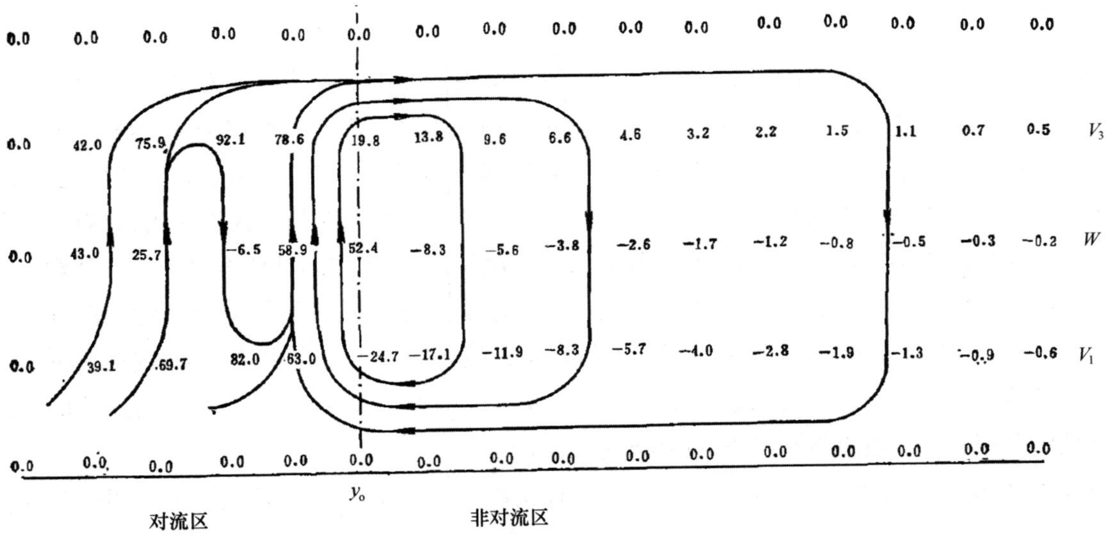

图 8　ITCZ 中的经向运动

（$y_o=1.5\times10^8$ m，$\varphi=10°$，$\sigma=(0.8+0.4i)\times10^{-7}$/s）

参考文献

[1]　中国科学院地理研究所长期天气预报组. 科学通报，1977，22(7)：313-317.

[2]　潘怡航. 大气科学，1978(3)：246-252.

[3]　符淙斌，等. 赤道海温异常与大气的赤道环流圈. 大气科学，1979，3(1).

[4]　长期数值天气预报研究小组. 中国科学，1977(2)：162-172.

[5]　Charney J G. Planetary Fluid Dynamics Dynamic Meteorology. 1973，D.Reidel Publishing company.

[6]　Kuo，H. L.，*J. Atmos. Sci.* Vol.，1973，30(6)：969-983.

[7]　巢纪平，伍荣生. 热带大尺度运动的尺度分析. 大气科学，1979.

热带运动的尺度分析*

巢纪平[1] 伍荣生[2]

(1. 中国科学院大气物理研究所;2. 南京大学气象系)

摘要:本文对热带大尺度运动的动力学特征作了尺度分析,与 Charney[1] 不同,在文中首先引进热带经向宽度动力学定义,分析表明,在这种情况下,热带运动的性质将因其纬向尺度与这个经向宽度之比而异。当这个比值的量级为 $O(1)$ 时,运动是层结三维的,当这个比值的量级甚大于 $O(1)$ 时,类似于 Charney 的结果,运动将趋于正压化。

1 引言

Charney 对热带大尺度运动的尺度分析指出,运动基本上是准水平和无辐散的,即是准正压的。沿用 Charney 提出的观点,Matsuno[2] 对热带波动的水平特征作过分析。虽然 Charney 也指出发生在赤道辐合带中的上升运动所造成的潜热释放过程对热带运动的重要性,但认为由于它们的尺度很窄,可以看成是热带大尺度准水平运动的内边界层。然而,宽窄的含义是相对的。我们认为,只有当对热带区域的经向尺度给出确切的定义后,与这个经向尺度相比较,才能确定运动的宽窄程度。从观测事实看,发生在热带地区的如东风波、赤道波或热带低压等扰动,都是有垂直结构的,而很多研究指出(见参考文献[3]),热带扰动在垂直方向上的能量传播,将直接影响着平流层中的运动,进而有可能使全球的大气环流产生影响。这种现象显然不是正压过程所能解释的。在另一方面,目前成功的低纬度区域性数值预报模式一般也不是正压的(如参考文献[4])。

由此来看,热带大尺度运动的特征还有进一步探索的必要。在本文中,将首先对热带的经向范围给出一个动力学意义下的定义,与这个经向尺度相比,我们将指出,热带大尺度运动的动力学特征,因其纬向尺度的大小而异。

2 运动方程和热带范围的定义

在邻近赤道地区,如不考虑摩擦和加热过程,在 β 平面近似下,运动方程的标量式可以写成

$$\frac{\mathrm{d}u}{\mathrm{d}t} - \beta yv = - \frac{1}{\bar{\rho}}\frac{\partial p'}{\partial x} \tag{1a}$$

$$\frac{\mathrm{d}v}{\mathrm{d}t} + \beta yu = - \frac{1}{\bar{\rho}}\frac{\partial p'}{\partial y} \tag{1b}$$

* 大气科学,1980,4(2):103-110.

$$0 = -\frac{1}{\bar{\rho}}\frac{\partial p'}{\partial z} + \frac{g}{\bar{\theta}}\theta' \tag{1c}$$

$$\frac{\mathrm{d}\theta'}{\mathrm{d}t} + \frac{\partial\bar{\theta}}{\partial y}v + \frac{\partial\bar{\theta}}{\partial z}w = 0 \tag{1d}$$

$$\frac{\partial u}{\partial x} + \frac{\partial v}{\partial y} + \frac{\partial w}{\partial z} = \frac{1}{\mathrm{H}_s}w \tag{1e}$$

其中，压力 p'、位温 θ' 都是相对于标准大气某一基准 $\bar{p}$，$\bar{\theta}$ 值的偏差，标准大气中 $\bar{p}$，$\bar{\theta}$ 和密度 $\bar{\rho}$ 的形成是气候学研究的问题，在这里把它们看成是已知的背景场。符号

$$\frac{\mathrm{d}}{\mathrm{d}t} \equiv \frac{\partial}{\partial t} + u\frac{\partial}{\partial x} + v\frac{\partial}{\partial y} + w\frac{\partial}{\partial z}$$

而 $H_s = -(\partial\ln\bar{\rho}/\partial z)^{-1} \approx c^2/g$，$c = \sqrt{RT}$ 为声速，H_s 是由密度递减率所确定的标准大气的标高。如取运动的垂直特征尺度为 H，则当 $H \ll H_s$ 时，方程(1e)的右端可以忽略，这时的运动可以取 Boussinesq 近似，或称准不可压缩近似[5]。由于 $H_s \approx 10$ km，我们所研究的运动其垂直尺度 $H \leqslant H_s$，所以方程(1e)的右端项保留。方程(1)中其他符号为常用。

旋转大气中的运动特征，将因水平加速度和柯里奥利加速度之比的大小而异。如设 V 和 L 是运动的水平特征速度和特征尺度，则这两个加速度之比的量级为

$$R_0 = \frac{V}{2\Omega\sin\varphi L}$$

称为 Rossby 数。在中纬度 $f = 2\Omega\sin\varphi \approx 10^{-4}/\mathrm{s}$，如 V 取 10 m/s，则对 $L \geqslant 10^3$ km 的运动，$R_0 \sim O(10^{-1})$。但在低纬，当 f 已小到 $10^{-5}/\mathrm{s}$ 时，R_0 将增大到 $O(1)$ 的量级。因此可以把 $R_0 \gg O(10^{-1})$（例如 $R_0 > 0.5$）的大尺度运动，定义成是热带运动。

如令 $y = l\tilde{y}$，$\tilde{y}$ 为无量纲经向坐标，$u = U\tilde{u}$，而特征纬向速度 U 的水平变化 $\delta U \sim U$，则由方程(1a)，定义

$$\beta yv / v\frac{\partial u}{\partial y} \sim \frac{\beta l^2}{U} \equiv R_l^{-1} \tag{2}$$

R_l 称为经向 Rossby 数。如按上面的讨论，把 $R_l \geqslant 0.5$ 的运动定义成是发生在热带的，则由此可以定义出热带最大经向宽度为

$$ls \approx 1.4\sqrt{\frac{U}{\beta}} \tag{3}$$

如取 $U \sim 10$ m/s，则 ls 约为 15 个纬距左右。除在赤道这一奇异线附近外，在南北半球这一低纬度带中，R_l 的量级为 $O(1)$。

3　热带运动的尺度分析

3.1　经向速度

令 $x = L\tilde{x}$，$v = V\tilde{v}$，由方程(1a)

$$u\frac{\partial u}{\partial x} \sim \beta yv$$

得到

$$V/U \sim \left(\frac{l}{L}\right) R_l \tag{4}$$

可见,运动的经向速度,将因纬向特征尺度的增加而减小,对于全球尺度的低纬环流,如 L 已大于 150 个经度,这时经向速度只有纬向速度的 1/10 量级。

3.2　水平辐散与垂直涡度分量之比

由方程(1a)和方程(1b)消去 p',得到涡度方程

$$\frac{\partial \zeta}{\partial t} + \left(u\frac{\partial \zeta}{\partial x} + v\frac{\partial \zeta}{\partial y}\right) + (\beta y + \zeta) D + \cdots = 0 \tag{5}$$

式中

$$\zeta = \frac{\partial v}{\partial x} - \frac{\partial u}{\partial y},\quad D = \frac{\partial u}{\partial x} + \frac{\partial v}{\partial y}$$

如令 $\zeta = \tilde{\zeta}\xi$, $D = \mathscr{D}\tilde{D}$,以及 $\delta\xi \sim \xi$,则由方程(5)中第二、第三两项,在考虑了式(4)后,估计得出

$$\mathscr{D}/\xi = \left(\frac{U}{L} + \mathscr{D}\right)/\beta l$$

由于一般来说,$\mathscr{D} \lesssim U/L$,所以有

$$\mathscr{D}/\xi \sim \frac{l}{L} R_l \tag{6}$$

由此可见,对于 $L \approx l$ 的天气系统尺度的扰动(如东风波、热带低压),有 $\mathscr{D} \sim \xi$,即水平辐散与垂直涡度分量同级,这与在中纬度的天气系统中一般 $\mathscr{D} < \xi$ 有所不同。但对 $L \gg l$ 的低纬纬向环流,仍和中纬度的环流一样,有 $\mathscr{D} < \xi$。

3.3　垂直运动和扰动温度

取 $z = Hs\,\tilde{z}$, $w = W\,\tilde{w}$,因此 $\partial w/\partial z \sim W/H_s$,由方程(1e),并考虑到式(6)后,得到

$$W/H_s \sim \mathscr{Z} \sim \frac{l}{L} R_l \cdot \xi \tag{7}$$

如取 $\xi \sim \partial v/\partial x \sim V/L$,则考虑到式(4)后有

$$W/H_s \sim \left(\frac{l}{L} R_l\right)^2 \frac{U}{L}, \tag{8a}$$

如取 $\xi \sim \partial u/\partial y \sim U/l$,则有

$$W/H_s \sim R_l \frac{U}{L} \tag{8b}$$

联合式(8a)、式(8b)得到

$$W/H_s \sim \left(\frac{l}{L} R_l\right)^2 \frac{U}{L} \lesssim R_l \frac{U}{L} \tag{9}$$

这是因为研究的是大尺度运动,$L \gtrsim l$,即 L 也至少取 10^3 km 的量级。由此得到

$$W/U \sim \left(\frac{l}{L} R_l\right)^2 \frac{H_s}{L} \lesssim R_l \frac{H_s}{L} \tag{10}$$

可见,对于天气系统尺度的扰动,即 $L \sim l$,垂直运动将取 $W \sim R_l \dfrac{H_s}{L} U$,而当运动的纬向尺度增大时,垂直

运动将迅速小于此值。

在另一方面，由 Ooyama 给出的热带大气典型探空资料，在 1000 mbar 到 500 mbar 之间的位温差可达 25℃左右（见参考文献[6]），由算得的 Brunt-Väisälä 频率虽比温带小，但也可达 1.0×10^{-2}/s 左右，因此层结的作用不能事先省略，应视分析结果而定。如取 $\theta'=\theta\,\tilde{\theta}'$，以及 $\delta\theta\sim\theta$，则由方程（1d）估计出

$$\theta \sim \left(\frac{\partial\bar{\theta}}{\partial z}\right)\frac{L}{U}W \tag{11}$$

将式（10）代入，得到

$$\theta \sim \left(\frac{l}{L}R_l\right)^2\left(\frac{\partial\bar{\theta}}{\partial z}\right)H_s \lesssim R_l\left(\frac{\partial\bar{\theta}}{\partial z}\right)H_s \tag{12}$$

可见，当 $L\sim l$ 时，θ 的量级为 $R_l\left(\frac{\partial\bar{\theta}}{\partial z}\right)H_s$，和一般情况下，与垂直运动一样，将随着 L 的增大而迅速减小。

由此可见，只有对纬向尺度甚大的热带大气环流，垂直运动和扰动温度才是不重要的，这时运动将趋于正压化。因此 Charney 的结果只适用于 $L\gg l$ 的热带大尺度运动，而对于 $L\approx l$ 的热带天气扰动，准正压性是难于成立的。

3.4　气压场水平变化的均匀性

在热带天气分析中，由于风场变化明显，气压场（指扰动量）的水平变化微弱，所以一般分析流线而不分析等压线。这一特征是热带大尺度运动与温带准地转运动的一个重要差异之点，能否从尺度分析得出气压场较均匀的结果，是衡量所建立的尺度分析理论是否正确的一个标志。

Holton[6] 在对热带运动作尺度分析时，取扰动压力的垂直变化 ΔP 与水平变化 δP 同级①，即 $\Delta P\approx\delta P$，其中 $p'=P\,\tilde{p}'$。在这里 P 为 p' 的特征量。在旋转流体中，由于转动的作用，运动将趋于水平化（Taylar-Proudman 定理），因此对某些物理量，其水平和垂直变化不一定具有各向同性的性质。所以一般来讲，这样的取法不一定合适。因为现在考虑的是压力的扰动量，例如在一个垂直波长内，倒可以取 $\Delta P\sim P$，这样由方程（1c）估计出

$$P \sim \bar{\rho}H_s\frac{g}{\bar{\theta}}\theta \tag{13}$$

将式（12）代入，得到

$$P \sim \bar{\rho}\left(\frac{l}{L}R_l\right)^2N^2H_s^2 \lesssim \bar{\rho}R_lN^2H_s^2 \tag{14}$$

式中，$N=\left(\frac{g}{\bar{\theta}}\frac{\partial\bar{\theta}}{\partial z}\right)^{\frac{1}{2}}$ 为 Brunt-Väisälä 频率。

另一方面，在热带大尺度运动中，水平气压梯度需与惯性力平衡[6]，由方程（1a）得到

$$\delta P \sim \bar{\rho}U^2 \tag{15}$$

由式（14）、式（15）两式，有

---

① Holton 用的是扰动重力位势 ϕ'。

$$\delta P/P \sim \left(\frac{l}{L}R_l\right)^{-2} F_0 \tag{16}$$

式中

$$F_0 \equiv U^2/C_g^2$$

为内 Froude 数，而 $C_g = \left(\frac{g}{\bar{\theta}}\frac{\partial \bar{\theta}}{\partial z}H_s^2\right)^{1/2}$ 为重力内波波速，由于 $N \approx 10^{-2}/\mathrm{s}$，$H_s \approx 10^4$ m，所以 $C_g \approx 10^2$ m/s，如取 $U \sim 10$ m/s，则 $F_0 \sim O(10^{-2})$。

由此可见，对于 $L \sim l$ 的天气系统尺度的扰动，由式(16)估计出

$$\delta P/P \sim F_0 \sim O(10^{-2}) \ll 1 \tag{17}$$

这表明，在这样尺度的天气系统中，扰动气压的水平变化比起扰动气压本身是很小的，也即扰动气压在水平方向上是十分均匀的。

对于全球尺度的纬向环流，这时 $L/l \sim O(10^1)$，所以有

$$\delta P/P \sim O(1) \tag{18}$$

在这类运动中，扰动气压在水平方向的变化才能与扰动气压本身同级，这是很自然的结果，如沿纬向为一个波，则绕地球一圈后，扰动压力的变化将与扰动压力同级。

3.5 斜压性的作用

在中纬度，由热成风关系

$$\frac{\partial \bar{\theta}}{\partial y} = \frac{\bar{\theta}}{g} f \frac{\partial u}{\partial z}$$

可知在方程(1d)中 $v\frac{\partial \bar{\theta}}{\partial y}$这一项反映了斜压性的作用。在热带上式不能用，但仍可估计出一这项的相对重要性，为

$$v\frac{\partial \bar{\theta}}{\partial y}\Big/ u\frac{\partial \theta'}{\partial x} \sim \left(\tan\alpha \cdot \frac{l}{H_s}\right) R_l^{-1} \left(\frac{l}{L}\right)^{-2} \tag{19}$$

式中

$$\tan\alpha \equiv \left(\frac{\partial \bar{\theta}}{\partial y}\right) \Big/ \left(\frac{\partial \bar{\theta}}{\partial z}\right) \tag{20}$$

α 为等熵面的坡度。在热带的气候背景场中，等熵面接近与地面平行，即 α 值很小，一般都有$\tan\alpha \cdot \frac{l}{H_s} \ll 1$。

上面的分析表明，对 $L \sim l$ 的运动，斜压性的作用很小，只有对全球尺度的运动，这一项才变得重要起来，但这时如分析已表明的，温度场的影响已可以忽略了。

4 热带大尺度运动的控制方程

除以上已决定的各物理量的特征值外，并取 $t = \frac{L}{U}\tilde{t}$，于是方程组(1)的无量纲形式为(略去“~”号)：

$$\frac{\partial u}{\partial t} + u\frac{\partial u}{\partial x} + R_l v\frac{\partial u}{\partial y} + \left(\frac{1}{L}R_l\right)^2 w\frac{\partial u}{\partial z} - yv = -\frac{\partial p'}{\partial x} \tag{21a}$$

$$\left(\frac{l}{L}R_l\right)^2\left[\frac{\partial v}{\partial t}+u\frac{\partial v}{\partial x}+R_l v\frac{\partial v}{\partial y}+\left(\frac{l}{L}R_s\right)^2 w\frac{\partial v}{\partial z}\right]+yu=-R_l\frac{\partial p'}{\partial y} \tag{21b}$$

$$0=-\frac{\partial p'}{\partial z}+\theta' \tag{21c}$$

$$\left(\frac{l}{L}R_l\right)^2\left[\frac{\partial \theta'}{\partial t}+u\frac{\partial \theta'}{\partial x}+R_l v\frac{\partial \theta'}{\partial y}+\left(\frac{l}{L}R_l\right)^2 w\frac{\partial \theta'}{\partial z}+\frac{\partial \bar{\theta}}{\partial z}w\right]+\frac{\partial \tilde{\theta}}{\partial y}\cdot \tan\alpha\cdot\frac{l}{H_s}R_l v=0 \tag{21d}$$

$$\frac{\partial u}{\partial x}+R_l\frac{\partial v}{\partial y}+\left(\frac{l}{L}R_l\right)^2\left(\frac{\partial w}{\partial z}-w\right)=0 \tag{21e}$$

式中，$\widetilde{\partial\bar{\theta}/\partial y}$，$\widetilde{\partial\bar{\theta}/\partial z}$ 均为无量纲量。

4.1　对于天气系统尺度的运动($L\approx l$)

由于 $R_l\sim O(1)$，这时除由于 $\tan\alpha\cdot\frac{l}{H_s}<1$，斜压性的作用可以忽略外，不能再作进一步简化，因此控制这类运动的方程组为

$$\frac{\mathrm{d}u}{\mathrm{d}t}-yv=-\frac{\partial p'}{\partial x} \tag{22a}$$

$$\frac{\mathrm{d}v}{\mathrm{d}t}+yu=-\frac{\partial p'}{\partial y} \tag{22b}$$

$$\frac{\mathrm{d}}{\mathrm{d}t}\left(\frac{\partial p'}{\partial z}\right)+\widetilde{\frac{\partial\bar{\theta}}{\partial z}}w=0 \tag{22c}$$

$$\frac{\partial u}{\partial y}+\frac{\partial v}{\partial y}+\frac{\partial w}{\partial z}=w \tag{22d}$$

可见运动是层结、三维的，这表明对于天气系统尺度的扰动，具有垂直结构，并可以预料，它们在高低空运动的耦合中将起重要的作用。

4.2　对于纬向尺度更大的运动($L\gg l$)

这时$\left(\frac{l}{L}\right)^2\sim O(10^{-1}-10^{-2})$，在方程组(21)中略去量级为 $O\left[\left(\frac{l}{L}\right)^2\right]$的项后，保持同量级项的方程组为

$$\frac{\mathrm{D}u}{\mathrm{D}t}-yv=-\frac{\partial p'}{\partial x} \tag{23a}$$

$$yu=-\frac{\partial p'}{\partial y} \tag{23b}$$

$$\frac{\partial u}{\partial x}+\frac{\partial v}{\partial y}=0 \tag{23c}$$

式中

$$\frac{\mathrm{D}}{\mathrm{D}t}\equiv\frac{\partial}{\partial t}+u\frac{\partial}{\partial x}+v\frac{\partial}{\partial y}$$

由此可见，对于这一类运动，不仅是正压、水平无辐散的，而且纬向风是地转的，同时，由方程(21b)可见，在赤道这一奇异线($y=0$)上，需要有 $\partial p'/\partial y=0$，即通过赤道扰动气压具有经向对称性。

如不具有这种对称性，则意味着在赤道两侧存在一窄的黏性边界层，这一层内，需考虑湍流的侧向混合作用。

或者，由方程（23c）引进流函数 ψ，定义成

$$u = -\frac{\partial \psi}{\partial y}, \quad v = \frac{\partial \psi}{\partial x} \tag{24}$$

由式（23a）、式（23b）得到涡度方程为

$$\frac{D}{Dt}\left(\frac{\partial^2 \psi}{\partial y^2}\right) + \frac{\partial \psi}{\partial x} = 0 \tag{25}$$

这基本上类似于 Charney 的结果，不同者，在这里对运动的纬向尺度有个要求：$L>l$，同时在涡度的流函数表达式中，只保留了对 y 的二阶微商项。

以上的结果，是在绝热条件下获得的，这并不意味着在热带像积云加热等非绝热过程可以忽略，在这里只分析了热带大尺度运动动力学的基本特征。

5 热带运动的双时态特征

叶笃正等[7]揭露了在大气运动中普遍存在多时间尺度特征的现象，本文作者之一[8]曾对出现这种现象的物理意义作过一个简单的注释。同样地，在热带运动中也存在类似的现象。

根据上面引进的特征量，热带运动也存在两个时间尺度，即

$$t_1 = (\beta l)^{-1}, \quad t_2 = R_l \frac{l}{V} = \frac{l}{U} \tag{26}$$

前者是经向 Rossby 波的特征周期，后者是平流过程的特征时间。

这两个时间尺度之比为

$$t_1/t_2 = U/\beta l L = \left(\frac{l}{L}\right) R_l \tag{27}$$

由此可以推论，对 $L \approx l$ 的天气系统尺度的扰动运动，当副热带一个经向 Rossby 波进入热带后（或过程反过来），热带运动将发生一次调整，这个调整的特征时间为 t_1，当调整时段基本结束后，热带扰动将进入以平流为主的第二时态，其特征时间为 t_2，由于现在 $t_1 \approx t_2$，即第一时态和第二时态同级，这表明，这个副热带波动进入热带后（或过程反过来），热带区域中的运动需要一段很长的调整、适应时间，因此，对于这类尺度的运动，热带和副热带之间的相互作用，或动力学耦合，是十分重要的。

在另一方面，对于 $L \gg l$ 的纬向运动，其 $t_1 \ll t_2$，这表明热带运动对来自副热带的波动，其调整或响应时间很短，运动将很快按本区域中内在的规律而演变。由此也似乎表明，热带全球尺度的纬向环流，受副热带天气扰动的影响并不重要，运动的能源主要在热带区域内部。

致谢：叶笃正、周晓平对本文的初稿提出了宝贵的意见，在此谨表谢意。

参考文献

[1] Charney, J. G. *J. Aimos.Sci.* 1963, 20: 607-609.
[2] Matsuno T., *J. Meteor. Soc.* 1966, Japan, 44:25-43.
[3] Holton J. R., *J. Atmot. Sci.* 1972, 29:365-375.
[4] 陈隆勋，等．一个四层初始方程热带数值预报模式的初步结果．第二次数值天气预报会议论文集．科学出版社，

1980:127-136.
[5]　巢纪平. *Scientia Sinica*. 1962, 11:1789-1706.
[6]　Holton J. R. Introduction of Dynamic Meteorology, 1972.
[7]　叶笃正,李麦村. 大气各类运动的多时间尺度特征. 同[4]论文集,181-192.
[8]　巢纪平. 关于大气各类运动的多时间尺度特征的讨论. 同[4]论文集,193-195.

The Air-sea Interaction Waves in the Tropics and Their Instabilities*

Chao Jiping(巢纪平)[1] Zhang Renhe(张人禾)[2]

(1. National Research Center For Marine Environmental Forecasts, State Oceanic Administration, Beijing; 2. Institute of Atmospheric Physics, Academia Sinica, Beijing)

ABSTRACT: By using a simple air-sea coupled model. the interaction of Rossby waves between the air and sea in the tropics is discussed. It is shown that the coupling of Rossby waves in the two media produces not only the westward propagating waves, but also a type of new wave which moves eastward. The eastward propagating waves exist in the scope of comparatively long wavelengths and this scope is governed by the intensity of the air-sea interaction. In addition, instability may appear in both the eastward and westward propagating waves, and the wave amplifying rates are also governed by the intensity of the air-sea interaction. In the end, a possible explanation to ENSO events is given in terms of the air-sea interaction waves.

I INTRODUCTION

Since Bjerkness (1966, 1969) pointed out the internal relationship between Southern Oscillation and El Niño, Scientists have realized more and more that these two anomalous phenomena occurring in the atmosphere and ocean respectively, are the results of the air-sea interaction. So they are referred to as ENSO (Philander, 1983a).

Some theories have been proposed by oceanographers (Wyrtki, 1975, for example) to explain El Niño events. They believe that the weakening of southeast trade winds, on the one hand, will slow down the upwelling along the coast of Peru and near the eastern equatorial Pacific Ocean, which will weaken the development of cool water or even lead to the appearance of a warm sea surface temperature (SST); and on the other hand, Kelvin waves can be excited in the western equatorial Pacific because of the weakening of trade winds or the appearance of westerlies. The transport of warm water from west to east by Kelvin waves makes the warm SST in the eastern equatorial Pacific expand. Moreover, when Kelvin waves reach eastern coast, the Rossby waves can be excited as reflecting waves, which make the positive SST anomalies expand westward. Thus, a large amount of warm water is formed and an El Niño phenomenon appears.

However, the 1982–1983 ENSO event is in contradiction with the theory mentioned above because the

* Acta Meteorologica Sinica, 1988, 2(3): 275–287.

positive SST anomalies first appeared near 160°E in the middle equatorial Pacific and then propagated eastward. Together with the eastward propagating of the positive SST anomalies, the anomalies of winds at 850 hPa and of the outgoing longwave radiation representative of convection activity were all going eastward (see Fig.1). The unification of these physical quantities in the eastward propagating both in the atmosphere and ocean further shows that ENSO events are caused by the large-scale air-sea interaction. The 1982—1983 ENSO event also shows the variety of ENSO events. They are not all in accordance with the developing phases given by Rasmusson et al. (1983).

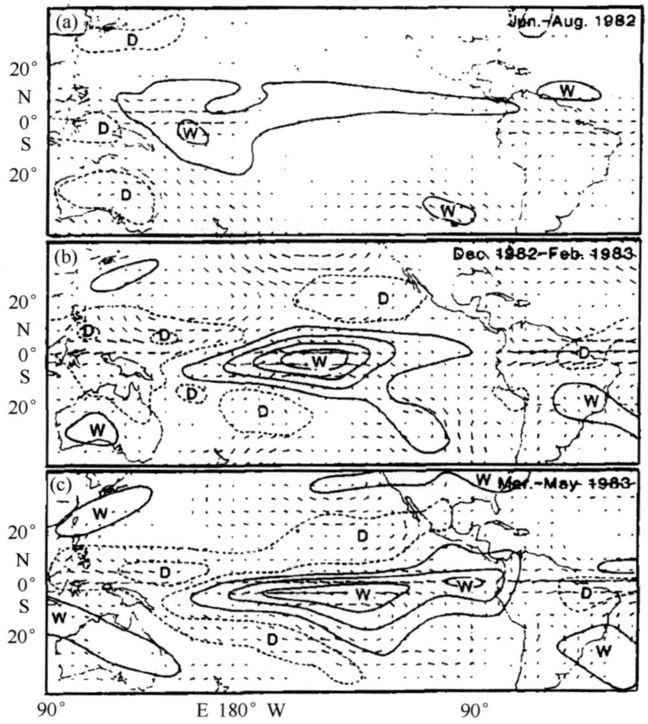

Fig.1 Anomalies in satellite-sensed outgoing longwave radiation (OLR) (contours) and winds at 850 hPa (arrow) for three seasons during the 1982-1983 episode. Negative anomalies in OLR, indicated by the solid contours and labeled W for "wet", correspond to regions of enhanced precipitation, and vice versa (D, "dry") (Rasmusson, 1983)

Since the 1982-1983 ENSO event, a lot of work has been done to study the ENSO theoretically and numerically. Among them the most striking is the unstable air-sea interaction theory given by Philander et al. (1984). This theory indicates that the perturbation can propagate eastward due to the air-sea interaction. Because the propagating velocity is slower than that of Kelvin waves, it can be inferred that there exists a type of air-sea interaction wave besides Kelvin waves in the tropical region.

In our work here, we shall use shallow water model both in the atmosphere and ocean to further study the properties of the tropical air-sea interaction waves. It is different from Philander et al. in that a higher approximate relationship between the speed field and pressure (height) field is used. In such a way, there only exist Rossby waves in each medium. But our results show that a type of eastward propagating wave appears because of the air-sea interaction between the Rossby waves in the two media. This proves that the eastward propagating ENSO event in 1982-1983 is probably caused by the air-sea interaction waves.

Ⅱ MODEL

1. Atomspheric Equations

For simplicity, a one-layer model in the atmosphere (Anderson et al., 1975)

$$fu_a = -g\frac{\partial h_a}{\partial y} - \frac{g}{f}\frac{\partial^2 h_a}{\partial x \partial t} - \frac{\beta g}{f}h_a \tag{1}$$

$$fv_a = g\frac{\partial h_a}{\partial x} - \frac{g}{f}\frac{\partial^2 h_a}{\partial y \partial t} \tag{2}$$

$$\frac{\partial h_a}{\partial t} + H_a\left(\frac{\partial u_a}{\partial x} + \frac{\partial v_a}{\partial y}\right) = -Q \tag{3}$$

is used, where u_a and v_a are velocities in the air in the x and y directions, respectively; h_a is the perturbed height of certain eigen layer; f is the Coriolis parameter; H_a is the equavalent height or the eigenvalue corresponding to some vertical mode, in Gill's (1980) study, H_a was taken to be 400 m, and Q is the mass source or sink, depending on the heating rate.

2. Oceanic Equations

In the ocean, the shallow water model

$$fu_s = -g\frac{\partial h_s}{\partial y} - \frac{g}{f}\frac{\partial^2 h_s}{\partial x \partial t} - \frac{\beta g}{f}h_s + \frac{\tau_s^y}{D} \tag{4}$$

$$fv_s = g\frac{\partial h_s}{\partial x} - \frac{g}{f}\frac{\partial^2 h_s}{\partial y \partial t} - \frac{\tau_s^x}{D} \tag{5}$$

$$\frac{\partial h_s}{\partial t} + \Delta \overline{T}_z D a\left(\frac{\partial u_s}{\partial x} + \frac{\partial v_s}{\partial y}\right) = 0 \tag{6}$$

is also used, where u_s and v_s are velocities in the x and y directions, respectively; h_s is the depth of perturbation; $\Delta \overline{T}_z$ is the difference of climatological sea temperature between the sea surface and the top of the thermocline; α is the thermal expansion coefficient of sea water; D is the depth of mixed layer; and τ_s^x and τ_s^y are the body forces in the x and y directions, respectively.

3. The Air-Sea Coupled Model

Considering $h_s \propto T'_s$, where T'_s is the SST anomaly, we can assume that the heating of the ocean to the atmosphere is proportional to the T'_s (Philander et al., 1984; Zebiak, 1982). Thus

$$Q = Ah_s \tag{7}$$

Winds are assumed to act as body force on the ocean

$$(\tau_s^x, \tau_s^y) = \gamma(u_a, v_a) \tag{8}$$

where A and γ are constants, their values are taken to be $A = 10^{-2}\ \mathrm{s}^{-1}$ and $\gamma = 5\times10^{-5}$ m/s (Philander et al., 1984).

Substituting Eqs. (7) and (8) into Eqs. (3)-(5) and eliminating u_a and v_a in Eqs. (1)-(3), we obtain

$$\left(\frac{\partial^2}{\partial x^2}+\frac{\partial^2}{\partial y^2}-\frac{f^2}{C_a^2}\right)\frac{\partial h_a}{\partial t}+\beta\frac{\partial h_a}{\partial x}=\frac{f^2}{C_a^2}Ah_s \tag{9}$$

where $C_a=(gH_a)^{1/2}$ is the speed of gravity wave in the atmosphere. Eliminating u_s and v_s is Eqs. (4)-(6), we get

$$\left(\frac{\partial^2}{\partial x^2}+\frac{\partial^2}{\partial y^2}-\frac{f^2}{C_s^2}\right)\frac{\partial h_s}{\partial t}+\beta\frac{\partial h_s}{\partial x}=\frac{f\gamma}{Dg}\left(\frac{\partial v_a}{\partial x}-\frac{\partial u_a}{\partial y}\right) \tag{10}$$

where $C_s=(g\Delta\overline{T}_z Da)^{1/2}$, the speed of gravity wave in the ocean. By introducing the equatorial β-plane approximation $f=\beta y$, with the help of the approximate relationship

$$f\left(\frac{\partial v_a}{\partial x}-\frac{\partial u_a}{\partial y}\right)=g\left(\frac{\partial^2}{\partial x^2}+\frac{\partial^2}{\partial y^2}\right)h_a \tag{11}$$

Eqs.(9) and (10) become (Leghthill, 1969)

$$\left(\frac{\partial^2}{\partial x^2}+\frac{\partial^2}{\partial y^2}-\frac{\beta^2}{C_a^2}y^2\right)\frac{\partial h_a}{\partial t}+\beta\frac{\partial h_a}{\partial x}=\frac{\beta^2}{C_a^2}Ay^2h_s \tag{12}$$

$$\left(\frac{\partial^2}{\partial x^2}+\frac{\partial^2}{\partial y^2}-\frac{\beta^2}{C_s^2}y^2\right)\frac{\partial h_s}{\partial t}+\beta\frac{\partial h_s}{\partial x}=\frac{\gamma}{D}\left(\frac{\partial^2}{\partial x^2}+\frac{\partial^2}{\partial y^2}\right)h_a \tag{13}$$

Thus, Eqs. (12) and (13) constitute an air-sea coupled model which governs the motion in the air and ocean respectively.

Ⅲ　NON-DIMENSIONAL EQUATIONS AND THE METHOD OF SOLUTION

Introducing non-dimensional variables

$$t=(2\beta C_s)^{-1/2}t^*,\qquad (x,y)=\left(\frac{C_a}{2\beta}\right)^{1/2}(x^*,y^*) \tag{14}$$

$$h_a=\left(\frac{C_a^2}{2g}\right)h_a^*,\qquad h_s=\left(\frac{C_aC_s}{2g}\right)h_s^*$$

substituting them into Eqs. (12) and (13) and dropping asterisks, we obtain the nondimensional forms of Eqs. (12) and (13)

$$\left(\frac{\partial^2}{\partial x^2}+\frac{\partial^2}{\partial y^2}-\frac{y^2}{4}\right)\frac{\partial h_a}{\partial t}+\frac{1}{2\epsilon^{1/2}}\frac{\partial h_a}{\partial x}=\frac{\epsilon AT}{4}y^2h_s \tag{15}$$

$$\left[\epsilon\left(\frac{\partial^2}{\partial x^2}+\frac{\partial^2}{\partial y^2}\right)-\frac{1}{\epsilon}\frac{y^2}{4}\right]\frac{\partial h_s}{\partial t}+\frac{\epsilon^{1/2}}{2}\frac{\partial h_s}{\partial x}=\frac{\gamma T}{D}\left(\frac{\partial^2}{\partial x^2}+\frac{\partial^2}{\partial y^2}\right)h_a \tag{16}$$

where $\epsilon=C_s/C_a$ and $T=(2\beta C_s)^{-1/2}$. Eqs. (15) and (16) are the basic equations to be studied in this paper.

Take the solutions of Eqs. (15) and (16) as

$$(h_a,h_s)=[h_a^{(0)}(y),h_s^{(0)}(y)]\cdot\exp[\mathrm{i}(kx-\sigma t)] \tag{17}$$

and also set

$$h_a^{(0)}(y)=\sum_{n=0}^{\infty}h_{an}^{(0)}D_n(y) \tag{18}$$

$$h_s^{(0)}(y)=\sum_{n=0}^{\infty}h_{sn}^{(0)}D_n(y/\sqrt{\epsilon})=\sum_{n=0}^{\infty}h_{sn}^{(0)}D_n(z) \tag{19}$$

where $z=y\sqrt{\epsilon}$, $D_n(y)$ is the nth-order parabolic cylinder function (Wang and Guo, 1979) which satisfies the following Weber equation

$$\frac{d^2}{dy^2}D_n(y)+\left(n+\frac{1}{2}-\frac{y^2}{4}\right)D_n(y)=0 \tag{20}$$

and has the recursive relation

$$y^2D_n(y)=D_{n+2}(y)+(2n+1)D_n(y)+n(n-1)D_{n-2}(y) \tag{21}$$

By expanding $D_n(y/\sqrt{\epsilon})$ in terms of the parabolic cylinder function of variable y

$$D_n(y/\sqrt{\epsilon})=\sum_{m=0}^{\infty}a_{mn}D_m(y) \tag{22}$$

where

$$a_{mn}=\frac{1}{m!\ \sqrt{2\pi}}\int_{-\infty}^{\infty}D_n(y/\sqrt{\epsilon})D_m(y) \tag{23}$$

and taking the first three terms in Eq. (22) as the approximation

$$D_n(y/\sqrt{\epsilon})=a_{0n}D_0(y)+a_{1n}D_1(y)+a_{2n}D_2(y) \tag{24}$$

Eqs. (15) and (16) then become

$$\begin{aligned}
&-\sum_{n=0}^{\infty}h_{an}^{(0)}\left[\left(k^2+\frac{1}{2\epsilon^{1/2}}\frac{k}{\sigma}\right)+\left(n+\frac{1}{2}\right)\right]D_n(y)\\
&=i\frac{\epsilon^2AT}{4\sigma}\sum_{n=0}^{\infty}h_{sn}^{(0)}\{[a_{0n+2}D_0(y)+a_{1n+2}D_1(y)+a_{2n+2}D_2(y)]\\
&\quad+(2n+1)[a_{0n}D_0(y)+a_{1n}D_1(y)+a_{2n}D_2(y)]\\
&\quad+n(n-1)[a_{0n-2}D_0(y)+a_{1n-2}D_1(y)+a_{2n-2}D_2(y)]\}
\end{aligned} \tag{25}$$

$$\begin{aligned}
&\sum_{n=0}^{\infty}h_{sn}^{(0)}\left[\left(\epsilon k^2+\frac{\epsilon^{1/2}}{2}\frac{k}{\sigma}\right)+\left(n+\frac{1}{2}\right)\right]\cdot[a_{0n}D_0(y)+a_{1n}D_1(y)+a_{2n}D_2(y)]\\
&=-i\frac{\gamma T}{D\sigma}\sum_{n=0}^{\infty}h_{an}^{(0)}\left[\frac{1}{4}D_{n+2}(y)-\left(k^2+\frac{n}{2}+\frac{1}{4}\right)D_n(y)+\frac{1}{4}n(n-1)D_{n-2}(y)\right]
\end{aligned} \tag{26}$$

Now truncating the series (25) and (26) at $n=2$, considering the orthogonality of parabolic cylinder functions, we obtain

$$h_{a0}^{(0)}\left(k^2+\frac{1}{2\epsilon^{1/2}}\frac{k}{\sigma}+\frac{1}{2}\right)=-h_{s0}^{(0)}i\frac{\epsilon^2AT}{4\sigma}(a_{02}+a_{00})-h_{s2}^{(0)}i\frac{\epsilon^2AT}{4\sigma}\times(a_{04}+5a_{02}+2a_{00}) \tag{27}$$

$$h_{a2}^{(0)}\left(k^2+\frac{1}{2\epsilon^{1/2}}\frac{k}{\sigma}+\frac{5}{2}\right)=-h_{s0}^{(0)}i\frac{\epsilon^2AT}{4\sigma}(a_{22}+a_{20})-h_{s2}^{(0)}i\frac{\epsilon^2AT}{4\sigma}\times(a_{24}+5a_{22}+2a_{20}) \tag{28}$$

$$\begin{aligned}
&h_{s0}^{(0)}\left(\epsilon k^2+\frac{\epsilon^{1/2}}{2}\frac{k}{\sigma}+\frac{1}{2}\right)a_{20}+h_{s2}^{(0)}\left(\epsilon k^2+\frac{\epsilon^{1/2}}{2}\frac{k}{\sigma}+\frac{5}{2}\right)a_{22}\\
&=-h_{a0}^{(0)}i\frac{\gamma T}{4D\sigma}+h_{a2}^{(0)}i\frac{\gamma T}{D\sigma}\left(k^2+\frac{5}{4}\right),
\end{aligned} \tag{29}$$

$$\begin{aligned}
&h_{s0}^{(0)}\left(\epsilon k^2+\frac{\epsilon^{1/2}}{2}\frac{k}{\sigma}+\frac{1}{2}\right)a_{00}+h_{s2}^{(0)}\left(\epsilon k^2+\frac{\epsilon^{1/2}}{2}\frac{k}{\sigma}+\frac{5}{2}\right)a_{02}\\
&=h_{a0}^{(0)}i\frac{\gamma T}{D\sigma}\left(k^2+\frac{1}{4}\right)-h_{a2}^{(0)}i\frac{\gamma T}{2D\sigma}
\end{aligned} \tag{30}$$

$$h_{a1}^{(0)}\left(k^2+\frac{1}{2\epsilon^{1/2}}\frac{k}{\sigma}+\frac{3}{2}\right)=-h_{s1}^{(0)}\mathrm{i}\frac{\epsilon^2AT}{4\sigma}(a_{13}+3a_{11}) \tag{31}$$

$$h_{s1}^{(0)}\left(\epsilon k^2+\frac{\epsilon^{1/2}}{2}\frac{k}{\sigma}+\frac{3}{2}\right)a_{11}=h_{a1}^{(0)}\mathrm{i}\frac{\gamma T}{D\sigma}\left(k^2+\frac{3}{4}\right) \tag{32}$$

Here, considering

$$(D_0,D_1,D_2,D_3,D_4)(y)=(1,y,y^2-1,y^3-3y,y^4-6y^2+3)\cdot\exp(-y^2/4) \tag{33}$$

from Eq. (23), we have

$$a_{01}=a_{03}=a_{10}=a_{12}=a_{14}=a_{21}=a_{23}=0 \tag{34}$$

$$\begin{cases} a_{00}=\left(\dfrac{2}{1/\epsilon+1}\right)^{1/2}, \quad a_{02}=\left(\dfrac{2}{\epsilon+1}-1\right)\left(\dfrac{2}{1/\epsilon+1}\right)^{1/2}, \\ a_{04}=3\left[\dfrac{4}{(1+\epsilon)^2}-\dfrac{4}{1+\epsilon}+1\right]\left(\dfrac{2}{1/\epsilon+1}\right)^{1/2}, \\ a_{11}=\dfrac{2}{1/\epsilon+1}\left(\dfrac{2}{1+\epsilon}\right)^{1/2}, a_{13}=6\left[\dfrac{2\epsilon}{(1+\epsilon)^2}-\dfrac{1}{1/\epsilon+1}\right]\left(\dfrac{2}{1+\epsilon}\right)^{1/2}, \\ a_{20}=\left(\dfrac{1}{1/\epsilon+1}-\dfrac{1}{2}\right)\left(\dfrac{2}{1/\epsilon+1}\right)^{1/2}, \\ a_{22}=\left[\dfrac{6\epsilon}{(1+\epsilon)^2}-\dfrac{1}{2}\right]\left(\dfrac{2}{1/\epsilon+1}\right)^{1/2}, \\ a_{24}=\left[\dfrac{60\epsilon}{(1+\epsilon)^3}-\dfrac{6(6+1/\epsilon)\epsilon}{(1+\epsilon)^2}+\dfrac{3(1+2/\epsilon)}{1+1/\epsilon}-\dfrac{3}{2}\right]\left(\dfrac{2}{1/\epsilon+1}\right)^{1/2} \end{cases} \tag{35}$$

Thus, we can see that Eqs. (27)-(30) are an eigenvalue problem about $h_{a0}^{(0)}, h_{a2}^{(0)}, h_{s0}^{(0)}$ and $h_{s2}^{(0)}$, and Eqs. (31) and (32) are those about $h_{a1}^{(0)}$ and $h_{s1}^{(0)}$. Therefore, variables with even subscripts are independent of those with odd subscripts. We shall discuss them respectively in the following.

Ⅳ THE BASIC SOLUTION WITHOUT CONSIDERATION OF THE AIR-SEA INTERACTION

Take $A=0$ and $\gamma=0$ in Eqs. (15) and (16). Now the air-sea interaction is cut off and the equations governing the air and sea motions become

$$\left(\frac{\partial^2}{\partial x^2}+\frac{\partial^2}{\partial y^2}-\frac{y^2}{4}\right)\frac{\partial h_a}{\partial t}+\frac{1}{2\epsilon^{1/2}}\frac{\partial h_a}{\partial x}=0 \tag{36}$$

$$\left[\epsilon\left(\frac{\partial^2}{\partial x^2}+\frac{\partial^2}{\partial y^2}\right)-\frac{1}{\epsilon}\frac{y^2}{4}\right]\frac{\partial h_s}{\partial t}+\frac{\epsilon^{1/2}}{2}\frac{\partial h_s}{\partial x}=0 \tag{37}$$

Considering Eqs. (17), (18) and (19), we get two frequency-equations corresponding to the atmosphere and ocean respectively

$$\sigma_a=-\frac{1}{2\epsilon^{1/2}}\frac{k}{k^2+n+\dfrac{1}{2}} \tag{38}$$

$$\sigma_s=-\frac{\epsilon^{1/2}}{2}\frac{k}{\epsilon k^2+n+\dfrac{1}{2}} \tag{39}$$

Obviously, they are the westward propagating Rossby waves in the atmosphere and ocean, respectively (Matsuno, 1966).

Taking the following parameters: $\Delta\bar{T}_z = 8.0$ K, $D = 100$ m, $H_a = 400$ m, $\alpha = 3.413\times10^{-4}$ K^{-1}, we get $C_s =$ 1.64 m/s, $C_a = 62.63$ m/s and $\epsilon = 0.026$. The dispersion relation-ship of Rossby waves in the atmosphere and ocean at $n = 0, 1$ and 2, according to Eqs. (38) and (39), is given in Fig.2.

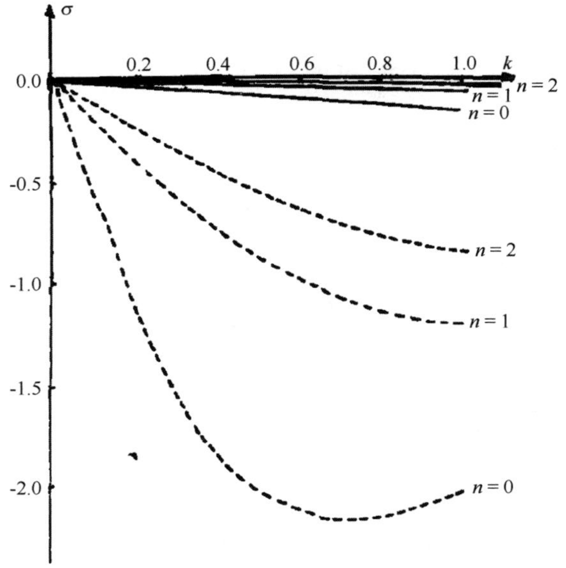

Fig.2 Dispersion relationships of Rossby waves in the atmosphere (dotted lines) and ocean (solid lines)

V INTERACTION BETWEEN ANTI-SYMMETRIC WAVES

Corresponding to Eqs. (31) and (32), there exist anti-symmetric waves about the equator for $n = 1$ in the air-sea coupled system. If $h_{a1}^{(0)}$ and $h_{s1}^{(0)}$ do not equal zero simultaneously, we have

$$\left(k^2+\frac{3}{2}\right)\left(\epsilon k^2+\frac{3}{2}\right)\sigma^2+\left[\left(k^2+\frac{3}{2}\right)\epsilon^{1/2}+\left(\epsilon k^2+\frac{3}{2}\right)/\epsilon^{1/2}\right]\cdot\frac{k}{2}\sigma + \frac{1}{4}\left[k^2-\frac{\epsilon^2 A\gamma T^2}{Da_{11}}(a_{13}+3a_{11})\left(k^2+\frac{3}{4}\right)\right]=0 \tag{40}$$

Thus we get two waves with different frequencies:

$$\sigma_{1,2}=\frac{1}{2[k^2+(3/2)][\epsilon k^2+(3/2)]}\left\{-\left[\left(k^2+\frac{3}{2}\right)\epsilon^{1/2}+\left(\epsilon k^2+\frac{3}{2}\right)/\epsilon^{1/2}\right]\cdot\frac{k}{2}\right.$$
$$\pm\left[\left[\left(k^2+\frac{3}{2}\right)\frac{\epsilon^{1/2}}{2}k+\left(\epsilon k^2+\frac{3}{2}\right)\frac{k}{2\epsilon^{1/2}}\right]^2-\left(k^2+\frac{3}{2}\right)\left(\epsilon k^2+\frac{3}{2}\right)\right.$$
$$\left.\left.\times\left[k^2-\frac{\epsilon^2 A\gamma T^2}{Da_{11}}(a_{13}+3a_{11})\left(k^2+\frac{3}{4}\right)\right]\right]^{1/2}\right\} \tag{41}$$

If the air-sea interaction is cut off, we obtain two roots

$$\sigma_1=-\frac{1}{2\epsilon^{1/2}}\frac{1}{k^2+3/2}, \tag{42}$$

$$\sigma_2 = -\frac{\epsilon^{1/2}}{2}\frac{1}{\epsilon k^2 + 3/2}. \tag{43}$$

Apparently, they are the Rossby waves in the atmosphere and ocean respectively, the same as those given in Eqs. (38) and (39) when $n=1$.

Eigenfrequencies are shown in Fig.3, together with frequencies of Rossby waves in the atmosphere and ocean when $n=1$. From Fig.3 we can see that the fast wave, corre-sponding to σ_1, is much the same as the Rossby wave in the atmosphere, meaning that the air-sea interaction has only a slight effect on this air-sea coupled wave. The wave corresponding to σ_2 is close to the Rossby wave in the ocean, but now the former becomes eastward propagating when its wavenumber k is small, and at about $k=0.52$, this wave changes its direction and becomes westward propagating again.

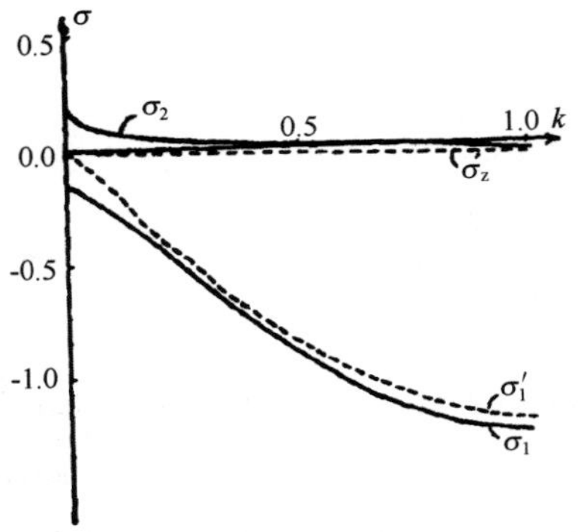

Fig. 3　Dispersion relationships for $\sigma_{1,2}$ (solid lines) and for $\sigma'_{1,2}$ (dotted lines), in which the air-sea interaction is cut off

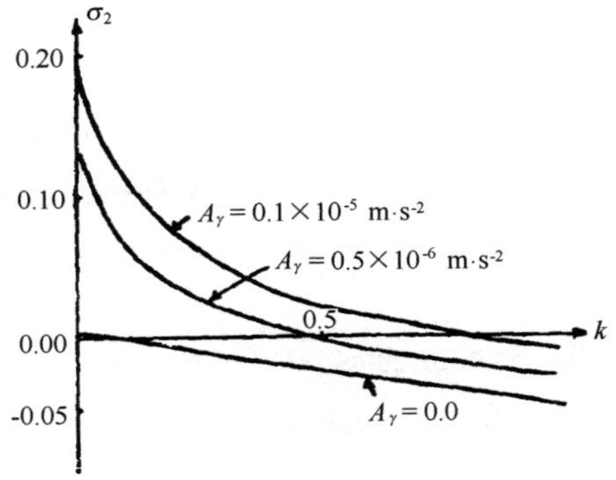

Fig.4　Variation of σ_2 with A_γ ($\epsilon=0.024$)

The effect of the air-sea interaction coefficient A_γ on σ_2 is given in Fig.4. It is shown that the eastward propagating wave moves towards higher wavenumbers as the air-sea interaction becomes stronger.

VI　INTERACTION BETWEEN SYMMETRIC WAVES

From Eqs. (27)-(30), we obtain four eigenfrequencies $\sigma_3, \sigma_4, \sigma_5$ and σ_6, which are shown in Fig.5, together with the eigenfrequencies of Rossby waves in the atmosphere (σ_{a0}, σ_{a2}, subscripts 0 and 2 represen-ting $n=0$ and 2, respectively.) and in the ocean (σ_{s0}, σ_{s2}, subscripts 0 and 2 as in the atmosphere). Fig.5 shows that the frequencies of fast waves, corresponding to σ_3 and σ_4, are close to those of Rossby waves in the atmosphere, but here the air and sea are a coupled system and the perturbations in the two media propa-gate westward simultaneously. Obviously, the waves corresponding to σ_5 and σ_6 are the air-sea interaction waves in which the Rossby waves in the ocean play a very important role. The difference is that the two waves here become eastward when the wavenumber k is small and at about $k=0.35$, the waves turn their directions and become westward again.

It is worth pointing out that when $k<0.03$, the waves corresponding to σ_5 and σ_6 are stable, while when $k \geqslant 0.03$, these two waves become unstable. The wave corresponding to σ_5 decays and that corresponding to

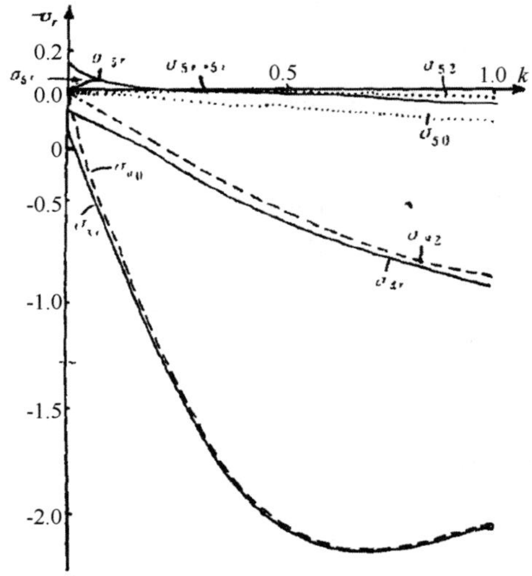

Fig.5 The real part of eigenfrequencies σ_3, σ_4, σ_5 and σ_6 given by Eqs. (27)-(30) (solid lines) and the eigenfrequencies of Rossby waves in the atmosphere (dashed lines) and in the ocean (dotted lines). Subscript r represents real part

σ_6 amplifies as time goes on. Fig.6 shows the decaying rate and amplifying rate. Here we see that the stable Rossby waves in the atmosphere and ocean can produce unstable westward and eastward propagating waves because of the air-sea interaction. Fig.7 shows the effect of the air-sea interaction coefficients on the unstable wave. When the air-sea interaction becomes stronger, the unstable eastward propagating wave moves towards higher wavenumbers and the amplifying rate becomes large.

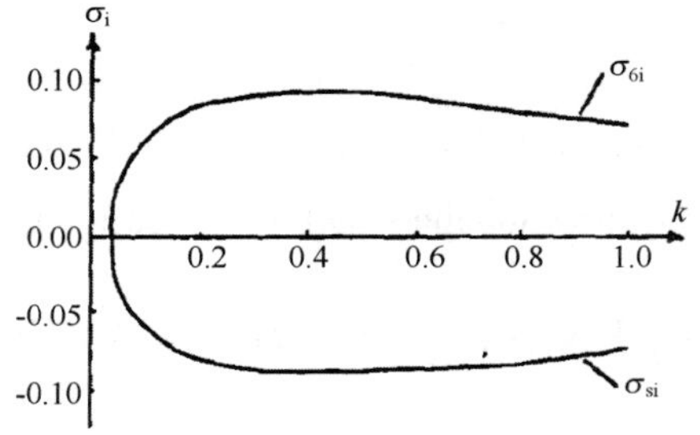

Fig.6 The imaginary part (denoted by subscript i) of σ_5 and σ_6

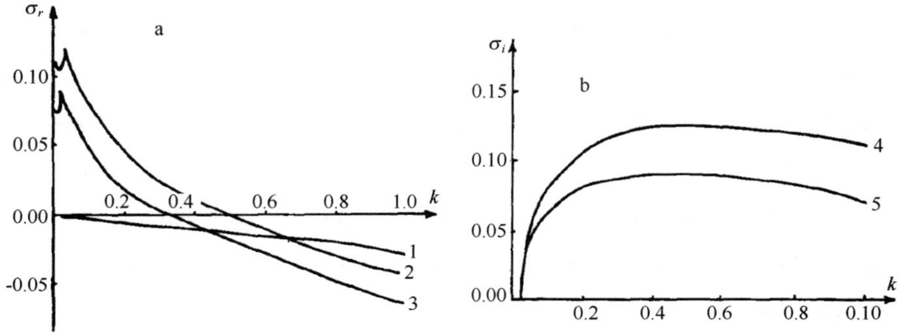

Fig.7 The effect of the air-sea interaction coefficient on σ_5; (a) for σ_{6r}, and (b) for σ_{6i}. Curves 1,2,3,4 and 5 are for $A_\gamma = 0.0, 0.1\times10^{-5}, 0.5\times10^{-6}, 0.1\times10^{-5}$ and 0.5×10^{-6} m · s^{-2}, respectively

Fig.8 shows the eigenfunctions of the westward propagating waves corresponding to $\sigma_3 \sim \sigma_6$. We can see that there are only very weak responses in the ocean corresponding to fast waves σ_3 and σ_4(Fig.8a, b). As for the unstable decaying wave σ_5(Fig.8c), the phase of the height field in the atmosphere is almost inverse to that in the ocean and so is the phase of velocity field, while for the unstable amplifying wave σ_5(Fig.8d), the phases of the height field and the velocity field are almost the same in the two media. We can give the following explanation for this instability. When there is a deepening of the thermocline somewhere in the ocean, the atmosphere is heated and winds converge near the equator. Because of the convergence of winds, the convection is intensified in the atmosphere. Since the winds and the ocean currents are of same phase, the convergence of the ocean currents is also intensified through the convergent stress, which in turn makes the thermocline deeper. In this case, the SST becomes warmer. So this is a process of positive feedback. In the case of unstable decaying, the direction of the wind stress is opposite to that of the ocean currents, so the convergence of the ocean currents will become weaker, which is unfavourable for the deepening of the thermocline, that is, unfavourable for the SST becoming warmer.

Ⅶ CONCLUSIONS

According to the above discussion, we obtain some important results as below:

(1) The coupling of the westward propagating Rossby waves between the atmosphere and ocean can produce eastward propagating waves in the scope where the wavelength is comparatively long. This scope is governed by the intensity of the air-sea interaction. The stronger the interaction, the larger the scope becomes. There eastward propagating waves are different from Kelvin waves, because they are produced only by the interaction of Rossby waves between the atmosphere and ocean.

(2) There exists instability for both the westward and eastward propagating waves, and the unstable amplifying rate is also governed by the intensity of the air-sea interaction. The stronger the interaction, the larger the unstable amplifying rate. In the air-sea coupled system, it is stable for the coupling waves with their frequencies close to those of the Rossby waves in the atmosphere, while it is unstable only for the coupling waves with their frequencies close to those of the Rossby waves in the ocean. Hence, this indicates that the Rossby waves in the ocean play a very important role in the air-sea coupled system. Without the Rossby waves in the ocean, neither appear the eastward propagating waves, nor exists the instability.

(3) In the past, the eastward propagating waves in ENSO events were all owed to Kelvin waves. However, not only the occurrence of Kelvin waves needs very strict conditions, that is, the meridional velocities must disappear, but the eastward propagating velocity of the 1982–1983 ENSO event (see Fig.1) is slower than that of Kelvin waves. As for ENSO events, warm water usually appears near the coast of Peru at first, then propagates westward. But in 1982–1983, warm water appeared in the middle equatorial Pacific first, then propagared eastward. For the former case, according to our model results, we can regard it as the case that the air-sea interaction is comparatively weak. In this situation, the waves mainly propagate westward and the unstable amplifying rate is small. Taking the values of A and γ as given is Section Ⅳ, that is, $A\gamma = 0.5 \times 10^{-5}$, for the unstable wave with a wavelength about 10 000 km, the westward propagating velocity is about 15.5 deg. long./month. Thus, in about 7 months, the unstable wave can arrive at 180°W from the coast of

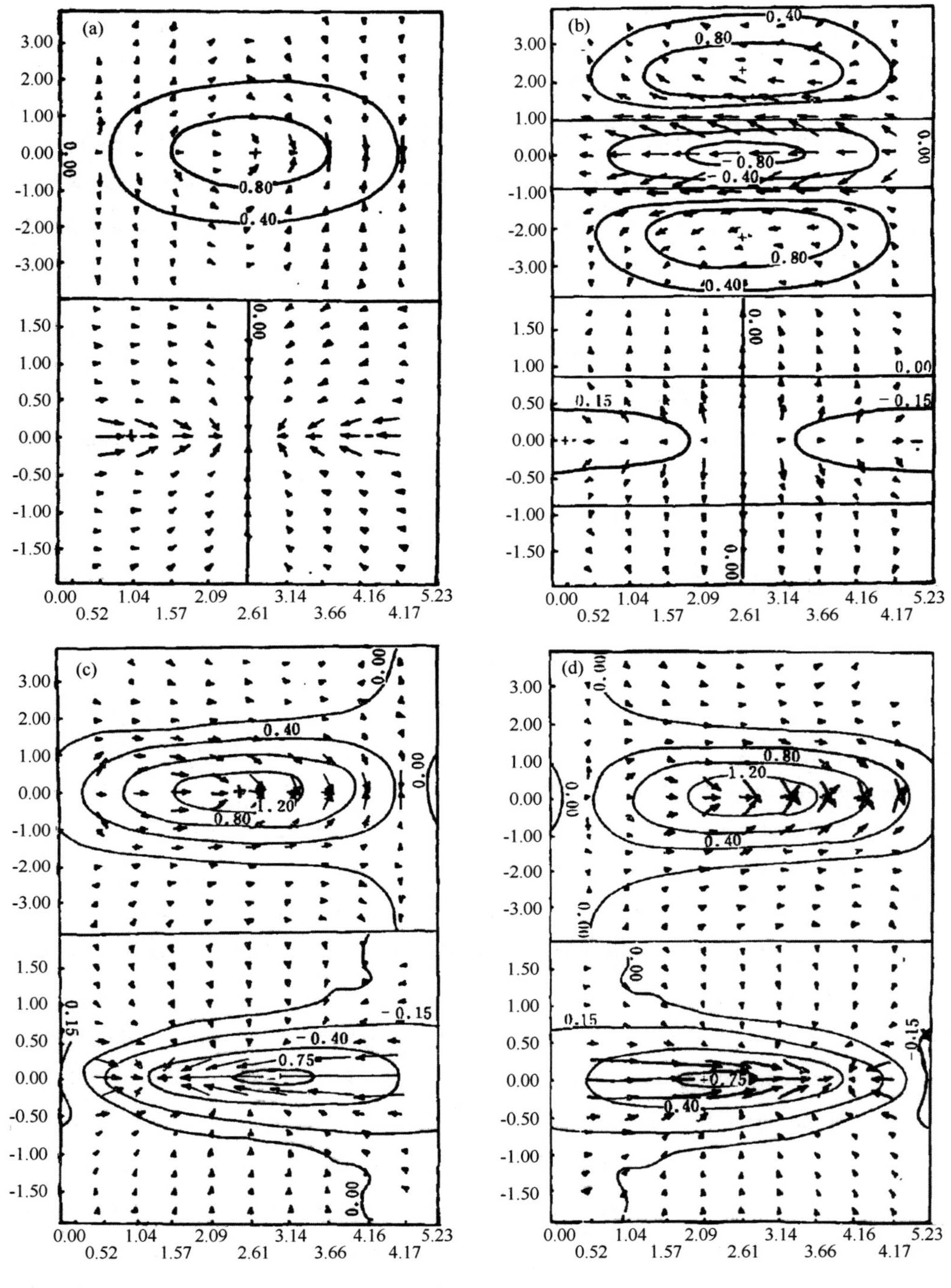

Fig.8 The eigenfunctions of westward propagating waves in the atmosphere and ocean corresponding to σ_3(a), σ_4(b), σ_5(c) and σ_6(d) when $k=0.6$. Solid lines and arrows stand for height field and wind (ocean current) field respectively. The upper half and the lower half represent atmosphere and ocean respectively.

Peru. For the latter case, we can regard it as the case that the air-sea interaction is comparatively strong. In this situation, waves with wavelengths comparatively long mainly propagate eastward and the unstable amplifying rate is larger. By taking $A\gamma=0.25\times10^{-5}$, and also for the unstable wave with a wavelength about 10 000 km, it propagates eastward at a velocity about 10.5 deg.long./month, which is close to the eastward propagating velocity of the 1982–1983 ENSO event. Let us take two actual examples. For the 1982–1983 ENSO event, up to September 1982, the SST in the eastern equatorial Pacific was 6℃ higher than that of the mormal and anomalies lasted until June 1983 (Philander, 1983b). But usually, in 1975 for example, the SST was only about 4℃ above the normal and the total duration of the event was short. Therefore, when an ENSO event appears, whether from our model results or the actual situations, the propagating direction of anomalies is determined by the intensity of air-sea interaction.

Of course, these are primary conclusions, because the truncating model method is used in the process of solution, which will undoubtedly affects the results. Nevertheless, we believe that at least in the case that the physical quantities are comparatively uniform in the meridional direction, our results are valuable, since the modes of higher n are not very important. In fact, the air-sea coupled model is rather simple. As is shown above, it is because the approximate relationship (11) is introduced and the advection terms are neglected that we can obtain the interactions between the anti-symmetric waves and symmetric waves respectively. The air-sea interaction waves here are linear and the instability in discussion is also linear. Further results without considering the approximate relation-ship (11) will be given in another paper. Like Paper et al., we are also able to use numerical method to prove our results.

REFERENCES

Anderson, D. T. and Gill, A. E. 1975. Spin-up of a stratified ocean, with applications to upwelling. *Deep-Sea Res.*, 22: 283–596.

Bjerkness, J. 1966. A possible response of the atmospheric Hadley Circulation to equatorial anomalies of ocean temperature, *Tellus*, 18:820–829.

Bjerkness, J. 1969. Atmospheric teleconnections from the equatorial Pacific. *Mon. Wea. Rev.*, 97:163–172.

Gill, A. E. 1980. Some simple solutions for heat-induced tropical circulation. *Quort. J. R. Met. soc.*, 106:447–462.

Lighthill, M. J. 1969. Dynamical response of the Indian Ocean to onset of the southwest monsoon, *Phil. Trans. Roy. Soc.*, A265:45–92.

Matsuno, T. 1966. Quasi-geostrophic motions in the equatorial area. *J. Meteor. Soc. Japan*, 44:25–43.

Philander, S.G.H. 1983a. El Niño Southern Oscillation phenomena. *Nature*, 302:295–301.

Philander, S.G.H. 1983b. Anomalous El Niño of 1982–83. *Nature*, 305:16.

Philander, S.G.H., Yamagata, T. and Pacanowski, R. C. 1984. Unstable air-sea interaction in the tropics. *J.Atmos. Sci.*, 41: 604–613.

Rasmusson, E. M. and Wallace, J. M. 1983. Meteorological aspects of the El Niño/Southern Oscillation. *Science*, 222:1195–1202.

Wang Zhuxi and Guo Dunren. 1979. *An Introduction to Special Functions.* Science Press, Beijing, 762 pp(in Chinese).

Wyrtki, K. 1975. El Niño-the dynamical response of the equatorial Pacific Ocean to atmospheric forcing. *J. Phys. Oceanogy*, 5:572–584.

Zebiak, A. E. 1982. A simple atmospheric model of rclevance to El Niño. *J. Atinos. Sci.*, 39:2017–2027.

Instability of the Oceanic Waves in the Tropical Region*

Ji Zhengang (李振刚), Chao Jiping (巢纪平)

(National Research Center For Marine Environmental Forecasts, State Oceanic Administration, Beijing)

ABSTRACT: In this paper, the effects of the large-scale mean sea temperature fields of the tropical ocean and the zonal current field (southern equatorial current) have been comprehensively entered in consideration on the basis of Chao and Ji (1985), and Ji and Chao (1986), the equatorial oceanic waves of the tropical ocean have been discussed by use of linearized primitive equations, then, the significant influence of the climatic background fields of the tropical ocean upon the oceanic waves of this region has been further testified. When very cold water appears in the tropical region, and the southern equatorial current is also relatively strong, the effect of the Rossby wave weakens, as a consequence, there are substitutive slow waves (i.e. thermal waves) which travel in opposite direction (eastward) to the Rossby wave. The characteristics of the slow wave are similar to those of Rossby waves, only the travelling direction is opposite. Under a certain environmental background field, the slow wave and the modified Rossby wave may be instable. With this conclusion, the mechanism of the occurrence, development and propagation of El Niño events has been studied. It is pointed out that the opposite travelling direction of the thermal wave and Rossby wave will bring repectively into action under different marine environmental background fields. The physical causes for that the abnormal warm water inclines to occur along the South American coast have also been explored in this paper.

I INTRODUCTION

At present, the studies on the air-sea interaction and El Niño/Southern Oscillation (ENSO) in the tropical areas have widely interested and drawn much attention in the community of meteorlogy and oceanography of the world. A lot of discussions about the propagation, extension of SST anomaly (SSTA) and the change of the thermocline have been undertaken. For example, Wyrtki (1975) pointed out that if trade winds weaken the Kelvin wave would be generated along the west coast of the Pacific, and when the Kelvin wave travels to the east coast, it would change the marine status of this region, i.e. the cold water upwelling weakens, then, SSTA increases correspondingly, as a consequence, the signal of the El Niño event appears. However,

* Acta Meteorologica Sinica, 1990, 4(2): 135-145.

more scientists think (including Wyrtki, 1984) that ENSO is the results of the air-sea interaction.

Beyond doubt, it is very important to study the ENSO events from the point of view of air-sea coupling interaction. However, it is also necessary to study the dynamic characteristics of the marine itself and its effect in the El Niño events, because the marine dynamic process, in fact, has not been made completely clear. For instance, in the specific marine and mean field of the Pacific, the equatorial wave system is different from the Matsuno's results (1966). Using a filtered model of the shallow water wave of mixed layer with inertia-gravitational waves filtered out Chao and Ji (1985) and Ji and Chao (1986) have discussed the oscillation period and instability of the modified equatorial Rossby wave affected by a specific SST field in the Pacific through theoretical analyses and numerical experiments. They have pointed out that a group of thermal wave, which travels in opposite direction (eastward) to that of the Rossby wave may be generated in a certain conditon. The generation of this group of wave is caused by the horizontal inhomogeneity of SST, particularly by meridional gradicnts. The results of numerical experiment show that SSTA may propagate and increase with the aid of the castward thermal waves and westward Rossby waves under the influence of the distribution of the specific equatorial mean SST field.

Based on the previous research (Chao and Ji, 1985; Ji and Chao, 1986), under the comprchensive consideration of the large-scale temperature fields and the equatorial current field, the equatorial wave and its instability have been further discussed theoretically with the shallow water model of the linearized primitive equation, and the physical factors of the occurrence and development of the El Niño events have bcen preliminarily studied with these results in this paper.

Ⅱ MODEL

The observational results show that the surface oceanic current of the Pacific in the equatorial region is the westward southern equatorial current. The position of this current has less change at the equator all the year round, only its strength changes with seasons. In this paper, we have not discussed the climatic causes for the formation of the equatorial current system, but, we only regard the equatorial current as a constant and basic current and as a background field to study the characteristics of the equatorial waves.

Fig.2 in the paper by Chao and Ji (1985) shows that in the tropical Pacific regions, particularly the eastern Pacific, there is a very cold water area, SST of the tropical region increases with the increase of latitudes, and the temperature gradients exist in the east-west direction. The discussion in the paper (Chao and Ji, 1985) has shown that the temperature gradients in the east-west direction in the Pacific have little influence on the period of the slow wave (modified Rossby wave). Thus, only the influence of the gradients in the south-north direction of SST upon the tropical waves has been considered, and for the convenience of mathematical treatment, the mean gradient field of SST is assumed not to change with time.

Let us take x-axis of the coordinates system along the equator and positive eastward, y-axis is vertical to the x-axis and positive northward. The z-axis is positive upward. Analogous to the treatment in the paper (Chao and Ji, 1985), the sea surface is regarded as the upper surface of the model, and the top of the thermocline as the lower surface of the model.

The equation of continuity for the imcompressible fluid can be written as

$$\frac{\partial u}{\partial x}+\frac{\partial v}{\partial y}+\frac{\partial w}{\partial z}=0 \tag{1}$$

then, we obtain approximately:

$$w(x,y,z,t)=-z\left(\frac{\partial u}{\partial x}+\frac{\partial v}{\partial y}\right)+\widetilde{w}(x,y,t) \tag{2}$$

where $\widetilde{w}(x,y,t)$ is an arbitrary function independent of z. The other symbols in this paper represent the same meaning as usual, except for special definition. On the sea surface at $z=h$, the normal velocity to this surface should equal zero because of the condition of vertical boundary. We thus obtain

$$w\big|_{z=h}=u\frac{\partial h}{\partial x}+v\frac{\partial h}{\partial y} \tag{3}$$

Inserting (3) into (2) obtains

$$\hat{w}_S(x,y,z,t)=(h-z)\frac{\partial u}{\partial x}+\frac{\partial v}{\partial y}+u\frac{\partial h}{\partial x}+v\frac{\partial h}{\partial y} \tag{4}$$

Thus, the perpendicularly-averaged vertical velocity $\hat{w}_s(x,y,t)$ is given by

$$\hat{w}_S(x,y,t)=\frac{D}{2}\left(\frac{\partial u}{\partial x}+\frac{\partial v}{\partial y}\right)+u\frac{\partial h}{\partial x}+v\frac{\partial h}{\partial y} \tag{5}$$

where D is the chaactcristic depth from sea surface to the bottom of the mixed layer.

As an approximation, it may be assumed that the basic current is gcostrophic. Although, strictly speaking this relation is not proper in some degree for the tropical regions, the diagnostic analysis of the dynamic height shows that in spite of the disturbed fluid field being a geostrophic, the mean ocean current in the tropical regions sitll have geostrophic features (Lukas and Firing, 1984). (Here, only the effect of the southern equatorial current is considered) Therefore, we obtain

$$g\frac{\partial \bar{p}_S}{\partial y}=g'\frac{\partial \bar{h}}{\partial y}=-u_0 f_0 \tag{6}$$

where $g'=\Delta\rho/\rho g$ is apparent gravitational acceleration, h is mean depth of mixed layer, $\partial \bar{p}_y/\partial y$ is mean meridional pressure gradient, and f_0 is mean value of Coriolis parameter near the equator. In fact, it is rational to take $f=\beta y$, however, there is no substantial influence on the results (Chao and Zhang, 1988).

Splitting disturbed and mean components, we have

$$u=u_g+u',\quad v=v',\quad \hat{w}_S=w'_S,\quad h=\bar{h}+h' \tag{7}$$

Inserting (6),(7) into (5) obtains

$$w'_S(x,y,t)=\frac{D}{2}\left(\frac{\partial u'}{\partial x}+\frac{\partial v'}{\partial y}\right)-\frac{u_g f_0}{g'}v' \tag{8}$$

If there is no external heat sources, then the thermodynamic equation is

$$\frac{\mathrm{D}T'}{\mathrm{D}t}+v'\frac{\partial \bar{T}}{\partial y}+w'_S\frac{\Delta \bar{T}_z}{D}=0 \tag{9}$$

where T' is disturbed temperature, $\partial \bar{T}/\partial y$ is gradient in the north-south direction of the mean sea temperature field of the equatorial Pacific Ocean, $\Delta \bar{T}_z$ is temperature difference between sea surface and bottom of the mixed layer, and

$$\frac{\mathrm{D}}{\mathrm{D}t}=\frac{\partial}{\partial t}+u_g\frac{\partial}{\partial x} \tag{10}$$

In addition, the horizontal equations of motion are

$$\frac{\mathrm{D}u'}{\mathrm{D}t}-\beta yv'=-G\frac{\partial T'}{\partial x} \tag{11}$$

$$\frac{\mathrm{D}v'}{\mathrm{D}t}+\beta yu'=-G\frac{\partial T'}{\partial y} \tag{12}$$

where $G=g\Delta\alpha$, and α is the coefficient of heat expansion.

Eeq. (8), (9), (11) and (12) constitute a completely closed equation system related to u', v', w' and T'.

Ⅲ SOLUTIONS OF THE EQUATIONS

Assuming that the form of the solutions are as follows

$$\begin{bmatrix} u' \\ v' \\ w' \\ T' \end{bmatrix}=\begin{bmatrix} u(y) \\ v(y) \\ w(y) \\ T(y) \end{bmatrix}\mathrm{e}^{\mathrm{i}(kx-\sigma t)} \tag{13}$$

where k is wavenumber, σ is frequency. From Eqs. (8)-(13), we obtain

$$-\mathrm{i}(\sigma-ku_g)u=-GikT+\beta yv \tag{14}$$

$$-\mathrm{i}(\sigma-ku_g)v=-G\frac{\mathrm{d}T}{\mathrm{d}Y}-\beta yu \tag{15}$$

$$-\mathrm{i}(\sigma-ku_g)T+\frac{\Delta\overline{T}_z}{2}\left(iku+\frac{\mathrm{d}v}{\mathrm{d}y}\right)+S_yv=0 \tag{16}$$

where

$$S_y=\frac{\partial\overline{T}}{\partial y}-\frac{u_gf_0}{2Dg} \tag{17}$$

The expression $\Delta\rho=\alpha\Delta\overline{T}_z\rho$ is used here. Putting

$$\overline{S}_y=\frac{2S_y}{\Delta\overline{T}_z},\quad C_S^2=\frac{1}{2}gD\alpha\Delta\overline{T}_z \tag{18}$$

From Eqs. (14)-(16) we obtain

$$\frac{\mathrm{d}^2v}{\mathrm{d}y^2}+b\frac{\mathrm{d}v}{\mathrm{d}y}+(c-dy^2)v=0 \tag{19}$$

where

$$\begin{cases} b=\widetilde{S}_y, \\ c=\dfrac{(\sigma-ku_g)^2}{C_S^2}-\dfrac{ku_g(\sigma-ku)}{C_S^2}-\dfrac{[k(\sigma-ku_g)^2+\beta]k}{(\sigma-ku_g)}+\dfrac{k\widetilde{S}_yf_0}{(\sigma-ku)} \\ d=\dfrac{\beta^2}{C_S^2} \end{cases} \tag{20}$$

By taking the following forms of transformation

$$v(y) = \tilde{v}(y) e^{-\frac{1}{2}\tilde{S}_y y} \tag{21}$$

$$\left(\frac{\beta}{C_S}\right)^{1/2} y = y^* \tag{22}$$

and omitting tildes and asterisks over variables, Eq. (19) then becomes

$$\frac{d^2 v}{dy^2} + \left[\frac{c - \frac{1}{4}b^2}{d^{1/2}} - y^2\right] v = 0 \tag{23}$$

To obtain Weber equation, the solution with boundary when $|y| \to \infty$ requires

$$\frac{c - \frac{b^2}{4}}{dy^2} = 2n + 1, \quad n = 0,1,2,\cdots \tag{24}$$

The eigenfunction is as below

$$\tilde{v}(y^*) = e^{-\frac{1}{2}y^{*2}} H_n(y^*) \tag{25}$$

where H_n is Hermite's polynomial of n order.

The solution of the disturbed variable $v'(x,y,t)$ is

$$v'(x,y,t) = \mathrm{e}^{-\frac{1}{2}\left(\frac{\beta}{C_S}y^2 + \tilde{S}_y y\right)} \mathrm{e}^{\mathrm{i}(kx-\sigma t)} Hn\left[\left(\frac{\beta}{C_S}\right)^{1/2} y\right] \tag{26}$$

At the same time, we can easily obtain the solutions of disturbed variables $fu'(x,y,t)$, $\hat{w}_s(x,y,t)$ and $T'(x,y,t)$. Here, we do not express one by one.

Ⅳ INSTABILITY WAVES IN THE TROPLCAL OCEAN

Now, let us discuss the period and the instability of the tropical wave. Erom Eq. (24) we can obtain

$$\omega^3 - \left[k^2 C_S^2 + \frac{1}{4}\tilde{S}_y^2 C_S^2 + (2n+1)\beta C_S\right]\omega - (\beta - \tilde{S}_y f_0) C_S k = 0 \tag{27}$$

where

$$\omega = \sigma - k u_g \tag{28}$$

1. Brief Analysis of the Equation of Frequency

If the effects of the mean current and large-scale sea temperature fields of the equatorial zone are ignored, then

$$\tilde{S}_y = 0 \tag{29}$$

and Eq. (27) can be written as

$$\omega^3 - \left[k^2 C_S^2 + (2n+1)\beta C_S\right]\omega - \beta C_S^2 k = 0 \tag{30}$$

This is equation of dimensional form in the paper by Matsuno (1966). Similar to Matsuno's analysis, two fast waves (one traveling eastward and the other westward, i.e. inertia-gravitational waves) and the third one, equatorial Rossby wave traveling westward, have bcen obtained from Eq. (30). These three waves are all stable waves.

If $\widetilde{S}_y \neq 0$, for the waves of relatively high wave numbers and high frequencies, we can approximately derivc the following expression from Eq. (27)

$$\omega \doteq \sqrt{k^2 C_S^2 + \widetilde{S}_y C_S^2/4 + (2n+1)\beta C_S} \tag{31}$$

It may be obviously seen that these are still two incrtia-gravitational waves with different traveling directions, only the frequencies are modifred owing to the existence of $\widetilde{S}_y$.

For the waves of low frequency, we approximately derive from Eq. (27) the following equation

$$\omega \doteq -\frac{(\beta - \widetilde{S}_y f_0) C_S^2 k}{k^2 C_S^2 + \widetilde{S}_y^2 C_S^2/4 + (2n+1)\beta C_S} \tag{32}$$

The results are similar to those of Chao and Ji (1985). When the parameter $\widetilde{S}_y$ is relatively small, thus $\omega<0$, the wave is the modified Rossby wave of the tropical region. When $\widetilde{S}_y$ relatively large, $\omega>0$, this slow wave travels eastward, and we call it thermal wave.

From the discriminant of a cubic algebraic equation. we obtain the condition for conjugate complex root of Eq. (27) as

$$\Delta = \frac{(\beta - \widetilde{S}_y f_0)^2}{4} K^2 - \frac{[Kk + \bar{S}_S C_S/4 + (2n+1)\beta C_S]^3}{27} > 0 \tag{33}$$

where

$$K = kC_S^2 \tag{34}$$

For the high wavenumber and relatively large value of $|\widetilde{S}_y|$. we have obtained approxi-mately the following expression

$$27 |\widetilde{S}_y| f_0/4 > k^2 C_S \tag{35}$$

By use of wavelength $L = 2\pi/k$, we obtain

$$L > \frac{C}{\sqrt{|\widetilde{S}_y|}} \tag{36}$$

where C is a constant greater than zero. Thus, when $|\widetilde{S}_y|$ is relatively big and wavelength L satisfies Eq. (36), inevitably, there is one instable wave with growing amplitude among three roots of Eq. (27). It can be seen from Eq. (36) that when $\widetilde{S}_y \ll 0$ and $\widetilde{S}_y \gg 0$, two instable areas would appear correspondingly.

2. Dispersion Relation and Growing Rate of Instable Wave

When effects of the zonal basic circulation and the large-scale sea temperature field of the tropical regions are taken into consideration, usually, $\widetilde{S}_y \neq 0$. From the numerical calculation of Eq. (27), we obtain dispersion relation and relation between growing rate and wavenumber.

Taking $\alpha = 3\times10^{-4}\ \mathrm{K}^{-1}$, $D = 100$ m, $g = 9.8\ \mathrm{m/s^2}$, $f_0 = 1.5\times10^{-5}$/s, $u_g = -0.1$ m/s, $\partial T/\partial y = 2.2\times10^{-6}$℃/m, and $\Delta T_z = 4.3$℃, from Eq.(18), we can obtain $\widetilde{S}_y = 3.3\times10^{-6}$/m. At this time, there is a very cold water area in the equatorial zone. From Eq. (27), we obtain Fig.1 of dispersion relation and Fig.2 of relation between growing (decaying) rate and wavenumber. Comparing Fig.1 with the results of Cane and Sarachik (1976) shows that when $n \geqslant 1$, though $\widetilde{S}_y \neq 1$, the frequencies of two branches of inertia-gravitational waves have little changes. However, there is no Rossby wave traveling eastward (Cane and Sarachik, 1976), alternately, there is another slow wave traveling in the opposite direction to the Rossby wave, We call it thermal wave (Chao and Ji, 1985).

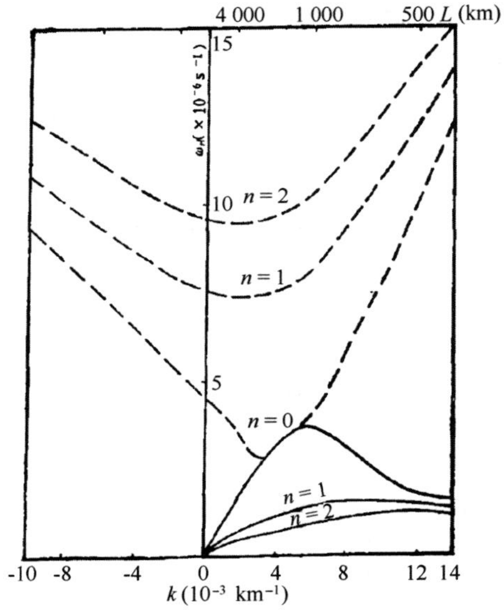

Fig.1 Dispersion relation of the tropical oceanic wave when $\widetilde{S}_y = 3.3\times10^{-6}\ m^{-1}$. The dashed lines represent the inertia-gravitational wave, while the solid lines represent slow wave

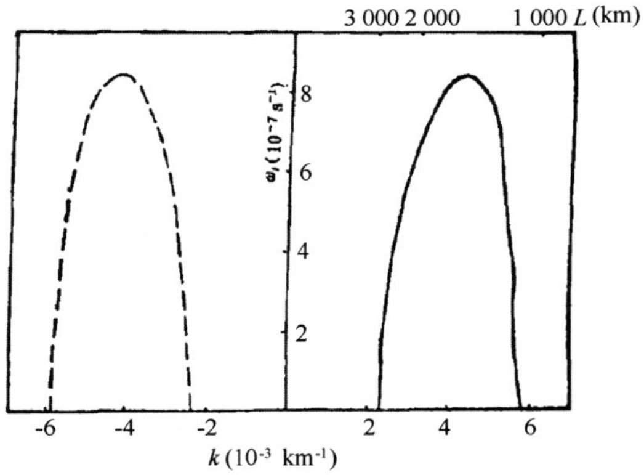

Fig.2 Growing (decaying) rate, represented by solid (dashed) line, of the tropical oceanic wave when $\widetilde{S}_y = 3.3\times10^{-6}\ m^{-1}$

The generation of this slow wave is just due to the consideration of large-scale mean temperature field and mean current field of the tropical regions. When $n=0$, the mixed thermal-inertia-gravitational wave may be seen from Fig.1. Fig.2 shows that within the range of coincided frequencies of the inertia-gravitational wave with thermal wave, the waves are instable.

In Fig.2, solid (dashed) lines represent the growing (decaying) rate of thermal-inertia-gravitational waves. The calculations show that when $n \geqslant 1$, the instability of the wave may appear if $|\widetilde{S}_y|$ is certainly umch greater. While the actual large-scale mean current field in the ocean is difficult to satisfy this condition. Therefore, only the case is given when $n = 0$, in Fig. 2. Two branches of conjugate thermal-inertia-gravitaitonal mixed waves can be seen in Fig.2, one is decaying wave, and the other intensified. The length of unstable waves lies between 2200−1280 km.

When the southern equatorial current of the equatorial zone weakens, even the east-ward oceanic current appears (i.e. $u_g>0$), and sea temperature has relatively large positive anomalies in this area, then it is known from Eq. (8) that $\widetilde{S}_y<0$. Fig.3 shows dispersion relation of the waves in the tropical region when $\widetilde{S}_y=-0.07\times10^{-6}/\text{m}$. Comparing Fig.3 with the results of Cane and Sarachik (1976), it may be seen that when $n \geqslant 1$, frequencies of the Rossby wave and two branches of inertia-gravitational waves have little change. Bue, when $n=0$, it is dnown from Eq. (27) that the frequency relation of $\omega=-k$ does not exist if $\widetilde{S}_y\neq0$. Therefore, we have not taken away any root in Exp. (2), still, we can obtain the frequencies of three branches of waves. It may be seen from Fig.3 that within the wave-length of 1.3×10^3 km -2.2×10^3 km, there are two waves i.e. Rossby-inertia-gravitational mixed waves, with conjugate frequcncies. The relation be-twecn their growing (decaying) rate and wave number k is similar to Fig.2, we will not explain in this paper.

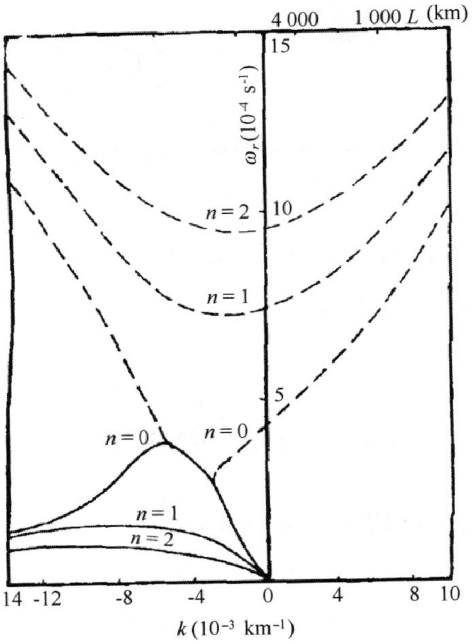

Fig.3　Dispersion relation of the tropical oceanic wave when $\widetilde{S}_y=-0.07\times10^{-6}\ \text{m}^{-1}$

The calculations show that the frequency of the modified Rossby wave is gradually decreasing whth the increase of $\widetilde{S}_y$, when $\widetilde{S}_y \geqslant S_{yC_1}=1.52\times10^{-6}/\text{m}$, its frequency trends to be zero. The alternative is thermal wave travelling eastward. Thus, when $\widetilde{S}_y<S_{yC_1}$, the relation of dispersion obtained from Eq. (27) is similar to Fig.3, while if $\widetilde{S}_y>S_{yC_1}$, this relation is similar to Fig.1, only with frequencies different (particularly for slow waves).

3. Relation between Parameter $\widetilde{S}_y$ and Instable Wavelength

Fig.4 shows the relation between $\widetilde{S}_y$ and instable wavelength for $n=0$, which can be obtained by setting $\Delta=0$ in (33). Within the range of short wavelengths the two instable areas (hatched in Fig.4) are generally similar to (36) approximately obtained above. From Fig.4 we can see that when $\widetilde{S}_y>3.12\times10^{-6}$/m or $\widetilde{S}_y<0$, it is easy to generate instable waves near wavelength $L=1\ 500$ km. Because the oceanic currents of the upper layer of the equatorial oceans are generally the southern equatorial currents travelling westward ($u_g<0$), when there is a cold water in the equatorial Pacific, the waves incline to enter into the right instable area in Fig. 4 according to (18). When a warm water area apears in the equatorial Pacific and the southern equatorial current weakens, the oceanic background field may satisfy the condition of $\widetilde{S}_y<0$, then the waves enter into the left instable area in Fig.4.

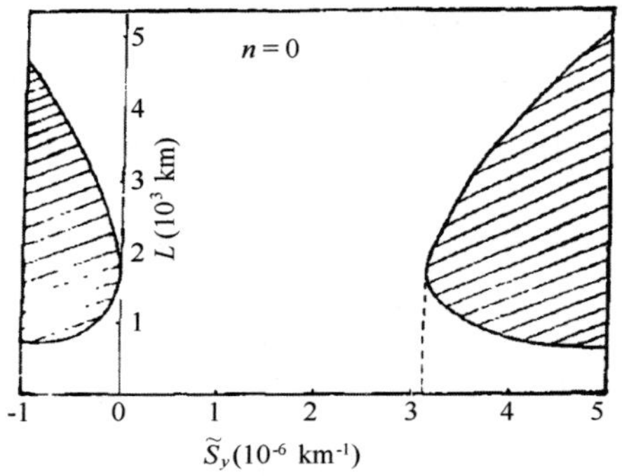

Fig.4 Relation between parameter $\bar{S}_y$ and instable length of oceanic wave. The area with the oblique lines is instable region

V INSTABLE WAVES AND EL NINO EVENTS

The occurrence and development of El Nino events and the propagation of SSTA always attract much interests and attention. Because ENSO is the result of very complicated air-sea coupling interaction, the detailedly theoretical studies on ENSO are not enough now. For the SSTA travelling westward, it is well known that this is owing to the effect of the Rossby wave, while for the SSTA travelling eastward, many researchers think this is the effect of the Kelvin wave that makes SSTA of the western Pacific travel eastward according to Wyrtki's theory (1975).

However, as Chao and Ji (1985) and Ji and Chao (1986) pointed out, the equatorial Kelvin waves certainly have influence on the El Niño events, but the dynamic condition of the existence of the equatorial Kelvin wave demands the current components in the south-north direction to be zero, in fact this condition could not be strictly satisfied in the current field of the equatorial Pacific, and certainly weakens the effect of the Kelvin wave. Ji (1987) also discussed this problem in his research on the response of the tropical atmosphere to the sea surface heat source. It is known from the above discussion that when $\widetilde{S}_y > S_{yC_1}$, we may obtain eastward travelling thermal wave. In the El Niño events, the thermal wave may play the similar role to the Kavlin wave, and it makes SSTA propagate from the west to the east. In Fig.1, if $L=2\ 100$ km is given, the phase speed of the thermal wave is 0.95 m/s, i.e. it travels 10 000 km in 130 days. This is approximated to the observational propagating speed of SSTA in the El Niño events. It should be pointed out that this speed is not sensitive to the change of parameters.

Although the air-sea interaction is not directly taken into account in this paper, only the physical characteristics of the ocean itself, i.e. the influence of the particular distribution of the large-scale mean SST field and current field in the equatoriai Pacific on the oceanic waves (of course, these fields are the consequences of the long-range air-sea interaction) have been considered. Because instability exists in oceanic waves, it seems that this kind of instability may give a qualitative physical explanation to the mechanism of occurrence, development and propagation of the El Niño events. Fig. 4, shows that the equatorial waves are stable within the range of parameter $0<\widetilde{S}_y<3.1\times10^{-6}$/m. Normally, actual large-scale mean current field (souhtern equatorial current) and SST field of the equatorial oceans confine the change of $\widetilde{S}_y$ in this range. It scems that this may explain why large anomaly of SST does not exist. However if the equatorial trade wind persistently intensifies, it causes the southern equatorial current and the equatorial water upwelling to strengthen, and SST to decrease, thus leading to increase of parameter $\widetilde{S}_y$. When $\widetilde{S}_y$ exceeds the critical value, i.e. $S_{yC_2}=3.1\times10^{-6}$/m, the equatorial oceanic waves become instable, and the disturbed SST field increases instably. It may be seen from the curves of the growing ratc in Fig.2 that if $L=2\ 100$ km, the *e*-fold growing time is 25 days. At the early stage of the El Niño events, SSTA may grow at this rate, at this moment, the thermal waves would play a dominant role and cause SSTA to propagate and to extend from west to east. While at the mature stage of the El Niño events, large areas of the warm water appear in the equatorial zone, the cold water areas even disappear. It is known from the analysis of Section Ⅳ, the modified Rossby waves play a dominant role at that time. Because $\widetilde{S}_y<0$, the instable Rossby waves would travel westward. In this way, SSTA may complcte an entire process, i.e. first, it propagates from west to east, then from east to west during El Niño events. The evolution and development of SSTA in the El Niño event of 1982–1983 are similar to this process.

During El Niño events, generally the positive SSTA appcars first along the Southern American coast, i.e. the instably increasing of SST occurs first. The observations show that the cold water of the Southern American coast is the coldest. It is easy to satisfy the instable condition, $\widetilde{S}_y>S_{yC_2}=3.0\times10^{-6}\ \mathrm{m}^{-1}$ in Fig.4 of this paper. It seems to explain qualitatively the physical reasons why abnormal SST inclines to occur along the Southern American coast at first. The conclusion drawn by Ji (1987) shows that the response of the atmosphere to the positive SSTA in Eastern Pacific may generate anomalous westerlies in the Middle and Western Pacific, furthermore, the anomalous westerlies are favourable to the further increase of SSTA.

Ji and Chao (1987) pointed out from their studies on the teleconnection of SST that during the early stage for each of El Niño events of 1975, 1963, 1969, 1972 and 1976, relatively strong negative SSTA may appear in the Eastern Pacific. We think that persistent decreasing of SST of the equatorial Eastern Pacific (i. e. the relatively large negative anomaly of SST appears) would tend to the increment of the gradients of the large scale sea temperature and makes $\widetilde{S}_y$ increasing, thus the ocean waves enter into the left instable area in Fig.4.

Ⅵ DISCUSSION AND CONCLUSION

On the basis of the studies of Chao and Ji (1985) and Ji and Chao (1986), the effect of the large-scale mean temperature field of the equatorial ocean and zonal mean oceanic current have been taken into consideration, and the equatorial oceanic waves have been studied by use of linearized primitive equations in this paper. And the important influence of the climate background of the equatorial ocean on the tropical waves has been further verified. Under the effect of those climatic mean fields, the tropical waves may become instable. When $\widetilde{S}_y > S_{yC_1} = 1.52\times10^{-6}\ \mathrm{m}^{-1}$, the Rossby waves disappear, the subsitutive eastward slow waves appear. We call them thermal waves (Chao and Ji, 1985). The characteristics of the thermal wave are similar to those of Rossby waves, only the traveling direction is opposite. When the cold water of the equatorial zone becomes colder and the southern equatorial current strengthens, it leads to $\widetilde{S}_y > \widetilde{S}_{yC} = 3.1\times10^{-6}\ \mathrm{m}^{-1}$, at this time the castward thermal waves play a dominant role, and the instable phenomenon appears around the wavchength $L = 1\ 500$ km. If the warmer water appears in the equatorial zone and the southern equatorial current weakens, then, $\widetilde{S}_y < 0$, at that time the westward Rossby waves play a dominant role, and the instable phenomenon occurs around the wavelength $L = 1\ 500$ km.

According to the conclusion above, the mechanism of the occurrence, development and propagation of the El Niño events have been discussed in this paper. It is pointed out that under the background fields of the different oceanic environments, the thermal wave and Rossby wave which travel in the opposite direction may play different roles respectively. The effect of the thermal wave is similar to that of Kelvin wave, i.e. it causes the instable SSTA to propagate from west to east. While the modified Rossby wave causes instable SSTA to propagate from east to west. Under the influence of these two slow waves, the whole propagating process of SSTA in an El Niño event may complete.

The physical causes for the abnormal warm water which tends to appear along the Southern American coast have been also discussed in this paper. There is a very strong equatorial cold water zone along the Southern American coast. When the trade wind intensifies, it causes the ocean water upwelling and the Southern Equatorial Current to strengthen and the gradients of the large-scale ocean temperature to increase, furthermore, $\widetilde{S}_y$ to increase, At this moment, the oceanic environmental background fields enter into instable area of $\widetilde{S}_y > \widetilde{S}_{yC}$ of Fig.3. Fig.1 in the paper by Ji and Chao (1987) also illustrated that generally SST may greatly decreases at the early stage of the El Niño event.

El Niño event is the result of the complicated nonlinear air-sea coupling interaction. It is impossible to explain its whole process by linearized shallow awater model in this paper. Discussed in this paper are only the instable triggering mechanism which may be generated by the large-scale oceanic background field and the influence of the field on the propagating process of the SSTA during the El Niño event. As for the problem of the interannual change of the large-scale oceanic current and SST, it is not the concernment in this paper. The further study of the equatorial waves and El Niño events should use the coupling air-sea interaction mode. Now, we have already set up a simple analytic model of this kind for tropical regions (Ji and Chao, 1988), and will further develop and perfect coupling air-sea interaction models.

REFERENCES

Cane, M.A. and Sarachik, E.S. 1976. Forced baroclinic ocean motions, I: The linear equatorial boundary case, *J. Mar. Res.*, 35:395–432.

Chao Jiping and Ji Zhengang. 1985. On the influence of large-scale inhomogencity of sea temperature upon the oceanic waves in the tropical regions, Part Ⅰ: Linear theoretical analysis. *Adrances in Atmos. Sci.*, 2: 295–306.

Chao Jiping and Zhang Renhe. 1988. The air-sea interaction waves in the tropical ocean and their instabilities, *Acta Meteor. Sinica*, 2:275–287.

Ji Zhengang. 1987. Dynamic studies on the tropical large-scale ocean and atmosphere, Dcctorial thesis of the Institute of the Atmospheric Physics, Academia Sinica.

Ji Zhengang and Chao Jiping. 1986. On the influence of large-scale inhomogeneity of sea temperature upon the oceanic waves in the tropical regions, Part Ⅱ: Linear numerical experiments. *Advances in Atmos. Sci.*, 3:238–244.

Ji Zhengang and Chao Jiping. 1987. Teleconnections of the sea surface temperature in the Eastern Equatorial Pacific, and with 500 hPa geopotential height field in the Northern Hemisphere, *Advances in Atmos. Sci.*, 4:343–348.

Ji Zhengang and Chao Jiping. 1988. An analytic model of the tropical air-sea coupling interaction (to be published).

Lukas, K. and Firing, E. 1984. The geostrophic balance of the Pacific equatorial undercurrent. *Deep-Sea Res.*, 31(A):61–66.

Maatsuno, T. 1966. Quasi-geostrophic motions in the equatorial area. *J. Metcor. Soc. Jan.*, 44:25–42.

Wyrthi, K. 1975. El Niño-the dynamic response of the Equatorial Pacific Ocean to atmospheric forcing. *J. Phys*, *Oceanogr.*, 5: 572–584.

Wyrtki, K. 1984. The slope of sea level along the Equator during the 1982/83 El Niño. *J. Geophys. Res.*, 89:10419–10424.

An Analytical Coupled Air-sea Interaction Model*

Zhen-Gang Ji, Ji-Ping Chao

(National Research Center for Marine Environmental Forecasts, Beijing 100081, China)

ABSTRACT: Using time-dependent, linear, reduced gravity equations (gravity waves are filtered out), an analytical coupled air-sea interaction model is constructed. Parabolic cylinder functions $D_n(y)$ are used to solve partial differential equations. The variables are expanded into the first three terms of $D_n(y)$ series in the y-direction. Then the approximative wave solutions of the coupled atmophere-ocean system (AOS) are obtained. According to the dispersion relationship of the coupled AOS, for coupling parameters in a certain interval, the waves which are symmetric to the equator, are unstable. If the coupling parameters are too large or too small, there is no unstable phenomenon in the coupled AOS. Under certain conditions, the AOS has an eastward propagating wave caused by air-sea interactions. The physical explanation for the air-sea interaction wave is given. Horizontal distributions of the coupled AOS are analysed and depicted also. The unstable feedback process and the possible relationship of the above results with ENSO events are explored.

Introduction

Many investigations of ENSO highlight these two questions: how the ocean responds to anomalous wind stress forcing (e.g., Wyrtki, 1975; McCreary, 1976 and Philander, 1981) and how the atmosphere responds to the anomalous heating of the ocean (e.g., Webster, 1972; Gill, 1980 and Ji and Chao, 1989). Although these investigations can not uncover the development of ENSO completely, they help us understanding ENSO events. Since 1982/83 ENSO, more and more authors have researched coupling mechanisms between the atmosphere and ocean. Some authors (e.g., Gill, 1985 and Philander et al., 1984) developed simple coupled air-sea interaction models in which the atmospheric equations are steady, but the oceanic equations are time-dependent. In the present paper, we construct a linear coupled model in which the atmospheric and oceanic equations are all time-dependent and discuss some basic features of large-scale motion in the atmosphere and ocean system (AOS) and explore their possible roles during ENSO events.

* Journal of Marine Systems, 1991, 1(3): 263-270.

Model

The basic equations

The oceanic motion is described by a mixed layer shallow water model. Basic equations are similar to the ones discussed by Anderson and Gill (1975) and the gravity wave is filtered out. The ocean is only forced by wind stress.

$$fu_s = -g\frac{\partial h_s}{\partial y} - \frac{g}{f}\frac{\partial^2 h_s}{\partial x \partial t} - \frac{\beta g}{f}h_s + \tau^y \tag{1}$$

$$fv_s = g\frac{\partial h_s}{\partial x} - \frac{g}{f}\frac{\partial^2 h_s}{\partial y \partial t} - \tau^x \tag{2}$$

$$\frac{\partial h_s}{\partial t} + H_s\left(\frac{\partial u_s}{\partial x} + \frac{\partial v_s}{\partial y}\right) = 0 \tag{3}$$

H_s is the equivalent depth of the shallow water model. Except for special description, other notations have common meteorological meanings. We (Chao and Ji, 1985 and Ji and Chao, 1986) used this group of equations to investigate equatorial waves.

The atmosphere is only forced by surface heating Q. Its equations are similar to the ones of the ocean:

$$fu_a = -g\frac{\partial h_a}{\partial y} - \frac{g}{f}\frac{\partial^2 h_a}{\partial x \partial t} - \frac{\beta g}{f}h_a \tag{4}$$

$$fv_a = g\frac{\partial h_a}{\partial x} - \frac{g}{f}\frac{\partial^2 h_a}{\partial y \partial t} \tag{5}$$

$$\frac{\partial h_a}{\partial t} + H_a\left(\frac{\partial u_a}{\partial x} + \frac{\partial v_a}{\partial y}\right) = -Q \tag{6}$$

H_a is the equivalent depth of the model atmosphere.

Coupled air-sea interactions and parameters

A simple and convenient way to couple the AOS is to consider that anomalous heating caused by the variation of sea surface temperature (SST) is proportional to the variation of the depth of the mixed layer, i.e., in eqn. (6) heating function Q is expressed in this way:

$$Q = Ah_s \tag{7}$$

The atmospheric wind field acts upon the mixed layer as a body force

$$(\tau^x, \tau^y) = \gamma(u_a, v_a) \tag{8}$$

Philander and many other authors obtained meaningful results by using this method to couple the AOS. They also discussed its physical meaning.

Putting eqns. (7) and (8) into eqns. (1)-(6), the governing equations are deduced:

$$\left(\frac{\partial^2}{\partial x^2} + \frac{\partial^2}{\partial y^2} - \frac{f^2}{C_a^2}\right)\frac{\partial h_a}{\partial x} + \beta\frac{\partial h_a}{\partial x} = \frac{f^2}{C_a^2}Ah_s \tag{9}$$

$$\left(\frac{\partial^2}{\partial x^2} + \frac{\partial^2}{\partial y^2} - \frac{f^2}{C_s^2}\right)\frac{\partial h_s}{\partial t} + \beta\frac{\partial h_s}{\partial x} = \gamma\left(\frac{\partial^2}{\partial x^2} + \frac{\partial^2}{\partial y^2}\right)h_a \tag{10}$$

where $C_s = \sqrt{gH_s}$, $C_a = \sqrt{gH_a}$ are phase speeds of reduced oceanic and atmospheric gravity waves, respectively. A and γ are coupling parameters of the AOS. Our paper is concerned with features of the atmosphere and ocean which are described by time-dependent equations. It is not our aim to discuss the variations of coupling parameters in detail. For the sake of convenience, all of the parameters in the present paper are the same as the ones given by Philander et al. (1984), except where specially stated.

In the model, the atmospheric (oceanic) Rossby deformation radius is used as a length scale in zonal (meridional) direction and the atmospheric Rossby wave period is taken to be a time scale, viz.:

$$x = (C_a/\beta)^{1/2}x^*, \quad y = (C_s/2\beta)^{1/2}y^*,$$

$$t = (C_a/\beta)^{-1/2}t^* \tag{11}$$

$$h_s(x,y,t) = h_s^*(y)\exp[i(kx - \sigma t)] \tag{12}$$

$$h_a(x,y,t) = h_a^*(y)\exp[i(kx - \sigma t)] \tag{13}$$

Putting eqns. (11)-(13) into eqns. (9) and (10) and omitting the notation " * ", we have

$$\frac{d^2h_s}{dy^2} + \left(\epsilon a - \frac{y^2}{4}\right)h_s = Q_A\left(2\frac{d^2h_a}{dy^2} - \epsilon k^2 h_a\right) \tag{14}$$

$$\frac{d^2h_a}{dy^2} + \left(\epsilon a - \frac{\epsilon^2 y^2}{4}\right)h_a = Q_S y^2 h_s \tag{15}$$

where

$$T_a = (C_a\beta)^{1/2}, \quad Q_S = -\frac{AT_a\epsilon^2}{i4\sigma},$$

$$Q_A = -\frac{\gamma T_a}{i2\sigma}, \quad a = \frac{1}{2}\left(-\frac{k}{\sigma} - k^2\right), \quad \epsilon = \frac{C_s}{C_a}$$

To the present stage, we obtained the governing ordinary differential eqns. (14) and (15) of the AOS.

The solutions of the coupled equations

Expanding $h_a(y)$ and $h_s(y)$ in terms of parabolic cylinder functions, viz.:

$$h_a(y) = \sum_{m=0}^{\infty} A_m D_m(y) \tag{16}$$

$$h_s(y) = \sum_{n=0}^{\infty} S_n D_n(y) \tag{17}$$

and taking the first three terms of eqns. (16) and (17) (i.e., truncating the series at $m,n=2$), yields

$$h_s(y) = A_0D_0(y) + A_1D_1(y) + A_2D_2(y) \tag{18}$$

$$h_s(y) = S_0D_0(y) + S_1D_1(y) + S_2D_2(y) \tag{19}$$

Strictly speaking, to get accurate solutions of eqns. (14) and (15), we should take infinite terms of $D_n(y)$ to express $h_s(y)$ and $h_a(y)$. However, Yamagata (1985) and Ji's (1987) investigations show that if one is concerned only with the largescale dynamics of the tropical atmosphere and ocean, the small scale [high order terms of $D_n(y)$] can be neglected. The approximative solutions of eqns. (14) and (15) obtained through eqns. (18) and (19) are still able to describe some large-scale features of the AOS qualitatively.

Substituting eqns. (18) and (19) into eqns. (14) and (15) by using the orthogonality of $D_n(y)$, we obtain equations for the coefficients S_0, S_2, A_0, A_2, A_1 and S_1:

$$(\epsilon a - \frac{1}{2})S_0 + (\frac{1}{2} + \epsilon k^2)Q_A A_0 - Q_A A_2 = 0 \tag{20}$$

$$(\epsilon a - \frac{5}{2})S_2 - \frac{Q_A}{2}A_0 + (\frac{5}{2} + \epsilon k^2)Q_A A_2 = 0 \tag{21}$$

$$Q_S S_0 + 2Q_S S_2 - \left(\epsilon a - \frac{1}{2} + \frac{B}{4}\right)A_0 - \frac{B}{2}A_2 = 0 \tag{22}$$

$$Q_S S_0 + 5Q_S S_2 - \frac{B}{4}A_0 - (\epsilon a - \frac{5}{2} + \frac{5}{4}B)A_2 = 0 \tag{23}$$

and

$$(\epsilon a - \frac{3}{2})S_1 + (\frac{3}{2} + \epsilon k^2)Q_A A_1 = 0 \tag{24}$$

$$3Q_S S_1 - \left(\epsilon a - \frac{3}{2} + \frac{3B}{4}\right)A_1 = 0 \tag{25}$$

where $B = 1 - \epsilon^2$. As parabolic cylinder functions $D_0(y)$ and $D_2(y)$ are symmetric to the equator, and $D_1(y)$ is asymmetric to the equator, eqns. (20)-(25) show that the symmetric (asymmetric) perturbation height of the atmosphere only interacts with the symmetric (asymmetric) perturbation height of the ocean.

Equations (20)-(23) can be solved only if the determinant of the coefficients is equal to zero, i.e., if

$$C_0\sigma^4 + C_1\sigma^3 + C_2\sigma^2 + C_3\sigma + C_4 = 0 \tag{26}$$

The coefficients C_0, C_1, C_2, C_3 and C_4 are complicated and are not shown here. It should be noticed that we can not distinguish clearly that the four waves of eqn. (26) are atmospheric Rossby waves or oceanic Rossby waves, because the four frequencies of eqn. (26) are obtained from the coupled model. What we know is that they are coupled Rossby waves of the AOS as $m, n = 0, 2$.

Similarly, from eqns. (24) and (25) we have

$$(\epsilon k^2 + 3)(\epsilon k^2 + 3 - \frac{3}{2}B)\sigma^2 + \epsilon k(2\epsilon k^2 + 6 - \frac{3}{2}B)\sigma + \epsilon^2 k^2 - \frac{3}{4}\gamma A T_a^2 \epsilon^2\left(\frac{3}{2} + \epsilon k^2\right) = 0 \tag{27}$$

Therefore, the dispersion diagrams can be obtained through eqns. (26) and (27) and the coefficients S_0, S_2, A_0, A_2, A_1 and S_1 are determined through eqns. (20)-(25). Furthermore, approximation solutions of ordinary differential eqns. (14) and (15) can be given.

Unstable waves in the coupled AOS

According to frequency eqns. (26) and (27), unstable waves in the coupled AOS can be analysed. If the coefficients of σ^2, σ^1 and σ^0 in eqn. (27) are denoted by a, b and c respectively, we have

$$b^2 - 4ac = \frac{9}{4}\epsilon^2 k^2 B^2 + 3(\epsilon^2 k^4 + 6\epsilon k^2 + 9 - \frac{3}{2}\epsilon k^2 B - \frac{9}{2}B) \times \gamma A T_a^2 \epsilon^2\left(\frac{3}{2} + \epsilon k^2\right) \geqslant 0 \tag{28}$$

Hence, there are no complex roots in eqn. (27), i.e., the asymmetric modes of the AOS in the present paper are always stable. Perhaps this result can help us to understand the cause of which the anomalous fields during ENSO are mainly symmetric about the equator. When $\gamma A = 0$, the two roots of eqn. (27) correspond to oceanic and atmospheric equatorial Rossby waves at $n = 1$ (Matsuno, 1966). By using eqn. (27) it can be deduced that the condition for positive frequencies is

$$\gamma A > \frac{8}{9T_a^2}k^2 \tag{29}$$

When the coupling coefficients γ and A are large enough to satisfy the inequality (29), there is a kind of slow eastward propagating wave in the AOS.

The dispersion relationship of the coupled AOS, according to eqns. (26) and (27), is shown in Fig.1. Figure la represents the real part of roots as $\gamma A = 5\times10^{-9}\ \mathrm{s}^{-2}$ and Fig.1b shows the corresponding imaginary part (amplifying and decaying rates). When the two waves of $m, n=1$ (the dashed lines shown in Fig.1a) are compared with Matsuno's (1966) results, one finds that the higher frequency wave is not modified much by the interactions. The lower frequency component involves slow eastward propagation in the low wave number section, which is consistent with form (29). Among the four waves of $m, n=0,2$ (the solid lines shown in Fig. 1a), the higher frequency one is not modified much by coupling, which shows that interactions between the atmosphere and ocean do not have much influence upon the high frequency waves with periods of one day or less. At $m, n=0,2$, there is also slow eastward propagation in the low wave number section with a period of about three months.

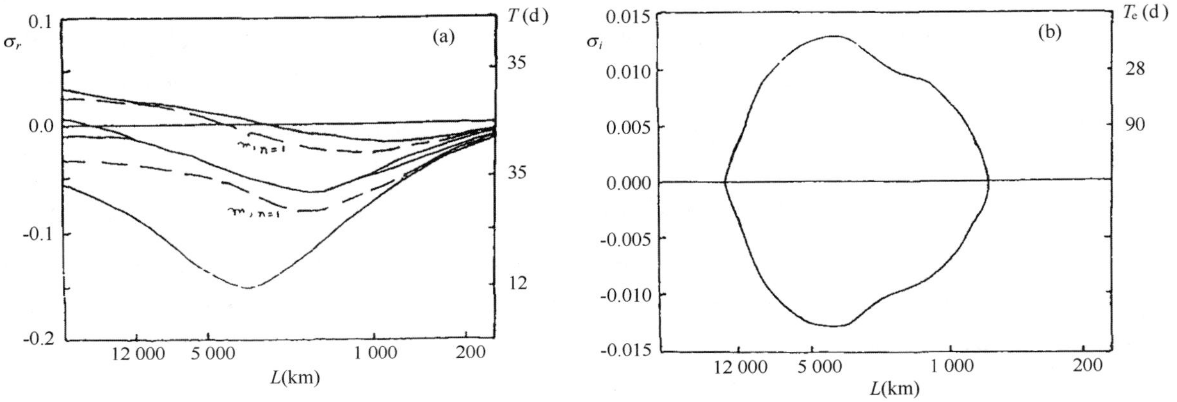

Fig.1 As the coefficient $\gamma A = 5\times10^{-9}\ \mathrm{s}^{-2}$ (a) the dispersion relationship between the real part of wave frequencies and wave lengths and (b) the corresponding amplifying (or decaying) rates. Solid (dashed) lines represent the wave frequencies as $m, n=0,2$ ($m, n=1$)

It is worth pointing out that gravity waves (including Kelvin wave) have been filtered from our atmospheric and oceanic equations. Only westward Rossby waves remain. Therefore, the slow eastward propagating wave obtained here must be the result of coupling within the AOS. Rasmusson (1985) analysed cloudiness, wind components, radiation, SST and other parameters in the equatorial Pacific during 1982/83 ENSO and found that all these anomalies traveled eastward consistently. Though, at present, we can not verify that the eastward propagating wave in the present paper is the physical cause of the eastward movement of anomalies during ENSO, we believe that this wave has some influence upon the propagation of anomalies. Gill (1981) pointed out that the rotation of the earth causes the tropical circulation of the atmosphere to be asymmetric in the west-east direction. Ji (1987) used Gill's (1980) model to investigate the response of atmospheric circulation to steady heating, and found that as the area of heating expands (i.e., as the wave number decreases), the induced atmospheric low pressure center (ALPC) moves eastward gradually. The larger the heating area, the greater the eastward displacement of the ALPC. Figure 2a (from Ji, 1987) shows the distribution of tropical steady heating. Figure 2b (from Ji, 1987) depicts the response of atmospheric circulation to this heating. Comparing the ALPC in Fig.2b with the heating in Fig. 2a, it can be seen that the induced ALPC in Fig. 2b has an apparent eastward displacement. If the feedback process in the AOS is strong enough, the induced ALPC will force sea water to converge and SST to increase. The response of the atmosphere to this eastward moved heating source causes the ALPC to move further eastward. This is the mechanism which produces the eastward wave in the coupled AOS. In the meanwhile, it also helps us to understand the result that in Fig. 1a, the eastward waves appear first in the long wave length section. The eastward propagating wave obtained in the present paper is the comprehensive result of the earth rotation and the coupling between the atmosphere and ocean.

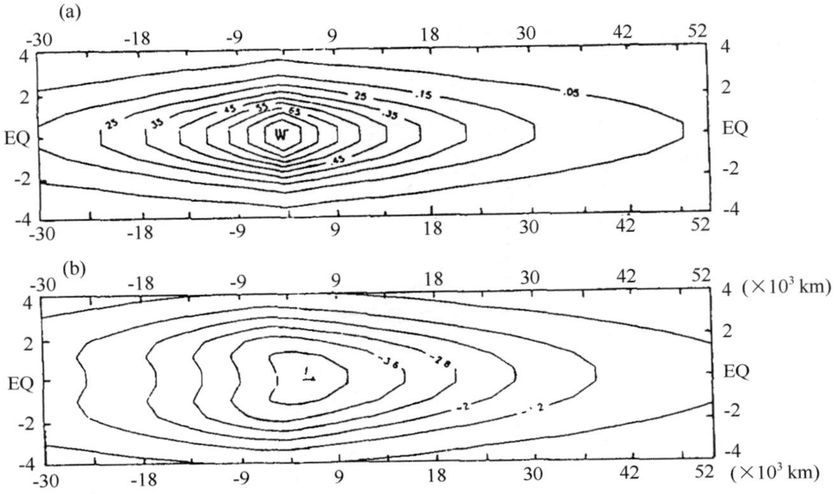

Fig.2 (a) The distribution of tropical steady heating and (b) the response of atmospheric circulation to this heating. Warm (low pressure) center is represented by $W(L)$. (from Ji, 1987)

As $\gamma A \neq 0$, another striking feature of the coupled AOS is the instability. Fig. 1b indicates an e-folding time of about one month. The range of unstable wave lengths is 800–11 000 km and the most unstable wave length is about 4 000 km. Letting $\gamma A = 5\times10^{-8}\ s^{-2}$, which is one order larger than the common value, the ocean heating the atmosphere and the atmosphere forcing the ocean are both very strong. The resulting dispersion relationship is shown in Fig.3. In this case, the waves of the AOS become stable.

Fig. 4 shows the relationship between the product γA of the coupling coefficients γ and A and the unstable wave lengths. The unstable area is hatched. From Fig. 4 it can be seen that the AOS is easy to be unstable at small and large wave lengths. As γA increases, the range of instability decreases. When γA is greater than the critical value of $3.8\times10^{-8}\ s^{-2}$, the AOS becomes stable again. This illustrates that the AOS can be unstable, only if the coupling coefficients γ and A vary within a certain range.

According to the above discussions, the unstable waves obtained here can cause the initial perturbation fields of the AOS to increase rapidly. This kind of instability is the result of coupling within the AOS. We can use this to investigate the trigger mechanisms of ENSO. The relationship between the product γA and unstable wave lengths shown in Fig. 4, indicates the "efficiency" of interactions between the ocean and the atmosphere. If there is no interaction ($\gamma A = 0$), there is no unstable increase of the perturbation fields, When the

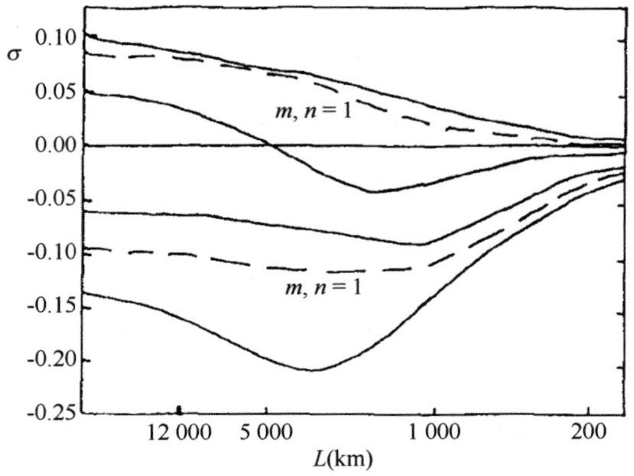

Fig.3 The same as in Fig.1a except for $\gamma A = 5\times10^{-8}\ s^{-2}$

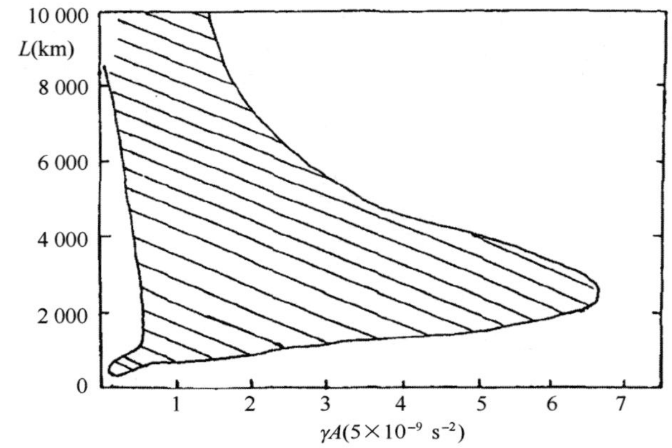

Fig.4 The relationship between coefficient γA and unstable wave lengths. The unstable area is hatched

interactions are efficient (e.g., for the values used by Philander et al., $\gamma = 5\times10^{-7}\ s^{-1}$, $A = 10^{-2}\ s^{-1}$), the atmospheric and oceanic flows cooperate with each other (as discussed below). The perturbation fields can develop rapidly with help of the unstable waves mentioned above, This may represent the developing phase of ENSO. When the interactions are very strong (the values of γ and A are very large), this kind of coupfing will be unfavorable for the development of perturbations (as shown in Fig. 3). If diffusion and other effects are taken into account, it may cause the perturbation fields to decay.

Utilizing these results, the perturbation fields h_s, u_s, v_s, h_a, u_a and v_a of the AOS can be calculated. Fig 5a gives the distribution of oceanic perturbation height h_s in the tropics for $\gamma = 5\times10^{-7}\ s^{-1}$, $A = 10^{-2}\ s^{-1}$, nondimensional frequency $\sigma = -0.035 + i0.013$ and wave length $L = 5\ 000$ km. Figure 5b shows the corresponding perturbation height h_a of the atmosphere. Black arrows indicate the directions of atmospheric winds and oceanic currents sketchily. Comparing Figs. 5a and 5b, it can be seen that the convergence (divergence) of the oceanic currents corresponds to the convergence (divergence) of the wind field. In this way, the AOS can be organized to form positive feedback processes which, in turn, cause the perturbation fields of the AOS to develop even more. By using linear primitive equations, Yamagata (1985) deduced that a necessary condition for the AOS instability was: a positive correlation between atmospheric and oceanic flows. Our conclusion is consistent with Yamagata's. And we give some physical explanations. It is also an illustration that though the present truncated filtered model is a quite simple approximation, it is still able to describe some important features of the large-scale atmosphere-ocean system.

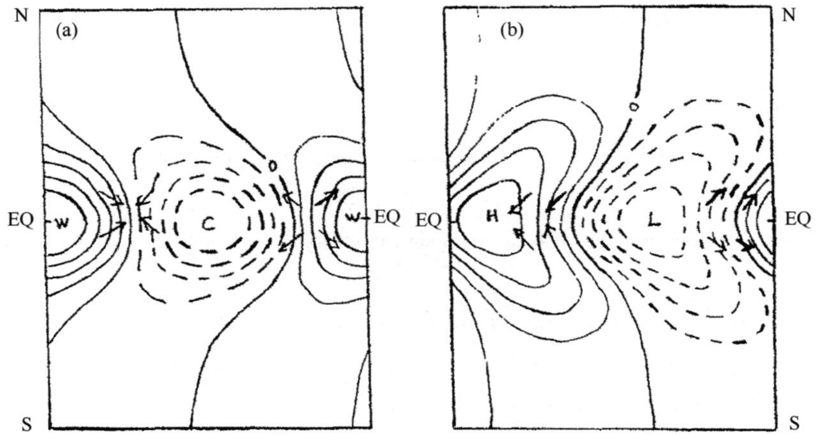

Fig.5 Charts (a) and (b) represent the perturbation height of the model ocean and atmosphere, respectively. Warm and cold center are represented by W and C, and high and low pressure center are represented by H and L, respectively. Black arrows indicate the directions of the wind and current sketchily. Contour intervals are 2.0 for chart (a) and 0.8 for chart (b)

Conclusions

In the present paper, we obtained the following results:

(1) Using time-dependent equations to describe the atmosphere and ocean system, an analytical coupled air-sea interaction model was constructed. Approximate analytical solutions and wave dispersion relationship were obtained by expanding expressions for the perturbation height h_s and h_a in the first three terms of a series of parabolic cylinder functions $D_n(y)$.

(2) Although gravity and Kelvin waves are filtered out and only westward Rossby waves remain in the oceanic and atmospheric equations, a slow eastward propagating wave is produced as the result of coupling air-sea interactions. Its existence and eastward propagating mechanisms, as well as the possible relationship with ENSO, are investigated.

(3) Only the symmetric modes of the AOS can be unstable in the present model. The asymmetric modes are always stable. The mechanisms of instability are discussed by analyzing the perturbation fields of the atmosphere and ocean. Their possible role in ENSO is explored.

(4) To a certain degree, the coupling coefficients γ and A represent the interaction efficiency of the AOS. In different phases of ENSO, the values of γ and A should be different.

The present model is quite simple. Latent heat release, diffusion and nonlinearity are not considered directly. How to determine appropriate values of γ and A is actually an unsolved problem. ENSO is a very complicated nonlinear process, it is impossible to describe its whole evolution with the present simple and linear model. The possible unstable interactions and eastward propagating process of ENSO discussed in this paper need to be verified by observation and further numerical experiments. We are currently developing more comprehensive air-sea interaction models.

Acknowledgment

We would like to thank the anonymous reviewer for his helpful comments and Prof. S.G.H. Philander for reading the manuscript and correcting its English.

References

Anderson, D.L.T. and Gill, A.E. 1975. Spin-up of a stratified ocean, with application to upwelling. Deep-Sea Res., 22: 583-596.

Bjerknes, J. 1966. A possible response of the atmospheric Hadley circulation to equatorial anomalies of ocean temperature. Tellus, 18: 820-829.

Bjerknes, J. 1969. Atmospheric teleconnections from the equatorial Pacific. Mort. Weather Rev., 97: 163-172.

Chao, Ji-Ping and Ji, Zhen-Gang. 1985. On the influences of large-scale inhomogeneity of sea temperature upon the oceanic waves in the tropical regions. Part 1: Linear theoretical analysis. Adv. Atmos. Sci., 2: 295-306.

Gill, A.E. 1980. Some simple solutions for heat-induced tropical circulation. Q.J.R. Meteorol. Soc., 106: 447-462.

Gill, A.E. 1985. Elements of coupled ocean-atmosphere models for the tropics. In.: J.C.J. Nihoul (Editor), Coupled Ocean-Atmosphere Models. Elsevier Oceanogr. Ser., 40. Elsevier, Amsterdam, pp. 303-327.

Ji, Zhen-Gang. 1987. Some researches on large-scale dynamics of the tropical atmosphere and ocean. Diss. Inst. Atmos. Phys. Acad. Sin., 112 pp.

Ji, Zhen-Gang and Chao, Ji-Ping. 1986. On the influences of large-scale inhomogeneity of sea temperature upon the oceanic waves in the tropical regions. Part Ⅱ: Linear numerical experiments. Adv. Atmos. Sci., 3: 238-244.

Ji, Zhen-Gang and Chao. Ji-Ping, 1989. Response of tropical atmospheric circulations to the ocean-land surface heating. Acta Meteorol. Sin., 3: 119-131.

Matsuno, T. 1966. Quasi-geostrophic motions in the equatorial area. J. Meteorol. Soc. Jpn., 44: 25-42.

McCreary, J. 1976. Eastern tropical ocean response to changing wind systems: With application to El Niho. J. Phys. Oceanogr., 6: 632-645.

Philander, S.G.H. 1981. The response of equatorial oceans to a relaxation of the trade winds. J. Phys. Oceanogr., 11: 176-189.

Philander, S.G.H., Yamagata, T. and Pacanowski, R.C., 1984. Unstable air-sea interactions in the tropics. J. Atmos. Sci., 41: 604-613.

Rasmusson, E.M. 1985. In: WMO, Scientific plan for the tropical atmosphere and global atmospheric programme. WMO/TD-No. 64, 1985.

Webster, P.J., 1972. Response of the tropical atmosphere to local steady forcing. Mon. Weather Rev., 100, 518-541.

Wyrtki, K. 1975. El Niño—the dynamic response of the equatorial Pacific Ocean to atmospheric forcing. J. Phys. Oceanogr., 5:572-584.

Yamagata, T. 1985. Stability of a simple air-sea coupled model in the tropics. In: J.C.J. Nihoul (Editor), Coupled Ocean-Atmosphere Models. Elsevier Oceanogr. Set. 40. Elsevier, Amsterdam, pp. 637-657.

Numerical Experiments on the Tropical Air-sea Interaction Waves*

Zhang Renhe(张人禾)[1]　Chao Jiping(巢纪平)[2]

(1. Institute of Atmospheric Physics, Academia Sinica, Beijing 100080; 2. National Research Center for Marine Environmental Forecasts, State Oceanic Administration, Beijing 100081)

ABSTRACT: By means of the numerical method, the tropical air-sea interaction waves are studied. The results show that when the Kelvin waves are filtered out and only the equatorial Rossby waves are reserved both in the atmosphere and in the ocean, the disturbances can also propagate eastward because of the air-sea interaction. The critical wavelength of the eastward propagating waves is related to the intensity of the air-sea interaction. The stronger the air-sea interaction, the larger the eastward propagating components of the air-sea interaction waves. The results of the numerical experiments are in good agreement with those of the theoretical analysis (Chao and Zhang, 1988).

Key words: equatorial Rossby waves, Kelvin waves, tropical air-sea interaction waves

I　Introduction

Philander et al. (1984) and Yamagata (1985) have discussed the unstable air-sea interaction in the tropics. Their calculating results showed that the disturbances can propagate eastward because of the air-sea interaction, and they believed that the eastward propagating disturbances were caused by the Kelvin waves after the air-sea interaction. Chao and Zhang (1988) pointed out that there also exist unstable air-sea interaction waves which propagate eastward by means of a simple tropical air-sea coupled model, in which the Kelvin waves are filtered out in both the atmosphere and oceans, and the roles of the tropical air-sea interaction waves in the ENSO events were discussed. In Chao and Zhang's work, the truncating model method was used in the process of solution, which would undoubtedly affect the results. Here we will solve the tropical air-sea interaction model by using the numerical method instead of the truncating model method, to further study the properties of tropical air-sea interaction waves.

* Acta Meteorologica Sinica, 1992, 6(2): 148-158.

II Numerical Model

1. Model Equations

Chao and Zhang (1988) have described the model equations in detail in both the atmosphere and oceans, so we will not reiterate them here. According to Chao and Zhang, the nondimensional equations on the equatorial β-plane in the atmosphere and oceans are given respectively as follows:

$$\left(\frac{\partial^2}{\partial x^2}+\frac{\partial^2}{\partial y^2}-\frac{y^2}{4}\right)\frac{\partial h_a}{\partial t}+\frac{1}{2\varepsilon^{1/2}}\frac{\partial h_a}{\partial x}=\frac{\varepsilon AT}{4}y^2 h_s \tag{1}$$

$$\left[\varepsilon\left(\frac{\partial^2}{\partial x^2}+\frac{\partial^2}{\partial y^2}\right)-\frac{1}{\varepsilon}\frac{y^2}{4}\right]\frac{\partial h_s}{\partial t}+\frac{\varepsilon^{1/2}}{2}\frac{\partial h_s}{\partial x}=\frac{\gamma T}{D}\left(\frac{\partial^2}{\partial x^2}+\frac{\partial^2}{\partial y^2}\right)h_a \tag{2}$$

where t, x and y are time, zonal and meridional coordinates respectively; h_a and h_s are disturbed height fields in the atmosphere and oceans respectively; $\varepsilon=C_s/C_a$, $C_s=(g\Delta\overline{T}_z D\alpha)^{1/2}$, $C_a=(gH_a)^{1/2}$, where C_a and C_s are gravity velocities in the atmosphere and oceans respectively; $\Delta\overline{T}_z$, α, D and H_a are the difference of climatological sea temperature between the sea surface and the thermocline, thermal expansion coefficient of sea water, depth of mixed layer and equivalent height in the air, respectively; A and γ are the affecting coefficients of sea to air and air to sea respectively (see Chao and Zhang, 1988); $T=(2\beta C_s)^{-1/2}$, the typical time scale of the motion in the sea. Here we take $\Delta\overline{T}_z=8.0\text{K}$, $D=100$ m, $H_a=400$ m and $\alpha=3.413\times 10^{-4}\ \text{K}^{-1}$.

2. The Method of Solution

Rectangular areas are taken both in the atmosphere and oceans. The nondimensional distance in the zonal direction is taken from −6.45 to 6.45, which corresponds to the dimensional distance from -0.75×10^4 km to 0.75×10^4 km; while nondimensional distance in the meridional direction is taken from −2.0 to 2.0, which corresponds to the dimensional distance from -2.35×10^3 km to 2.35×10^3 km. Grid points in the zonal and meridional directions are taken to be 129 and 41 respectively, with the intervals being nondimensional distance 0.1 for both zonal and meridional directions, corresponding to the dimensional distance 117 km, e.g., $\Delta x=\Delta y=d=0.1$. For $\partial h_a/\partial t$, and $\partial h_s/\partial t$, Eqs. (1) and (2) are two elliptic equations, which can be solved by the Gauss-Seidel iterative method with overrelaxation. Setting $Z_a=\partial h_a/\partial t$ and $Z_s=\partial h_s/\partial t$, then we have

$$Z_{aij}^{k,v+1}=Z_{aij}^{k,v}+\frac{\alpha_a}{\mu_{aj}}\left[Z_{ai+1j}^{k,v}+Z_{ai-1j}^{k,v+1}+Z_{aij+1}^{k,v}+Z_{aij-1}^{k,v+1}-\mu_{aj}^2 Z_{aij}^{k,v}-F_{aij}^k\right] \tag{3}$$

$$Z_{sij}^{k,v+1}=Z_{sij}^{k,v}+\frac{\alpha_s}{\mu_{sj}}\left[Z_{si+1j}^{k,v}+Z_{si-1j}^{k,v+1}+Z_{sij+1}^{k,v}+Z_{sij-1}^{k,v+1}-\mu_{sj}^2 Z_{sij}^{k,v}-F_{sij}^k\right] \tag{4}$$

$$(i=1,2,\cdots,129;j=1,2,\cdots,41)$$

where

$$\mu_{aj}^2=4+\frac{d^2}{4}[(j-21)d]^2$$

$$\mu_{sj}^2 = 4 + \frac{d^2}{4\varepsilon^2}[(j-21)d]^2$$

$$F_{aij}^k = \frac{\varepsilon ATd^2}{4}[(j-21)d]^2 - \frac{d}{4\varepsilon^{1/2}}(h_{ai+1j}^k - h_{ai-1j}^k)$$

$$F_{sij}^k = \frac{\gamma T}{\varepsilon D}(h_{ai+1j}^k + h_{ai-1j}^k + h_{aij+1}^k + h_{aij-1}^k - 4h_{aij}^k) - \frac{d^2}{4\varepsilon^2}(h_{sj+1j}^k - h_{si-1j}^k)$$

α_a and α_s are coefficients of overrelaxation. In order to get the fastest convergent rate in the iterative process, according to actual calculating test, the α_a and α_s are taken to be 1.86 and 1.70 respectively. k is the number of time step and v is the time of the iteration.

For the time integration, frog-leap scheme is used, that is,

$$h_{ai,j}^{k+1} = h_{ai,j}^{k-1} + 2\Delta t_a Z_{ai,j}^k \qquad (5)$$

$$h_{si,j}^{k+1} = h_{si,j}^{k-1} + 2\Delta t_s Z_{si,j}^k \qquad (6)$$

$$(i = 1,2,\cdots,129; j = 1,2,\cdots,41)$$

Δt_a and Δt_s are the time steps of atmospheric equation (1) and oceanic equation (2), respectively. Because the velocity of gravity wave in the atmosphere (C_a) is much larger than that in oceans (C_s), we take $\Delta t_a = 0.05$ and $\Delta t_s = 0.5$, which correspond to dimensional time 1.6 h and 16 h respectively. During the time integration, we integrate the oceanic equation for one step and then integrate the atmospheric equation for ten steps. This integrating method means that the process of the air-sea interaction is taken approximately as follows: After the heating of the atmosphere by oceans for 16 h, the wind stress of the atmosphere then acts on the ocean. The ocean is integrated for 16 h under the action of this wind stress, then a new heating field is given to the atmosphere by the ocean.

For frog-leap schemes (5) and (6), Matsuno scheme (Matsuno, 1966) is used for the beginning of the numerical calculating, that is,

$$\begin{cases} h_{ai,j}^{*2} = h_{ai,j}^1 + \Delta t_a Z_{ai,j}^1, \\ h_{ai,j}^2 = h_{ai,j}^1 + \Delta t_a Z_{ai,j}^{*2} \end{cases} \qquad (7)$$

$$\begin{cases} h_{si,j}^{*2} = h_{si,j}^1 + \Delta t_s Z_{si,j}^1, \\ h_{si,j}^2 = h_{si,j}^1 + \Delta t_s Z_{si,j}^{*2} \end{cases} \qquad (8)$$

$$(i = 1,2,\cdots,129; j = 1,2,\cdots,41)$$

where superscripts 1 and 2 are the time steps; $Z_{ai,j}^{*2}$ and $Z_{si,j}^{*2}$ are calculated by $h_{ai,j}^{*2}$ and $h_{si,j}^{*2}$, respectively.

Boundary conditions are taken to be zero, i.e., when $x = \pm 6.45$ and $y = \pm 2.0$, set $h_a = h_s = 0$. In the grid points, we have

$$\begin{cases} h_{a1,j} = h_{a129,j} = h_{s1,j} = h_{s129,j} = 0 \\ h_{ai,1} = h_{ai,41} = h_{si,1} = h_{si,41} = 0 \end{cases} \qquad (9)$$

$$(i = 1,2,\cdots,129; j = 1,2,\cdots,41)$$

Ⅲ　Calculating Results

1. The Basic Solution without the Air-Sea Interaction

For the purpose of the physical comparison and also for examining the stability of the calculating scheme, first of all, we numerically solve the tropical atmospheric equation (1) and oceanic equation (2) without considering the air-sea interaction. Setting $A=\gamma=0$ in Eqs. (1) and (2), then we have

$$\left(\frac{\partial^2}{\partial x^2}+\frac{\partial^2}{\partial y^2}-\frac{y^2}{4}\right)\frac{\partial h_a}{\partial t}+\frac{1}{2\varepsilon^{1/2}}\frac{\partial h_a}{\partial x}=0 \tag{10}$$

$$\left[\varepsilon\left(\frac{\partial^2}{\partial x^2}+\frac{\partial^2}{\partial y^2}\right)-\frac{1}{\varepsilon}\frac{y^2}{4}\right]\frac{\partial h_s}{\partial t}+\frac{\varepsilon^{1/2}}{2}\frac{\partial h_s}{\partial x}=0 \tag{11}$$

According to the theoretical analysis (Chao and Zhang, 1988), now in the atmosphere and in oceans are the equatorial Rossby waves which propagate westward. At $t=0$, the initial disturbed height field in the atmosphere and oceans are given as below

$$h_s^0=h_a^0=\exp[-2(x^2+y^2)]. \tag{12}$$

Fig.1 shows the evolutions of the disturbed height fields h_a and h_s in the atmosphere and oceans along the equator. Because only the equatorial Rossby waves can be found in both the media, then the disturbed height fields are all moving westward. The westward propagating velocity of the disturbed height field in the air is about −18.7 m/s and that in the sea is about −1.3 m/s, which is much smaller than that in the air. The oscillation of the disturbed height field in the air is of high frequency with the period about 4.5 d, while that in the sea is of slowly changing, with the period about 48 d. Fig. 2 shows the initial and the 26.7th day disturbed fields in the sea. We can see that the disturbed field goes westward and at 26.7 d, the structure of disturbed field is much different from the initial field.

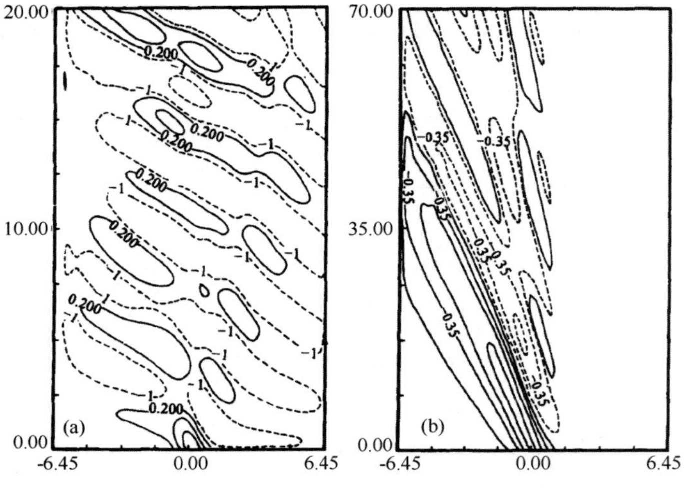

Fig.1 The evolution of the disturbed height field in the atmosphere (a) and that in the ocean (b) along the equator (the ordinate denoting nondimensional time with 1 corresponding to 32 h, solid lines and dotted lines standing for positive and negative disturbed height respectively)

2. Solutions to General Cases

At the initial time, we take the initial field of Eq. (12) at $t=0$ in the sea (see Fin.2a), that is, there only exists the initial disturbed field in the sea, while nothing exists in the air, which means that at this time the sea gives a heating field to the air (see Chao and Zhang, 1988). With $A=10^{-2}$ s^{-1} and $\gamma=5.0\times10^{-5}$ m/s. Fig.3 shows the disturbed fields in the atmosphere and oceans respectively at time 26.7 d. Compared with the

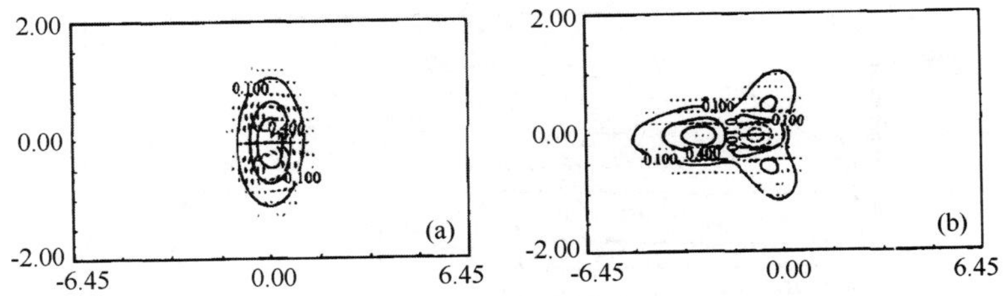

Fig.2　The initial disturbed field (a) and the disturbed field at 26.7 d (b) in the sea (solid and dotted lines being the same as in Fig.1, arrows representing the ocean current field)

case without the air-sea interaction (see Fig.2b), it can be seen that there are great changes in the structure of disturbed field between two cases. When the air-sea interaction is cut off, in the ocean the disturbed field near the equator has a positive disturbed height area in the western part and a negative one in the eastern part, while to the higher latitudes both in the north and south directions of the negative disturbed height area there exists a positive disturbed height area respectively; When the air-sea interaction is taken into account, in the ocean the phase of the disturbed height field in the west-east direction near the equator is just opposite to that without the air-sea interaction. Namely, there is a positive disturbed height area in the eastern part and a negative one in the western part, while the positive disturbed height areas to the south and north of the negative disturbed height area in the higher latitudes disappear. There indicate that the structure of equatorial Rossby waves in the ocean is changed by the air-sea interaction.

Fig.4 shows the evolution of the disturbed height fields in the atmosphere and oceans along the equator. It can be seen that either in the atmosphere or in the ocean there exist two kinds of disturbances, one is of high frequency and the other is of low frequency. For the low-frequency disturbances, they go westward with a velocity about −0.73 m/s, which is smaller than that in the ocean without the air-sea interaction. From Figs.3 and 4, we can also see that the low-frequency disturbances in the atmosphere and ocenas are mainly negatively correlated. Fig.4 indicates that the amplitudes of the low-frequency disturbances in both the atmosphere and oceans decay with time, which is in good agreement with the results of the theoretical analysis (Chan and Zhang, 1988).

Fig.4 also shows that for the high-frequency disturbance in the tropical air-sea coupled system, the period is about 4.5 d, which is very close to that in the atmosphere without the air-sea interaction. From the evolution of the disturbed height field in the atmosphere along the equator (Fig.4a), we can see that the high-frequency disturbances propagate westward. In the theoretical analysis (Chao and Zhang, 1988), we have already known that the high-frequency tropical air-sea interaction waves are very close to the equatorial Rossby waves in the atmosphere, which is also proved by the numerical calculating results here.

Fig.5 shows the evolution of the disturbed height fields in the atmosphere and oceans along the equator when the air-sea interaction is intensified. Setting $A\gamma=2.0\times10^{-6}$ m/s, from Fig.5a we can see that the westward propagating component of the low-frequency disturbances reduces, while the eastward propagating component of the low-frequency disturbances enlarges. When the air-sea interaction is further intensified, e.g., setting $A\gamma=4.5\times10^{-6}$ m/s, Fig.5b shows that the westward propagating component of the low-frequency disturbances disappears. Now all the low-frequency disturbances both in the atmosphere and in the ocean move

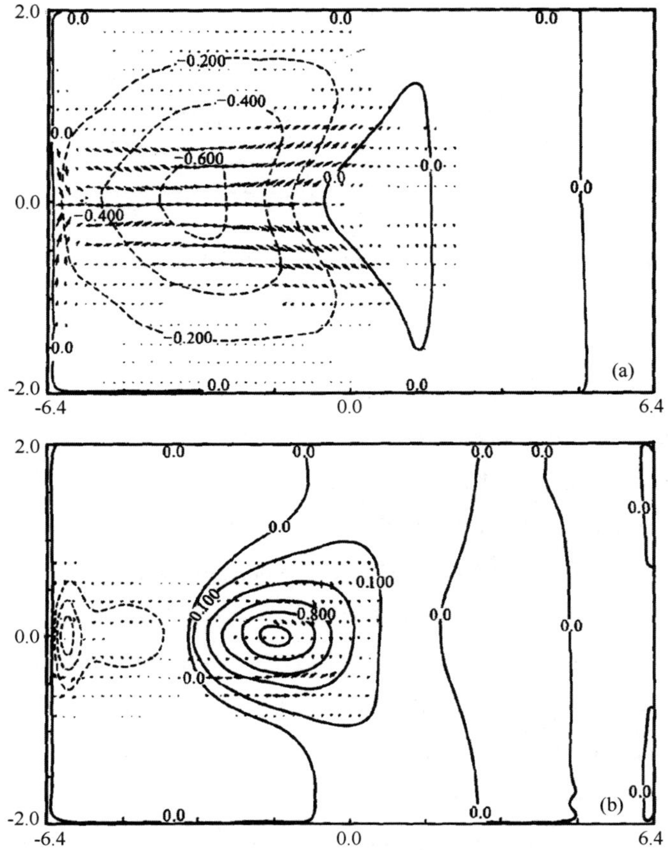

Fig.3 The disturbed fields in the atmosphere (a) and oceans (b) at time 26.7 d (explanations as Fig.2, but the arrows in the upper part standing for the wind field)

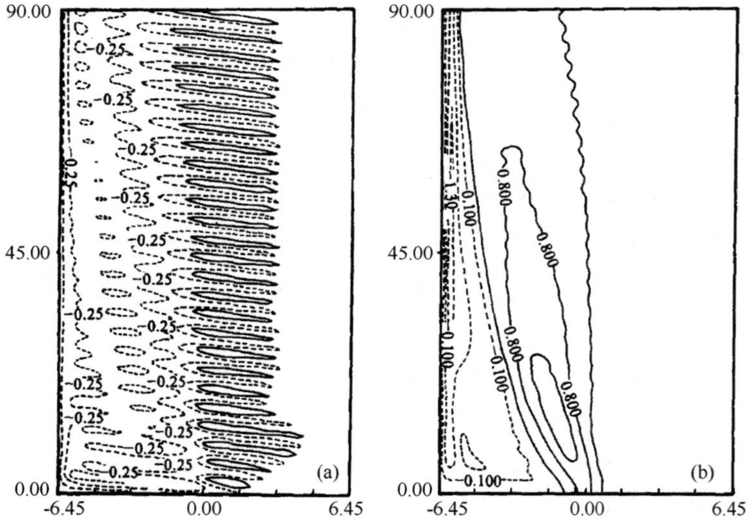

Fig.4 The evolution of the disturbed height fields in the atmosphere (a) and that in the ocean (b) along the equator (explanations as Fig.1)

eastward. Here we can see that the stronger the air-sea interaction, the larger the eastward propagating component of the low-frequency disturbances in the air-sea coupled system. Fig.5 also indicates that when the air-sea interaction is intensified, the low-frequency disturbances propagate eastward in the form of a wave train. In the process of propagating, the disturbances in the eastern part is decaying, while those in the western part is intensifying. Thus we can suppose that although the phase speed of the low-frequency disturbances is propagating eastward, the energy is dispersing westward. In addition, Fig.5 shows that the intensification of the air-sea interaction has little effects on the high-frequency disturbances. Their periods are also about 4. 5 d and from the evolution of the disturbed height field in the atmosphere along the equator, we can see that the high-frequency disturbances also propagate westward.

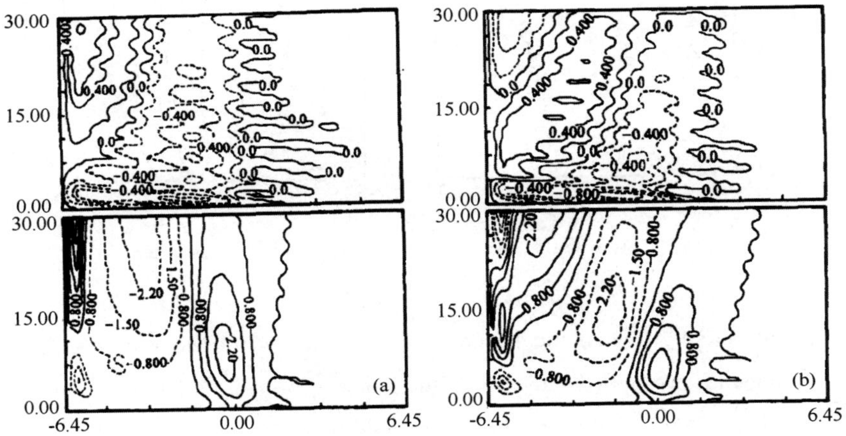

Fig.5　The evolution of the disturbed height fields in the atmosphere (upper part) and that in the ocean (lower part) along the equator. (a) and (b) correspond to $A\gamma = 2.0\times10^{-6}$ m/s and 4.5×10^{-6} m/s, respectively (explanations as Fig.1)

3. Effect of the Initial Fields on the Tropical Air-Sea Interaction Waves

According to Chao and Zhang (1988), the eastward propagating waves appear in the scope of comparatively long wavelengths. In order to prove this, set $A\gamma = 0.5\times10^{-6}$ m/s and take the initial field as follows

$$h_s'^0 = \exp[-0.4x^2 - 2y^2]. \tag{13}$$

Compared with Eq.(12), Eq. (13) is equivalent to enlarging the scope of the initial field in the zonal direction. $h_s'^0$ in Eq. (13) and h_s^0 in Eq. (12) correspond to e-fold scales in zonal direction about 1850 km and about 827 km, respectively, i.e., the initial field (13) gives prominence to the effect of the long wave components in fact. In order to reduce the effect of the short wave components, we enlarge the intervals between the grid points in the zonal direction by taking $\Delta x = 0.5$, corresponding to dimensional distance about 585 km.

Fig.6 shows that when the initial field (13) is used, the evolution of the disturbed height fields in the atmosphere and oceans along the equator. We can see that the disturbed height fields in both the atmosphere and oceans move toward eastward simultaneously. Thus, we know that if the initial field is composed of the waves with comparatively long wavelengths, the westward propagating disturbances (see Fig.4) can become eastward propagating. From the above we can see that the eastward propagating waves mainly appear in the scope of comparatively long wavelengths, which is consistent with the results of the theoretical analysis (Chao and Zhang, 1988).

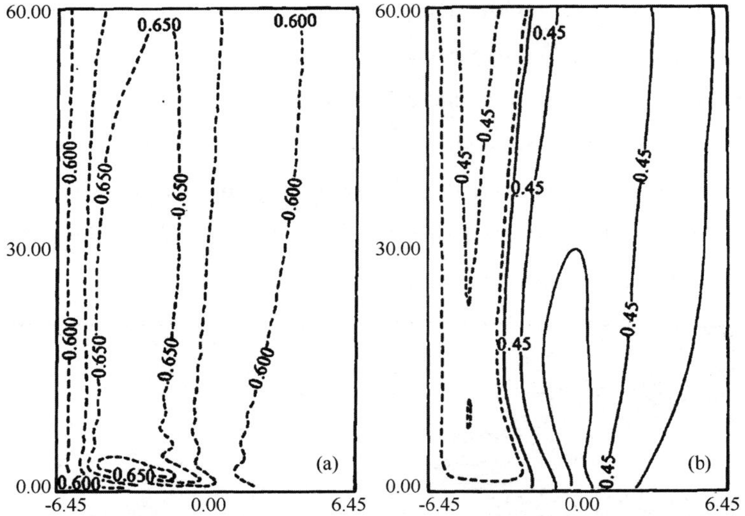

Fig.6 As in Fig.4, except that the initial field $h_s'^0$, is taken

We have known the situation when the initial field is composed of the waves with comparatively long wavelenghts. In the following, we will discuss the situation when the initial field is composed of the waves with comparatively short wavelengths. Setting the initial field as

$$h''^0_s = \exp(-200x^2 - 2y^2), \tag{14}$$

it can be seen that Eq. (14) has much less scope in the zonal direction than that of Eq. (12). The e-fold scale that h''^0_s corresponds to is about 83 km. In order to take more short wave components into account, in calculating we reduce the intervals between the grid points in the zonal direction by taking $\Delta x = 0.01$, corresponding to the dimensional distance about 11.7 km.

Fig.7 shows that when the initial field h''^0_s is taken, the evolution of the disturbed height fields in the atmosphere and oceans along the equator. From Fig.7 we can see that although the phase speed of the disturbed height fields propagate westward, the energy of the disturbances disperses eastward. During the process of the eastward dispersing of the energy, the amplitudes of the disturbances are unstably amplified with the increasing of the time, meaning that the energy of the short wave component in the tropical air-sea interaction waves has the characteristics of unstably eastward dispersing.

Ⅳ Conclusions

According to the analysis above, we can see that when the equatorial Rossby waves in the atmosphere and oceans are coupled together, the coupled waves have the eastward propagating components, which are different from the eastward propagating Kelvin waves. The structure of the tropical air-sea interaction waves is different from that of the equatorial Rossby waves. The propagating direction of the tropical air-sea interaction waves is governed by the intensity of the air-sea interaction. In the general case, the tropical air-sea interaction waves propagate westward. But when the air-sea interaction is intensified, the propagating direction of

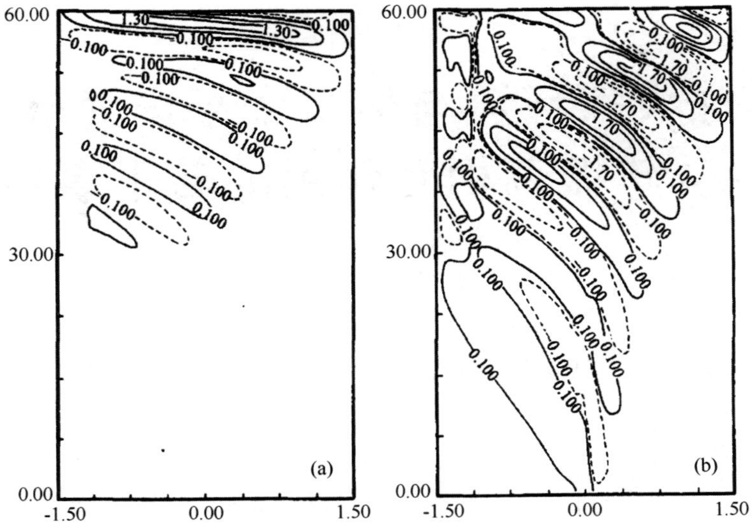

Fig.7　As in Fig.4, except that the initial field h''^0_s is taken

the tropical air-sea interaction waves can become eastward, and the eastward propagating air-sea interaction waves mainly appear in the scope of comparatively long wavelengths. For the eastward propagating tropical air-sea interaction waves, though the phase speed is eastward, the energy disperses westward. And for the short wave components of the tropical air-sea interaction waves, the phase speed is westward, while the energy unstably disperses eastward.

The results of the numerical experiments also indicate that the air-sea interaction has little effects on the propagating direction and the period of the high-frequency tropical air-sea interaction waves, which are always very close to the equatorial Rossby waves in the atmosphere. Thus we can see that in the long period process of the tropical air-sea coupled system, the equatorial Rossby waves in the ocean play a very important part.

For the convenience of theoretical analysis, truncating model method was used by Chao and Zhang (1988). By using the numerical method to solve Eqs.(1) and (2) directly, the same results as Chao and Zhang (1988) are obtained, which further demonstrate the importance of the air-sea interaction waves in the tropical air-sea coupled system.

REFERENCES

Chao Jiping and Zhang Renhe. 1988. The air-sea interaction waves in the tropics and their instabilities. *Acta Meteorologica Sinica*, 2:275-287.

Matsuno, T. 1966. Numerical integration of the primitive equations by a simulated backward difference method. *J. Meteor. Soc. Japan*, 44:76-84.

Philander, S. G. H., Yamagata, T. and Pacanowski, R. C. 1984. Unstable air-sea interaction in the tropics. *J. Atmos. Sci*, 41:604-613.

Yamagata, T. 1985. Stability of a simple air-sea coupled model in the tropics. *Coupled Ocean-Atmosphere Model*, ed. J. C. J. Nihoul, Amsterdam, 767.

北半球大气环流与热带太平洋海表温度 3~4 年振荡相互作用的若干事实*

巢纪平[1] 王彰贵[2]

(1. 国家海洋环境预报研究中心,100081;2. 中国科学院大气物理研究所)

摘要:本文应用统计学上的带通滤波方法,对 COADS 资料集中的热带海表温度资料以及大气环流资料(500 hPa 高度场)进行了处理,使处理后的资料只保留 3~4 年周期变化的信息。由此计算了海温和高度场之间的时滞相关场,并分析了相关系数的时间经度和时间纬度剖面图,从而对东、中和西太平洋热带海表温度与大气环流之间的相互作用,揭示了一些有意义的现象。作者认为大尺度海气相互作用是复杂的,它依赖于信号的时间尺度以及海洋加热(或冷却)的地理位置

1 引言

许多研究表明,海洋中以赤道东太平洋(EEP)的增温为表征的 El Niño 现象和以 Walker[1] 早期发现的以 Darwin(12°S,130°E)和 Tahiti(17°S,150°W)的海平面气压差为表征的南方涛动(Southern Oscillation)对全球的气候异常有着重要的影响。Bjerknes[2,3] 指出,El Niño 和南方涛动之间存在着内在的联系,实际上是热带大尺度海气相互作用的结果。近年来,不少气象学家把这两种分别发生在海洋和大气中的现象联合起来称之为 ENSO。当前,ENSO 已经成为世界性海洋和气象学家研究的一个重要课题。

Pan 和 Oort[4]、Oort 和 Pan[5] 应用不同时间尺度的资料,详细地分析了 EEP(5°N 至 5°S,130°W)的海表温度与全球大气物理参数之间的统计关系,得到了一些十分有意义的结果。然而,由于西太平洋海洋资料的缺乏,过去很少有人研究过热带西太平洋海表温度与全球气候变化之间的关系,对热带不同经度上的海表温度与大气环流之间的相互作用及其性质上有哪些共同点和差异点,也很少进行过详细的比较。在另一方面,很多的研究(包括参考文献[4,5])都是以实际的热带海表温度作为研究对象的。然而,我们对赤道东太平洋海表温度进行了能谱分析,发现除年变化外,在能谱上存在两个不同周期的峰值,一个周期为 2.3 年,另外一个是 3~4 年(图 1)。因此,我们有理由认为,不同时间尺度的热带海表温度振荡,它们与大气环流之间的相互作用在物理过程以及造成短期气候变化上可能是不同的。因此在参考文献[6]我们采用了统计学上的带通滤波技术,使得海温资料中只包含 2 年左右的周期,并研究它们与北半球冬季大气环流之间的统计关系,由此发现了一些很有意义的现象。虽然这些现象在物理学和动力学方面尚未给出解释,但作为揭示现象和事实来讲,这些结果是值得参

* 气象学报,1992,50(3):372-378.

考的,也可以作为理论研究的参考。为了与参考文献[6]2 年振荡结果比较,在本文中我们仍应用带通滤波方法,将海温和大气资料(500 hPa)高度场进行处理,使其只保留 3~4 年的振荡信息,以分析它们之间的关系。统计上的处理和分析方法仍同参考文献[6],在此不再重述,只给出分析结果,但 500 hPa高度场是全年的,而不是冬季。

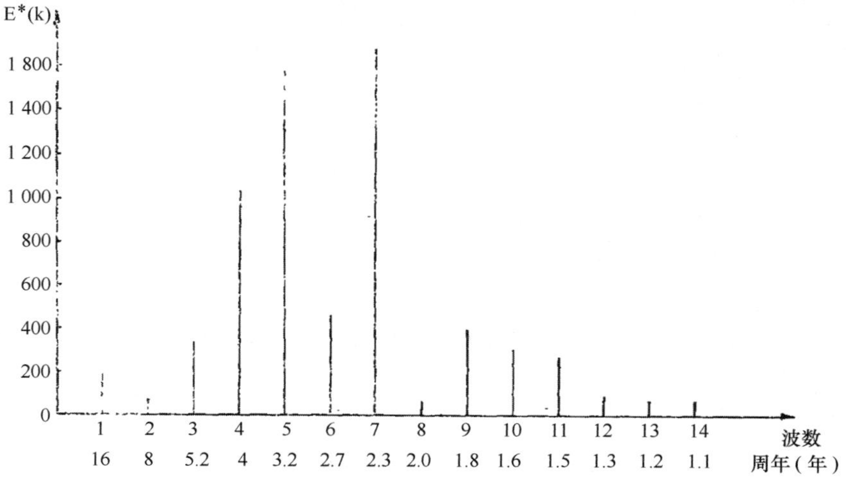

图 1　赤道东太平洋海表温度能谱(1°S,130°W)

2　东太平洋海温变化与北半球 500 hPa 高度场之间的关系

在图 2(a)和图 2(b)中给出了赤道东太平洋(1°S,130°W)海表温度与北半球 500 hPa 高度场之间的统计相关。图中纵坐标是时间 τ,当 $\tau<0$ 时表示海温超前于高度场,相反 $\tau>0$ 时表示高度场超前于海温场。这种时滞相关分析,既反映了海温对大气的影响,也反映了大气对海温的影响。为了使结果易于分析,虽然我们计算了一系列相关场,但在图 2 中只给出中纬度(20°—40°N)和中高纬度(40°—70°N)的经度时间剖面图。剖面图的作法同参考文献[6]。从中纬度的情况看(图 2(a)),相关区沿纬圈的演变可以分为两种类型。从中太平洋直到大西洋是振荡型,而从大西洋的东岸通过欧亚大陆直到中太平洋,虽然在时间演变上相关区也是呈正-负相间的振荡形式,但出现了向东的演变。平均来讲,约 6 个季节东移了 240 个经度。同时这两个不同的区域,相关系数沿纬圈的分布是相反的,即在东移区 90°E 上,当海温场超前或落后高度场多时,它们之间是负相关,而正相关出现在同期附近。但在另一个振荡区,相关场的演变几乎完全是相反的。这意味着在这两个不同区域,海气之间相互作用的物理过程可能是不同的。

在中高纬度(40°—70°N)(图(2)b),情况与中纬度的不同。从东太平洋通过美洲大陆到大西洋仍然和中纬度的情况相似,在时间上相关区呈振荡型,但沿纬圈的波长要短些。然而从欧亚大陆到中西太平洋,相关区明显地呈自东向西移动或传播。

把上述这种现象与 Yasumari 和 Krishnamurti[7] 对赤道高、低空风场的时间经度演变的结果相比较,现象很不相同(图 3a),但与我们赤道地区(10°S 至 10°N)的计算结果相一致,即在印度洋和大西洋,相关区随时间的演变呈振荡型,只有在太平洋地区才出现向东传播(图 3b)。这表明不同纬度带,大气对热带海洋加热的响应,或反过来大气对热带海温变化的作用是不同的。可能也正说明大尺度海气相互作用以及它们之间的遥相关具有质的多样性,是很复杂的,很难用一个模型来概括。也说明

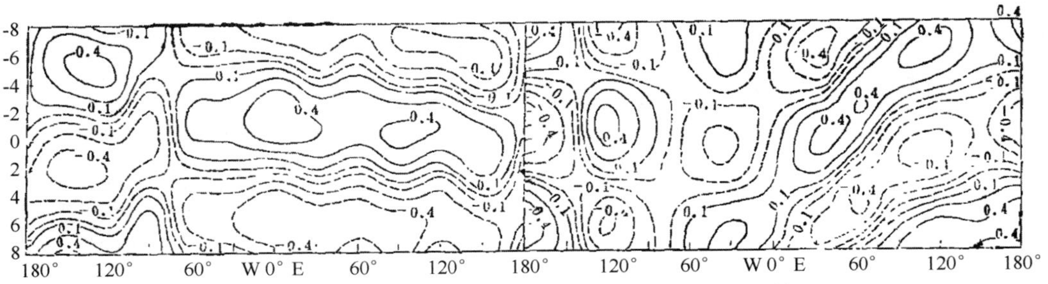

(a) (b)

图 2 相关系数的经度时间剖面

(海温参考点位于(1°S,130°W);a. 中纬度(20°—40°N);b. 中高纬度(40°—70°N);实(虚)线表示正(负)相关,相关系数≥0.32,置信度超过 1%;时间单位为季)

大尺度海气相互作用的研究处于开始阶段,很难说问题已经基本清楚。正是因为这样,El Niño 现象的长期预报是一个较为困难的问题。

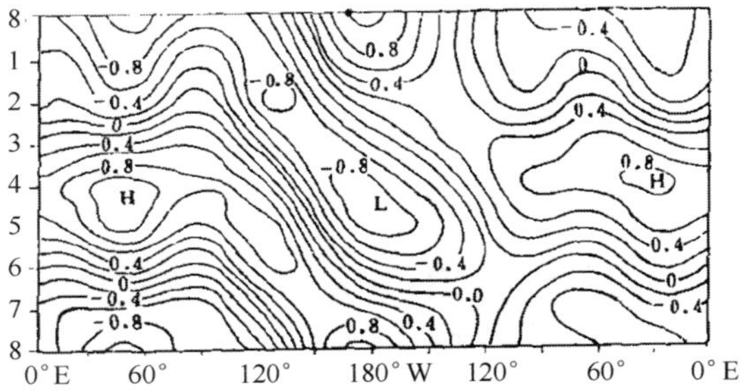

图 3a 赤道纬带上(10°S 至 10°N)纬向风的时间经度剖面

(200 hPa,时间尺度为 5~7 个月[7])

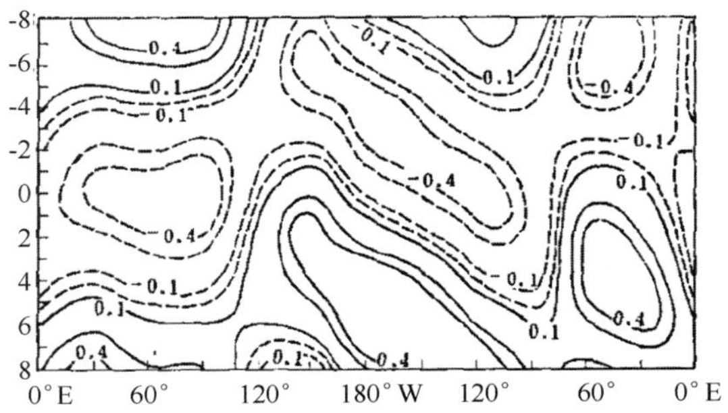

图 3b 东太平洋(1°S,130°W)海表温度与赤道地区(10°S 至 10°N)500 hPa 纬向风之间时滞相关系数的时间经度剖面(时间尺度为季)

与参考文献[6]中的图 4 和图 5 相比较,在中高纬度地区,3~4 年和 2 年时间尺度的结果正好相反。这表明海气相互作用具有时间尺度的特征。最近巢纪平等对赤道太平洋海表温度进行了分析,发现 2 年和 3~4 年时间尺度的海温向相反方向传播。实际上在 ENSO 事件中,海温有东传和西传两种现象[8,9]。用时间尺度不同来解释,这有待于进一步研究。

转向对不同经度上相关系数时间纬度演变的剖面图,即研究它们南北方向的变化。

为便于比较，图 4 给出了相关区经向输送的示意图。图中的箭号表示相关区的传播方向，Ⅰ、Ⅱ表示相关系数分布类型，$Ⅰ_A$、$Ⅱ_A$ 分别表示与Ⅰ、Ⅱ相反的分布类型。

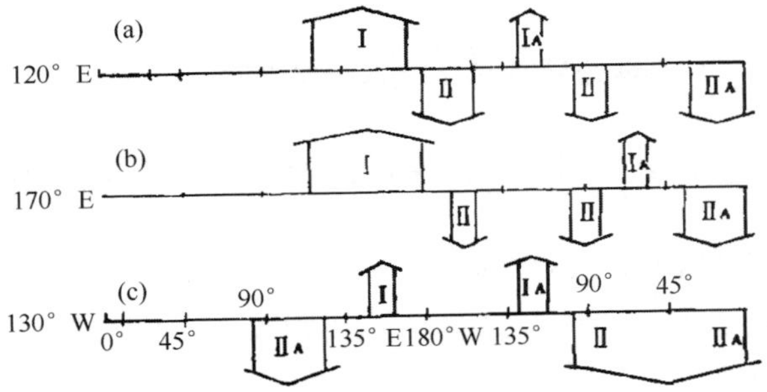

图 4　相关区经向传播的比较示意图

（⇧（⇩）表示相关区从低（高）纬向高（低）纬传播）

由图 4(c) 可以看到，相关区从低纬向高纬传播主要出现在 150°E 至 120°W 经度上，而中太平洋地区传播较弱（图中未表示）。相关区向低纬传播区为 80°—120°E 和 0°—100°W。为叙述方便，相关区在南北方向的传播宽度（指经度范围）简称为传播区。各主要传播区上的时间纬度剖面见图 5。在美洲西海岸（图 5(a)），当气压场和海温场接近同期前后，低纬为负相关，高纬为正相关。当海温远超前或落后于气压场时，相关的符号反了过来，这种分布型记为Ⅱ（图 4(c)）。到了大西洋，尽管信号的传播方向与美洲西海岸相同，但相关系数的分布类型相反。在印度洋到西太平洋经度上，另一支向低纬传播的相关系数分布与美洲西海岸的分布相反（图 5(d)）。对于东太平洋的传播区（图 5(b)），当海温场超前气压场或落后于气压场较多时，低纬地区为正相关，高纬度地区为负相关。当海温场与气压场同时或落后气压场，低纬为负相关，高纬为正相关。这种相关系数分布类型记为 $Ⅰ_A$。到了西太平洋经度所在区域，传播区仍指向高纬，但其分布类型与东太平洋相关系数分布类型相反（图 5(c)）。

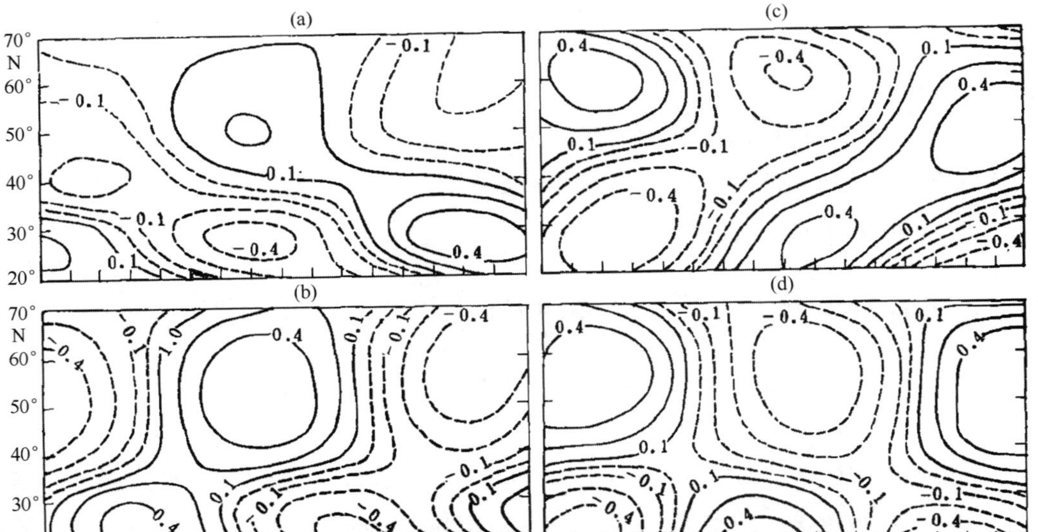

图 5　相关系数的时间纬度剖面

（参考点海温位于东太平洋（1°S，130°W）；(a) 90°W；(b) 120°W；(c) 150°E；(d) 120°E，其余说明同图 2）

总之,在海气相互作用过程中,高低纬之间存在着相互输送信息的区域,这在直观上可以理解,尤其是在海温异常区,存在着高低纬能量的输送,这已被许多数值试验所证实(即波迹的传播)[10]。为什么还在其他地区存在高低纬输送相互作用信息,作者还不能解释。笔者认为如果能说明西太平洋传播区存在的原因,赤道东太平洋海温异常对东亚大气环流的遥相关影响就不难作出解释。

3 中太平洋海温变化与北半球 500 hPa 高度场之间的相关

与上节一样,我们计算了海温参考点位于中太平洋(1°S,170°E)时,它与高度场之间的统计结果。

对于相关系数的时间经度演变(图略),基本上与海温参考点位于东太平洋没有明显的差别(图 2 和图 3b)。实际上如参考文献[4]的分析表明,从东太平洋到日界线附近,海表温度 3~4 年的变化趋势是相当一致的,为节省篇幅在此不再加以说明。

相关区在南北方向传播见图 4(b)。与海温参考点位于东太平洋的情况相比较(图 4(c)),120°E 至日界线,与东太平洋一样,相关区仍然从低纬向高纬传播,并且相关系数的类型也相同(图略)。从 0°W 至 90°W,除了在 60°W 附近存在一个较弱的指向高纬的传播外(图 6),同样有两个指向低纬的传播区,其类型也相同。不同的是在东太平洋上,有一个指向低纬度的传播区,而 120°E 以西则没有(见图 4(b))。

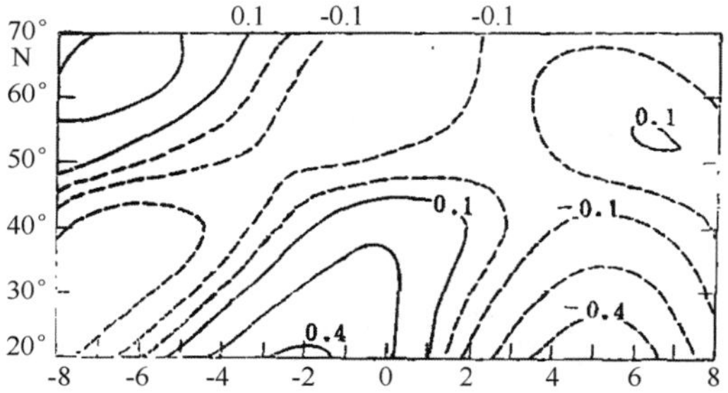

图 6 60°W 经圈上相关系数的时间纬度剖面

(参考点海温位于中太平洋,其余说明同图 2)

图 4(b)中有一个明显不同于东太平洋情况的特点,即从高纬向低纬传播区位于西半球,而指向高纬的传播区主要在中西太平洋。东、中太平洋海温异常产生的这种差异,在一定程度上反映了大气环流对不同地区海温异常响应的不同(经圈方向)[11]。

4 西太平洋海温变化与北半球 500 hPa 高度场之间的相关

现在来分析西太平洋海温(1°S,120°E)与 500 hPa 高度场之间的统计相关,并与中、东太平洋的情况进行比较。

从相关系数的时间经度剖面图来看(图略),从东太平洋到欧洲大陆是振荡型,而在其以东至日界线,在中纬度地区的相关区是向东传播的,速率很快。中高纬度的相关区是向西传播的,其速率很慢,每两年约为 150 个经度。这与中、东太平洋相比,没有实质性的差异,但在赤道地区,相关系数的正负号分布与中、东太平洋的情况正好相反(图略)。这与东、西太平洋 3~4 年时间尺度变化的海温呈反相变化是一致的[4]。

相关区在经向上的传播,除了在西半球向高纬传播的相关区移到 120°W 附近外,其余与中太平洋的情况一致(比较图4中的(a)与(b)),不但传播区的指向和位置相同,而且它的类型也相同。

从图4中还可以注意到,在120°—180°E区域,从低纬向高纬传播的相关区与海温在中、东、西太平洋的位置无关,但传播区的宽度与位置有关。其原因有待于研究。

5　结论和讨论

与参考文献[6]2年振荡的结果相比较,3~4年振荡热带海温场与北半球500 hPa高度场之间的遥相关,在时间经度演变方面看,相关区的传播要简单些。在中纬度地区,从60°W到中太平洋,相关区很快向东传播,其他地方基本上是振荡型。在中高纬度,除振荡型外,在30°W直至180°附近,相关区是自东向西传播的。在赤道地区(10°S至10°N),太平洋区域(120°E至120°W),相关区向东传播,其速率约为每两年130个经度,而印度洋和大西洋区域为振荡型。

当热带海温异常位于不同太平洋区域时,其高低纬度之间相关区的传播方向及位置是不同的。

结合参考文献[6]的结果可以看到,海气之间的大尺度相互作用,不但依赖于热带SST异常变化的周期,同时也依赖于热带海表温度异常的位置(即经度)。尽管这样,在3~4年时间尺度海气相互作用中,相关区的经向传播也存在一些相同的规律。在120°—180°E区域上,从低纬指向高纬的传播区以及在0°—90°W区域上,从高纬指向低纬的传播区则与海温异常的位置无关。与Simmons[12]关于遥相关型只取决于气候基本态,而与海温异常位置无关的结论相比较,其结论是不一致的。原因可以有以下两个方面:① 参考文献[12]中的遥相关型与海温异常位置无关,指的是与PNA和EA(East Atlantic)型无关。在实际大气-海洋系统中,大气对海温异常的响应是多种类型的,不只是上面两种类型;② 正如参考文献[6]所指出的,大尺度热带海温与全球大气环流之间的时滞相关是复杂的。目前研究中所采用的模式大都还没有反映出这种复杂性。

参考文献

[1] Walker, C.T., and E.M.Bliss. World Weather, Memoirs. *R. Met. Soc.* 1932. 4(36)52-84.

[2] Bjerknes, J., A possible response of the atmospheric Hadley circulation to equatorial anomalies of ocean temperature. *Tellus*, 1966, 18:820-829.

[3] Bjerknes, J. Atmospheric teleconnections from the equatorial Pacific. *Mon. Wea. Rev.* 1969, 97:163-172.

[4] Pan, Y. H., and A. H. Oort. Global climate variations connected with sea surface temperature anomalies in the eastern equatorial Pacific Ocean for the 1958-1973 period. *Won. Mea. Rev.* 1983, 111:1244-1258.

[5] Oort, A. H., and Y. H. Pan. Diagnosis of historical ENSO events, Proc. 1st WMO Workshop on the diagnosis and prediction of monthly and seasonal atmospheric variations over the global. College park, U. S. A. 1986, 29 July-2 August: 249-258.

[6] 王彰贵,巢纪平. 北半球冬季大气环流与热带太平洋海表温度二年振荡相互作用的若干事实. 气象学报,1990,48(4):438-449.

[7] Yasumari, T., and T. N. Krishnamurti. Glodal structure of the southern oscillation. *Same as* [5],1986:259-263.

[8] Philander, S. G. H. The southern oscillation and El Niño. *Advances in Geophysics*, 1985, 28:197-215.

[9] Wyrtki, K., El Niño-The dynamic response of the equatorial Pacific Ocean to atmospheric forcing, *J. Phy. Oceanogr*, 1975, 5:572-582.

[10] 黄荣辉,李维京. 夏季热带西太平洋上空的热流异常对东亚上空副热带高压的影响及其物理机制. 大气科学(特刊). 1988:107-116.

[11] Palmer, T. N. Response of two atmospheric general circulation models to sea surface temperature anomalies in tropical East and West Pacific, *Nature*, 1984,310:483-485.

[12] Simmons, A. J., J. M. Wallace and G. W. Branstator. Barotropic wave propagation and instability, and atmospheric teleconnection patterns, *J. Atmos. Sci.* 1983, 42:1361-1392.

简单的热带海气耦合波
——Rossby 波的相互作用*

巢纪平　王彰贵

(国家海洋环境预报研究中心，北京 100081)

摘要：在本文中分析了当大气和海洋中未经耦合前的自由波均为 Rossby 模时，经相互作用后所激发出的耦合波的物理性质。结果表明，由于大气和海洋的背景状态不同，可以激发出两类不稳定耦合 Rossby 波。一类波要求大气的背景场是斜压的，而海洋的混合层较深，即热容量较大。这是一类弱相互作用的不稳定波。另一类要求大气的背景场趋于正压性，而海洋的混合层较浅，即热容量较小。这是一类强相互作用的不稳定波。色散关系的计算表明，这两类不稳定波产生的物理机制也不相同。文中对解不同截断模的本征值问题提出了几种数学方法，同时还进一步提出了一种使大气和海洋自由 Rossby 模的色散关系不受歪曲的处理方法。

关键词：海气耦合波；Rossby 波；不稳定耦合波

1　引言

Philander 等[1]在热带不稳定海气相互作用的文章中指出，Kelvin 波和 Rossby 波都具有不稳定性。由于他们用的是原始运动方程，在海气耦合系统中，Kelvin 模和 Rossby 模是并存的，因此必然存在两种不同物理性质模态之间的相互作用。为此，巢纪平、张人禾[2]和季振刚、巢纪平[3]用赤道 β 平面近似下的位势涡度方程来研究海气相互作用。在这种情况下，大气和海洋中的自由模只为 Rossby 波。计算结果表明，由大气和海洋中的 Rossby 模，经相互作用后所激发出的耦合波，在长波波段可以是向东传播的，同时在一定的海气相互作用强度下，波在此波段是不稳定的。然而，仔细比较他们的结果，可以看到耦合波产生的物理本质是不同的。前者的耦合波是由两个不同的经圈海洋 Rossby 波模经相互作用后激发出来的，这接近 Philander 等的结果。因为在 Philander 的海气耦合模式中，大气取了定常近似，大气中的瞬变波已自然地被过滤掉。在另一方面，后者的耦合波则由一个低频大气自由 Rossby 波模和一个高频海洋自由 Rossby 波模经相互作用后产生。同时，两者在经圈方向所取的特征尺度也不一样，即分别取大气和海洋的赤道 Rossby 变形半径。为了使耦合系统中海洋和大气的运动能用同一的经圈尺度来量度，他们采取了不同的处理方法。由于本征值问题的解在经圈方向是用抛物圆柱函数的级数和来表示的，因此除了解截断到某一经圈模态会使强迫函数受到歪曲外，如用大气赤道 Rossby 变形半径作为经圈运动的特征尺度时，海洋的自由 Rossby 模的色散关系会受到一

* 气象学报，1993，51(4)：385-393.

定的歪曲。反过来，大气的自由 Rossby 模的色散关系也会受到一定的歪曲。这两种误差都会对耦合波的物理性质产生影响，其影响程度是需要讨论的。

另外，Hirst[4,5]用差分方法和高阶截断模方法，分析了海气耦合波的性质。他指出当海洋对大气的加热取热力局地平衡近似，即海表温度距平正比于温跃层的扰动高度时，Rossby 模是稳定的，且没有发现向东传播的波段，这与参考文献[2,3]的结果是不一致的。

在本文中，我们试图对上面指出的这些矛盾做一些分析讨论，同时改进对截断模本征值问题的数学处理方法。在这基础上进一步分析由大气与海洋 Rossby 模经相互作用后耦合波的物理性质。

2 海气相互作用模式

当海洋以热力局地近似的方式加热大气，而大气给海洋以风应力，则在赤道 β 平面近似下，大气和海洋运动的位势涡度方程分别为

$$\left(\frac{\partial^2}{\partial x^2}+\frac{\partial^2}{\partial y^2}-\frac{\beta^2}{C_a^2}y^2\right)\frac{\partial h_a}{\partial t}+\beta\frac{\partial h_a}{\partial x}=\frac{\beta^2}{C_a^2}\alpha y^2 h_s \tag{1}$$

$$\left(\frac{\partial^2}{\partial y^2}+\frac{\partial^2}{\partial y^2}-\frac{\beta^2}{C_s^2}y^2\right)\frac{\partial h_s}{\partial t}+\beta\frac{\partial h_s}{\partial x}=\gamma\left(\frac{\partial^2}{\partial x^2}+\frac{\partial^2}{\partial y^2}\right)h_a \tag{2}$$

式中，$C_a=(gH)^{1/2}$，H 可以看成是大气运动在垂直方向某一特征模的未扰动厚度；$C_s=(gD)^{1/2}$，D 是未扰动的海洋混合层的深度，或温跃层顶离海表的厚度；h_a 是大气某一等压面的扰动高度；h_s 是海洋温跃层的扰动高度。

取特征量

$$\begin{cases}t=(2\beta C_s)^{-\frac{1}{2}}t^*\equiv Tt^*,\quad x=(C_a/2\beta)^{1/2}x^*\\ h_a=\left(\dfrac{C_a^2}{2g}\right)h_a^*,\quad h_s=\left(\dfrac{C_s^2}{2g}\right)h_s^*\end{cases} \tag{3}$$

以及如经圈方向的特征尺度取大气赤道 Rossby 变形半径，即

$$y=\left(\frac{C_a}{2\beta}\right)^{1/2}y^* \tag{4}$$

则方程(1)、方程(2)的无量纲形式为(已略去“ * ”号)

$$\left(\frac{\partial^2}{\partial x^2}+\frac{\partial^2}{\partial y^2}-\frac{y^2}{4}\right)\frac{\partial h_a}{\partial t}+\frac{\varepsilon^{-\frac{1}{2}}}{2}\frac{\partial h_a}{\partial x}=\frac{\varepsilon^2\alpha T}{4}y^2 h_s \tag{5}$$

$$\left[\varepsilon\left(\frac{\partial^2}{\partial x^2}+\frac{\partial^2}{\partial y^2}\right)-\frac{y^2}{4\varepsilon}\right]\frac{\partial h_s}{\partial t}+\frac{\varepsilon^{-\frac{1}{2}}}{2}\frac{\partial h_s}{\partial x}=\frac{\gamma T}{\varepsilon}\left(\frac{\partial^2}{\partial x^2}+\frac{\partial^2}{\partial y^2}\right)h_a \tag{6}$$

这是参考文献[2]的模式，其中 $\varepsilon=C_s/C_a$。

容易看出，如取解的形式为

$$h_a=\sum_n h_{an}^{(0)}D_n(y)\mathrm{e}^{\mathrm{i}(kx-\sigma t)} \tag{7}$$

$$h_s=\sum_n h_{sn}^{(0)}D_n(y_s)\mathrm{e}^{\mathrm{i}(kx-\sigma t)} \tag{8}$$

式中，$y_s=y/\sqrt{\varepsilon}$，则 D_n 满足抛物圆柱函数所定义的方程

$$\frac{\mathrm{d}^2}{\mathrm{d}Z^2}D_n(Z)+\left(n+\frac{1}{2}-\frac{y^2}{4}\right)D_n(Z)=0 \tag{9}$$

式中 $Z=y$ 或 y_s。在这种情况下，对每一个经圈模 n，大气与海洋自由 Rossby 波的色散关系没有歪曲。但为了使这一耦合系统中的运动能用同一个经圈变量或同样的结构函数来表示，他们把 $D_n(y_s)$ 进一步展开，即

$$D_n(y_s) = \sum_{m=0}^{\infty} a_{mn} D_m(y) \tag{10}$$

式中

$$a_{mn} = \frac{1}{m!\ \sqrt{2\pi}} \int_{-\infty}^{\infty} D_n(y_s) D_m(y) \mathrm{d}y \tag{11}$$

在另一方面，在方程(5)和方程(6)右端的强迫项中都会出现如 $D_n(y)y^2$ 的项。为此又应用了下面的递推公式

$$y^2 D_n(y) = D_{n+2}(y) + (2n+1) D_n(y) + n(n-1) D_{n-2}(y) \tag{12}$$

如果 n, m 都取无穷多个经圈模，则这样的展开方法不会对问题的解带来误差。然而，为了求得本征值问题的解析解，只取 $n, m=2$，由于这一截断，使强迫函数和海洋的自由 Rossby 模都受到了歪曲。

另一种情况，如取海洋赤道 Rossby 变形半径为运动的经圈尺度，即

$$y = (C_s/2\beta)^{1/2} y_s^* \tag{13}$$

而其他的特征量同式(3)，则无量纲方程的形式为

$$\left(\frac{\partial^2}{\partial x^2} + \frac{1}{\varepsilon}\frac{\partial^2}{\partial y_s^2} - \frac{\varepsilon}{4}y_s^2\right)\frac{\partial h_a}{\partial t} + \frac{\varepsilon^{-\frac{1}{2}}}{2}\frac{\partial h_a}{\partial x} = \frac{\varepsilon^3 \alpha T}{4} y_s^2 h_s \tag{14}$$

$$\left(\varepsilon\frac{\partial^2}{\partial x^2} + \frac{\partial^2}{\partial y_s^2} - \frac{y_s^2}{4}\right)\frac{\partial h_s}{\partial t} + \frac{\varepsilon^{\frac{1}{2}}}{2}\frac{\partial h_s}{\partial x} = \frac{\gamma T}{\varepsilon^2}\left(\varepsilon\frac{\partial^2}{\partial x^2} + \frac{\partial^2}{\partial y^2}\right) h_a \tag{15}$$

这是参考文献[3]的模式。为了使解的结构函数均用 $D_n(y_s)$ 展开，可将式(14)改写成

$$\left[\frac{\partial^2}{\partial x^2} + \frac{1}{\varepsilon}\left(\frac{\partial^2}{\partial y_s^2} - \frac{y_s^2}{4}\right)\right]\frac{\partial h_a}{\partial t} + \frac{\varepsilon^{-\frac{1}{2}}}{2}\frac{\partial h_a}{\partial x} + \frac{1-\varepsilon^2}{4\varepsilon} y_s \frac{\partial h_a}{\partial t} = \frac{\varepsilon^3 \alpha T}{4} y_s^2 h_s \tag{16}$$

当考虑了式(12)后，也可以求得问题的截断解。容易看出，在现在的情况下，海洋的自由 Rossby 模对任一经圈模 n，其色散关系是正确的，但大气的自由 Rossby 模的色散关系，将因解的截断而受到歪曲。

当取 $C_s=1.64$ m/s，$C_a=62.63$ m/s，相当 $\varepsilon=0.026$，$\alpha\gamma=5\times10^9/\mathrm{s}^2$ 时，巢纪平、张人禾模式的耦合波是不稳定的。如取 $C_s=1.4$ m/s，$C_a=66$ m/s，相当 $\varepsilon=0.021$，$\alpha\gamma=5\times10^{-9}/\mathrm{s}^2$ 时，季振刚、巢纪平模式的耦合波也是不稳定的。但比较这两个模式的不同处理方法，可以看到不稳定耦合波产生的物理机制是不同的。参考自由模的频率后，容易判断出在前者的模式中，耦合波是由两个不同经圈模的海洋自由 Rossby 波，经相互作用后激发出来的，而在后者的模式中，耦合波是由一个“低频”的大气自由 Rossby 模和一个“高频”的海洋自由 Rossby 模，经相互作用后激发出来的。

为了进一步了解耦合波的上述差异是由于物理原因造成的，还是由于数学上的处理方法不同所带来的误差不一致而造成。为此，在下面我们将用同样的方法来求问题的解。另外，若不用展开式(10)，则可以将方程(6)改写成类似于方程(16)的形式

$$\left[\varepsilon\frac{\partial^2}{\partial x^2} + \varepsilon\left(\frac{\partial^2}{\partial y^2} - \frac{y^2}{4}\right)\right]\frac{\partial h_s}{\partial t} + \frac{\varepsilon^{\frac{1}{2}}}{2}\frac{\partial h_s}{\partial x} - \frac{1-\varepsilon^2}{4\varepsilon} y^s \frac{\partial h_s}{\partial t} = \frac{\gamma T}{\varepsilon}\left(\frac{\partial^2}{\partial x^2} + \frac{\partial^2}{\partial y^2}\right) h_a \tag{17}$$

3 两类耦合 Rossby 波

如用大气赤道 Rossby 变形半径为经圈运动的特征尺度，则耦合系统的控制方程为式(5)和式(17)。取解的形式为

$$(h_a, h_s) = \sum (h_{an}^{(0)}, h_{sn}^{(0)}) D_n(y) \mathrm{e}^{\mathrm{i}(kx-\sigma t)} \tag{18}$$

由此得到

$$\sum_n h_{an}\left[\left(k^2+n+\frac{1}{2}\right)\sigma+\frac{\varepsilon^{-\frac{1}{2}}k}{2}\right]D_n(y)=-\frac{\mathrm{i}\varepsilon^2\alpha T}{4}\sum h_{sn}y^2D_n(y) \tag{19}$$

$$\sum_n h_{sn}\left[\varepsilon\left(k^2+n+\frac{1}{2}\right)\sigma+\frac{\varepsilon^{\frac{1}{2}}k}{2}\right]D_n(y)+\frac{1-\varepsilon^2}{4\varepsilon}\sigma\sum h_{sn}y^2D_n(y)$$

$$=-\mathrm{i}\frac{\gamma T}{\varepsilon}\sum h_{an}\left[\frac{y^2}{4}-\left(k^2+n+\frac{1}{2}\right)\right]D_n(y) \tag{20}$$

现称此为模式Ⅰ。

如用海洋赤道 Rossby 变形半径为经圈运动的特征尺度，则耦合系统的控制方程为式(16)和式(15)。取解的形式为

$$(h_a, h_s) = \sum_n (h_{an}^{(0)}, h_{sn}^{(0)}) D_n(y_s) \mathrm{e}^{\mathrm{i}(kx-\sigma t)} \tag{21}$$

由此得到

$$\sum_n h_{an}^{(0)}\left[\left(k^2+\frac{1}{\varepsilon}\left(n+\frac{1}{2}\right)\right)\sigma+\frac{\varepsilon^{-\frac{1}{2}}k}{2}\right]D_n(y_s)-\frac{1-\varepsilon^2}{4\varepsilon}\sigma\sum_n h_{an}^{(0)}y_s^2D_n(y_s)$$

$$=-\mathrm{i}\frac{\varepsilon^3\alpha T}{4}\sum_n h_{sn}^{(0)}y_s^2D_n(y_s) \tag{22}$$

$$\sum_n h_{sn}^{(0)}\left[\left(\varepsilon k^2+n+\frac{1}{2}\right)\sigma+\frac{\varepsilon^{\frac{1}{2}}k}{2}\right]D_n(y_s)$$

$$=-\mathrm{i}\frac{\gamma T}{\varepsilon^2}\sum_n h_{an}^{(0)}\left[\frac{y_s^2}{4}-\left(\varepsilon k^2+n+\frac{1}{2}\right)\right]D_n(y_s) \tag{23}$$

现称此为模式Ⅱ。

考虑到低阶模时 Weber 函数的表达式分别为

$$(D_0, D_1, D_2, D_3)(Z) = (1, Z, Z^2-1, Z^3-3Z)\exp\left(-\frac{Z^2}{4}\right) \tag{24}$$

因此，只要应用积分

$$\int_{-\infty}^{\infty} \mathrm{e}^{-az^2}\mathrm{d}Z = \sqrt{\frac{\pi}{a}} \tag{25}$$

就不难算出形如

$$\int_{-\infty}^{\infty} Z^m \mathrm{e}^{-az^2}\mathrm{d}Z \quad (m\ 为正整数) \tag{26}$$

的积分值，其中 $Z=(y, y_s)$。

当应用上面的积分,并考虑到 Weber 函数的正交性

$$\int_{-\infty}^{\infty} D_n(Z) D_m(Z) \mathrm{d}Z = n! \ \sqrt{2\pi}\delta_{mn} \tag{27}$$

式中　$\delta_{mn}\begin{cases} = 0 & m \neq 1 \\ = 1 & m = n \end{cases}$

在不用递推公式(12)的情况下，当将解截断到 $n=2$ 时，容易算得模式 I、模式 II 对赤道对称运动的本征值和本征函数。

当 $k=0.5$，1.0，并本征值即频率 σ 出现虚部时，ε 和 $\alpha\gamma$ 的依赖关系见图 1。事实上，这是耦合波的不稳定区。由图可见，对模式 I，解的不稳定性只有当 $\varepsilon>0.20$ 且 $\alpha\gamma<10^{-10}/\mathrm{s}^2$ 的条件下才能出现。而对于模式Ⅱ，在通常取的海气相互作用强度，即 $10^{-10}/\mathrm{s}^2<\alpha\gamma<10^{-8}/\mathrm{s}$ 的范围内，只要 ε 值很小，例如 $\varepsilon<0.05$，这时解就能出现不稳定性。当 ε 值较大时，出现不稳定性要求 $\alpha\gamma$ 较小，这时模式Ⅱ和模式 I 的不稳定区基本上接近。

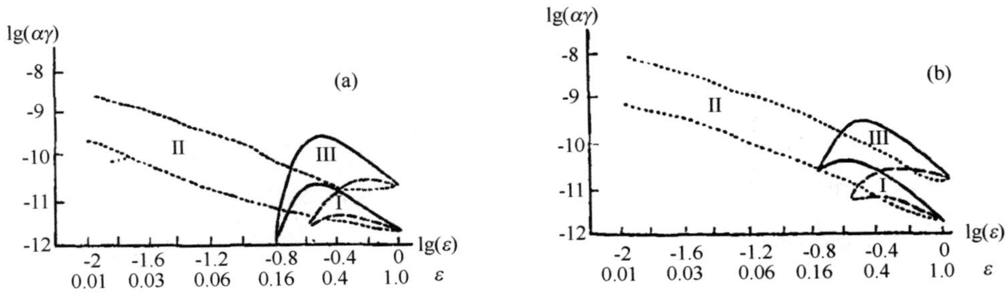

图 1　不稳定区对 ε 和 $\alpha\gamma$ 的依赖关系

(a) $k=0.5$；(b) $k=1.0$

不稳定区依赖于 ε 和 $\alpha\gamma$ 的值，这在物理上是一个有意思的问题。考虑到

$$\varepsilon = \frac{C_s}{C_a} = \left(\frac{D}{H}\right)^{1/2} \tag{28}$$

在这里 D 可以看成是海洋混合层的厚度，或温跃层顶离海表的深度。H 可以看成是大气在垂直方向某一模态所对应的等值厚度，H 值小表示大气运动在垂直方向有多个模，亦即层结性或斜压性强。因此，在模式 I 中要求 ε 值大时才能出现不稳定区，这意味着要求海洋的混合层很厚即热容量很大，或大气的斜压性很强，这在物理上是可以理解的。因为要求耦合系统中的运动，在大气赤道 Rossby 变形半径这样大的经圈尺度使运动不稳定，而海洋的加热主要集中在比它小一个量级的海洋赤道 Rossby 变形半径上，因而必然要求海洋的热容量很大，即海洋的混合层很深，或者只能使等值高度较小的斜压大气不稳定地运动起来。由于海洋的热容量很大，大气又很薄，因此只要适度大小的海气相互作用强度就足够使耦合系统中的运动产生不稳定。

在另一方面，对于模式Ⅱ，不稳定运动出现时 ε 值的范围很宽，在 ε 大值的一端，其物理解释与模式 I 类似，事实上这时 $y_s \approx y$。有兴趣的是 ε 值很小的情况，这意味着不稳定出现的背景状态是海洋的混合层很薄，即相应的热容量很小，或者大气的等值厚度很大，即大气状态趋于正压性。因此在这样的背景状态下，要求耦合系统中的运动不稳定地发展起来，必然要求海气相互作用强度很大。不大的海洋热容量也可以使深厚的大气运动和海洋运动一起不稳定地增长，这在物理上也是容易理解的。这是因为耦合运动的经圈特征尺度主要集中在海洋赤道 Rossby 变形半径这样一个只有百千米量级的范围内。

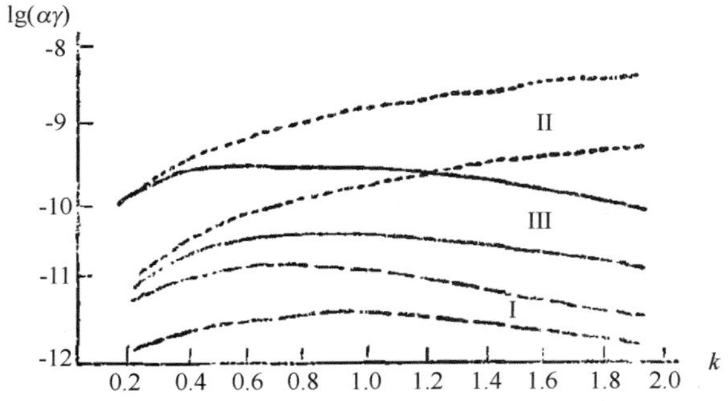

图 2 不稳定波数与 $\alpha\gamma$ 的关系

（图中标记Ⅰ、Ⅱ、Ⅲ分别代表模式类型，同时对应的 ε 分别为 0.3，0.03 和 0.25）

在图 2 中，对模式Ⅰ取 $\varepsilon=0.3$，对模式Ⅱ取 $\varepsilon=0.03$。由图可见，对于模式Ⅰ，当 $\alpha\gamma$ 在 $10^{-12}\sim10^{-11}/s^2$ 范围内，以及对于模式Ⅱ，当 $\alpha\gamma$ 在 $(10^{-10}\sim10^{-8})/s^2$ 范围内，波数 k 从 0.2 的长波到 2.0 的短波很宽的波段内，耦合 Rossby 波都可以是不稳定的。

由以上的分析可以看到，当大气和海洋中的自由波均为 Rossby 波时，其经相互作用激发出的耦合 Rossby 波，可以在两种不同的大气和海洋背景状态下发生不稳定性。若以大气的背景场为参考，一类是斜压的，另一类是正压的。若以海洋的背景场为参考，则一类海洋混合层是深厚的，另一类混合层是浅薄的。对于前者，不稳定波发生在弱的海气相互作用下。对于后者，不稳定波发生在强的海气相互作用下。由于大气的斜压性和正压性，海洋混合层的深厚和浅薄有地理性和季节性，同时在一个 ENSO 事件发展的各个阶段（位相），混合层的深浅和大气的背景场均有变化。因此，通过海气相互作用能激发出哪一类的不稳定耦合 Rossby 波，自然也因地区、季节和 ENSO 事件处于的不同阶段而不同。

4 不稳定耦合波的色散关系

对模式 I，当 $\varepsilon=0.4$，$\alpha\gamma=2\times10^{-11}/s^2$ 时，所算得的 4 个根的频率见图 3。在图中同时给出两个大气自由 Rossby 波频率（σ_{a0}，σ_{a2}）和两个海洋自由 Rossby 模的频率（σ_{s0}，σ_{s2}）。由图可见，耦合波 σ_1 的色散关系基本上接近 σ_{aR0} 海气相互作用对这一高频自由 Rossby 模几乎没有影响。耦合波 σ_4 的色散关系接近 σ_{sR2}，但由于海气相互作用，使频率比原来的自由模有所减小，且在长波波段转为向东传播。有意义的是频率接近 σ_{aR2} 和 σ_{sR0} 的两支耦合波 σ_3 和 σ_2 在波长约为 1800 km 到 5000 km 之间两者重合，即两者变成共轭复根，于是波变成不稳定，其增长和阻尼率见图中的 σ_{i23}。最不稳定的波长为 3100 km，其增长到 2.72 倍（e-fold）所需的时间约 40 天，这时向西传播的相速度约为 23 经度/月。

对于模式 II，当取 $\varepsilon=0.03$，$\alpha\gamma=10^{-9}/s^2$ 时，所算得的耦合波的色散关系和增长（阻尼）率见图 4。可见，与模式 I 不同，这时不稳定的耦合 Rossby 波，是由频率接近大气和海洋两个慢的自由模 σ_{aR2} 和 σ_{sR2} 经相互作用后激发出来的。在不稳定波段，波可以向西传播，也可以在长波段向东传播。最不稳定波长约为 5200 km，其增长到 e 倍的时间约为 55 天，向西的相速度约为 5 经度/月。

5 模式的改进

在以上的模式和处理方法中，或者是海洋的自由 Rossby 模受到歪曲，或者是大气的自由 Rossby

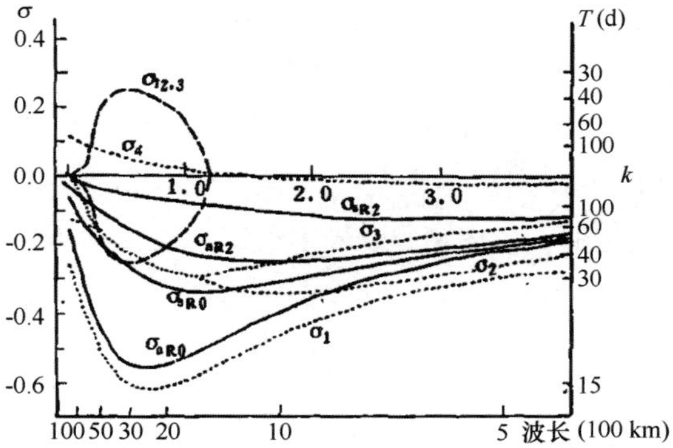

图 3　模式 I 当取 $\varepsilon=0.4,\alpha\gamma=2\times10^{-11}/s^2$ 时，耦合波的色散关系和增长(阻尼)率

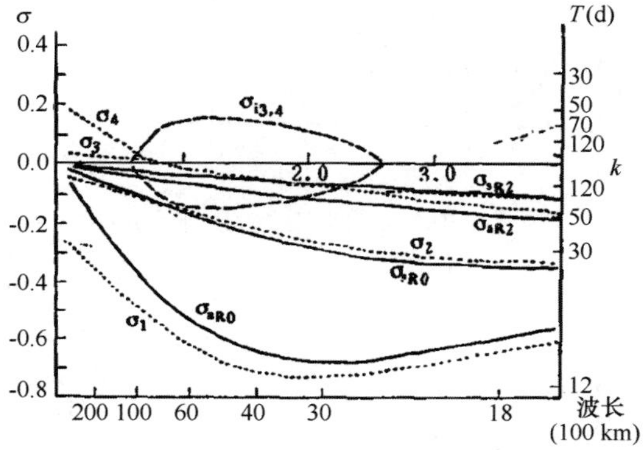

图 4　模式 II 当取 $\varepsilon=0.03,\alpha\gamma=10^{-9}/s^2$ 时，耦合波的色散关系和增长(阻尼)率

模受到歪曲。为了使自由 Rossby 模都不受到歪曲，我们对方程(5)和方程(6)取下面形式的解

$$h_a(x,y,t)=\sum_n h_{an}^{(0)}D_n(y)e^{i(kx-\sigma t)} \tag{29}$$

$$h_s(x,y,t)=\sum_m h_{sm}^{(0)}D_m(y_s)e^{i(kx-\sigma t)} \tag{30}$$

由此得到

$$\sum_n h_{an}^{(0)}\left[\left(k^2+n+\frac{1}{2}\right)\sigma+\frac{\varepsilon^{-\frac{1}{2}}k}{2}\right]D_n(y)=-i\frac{\varepsilon^2\alpha T}{4}\sum_m h_{sm}^{(0)}y^2D_n(y_s) \tag{31}$$

$$\sum_m h_{sm}^{(0)}\left[\left(\varepsilon k^2+n+\frac{1}{2}\right)\sigma+\frac{\varepsilon^{\frac{1}{2}}k}{2}\right]D_m(y_s)=-i\gamma T\sum_n h_{an}^{(0)}\left[\frac{y^2}{4}-\left(k^2+n+\frac{1}{2}\right)\right]D_n(y) \tag{32}$$

我们称之为模式Ⅲ。两个 Weber 函数的自变量虽然不同(y 和 y_s)，但应用上面的积分方法，容易得到截断模所构成的方程组。现在将 n,m 分别截断到 3，这样使有两个对赤道偶对称的模 $m,n=0,2$ 和两个对赤道奇对称的模 $n,m=1,3$。

计算所得，当 $k=0.5$，和 $k=1.0$ 时，耦合 Rossby 波当对赤道对称时的不稳定区，其与 $\varepsilon,\alpha\gamma$ 的依赖关系见图 1 中的区域Ⅲ。而当 $\varepsilon=0.25$ 时，不稳定波长(波数)对 $\alpha\gamma$ 的依赖关系见图 2 中的Ⅲ区。由图可见，不稳定区出现时的 ε 值要小于模式 I，即所要求的背景条件大气的斜压性或海洋的混合层深度，要比模式Ⅰ小，而海气相互作用强度在 $10^{-11}\sim10^{-10}/s^2$之间是中等强度的相互作用，不稳定波长对海气相互作用强度的依赖性不大。

当取 $\varepsilon=0.3$，$\alpha\gamma=10^{-10}/s^2$时，对赤道偶对称耦合波的色散关系和增长(阻尼)率分别见图 5。若与自由波的频率相比较，现在的不稳定波由慢的一支大气 Rossby 模 $\sigma_{\alpha R2}$和快的一支海洋 Rossby 模 σ_{sR0}耦合而成，与模式Ⅰ的情况相似。由图可见，最不稳定的波长约为 3800 km，增长到 e 倍的时间需 30 天，向西传播的相速度约 26 经度/月。

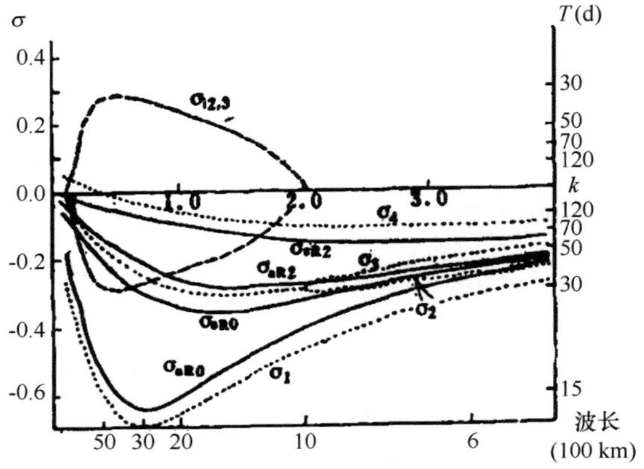

图 5　模式Ⅲ中对赤道偶对称的耦合波色散关系

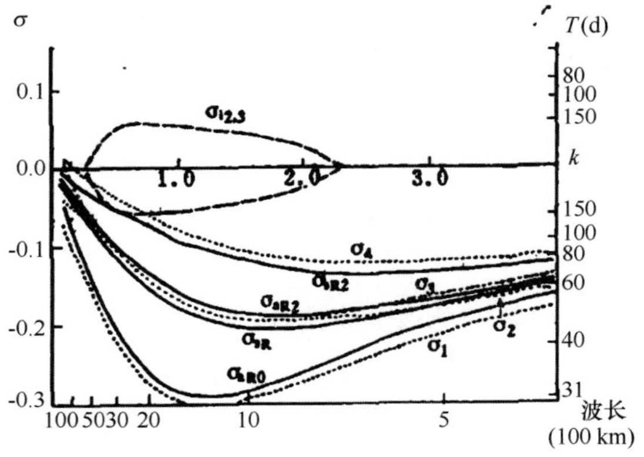

图 6　模式Ⅲ中对赤道奇对称的耦合波色散关系

由于解截断到 $n=3$，因此有两个对赤道奇对称的模，当取 $\varepsilon=0.5$，$\alpha\gamma=2\times10^{-12}/s^2$时，对赤道奇对称的耦合波也出现了不稳定，其色散关系和增长率见图 6 。与赤道对称耦合波的色散关系(图 5)比较，两者十分类似。但不同的地方在于，图 6 中的最不稳定波的波长变短(2 300 km)、增长到 e 倍所需的时间变长(150 天)，以及向西的相速度变慢(9.6 经度/月)。

计算还表明，在运动对赤道奇对称的情况下，要激发出不稳定的耦合波，其最小的 ε 值要大于偶对称情况，也即要求大气有更强的斜压性，或海洋有更深的混合层。

6　结论

经过以上较为详细的分析可以认为，大气和海洋的自由 Rossby 波，在海洋对大气加热取热力局地平衡近似下，通过相互作用是可以激发出不稳定的耦合 Rossby 波来的。但不稳定耦合波的出现，依赖于海洋和大气赤道 Rossby 变形半径之比和海气相互作用强度。由此可以分成两类：一类不稳定波出现的条件是大气的背景状态具有较强的斜压性，海洋的混合层较深，或所含的热容量较大；另一类不稳定波出现的条件，要求大气的基本状态趋于正压的，而海洋的混合层较浅，或所含的热容量较小。对于前者，不稳定耦合波出现在弱的海气相互作用情况下；对于后者，则要求有强的海气相互作用。

在另一方面，这两类不稳定耦合波是由不同的大气与海洋自由 Rossby 波经相互作用后激发出来，亦即物理过程不同。

Hirst 认为，在热力局地加热平衡条件下，由 Rossby 模的耦合波是不稳定的，而由 Kelvin 模的耦合波才具有不稳定性。对此，比较 Hirst 所取的参数值后，其结果是不矛盾的。因为若取他所用的参数值，γ（即 K_s）$= 8\times10^{-8}/\mathrm{s}$，$\alpha$（即 K_aK/g）$= 0.21\times10^{-4}/\mathrm{s}$，则有 $a\gamma = 1.68\times10^{-12}/\mathrm{s}^2$。在这样弱的相互作用下，模式Ⅰ、模式Ⅱ、模式Ⅲ，耦合波均不可能出现不稳定性。

参考文献

[1] Philander S G H, et al. Unstable air sea interaction in the tropics. J Atmos Sci, 1984,41:604-613.

[2] Chao Jiping, Zhang Renhe. The air-sea interaction waves in the tropics and their instabilities. Acta Meteor Sinica, 1988, 2:275-287.

[3] Ji Zhengang, Chao Jiping. An analysitcal coupled air-sea interaction model. J Marine System, 1991, 1:263-270.

[4] Hirst A C. Free Equatorial Instabilites in Simple Coupled Aimosphere-Ocean Model, Coupled Ocean-Atmosphere Models, J. C. J. Nihoul, Ed. Elsevier Oceanography Series, 40, Elseveier, 1985:153-165.

[5] Hirst A C. Unstable and demped equatorial modes in simple coupled ocean-atmosphere mode. J Atmos Sci, 1986, 43: 606-630.

On the Instability of Axisymmetric Vortex in the Tropical Air-sea Coupled System*

Wang Zhanggui Chao Jiping

(National Research Center for Marine Environment Forecasts. Beijing, China)

Abstract: Based on a viewpoint of the air-sea interaction, it is discussed the instability on an axisymmetric vortex in the tropics. Atmospheric and oceanic motions are respectively desctribed by shallow-water equations. It is shown that the tropical vartex is unstable under some conditions. The growth rate depends on tatitude and ait-sea coupling coefficients. There exists a structure similar to typhoon eye and the cold water upwelling in the central region of vortex when the unstable vortex develops for 1.5 d. It cuts off the hear supply coming from the sea surface. Results also show that the atmosphere besic flow is bencficial to the development of coupted vortx, but the ocean tasic current is not beneficial to.

I INTRODUCTION

The vortex motion is one of the most commonest phenomena in the atmosphere and the ocean, for example, typhoon, the tropical vortex, etc. After Palmen (1949) studied the convective instability in the tropical atmosphere, he put forward that the tropical vortex developed only over regions where the sea surface temperature (SST) exceeded 26℃. It means that the energy of vortex development is mainly derived from the release of latent heat. In addition, the development of tropical vortex also bears a relation to atmospheric environment conditions, such as the inertial stability, stratified stability (Liu and Ni, 1983), the vertical and horizontal wind distribution (Gray, 1968). Charney and Eilassen (1964) contributed the development of cyclone to the conditional instability of second kind (CISK), that is, interaction between the cumulus and cyclone scale motions. Chen (1984) proposed that the vortex development is due to the existence of two circulation cells in its central region.

The development of tropical vortex depends on the atmospheric conditions, the sensible heat and latent heat energy supplied by the ocean. But its delvelopment will cause SST to be anomalous and the current field to change. Figure 1 is a schematic diagram of the observed current structure when hurricane crossed the ocean (Leipper, 1967). The warm ocean surface layers were transported outward from the hurricane center, and there waters converged outside of the central storm area. The corresponding downwelling to some 80 to 100 m in

* The symposium on the Physical and Chemical Oceanography of the China Seas, Hangzhou, China Ocean Press, 1993.

depth took place there. The cold water upwelled along the hurricane phat from depth of approximately 60 m. SST decreased by more than 5℃ within a radius of some 70 to 200 miles (1 mile = 1 609.344 m).

The atmosphere and ocean have not been regarded as a united system in the above-mentioned studies. It ignores that the interaction of the atmosphere with the ocean exists in the process of vortex development. In this paper, we will use a shallow-water coupled model. which contains the feedback processes between the atmosphere and the ocean, to study the instability of axisymmetric tropical vortex and the influence of basic currents on the vortex instability.

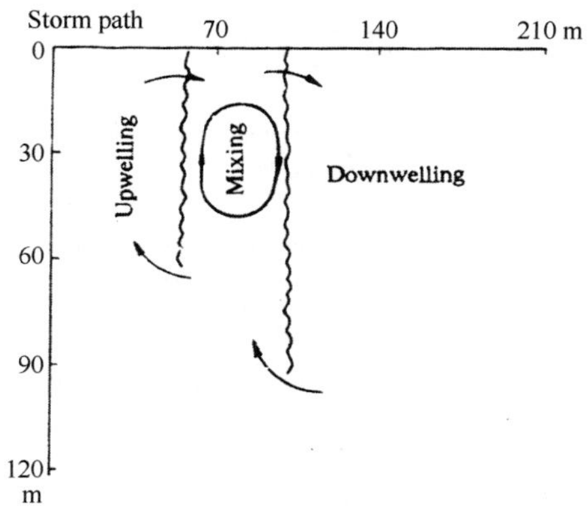

Fig.1　Schematic diagram showing the influence of hurricane on the ocean

Ⅱ　AN ATMOSPHERE-OCEAN COUPLED MODEL

Atmosphere

For the purpose of simplicity, the vortex motion is described by shallow-water equations, and the oceanic heating is directly proportional to the mixed-depth perturbation. So atmospheric motion equations in the cylindrical coordinate are

$$\frac{\mathrm{d}U_a}{\mathrm{d}t} - \left(\frac{V_a}{r} + f\right)V_a = -g\frac{\partial h_a}{\partial r} \tag{1}$$

$$\frac{\mathrm{d}V_a}{\mathrm{d}t} + \left(\frac{V_a}{r} + f\right)U_a = -g\frac{\partial h_a}{r\partial \varphi} \tag{2}$$

$$\frac{\mathrm{d}h_a}{\mathrm{d}t} + H\left[\frac{\partial(rU_a)}{r\partial r} + \frac{\partial V_a}{r\partial \varphi}\right] = -\alpha h'_s \tag{3}$$

where $\frac{\mathrm{d}}{\mathrm{d}t} = \frac{\partial}{\partial t} + U_a\frac{\partial}{\partial r} + V_s\frac{\partial}{r\partial\varphi}$; U_a and V_a are the radial and tangential velocities; r is the radial coordinate; subscripts a and s represent the atmosphere and the ocean respectively; α is the air-sea coupling coefficient; other symbols are commonly used.

It is assumed that the basic flows in the atmosphere and the ocean satisfy the static equilibrium and gradient wind equilibrium

$$-\frac{1}{\bar{\rho}_j}\frac{\partial \bar{P}_j}{\partial z} = \bar{g}_j;\quad \frac{\bar{V}_j^2}{r} + f\bar{V}_j = -\bar{g}_j\frac{\partial \bar{h}_j}{\partial r} \tag{4}$$

where j stands for a or s. The bar refers to the basic flow. It is convenient to decompose U_a and the corresponding variables into basic flows and disturbances as follows

$$U_j = U'_j,\quad V_j = \overline{V}_j + V'_j,$$
$$h_a = H + h'_a,\quad h_s = D + h'_s \tag{5}$$

So the linear equations derived from Eq. (1) to Eq. (3) are

$$\left(\frac{\partial}{\partial t} + \Omega_a \frac{\partial}{\partial \varphi}\right) U'_a - f'_a V'_a = - g \frac{\partial h'_a}{\partial r} \tag{6}$$

$$\left(\frac{\partial}{\partial t} + \Omega_a \frac{\partial}{\partial \varphi}\right) V'_a + f'_a U'_a = - g \frac{\partial h'_a}{r\partial \varphi} \tag{7}$$

$$\left(\frac{\partial}{\partial t} + \Omega_a \frac{\partial}{\partial \varphi}\right) h'_a + H\left[\frac{\partial (rU'_a)}{r\partial r} + \frac{\partial V'_a}{r\partial \varphi}\right] = - \alpha h'_s \tag{8}$$

where $V_{ao} = r\Omega_a$, Ω_a is the angular speed of the atmospheric basic flow; $f'_a = f+2\Omega_a$, f is the Coriolis perameter.

Cteras

The oceanic motion in response to wind stress which act as a body force on the mixed layer is given by the follwing equations

$$\frac{\mathrm{d}U_s}{\mathrm{d}t} - \left(\frac{V_s}{r} + f\right) V_s = - g^* \frac{\partial h_s}{\partial r} + \gamma U'_a \tag{9}$$

$$\frac{\mathrm{d}V_s}{\mathrm{d}t} + \left(\frac{V_s}{r} + f\right) U_s = - g^* \frac{\partial h_s}{r\partial \varphi} + \gamma V'_a \tag{10}$$

$$\frac{\mathrm{d}h_s}{\mathrm{d}t} + D\left[\frac{\partial (rU_s)}{r\partial r} + \frac{\partial V_s}{r\partial \varphi}\right] = 0 \tag{11}$$

By using Eqs (4) and (5), we will derive linear equations from Eqs (9) to (11)

$$\left(\frac{\partial}{\partial t} + \Omega_s \frac{\partial}{\partial \varphi}\right) U'_s - f'_s V'_s = - g^* \frac{\partial h'_s}{\partial r} + \gamma U'_a \tag{12}$$

$$\left(\frac{\partial}{\partial t} + \Omega_s \frac{\partial}{\partial \varphi}\right) V'_s + f'_s U'_s = - g^* \frac{\partial h'_s}{r\partial \varphi} + \gamma V'_a \tag{13}$$

$$\left(\frac{\partial}{\partial t} + \Omega_s \frac{\partial}{\partial \varphi}\right) h'_s + D\left[\frac{\partial (rU'_s)}{r\partial r} + \frac{\partial V_s}{r\partial \varphi}\right] = 0 \tag{14}$$

where $V_{so} = r\Omega_s$, Ω_s is the angular speed of the ocean basic current; $f_s = f+2\Omega_s$; $g^* = g\Delta\rho/\rho_s$, g is the gravitational acceleration, ρ_s the density of sea water, $\Delta\rho$ is the density difference, between thermocline layer and mixed layer.

Ⅲ THE FREQVENCY AND INSTABILITY OF THE TROPICAL VORTEX

Introducing the non-dimensional variables:

$$t_1 = \frac{t}{T},\quad r_1 = \frac{r}{r_0},\quad (U'_{a1}, V'_{a1}) = \frac{(U'_a, V'_a)}{C_a},$$
$$h'_{a1} = \frac{h'_a}{H},\quad h'_{s1} = \frac{h'_s}{D},\quad (U'_{s1}, V'_{s1}) = \frac{(U'_s, V'_s)}{C_s} \tag{15}$$

where r_0 represents the characteristic horizontal scale; $T = r_0/\sqrt{C_a C_s}$ time scale; $C_s = \sqrt{gH}$ and $C_s = \sqrt{g^* D}$ are the gravity wave speeds in the atmosphere and the ocean, respectively. By using the axisymmetric approximation and Eq. (15), the non-dimensional equations are derived from Eq. (6) to Eq. (8) and Eq. (12) to Eq. (14) (subscripts 1 omited)

$$\varepsilon^{1/2}\frac{\partial U'_a}{\partial t} - \varepsilon_2 V'_a = -\frac{\partial h'_a}{\partial r} \tag{16}$$

$$\varepsilon^{1/2}\frac{\partial V'_a}{\partial t} + \varepsilon_2 U'_a = 0 \tag{17}$$

$$\varepsilon^{1/2}\frac{\partial h'_a}{\partial t} + \frac{\partial(rU'_a)}{r\partial r} = -\varepsilon_3 h'_s \tag{18}$$

$$\varepsilon^{-1/2}\frac{\partial U'_s}{\partial t} - \lambda_2 V'_s = -\frac{\partial h'_s}{\partial r} + \lambda_3 U'_a \tag{19}$$

$$\varepsilon^{-1/2}\frac{\partial V'_s}{\partial t} + \lambda_2 U'_a = \lambda_3 V'_a \tag{20}$$

$$\varepsilon^{-1/2}\frac{\partial h'_s}{\partial t} + \frac{\partial(rU'_s)}{r\partial r} = 0 \tag{21}$$

hwere $\varepsilon = C_s/C_a, \varepsilon_2 = f'_a\varepsilon^{1/2}T, \varepsilon_3 = \alpha T\varepsilon^{3/2}g/g^*, \lambda_2 = f'_a\varepsilon^{-1/2}T, \lambda_3 = \gamma T\varepsilon^{-3/2}$.

We now seek solutions to Eqs (16) to (21) of the form

$$(h'_a, h'_s) = \sum_{n=1}^{\infty}(h_{as}, h_{ss})J_0(\sqrt{\mu_n}r)e^{-i\sigma t}$$

$$(U'_a, V'_a, U'_s, V'_s) = \sum_{n=1}^{\infty}(U_{as}, V_{as}, U_{ss}, V_{ss})J_1(\sqrt{\mu_n}r)e^{-i\sigma t} \tag{22}$$

where $J_m(x_s)$ is the Bessel function of the first kind and order m. Substituting Eq. (22) into Eq. (16) to Eq. (21) and using the orthogonality of Bassel function

$$\int_0^{\infty} J_m(\sqrt{\mu_n}r)J_m(\sqrt{\mu_t}r)r\mathrm{d}r = [N_n^{(m)}]^2\delta_{nt} \tag{23}$$

where

$$\delta_{nl} = \begin{cases} 0, & n \neq l \\ 1, & n = l \end{cases}$$

we have the frequency equation on the tropical vortex

$$\sigma^6 - \left[\frac{\varepsilon_2^2 + \mu}{\varepsilon} + \varepsilon(\lambda_2^2 + \mu_n)\right]\sigma^4 + [(\mu_n + \lambda_2^2)(\mu_a + \varepsilon_2^2) - \varepsilon_3\lambda_3]\sigma^2 - \varepsilon_2\lambda_2\varepsilon_3\lambda_3 = 0. \tag{24}$$

when $\varepsilon_3 = \lambda_3 = 0$, this is, there is no air-sea interaction, from Eq. (24) we get

$$\sigma_{a\pm} = \pm\varepsilon^{-1/2}\sqrt{\mu_n + f'^2_a\varepsilon T^2} \tag{25}$$

the inertia-gravitational oscillation in the atmosphere, and

$$\sigma_{s\pm} = \pm\varepsilon^{1/2}\sqrt{\mu_a + f'^2_s\varepsilon^{-1}T^2} \tag{26}$$

the inertia-gravitational oscillation in the ocean.

The instability of perturbation vortex is discussed under four kinds of basic currents as follows:

Ⅰ: There exists a cyclonic vortex in the atmosphere and the ocean respectively;

II: the atmosphere is stationary, and there exists a cyclonic vortex in the ocean;

III: there is a cyclonic vortex in the atmosphere and the ocean is stationary;

IV: both the atmosphere and the ocean are stationary.

Taking $\Omega_a = 5\times10^{-5}\ s^{-1}$, $\Omega_s = 5\times10^{-6}\ s^{-1}$, $r_0 = 10^5$ m, $C_s = 1.0$ m/s, $C_a = 10$ m/s, $\Delta\rho/\rho_s = 10^{-3}$, $\alpha r = 3.5\times 10^{-11}\ s^{-2}$, Fig.2 are computed from Eq. (24) of σ as a function of $\sqrt{\mu_n}$ for $f = 3.8\times10^{-5}\ s^{-1}$. The air-sea interaction has a weak influence upon the frequency of the inertia-gravitational oscillation of the atmosphere, but it causes the frequency of the inertia-gravitational oscilltiam of the ocean to diminish and produce new low frequency oscillations. The new oscillations and the inertia-gravitational oscillation in the ocean will couple and become unstable in the range $\sqrt{\mu_a} = 1.5$ to 2.7. The maximum growth rate is about 0.14, that is, *e*-folding time of 2.6 d. Various coupling coeaxients will be used to study the evolution of unstable vortex. We find that the inertia-grvitational mcillation in the ocean and the new oscillations first approach at $\sqrt{\mu_a} = 2.5$ and come about the uncidle coupling as the air-sea coupling coefficients increase.

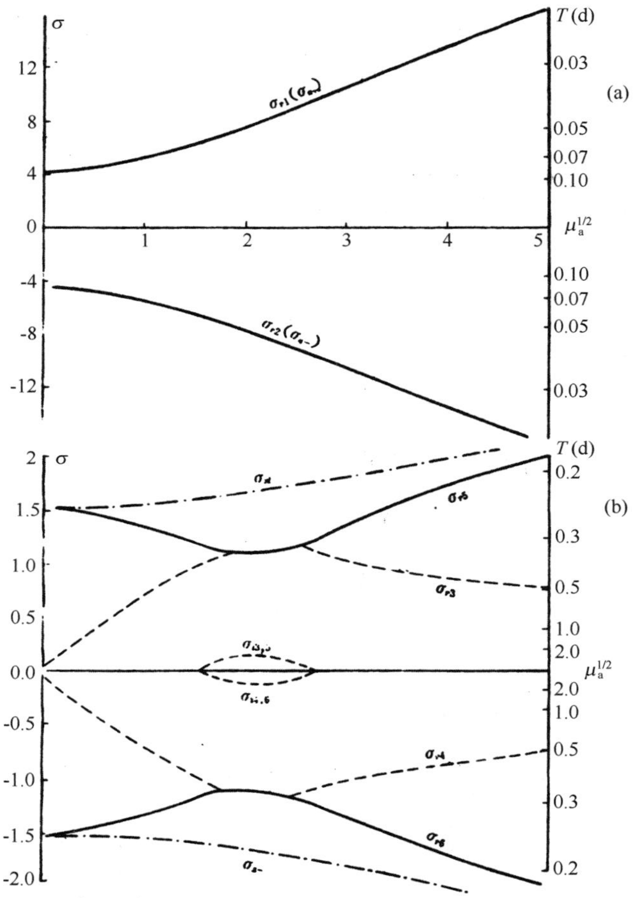

Fig.2 Real part (σ_r) and imaginary part (σ_i) of frequency as a function of eigenvalue ($\sqrt{\mu_s}$) for the basic current I. (a) Atmosphere and (b) ocean. $\sigma_{a\pm}$ and $\sigma_{s\pm}$ stand for the inertia-gravitational oscillations in the atmosphere and the ocean respectively, $\sqrt{\mu_1} = 2.18$ corresponds the dimensional scale of $r = 100$ km

Results mentioned above show that the unstable coupling frequency is not only related to air-seacaopling coefficients but also to the horizontal scale of tropical vortex ($\sqrt{\mu_n}$). Figure 3 gives the unstabe regions of tropical vortex. The unstable regions are located on the right of curves. The marks Ⅱ ~ Ⅳ in the curves represent the basic currents Ⅰ ~ Ⅳ. It is seen from the figure that the most unstabe vortex lies in the range $\sqrt{\mu_n}$ = 1.0 to 2.15. The column two in the Table 1 lists the minimum values of coupling coefficients for which the vortex becomes unstable under the various basic currents. The existence of the atmosphere basic flow is beneficial to the development of perturbation vortex, but the ocean basic current is not beneficial. These results are easily understood on the viewpoint of physics. At the initial period of the vortex development, the thermal structure in the ocean plays a main role in the air-sea coupled system. So the existence of atmosphere basic flow is helpful to the convective activity over the warm sea surface by release of latent heat in the atmosphere. The air in the low layer converges, and a perturbation cyclone forms under the action of Coriolis force. At last, it makes the perturbation vortex in the tropics enhance. So the atmosphere basic flow plays a positive feedback role in the perturbation vortex. For the ocean, by the action of Ekman transport, the existence of the ocean basic current will cause the warm water in the central areas to flow outside. As a result, the cold water upwells and SST decreases. The effect of oceanic heating on atmosphere is reduced, that is, the ocean basic current plays a role of negative feedback. The effects are easily seen from Fig.4. The basic currents in the atmosphere (ocean) make the growth rate of the perturbation vortex in crease (decrease).

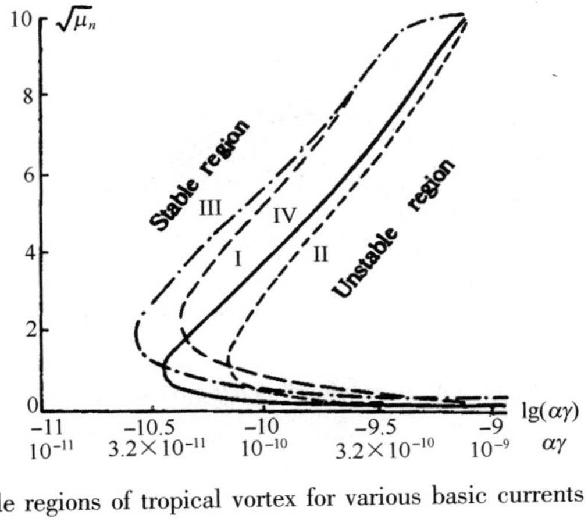

Fig.3　The unstable regions of tropical vortex for various basic currents with $f = 3\times10^{-11}\ s^{-1}$

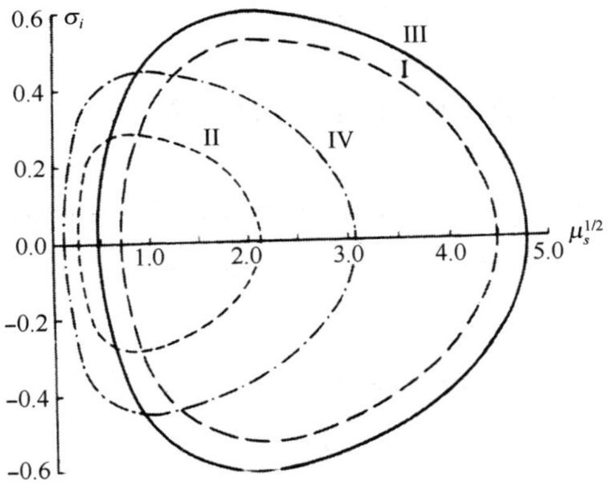

Fig.4　Relation of the growth rate to the various basic currents for $\alpha\gamma = 5\times10^{-11}\ s^{-2}$, $f = 3\times10^{-5}\ s^{-1}$

Table 1 The maximum growth rate (σ_{imax}), the minimum coupling coefficients ($\alpha\gamma_{min}$) and the critical latitude (φ_c) for various basic currents

Basic current	σ_{imax}	$\alpha\gamma_{min}(s^{-2})$	φ_c(°N)
I	0.52	4×10^{-11}	20
II	0.28	5.5×10^{-11}	15
III	0.6	2.2×10^{-11}	24
IV	0.44	3.6×10^{-1}	20

Most tropical hurricanes, as we know, form in regions equatorward of 20° latitude and poleward side of doldrum equatorial troughs. Besides a fact that the ocean supplies a great amount of water vapour to the atmosphere, the Coriolis force is another main factor. For the basic current I, the relation of the growth rate of perturbation vortex to the latitude is given in Fig.5. The growth rate of vortex increases in the latitude range of 0° to 5° N, but it decreases beyond 5°N and the unstable region becomes small as the latitude increases. The growth rate, however, diminishes when the latitude approaches to 20°N. The reason is that both the Coriolis force and the Ekman transport result in that the cold water upwelling is enhanced and SST decreases so that the effect of the oceanic heating upon atmosphere is weakened.

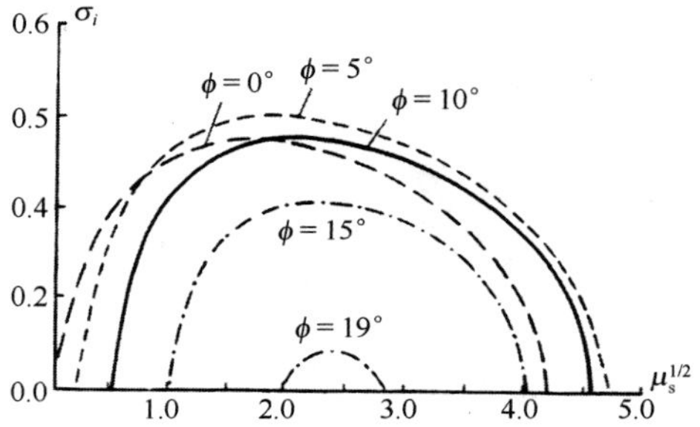

Fig.5 Variation of the growth rate with latitude for the basic current I with $\alpha\gamma=5\times10^{-11}$ s^{-2} (φ in most internal circle should be 19°)

It is seen from the column three in the Table 1 that the ocean basic current makes the critical latitude decrease, but the atmosphere basic flow makes it increase.

Ⅳ THE STRUCTURE OF PHYSICAL FIELD

In the last section we discussed the instability of tropical vortex. The structure of physical field corresponding to the maximum growth rate will be given in the section in order to understand the developing process of axisymmetric tropical vortex.

The eigenfunctions of U_a and other variables can be obtained from Eqs (16) to (22). Taking $\sigma=1.1+0.14\,i$ (e-folding time of 2.6 d), $f=3.8\times10^{-5}$ s^{-1}, $\alpha\gamma=3.5\times10^{-11}$ s^{-2}, the eigenfunctions belonging to the basic current I are shown in Fig.6. The air converges toward the vortex central area from outside and rises here. The perturbation height decreases. There is a maximum tangential velocity of the cyclonic vortex away from center about 100 km. The atmospheric structure bears analogs with the initial structure of tropical hurricane. For the ocean, the distribution of current field is similar to that of the atmospheric wind field, that is, there exists a cyclonic vortex which the sea water cocverges to its center region. It results in that the sea surface water sinks ($W'_s<0$) , and the mixed depes and SST increase. So the ocean continues to supply the sensible heat and tatent hear to the atmosphere so that the perturbation vortex can develop further.

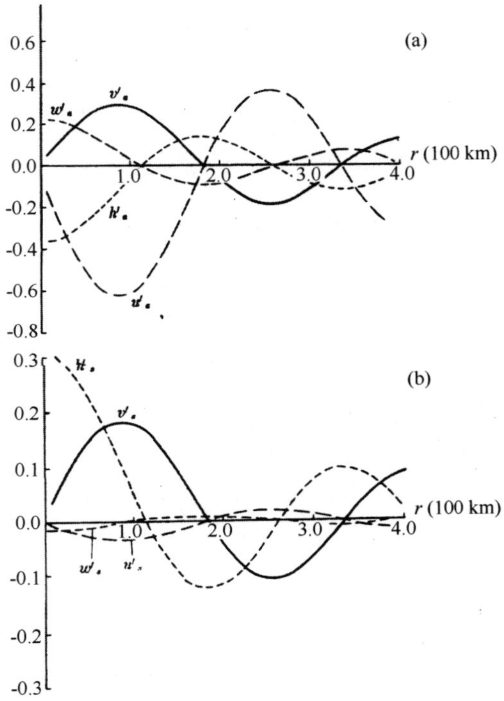

Fig.6　The structure of physical field when $t=0$. (a) Atmosphere and (b) Ocean

Figure 7 is the variation of variables, which are standarized, with time. The cyclonic vortein the atmosphere is enhanced further and the sinking motion begins to occur in the center region wihe the perturbation vortex develops for 1.5 d. The structure of atmospheric flow is analogous to that typhoon eye. At the same time, the sea surface water diverges outward from the central regions with the result that upwelling and cooling ($h'_s<0$) take place there.

V　CONCLUTION

We have analysed the instability of tropical vortex in the air-sea coupled system, and illustrate the vital role of the basic currents, latitude and the coupling coefficients in the leading to the instability of votex. The main results from this study are:

(1) The axisymmetric vortex in the air-sea coupled system can be developed unstably on certain conditions. For the atmosphere, the development of vortex is similar to that of the tropical hurricane, but its mechanism is different from CISK. The development of unstable vortex, on the other hand, is mestricted within the low latitude regions.

(2) The sturcture of perturbation vortex in the initial period differs from that in the mature periat.

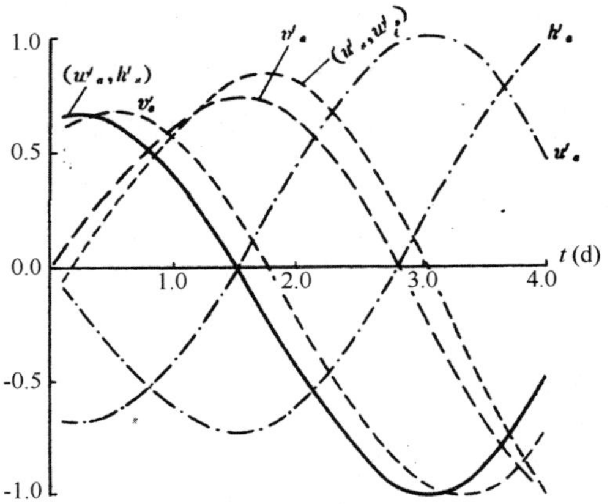

Fig.7 Variation of standarized variabies with time at $r=10^5$ m

Typhoon eye only comes about in the mature period of unstable vortex.

(3) As the unstable vortex develops further, the positive SST anomaly in the initial period is changed into the negative one in the mature period.

Because of using a linear shallow-water coupled model, the radial scale of sinking motion in the wortex center is 2 times larger than that of typhoon eye. So it is necessary to introduce the nonlinear effects into the air-sea coupled model.

References

Ctarney J. G. and A. Eliassen. 1964. On the growth of the hurricane depression. *J. Atmos. Sci.*, 21:68-75.

Chen Yingyi. 1984. Conservation of the wave action density and instability in the vortex motion. *Scientia Sinica*, Series, B. 5: 476-483.

Cnay W. M. 1968. Global view of the origin of tropical disturbances and storms. *Man. Wea. Rev.*, 96:669-700.

Leipper D. F. 1967. Observed ocean conditions and hurricane Hilda, 1694. *J. Atmos. Sci.*, 24:182-196.

Liu Shikuo and Ni Bingjing. 1983. Influence of the inertial stability and stratified stability on the development of typhoon. In: *Collecttioa of Typhoon Meeting*, Shanghai Scientific Publishing House, Shanghai.

Pwimen E. 1948. On the formation and structure of tropical hurricanes. *Geophysica*, 3:26-38.

On the Instability of Tropical Vortex in an Air-sea Coupled Model: Ⅱ. a Numerical Experiment

Chao Jiping, Wang Zhanggui

(National Research Center for Marine Environment Forecasts, Beijing 100081, China)

ABSTRACT: In this paper, a numerical method is used to study the instability of a tropical vortex in the linear and nonlinear air-sea coupled models. For the linear coupled model, the numerical results are consistent with those of the theoretical analysis in paper Ⅰ (Wang et al., 1991). The instability of the vortex depends on the air-sea coupling coefficients and latitude, but those coefficients have little effect on oscillation periods. For the nonlinear coupled model, there are two aspects different from the linear model. One is that the nonlinear effect makes the critical couling coefficients increase. The other is that there exists a mulit-period oscillation which consists of two main periods about 2 days and 40 days when a suitable coupling coefficient is taken. The CISK mechanism is not considered in this paper. So the suggestion of a 40 day oscillation indicates that the low frequency oscillation in the tropics may be relevant to the mesoscale air-sea interaction.

Ⅰ INTRODUCTION

The motion and development of a tropical vortex have been studied by many meteorologists (palmen, Gray and Charney). Most researches, however, are focused on the atmospheric conditions (Liu, 1983). It is, on the other hand, considered that the tropical vortex is a kind of strong mesoscale system. Its development depends on not only atmospheric conditions, but also the sensible heat and latent heat supplied by the ocean. Owing to the strong wind in the tropical vortex, the windstress in the lower layer makes the oceanic state change, e. g., the cold water upwelling caused by Ekman pumping results in such a thermodynamic structure change as to modify the heat flow from the ocean to the atmosphere. So, in this paper, the development of a tropical vortex is discussed in an air-sea coupled system.

Ⅱ AIR-SEA COUPLED MODELS

As with the paper I, it is assumed that the vortex motion is axisymmetric, and it is described by shallow-water equations. The atmospheric motion is coupled with the ocean by using a simple parametric method, that is, the atmospheric windstress acts on the ocean as a body force and the oceanic heating is directly proportional to the mixed layer depth. So air-sea coupled models in the cylindrical coordinate are as follows:

1) A linear air-sea coupled model

$$\frac{\partial u'_a}{\partial t} - f_a v'_a = - g\frac{\partial h'_a}{\partial r} - a_1 u'_a \tag{1}$$

$$\frac{\partial v'_a}{\partial t} + f_a u'_a = - a_1 v'_a \tag{2}$$

$$\frac{\partial h'_a}{\partial t} + H\frac{\partial (ru'_a)}{r\partial r} = - \alpha h'_s - b_1 h'_a \tag{3}$$

$$\frac{\partial u'_s}{\partial t} - f_s v'_s = - g^*\frac{\partial h'_s}{\partial r} + \gamma u'_a - a_2 u'_s \tag{4}$$

$$\frac{\partial v'_s}{\partial t} + f_s u'_s = \gamma v'_a - a_2 v'_s \tag{5}$$

$$\frac{\partial h'_s}{\partial t} + D\frac{\partial (ru'_s)}{r\partial r} = - b_2 h'_s \tag{6}$$

where U and V are the radial and tangential velocities; r is the radial coordinate; subscripts a and s represent the atmosphere and ocean respectively; α and γ are the air-sea coupling coefficients; $f_a = 2\Omega_a + f, f_s = 2\Omega_s + f$. Ω_a and Ω_s are the angular speeds of the atmosphere and ocean basic flows, f the coriolis parameter $g^* = g \cdot \Delta\rho/\rho_s$ g is the gravitational acceleration, ρ_s the density of sea water, $\Delta\rho$ the density difference between thermocline layer and the mixed layer; a_i and $b_i (i=1,2)$ are the damping coefficents. Other symbols are as commonly used.

2) A nonlinear air-sea coupled model

$$\frac{\partial u_a}{\partial t} + u_a\frac{\partial u_a}{\partial r} - \left(\frac{v_a}{r} + f\right)V_a = - g\frac{\partial h_a}{\partial r} - A_1 u_a \tag{7}$$

$$\frac{\partial v_a}{\partial t} + u_a\frac{\partial v_a}{\partial r} + \left(\frac{v_a}{r} + f\right)u_a = - A_1 v_a \tag{8}$$

$$\frac{\partial h_a}{\partial t} + u_a\frac{\partial h_a}{\partial r} + (H + h_a)\frac{\partial ru_a}{r\partial r} = - \alpha h_s - B_1 h_a \tag{9}$$

$$\frac{\partial u_s}{\partial t} + u_s\frac{\partial u_s}{\partial r} - \left(\frac{v_s}{r} + f\right)v_s = - g\frac{\partial h_s}{\partial r} + \gamma u_a - A_2 u_s \tag{10}$$

$$\frac{\partial v_s}{\partial t} + u_s\frac{\partial v_s}{\partial r} + \left(\frac{v_s}{r} + f\right)u_s = - \gamma v_a - A_2 v_s \tag{11}$$

$$\frac{\partial h_s}{\partial t} + u_s\frac{\partial h_s}{\partial t}(D + h_s)\frac{\partial ru_s}{r\partial r} = - B_2 h_s \tag{12}$$

where A_i and $B_i (i=1,2)$ are the damping coeffecients.

We use the boundary conditions:

$$(u,v)\mid_{r=0} = 0,$$

$$(u,v,h)\mid_{x\to\infty} = 0,$$

$$\frac{\partial h_a}{\partial t}\Big|_{r=0} = \left[-(H + h_a)\frac{\partial (ru_a)}{r\partial r} - \alpha h_s - B_1 h_a\right]_{r=0} \tag{13}$$

$$\frac{\partial h_s}{\partial t}\Big|_{r=0} = \left[(-D + h_s)\frac{\partial(ru_s)}{r\partial r} - B_2 h_s \right]_{r=0}$$

Ⅲ　THE NUMERICAL RESULTS

1) The linear model

Taking $\Omega_a = 5\times10^{-5}$ s, $\Omega_s = 5\times10^{-6}$ s, $f = 3.8\times10^{-5}$ s (about 15 N), Figs 1-4 are the variation of physical variables of a tropical vortex with time in the 200 km from the vortex center. For the weak air-sea interaction, the vortex shows a damped oscillation with a period of 15 days. As the coupling coefficients increase, the vortex tends towards an equal amplitude oscillation, and the period decreases rapidly and reaches about 2 days. They become unstable when the coupling coefficients exceed 7×10^{-10} s^2, but the period is only modified a little (Fig.3). As compared with paper Ⅰ, there is a similar relationship between the vortex instability and the coupling coefficients, but the critical coupling coefficients rise about 10 times. One reason is that the model includes a damping effect. It is seen from Fig.1 and Fig.3 that the damped oscillation is different from the unstable oscillation. In the damped oscillation, a cyclone in the atmosphere directly produces a cold cyclonic vortex in the ocean by action of windstress. The air divergence over the cold water causes an atmospheric anticyclone to develop. On the other hand, the anticyclone will excite a warm anticyclonic vortex in the ocean. The oceanic heating makes the air in the low layer converge so that the atmospheric cyclone develops. Such a cycle makes up a oscillation process in the air-sea coupled system. In fact, this oscillation is a direct response of the atmosphere to the ocean heating and the ocean to the atmospheric windstress. For the unstable oscillation, thd former is same, but the latter is different. Under the action of the Ekman pumping, the cyclone in the atmosphere results in the cold water upwelling and divergence outside the vortex center. As a results, a cold anticyclone vortex is developed in the ocean. On the contary, the warm cyclonic vortex in the ocean comes to pass again. It implies that the unstable oscillation contains a current adjustment procedure in the ocean.

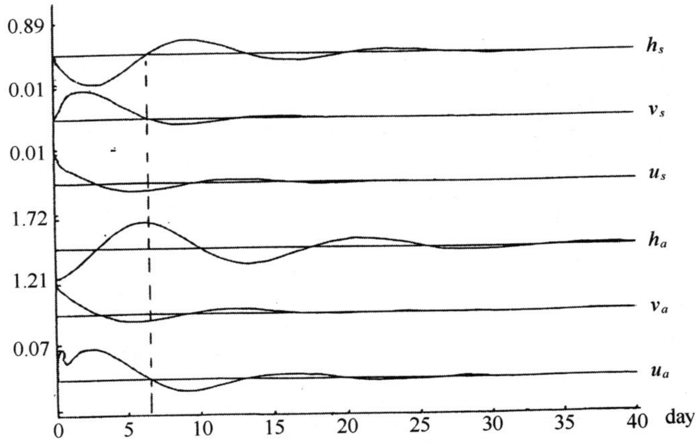

Fig.1　Variation of perturbation variables with time in the linear model for the coupling coefficient $5\times10^{-11}/s^2$

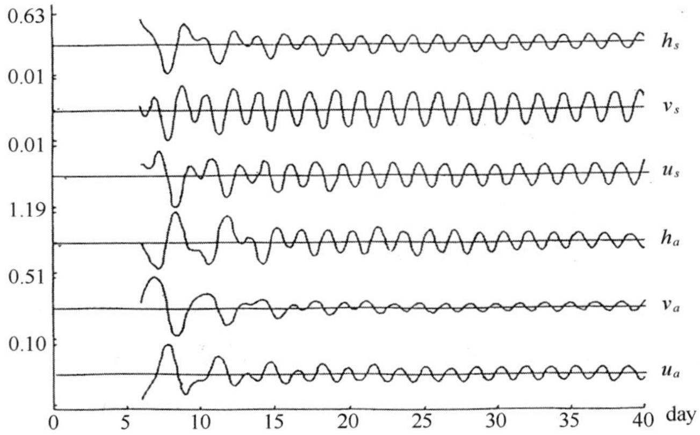

Fig.2 As in Fig.1, except for the coupling coefficient $5.5\times10^{-10}/s^2$

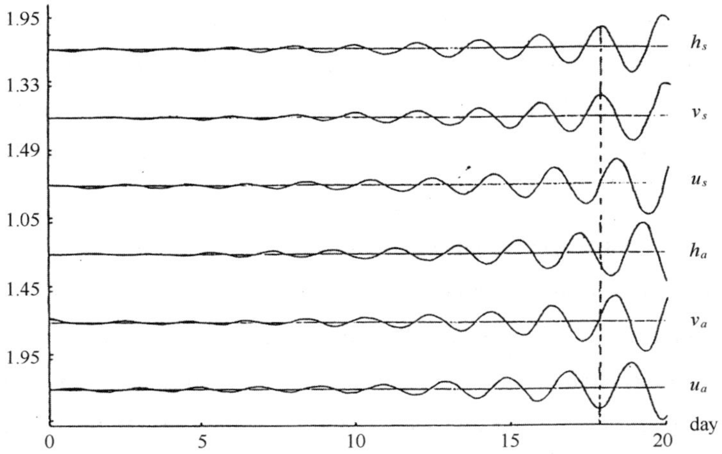

Fig.3 As in Fig.1, except for the coupling coefficient $8\times10^{-10}/s^2$

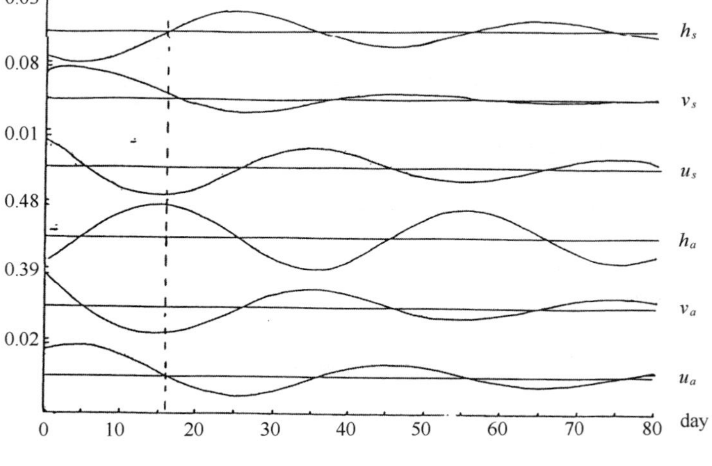

Fig.4 Variation of perturbation variables with time in the nonlinear model for the coupling coefficient $5\times10^{-9}/s^2$

Results also show that the instability of tropical vortex is sensitive to latitude. Under conditions of the most probable values of the coupling coefficients, the vortex instabilty only occurs in the regions of low latitude. Reason is the same as that mentioned in paper I.Both the coriolis force and the Ekman pumping make the cold water upwelling enhance and SST decrease so that an effect of oceanic heating upon the atmosphere is weakened.

2) The nonlinear model

The numerical results calculated from the nonlinear coupled model are given in Figs 4–6. There are two different characteristics in the nonlinear model by comparision with the linear model. One is a multi-period oscillation which consists of two main periods for 2 days and 40 days. The multi-period exists only under a condition of suitable coupling coefficients. As the coupling coefficients increase, the first period is changed a little, but the second will vanish. At the same time, the vortex motion becomes unstable (see Fig.6). The other difference is that the nonlinear effect leasd to an increase of the critical coupling coefficients. It results from the advective effect in the nonlinear model.

The influence of the coupling coefficients and latitude on the instability of a tropical vortex is the same in the nonlinear model as in the linear model.

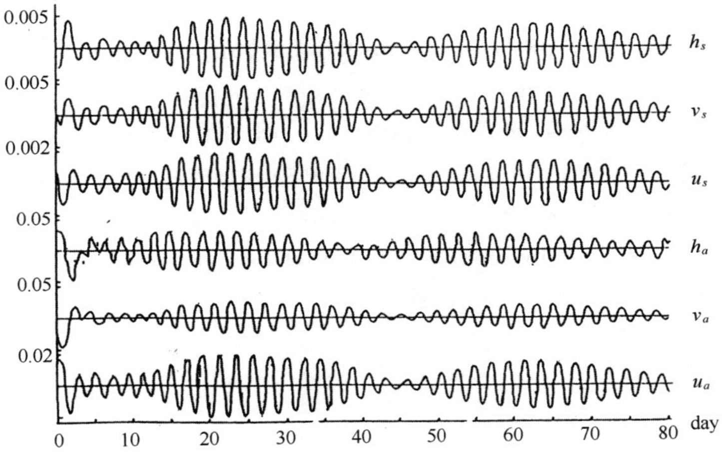

Fig.5　As in Fig.4, except for the coupling coefficient $5\times10^{-8}/s^2$

Ⅳ　CONCLUSIONS

The relationship of the instability of a tropical vortex to various parameters is discussed in the linear and nonolinear air-sea coupled models. The main results from this study are

1) In the linear and nonlinear models, the instability of tropical vortex depends on the coupling coefficients and latitude, but those parameters have little effect on the unstable period.

2) There exists a multi-period oscillation in the nonlinear model. Main periods are approximately 2 days and 40 days.

3) The mechanism of the damped oscillation is different from that of the unstable oscillation. The former is a direct response of the atmosphere to the oceanic heating and the ocean to the atmopheric windstress. The

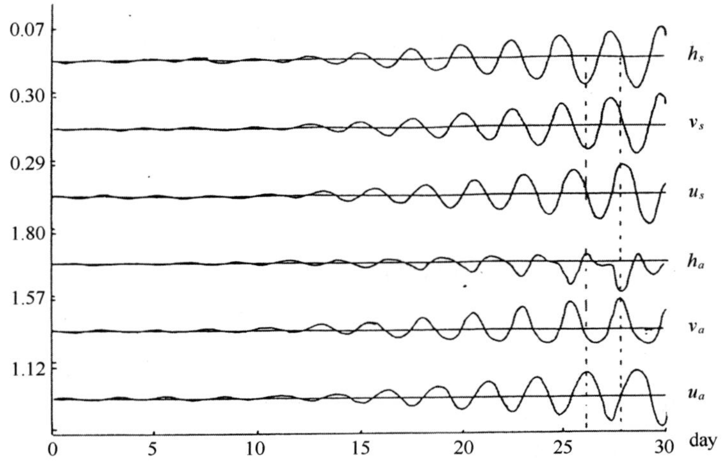

Fig.6 As in Fig.4, except for the coupling coefficient $7\times10^{-8}/s^2$

latter involves a current adjustment process in the ocean besides the direct response.

References

[1] Charney, J. G. and Eliassen, A. On The Growth of The Hurricane Depression, J. Atmos. Sci, 1964:2168–2175.

[2] Gray, W. M. Grobal View of The Origin of Tropical Disturbances And Storms. Mon. Wea. Rew., 1968, 96:669–700.

[3] Liu, Shikuo and Ni Bingjing. Influence of The Inertial Stability And Stratified Stability on The Development of Typhoon. Collection of Typhoon Meeting, Shanghai Scientific Publishing House, 1983.

[4] Palmen, E. On The Formation And Structure of Tropical Hurricanes. Geophysican, 1949, 3:26–38.

[5] Wang Zhanggui and Chao Jiping. On The Instability of A Tropical Vortex in An Air-Sea Coupled Model: Ⅰ. The Theoretical Analysis. Submitted to ACTA Meteorology, 1991.

热带地转适应运动的动力学基础*

巢纪平

（国家海洋环境预报研究中心，北京 100081）

摘要： 文中讨论了热带斜压大气地转适应过程中的若干动力学约束关系，在不考虑行星位势涡度梯度的前提下给出了三维重力惯性波的频散方程、位势涡度时间不变式。在这基础上指出由于 Taylor-Proudman 定理成立，运动将趋于水平化。同时指出，在热带纬圈半地转平衡更易出现。地转适应后的运动，一般是水平无辐散的，虽然垂直运动趋于零，但物理场随高度仍然有变化，即是层结的。

关键词： 热带斜压大气；地转适应；动力学约束

1　引言

在重力场和旋转力场作用下的大尺度大气（包括海洋）运动，其基本状态是静力平衡和地转平衡的。当地转平衡受到破坏出现非地转风后，将激发出重力惯性波，随着重力惯性波在无界空间中的频散，运动的非地转分量消失，而重新建立起风、压场之间的地转关系。这一过程即为地转适应，最早是由 Rossby[1,2] 提出的。

在地转适应过程中，科氏力的作用是不可缺少的，但不需要考虑由于球面而引起的科氏参数随纬度的梯度，即行星涡度梯度。因为若计入行星涡度梯度，将激发出低频的行星波（长波或 Rossby 波）。这样运动将达不到地转的平衡状态，而将较慢地随时间演变。在地转适应过程中，由于行星涡度的存在，对运动有一个重要的约束关系，称为位势涡度的时间不变式，它最早由 Oboukhov[3] 提出。由于位势涡度时间不变式，使我们不必去研究地转平衡的建立过程，而只需要研究过程的最终状态，这给研究带来了极大的方便。

关于中、高纬度地转适应中的一些动力学约束关系，经叶笃正、曾庆存等[4,5] 的研究后已十分清楚了（见参考文献[6]以及最近的评述文章[7]）。对于热带地区，过去由于认为科氏参数 $f(=2\Omega\sin\psi$，Ω 为地转角速度，ψ 为纬度）很小，是否仍然存在地转适应过程是一个可值得质疑的问题，很少有人探讨。但注意到，实际的热带大气和海洋状态表明，其纬圈风（如信风带或 Walker 环流中的高、低空的东、西风带）或纬圈流（如南、北赤道洋流）是不小的，可达到 10^0 m/s 或 10^0 cm/s 的量级，因此其科氏力可以不小，且压力在经圈方向的梯度并不大，因此至少在纬圈方向出现地转平衡是可能的。事实上，Gill[8] 研究在低频演变运动的模式中，已用了纬圈地转平衡的长波近似。鉴于此，最近巢纪平和林永辉[9] 研究了热带大气和海洋的纬圈半地转平衡的建立，同时也指出，在海洋经圈边界附近，经圈流

* 气象学报，2000，58(1)：1-10.

速很大,因此在这样特殊的地区,建立经圈地转平衡也是可能的。

在此后的一些论文中,将进一步研究热带大气和海洋的地转适应运动。而文中先给出适应过程中动力学的一些约束关系。至于由于行星涡度梯度的存在,地转适应完成后,运动进入到较缓慢变化的发展(或称演变)阶段的情况,以及非线性对流项的作用将另文再讨论。

2　基本方程

在赤道 $\beta(f=f_0\beta y)$ 平面近似下,Boussinesq 流体的基本运动方程为

$$\frac{\partial u}{\partial t}-\beta yv=-\frac{\partial}{\partial x}\left(\frac{p}{\rho_0}\right) \tag{1}$$

$$\frac{\partial v}{\partial t}+\beta yu=-\frac{\partial}{\partial y}\left(\frac{p}{\rho_0}\right) \tag{2}$$

$$\frac{\partial u}{\partial x}+\frac{\partial v}{\partial y}+\frac{\partial w}{\partial z}=0 \tag{3}$$

$$\frac{g}{\theta_0}=\frac{\partial}{\partial z}\left(\frac{p}{\rho_0}\right) \tag{4}$$

$$\frac{\partial}{\partial t}\theta'+\frac{\mathrm{d}\theta_0}{\mathrm{d}z}w=0 \tag{5}$$

式中,ρ_0,θ_0 分别为背景场的密度和位温,其他符号同常用。

设大气的特征厚度为 H,则可引进重力内波波速,为

$$C=\left(\frac{g}{\theta_0}\frac{\mathrm{d}\theta_0}{\mathrm{d}z}H^2\right)^{\frac{1}{2}} \tag{6}$$

引进特征量

$$(x,y)=\left(\frac{C}{2\beta}\right)^{\frac{1}{2}}(x',y'),\quad z=Hz',\quad t=(2\beta C)^{-\frac{1}{2}}t',\quad (u,v)=C(u',v'),$$

$$w=(2\beta C)^{-\frac{1}{2}}Hw',\quad \frac{p}{\rho_0}=C^2\varphi',\quad \vartheta=\frac{\mathrm{d}\theta_0}{\mathrm{d}z}H'\vartheta \tag{7}$$

式中,特征长度为赤道 Rossby 变形半径;特征时间为扰动以速度 C 传过这一变形半径时所需的时间,无量纲方程为(略去“′”号)

$$\varepsilon_1\frac{\partial u}{\partial t}-\frac{1}{2}yv=-\frac{\partial\varphi}{\partial x} \tag{8}$$

$$\varepsilon_2\frac{\partial v}{\partial t}+\frac{1}{2}yu=-\frac{\partial\varphi}{\partial y} \tag{9}$$

$$\varepsilon_3\frac{\partial}{\partial t}+w=0 \tag{10}$$

$$\vartheta=\frac{\partial\varphi}{\partial z} \tag{11}$$

$$\frac{\partial u}{\partial x}+\frac{\partial v}{\partial y}+\frac{\partial w}{\partial z}=0 \tag{12}$$

式中,$\varepsilon_{i=1,2,3}$为引进的标识符,其值取 1 或 0,当取 0 值时,式(8)至式(10)简化成

$$\frac{1}{2}yv = \frac{\partial \varphi}{\partial x} \tag{13}$$

$$\frac{1}{2}yu = \frac{\partial \varphi}{\partial y} \tag{14}$$

$$w = 0 \tag{15}$$

即运动是地转的和水平的，而式(11)表明运动又是静力平衡的，如果边界上垂直运动为零，则由式(15)导出

$$\frac{\partial u}{\partial x} + \frac{\partial v}{\partial y} = 0 \tag{16}$$

即平衡状态下的运动是水平无辐散的。适应过程要研究的是怎样达到这样的平衡状态。

如果设运动的垂直分布为

$$(w, \vartheta) = \sum_m (W(x,y,t), \theta(x,y,t)) \sin m\pi z \tag{17}$$

$$(u, v, \varphi) = \sum_m (U(x,y,t), V(x,y,t), \Phi(x,y,t)) \cos m\pi z \tag{18}$$

则以上各式给出

$$\varepsilon_1 \frac{\partial U}{\partial t} - \frac{1}{2}yV = -\frac{\partial \Phi}{\partial x} \tag{19}$$

$$\varepsilon_2 \frac{\partial V}{\partial t} + \frac{1}{2}yU = -\frac{\partial \Phi}{\partial y} \tag{20}$$

$$\varepsilon_3 \frac{\partial \Phi}{\partial t} + \frac{1}{m^2\pi^2}\left(\frac{\partial U}{\partial x} + \frac{\partial V}{\partial y}\right) = 0 \tag{21}$$

因此，只要将重力内波波速改写成

$$C = \left[g\left(\frac{1}{\theta_0}\frac{\mathrm{d}\theta_0}{\mathrm{d}z}\frac{H^2}{m^2\pi^2}\right)\right]^{-\frac{1}{2}} \tag{22}$$

式(21)即可写成

$$\varepsilon_3 \frac{\partial \Phi}{\partial t} + \frac{\partial U}{\partial x} + \frac{\partial V}{\partial y} = 0 \tag{23}$$

式(19)，式(20)和式(23)称为等值浅水运动方程，而

$$h_m = \left(\frac{1}{\theta_0}\frac{\mathrm{d}\theta_0}{\mathrm{d}z}\frac{H^2}{m^2\pi^2}\right) \tag{24}$$

称为等值高度。这表明，等值浅水模式并不一定是正压的，而只是说，对任一个垂直本征模来讲，其水平结构方程很像正压运动[10]。文中将直接分析方程(8)至方程(12)，但如果在某些情况下应用式(19)，式(20)和式(23)并不意味着讨论的一定是正压运动，而只是讨论了在垂直方向的某一模态。

如运动不处在平衡状态，则 u,v,w 可以用 ψ 表示：

$$\varepsilon_1\varepsilon_2 \frac{\partial^2 u}{\partial t^2} - \frac{1}{4}y^2 u = -\left(\varepsilon_2 \frac{\partial^2 \psi}{\partial t \partial x} + \frac{1}{2}y\frac{\partial \psi}{\partial y}\right) \tag{25}$$

$$\varepsilon_1\varepsilon_2 \frac{\partial^2 v}{\partial t^2} - \frac{1}{4}y^2 v = -\left(\varepsilon_2 \frac{\partial^2 \psi}{\partial t \partial y} - \frac{1}{2}y\frac{\partial \psi}{\partial x}\right) \tag{26}$$

$$\varepsilon_1\varepsilon_2 \frac{\partial^2 w}{\partial t^2 \partial z} - \frac{1}{4}y^2 \frac{\partial w}{\partial z} = \frac{\partial}{\partial t}\left(\varepsilon_2 \frac{\partial^2 \psi}{\partial x^2} + \varepsilon_1 \frac{\partial^2 \psi}{\partial y^2}\right) \tag{27}$$

由此可以得到对变量 v 的单一方程，为

$$\frac{\partial}{\partial t}\left[\varepsilon_1\varepsilon_2\varepsilon_3\frac{\partial^4 v}{\partial t^2\partial z^2}+\varepsilon_2\frac{\partial^2 v}{\partial x^2}+\varepsilon_1\frac{\partial^2 v}{\partial y^2}+\varepsilon_3\frac{1}{4}y^2\frac{\partial^2 v}{\partial z^2}\right]+\frac{1}{2}\frac{\partial v}{\partial x}=0 \tag{28}$$

显而易见，这是斜压大气的 Matsuno 方程[11]。最后不带 ε_i 的项是由于考虑了行星涡度梯度而得到的，由于有这一项，在式(28)中除包含有高频的重力惯性波外，尚有低频的 Rossby 波，如前述，在讨论地转适应过程时，将不考虑由于行星涡度梯度而激发出的低频 Rossby 波。我们将遵循这一约定：即保留行星涡度的经圈不均匀性(赤道 β 平面的几何效应)，但不引进由其梯度产生的动力效应，事实上这是过滤掉低频波的一种方法。不作这样的假定，把快的适应过程和慢的演变过程统一起来处理的方法将在另文中给出。

3 三维重力惯性波动方程

略去式(28)不带 ε_i 的项，得到

$$\frac{\partial}{\partial t}\left[\varepsilon_1\varepsilon_2\varepsilon_3\frac{\partial^4 v}{\partial t^2\partial z^2}+\varepsilon_2\frac{\partial^2 v}{\partial x^2}+\varepsilon_1\frac{\partial^2 v}{\partial y^2}+\varepsilon_3\frac{1}{4}y^2\frac{\partial^2 v}{\partial z^2}\right]=0 \tag{29}$$

对时间积分一次，给出

$$\begin{aligned}&\varepsilon_1\varepsilon_2\varepsilon_3\frac{\partial^4 v}{\partial t^2\partial z^2}+\varepsilon_2\frac{\partial^2 v}{\partial x^2}+\varepsilon_1\frac{\partial^2 v}{\partial y^2}+\varepsilon_3\frac{1}{4}y^2\frac{\partial^2 v}{\partial z^2}\\&=\left[\varepsilon_1\varepsilon_2\varepsilon_3\frac{\partial^4 v}{\partial t^2\partial z^2}+\varepsilon_2\frac{\partial^2 v}{\partial x^2}+\varepsilon_1\frac{\partial^2 v}{\partial y^2}+\varepsilon_3\frac{1}{4}y^2\frac{\partial^2 v}{\partial z^2}\right]_{t=0}\end{aligned} \tag{30}$$

将式(26)及式(10)至式(12)应用到上式右边，可改写成

$$\varepsilon_1\varepsilon_2\varepsilon_3\frac{\partial^4 v}{\partial t^2\partial z^2}+\varepsilon_2\frac{\partial^2 v}{\partial x^2}+\varepsilon_1\frac{\partial^2 v}{\partial y^2}+\varepsilon_3\frac{1}{4}y^2\frac{\partial^2 v}{\partial z^2}=\frac{\partial}{\partial x}\left[\left(\varepsilon_2\frac{\partial v}{\partial x}-\varepsilon_1\frac{\partial u}{\partial y}\right)+\varepsilon_3\frac{1}{2}y\frac{\partial^2\varphi}{\partial z^2}\right]_{t=0} \tag{31}$$

显而易见，右端括号中的量是一位势涡度，该式表明，在斜压大气中以经圈速度为表征的波动，可以是在初始时刻位势涡度的纬圈梯度作用下被激发出来的，而这一波动即为赤道 β 平面中的斜压重力惯性波。事实上，这可以做下面简单的处理后看出，令式(31)右端为

$$F(x,y,z,0)=\frac{\partial}{\partial x}\left[\left(\varepsilon_2\frac{\partial v}{\partial x}-\varepsilon_1\frac{\partial u}{\partial y}\right)+\varepsilon_3\frac{1}{2}y\frac{\partial^2\varphi}{\partial z^2}\right]_{t=0}=\sum_m F_m(x,y,0)\cos m\pi z \tag{32}$$

而由式(18)

$$v(x,y,t)=\sum_m V_m(x,y,t)\cos m\pi z \tag{33}$$

则式(31)给出

$$\varepsilon_1\varepsilon_2\varepsilon_3 m^2\pi^2\frac{\partial^2 V_m}{\partial t^2}-\left(\varepsilon_2\frac{\partial^2 V_m}{\partial x^2}+\varepsilon_1\frac{\partial^2 V_m}{\partial y^2}-\frac{1}{4}m^2\pi^2y^2V_m\right)=F_m(x,y,0) \tag{34}$$

方程左端是 Klein 波动算子，在这里它描写的是赤道 β 平面中的重力惯性波。

类似地，容易得到下面的波动方程

$$\varepsilon_1\varepsilon_2\varepsilon_3\frac{\partial^4 u}{\partial t^2\partial z^2}+\varepsilon_2\frac{\partial^2 u}{\partial x^2}+\varepsilon_1\frac{\partial^2 u}{\partial y^2}+\varepsilon_3\frac{1}{4}y^2\frac{\partial^2 u}{\partial z^2}=-\frac{\partial}{\partial y}\left[\left(\varepsilon_2\frac{\partial v}{\partial x}-\varepsilon_1\frac{\partial u}{\partial y}\right)+\varepsilon_3\frac{1}{2}y\frac{\partial^2\varphi}{\partial z^2}\right]_{t=0} \tag{35}$$

这表明，重力惯性波中的纬圈风受初始时刻位势涡度的经圈梯度制约。对 φ 有

$$\varepsilon_1\varepsilon_2\varepsilon_3\frac{\partial^4 \varphi}{\partial t^2\partial z^2}+\varepsilon_2\frac{\partial^2 \varphi}{\partial x^2}+\varepsilon_1\frac{\partial^2 \varphi}{\partial y^2}+\varepsilon_3\frac{1}{4}y^2\frac{\partial^2 \varphi}{\partial z^2}=\frac{1}{2}y\left[\left(\varepsilon_2\frac{\partial v}{\partial x}-\varepsilon_1\frac{\partial u}{\partial y}\right)+\varepsilon_3\frac{1}{2}y\frac{\partial^2\varphi}{\partial z^2}\right]_{t=0} \tag{36}$$

而波动中的重力位势高度场，直接受位势涡度制约。对 w 有

$$\varepsilon_1\varepsilon_2\varepsilon_3\frac{\partial^4 w}{\partial t^2\partial z^2}+\varepsilon_2\frac{\partial^2 w}{\partial x^2}+\varepsilon_1\frac{\partial^2 w}{\partial y^2}+\varepsilon_3\frac{1}{4}y^2\frac{\partial^2 w}{\partial z^2}=0 \tag{37}$$

注意到，垂直运动的频散特征与其他物理量不同，它没有初始时刻的位势涡度梯度支持，能激发它的只能是初始时刻的垂直运动或垂直运动的时间变化，或者边界上的垂直运动。

4　位势涡度时间不变式

将式(26)及式(10)至式(12)应用到式(31)左边，并对 x 积分一次，给出

$$\varepsilon_2\frac{\partial v}{\partial x}-\varepsilon_1\frac{\partial u}{\partial y}+\varepsilon_3\frac{1}{2}y\frac{\partial^2\varphi}{\partial z^2}=\left[\varepsilon_2\frac{\partial v}{\partial x}-\varepsilon_1\frac{\partial u}{\partial y}+\varepsilon_3\frac{1}{2}y\frac{\partial^2\varphi}{\partial z^2}\right]_{t=0} \tag{38}$$

即为位势涡度的时间不变式，它在研究适应过程的最终状态时有重要的作用。同样也可由式(35)或式(36)导出此式。

5　Taylor-Proudman 定理

在旋转力场作用下的流体运动，沿旋转轴方向的速度分量消失，即为 Taylor-Proudman 定理[12]。对于地球大气旋转力表现为科氏力，在科氏力作用下大尺度运动的垂直速度很小，即运动将趋于水平化，为 Taylor-Proudman 在大气运动中的具体表现。

在热带，虽然科氏力很小，但仍受 Taylor-Proudman 定理的制约。

$$\text{设}\qquad z=0,1\qquad w=0 \tag{39}$$

w 可写成

$$w=\sum_m W_m(x,y,t)\sin m\pi z \tag{40}$$

由此对任一个 m，方程(37)给出(略去标识符)

$$m^2\pi^2\frac{\partial^2 W_m}{\partial t^2}-\frac{\partial^2 W_m}{\partial x^2}-\frac{\partial^2 W_m}{\partial y^2}+m^2\pi^2\frac{1}{4}y^2W_m=0 \tag{41}$$

作变换

$$\tau=\frac{t}{\overline{m\pi}},X=\overline{m\pi}x,Y=\overline{m\pi}y \tag{42}$$

式(41)可改写成

$$\frac{\partial^2 W_m}{\partial\tau^2}-\frac{\partial^2 W_m}{\partial X^2}-\frac{\partial^2 W_m}{\partial Y^2}+\frac{1}{4}Y^2W_m=0 \tag{43}$$

初始条件为

$$\tau=0,\quad W_m=\Phi_{1m}(X,Y),\quad \frac{\partial W_m}{\partial\tau}=\Phi_{2m}(X,Y) \tag{44}$$

边界条件为

$$|X|\to\infty,\qquad W_m\to 0 \tag{45}$$

将变量在 Y 方向用 Weber 函数展开，即

$$W_m(X,Y,\tau)=\sum_n W_{mn}(X,\tau)D_n(Y) \tag{46}$$

$$\Phi_{1m}(X,Y)=\sum_n \Phi_{1mn}(X)D_n(Y) \tag{47}$$

$$\Phi_{2m}(X,Y)=\sum_n \Phi_{2mn}(X)D_n(Y) \tag{48}$$

于是

$$\frac{\partial^2 W_{mn}}{\partial \tau^2}-\frac{\partial^2 W_{mn}}{\partial X^2}+\left(n+\frac{1}{2}\right)W_{mn}=0 \tag{49}$$

$$\tau=0,\quad W_{mn}=\Phi_{1mn}(X),\quad \frac{\partial W_{mn}}{\partial f}=\Phi_{2mn}(X) \tag{50}$$

$$|X|\to\infty,\qquad W_{mn}\to 0 \tag{51}$$

其解为

$$W_{mn}=\frac{1}{2}\int_{-\infty}^{\infty}\Phi_{1mn}(X')\frac{\mathrm{J}_1\left(\sqrt{n+\frac{1}{2}}\sqrt{\tau^2-(X-X')^2}\right)}{\sqrt{\tau^2-(X-X')^2}}\mathrm{d}X'+\frac{1}{2}\int_{-\infty}^{\infty}\Phi_{2mn}(X')\mathrm{J}_0\left(\sqrt{n+\frac{1}{2}}\sqrt{\tau^2-(X-X')^2}\right)\mathrm{d}X' \tag{52}$$

式中 J_ν 为贝塞尔函数。

若初始扰动只局限一有限的区间内，则根据贝塞尔函数的性质，容易得到，当 $\tau\to\infty$ 时，$W_{mn}\to 0$，即运动趋于水平化，此即 Taylor-Proudman 定理。如果时间是有限的（但很大），虽然 W_{mn} 仍有值，但其值已很小，因此运动仍然是准水平的。就这一点来讲，在科氏力场作用下热带大尺度运动，具有和中、高纬度大尺度运动接近水平的动力学性质。

6 重力惯性波的频散

上节已研究了重力惯性波的频散，但由于垂直运动场没有初始位势涡度梯度的支持，因此初始垂直运动的影响将很快消失，对其他的场因有位势涡度的支持，其动力学行为将变得不同。

如对经圈风，由方程（34）可得到（略去标识符）

$$\frac{\partial^2 V_m}{\partial \tau^2}-\frac{\partial^2 V_m}{\partial X^2}-\frac{\partial^2 V_m}{\partial Y^2}+\frac{1}{4}Y^2 V_m=F_m/m\pi \tag{53}$$

初始条件和边界条件分别为

$$\tau=0,\quad V_m=\Phi_{1m}(X,Y),\quad \frac{\partial V_m}{\partial \tau}=\Phi_{2m}(X,Y) \tag{54}$$

$$|X|\to\infty,\quad V_m\to 0 \tag{55}$$

将变量在 Y 方向用 Weber 函数展开成

$$V_m(\tau,X,Y)=\sum_n V_{mn}(\tau,X)D_n(Y) \tag{56}$$

则问题变为

$$\frac{\partial^2 V_{mn}}{\partial t^2}-\frac{\partial^2 V_{mn}}{\partial x^2}+\left(n+\frac{1}{2}\right)V_{mn}=F_m/m\pi \tag{57}$$

$$\tau=0,\quad V_{mn}=\Phi_{1mn}(X),\quad \frac{\partial V_m}{\partial \tau}=\Phi_{2mn}(X) \tag{58}$$

$$|X|\to\infty,\qquad V_{mn}\to 0 \tag{59}$$

其解为

$$V_{mn}=\frac{1}{2}\int_{-\infty}^{\infty}\Phi_{1mn}(X')\frac{J_1\left(\sqrt{n+\frac{1}{2}}\sqrt{\tau^2-(X-X')^2}\right)}{\sqrt{\tau^2-(X-X')^2}}\mathrm{d}X'+$$

$$\frac{1}{2}\int_{-\infty}^{\infty}\Phi_{2mn}(X')J_0\left(\sqrt{n+\frac{1}{2}}\sqrt{\tau^2-(X-X')^2}\right)\mathrm{d}X'+$$

$$\frac{1}{2m\pi}\int_0^{\tau}\int_{-\infty}^{\infty}F_{mn}(X')J_0\left(\sqrt{n+\frac{1}{2}}\sqrt{\tau^2-(X-X')^2}\right)\mathrm{d}X'\mathrm{d}\tau' \tag{60}$$

注意到，当时间充分长后，初值的影响消失很快，解的主要贡献部分来自位势涡度梯度的影响，为

$$V_{mn}=\frac{1}{2m\pi}\int_0^{\tau}\int_{-\infty}^{\infty}F_{mn}(X')J_0\left(\sqrt{n+\frac{1}{2}}\sqrt{\tau'^2-(X-X')^2}\right)\mathrm{d}X'\mathrm{d}\tau' \tag{61}$$

其时间导数为

$$\frac{\partial V_{mn}}{\partial\tau}=\frac{1}{2m\pi}\int_{-\infty}^{\infty}F_{mn}(X')J_0\left(\sqrt{n+\frac{1}{2}}\sqrt{\tau^2-(X-X')^2}\right)\mathrm{d}X' \tag{62}$$

设 F_{mn} 只集中在原点附近，则可近似地表示成

$$F_{m,n}=A_{m,n}\delta(X) \tag{63}$$

式中，$\delta(X)$ 为 Delta 函数，于是式(61)，式(62)分别为

$$V_{mn}=\frac{A_{mn}}{2m\pi}\int_0^{\tau}J_0\left(\sqrt{n+\frac{1}{2}}\sqrt{\tau^2-X^2}\right)\mathrm{d}\tau' \tag{64}$$

$$\frac{\partial V_{mn}}{\partial\tau}=\frac{A_{mn}}{2m\pi}J_0\left(\sqrt{n+\frac{1}{2}}\sqrt{\tau^2-X^2}\right)\mathrm{d}\tau' \tag{65}$$

由此可见，在 $\tau>x$ 的区域中，当 $\tau\to\infty$ 时，式(65)的渐近性态为

$$\frac{\partial V_{mn}}{\partial\tau}\sim O(\tau^{-\frac{1}{2}}) \tag{66}$$

即经圈风的时间导数按 $\tau^{-\frac{1}{2}}$ 次幂衰减。但这时在原点($X=0$)式(64)给出

$$V_{mn}(0,\tau\to\infty)=\frac{A_{mn}}{2m\pi}\int_0^{\infty}J_0\left(\sqrt{n+\frac{1}{2}}\tau'\right)\mathrm{d}\tau'=\frac{A_{mn}}{2m\pi\sqrt{n+\frac{1}{2}}} \tag{67}$$

可见，其渐近值为有限，这表明在初始时刻位势涡度的纬圈梯度作用下，随着重力惯性波的频散，经圈流将被保留下来。巢纪平和林永辉曾计算当 $n=1$ 时在原点经圈流随时间的演变[13]。计算表明，约当 10 个无量纲时间后，由式(67) 算出的估计值已接近计算值。在正常的参数值下，对热带大气一个特征时间约为一周，10 个特征时间约相当于 10 周，这表明重力惯性波的频散速度并不是太快的。但注意到计算是对 $n=1$ 的扰动而言的，对 $n>1$ 的扰动频散自然要更快些。这表明由于重力惯性波的频散，使物理场在经圈方向趋于大尺度化。

7　准地转平衡和水平无辐散

设纬圈非地转平衡分量为

$$X_u = \frac{1}{2}yu + \frac{\partial \varphi}{\partial y} \tag{68}$$

则在不考虑行星涡度梯度的约定下,由式(35)和式(36)可得到

$$\varepsilon_1\varepsilon_2\varepsilon_3 \frac{\partial^4 X_u}{\partial t^2 \partial z^2} + \varepsilon_2 \frac{\partial^2 X_u}{\partial x^2} + \varepsilon_1 \frac{\partial^2 X_u}{\partial y^2} + \varepsilon_3 \frac{1}{4}y^2 \frac{\partial^2 X_u}{\partial z^2} = 0 \tag{69}$$

由以上的讨论可见,由于纬圈非地转风分量没有初始位势涡度梯度支持,如果也没有边界效应,则初始时刻的非地转风将以重力惯性波的方式被频散掉,而建立起新的纬圈地转平衡。

类似地,可设经圈非地转平衡分量为

$$X_v = \frac{1}{2}yv - \frac{\partial \varphi}{\partial x} \tag{70}$$

由式(31)和式(36)给出

$$\varepsilon_1\varepsilon_2\varepsilon_3 \frac{\partial^4 X_v}{\partial t^2 \partial x^2} + \varepsilon_2 \frac{\partial^2 X_v}{\partial x^2} + \varepsilon_1 \frac{\partial^2 X_v}{\partial y^2} + \varepsilon_3 \frac{1}{4} \frac{\partial X_v}{\partial z^2} = 0 \tag{71}$$

同样地,由于没有初始位势涡度梯度的支持,初始经圈非地转风分量也将以重力惯性波的方式被频散掉,而重新建立起新的经圈地转平衡。

在另一方面,如设水平辐散为

$$D = \frac{\partial u}{\partial x} + \frac{\partial v}{\partial y} \tag{72}$$

则由式(30)和式(35)得到

$$\varepsilon_1\varepsilon_2\varepsilon_3 \frac{\partial^4 D}{\partial t^2 \partial z^2} + \varepsilon_2 \frac{\partial^2 D}{\partial x^2} + \varepsilon_1 \frac{\partial^2 D}{\partial y^2} + \varepsilon_3 \frac{1}{4} \frac{\partial^2 D}{\partial z^2} = 0 \tag{73}$$

由此可见,如果没有边界效应,当时间充分长后将有 $t\to\infty$,$D\to 0$,即运动将趋于水平无辐散。

由这些讨论可知,如果不考虑行星涡度梯度的动力作用,运动的最终状态即为由式(13)至式(16)所表示的,是地转的、水平的和无辐散的。

然而,物理场可以在垂直方向分布为层结,这一方面是因为当 Taylor-Proudman 定理成立时,有

$$\frac{\partial}{\partial t} = \frac{\partial^2 \varphi}{\partial t \partial z} = 0 \tag{74}$$

即运动的最终状态,位温将趋于定常,而如果初始时刻位温随高度有分布,则最终状态时的位温将具有相同的垂直分布,或者对气压而言有

$$\frac{\partial \varphi}{\partial z} = \left.\frac{\partial \varphi}{\partial z}\right|_{t=0} \tag{75}$$

即气压随高度的分布同初始时刻。在另一方面,由于 u,v,φ 在适应过程完成后将保留下来,而这些物理量随高度是有分布的(展式(18)表明)。

8 结论

在赤道 U 平面近似下,如果不考虑行星涡度梯度的动力影响,即不考虑低频及 Rossby 波的作用,则热带运动的最基本状态是地转的、水平的、无辐散的,但可以是层结的。所有这些特征将通过重力惯性波的频散来实现,而初始位势涡度梯度为支持这样的运动起了重要的作用,否则最终状态是静态。

文中给出了地转适应运动中若干动力学约束，应用这些动力学基础可以对热带地转或半地转的适应后的状态作进一步讨论。

参考文献

[1] Rossby C G. On the mutual adjus tment of pressure and velocity distribution in certain simple current system Ⅰ. J Mar Res, 1937, 1:15-28.

[2] Rossby C G. On the mutual adjus tment of pressure and velocity distribution in certain simple current systems Ⅱ. J Mar Res, 1938, 2:239-263.

[3] Oboukhov A M. The problem of the geos trophic adaptation. Izvestiya of Academy of Science U SSR, Ser Geography and Geophysics, 1949, 13:281-189.

[4] Yeh T C. On the formation of quasi-geost rophic motion in the atmosphere. J Meteor Soc Japan, 1957:130-134.

[5] 曾庆存. 大气中的适应过程和发展过程(一)和(二). 气象学报, 1963, 33:163-174, 281-189.

[6] 叶笃正, 李麦村. 大气运动的适应问题. 北京：科学出版社, 1965.

[7] 叶笃正, 巢纪平. 论大气运动的多时态特征——适应、发展和准定常演变. 大气科学, 1998, 22:385-398.

[8] Giil A E. Some simple solution of the heat-induced tropical circulation. Quart J Roy Meteor Soc, 106:447-462.

[9] Chao J P, Lin Y H. The foundati on and movement of tropical semi-geostrophic adaptation. Acta Meteor Sinica, 1996, 10: 129-141.

[10] Moore D W, Philander S G M. Modelling of the tropical ocean circulation. In "The Sea" Goldberg E D, etc eds. New York: Wiley (Interscience), 1977, 6:319-362.

[11] Matsuno T. Quasi-geost rophic motion in the equatorial area. J M eteor Soc Japan, 1966, 44:25-43.

[12] Pedlosky J. Geoph ysical Fluid Dynamics. New York, Heid elberg, Berlin: Springer-Verlag, 1979.

论热带纬圈半地转运动的建立*

巢纪平

(国家海洋环境预报研究中心,北京 100081)

摘要: 文中在赤道 U 平面上,在滤掉低频 Rossby 波的情况下,研究了纬圈半地转运动的建立。指出,只有当运动的纬圈尺度很大时,非地转风分量才能随着重力惯性波的频散而消失,从而建立起纬圈半地转平衡。应用位势涡度不变式,给出了纬圈半地转适应后物理场的解。同时指出,Kelvin 波(对赤道对称情况)和混合波的 Rossby 波波段(对赤道反对称情况)将不参与适应运动,它们属于发展运动中的角色。

关键词: 纬圈半地转平衡;重力惯性波;位势涡度时间不变式

1 引言

在中、高纬度大尺度大气(也包括海洋)最基本的平衡状态是地转运动。当风、压场之间的地转平衡遭到破坏后,在 $f(=2\Omega\sin h$, Ω 为地转角速度, h 为纬度)平面上将激发出重力惯性波,当重力惯性波频散后,运动的势量消失,而管量保留下来,并重新建立起地转关系。场的这一过程称地转适应,最早由 Rossby 提出[1,2]。以后 Oboukhov[3]、叶笃正[4]和曾庆存[5]等对中、高纬度的地转适应过程进行了深入的研究,其物理机制已基本上清楚了[6,7]。

在热带地区,由于科氏参数很小,对是否还存在准地转运动是可以质疑的,因此很少有人来研究热带地区的地转适应过程。但我们注意到,在热带地区的大气中盛行东北或东南信风,风速可达 10^0 m/s,在海洋中是南、北赤道洋流,其流速可达 10^1 cm/s,因此除赤道是条奇异线外,离开赤道后,科氏力仍然可以达到与经圈压力梯度相平衡的量级。而理论上在广泛应用的 Gill 模式[8]中就假定运动是纬圈地转平衡的(长波近似),而即使在热带动力学中起重要作用的 Kelvin 波,其存在的条件也是纬圈地转平衡。因此研究纬圈地转平衡是如何建立起来的,是一个很有趣的热带运动的动力学问题。

由于地转平衡只在一个方向成立(纬圈的或经圈的),因此可以称为半地转运动,而它建立的适应过程可称为半地转适应。最近笔者等就这个问题写了几篇文章[9-11],得到了与中纬度地转适应相类似的结果。这自然是地转适应问题向热带地区的拓展,是一个值得进一步研究的问题。我们将在文中用另一种方法来进一步分析纬圈半地转适应过程。

2 基本方程

引进重力波波速,为

* 气象学报,2000,58(2):129-136.

$$C = (gh)^{\frac{1}{2}} \tag{1}$$

其中 h 为等值厚度，为三维方程在垂直方向按本征模展开时的本征值。取特征量为：长度为 $(C/2U)^{1/2}$，时间为 $(2UC)^{-1/2}$，速度为 C，重力位势高度为 C^2，于是对任一本征模的水平运动方程的无量纲方程为

$$\frac{\partial u}{\partial t} - \frac{1}{2}yv + \frac{\partial h}{\partial x} = 0 \tag{2}$$

$$\frac{\partial v}{\partial t} + \frac{1}{2}yu + \frac{\partial h}{\partial y} = 0 \tag{3}$$

$$\frac{\partial h}{\partial t} + \frac{\partial u}{\partial x} + \frac{\partial v}{\partial y} = 0 \tag{4}$$

按 Gill 和 Clarke[12] 引进变量

$$q = h + u, \qquad r = h - u \tag{5}$$

由此有

$$h = \frac{1}{2}(q + r), \qquad u = \frac{1}{2}(q - r) \tag{6}$$

于是有方程

$$\frac{\partial q}{\partial t} + \frac{\partial q}{\partial x} + \left(\frac{\partial v}{\partial y} - \frac{1}{2}yv\right) = 0 \tag{7}$$

$$\frac{\partial r}{\partial t} - \frac{\partial r}{\partial x} + \left(\frac{\partial v}{\partial y} + \frac{1}{2}yv\right) = 0 \tag{8}$$

$$\frac{\partial v}{\partial t} + \frac{1}{2}\left(\frac{\partial q}{\partial y} + \frac{1}{2}yq\right) + \frac{1}{2}\left(\frac{\partial r}{\partial y} - \frac{1}{2}yr\right) = 0 \tag{9}$$

对所有的物理量用抛物圆柱函数即 Weber 函数展开，例如对 q 有

$$q = \sum_n q_n(x,t)D_n(y) \tag{10}$$

考虑到抛物圆柱函数的循环公式

$$\frac{\mathrm{d}D_n}{\mathrm{d}y} + \frac{1}{2}yD_n = nD_{n-1} \tag{11}$$

$$\frac{\mathrm{d}D_n}{\mathrm{d}y} - \frac{1}{2}yD_n = -D_{n+1} \tag{12}$$

并利用函数的正交性，于是得到 Weber 函数的系数方程，为

$$\frac{\partial q_n}{\partial t} + \frac{\partial q_n}{\partial x} - v_{n-1} = 0 \tag{13}$$

$$\frac{\partial r_n}{\partial t} - \frac{\partial r_n}{\partial x} + (n+1)v_{n+1} = 0 \tag{14}$$

$$\frac{\partial v_n}{\partial t} + \frac{1}{2}(n+1)q_{n+1} - \frac{1}{2}r_{n-1} = 0 \tag{15}$$

这是模式的基本方程。

进而，如果消去 v_n，则给出[13,14]

$$\frac{\partial q_0}{\partial t} + \frac{\partial q_0}{\partial x} = 0 \tag{16}$$

$$\frac{\partial^2 q_1}{\partial t^2}+\frac{\partial^2 q_1}{\partial t\partial x}+\frac{1}{2}q_1=0 \tag{17}$$

$$\frac{\partial^3 q_{n+2}}{\partial t^3}-\frac{\partial^3 q_{n+2}}{\partial t\partial x^2}+\frac{1}{2}(2n+3)\frac{\partial q_{n+2}}{\partial t}-\frac{1}{2}\frac{\partial q_{n+2}}{\partial x}=0 \tag{18}$$

$$\frac{\partial^3 r_n}{\partial t^3}-\frac{\partial^3 r_n}{\partial t\partial x^2}+\frac{1}{2}(2n+3)\frac{\partial r_n}{\partial t}-\frac{1}{2}\frac{\partial r_n}{\partial x}=0 \tag{19}$$

此外尚有

$$\frac{\partial^2 q_{n+2}}{\partial t^2}+\frac{\partial^2 q_{n+2}}{\partial t\partial x}+\frac{1}{2}(n+2)q_{n+2}-\frac{1}{2}r_n=0 \tag{20}$$

$$\frac{\partial^2 r_n}{\partial t^2}-\frac{\partial^3 r_n}{\partial t\partial x}+\frac{1}{2}(n+1)r_n-\frac{1}{2}(n+1)(n+2)q_{n+2}=0 \tag{21}$$

$$(n+1)\left(\frac{\partial q_{n+2}}{\partial t}+\frac{\partial q_{n+2}}{\partial x}\right)+\frac{\partial r_n}{\partial t}-\frac{\partial r_n}{\partial x}=0 \tag{22}$$

在导出上面这些方程时未引进任何的简化。

注意到，方程(16)是 Kelvin 波方程，方程(17)是 Rossby 重力混合波方程，由色散公式可知式(18)，式(19)是包含重力惯性波和 Rossby 波的 Matsuno 方程[15]。方程(15)如略去经圈速度的变化，则简化成

$$r_n=(n+2)q_{n+2} \tag{23}$$

这是纬圈地转平衡，也即

$$\frac{1}{2}yu=-\frac{\partial h}{\partial y} \tag{24}$$

3 重力惯性波方程

注意到，在方程(18)和方程(19)中，包含了高频的重力惯性波和低频的 Rossby 波，如果略去方程中不带时间的项，而这一项众所周知是由于行星涡度梯度所造成，由它而产生 Rossby 波，于是方程中的高频的重力惯性波方程为

$$\frac{\partial}{\partial t}\left[\frac{\partial^2 q_{n+2}}{\partial t^2}-\frac{\partial^2 q_{n+2}}{\partial x^2}+\frac{1}{2}(2n+3)q_{n+2}\right]=0 \tag{25}$$

$$\frac{\partial}{\partial t}\left[\frac{\partial^2 r_n}{\partial t^2}-\frac{\partial^2 r_n}{\partial x^2}+\frac{1}{2}(2n+3)r_n\right]=0 \tag{26}$$

对时间积分，得到

$$L_k(q_{n+2},r_n)=\begin{cases}\left[\dfrac{\partial^2 q_{n+2}}{\partial t^2}-\dfrac{\partial^2 q_{n+2}}{\partial x^2}+\dfrac{1}{2}(2n+3)q_{n+2}\right]_{t=0}\\[2ex]\left[\dfrac{\partial^2 r_n}{\partial t^2}-\dfrac{\partial^2 r_n}{\partial x^2}+\dfrac{1}{2}(2n+3)r_n\right]_{t=0}\end{cases} \tag{27}$$

式中

$$L_k=\frac{\partial^2}{\partial t^2}-\frac{\partial^2}{\partial x^2}+\frac{1}{2}(2n+3)$$

为 Klein 波动算子,在这里描写的是重力惯性波。

应用方程(20)至方程(23)到方程(27)的右端,给出

$$L_k(q_{n+2}) = \left[-\frac{\partial v_{n+1}}{\partial x} + \frac{1}{2}(n+1)q_{n+2} + \frac{1}{2}r_n \right]_{t=0} \equiv F_q(x,0) \tag{28}$$

$$L_k(r_n) = \left[-(n+1)\left(\frac{\partial v_{n+1}}{\partial x} - \frac{1}{2}(n+2)q_{n+2} - \frac{1}{2}r_n \right) + \frac{1}{2}r_n \right]_{t=0} \equiv F_r(x,0) \tag{29}$$

容易看出,这两个方程的右端是位势涡度的一种形式,于是方程(28),方程(29)表明,除初始条件外,重力惯性波是在初始时刻的位势涡度作用下激发出来的。

4　位势涡度时间不变式

将式(15)写成

$$\frac{\partial v_{n+1}}{\partial t} + \frac{1}{2}(n+2)q_{n+2} - \frac{1}{2}r_n = 0 \tag{30}$$

取对 x 的微商,给出

$$\frac{\partial^2 v_{n+1}}{\partial t \partial x} + \frac{1}{2}(n+2)\frac{\partial q_{n+2}}{\partial x} - \frac{1}{2}\frac{\partial r_n}{\partial x} = 0 \tag{31}$$

应用式(13)和式(14)消去$\frac{\partial q_{n+2}}{\partial x}$和$\frac{\partial r_n}{\partial x}$,得到

$$\frac{\partial}{\partial t}\left[\frac{\partial v_{n+1}}{\partial x} - \frac{1}{2}(n+2)q_{n+2} - \frac{1}{2}r_n \right] = -\frac{1}{2}v_{n+1} \tag{32}$$

对时间积分,给出

$$\frac{\partial v_{n+1}}{\partial x} - \frac{1}{2}(n+2)q_{n+2} - \frac{1}{2}r_n = \left[\frac{\partial v_{n+1}}{\partial x} - \frac{1}{2}(n+2)q_{n+2} - \frac{1}{2}r_n \right]_{t=0} - \frac{1}{2}\int_0^t v_{n+1}\mathrm{d}t \tag{33}$$

考虑到经圈速度将以重力惯性波的形式变化,由于运动是振荡的,因此当积分时间充分长时,其积分值会变得很小,由此有近似式

$$\frac{\partial v_{n+1}}{\partial x} - \frac{1}{2}(n+2)q_{n+2} - \frac{1}{2}r_n = \left[\frac{\partial v_{n+1}}{\partial x} - \frac{1}{2}(n+2)q_{n+2} - \frac{1}{2}r_n \right]_{t=0} \tag{34}$$

即为位势涡度的时间不变式。事实上,由位势涡度不变式[9,11]

$$\left[\frac{\partial v}{\partial x} - \frac{\partial u}{\partial y} - \frac{1}{2}yh \right] = \left[\frac{\partial v}{\partial x} - \frac{\partial u}{\partial y} - \frac{1}{2}yh \right]_{t=0} \tag{35}$$

应用 Weber 函数展开及循环公式(11),式(12)后即可得到式(34)。

5　非地转风的频散

设非地转风或地转偏差为

$$X_{n+1} = \frac{1}{2}[r_n - (n+2)q_{n+2}] \tag{36}$$

事实上由式(15),此即为 $\partial v_{n+1}/\partial t$,现在分析地转偏差随时的变化,如一旦地转偏差趋于零,即意味着纬圈地转平衡建立。

将方程(29)乘以$\frac{1}{2}$,减去乘以$\frac{1}{2}(n+2)$后的方程(28),于是得到

$$L_k(X_{n+1}) = \frac{1}{2}\frac{\partial v_{n+1}}{\partial x}\bigg|_{t=0} \equiv F_{v,n+1}(x,0) \tag{37}$$

如运动的纬圈尺度很大,则经圈风的纬圈梯度很小,这时上式简化成

$$L_k(X_{n+1}) = 0 \tag{38}$$

设方程(37)的初始条件为

$$t = 0, \quad X_{n+1} = H_{1,n+1} \tag{39}$$

$$t = 0, \quad \frac{\partial X_{n+1}}{\partial t} = H_{2,n+1} \tag{40}$$

边界条件为

$$x \to \infty, \quad X_{n+1} \to 0 \tag{41}$$

应用傅氏方法或拉普拉斯变换,容易求得问题的解为

$$X_{n+1} = \frac{1}{2}\int_{-\infty}^{\infty} H_{1,n+1}(x')\frac{\mathrm{J}_1\left(\sqrt{\frac{1}{2}(2n+3)}\sqrt{t^2-(x-x')^2}\right)}{\sqrt{t^2-(x-x')^2}}\mathrm{d}x' +$$
$$\frac{1}{2}\int_{-\infty}^{\infty} H_{2,n+1}(x')\mathrm{J}_0\left(\sqrt{\frac{1}{2}(2n+3)}\sqrt{t^2-(x-x')^2}\right)\mathrm{d}x' +$$
$$\frac{1}{2}\int_0^t\int_{-\infty}^{\infty} F_{v,n+1}(x')\mathrm{J}_0\left(\sqrt{\frac{1}{2}(2n+3)}\sqrt{t'^2-(x-x')^2}\right)\mathrm{d}x'\mathrm{d}t' \tag{42}$$

注意到,由 Bessel 函数 J_1 和 J_0 的性质,当时间充分长时,上式前面两项很快消失,其主要贡献部分为

$$X_{n+1} \approx \frac{1}{2}\int_0^t\int_{-\infty}^{\infty} F_{v,n+1}(x,0)\mathrm{J}_0\left(\sqrt{\frac{1}{2}(2n+3)}\sqrt{t'^2-(x-x')^2}\right)\mathrm{d}x'\mathrm{d}t' \tag{43}$$

在参考文献[9]已讨论过,式(43)的值即使当 $t\to\infty$ 时仍趋于有限。

由此得到一个重要的结论,只有当初始扰动中经圈风的纬圈梯度很小时,地转偏差才会随着重力惯性波的频散而消失,并建立起纬圈地转平衡。事实上,这正是 Gill 模式成立的条件,因为是长波,因此物理量的纬圈梯度是不大的,这时方程(37)可用式(38)来近似,而方程(38)表明当时间充分大时纬圈地转平衡将建立。

6 纬圈地转适应后的运动

当地转偏差频散后,纬圈地转平衡式(23),重新写为

$$r_n = (n+2)q_{n+2} \tag{44}$$

将此式代入位势涡度时间不变式(34),得出

$$\frac{\partial v_{n+1}}{\partial x} - r_n = \left[\frac{\partial v_{n+1}}{\partial x} - \frac{1}{2}(n+2)q_{n+2} - \frac{1}{2}r_n\right]_{t=0} \tag{45}$$

在另一方面,式(43)表明,当时间充分大时,$\partial X_{n+1}/\partial t\to 0$,将此条件应用到式(13)和式(14),立即得到

$$\frac{\partial}{\partial x}[(n+2)q_{n+2} + r_n = (2n+3)v_{n+1} \tag{46}$$

将此式与式(45)消去 v_{n+1}，再用式(44)消去 q_{n+2}，最后得到

$$\frac{\partial^2 r_n}{\partial x^2} - \left(n + \frac{3}{2}\right) r_n = \left(n + \frac{3}{2}\right)\left[\frac{\partial v_{n+1}}{\partial x} - \frac{1}{2}(n+2)q_{n+2} - \frac{1}{2}r_n\right]_{t=0} \equiv \Omega_n(x,0) \tag{47}$$

右端为初始时刻的位势涡度。

设方程(47)的边界条件为

$$|x| \to \infty, \qquad r_n \to 0 \tag{48}$$

于是解为

$$r_n = \int_{-\infty}^{\infty} G(x,a)\Omega_n(a,0)\,\mathrm{d}a \tag{49}$$

其中 Green 函数为

$$-\infty < x < a, \qquad G(x,a) = -\sqrt{n+\frac{3}{2}}\,\mathrm{e}^{-n+\frac{3}{2}(a-x)}$$

$$a < x < \infty, \qquad G(x,a) = -\sqrt{n+\frac{3}{2}}\,\mathrm{e}^{-n+\frac{3}{2}(x-a)} \tag{50}$$

当 r_n 求得后，由纬圈地转平衡可求得 q_{n+2}，而由式(46)算出 v_{n+1}，于是纬圈地转适应后的场全部算出。由于应用了位势涡度时间不变式(34)，这样就不需求解方程(28)和方程(29)，这就是位势涡度不变式带来的方便。

7　Kelvin 波和 Rossby-重力混合波

至此，在 Weber 函数展开式中，尚有两个模需要处理。对赤道对称的运动要处理 q_0，对赤道反对称运动要处理 q_1。前者由方程(16)控制，它描写的是 Kelvin 波，后者由方程(17)控制，它描写的是 Rossby-重力混合波。

先讨论 Kelvin 波。设初始扰动为

$$t = 0, \qquad q_0 = q_0^0(x) \tag{51}$$

则方程(16)的解为

$$q_0 = q_0^0(t - x), \qquad t > x \tag{52}$$

这是一个非色散的向东(即下游)传播的行波。注意到，如 $q_0^0(x)$ 只在一个有限区域中有值，则在这个区域的上游(区域西侧)q_0 为零，因为没有信号从西边传过来。在区域以东，即使在超出 $q_0^0(x)$ 所在的区域，这时仍有解，但解只存在 $t>x$ 的区域中，因为超过这个区域 Kelvin 波的信号尚未传到。

但是注意到，Kelvin 波要求纬圈地转平衡，因此它已属于纬圈半地转适应后的发展或演变运动，将不参与半纬圈地转适应过程。

对于 Rossby-重力混合波，设方程(17)的初始条件为

$$t = 0, \qquad q_1^0 = H_{1(x)}, \qquad \frac{\partial q_1^0}{\partial t} = H_{2(x)} \tag{53}$$

边界条件为

$$x \to -\infty, \qquad q_1 \to 0 \tag{54}$$

应用拉普拉斯变换，方程(17)给出

$$\frac{\mathrm{d}\hat{q}_1}{\mathrm{d}x}+\left(s+\frac{1}{2s}\right)\hat{q}_1=\frac{1}{s}\left(sH_1+\frac{\mathrm{d}H_1}{\mathrm{d}x}+H_2\right) \tag{55}$$

解为

$$\hat{q}=\int_{-\infty}^{x}\mathrm{e}^{-\left(s+\frac{1}{2s}(x-x')\right)}\frac{1}{s}\left(sH_1+\frac{\mathrm{d}H_1}{\mathrm{d}x}+H_2\right)\mathrm{d}x' \tag{56}$$

当时间很小即 s 很大时,上式简化成

$$\hat{q}_1=\int_{-\infty}^{x}e^{-s(x-x')}H_1(x')\mathrm{d}x' \tag{57}$$

反变换给出

$$q_1=\int_{-\infty}^{x}\{W[t-(x-x')]H_1(x')\}\mathrm{d}x' \tag{58}$$

即为

$$q_1=H_1(t-x),\qquad t>x \tag{59}$$

显然,这是混合波中的重力波波段。自然,这一波段将参与地转适应过程,因此只有当重力波传过的区域,场才完成适应过程。

当时间很大即 s 很小时,可略去式(56)括号中的 sH_1,于是有近似式

$$\hat{q}_1=\int_{-\infty}^{x}\mathrm{e}^{-\frac{1}{2s}(x-x')}\frac{1}{s}\left(\frac{\mathrm{d}H_1}{\mathrm{d}x'}+H_2\right)\mathrm{d}x' \tag{60}$$

反变换给出

$$q_1=\int_{-\infty}^{x}\mathrm{J}_0(\overline{2(x-x')t})\left(\frac{\mathrm{d}H_1}{\mathrm{d}x'}+H_2\right)\mathrm{d}x' \tag{61}$$

这是混合波中的 Rossby 波波段。由 Bessel 函数的性质,这个解虽然当时间充分大时其值趋于零,但在物理上它应归到发展运动中去。由于适应后的发展运动是纬圈半地转平衡的长波,而混合波中的 Rossby 波波段,属于短波性质,因此并不一定需要考虑。

Kelvin 波和 Rossby-重力混合波的 Rossby 波波段只应参加到发展过程中去,这是热带大气和海洋动力学要讨论的特殊问题,我们将在另文中讨论。

参考文献

[1] Rossby C G. On the mutual adjustment of pressure and velocity distribution in certain simple current system Ⅰ. J Mar Res, 1937, 1: 15-28.

[2] Rossby C G. On the mutual adjustment of pressure and velocity distribution in certain sample current system Ⅱ. J Mar Res, 1938, 2: 239-263.

[3] Oboukhov A M. The problem of the geostrophic adaptation, Izvestiya of Academy of Science USSR, Ser. Geography and Geophysics, 1949, 13: 281-289.

[4] Yeh T C. On the formation of quasi-geostrophic motion in the atmosphere. J Meteor Soc Japan: The 75th Anniversary, 1957: 130-134.

[5] 曾庆存. 大气中的适应过程和发展过程(一)和(二). 气象学报, 1963, 33: 163-174, 281-289.

[6] 叶笃正,李麦村. 大气运动的适应问题. 北京: 科学出版社, 1956.

[7] 叶笃正,巢纪平. 论大气运动的多时态特征——适应、发展和准定常演变.大气科学, 1998, 22: 385-398.

[8] Gill A E. Some simple solutionfor heatinduced tropical circulation. Quart, J Roy Meteor Soc, 1980, 106: 447-462.

[9]　Chao J P, Lin Y H. The foundation and movement of tropical semi-geostrophic adaptation. Acta Meteor Sinica, 1996, 10: 129-141.

[10]　Chao J P, Lin Y H. The motion of tropical semi-geostrophic adaptation. In: IAP CAS, eds. From Atmos pheric Circul ation to Global Change. Beijing: Chins Meteorelogical Press, 1996: 237-246.

[11]　林永辉,巢纪平. 热带半地转适应过程. 中国科学(D 辑), 1997, 27: 566-573.

[12]　Gill A E, Clarke A J. Wind-inducing upwelling, coastal current and sea-level changs. Deep Sea Res, 1974, 21: 325-345.

[13]　Aderson D L J, Rowlands P B. The role of inertia-gravity and planetary waves in the response of at ropical ocean to the incidence of an equatorial Kelvin wave on a meridional boundary. J Mar Res, 1976, 34: 295-312.

[14]　Anderson D L J, Rowlands P B. The Somali Current response to southwest monsoon: the relativeim portance of local and remote forcing. J Mar Res, 1976, 34: 395-417.

[15]　Matsuno T. Quasi-geostrophic motion in the equatorial area. J Meteor Soc, Japan, 1966, 44: 25-43.

热带大气和海洋的半地转适应和发展运动*

巢纪平

(国家海洋环境预报研究中心,北京 100081)

摘要:半地转适应和半地转发展是热带大气和海洋运动的两种基本形态,它们在时间上是可分的,反映了不同的物理过程。当初始扰动作用于大气或海洋时,首先将激发出重力惯性波,当重力惯性波频散后,建立起半地转的平衡状态,此后运动进入到以 Rossby 波(长波或短波)、Kelvin 波和混合波中的 Rossby 短波为主导的发展状态。文中研究的纬圈半地转适应和发展运动,是 Gill 长波近似模式的理论基础。同时研究了经圈半地转适应和发展运动,实际上这相当于短波近似模式,它可以应用到研究海洋经圈边界附近的一类问题。

关键词:半地转运动;适应过程;发展过程

1 引言

地转适应运动和准地转发展运动是中纬度大气和海洋中两种基本的运动形态[1],这两种形态在时间上是可分的,由不同的动力学过程制约[2,3],前者反映了重力惯性波的频散,后者是 Rossby 波的表现。把运动区分出适应阶段和发展阶段,是中纬度大气和海洋动力学研究的一大贡献,在气象学中,它对早期数值天气预报的发展起了积极的作用。

早在 20 世纪 60 年代 Matsuno[41] 就指明了热带运动的最基本形态,并表明除重力惯性波、Rossby 波外,Kelvin 波和 Rossby-重力混合波在调节热带大气和海洋环流中同样也起着重要的作用。这是热带运动不同于中纬度运动的特点之一。

另一方面,Gill[5] 引进长波近似,即纬圈半地转近似,建立了一个类似于中纬度的准地转模式,不同之处除重力惯性波被过滤以及 Rossby 波被非色散外,尚保存了 Kelvin 波和 Rossby-重力混合波的作用。这个模式在热带大气和海洋动力学的研究中,曾有过广泛的应用。然而,Gill 模式的理论基础是不够充实的,例如纬度半地转平衡为什么能够建立、通过什么过程建立等,在物理上都没有像中纬度准地转模式的建立分析得那么清楚。

最近,巢纪平等[6] 提出了热带半地转适应的概念,并指出,当初始场存在位势涡度的纬圈梯度时,随着重力惯性波的频散,经圈流将保留下来,但经圈流的时间变化消失,从而纬圈半地转平衡建立。这样就为 Gill 模式的建立提供了一个物理基础。但是,从纬圈半地转适应状态是如何过渡到纬圈半地转发展状态的,尚需要用统一的观点和方法做出较好的处理,这正是本文的目的之一。

在另一方面,长波近似下的 Gill 模式虽然可以用来研究热带纬圈大尺度的大气和海洋运动,但对

* 气象学报,2000,58(3):259-264.

有些运动,例如海洋经圈边界附近的运动,其纬圈尺度很小,这时 Gill 模式就失效了。为此,在本文中又研究了经圈半地转的适应过程,以及在经圈半地转平衡下的发展运动,而经圈半地转近似实质上相当于 Rossby 波的短波近似。这样短波近似模式和 Gill 的长波近似模式就能相互补充了,成为热带动力学中的一个较为完整的体系。

2　基本方程

当运动的物理量在垂直方向用本征模展开时,其本征值相当于等值厚度,记以 h,于是可定义重力波波速为

$$C = (gh)^{\frac{1}{2}} \tag{1}$$

在赤道 β 平面上,引进特征量:长度为 $(C/2\beta)^{1/2}$,时间为 $(2\beta C)^{-1/2}$,速度为 C,重力位势高度为 C^2,则对任一本征模其水平运动方程相当于浅水运动方程,为

$$\frac{\partial u}{\partial t} - \frac{1}{2}yv = -\frac{\partial \varphi}{\partial x} \tag{2}$$

$$\frac{\partial v}{\partial t} + \frac{1}{2}yu = -\frac{\partial \varphi}{\partial y} \tag{3}$$

$$\frac{\partial \varphi}{\partial t} + \frac{\partial u}{\partial x} + \frac{\partial v}{\partial y} = 0 \tag{4}$$

引进变量

$$q = \varphi + u, \quad r = \varphi - u \tag{5}$$

由此有

$$\varphi = \frac{1}{2}(q + r), \quad u = \frac{1}{2}(q - r) \tag{6}$$

于是由方程式(2)至式(4)给出

$$\frac{\partial q}{\partial t} + \frac{\partial q}{\partial x} + \frac{\partial v}{\partial y} - \frac{1}{2}yv = 0 \tag{7}$$

$$\frac{\partial r}{\partial t} - \frac{\partial r}{\partial x} + \frac{\partial v}{\partial y} + \frac{1}{2}yv = 0 \tag{8}$$

$$\frac{\partial v}{\partial t} + \frac{1}{2}\left(\frac{\partial q}{\partial y} + \frac{1}{2}yq\right) + \frac{1}{2}\left(\frac{\partial r}{\partial y} - \frac{1}{2}yr\right) = 0 \tag{9}$$

对所有变量在 y 方向用抛物圆柱函数即 Weber 函数展开成

$$(q, r, v) = \sum_n (q_n, r_n, v_n) D_n(y) \tag{10}$$

Weber 函数有循环公式

$$\frac{\mathrm{d}D_n}{\mathrm{d}y} + \frac{1}{2}yD_n = nD_{n-1} \tag{11}$$

$$\frac{\mathrm{d}D_n}{\mathrm{d}y} - \frac{1}{2}yD_n = -nD_{n+1} \tag{12}$$

考虑到函数 $D_n(y)$ 的正交性,由方程式(7)至式(9)给出展开式的系数方程为

$$\frac{\partial q_n}{\partial t} + \frac{\partial q_n}{\partial x} - v_{n-1} = 0 \tag{13}$$

$$\frac{\partial r_n}{\partial t}-\frac{\partial r_n}{\partial x}+(n+1)v_{n+1}=0 \tag{14}$$

$$\frac{\partial v_n}{\partial t}+\frac{1}{2}(n+1)q_{n+1}-\frac{1}{2}r_{n-1}=0 \tag{15}$$

或者,可写成

$$\frac{\partial q_0}{\partial t}+\frac{\partial q_0}{\partial x}=0 \tag{16}$$

$$\frac{\partial^2 q_1}{\partial t^2}+\frac{\partial^2 q_1}{\partial t\partial x}+\frac{1}{2}q_1=0 \tag{17}$$

$$\frac{\partial^3 q_{n+2}}{\partial t^3}-\frac{\partial^3 q_{n+2}}{\partial t\partial x^2}+\frac{1}{2}(2n+3)\frac{\partial q_{n+2}}{\partial t}-\frac{1}{2}\frac{\partial q_{n+2}}{\partial x}=0 \tag{18}$$

$$\frac{\partial^3 r_n}{\partial t^3}-\frac{\partial^3 r_n}{\partial t\partial x^2}+\frac{1}{2}(2n+3)\frac{\partial r_n}{\partial t}-\frac{1}{2}\frac{\partial r_n}{\partial x}=0 \tag{19}$$

注意到,式(16)是 Kelvin 波方程,式(17)是 Rossby-重力混合波方程,式(18)和式(19)是描写重力惯性波和 Rossby 波的方程。

考虑到式(18)和式(19)后,由式(15)可以得到对 v_n 的方程,为

$$\left[\frac{\partial^2}{\partial t^2}-\frac{\partial^2}{\partial x^2}+\left(n+\frac{1}{2}\right)\right]\frac{\partial v_n}{\partial t}-\frac{1}{2}\frac{\partial v_n}{\partial x}=0 \tag{20}$$

这是 Matsuno 方程。

3 纬圈半地转适应过程

引进拉普拉斯变换

$$\hat{v}_n=\int_0^\infty v_n e^{-st}dt \tag{21}$$

则方程(20)给出

$$s\frac{d^2\hat{v}}{dx^2}+\frac{1}{2}\frac{d\hat{v}_n}{dx}-\left(s^2+n+\frac{1}{2}\right)s\hat{v}_n$$

$$=\left[\left(\frac{d^2v_n^0}{dx^2}-\left(n+\frac{1}{2}\right)v_n^0\right)-s^2v_n^0-sv'^0_n-v''^0_n\right]\equiv F_n^0(x) \tag{22}$$

式中,v_n^0 表示初始值,撇号表示对时间的微商。

方程(22)的自由方程的色散关系为

$$sk_n^2+\frac{1}{2}k_n-\left(s^3+\left(n+\frac{1}{2}\right)s\right)=0 \tag{23}$$

其根为

$$k_n^+=-\frac{1}{4s}\left[1-\sqrt{1+16\left(n+\frac{1}{2}\right)s^2+16s^4}\right] \tag{24}$$

$$k_n^-=-\frac{1}{4s}\left[1+\sqrt{1+16\left(n+\frac{1}{2}\right)s^2+16s^4}\right] \tag{25}$$

注意到,当 s 很大时,上两式的渐近式为

$$k_n^+ \approx \sqrt{\left(n+\frac{1}{2}\right)+s^2} \tag{26}$$

$$k_n^- \approx -\sqrt{\left(n+\frac{1}{2}\right)+s^2} \tag{27}$$

容易看出，如方程(20)略去由行星涡度梯度而造成的最后一项（即不带时间变化的项），也即相当于过滤掉 Rossby 波，这时方程退化成

$$\frac{\partial^2 v_n}{\partial t^2}-\frac{\partial^2 v_n}{\partial x^2}+\left(n+\frac{1}{2}\right)v_n=0 \tag{28}$$

这一 Kelvin 方程描写的是重力惯性波，其色散关系即为式(26)和式(27)。

由此可见，主导适应过程的重力惯性波和主导发展过程之一的 Rossby 波，在时间上是可分的，一是在初值附近的过程，二是离开初值很长时间后的过程。事实上，我们知道重力惯性波的最小频率和 Rossby 波的最大频率分别为[7]

$$\sigma_{\min}=C_m^{\frac{1}{2}}\left[\left(\frac{n+1}{2}\right)^{\frac{1}{2}}+\left(\frac{n}{2}\right)^{\frac{1}{2}}\right] \tag{29}$$

$$\sigma_{\max}=C_m^{\frac{1}{2}}\left[\left(\frac{n+1}{2}\right)^{\frac{1}{2}}-\left(\frac{n}{2}\right)^{\frac{1}{2}}\right] \tag{30}$$

它们是不相重合的。这表明适应过程和发展过程反映的是两个不同时间阶段上的过程。

现在用格林函数方法来求解方程(22)，边界条件为

$$|x|\to\infty,\quad \hat{v}_n\to 0 \tag{31}$$

在这一条件下容易求得问题的格林函数为

$$-\infty\leqslant x<\xi,\quad G(x,\xi)=-\frac{1}{s(k_n^+-k_n^-)}e^{-k_n^+(\xi-x)} \tag{32}$$

$$\xi\leqslant x<\infty,\quad G(x,\xi)=-\frac{1}{s(k_n^+-k_n^-)}e^{-k_n^-(\xi-x)} \tag{33}$$

方程(22)的解为

$$\bar{v}_n=\int_{-\infty}^{\infty}F_n^0(\xi)G(x,\xi)\mathrm{d}\xi \tag{34}$$

或者写成

$$\hat{v}_n=-\frac{1}{s(k_n^+-k_n^-)}\left[\int_{-\infty}^{x}e^{k_n^-(x-\xi)}F(\xi)\mathrm{d}\xi+\int_{x}^{\infty}e^{-k_n^+(\xi-x)}F(\xi)\mathrm{d}\xi\right] \tag{35}$$

当时间很小即 s 很大时，式(34)的渐近式为

$$\hat{v}_n=-\frac{1}{2s\sqrt{n+\frac{1}{2}+s^2}}\int_{-\infty}^{\infty}e^{-|x-\xi|\sqrt{n+\frac{1}{2}+s^2}}F^0(\xi)\mathrm{d}\xi \tag{36}$$

由反变换给出

$$v_n=-\frac{1}{2}\left[\int_{-\infty}^{\infty}\int_0^t\left(\frac{\mathrm{d}^2v_n^0(\xi)}{\mathrm{d}\xi^2}-\left(n+\frac{1}{2}\right)v_n^0(\xi)-v''^0_n(\xi)\right)\times\right.$$

$$J_0\left(\sqrt{n+\frac{1}{2}}\sqrt{\tau^2-(x-\xi)^2}\right)\mathrm{d}\tau\mathrm{d}\xi-\int_{-\infty}^{\infty}v'^0_n(\xi)J_0\left(\sqrt{n+\frac{1}{2}}\sqrt{t^2-(x-\xi)^2}\right)\mathrm{d}\xi+$$

$$\sqrt{n+\frac{1}{2}}\int_{-\infty}^{\infty}v_n^0(\xi)\frac{t}{\sqrt{t^2-(x-\xi)^2}}\mathrm{J}_1\left(\sqrt{n+\frac{1}{2}}\sqrt{t^2-(x-\xi)^2}\right)\mathrm{d}\xi\Bigg] \tag{37}$$

现在来分析公式(37)的性质。由贝塞尔函数 J_v 的渐近性态,当时间充分大时,式中第2、第3两项的贡献趋向于零,只有第一项保留下来,即当时间充分大时,由初始扰动的影响,尚存的经圈速度为

$$v_n=-\frac{1}{2}\int_{-\infty}^{\infty}\int_0^t\left(\frac{\mathrm{d}^2v_n^0(\varepsilon)}{\mathrm{d}\xi^2}-\left(n+\frac{1}{2}\right)v_n^0(\xi)-v''^0_n(\xi)\right)\times$$

$$\mathrm{J}_0\left(\sqrt{n+\frac{1}{2}}\sqrt{\tau^2-(x-\xi)^2}\right)\mathrm{d}\tau\mathrm{d}\xi \tag{38}$$

其时间微分为

$$\frac{\partial v_n}{\partial t}=-\frac{1}{2}\int_{-\infty}^{\infty}\left(\frac{\mathrm{d}^2v_n^0(\xi)}{\mathrm{d}\xi^2}-\left(n+\frac{1}{2}\right)v_n^0(\varepsilon)-v''^0_n(\xi)\right)\times$$

$$\mathrm{J}_0\left(\sqrt{n+\frac{1}{2}}\sqrt{t^2-(x-\xi)^2}\right)\mathrm{d}\xi \tag{39}$$

由此可见,当时间充分大时,其值趋于零,于是由式(15)给出

$$r_n=(n+2)q_{n+2}\quad n\geqslant 0 \tag{40}$$

即建立起纬圈地转平衡。这是 Gill 长波近似成立的理论基础。

4 纬圈半地转发展过程

由于式(40)的约束,由式(13)和式(14)给出方程

$$\frac{\partial q_{n+2}}{\partial t}-\frac{1}{2n+3}\frac{\partial q_{n+2}}{\partial x}=0\quad n\geqslant 0 \tag{41}$$

当初值为 $q_{n+1}^0(x)$,则解为

$$q_{n+2}=q_{n+2}^0\left(x+\frac{1}{2n+3}t\right) \tag{42}$$

这是以速度为 $1/(2n+3)$ 的向西传播的 Rossby 长波。

注意到,对于赤道对称的运动,由于 Kelvin 波中的速度和压力之间要求地转平衡,因此它将参与发展运动。如果 q_0 的初值为 $q_0(x)$,则方程(16)的解为

$$q_0=q_0^0(x-t),\quad t>x \tag{43}$$

这是以速度为1的向东传播的 Kelvin 波。

在另一方面,当初值为

$$t=0,\quad q_1=q_1^0,\quad \frac{\partial q_1}{\partial t}=q'^0_1 \tag{44}$$

则方程(17)由拉普拉斯变换得出

$$\frac{\mathrm{d}\hat{q}_1}{\mathrm{d}x}+\left(s+\frac{1}{2s}\right)\hat{q}_1=\frac{1}{s}\left(\frac{\mathrm{d}q_1^0}{\mathrm{d}x}+sq_1^0+q'^0_1\right) \tag{45}$$

在 $x\to\infty$, $q_1\to 0$ 的条件下,其解为

$$\hat{q}_1=\int_{-\infty}^{x}\mathrm{e}^{-\left(s+\frac{1}{2s}\right)(x-\xi)}\left[\frac{1}{s}\left(\frac{\mathrm{d}q_1^0(\xi)}{\mathrm{d}\xi}+q_1^0sq_1^0(\xi)+q'^0_1(\xi)\right)\right]\mathrm{d}\xi \tag{46}$$

当 s 很大时,式(46)的近似解为

$$\hat{q}_1 = \int_{-\infty}^{x} q_1^0(\xi)\delta(t-(x-\xi))\mathrm{d}\xi = q_1^0(x-t) \quad t \geqslant x \tag{47}$$

式中,$\delta(t)$ 为 Deha 函数,这是混合波中的重力波波段的行为,它将参与适应过程。当 s 很小时,式(46)的近似解为

$$q_1 = \int_{-\infty}^{x} \left(\frac{\mathrm{d}q_1^0(\xi)}{\mathrm{d}\xi} + q'^0_1(\xi)\right) J_0(\sqrt{2(x-\xi)t})\mathrm{d}\xi \tag{48}$$

这是混合波中的 Rossby 波波段的行为。

但注意到,混合波中的 Rossby 波相当于一类短波,因此为了与长波近似在逻辑上一致,混合波中的 Rossby 短波不应当参加到发展过程中去,事实上如将 $\partial v_n/\partial t$ 的约束,也用到 $n=0$ 的情况,即用到 Matsuno 方程中混合波的情况,则由式(15)给出

$$q_1 = 0 \tag{49}$$

即 Rossby 短波将不参与纬圈半地转适应后的发展过程。在有外源时,Gill[5] 令 q_1 与外源相平衡。由于 Gill 的模式是长波近似下的模式,而 q_1 保留下来与外源平衡的处理方法多少与长波假定不相一致。

5 经圈半地转适应和发展过程

如前所述,长波近似下的 Gill 模式,在对热带大气和海洋运动的研究中,有一定的局限性。因此作为相互补充,现研究经圈的半地转适应和发展过程。

注意到,方程(18)和方程(19)的算子是一样的,可以合在一起写成

$$L(q_{n+2}, r_n) = 0 \tag{50}$$

其特征值为式(24)和式(25)。当 s 很大时,其渐近值分别为式(26)和式(27),它们相当于方程

$$L_k(q_{n+2}, r_n) = 0 \tag{51}$$

的特征根,其中算子

$$L_k = \frac{\partial^2}{\partial t^2} - \frac{\partial^2}{\partial x^2} + \left(n + \frac{3}{2}\right) \tag{52}$$

方程(50)略去不带时间的项,并对时间积分一次,由此得到 s 很大时的解同式(37),除将 v_n 改写成 q_{n+2} 或 r_n,以及将 $n+1/2$ 改写成 $n+3/2$。于是得到结论,当时间充分大时 $\frac{\partial q_{n+2}}{\partial t}$ 及 $\frac{\partial r_n}{\partial t}$ 消失,而 q_{n+2} 及 r_n 则保留下来。并由下面的方程决定,即

$$\frac{\partial q_n}{\partial x} = v_{n-1} \tag{53}$$

$$\frac{\partial r_n}{\partial x} = (n+1)v_{n+1} \tag{54}$$

或者有

$$\frac{\partial \varphi_n}{\partial x} = \frac{1}{2}[(n+1)v_{n+1} + v_{n-1}] \tag{55}$$

这是经圈半地转平衡的条件。即

$$\frac{1}{2}yv = \frac{\partial \varphi}{\partial x} \tag{56}$$

将条件式(53)和式(54)代入式(15),得到

$$\frac{\partial^2 v_n}{\partial t \partial x} + \frac{1}{2} v_n = 0 \tag{57}$$

这是经圈半地转平衡运动的发展方程,显然这是描写 Rossby 短波行为的方程。设条件为

$$t = 0, \quad \frac{\partial v_n}{\partial x} = \Phi_n(x) \tag{58}$$

$$x \to \infty, \quad v_n \to 0 \tag{59}$$

方程(56)的解为

$$n \geqslant 1, \quad v_n = \int_{-\infty}^{x} J_0(\sqrt{2(x-\xi)t}\,\Phi_n(\xi)\,\mathrm{d}\xi \tag{60}$$

当 v_n 求得后,可由式(53)和式(54)算出 q_{n+2} 和 r_n 以及进而可算出相应的 u_n 和 φ_n。

但要注意到,当 $n=0$ 时,由方程(15)得到

$$\frac{\partial v_0}{\partial t} + \frac{1}{2} q_1 = 0 \tag{61}$$

而 q_1 是 Rossby-重力混合波方程(17)的解,即式(17)当 s 很小时的解即式(48),当 q_1 算出后可直接算出 v_0。这表明混合波中的 Rossby 波波段将参与发展运动。事实上,只要表明式(57)中的 n 从 $n=0$ 算起,则包括整个的短波活动。因为由式(53)和式(61)即可得到

$$\frac{\partial^2 v_0}{\partial t \partial x} + \frac{1}{2} v_0 = 0 \tag{62}$$

注意到,Kelvin 波是长波近似下的产物,因此它不应参与到短波近似下的发展运动。事实上,在式(2)中如令 $\partial u/\partial t=0$(短波近似)就得到式(56),而由于 Kelvin 波中经圈速度滤失,因此这相当于把 Kelvin 波过滤掉。

6 结语

由以上的讨论可见,即使在热带,运动也存在适应过程和发展过程,这两个过程在时间上是可分的,这一点是和中纬度大气运动的动力学特征相似的,但在这里发展运动是在纬圈或经圈半地转近似下进行的,而不是像中纬度的运动那样,发展运动是在准地转近似下进行的。在另一方面,在两个不同的纬度带,适应过程虽然都是通过重力惯性波的频散来实现的,但在发展过程中,在热带除了有 Rossby 波(退化成 Rossby 长波或 Rossby 短波)作用外,在纬圈半地转近似下尚有 Kelvin 波参与,而在经圈半地转近似下尚有重力混合波中的 Rossby 波波段参与,这在动力学行为上和中纬度是不同的。

注意到,纬圈半地转近似或长波近似虽然对热带大多数运动都是一个很好的近似,但由于长波近似下的 Rossby 波和 Kelvin 波都是非色散波,因此像波能量的传播速度不同于波信号的传播速度或反向于波信号的传播速度而造成的一些重要现象就被过滤掉了。但经圈半地转平衡下的 Rossby 短波是色散波,而且其能量是向东传播的,因此可以用来研究经圈边界附近的一些现象,这两种近似显然可以起到互补的作用。当然,用来描写热带发展运动的更好的近似,是应把长波和短波近似统一地包含在一个模式中,这我们将在另一篇文章中讨论。

参考文献

[1] 曾庆存. 大气中的适应过程和发展过程(一)和(二).气象学报,1963,33:163-174,281-289.

[2] Yeh T C,Li M C. On the characteristics of scale of the atmospheric motion. J Meteor Soc Japan, 1982,60:16-23.

[3] 叶笃正,巢纪平.论大气运动的多时态特征——适应、发展和准定常演变.大气科学,1998,22:385-398.

[4] Matsuno T. Quasi geostrophic motions in the equatorial area. J Meteor Soc Japan, 1966,44:25-43.

[5] Gill A E. Some simple solution for heat-induced tropical circulation. Qusrt J Roy Meteor Soc, 1980, 106:447-462.

[6] Chao J P, Lin Y H. The foundation and movement of tropical semi-geostrophic adsptation. Acta Meteor Sinica, 1996.20: 129-141.

[7] Moore D W. Philander S G H. Modeling of the ocean circulation, In: Goldber E D. et al. Eds. The Sea: Vol 6. New York: Wiley (Interscience), 1977:319-362.

热带大气发展运动的低频模式*

巢纪平

（国家海洋环境预报研究中心，北京 100081）

摘要：当热带大气运动的适应过程完成后，即进入到缓变的发展运动，描写发展运动的模式有 Gill 的长波近似模式以及作者提出的短波近似模式。文中在分析这两个模式的特点和不足之处后，提出了低频近似下的发展模式。如 f_0 是热带地区典型的科氏参数值或惯性振荡频率，设运动的特征频率为 e，则长波和短波近似的成立条件之一，要求 $e \ll f_0$。而低频近似模式要求 $e^2 \ll f_0^2$，即对运动频率的要求比长波或短波近似模式降低一个量级，因此适用性更广。

关键词：发展过程；低频近似；位势涡度方程

1 引言

像中纬度的大气运动一样，热带大气运动也可以分成适应阶段和发展阶段[1]，不管是纬圈半地转的或经圈半地转的适应过程都是以重力惯性波的频散为特征的[2]。但进入到发展运动后，控制方程（或模式）可以有不同形式，其中常用范围较广的是 Gill 的长波近似模式[3]，但这一模式不能用到纬圈尺度小的运动，例如在海洋经圈边界附近的运动，这时就需要用短波近似模式[1]。这两种模式，除对运动的纬圈尺度或纬圈波长有所限制外，实际上对运动的特征时间或频率也有所限制。如取热带的惯性运动频率为 f_0，则两个模式都要求发展运动的频率 $e \ll f_0$，即要求运动的频率甚低。除此之外，长波近似模式使 Rossby 波非色散化，这样使相速度和群速度取同一个值，限制了波能量的传播速度和方向。

鉴于以上这些问题，在本文中将提出一个精度更高、限制更少的发展运动模式，而上面的长波近似模式和短波近似模式将是现在所建立的模式在一定条件下的特例。

2 基本方程

对于斜压等值浅水模式，对任一个垂直本征模其无量纲的水平地运动方程为

$$\frac{\partial u}{\partial t} - \frac{1}{2}yv = -\frac{\partial h}{\partial x} \tag{1}$$

$$\frac{\partial v}{\partial t} + \frac{1}{2}yu = -\frac{\partial h}{\partial y} \tag{2}$$

* 气象学报，2000，58（4）：385-390.

$$\frac{\partial H}{\partial t}+\frac{\partial u}{\partial x}+\frac{\partial v}{\partial y}=0 \tag{3}$$

所取特征量为：时间$(2U_c)^{-\frac{1}{2}}$，速度C，位势高度C^2，其中$C=(gh)^{\frac{1}{2}}$为重力波波速，h为本征模的垂直本征值，或等值厚度。

引进变量

$$q=h+u,\qquad r=h-u \tag{4}$$

由方程式(1)至式(3)给出

$$\frac{\partial q}{\partial t}+\frac{\partial q}{\partial x}+\left(\frac{\partial v}{\partial y}-\frac{1}{2}yv\right)=0 \tag{5}$$

$$\frac{\partial r}{\partial t}-\frac{\partial r}{\partial x}+\left(\frac{\partial v}{\partial y}+\frac{1}{2}yv\right)=0 \tag{6}$$

$$\frac{\partial v}{\partial t}+\frac{1}{2}\left(\frac{\partial q}{\partial y}+\frac{1}{2}yq\right)+\frac{1}{2}\left(\frac{\partial r}{\partial y}-\frac{1}{2}yr\right)=0 \tag{7}$$

对物理量用抛物圆柱函数或Weber函数展开成

$$(q,r,v)=\sum_n (q_n,r_n,v_n)D_n(y) \tag{8}$$

考虑到函数$D_n(y)$的性质

$$\frac{\mathrm{d}D_n}{\mathrm{d}y}+\frac{1}{2}yD_n=nD_{n-1} \tag{9}$$

$$\frac{\mathrm{d}D_n}{\mathrm{d}y}-\frac{1}{2}yD_n=-D_{n+1} \tag{10}$$

以及函数的正交性，于是展开式的系数方程为

$$\frac{\partial q_n}{\partial t}+\frac{\partial q_n}{\partial x}-v_{n-1}=0 \tag{11}$$

$$\frac{\partial r_n}{\partial t}-\frac{\partial r_n}{\partial x}+(n+1)v_{n+1}=0 \tag{12}$$

$$\frac{\partial v_n}{\partial t}+\frac{1}{2}(n+1)v_{n+1}-\frac{1}{2}r_{n-1}=0 \tag{13}$$

方程式(11)至式(13)如消去v_n则有

$$\frac{\partial q_0}{\partial t}+\frac{\partial q_0}{\partial x}=0 \tag{14}$$

$$\frac{\partial^2 q_1}{\partial t^2}+\frac{\partial^2 q_1}{\partial t\partial x}+\frac{1}{2}q_1=0 \tag{15}$$

以及

$$L_n(q_{n+2},r_n)=0 \tag{16}$$

式中算子

$$L_n\equiv\frac{\partial^3}{\partial t^3}-\frac{\partial^3}{\partial t\partial x^2}+\frac{1}{2}(2n+3)\frac{\partial}{\partial t}-\frac{1}{2}\frac{\partial}{\partial x} \tag{17}$$

另外，尚有

$$\frac{\partial^2 q_{n+2}}{\partial t^2}+\frac{\partial^2 q_{n+2}}{\partial t\partial x}+\frac{1}{2}(n+2)q_{n+2}-\frac{1}{2}r_n=0 \tag{18}$$

$$\frac{\partial^2 r_n}{\partial t^2}+\frac{\partial^2 r_n}{\partial t\partial x}+\frac{1}{2}(n+1)r_n-\frac{1}{2}(n+1)(n+2)q_{n+2}=0 \tag{19}$$

以上是本文用到的基本方程。

3　适应过程和发展过程

注意到，方程(14)是 Kelvin 波方程，方程(15)是 Rossby-重力混合波方程，这两种是热带动力学中特有的运动。

为讨论方便，引进拉普拉斯变换，变换参数为 s，如不考虑初始条件，则式(16)给出

$$\hat{L}_n(\hat{q}_{n+2},\hat{r}_n)=0 \tag{20}$$

式中

$$\hat{L}_n=s^3-s\frac{\mathrm{d}^2}{\mathrm{d}x^2}+\frac{1}{2}(2n+3)s-\frac{1}{2}\frac{\partial}{\partial x} \tag{21}$$

这个算子的特征方程为

$$sk_n^2+\frac{1}{2}k_n-\left(s^3+\frac{1}{2}(2n+3)s\right)=0 \tag{22}$$

根为

$$k_n^+=-\frac{1}{4s}\left[1-\sqrt{1+16\left(n+\frac{3}{2}\right)s^2+16s^4}\right] \tag{23}$$

$$k_n^-=-\frac{1}{4s}\left[1+\sqrt{1+16\left(n+\frac{3}{2}\right)s^2+16s^4}\right] \tag{24}$$

对于适应过程，因发生在初值附近，可以取 s 大时的渐近值，分别为

$$k_n^+=\sqrt{\left(n+\frac{3}{2}\right)+s^2} \tag{25}$$

$$k_n^-=-\sqrt{\left(n+\frac{3}{2}\right)+s^2} \tag{26}$$

这两个根式对应的方程为

$$L_k(q_{n+2},r_n)=0 \tag{27}$$

算子

$$L_k\equiv\frac{\partial^2}{\partial t^2}-\frac{\partial^2}{\partial x^2}+\frac{1}{2}(2n+3) \tag{28}$$

这是描写重力惯性波的 Klein 方程。热带半地转适应正是通过重力惯性波的频散来实现的，这已在参考文献[1]中已作过详细的分析，在此不再作讨论。

当时间很大或 s 很小时，式(23)和式(24)的近似式为

$$k_n^+=-\frac{1}{4s}\left[1-\sqrt{1+16\left(n+\frac{3}{2}\right)s^2}\right] \tag{29}$$

$$k_n^-=-\frac{1}{4s}\left[1+\sqrt{1+16\left(n+\frac{3}{2}\right)s^2}\right] \tag{30}$$

它所对应的方程为

$$L_R(q_{n+2},r_n)=0 \tag{31}$$

算子

$$L_R\equiv\left(\frac{\partial^2}{\partial x^2}-\frac{1}{2}(2n+3)\right)\frac{\partial}{\partial t}+\frac{1}{2}\frac{\partial}{\partial x} \tag{32}$$

这是描写发展过程的控制方程。与式(16)的算子 L_n 相比,略去了与时间三次导数有关的项,实际上这相当于过滤了重力惯性波,而只保留了慢的过程。

4　低频近似

在参考文献[2]中对长波近似及短波近似下发展方程成立的条件已做了讨论,现在来讨论发展方程(32)成立时对运动的要求。

设运动的频率已低到可略去方程(18)和方程(19)中对时间的二阶导数项,于是有

$$\frac{\partial^2 q_{n+2}}{\partial t\partial x}+\frac{1}{2}(n+2)q_{n+2}-\frac{1}{2}r_n=0 \tag{33}$$

$$\frac{\partial^2 r_n}{\partial t\partial x}+\frac{1}{2}(n+1)r_n-\frac{1}{2}(n+1)(n+2)q_{n+2}=0 \tag{34}$$

另一方面,由式(11)和式(12)有

$$\frac{\partial r_n}{\partial t}-\frac{\partial r_n}{\partial x}=-(n+1)\left(\frac{\partial q_{n+2}}{\partial t}+\frac{\partial q_{n+2}}{\partial x}\right) \tag{35}$$

将式(35)代入式(33)消去 r_n,并略去带有时间二阶导数的项,得到

$$\left(\frac{\partial^2}{\partial x^2}-\frac{1}{2}(2n+3)\right)\frac{\partial q_{n+2}}{\partial t}+\frac{1}{2}\frac{\partial q_{n+2}}{\partial x}=0 \tag{36}$$

或者,将式(35)代入式(34)消去 q_{n+2},并略去带有时间二阶导数的项,得到

$$\left(\frac{\partial^2}{\partial x^2}-\frac{1}{2}(2n+3)\right)\frac{\partial r_n}{\partial t}+\frac{1}{2}\frac{\partial r_n}{\partial x}=0 \tag{37}$$

方程(36)和方程(37)即方程(31)。

考虑到 $f=U_y$,如取 f 在热带的平均值为 f_0,同时设运动的特征频率为 e,由于长波近似的发展方程要求$\frac{\partial v_n}{\partial t}$很小而短波近似的发展要求$\frac{\partial q_{n+2}}{\partial t}$和$\frac{\partial r_n}{\partial t}$很小,这表明这两个近似要求

$$e\ll f_0 \tag{38}$$

而低频近似的发展模式是在略去 $q_{(n+2)}$ 或 r_n 的二阶时间导数项后建立的,亦即其近似度为

$$e^2\ll f_0^2 \tag{39}$$

这表明,比起长波近似或短波近似来,低频近似的发展方程可容纳频率较高的运动,因此,是一个精度高一级的模式。

在另一方面,如果运动的纬向尺度很大,或波长很长,则在式(31)中可略去对 x 的二阶导数项,而方程退化成

$$\frac{\partial(q_{n+2},r_n)}{\partial t}-\frac{1}{2n+3}\frac{\partial(q_{n+2},r_n)}{\partial x}=0 \tag{40}$$

此即为长波近似下的发展方程。另一个特殊情况，如运动的纬向尺度很小，或波长很短，则在式(31)中可略去$\frac{(2n+3)}{2}$这一项(在 n 不大时)，于是得到

$$\frac{\partial^2(q_{n+2},r_n)}{\partial t\partial x}-\frac{1}{2}(q_{n+2},r_n)=0 \tag{41}$$

这是短波近似下发展方程。由此可见，低频近似下的发展方程(31)可容纳纬向各种尺度的运动。

注意到 Kelvin 波将参与到发展过程中去，由于它满足纬圈地转关系，因此在长波近似的发展模式中是重要的角色，同时它也将包含在低频近似的发展方程中。在另一方面，对于 Rossby-重力混合波，如对 $q_{n+2},r_n(n\geqslant 0)$ 的频率要求也用到 q_1，则式(15)退化成

$$\frac{\partial^2 q_1}{\partial t\partial x}+\frac{1}{2}q_1=0 \tag{42}$$

即混合波是 Rossby 短波形式参加到发展运动中去。

5 频散性质

设方程(31)的解正比于 exp(i($kx-et$))，则色散公式为

$$e=-\frac{\frac{1}{2}k}{k^2+\frac{1}{2}(2n+3)} \tag{43}$$

这是考虑了辐散后的 Rossby 波速公式，当 k 很小时(长波)就退化到长波近似下的色散公式，当 k 很大时(短波)就退化到短波近似下的色散公式。

由式(43)算得群速度为

$$C_g=\frac{\mathrm{d}e}{\mathrm{d}k}=\frac{k^2-\frac{1}{2}(2n+3)}{2\left(k^2+\frac{1}{2}(2n+3)\right)^2} \tag{44}$$

由此可见，对于

$$k>\sqrt{\frac{1}{2}(2n+3)}=k_c \tag{45}$$

的短波，群速度是正的(向东的)，而对于

$$k<\sqrt{\frac{1}{2}(2n+3)}=k_c \tag{46}$$

的长波，群速度是负的(向西的)，因此在低频的发展模式中，能量可以不同于相速的速度传播，并可向正、负两个方向传播(依赖于波长)。这表明低频近似发展模式中的运动，其动力学性质比起其他两个模式来要丰富多彩一些。

由于这些特点，低频近似下的发展模式在热带大气(和海洋)动力学中能得到更广泛的应用。

6 结论

在分析发展运动的长波近似模式(即 Gill 模式)和短波近似模式后，指出了这两种模式的局限性，

于是提出了低频近似下的发展模式,指出它优于长波近似模式和短波近似模式。因为可容纳频率较低和波长更宽的运动,因此是研究热带大气(和海洋)动力学问题的一个性能较好的模式。

参考文献

[1] 巢纪平. 热带大气和海洋半地转的适应运动和发展运动. 气象学报,2000,58(3):257-264.

[2] Chao J P,Lin Y H. The foundation and mov ement of tropical semi-geostrophic adap tation. Acta Metqeor Sinica,1996,10:129-141.

[3] Gill A E. Some simple solution for heat-induced tropical circulation. Quart J Roy Meteor. Soc,1980,106:447-462.

热带半地转适应过程*

林永辉[1],巢纪平[2]

(1.中国气象科学研究院,北京 100081;2.国家海洋环境预报研究中心,北京 100081)

摘要:在热带地区,当纬圈或经圈方向上的地转平衡遭到破坏后,非地转运动将激发出重力惯性波。随着重力惯性波的频散,纬圈或经圈方向的地转平衡将重新建立,且遵循半位势涡度不变式。对半位势涡度不变式的讨论指出,纬圈或经圈半地转适应过程的方向主要依赖于初始扰动的经圈特征尺度。对于纬(经)圈半地转适应运动来说,只要初始扰动的纬圈特征尺度足够大(小),则适应场的特点总是压力场和纬(经)圈流的相互适应。

关键词:重力惯性波的频散;半地转适应;尺度准则

适应过程和演变过程是大气和海洋中两种最基本的运动形式。地转适应概念最初由 Rossby[1] 提出,在以后的 30 多年里,地转适应问题有了很大的进展[2-5]。可以说在中、高纬地区地转适应问题已基本清楚了[6]。

在热带地区,由于柯氏参数太小,一般认为流场和压力场之间难以建立地转平衡关系。但实际上,不论是大气还是海洋运动,物理场的变化主要表现在赤道两侧 Rossby 变形半经这一狭窄的区域内,运动沿纬圈方向的变化比较均匀且强度大,在这种形势下运动呈纬圈方向的地转平衡是完全可能的,这样的平衡称为纬圈半地转平衡,一般也称为长波近似或低频近似[7]。而在海洋经圈边界附近满足的地转平衡称为经圈半地转平衡,也称为短波近似。半地转平衡的建立过程称为半地转适应过程。最近,巢纪平和林永辉[8] 对此问题进行了研究。在本文中,将进一步研究半地转适应过程,及在半地转平衡下流场和压力场之间相互适应的尺度准则。

1 基本运动方程

在赤道 β 平面中线性化的浅水无量纲方程组为

$$\varepsilon_1 \frac{\partial u}{\partial t} - \frac{1}{2} y v + \frac{\partial \varphi}{\partial x} = 0 \tag{1.1}$$

$$\varepsilon_2 \frac{\partial v}{\partial t} - \frac{1}{2} y u + \frac{\partial \varphi}{\partial y} = 0 \tag{1.2}$$

$$\varepsilon_3 \frac{\partial \varphi}{\partial t} + \frac{\partial u}{\partial x} + \frac{\partial v}{\partial y} = 0 \tag{1.3}$$

式中 $\varepsilon_1(i=1,2,3)$ 为引进的标识符,其值取 1 或 0。当取 0 值时,分别表示运动处于经圈地转平衡、纬

* 中国科学(D 辑),1997,27(6):566-573.

圈地转平衡和无辐散状态。

由方程式(1.1)至式(1.3)得到

$$\varepsilon_1\varepsilon_3\frac{\partial^2 u}{\partial t^2}-\frac{\partial^2 u}{\partial x^2}=\varepsilon_3\frac{1}{2}y\frac{\partial v}{\partial t}+\frac{\partial^2 v}{\partial x\partial y} \tag{1.4}$$

$$\varepsilon_2\varepsilon_3\frac{\partial^2 v}{\partial t^2}-\frac{\partial^2 v}{\partial y^2}=-\varepsilon_3\frac{1}{2}y\frac{\partial u}{\partial t}+\frac{\partial^2 u}{\partial x\partial y} \tag{1.5}$$

由以上两个方程消去 u 可以得到对单一变量 v 的方程，即

$$\varepsilon_1\varepsilon_2\varepsilon_3\frac{\partial^3 v}{\partial t^3}-\left(\varepsilon_2\frac{\partial^2}{\partial x^2}+\varepsilon_1\frac{\partial^2}{\partial y^2}-\varepsilon_3\frac{1}{4}y^2\right)\frac{\partial v}{\partial t}-\frac{1}{2}\frac{\partial v}{\partial x}=0 \tag{1.6}$$

方程(1.6)即是研究赤道自由波运动的方程[9]。在赤道自由波系中包括了高频的重力惯性波、低频的 Rossby 波、重力惯性——Rossby 混合波和非频散的 Kelvin 波。

考虑到 Rossby 波的低频特性时，可将式(1.6)中的第 3 项忽略掉(即将 Rossby 波过滤掉)，则方程(1.6)成为

$$\varepsilon_1\varepsilon_2\varepsilon_3\frac{\partial^3 v}{\partial t^3}-\left(\varepsilon_2\frac{\partial^2}{\partial x^2}+\varepsilon_1\frac{\partial^2}{\partial y^2}-\varepsilon_3\frac{1}{4}y^2\right)\frac{\partial v}{\partial t}=0 \tag{1.7}$$

上式对时间积分后给出

$$\varepsilon_1\varepsilon_2\varepsilon_3\frac{\partial v^2}{\partial t^2}-\left(\varepsilon_2\frac{\partial^2}{\partial x^2}+\varepsilon_1\frac{\partial^2}{\partial y^2}-\varepsilon_3\frac{1}{4}y^2\right)v=$$
$$\left[\varepsilon_1\varepsilon_2\varepsilon_3\frac{\partial v^2}{\partial t^2}-\left(\varepsilon_2\frac{\partial^2}{\partial x^2}+\varepsilon_1\frac{\partial^2}{\partial y^2}-\varepsilon_3\frac{1}{4}y^2\right)v\right]_{t=0} \tag{1.8}$$

方程(1.8)右边描述的是初始时刻的运动，它可以是非地转和有辐散的，考虑了方程(1.5)和式(1.1)后得到

$$\varepsilon_1\varepsilon_2\varepsilon_3\frac{\partial v^2}{\partial t^2}-\left(\varepsilon_2\frac{\partial^2}{\partial x^2}+\varepsilon_1\frac{\partial^2}{\partial y^2}-\varepsilon_3\frac{1}{4}y^2\right)v=$$
$$\frac{\partial}{\partial x}\left[\varepsilon_3\frac{1}{2}y\varphi-\left(\varepsilon_2\frac{\partial v}{\partial x}-\varepsilon_1\frac{\partial u}{\partial y}\right)\right]_{t=0} \tag{1.9}$$

略去标识符号后有

$$\frac{\partial^2 v}{\partial t^2}-\left(\frac{\partial^2}{\partial x^2}+\frac{\partial^2}{\partial y^2}-\frac{1}{4}y^2\right)v=F(x,y,0) \tag{1.10}$$

其中

$$F(x,y,0)=\frac{\partial}{\partial x}\left[\frac{1}{2}y\varphi-\left(\frac{\partial v}{\partial x}-\frac{\partial u}{\partial y}\right)\right]_{t=0} \tag{1.11}$$

注意到，方程(1.10)左端描写的是重力惯性波，因此方程(1.10)研究的是在初始位势涡度梯度作用下重力惯性波的频散，其定解条件设为

$$t=0, v=\varphi_1(x,y),\quad \frac{\partial v}{\partial t}=\varphi_2(x,y) \tag{1.12}$$

$$|x|\to\infty,\quad v\to 0 \tag{1.13}$$

2　半位势涡度不变式

巢纪平和林永辉[8]对方程(1.10)定解问题研究指出，当时间充分大时，经圈流的时间导数将趋近

于零，即重力惯性波将在全场发生频散，但此时经圈流并不等于零。此时，式(1.2)退化成

$$\frac{1}{2}yu = -\frac{\partial \varphi}{\partial y} \tag{2.1}$$

即热带运动建立起纬圈的地转平衡。事实上，纬圈地转平衡的建立，相当于取 $\varepsilon_2=0$。参照式(1.9)的标识符号，则场适应后的经圈流满足方程

$$\left(\frac{\partial^2}{\partial y^2} - \frac{1}{4}y^2\right)v = \frac{\partial}{\partial x}\left[\left(\frac{\partial v}{\partial x} - \frac{\partial u}{\partial y}\right) - \frac{1}{2}y\varphi\right]_{t=0} \tag{2.2}$$

注意到，初始时刻的运动可以是非地转及有辐散的，因此式(1.9)右端的标识符号取为 1。上式在物理上表现为适应后的经圈流将由初始场中的位势涡度沿纬圈方向的梯度决定。

由方程式(1.2)及式(1.3)并考虑到式(2.1)和式(2.2)，可以推知纬圈半地转适应运动的压力场满足方程

$$\frac{\partial^2 \varphi}{\partial y^2} - \frac{1}{4}y^2\varphi = \frac{1}{2}y\left[\left(\frac{\partial v}{\partial x} - \frac{\partial u}{\partial y}\right) - \frac{1}{2}y\varphi\right]_{t=0} \tag{2.3}$$

由于上式右端括号中的量是初始的位势涡度，因此，上式即是位势涡度守恒式，但由于涡度算子中缺省了对 x 的二阶导数项，故称式(2.3)为半位势涡度不变式。

注意到，由方程(1.9)容易看出，当流场和压力场之间取经圈的地转平衡(即 $\varepsilon_1=0$)时，重力惯性波同样被过滤掉，流场和压力场达到相互适应的状态。此时，经圈流满足

$$\frac{1}{2}yv = \frac{\partial \varphi}{\partial x} \tag{2.4}$$

类似于前面的推导，容易推知经圈半地转适应后的纬圈流和压力场满足

$$\frac{\partial^2 u}{\partial x^2} - \frac{1}{4}y^2 u = \frac{\partial}{\partial y}\left[\frac{1}{2}y\varphi - \left(\frac{\partial v}{\partial x} - \frac{\partial u}{\partial y}\right)\right]_{t=0} \tag{2.5}$$

$$\frac{\partial^2 \varphi}{\partial x^2} - \frac{1}{4}y^2\varphi = \frac{1}{2}y\left[\left(\frac{\partial v}{\partial x} - \frac{\partial u}{\partial y}\right) - \frac{1}{2}y\varphi\right]_{t=0} \tag{2.6}$$

方程式(2.6)即为经圈半地转适应运动所遵循的半位势涡度不变式。

3 半地转适应过程的尺度准则

中高纬的地转适应研究表明，场的相互适应过程依赖于运动的尺度，即当初始扰动尺度大时，流场向压力场适应；反之，压力场向流场适应。现在来分析热带半地转适应过程中场适应的相应尺度准则。

3.1 纬圈半地转适应运动

对于纬圈半地转适应运动而言，当初始的位势涡度场给定后，适应后的压力场由式(2.3)决定，纬圈流和经圈流分别由式(2.1)和式(2.2)决定。为简单起见，设初边值条件为

$$t = 0, \quad u = v = 0, \quad \varphi = \varphi^0 e^{\alpha_1 x^2 \quad \alpha_2 y^2} \tag{3.1}$$

$$|x| \to \infty, \quad u, v, \varphi \to 0, \tag{3.2}$$

其中

$$\alpha_1 = (L_0/L_1)^2, \quad \alpha_2 = (L_0/L_2)^2 \tag{3.3}$$

L_0 为赤道 Rossby 变形半经，L_1 和 L_2 分别为初始扰动的纬圈和经圈特征尺度，α_1 和 α_2 很小时，即表示

纬圈和经圈特征尺度很大。

将式(3.1)代入式(2.1)至式(2.3)中,并将 u,v 及 φ 用抛物圆柱函数(即 Weber 函数)展开,即

$$(u,v,\varphi)=\sum_{n=0}^{\infty}(u_n,xu_n,\varphi_n)e^{a_1x^2}\cdot D_n(y) \tag{3.4}$$

式中,$D_n(y)$为 n 阶 Weber 函数,利用 Weber 函数的性质,取最低阶近似有:

$$\begin{cases}u_0=\varphi_0,\\ v_1=-\dfrac{\sqrt{2}}{6}\varphi^0\alpha_1(\alpha_2+1/4)^{-3/2},\\ \varphi_0=\dfrac{\sqrt{2}}{8}\varphi^0(\alpha_2+1/4)^{-3/2}\end{cases} \tag{3.5}$$

由上式知,对纬圈半地转适应运动而言,当取最低阶近似时,适应后的等压线和等纬圈流重合。当 $a_2\ll1$(即初始扰动的经圈特征尺度远大于赤道 Rossby 变形半径)时,适应后的压力场和初始的压力场差不多,但纬圈流的变化较大(因初始纬圈流是静止的)。适应后的经圈流强度依赖于 α_1,当 α_1 很小(即初始扰动的纬圈特征尺度很大)时,经圈流强度的变化较小,即穿越等压线的运动较弱,此时是纬圈流向初始的压力场适应。当 $\alpha_2\gg1$(即初始扰动的经圈特征尺度很小)时,适应后的压力场较初始压力场变化较大,而纬圈流变化不大,适应后的经圈流强度仍然依赖于 α_1,此时当纬圈特征尺度很大(即 α_1 很小时),压力场向初始的纬圈流适应。

在以后的讨论中,为方便将无量纲初始值 φ^0 取为 1,并引入如下定义:$\alpha_1\ll\alpha_2$(即初始扰动的纬圈特征尺度远大于经圈特征尺度)的情形称为纬圈型初始扰动;$\alpha_1\gg\alpha_2$ 的情形称为经圈型初始扰动;$\alpha_1=\alpha_2\ll1$的情形称为大尺度均衡型初始扰动,反之,称为小尺度均衡型初始扰动。

图 1 给出纬圈半地转适应后的压力场和初始压力场的东西向剖面图。由图可见,对于经圈型和大尺度均衡型初始扰动(即 α_2 很小)而言,适应后的压力场和初始压力场差不多(图 1(a)、图 1(b))。但对于纬圈型和小尺度均衡型初始扰动(即 α_2 很大)来说,适应后的压力场较初始压力场的变化是非常激烈的(图 1(c)和图 1(d)),此时纬圈半地转适应过程是以压力场调整为主。

对于纬圈流来说,图 2 给出纬圈半地转适应后的纬圈流随初始扰动特征尺度变化的南北向剖面图。对于经圈型和大尺度均衡型初始扰动(即 α_2 很小)而言,适应后的纬圈流较初始的纬圈流变化很大,即此时以纬圈流的调整为主。而对于纬圈型和小尺度均衡型初始扰动(即 α_2 很大)来说,适应后的纬圈流趋近于零,即适应后的纬圈流较初始纬圈流变化不大。

对经圈流而言,纬圈半地转适应运动中的经圈流变化不同于压力场和纬圈流。图 3 给出适应后的经圈流随初始扰动特征尺度变化的南北向剖面图。从图可知,除经圈型初始扰动外,对于其他型的扰动,经圈流变化都趋近于零,即适应后的经圈流较初始经圈流变化不大。因此可以说,对于纬圈半地转适应运动而言,只要初始扰动的经圈特征尺度足够大(即 α_1 较小)则适应场的性质特点是压力场和纬圈流的相互适应,适应过程的方向依赖于初始扰动的经圈特征尺度,即当初始扰动的经圈特征尺度很大(此时为大尺度均衡型扰动)时,纬圈流向初始的压力场适应;而当初始扰动的径圈特征尺度很小(此时为纬圈型初始扰动)时,压力场向初始的纬圈流适应。

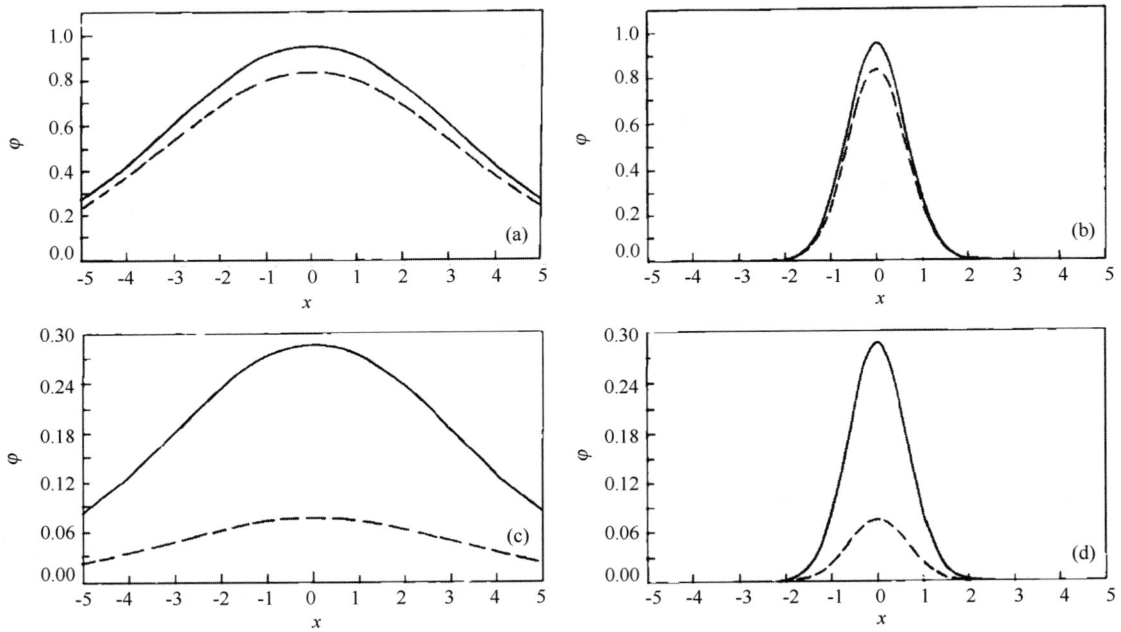

图 1 纬圈半地转适应后的压力场与初始压力场沿 $y=1.0$ 的东西向剖面图

纵轴为压力值，其中，(a) $\alpha_1=\alpha_2=0.05$；(b) $\alpha_1=1.25,\alpha_2=0.05$；(c) $\alpha_1=0.05,\alpha_2=1.25$；(d) $\alpha_1=\alpha_2=0.025$。实线对应初始值，虚线对应适应后的值

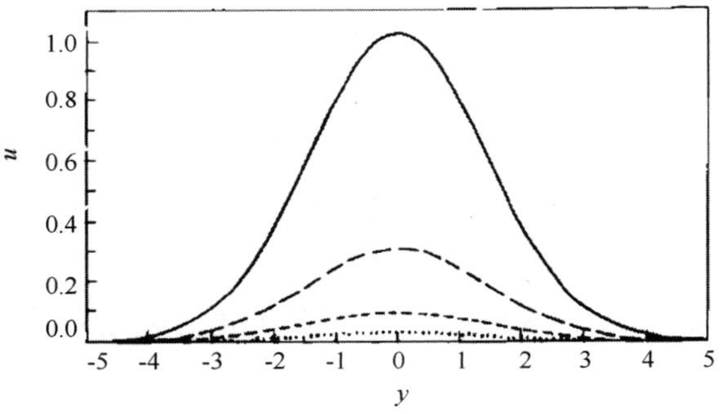

图 2 纬圈半地转适应后的纬圈流随初始扰动特征尺度变化沿 $x=1.0$ 的南北向剖面图

纵轴为纬圈流。其中，实线对应 $\alpha_1=\alpha_2=0.05$；长虚线对应 $\alpha_1=1.25,\alpha_2=0.05$；短线对应剖面 $\alpha_1=0.05,\alpha_2=1.25$；点线对应 $\alpha_1=\alpha_2=1.25$

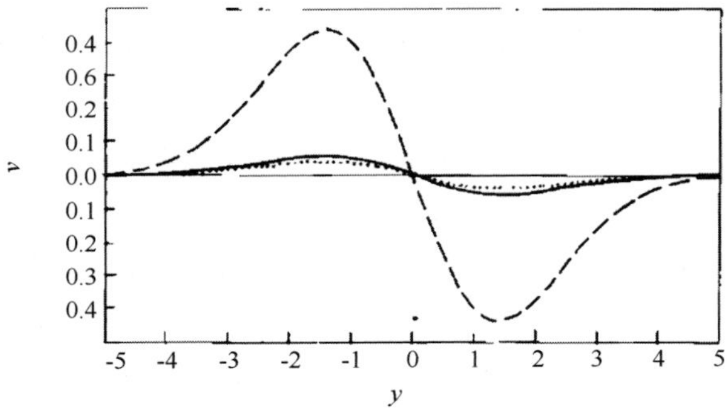

图 3 纬圈半地转适应后的径圈流随初始扰动特征尺度变化沿 $x=1.0$ 的南北向剖面图

纵轴为纬圈流。其中，实线对应 $\alpha_1=\alpha_2=0.05$；长虚线对应 $\alpha_1=1.25,\alpha_2=0.05$；短线对应剖面，$\alpha_1=0.05,\alpha_2=1.25$；点线对应 $\alpha_1=\alpha_2=1.25$

3.2 经圈半地转适应运动

同样地，当初始场的位势涡度给定后，由式(2.4)、式(2.5)及式(2.6)可得经圈半地转适应后的经圈流、纬圈流和压力场分布。

将 u,v 和 φ 用抛物圆柱函数(即 Weber 函数)展开，即

$$(u,v,\varphi)=\sum_{n=0}^{\infty}(u_n,v_n,\varphi_n)D_n(y) \tag{3.6}$$

式中，$D_n(y)$为n阶Weber函数，将式(3.6)代入式(2.4)至式(2.6)中，并考虑到式(3.1)和式(3.2)，利用Weber函数的性质，取最低阶近似，并用Weber函数法求解可得：

$$\begin{cases} u_0(x) = -\beta_2[(1+\mathrm{erf}(x_1))e^{x/2} + (1-\mathrm{erf}(x_2))e^{x/2}], \\ \varphi_0(x) = -\beta_1[(1+\mathrm{erf}(x_1))e^{x/2} + (1-\mathrm{erf}(x_2))e^{x/2}], \\ v_1(x) = \beta_1[(1+\mathrm{erf}(x_1))e^{-x/2} + (1-\mathrm{erf}(x_2))e^{x/2}]. \end{cases} \tag{3.7}$$

式中：

$$\begin{cases} \beta_1 = -\dfrac{\sqrt{2\pi}}{64}\varphi^0 e^{1/(16\alpha_1)}\alpha_1^{-1/2}(\alpha_2+1/4)^{-3/2}, \\ \beta_2 = \dfrac{\sqrt{2\pi}}{16}\varphi^0 e^{1/(16\alpha_1)}\alpha_1^{1/2}[(\alpha_2+1/4)^{1/2} - \alpha_2(\alpha_2+1/4)^{3/2}], \\ x_1 = \sqrt{\alpha_1}[x-1/(4\alpha_1)], \\ x_2 = \sqrt{\alpha_1}[x+1/(4\alpha_1)], \\ \mathrm{erf}(\xi) = \dfrac{2}{\sqrt{\pi}}\displaystyle\int_0^{\varepsilon} e^{-\tau^2}d\tau \end{cases} \tag{3.8}$$

图4给出经圈半地转适应后的压力场和初始压力场的南北向剖面图（在计算中仍然取无量纲初始值φ^0为1，下同）。从图可知，对于经圈型和大尺度均衡型初始扰动（即α_2较小）而言，适应后的压力场和初始的压力场差不多（图4(a)、图4(b)）。但对于纬圈型和小尺度均衡型初始扰动（即α_2较大）来说，适应后的压力场较初始压力场变化非常激烈（图4(c)、图4(d)），即此时适应过程以压力场的调整为主。

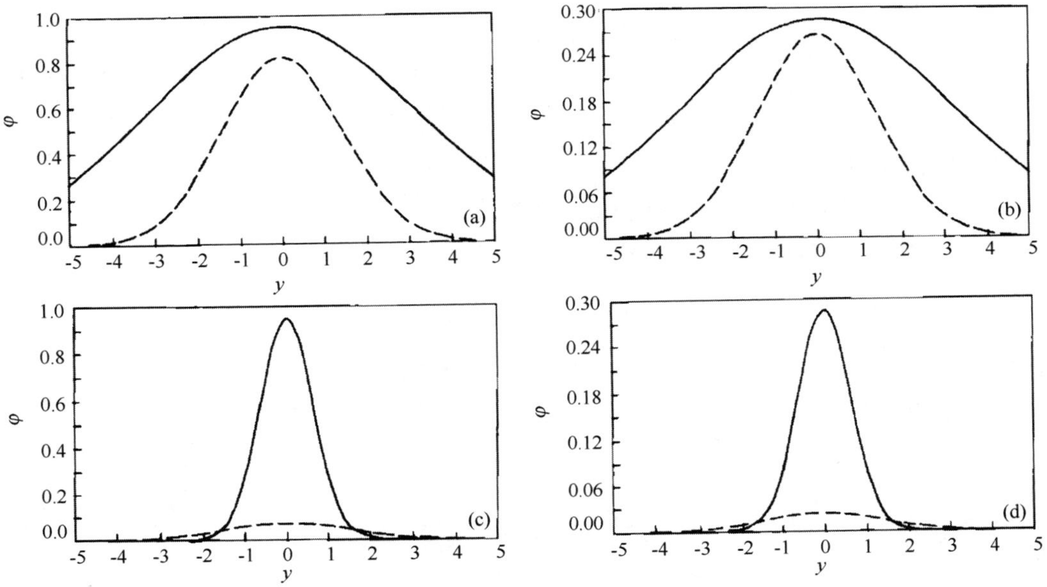

图4　经圈半地转适应后的压力场与初始压力场沿$x=1.0$的南北向剖面图

纵轴为压力值。其中，(a)$\alpha_1=\alpha_2=0.05$；(b)$\alpha_1=1.25,\alpha_2=0.05$；(c)$\alpha_1=0.05,\alpha_2=1.25$；(d)$\alpha_1=\alpha_2=0.025$。实线对应初始值，虚线对应适应后的值

图5是经圈半地转适应后的纬圈流随初始扰动特征尺度变化的南北向剖面图。对于经圈型和大

尺度均衡型初始扰动(即 α_2 很小)而言,适应后的纬圈流较初始纬圈流变化很大,此时,适应过程以纬圈流的调整为主。对于纬圈型和小尺度均衡型初始扰动(即 α_2 很大)来说,适应后的纬圈流趋近于零,说明适应后的纬圈流较初始纬圈流变化不大。从图 4 和图 5 也可看出,对于经圈半地转适应运动而言,压力场和纬圈流的变化仍然主要依赖于初始扰动的经圈特征尺度。同样地,经圈半地转适应运动中的经圈流变化异于纬圈流变化。图 6 所显示的经圈流变化特征类似于图 5 中的纬圈流变化,即对于经圈型和大尺度均衡型初始扰动(即 α_2 很小)来说,适应后的经圈流变化很大;而对于纬圈型和小尺度均衡型初始扰动(即 α_2 很大),适应后的经圈流较初始的经圈流变化不大。值得注意的是,当初始扰动的纬圈特征尺度不断减小(即 a_1 不断增加)时,经圈半地转适应后的纬圈流不断减小(图 7(a)),但经圈流强度几乎不变(图 7(b))。因此,在经圈半地转适应运动中,对经圈型初始扰动,适应场的特点是经圈流向初始压力场适应;对大尺度均衡型初始扰动而言,经圈流和纬圈流向初始压力场适应;而对于纬圈型和小尺度均衡型初始扰动来说,适应场的特点是压力场向初始流场适应。换句话说,只要初始扰动的纬圈特征尺度足够小,则适应场的性质特点总是压力场和经圈流相互适应,适应过程的方向依赖于初始扰动的经圈特征尺度,即当初始扰动的经圈特征尺度很大时,经圈流向初始的压力场适应;反之,压力场向初始的经圈流适应。

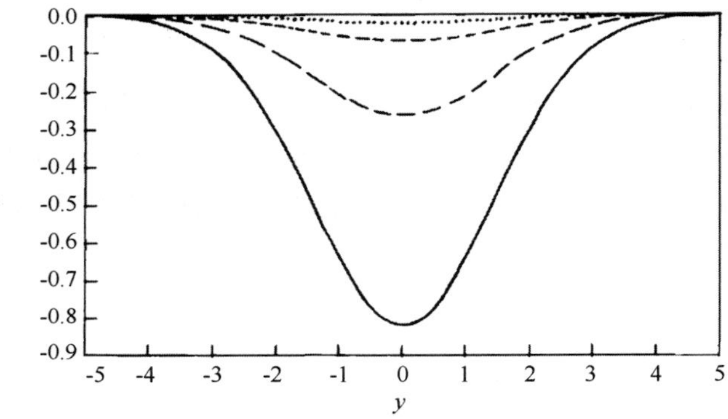

图 5 经圈半地转适应后的经圈流随初始扰动特征尺度变化沿 $x=1.0$ 的南北向剖面图

纵轴为纬圈流。其中,实线对应 $\alpha_1=\alpha_2=0.05$;长虚线对应 $\alpha_1=1.25$, $\alpha_2=0.05$;短线对应剖面,$\alpha_1=0.05,\alpha_2=1.25$;点线对应 $\alpha_1=\alpha_2=1.25$

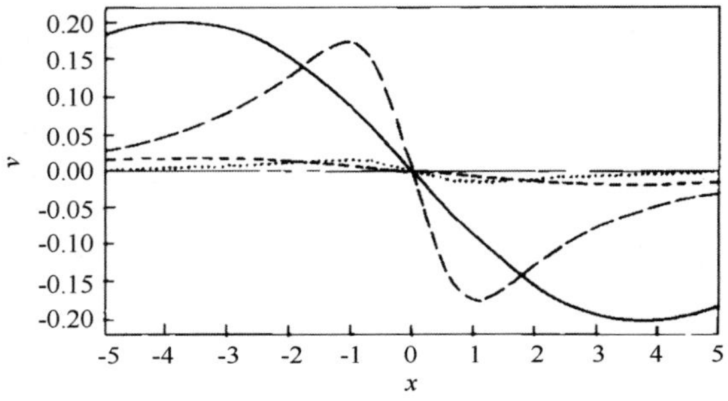

图 6 经圈半地转适应后的经圈流随初始扰动特征尺度变化沿 $y=1.0$ 的东西向剖面图

纵轴为纬圈流。其中,实线对应 $\alpha_1=\alpha_2=0.05$;长虚线对应 $\alpha_1=1.25,\alpha_2=0.05$;短线对应剖面,$\alpha_1=0.05,\alpha_2=1.25$;点线对应 $\alpha_1=\alpha_2=1.25$

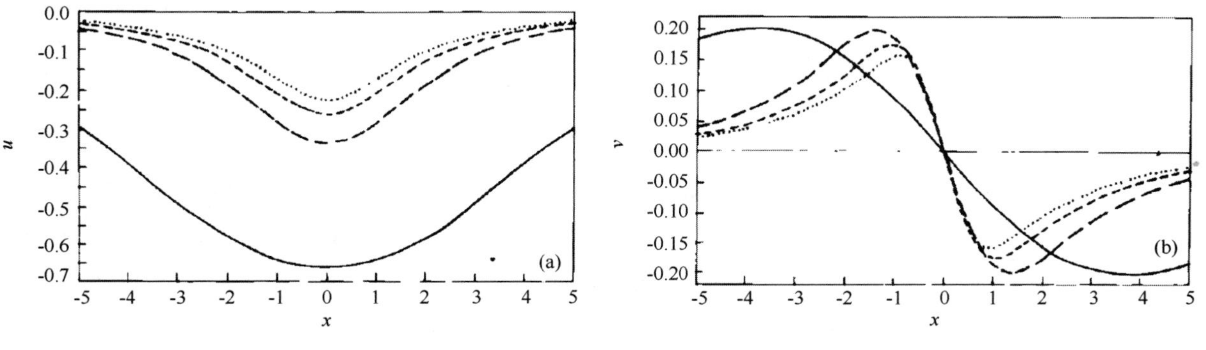

图 7 经圈半地转适应后的纬圈流(a)和经圈流(b)随初始扰动特征尺度变化沿 $y=0.1$ 的东西向剖面图

其中,实线对应 $\alpha_1=\alpha_2=0.05$;长虚线对应 $\alpha_1=0.65,\alpha_2=0.05$;短虚线对应 $\alpha_1=1.25,\alpha_2=0.05$;点虚线对应 $\alpha_1=1.85,\alpha_1=0.05$

4　小结和讨论

在热带地区，当纬圈或经圈方向的地转平衡遭到破坏后，非地转运动将激发出重力惯性波，伴随着重力惯性波的频散，将重新建立起纬圈或经圈方向上的地转平衡关系，且这种运动遵循半位势涡度不变式。对半位势涡度不变式的讨论指出，不论对纬圈或经圈半地转适应运动而言，适应过程的方向主要依赖于初始扰动的经圈特征尺度。但对于纬圈半地转适应运动来说，只要初始扰动的纬圈特征尺度足够大，则适应场的特点总是压力场和纬圈流的相互适应。而对于经圈半地转适应运动而言，只要初始扰动的纬圈特征尺度足够小，则总是压力场和经圈流的相互适应。

纬圈和经圈半地转适应过程所表现出来的这种不同特点有其各自的物理原因。对纬圈半地转适应过程而言，当初始扰动的纬圈特征尺度足够大，而经圈特征尺度较小时，由南北压力梯度力产生的经圈流一方面向低压一方输送质量，另一方面在柯氏力作用下产生纬圈流以平衡压力梯度力，但由于经圈范围较小，故纬圈流还来不及发展到平衡压力梯度力时，压力场就被填塞掉了，即此时主要是压力场向纬圈流场适应；当经圈尺度很大时，经圈流在柯氏力作用下产生的纬圈流有足够的时间来平衡压力梯度力，即此时纬圈流向初始的压力场适应。因此，在赤道两侧常常可以看到比较均匀而且强大的大气或海洋运动，它们均满足纬圈方向的地转平衡。而对经圈半地转适应过程来说，虽然在理论上得到经圈型扰动有利于经圈流的调整变化，但由于纬圈范围较小，故对热带大气运动而言，经圈流还不足以发展到平衡东西向压力梯度力时压力场就被填塞了。因此，在热带大气中纯粹的经圈半地转运动是非常少见的。但是，在海洋的西边界附近，情况就完全不一样了，纬圈压力梯度力产生的纬圈流驱使的质量输送并没有在西边界附近大量堆集，而是沿西边界向北输送（北半球），因此，当经圈特征尺度很大时，经圈流有足够的时间来平衡纬圈方向的压力梯度力；但当经圈尺度很小时，质量输送也易在北边堆集，导致压力场的填塞。

需要指出的是，在本文中取了低阶截断解，这必然会对结果产生一定的影响，但考虑到热带大气主要以低阶模运动为主[10]，因此，从定性讨论的角度出发，这样的方法是可行的。

参考文献

[1] Rossby C G. On the mutual adjustment of presure and velocity distributions in certain simple current systems. J Mar Res, 1973, 1: 15-28.

[2] Oboukhov A M. The problem of the geostrophic adaptaion, Izvestiya of Academy of Science USSR. Series Geography and Geophysics, 1949, 13: 281-289.

[3] Yeh T C. On the formation of quasi-geostrophic motion in the atmosphere. J Met Sec Japan, 1957, 35: 130-134.

[4] 曾庆存．大气中的适应过程和发展过程（一）和（二）．气象学报，1993，33：163-174.

[5] Blumen W. Geostrophic adjustment. Reviews of Geophysics and Space Physics, 1972, 10: 485-528.

[6] 叶笃正，李麦村．大气运动中的适应问题．北京：科学出版社，1965.

[7] Gill A E. Atmosphere-Ocean Dynamics. New York: Academic Press, 1982.

[8] Chao J P, Lin, Y H. The foundarion and movement of tropical semigeostrophic adaptation. Acta Meteor Sinica, 1996, 10: 129-141.

[9] Matsuno T. Quasi-geostrophic motions in the equatorial area. J Meteor Soc Japan, 1966, 44: 25-43.

[10] 巢纪平．厄尔尼诺和南方涛动动力学．北京：气象出版社，1993.

The Foundation and Movement of Tropical Semi-geostrophic Adaptation*

Chao Jiping(巢纪平)[1] Lin Yonghui(林永辉)[2]

(1. National Research Center for Marine Environment Forecasts, Beijing 100081;2. Institute of Atmospheric Physics. Chinese Academy of Sciences, Beijing 100029)

ABSTRACT: The breakdown and foundation of geostrophic balance is one of the important movements in the mid-and high-latitude atmosphere and oceans. In the tropical area, the value of Coriolis parameter is so small that it is difficult to satisfy the bi-geostrophic equilibrium between the pressure and velocity fields. However. in the tropical area, the zonal velocity of some motions in the atmo-sphere and oceans is large, so the Coriolis force is not small. geostrophic balance can exist in zonal direction, i. e. semi-geostrophic balance. Furthermore, in the dominant area of Hadley circulation in the atmosphere or the area near the ocean meridional boundary, the meridional velocity is large, so geostrophic balance can also exist in meridional direction. In this paper, the process of the dis-persion of inertial gravity wave and the foundation of semi-geostrophic balance are first discussed. Second, the adjustment process between the velocity and pressure fields after adaptation is also viewed. and the scale criterion of the semi-geostrophic adaptation is discussed, i.e. for the motion with meridional scale greater than the equatorial Rossby radius of deformation. the velocity and pressure fields after adaptation change to fit the initial pressure field; on the contrary, the fields change to fit the initial zonal velocity field, and the strength of the fields after adaptation depends on the zonal scale.

Key words: dispersion of inertial gravity wave; semi-geostrophic adaptation; scale criterion

I INTRODUCTION

For the motions of atmosphere and ocean of a rotating earth, if the change of Coriolis parameter $f(=2\Omega\sin\varphi, \Omega$ is the angular velocity of the rotating earth, φ is the latitude) with the latitude (i.e. β-effect) is ignored, and the initial equilibrium is disterbed, then inertial gravity wave will be stimulated, with the dispersion of the wave, a new geostrophic balance will be founded founded through the mutual adjustment and adaptation between the pressure and velocity fields. The process of the dispersion of inertial gravity wave and the foundation of geostrophic balance is known as the geostrophic adaptation (the field of adaptation follows

* Acta meteorologica Sinica, 1996(2):129-141.

an invariant of potential vorticity). The movement under the geostrophic balance is called evolution movement. Since Rossby (1937;1938) first put forward the concept, there have been many advances, including the contributions of Yeh (1957) and Zeng (1963). So to speak, the geostrophic adaptation of mid- and high-latitudes is basically clear (Yeh and Li 1965).

In the tropical area, the Coriolis parameter is so small that it is very difficult to satisfy the bi-geostrophic balance in both zonal and meridional directions, simultaneously. The geostrophic balance often exists in one direction, for example, the changes of physical field of atmosphere and ocean mainly emerge in a narrow equatorial area whose width is about one Rossby radius of deformation, and the movements along the zonal direction are uniform and the strength is strong, so it is possible for the geostrophic balance to exist in zonal direction, and generally we name it the long-wave approximation or low-frequency approximation. On the other hand, in the area near ocean meridional boundary, the merid-ional velocity along the boundary, and the pressure gradient vertical to the boundary are large, they approximately satisfy geostrophic balance. Similarly, in the dominant area of Hadley circulation in the atmosphere. the movement also satisfies geostrophic balance in meridional direction, the change of physical field of this movement along the zonal direction is large or the wave numbers are big. so it is called short wave approximation. Since the geostrophic balance generally exists in one direction in the tropical area, the balance is named semi-geostrophic balance, the founding process of semi-geostrophic balance is known as semi-geostrophic adaptation, the movement after semi-geostrophic adaptation is named semi-geostrophic evolution moventent.

In this paper, the process of the dispersion of inertial gravity wave and the foundation of semi-geostrophic balance in the tropical area is viewed, and at the same time, the mutual adjustment of the fields under the semi-geostrophic adaptation is also discussed.

Ⅱ THE EQUATIONS OF MOTION

The linearized shallow-water equations in the equatorial β-plane are

$$\frac{\partial u}{\partial t} - \beta y v + \frac{\partial \varphi}{\partial x} = 0 \tag{1}$$

$$\frac{\partial v}{\partial t} + \beta y u + \frac{\partial \varphi}{\partial y} = 0 \tag{2}$$

$$\frac{\partial \varphi}{\partial t} + C^2\left(\frac{\partial u}{\partial x} + \frac{\partial v}{\partial y}\right) = 0 \tag{3}$$

where u and v represent the eastward (x) and northward (y) components of velocity, respectively. φ is the geopotential height of the atmosphere, for the ocean $\varphi = g'\eta$, η is the disturbed depth of the thermocline, and

$$g' = \frac{\rho_2 - \rho_1}{\rho_2} g \tag{4}$$

where g is the acceleration of gravity, g' is the reduced gravity. The ocean is separated into two layers with density ρ_1 and ρ_2, respectively; the lower-layer with density ρ_2 is motionless. The corresponding speed of gravity wave is

$$C = (g'H)^{1/2} \tag{5}$$

where H is the depth of the upper-layer. Generally, for the baroclinic fluid, the motion may be expanded in terms of eigenmodes in vertical direction. For any mode, we have

$$C=(gh)^{1/2} \tag{6}$$

where h is the equivalent depth of the eigenmode, $\varphi = gh$. In this case, Eqs. (1)-(3) are the horizontal structure equations of the baroclinic fluid corresponding to certain vertical eigenmode.

Taking the time scale $(2\beta C)^{-1/2}$, the horizontal scale $(C/2\beta)^{1/2}$, the velocity scale C. and the geopotential height scale C^2. the forms of non-dimensional equations can be written as

$$\epsilon_1 \frac{\partial u}{\partial t} - \frac{1}{2} yv + \frac{\partial \varphi}{\partial x} = 0 \tag{7}$$

$$\epsilon_2 \frac{\partial v}{\partial t} + \frac{1}{2} yu + \frac{\partial \varphi}{\partial y} = 0 \tag{8}$$

$$\epsilon_3 \frac{\partial \varphi}{\partial t} + \frac{\partial u}{\partial x} + \frac{\partial v}{\partial y} = 0 \tag{9}$$

where $\epsilon_i (i=1,2,3)$ is the sign symbol, it takes 1 or 0. When taking 0, Eqs. (7)-(9) represent the meridional geostrophic balance, zonal geostrophic balance and non-divergent motion, respectively.

Eqs. (7)-(9) give

$$\epsilon_1\epsilon_3 \frac{\partial^2 u}{\partial t^2} - \frac{\partial^2 u}{\partial x^2} = \epsilon_3 \frac{1}{2} y \frac{\partial v}{\partial t} + \frac{\partial^2 v}{\partial x \partial y} \tag{10}$$

$$\epsilon_2\epsilon_3 \frac{\partial^2 v}{\partial t^2} - \frac{\partial^2 v}{\partial y^2} = -\epsilon_3 \frac{1}{2} y \frac{\partial u}{\partial t} + \frac{\partial^2 u}{\partial x \partial y} \tag{11}$$

Eliminating u from Eqs. (10) and (11) gives

$$\epsilon_1\epsilon_2\epsilon_3 \frac{\partial^3 v}{\partial t^3} - \left(\epsilon_2 \frac{\partial^2}{\partial x^2} + \epsilon_1 \frac{\partial^2}{\partial y^2} - \epsilon_3 \frac{1}{4} y^2\right) \frac{\partial v}{\partial t} - \frac{1}{2} \frac{\partial v}{\partial x} = 0 \tag{12}$$

Eq. (12) is Matsuno (1966) equation describing the tropical trapped waves, including high-frequency inertial gravity wave, low-frequency Rossby wave, inertial gravity-Rossby mixed wave and non-dispersion Kelvin wave. Figure 1 gives the dispersion curves of various waves, including the westward propagating boundary-wave with damping amplitude.

Ⅲ THE DISPERSION OF LNERTIAL GRAVITY WAVE

Ignoring β-effect, Eq. (12) becomes

$$\epsilon_1\epsilon_2\epsilon_3 \frac{\partial^3 v}{\partial t^3} - \left(\epsilon_2 \frac{\partial^2}{\partial x^2} + \epsilon_1 \frac{\partial^2}{\partial y^2} - \epsilon_3 \frac{1}{4} y^2\right) \frac{\partial v}{\partial t} = 0 \tag{13}$$

Integrating (13) over the time gives

$$\epsilon_1\epsilon_2\epsilon_3 \frac{\partial^2 v}{\partial t^2} - \left(\epsilon_2 \frac{\partial^2}{\partial x^2} + \epsilon_1 \frac{\partial^2}{\partial y^2} - \epsilon_3 \frac{1}{4} y^2\right) v$$

$$= \left[\epsilon_1\epsilon_2\epsilon_3 \frac{\partial^2 v}{\partial t^2} - \left(\epsilon_2 \frac{\partial^2}{\partial x^2} + \epsilon_1 \frac{\partial^2}{\partial y^2} - \epsilon_3 \frac{1}{4} y^2\right) v\right]_{t=0} \tag{14}$$

Since the initial motion may be non-geostrophic and divergent, the motion is still controlled by Eqs. (7), (8), (10) and (11). Considering Eqs. (11) and (7), (14) becomes

$$\epsilon_1\epsilon_2\epsilon_3 \frac{\partial^2 v}{\partial t^2} - \left(\epsilon_2 \frac{\partial^2}{\partial x^2} + \epsilon_1 \frac{\partial^2}{\partial y^2} - \epsilon_3 \frac{1}{4}y^2\right) v$$

$$= \frac{\partial}{\partial x}\left[\epsilon_3 \frac{1}{2} y\varphi - \left(\epsilon_2 \frac{\partial v}{\partial x} - \epsilon_1 \frac{\partial u}{\partial y}\right)\right]_{t=0} \tag{15}$$

Ignoring the sign symbol, Eq. (15) becomes

$$\frac{\partial^2 v}{\partial t^2} - \left(\frac{\partial^2}{\partial x^2} + \frac{\partial^2}{\partial y^2} - \frac{1}{4}y^2\right) v = F(x,y,0) \tag{16}$$

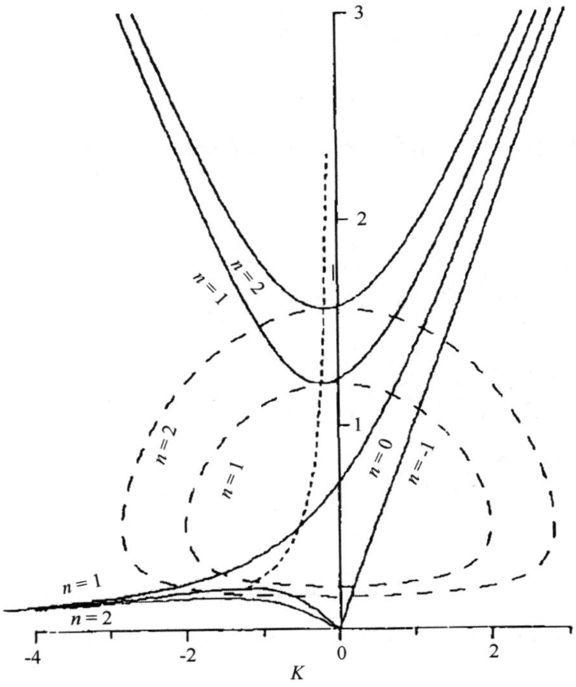

Fig.1 Dispersion curves for equatorial waves. The vertical axis is the frequency and the horizontal axis is the east-west wavenumber. The curve labeled 0 corresponds to the mixed Rossby-gravity wave. The upper curves labeled 1 and 2 are the first two gravity wave modes and the corresponding lower curves are the first two Rossby wave modes. The straight line labeled −1 corresponds to the Kelvin wave. The dashed curves are the boundary waves

$$F(x,y,0) = \frac{\partial}{\partial x}\left[\frac{1}{2} y\varphi - \left(\frac{\partial v}{\partial x} - \frac{\partial u}{\partial y}\right)\right]_{t=0} \tag{17}$$

Eq. (16) describing the dispersion of inertial gravity wave is resolved under the following conditions

$$t = 0, \quad v = \Phi_1(x,y), \quad \frac{\partial v}{\partial t} = \Phi_2(x,y) \tag{18}$$

$$|x| \to \infty, \quad v \to 0 \tag{19}$$

Variables v, F, Φ_1, Φ_2 are expanded in terms of the parabolic cylinder functions (i.e. Weber functions):

$$v = \sum_n v_n(x,t) D_n(y) \tag{20a}$$

$$F = \sum_n F_n(x) D_n(y) \tag{20b}$$

$$\Phi_1 = \sum_n \Phi_{1n}(x) D_n(y) \tag{20c}$$

$$\Phi_2 = \sum_n \Phi_{2n}(x) D_n(y) \tag{20d}$$

Therefore the coefficients of the Weber functions satisfy

$$\frac{\partial^2 v_n}{\partial t^2} - \frac{\partial^2 v_n}{\partial x^2} + \left(n + \frac{1}{2}\right) v_n = F_n \tag{21}$$

and

$$t = 0, \quad v_n = \Phi_{1n}, \quad \frac{\partial v_n}{\partial t} = \Phi_{2n} \tag{22}$$

$$| x | \to \infty, \quad v_n \to 0 \tag{23}$$

Using Fourier method or Laplance transformation. the solution of Eq. (21) satisfying conditions (22) and (23) is

$$\begin{aligned} v_n = & \frac{1}{2}\int_{-\infty}^{\infty} \Phi_{1n}(x') \frac{J_1\left(\sqrt{n + \frac{1}{2}}\sqrt{t^2 - (x - x')^2}\right)}{\sqrt{t^2 - (x - x')^2}} dx' \\ & + \frac{1}{2}\int_{-\infty}^{\infty} \Phi_{2n}(x') J_0\left(\sqrt{n + \frac{1}{2}}\sqrt{t^2 - (x - x')^2}\right) dx' \\ & + \frac{1}{2}\int_0^t\int_{-\infty}^{\infty} F_n(x') J_0\left(\sqrt{n + \frac{1}{2}}\sqrt{t'^2 - (x - x')^2}\right) dx' dt' \end{aligned} \tag{24}$$

where J_v is Bessel function of order v. Using the properties of Bessel function, i. e., when time is longer, the influence due to the initial conditions disappears quickly, then the main contribution of (24) is

$$v_n = \frac{1}{2}\int_0^t\int_{-\infty}^{\infty} F_n(x') J_0\left(\sqrt{n + \frac{1}{2}}\sqrt{t'^2 - (x - x')^2}\right) dx' dt' \tag{25}$$

Derivation of (25) with respect to the time is

$$\frac{\partial v_n}{\partial t} = \frac{1}{2}\int_{-\infty}^{\infty} F_n(x') J_0\left(\sqrt{n + \frac{1}{2}}\sqrt{t^2 - (x - x')^2}\right) dx' \tag{26}$$

Now, continuously evaluate the solution as the time increases greatly. If the value of F_n focuses on the original point, then assume

$$F_n(x) = A_n \delta(x) \tag{27}$$

where $\delta(x)$ is Delta function. Considering (27), Eqs. (25) and (26) respectively become

$$v_n = \frac{A_n}{2}\int_0^t J_0\left(\sqrt{n + \frac{1}{2}}\sqrt{t'^2 - x^2}\right) dt' \tag{28}$$

$$\frac{\partial v_n}{\partial t} = \frac{A_n}{2} J_0\left(\sqrt{n + \frac{1}{2}}\sqrt{t^2 - x^2}\right) \tag{29}$$

Note that in the area of $t>x$, when $t \to \infty$, Eq. (29) gives

$$\frac{\partial v_n}{\partial t} \sim O(t^{-\frac{1}{2}}) \tag{30}$$

i.e. the time derivative of meridional velocity decays according to $t^{-1/2}$, but the meridional velocity does not equal zero. In fact, near the original point ($x=0$), Eq. (28) gives

$$v_n(0,t\rightarrow\infty)=\frac{A_n}{2}\int_0^{\infty}J_0\left(\sqrt{n+\frac{1}{2}}t'\right)\mathrm{d}t'=\frac{A_n}{2\sqrt{n+\frac{1}{2}}} \tag{31}$$

which denotes that the meridional velocity is definite, and the value is small with higher n, i. e. the lower-order mode plays an important role. When $n=1$, the change of v_1/A_1 with the time is given in Fig.2, showing that the valus is almost the same as the theoretical valus 0.408 of Eq. (31) as the fime equals 20.

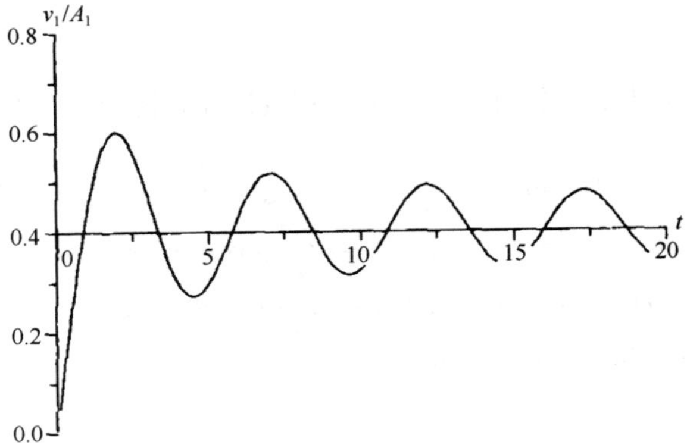

Fig.2　Solution v_1 of Eq. (28) at $n=1$. The Vertical axis is the meridional velocity in units of A_1. The horizontal axis is the time

Ⅳ　THE FOUNDATION OF SEMI-GEOSTROPHIC BALANCE

As the time is longer, after the dispersion of inertial gravity wave, since the time derivative of meridional velocity almost equals zero, then Eq. (8) becomes

$$\frac{1}{2}yu=-\frac{\partial\varphi}{\partial y} \tag{32}$$

i.e. the motion satisfies the zonal geostrophic balance. In fact, the foundation of zonal geostrophic balance is equavelent to $\epsilon_2=0$. Ignoring the sign symbol. the meridional velocity after adaptation satisfies

$$-\left(\frac{\partial^2}{\partial y^2}-\frac{1}{4}y^2\right)v=\frac{\partial}{\partial x}\left[\frac{1}{2}y\varphi-\left(\frac{\partial v}{\partial x}-\frac{\partial u}{\partial y}\right)\right]_{t=0} \tag{33}$$

Noticeably, the initial motion may be non-geostrophic and divergent, so the sign symbol of the right-hand side of (15) should be taken as 1.

Eq. (33) shows that the meridional velocity is directly given by (33), when the geostrophic balance is founded and the inertial gravity wave completely dispersed. Eq. (33) also denotes that the meridional velocity after adaptation is determined by the gradient of initial potential vorticity along the zonal direction.

Note that the zonal geostrophic balance can remove the inertial gravity wave, and make the fluid fields mutual adaptation. The zonal geostrophic balance is not a unique method to exclude the inertial gravity wave. In fact, from Eq. (15), when meridional geostrophic balance exists between the velocity and pressure fields, i.e.

$$\frac{1}{2}yv = \frac{\partial \varphi}{\partial x} \tag{34}$$

the inertial gravity wave is also filtered, and the fields are in mutul adaptation. Similar to the previous deduction, the zonal velocity after adaptation is given by

$$-\left(\frac{\partial^2}{\partial x^2} - \frac{1}{4}y^2\right)u = \frac{\partial}{\partial y}\left[\left(\frac{\partial v}{\partial x} - \frac{\partial u}{\partial y}\right) - \frac{1}{2}y\varphi\right]_{t=0} \tag{35}$$

showing that the zonal velocity after adaptation is determined by the gradient of initial potential vorticity along meridional direction.

V THE INVARIANT OF SEMI-POTENTIAL VORTICITY

In the motion of geostrophic adaptation of the mid-and high-latitudes, the physical field after adaptation follows an invariant of potential vorticity (Obukhow 1949) governing the motion after adaptation. Now, the corresponding laws in tropical semi-geostrophic adaptation are discussed.

For the zonal semi-geostrophic motion, based on Eqs. (7) and (9), and considering Eq. (32), we obtain

$$\frac{\partial^2 u}{\partial x \partial y} + \frac{1}{2}y\frac{\partial \varphi}{\partial x} = -\left(\frac{\partial^2 u}{\partial y^2} - \frac{1}{4}y^2 v\right) \tag{36}$$

Substituting Eqs. (32) and (33) into Eq. (36), and ignoring the term associated with the change of y in order to exclude β-effect. we yield

$$\frac{\partial}{\partial x}\left[\frac{\partial^2 \varphi}{\partial y^2} - \frac{1}{4}y^2\varphi\right] = \frac{\partial}{\partial x}\left[\frac{1}{2}y\left(\frac{\partial v}{\partial x} - \frac{\partial u}{\partial y}\right) - \frac{1}{4}y^2\varphi\right]_{t=0} \tag{37}$$

Integrating (37) with respect to x becomes

$$\frac{\partial^2 \varphi}{\partial y^2} - \frac{1}{4}y^2\varphi = \frac{1}{2}y\left[\left(\frac{\partial v}{\partial x} - \frac{\partial u}{\partial y}\right) - \frac{1}{2}y\varphi\right]_{t=0} \tag{38}$$

denoting that the inside bracket of the right-hand side is initial potential vorticity, so the adaptation field of the left-hand side is potential vorticity under the long-wave approximation. Due to the lack of x-derivatives of order two in the vorticity. Eq. (38) is named the invariant of semi-potential vorticity.

Similarly, for the meridional semi-geostrophic motion, from Eqs. (8), (9) and (34), we obtain

$$\frac{\partial^2 u}{\partial x^2} - \frac{1}{4}y^2 u = -\frac{\partial^2 v}{\partial x \partial y} + \frac{1}{2}y\frac{\partial \varphi}{\partial y} \tag{39}$$

Substituting Eq. (35) into Eq. (39) becomes

$$\frac{\partial}{\partial y}\left[\frac{\partial v}{\partial x} - \frac{1}{2}y\varphi\right] = \frac{\partial}{\partial y}\left[\left(\frac{\partial v}{\partial x} - \frac{\partial u}{\partial y}\right) - \frac{1}{2}y\varphi\right]_{t=0} \tag{40}$$

Integrating (40) with respect to y yields

$$\frac{\partial v}{\partial x} - \frac{1}{2}y\varphi = \left[\left(\frac{\partial v}{\partial x} - \frac{\partial u}{\partial y}\right) - \frac{1}{2}y\varphi\right]_{t=0} \tag{41}$$

Substituting Eq. (34) into Eq. (41) gives

$$\frac{\partial^2 \varphi}{\partial x^2} - \frac{1}{4}y^2\varphi = \frac{1}{2}y\left[\left(\frac{\partial v}{\partial x} - \frac{\partial u}{\partial y}\right) - \frac{1}{2}y\varphi\right]_{t=0} \tag{42}$$

Ⅵ THE SCALE CRITERION OF ADAPTATION FIELD

The study of geostrophic adaptation of mid-and high-latitudes shows that the process of mutual adaptation between the fields depends on the scale of motion, for the large-scale initial disturbance, the velocity field changes to fit the pressure field; for the small-scale initial disturbance, the pressure field changes to fit the velocity field. Now, the scale-criterion in tropical semi-geostrophic adaptation is viewed.

For the zonal semi-geostrophic motion, if knowing the initial potential vorticity, then the pressure field is given by (38), the zonal velocity is given by (32), and the meridional velocity is given by (33).

Let

$$t = 0, \quad \begin{cases} v = 0, \\ u = u^0 e^{-\alpha_1 x^2} e^{-\alpha_2 y^2}, \\ \varphi = \varphi^0 e^{-\alpha_1 x^2} e^{-\alpha_2 y^2}, \end{cases} \tag{43}$$

where

$$\alpha_1 = (L_0/L_1)^2, \quad \alpha_2 = (L_0/L_2)^2 \tag{44}$$

$L_0 = (C/2\beta)^{1/2}$ is the equatorial Rossby radius of deformation. L_1 and L_2 are zonal and meridional characteristic scale of initial perturbation. respectively.

Substituting (43) into (33) and (38) gives

$$\left(\frac{\partial^2}{\partial y^2} - \frac{1}{4}y^2\right) v = \alpha_1(\varphi^0 - 4\alpha_2 u^0) x e^{-\alpha_1 x^2} y e^{-\alpha_2 y^2} \tag{45}$$

$$\left(\frac{\partial^2}{\partial y^2} - \frac{1}{4}y^2\right) \varphi = \frac{1}{4}(4\alpha_2 u^0 - \varphi^0) e^{-\alpha_1 x^2} y^2 e^{-\alpha_2 y^2} \tag{46}$$

Variables u, v, φ are expanded in terms of the parabolic cylinder functions, i.e. Weber functions:

$$(u, v, \varphi) = \sum_{n=0}^{\infty} (u_n, x v_n, \varphi_n) e^{-\alpha_1 x^2} D_n(y) \tag{47}$$

Substituting (47) into (45), (46), (32), using the properties of Weber functions, and taking the lowest order, we obtain

$$\begin{cases} \varphi_0 = \dfrac{\sqrt{2}}{8}(\varphi^0 - 4\alpha_2 u^0)(\alpha_2 + \dfrac{1}{4})^{-3/2}, \\ v_1 = \dfrac{\sqrt{-2}}{6}(\varphi^0 - 4\alpha_2 u^0)\alpha_1(\alpha_2 + \dfrac{1}{4})^{-3/2}, \\ u_0 = \varphi_0 \end{cases} \tag{48}$$

From (48), as $\alpha_2 = (L_0/L_2)^2 \ll 1$, the pressure field is more important than the zonal velocity field for the initial disturbance, i. e. the pressure, zonal and meridional velocity fields after adaptation change to fit the initial pressure field. As $\alpha_2 = (L_0/L_2)^2 \gg 1$, the zonal velocity field is more important than the pressure field for the initial disturbance, and the fields after adaptation change to fit the initial zonal velocity field. Furthermore, the strength of the meridional velocity after adaptation depends on α_1. As α_1 is small (i.e. the initial zonal scale is large), the strength of the meridional velocity is small.

In computation, let the non-dimensional variable $\varphi^0 = u^0 = 1$. Substituting (48) into (47), we obtain the structure fields after zonal semi-geostrophic adaptation.

From Fig.3, for the large scale initial disturbance (i.e. α_2 is small), the value of the pressure after zonal semi-geostrophic adaptation changes a little, but the velocity changes greatly, i.e. the velocity field changes to fit the initial pressure field. On the other hand, for the small scale initial disturbance (i.e. α_2 is large), with the increase of the value of α_2, the pressure and velocity fields after adaptation are close to zero, the value of the pressure changes a lot, and the velocity changes a little, i.e. the pressure field changes to fit the initial velocity field. When the scale of the initial disturbance equals the Rossby radius of deformation (i.e. α_2 equals 1), the values of the pressure and velocity after adaptation both change greatly. The conclusions agree with the scale-criterion of the geostrophic adaptation in mid-and high-latitude area.

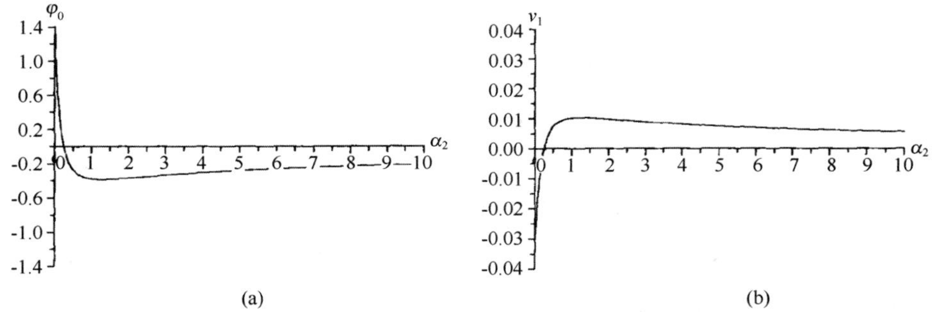

Fig.3 The relation of the variations of the fields after zonal semi-geostrophic adaptation with the scale α_2 of initial disturbance. (a) The variations of the pressure. (b) The variations of the meridional velocity. In the computation. $\alpha_1 = 0.02$

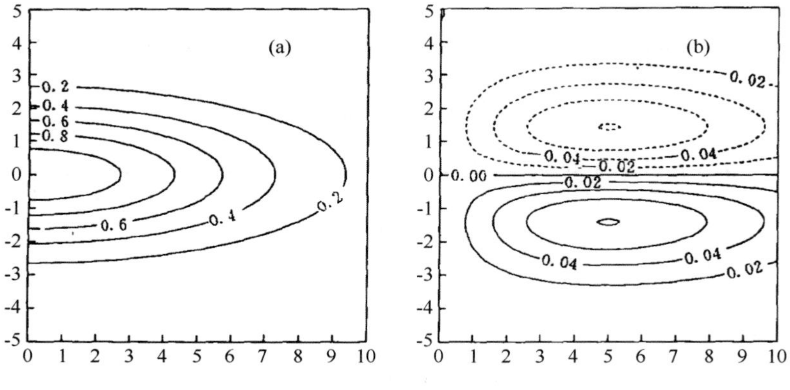

Fig.4 The structure of the physical fields after zonal semi-geostrophic adaptation. The vertical axis is meridional length, the horizontal axis is zonal length. (a) The pressure field. (b) The meridional velocity field. In the computation, $\alpha_1 = \alpha_2 = 0.02$

Figures 4 and 5 give the structure fields of the velocity and pressure fields after zonal semi-geostrophic adaptation. For the large-scale initial disturbance, the velocity field and the pressure field exist in large area, and the pressure field changes a little, i.e. the velocity field changes to fit the pressure field (Fig.4). For the

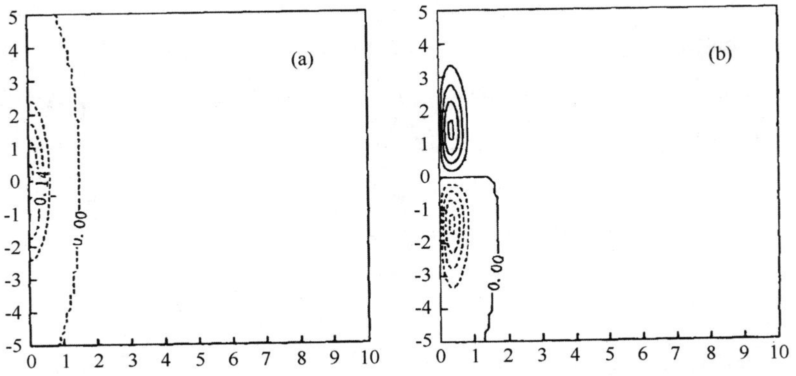

Fig.5　As in Fig.4 but $\alpha_1=\alpha_2=4.0$. The interval is 0.07 in diagram b

small-scale initial disturbance, the velocity field and the pressure field exist in the local area, the pressure field changes greatly, i.e. the pressure field changes to fit the velocity field (Fig.5).

If the motion is meridional geostrophic balance, then after adaptation, the pressure field is given by (42), the meridional velocity is given by (34) and the zonal velocity is given by (35).

Substituting (43) into (42) and (35) gives

$$\frac{\partial^2\varphi}{\partial x^2}-\frac{1}{4}y^2\varphi=\left(\alpha_2u^0-\frac{1}{4}\varphi^0\right)y^2e^{-\alpha_1x^2}e^{-\alpha_2y^2} \tag{49}$$

$$\frac{\partial^2 u}{\partial x^2}-\frac{1}{4}y^2u=\left(\frac{1}{2}\varphi^0-2\alpha_2u^0\right)(1-2\alpha_2y^2)e^{-\alpha_1x^2}e^{-\alpha_2y^2} \tag{50}$$

Variables u,v,φ are expanded in terms of Weber functions, i.e.

$$(u,v,\varphi)=\sum_{n=0}^{\infty}(u_n,v_n,\varphi_n)D_n(y) \tag{51}$$

Substituting (51) into (49), (50), (34), considering the properties of Weber functions, taking the lowest order and using Green functions, the solutions of Eqs. (49), (50) and (34) become

$$\begin{cases}\varphi_0(x)=-\beta_1[(1+\mathrm{erf}(x_1))\mathrm{e}^{-\frac{1}{2}x}+(1-\mathrm{erf}(x_2))\mathrm{e}^{\frac{1}{2}x}]\\ u_0(x)=-\beta_2[(1+\mathrm{erf}(x_1))\mathrm{e}^{-\frac{1}{2}x}+(1-\mathrm{erf}(x_2))\mathrm{e}^{\frac{1}{2}x}]\\ v_1(x)=\beta_1[(1+\mathrm{erf}(x_1))\mathrm{e}^{\frac{1}{2}x}-(1-\mathrm{erf}(x_2))\mathrm{e}^{\frac{1}{2}x}]\end{cases} \tag{52}$$

where

$$\begin{cases}\beta_1=\frac{\sqrt{2\pi}}{16}\left(\alpha_2u^0-\frac{1}{4}\varphi^0\right)\mathrm{e}^{\frac{1}{16\alpha_1}}\cdot\alpha_1^{-1/2}\cdot\left(\alpha_2+\frac{1}{4}\right)^{-3/2}\\ \beta_2=\frac{\sqrt{2\pi}}{16}(\varphi^0-4\alpha_2u^0)\mathrm{e}^{\frac{1}{16\alpha_1}}\cdot\alpha_1^{-1/2}\left[\left(\alpha_2+\frac{1}{4}\right)^{-1/2}-\alpha_2\left(\alpha_2+\frac{1}{4}\right)^{-3/2}\right]\\ x_1=\sqrt{\alpha_1}\left(x-\frac{1}{4\alpha_1}\right)\\ x_2=\sqrt{\alpha_1}\left(x+\frac{1}{4\alpha_1}\right)\\ \mathrm{erf}(\xi)=\frac{2}{\sqrt{\pi}}\int_0^{\xi}e^{-\tau^2}\mathrm{d}\tau\quad\text{is error function.}\end{cases} \tag{53}$$

From Eqs. (52) and (53), as $\alpha_2=(L_0/L_2)^2 \ll 1$, the pressure field is more important than the zonal velocity field for the initial disturbance, i.e. the pressure, zonal and meridional velocity fields after adaptation change to fit the initial pressure field. As $\alpha_2=(L_0/L_2)^2 \gg 1$, the zonal velocity field is more important than the pressure field for the initial disturbance, and the fields after adaptation change to fit the initial zonal velocity field. Furthermor, the strength of the fields after adaptation depends on α_1, as α_1 is small (i.e. the initial zonal scale is large), and the strengtn of the fields is large.

In the computation, let $\varphi^0=u^0=1$, Substituting (52) into (51) obtains the structure fields after meridional semi-geostrophic adaptation.

From Fig.6, for the large scale initial disturbance (i.e. α_2 is small), the value of the pressure after meridional semi-geostrophic adaptation changes a little, but the velocity changes greatly, i.e. the velocity field changes to fit the pressure field. On the contrary, for the small scale initial disturbance (i.e. α_2 is large), with the increase of α_2. the pressure and velocity fields after adaptation are chose to zero, the value of the pressure changes greatly, and the velocity changes a little, i.e. the pressure field changes to fit the initial velocity field. When the scale of the initial disturbance equals the Rossby radius of defor-mation (i.e. α_2 equals 1), the values of the pressure and vilocity after adaptation both change greatly. The conclusions agree with the scale-criterion of the geostrophic adaptation in mid-and high-latitude area.

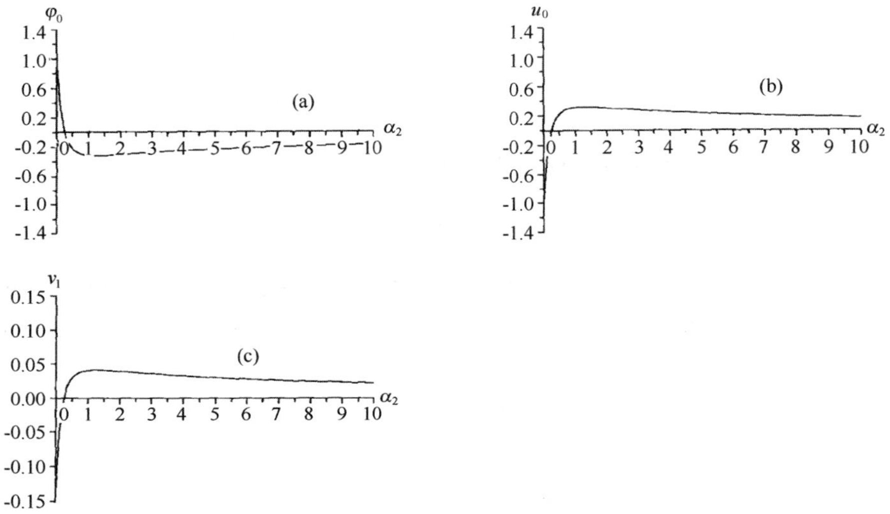

Fig. 6 The relation of the variations of the fields after meridional semi-geostrophic adaptation with the scale α_2 of initial distur-bance. (a) The variations of the pressure. (b) The variations of the zonal velocity. (c) The variations of the meridional velocity. In the computation, $x=2.0$, $\alpha_1=0.02$

Figures 7 and 8 give the pressure field and meridional velocity field after meridional semi-geostrophic adaptation. For the large-scale initial disturbance, the velocity field and the pressure field also exist in large area, and the velocity field changes a lot, i.e. the velocity field changes to fit the pressure field (Fig.7). For the small-scale initial disturbance, the velocity and pressure fields both exist in the local area, and the pressure field changes greatly, i.e. the pressure field changes to fit the velocity field (Fig.8).

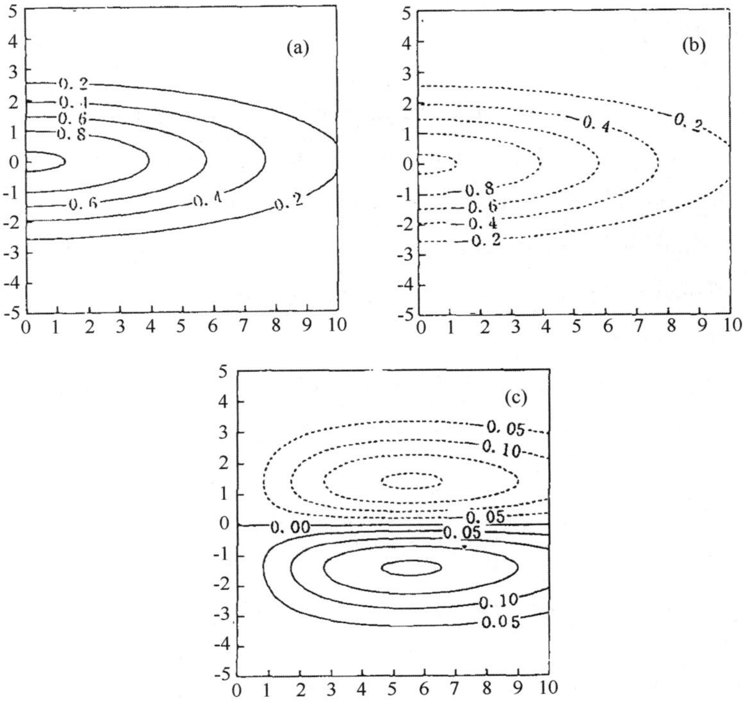

Fig.7　The sturcture of the physical fields after meridional semi-geostrophic adaptation, The vertical axis is meridional length, the horizontal axis is zonal length. (a) The pressure field. (b) The zonal velocity field. (c) The meridional velocity field. In the computation, $\alpha_1=\alpha_2=0.02$

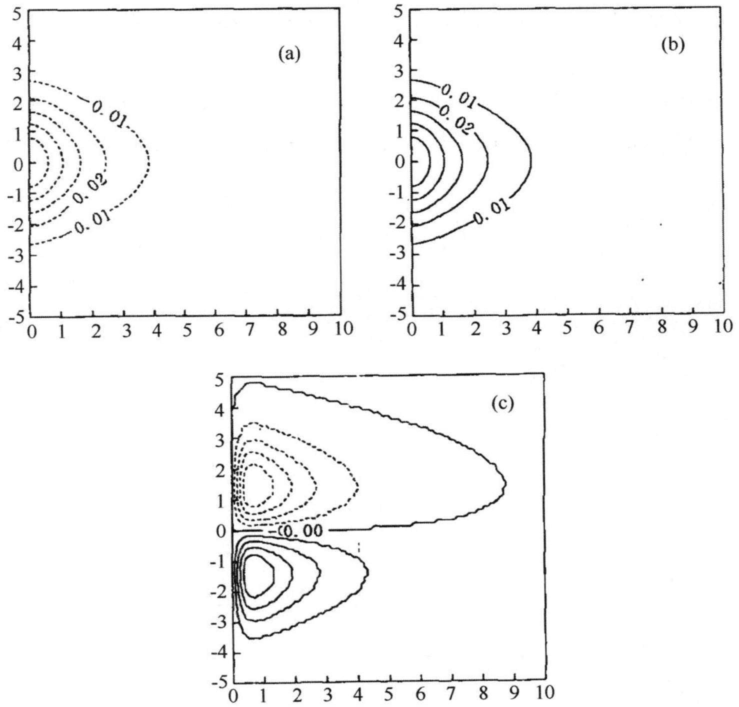

Fig.8　As in Fig.7 but $\alpha_1=\alpha_2=4.0$. The interval is 0.000 5 in diagram c

Ⅶ CONCLUSIONS

In the tropical area, after the dispersion of an inertial gravity wave, the zonal or meridional semi-geostrophic balance can be founded, it follows an invariant of semi-potential vorticity. Based on the invariant, the author points out that the velocity and pressure fields after adaptation change to fit the initial pressure field for the large meridional-scale initial disturbance, and the fields change to fit the initial zonal velocity field for the small meridional-scale initial disturbance, and the strength of the fields after adaptation depends on the zonal scale.

In this article, taking zero as the value of the initial meridional velocity field; for the case that meridional velocity is not zero, another paper will discuss it.

REFFRECNES

Matsuno. T. 1966. Quasi-geostrophic motions in the equatorial area, *J. Meteor. Soc. Japan.* 44:25–43.

Obukhow, A. M. 1949. The problem of the geostropic adaptation. *Izvestiya of Academy of Science USSR*, Series Geography and Geophysics, 13:281–28

Rossby, C.G. 1937. On the mutual adjustment of pressure and velocity distributions in certain simple current systems, Ⅰ. *J. Mar. Res.*, 1:15–28.

Rossby, C. G. 1938. On the mutual adjustment of pressure and velocity distributions in certain simple current systems, Ⅱ. *J. Mar. Res.*, 2:239–263.

Wu Rongsheng and Chan Jiping. 1978. Characteristics of multi-time scale of motion and temporal boundary layers in rotating atmosphere, *Chinese J. Atmos, Sci.*, 2:267–275(in Chinese).

Yeh, T.C. 1957. On the formation of quasi-geostrophic motion in the atmosphere, *J. Met. Soc. Japan.* The 75th Anniversary volume:130–134.

Yeh, T.C. and Li, M. T. 1965. On the Adaptation of the Atmospheric Motion. Science Press, Beijing (in Chinese).

Zeng Qingcun. 1963. The adjustment and evolutional process in atmosphere, *Acta Meteor. Sin.*, 33:163–174,281–289(in Chinese).

热带斜压大气的适应运动和发展运动*

巢纪平

(国家海洋环境预报研究中心,北京 100081)

摘要:研究了热带斜压大气中的适应运动和发展运动。当以重力惯性波为特征的适应过程基本完成后,运动进入缓变的发展或演变阶段。发展运动可以是纬圈半地转平衡的(即长波近似的 Gill 模式),也可以是经圈半地转平衡的(短波近似)。分析了这两种半地转发展模式的特点后,提出了低频近似的发展模式。在低频动力系统中,包括了除重力惯性波外,由 Mastuno 指出的所有热带基本运动,即 Kelvin 波、Rossby 波和混合波中的 Rossby 短波。因此,这个模式能反映热带运动更多的动力学行为。

关键词:适应和发展过程;长波和短波近似;低频近似

重力场中的大尺度大气运动是静力平衡的,表明垂直加速度很小,而由于地球边界无垂直速度的约束(如不计地形影响),因而大气中的垂直运动也很小,即大尺度运动基本上是水平的。大尺度的水平运动,在旋转 Coriolis 力作用下,在中纬度风场与压力场之间,一般处在地转平衡的状态下。如由于某种原因,使地转平衡遭到破坏,则地转偏差就会以重力惯性波的形式频散,并最后重建地转平衡。这一过程称为地转适应过程,或简称适应过程[1,2]。相对于这类快变过程,是在地转平衡附近进行的运动,称准地转运动或发展运动。在没有非线性过程时,中纬度发展过程一般表现为 Rossby 波的频散,Rossby 波的频率相对重力惯性波来要慢得多,因此发展过程是一类慢过程。由于适应运动和发展运动由不同的物理过程造成,因此它们在时间上是可分的[3-5]。过去半个世纪中人们对中纬度大气运动中的适应过程和发展过程已研究得相当清楚,并已成为 20 世纪 50 年代发展数值天气预报的动力学基础。

对于热带大气,由于科氏力参数很小,其基本运动形式与中纬度有所不同。Mastuno[6] 指出,在热带除重力惯性波和 Rossby 波外,尚存在 Kelvin 波和 Rossby-重力混合波,后者在调节热带大气环流中也起着重要作用。这是热带大气运动不同于中纬度大气运动的特色之一。另一方面,在科氏力参数小的情况下,还是否存在地转适应过程,这也是热带动力学需要研究的一个基本问题。Chao[7] 指出,科氏力参数在热带虽然很小,但沿纬圈的运动速度不小(如信风、Walker 环流等),且沿经圈方向的气压梯度相对中纬度来讲也较小,因此在纬圈方向风、压场之间存在地转平衡是可能的,并可称为纬圈半地转平衡。相应地,在海洋经圈边界附近由于经圈方向流速很大,也可以存在经圈半地转平衡。这样在纬圈或经圈方向,运动也存在适应过程,并同样是通过重力惯性波的频散来实现的。当半地转适应过程完成后,运动将进入较缓变的发展阶段。巢纪平指出,Gill 的长波近似模式[8] 实际上正是纬圈

* 中国科学(D 辑),1999,29(3):279-288.

半地转近似下的发展运动模式。在线性情况下,运动的主要特征表现在 Kelvin 波和 Rossby 长波的传播上。作为对长波近似模式的补充,巢纪平指出可以建立在短波近似下的经圈半地转发展模式,这个模式中的运动形式主要是 Rossby 短波和 Rossby-重力惯性波中的 Rossby 波波段(实际上也是 Rossby 短波)。巢纪平又指出①,纬圈半地转发展运动的 Gill 模式虽然是一个很好用的模式,但由于把原来色散的 Rossby 波非色散化了,因此使波能量的传播受到了限制。同时,对一类纬圈尺度不是很大的运动,Gill 模式失效。经圈半地转发展运动的短波模式,虽然其中 Rossby 短波仍然是色散波,但其波能量只是向东传播的,因此也影响对一些发生在扰动源西侧的现象的研究。为此巢纪平提出了一个近似度高于长波和短波近似的低频近似模式。在低频近似下的发展运动模式中,波长是不受限制的。波能量的传播将因波长而异,可以是向东的,也可以是向西的,而能容纳的运动频率也要比长波或短波近似下的频率高,即所包含的运动的时间尺度更宽,低频近似模式中的运动形式包含有 Kelvin 波、Rossby 波以及 Rossby-重力混合波中的 Rossby 波波段,也即包含了 Mastuno 赤道波系中除重力惯性波外的所有波动,而重力惯性波是在建立低频发展运动的过程中被频散掉的。

但是,上述工作都是在等值浅水模式中进行的,即是对某一特定的垂直本征模进行的。在本文中,将对三维斜压大气进一步研究上面论述的一些问题。

1 基本方程

在赤道 β 平面近似下,Boussinesq 流体的线性运动方程为

$$\frac{\partial u}{\partial t} - \beta y v = -\frac{1}{\rho_0}\frac{\partial p}{\partial x} \tag{1}$$

$$\frac{\partial v}{\partial t} + \beta y u = -\frac{1}{\rho_0}\frac{\partial p}{\partial y} \tag{2}$$

$$\frac{\partial u}{\partial x} + \frac{\partial v}{\partial y} + \frac{\partial w}{\partial z} = 0 \tag{3}$$

$$\frac{g}{\theta_0}\vartheta = \frac{1}{\rho_0}\frac{\partial p}{\partial z} \tag{4}$$

$$\frac{\partial \vartheta}{\partial t} + \frac{\mathrm{d}\theta_0}{\mathrm{d}z}w = 0 \tag{5}$$

式中,ρ_0 和 θ_0 为背景场的密度和位温;ϑ 为位温偏差,其他符号均同常用符号。

设大气的垂直特征厚度为 H,引进重力内波波速

$$C = \sqrt{\frac{g}{\theta_0}\frac{\mathrm{d}\theta_0}{\mathrm{d}z}}H \tag{6}$$

及 $z = Hz'$, $(x,y) = (c/2\beta)^{1/2}(x',y')$, $t = (2\beta c)^{-1/2}t'$, $(u,v) = C(u',v')$, $w = (2\beta c)^{1/2}Hw'$, $p/\rho^0 = c^2\varphi'$, $\vartheta = (\mathrm{d}\theta_0/\mathrm{d}zH)\vartheta'$,由此得到无量纲方程组为(略去“′”号):

$$\varepsilon_1\frac{\partial u}{\partial t} - \frac{1}{2}yv + \frac{\partial \varphi}{\partial x} = 0 \tag{7}$$

$$\varepsilon_2\frac{\partial v}{\partial t} + \frac{1}{2}yu + \frac{\partial \varphi}{\partial y} = 0 \tag{8}$$

① 巢纪平. 低频近似下的热带大气的发展运动. 气象学报.

$$\frac{\partial u}{\partial x}+\frac{\partial v}{\partial y}+\frac{\partial w}{\partial z}=0 \tag{9}$$

$$\vartheta=\frac{\partial \varphi}{\partial z} \tag{10}$$

$$\varepsilon_3\frac{\partial \vartheta}{\partial t}+w=0 \tag{11}$$

式中，ε_i 为标识符，是加上去的，其值取 0 或 1，当取 0 时，上述方程组给出

$$\frac{1}{2}yv=\frac{\partial \varphi}{\partial x} \tag{12}$$

$$\frac{1}{2}yu=-\frac{\partial \varphi}{\partial y} \tag{13}$$

$$w=0 \tag{14}$$

以及式(10)，这表明大气的平衡状态是地转的、水平的和静力平衡的。

由方程(9)至方程(11)给出

$$\varepsilon_3\frac{\partial^3 \varphi}{\partial t\partial z^2}-\left[\frac{\partial u}{\partial x}+\frac{\partial v}{\partial y}\right]=0 \tag{15}$$

这样由方程(7)，方程(8)和方程(15)组成对变量 u,v,φ 的闭合方程组。

由基本方程可以导出关系式

$$\varepsilon_1\varepsilon_2\frac{\partial^2 u}{\partial t^2}+\frac{1}{4}y^2u=-\varepsilon_2\frac{\partial^2 \varphi}{\partial t\partial x}-\frac{1}{2}y\frac{\partial \varphi}{\partial y} \tag{16}$$

$$\varepsilon_1\varepsilon_2\frac{\partial^2 v}{\partial t^2}+\frac{1}{4}y^2v=-\varepsilon_1\frac{\partial^2 \varphi}{\partial t\partial y}+\frac{1}{2}y\frac{\partial \varphi}{\partial x} \tag{17}$$

以及

$$\varepsilon_1\varepsilon_3\frac{\partial^4 u}{\partial t^2\partial z^2}+\frac{\partial^2 u}{\partial x^2}=\varepsilon_3\frac{1}{2}y\frac{\partial^3 v}{\partial t\partial z^2}-\frac{\partial^2 v}{\partial x\partial y} \tag{18}$$

$$\varepsilon_2\varepsilon_3\frac{\partial^4 v}{\partial t^2\partial z^2}+\frac{\partial^2 v}{\partial x^2}=-\varepsilon_3\frac{1}{2}y\frac{\partial^3 u}{\partial t\partial z^2}-\frac{\partial^2 u}{\partial x\partial y} \tag{19}$$

2　斜压 Mastuno 方程

由式(18)和式(19)消去 u 后得到

$$L(v)\equiv\left\{\left[\varepsilon_1\varepsilon_2\varepsilon_3\frac{\partial^4}{\partial t^2\partial z^2}+\left(\varepsilon_2\frac{\partial^2}{\partial x^2}+\varepsilon_1\frac{\partial^2}{\partial y^2}\right)+\varepsilon_3\frac{1}{4}y^2\frac{\partial^2}{\partial z^2}\right]\frac{\partial}{\partial t}+\frac{1}{2}\frac{\partial}{\partial x}\right\}\frac{\partial v}{\partial t}=0 \tag{20}$$

对时间积分，如初始状态为静止，则积分后的方程为热带斜压大气的 Mastuno 方程。

将 v 在垂直方向用余弦函数展开

$$v(x,y,z,t)=\sum_m v_m(x,y,t)\cos m\pi z \tag{21}$$

作变换

$$\tau=\frac{t}{\sqrt{m\pi}},\quad (X,Y)=\sqrt{m\pi}(x,y) \tag{22}$$

方程(20)给出(略去标识符)

$$\left\{\left[\frac{\partial^2}{\partial\tau^2}-\frac{\partial^2}{\partial X^2}-\left(\frac{\partial^2}{\partial Y^2}-\frac{1}{4}Y^2\right)\right]\frac{\partial}{\partial\tau}-\frac{1}{2}\frac{\partial}{\partial X}\right\}\frac{\partial v_m}{\partial\tau}=0 \tag{23}$$

将 v_m 在 Y 方向用抛物圆柱函数即 Weber 函数展开成

$$v_m=\sum_n v_{m,n}(X,\tau)D_n(Y) \tag{24}$$

则方程(23)给出

$$L_{m,n}(v_{m,n})\equiv\left\{\left[\frac{\partial^2}{\partial\tau^2}-\frac{\partial^2}{\partial X^2}+\left(n+\frac{1}{2}\right)\right]\frac{\partial}{\partial\tau}-\frac{1}{2}\frac{\partial}{\partial X}\right\}v_{m,n}$$

$$=\left\{\left[\frac{\partial^2}{\partial\tau^2}-\frac{\partial^2}{\partial X^2}+\left(n+\frac{1}{2}\right)\right]\frac{\partial}{\partial\tau}-\frac{1}{2}\frac{\partial}{\partial X}\right\}v_{m,n}\big|_{t=0} \tag{25}$$

如初始状态是静止的,则这是对某一斜压模的 Mastuno 方程。

注意到,由于行星涡度的纬度变化,基本方程(7)和方程(8)是变系数的,对变量 u 和 φ 得不出像方程(20)那样简洁的形式。为了得到类似的方程,需要采取另外的处理方法。

设

$$(u,\varphi)=\sum_m(u_m,\varphi_m)\cos m\pi z \tag{26}$$

并采用式(22)的坐标变换,于是有方程

$$\varepsilon_1\frac{\partial u_m}{\partial\tau}-\frac{1}{2}Yv_m+\frac{\partial m\pi\varphi_m}{\partial X}=0 \tag{27}$$

$$\varepsilon_2\frac{\partial v_m}{\partial\tau}+\frac{1}{2}Yu_m+\frac{\partial m\pi\varphi_m}{\partial Y}=0 \tag{28}$$

$$\varepsilon_3\frac{\partial m\pi\varphi_m}{\partial\tau}+\frac{\partial u_m}{\partial X}+\frac{\partial v_m}{\partial Y}=0 \tag{29}$$

这组方程和等值浅水运动方程在形式上是一样的。引进变量

$$q_m=m\pi\varphi_m+u_m,\qquad r_m=m\pi\varphi_m-u_m \tag{30}$$

于是有(标识符 $\varepsilon_i=1$)

$$\frac{\partial q_m}{\partial\tau}+\frac{\partial q_m}{\partial X}+\left(\frac{\partial v_m}{\partial Y}-\frac{1}{2}Yv_m\right)=0 \tag{31}$$

$$\frac{\partial r_m}{\partial\tau}-\frac{\partial r_m}{\partial X}+\left(\frac{\partial v_m}{\partial Y}-\frac{1}{2}Yv_m\right)=0 \tag{32}$$

将 q_m 和 r_m 用抛物圆柱函数展开

$$(q_m,r_m)=\sum_n(q_{m,n},r_{m,n})D_n(Y) \tag{33}$$

得到

$$\frac{\partial q_{m,n}}{\partial\tau}+\frac{\partial q_{m,n}}{\partial X}-v_{m,n-1}=0 \tag{34}$$

$$\frac{\partial r_{m,n}}{\partial\tau}+\frac{\partial r_{m,n}}{\partial X}+(n+1)v_{m,n+1}=0 \tag{35}$$

而式(28)可写成

$$\frac{\partial v_{m,n}}{\partial\tau}+\frac{1}{2}(n+1)q_{m,n-1}-\frac{1}{2}r_{m,n-1}=0 \tag{36}$$

这组方程组如消去 $v_{m,n}$ 则有

$$\frac{\partial q_{m,0}}{\partial \tau}+\frac{\partial q_{m,0}}{\partial X}=0 \tag{37}$$

$$\frac{\partial^2 q_{m,1}}{\partial \tau^2}+\frac{\partial^2 q_{m,1}}{\partial \tau \partial X}+\frac{1}{2}q_{m,1}=0 \tag{38}$$

$$\frac{\partial^3 q_{m,n+2}}{\partial \tau^3}-\frac{\partial^3 q_{m,n+2}}{\partial \tau \partial X^2}+\frac{1}{2}(2n+3)\frac{\partial q_{m,n+2}}{\partial \tau}-\frac{1}{2}\frac{\partial q_{m,n+2}}{\partial X}=0 \tag{39}$$

$$\frac{\partial^3 r_{m,n}}{\partial \tau^3}-\frac{\partial^3 r_{m,n}}{\partial \tau \partial X^2}+\frac{1}{2}(2n+3)\frac{\partial q_{m,n}}{\partial \tau}-\frac{1}{2}\frac{\partial q_{m,n}}{\partial X}=0 \tag{40}$$

其中式(37)是 Kelvin 波的控制方程，式(38)是 Rossby-重力混合波的控制方程，而式(39)和式(40)都是重力惯性波和 Rossby 波的控制方程。显然

$$L_{m,n}(q_{m,n+1},\partial r_{m,n-1})=0 \tag{41}$$

其算子同方程(25)。事实上，考虑到式(36)后，由式(41)立即可得到式(25)。这样对变量 $v_{m,n}$，$q_{m,n+1}$，$r_{m,n-1}$ 都可以用同一算子表示。

3　适应运动

重力惯性波是一类高频运动，而 Rossby 是一类低频运动，重力惯性波的最小频率和 Rossby 波的最大频率是不重合的，表明这两种运动在时间上是可分的，为此在研究以重力惯性波为特征的适应过程时，可以不考虑 Rossby 波的作用。在方程(23)或方程(25)，方程(39)及方程(40)中过滤掉 Rossby 波的最简单方法，是略去与时间变化无关的项，而在物理上可以理解成只保留了高频运动，也可以理解成略去这一项相当于不考虑行星涡度梯度的作用，这自然过滤了 Rossby 波。

用积分后的方程(23)即 Mastuno 方程来研究适应过程。略去该方程最后一项，并对时间积分，给出

$$\left[\frac{\partial^2}{\partial \tau^2}-\frac{\partial^2}{\partial X^2}-\left(\frac{\partial^2}{\partial Y^2}-\frac{1}{4}Y^2\right)\right]v_m=\left[\frac{\partial^2}{\partial \tau^2}-\frac{\partial^2}{\partial X^2}-\left(\frac{\partial^2}{\partial Y^2}-\frac{1}{4}Y^2\right)\right]v_m\Big|_{\tau\geqslant 0} \tag{42}$$

考虑到式(17)和式(15)两式，式(42)右端可改写成

$$\left[\frac{\partial^2}{\partial \tau^2}-\frac{\partial^2}{\partial X^2}-\left(\frac{\partial^2}{\partial Y^2}-\frac{1}{4}Y^2\right)\right]v_m=\frac{\partial}{\partial x}\left[\frac{1}{2}Y(m\pi\varphi_m)-\left(\frac{\partial v_m}{\partial X}-\frac{\partial u_m}{\partial Y}\right)\right]_{\tau=0}\equiv F_m(X,Y,0) \tag{43}$$

式中

$$\Omega_m=\left(\frac{\partial v_m}{\partial X}-\frac{\partial u_m}{\partial Y}\right)-\frac{1}{2}Y(m\pi\varphi_m) \tag{44}$$

是初始时刻的位势涡度。这表明，除任意的初始扰动外，初始时刻的位势涡度的纬圈梯度同样可以激发出重力惯性波。

应用抛物圆柱函数后，方程(43)为

$$\left[\frac{\partial^2}{\partial \tau^2}-\frac{\partial^2}{\partial X^2}+\left(n+\frac{1}{2}\right)\right]v_{m,n}=F_{m,n}(X,0) \tag{45}$$

将初值取成

$$\tau=0,\quad v_{m,n}=\Phi_{m,n}^{(1)}(X),\quad \frac{\partial v_{m,n}}{\partial \tau}=\Phi_{m,n}^{(2)}(X) \tag{46}$$

其解为[7]

$$v_{m,n}=\frac{1}{2}\int_{-\infty}^{\infty}\Phi_{m,n}^{(1)}(X')\frac{\mathrm{J}_1\left(\sqrt{n+\frac{1}{2}}\sqrt{\tau^2-(X-X')}\right)}{\sqrt{\tau^2-(X-X')^2}}\mathrm{d}X'+$$

$$\frac{1}{2}\int_{-\infty}^{\infty}\Phi_{m,n}^{(2)}(X')\mathrm{J}_0\left(\sqrt{n+\frac{1}{2}}\sqrt{\tau^2-(X-X')^2}\mathrm{d}X'\right)+$$

$$\frac{1}{2}\int_0^{\tau}\int_{-\infty}^{\infty}F_{m,n}(X',0)\mathrm{J}_0\left(\sqrt{n+\frac{1}{2}}\sqrt{\tau'^2-(X-X')^2}\right)\mathrm{d}X'\mathrm{d}\tau' \tag{47}$$

式中，J_ν 为 Bessel 函数。由 Bessel 函数的性质可知，当时间充分大时，任意初值的影响消失，而解的主要贡献来自初始的位势涡度纬圈梯度的作用，即

$$v_{m,n}=\frac{1}{2}\int_0^{\tau}\int_{-\infty}^{\infty}F_{m,n}(X',0)\mathrm{J}_0\left(\sqrt{n+\frac{1}{2}}\sqrt{\tau'^2-(X-X')^2}\right)\mathrm{d}X'\mathrm{d}\tau \tag{48}$$

其时间导数为

$$\frac{\partial v_{m,n}}{\partial\tau}=\frac{1}{2}\int_{-\infty}^{\infty}F_{m,n}(X',0)\mathrm{J}_0\left(\sqrt{n+\frac{1}{2}}\sqrt{\tau'^2-(X-X')^2}\right)\mathrm{d}X' \tag{49}$$

现对解在时间充分大时的渐进性态做一分析。如果 $F_{m,n}(x,0)$ 的值集中在原点附近，则可设

$$F_{m,n}=A_{m,n}\delta(X) \tag{50}$$

式中，$\delta(X)$ 为 Delta 函数，于是式(48)和式(49)为

$$v_{m,n}=\frac{A_{m,n}}{2}\int_0^{\tau}\mathrm{J}_0\left(\sqrt{n+\frac{1}{2}}\sqrt{\tau'^2-X^2}\right)\mathrm{d}\tau' \tag{51}$$

$$\frac{\partial v_{m,n}}{\partial\tau}=\frac{A_{m,n}}{2}\mathrm{J}_0\left(\sqrt{n+\frac{1}{2}}\sqrt{\tau'^2-X^2}\right) \tag{52}$$

在原点，当时间充分大时有

$$v_{m,n}=\frac{A_{m,n}}{2}\int_0^{\infty}\mathrm{J}_0\left(\sqrt{n+\frac{1}{2}}\tau'\right)\mathrm{d}\tau'=\frac{A_{m,n}}{2\sqrt{n+\frac{1}{2}}} \tag{53}$$

$$\frac{\partial v_{m,n}}{\partial\tau}=\frac{A_{m,n}}{\sqrt{2\pi}}\left(n+\frac{1}{2}\right)^{-\frac{1}{4}}\sqrt{\frac{1}{\tau}}\cos\left(\sqrt{n+\frac{1}{2}}\tau-\frac{\pi}{4}\right) \tag{54}$$

由此可见，当时间充分大时，经圈速度为有限值，而其时间变化按 $\mathrm{O}(\tau^{-1/2})$ 的速度趋于零。另外可估计出

$$\frac{\partial^2 v_{m,n}}{\partial\tau^2}\sim\mathrm{O}(\tau)^{-\frac{3}{2}} \tag{55}$$

可见经圈速度振幅的二阶时间导数按 $\mathrm{O}(\tau^{-\frac{3}{2}})$ 速度趋于零，快于一阶时间导数的衰减速度。

这一讨论同样适用于 $q_{m,n+2}$ 和 $r_{m,n}$。由于这些物理量当时间充分大后，其一阶时间导数趋于零，因此式(12)和式(13)建立，即场完成地转适应。

关于适应后的场，可以不用解式(48)，而将式(17)和式(15)用到式(43)左端而得到

$$\left[\varepsilon_1\frac{\partial v_m}{\partial X}-\varepsilon_1\frac{\partial u_m}{\partial Y}-\varepsilon_3\frac{1}{3}Y(m\pi\varphi)\right]=\Omega_m(X,Y,0) \tag{56}$$

这是位势涡度的时间不变式，这一守恒方程给出了适应后物理场之间的联系。

4　半地转发展运动

当适应过程完成后，场进入到缓变的发展阶段，这时重力惯性波的作用自然已消失。下面分别对纬圈半地转和经圈半地转的发展运动进行讨论。

4.1　纬圈半地转发展运动

上面的分析表明，当时间充分长时，$\partial v/\partial t$ 消失，而 v 保留，$\partial v/\partial t$ 消失相当于在基本方程中取 $\varepsilon_2=0$，于是控制这类运动的方程为

$$\frac{\partial u}{\partial t}-\frac{1}{2}yv=-\frac{\partial \varphi}{\partial x} \tag{57}$$

$$\frac{1}{2}yu=-\frac{\partial \varphi}{\partial y} \tag{58}$$

$$\frac{\partial^3 \varphi}{\partial t \partial z^2}-\left(\frac{\partial u}{\partial x}+\frac{\partial v}{\partial y}\right)=0 \tag{59}$$

这一控制方程组如对 v 写出，只需在式(20)中令 $\varepsilon_2=0$ 即可，为

$$\left(\frac{\partial^2}{\partial y^2}+\frac{1}{4}y^2\frac{\partial^2}{\partial z^2}\right)\frac{\partial v}{\partial t}+\frac{1}{2}\frac{\partial v}{\partial x}=0 \tag{60}$$

在物理量于垂直方向用余弦函数展开的情况下，式(58)为

$$\frac{1}{2}Yu_m=-\frac{\partial m\pi\varphi_m}{\partial Y} \tag{61}$$

如令式(36)中 $\partial v_{m,n}/\partial \tau=0$，则给出

$$r_{m,n-1}=(n+1)q_{m,n+1} \tag{62}$$

应用这一关系式，由式(34)和式(35)给出

$$\frac{\partial q_{m,n+2}}{\partial \tau}-\frac{1}{2n+3}\frac{\partial q_{m,n+2}}{\partial X}=0 \tag{63}$$

这是向西传播的非色散 Rossby 长波方程。

这一发展运动系统中尚应包括 Kelvin 波方程(37)。可注意到，当适应过程结束后，Rossby-重力混合波中的重力波波段已完成任务，剩下的只有其中的 Rossby 波波段，其控制方程为

$$\frac{\partial^2 q_{m,1}}{\partial \tau \partial X}+\frac{1}{2}q_{m,1}=0 \tag{64}$$

但由于混合波中的 Rossby 波是一类短波，从逻辑上讲是不应该参与长波近似或纬圈半地转近似下的发展运动的，但 Gill[8] 把这一部分运动保留下来并使之与外源平衡。

纬圈半地转近似下的发展运动，对垂直方向某一本征模来讲，它即是 Gill 的长波近似模式。这样上面的讨论把 Gill 的长波近似模式建立的物理过程说清楚了。

4.2　经圈半地转发展运动

如运动是在经圈半地转平衡

$$\frac{1}{2}yv=\frac{\partial \varphi}{\partial X} \tag{65}$$

下进行的,可称为短波近似[7],式(65)相当于在方程(7)中取 $\partial u/\partial t=0$ 或 $\varepsilon_1=0$。这类运动的控制方程除(65)外,尚有式(8)和式(15),或者对单一变量 v 写出时,可令式(20)中 $\varepsilon_1=0$,即为

$$\left(\frac{\partial^2}{\partial x^2}+\frac{1}{4}y^2\frac{\partial^2}{\partial z^2}\right)\frac{\partial v}{\partial t}+\frac{1}{2}\frac{\partial v}{\partial x}=0 \tag{66}$$

当物理量在垂直方向用余弦函数展开时,短波近似要求

$$\frac{1}{2}Yv_m=\frac{\partial m\varphi_m}{\partial X} \tag{67}$$

考虑到式(28)和式(29)后,给出

$$\left(\frac{\partial^2}{\partial X^2}-\frac{1}{4}Y^2\right)\frac{\partial v_m}{\partial \tau}+\frac{1}{2}\frac{\partial v_m}{\partial X}=0 \tag{68}$$

当 v_m 在 Y 方向用 Weber 函数展开时,考虑到

$$Y^2D_n(Y)=D_{n+2}(Y)+(2n+1)D_n(Y)+(n-1)nD_{n-2}(Y) \tag{69}$$

方程(68)给出

$$\frac{\partial^3 v_{m,n}}{\partial\tau\partial X^2}+\frac{1}{2}\frac{\partial v_{m,n}}{\partial X}=\frac{1}{4}\left[\frac{\partial v_{m,n-2}}{\partial\tau}+(2n+1)\frac{\partial v_{m,n}}{\partial\tau}+(n+1)(n+2)\frac{\partial v_{m,n+2}}{\partial\tau}\right] \tag{70}$$

这是对 n 的联立方程组。

设热带的最高纬度为 Y_c,运动沿纬圈方向的波数为 k,如波数大到(或波长短到)满足条件

$$k^2\gg\frac{1}{4}Y_c^2\quad 或\ k>\frac{1}{2}Y_c \tag{71}$$

这时式(68)简化成

$$\frac{\partial^2 v_m}{\partial\tau\partial X}+\frac{1}{2}v_m=0 \tag{72}$$

这是 Rossby 短波的控制方程。

注意到式(68)中与 Y^2 有关的项由水平辐散引起,事实上除令 $\varepsilon_1=0$ 外,如再令 $\varepsilon_3=0$(水平无辐散),则由式(27)至式(29)当取 $\varepsilon_1=\varepsilon_3=0$ 时立即得到式(72)。因此,式(68)是有辐散作用时的 Rossby 短波方程,而式(72)是无辐散作用时的 Rossby 短波方程。

对经圈半地转并在无辐散条件下的发展运动的建立,还可以进一步从适应过程的角度加以论证。如在前文中对 $v_{m,n}$ 的讨论用到对 $q_{m,n}$ 和 r_m(方程(39)和方程(40))的情况,则当时间充分大时 $\partial q_{m,n}/\partial\tau$,$\partial r_{m,n}/\partial\tau$ 消失,但 $q_{m,n}$ 和 $r_{m,n}$ 保留。考虑到

$$u_{m,n}=\frac{1}{2}(q_{m,n}+r_{m,n}),\quad m\pi\varphi_{m,n}=\frac{1}{2}(q_{m,n}-r_{m,n}) \tag{73}$$

$q_{m,n}$ 和 $r_{m,n}$ 时间导数的消失相当于 $u_{m,n}$ 和 $m\pi\varphi_{m,n}$ 的时间导数均消失,即经圈半地转平衡和无辐散状态将同时建立。这时方程(34)和方程(35)变成

$$\frac{\partial q_{m,n+1}}{\partial X}=v_{m,n} \tag{74}$$

$$\frac{\partial r_{m,n-1}}{\partial X}=nv_{m,v} \tag{75}$$

将此式代入式(36)给出

$$\frac{\partial^2 v_{m,n}}{\partial\tau\partial X}+\frac{1}{2}v_{m,n}=0 \tag{76}$$

此即对任一经圈模 n 的方程(72)。另外尚有

$$\frac{\partial^2 q_{m,n+1}}{\partial\tau\partial X}+\frac{1}{2}q_{m,n-1}=0 \tag{77}$$

$$\frac{\partial^2 r_{m,n-1}}{\partial\tau\partial X}+\frac{1}{2}r_{m,n-1}=0 \tag{78}$$

注意到,当 $n=0$ 时,式(77)即为式(64),这是混合波中的 Rossby 波,因此只要将式(77)的下标改成 $n\geqslant 0$,则式(77)就包括了式(64)。这表明,混合波中的 Rossby 波波段是参与经圈半地转发展运动的。

5　低频近似的发展运动

纬圈半地转和经圈半地转模式构成了热带大气(或热带海洋)发展运动的两个互为补充的模式。这两个动力学模式虽然有简洁、好用的长处,但也存在短处。在空间尺度上,纬圈半地转模式不适用于研究纬圈尺度小的现象,经圈半地转模式不适用于研究经圈尺度大的现象。在时间尺度上,两者都是在适应过程的末期以经圈速度或纬圈速度的一级时间导数消失为条件的,这意味着要求发展运动的频率很低,这就对发展运动的时间尺度给了较严格的限制。在动力学上,纬圈半地转模式中,Kelvin 波和 Rossby 波都是非色散波,这样波能量就不能以不同于信号传播速度的速度传播(即群速度等于相速度)。在经圈半地转模式中,Rossby 短波的群速度虽然不同于相速度,但短波能量也只存在单方向的,即向东的传播。因此,建立热带发展运动更好的模式是必要的。在这方面,巢纪平①最近提出了低频近似下的发展运动模式。

设热带惯性运动的平均频率或参考频率为 f_0,运动的特征频率为 σ,如果运动的频率满足条件

$$\sigma<f_0,\text{或}\ \sigma^2\ll f_0^2 \tag{79}$$

则这样的近似称为低频近似。可以看到,这对运动频率或时间尺度的要求要比纬圈半地转模式宽,在这样的条件下,式(16)和式(17)两式中可略去时间二阶导数项,即

$$\frac{1}{4}y^2u=-\varepsilon_2\frac{\partial^2\varphi}{\partial t\partial x}-\frac{1}{2}y\frac{\partial\varphi}{\partial y} \tag{80}$$

$$\frac{1}{4}y^2v=-\varepsilon_1\frac{\partial^2\varphi}{\partial t\partial y}+\frac{1}{2}y\frac{\partial\varphi}{\partial x} \tag{81}$$

将此两式代入式(15),给出

$$\left[\varepsilon_2\frac{\partial^2}{\partial x^2}+\varepsilon_1\frac{\partial^2}{\partial y^2}-\varepsilon_1\frac{2}{y}\frac{\partial}{\partial y}+\varepsilon_3\frac{1}{4}y^2\frac{\partial^2}{\partial z^2}\right]\frac{\partial\varphi}{\partial t}+\frac{1}{2}\frac{\partial\varphi}{\partial x}=0 \tag{82}$$

此即低频近似下的斜压发展运动方程。或者,在条件(79)下由方程(20)给出

$$\left\{\left[\left(\varepsilon_2\frac{\partial^2}{\partial x^2}+\varepsilon_1\frac{\partial^2}{\partial y^2}\right)+\varepsilon_3\frac{1}{4}y^2\frac{\partial^2}{\partial z^2}\right]\frac{\partial}{\partial t}+\frac{1}{2}\frac{\partial}{\partial x}\right\}\frac{\partial v}{\partial t}=0 \tag{83}$$

如初始值为零,对时间积分一次后给出

① 巢纪平. 低频近似下的热带大气的发展运动. 气象学报(待刊).

$$\left[\left(\varepsilon_2 \frac{\partial^2}{\partial x^2} + \varepsilon_1 \frac{\partial^2}{\partial y^2}\right) + \varepsilon_3 \frac{1}{4} y^2 \frac{\partial^2}{\partial z^2}\right] \frac{\partial v}{\partial t} + \frac{1}{2} \frac{\partial v}{\partial x} = 0 \tag{84}$$

容易看出,式(84)是对波长无限制的 Rossby 波方程,并当 $\varepsilon_2 = 0$ 时,退化为长波近似模式式(60),当 $\varepsilon_1 = 0$ 时,退化成短波近似模式式(66)。

可注意到,在对适应过程的分析中,由式(54)和式(55)表明,物理量二阶时间导数要比一阶时间导数消失得更快,这意味着低频近似的发展运动要比半地转近似的发展运动建立更早。考虑到这一特征,或者考虑到条件(79),在适应运动进行到一定时间后,就会由式(39)和式(40)得到

$$L(q_{m,n+2}, r_{m,n}) = \left[\frac{\partial^3}{\partial \tau \partial X^2} + \frac{1}{2}(2n+3)\frac{\partial}{\partial \tau} - \frac{1}{2}\frac{\partial}{\partial X}\right](q_{m,n+2}, r_{m,n}) = 0 \tag{85}$$

此即对于斜压本征模的低频发展运动方程。在考虑到式(36),有

$$L\left(\frac{\partial v_{m,n+1}}{\partial \tau}\right) = 0 \tag{86}$$

当然,由于频近似对发展运动的波长无限制,因此除 Rossby 波外,在运动系统中尚应包括式(37)的 Kelvin 波和式(64)的混合波中的 Rossby 短波。

6 结论

本文讨论了热带斜压大气的适应运动和发展运动,指出这两种运动是不同的物理过程的表现,它们在时间上是可分的,当地转平衡被破坏后,运动即进入到以重力惯性波频散的适应阶段。发展运动的形式可以不同,依赖于激发适应过程的初始位势涡度的情况①。在一般情况下是由 Gill[8] 建立的纬圈半地转发展运动,但也可以建立起经圈半地转发展运动。文中讨论了这两种互补的发展运动模式的长处和不足之处后,提出了低频近似的发展模式,它包括了作为特殊情况的长波近似(纬圈半地转)模式和短波近似(经圈半地转)模式。参加到这一发展模式中的基本运动有 Rossby 波、Kelvin 波和混合波中的 Rossby 短波。在这一模式中,对运动的空间尺度的限制只要求是大尺度,对时间尺度的要求要比另两种模式宽,且波能量可以因波长不同而向两个方向传播,这样该模式可以用来研究热带大气和海洋中更多的动力学现象。

参考文献

[1] Rossby C G. On the mutual adjustment of pressure and velocity distribution in certain simple current system. Ⅰ. JMar Res,1937,1:15-18.

[2] Rossby C G. On the mutual adjustment of pressure and velocity distribution in certain simple current system. Ⅱ. J Mar Res,1938,2:239-263.

[3] 曾庆存. 大气中的适应过程和发展过程(一)和(二). 气象学报,1963,33:163-174,281-189.

[4] 叶笃正,李麦村. 大气运动的适应问题. 北京:科学出版社,1965.

[5] 叶笃正,巢纪平. 论大气运动的多时态特征——适应、发展和准定常演变. 大气科学,1998,22:385-398.

[6] Mastuno T. Quasi-geostrophic motions in equatorial area. J Meteor Soc,Japan. 1966,44:25-43.

[7] Chao Jiping, Lin Yonghui. The foundation and movement of tropical semigenostrophic adaptation. Acta Meteor Sinica, 1996,10:129-141.

[8] Gill A E. Some simple solution for heat-induced tropical circulation. Quart J R Meteor Soc,1980,106:447-562.

① 巢纪平. 热带地转适应运动的动力学基础. 气象学报,2000,58(1):1-10.

A Data Analysis Study on the Evolution of the El Niño/La Niña Cycle*

Chao Jiping (巢纪平)[1], Yuan Shaoyu(袁绍宇)[2], Chao Qingchen(巢清尘)[3], Tian Jiwei(田纪伟)[2]

(1. *National Marine Environmem Forecast Center*, *Beijing* 100081; 2. *Ocean University of Qintdao*, *Qingdao* 266003; 3. *National Climate Center*, *Beijing* 100081)

ABSTRACT: The curved surface of the maximum sea temperature anomaly (MSTA) was created from the JEDAC subsurface sea temperature anomaly data at the tropical Pacific between 1955 and 2000. It is quite similar to the depth distribution of the 20℃ isotherm, which is usually the replacement of thermocline. From the distribution and moving trajectory of positive or negative sea temperature anomalies (STA) on the curved surface we analyzed all the El Niño and La Niña events since the later 1960s. Based on the analyses we found that using the subsurface warm pool as the beginning point, the warm or cold signal propagates initially eastward and upward along the equatorial curved surface of MSTA to the eastern Pacific and stays there several months and then to turn north, usually moving westward near 10°N to western Pacific and finally propagates southward to return to warm pool to form an off-equator closed circuit. It takes about 2 to 4 years for the tempcrature anomaly to move around the cycle. If the STA of warm (cold) water is strong enough, there will be two successive El Niño (La Niña) events during the period of 2 to 4 years. Sometime, it becomes weak in motion due to the unsuitable oceanic or atmospheric condition. This kind process may not be considered as an El Niño (La Niña) event, but the moving trajcctory of warm (cold) water can still be recognized. Because of the alternatc between warm and cold water around the circuits, the positive (negative) anomaly signal in equatorial western Pacific coexists with negative (positive) anomaly signal near 10°N in eastern Pacific before the outbreak of El Niño (La Niña) event. The signals move in the opposite directions. So it appears as El Niño (La Niña) in equator at 2–4 years intervals. The paper also analyzed several excсptional cases and discussed the effect and importance of oceanic circulation in the evolution of El Niño/La Niña event.

Key words: El Niño (La Niña) events; curved surface of maximum sea temperature anomaly; Kelvin wave and Rossby wave; air-sea interaction

* Advances in Atmospheric Sciences, 2002, 19(5): 837–845.

1 Introduction

The El Niño phenomenon is characterized as the interannual appearance of unusual oceanographic conditions-exceptionally high sea surface temperature (SST) in the eastern equatorial Pacific. Because it can cause the abnormal variability of the global climate by the way of air-sea interactions, a great deal of work has been done to try to make clear the physical process of its appearance and development in the past fifty years. Before the 1980s, it was believed that the original positive sea surface temperature anomalies (SSTA) originated from near the coast of Peru in South Americal and propagated to equator and then westward along the equator. And the composite model synthesized by Rasmusson and Carpenter (1982), according to SST evolvement, gave a description of such an analysis at the beginning of 1980s. However the El Niño of 1982-1983 which attained a very large amplitude (Gill and Rasmusson 1983) and the El Niño of 1986-1987 shich attained a moderate amplitude (Mcphaden et al. 1990) were exceptional because their initial SSTA came into being in the central and western equatorial Pacific and then the positive SSTA propagated eastward along the equator. At that time, according to the evolution of SSTA, it was taken for granted that the development of El Niño may have two different processes, one from the east to west and the other from the west to east. One of the major accomplishments of the 10-year (1985-1994) Tropical Ocean Global Atmosphere (TOGA) program is to improve the in-suit data and the historical data, and offer the data basis to study the physical process of development.

The El Niño of 1997-1998 was, by some measures, the strongest on record. Analysis made by Li and Mu (1999), McPhaden (1999) show that warm water of 1997/1998 El Niño event originates in the subsurface warm pool in the western Pacific, then spreads toward the east and upward along the equatorial thermocline. Further investigation made by Chao and Chao (2001), they indicated that the warm water of eleven El Niño events and also the cold water of eleven La Niña events are appearance in the subsurface warm pool, then propagate eastward along the equatorial curved surface of MSTA, when they reach the Nino 3 region the positive/negative temperature anomalies have already ascended to the surface and then appear as El Niño/La Niña events we usually consider.

The important question caused by the previous research is where the warm/cold water as the source of El Niño/La Niña event comes from and where it goes after they reach the Nino 3 region. For this question, Chao et al. (2002) recently studied the case of 1997/1998 El Niño and gave the moving path of warm/cold water mass. In this paper, the systematic analyses for all El Niño/La Niña events since late 1960s have been presented.

Monthly sea surface temperature and sea subsurface temperature adta are used in this study. They are taken from Scripps Institute of Oceanography (JEDAC) on a 5° (lat.)×2° (long.) grid. This study emphasizes the region of tropical and subtropical Pacific between 20°N and 20°S.

2 Seasonal variation of MSTA

The varied characteristic of sea temperature over thermocline is firstly to be analyzed. The position of 20℃ isotherm is usually regarded as the thermocline depth. However, the SST over the equator and the southern part of the Eastern Pacific is less than 20℃ and there still exists thermochine there. So another method needs to be developed. In this study, based on the previous research along equator, we point out that the variation of temperature near the thermocline is largest, then the historical data is used to consturct the climatological curved surface of maximum sea temperature anomaly (MSTA) for approximately taking the place of the traditional thermocline curved surface in the whole tropical region. Figure 1 is the result based on the above two methods, and a very similar characteristic is shown in most regions of the tropical ocean, except for the smaller difference over the eastern Pacific. How does one construct the surface of MSTA? The first step is to find the depth over the region of MSTA during each month in each grid. Using the weighted averages method for the same month of each year, twelve climatic-averaged curved surface of MSTA are obtained.

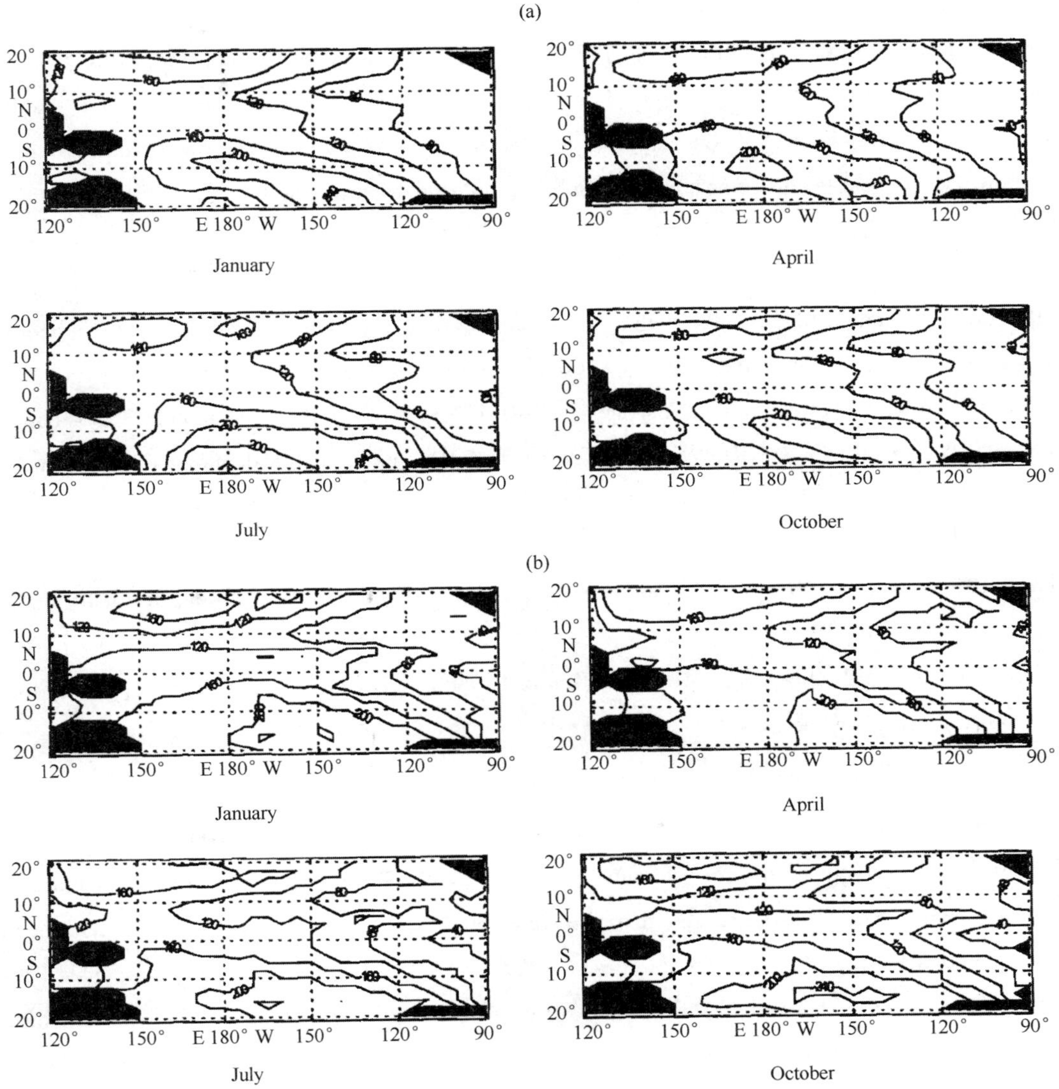

Fig.1　(a) Climatically-averaged curved surface of maximum sea temperature anomaly (MSTA) (units: m)

(b) Depth distribution of the 20℃ isotherm (units: m)

The curved surface of MSTA of January. April, July and October is shown in Fig. 1a. Analysis shows little variation in different months. Among them, the depth is shallowest in April, becoming deeper as the month goes. It reaches the deepest level in December. The general depth distribution along the equator is that the thermocline reaches its deepest position beyond 120 m in the warm pool, and gradually becomes shallower to the east. It almost reaches the sea surface in the eastern Pacific. The deepest is not in the warm pool, but near 15°S. That phenomenon is perhaps related to the South Equatorial Current. Comparing Figs. 1a and 1b (20℃ isotherm curved surface, i.e., thermocline curved surface), the trend is very similar.

3 The evolution characteristics of sea temperature anomaly in the MSTA curved surface

If defined by the index of SSTA in the Nino 3 region, there are totally seven El Niño events, which are in 1972/1793, 1976/1977, 1982/1983, 1986/1987, 1991/1992, 1994/1995, and 1997/1998. Moreover, one finds six La Niña events, in 1971, 1973/1974, 1978, 1983/1984, 1988/1989, and 1998. Observing the development of subsurface cold/warm water (Fig.2), the appearing time of El Niño/La Niña may be not in the above years. The reason is that the values of sea temperature anomaly in the MSTA curved surface and the SSTA may be not totally the same, even over the equatorial eastern Pacific.

The abscissa in Fig.2 is divided into five parts. For the left panels, the first part covers 140°E to 95°W along the equator. the second 2°N to 10°N along 95°W, the third 100°W to 145°E along 10°N, and the fourth 10°N to 2°N along 140°E. These four parts form a closed circuit in the Northern Hemisphere, being completed by the fifth part couvering 140°E to 160°E along the equator.

For the right panels, the first part also covers 140°E to 90°W along the equator, the second 2°S to 10°S along 95°W, the third part 100°W to 170°E along 10°S, and the fourth along the grid of (165°E, 10°S), (160°E, 8°S), (155°E, 6°S), (150°E, 4°S), and (145°E, 2°S). The reason that we chose such a route in the fourth part is that there are many islands near the warm pool in the south of 140°E and no data there. The fifth part covers 140°E to 180° along the equator and hence forms a closed circuit in the Southern Hemisphere.

The ordinate in Fig.2 shows time. The evolution of El Niño/La Niña, demonstrated in the figure, is explained in the following subsections.

3.1 The development of typical warm water enents

Judged by the SSTA index in the Nino 3 zone, the event of 1968/1969 is not considered and El Niño event. However, if it is judged by the intensity and duration of sea subsurface temperature anomaly, it is indeed a warm water event. The warm signal in the warm pool of the western Pacific propagated eastward in 1968 and reached the equatorial eastern Pacific in 1969. It then left the equator and propagated westward along both 10°S and 10°N. It reached the warm pool in 1971 and propagated eastward to the equatorial eastern Pacific to become the El Niño event of 1972/1973. Another strong El Niño is the event of 1997/1998, which was analyzed by Chao and Chao (2001) and Chao et al. (2002). In Fig.2 it is seen that the warm water in the warm pool subsurface comes from the warm water event of 1995. This signal of positive sea temperature anomaly of 1995 in the Nino 3 zone propagated westward along the closed circuit from 10°S and 10°N to the warm pool subsurface, and stayed and developed there to be a warm water mass. It then moved east-

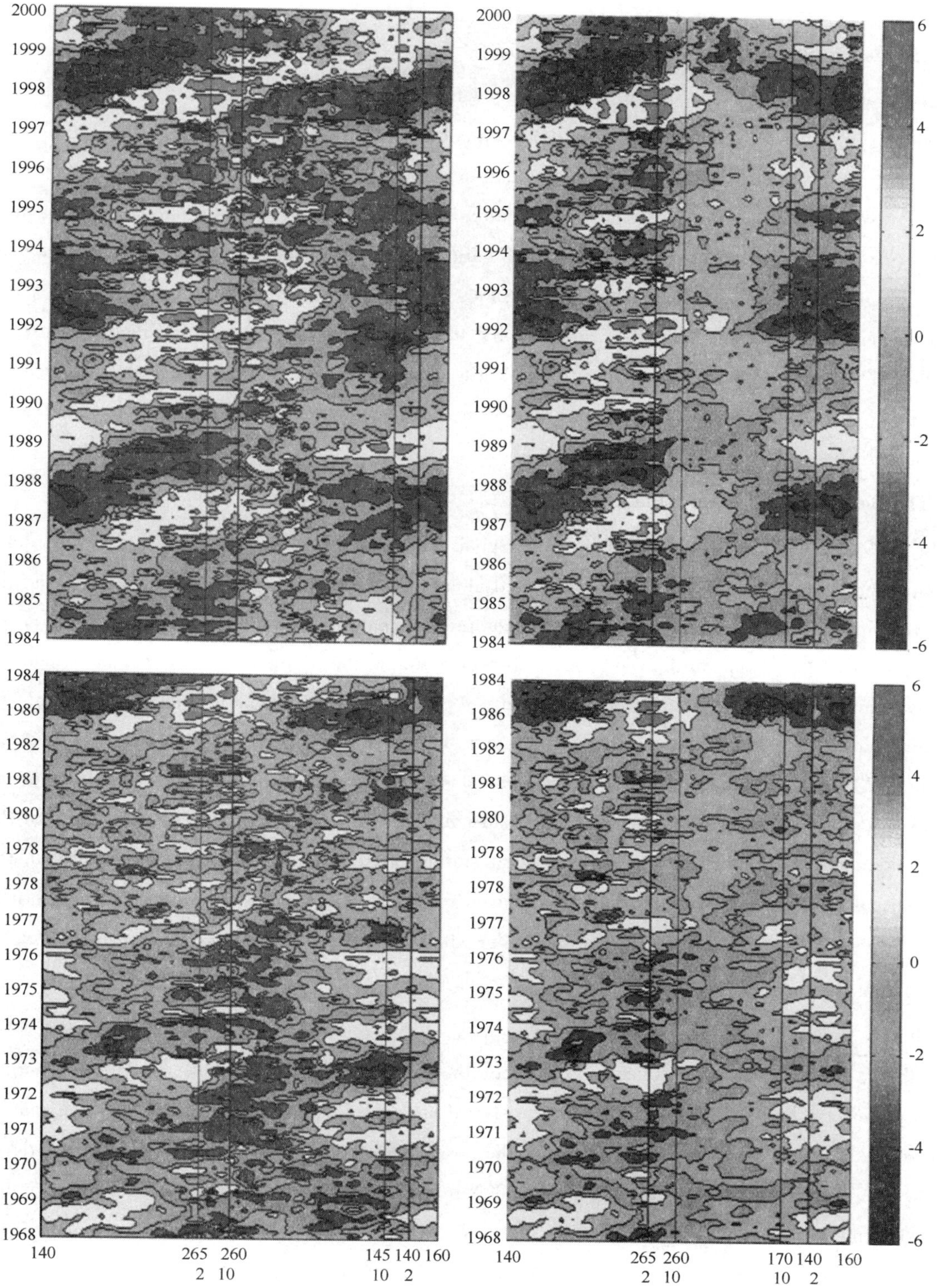

Fig.2　The evolution of the El Niño (La Niña) events(见书后彩插)

ward along the equatorial thermocline to the equatorial eastern Pacific in 1997 to become the event of 1997/1998. Both cases show that the warm water in the warm pool subsurface usually comes from the positive sea temperature anomaly in the equatorial eastern Pacific and moves westward along the respective off equator circuits.

3.2 The development of weak El Niño events

The event of 1976/1977 was a weak one. Following the origin of its warm water, it was found that the positive sea temperature anomaly of 1973 in the equatorial eastern Pacific moved to the warm pool from 10°N. The signal of the weaker positive sea temperature anomaly moved to the Nino 3 zone along the equator and then westward along the north branch of the circuits. During the period, the temperature anomaly was negative. However, when it moved to the central Pacific in 1975, it changed to positive. After if reached the warm pool and propagated eastward along the equator, it became stronger. The process in finally developed into a weaker El Niño event. This case indicates that the intensity of sea temperature anomaly varies during its evolution, even totally changing sign on one occasion, while usually propagating westward along the circuit of 10°N. If the intensity of the temperature anomaly cannot be strengthened for some time, the warm water event will disappear.

3.3 The development of La Niña events

In 1969, the cold water stayed in the warm pool subsurface after the warm water of 1968 propagated eastward. Similarly, the cold water moved eastward along the equatorial thermocline and developed into a La Niña event in 1971. Then, the negative sea temperature anomaly near the surface of the equatorial eastern Pacific propagated westward to the warm pool subsurface around the northern (stronger signal) and southern (weaker signal) branch of the circuits respectively. A strong negative anomaly formed in the warm pool in 1973 and moved eastward to the eastern Pacific along the equatorial thermocline to form the strong La Niña event of 1973/1974. Another case is the development of cold water after the strong El Niño event of 1998 came to an end. Notice that when the sea temperature anomaly was positive in the equatorial eastern Pacific, the negative anomaly occurring in the warm pool subsurface started to move eastward along the equator to the equatorial eastern Pacific in late 1996. It then moved westward to the warm pool along the northern branch and moved eastward to the eastern Pacific along the equator to form the strong cold event of 1998.

3.4 Alternative occurrence

The positive and negative anomalies always appear at the same time at the equator and at 10°N or 10°S before the El Niño or La Niña event begins. Maybe, the positive anomaly moves westward to the north of the negative signal as the negative anomaly moves eastward along, the equator. Or, the negative anomaly moves westward to the north of the positive anomaly as the positive anomaly moves eastward. In other words, before El Niño occurs, the negative signal moves near 10°N to the warm pool when the positive anomaly propagates eastward along the thermocline. The situation for La Niña is similar, but with the negative anomaly instead of the positive one. Therefore, the La Niña (El Niño) event is growing when El Niño (La Niña) develops. The El Niño and La Niña events form a cycle that alternates between a warm phase and a cold phase. Of course, the period between two warm (cold) events in succession may be different. For example, the interval between El Niño events of 1976/1977 and 1982/1983 is greater than six years but only two years between the La Niña events of 1971 and 1973/1974. The average interval, however, is 3 to 4 years in the last 30 years.

Of course, there are some unique characteristics for each El Niño or La Niña event beyond the above basic ones. The El Niño event of 1982/1983 is a rare case in which the warm water did not originate from the warm pool but from the central Pacific. In fact, a portion of the warm water came from the western Pacific, and the event's position was to the north of the equator and the intensity was weaker. So, the developing process has not been shown in the figure. This phenomenon also indicates external influences on El Niño events.

4　Summary and discussion

The above analysis shows that, corresponding to the warm event or cold event, the positive or negative anomaly propagates along the two closed circuits above the thermochline curved surface in the tropical off equator ocean. The signal propagates eastward along the equator and westward along 10°N and 10°S respectively. Generally speaking, the warm event appears alternating with cold event. The average interval between two warm (cold) events in succession is about 3 to 4 years. The period is related to the propagating time along the circuits.

The following questions arise from the above analysis.

(1) Why does the signal propagate along the circuit above the thermocline curved surface? Can we consider that the variation in the temperature anomaly mainly reflects the depth variation of the thermocline in the process of internal ocean and atmospheric forcing?

(2) Why does the signal propagate westward along 10°N and 10°S respectively? Is this related to the distribution of equatorial current or to the pattern of atmospheric circulation in the tropical region?

(3) Does the Kelvin wave or another process reflect the eastward propagation of temper-ature anomaly along the equatorial thermocline?

(4) In many cases, it is difficult to explain with Rossby wave theory the westward propagation of the temperature anomaly along 10°N or 10°S, because the signal propagates westward too fast sometimes. For example, the speed of the westward propagation between 5°N and 15°N is faster than the warm signal speed of the eastward propagation along the equator in the event of 1997/1998. Even if the speed of the current is considered, the speed cannot reach that fast. So, it may be due to the sea temperature change responding to air-sea interaction. Thus, this needs further investigation.

In a sense, the results shown in this paper may change the traditional understanding on the evolution of the El Niño/La Niña event. So we should reconsider some past theory and develop new theoretical framework as well as conduct prediction simulations.

Acknowledgments

This work was supported by the National Natural Science Foundation of China under Grant No.40126002.

REFERENCES

Chao Qingchen, and Chao Jiping. 2001. The influence of western tropical Pacific and eastern Indian Ocean on the development of ENSO event. *Progress in Natural Science*, 11(12):1293–1300 (in Chinese).

Chao Jiping, Yuan Shaoyu, Chao Qingchen, and Tian Jiwei. 2002. The source of the subface warm water of the warm pool in the equatorial western Pacific-the analysis on the El Niño event in 1997/1998. *Chinese J. Atmos. Sci.* (in Chinese), (to be published).

Gill. A. E., and E. M. Rasmusson. 1983. The 1982-1983 climate anomaly in equatorial Pacific. *Nature*, 306:229-234.

Li Chongyin, and Mu Mingquan. 1999. The occurrence of the El Niño event and the subsurface temperature anomaly of warm pool in the equatorial western Pacific. *Chinese J. Atmos. Sci.*, 23:513-521 (in Chinese).

McPhaden, M. J., S. P. Havers, L. J. Maugum, and J. M. Tool. 1990. Variability in the weastern equatorial Pacific ocean during 1986-1987 El Niño/Southern Oscillation event. *J. Phys. Oceanogr.*, 20(2):190-208.

McPhaden, M. J. 1999. Genesis and evolution of the 1997-1998 El Niño. Science, 283:950-953.

Rasmusson, E. M., and T. H. Carpenter. 1982. Variations in tropical sea surface temperature and surface wind fields associated with the Southern Osccillation/El Niño.*Mon. Wca. Rew.*, 110:354-384.

ENSO 事件中次表层海温距平在 10°N 附近向西传播的机理*

巢纪平[1,2]，蔡　怡[1]

（1.国家海洋环境预报研究中心，北京，100081；2.国家海洋局第一海洋研究所，青岛，266061）

摘要：在最大温度距平的极值曲面上，对观测资料的分析表明，在这个曲面上的次表层海温距平，一般从西太平洋暖池附近沿赤道向东传播，然后在东太平洋 95°W 附近向两极传播，并在 10°N 附近（北半球比南半球清楚）向西传播，再在 140°E 暖池海域传向赤道，形成一个信号传播的回路。文章试图研究东太平洋次表层海温距平信号在 10°N 附近向西传播的可能机制。低空 850 hPa 风场的资料分析表明，当 ENSO 处在暖（冷）位相时，东太平洋沿岸附近将出现经向风，首先在经向风的吹引下，将产生沿岸的 Kelvin 波，进而在经向风的辐散（辐合）作用下，通过沿岸的上升（下沉）流在各个纬度激发出向西的 Rossby 波，但理论表明在与观测接近的周期性经向风作用下由 Kelvin 波产生的沿岸上升（下沉）流在 10°N 附近最大，因此在那个纬度附近 Rossby 波的振幅最大，更易将距平意义下的冷（暖）水传向西太平洋。

关键词：ENSO；Kelvin 和 Rossby 波；扰动向西传播；海温距平

1　引言

自从“中、美西太平洋海气相互作用试验”（1985—1991 年）观测到 1986/1987 年的 El Niño 事件的正的海温距平首先出现在西太平洋暖池的次表层以来[1]，中国的气象和海洋学家十分注意西太平洋在 ENSO 事件形成、发展中的作用。在理论分析方面曾指出，赤道西太平洋海温对风应力的响应强度远大于赤道东太平洋[2~6]，而温跃层海温响应的振幅也远大于其上面的混合层[7,8]。在资料分析方面，李崇银、穆明权[9]就 1997 年的 ENSO 事件，分析了次表层海温距平的演变[9]。当分析次表层温度距平演变时，需要对次表层给出一个能对所有例子进行比较的参考面，为此，巢清尘、巢纪平[10]提出最大海温距平曲面（MSTA）的概念，这是海温距平极值所在的曲面，它的深度分布接近于 20℃ 为参考的温跃层曲面的深度分布，图 1（彩图）是年平均 MSTA 的深度分布。在这个曲面上巢纪平等曾分析了 1997/1998 年 El Niño 事件中正、负海温距平发展和演变的全过程[11]，进而又分析了 20 世纪 60 年代到 2000 年所有暖、冷事件的海温距平在这个曲面上的行为[12]。图 2（彩图）即为近年来海温距平在 MSTA 面上的演变，这张图取自参考文献[12]但经过 9 个月的滑动平均，由于过滤掉了一些弱的信号，所以要比原图清楚。横坐标第 1 分区是沿赤道从西太平洋到东太平洋（图 2（a）），第 2 分区是沿北美海岸附近北上到 10°N（图 2（b）），然后再沿 10°N 向西到西太平洋（图 2（c）），再在 140°E 附近南

* 气象学报，2005，63（4）：385-390.

下到赤道(图2(d)),最后是第一分区的重复(图2e),这样在北半球从赤道到副热带形成一个回路。由图2特别注意到,在赤道有一个正(负)温度距平向东传播时,在副热带10°N附近有一个负(正)温度距平向西太平洋传播,再南下传到赤道就形成一个近似的ENSO正、负位相的循环。

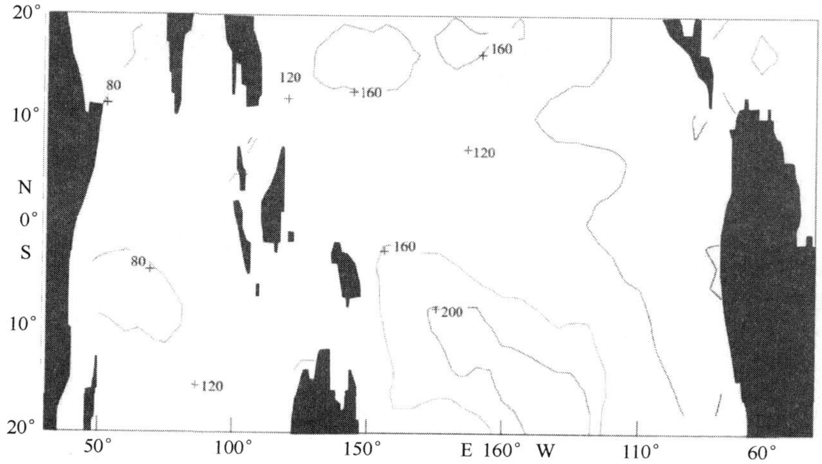

图1 年平均MSTA的深度分布(单位:m)(见书后彩插)

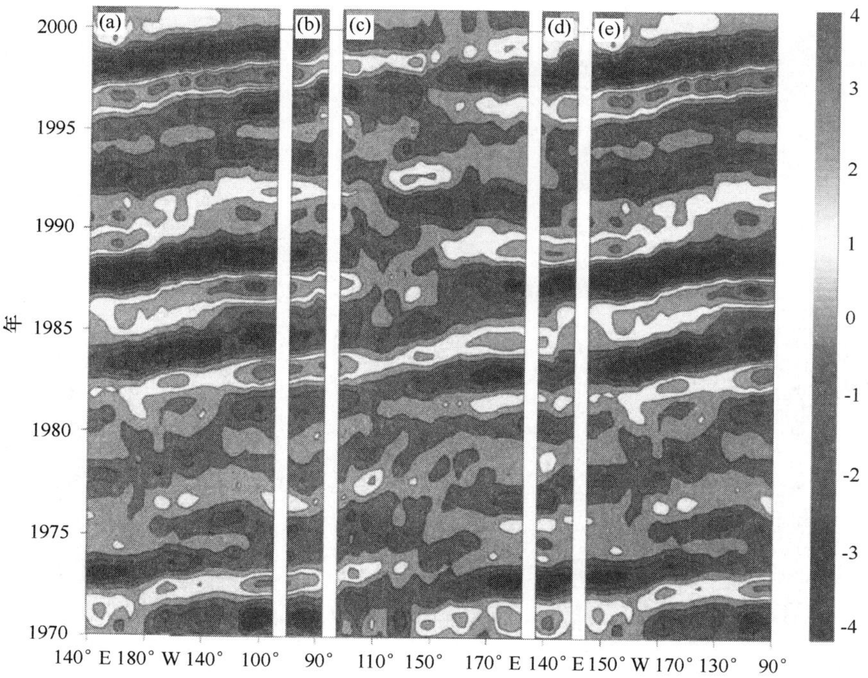

图2 1970—2000年海温距平在MSTA面上的演变(见书后彩插)

(a-e分别为第1分区至第5分区)

本文试图对图2中为什么信号在10°N附近向西传播的现象给出初步的物理解释(如参考文献[11]的个例分析表明,在南半球也存在相似的回路但不如北半球清楚,因此下面的分析对南半球也同样适合)。首先分析东边界附近Kelvin波的经向传播特征,确定最易产生向西传播的Rossby波所在的纬度,进而分析Rossby波的特征。考虑到Kelvin波是在经向风应力强迫下激发出来的,在图3给出东边界附近经向风距平的周期和强度的分布。

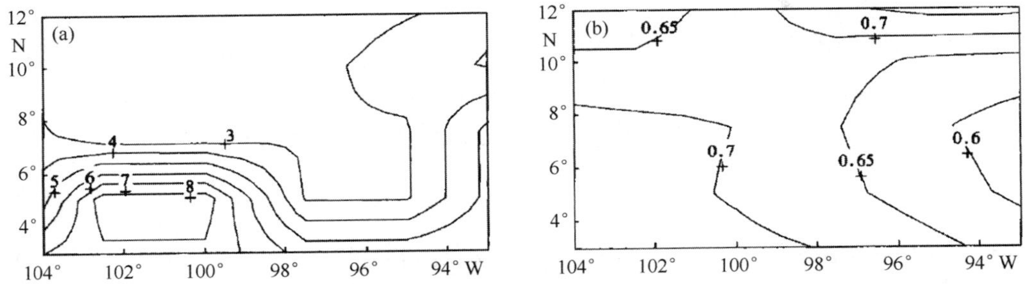

图 3　经向风距平的周期(图(a),单位:月)和均方根强度值(图(b),单位:m/s)

2　基本方程

应用 $1\frac{1}{2}$层模式,即海表下是一薄的混合层,密度为 ρ_1,其下是温跃层,密度为 ρ_2,视重力为 $g'=g(\rho_2-\rho_1)/\rho_2$,温跃层的气候深度分布若只考虑其经向分布,可设成

$$H(y)=H_0\left(1-\frac{h^*(y)}{H_0}\right) \tag{1}$$

式中 H_0 可看成温跃层在赤道的气候深度。重力波波速为 $c=\sqrt{g'H}$,其在赤道的速度为 $c_0=\sqrt{g'H_0}$。与式(1)对应的气候纬向流速为

$$u^*(y)=\frac{g'}{f}\frac{\mathrm{d}h^*}{\mathrm{d}y} \tag{2}$$

对气候偏差的非定常、线性化方程组为

$$\frac{\mathrm{D}v}{\mathrm{D}t}+fu=-g'\frac{\partial h}{\partial y}+\frac{\tau^y}{H_0} \tag{3}$$

$$\delta\frac{\mathrm{D}u}{\mathrm{D}t}-fv=-g'\frac{\partial h}{\partial x} \tag{4}$$

$$\frac{\mathrm{D}h}{\mathrm{D}t}+H_0\left(1-\frac{h^*(y)}{H_0}\right)\left(\frac{\partial u}{\partial x}+\frac{\partial v}{\partial y}\right)-v\frac{\partial h^*}{\partial y}=0 \tag{5}$$

式中

$$\frac{\mathrm{D}}{\mathrm{D}t}=\frac{\partial}{\partial t}+u^*\frac{\partial}{\partial x}$$

而 $\tau^y(t,x,y)$ 为东边界附近的经向风应力,可认为 $h^*/H_0=\varepsilon$ 是个小量,因此一般可取 $c=\sqrt{g'H_0(1-h^*/H_0)}\approx c_0$。

3　Kelvin 波在东边界附近的传播

考虑到 Kelvin 波需满足经向的地转平衡,可取标识符 $\delta=0$,并令

$$v=\hat{v}(y,t)\mathrm{e}^{\frac{f}{c}x} \tag{6}$$

有

$$h \sim \frac{c}{g'}v \tag{7}$$

对 Kelvin 波,$u\approx 0$,并且由于它是边界附近的波,其存在区域为 $x<c/f=l_R$-Rossby 变形半径,在另一方面由图 1(彩图)表明,在东边界附近,从赤道以北直到副热带,纬度背景场的不均匀性并不明显,因此在分析 Kelvin 波时可以不考虑背景场的不均匀性。考虑到式(5),式(1)可写成

$$\frac{\partial v}{\partial t} + c\frac{\partial v}{\partial y} = \frac{\tau^y}{H} \tag{8}$$

由于 Kelvin 波传播区域很窄,如图 3 由观测资料的分析所示可以认为由经向风距平造成的 τ^y 在赤道以北在周期上相对来讲是均匀的(3 个月左右),强度虽然向北增大,但变化只在 0.6~0.8 m/s,可以近似地取其平均值,这样可将式(8)写成

$$\frac{\partial v}{\partial t} + c\frac{\partial v}{\partial y} = \frac{\tau_0}{H_0}\sin\left(\frac{2\pi}{P}t\right) \tag{9}$$

这里设 $c\approx c_0$,近似解为

$$v(y,t) = F(y-ct) - \frac{\tau_0}{H_0}\frac{P}{2\pi}\cos\left(\frac{2\pi}{P}t\right) \tag{10}$$

其中第 1 项是自由波。条件为

$$y\approx 0,\qquad v=0 \tag{11}$$

有

$$F(-ct) = \frac{P\tau_0}{2\pi H_0}\cos\left(\frac{2\pi}{P}t\right) \tag{12}$$

得到解为

$$v(y,t) = -\frac{P\tau_0}{\pi H_0}\sin\left(\frac{\pi}{Pc}y\right)\cdot\sin\left(\frac{\pi}{Pc}(y-2ct)\right) \tag{13}$$

给出半个周期的积分,为

$$\tilde{v} \equiv \int_0^{P/2} v(y,t)\,\mathrm{d}t = -\frac{1}{H_0}\left(\frac{P}{\pi}\right)^2\sin\left(\frac{\pi}{Pc}y\right)\cos\pi\left(1-\frac{2}{Pc}y\right) \equiv -\frac{1}{H_0}\left(\frac{P}{\pi}\right)^2 F(P,y) \tag{14}$$

由此可求得 $\tilde{v}|_{max}$ 时,P,y 的关系,即由

$$\frac{\partial F}{\partial y} = 0 \tag{15}$$

得到

$$P = G(y_{max}) \tag{16}$$

由式(14),$y=Pc/2$,$F(P,y)=1$。如 $y_{max}=3\ 000$ km,$c=50$ cm/s,则 $P\approx 120$ d,接近资料给出的 3~4 个月。

在 y_{max} 以南,经圈风是辐散的,由此可引起沿岸涌升流,近似地估计为

$$h|_{P/2} \sim -H\frac{\partial\tilde{v}}{\partial y} = \left(\frac{P}{\pi}\right)^2\frac{\partial F}{\partial y} \tag{17}$$

注意到涌升流极值位置可近似地由下式决定,即

$$\frac{\partial h|_{P/2}}{\partial y} = \left(\frac{P}{\pi}\right)^2\frac{\partial^2 F}{\partial y^2} = 0 \tag{18}$$

考虑到

$$F = \sin\left(\frac{\pi}{Pc}y\right)\cos\left(\pi - \frac{2\pi}{Pc}y\right) = -\sin\left(\frac{\pi}{Pc}y\right)\cos\left(\frac{2\pi}{Pc}y\right)$$

利用倍角公式有

$$F - \sin\left(\frac{\pi}{Pc}y\right)\left[1 - 2\sin^2\left(\frac{\pi}{Pc}y\right)\right] = -\sin\left(\frac{\pi}{Pc}y\right) + 2\sin^3\left(\frac{\pi}{Pc}y\right) \tag{19}$$

由此得

$$\frac{\partial^2 F}{\partial y^2} = 2\left(\frac{\pi}{Pc}\right)^2 \sin\left(\frac{\pi}{Pc}y\right)\cdot\left[1 - 4\sin^2\left(\frac{\pi}{Pc}y\right)\right] = 0 \tag{20}$$

考虑到由 $\sin\left(\frac{\pi}{Pc}y\right) = 0$，得到 $y_c = Pc$，大于 $y_{\max}$，此根不可取。而由式(20)第 2 个括号等于零，有

$$\sin\left(\frac{\pi}{Pc}y\right) = \frac{1}{2} \tag{21}$$

由此得

$$y_c = \frac{Pc}{6} \tag{22}$$

可以估算，如取 $y_{\max} = 3\ 000$ km，有 $y_c = 1\ 000$ km。亦即在 10°N 附近由涌升流产生的温跃层距平较大，图 4 给出资料分析结果，可以看到东南角有一大值区之外，距平大的区在 8°–10°N 的一条带上。

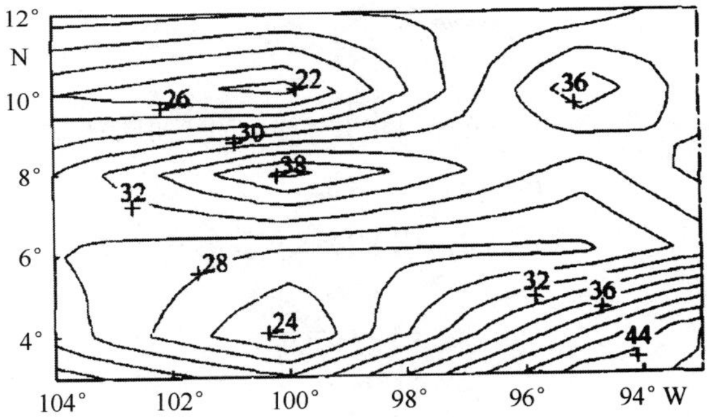

图 4　东太平洋边界附近温跃层距平的分布(单位:m)

4　Rossby 波的特征

在离开东边界一段距离，经向风应力已变得很弱，可以忽略不计，由式(3)，式(4)略去高频振荡给出

$$u = -\frac{g'}{f}\frac{\partial h}{\partial y} - \frac{g'}{f^2}\frac{\mathrm{D}}{\mathrm{D}t}\left(\frac{\partial h}{\partial x}\right)$$

$$v = \frac{g'}{f}\frac{\partial h}{\partial x} - \frac{g'}{f^2}\frac{\mathrm{D}}{\mathrm{D}t}\left(\frac{\partial h}{\partial y}\right) \tag{23}$$

代入式(5)得到位势涡度方程，为

$$\left(\frac{\partial^2}{\partial x^2} + \frac{\partial^2}{\partial y^2} - \frac{1}{\lambda^2} + \frac{1}{H_0}\frac{\mathrm{d}h^*}{\mathrm{d}y}\frac{\partial}{\partial y}\right)\frac{\mathrm{D}h}{\mathrm{D}t} + \beta_M\frac{\partial h}{\partial x} = 0 \tag{24}$$

式中 $\lambda=c/f$ 为 Rossby 变形半径，而

$$\beta_M=\beta-\frac{f}{H_0}\frac{\mathrm{d}h^*}{\mathrm{d}y} \tag{25}$$

作变换

$$h=\widetilde{h}(x,y,t)\exp\left(-\frac{1}{2H_0}\frac{\mathrm{d}h^*}{\mathrm{d}y}y\right) \tag{26}$$

给出

$$\left(\frac{\partial}{\partial t}+u^*(y)\frac{\partial}{\partial x}\right)\left(\frac{\partial^2}{\partial x^2}+\frac{\partial^2}{\partial y^2}-\alpha^2\right)\tilde{h}-\beta_M\frac{\partial\tilde{h}}{\partial x}=0 \tag{27}$$

式中

$$\alpha^2=\frac{1}{\lambda^2}+\left(\frac{1}{2H_0}\frac{\mathrm{d}h^*}{\mathrm{d}y}\right)^2+\frac{1}{2H_0}\frac{\mathrm{d}^2h^*}{\mathrm{d}y^2} \tag{28}$$

应用 WKB 方法，设波振幅是空间和时间的缓变函数，即

$$\tilde{h}=A(T,X,Y)\mathrm{e}^{i\varphi} \tag{29}$$

其中位相为

$$\varphi=kx+ly-\sigma t \tag{30}$$

而

$$\sigma=-\frac{\partial\varphi}{\partial t},\quad k=\frac{\partial\varphi}{\partial x},\quad l=\frac{\partial\varphi}{\partial y} \tag{31}$$

在这里假定 σ,k,l 在整个定义域中是可变的，但也是空间和时间的缓变函数。缓变函数取成

$$T=\varepsilon t,\quad X=\varepsilon x,\quad Y=\varepsilon y \tag{32}$$

容易算得修正(因有背景场的不均匀性在内)Rossby 波的色散关系为

$$\sigma=u^*k-\frac{\beta_M k}{K^2} \tag{33}$$

式中 $K^2=k^2+l^2+\alpha^2$。群速度为

$$C_{gx}=\frac{\sigma}{k}+\frac{2\beta_M k^2}{K^4}=u^*+\frac{\beta_M}{K^4}(k^2-l^2) \tag{34}$$

$$C_{gy}=\frac{2\beta_M kl}{K^4} \tag{35}$$

注意到，当 $u^*\approx 0$，而 k 很小(长波)时，沿 x 方向的群速度是向西的。

引进波作用密度，定义为

$$E=\frac{K^4}{\beta_M}A^2 \tag{36}$$

则有守恒方程[13]

$$\frac{\partial E}{\partial T}+\frac{\partial C_{gx}E}{\partial X}+\frac{\partial C_{gy}E}{\partial Y}=0 \tag{37}$$

这表明，波作用密度或波能量以群速度传播，传播的路径称射线。对于缓慢的运动，上式可近似地写成

$$\frac{\partial}{\partial s}(s \cdot C_{gs}E)=0 \quad (38)$$

式中 s 为沿射线的单位向量。由此沿射线

$$C_{gs}E_s = \text{常数} \quad (39)$$

假设在射线的有限线段上群速度不变,则在这有限线段上波作用密度也将不变。由图 1 看到,在 10°N 附近,气候的温跃层深度最浅,按定义 $h^*(y)$,在该纬度附近应取极大值,即 $dh^*/dy=0$,而 $d^2h^*/dy^2<0$,因此如 $E_s \approx E$,则由式(28)α^2 变小,相应的 K^2 也变小,而 $\beta_M \to \beta$,但一般来论,在极值区附近这一变化并不重要,也即 K^2 的变化相对来讲要重要些,于是为了保持 E 的不变,A^2 需增加。这表明在 10°N 附近向西传播的长波信号,要比其他纬度更明显。

5　结论

资料分析表明,在 ENSO 事件中冷、暖次表层海温距平,可在接近温跃层的最大海温距平曲面上,形成一个传播的回路,海温距平先从暖池沿赤道向东传播,到东太平洋后转向极方向传播,一般在 8°N~12°N 范围内转向西传播,然后再在西太平洋向赤道传播。一般认为,信号在赤道的向东传播,Kelvin 波是载体,但为什么在 10°N 附近转向西传播虽然有一些工作讨论[14~17],但研究的时间尺度都是 10 年左右年代际短,长于 ENSO 的特征时间尺度。本文结合资料分析从另一个角度初步研究了这一问题,指出沿大洋东边界附近向极传播的强迫 Kelvin 波在 10°N 附近造成的沿岸涌升流最强,表明在这一纬度附近激发向西传播的 Rossby 波的动力最强。进一步分析指出,在这个纬度向西传播的 Rossby 波受背景气候温跃层深度分布的影响,波的振幅也强于其他纬度。当然这只是初步的启示性的理论分析,进一步的研究尚待进行。

参考文献

[1] Wang Zongshan, et al, Air-sea Interaction in Tropical Western Pacific. Proc. US-PRC International TOGA Symposiun, Nov. 15-17, 1988, Beijing. Editor in Chief Chao Jiping and J. A. Young. Beijing: China Ocean Press, 1990. 15-26
[2] 巢纪平,张丽．赤道不同海域对信风张弛的响应特征——对 El Niño 的研究的启示．大气科学,1998,22(4):428-442.
[3] 巢纪平,陈峰．热带大洋东、西部对风应力经圈不对称的响应．大气科学,2000,24:723-738.
[4] 巢纪平,巢清尘．热带西太平洋对风应力响应的动力学．大气科学,2002,26:145-160.
[5] 林永辉．风应力强迫变化对赤道太平洋斜压扰动的影响．见:ENSO 循环机理和预测研究．北京:气象出版社,2003:209-216.
[6] 林永辉,布和朝鲁．强迫耗散作用下 Kelvin 波的解析求解．见:ENSO 循环机理和预测研究,北京:气象出版社,2003:217-222.
[7] 巢纪平,陈鲜艳,何金海．热带西太平洋对风应力的斜压响应．地球物理学报,2001,45:176-187.
[8] 巢纪平,陈鲜艳,何金海．风应力对热带斜压海洋的强迫．大气科学,2002,26:578-594.
[9] 李崇银,穆明权．厄尔尼诺的发生与赤道西太平洋暖池次表层海温异常．大气科学,1999,23:513-521.
[10] 巢清尘,巢纪平．热带西太平洋和东印度洋对 ENSO 发展的影响．自然科学进展,2001,11:1293-1300.
[11] 巢纪平,袁绍宇,巢清尘等．热带西太平洋暖池次表层暖水的起源——对 1997/1998 年事件的分析．大气科学,2003,27:145-150.
[12] Chao Jiping, Yuan Shaoyu, Chao Qingchen, et al. A data analysis study on the evolution of the El Niño/La Niña cycle

. Adv Atmos Sci,2002,19:837-844.

[13] 陈英仪,巢纪平. 螺旋 Rossby 波的波作用密度守恒和稳定性. 中国科学(B 辑),1983,7:663-672.

[14] Huang R X(黄瑞新),Wang Qi. Interior communication from subtropical to the tropical oceans. J Phys Oceanogr,2001,31:3538-3550.

[15] Capotondi A,Alexander M A,et al. Why are there Rossby wave maxima in the Pacific at and 13°N. J Phys Oceanogr,2003,33:1549-1563.

[16] Capotondi A,Alexander M A. Rossby waves in the tropicl north Pacific and their role in decadal thermocline variability. J Phys Oceangr,2001,31:3496-3515.

[17] Galanti E,Tziperman E. A midlatitude-ENSO teleconnection mechanism viabarovlinically unstable long Rossby waves. J Phys Oceangr,2003,33:1877-1888.

热带印度洋的大尺度海气相互作用事件*

巢纪平[1]　袁绍宇[2]　蔡怡[1]

(1. 国家海洋环境预报研究中心,北京 100081;2. 中国海洋大学,青岛 266003)

摘要: 分析了热带温跃层上海温距平资料后指出,在印度洋东西方向的海温距平分布呈现出距平符号相反的偶极子现象,在大气中的纬圈环流即 Walker 环流上也呈现出与海温距平相协调的或匹配的上升和下沉分支(距平意义下)分布。这一分析表明,印度洋也存在着与太平洋类似 ENSO 的大尺度海气相互作用事件。

关键词: 印度洋;偶极子;海气相互作用

1　引言

Webster 等[1]报告在 1997—1998 年热带印度洋发生了 40 年纪录中未发生的特殊事件,西部赤道附近的海表温度是正距平,而东南部的海表温度是负距平,认为这一海洋事件是与上空大气中环流相互作用的结果。Saji 等[2]把海洋的这一现象定名为偶极子(dipole),进而将印度洋西部赤道附近的海表温度距平减去印度洋东部赤道附近的海表温度距平的值称为偶极子模态指数(DMI),DMI 可呈有正、负值的年际时间变化形式,通过对大气中风和降水等资料分析后指出,这是热带印度洋的海气相互作用现象。李崇银等[3]也给出了 100 年以来 DMI 的时间序列。但需要指出,DMI 的正、负值在大多数年份表现为西、东印度洋赤道附近的海表温度在相同符号下强度的差异,很少出现像 1997—1998 年的海温分布,海表温度距平在符号上也是反的。

我们认为,严格物理意义下的偶极子,在海洋上应表现为像 1997—1998 年那样的海表温度距平在赤道印度洋东、西两个部位呈符号相反的分布,而不只是海温距平符号相同下强度上的差异;表现在海温距平分布上的偶极子现象,是一种海气相互作用现象,尚需进一步用长时间序列资料来分析和印证。

文中将用海温距平,高、低空纬向风来分析从 20 世纪 60 年代以来在印度洋赤道海域所发生的与海洋中偶极子现象相关联的大气中的现象。

文中所用的海洋资料为美国 Scripps 海洋研究所的海温再分析资料,时间为 1961 年 1 月至 2001 年 12 月,深度为 0~400 m;大气资料为 NCEP 的全球 17 层格点资料,时间同海温。

* 气象学报,2003,61(2):251-256.

2 热带印度洋海温特征分析

2.1 海表温度距平的特征

由热带印度洋40年的海表温度距平的绝对值(即强度或振幅值)的平均值分布,可以看到在印度洋西部(10°S 至 10°N,50°—70°E)和东部(10°S 至 EQ,90°—110°E)各有一个强度达到0.5℃和0.4℃的区域,从这两个区域中的海表温度距平逐年变化(图1)可以看到,除1997—1998年有明显的符号相反外,多数年份温度距平的符号是相同的,这两条曲线的相关系数为0.23,计算了两者之间的时滞相关系数,在前后6个月的时滞内,均未发现负相关。由此而构造的DMI,主要反映的是赤道附近西、东印度洋海表温度距平在强度上的差异。

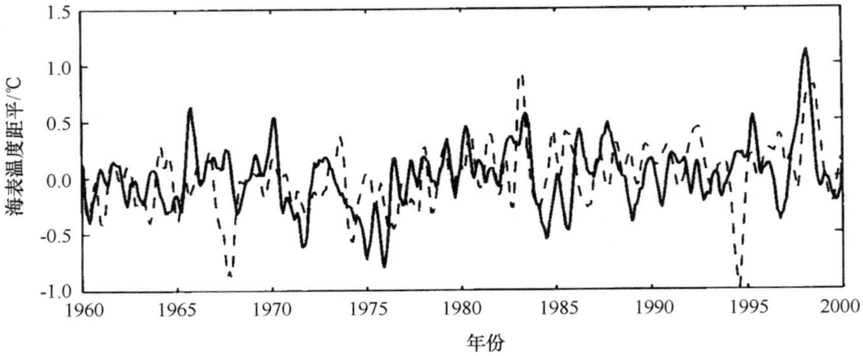

图1 印度洋海表温度距平的时间变化
(实线:印度洋西部;虚线:印度洋东部)

2.2 次表层海温距平的特征

考虑到海表温度信号弱,噪音也大,影响海表温度变化的因子和物理过程也比较复杂,因此巢清尘和巢纪平[4]及Chao Jiping等[5]指出,改用垂直方向海温变化的极值面(接近温跃层曲面)上的温度距平来分析海洋中温度的变化能更好地反映海气相互作用。极值曲面是这样构造的,在每一个网格点上取次表层中的最大海温距平值(绝对值),由此得到极值深度分布的气候曲面,在热带太平洋这个曲面与20℃为标准的温跃层曲面基本符合。

图2(a)是极值面上赤道印度洋西部(10°S 至 10°N,50°—70°E)和东部(10°S 至 EQ,90°—110°E)海温距平的时间变化曲线(取5个月的滑动平均,两条曲线的起始月份不同,差了4个月,东部在前)。可以看到这两条曲线的变化趋势基本上是反向的,其同期的相关系数为负的(超过信度99.9%),最大的时滞相关系数为-0.46,出现在东部早于西部2个月以上(图2(b))。

如果将西部指标区值减去东部指标区值,也可以构成一条DMI曲线,但这时DMI的正值主要反映西部为海温正距平,东部为海温负距平,DMI的负值,正、负海温距平的分布形势反过来,所以这是海温距平偶极子模态的更好的表示方法。

3 与海温偶极子模态相耦合的热带印度洋Walker环流

在气候图上与热带太平洋的Walker环流相毗邻的是,在印度洋上空是一个反方向转动的垂直环流,或可称这是印度洋的Walker环流。对于印度洋的Walker环流可以用850 hPa和200 hPa的纬向风

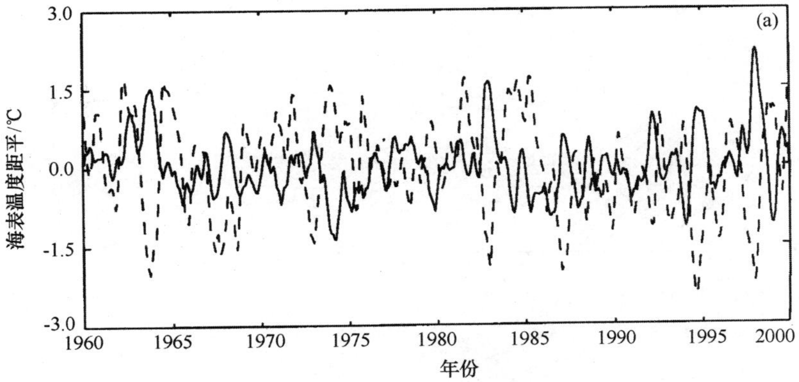

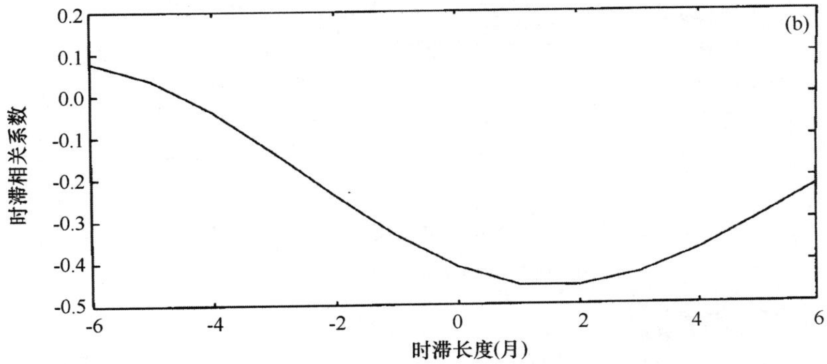

图 2　印度洋次表层海温距平的时间变化

(a)实线:西部(10°S 至 10°N,50°—70°E),虚线:东部(10°S 至 EQ,90°—110°E)

(b)西部与东部两区海表温度时滞相关)

来表示。为探索到与海温偶极子模态相耦合的 Walker 环流,分别作西部指标区的海温距平与 850 hPa 和 200 hPa 纬向风距平的相关图(图 3(a),图 3(b))。可见,与 850 hPa 纬向风距平的主要分布负相关区在(5°S 至 2.5°N,70°—90°E),相关系数超过-0.50,与 200 hPa 纬向风距平的正相关区主要在(12.5°S 至 EQ,70°—90°E),相关系数也超过 0.4。

为了更确切地说明上面高、低空反向的纬向风距平是 Walker 环流的组成部分,可以从图 3 看到,在西部海温距平区的两侧,相关系数的符号相反,这表明当指标区的海温距平为正(负)时,低空纬向距平风是辐合(辐散),而高空是辐散(辐合),也即是说在指标区附近,当海温距平为正(负)时是上升(下沉)运动,与海温距平模态相对应,的确是距平意义下的 Walker 环流。

4　热带印度洋海气相互作用事件

偶极子模态是海洋事件,Walker 环流是大气事件,要构成一个海气相互作用事件,两者之间要有协调的耦合。在图 4(a)中给出西部指标区中温度距平的年际变化(温度距平值增大 2 倍)和 850 hPa 的纬向风距平的年际变化。对于一个相互协调的海气耦合事件,当西部为温度正距平时,相应的低空纬向风应是距平东风,即它们之间应是负相关,而图中两条曲线的相关系数为-0.65,这个负相关系数是相当高的。图 4(b)是西部指标区中温度距平的年际变化和 200 hPa 纬向风距平的年际变化,对一个协调的海气耦合事件,温度正距平应对应为西风距平,这两个曲线的相关系数为 0.44。

由此可见,在热带印度洋也存在类似热带太平洋的 ENSO 事件,为简便称为 D-W 事件。当西部

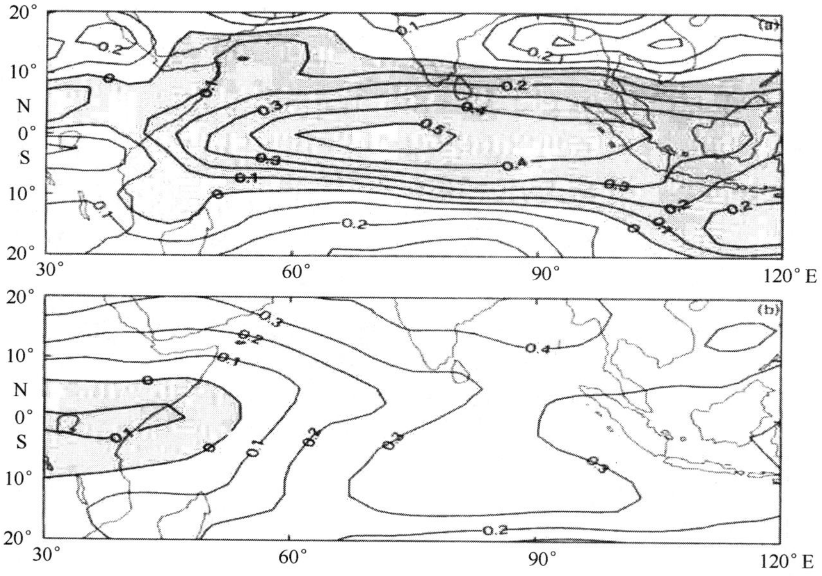

图 3　西部海温距平与印度洋 850 hPa 和 200 hPa 距平纬向风的相关分布

(a) 850 hPa; (b) 200 hPa; 阴影区表示负相关

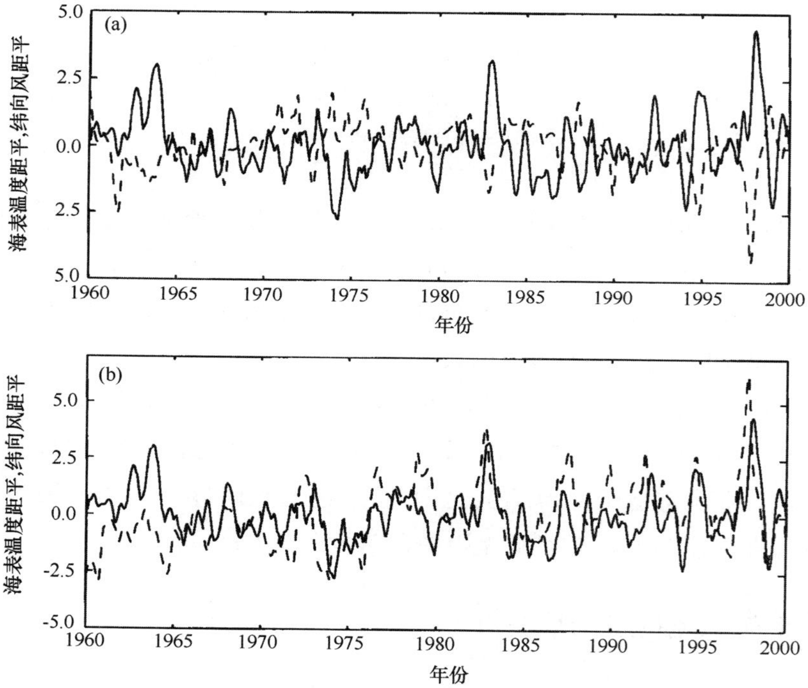

图 4　西部温度距平(增大 2 倍,单位℃)和纬向风距平的年际变化

(a) 850 h Pa; (b) 200 hPa; 实线:温度线;虚线:纬向风距平

的温度距平为正(负)时,称正(负)D-W 事件。对照图 4(a),图 4(b),可以看到,除 1997—1998 年是次强的正 D-W 事件外,1962 年,1967—1968 年,1972—1973 年,1982—1983 年,1987 年,1994 年,都出现正的 D-W 事件。出现负的(或反的)D-W 事件的年份为 1971 年,1973—1975 年,1984—1985 年,

1988 年,1999 年。当然究竟哪些年份是 D-W 事件,像定义 ENSO 事件那样,需在强度上,持续时间给出一定的要求,在这里只初步地说,海气相互作用事件在热带印度洋也是存在的。

5　结语

文中分析表明,在热带印度洋若不用传统的分析海表温度距平的分布和变化,而改用次表层海温资料,在海温距平的极值面上来分析海温距平的分布和变化,则可以清楚地看到,赤道印度洋西部和东部的海温距平分布,在很多年份它们的距平符号是相反的,即在海温距平的西、东方向分布上存在物理意义下的偶极子模态,而海洋中的这种偶极子模态,与上空大气中的距平纬圈环流或距平 Walker 环流,有着很好的耦合关系,亦即这是一类大尺度海气相互作用事件,本质上与热带太平洋的 ENSO 事件是一致的。热带印度洋的海气相互作用事件与热带太平洋的 ENSO 事件也会有一定的联系[6],这将在另文讨论。

参考文献

[1]　Wedster P T, Moore A M, Loschning J P, et al. Coupled ocean-atmosphere dynamics in the Indian ocean during 1997-1998. Nature, 1999, 401: 337-339.

[2]　Saji N H, Goswami B N, Viayachandrom P N, et al. A dipole mode in the tropical Indian Ocean. Nature, 1999, 401: 360-363.

[3]　李崇银,穆明权.赤道印度洋海温偶极子型振荡及其气候影响.大气科学,2001,25(4):433-443.

[4]　巢清尘,巢纪平.热带西太平洋和东印度洋对 ENSO 发展的影响. 自然科学进展,2001,11:1293-1300.

[5]　Chao Jiping, Yuan Shaoyu, Chao Qingchen, et al. A data analysis study on the evolution of the El Niño/La Niña cycle. Adv Atmos Sci, 2002, 19: 837-844.

[6]　Ji Zhengang, Chao Jiping. Teleconnections of sea surface temperature in the Indian ocean with Pacific, and with the 500 hPa geopotential height field in the northern hemisphere. Adv Atmos Sci, 1987, 4: 343-348.

热带印度洋和太平洋海气相互作用事件间的联系*

巢纪平[1] 袁绍宇[2]

(1. 国家海洋局第一海洋研究所,青岛 266071;2. 中国海洋大学,青岛 266003)

摘要:在次表层海温距平极值面上海温距平的分布和变化表明,热带西、东印度洋的海温距平呈偶极子模态,即当西印度洋海温距平为正(负)时,东印度洋海温距平为负(正),偶极子模态的海温距平分布在热带太平洋同样存在,两大洋海温距平的偶极子模态间有密切的联系,在分析它们和850 hPa纬向风距平后指出,不仅这种海温距平偶极子模态的形成、发展是与这两大洋热带上空Walker环流相互作用的结果,同时正是Walker环流异常把两大洋的海温距平变化联系起来。

关键词:印度洋、太平洋海温距平偶极子模态;Walker环流;海气相互作用;次表层海温距平极值曲面

在热带太平洋,自从Bjerknes[1]提出以赤道东太平洋海表温度距平的正、负为表征的年际变化即El Niño/La Niña现象,和以海平面气压场在东、西方向以"跷跷板"形式呈年际变化的南方涛动现象,或纬圈环流即Walker环流的年际变化现象,是一类大尺度海气相互作用现象并称为ENSO事件以来,对ENSO的研究,在观测资料的分析、理论研究和数值模拟等方面均有较大的进展(见参考文献[2,3])。近年来,巢清尘、巢纪平[4]和巢纪平等[5,6]对20世纪60年代以来所有的El Niño和La Niña事件的分析表明,若在次表层构造一个最大温度距平极值曲面(这个曲面接近气候的温跃层面),并在这个面上来分析海温距平的动态行为,则可以对El Niño和La Niña事件的形成、发展过程看得更清楚,特别是热带西太平暖池次表层的温度距平往往是以赤道东太平洋海表温度距平为表征的El Niño和La Niña事件的源地。李崇银等[7]对1997—1998年那次强El Niño事件的分析也指出了这一点。

过去对热带印度洋海气相互作用事件的研究很少。1997—1998年在热带太平洋出现了历史上最强的一次El Niño事件,而在赤道印度洋则出现西高东低的海面温度分布形势。Webster等[8]首先报道了印度洋这一异常事件,Saji等[9]指出,热带印度洋西部的海表温度距平和东部的海表温度距平呈偶极子模态的分布,并定义了一个偶极子模指数*DMI*(西部赤道附近的海表温度距平减去东部赤道附近的海表温度距平),40年海表温度的时间序列表明,*DMI*有正、负相间的年际变化。他们分析了大气资料后又指出,热带印度洋海表温度的偶极子模态是海气相互作用的结果,并是与热带太平洋的ENSO并无关系的独立过程。李崇银等[10]也给出了100年来*DMI*的年际变化,但认为它们与ENSO有关。最近,巢纪平等[11]分析了1961—2001年印度洋海表温度距平后指出,Saji等给出的*DMI*正、负

* 自然科学进展,2003,13(12):1280-1285.

相间的年际变化,实际上在大多数年份是在热带印度洋西部和东部海表温度距平符号相同的情况下由强度不同而造成,像 1997—1998 年那次强事件热带印度洋西、东部海表温度距平符号也是相反的真正偶极子模态是不多的。

另一方面,1987 年季振刚等[12]注意到,印度洋赤道两侧的海表温度的年际变化与赤道东太平洋 130°E 的海表温度的年际变化有着很好的时滞相关关系,吴国雄等[13]和孟文等[14]指出,沿赤道印度洋的季风环流和太平洋上空 Walker 环流间存在"齿轮"式的耦合并致使海洋产生相应的响应使两大洋海表温度变化产生有机的联系。王东晓等[15]最近分析了 1997—1998 年 El Niño 期间热带东印度洋和西太平洋海洋上层热含量之间的联系后,认为印度尼西亚贯穿流在连接这两大洋海温变化中起着作用。

由此可见,在已有的论文中,对印度洋海气相互作用以及与 ENSO 间的关系的认识不尽相同,为此本文将对热带印度洋和太平洋用统一的物理指标,首先来分析发生在这两大洋中的海温距平的年际变化的相关性,进而从海气相互作用的观测探索这两大洋中海气相互作用事件的联系的可能的物理过程。

分析所用资料是 1960—2000 年 Scripps 海洋研究所 0~400 m 5°×2°(纬度×经度)海温再分析资料,及同期 NCEP/NCAR17 层 2.5°×2.5°(纬度×经度)网格点大气资料。

1　热带印度洋的海气相互作用

对于分析热带印度洋的海气相互作用,海表温度距平并不是一个很好的物理量,因为 Saji 等提出的用 *DMI* 表示的偶极子模态,在多数年份是在西、东印度洋的海表温度距平符号相同下振幅的差异。

但如果在次表层海温距平极值曲面来分析海温距平(本文称海温距平极值面上的海温距平为 SM-TA)的分布和演变,结果会有大的改变。现构造一个海温距平极值曲面,它是这样构造的,取每一点次表层最大的温度距平(绝对值)所对应的深度,由这些深度的多年平均值就构成一个深度曲面,此即海温距平极值曲面。图 1 是热带印度洋 4 个季节的海温距平极值深度的分布。若与热带太平洋的温度距平极值曲面相比(见参考文献[5,6]),印度洋温度距平极值所在的深度分布形势恰好与热带太平洋反过来,浅的海域是在热带西印度洋。

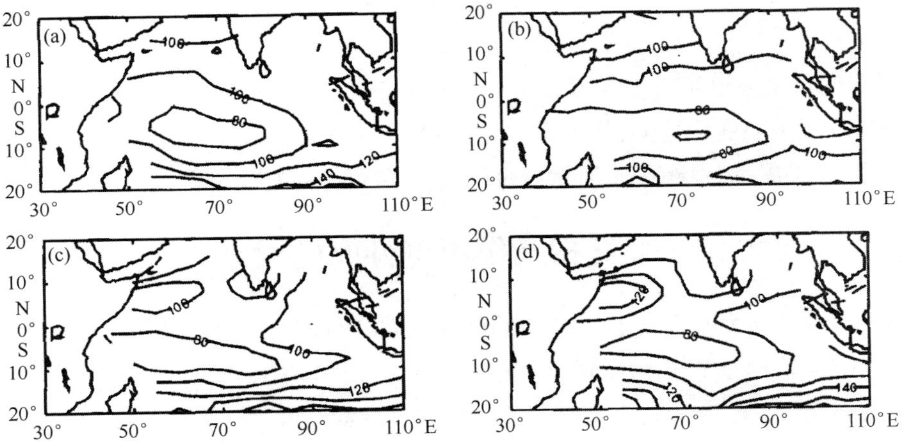

图 1　热带印度洋海温距平极值曲面的深度分布

(a)1 月;(b)4 月;(c)7 月;(d)10 月

在图2((a))中给出西、东印度洋海表温度距平的年际变化,它们基本上呈同向的变化趋势,相关系数为0.24,可见Saji等的资料分析并没有给出真正物理意义下的偶极子模态。但如在极值面上取赤道附近西、东印度洋海温距平的年际变化(图2(b)),它们的变化趋势总体上是反向的,相关系数为-0.46,由此可见,赤道附近西、东印度洋的海温距平在统计上呈偶极子模态的年际变化。应用极值曲面上东、西印度洋在上述指标区的温度差,也构造一条年际变化曲线,这是真正物理含义下的偶极子模态指数,称IDMI(图2(c))。

图2(d)是赤道西印度洋海温距平的年际变化和赤道印度洋中部上空850 hPa纬向风距平的年际变化,这两条曲线的变化趋势是反向的,最大相关系数为-0.66,即当赤道西印度洋为正(负)的海温距平时,赤道印度洋上空850 hPa为距平东(西)风。图2(e)是和200 hPa纬向风距平年际变化间的相关性,它们的变化趋势呈正相关,最大相关系数为0.44。可见印度洋海温距平的年际变化与其上空环流的年际变化有着有机的联系,这时气候的Walker环流减弱(加强),这表明印度洋也存在大尺度海气相互作用现象。为了进一步说明这是一类海气相互作用现象,在图2(f)中给出西印度洋海温距平年际变化和850 hPa及200 hPa纬向风距平年际变化的时滞相关,可见温度场和风场的最大相关均发生在风场超前温度场3个月,而且两者的时滞相关曲线也是反位相的,进一步说明与海温距平模态相对应的的确是距平意义下的Walker环流。

2 热带太平洋的海气相互作用

在众多的研究中,一般都用Niño 3区的海表温度距平来表示El Niño现象,Niño 3区海表温度距平和该区极值面上的海温距平的年际变化,有明显相同的变化趋势,它们的最大相关系数达0.83,因此对于赤道东太平洋用海表温度距平或用次表层海温距平来分析El Niño现象不会有大的区别。

图3(a)是赤道东太平洋极值面上的海温距平的年际变化和西太平洋暖池次表层极值面上的海温距平的年际变化,可见这两条曲线的变化趋势是反向的,相关系数为-0.77。这自然也是海温距平的偶极子模态的表现,把赤道东太平洋极值面上的海温距平减去西太平洋极值面上的海温距平,也构造一个偶极子模态指数,称PDMI(图3(b)),强的正指数是El Niño事件,大的负指数是La Niña事件。如果以温度距平超过4℃作为指标,则1973年,1983年,1987年,1992年,1994年,1997年均为强的暖事件,另外1964,1966,1969,1977,1993也为比较强的暖事件,其发生的时间与印度洋的暖事件具有很强的相关性。

图3(c),图3(d)是赤道东太平洋海温距平和850 hPa,200 hPa纬向风距平的年际变化,它们之间分别呈正相关和负相关的变化趋势,相关系数为0.83及-0.55。这表明,当赤道东太平洋为海温正距平时,低空距平西风发展,高空距平东风发展也即气候的Walker环流减弱。

3 热带印度洋与太平洋海气相互作用事件间的联系

现分析热带印度洋与太平洋海温距平间的联系。对印度洋和太平洋各取两个指标区,西印度洋为(50°—70°E,10°S至10°N),东印度洋为(90°—110°E,10°S至EQ),与Saji所取的位置相同,西太平洋为(130°—155°E,4°至10°N),东太平洋为(90°—150°W,4°S至4°N),对应于Niño 3区。

图4(a)是西印度洋和东太平洋温度距平的时间变化,它们的变化趋势基本同相,相关系数当东太平洋超前西印度洋2~3个月时最大,为0.37。这似乎表明,有一种过程把它们的变化联系起来,由于这两个区相隔甚远,没有一种海洋过程可以把它们联系起来,因此最有可能的是它们对大气环流异

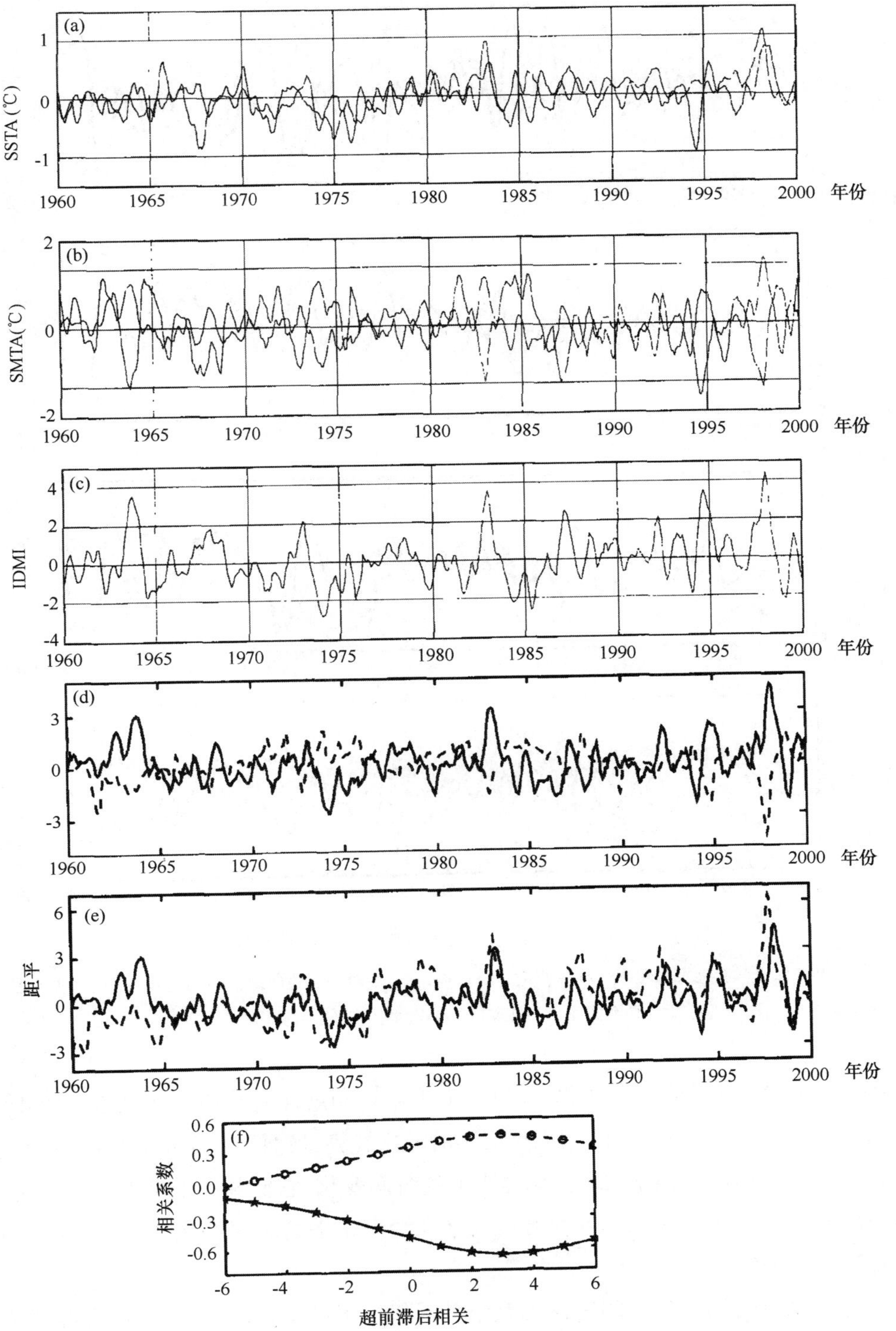

图 2　赤道西、东印度洋海温距平变化

(a)赤道西、东印度洋次表层海温距平年际变化；(b)赤道西、东印度洋海表温度距平的年际变化；(c) IDMI(℃)的时间变化曲线；(d)赤道西印度洋海温距平及 850 hPa 纬向风距平的年际变化；(e)赤道西印度洋海温距平及 200 hPa 纬向风距平的年际变化；(f)西印度洋海温距平年际变化和 850 hPa 及 200 hPa纬向风距平年际变化的时滞相关

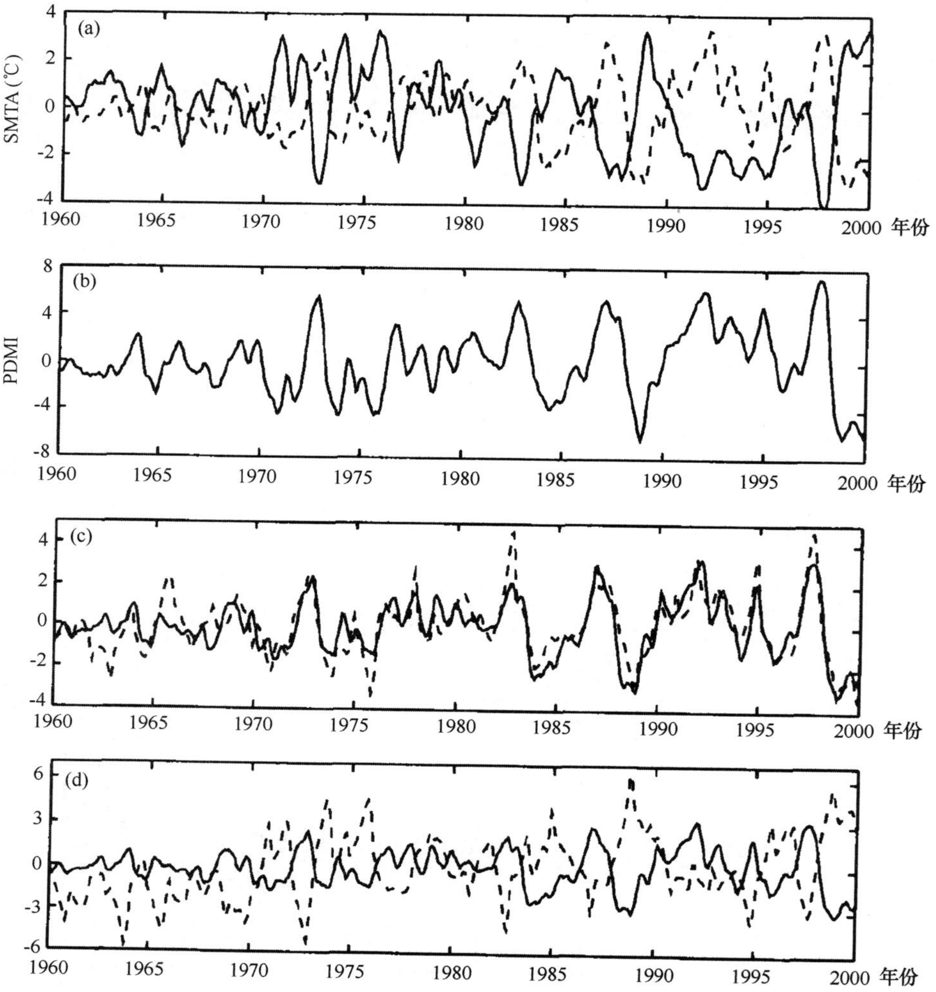

图 3　赤道东太平洋海温距平和纬向风距平的年际变化

(a)赤道东太平洋和西太平洋暖池海温距平的年际变化;(b)PDMI(℃)的时间变化曲线;(c)赤道东太平洋海温距平和 850 hPa 纬向风距平的年际变化;(d)赤道东太平洋海温距平和 200 hPa 纬向风距平的年际变化

常变化的响应,而这种大气环流异常变化,在热带太平洋和印度洋有着内在的联系。图 4(b)是西印度洋和西太平洋温度距平的时间变化,它们呈反相的变化趋势,当西太平洋超前 2~3 个月时相关系数为-0.33。图 4(c)是东印度洋和东太平洋海温距平的时间变化,它们基本上呈反向的变化趋势,同期的相关系数为-0.45。图 4(d)是东印度洋和西太平洋温度距平的时间变化,它们的变化趋势是同相的,同期相关系数为 0.53。

这些相关系数表明,热带西、东印度洋海温距平的偶极子模态和西、东太平洋海温距平的偶极子模态间有着密切的联系,而联系它们的可能是两大洋热带上空 Walker 环流的异常变化。由于这 4 个区中海温距平变化相互的统计规律已经清楚,因此只要分析其中一个海域的海温距平变化和大气环流的变化间的关系就可以了。

图 5 是赤道东太平洋的海温距平变化与 850 hPa 纬向风距平变化之间同期的相关系数分布。以赤道东太平洋海温距平为正值时讨论,可以看到,在热带太平洋很大范围内是距平西风,这正是维持 El Niño 所需要的大气风场条件。而在热带印度洋是距平东风,这也正是维持西印度洋为正值海温距

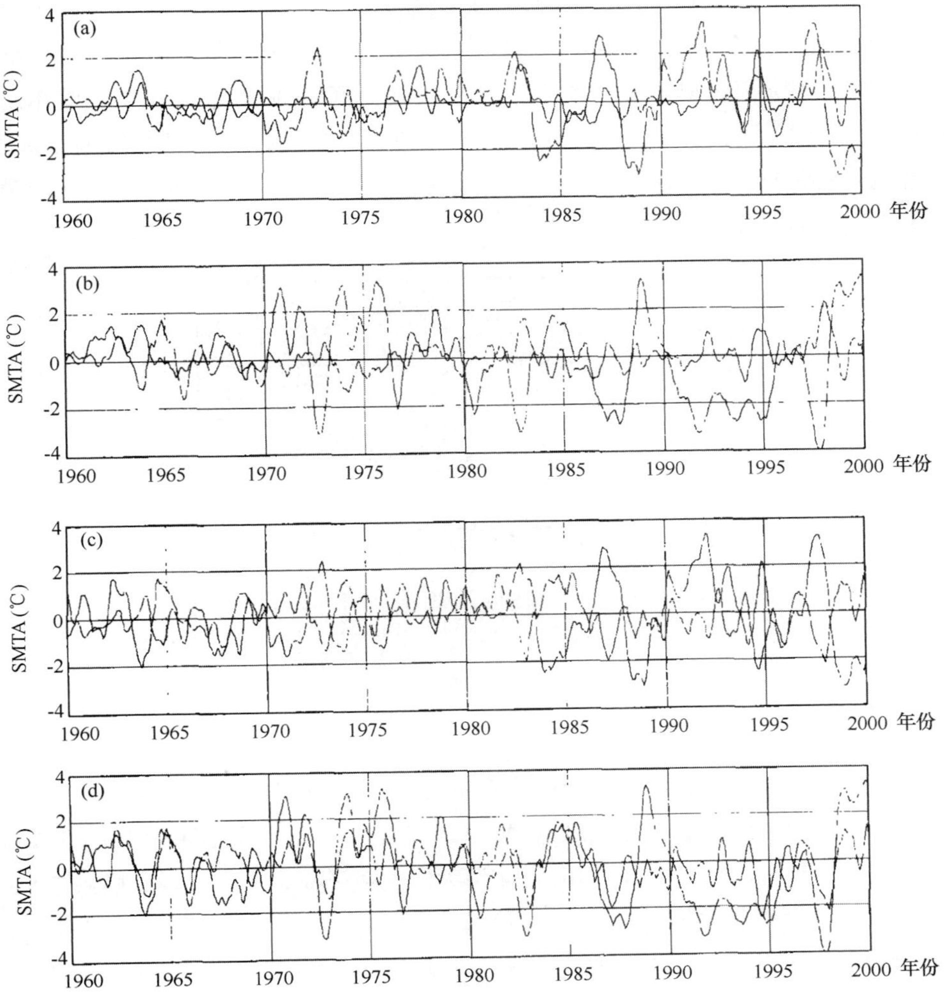

图 4　印度洋和太平洋温度距平的年际变化

(a)西印度洋和东太平洋温度距平的年际变化;(b)西印度洋和西太平洋温度距平的年际变化;(c)东印度洋和东太平洋温度距平的年际变化;(d)东印度洋和西太平洋温度距平的年际变化

平所需的大气风场条件。由纬向风距平在两大洋上空的分布可以看到,这时在太平洋和印度洋都是与气候 Walker 环流反向的距平 Walker 环流。这表明,赤道东太平洋和西印度洋都处在正的海温距平状态,是借减弱气候的 Walker 环流来实现的。因此必然有从副热带纬度来的气流,在赤道附近流向太平洋加强那里的距平西风,同时也有气流流向印度洋加强那里的东风距平。

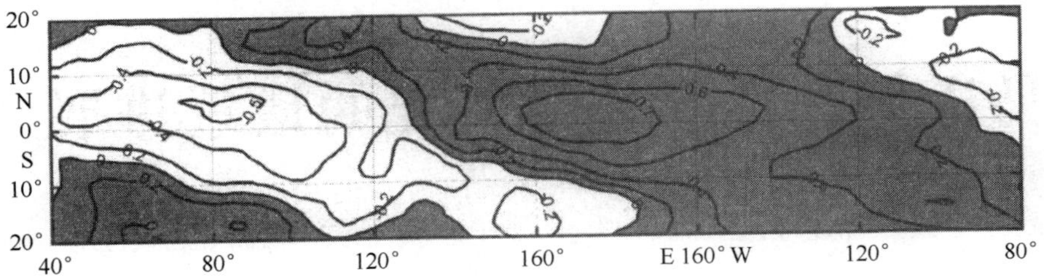

图 5　赤道东太平洋的海温距平变化与 850 hPa 纬向风距平变化之间同期的相关系数分布

图6是1997年7月850 hPa的距平风场,可以看到,在西、中太平洋赤道两侧的南、北半球各有一个气旋性环流,它们的气流在西太平洋向赤道辐合,并共同流向中太平洋,使那里的距平西风大大加强。类似的,在东、中印度洋赤道两侧分别有一个反气旋环流,它们的气流在东印度洋向赤道辐合后流向中印度洋,使那里的距平东风加强。这两大洋的环流,太平洋强于印度洋,南半球强于北半球。这样的低空环流形势是对上面相关场的一个很好的注释。这种低空异常环流形势在赤道东太平洋和西印度洋都处在暖态时是带有共性的。

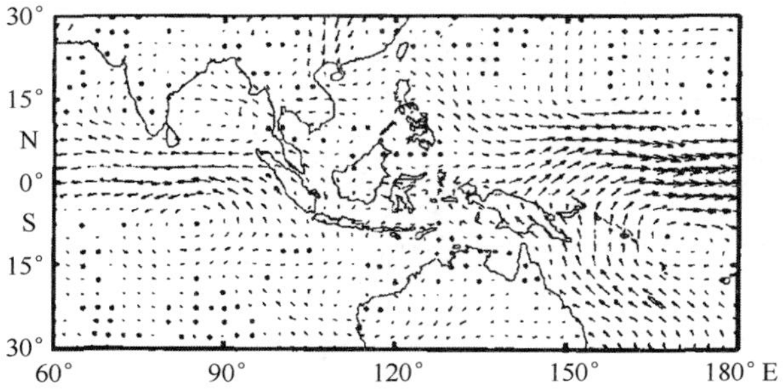

图6 1997年7月太平洋,印度洋850 hPa的距平风场

4 结论

应用次表层海温极值面上海温距平的分布和变化,指出热带西、东印度洋和热带东、西太平洋都存在偶极子模态的海温距平分布,它们和上空 Walker 环流的异常变化有着密切的联系,也正是通过减弱热带太平洋和印度洋气候的 Walker 环流,使两大洋偶极子模态的海温距平联系起来的。

需要指出,在文中制作相关场时,是固定某一海域的海温距平后对风场作出不同时间的相关场的(时滞相关),但这并不意味在海温与风场之间海温的变化总是主导的,而相隔距离这么远的两大洋海温距平会有密切联系的变化,说明在相当程度上海温变化与大气环流变化间存在大尺度相互响应的过程。

参考文献

[1] Bjerknes J. Atmospheric teleconnections from the equatorial Pacific. Mon Wea Rev, 1969,97:163.

[2] Neelin J D. et al. ENSO theory. J Geophys Res, 1998,103:14261.

[3] 巢纪平．厄尔尼诺和南方涛动动力学. 北京．气象出版社,1993:309.

[4] 巢清尘,等．热带西太平洋和东印度洋对ENSO发展的影响．自然科学进展,2001,11(12):1293.

[5] 巢纪平,等．热带西太平洋暖池次表层暖水的起源——对1997/1998年ENSO事件的分析．大气科学,2003,27(2):145.

[6] Chao Jiping, et al. A data analysis study on the evolution of the El Niño/La Niña cycle. Adv Atmos Sci,2002,19(5):837.

[7] 李崇银,等．赤道印度洋海温偶极子型振荡及其气候影响．大气科学,2001,25(4):433.

[8] Webster P T, et al. Coupled ocean-atmosphere dynamics in the Indian Ocean during 1997 - 1998. Nature, 1999, 401:337.

[9] Saji N H, et al. A dipole mode in the tropical Indian Ocean. Nature, 1999, 401:360.

[10] 李崇银,等．印度洋海温偶极子和太平洋海温异常．科学通报,2001,46(20):1747.

[11] 巢纪平,等. 热带印度洋的大尺度海气相互作用事件. 气象学报,2003,61(2):251.

[12] Ji Zhengang, et al. Teleconnections of sea surface temperature in the Indian ocean with Pacific, and with the 500 hPa geopotential height field in the northern hemisphere. Adv Atmos Sci, 1987,4:343.

[13] 吴国雄,等. 赤道印度洋-太平洋地区海气系统的齿轮式耦合和 ENSO 事件. I 资料分析. 大气科学,1998,22(4):470.

[14] 孟文,等,赤道印度洋-太平洋地区海气系统的齿轮式耦合和 ENSO 事件. Ⅱ数值模拟. 大气科学,2000,24(1):15.

[15] 王东晓,等. 1997—1998 年厄尔尼诺期间印度洋和西太平洋上层海洋的联系. 自然科学进展,2003,13(9):957.

热带西太平洋和东印度洋对 ENSO 发展的影响*

巢清尘[1] 巢纪平[2]

（1. 国家气候中心,北京 100081;2. 国家海洋环境预报中心,北京 100081）

摘要:分析了从 1955—1998 年 44 年热带西太平洋次表层海温距平资料,指出在这一期间共发生了 11 次 El Niño 事件,这些事件的初始海温正距平无一例外地都在“暖池”次表层 160 m 附近生成,发展到一定强度,到西太平洋出现异常的持续西风距平后,暖的海温距平沿着气候的温跃层向东向海表传播,经过 1 年左右的时间,传播到赤道东太平洋海表的 El Niño 3 或 1、2 海域形成通常认为的 El Niño 事件。进一步分析 1960—1999 年的 850 hPa 风场指出,在热带西太平洋西风距平暴发前,赤道东印度洋已经有距平西风持续维持了 2 年左右(西南季风),然后距平西风快速地向东发展产生热带西太平洋的距平西风,或真正的强西风,致使西太平洋出现 El Niño 暴发不可缺少的条件之一,在这 40 年中,这一过程对所有的 El Niño 事件无一例外。同样的分析方法也用到这 44 年中的 La Niña 事件,冷的海温距平也都首先在“暖池”的次表层形成,然后沿着温跃层向东向表层传播,在热带西太平洋出现东风前,印度洋也早已是持续的东风。

关键词:El Niño/La Niña 事件;暖池次表层海温距平;印度洋西风

近年来,科学家普遍认为发生在大气-海洋这个耦合系统中的物理过程,是对全球气候影响最为重要的物理过程之一。越来越多的资料分析和数值模拟都已表明 El Niño/南方涛动事件(ENSO)的发生不仅直接影响到热带地区的大气环流和气候,而且通过遥相关过程随后也影响到全球范围的大气环流和气候变化,从而造成世界很多地方严重的旱涝与异常的温度变化,给许多国家和地区的工农业生产带来重大损失。

近半个世纪以来,由于海洋观测资料在质量和数量上都有重大的改善,使人们可以进一步来分析 El Niño/La Niña 发生时的海洋实际状态,为此本文将对 1955—1998 年热带太平洋的表层和次表层海温资料作一全面分析,希望能对 El Niño 初始正、负温度的源地和发展过程有一个较为全面的、规律性的认识。类似地,通过对全球 850 hPa 风资料的分析将对 El Niño 事件暴发时西(东)风的发生、发展过程有一个较为全面的、规律性的认识。

1 资料

本文所用的资料为国家气候中心资料室提供的美国 Scripps 海洋研究所联合环境分析中心(JEDCA)的海表和次表层海温观测资料。这是根据投弃式温深计(XBT)、船舶和浮标观测资料经过最优插值

* 自然科学进展,2001,11(12):1293-1300.

方法得到的，包括 11 个标准层上的海温距平（0，20 m，40 m，60 m，80 m，120 m，160 m，200 m，300 m，400 m），水平分辨率为 5°×2°（纬度×经度），范围包括 30°E—180°—30°E（360°），60°N 至 60°S。资料的时间长度从 1955 年 1 月到 1998 年 12 月共 44 年。

风的资料取自国家气候中心资料室提供的美国国家大气研究中心国家环境计划委员会（NCEP/NCAR）全球月平均再分析数据集，包括全球范围的 17 层三维风场资料，分辨率为 2.5°×2.5°，资料年代为 1960 年 1 月至 1999 年 12 月，为消除年际观测误差将上述资料做了距平计算。为消除高频观测误差或高频振荡，对上述所有资料在时间演变分析时做了 5 个月的滑动平均。

2　热带太平洋次表层海温演变特征

首先对热带太平洋沿赤道各个经度的海温距平的垂直分布做了分析。在图 1（a）中给出暖池（140°—160°E，4°S 至 4°N）从 1955 年到 1998 年 1 月，4 月，7 月和 10 月海温距平的均方根值（$\sqrt{\sum_j (T_j-\overline{T})^2/N}$）的垂直分布。可见 1 月，10 月次表层温度距平均方根最大值出现在 120 m 处，4 月、7 月最大值出现在 160 m 附近。为了比较，同时给出 Niño 3 区相类似的分布（图 1（b）），从这一组图中看到，4 个月份的温度距平最大值均出现在次表层 100 m 以上，大部分出现在次表层 40 m 附近。实际上这说明海温距平最大的深度在气候的温跃层附近，由此可见，在热带西太平洋分析次表层 150 m 附近的海温变化特征远比分析海表温度变化来得重要；而在热带东太平洋，次表层 40 m 附近的海温距平值和该处的海表温度值几乎差不多，因此只要分析海表温度就可以了。

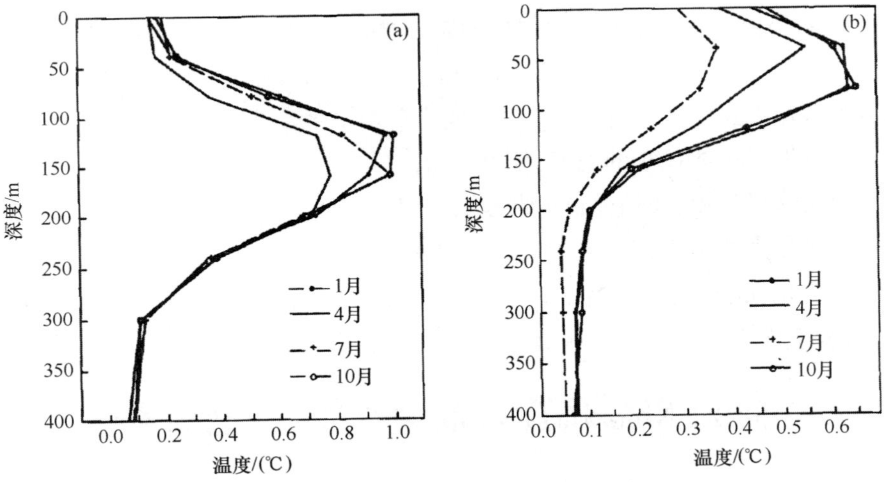

图 1　1995—1998 年海温距平的均方根值的垂直分布

（a）暖池；（b）Niño 3 区

这两个区域温度的年际变化表明（图略），对于 El Niño 年，西太平洋暖池的次表层 160 m 深处温度的最大值与 Niño 3 区海表温度的最大值具有 0.5～1 年左右的时滞。当暖池次表层温度超前时，这两个海域温度的正相关系数达到置信度 99%要求（置信度 99%的相关系数为 0.114 5）的时间约为 10 个月以后到 2 年左右，最大相关系数值出现在暖池超前 Niño 3 约 1.5 年左右。另一方面，分析还表明，在赤道东太平洋正（负）的极大值出现 2 年左右后，暖池次表层海温又将出现正（负）的极大值，如果这些相关系数的极值主要由 El Niño/La Niña 事件造成，则表明一个 ENSO 循环的时间为 3～4 年。

1955—1998 年赤道东太平洋 Niño 3 区域平均的海表温度距平的年际变化表明，从 1955—1998 年的44年中，赤道中、东太平洋发生了11次很明显的增暖事件，即1957/1958年，1972/1973年，1976/1977年，1982/1983年，1986/1987年，1991/1992年，1994/1995年和1997/1998年，以及大家认可的1963/1964年，1965/1966年和1969/1970年，而这些海表温度增暖年大部分都已达到El Niño事件的标准，其中1982/1983年和1997/1998年的过程为近50年来最强的两次过程。由Niño 3区的温度年际变化同时可知，还出现了11次很明显的冷事件，即1955/1956年，1961/1962年，1964/1965年，1967/1968年，1971年，1973/1974年，1978年，1984/1985年，1988/1989年，1995/1996年和1998年冬季，这些年也都发生了El Niño事件。注意到在这些El Niño事件发生0.5~1年，赤道西太平洋暖池次表层160 m深处的温度也都有明显的增暖。而在上述大部分La Niña事件发生前1~1.5年多的时间之前，赤道西太平洋暖池次表160 m深处的温度也都有明显的变冷。由此证实西太平洋暖池次表温度的变化早于东太平洋表层温度的变化。

3 1997/1998年次表海温距平的演变特征

1997/1998年是10年来最强的一次El Niño过程，研究并分析格点海温资料的结果表明，早在1996年就有异常暖水在赤道西太平洋滞留，到1997年春，最大值达到3.4℃，而这时东太平洋则还残留着一个负海温距平。这时暖池次表层的海温正距平沿着气候的温跃层向东传播。1997年夏，正距平已经移到了中东太平洋，并且强度达到了8℃，这时20℃等温线下降到了100 m左右，几乎使东太平洋的混合层厚度增加了一倍，海表温度剧烈上升，形成了一次异常猛烈的El Niño事件。

随之，中、西太平洋次表层有一个冷距平发展，并跟随着海温正距平向东移动并发展，即一次La Niña正在形成。随着负海温距平沿着斜温层的向东传播，1998年夏季整个次表层海温距平都出现了负距平，中心最大值为-7℃，位于170°—110°W之间，使La Niña那里的斜温层抬升到50 m的深度。结果在这一地区海表温度正距平迅速减弱，并出现了负距平的海表温度。直到1999年初，次表层海温距平仍然维持在日期变更线以东的地区，使事件得以发展。

为了进一步说明这个过程，我们分析了次表层海温在不同深度的发展。分析表明，1997年3月次表层160 m海温距平最大值出现在赤道170°E至170°W附近区域，之后迅速向东向表层移动，4月最大值已移到了次表层80 m，即对应赤道120°W附近和次表层40 m，即对应赤道110°W附近，此时表示一次El Niño过程全面生成。

4 1955—1999年间El Niño和La Niña过程的次表海温演变规律

1997/1998年的El Niño事件表明，海表温度的正距平是先在西、中太平洋的暖池次表层150 m深附近的温跃层发生，在那里发展后，沿着气候的温跃层向东传播，并在东海岸附近上升至海表。为此，我们利用1955—1998年44年的海洋资料分析了11次El Niño事件的次表层海温变化。结果表明，所有个例都是先在次表层160 m赤道西太平洋处生成暖水，之后，暖水不断向东、向表层传播，最后到达东太平洋表层，形成一次El Niño事件。图2为11个个例的合成图（将11个例子各次表层的最大温度距平值加在一起再平均），这组图很清楚地反映了El Niño事件中温度正距平的上述变化规律。当然，各次事件从暖池次表层160 m到达东边各层所需的时间是不同的。平均来讲，从160 m传到120 m的时间比较快，为1.64个月，再传到次表层80 m用了2.18个月，传到次表层40 m用了1.27个月，而后又用了3.46个月传到海洋东部的海表层。同样我们也依此方法做出11个La Niña事件相应的图，可看到它们初始的冷水也是在热带西太平洋次表层出现后，沿温跃层向东传播，并最终在赤道东太平洋表层出现大面积冷水。东传到上述各层的平均传播时间分别为:1.64个月,0.91个月,3个月和3.55个月。

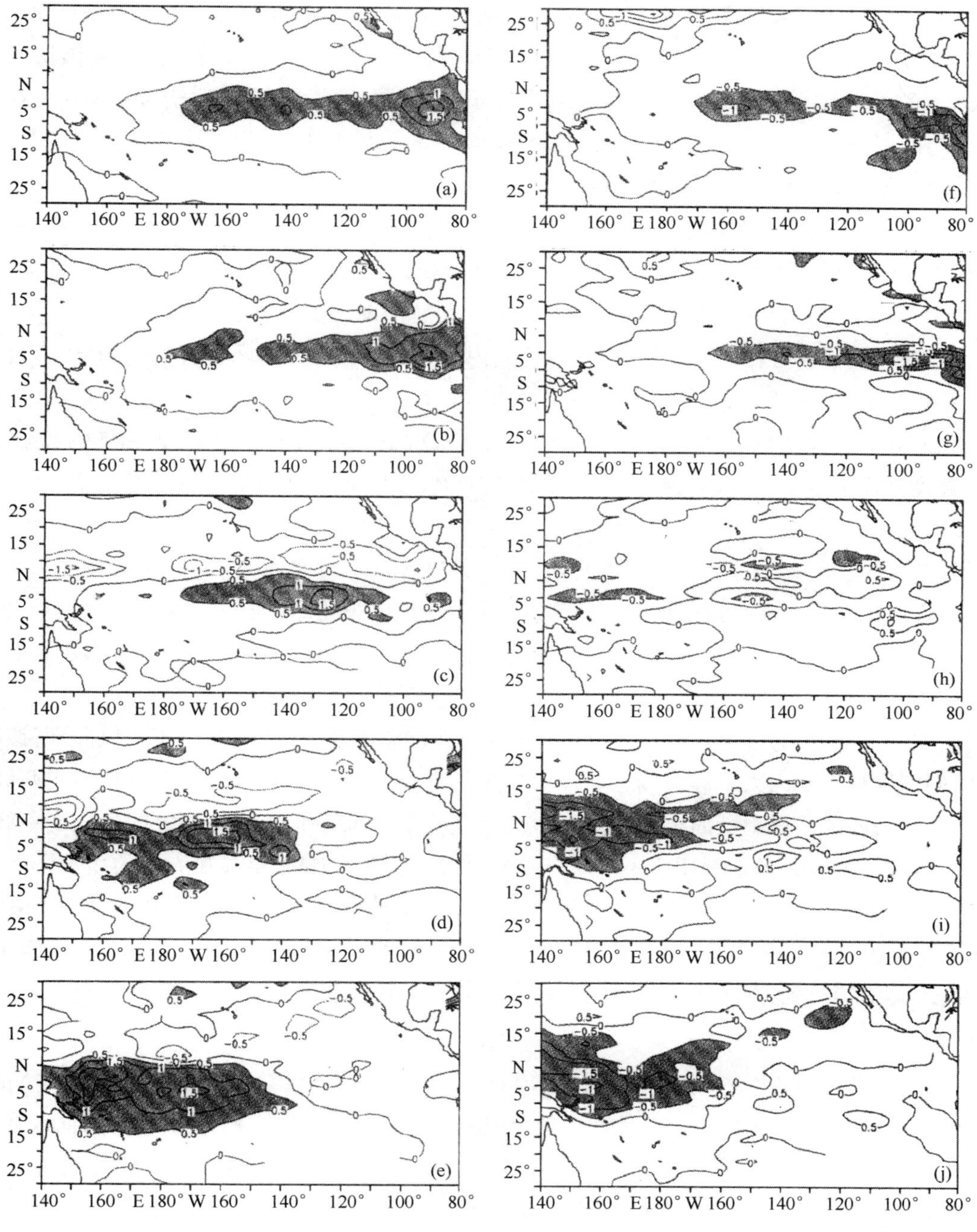

图2　11个El Niño(a—e)和11个La Niña(f—j)事件次表层海温变化合成图

El Niño事件是以赤道东太平洋海表温度增暖来定义的,分析表明,El Niño和La Niño生成的初始正、负距平温度首先出现在西太平洋次表层160 m,因此可以从那里的海温变化中发现先期信号。

5　ENSO事件与前期印度洋西风异常的关系

国内外许多专家的研究都表明西风异常与ENSO的暴发有着密切的关系。在Wyrtki[2]关于El

Niño 发展的信风张弛理论中已提出了西风暴发的必要性。Rasmusson 和 Carpenter[3] 发现每次 El Niño 发生前冬春季西太平洋有西风异常。黄荣辉、张人禾[4] 分析了 1980—1994 年发生的 3 次 ENSO 事件，指出它们的发生与赤道太平洋上空对流层下层纬向风的出现有关。丁一汇[5] 分析了 TOGA-COARE IOP 时期(1992 年 11 月至 1993 年 2 月)大尺度流场的特点，提出强西风大范围的连续出现或暴发是 El Niño 事件发生的关键条件。看来赤道西风异常与西太平洋暖池温跃层加深是 El Niño 发生前两个必不可少的前提条件。本文对 44 年 10 次 El Niño 事件的分析表明，它们形成的前期热带中、西太平洋低空均出现了 3 个月以上的持续距平西风。

在本文中，分析的重点在热带东印度洋风场对热带中、西太平洋距平西风出现的诱导作用。

分析 1960—1999 年纬向风场沿 5°N、赤道、5°S 的时间-经度剖面图知道，一个最明显的特点就是西(东)风异常在东印度洋-太平洋范围内具有很好的交替出现规律性，其“周期”一般是 3~4 年，这种规律性清楚地表明了只有在东印度洋-太平洋范围内才会出现纬向风年际交替出现的异常，可见东印度洋-太平洋为 El Niño 事件的产生提供了很好的环境。

对比 El Niño 事件和 La Niña 事件发生的年份还发现，它们在沿赤道附近，即沿 5°N、赤道和 5°S 的时间-经度剖面上西(东)风加强、发展和成熟的年份与 El Niño 和 La Niña 事件出现的年份完全一致。

同时我们还发现一个重要的、新的事实是，在每次 El Niño 发生的前 1~2 年，在东印度洋均已有西风存在，当该地区西风存在一段时期如 1~2 年后，在 140°E 以东的赤道西太平洋西风突然暴发。暴发后的西风基本上是随时间由西向东移动的。为此我们相信东印度洋在触发西(东)风的发展上具有重要作用。

分析 1997/1998 年的过程表明，早在 1996 年春天在东印度洋 60°E 就酝酿了西风异常，该西风异常随着时间的推移不断向太平洋移动，到 1996 年冬季在赤道西太平洋 120°E 上空已形成西风距平。热带西太平洋的西风距平不断向东扩展，到 1997 年夏秋季西风距平已越过 180°的日期变更线，一次大的 El Niño 过程暴发。该西风距平不断向东发展，1997 年冬至 1998 年春在赤道的太平洋上形成了最强的西风距平中心。这次的西风距平范围和强度都大大超过了 1982/1983 年的过程，El Niño 衰减后，西风距平依然在赤道东太平洋存在。从 1998 年初热带西太平洋地区已开始出现东风距平，东风距平不断发展并东移，到 1998 年秋冬季已跨过 180°日期变更线到达热带东太平洋，这时一次 La Niña 过程也开始了。

事实上上述特征对 1960—1999 年发生的 10 次 El Niño 事件都很类似。为此我们对 El Niño 暴发前后各 24 个月，做了 10 次 El Niño 事件发生的合成图(图 3)。通过分析 El Niño 事件纬向风在不同纬度的合成变化，可以看到在南北纬 10°范围内，从 El Niño 事件发生的前两年西风异常就在东印度洋开始出现并维持，并在一定月份突发性地向赤道西太平洋推进，即在很短的时间内，西太平洋突然吹起西风，然后西太平洋的西风不断向东发展，最后在日界线附近西风异常达到最大，使得 El Niño 事件全面暴发。为此，我们有理由认为，赤道附近西太平洋的西风异常主要是由于印度洋西风异常迅速东移引起。由分析经向风变化可知，东印度洋的西风实际上是西南季风的反映。

下面分析印度洋在 La Niña 事件发展中的作用。以同样的方法，我们确定了 10 次 La Niña 事件暴发的有效月份的前后 24 个月，做了 10 次 La Niña 事件发生的合成图(图 3)。可以看到 La Niña 事件发生前后的风场变化与 La Niña 的风场变化有类似的特征，只是符号相反。在南北纬 10°范围内，在 La Niña 事件发生前 1.5 年，东风异常就在东印度洋酝酿出现，东风异常随时间的推移不断向东发展，也基本在日界线附近东风异常达到最大，这时一次 La Niña 事件发生。

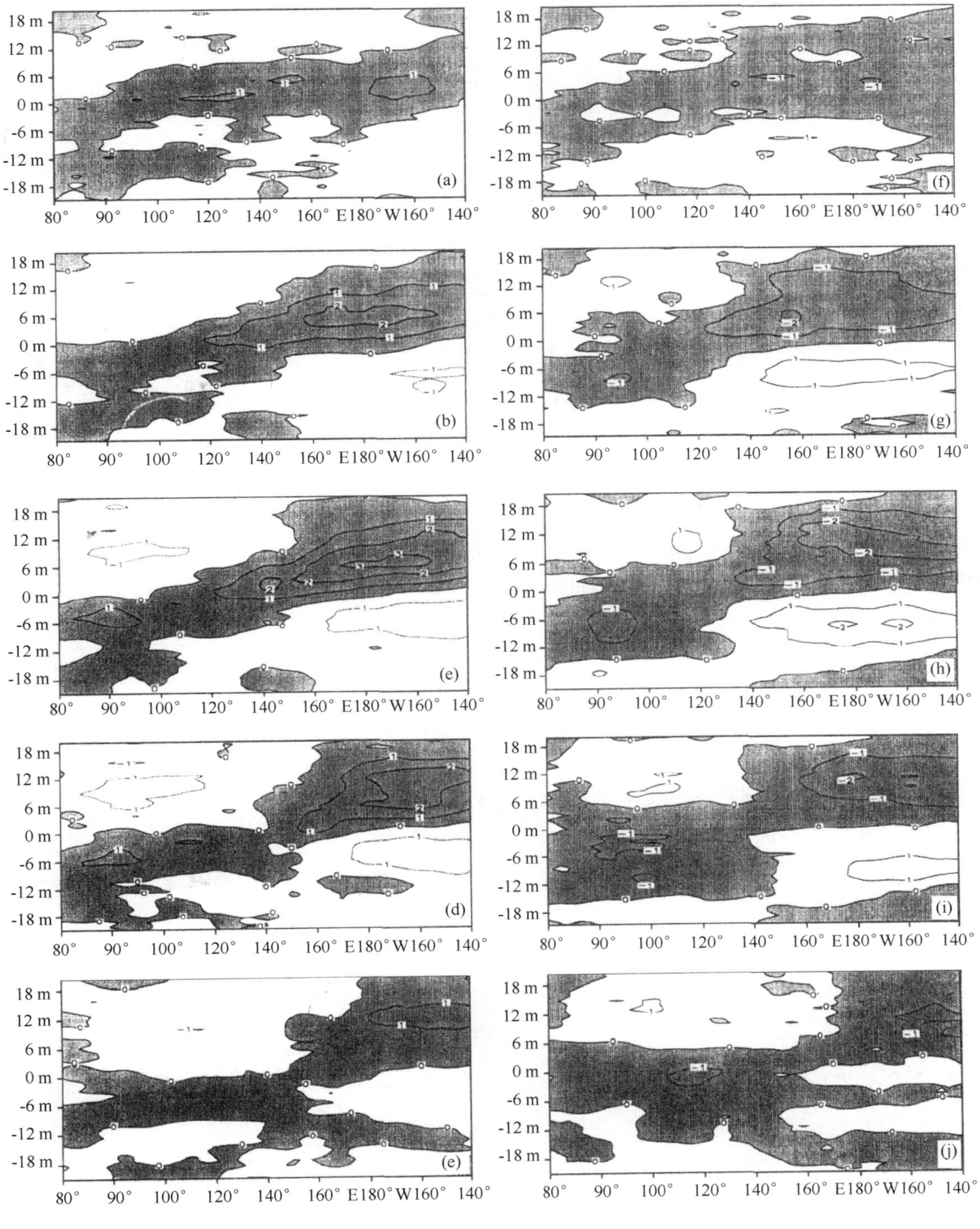

图 3　El Niño 和 La Niña 暴发前后 24 个月纬向风场的合成图

6　总结

本文的分析表明：

(1)近半个世纪发生的 11 次 El Niño 事件和 11 次 La Niña 事件，其初始的正、负海温距平都首先在暖池 160 m 深的温跃层附近生成并发展。

(2)当热带西(东)风或距平西(东)风出现后,这一温度距平沿着气候的温跃层向东、向表层传播,使赤道东太平洋的表层出现暖/冷水,于是一次 El Niño/La Niña 形成。

(3)印度洋对热带太平洋西风的暴发具有重要的作用,引起 El Niño/La Niña 发生的西(东)风的源地均在热带东印度洋。

(4)在 El Niño/La Niña 事件中,印度洋东部西/东风的东传不完全是渐变的,还存在一个突变的过程。在 El Niño 事件暴发约 2 年前,东印度洋的西风异常出现,经历约 1 年的时间后,在西风传播到 120°—140°E 时,发生一个向北的突变,西风距平增大并向东北传播。La Niña 事件中的东风传播同样有这样的现象,只是东风异常出现的时间要略短于 2 年,约为 1.5 年。

本文认为需进一步研究为什么所有的 El Niño/La Niña 事件的初始海温距平都起源于暖池次表层,物理过程和条件是什么?为什么在热带西太平洋西(东)风暴发前,在东印度洋已有西(东)风存在,那里的异常风的起因是什么,它与热带西(东)风又是如何联系的,这些都需做更多的研究。

参考文献

[1] Pan Y H, et al. Global climate variations connected with sea surface temperature anomalies in the Eastern Equatorial Pacific Ocean for the 1958-73 period. Mon Wea Rev, 1983, 111:1244.

[2] Wyrtki K. El Niño-the dynamic response of the Equatorial Pacific Ocean to atmosphere forcing. J Phys Oceanogr, 1975, 5:572.

[3] Rasmusson E, et al. Variations in tropical sea surface temperature and surface wind fields associated with the Southern Oscillation/El Niño. Mon Wea Rev, 1982, 110:354.

[4] 黄荣辉,等.ENSO 循环与东亚季风环流相互作用过程的诊断研究.赵九章纪念文集,叶笃正主编. 北京:科学出版社,197,93-109.

[5] 丁一汇.TOGA-COARE IOP 时期大尺度流场的分析.气象学报,1998,56:284.

热带太平洋 ENSO 期间的海气相互作用分析*

——大气环流无旋和无辐散分量的年际变化

于卫东[1] 巢纪平[1,2]

(1. 国家海洋局第一海洋研究所,海洋环境科学和数值模拟国家海洋局重点实验室,青岛 266061;2. 国家海洋局海洋环境预报中心,北京 100081)

摘要:利用 NCEP/NCAR 再分析资料,将风场分解为无辐散分量和无旋分量两部分,通过求解球面 Poisson 方程得到大气的扰动流函数和扰动速度势。利用 850 hPa,200 hPa 扰动流函数和速度势分别与 Niño-3 指数做相关分析,研究了 Walker 环流、Hadley 环流水平分量的变化和上升/下沉分支的相应位置变化,从而得到了 ENSO 期间大气环流变化的完整图像。

关键词:ENSO;Walker 环流;Hadley 环流;流函数和速度势

El Niño 是发生在热带太平洋的年际变化现象,南方涛动(southern oscillation)是热带大气的年际变化现象,Bjerkines[1,2]最早将两者通过大尺度海气相互作用联系起来,从而形成了 ENSO 的概念。自此之后,大气和海洋科学家围绕 ENSO 的观测、资料分析、理论和数值模拟做了大量的工作[3],对 ENSO 建立了比较清晰的概念性框架,并成为 20 世纪 80 年代以来海洋科学取得的最重大成果之一。

Walker 环流和 Hadley 环流变化是 ENSO 循环中海气相互作用的重要组成部分,目前对于它们已有充分的定性认识[4],但是从资料中揭示 ENSO 循环中 Walker 环流和 Hadley 环流变异过程的工作仍然不足,而且以前的海气相互作用资料分析多关注大气纬向风分量的变化[5]。由于 Walker 环流和 Hadley 环流在本质上是热力驱动环流,与大气的辐合辐散密切联系,因此有必要将大气运动分解成无辐散(或有旋)和有辐散(或无旋)两部分来研究。最近 Wang[6]采用有辐散大气运动分量对与 ENSO 相关的大气环流分量,包括 Walker 环流、Hadley 环流、Ferrel 环流以及中纬度西风急流给出了一个较为全面的描述。虽然无辐散分量并不直接对应着大气对海洋加热的响应,但是它是 ENSO 循环中大气环流调整的重要组成部分,而且在大气对海洋的反馈过程中通过 Ekman 抽吸起到重要作用,因此本文针对太平洋区域大气运动有旋和有辐散分量(以流函数和速度势表征)在 ENSO 期间的变化以及和 Niño-3 区海温异常的关系给出一个全面的描述,从而给出了 ENSO 循环中大气环流变化的清晰、全面的图像。

1 资料及其处理

本文所使用的资料包括 NCAR/NCEP 再分析大气资料[7]中的 850 hPa 和 200 hPa 高度上的月平均风场和太平洋 Niño-3 区月平均海表温度(SST),首先根据原始资料计算风场和 SST 的气候平均值,

* 自然科学进展,2004,14(8):917-924.

然后将逐月资料减去对应月份的气候平均值得到月距平资料。

1.1 风场资料分解的原理

根据 Helmholtz 定理[8]，速度场 $\boldsymbol{V}$ 可以分解为无辐散分量 $\boldsymbol{V}_\psi$ 和无旋分量 $\boldsymbol{V}_\phi$ 两部分，即

$$\boldsymbol{V} = \boldsymbol{V}_\psi + \boldsymbol{V}_\phi \tag{1}$$

其中，$\nabla \cdot \boldsymbol{V}_\psi = 0$，$\nabla \times \boldsymbol{V}_\phi = 0$。

对于二维速度场无辐散分量可以用流函数 ψ 表达，即

$$\boldsymbol{V}_\psi = k \times \nabla \psi \tag{2}$$

在 Cartesian 坐标系下的表达式为

$$u_\psi = -\frac{\partial \psi}{\partial y}, \quad v_\psi = \frac{\partial \psi}{\partial x} \tag{3}$$

类似的，无旋部分可以用速度势 ϕ 表达，即

$$\boldsymbol{V}_\phi = \nabla \phi \tag{4}$$

在 Cartesian 坐标系下的表达式为

$$u_\phi = \frac{\partial \phi}{\partial x}, \quad v_\phi = \frac{\partial \phi}{\partial y} \tag{5}$$

容易验证

$$\zeta = \Delta \varphi \tag{6}$$

$$D = \Delta \phi \tag{7}$$

其中上式中的 ζ 为旋度，D 为辐散。

式(6)，式(7)的左端量可以通过 NCAR/NCEP 再分析大气资料计算得到，从而问题变成了求解两个 Poisson 方程。

1.2 Poisson 方程的球面解

虽然本文只关心热带区域，但是由于有限区域的风场分解问题需要求解两个耦合的 Poisson 方程，解不具有唯一性，虽然 Bijlsma 等[9]对此问题提出了解决办法，但是本文没有采用该方法（原因参考附录），转而直接在整个球面上处理风场的分解和重构。因为对于球面情况，风场分解具有唯一性，重构也非常直接，关于这个问题的详细讨论参见附录。另外值得注意的是，如果解上述 Poisson 方程时简单地取零边界条件，将会得到一个错误的解，这是因为椭圆型方程对边值具有很强的敏感性，边界误差会入侵到整个区域。对于球面 Poisson 方程求解算法本文采用经典的 Fishpack① 算法软件包。

2 结果与分析

2.1 大气环流和 SST 的相关分析

图版 A 是 850 hPa，200 hPa 高度上大气扰动流函数和 Niño-3 指数（定义为 Niño 3 区 SST 的归一化距平）的相关系数。可以看出，太平洋区域的低空大气在距平意义上被一对气旋式环流（应当注意，由于科氏参数在南北半球反号，所以南半球的反气旋流函数，在动力学效应上它相当于北半球的气旋

① Adams J, et al.Fishpack—A package of Fortran subprograms for the solution of separable elliptic partial differential equations.1980, available from: http://www.netlib.org/fishpack.

性环流使海洋温跃层产生抬升，因此本文称其为气旋式结构）所控制，它们分别位于赤道南北两侧。类似的，太平洋区域的高空大气在距平意义上被一对反气旋式环流所控制。在 850 hPa 高度上，北半球的最大相关系数超过 0.4，南半球的最大相关系数超过 0.3；在 200 hPa 高度上，北半球的相关较弱，最大相关系数超过 0.15，而南半球的相关系数超过了 0.4。

首先，从空间结构的对称性上分析，在 850 hPa 高度上的相关图上可以看出，在赤道东太平洋表层为暖水时期赤道两侧的一对气旋在结构上并不对称，在北半球上呈现基本水平的结构，而在南半球最大相关系数区域的主轴呈现明显的西北—东南倾斜，这是和北半球赤道辐合带（ITCZ）较为水平而南太平洋辐合带（SPCZ）具有西北-东南走势一致的。反观 200 hPa 相关场分布，赤道两侧的一对反气旋基本上对称。另外，低空、高空最大相关区域的位置基本对应。

其次，由这一相关场的分布可以推测流函数所表征的距平大气环流的结构，它们很好地描述了 Walker 环流和 Hadley 环流在 ENSO 期间的变化。当赤道东太平洋海温为正距平即 El Niño 时期，在赤道附近，850 hPa 高度上出现强的距平西风，这是 ENSO 期间大气环流的典型特征。相应的，在 200 hPa 高度上出现强的距平东风，这清楚地表明了整个 Walker 环流两个水平分支的逆转。La Niña 时期流场的结构反过来。由于赤道低空西风/高空东风的出现，质量守恒关系要求必然存在其他地区向赤道的质量输送，在图版 A 中可以清楚地看到这一要求引导出西侧低空的风场向赤道辐合/高空的风场离赤道辐散，在距平环流东侧低空的风场离赤道辐散/高空的风场向赤道辐合。这一经向风的结构反映了 Hadley 环流在 ENSO 循环中的作用。表明中纬度和热带之间的信号通道是通过东西太平洋上空的 Hadley 环流所建立。

再者，从速度势所表征的大气环流辐合/辐散特征上进一步支持和补充了流函数的分析结果。图版 B 是 850 hPa 和 200 hPa 高度上大气扰动速度势和 Niño-3 指数的相关系数，可以清楚地看到，伴随着 El Niño 的发生，在距平意义下 Walker 环流的上升分支移到了中东赤道太平洋，中心位于西经 120°，而下沉分支出现在印度尼西亚海域。相应的，在 200 hPa 高度上辐合/辐散的位置正好与低空相反。因此，图版 A 和图版 B 完整地说明了 ENSO 期间在距平意义上 Walker 环流发生了与其气候态的偏离甚至逆转，而且大气的这种对于 SST 异常的响应，在垂直方向基本上以第一斜压模为主，或者说，热带大气的垂直环流基本上是热力性环流。

2.2　大气环流对于 SSTA 的响应过程分析

按照 Bjerkines[1,2] 的观点，大尺度的海-气相互作用在 ENSO 框架中占据中心位置。遵循这一观点，我们可以把这种海-气相互作用分解成两个方面来阐述，即大气环流如何对海洋变异（以 SSTA 代表）响应和海洋如何对大气环流变异响应，本文首先通过流函数、速度势和 SSTA 的延时相关来着重分析大气环流如何对 SSTA 响应。

首先，定义几个表征特定区域流函数和速度势变化的指数，其中北太平洋扰动流函数指数定义为范围（170°E 至 130°W，5°—20°N）的扰动流函数面积平均，相应的，南太平洋扰动流函数指数定义为（160°—140°W；10°—25°S）区域的扰动流函数面积平均，东太平洋扰动速度势指数定义为（120°—80°W；15°S 至 10°N）区域的扰动速度势面积平均，西太平洋扰动速度势指数定义为（120°—160°E；10°S 至 10°N）区域的扰动速度势面积平均。上述 4 个指数分别定义在 850 hPa 和 200 hPa 的高度层上。把上述定义的流函数和速度势变化指数分别与 Niño-3 SSTA 指数做滞后相关（流函数和速度势相对于 SSTA 的滞后），可以看出大气环流如何对于 SST 的变化进行响应。图 1 表示了南、北太平洋流函数在 850 hPa 和 200 hPa 两个高度层上相对于 Niño-3 SSTA 指数的滞后相关系数，图 2 表示了东、西太平洋速度势在 850 hPa 和 200 hPa 两个高度层上 Niño-3 SSTA 指数的滞后相关系数。

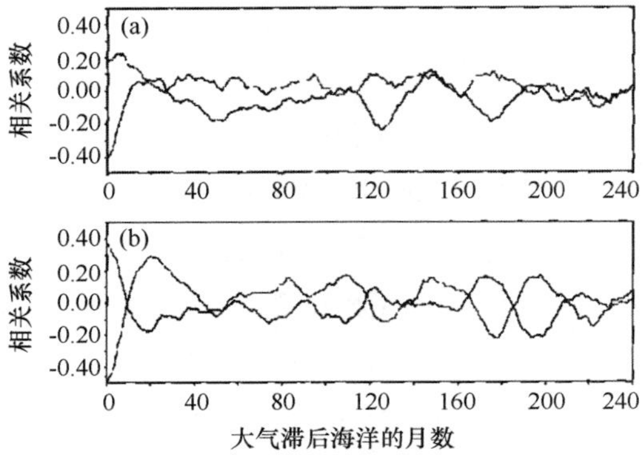

图 1 流函数指数与 Niño 3 指数的滞后相关系数

实线:850 hPa;虚线:200 hPa

(a)北太平洋;(b)南太平洋

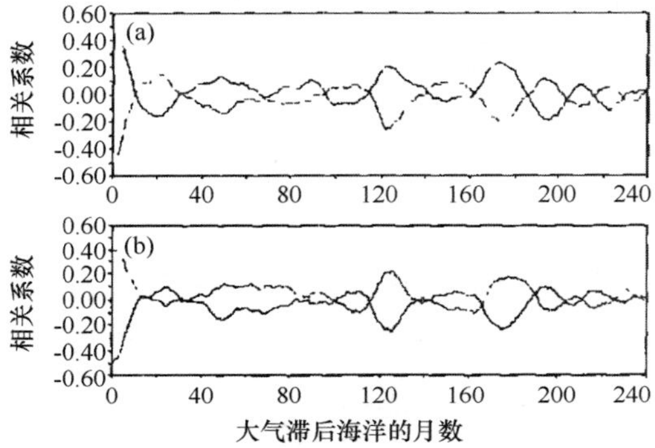

图 2 速度势指数与 Niño 3 指数的滞后相关系数

实线:850 hPa,虚线:200 hPa

(a)东太平洋;(b)西太平洋

图 1 中,在北太平洋负的相关系数对应着气旋式距平风场,在南太平洋正的相关系数对应着气旋式距平风场。图中可以看出如下几个特点。

(1)在每个高度层上南、北太平洋区域的相关系数反号,意味着气旋式或反气旋式距平风场总是在赤道两侧相伴出现。

(2)在南、北太平洋区域 850 hPa 和 200 hPa 高度上的相关系数反号,而且这种反位相关系在南太平洋表现的更加明显,意味着 Walker 环流的低空和高空分支协调一致的变化,即当低空出现西风距平时必然伴随着高空东风距平的发生。

(3)在 850 hPa 高度上,北太平洋的相关强于南太平洋,而在 200 hPa 高度上南太平洋的相关强于北太平洋,这种非对称性值得进一步关注。

(4)4 个滞后相关系数分别在 4 年、10 年和 15 年左右出现峰值,并且 10 年和 15 年的滞后相关较为显著,这表明 ENSO 过程一方面具有 3~5 年的年际变化周期,同时也有年代际的变化周期,在年代际上的较强相关也包含了 3~5 年周期信号的贡献。

在图 2 中,正/负相关系数分别代表了风场的辐合/辐散,从中可以发现与图 1 协调一致的几个特征。

(1)在每个高度层上东、西太平洋区域的相关系数反号,意味着辐合和辐散总是在热带海区两侧相伴出现。

(2)在东、西太平洋区域 850 hPa 和 200 hPa 高度上的相关系数反号,意味着 Walker 环流的上升和下沉分支协调一致的变化,即低空的辐合必然伴随着高空的辐散,反之亦然。

(3)在太平洋的东、西以及低空、高空速度势与 Niño-3 SSTA 的滞后相关系数在强度上与流函数相比较为一致。

(4)4 个滞后相关系数分别在 4 年、10 年和 15 年左右出现峰值,并且 10 年和 15 年的滞后相关较为显著,这一特征与流函数的情况相同。

图 1,图 2 说明了 ENSO 循环中大气环流变化的总体形势,即存在 3~5 年的重复周期在大气环流上表现为在低空出现关于赤道对称的气旋式距平风场,在高空出现反气旋式距平风场,同时在低空出现东/西太平洋的辐合/辐散和高空出现东/西太平洋的辐散/辐合距平风场,从而表明了 El Niño 发生时 Walker 环流在距平意义上发生反转,与此相匹配的 Hadley 环流也发生相应的调整,在西/东太平洋的低空向赤道产生辐合/辐散,在高空向赤道产生辐散/辐合。

为了说明 ENSO 循环期间大气环流的变化过程,进一步给出流函数(图 3)、速度势(图 4)和 Niño-3 指数的延时相关图,延时时间从大气超前海洋 8 个月到大气落后海洋 8 个月。从图中可以看出,在超前海洋 6 个月时大气出现显著的变化,包括西/东太平洋出现显著的辐散/辐聚,代表了 Walker 环流下沉/上升分支的位置发生变化,其强度减弱。Hadley 环流在西/东太平洋发生不同的响应,西侧加强东侧减弱。同时,大气的无辐散分量出现协调变化,沿赤道的距平西风和关于赤道对称的气旋式环流出现,在动力学上对应着抬升西太平洋的温跃层深度,加强了西太平洋温度负距平和沿赤道东传的暖 Kelvin 波发展,因此在整个 El Niño 的发展期,大气海洋之间存在着相互加强的正反馈。在 El Niño 的成熟期之后,Walker 环流逐渐向着气候平均态恢复,同时关于赤道对称的气旋式环流位置东移,甚至在西太平洋出现气旋式环流,上述过程都减弱 El Niño 事件,因此在 El Niño 消衰期大气海洋之间存在着相互减弱的负反馈。

2.3　ENSO 循环中大气环流变异的个例分析

为了进一步说明大气环流在 ENSO 期间的变异,以 20 世纪最强的一次 El Niño 事件,即 1997/1998 年的 El Niño 为例,给出大气环流的各种变异特征的描述(图版 C)。该图给出了 1997 年 10 月在 850 hPa 和 200 hPa 高度上相对于气候平均值的扰动流函数和速度势(为了显示方便,流函数、速度势的数值分别除以了 10^6)以及它们所代表风场的无辐散分量和无旋分量。

从中可以清晰地看出,对应着 Niño-3 指数达到最大值,即 El Niño 的最盛期,Walker 环流发生了距平意义上的反转,即低空为西风距平,高空为东风距平,上升分支位于东太平洋 120°W 附近,下沉分支位于印度尼西亚海域。同时,850 hPa 上的 Hadley 环流分支在西/东太平洋向赤道辐聚/辐散,在 200 hPa 上分别在西/东太平洋向赤道辐散/辐合,以平衡 Walker 环流的变化而保持质量守恒。

3　讨论与结论

在 ENSO 循环中大气环流变化的资料分析中,前人的工作多采用纬向风分量作为 Walker 环流变化的指标,在本文的工作中没有沿用这一做法,而是将风场分解为无旋分量和无辐散分量两部分,从而引入了流函数和速度势,这种做法有它的优点:首先,从海气相互作用的动力学上来讲,SST 变化由

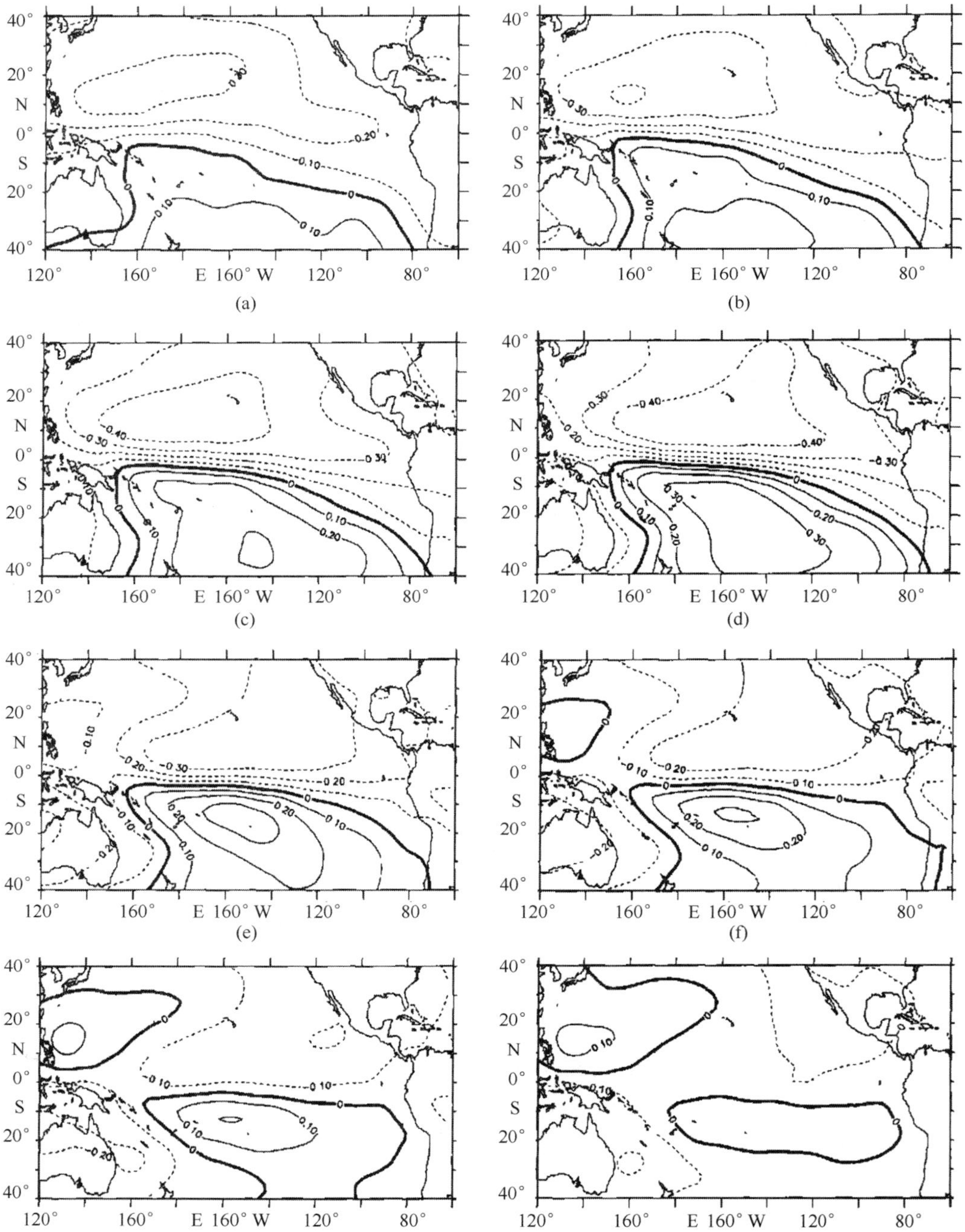

图 3　流函数和 Niño-3 指数的延时相关系数空间分布

(a)超前 8 个月;(b)超前 6 个月;(c)超前 4 个月;(d)超前 2 个月;(e)滞后 2 个月;

(f)滞后 4 个月;(g)滞后 6 个月;(h)滞后 8 个月

海洋温跃层的变化引起,而温跃层变化的强迫力正是风场的旋度部分,即对应着风场的流函数,采用这种方式的风场分解有利于逐步深入地开展海气相互作用的资料分析;其次,Walker 环流和 Hadley 环流是热力驱动环流,风场的无旋分量是描述大气对海洋热强迫响应的最直接的指标。因此,采用这种分析方法能够获得 ENSO 循环期间大气环流变异的更全面描述。本文利用 NCEP/NCAR 再分析资

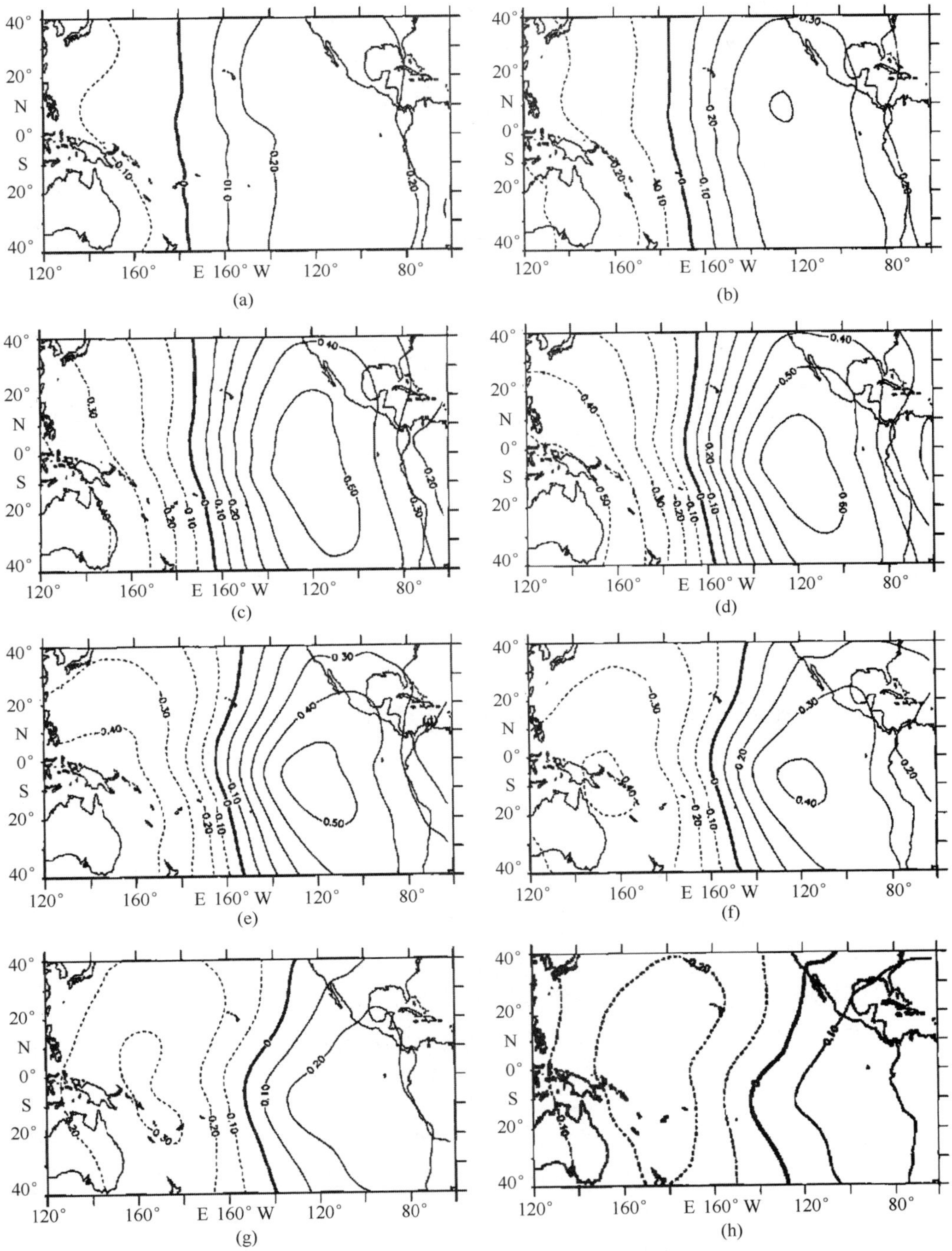

图4　速度势和Niño-3指数的延时相关空间分布

(a)超前8个月;(b)超前6个月;(c)超前4个月;(d)超前2个月;(e)滞后2个月;
(f)滞后4个月;(g)滞后6个月;(h)滞后8个月

料(1948—2002年),求解出大气运动的扰动流函数和扰动速度势,并进一步把流函数、速度势与Niño-3指数进行相关分析,利用所定义的流函数指数研究了低空(850 hPa)和高空(200 hPa)Walker环流、Hadley环流水平分量的变化,利用所定义的速度势指数分析了Walker环流的上升/下沉分支的相应位置变化,从而得到了ENSO期间大气环流变化的清晰图像。

通过资料分析可以清晰地得到 ENSO 循环期间大气环流变化的如下特征。

(1)伴随 El Niño 的发生,在南、北太平洋低空出现关于赤道对称的一对气旋式涡旋,高空出现一对反气旋式涡旋,它们表征了 Walker 环流、Hadley 环流水平分支的变化。

(2)同时,在东(120°W 附近)/西(印度尼西亚附近)太平洋低空出现风场辐合/辐散,高空出现对应的辐散/辐合,它们表征了 Walker 环流的上升/下沉分支,与上一条特征共同表明了 ENSO 期间 Walker 环流在距平意义上发生了反转。

(3)滞后相关分析表明大气环流的这种变化具有 4 年、10 年和 15 年左右的 3 个显著周期,分别对应着 ENSO 的年际和年代际变化特性。关于这种大气环流的 4 年振荡周期最早由中国科学院地理研究所长期天气预报组[9]在研究热带海洋对于副热带高压长期变化的影响时指出,他们发现东太平洋 SSTA 和副高强度都存在 3.5 年的振荡周期,与 Bjerknes[1]提出的热带海气相互作用的准两年周期不同。由于相应的 Hardley 环流变化,中纬度和热带在 4 年周期上的相互作用值得进一步关注。

附录:关于风场分解的讨论

对于本文所关心的热带区域,似乎直接求解上述方程就可以得到流函数 ψ 和速度势 ϕ,但是这会导致错误的结果,数值试验发现所得到的流函数 ψ 和速度势 ϕ 并不能正确反映热带大气的运动(图略)。正如 Miyakoda① 所指出,对于有限区域上述分解问题的解并不唯一,Bijlsma 等[9]对这个问题给出了全面的讨论,为了清楚起见,本文按照其思路对这个问题作一个简要说明。

给定球面有限区域 Ω,文中方程(6),方程(7)在边界 Γ 上满足下列条件:

$$-\frac{\partial\psi}{R\partial\theta}+\frac{1}{R\cos\theta}\frac{\partial\phi}{\partial\lambda}=u \tag{1}$$

$$\frac{1}{R\cos\theta}\frac{\partial\psi}{\partial\lambda}+\frac{\partial\phi}{R\partial\theta}=v \tag{2}$$

其中,λ 和 θ 是经度和纬度。

由于每个边界条件中均包含了两个变量,因此使流函数 ψ 和速度势 ϕ 耦合在一起,需要把文中方程(6)、方程(7)联立求解,它们有唯一解的充分必要条件是相应的齐次系统只有平凡解,但事实并非如此,可以举例说明,令:

$$\psi = A\log(\sec\theta+\tan\theta)+B\lambda \tag{3}$$

$$\phi = -B\log(\sec\theta+\tan\theta)+A\lambda \tag{4}$$

其中,A 和 B 为常数。

可以验证式(3),式(4)是满足齐次系统的解,相应的风场为零,(u_ψ, v_ψ)代表沿固定方向但是流速反比于 $\cos\theta$ 的流动,(u_ψ, v_ψ)恰好是它的反向流。因此可以知道,对于有限区域将风场分解为流函数 ψ 和速度势 ϕ 问题的解不唯一。为了解决这一问题,Bijlsma 等[9]提出了如何在无限多个解中寻求符合物理意义解的方法,正如 Sangster[11]指出的那样,这种做法意味着将辐散风场部分的动能最小化,而将有旋风场部分的动能最大化。一般来讲,这在中纬度气象学上是可以接受的方式,但是对于热带过程并不是一种理想的处理方式。

① Miyakoda K.Numerical solution of the balance equation. Japan Meteor Agency, Tech Rep, No.3:15

参考文献

[1] Bjerknes J. A possible response of the atmospheric Hadley circulation to equatorial anomalies of ocean temperature. Tellus, 1966, 18:820.

[2] Bjerknes J.Atmospheric teleconnection from the equatorial Pacific.Mon Weather Rev,1969,97:163.

[3] Philander S G.El Niño,La Niña,and the Southern Oscillation.London:Academic Press,1990.

[4] McPhaden M J,et al.The tropical ocean-global atmosphere observing system:A decade of progress.J Geophys Res,103(C7):14169.

[5] Wallace J M,et al.On the structure and evolution of ENSO-related climate variability in the tropical Pacific:Lessons from TOGA.J Geophys Res,1998,103(C7):14,241.

[6] Wang C.Atmospheric circulation cells associated with the El Niño-Southern Oscillation .J Climate,2002,15(4):399.

[7] Kalnay E,et al.The NCEP/NCAR reanalysis 40-year project.Bull Amer Meteor Soc,1996,77:437.

[8] Holton J R.An introduction to dynamic meteorology.3rd ed.London:Academic Press,1992.

[9] Bijlsma S J,et al.Computation of the streamfunction and velocity potential and reconst ruction of the wind field.J Atmos Sci,1986,114:1547.

[10] 中国科学院地理研究所长期天气预报组.热带海洋对于副热带高压长期变化的影响.科学通报,1977,7:313.

[11] Sangster W E.A method of representing the horizontal pressure force without reduction of pressure to sea level.J Meteor,1960,17:166.

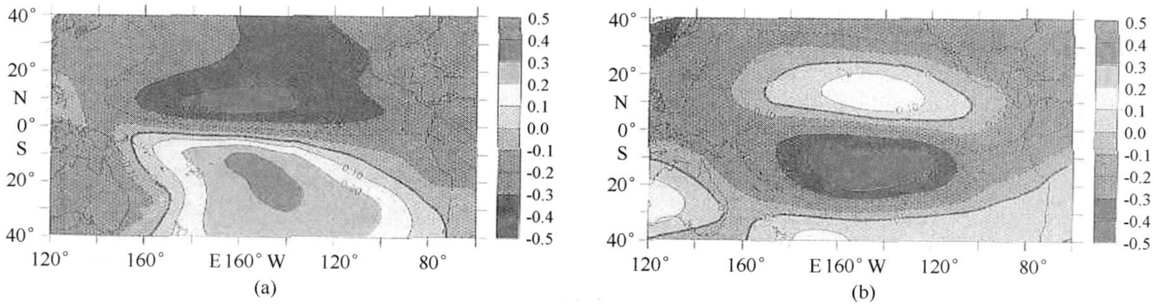

版图 A　流函数和 Niño-3 指数的同期相关系数空间分布(见书后彩插)

(a)850 hPa 高度;(b)200 hPa 高度

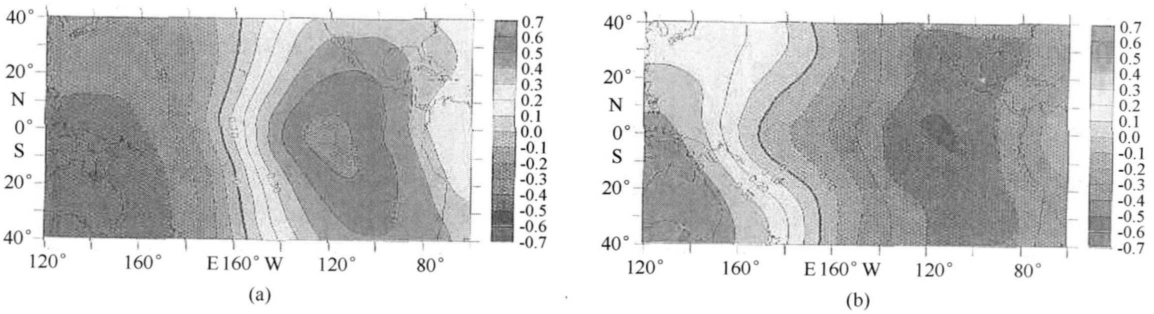

版图 B　速度势和 Niño-3 指数的同期相关系数空间分布(见书后彩插)

(a)850 hPa 高度;(b)200 hPa 高度

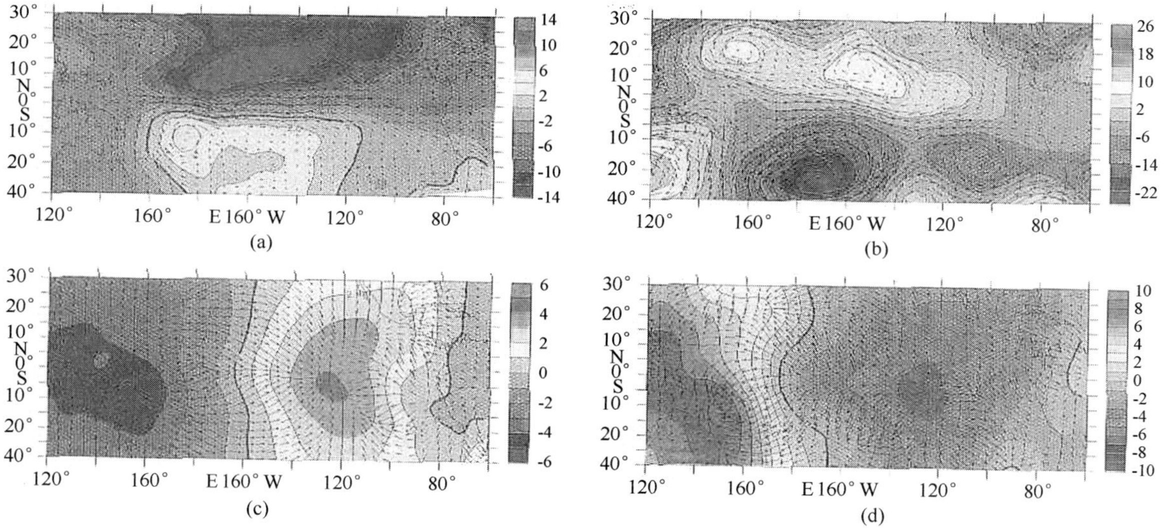

版图 C　1997 年 10 月的扰动流函数、速度势及相应的风场分量(见书后彩插)

(a)850 hPa 流函数及无辐散风场;(b)200 hPa 流函数及无辐散风场;

(c)850 hPa 速度势及无旋风场;(d)200 hPa 速度势及无旋风场

热带太平洋 ENSO 事件和印度洋的 DIPOLE 事件*

巢纪平[1,3]　巢清尘[2]　刘琳[3,4]

（1. 国家海洋环境预报中心，北京 100081；国家海洋局；2. 国家气候中心，北京 100081；3. 国家海洋局第一海洋研究所，青岛 266061；4. 中国海洋大学，青岛 266003）

摘要：在热带太平洋和印度洋的次表层构造了一个气候上的海温距平极值曲面（接近由20℃定义的温跃层曲面），分析了 1960—2000 年海温距平在这一曲面上演变的统计行为，指出，在这个曲面上分析海温距平的演变要比分析海表温度距平的演变规律更清楚，例如热带太平洋的 ENSO 事件，海温距平信号在赤道和南北 10°左右的纬带附近呈逆时针方向传播，在传播过程中其强度产生变化甚至变号；在热带印度洋的 Dipole 若在最大海温距平曲面上来分析，则西、东印度洋的海温距平在统计上呈负相关（真正物理意义下的 Dipole），而不像用海表温度距平分析那样只在西、东温度距平梯度上呈现年际的正、负号变化。进一步的分析表明，ENSO 和 Dipole 的发展，在统计上呈现出时滞的相互关系，一般赤道东太平洋的海温距平变化在前（一个季度左右），联系这两者变化之间的纽带是赤道太平洋和印度洋的一对反相转动的 Walker 环流的耦合演变。

关键词：最大海温距平曲面；ENSO；Dipole；Walker 环流；耦合演变

1　引言

自 20 世纪 60 年代 Bjerknes[1,2]指出发生在热带太平洋的 El Niño 和发生在大气中的南方涛动实质上是相互联系的大尺度海气相互作用事件（以后称为 ENSO）以来，ENSO 的研究已被众多的气象学家和海洋学者所重视，普遍认为 ENSO 是引起年际全球气候异常的重要信号之一。一般对 ENSO 发展的资料分析研究中，用海表温度距平（SSTA）来表征海洋的变化，用 850 hPa 纬向距平风来表征大气的变化。应用这两个指标，Rasmusson 和 Carpenter[3]分析了 ENSO 发展过程中各个位相海洋和大气的特征。然而在“中、美西太平洋海气相互作用试验”中，观测到 1986/1987 年那次 ENSO 中暖的海温异常首先出现在 147°—175°E 的 80~115 m 的次表层（温跃层附近）[4]。这一观测事实启示人们，不仅西太平洋在 El Niño 发展中具有重要作用（如 1982/1983 年的正的 SSTA 首先在赤道中西太平洋出现），而强的海温指示性信号也可能首先在次表层出现。李崇银和穆明权[5]就分析了 1997/1998 年 El Niño 事件的发展和西太平洋次表层海温异常的关系。

在另一方面，Saji 等[6]指出，在热带印度洋的海表温度距平在西、东方向的梯度呈年际的正、负交替变化，他们把这一现象称为 Dipole。但如按物理学的 Dipole 定义，西、东印度洋的海温距平应在符

* 气象学报，2005，63(5)：594-602.

号上是相反的,像 1997/1998 年和 1994 年西、东印度洋的海表温度距平是反号的除外,其他年份海表温度这种相反符号的分布并不多见,但这也促使人们去探索新的分析方法。

2 海温距平极值曲面及应用

事实上若用温跃层为次表层的代表层是合理的,但是考虑到若以 20℃ 为参考温度来定温跃层的深度,则在赤道东太平洋的冷舌区其温度通常低于 20℃,这给分析带来一定的困难,而在印度洋以什么温度作为温跃层的参考温度也未取得共识。为此巢清尘等在参考文献[7]中提出构造一个次表层海温距平极值曲面(MSTA)用它的深度来代替温跃层深度的想法。MSTA 是这样构造的:在每一点取表层以下温度距平的极值(不计符号),把所有极值对应的深度做气候平均,得到深度曲面,此即海温距平极值曲面。Chao 等在参考文献[8]中比较了 MSTA 的深度分布和以 20℃ 为参考的温跃层曲面的深度,指出它们的深度分布形势十分相似。

在以下的分析中将用 MSTA 曲面上的次表层海温距平来代替通常用的海表温度距平(SSTA)。作这种替换在物理上是考虑到年际时间尺度上大气对海洋的作用,主要通过低空风应力的辐合、辐散或旋度影响了温跃层的深度,而温跃层深度变化,在一定程度上可用气候温跃层上的温度变化来表示,例如当温跃层变浅(深)时,气候温跃层面上的海温变冷(暖)。在另一方面,海表温度的变化除受风应力影响外还受到感热、潜热及降水、蒸发等影响,使海表温度变化比较复杂,对一定时间尺度的信号来讲,噪音较多,且海表温度变化的强度也较弱。

作为例子文中分析了从 1968 年到 2000 年海温距平在 MSTA 上的演变特征。图 1 是参考文献[8]中的图 2 对信号做了 9 个月的滑动平均处理后给出的新图。

图 1 明显地呈现出,当一个暖(冷)的海温信号沿赤道从西太平洋暖池向东传播到东太平洋北上的过程中,在这期间一般有一个冷(暖)的海温距平信号在 10°N 附近(一般为 8°—14°N)从太平洋的东侧向西传向西太平洋后并南下到暖池的赤道附近。这样冷、暖海温距平信号在不同纬度上的传播,在赤道某一指定的经度上构成冷、暖位相的交替出现,即形成一个 ENSO 循环。一般来说,在南半球也存在类似的海温距平传播回路,但不如北半球明显,这可能因为赤道以南海温距平极值曲面较深,在那里海温距平强度不强,不易分析,也可能易受干扰。海温信号在 MSTA 曲面上的这种有规则的传播特征,用 SSTA 是分析不出来的。

3 热带太平洋的海气相互作用事件(ENSO)

从图 1 可以看到,在众多的例子中,同一时间在赤道东太平洋为暖(冷)的温度距平时,西太平洋暖池区为冷(暖)的温度距平,图 2 分别给出 MSTA 曲面上太平洋西部 10°N 至 4°S,130°—155°E 和东部 4°N 至 4°S,90°—150°W 海温距平的年际变化曲线(资料取自 Scripps 海洋研究所),可见两者呈反相关趋势,相关系数为-0.43(最大相关系数出现在东太平洋海温距平超前 2 个月时,为-0.45)。可见,在 MSTA 曲面上大洋东、西两部海温距平呈偶极子(Dipole)分布。

图 3(a)给出西太平洋上述指标区 MSTA 曲面上的海温距平及相应区的 MSTA 面深度距平的年际变化曲线,这两条曲线呈正相关趋势,相关系数为 0.31,可见,MSTA 曲面上的海温距平变化反映了温跃层深度的变化。图 3(b)给出太平洋中部(10°N 至 10°S,180°—145°W)850 hPa 纬向风距平和西太平洋指标区次表层海温距平的年际变化曲线(纬向风超前 3 个月),它们之间呈反相关变化趋势,相关系数为-0.51。图 3(c)给出东太平洋的情况,海温距平与 MSTA 深度变化之间的相关系数为 0.28,而

海温距平和纬向风(纬向风超前4个月)之间为正相关,相关系数为0.41(图3(d))。这几张图表明用MSTA曲面上的海温距平可以很清楚地表明ENSO事件。

图中纬向风取超前的月份,因那时的相关系数值最大,这表明在这一海气相互作用的位相中大气起主导作用。关于海洋影响大气的过程待分析。

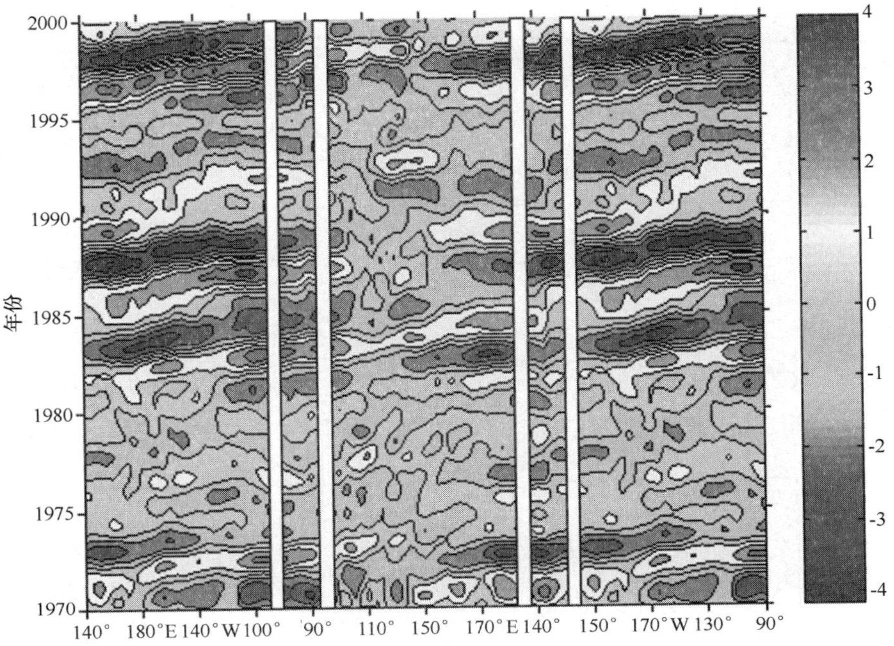

图1　在MSTA上海温距平的演变(见书后彩插)

虚线表示变化趋势;横坐标第一分格是沿赤道从西太平洋140°E到东太平洋115°W;第二分格是在115°W上从赤道到10°N(由于空间较小纬度标度未给出);第三分格是10°N从115°W回到140°E;第四分格在140°E上从10°N回到赤道(由于空间较小纬度未标出);第五分格同第一分格

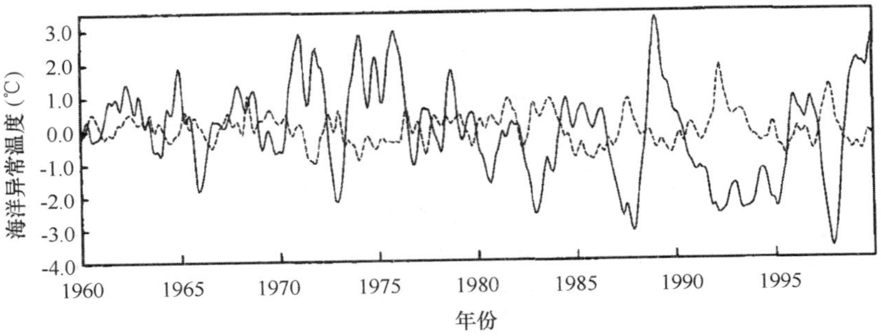

图2　MSTA曲面上太平洋西部(10°N至4°S,130°—155°E,实线)和东部(4°N至4°S,90°—150°W,虚线)海温距平的年际变化曲线

4　热带印度洋海气相互作用事件(Dipole)

在热带印度洋,Saji等[6]用SSTA提出的Dipole指数(DMI)实际上反映了西、东印度洋海表温度距平梯度的年际变化,在相当多的情况下,西、东印度洋海温距平的符号是一样的,只不过强度不同。

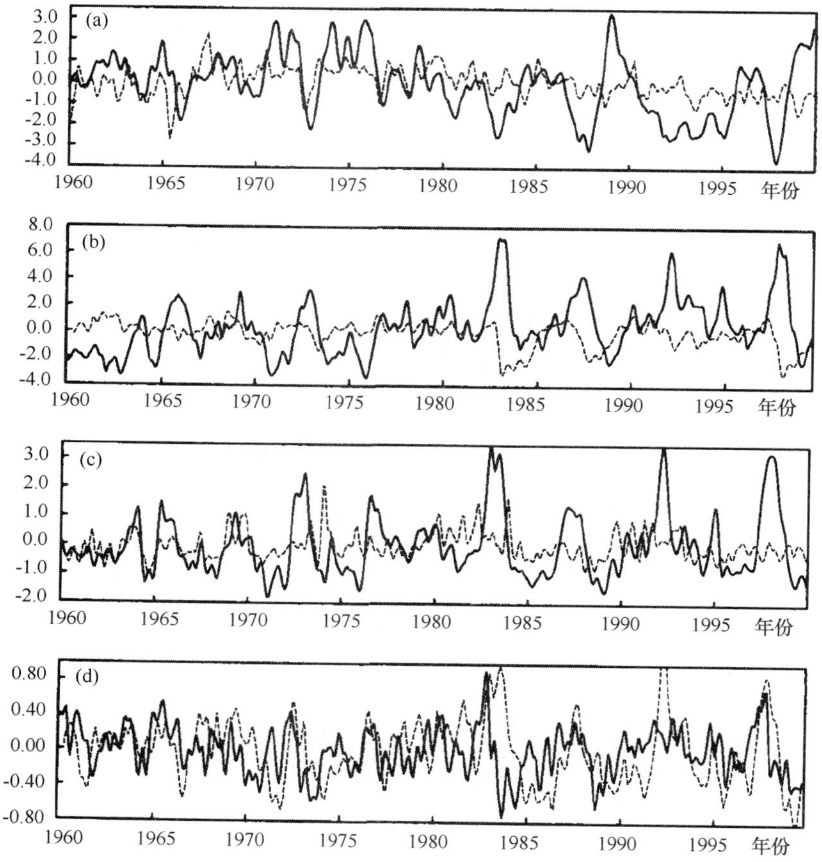

图 3　MSTA 曲面上的海温距平(实线)和该曲面的深度距平(虚线)的时间序列

(a)西太平洋;(c)东太平洋;太平洋中(b)、东部(d)850 hPa 纬向风距平(实线)分别与西(b)、东(d)太平洋 MSTA 曲面上的温度距平(虚线)的年际变化曲线(以上各变化曲线经过了 5 个月的滑动平均)

巢纪平等[9]指出,若在印度洋的 MSTA 曲面上来分析,则西印度洋(10°N 至 10°S,50°—70°E)和东印度洋(10°N 至 10°S,90°—110°E)的海温距平年际变化(图 4)呈反相关变化趋势,相关系数为-0.38。

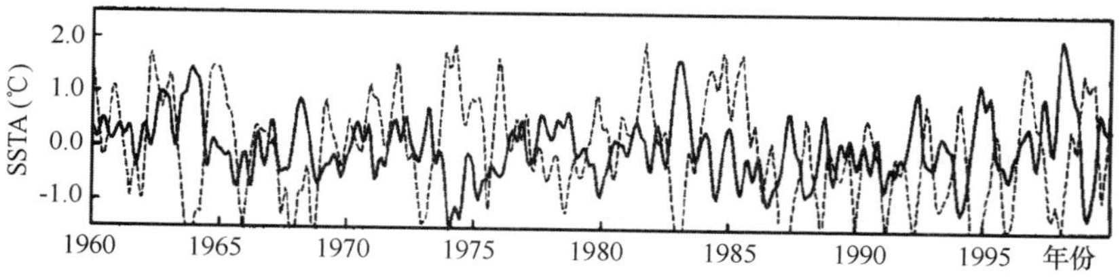

图 4　西印度洋(10°N 至 10°S,50°—70°E,实线)和东印度洋(10°N 至 10°S,90°—110°E,虚线)的海温距平年际变化

图 5(a)是西印度洋的海温距平年际变化及相应的 MSTA 曲面深度的年际变化,两者之间的相关系数为 0.32 。这同样地表明在印度洋也是一样,MSTA 曲面上的温度距平变化是该曲面深度变化的反映。在图 5(b)中也给出印度洋中部(4°N 至 10°S,60°—80°E)的纬向风变化和西印度洋的次表层的温度距平曲线,温度距平和纬向风距平呈反相关的变化趋势,相关系数为-0.34。图 5(c)是东印度洋的情况,这时海温距平与 MSTA 深度变化之间的相关系数为 0.23,而纬向风(超前 2 个月)和温度距

平是正相关变化,相关系数为 0.34(图 5(d))。可见印度洋的 Dipole 事件同样是海气相互作用事件。

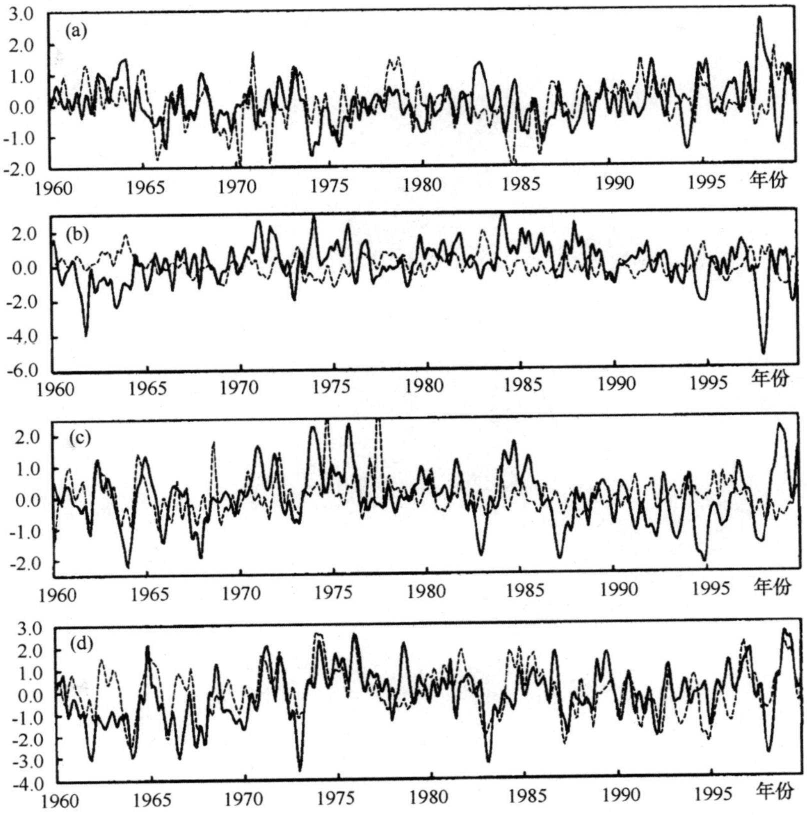

图 5　印度洋情况(说明同图 3)

5　ENSO 事件与 Dipole 发展的相关

Saji 等[6]认为印度洋的 Dipole 是独立于太平洋 ENSO 的事件,但 Webster 等[10]指出就 1997/1998 年印度洋的海气相互作用事件而言,它与 ENSO 事件是有关联的,李崇银和穆明权[11]认为用海表温度距平表示的印度洋 Dipole 与太平洋的 ENSO 是有联系的。巢纪平和袁绍宇[12]分析了两大洋次表层的海温距平与纬向风的关系后认为 Dipole 与 ENSO 是有联系的。这里用资料进一步分析它们的联系。

图 6(a)是赤道东太平洋(4°N 至 4°S,90°—150°W)次表层海温和两大洋 MSTA 面上海温距平的同期相关系数的分布,可以看到与西太平洋暖池的海温呈负相关(信度 99.9%的相关系数值为 0.1),这是 El Niño 形势,而在印度洋与东部呈负相关,与西部呈正相关,后者是 Dipole 形势,这表明 El Niño 和 Dipole 两者的发展,在海洋 MSTA 曲面上的海温距平有着明显的联系。图 6(b)是西印度洋上述指标区中次表层海温距平与 MSTA 面上热带两大洋海温距平的同期相关,可以看到它和东印度洋的海温是负相关,这是印度洋的 Dipole,与西太平洋的海温是负相关(相关系数超过-0.2),与东太平洋是正相关(相关系数超过 0.2),这表明,就统计上看(不是哪一个别年份),Dipole 与 El Niño 两者的发展在 MSTA 曲面的海温距平上有着明显的相关联系。

参考文献[12]指出,联系 El Niño 与 Dipole 发展的纽带是热带太平洋和印度洋那一对相互依存的 Walker 环流的变化,本文在这里作进一步分析。对于 Walker 环流的演变可用垂直运动来表示,而垂直运动可以通过与风的辐合辐散相联系的速度势来计算。

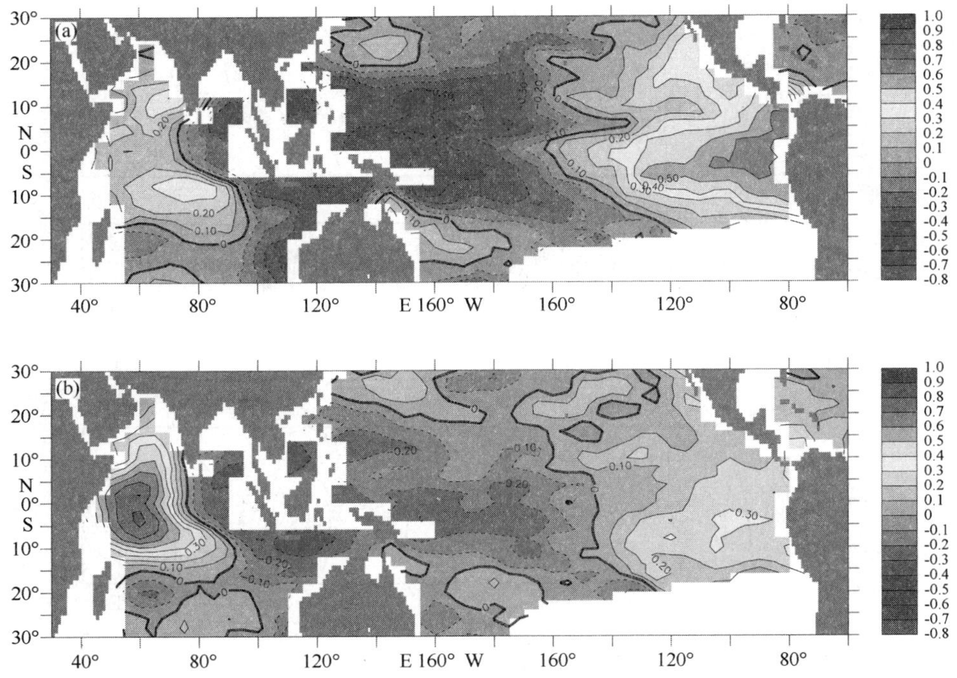

图 6　东太平洋(a)和西印度洋(b)次表层海温距平分别与两大洋 MSTA 面上海温距平的相关系数(见书后彩插)

图 7 是西印度洋指标区 MSTA 曲面上的海温距平与热带上空两大洋速度势的相关系数分布。由这一相关系数场的分布容易推断出当西印度洋次表层温度是正(负)距平时,在中、西印度洋是距平意义下的上升(下沉)气流,从印度洋中部到太平洋中部是大范围的下沉(上升)气流,而在东太平洋是上升(下沉)气流,从而在低空将有距平东(西)风从西太平洋—东印度洋吹向西(东)印度洋,距平西(东)风从西、中太平洋吹向东(西)太平洋。这反映了跨越印度洋和太平洋两个纬圈环流(在太平洋即为 Walker 环流)在西印度洋次表层海温出现异常时的结构。

图 8 是东太平洋 MSTA 曲面上的海温距平和距平速度势的相关系数分布。显然,相关系数场的分布形势基本同图 7,因此上升和下沉气流的分布基本上也是一样的。

这两张图一方面反映了印度洋和太平洋海气相互作用的形势,进而也表明这种海气相互作用形势的发展,在统计意义上,两大洋的海气相互作用事件是有关系的。

前面关于海温距平和低空纬向风的相关分析表明,最大的相关系数出现在风场超前海温,因此可以认为热带两大洋纬圈环流的耦合变化,制约并联系着两大洋海温距平的变化。

6　结论

最近一系列的资料分析表明,应用最大海温距平曲面(其深度在热带太平洋接近以 20℃ 为参考的温跃层深度)上的次表层海温距平能更好地分析 El Niño 和 Dipole 的发展,分析表明,印度洋的 Dipole 事件,在多数情况下不是独立的,它们与 ENSO 的发展有着紧密的联系,联系它们发展的是两大洋上空的距平 Walker 环流的耦合变化。

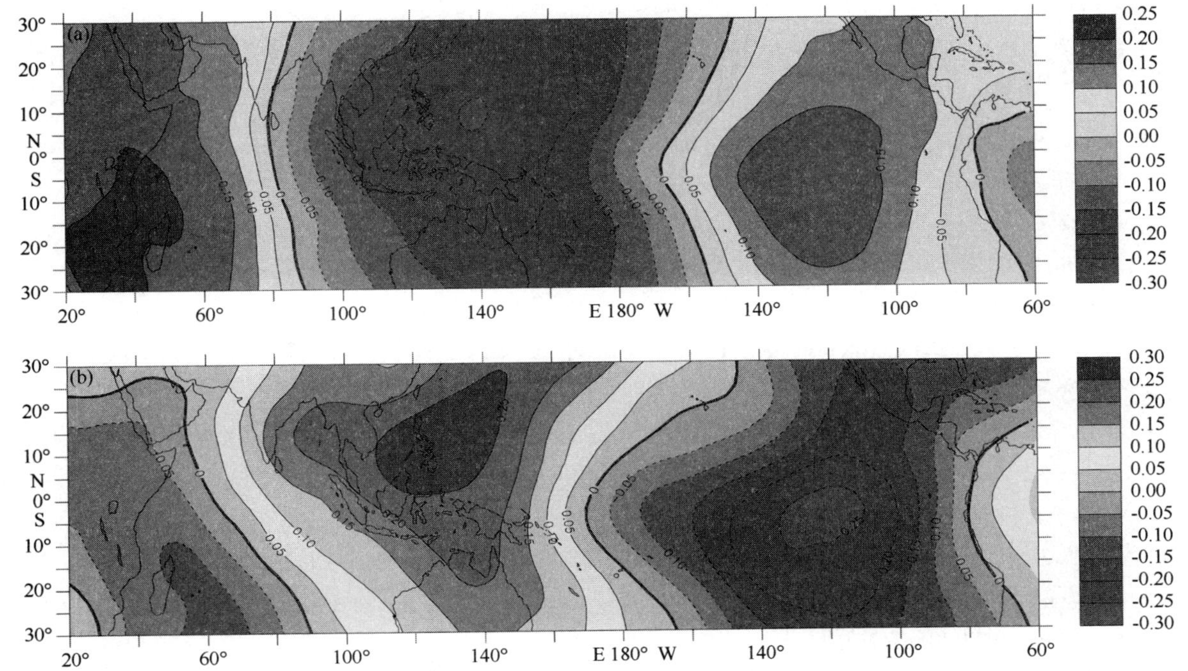

图 7　西印度洋 MSTA 曲面上的海温距平分别和 200 hPa(a),850 hPa(b)速度势的相关系数(见书后彩插)

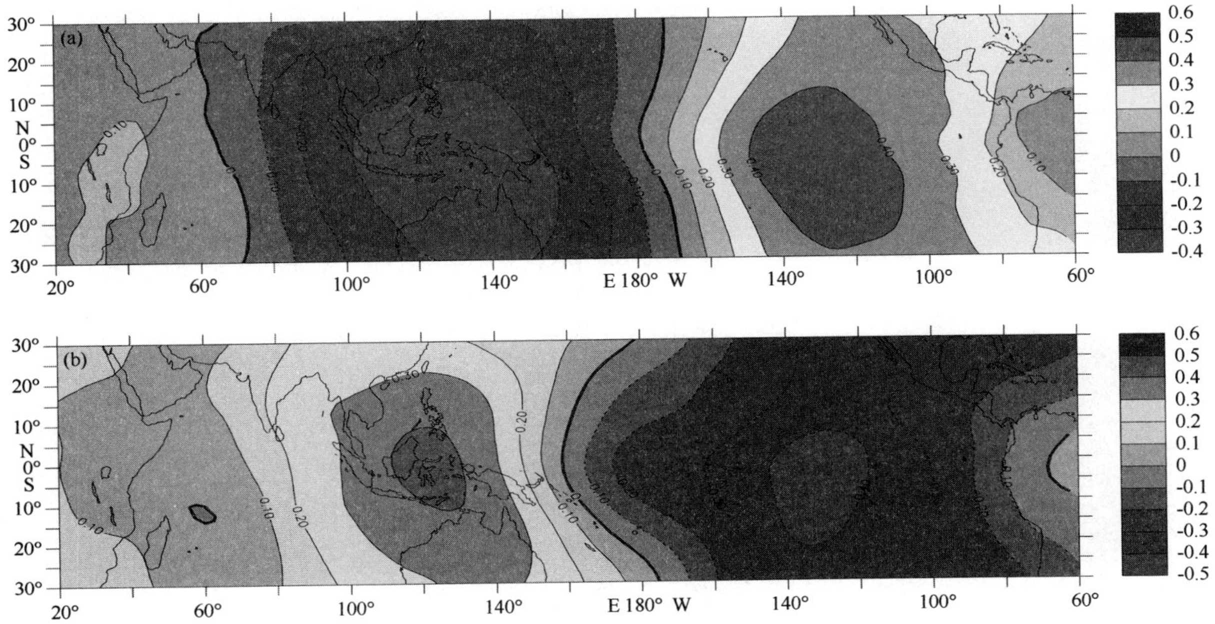

图 8　东太平洋 MSTA 曲面上的海温距平分别和 200 hPa(a),850 hPa(b)速度势的相关系数(见书后彩插)

参考文献

[1] Bjerknes J. A possible response of the atmospheric Hadley circulation to equatorial anomalies of ocean temperature. Tellus,1966,18:820-829.

[2] Bjerknes J. Atmospheric teleconnection from the equatorial Pacific. Mon Wea Rev,1969,97:163-172.

[3] Rasmusson E M,Carpenter T H. Variation in tropical sea surface temperature and surf ace wind fields associated with the Southern Oscillation/El Niño. Mon Wea Rev,1982,110:354-384.

[4] Wang Zongshan,Zou Emei,Tool J M,et al.Air sea interaction in tropical Western Pacifi c.Beijing.Chao Jiping,Young J Aed. In:Proc.US-PRC International TOGA Sym.,Beijing:China Ocean Press,1990:15-26.

[5] 李崇银,穆明权.厄尔尼诺的发生与赤道西太平洋暖池次表层海温异常.大气科学,1999,23:513-526.

[6] Saji N H, Goswami B N, Viayachandrom P N, et al. A dipole mode in the tropical Indian Ocean. Nature, 1999, 401: 360-363.

[7] 巢清尘,巢纪平.热带西太平洋和东印度洋对 ENSO 发展的影响.自然科学进展,2001,11:1293-1300.

[8] Chao Jiping,Yuan Shaoyu,Chao Qingchen,et al.A data analysis study on the evolution of the El Niño/La Niña cycle.Adv Atmos Sci,2003,19:837-844.

[9] 巢纪平,袁绍宇,蔡怡.热带印度洋的大尺度海气相互作用事件.气象学报,2003,61:251-256.

[10] Webster P T,Moore A M,Loschning J P,et al.Coupled ocean-atmosphere dynamics in the Indian Ocean during 1997-98. Nature,1999,401:337-339.

[11] 李崇银,穆明权.赤道印度洋海温偶极子型振荡及其气候影响.大气科学,2001,25:433-443.

[12] 巢纪平,袁绍宇.热带太平洋和印度洋海气相互作用事件的相互联系.自然科学进展,2003,13(12):1280-1285.

风应力对热带斜压海洋的强迫*

巢纪平[1]，陈鲜艳[2]，何金海[2]

(1.国家海洋局海洋环境科学和数值模拟重点实验室，青岛 266003；2.南京气象学院，南京 210044)

摘要：利用一个线性的具奋不同密度、温度的热带海洋两层模式，分析了热带西太平洋对纬圈风应力的响应。解析地求得热带西太平洋温跃层厚度、洋流及海温分布。结果表明次表层温度变化明显要比表层海温变化大，同时在大洋西部次表层发展起来的扰动向东传播能引起海温分布形态的异常。理论结果支持观测已表明的热带西太平洋物理量的变异在 EI Niño/La Niña 事件中起着重要作用的事实。

关键词：斜压热带海洋；风应力强迫；厄尔尼诺

1　引言

El Niño/La Niña 事件和南方涛动结合起来称为 ENSO，这种海-气相互作用一旦发生异常就会在全世界包括东亚在内的相当多的地区造成气候反常，因此不少科学家们致力于 ENSO 事件的研究。然而由于大气、海洋运动的复杂性，以及海上资料的缺乏，使得对 ENSO 发生机制的认识受到限制。Rasmusson 和 Carpenter[1]曾把 20 世纪 50 年代以来的 ENSO 事件做过综合分析，认为热带东太平洋是发生此类事件的先兆所在。然而 20 世纪 80 年代后，如 1982/1983 年、1986/1987 年、1997/1998 年几次 El Niño 事件其先兆均发生在热带中、西太平洋，而且又都比较强。最近的资料诊断结果表明，在 1960—1999 年这 40 年中，共发生了 11 次 E1 Niño 事件，它们的初始正的温度扰动距平，都首先在热带西太平洋“暖池”160 m 附近的次表层出现，当发展到一定强度后沿着气候的温跃层向东传播，并在东太平洋影响到海表。根据这一事实，巢纪平、张丽[2]和巢纪平、巢清尘[3]用热带海洋的等值浅水模式解析地求得了大洋西部对纬圈风应力的响应解，研究了扰动的发展和向东传播的物理过程。进而巢纪平、陈鲜艳[4]研究了热带二层层结海洋对纬圈风应力的响应，结果表明次表层海温的响应强度要远大于表层，这和观测相符。然而在那个工作中，为了得到解析解作了较多的假定，即当时为使问题简化，并考虑到正压辐散项很小，因此只给出了去除正压辐散对斜压模强迫情况下西太平洋对风应力的斜压响应。为得到更准确的响应结果，在本文中，作为对参考文献[4]的延拓，继续采用参考文献[4]的方法，加上了正压辐散项对斜压连续方程的强迫，重新计算西太平洋边界影响下的解。

2　模式简介

模式结构如图 1 所示，即将海洋分为上、下两层，厚度、密度和温度各不相同。

* 国家自然科学基金资助项目 49976001、49975025 和国家重点基础发展规划项目 G1998040900 第 1 部分共同资助

大气科学，2002，26(5)：577-594.

设 $z_1=H_1+\eta_1$，$z_2=H_2+\eta_2$，其中 H_1 和 H_2 分别是海洋静止时海-气界面和混合层-温跃层界面，D 是海洋处于静止状态下厚度，而 η_l 及 η_2 则是海洋在运动状态下对这两层平均界面的扰动。

线性控制方程组为

$$\frac{\partial \boldsymbol{u}_s}{\partial t}+f\boldsymbol{k}\times\boldsymbol{u}_s=-g'\nabla h+\frac{\eta}{2}g\alpha\nabla T_s+(\eta_2-\eta_1)g\alpha\nabla T_b+\frac{\tau}{\eta} \tag{1}$$

$$\frac{\partial u_b}{\partial t}+f\boldsymbol{k}\times\boldsymbol{u}_b=-g'\nabla h+\eta g\alpha\nabla T_s+(\frac{h^*}{2}+\eta_2-\eta_1)g\alpha\nabla T_b \tag{2}$$

$$\frac{\partial h}{\partial t}+(\eta\nabla\cdot\boldsymbol{u}_s+h^*\nabla\cdot\boldsymbol{u}_b)=0 \tag{3}$$

$$\frac{\partial T_s}{\partial t}+(\frac{1}{2}\Delta\bar{T}_s)\nabla\cdot\boldsymbol{u}_s=0 \tag{4}$$

$$\frac{\partial T_b}{\partial t}+\frac{1}{2}\left[\left(\frac{\eta}{h^*}\right)\nabla\cdot\boldsymbol{u}_s+\nabla\cdot\boldsymbol{u}_b\right]\Delta\bar{T}_b=0 \tag{5}$$

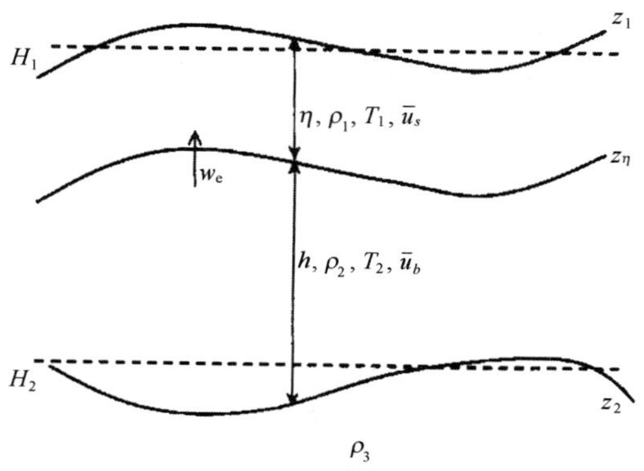

图 1 模式的两层结构

其中下标 s、b 分别表示表层和次表层运动，η 为混合层厚度，h 为温跃层厚度，h^* 为 h 的平均值，α 为海水热膨胀系数，τ 为直接作用于表层洋面的风应力，$\Delta\bar{T}_s$、$\Delta\bar{T}_b$ 分别是上下两层流体中的气候温度垂直差，γ 是 η 和 D 的比值。由于是线性模式，在模式化简时已经去掉了两层之间的非线性垂直交换（挟卷速度），而表层和次表层之间就可由 η_1、η_2 联系起来，即 $\eta_2-\eta_1=D-\eta-h$。定义运动的正压模和斜压模

$$\begin{cases}\hat{\boldsymbol{u}}=\frac{1}{D}(h^*\boldsymbol{u}_b+\eta\boldsymbol{u}_s)=\gamma\boldsymbol{u}_s+(1-\gamma)\boldsymbol{u}_b,\\ \tilde{\boldsymbol{u}}=\gamma(\boldsymbol{u}_s-\boldsymbol{u}_b)\end{cases} \tag{6}$$

可见，不论是正压模还是斜压模，都受到上、下层流体运动共同作用，并将两层之间的运动互相联系起来。考虑到 $\eta\ll D$，$(\eta_2-\eta_1)\ll\frac{h^*}{2}$，由此得简化的正压模和斜压模的方程分别为

$$\frac{\partial\hat{\boldsymbol{u}}}{\partial t}+f\boldsymbol{k}\times\hat{\boldsymbol{u}}=-g'\nabla h+\frac{1}{2}g\alpha(2\eta\nabla T_s+h^*\nabla T_b)+\gamma\frac{\tau}{\eta} \tag{7}$$

$$\frac{\partial\tilde{\boldsymbol{u}}}{\partial t}+f\boldsymbol{k}\times\tilde{\boldsymbol{u}}=-\frac{\gamma}{2}g\alpha(\eta\nabla T_s+h^*\nabla T_b)+\gamma\frac{\tau}{\eta} \tag{8}$$

$$\frac{\partial T_s}{\partial t} + \frac{\Delta \bar{T}_2}{2\eta} \nabla \cdot (\eta \hat{\boldsymbol{u}} + h^* \tilde{\boldsymbol{u}}) = 0 \tag{9}$$

$$\frac{\partial T_b}{\partial t} + \frac{1}{2} \Delta \bar{T}_b \left[\left(1 + \frac{\eta}{h^*} \right) \nabla \cdot \hat{\boldsymbol{u}} + \nabla \cdot \tilde{\boldsymbol{u}} = 0 \right] \tag{10}$$

取特征量

$$(x, y) \sim (c/2\beta)^{1/2}, \quad h \sim D, \quad (\hat{\boldsymbol{u}}, \tilde{\boldsymbol{u}}) \sim c, \quad T_s \sim \Delta \bar{T}_s,$$

$$T_b \sim \Delta \bar{T}_b, \quad \tau \sim \tau_0, \quad t \sim (2\beta c)^{-1/2}$$

则在赤道 β 平面上无量纲方程组为

$$\frac{\partial \hat{\boldsymbol{u}}}{\partial t} + \frac{1}{2} y \boldsymbol{k} \times \hat{\boldsymbol{u}} = - \nabla h + \left(\frac{c_1}{c} \right)^2 2 \nabla T_s + \left(\frac{c_2}{c} \right)^2 \nabla T_b + F\tau \tag{11}$$

$$\frac{\partial \tilde{\boldsymbol{u}}}{\partial t} + \frac{1}{2} y \boldsymbol{k} \times \tilde{\boldsymbol{u}} = - \gamma \left(\frac{c_1}{c} \right)^2 \nabla T_s - \left(\frac{c_2}{c} \right)^2 \nabla T_b + F\tau \tag{12}$$

$$\frac{\partial h}{\partial t} + \nabla \cdot \hat{\boldsymbol{u}} = 0 \tag{13}$$

$$\frac{\partial T'_1}{\partial t} + \frac{1}{2} \nabla \cdot \left(\hat{\boldsymbol{u}} + \frac{h^*}{\eta} \tilde{\boldsymbol{u}} \right) = 0 \tag{14}$$

$$\frac{\partial T'_2}{\partial t} + \frac{1}{2} \left[\left(1 + \frac{\eta}{h^*} \right) \hat{\boldsymbol{u}} + \tilde{\boldsymbol{u}} \right] = 0 \tag{15}$$

式中 $F = \gamma \tau_0 T / \eta c$，$T = (2\beta c)^{-1/2}$，而

$$\begin{cases} c_1 = \left(\frac{1}{2} g D \alpha \nabla \bar{T}_s \right)^{1/2}, \\ c_2 = \left(\frac{1}{2} g D \alpha \nabla \bar{T}_b \right)^{1/2} \end{cases} \tag{16}$$

为混合层和温跃层的重力内波波速。引进量 $T' = \left(\frac{c_1}{c} \right)^2 T_s + \left(\frac{c_2}{c} \right)^2 T_b$，得到

$$\frac{\partial T'}{\partial t} + A \nabla \cdot \hat{\boldsymbol{u}} + B \nabla \cdot \tilde{\boldsymbol{u}} = 0 \tag{17}$$

式中 A，B 为系数。并可将方程式(11)、式(12)改写成

$$\frac{\partial \hat{\boldsymbol{u}}}{\partial t} + \frac{1}{2} y \boldsymbol{k} \times \hat{\boldsymbol{u}} = - \nabla h + \nabla T' + F\tau \tag{18}$$

$$\frac{\partial \tilde{\boldsymbol{u}}}{\partial t} + \frac{1}{2} y \boldsymbol{k} \times \tilde{\boldsymbol{u}} = - \gamma \nabla T' + F\tau \tag{19}$$

如取 $D = 20\ 000$ cm，$\eta = 5\ 000$ cm，$\Delta \bar{T}_s = 0.5$℃，$\Delta \bar{T}_b = 5$℃则对于正压模重力波波速 $c \approx 200$ cm/s，而 c_2、c_1 值分别约为 98.9 cm/s、31.3 cm/s，故有 $c_1 < c_2 \ll c$。因此式(11)和式(12)可简化成

$$\frac{\partial \hat{\boldsymbol{u}}}{\partial t} + \frac{1}{2} y \boldsymbol{k} \times \hat{\boldsymbol{u}} = - \nabla h + F\tau \tag{20}$$

$$\frac{\partial \tilde{\boldsymbol{u}}}{\partial t'} + \frac{1}{2} y' \boldsymbol{k} \times \tilde{\boldsymbol{u}} = - \nabla' \tilde{T} + (B\gamma)^{-1/4} F\tau \tag{21}$$

这可称之为快波近似。为方便起见，重新写出式(13)、式(17)如下

$$\frac{\partial h}{\partial t} + \nabla \cdot \hat{\boldsymbol{u}} = 0 \tag{22}$$

$$\frac{\partial \tilde{T}}{\partial t'} + \frac{\partial \tilde{u}}{\partial x'} + \frac{\partial \tilde{v}}{\partial y'} = - AB^{-3/4}\gamma^{1/4} \nabla \cdot \hat{\boldsymbol{u}} \tag{23}$$

由此，可以看到正压模方程与浅水运动方程相似，而斜压模方程除了在连续方程的右端多了由正压模造成的强迫之外，其余也同浅水方程。在前文中我们已经分析了正压模方程组和无正压辐散强迫时的斜压模方程组对风应力响应的解。

3 边界条件

将正压模方程组式(20)、式(22)写成

$$\frac{\partial \hat{u}}{\partial t} - \frac{1}{2}y\hat{v} = - \frac{\partial h}{\partial x} + F\tau^{x} \tag{24}$$

$$\frac{\partial \hat{v}}{\partial t} + \frac{1}{2}y\hat{u} = - \frac{\partial h}{\partial y} + F\tau^{y} \tag{25}$$

$$\frac{\partial h}{\partial t} + \frac{\partial \hat{u}}{\partial x} + \frac{\partial \hat{v}}{\partial y} = 0 \tag{26}$$

引进变量 $\hat{q}=h+\hat{u}$，$\hat{r}=h-\hat{u}$ 代入式(24)至式(26)并将物理量用抛物柱函数（Weber 函数）展开成 $(\hat{q},\hat{r},\hat{v},\tau)= \sum\limits_{n}(\hat{q}_n,\hat{r}_n,\hat{v}_n,\tau_n)D_n(y)$，有

$$\frac{\partial \hat{q}_0}{\partial t} + \frac{\partial \hat{q}_0}{\partial x} - \hat{v}_{n-1} = F\tau_{n}^{x} \tag{27}$$

$$\frac{\partial^2 \hat{q}_1}{\partial t^2} + \frac{\partial^2 \hat{q}_1}{\partial x \partial t} + \frac{1}{2}\hat{q}_1 = F\frac{\partial \tau_{n}^{x}}{\partial t} + F\tau_{0}^{y} \tag{28}$$

$$\frac{\partial^2 \hat{q}_{n+2}}{\partial t^2} + \frac{\partial^2 \hat{q}_{n+2}}{\partial x \partial t} + \frac{1}{2}(n+2)\hat{q}_{n+2} = F\frac{\partial \tau_{n+2}^{x}}{\partial t} + F\tau_{n+1}^{y} \tag{29}$$

$$\frac{\partial^2 \hat{r}_n}{\partial t^2} - \frac{\partial^2 \hat{r}_n}{\partial t \partial x} + \frac{1}{2}(n+1)\hat{r}_n - \frac{1}{2}(n+1)(n+2)\hat{q}_{n+2} = - \frac{\partial F\tau_{n}^{x}}{\partial t} - (n+1)F\tau_{n+1}^{y} \tag{30}$$

$$\begin{aligned}\frac{\partial^3 \hat{q}_{n+2}}{\partial t^3} - \frac{\partial^3 \hat{q}_{n+2}}{\partial t \partial x^2} + \frac{1}{2}(2n+3)\frac{\partial \hat{q}_{n+2}}{\partial t} - \frac{1}{2}\frac{\partial \hat{q}_{n+2}}{\partial x} &= \frac{\partial^2 F\tau_{n+2}^{x}}{\partial t^2} - \frac{\partial^2 F\tau_{n+2}^{x}}{\partial t \partial x} + \\ &\frac{1}{2}(n+1)F\tau_{n+2}^{x} - \frac{1}{2}F\tau_{n}^{x} + \frac{\partial F\tau_{n+1}^{y}}{\partial t} - \frac{\partial F\tau_{n+1}^{y}}{\partial x}\end{aligned} \tag{31}$$

$$\begin{aligned}\frac{\partial^3 \hat{r}_n}{\partial t^3} - \frac{\partial^3 \hat{r}_n}{\partial t \partial x^2} + \frac{1}{2}(2n+1)\frac{\partial \hat{r}_n}{\partial t} - \frac{1}{2}\frac{\partial \hat{r}_n}{\partial x} &= - \frac{\partial^2 F\tau_{n}^{x}}{\partial t^2} - \frac{\partial^2 F\tau_{n}^{x}}{\partial x \partial t} + \frac{1}{2}(n+1)(n+2)F\tau_{n+2}^{x} - \\ &\frac{1}{2}(n+2)F\tau_{n}^{x} - (n+1)\frac{\partial F\tau_{n+1}^{y}}{\partial t} - (n+1)\frac{\partial F\tau_{n+1}^{y}}{\partial x}\end{aligned} \tag{32}$$

$$\frac{\partial \hat{v}_{n+1}}{\partial t} + \frac{1}{2}(n+2)\hat{q}_{n+2} - \frac{1}{2}\hat{r}_n = F\tau_{n+1}^{y} \tag{33}$$

引进拉普拉斯变换 $\hat{G}_n = \int_0^{\infty} \mathrm{e}^{-st}\hat{g}_n \mathrm{d}t$，这里 g 表示任一物理量，其中 s 在物理意义上相当于频率 w。

以上方程可写成

$$\frac{\mathrm{d}\hat{Q}_0}{\mathrm{d}x} + s\hat{Q}_0 = F\hat{\tau}_0^x \tag{34}$$

$$s\frac{\mathrm{d}\hat{Q}_1}{\mathrm{d}x} + \left(s^2 + \frac{1}{2}\right)\hat{Q}_1 = sF\hat{\tau}_1^x + F\hat{\tau}_0^y \tag{35}$$

$$s\frac{\mathrm{d}\hat{Q}_{n+2}}{\mathrm{d}x} + \left[\frac{1}{2}(n+2) + s^2\right]\hat{Q}_{n+2} - \frac{1}{2}\hat{R}_n = sF\hat{\tau}_{n+2}^x + F\hat{\tau}_{n+1}^y \tag{36}$$

$$s\frac{\mathrm{d}\hat{R}_n}{\mathrm{d}x} - \left[\frac{1}{2}(n+1) + s^2\right]\hat{R}_n + \frac{1}{2}(n+1)(n+2)\hat{Q}_{n+2} = sF\hat{\tau}_n^x + (n+1)F\hat{\tau}_{n+1}^y \tag{37}$$

$$\begin{aligned} & s\frac{\mathrm{d}^2\hat{Q}_{n+2}}{\mathrm{d}x^2} + \frac{1}{2}\frac{\mathrm{d}\hat{Q}_{n+2}}{\mathrm{d}x} - \left[\frac{1}{2}(2n+3)s + s^3\right]\hat{Q}_{n+2} \\ &= F\left\{s\frac{\mathrm{d}\hat{\tau}_n^x}{\mathrm{d}x} - \left[\frac{1}{2}(n+1) + s^2\right]\hat{\tau}_{n+2}^x + \frac{1}{2}\hat{\tau}_n^x\right\} + F\left(\frac{\mathrm{d}\hat{\tau}_{n+1}^y}{\mathrm{d}x} - s\hat{\tau}_{n+1}^y\right) \end{aligned} \tag{38}$$

$$\begin{aligned} & s\frac{\mathrm{d}^2\hat{R}_n}{\mathrm{d}x^2} + \frac{1}{2}\frac{\mathrm{d}\hat{R}_n}{\mathrm{d}x} - \left[\frac{1}{2}(2n+3)s + s^3\right]\hat{R}_n \\ &= F\left\{s\frac{\mathrm{d}\hat{\tau}_n^x}{\mathrm{d}x} + \left[\frac{1}{2}(n+2) + s^2\right]\hat{\tau}_n^x - \frac{1}{2}(n+1)(n+2)\hat{\tau}_{n+2}^x\right\} + F\left[(n+1)\frac{\mathrm{d}\hat{\tau}_{n+1}^y}{\mathrm{d}x} + (n+1)s\hat{\tau}_{n+1}^y\right] \end{aligned} \tag{39}$$

$$s\hat{V}_{n+1} + \frac{1}{2}(n+2)\hat{Q}_{n+2} - \frac{1}{2}\hat{R}_n = F\hat{\tau}_{n+1}^y \tag{40}$$

由于在动力学上，纬圈半地转模式中，Kelvin 波和 Rossby 波都是非色散的，这样波能量就不能以不同于信号传播速度传播。而在经圈半地转模式中，Rossby 短波的群速度虽然不同于相速度，但短波能量只能单向(向东)传播。基于这种前提下，巢纪平[5]提出了低频近似下的发展运动。设热带惯性运动的平均频率或参考频率为 f_0，运动的特征频率为 ω，若运动的特征频率满足条件 $\omega < f_0$ 或 $\omega^2 \ll f_0^2$，则称此为低频近似，即相当于滤去频率较高的重力波。将低频近似运用到本模式，可以在模式中略去与时间运动有关特征的二次项或更高次项。由此可得

$$\frac{\mathrm{d}\hat{Q}_0}{\mathrm{d}x} + s\hat{Q}_0 = F\hat{\tau}_0^x \tag{41}$$

$$s\frac{\mathrm{d}\hat{Q}_1}{\mathrm{d}x} + \frac{1}{2}\hat{Q}_1 = sF\hat{\tau}_1^x + F\hat{\tau}_0^y \tag{42}$$

$$\begin{aligned} & s\frac{\mathrm{d}^2\hat{Q}_{n+2}}{\mathrm{d}x^2} + \frac{1}{2}\frac{\mathrm{d}\hat{Q}_{n+2}}{\mathrm{d}x} - \frac{1}{2}(2n+3)s\hat{Q}_{n+2} = F\left[s\frac{\mathrm{d}\hat{\tau}_{n+2}^2}{\mathrm{d}x} - \frac{1}{2}(n+1)\hat{\tau}_{n+2}^x + \frac{1}{2}\hat{\tau}_n^x\right] + \\ & F\left(\frac{\mathrm{d}\hat{\tau}_{n+1}^y}{\mathrm{d}x} - s\hat{\tau}_{n+1}^y\right) \end{aligned} \tag{43}$$

$$s\frac{\mathrm{d}^2\hat{R}_n}{\mathrm{d}x^2} + \frac{1}{2}\frac{\mathrm{d}\hat{R}_n}{\mathrm{d}x} - \frac{1}{2}(2n+3)s\hat{R}_n = F\left[s\frac{\mathrm{d}\hat{\tau}_n^x}{\mathrm{d}x} + \frac{1}{2}(n+2)\hat{\tau}_n^x + \frac{1}{2}(n+1)(n+2)\hat{\tau}_{n+2}^x\right] +$$

$$F\left[(n+1)\frac{\mathrm{d}\hat{\tau}^y_{n+1}}{\mathrm{d}x}+(n+1)s\hat{\tau}^y_{n+1}\right] \tag{44}$$

容易看出，方程(41)的自由方程描写的是 Kelvin 波，方程式(43)、式(44)相当位势涡度方程，描写的是 Rossby 波。

方程式(36)、式(37)对 x 的微商是一阶的，其可作为对式(38)、式(39)的边界条件。在边界上法向速度为 0，若边界是南北走向的，则有 $\hat{u}_n=0$，于是根据前面定义，有 $\hat{q}_n=\hat{r}_n=h_n$。为与模式的近似度协调，现考虑在边界上也无高频运动作用。则由式(36)、式(37)当 $x=0$ 时，有

$$s\frac{\mathrm{d}\hat{Q}_{n+2}}{\mathrm{d}x}+\frac{1}{2}(n+2)\hat{h}_{n+2}-\frac{1}{2}\hat{h}_n=Fs\hat{\tau}^x_{n+2}+F\hat{\tau}^y_{n+1} \tag{45}$$

$$s\frac{\mathrm{d}\hat{R}_n}{\mathrm{d}x}-\frac{1}{2}(n+1)\hat{h}_n+\frac{1}{2}(n+1)(n+2)\hat{h}_{n+2}=Fs\hat{\tau}^x_n+(n+1)F\hat{\tau}^y_{n+1} \tag{46}$$

由于文章只研究半无界海洋解，另一个条件要求

$$\text{当 } x\to\infty,\qquad \hat{Q}_n,\hat{R}_n\to\text{有限} \tag{47}$$

经圈流可表示为

$$s\hat{V}_{n+1}+\frac{1}{2}(n+2)\hat{Q}_{n+2}-\frac{1}{2}\hat{R}_n=F\hat{\tau}^y_{n+1} \tag{48}$$

4 正压模对纬圈风应力的响应

由于在赤道地区以西风为主，因此本文主要研究西边界对纬圈风应力 τ^x 的响应，如风应力集中在赤道附近，则可取风应力在 y 方向用 Weber 函数的 0 阶模表示，即 $\tau^x(y)=D_0(y)=\mathrm{e}^{-\frac{y^2}{4}}$。并给出风应力随时间变化为 $\hat{X}=\frac{1-2\mathrm{e}^{-t_0 s}}{s}$，如 $\tau_0>0$，则相当于洋面上先吹东风，到 $t=t_0=10$ 时刻转成西风。其他变量截取到 4 阶(n 取 0,2,4)，可得到以上方程的截断模解如下：

$$\hat{h}_0=\frac{1}{2}(\hat{Q}_0+\hat{R}_0)=\frac{F\hat{X}}{6s}(1-5\mathrm{e}^{-sx}) \tag{49}$$

$$\hat{U}_0=\frac{1}{2}(\hat{Q}_0-\hat{R}_0)=\frac{5F\hat{X}}{6s}(1-\mathrm{e}^{-sx}) \tag{50}$$

$$\hat{h}_2=\frac{1}{2}(\hat{Q}_2+\hat{R}_2)=-\frac{F\hat{X}}{6s}[1+\mathrm{e}^{-(1/2s+7s)x}] \tag{51}$$

$$\hat{U}_2=\frac{1}{2}(\hat{Q}_2-\hat{R}_2)=-\frac{F\hat{X}}{6s}[1-\mathrm{e}^{-(1/2s+7s)x}] \tag{52}$$

$$\hat{h}_4=\frac{1}{2}(\hat{Q}_4+\hat{R}_4)=-\frac{F\hat{X}}{18s}[\mathrm{e}^{-(1/2s+7s)x}+\mathrm{e}^{-(1/2s+11s)x}] \tag{53}$$

$$\hat{U}_4=\frac{1}{2}(\hat{Q}_4-\hat{R}_4)=-\frac{F\hat{X}}{18s}[\mathrm{e}^{-(1/2s+7s)x}-\mathrm{e}^{-(1/2s+11s)x}] \tag{54}$$

5 斜压模对赤道纬圈风应力的响应

由斜压模方程组

$$\frac{\partial \tilde{u}}{\partial t'} - \frac{1}{2}y'\tilde{v} = -\frac{\partial \tilde{T}}{\partial x'} + B^* F\tau^x \tag{55}$$

$$\frac{\partial \tilde{v}}{\partial t'} + \frac{1}{2}y'\tilde{u} = -\frac{\partial \tilde{T}}{\partial y'} \tag{56}$$

$$\frac{\partial \tilde{T}}{\partial t'} + \frac{\partial \tilde{u}}{\partial x'} + \frac{\partial \tilde{v}}{\partial y'} = -A^*\left(\frac{\partial \hat{u}}{\partial x} + \frac{\partial \hat{v}}{\partial y}\right) = A^* \frac{\partial h}{\partial t} \tag{57}$$

其中 $A^* = AB^{-3/4}\gamma^{1/4}, B^* = (B\gamma)^{-1/4}$。

同正压模方法，引进变量 $\tilde{q}=\tilde{T}+\tilde{u}, \tilde{r}=\tilde{T}-\tilde{u}$，并经用抛物圆柱函数、拉普拉斯变换以及低频近似后，可得

$$\frac{d\hat{Q}_0}{dx'} + s\tilde{Q}_0 = \frac{1}{\sqrt{2\pi}}[A^*(B\gamma)^{3/4}s\hat{h}_m + B^*(B\gamma)^{1/4}F\hat{\tau}^x_m]\int_{-\infty}^{+\infty} D_m(y)D_0(y')\,dy' \tag{58}$$

$$\begin{aligned} s\frac{d^2\tilde{Q}_{n+2}}{dx'^2} + \frac{1}{2}\frac{d\tilde{Q}_{n+2}}{dx'} - \frac{1}{2}(2n+3)s\tilde{Q}_{n+2} = &-\frac{1}{2}\frac{n+1}{(n+2)!\sqrt{2\pi}}[A^*(B\gamma)^{3/4}s\hat{h}_m \\ &+ B^*F(B\gamma)^{1/4}\hat{\tau}^x_m]\int_{-\infty}^{+\infty} D_m(y)D_{n+2}(y')\,dy' - \frac{1}{2n!\sqrt{2\pi}}[A^*(B\gamma)^{3/4}s\hat{h}_m \\ &-(B\gamma)^{1/4}B^*F\hat{\tau}^x_m]\int_{-\infty}^{+\infty} D_m(y)D_n(y)\,dy' \end{aligned} \tag{59}$$

$$\begin{aligned} s\frac{d^2\tilde{R}_n}{dx'^2} + \frac{1}{2}\frac{d\tilde{R}_n}{dx'} - \frac{1}{2}(2n+3)s\tilde{R}_n = &-\frac{n+2}{2n!\sqrt{2\pi}}[A^*(B\gamma)^{3/4}s\hat{h}_m \\ - B^*F(B\gamma)^{1/4}\hat{\tau}^x_m]\int_{-\infty}^{+\infty} D_m(y)D_0(y')\,dy' &- \frac{1}{2}\frac{(n+1)(n+2)}{(n+2)!\sqrt{2\pi}}[A^*(B\gamma)^{3/4}s\hat{h}_m \\ &+ B^*(B\gamma)^{1/4}F\hat{\tau}^x_m]\int_{-\infty}^{+\infty} D_m(y)D_{n+2}(y')\,dy' \end{aligned} \tag{60}$$

$$s\tilde{V}_n + \frac{1}{2}(n+1)\tilde{Q}_{n+1} - \frac{1}{2}\tilde{R}_{n-1} = 0 \tag{61}$$

此时方程(58)描写的是斜压 Kelvin 波，方程式(59)、式(60)描写的是斜压 Rossby 波。需注意的是在方程组的右端正压模和斜压模具有不同的经向尺度。

同样有边界条件表示为，当 $x=0$，

$$s\frac{d\tilde{Q}_n}{dx'} + \frac{n}{2}\tilde{Q}_n - \frac{1}{2}\tilde{R}_{n-2} = \frac{1}{n!\sqrt{2\pi}}[(B\gamma)^{1/2}B^*Fs\hat{\tau}^x_m]\int_{-\infty}^{+\infty} D_m(y)D_n(y')\,dy' \tag{62}$$

$$\begin{aligned} &s\frac{d\tilde{R}_n}{dx'} - \frac{1}{2}(n+2)\tilde{R}_n + \frac{1}{2}(n+1)(n+2)\tilde{Q}_{n+2} \\ &= \frac{1}{n!\sqrt{2\pi}}[(B\gamma)^{1/2}B^*Fs\hat{\tau}^x_m]\int_{-\infty}^{+\infty} D_n(y)D_n(y')\,dy' \end{aligned} \tag{63}$$

及当 $x'\to\infty$，

$$\widetilde{Q}_n, \tilde{R} \to 有限 \tag{64}$$

对前述风应力,得截断模方程

$$\frac{\mathrm{d}\widetilde{Q}_0}{\mathrm{d}x'} + s\widetilde{Q}_0 = B^*(B\gamma)^{1/4} F\hat{\tau}_0^x a_{00} + A^*(B\gamma)^{3/4} s(\hat{h}_0 a_{00} + \hat{h}_2 a_{20} + \hat{h}_4 a_{40}) \tag{65}$$

$$s\frac{\mathrm{d}^2\widetilde{Q}_2}{\mathrm{d}x'^2} + \frac{1}{2}\frac{\mathrm{d}\widetilde{Q}_2}{\mathrm{d}x'^2} - \frac{3}{2}s\widetilde{Q}_2 = \frac{1}{2}B^* F(B\gamma)^{1/4}(\hat{\tau}_0^x a_{02} - \hat{\tau}_0^x a_{00})$$
$$-\frac{1}{2}A^*(B\gamma)^{3/4} s(\hat{h}_0 a_{02} + \hat{h}_2 a_{22} + \hat{h}_4 a_{42} + \hat{h}_0 a_{00} + \hat{h}_2 a_{20} + \hat{h}_4 a_{40}) \tag{66}$$

$$s\frac{\mathrm{d}\widetilde{Q}_4}{\mathrm{d}x'^2} + \frac{1}{2}\frac{\mathrm{d}\widetilde{Q}_4}{\mathrm{d}x'} - \frac{7}{2}s\widetilde{Q}_4 = -\frac{3}{2}B^* F(B\gamma)^{1/4}(\hat{\tau}_0^x a_{04} - \hat{\tau}_0^x a_{02})$$
$$-\frac{3}{2}A^*(B\gamma)^{3/4} s(\hat{h}_0 a_{04} + \hat{h}_2 a_{24} + \hat{h}_4 a_{44} + \hat{h}_0 a_{02} + \hat{h}_2 a_{22} + \hat{h}_4 a_{42}) \tag{67}$$

$$s\frac{\mathrm{d}^2\widetilde{R}_0}{\mathrm{d}x'^2} + \frac{1}{2}\frac{\mathrm{d}\widetilde{R}_0}{\mathrm{d}x'} - \frac{3}{2}s\widetilde{R}_0 = B^* F(B\gamma)^{1/4}(\hat{\tau}_0^x a_{00} - \hat{\tau}_0^x a_{02}) - A^*(B\gamma)^{3/4} s(\hat{h}_0 a_{00}$$
$$+ \hat{h}_2 a_{20} + \hat{h}_4 a_{40} + \hat{h}_0 a_{02} + \hat{h}_2 a_{22} + \hat{h}_4 a_{42}) \tag{68}$$

$$s\frac{\mathrm{d}^2\widetilde{R}_4}{\mathrm{d}x'^2} + \frac{1}{2}\frac{\mathrm{d}\widetilde{R}_2}{\mathrm{d}x'} - \frac{7}{2}s\widetilde{R}_2 = 2B^* F(B\gamma)^{1/4}(\hat{\tau}_0^x a_{02} - 3\hat{\tau}_0^x a_{00})$$
$$-2A^*(B\gamma)^{3/4} s(\hat{h}_0 a_{02} + \hat{h}_2 a_{22} + \hat{h}_{4a42} + 3\hat{h}_0 a_{04} + \hat{h}_2 a_{24} + \hat{h}_4 a_{44}) \tag{69}$$

$$s\frac{\mathrm{d}^2\widetilde{R}_4}{\mathrm{d}x'^2} + \frac{1}{2}\frac{\mathrm{d}\widetilde{R}_4}{\mathrm{d}x'} - \frac{11}{2}s\widetilde{R}_4 = 3B^* F(B\gamma)^{1/4}\hat{\tau}_0^x a_{04} - 3A^*(B\gamma)^{3/4} s(\hat{h}_0 a_{04} + \hat{h}_2 a_{24} + \hat{h}_4 a_{44}) \tag{70}$$

注意到由于在斜压连续方程中有正压模辐散项的强迫,因此在这组方程组中除了出现与前文相同的风应力0阶模外,另外非齐次项中还多了温跃层厚度强迫的项(其中系数见附录1)。

边条件写成($x=0$ 时)

$$s\frac{\mathrm{d}\hat{Q}_2}{\mathrm{d}x'} + \hat{\tilde{T}}_2 - \frac{1}{2}\hat{\tilde{T}}_0 = (B\gamma)^{1/2} B^* Fs\hat{\tau}_0^x a_{02} \tag{71}$$

$$s\frac{\mathrm{d}\hat{Q}_4}{\mathrm{d}x'} + 2\hat{\tilde{T}}_4 - \frac{1}{2}\hat{\tilde{T}}_2 = (B\gamma)^{1/2} B^* Fs\hat{\tau}_0^x a_{04} \tag{72}$$

$$s\frac{\mathrm{d}\widetilde{R}_0}{\mathrm{d}x'} - \frac{1}{2}\hat{\tilde{T}}_0 + \hat{\tilde{T}}_2 = (B\gamma)^{1/2} B^* Fs\hat{\tau}_0^x a_{00} \tag{73}$$

$$s\frac{\mathrm{d}\widetilde{R}_2}{\mathrm{d}x'} - \frac{3}{2}\hat{\tilde{T}}_2 + \hat{\tilde{T}}_4 = (B\gamma)^{1/2} B^* Fs\hat{\tau}_0^x a_{02} \tag{74}$$

$$s\frac{\mathrm{d}\widetilde{R}_4}{\mathrm{d}x'} - \frac{5}{2}\hat{\tilde{T}}_4 = (B\gamma)^{1/2} B^* Fs\hat{\tau}_0^x a_{04} \tag{75}$$

容易得到解为

$$\hat{\tilde{T}}_0 = [b_1 + b_2 e^{-(B\gamma)^{1/2}sx'}]\frac{\hat{X}}{s}e^{-sx'} + b_3\frac{\hat{X}}{s} + b_4\frac{\hat{X}}{s}e^{-(B\gamma)^{1/2}sx'} \tag{76}$$

$$\hat{U}_0 = [b_5 + b_6 e^{-(B\gamma)^{1/2}sx'}]\frac{\hat{X}}{s}e^{-sx'} + b_7\frac{\hat{X}}{s} + b_8\frac{\hat{X}}{s}e^{-(B\gamma)^{1/2}sx'} \tag{77}$$

$$\hat{\tilde{T}} = b_9\frac{\hat{X}}{s}e^{-(1/2s+7s)x'} + b_{10}\frac{\hat{X}}{s} \tag{78}$$

$$\tilde{U}_2 = b_{11}\frac{\hat{X}}{s}e^{-(1/2s+7s)x'} + b_{12}\frac{\hat{X}}{s} \tag{79}$$

$$\hat{\tilde{T}}_4 = b_{13}\frac{\hat{X}}{s}e^{-(1/2s+7s)x'} + b_{14}\frac{\hat{X}}{s}e^{-(1/2s+11s)x'} + b_{15}\frac{\hat{X}}{s}, \tag{80}$$

$$\tilde{U}_4 = b_{16}\frac{\hat{X}}{s}e^{-(1/2s+7s)x'} + b_{17}\frac{\hat{X}}{s}e^{-(1/2s+11s)x'} + b_{18}\frac{\hat{X}}{s}. \tag{81}$$

系数及拉普拉斯反变化的解见附录 2。

经圈流的表达式由 $\tilde{V}_{n+1}=\frac{1}{2s}[\tilde{R}_n-(n+2)]\tilde{Q}_{n+2}$ 给出，在此略。

上述解的形式中基本包含了三个部分，一是风应力直接作用，二是以 $e^{-sx'}$ 为表征的 Kelvin 波动，三是以 $e^{-\frac{1}{s}x'}$ 为表征的边界处 Rossby 短波。并且还可以注意到这些结果的形式和组成与没有辐散强迫时的解很相似，但是在系数上又增加了表征辐散强迫项的内容。

6　结果分析

文中有关的参数值同参考文献[4]，取 $\eta=5\ 000$ cm，$h^*=15\ 000$ cm，$D=20\ 000$ cm，$\gamma=0.25$，$c=200$ cm/s，$g=980$ cm/s^2，$g=2$ cm/s^2，$\alpha=2\times10^{-4}$/℃。根据以上参数可得量纲 $X=210$ km，$Y=210$ km，$t=29.3$ h。从斜压模解的表达式可以看出正压模对斜压模的辐散强迫主要由系数 A^* 决定。当 A^* 取 0 时的响应我们以前分析过，而 A^* 取不同值时，斜压模对风应力的响应显然也会不同。为便于比较，在此首先给出正压模瘟跃层沿赤道的传播（图 2）。最初东风作用时，大洋西边界处先激发出一个暖 Kelvin 波东传，$t>t_0=10$ 后风场变向转为西风，西边界处又激发出一个冷 Kelvin 波。由于海洋的“记忆”，注意到虽然温跃层深度在 Kelvin 波经过后就有变化，但边界处温跃层符号在 $t>20$ 后才开始由正转负。此前大洋温跃层厚度沿赤道呈东高西低的形势，相对海温而言是东暖西冷，而当 $t>20$ 之后，暖水逐渐传到大洋东部，而西边界也开始有了冷水出现，当 $t=40$ 时，海水分布相当于一个 El Niño 位相。正如 Fedorov 等[6]曾经指出的那样，无论定常风突然张弛还是突然转向，都会激发出一个可以影响到 El Niño 的 Kelvin 波动。Long 等[7]还分别用观测资料和数值试验结果来证明在赤道地区 Kelvin 波在 ENSO 或其他一些海洋事件中确有着重要作用。在这里还注意到，西风的盛行一方面推动由东风应力作用下激发出的暖的 Kelvin 波向东传播，造成 El Niño 位相；另一方面，西风应力在大洋西边界又激发出冷的 Kelvin 波，当这一个冷波向东传播后，就出现 El Niño 位相向 La Niña 位相的转变。

图 3 给出了上述风应力变化情况下沿赤道表层、次表层温度的传播（由于表层、次表层海水温度变化的无量纲数不是一个量级，为便于比较，在计算时已将海水温度的单位均用℃表示）。可以看到前期东风风场刚开始作用时，海洋表现出西暖东冷的形式。一定时间（$t>t_0$）后异常西风盛行于洋面

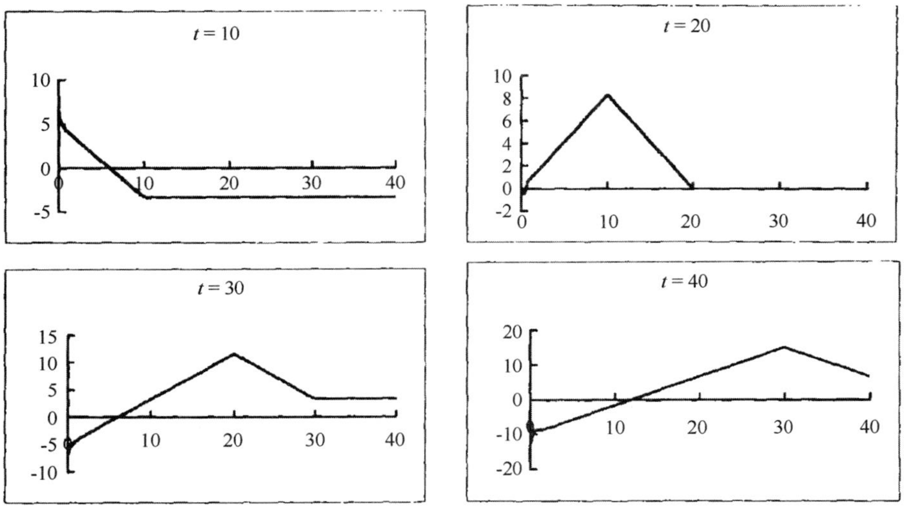

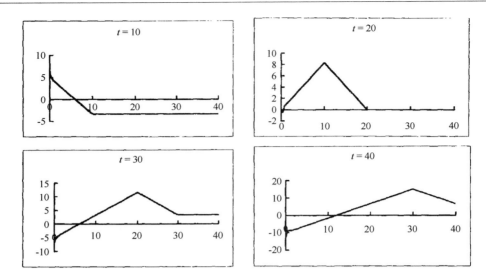

图 2 西太平洋边界附近温跃层在 t 时刻沿赤道随 x 的变化分布

y 轴为温跃层厚度(单位:10 m),x 轴为离开边界的距离(单位:1 km)

上,这时在西边界激发出的冷 Kelvin 波东传,波前沿的暖水被冷水波向东推进。当 $t>2t_0$ 后洋面上已经表现出西冷东暖的状态,出现了 El Niño 型的海水分布。西风的持续作用使东太平洋的海水继续增温,发展至 $t=4t_0$ 时已演变为 El Niño 成熟期,可以预计若西风继续作用,则西太平洋出现异常的冷水将东传至东太平洋,而整个洋面海水温度降低。事实上,从 $t=4t_0$ 时已经可以看到这时的冷水已向东传播了。此时若西风张弛,或东风出现(在西边界又重新激发出暖的 Kelvin 波东传)可能会形成 La Niña 型海水分布。不管是何种异常海温形态出现,我们都有理由认为事件的先兆是从西太平洋传至东太平洋的。而且,图中可看到次表层的温度变化的振幅比表层温度大了将近一个量级,也就是说,当有异常强迫条件发生时,次表层的物理变化比表层的变化更为显著。这个结果与近几年来的观测事实一致。温度的平面分布也说明了这个特征(图 4)。

图 5 是表层洋流和次表层洋流的传播情况。表层洋流在风应力的直接驱动下,其流速要比次表层大出一个量级,而且也直观地表现出了 Kelvin 波传播的特征。

图 6 是不考虑正压辐散对斜压模强迫时温度沿赤道的传播,可以看到,正压辐散的强迫使次表层温度的变化加大约 50%。

7 结论

用正压辐散来强迫斜压连续方程,计算得到在西太平洋边界处的两层海洋模式中不同层的温度、洋流的传播解。解析结果表明,当风场变化时,西太平洋次表层的温度变化可比表层温度变化大一个量级,在不同方向的纬圈风应力作用下,扰动的向东传播可能产生不同的海温异常分布型态。这也就是说,在大气强迫发生变化时,热带西太平洋在 EI Niño 或 La Niña 事件中都会起重要作用,一些 EI Niño / La Niña 事件,尤其是强事件发生前,其先兆可能先出现于热带西太平洋的暖池海域,而后才传播到热带东太平洋海区。

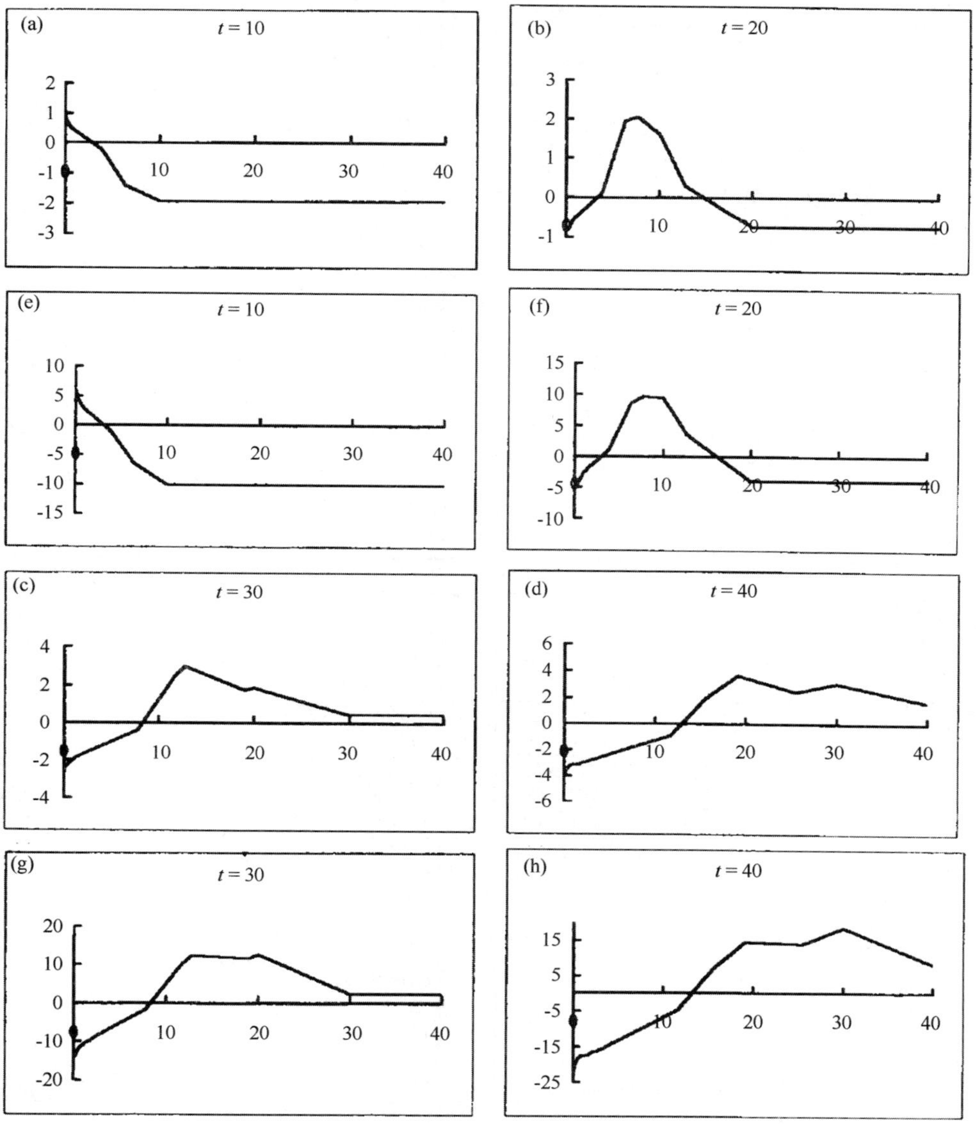

图 3　表层、次表层温度在不同时刻沿赤道随 x 的变化

y 轴为温度(单位:℃),x 轴为离开边界的距离(单位: km)

(a)至(d)为表层温度,(e)至(h)为次表层温度

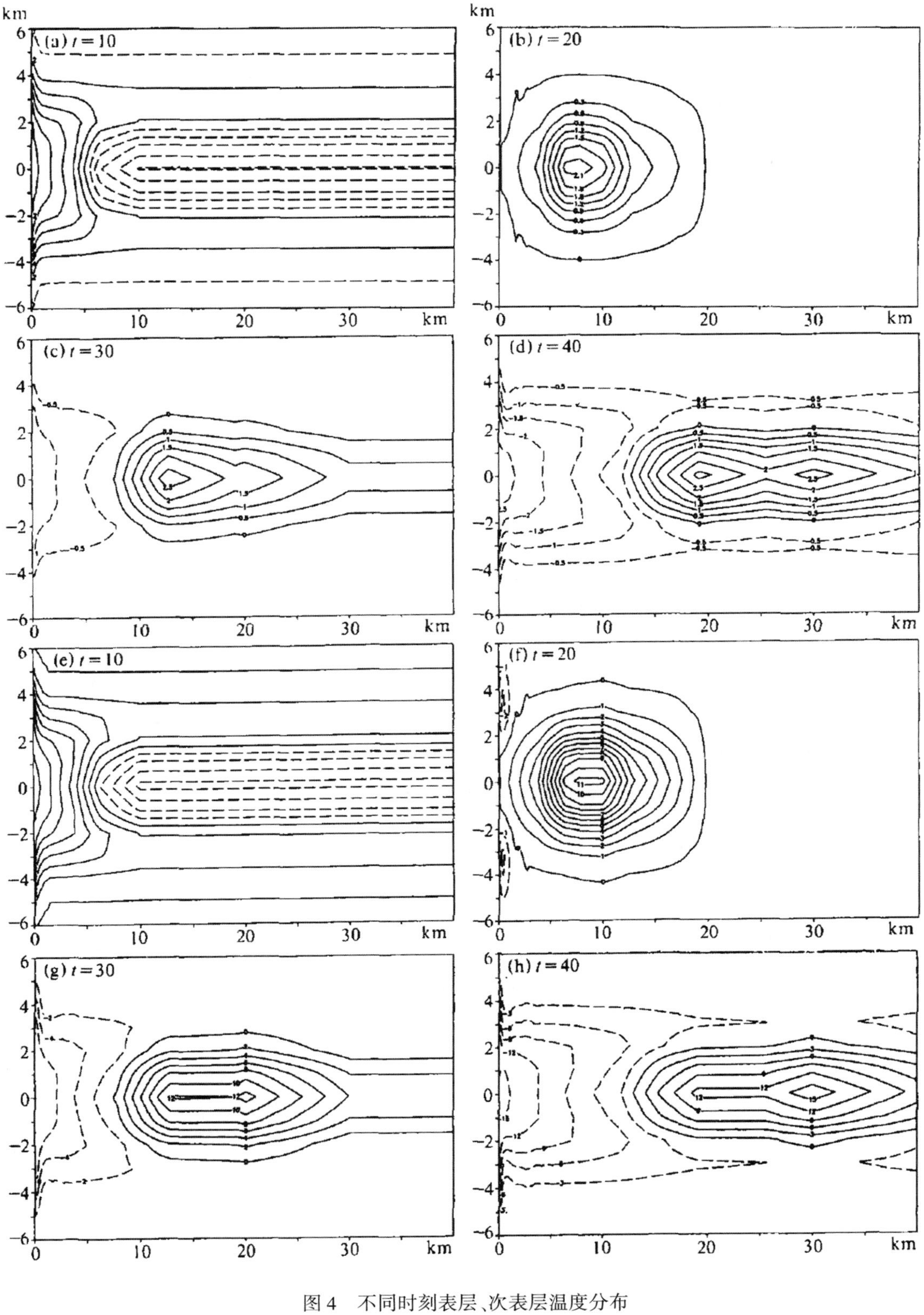

图 4 不同时刻表层、次表层温度分布

y 轴为经向距离(单位:km),x 轴为离开边界的距离(单位:km)

(a)至(d)为表层海水在各时刻的温度分布,(e)至(h)为次表层海水在各时刻的温度分布

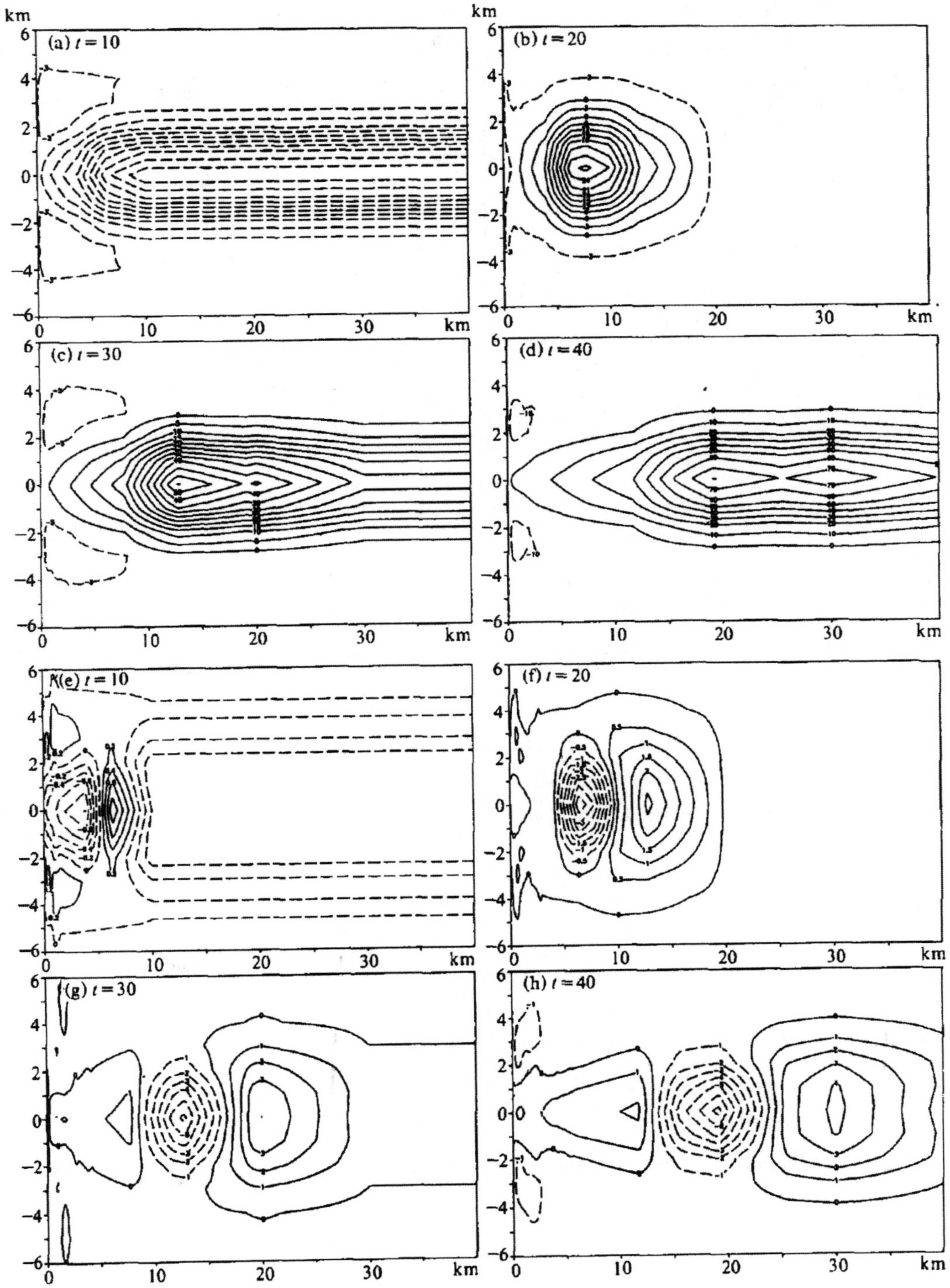

图 5　不同时刻表层、次表层洋流分布

y 轴为经向距离(单位:km),x 轴为离开边界的距离(单位:km)

(a)至(d)为表层洋流在各时刻的分布,(e)至(h)为次表层洋流在各时刻的分布

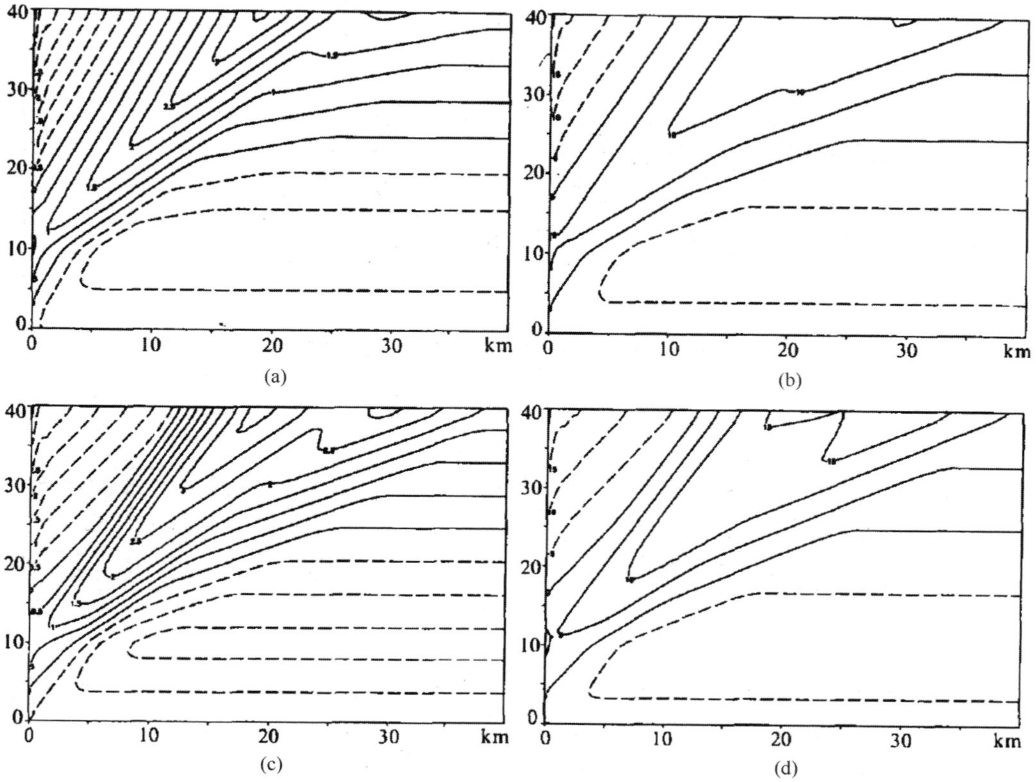

图6 表层、次表层在不同强迫条件下沿赤道传播

x 轴为离开西边界距离(单位:km),y 轴为时间 t(单位:d)

(a)至(b)无正压模项强迫斜压连续方程,(c)至(d)有正压模项强迫斜压连续方程

参考文献

[1] Rasmusson, S. G. H. EI Niño Southern Oscillation phenomena. *Nature*,1983,302:296-301.

[2] 巢纪平,张丽. 赤道不同海域对信风张弛的响应特征——对 EI Niño 研究的启示. 大气科学,1998,22:428-442.

[3] 巢纪平,巢清尘. 热带西太平洋对风应力响应的动力学. 大气科学,2002,26(2):145-160.

[4] 巢纪平,陈鲜艳,何金海. 热带西太平洋对风应力的斜压响应. 地球物理学报,2002,45(2):176-187.

[5] 巢纪平. 热带斜压大气的适应运动和发展运动. 中国科学(D 辑),1999,29:279-288.

[6] Fedorov, A. V. and W. K. Melville. Kelvin Fronts on the Equatorial Thermocline. *J. Phys. Oceanogr.*,2000,30(7):1692-1705.

[7] Long, B. and Ping Chang. Propagation of an Equatorial Kelvin Wave in a Varying Thermocline. *J. Geophys Res.*,1990,95:1826-1841.

附录

1　温跃层厚度强迫项的系数

$$a_{00}=\sqrt{\frac{2}{1+(B\gamma)^{1/2}}},$$

$$a_{02}=\left[\frac{1}{1+(B\gamma)^{1/2}}-\frac{1}{2}\right]a_{00},$$

$$a_{04}=\left\{\frac{1}{2[1+(B\gamma)^{1/2}]^{2}}-\frac{1}{2[1+(B\gamma)^{1/2}]}+\frac{1}{8}\right\}a_{00},$$

$$a_{20}=\left[\frac{2(B\gamma)^{1/2}}{1+(B\gamma)^{1/2}}-1\right]a_{00},$$

$$a_{22}=\left[\frac{6(B\gamma)^{1/2}}{[1+(B\gamma)^{1/2}]^{2}}-\frac{1}{2}\right]a_{00},$$

$$a_{24}=\left\{\frac{5(B\gamma)^{1/2}}{[1+(B\gamma)^{1/2}]^{3}}-\frac{6(B\gamma)^{1/2}+1}{2[1+(B\gamma)^{1/2}]^{2}}+\frac{(B\gamma)^{1/2}+2}{4[1+(B\gamma)^{1/2}]}-\frac{1}{8}\right\}a_{00},$$

$$a_{40}=\left\{\frac{12(B\gamma)^{1/2}}{[1+(B\gamma)^{1/2}]^{2}}-\frac{12(B\gamma)^{1/2}}{1+(B\gamma)^{1/2}}+3\right\}a_{00},$$

$$a_{42}=\left\{\frac{60(B\gamma)}{[1+(B\gamma)^{1/2}]^{3}}-\frac{6(B\gamma)^{1/2}[6+(B\gamma)^{1/2}]}{[1+(B\gamma)^{1/2}]^{2}}+\frac{6(B\gamma)^{1/2}+3}{1+(B\gamma)^{1/2}}-\frac{3}{2}\right\}a_{00},$$

$$a_{44}=\left\{\frac{70(B\gamma)}{[1+(B\gamma)^{1/2}]^{4}}+\frac{3-[1-8(B\gamma)^{1/2}+(B\gamma)]}{2[1+(B\gamma)^{1/2}]^{2}}-\frac{9}{8}\right\}a_{00}.$$

2　拉普拉斯反变化的解

记$\chi=t_0(B\gamma)^{1/4}$，

$$P(\xi)=\begin{cases}0 & t'<\xi\\ (B\gamma)^{1/4}(t'-\xi) & t'<\xi\end{cases}$$

$$Q(\xi)=\begin{cases}0 & t'<\xi\\ (B\gamma)^{-1/4}\left[\dfrac{2(t'-\xi)}{x'}\right]^{1/2}J_1\left[\sqrt{2x'(t'-\xi)}\right] & t'<\xi\end{cases}$$

文中解的表达式可写为

$$T_0=b_1[P(x')-2P(x'+\chi)]+b_4\{P[(B\gamma)^{1/2}x']-2P[(B\gamma)^{1/2}x'+\chi]\}+$$
$$b_2\{P[x'+(B\gamma)^{1/2}x']-2P[x'+(B\gamma)^{1/2}x'+\chi]\}+b_3[(B\gamma)^{-1/4}t'-2P(\chi)],\qquad(1)$$

$$\widetilde{u}_0=b_5[P(x')-2P(x'+\chi)]+b_8\{P[(B\gamma)^{1/2}x']-2P[(B\gamma)^{1/2}x'+\chi]\}+$$
$$b_6\{P[x'+(B\gamma)^{1/2}x']-2P[x'+(B\gamma)^{1/2}x'+\chi]\}+b_7[(B\gamma)^{-1/4}t'-2P(\chi)],\qquad(2)$$

$$\widetilde{T}_2=b_9[Q(7x')-2Q(\chi+7x')]+b_{10}[(B\gamma)^{-1/4}t'-2P(\chi)],\qquad(3)$$

$$\widetilde{U}_2=b_{11}[Q(7x')-2Q(\chi+7x')]+b_{12}[(B\gamma)^{-1/4}t'-2P(\chi)],\qquad(4)$$

$$\widetilde{T}_4=b_{13}[Q(7x')-2Q(\chi+7x')]+b_{14}[Q(7x')-2Q(\chi+11x')]+b_{15}[(B\gamma)^{-1/4}t'-2P(\chi)],\qquad(5)$$

$$\widetilde{u}_4 = b_{16}[Q(7x') - 2Q(\chi + 7x')] + b_{17}[Q(7x') - 2Q(\chi + 11x')] + b_{18}[(B\gamma)^{-1/4}t' - 2P(\chi)], \tag{6}$$

其中系数

$$b_1 = -\frac{1}{6}B^*F(B\gamma)^{1/4}(5a_{00} - 2a_{02}) + \frac{1}{36}A^*F(B\gamma)^{1/2}(-a_{00} + 2a_{02} - 2a_{22} + a_{20}),$$

$$b_2 = -\frac{5F}{12}\frac{A^*(B\gamma)^{1/2}a_{00}}{(B\gamma)^{1/2} - 1},$$

$$b_3 = \frac{F}{36}A^*(B\gamma)^{1/2}(5a_{00} - 5a_{20} + 2a_{02} - 2a_{22}) + \frac{1}{6}B^*(B\gamma)^{1/4}F(a_{00} + 2a_{02}),$$

$$b_4 = \frac{5}{12}\frac{A^*(B\gamma)^{1/2}Fa_{00}}{(B\gamma)^{1/2} - 1},$$

$$b_5 = -\frac{1}{6}B^*F(B\gamma)^{1/4}(5a_{00} - 2a_{02}) + \frac{1}{36}A^*F(B\gamma)^{1/2}(-a_{00} + 2a_{02} - 2a_{22} + a_{20}),$$

$$b_6 = -\frac{5F}{12}\frac{A^*(B\gamma)^{1/2}a_{00}}{(B\gamma)^{1/2} - 1},$$

$$b_7 = \frac{1}{6}B^*(B\gamma)^{1/4}F(5a_{00} - 2a_{02}) + \frac{F}{36}A^*(B\gamma)^{1/2}(a_{00} - a_{20} - 2a_{02} + 2a_{22}),$$

$$b_8 = \frac{5}{12}\frac{A^*(B\gamma)^{1/2}Fa_{00}}{(B\gamma)^{1/2} - 1},$$

$$b_9 = \frac{F}{42}B^*(B\gamma)^{1/4}(19a_{02} - 7a_{00} - 36a_{04}) + \frac{F}{252}A^*(B\gamma)^{1/2}(7a_{00} - 5a_{02} + 5a_{22} - 7a_{20} +$$
$$36a_{24} - 36a_{04}),$$

$$b_{10} = \frac{1}{42}B^*F(B\gamma)^{1/4}(36a_{04} - 5a_{02} - 7a_{00}) + \frac{A^*(B\gamma)^{1/2}F}{252}(19a_{02} - 19a_{22} + 7a_{00} - 7a_{20} +$$
$$36a_{04} - 36a_{24}),$$

$$b_{11} = -\left[\frac{F}{42}B^*(B\gamma)^{1/4}(19a_{02} - 7a_{00} - 36a_{04}) + \frac{F}{252}A^*(B\gamma)^{1/2}(7a_{02} - 5a_{02} + 5a_{22} - 7a_{20} + \right.$$
$$\left. 36a_{24} - 36a_{04})\right],$$

$$b_{12} = \frac{1}{42}B^*F(B\gamma)^{1/4}(19a_{02} - 7a_{00} - 36a_{04}) + \frac{A^*(B\gamma)^{1/2}F}{252}(-5a_{02} + 7a_{00} + 5a_{22} - 7a_{20} +$$
$$36a_{24} - 36a_{04}),$$

$$b_{13} = \frac{1}{126}B^*F(B\gamma)^{1/4}(-7a_{00} + 19a_{02} + 36a_{04}) + \frac{1}{756}A^*F(B\gamma)^{1/2}(7a_{00} - 5a_{02} + 5a_{22} -$$
$$7a_{20} + 36a_{24} - 36a_{04}),$$

$$b_{14} = \frac{1}{1386}B^*F(B\gamma)^{1/4}(-77a_{00} + 110a_{02} - 1890a_{06} + 279a_{04}) + \frac{F}{8316}A^*(B\gamma)^{1/2}(77a_{00} +$$
$$44a_{02} - 44a_{22} - 77a_{20} + 1890a_{26} + 477a_{24} - 1890a_{06} - 477a_{04}),$$

$$b_{15} = \frac{1}{154}B^*F(B\gamma)^{1/4}(-9a_{04} - 11a_{02} + 210a_{06}) + \frac{A^*(B\gamma)^{1/2}F}{924}(75a_{04} + 11a_{02} - 75a_{24} -$$

$11a_{22} + 210a_{06} - 210a_{26})$,

$$b_{16} = \frac{1}{126}B^*F(B\gamma)^{1/4}(-7a_{00} + 19a_{02} - 36a_{04}) + \frac{1}{756}A^*F(B\gamma)^{1/2}(7a_{00} - 5a_{02} + 5a_{22} - 7a_{20} + 36a_{24} - 36a_{04}),$$

$$b_{17} = -\left[\frac{1}{1386}B^*F(B\gamma)^{1/4}(-77a_{00} + 11a_{02}) - 1890a_{02} + 279a_{04} + \frac{1}{8316}A^*F(B\gamma)^{1/2}(77a_{00} + 44a_{02} - 44a_{22} - 77a_{20} + 1890a_{26} + 477a_{24} - 1890a_{06} - 477a_{04})\right],$$

$$b_{18} = \frac{1}{154}B^*F(B\gamma)^{1/4}(75a_{04} - 11a_{02} - 210a_{06}) + \frac{A^*(B\gamma)^{1/2}F}{924}(210a_{26} + 9a_{24} - 210a_{06} - 9a_{04} + 11a_{02} - 11a_{22}).$$

热带西太平洋对风应力的斜压响应*

巢纪平[1],陈鲜艳[2,3],何金海[2]

(1. 国家海洋局海洋环境科学和数值模拟重点实验室, 青岛 266000;2. 南京气象学院, 南京 210044;3. 北京市气象台, 北京 100089)

摘要:发展了一个线性热带西太平洋两层模式,分别为混合层和温跃层,其密度、温度各不相同。利用这一模式分析了热带西太平洋对纬圈风应力的响应,求出西边界的解,以及解析地求得热带西太平洋温跃层厚度、洋流及海温分布。结果表明,热带西太平洋物理量的变异在 El Niño/La Niña 事件中起着重要作用,在温跃层中海温变化的振幅明显大于混合层,这从理论上支持了近年来的观测事实。

关键词:热带西太平洋;厄尔尼诺;拉尼娜;斜压响应

1 引言

近年来,热带西太平洋对 El Niño 或 La Niña(厄尔尼诺或拉尼娜)形成和发展中的重要性在理论上受到关注[1],观测表明产生 El Niño 或 La Niña 的初始海温距平,主要出现在"暖池"150 m 左右的斜温层,当它发展到一定强度后,沿着气候的温跃层向热带东太平洋传播,并上升到海表[2]。注意到近年来在动力学上研究热带西太平洋对风应力的响应时,一般用的是等值浅水模式[3],浅水模式虽然可以研究扰动向东传播的物理过程,但不能用来研究海温距平在次表层温跃层附近响应最强的物理机制。要研究这样的问题至少需要两层斜压模式,甚至多层斜压模式。

Cane[4]最早提出一个简单的线性两层海洋模式,用以研究赤道地区的表层风生流的演变。余丰宁等[5]将 Cane 的模式作了改进,考虑了海温的作用,数值试验表明海温对风应力的响应是有一定强度的;为使模式简化,他们将温跃层和混合层密度、温度认为是相同的,因此仍不能用来研究为什么斜温层海温对风应力的响应强度要大于混合层这样的问题。为此,采用了巢纪平[6]最近提出的低频近似,使模式的近似度一致协调。并将混合层及温跃层的温度、密度取成不同值,这样更接近实际海况,并能用来研究所提出的现象。

2 模式

将海洋分为两层(如图 1),上层厚度为 η,参照 Cane[4]认为其值是恒定的,但具有自身的密度和温度,分别为 ρ_1、T_1,这一层可认为是混合层;下层厚度为 h,密度和温度分别为 ρ_2、T_2,可认为是温跃层。上标"s"和"b" 分别表示混合层和温跃层,并用"′" 表示变量扰动量,用" −" 表示变量平均量。

* 地球物理学报,2002,45(2):176-187.

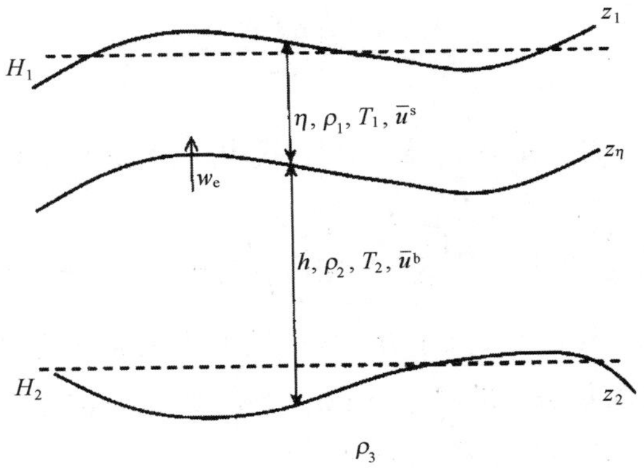

图 1　模式两层结构图

定义垂直平均量为

$$q^s = \eta^{-1}\int_{z_\eta}^{z_1} q\mathrm{d}z, \quad q^b = h^{-1}\int_{z_2}^{z_\eta} q\mathrm{d}z$$

$$\hat{q} = (h + \eta)^{-1}\int_{z_2}^{z_1} q\mathrm{d}z = (h + \eta)^{-1}(hq^b + \eta q^s) \tag{1}$$

其中 z_1 为海-气界面，z_η 为混合层-温跃层界面，z_2 为温跃层与准静止层的界面。在上、下两层的界面上

$$w(z_1) = \frac{\mathrm{d}z_1}{\mathrm{d}t} \tag{2}$$

设通过 $z=z_\eta$ 面另有一挟卷速度 w_e，于是有

$$w(z_\eta) = w(z_1) + w_e, \tag{3}$$

应用连续性方程后有

$$w_e = w(z_\eta) - w(z_1) = \int_{z_\eta}^{z_1} \Delta \cdot \boldsymbol{u}\mathrm{d}z = \eta\Delta \cdot \boldsymbol{u}^s \tag{4}$$

在另一方面，将连续方程从 z_1 到 z_2 积分得

$$\frac{\partial h}{\partial t} + \Delta \cdot (\eta\boldsymbol{u}^s + h\boldsymbol{u}^b) = 0 \tag{5}$$

设混合层的扰动温度为 T'^s，有方程

$$\frac{\partial T'^s}{\partial t} + w^s\left(\frac{\mathrm{d}\bar{T}^s}{\mathrm{d}z}\right)_s = 0 \tag{6}$$

近似地取

$$w^s = \frac{1}{2}w_e = \frac{1}{2}\eta\Delta \cdot \boldsymbol{u}^s \tag{7}$$

设 $\left(\frac{\mathrm{d}\bar{T}^s}{\mathrm{d}z}\right)_s \approx \frac{\Delta\bar{T}^s}{\eta}$，$\Delta\bar{T}^s$ 为混合层上、下的背景温度差，故有

$$\frac{\partial T'^s}{\partial t} + \left(\frac{1}{2}\Delta\bar{T}^s\right)\Delta \cdot \boldsymbol{u}^s = 0 \tag{8}$$

类似地，设温跃层的扰动温度为 T'^b，有方程

$$\frac{\partial T'^b}{\partial t} + w^b\left(\frac{\mathrm{d}\overline{T}^b}{\mathrm{d}z}\right)_b = 0 \tag{9}$$

式中 w^b 近似取 $\frac{1}{2}w(z_2) = \frac{1}{2}(\eta\Delta\cdot\boldsymbol{u}^s + h^*\Delta\cdot\boldsymbol{u}^b)$，式中 h^* 为 h 的平均值。设

$$\left(\frac{\mathrm{d}\overline{T}^b}{\mathrm{d}z}\right)_b \approx \frac{\Delta\overline{T}^b}{h^*}，于是有\frac{\partial T'^b}{\partial t} + \frac{1}{2}\left[\left(\frac{\eta}{h^*}\right)\Delta\cdot\boldsymbol{u}^s + \Delta\cdot\boldsymbol{u}^b\right]\Delta\overline{T}^b = 0 \tag{10}$$

设 $z_1 = H_1 + \eta_1$，$z_2 = H_2 + \eta_2$，$D = H_1 - H_2$，其中 H_1 和 H_2 分别是海洋无运动状态下海-气界面和混合层-准静止层界面，D 是海洋无运动状态下的厚度，而 η_1 及 η_2 则是海洋在运动状态下对这两层平均界面的扰动。设 $P(z_1) = P_0$，不难得出各层的静压力分布为

$$P_1 = P_0 + \rho_1 g(z_1 - z) \tag{11a}$$

$$P_2 = P_0 + \rho_1 g\eta + \rho_2 g(z_\eta - z) \tag{11b}$$

$$P_3 = P_0 + \rho_1 g\eta + \rho_2 gh + \rho_3 g(z_2 - z) \tag{11c}$$

其中，g 为重力加速度。

在 z_2 处，$P_2 = P_3$，由准静力平衡可假定 $\Delta P_3 = 0$，由此得

$$\Delta\eta_1 = \frac{\rho_3 - \rho_2}{\rho_3}\Delta h + \frac{\eta_2 - \eta_1}{\rho_3}\Delta\rho_2 \tag{12}$$

另外取准不可压缩近似，即

$$\frac{\Delta\rho}{\rho} \approx -\alpha\Delta T \tag{13}$$

式中 α 为海水的热膨胀系数。由此有

$$\frac{1}{\eta}\int_{z_\eta}^{z_1}\left(-\frac{\Delta P_1}{\rho_1}\right)\mathrm{d}z = -g'\Delta h + g(\eta_2 - \eta_1)\alpha\Delta T'_2 + \frac{1}{2}g\eta\alpha\Delta T'_1 \tag{14}$$

$$\frac{1}{h}\int_{z_2}^{z_\eta}\left(-\frac{\Delta P_2}{\rho_2}\right)\mathrm{d}z = -g'\Delta h + g\left(\eta_2 - \eta_1 + \frac{h}{2}\right)\alpha\Delta T'_2 + g\eta\alpha\Delta T'_1 \tag{15}$$

式中 $g' = g\left(\frac{\rho_3 - \rho_2}{\rho_3}\right)$ 为视重力。

3 控制方程

在线性情况下，模式的控制方程组为

$$\frac{\partial\boldsymbol{u}^s}{\partial t} + f\boldsymbol{k}\times\boldsymbol{u}^s = -g'\Delta h + \frac{\eta}{2}g\alpha\Delta T'_1 + (\eta_2 - \eta_1)g\alpha\Delta T'_2 + \frac{\tau}{\eta} \tag{16}$$

$$\frac{\partial\boldsymbol{u}^b}{\partial t} + f\boldsymbol{k}\times\boldsymbol{u}^b = -g'\Delta h + \eta g\alpha\Delta T'_1 + \left(\frac{h^*}{2} + \eta_2 - \eta_1\right)g\alpha\Delta T'_2 \tag{17}$$

$$\frac{\partial h}{\partial t} + (\eta\Delta\cdot\hat{u}^s + h^*\Delta\cdot\hat{\boldsymbol{u}}^b) = 0 \tag{18}$$

$$\frac{\partial T'_1}{\partial t} + \left(\frac{1}{2}\Delta T^s\right)\Delta\cdot\boldsymbol{u}^s = 0 \tag{19}$$

$$\frac{\partial T'_2}{\partial t} + \frac{1}{2}\left[\left(\frac{\eta}{h^*}\right)\Delta\cdot\boldsymbol{u}^s + \Delta\cdot\boldsymbol{u}^b\right]\Delta T^b = 0 \tag{20}$$

此为模式的基本方程组,其中,f 为地转参数,$\boldsymbol{k}$ 为 z 坐标单位矢量,τ 为风应力。

4　正压模和斜压模

现将运动分成正压模和斜压模,分别定义成

$$\hat{u} = \frac{1}{D}(h^* \boldsymbol{u}^b + \eta \boldsymbol{u}^s) = \gamma \boldsymbol{u}^s + (1-\gamma)\boldsymbol{u}^b, \quad \tilde{u} = \gamma(\boldsymbol{u}^s - \boldsymbol{u}^b) \tag{21}$$

式中 $\gamma=\eta/D$。考虑到 $\eta \ll D, (\eta_2-\eta_1) \ll \frac{h^*}{2}$,由此正压模和斜压模方程分别为

$$\frac{\partial \hat{u}}{\partial t} + fk \times \hat{u} = -g'\Delta h + \frac{1}{2}g\alpha(2\eta\Delta T'_1 + h^*\Delta T'_2) + \gamma\frac{\tau}{\eta} \tag{22}$$

$$\frac{\partial \tilde{u}}{\partial t} + fk \times \tilde{u} = -\frac{\gamma}{2}g\alpha(\eta\Delta T'_1 + h^*\Delta T'_2) + \gamma\frac{\tau}{\eta} \tag{23}$$

$$\frac{\partial T'_1}{\partial t} + \frac{\Delta T^s}{2\eta}\Delta \cdot (\eta\hat{\boldsymbol{u}} + h^*\tilde{\boldsymbol{u}}) = 0 \tag{24}$$

$$\frac{\partial T'_2}{\partial t} + \frac{1}{2}\Delta T^b\left[\left(1+\frac{\eta}{h^*}\right)\Delta \cdot \hat{\boldsymbol{u}} + \Delta \cdot \tilde{\boldsymbol{u}}\right] = 0 \tag{25}$$

引进重力波波速

$$c = (g'D)^{\frac{1}{2}} \tag{26}$$

取特征量

$$(x,y) \propto (c/2\beta)^{\frac{1}{2}}, h \propto D, (\hat{\boldsymbol{u}},\tilde{\boldsymbol{u}}) \propto c, T'_1 \propto \Delta\bar{T}^s, T'_2 \propto \Delta\bar{T}^b, \tau \propto \tau_0, t \propto (2\beta c)^{-\frac{1}{2}}$$

其中 β 为地转参数,则在赤道 β 平面上量纲为 1 的方程组为

$$\frac{\partial \hat{u}}{\partial t} + \frac{1}{2}y\boldsymbol{k} \times \hat{\boldsymbol{u}} = -\Delta h + \left(\frac{c_1}{c}\right)^2 2\Delta T'_1 + \left(\frac{c_2}{c}\right)^2 \Delta T'_2 + F\tau \tag{27}$$

$$\frac{\partial \tilde{u}}{\partial t} + \frac{1}{2}y\boldsymbol{k} \times \tilde{\boldsymbol{u}} = -\gamma\left(\frac{c_1}{c}\right)^2 \Delta T'_1 - \left(\frac{c_2}{c}\right)^2 \Delta T'_2 + F\tau \tag{28}$$

式中 $F=(\gamma\tau_0 T/\eta c)$,$T=(2\beta c)^{-\frac{1}{2}}$,而

$$c_1 = \left(\frac{1}{2}gD\alpha\Delta\bar{T}^s\right)^{\frac{1}{2}}, \quad c_2 = \left(\frac{1}{2}gD\alpha\Delta\bar{T}^{T^s}b\right)^{\frac{1}{2}} \tag{29}$$

为混合层和温跃层的重力内波波速。同时有方程

$$\frac{\partial h}{\partial t} + \Delta \cdot \hat{\boldsymbol{u}} = 0 \tag{30}$$

$$\frac{\partial T'_1}{\partial t} + \frac{1}{2}\Delta \cdot \left(\hat{\boldsymbol{u}} + \frac{h^*}{\eta}\tilde{\boldsymbol{u}}\right) = 0 \tag{31}$$

$$\frac{\partial T'_2}{\partial t} + \frac{1}{2}\left[\left(1+\frac{\eta}{h^*}\right)\hat{\boldsymbol{u}} + \tilde{\boldsymbol{u}}\right] = 0 \tag{32}$$

引进量 $T'=\left(\frac{c_1}{c}\right)^2 T'_1+\left(\frac{c_2}{c}\right)^2 T'_2$,由方程式(31),式(32)得到

$$\frac{\partial T'}{\partial t} + A\Delta \cdot \hat{\boldsymbol{u}} + B\Delta \cdot \tilde{\boldsymbol{u}} = 0 \tag{33}$$

式中

$$A = \frac{1}{2}\left[\left(\frac{c_1}{c}\right)^2 + \left(1 + \left(\frac{\eta}{h^*}\right)\right)\left(\frac{c_2}{c}\right)^2\right] \tag{34}$$

$$B = \frac{1}{2}\left[\left(\frac{h^*}{\eta}\right)\left(\frac{c_1}{c}\right)^2 + \left(\frac{c_2}{c}\right)^2\right] \tag{35}$$

方程式(27)、式(28)改写成

$$\frac{\partial \hat{u}}{\partial t} + \frac{1}{2}y\boldsymbol{k} \times \hat{\boldsymbol{u}} = - \Delta h + \Delta T' + F\tau \tag{36}$$

$$\frac{\partial \tilde{u}}{\partial t} + \frac{1}{2}y\boldsymbol{k} \times \tilde{\boldsymbol{u}} = - \gamma\Delta T' + F\tau \tag{37}$$

5 快波近似波速

设正压模(平均模)的重力波波速 c 满足条件

$$c \gg c_2 > c_1 \tag{38}$$

称此为快波近似。在快波近似下式(36)变成

$$\frac{\partial \hat{u}}{\partial t} + \frac{1}{2}y\boldsymbol{k} \times \hat{\boldsymbol{u}} = - \Delta h + F\tau \tag{39}$$

和

$$\frac{\partial h}{\partial t} + \Delta \cdot \hat{\boldsymbol{u}} = 0 \tag{40}$$

此两个方程构成对正压模的封闭方程组。

如果考虑在式(33)中 $B>A$,同时认为正压模的水平辐散很小,则斜压模方程为

$$\frac{\partial \tilde{u}}{\partial t} + \frac{1}{2}y\boldsymbol{k} \times \tilde{\boldsymbol{u}} = - \gamma\Delta T' + F\tau \tag{41}$$

$$\frac{\partial T'}{\partial t} + B\Delta \cdot \tilde{\boldsymbol{u}} = 0 \tag{42}$$

可见正压模方程和斜压模方程在形式上都和浅水运动方程相似。

再作变换,$T' \propto (B/\gamma)^{\frac{1}{2}}, t \propto (B\gamma)^{-\frac{1}{4}}, (x,y) \propto (B\gamma)^{\frac{1}{4}}$,于是有

$$\frac{\partial \tilde{u}}{\partial t'} + \frac{1}{2}y'\boldsymbol{k} \times \tilde{\boldsymbol{u}} = - \Delta' T + (B\gamma)^{-\frac{1}{4}}F\tau \tag{43}$$

$$\frac{\partial T}{\partial t'} + \Delta' \cdot \tilde{\boldsymbol{u}} = 0 \tag{44}$$

由于 $B\gamma \leqslant 1$,因此斜压模的特征时间要比正压模大,即过程慢,而特征尺度要比正压模小。

6 正压模对赤道纬圈风应力的响应

在低频近似下[4],对 $n=0,2,4$ 的截断模应用与参考文献[1]相同的解法。所用风应力形式为在

$(0,t_0)$这一时段内,其振幅为 1,当 $t>t_0$ 时,风应力消失。对于这样形式的风应力,其拉普拉斯变换为

$$X = \frac{1 - e^{-t_0 s}}{s} \tag{45}$$

风应力向东(西风),$F>0$,向西(东风),$F<0$ 。其在 y 方向的分布取单一模,即 $D_0(y)$。

在上述情况下正压模方程式(47)、式(48)的解同巢纪平、巢清尘文[2],除了现在增加了一个 $n=4$ 的经圈模。解法和结果的数学表达式略,只给出一些结果的图。

在以下的数值计算中,取 $t_0=10$,其他参数为:$\eta=5\ 000$ cm,$h^*=15\ 000$ cm,$D=20\ 000$ cm,$\gamma=0.25$,$c=200$ cm/s^2。以此计算得 $x=210$ km,$y=210$ km,$t=29.3$ h。

根据前面的风应力形式,得到沿赤道温跃层($h=h_0|_{y=0}-h_2|_{y=0}+3h_4|_{y=0}$)的传播如图 2(a)所示。当为西风异常时,大洋内部在西风应力作用下,温跃层是正距平,但在西边界处激发出来的 Kelvin 波是冷的(即负的温跃层距平)。另一方面,西边界还激发出 Rossby 短波,但 Rossby 短波在离开边界后不久其能量很快就频散掉了,而 Kelvin 波则沿着赤道向东传播过去。注意到当 $t=4t_0=40$ 时,整个洋面几乎都为冷水覆盖。即,尽管异常西风的作用时间在 $t=t_0=10$ 后很快就张弛了,但在西边界附近仍可以引起温跃层等物理量的异常,并且很快向大洋东部以 Kelvin 波速传播。然而,同样的风应力若作用于东边界,则场的信号向西传播要慢得多,而且振幅也要小于大洋西部的物理量[1]。因此,在热带西太平洋出现风应力异常或海洋状况异常时,更容易影响到大洋中、东部,从而可以引起较强的暖水/冷水事件,即 El Niño/La Niña。

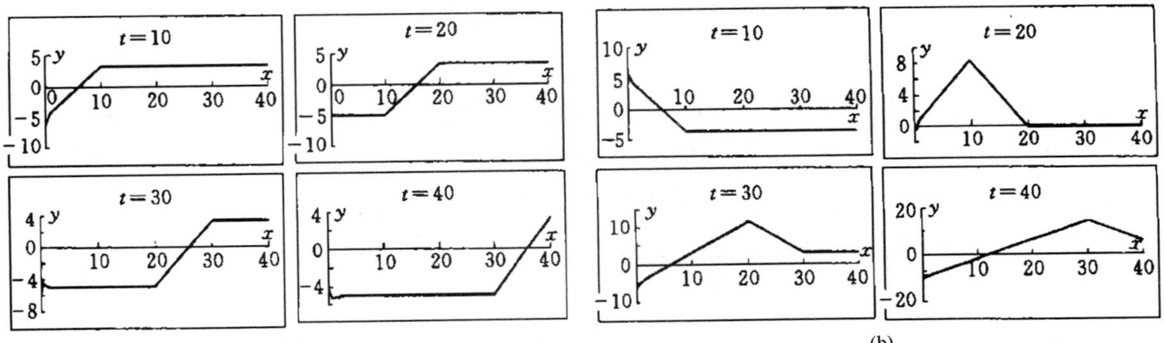

图 2　西太平洋边界附近温跃层在 t 时刻沿赤道随 x 的变化分布(t 的量纲为 1,下同)

(y 轴为温跃层厚度,x 轴为离开边界的距离,后同)

(a)洋面上先吹西风,在 $t=10$ 时停止;(b)洋面上先吹东风,在 $t=10$ 后改吹西风

若设风应力在$(0,t_0)$这一时段内,其振幅为-1,当 $t>t_0$ 时,风应力以相同强度转向,即洋面上先吹东风,到 $t=10$ 时刻转成西风。对于这样形式的风应力,其拉普拉斯变换为

$$X = \frac{1 - 2e^{-t_0 s}}{s} \tag{46}$$

得到沿赤道温跃层的变化如图 2(b)所示。这时,可看到当东风作用时,西边界处先激发出一个暖 Kelvin 波,一旦转为西风作用,又激发出一个冷 Kelvin 波向东传播,波动经过的地方,跃层由深变浅。注意到风场在 $t_0=10$ 时刻就已变化,然而边界处温跃层的符号一直到 $t>20$ 后才开始反号。也就是说,在吹了同样时间的反向风后海洋的变化才开始,这也就是通常所说的海洋的"记忆" 。另外值得注意的是,在 $t<20$ 时,由于暖的 Kelvin 波向东传播,大洋温跃层厚度沿赤道呈东高西低的形势,相对

海温而言是东暖西冷，当 $t>20$ 之后，暖水逐渐传到大洋东部，而西边界也开始有了冷水出现，当 $t=40$ 时，海水分布相当于一个 El Niño 位相。由此可见，赤道附近前期强劲的东风致使热带西太平洋出现距平意义下的暖水，Kelvin 波把暖水传向大洋东部；而当出现西风时只是由于海洋对风应力响应的“时滞”效应，西边界附近的海水仍保持正的距平，事实上，由于西风将产生离岸流，因此为西边界附近出现冷水创造了条件，当然在这时大洋东部的海水还是很暖的，即可以达到 El Niño 位相，但正是西风的盛行造成 El Niño 位相向 La Niña 位相的转变。

7　斜压模对纬圈风应力的响应

在只有均匀纬向风应力作用时斜压模方程的标量式为

$$\frac{\partial \tilde{u}}{\partial t'} - \frac{1}{2}y'\tilde{v} = -\frac{\partial T}{\partial x'} + (B\gamma)^{-\frac{1}{4}}F\tau^{x} \tag{47}$$

$$\frac{\partial \tilde{v}}{\partial t'} + \frac{1}{2}y'\tilde{u} = -\frac{\partial T}{\partial y'} \tag{48}$$

$$\frac{\partial T}{\partial t'} + \frac{\partial \tilde{u}}{\partial x'} + \frac{\partial \tilde{v}}{\partial y'} = 0 \tag{49}$$

为节省篇幅，方程求解方法详见附录。

8　计算结果

图 3 给出了斜压模温度沿赤道传播解，由斜压模温度定义就可计算混合层和温跃层温度，其中混合层单位为 0.1℃，温跃层温度单位为 1℃，在以下图中为便于比较，已将混合层温度值乘 0.1。值得注意的是，此时不管是混合层温度还是温跃层温度均受到正压模和斜压模的共同作用。当风应力的形成为先吹东风到 $t=10$ 时改吹西风时，混合层和温跃层温度的响应强度分别见图 4(a)和图 4(b)。

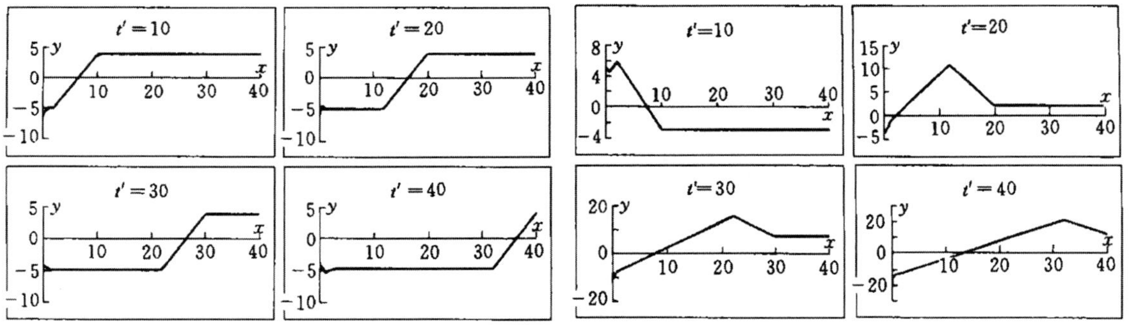

(a)　　　　(b)

图 3　西太平洋边界附近斜压模温度在 t'时刻沿赤道随 x 的变化分布

(a)洋面上先吹西风，在 $t'=10$ 时停止；(b)洋面上先吹东风，在 $t'=10$ 后改吹西风

由图 4 可见，当吹东风时，在大洋内部无论是混合层或温跃层温度距平都是负的，但在边界上却激发出一个暖的 Kelvin 波，波在温跃层的强度要比混合层大了约 5 倍，这个暖的 Kelvin 波向东传播，到 $t=30$ 时大洋内部在波阵面的前方也由冷水变成暖水，但在边界附近由于离岸流的作用，又转变为冷

水。洋流的响应强度则反过来,混合层的纬向流强度要比温跃层大出 5~6 倍(图 5)。

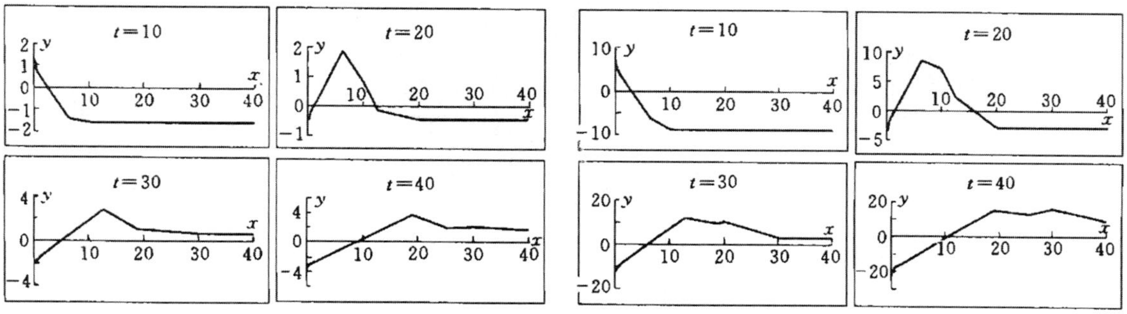

(a)　　(b)

图 4　西太平洋边界附近斜压模温度在 t 时刻沿赤道随 x 的变化分布

(a)混合层温度;(b)温跃层温度

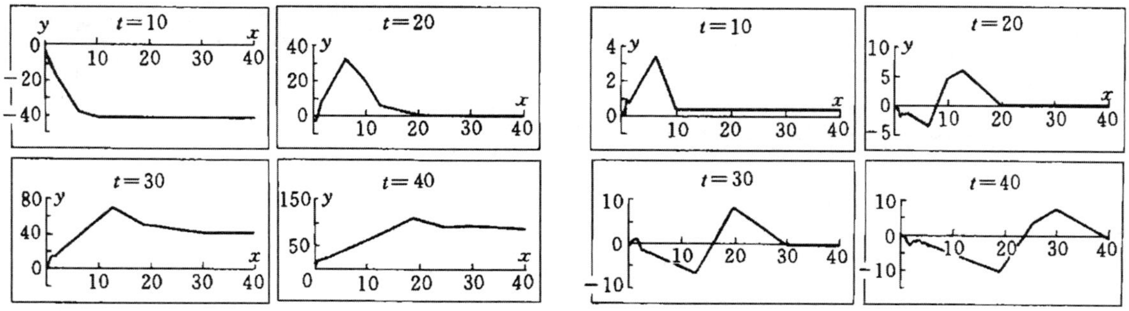

(a)　　(b)

图 5　西太平洋边界附近温跃层洋流在 t 时刻沿赤道随 x 的变化分布

(a)混合层洋流;(b)温跃层洋流

图 6(a)、图 6(b)分别是 $t=20$ 时表层和次表层的温度分布,可以看到这时在热带西太平洋边界附近是正温度距平区,但次表层的温度振幅要比表层几乎大了一个量级。图 7 是 $t=40$ 时的温度分布。这时正的温度距平区已东传到大洋中部甚至到东部(正温度距平前沿已超过 40 个 Rossby 变形半径),表明 El Niño 已发展到成熟期,这时次表层的温度距平仍大于表层,但只大了 1 倍左右。在温度正距平区的西边,直到边界已生成一负距平区,表明一个 La Niña 事件正在形成,这时次表层的负温度距平要比表层大了 5~6 倍。这一计算表明,无论是 El Niño 或者是 La Niña 事件,其初始的温度距平都主要在次表层发展起来。图 8 是纬向洋流在 $t=40$ 时的分布,可以看到表层和次表层洋流是反相的,同时在表层向东的洋流基本上对应着正的温度距平。

9　结论

本文发展了一个简单的线性两层海洋模式,用理论分析和解析结果分别说明热带西太平洋在 El Niño/La Niña 事件中的重要贡献,它们对于发生、发展 El Niño/La Niña 事件起主导作用。也就是说,在太平洋西边界大气强迫的异常可以引发次表层中大的温度异常,并主要随着东传的 Kelvin 波传播到大洋的东部,从而导致整个大洋的异常。事实上,回顾 El Niño 史,如 1982/1983 年、1997/1998 年等几次较为强的 El Niño 事件,它们的先兆均是先出现于西太平洋的次表层,而后才传播到中、东太平洋

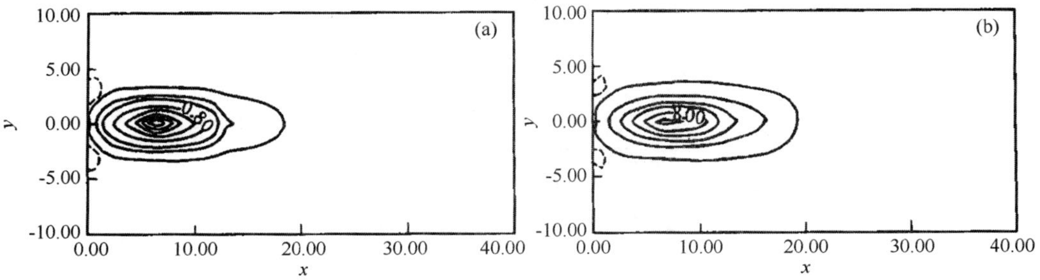

图 6　$t=20$ 时刻温度分布(虚线为负值)

(a)混合层温度;(b)温跃层温度

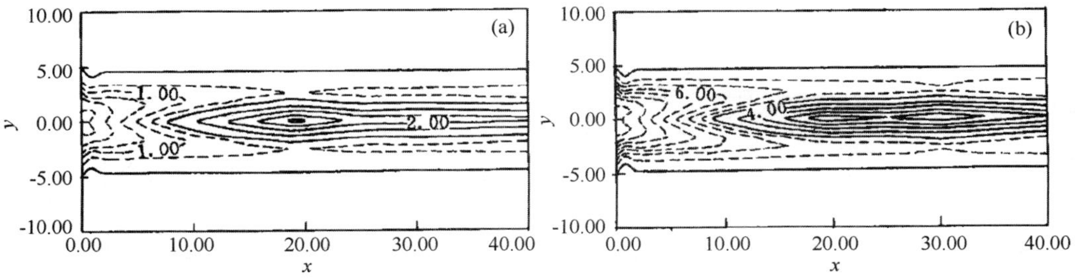

图 7　$t=40$ 时刻温度分布(虚线为负值)

(a)混合层温度;(b)温跃层温度

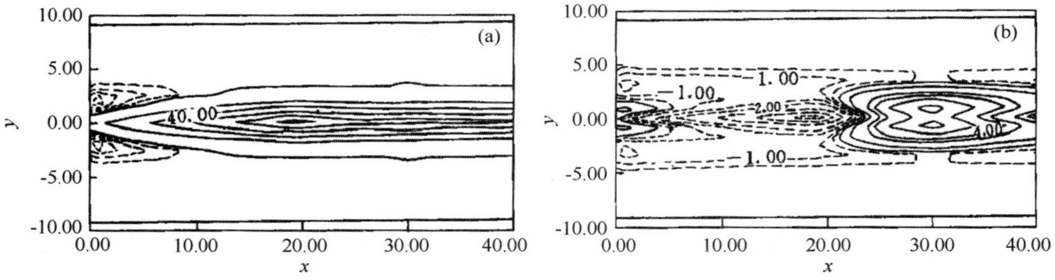

图 8　$t=40$ 时刻洋流分布(虚线为负值)

(a)混合层洋流;(b)温跃层洋流

而形成大家共识的 El Niño 事件的。

附　录

在只有均匀纬向风应力作用时斜压模方程的标量式为

$$\frac{\partial \widetilde{u}}{\partial t'} - \frac{1}{2} y' \widetilde{v} = -\frac{\partial \widetilde{T}}{\partial x'} + (B\gamma)^{-\frac{1}{4}} F_\tau^x \tag{1A}$$

$$\frac{\partial \widetilde{v}}{\partial t'} + \frac{1}{2} y' \widetilde{u} = -\frac{\partial \widetilde{T}}{\partial y'} \tag{2A}$$

$$\frac{\partial\widetilde{T}}{\partial t'}+\frac{\partial\widetilde{u}}{\partial x'}+\frac{\partial\widetilde{v}}{\partial y'}=0 \tag{3A}$$

引进变量

$$\widetilde{q}=\widetilde{T}+\widetilde{u},\quad \widetilde{r}=\widetilde{T}-\widetilde{u} \tag{4A}$$

代入方程组,并将物理量用 Weber 函数(抛物圆柱函数)展开成

$$(\widetilde{q},\widetilde{r},\widetilde{v})=\sum_n[\widetilde{q}_n(x',t'),\widetilde{r}_n(x',t'),\widetilde{v}_n(x',t')]D_n(y') \tag{5A}$$

同时,将正压物理量风应力也用抛物圆柱函数展开

$$\tau=\sum_m\tau_m(x,t)D_m(y) \tag{6A}$$

注意式(5A)和式(6A)的坐标尺度是不一样的。

引进拉普拉斯变换

$$(\widetilde{Q}_n,\widetilde{R}_n,\widetilde{V}_n)=\int_0^\infty e^{-st'}(\widetilde{q}_n,\widetilde{r}_n,\widetilde{v}_n)\,dt' \tag{7A}$$

由拉普拉斯变换的相似性可知

$$L[\tau_m]=L[\tau_m((B\gamma)^{-\frac{1}{4}}t')]=(B\gamma)^{\frac{1}{4}}\hat{\tau}_m[(B\gamma)^{-\frac{1}{4}}s] \tag{8A}$$

利用 Weber 函数的正交性及递推关系。则方程式(1A)至式(3A)化为

$$\frac{d\hat{Q}_0}{dx'}+s\hat{Q}_0=\frac{1}{\sqrt{2\pi}}[F\hat{\tau}_m^x]\int_{-\infty}^{+\infty}D_m(y)D_0(y')\,dy' \tag{9A}$$

$$s\frac{d\hat{Q}_1}{dx'}+\left(s^2+\frac{1}{2}\right)\hat{Q}_1=\frac{1}{\sqrt{2\pi}}(B\gamma)^{\frac{1}{4}}s[F\hat{\tau}_m^x]\int_{-\infty}^{+\infty}D_m(y)D_1(y')\,dy' \tag{10A}$$

$$\begin{aligned}&s\frac{d^2\widetilde{Q}_{n+2}}{dx'^2}+\frac{1}{2}\frac{d\widetilde{Q}_{n+2}}{dx'}-\left[\frac{1}{2}(2n+3)s+s^3\right]\widetilde{Q}_{n+2}=\\&-\frac{1}{(n+2)!\sqrt{2\pi}}(B\gamma)^{\frac{1}{2}}s^2[F\hat{\tau}_m^x]\int_{-\infty}^{+\infty}D_m(y)D_{n+2}(y')\,dy'-\\&\frac{1}{2}\frac{n+1}{(n+2)!\sqrt{2\pi}}[F\hat{\tau}_m^x]\int_{-\infty}^{+\infty}D_m(y)D_{n+2}(y')\,dy'+\\&\frac{1}{2}\frac{1}{n!\sqrt{2\pi}}[F\hat{\tau}_m^x]\int_{-\infty}^{+\infty}D_m(y)D_n(y')\,dy'\end{aligned} \tag{11A}$$

$$\begin{aligned}&s\frac{d^2R_n}{dx'^2}+\frac{1}{2}\frac{dR_n}{dx'}-\left[\frac{1}{2}(2n+3)s+s^3\right]\widetilde{R}_n=\\&\frac{1}{n!\sqrt{2\pi}}(B\gamma)^{\frac{1}{2}}s^2[F\hat{\tau}_m^x]\int_{-\infty}^{+\infty}D_m(y)D_n(y')\,dy'-\\&\frac{1}{2}\frac{(n+1)(n+2)}{(n+2)!\sqrt{2\pi}}[F\tau_m^x]\int_{-\infty}^{+\infty}D_m(y)D_{n+2}(y')\,dy'+\\&\frac{1}{2}\frac{n+2}{n!\sqrt{2\pi}}[F\tau_m^x]\int_{-\infty}^{+\infty}D_m(y)D_n(y')\,dy'\end{aligned} \tag{12A}$$

$$s\widetilde{V}_n+\frac{1}{2}(n+1)\widetilde{Q}_{n+1}-\frac{1}{2}\widetilde{R}_{n-1}=\frac{1}{n!\sqrt{2\pi}}F\hat{\tau}^y_m\int_{-\infty}^{+\infty}D_m(y)D_n(y')\mathrm{d}y' \tag{13A}$$

以及

$$s\frac{\mathrm{d}\hat{Q}_n}{\mathrm{d}x'}+\left[\frac{n}{2}+s^2\right]\hat{Q}_n-\frac{1}{2}R_{n-2}=\frac{1}{n!\sqrt{2\pi}}(B\gamma)^{\frac{1}{4}}s[F\hat{\tau}^x_m]\int_{-\infty}^{+\infty}D_m(y)D_n(y')\mathrm{d}y' \tag{14A}$$

$$s\frac{\mathrm{d}\hat{R}_n}{\mathrm{d}x'}-\left[\frac{1}{2}(n+1)+s^2\right]\hat{R}_n+\frac{1}{2}(n+1)(n+2)\hat{Q}_{n+2}=\frac{1}{n!\sqrt{2\pi}}(B\gamma)^{\frac{1}{4}}s[F\hat{\tau}^x_m]\int_{-\infty}^{+\infty}D_m(y)D_n(y')\mathrm{d}y' \tag{15A}$$

注意到式(14A)、式(15A)对x'是一阶的,它们可以作为二阶方程式(11A)、式(12A)的边界条件。

当风应力仍取$\tau^x=X(t)D_0(y)$,并采用低频近似后,得到截断模方程($n=0,2,4$)

$$\frac{\mathrm{d}\hat{Q}_0}{\mathrm{d}x'}+s\hat{Q}_0=F\widetilde{X}a_{00} \tag{16A}$$

$$s\frac{\mathrm{d}^2\hat{Q}_2}{\mathrm{d}x'^2}+\frac{1}{2}\frac{\mathrm{d}\hat{Q}_2}{\mathrm{d}x'}-\frac{3}{2}s\hat{Q}_2=-\frac{1}{2}F\widetilde{X}a_{02}+\frac{1}{2}FXa_{00} \tag{17A}$$

$$s\frac{\mathrm{d}^2\hat{Q}_4}{\mathrm{d}x'^2}+\frac{1}{2}\frac{\mathrm{d}\hat{Q}_4}{\mathrm{d}x'}-\frac{7}{2}s\hat{Q}_4=-\frac{3}{2}FXa_{04}+\frac{1}{2}F\widetilde{X}a_{02} \tag{18A}$$

$$s\frac{\mathrm{d}^2\widetilde{R}_0}{\mathrm{d}x'^2}+\frac{1}{2}\frac{\mathrm{d}\widetilde{R}_0}{\mathrm{d}x'}-\frac{3}{2}sR_0=-F\widetilde{X}a_{02}+F\widetilde{X}a_{00} \tag{19A}$$

$$s\frac{\mathrm{d}^2\widetilde{R}_2}{\mathrm{d}x'^2}+\frac{1}{2}\frac{\mathrm{d}\widetilde{R}_2}{\mathrm{d}x'}-\frac{7}{2}s\widetilde{R}_2=-6F\widetilde{X}a_{04}+2F\widetilde{X}a_{02} \tag{20A}$$

$$s\frac{\mathrm{d}^2\widetilde{R}_4}{\mathrm{d}x'^2}+\frac{1}{2}\frac{\mathrm{d}\widetilde{R}_4}{\mathrm{d}x'}-\frac{11}{2}sR_4=-15F\widetilde{X}a_{06}+3F\widetilde{X}a_{04} \tag{21A}$$

其中系数a_{mn}是利用积分公式$\int_{-\infty}^{+\infty}\mathrm{e}^{-ay^2}\mathrm{d}y=\sqrt{\frac{\pi}{a}}$及 Weber 函数的正交性计算而得,分别为

$$a_{00}=\sqrt{\frac{2}{1+(B\gamma)^{\frac{1}{2}}}},\quad a_{02}=\left(\frac{1}{1+(B\gamma)^{\frac{1}{2}}}-\frac{1}{2}\right)a_{00},$$

$$a_{04}=\left(\frac{1}{2[1+(B\gamma)^{\frac{1}{2}}]^2}-\frac{1}{2[1+(B\gamma)^{\frac{1}{2}}]}+\frac{1}{8}\right)a_{00}$$

边界条件方程($x=0$)写为

$$s\frac{\mathrm{d}\hat{Q}_2}{\mathrm{d}x'}+\hat{\widetilde{T}}_2-\frac{1}{2}\hat{\widetilde{T}}_0=(B\gamma)^{\frac{1}{4}}sF\hat{X}a_{02} \tag{22A}$$

$$s\frac{\mathrm{d}\hat{Q}_4}{\mathrm{d}x'}+2\hat{\widetilde{T}}_4-\frac{1}{2}\hat{\widetilde{T}}_2=(B\gamma)^{\frac{1}{4}}sF\hat{X}a_{04} \tag{23A}$$

$$s\frac{\mathrm{d}\hat{R}_0}{\mathrm{d}x'}-\frac{1}{2}\hat{\widetilde{T}}_0+\hat{\widetilde{T}}_2=(B\gamma)^{\frac{1}{4}}sFXa_{00} \tag{24A}$$

$$s\frac{d\hat{R}_2}{dx'} - \frac{3}{2}\hat{\tilde{T}}_2 + 6\hat{\tilde{T}}_4 = (B\gamma)^{\frac{1}{4}} sF\hat{X}a_{02} \tag{25A}$$

$$s\frac{d\hat{R}_4}{dx'} - \frac{5}{2}\hat{\tilde{T}}_4 = (B\gamma)^{\frac{1}{4}} sF\hat{X}a_{04} \tag{26A}$$

以及 $x\to\infty$ 时，$(\widetilde{Q}_0, \widetilde{Q}_2, \widetilde{Q}_4, \widetilde{R}_0, \widetilde{R}_2, \widetilde{R}_4)$有界。

注意到除非齐次项和坐标尺度不同外，斜压模方程和正压模方程在形式和结构上都和正压模方程一样，因此除系数外其解也和正压模一样，为节省篇幅在此略。

参考文献

[1] 巢纪平，张丽．赤道不同海域对信风张弛的响应特征——对 El Niño 研究的启示．大气科学，1998，22(4)：428-442.

[2] 巢纪平，巢清尘．热带太平洋对风应力响应的动力学．大气科学.

[3] 巢清尘，巢纪平．热带西太平洋和东印度洋对 ENSO 发展的影响．自然科学进展，2001，11(12)：1293-1300.

[4] Cane M. A. The response of an equatorial ocean to simple wind st ress patterns：Ⅰ. Model formulation and analytic results. *Journal of Marine Research*，1979，37(2)：233-252.

[5] She Fengning，Chao Jiping，Wang Lizhi. Numerical Experiment of the Influences of Heat and Wind Stress on the Equatorial Ocean Circulation. *Acta Meteorologica Sinica*，1991，5(2)：215-228.

[6] 巢纪平．热带斜压大气的适应运动和发展运动．中国科学(D 辑)，1999，29(3)：279-288.

热带大洋东、西部对风应力经圈不对称的响应*

巢纪平[1],陈峰[2]

(1.国家海洋环境预报研究中心,北京 100081;2. 中国科学院大气物理研究所, 北京 100029)

摘要:热带海洋,特别是热带太平洋,物理场在东、西两部分的经圈结构很不相同,西太平洋"暖池"的温度分布对赤道基本上是对称的,而东太平洋的"冷舌"偏在赤道以南,对赤道明显不对称。作者从波动性质解释了这种分布的特征,指出在西太平洋,由于对赤道对称的向东的Kelvin波具有较大的振幅,其权重明显大于Rossby短波,致使物理场具有对赤道的对称性;而在东太平洋,由边界激发出的偶次的和奇次的Rossby长波,振幅权重很相近,从而使物理场显不对称性。

关键词:对称和不对称结构;Kelvin 波;Rossby 波;风应力

1 引言

太阳辐射是使地球大气和海洋运动的最基本的能量。如果没有海陆分布,就年平均分布来讲,在太阳辐射作用下,大气和海洋的气候状态对赤道应该是对称的。海陆分布改变了气候状态在纬圈方向的对称性,同时使海洋物理场的不对称性要比大气来得更明显,这是因为发生在大气和海洋中的物理过程并不相同。例如,在热带太平洋盛吹偏东的信风,即使这一风场沿纬圈是均匀的,海洋对它的响应在大洋的东、西部分也不相同。在大洋东部,在信风作用下,洋流是向西的,并且是离岸的,为了补偿向西流,在海洋边界附近将产生向表层的涌升流,这样,使得那里的表层海水变冷。在大洋西部则相反,向西的洋流,在西边界附近将产生向下的垂直洋流,这样使得那里的表层海水变暖。同时由于这种过程,使东边界附近的温跃层厚度要比西边界附近浅。由此可见,即使是沿纬圈均匀的风应力也能使大洋东、西部形成物理场结构上的不对称。

另一方面,在热带海洋特别是热带太平洋,在东、西部分物理场在经圈方向上的结构也是不同的。在大洋西部的"暖池",海表温度的分布对赤道基本上是对称的,而在大洋东部,"冷舌"的最冷部分偏在南半球。这种东、西部分物理场对赤道的结构不同,很难用水平洋流的分布来解释。因为在理想情况下,大洋东部的离岸流,将使表层海水沿边界(及附近)向赤道辐合,大洋西部的向岸流,将使表层海水沿边界(及附近)向两极辐散,它们不会造成物理场对赤道的不对称性。Philander 等指出,海洋物理场对赤道的对称或不对称结构的形成,应有其他的原因,他们在一篇启发性但未见刊的文章中讨

* 大气科学,2000,24(6):723-738.

国家重点基础研究发展规划项目"我国重大气候和天气灾害机理和预测理论研究"第一部分:"我国重大气候灾害的形成机理和预测理论研究"项目(G1998040900)资助

论了几种可能的过程。

在 Philander 等提出这一问题后，Xie 和 Philander[1] 利用一个对纬圈平均的混合层海洋模式和一个简单的大气模式耦合，来研究一些过程对造成赤道不对称的贡献。他们指出，由混合过程造成赤道及两侧的涌升流，对海温的分布有重要的作用。由于在太平洋东部，温跃层较浅，海洋对风应力的响应快而强，因此对赤道不对称的风应力要比温跃层深的大洋西部更容易造成物理场对赤道的不对称分布。

Xie[2] 利用大气和海洋耦合的 GCM 来研究这种不对称现象，指出海气相互作用对形成不对称状态有重要作用。即使模式的气候状态对赤道对称，但一个对赤道不对称的初始扰动，由于不稳定的海气相互作用，可演变出一个不对称性物理场。也由于大洋东部温跃层较浅，因而在大洋东部更易形成不对称的物理场。这一结果支持了前面两维模式中不对称状态的发展机理。

由海气相互作用造成海洋状态的不对称性可能是一种重要的机制。但这种机制有一个重要基础，即考虑了大洋东、西部分的温跃层深度是不一样的。大洋东部浅的温跃层，由于海洋中垂直混合过程的发展，从而使海气相互作用强烈，使一个对赤道不对称的初始扰动更易得到发展，而造成对赤道不对称状态。如果温跃层在大洋的东、西部分是一样的，则即使是强烈的海气相互作用，也不可能使大洋东、西部分的物理场出现不同的经圈结构。

现在我们试图从纯海洋动力学的角度，即主要从分析热带海洋波动的性质来研究海洋物理场在大洋西部和东部将易形成的结构状态。

2　模式和启动海洋运动的风应力

如果大气和海洋的重力波波速分别为 C_a 和 C，一般来讲，$C/C_a \ll O(1)$，如大气运动和海洋运动的空间特征尺度分别为 $L_a=(C_a/2\beta)^{1/2}$ 和 $L=(C/2\beta)^{1/2}$，则这两个尺度显然是不一样的，并有 $L_a \ll L$。在另一方面，大气和海洋运动的特征时间分别为 $T_a=(2\beta C_a)^{-1/2}$ 和 $T=(2\beta C)^{-1/2}$，它们也是不一样的，并有 $T_a \ll T$，即大气运动的特征时间要比海洋运动来得快。这两个不同的空间尺度和时间尺度，给问题的研究带来一定的复杂性，但如研究的是一种慢过程，则可以取 T 作为大气运动的特征时间。进而，如果风应力是沿纬圈的（如信风），并且是均匀的（与 x 无关），则可以用 Weber 函数 D_n 写成

$$\tau^x = \sum_{n=0}^{\infty} \tau_n^x(t) D_n(y) \tag{1}$$

这是给定的启动海洋运动的风应力。为简单起见，在这里风应力在经圈方向的特征尺度也取成 L。

现用一个简单的模式来描写海洋运动。在这一模式中，温跃层是上、下两层流体的界面，上层流体的厚度为 D，流体只在上层运动，在下层是静止的，从而有

$$C = \sqrt{g\frac{\rho_2 - \rho_1}{\rho_2}D} = \sqrt{g'D} \tag{2}$$

C 为重力内波波速，g' 为视重力。如外界强迫只考虑纬圈风应力，在赤道 β 平面上小扰动的基本方程为

$$\frac{\partial u}{\partial t} - \beta y v + \frac{\partial \varphi}{\partial x} = \frac{\tau^x}{D} \tag{3}$$

$$\frac{\partial v}{\partial t} + \beta y v + \frac{\partial \varphi}{\partial y} = 0 \tag{4}$$

$$\frac{\partial \varphi}{\partial t}+C^{2}\left(\frac{\partial u}{\partial x}+\frac{\partial v}{\partial y}\right)=0 \tag{5}$$

取运动的特征时间为$(2\beta C)^{-1/2}$，特征尺度为$(C/2\beta)^{1/2}$，速度正比于C，重力位势正比于C^2。τ^x的特征值为τ_0，变化的特征时间同海洋，即为$(2\beta C)^{-1/2}$。方程式(3)、式(4)、式(5)的无量纲形式为

$$\frac{\partial u}{\partial t}-\frac{1}{2} y v+\frac{\partial \varphi}{\partial x}=\tau \tau^{x} \tag{6}$$

$$\frac{\partial v}{\partial t}-\frac{1}{2} y u+\frac{\partial \varphi}{\partial y}=0 \tag{7}$$

$$\frac{\partial \varphi}{\partial t}+\frac{\partial u}{\partial x}+\frac{\partial v}{\partial y}=0 \tag{8}$$

式中$\tau=\tau_0(T/CD)$，τ_0表示风应力的特征强度。

引进

$$q=\varphi+u, \quad r=\varphi-u \tag{9}$$

由式(6)加式(8)，得

$$\frac{\partial q}{\partial t}+\frac{\partial q}{\partial x}+\frac{\partial v}{\partial y}-\frac{1}{2} y v=\tau \tau^{x} \tag{10}$$

由式(8)减式(6)，得

$$\frac{\partial r}{\partial t}-\frac{\partial r}{\partial x}+\frac{\partial v}{\partial y}+\frac{1}{2} y v=-\tau \tau^{x} \tag{11}$$

注意到

$$\varphi=\frac{1}{2}(q+r), \quad u=\frac{1}{2}(q-r) \tag{12}$$

代入式(7)，得

$$\frac{\partial v}{\partial t}+\frac{1}{2}\left(\frac{\partial q}{\partial y}+\frac{1}{2} y q\right)+\frac{1}{2}\left(\frac{\partial r}{\partial y}-\frac{1}{2} y r\right)=0 \tag{13}$$

对式(10)、式(11)、式(13)中的各物理量用 Weber 函数展开成

$$F=\sum_{n=0}^{\infty} F_{n}(x, t) D_{n}(y) \tag{14}$$

利用 Weber 函数的正交性，得到系数方程

$$\frac{\partial q_{n}}{\partial t}+\frac{\partial q_{n}}{\partial x}-v_{n-1}=\tau \tau_{n}^{x} \tag{15}$$

$$\frac{\partial r_{n}}{\partial t}-\frac{\partial r_{n}}{\partial x}+(n+1) v_{n+1}=-\tau \tau_{n}^{x} \tag{16}$$

$$\frac{\partial v_{n}}{\partial t}+\frac{1}{2}(n+1) q_{n+1}-\frac{1}{2} r_{n-1}=0 \tag{17}$$

这组方程组消去v_n，则有

$$\frac{\partial q_{0}}{\partial t}+\frac{\partial q_{0}}{\partial x}=\tau \tau_{0}^{x} \tag{18}$$

$$\frac{\partial^{2} q_{1}}{\partial t^{2}}+\frac{\partial^{2} q_{1}}{\partial x \partial t}+\frac{1}{2} q_{1}=\tau \frac{\partial \tau_{1}^{x}}{\partial t} \tag{19}$$

$$\frac{\partial^3 q_{n+2}}{\partial t^3} - \frac{\partial^3 q_{n+2}}{\partial x^2 \partial t} + \frac{1}{2}(2n+3)\frac{\partial q_{n+2}}{\partial t} - \frac{1}{2}\frac{\partial q_{n+2}}{\partial x}$$

$$= \tau\left[\frac{\partial^2 \tau^x_{n+2}}{\partial t^2} + \frac{1}{2}(n+1)\tau^x_{n+2}\right] - \frac{1}{2}\tau\tau^x_n \tag{20}$$

$$\frac{\partial^3 r_n}{\partial t^3} - \frac{\partial^3 r_n}{\partial x^2 \partial t} + \frac{1}{2}(2n+3)\frac{\partial r_n}{\partial t} - \frac{1}{2}\frac{\partial r_n}{\partial x}$$

$$= -\tau\left[\frac{\partial^2 \tau^x_n}{\partial t^2} + \frac{1}{2}(n+2)\tau^x_n\right] + \frac{1}{2}(n+1)(n+2)\tau\tau^x_{n+2} \tag{21}$$

以及

$$\frac{\partial^2 q_{n+2}}{\partial t^2} + \frac{\partial^2 q_{n+2}}{\partial x \partial t} + \frac{1}{2}(n+2)q_{n+2} - \frac{1}{2}r_n = \tau\frac{\partial \tau^x_{n+2}}{\partial t} \tag{22}$$

$$\frac{\partial^2 r_{n-1}}{\partial t^2} - \frac{\partial^2 r_{n-1}}{\partial x \partial t} + \frac{1}{2}nr_{n-1} - \frac{1}{2}n(n+1)q_{n+1} = -\tau\frac{\partial \tau^x_{n-1}}{\partial t} \tag{23}$$

方程(18)的自由方程描写的是 Kelvin 波,而方程(19)的自由方程描写的是 Rossby-重力惯性混合波。方程式(20)和式(21)的自由方程在形式上是一样的,如对物理量取 Laplace 变换,即

$$(\hat{q}_n, \hat{r}_n) = \int_0^\infty e^{-st}(q_n、r_n)\,dt \tag{24}$$

则方程式(20)、式(21)的自由方程的色散关系为

$$k_n^+ = \frac{-\frac{1}{2} + \sqrt{\frac{1}{4} + 2(2n+3)s^2 + 4s^4}}{2s} \tag{25}$$

$$k_n^- = \frac{-\frac{1}{2} - \sqrt{\frac{1}{4} + 2(2n+3)s^2 + 4s^4}}{2s} \tag{26}$$

这与 $n \neq 0,1$ 时的 Matsuno 方程的色散关系相同。因此,方程式(20)和式(21)的自由方程描写的是 Rossby 波和重力惯性波。

注意到,方程式(22)和式(23)对 x 是一阶的,因此它们可作为方程式(20)和式(21)的边界条件。取 Laplace 变换,得到

$$s^2\hat{q}_{n+2} + s\frac{d\hat{q}_{n+2}}{dx} + \frac{1}{2}(n+2)\hat{q}_{n+2} - \frac{1}{2}\hat{r}_n = s\tau\tau^x_{n+2} \tag{27}$$

$$s^2\hat{r}_{n-1} - \frac{d\hat{r}_{n-1}}{dt} + \frac{1}{2}n\hat{r}_{n-1} - \frac{1}{2}n(n+1)\hat{q}_{n+1} = -s\tau\tau^x_{n-1} \tag{28}$$

3　纬圈风应力的作用

为简单起见,设风应力在经圈方向只取一个赤道对称的模 $D_0(y)$ 和一个赤道反对称的模 $D_1(y)$,且这两个模对应的振幅的时间变化是一样的,由此式(1)可简化成

$$\tau^x = X(t)\left[\tau^x_0 D_0(y) + \tau^x_1 D_1(y)\right] \tag{29}$$

式中 $\tau^x_0>0$,$\tau^x_1<0$,进而设风应力从零开始,线性加强到 t_0,这以后取常数值,即

$$X(t)=\widetilde{X}\begin{cases}t & \text{当}\ 0<t<t_0,\\ t_0 & \text{当}\ t>t_0.\end{cases} \tag{30}$$

引进 Laplace 变换后，有

$$\hat{X}(s)=\int_0^{\infty}\mathrm{e}^{-st}X(t)\,\mathrm{d}t=\frac{1-\mathrm{e}^{-t_0 s}}{s^2}\widetilde{X} \tag{31}$$

因为在赤道地区，纬向风一般以东风为主，所以令 $\widetilde{X}<0$。在另一方面，当时间很大，即 s 很小时，式(25)、式(26)可简化成

$$k_n^{+}=(2n+3)s \tag{32}$$

$$k_n^{-}=-\frac{1}{2s}-(2n+3)s \tag{33}$$

这相当于在频率方程中过滤掉重力惯性波。前者是向西的 Rossby 长波，后者是向东的 Rossby 短波。

对方程式(18)、式(27)、式(28)、式(29)及条件式(32)、式(33)取 Laplace 变换，并对经圈模截断到三阶，即取 $n=0,1,2,3$，然后分别讨论大洋西部解和大洋东部解。

4 西部海洋解

应用 Laplace 变换后的截断模方程为

$$\frac{\mathrm{d}\hat{q}_0}{\mathrm{d}x}+s\hat{q}_0=\tau\tau_0^x\hat{X} \tag{34}$$

$$\frac{\mathrm{d}\hat{q}_1}{\mathrm{d}x}+\left(s+\frac{1}{2s}\right)\hat{q}_1=\tau\tau_1^x\hat{X} \tag{35}$$

$$s\frac{\mathrm{d}^2\hat{q}_2}{\mathrm{d}x^2}+\frac{1}{2}\frac{\mathrm{d}\hat{q}_2}{\mathrm{d}x}-\left(s^2+\frac{3}{2}\right)s\hat{q}_2=\frac{1}{2}\tau\tau_0^x\hat{X} \tag{36}$$

$$s\frac{\mathrm{d}^2\hat{q}_3}{\mathrm{d}x^2}+\frac{1}{2}\frac{\mathrm{d}\hat{q}_3}{\mathrm{d}x}-\left(s^2+\frac{5}{2}\right)s\hat{q}_3=\frac{1}{2}\tau\tau_1^x\hat{X} \tag{37}$$

$$s\frac{\mathrm{d}^2\hat{r}_0}{\mathrm{d}x^2}+\frac{1}{2}\frac{\mathrm{d}\hat{r}_0}{\mathrm{d}x}-\left(s^2+\frac{3}{2}\right)s\hat{r}_0=\tau\tau_0^x\hat{X} \tag{38}$$

$$s\frac{\mathrm{d}^2\hat{r}_1}{\mathrm{d}x^2}+\frac{1}{2}\frac{\mathrm{d}\hat{r}_1}{\mathrm{d}x}-\left(s^2+\frac{5}{2}\right)s\hat{r}_1=\frac{3}{2}\tau\tau_1^x\hat{X} \tag{39}$$

$$s\frac{\mathrm{d}^2\hat{r}_2}{\mathrm{d}x^2}+\frac{1}{2}\frac{\mathrm{d}\hat{r}_2}{\mathrm{d}x}-\left(s^2+\frac{7}{2}\right)s\hat{r}_2=0 \tag{40}$$

$$s\frac{\mathrm{d}^2\hat{r}_3}{\mathrm{d}x^2}+\frac{1}{2}\frac{\mathrm{d}\hat{r}_3}{\mathrm{d}x}-\left(s^2+\frac{9}{2}\right)s\hat{r}_3=0 \tag{41}$$

方程(34)描写的是风应力强迫下向东传的 Klevin 波方程，其解为

$$\hat{q}_0=\hat{q}_0\Big|_{x=0}\mathrm{e}^{-sx}+\frac{1}{s}\tau\tau_0^x\hat{X}(1-\mathrm{e}^{-sx}) \tag{42}$$

考虑到 $x=0$，$u=0$，所以 $q=r=\varphi$，因此

$$\hat{q}_0=\hat{\varphi}_0\Big|_{x=0}\mathrm{e}^{-sx}+\frac{1}{s}\tau\tau_0^x\hat{X}(1-\mathrm{e}^{-sx}) \tag{43}$$

方程(35)描写的是 Rossby-重力惯性混合波，其解为

$$\hat{q}_1 = \hat{\varphi}_1 \big|_{x=0} e^{-(1/2s+2s)x} + \frac{1}{s+\frac{1}{2s}} \tau\tau_1^x \left(1 - e^{-(1/2s+2s)x}\right) \hat{X} \tag{44}$$

当时间很长即 s 很小时，近似得到

$$\hat{q}_1 = \varphi_1 \big|_{x=0} e^{-(1/2s+2s)x} + 2s\tau\hat{\tau}_1^x \left(1 - e^{-(1/2s+2s)x}\right) X \tag{45}$$

将方程(36) 的解取成正比于 $e^{\bar{k}_n x}$，考虑到当 s 很小时，$\bar{k}_n$ 的近似式为式(33)，则其解为

$$\hat{q}_2 = \hat{\varphi}_2 \Big|_{x=0} e^{-(1/2s+3s)x} - \frac{1}{2s^2+3} \tau\tau_0^x \left(1 - e^{-(1/2s+3s)x}\right) \hat{X} \tag{46}$$

近似得到

$$\hat{q}_2 = \hat{\varphi}_2 \Big|_{x=0} e^{-(1/2s+3s)x} - \frac{1}{3} \tau\tau_0^x \left(1 - e^{-1(1/2s+3s)x}\right) \hat{X} \tag{47}$$

类似地有

$$\hat{q}_3 = \hat{\varphi}_3 \Big|_{x=0} e^{-(1/2s+5s)x} - \frac{1}{5s} \tau\tau_1^x \left(1 - e^{-(1/2s+5s)x}\right) \hat{X} \tag{48}$$

$$\hat{r}_0 = \hat{\varphi}_0 \Big|_{x=0} e^{-(1/2s+3s)x} - \frac{2}{3s} \tau\tau_0^x \left(1 - e^{-(1/2s+3s)x}\right) \hat{X} \tag{49}$$

$$\hat{r}_1 = \hat{\varphi}_1 \Big|_{x=0} e^{-(1/2s+5s)x} - \frac{3}{5s} \tau\tau_0^x \left(1 - e^{-(1/2s+5s)x}\right) \hat{X} \tag{50}$$

$$\hat{r}_2 = \hat{\varphi}_2 \big|_{x=0} e^{-(1/2s+7s)x} \tag{51}$$

$$\hat{r}_3 = \hat{\varphi}_3 \big|_{x=0} e^{-(1/2s+9s)x} \tag{52}$$

由于 $x=0$ 时，$u=0$，$q=r=\varphi$，取 $n=0$，式(27)变为

$$s\frac{d\hat{q}_2}{dx} + (s^2+1)\hat{\varphi}_2 \Big|_{x=0} - \frac{1}{2}\hat{\varphi}_0 \Big|_{x=0} = 0 \tag{53}$$

取 $n=1$，式(28)变为

$$s\frac{d\hat{r}_0}{dx} + \hat{\varphi}_2 \Big|_{x=0} - \left(s^2 + \frac{1}{2}\right)\hat{\varphi}_0 \Big|_{x=0} = \tau\tau_0^x s\hat{X} \tag{54}$$

将式(47)代入式(53)，得

$$(1-4s^2)\hat{\varphi}_2 \big|_{x=0} - \hat{\varphi}_0 \big|_{x=0} = \frac{1}{3s}\tau\tau_0^x \hat{X}(1+6s^2) \tag{55}$$

将式(49)代入式(54)，得

$$\hat{\varphi}_2 \big|_{x=0} - (1+4s^2)\hat{\varphi}_0 \big|_{x=0} = \frac{1}{3s}\tau\tau_0^x \hat{X}(1+9s^2) \tag{56}$$

由式(55)和式(56)解得

$$\hat{\varphi}_0 \Big|_{x=0} = -\left(\frac{1}{48s^3} + \frac{3}{4s}\right)\tau\tau_0^x \hat{X} \tag{57}$$

$$\hat{\varphi}_2 \Big|_{x=0} = -\left(\frac{1}{48s^3} + \frac{1}{2s}\right)\tau\tau_0^x \hat{X} \tag{58}$$

取 $n=1$,式(27)变为

$$s\frac{\mathrm{d}\hat{q}_3}{\mathrm{d}x}+\left(\frac{3}{2}+s^2\right)\hat{\varphi}_3\Big|_{x=0}-\frac{1}{2}\hat{\varphi}_1\Big|_{x=0}=0 \tag{59}$$

取 $n=2$,式(28)变为

$$s\frac{\mathrm{d}\hat{r}_1}{\mathrm{d}x}-(s^2+1)\hat{\varphi}_1|_{x=0}+3\hat{\varphi}_3|_{x=0}=s\tau\tau_1^x\hat{X} \tag{60}$$

将式(48)代入式(59),得

$$(1-4s^2)\hat{\varphi}_3\Big|_{x=0}-\frac{1}{2}\hat{\varphi}_1\Big|_{x=0}=\frac{1}{5s}\tau\tau_1^x\hat{X}\left(\frac{1}{2}+5s^2\right) \tag{61}$$

将式(50)代入式(60),得

$$3\hat{\varphi}_3\Big|_{x=0}-\left(\frac{3}{2}+6s^2\right)\hat{\varphi}_1\Big|_{x=0}=\frac{1}{5s}\tau\tau_1^x\hat{X}\left(\frac{3}{2}+20s^2\right) \tag{62}$$

由式(61)和式(62)解得

$$\hat{\varphi}_1\Big|_{x=0}=-\left(\frac{1}{120s^3}+\frac{2}{3s}\right)\tau\tau_1^x\hat{X} \tag{63}$$

$$\hat{\varphi}_3\Big|_{x=0}=-\left(\frac{1}{240s^3}+\frac{1}{4s}\right)\tau\tau_1^x\hat{X} \tag{64}$$

考虑到边界值后,由以上有关各式得到

$$\hat{q}_0=-\tau\tau_0^x\left[\left(\frac{1}{48s^3}+\frac{7}{4s}\right)\mathrm{e}^{-sx}-\frac{1}{s}\right]\hat{X} \tag{65}$$

$$\hat{q}_1=-\tau\tau_1^x\left[\left(\frac{1}{120s^3}+\frac{2}{3s}+2s\right)\mathrm{e}^{-(x/2s+2sx)}-2s\right]\hat{X} \tag{66}$$

$$\hat{q}_2=-\tau\tau_0^x\left[\left(\frac{1}{48s^3}+\frac{1}{6s}\right)\mathrm{e}^{-(x/2s+3sx)}+\frac{1}{3s}\right]\hat{X} \tag{67}$$

$$\hat{q}_3=-\tau\tau_1^x\left[\left(\frac{1}{240s^3}+\frac{1}{20s}\right)\mathrm{e}^{-(x/2s+5sx)}+\frac{1}{5s}\right]\hat{X} \tag{68}$$

$$\hat{r}_0=-\tau\tau_0^x\left[\left(\frac{1}{48s^3}+\frac{1}{12s}\right)\mathrm{e}^{-(x/2s+3sx)}+\frac{2}{3s}\right]\hat{X} \tag{69}$$

$$\hat{r}_1=-\tau\tau_1^x\left[\left(\frac{1}{240s^3}+\frac{1}{20s}\right)\mathrm{e}^{-(x/2s+5sx)}+\frac{3}{5s}\right]\hat{X} \tag{70}$$

$$\hat{r}_2=-\tau\tau_0^x\left(\frac{1}{48s^3}+\frac{1}{12s}\right)\mathrm{e}^{-(x/2s+7sx)}\hat{X} \tag{71}$$

$$\hat{r}_3=-\tau\tau_1^x\left(\frac{1}{240s^3}+\frac{1}{4s}\right)\mathrm{e}^{-(x/2s+9sx)}\hat{X} \tag{72}$$

对于形如式(31)的风应力,以上各式经 Laplace 逆变换后,方程(65)给出

$$q_0=\tau\tau_0^x\widetilde{X}\left[\frac{1}{2}t^2-\begin{cases}\frac{1}{2}(t-t_0)^2 & t>t_0\\ 0 & t<t_0\end{cases}-\begin{cases}\frac{(t-x)^4}{1152}+\frac{7}{8}(t-x)^2 & t>x\\ 0 & t<x\end{cases}\right.$$
$$\left.+\begin{cases}\frac{(t-t_0-x)^4}{1152}+\frac{7}{8}(t-t_0-x)^2 & t>t_0+x\\ 0 & t<t_0+x\end{cases}\right] \tag{73}$$

容易看出式中前二项是风应力的直接作用，而后二项表示在 0 时刻和 t_0 时刻激发的 Kelvin 波。

方程(65)给出

$$
\begin{aligned}
q_1 = \tau\tau_1^x\widetilde{X}\Bigg[& \begin{cases} -\dfrac{1}{30}\left(\dfrac{t-2x}{x}\right)^2 J_4\sqrt{2x(t-2x)} \\ \quad -\dfrac{4(t-2x)}{3x}J_2\sqrt{2x(t-2x)} - 10J_0(\sqrt{2xt}) & \text{当 } t > 2x \\ 0 & \text{当 } t < 2x \end{cases} \\
& + \begin{cases} -\dfrac{1}{30}\left(\dfrac{t-t_0-2x}{x}\right)^2 J_4\sqrt{2x(t-t_0-2x)} \\ \quad -\dfrac{4(t-t_0-2x)}{3x}J_2\sqrt{2x(t-t_0-2x)} - 10J_0\sqrt{2x(t-t_0)} & \text{当 } t > t_0+2x \\ 0 & \text{当 } t < t_0+2x \end{cases} \\
& + 2 - \begin{cases} 2 & \text{当 } t > t_0 \\ 0 & \text{当 } t < t_0 \end{cases}\Bigg]
\end{aligned} \tag{74}
$$

式中的 $J_n(x)$ 为 Bessel 函数，与 $J_n(x)$ 有关的项是由 Rossby 短波造成的，由于它们的贡献主要在边界附近，这样使物理场在边界附近具有动力边界层结构。

方程(67)的逆变换给出

$$
\begin{aligned}
q_2 = \tau\tau_0^x\widetilde{X}\Bigg[& -\frac{1}{6}t^2 + \begin{cases} \dfrac{1}{6}(t-t_0)^2 & \text{当 } t > t_0 \\ 0 & \text{当 } t < t_0 \end{cases} \\
& + \begin{cases} -\dfrac{1}{12}\left(\dfrac{t-3x}{x}\right)^2 J_4\sqrt{2x(t-3x)} - \dfrac{t-3x}{3x}J_2\sqrt{2x(t-3x)} & \text{当 } t > 3x \\ 0 & \text{当 } t < 3x \end{cases} \\
& + \begin{cases} \dfrac{1}{12}\left(\dfrac{t-t_0-3x}{x}\right)^2 J_4\sqrt{2x(t-t_0-3x)} \\ \quad + \dfrac{t-t_0-3x}{3x}J_2\sqrt{2x(t-t_0-3x)} & t > t_0+3x \\ 0 & t < t_0+3x \end{cases}\Bigg]
\end{aligned} \tag{75}
$$

以及方程(68)的逆变换给出

$$
\begin{aligned}
q_3 = \tau\tau_0^x\widetilde{X}\Bigg[& -\frac{1}{10}t^2 + \begin{cases} \dfrac{1}{10}(t-t_0)^2 & \text{当 } t > t_0 \\ 0 & \text{当 } t < t_0 \end{cases} \\
& + \begin{cases} -\dfrac{1}{60}\left(\dfrac{t-5x}{x}\right)^2 J_4\sqrt{2x(t-5x)} - \dfrac{t-5x}{10x}J_2\sqrt{2x(t-5x)} & \text{当 } t > 5x \\ 0 & \text{当 } t < 5x \end{cases} \\
& + \begin{cases} \dfrac{1}{60}\left(\dfrac{t-t_0-5x}{x}\right)^2 J_4\sqrt{2x(t-t_0-5x)} \\ \quad - \dfrac{t-t_0-5x}{10x}J_2\sqrt{2x(t-t_0-5x)} & t > t_0+5x \\ 0 & t < t_0+5x \end{cases}\Bigg]
\end{aligned} \tag{76}
$$

此外，

$$
\begin{aligned}
r_0 = \tau\tau_0^x\widetilde{X}\Bigg[& -\frac{1}{3}t^2 + \begin{cases} \dfrac{1}{3}(t-t_0)^2 & \text{当 } t > t_0 \\ 0 & \text{当 } t < t_0 \end{cases} \\
& + \begin{cases} -\dfrac{1}{12}\left(\dfrac{t-3x}{x}\right)^2 J_4\sqrt{2x(t-3x)} - \dfrac{t-3x}{6x}J_2\sqrt{2x(t-3x)} & \text{当 } t > 3x \\ 0 & \text{当 } t < 3x \end{cases} \\
& + \begin{cases} \dfrac{1}{12}\left(\dfrac{t-t_0-3x}{x}\right)^2 J_4\sqrt{2x(t-t_0-3x)} \\ \quad + \dfrac{t-t_0-3x}{3x}J_2\sqrt{2x(t-t_0-3x)} & t > t_0+3x \\ 0 & t < t_0+3x \end{cases}\Bigg]
\end{aligned}
\tag{77}
$$

$$
\begin{aligned}
r_1 = \tau\tau_1^x\widetilde{X}\Bigg[& -0.3t^2 + \begin{cases} 0.3(t-t_0)^2 & \text{当 } t > t_0 \\ 0 & \text{当 } t < t_0 \end{cases} \\
& + \begin{cases} -\dfrac{1}{30}\left(\dfrac{t-5x}{x}\right)^2 J_4\sqrt{2x(t-5x)} - \dfrac{2(t-5x)}{15x}J_2\sqrt{2x(t-5x)} & \text{当 } t > 5x \\ 0 & \text{当 } t < 5x \end{cases} \\
& + \begin{cases} \dfrac{1}{60}\left(\dfrac{t-t_0-5x}{x}\right)^2 J_4\sqrt{2x(t-t_0-5x)} \\ \quad - \dfrac{t-t_0-5x}{10x}J_2\sqrt{2x(t-t_0-5x)} & t > t_0+5x \\ 0 & t < t_0+5x \end{cases}\Bigg]
\end{aligned}
\tag{78}
$$

$$
\begin{aligned}
r_2 = \tau\tau_0^x\widetilde{X}\Bigg[& \begin{cases} -\dfrac{1}{12}\left(\dfrac{t-7x}{x}\right)^2 J_4\sqrt{2x(t-7x)} - \dfrac{t-7x}{x}J_2\sqrt{2x(t-7x)} & \text{当 } t > 7x \\ 0 & \text{当 } t < 7x \end{cases} \\
& + \begin{cases} \dfrac{1}{12}\left(\dfrac{t-t_0-7x}{x}\right)^2 J_4\sqrt{2x(t-t_0-7x)} \\ \quad + \dfrac{t-t_0-7x}{x}J_2\sqrt{2x(t-t_0-7x)} & t > t_0+7x \\ 0 & t < t_0+7x \end{cases}\Bigg]
\end{aligned}
\tag{79}
$$

$$
\begin{aligned}
r_3 = \tau\tau_1^x\widetilde{X}\Bigg[& \begin{cases} -\dfrac{1}{60}\left(\dfrac{t-9x}{x}\right)^2 J_4\sqrt{2x(t-9x)} - \dfrac{t-5x}{2x}J_2\sqrt{2x(t-9x)} & \text{当 } t > 9x \\ 0 & \text{当 } t < 9x \end{cases} \\
& + \begin{cases} \dfrac{1}{60}\left(\dfrac{t-t_0-9x}{x}\right)^2 J_4\sqrt{2x(t-t_0-9x)} \\ \quad + \dfrac{t-t_0-9x}{2x}J_2\sqrt{2x(t-t_0-9x)} & \text{当 } t > t_0+9x \\ 0 & \text{当 } t < t_0+9x \end{cases}\Bigg]
\end{aligned}
\tag{80}
$$

在 q_2、q_3、r_0、r_1 中既有风应力强迫的作用，又有在风应力作用下 Rossby 短波的作用，而 r_2、r_3 则仅有风应力激发出的 Rossby 短波的作用。

图 1 给出了不同时刻 φ 场的分布情况。令 $|\tau_0^x|=|\tau_1^x|$,即初始风场中偶对称风场的权重和奇对称风场的权重是相等的,这时南半球的风应力较大。取 $t_0=5$,现取风应力定常后的四个时刻 $t=16,20,24,28$ 来看 φ 场的发展情况。图中的值都已除 $\tau\tau_n^x(n=0,1)$,所有量均为无量纲量。由于吹的是东风,温跃层是正的。同时由于风应力的不断强迫,温跃层振幅不断增大,但扰动却向东传播,由于 Rossby 短波只在边界附近是重要的,因此向大洋内部的传播主要是 Kelvin 波的作用。将大洋西部解减去大洋中部单纯由风应力强迫时的解,所得的 φ 场分布见图 2。这个场是去掉风应力强迫项只由赤道波动所造成的场的分布。这样就能更清楚的理解赤道波动的作用。从图 2 可以看出,在赤道南北纬平均一个无量纲长度范围内,赤道 Kelvin 波的振幅较大,从而有助于 φ 场在赤道地区呈现对称分布。另外也注意到,大洋边界附近的值较大,这是 Rossby 短波的作用,它表现出动力边界层的特征,而其结构在南、北半球是不对称的。

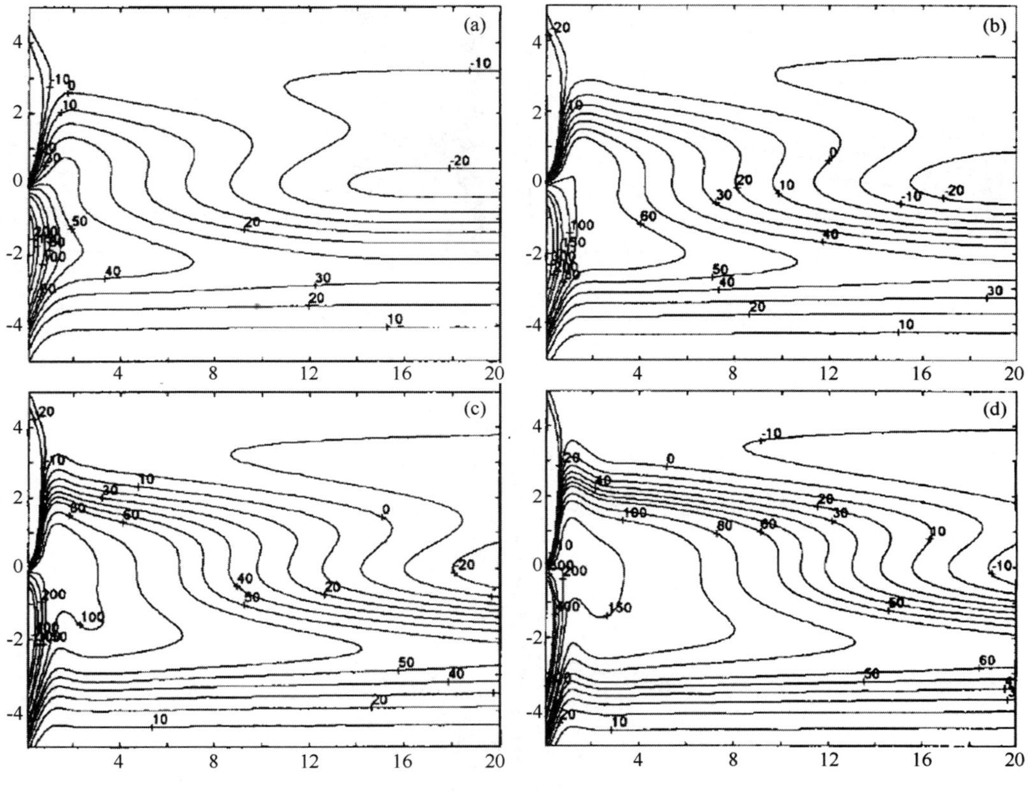

图 1　φ 场在四个时刻的分布情况

(a) $t=16$; (b) $t=20$; (c) $t=24$; (d) $t=28$

5　东部海洋解

设海洋的边界仍放在 $x=0$ 处,但 $x<0$,令 $x=-\xi,\xi>0$,式(43)改写成

$$\hat{q}_0=\left[\hat{\varphi}_0\bigg|_{x=0}-\frac{1}{s}\tau\tau_0^x\hat{X}\right]e^{s\xi}+\frac{1}{s}\tau\tau_0^x\hat{X}\tag{81}$$

为了得到 $\xi\to\infty$ 时的有限解,要求

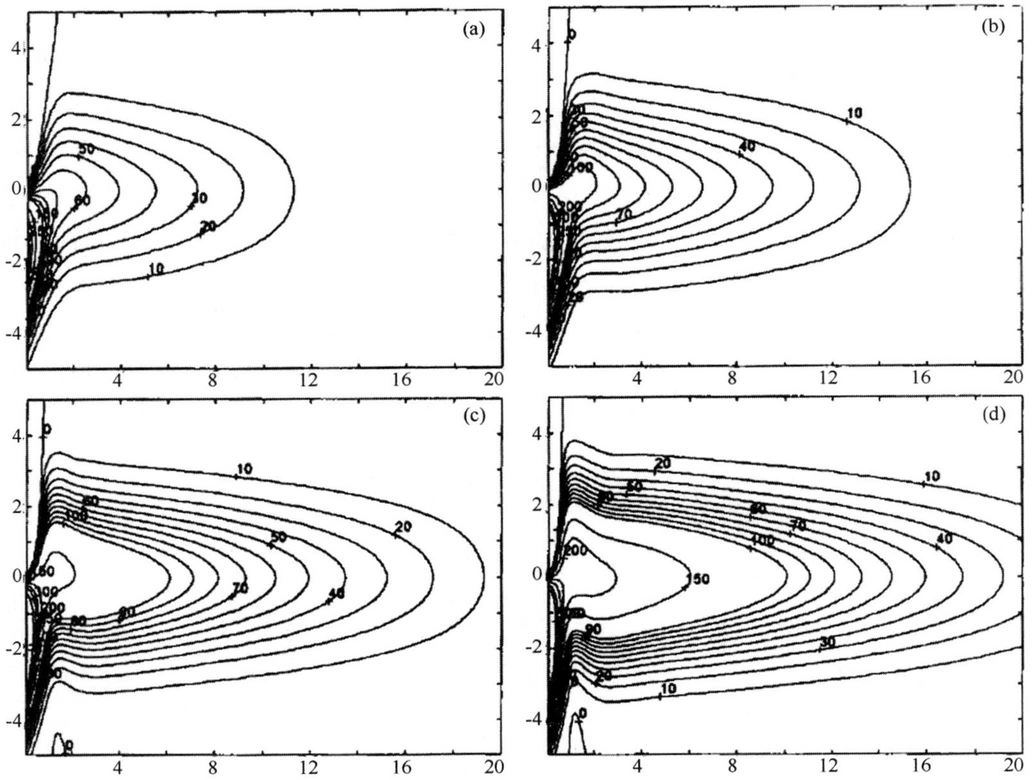

图 2　大洋西部去掉风应力强迫后 φ 场在四个时刻的分布情况,其余同图 1

$$\hat{q}_0\Big|_{\xi=0}=\frac{1}{s}\tau\tau_0^x\hat{X}=\varphi_0\Big|_{\xi=0}\tag{82}$$

由此得到

$$\hat{q}_0=\frac{1}{s}\tau\tau_0^x\hat{X}\tag{83}$$

同理,

$$\hat{\varphi}_1\mid_{\xi=0}=\hat{q}_1\mid_{\xi=0}=2s\tau\tau_1^x\hat{X}\tag{84}$$

$$\hat{q}_1=2s\tau\tau_1^x\hat{X}\tag{85}$$

对其他方程,取解的形式为 $\mathrm{e}^{-k_n^+\xi}$,考虑到 k_n^+ 的近似值取式(32),则有

$$\hat{r}_0=\tau\tau_0^x\left(\frac{5}{3}\mathrm{e}^{-3s\xi}-\frac{2}{3}\right)\frac{\hat{X}}{s}\tag{86}$$

$$\hat{r}_1=\tau\tau_1^x\left[\left(2s+\frac{3}{5s}\right)\mathrm{e}^{-5s\xi}-\frac{3}{5s}\right]\hat{X}\tag{87}$$

以及

$$\hat{q}_2=\hat{\varphi}_2\Big|_{\xi=0}\mathrm{e}^{-3s\xi}-\frac{1}{3s}\tau\tau_1^x(1-\mathrm{e}^{-3s\xi})\hat{X}\tag{88}$$

$$\hat{q}_3=\hat{\varphi}_3\Big|_{\xi=0}\mathrm{e}^{-5s\xi}-\frac{1}{5s}\tau\tau_1^x(1-\mathrm{e}^{-5s\xi})\hat{X}\tag{89}$$

考虑到边界条件，$\xi=0$ 处，$u=0$，当 $n=0$ 时，由式(88)代入方程(53)，得到

$$(1+3s^2)\hat{\varphi}_2\Big|_{\xi=0}-\frac{1}{2}\hat{\varphi}_0\Big|_{\xi=0}-\tau\tau_0^x s\hat{X}(t) \tag{90}$$

化简后，得到

$$\hat{\varphi}_2\Big|_{\xi=0}=\frac{1}{2}\hat{\varphi}_0\Big|_{\xi=0}=\frac{1}{2s}\tau\tau_0^x\hat{X}(t) \tag{91}$$

当 $n=1$ 时，由式(89)代入方程(59)，得到

$$\left(\frac{3}{2}+5s^2\right)\hat{\varphi}_3\Big|_{x=0}=\frac{1}{2}\hat{\varphi}_1\Big|_{x=0}-s\tau\tau_1^x\hat{X} \tag{92}$$

考虑到式(84)，得

$$\varphi_3=0 \tag{93}$$

所以

$$\hat{q}_2=\tau\tau_0^x\left(\frac{5}{6}e^{-3s\xi}-\frac{1}{3}\right)\frac{\hat{X}}{s} \tag{94}$$

$$\hat{q}_3=\tau\tau_1^x\left[-\frac{1}{5s}(1-e^{-5s\xi})\right]\hat{X} \tag{95}$$

$$\hat{r}_2=\frac{1}{2}\tau\tau_0^x e^{-7s\xi}\frac{\hat{X}}{s} \tag{96}$$

$$\hat{r}_3=0 \tag{97}$$

对形如式(31)的风应力，经 Laplace 逆变换给出

$$q_0=\tau\tau_0^x\widetilde{X}\left[\frac{1}{2}t^2-\begin{cases}\frac{1}{2}(t-t_0)^2 & t>t_0\\ 0 & t<t_0\end{cases}\right] \tag{98}$$

注意到在东部解的 q_0 中不包含 Kelvin 波解，仅有风应力的强迫作用，而

$$q_1=\tau\tau_1^x\widetilde{X}\begin{cases}0 & t>t_0\\ 2 & t<t_0\end{cases} \tag{99}$$

这个模也只有风应力的强迫作用，且当 $t>t_0$ 后，扰动消失。以及

$$q_2=\tau\tau_0^x\widetilde{X}\left[-\frac{1}{6}t^2+\begin{cases}\frac{1}{6}(t-t_0)^2 & \text{当 } t>t_0\\ 0 & \text{当 } t<t_0\end{cases}\right.$$
$$\left.+\begin{cases}\frac{5}{12}(t-3\xi)^2 & t>3\xi\\ 0 & t<3\xi\end{cases}-\begin{cases}\frac{5}{12}(t-t_0-3\xi)^2 & \text{当 } t>3\xi+t_0\\ 0 & \text{当 } t<3\xi+t_0\end{cases}\right] \tag{100}$$

其中前二项表示风应力的强迫，第三项表示0时刻激发的向西传播的波速为(1/3)C 的 Rossby 长波，第四项表示在 t_0 时刻激发的向西传播的波速为(1/3)C 的 Rossby 长波。此外

$$q_3=\tau\tau_1^x\widetilde{X}\left[-\frac{1}{10}t^2+\begin{cases}\frac{1}{10}(t-t_0)^2 & \text{当 } t>t_0\\ 0 & \text{当 } t<t_0\end{cases}\right.$$

$$+\begin{cases}\dfrac{1}{10}(t-5\xi)^2 & t>5\xi\\ 0 & t<5\xi\end{cases}-\left.\begin{cases}\dfrac{1}{10}(t-t_0-5\xi)^2 & \text{当}\ t>5\xi+t_0\\ 0 & \text{当}\ t<5\xi+t_0\end{cases}\right] \tag{101}$$

$$r_0=\tau\tau_0^x\widetilde{X}\left[-\frac{1}{3}t^2+\begin{cases}\dfrac{1}{3}(t-t_0)^2 & \text{当}\ t>t_0\\ 0 & \text{当}\ t<t_0\end{cases}\right.$$

$$+\begin{cases}\dfrac{5}{6}(t-3\xi)^2 & \text{当}\ t>3\xi\\ 0 & \text{当}\ t<3\xi\end{cases}-\left.\begin{cases}\dfrac{5}{6}(t-t_0-3\xi)^2 & \text{当}\ t>t_0+3\xi\\ 0 & \text{当}\ t<t_0+3\xi\end{cases}\right] \tag{102}$$

$$r_1=\tau\tau_1^x\widetilde{X}\left[-\frac{3}{10}t^2+\begin{cases}\dfrac{3}{10}(t-t_0)^2 & \text{当}\ t>t_0\\ 0 & \text{当}\ t<t_0\end{cases}\right.$$

$$+\begin{cases}2+\dfrac{3}{10}(t-5\xi)^2 & \text{当}\ t>5\xi\\ 0 & \text{当}\ t<5\xi\end{cases}-\left.\begin{cases}2+\dfrac{3}{10}(t-t_0-5\xi)^2 & \text{当}\ t>t_0+5\xi\\ 0 & \text{当}\ t<t_0+5\xi\end{cases}\right] \tag{103}$$

$$r_2=\tau\tau_0^x\widetilde{X}\begin{cases}(t-7\xi)^2 & \text{当}\ t>7\xi\\ 0 & \text{当}\ t<7\xi\end{cases}-\begin{cases}(t-t_0-7\xi)^2 & \text{当}\ t>t_0+7\xi\\ 0 & \text{当}\ t<7_0+7\xi\end{cases} \tag{104}$$

$$r_3=0 \tag{105}$$

在图 3 中和西部解一样，取 $t=16,20,24,28$ 四个时刻的 φ 场分布情况。在 Rossby 波没有传到的地方，也就是图中 φ 场纬向分布均匀的地方，和西部解相比较是一致的。这相当于大洋中部的解，也就是去掉了东、西部的各种波动作用，只有风应力的强迫在起作用时的解。大洋东部解和西部解相比，从图 3 可以看到，在 Rossby 波传过的地区，φ 场表现出对赤道的不对称性。同时可以看到，南半球有较大的梯度。

图 4 类似于图 2，也是大洋东部解减去大洋中部解，这相当于只保留各阶 Rossby 波后形成的 φ 场。在图中，各阶 Rossby 波的综合作用也是使 φ 场表现为关于赤道的不对称分布。这是因为南、北半球风应力强度不同，由此激发的 Rossby 长波的振幅在南、北半球也不相同。同时和图 2 相比，西部的值比东部的值大几倍。从这一点来说，西部以 Kelvin 波为表征的波动振幅明显强于东部 Rossby 波的振幅。可以这样说，大洋西部和大洋东部相比，由于多一个 Kelvin 波解，而且 Kelvin 波具有较大的振幅，对赤道对称的 Kelvin 波使大洋西部的赤道地区表现出对赤道的对称性，且振幅也大。大洋东部，各阶 Rossby 波的振幅相对较小，同时由于风应力对赤道的不对称性从而造成物理场对赤道的不对称性。

6　结论

由以上分析，可以得到如下结论：

由于赤道对称的 Kelvin 波的振幅比 Rossby 波的振幅大，使得物理场在西太平洋比在东太平洋更趋于对称性。在东太平洋由于边界不能直接激发出 Kelvin 波，而奇偶对称的风应力均能激发出 Rossby 波，而它们的振幅值相近，故呈现不对称分布。

这表明即使不考虑海气相互作用等过程，由于海洋本身的动力学性质，也可以使海洋的东西部状态分布有所不同，至少对距平状态是如此。在另一方面，大洋西部物理场的对称分布明显，响应场的

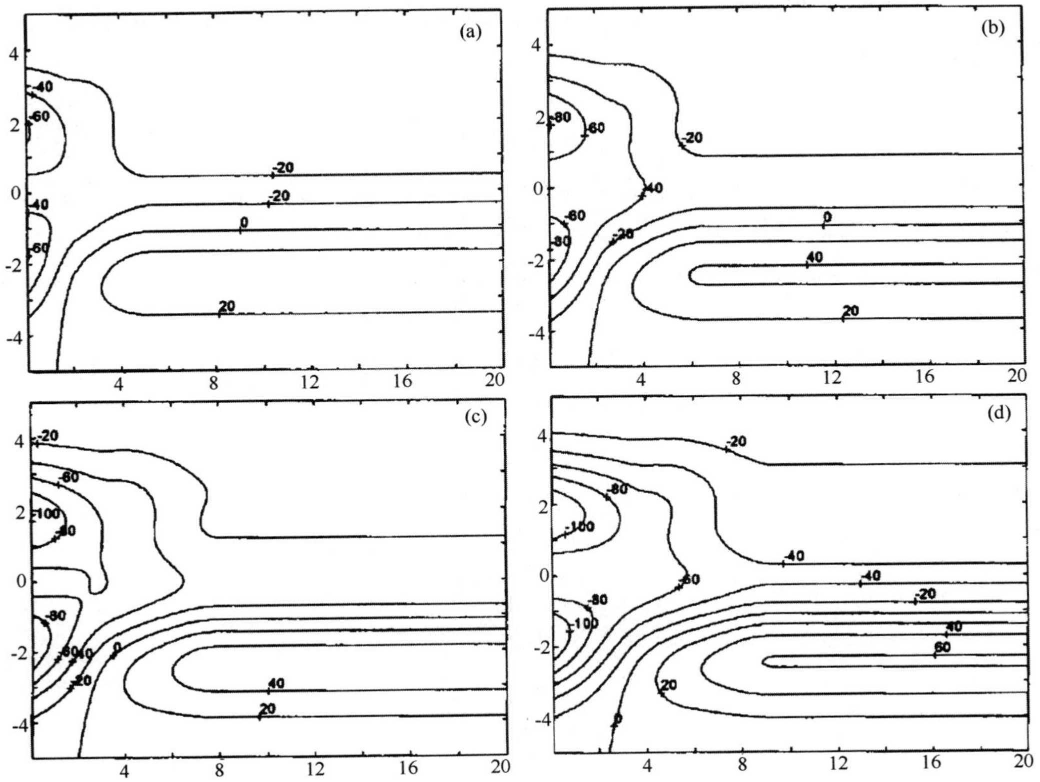

图 3　东部 φ 场在四个时刻的分布情况，其余同图 1

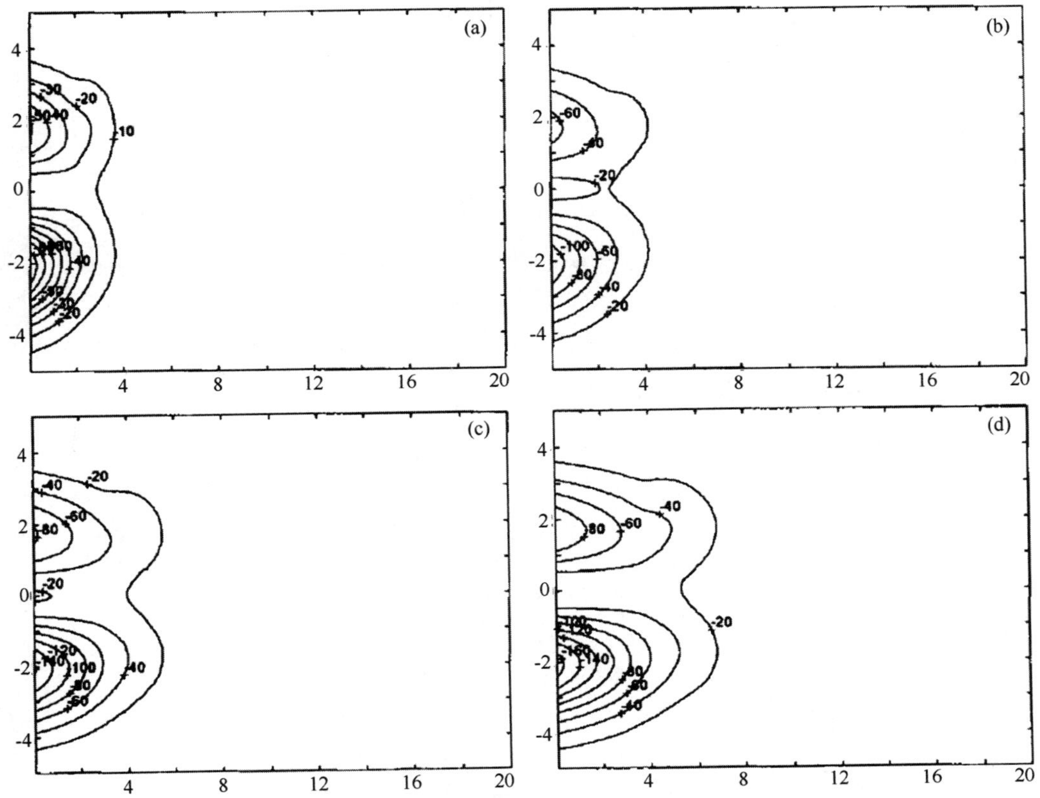

图 4　大洋东部去掉风应力强迫后 φ 场在四个时刻的分布情况，其余同图 1

振幅也大。这一特点,在研究 El Niño 事件的发生源地时是有参考意义的,即大洋西部更易成为 El Niño 发生的源地。

参考文献

[1] Shang-Ping Xie and S. George H. Philander, A coupled ocean-atmosphere model of relevance to the ITCZ in the eastern Pacific, *Tellus*, 1994,46A:340-350.

[2] Shang-Ping Xie. The maintenance of an equatorially asymmetric state in a hybrid soupled GCM, *J. Atmos*, *Sci.*,1994, 151:2602-2612.

热带海洋和大气中地形 Rossby 波和 Rossby 波的耦合不稳定*

巢纪平[1,2]　刘琳[1,3]　于卫东[1]

(1. 国家海洋局海洋环境科学和数值模拟国家海洋局重点实验室,青岛 266061; 2. 国家海洋局海洋环境预报中心,北京 100081;3. 中国海洋大学气象系,青岛 266003)

摘要:当大尺度背景场存在赤道急流时,由相应不均匀的温跃层(海洋)和高度场(大气)激发出的地形 Rossby 波和由 β 效应激发出的 Rossby 波,在一定条件下,通过相互作用后可产生一类新的不稳定,称为地形 Rossby 波和 Rossby 波的耦合不稳定。讨论了这类波系在 ENSO 发展中可能起的作用。

关键词:温跃层不均匀性;赤道急流;地形 Rossby 波;Rossby 波;耦合不稳定

从海洋方面考虑,在20世纪80年代以前,人们普遍认为厄尔尼诺(El Niño)是起源于南美洲秘鲁沿岸的海表增温现象,暖的海表温度信号到达一定强度后向赤道并沿赤道向西扩展,在赤道东太平洋形成大片距平意义下的暖水,这就是通称的 El Niño。Rasmusson 等[1]对 El Niño 事件中海表暖水的发展以及相应的大气风场的演变有过一个模型总结。但20世纪90年代以来一系列观测资料分析研究表明[2-9],无论在热带太平洋还是热带印度洋,暖/冷水事件中最强的海温距平多数都是沿温跃层曲面传播的。如果认为温度距平信号在温跃层曲面上的发展有波动传播的性质,则分析温跃层不均匀分布以及相应的地转赤道急流对热带海洋波动发展的影响就有它的重要性了。

从大气方面考虑,影响全球气候的 El Niño 现象的发展也需一定的大气条件来支持,如发展时低空信风减弱,甚至改吹西风。当西风暴发时,Madden-Julian 振荡即季节内振荡频繁,一般认为这类振荡发展的能量主要来自对流云团潜热释放而造成的对流不稳定(CISK),但它们向东的传播也需要一种"载体"或过程,其向东的传播速度也慢于理想状态下 Kelvin 波的波速。虽然对 Kelvin 波在季节内振荡传播中的作用已有不少研究[10],但在大尺度背景风的条件下,把 Kelvin 波放在整个赤道波系中来研究,仍然是一个值得关注的课题,甚至扰动向东的缓慢传播,其载体是否一定是 Kelvin 波,也需要进一步研究。

如果不计由水汽相变等非绝热过程激发的不稳定,可同时适用于解释热带大气和海洋中的波动发展的动力学不稳定理论,则尤其值得注意。Ji 等[11]参考巢纪平等[12]的早期分析,提出由于背景温度场的不均匀分布所产生的类似"地形 Rossby 波"[13]的"热力波",它与重力惯性波的相互作用可以出现波的不稳定。但这类不稳定的发生由于重力惯性波参与了,相速和增长率都很大,直接用来解释 ENSO 事件中的低频现象有一定困难。Marco De La Cruz-Heredia 等[14]参照层结流体中 Sakai[15]的

* 中国科学,2005,35(1):79-87.

Rossby 波-Kelvin 波形的不稳定性指出,即使流体是正压的,如果在经向两个相邻的有界区域中存在不同的切变流,则由于基本流位势涡度在两个区域连接处的不连续性可激发出 Rossby 波。另一方面,在区域的边界上可以激发出 Kelvin 波,而基本流的多普勒(Doppler)效应可改变两类波的相速度,在一定条件下,它们相互耦合而出现 Rossby 波-Kelvin 波型不稳定。在热带大气和海洋中,在赤道背景流场不存在相对涡度的不连续性,因此这一理论难以直接用来解释 ENSO 发展中的某些现象。注意到这两类不稳定都是由波与波的相互作用而产生的,可称为波的耦合不稳定。另一方面,Philander[16] 研究赤道潜流的作用后指出,对于相速度高的波动,如 Kelvin 波,潜流对它的影响很小,但对 Rossby 波,其相速度慢到已能与潜流的流速相比,潜流对它的作用变得显著,甚至可能使波产生不稳定性。注意到在地转情况下与潜流相适应的温跃层必然是纬向不均匀分布的,从而产生地形 Rossby 波,但在 Philander 的分析中我们没有看到这类波的作用。鉴于此,本文将用同时适合热带海洋和大气的浅水模式,在准长波近似下,用低阶截断模得到波的色散关系,从而提出一类新的低频不稳定理论,称它为地形 Rossby 波和 Rossby 波的耦合不稳定,并用这种不稳定性来解释 ENSO 发展中的一些观测现象。

1 基本方程

设 $H(y)=H_0(1+h_c(y))$,对海洋,$H(y)$ 为温跃层底距离海表的距离,是温跃层深浅的一个量度;对大气,$H(y)$ 可视为流体有自由表面的厚度。引进重力波波速:

$$C=\sqrt{g'H_0} \tag{1}$$

式中,g' 为约化重力,若 H_0 对海洋取 100 m,对大气取 1 km(相当行星边界层的平均厚度),取约化重力使 C 值对海洋和大气分别为 0.1 m/s 和 1 m/s。设基本流是地转的,即:

$$\beta yU=-g'\frac{\mathrm{d}H}{\mathrm{d}y} \tag{2}$$

在赤道 β 平面上,线性化的浅水运动方程为

$$\frac{\partial u}{\partial t}+U\frac{\partial u}{\partial x}+\frac{\partial U}{\partial y}v-\beta yv=-g'\frac{\partial h}{\partial x} \tag{3}$$

$$\frac{\partial v}{\partial t}+U\frac{\partial v}{\partial x}+\beta yv=-g'\frac{\partial h}{\partial y} \tag{4}$$

$$\frac{\partial h}{\partial t}+U\frac{\partial h}{\partial x}+H(y)\left(\frac{\partial u}{\partial x}+\frac{\partial v}{\partial y}\right)+\frac{\mathrm{d}H}{\mathrm{d}y}v=0 \tag{5}$$

取特征量为 $(u,v)=C(u^*,v^*)$,$t=(2\beta C)^{-\frac{1}{2}}t^*$,$(x,y)=\left(\frac{C}{2\beta}\right)^{\frac{1}{2}}(x^*,y^*)$,相对于气候场 $H(y)$ 的扰动高度为 $h=H_0h^*$。取 $\beta=2\times10^{-11}/(\mathrm{m}\cdot\mathrm{s})$,由此估计出特征时间对海洋约为 6 天,对大气约 2 天;特征空间尺度对海洋为约 50 km,对大气约为 150 km,这分别相当赤道 Rossby 变形半径的尺度。

运动方程的无量纲形式为(略去 * 号)

$$\frac{\partial u}{\partial t}-\frac{2}{y}\frac{\mathrm{d}h_c}{\mathrm{d}y}\frac{\partial u}{\partial x}-\frac{1}{2}yv+\frac{\partial h}{\partial x}=\left(\frac{2}{y}\frac{\mathrm{d}^2h_c}{\mathrm{d}y^2}-\frac{2}{y^2}\frac{\mathrm{d}h_c}{\mathrm{d}y}\right)v \tag{6}$$

$$\frac{\partial v}{\partial t}-\frac{2}{y}\frac{\mathrm{d}h_c}{\mathrm{d}y}\frac{\partial v}{\partial x}+\frac{1}{2}yu+\frac{\partial h}{\partial y}=0 \tag{7}$$

$$\frac{\partial h}{\partial t}-\frac{2}{y}\frac{\mathrm{d}h_c}{\mathrm{d}y}\frac{\partial h}{\partial x}+\left(\frac{\partial u}{\partial x}+\frac{\partial v}{\partial y}\right)=-\frac{\mathrm{d}h_c}{\mathrm{d}y}v-h_c\left(\frac{\partial u}{\partial x}+\frac{\partial v}{\partial y}\right) \tag{8}$$

引进变量 q,r 定义成

$$q = h + u, \quad r = h - u$$

于是有：

$$h = \frac{1}{2}(q + r), \quad u = \frac{1}{2}(q - r).$$

方程式(6)至式(8)变成：

$$\frac{\partial q}{\partial t} + \frac{\partial q}{\partial x} + \frac{\partial v}{\partial y} - \frac{1}{2}yv = \frac{2}{y}\frac{\mathrm{d}h_c}{\mathrm{d}y}\frac{\partial q}{\partial x} - h_c\left[\frac{1}{2}\left(\frac{\partial q}{\partial x} - \frac{\partial r}{\partial x}\right) + \frac{\partial v}{\partial y}\right] + \left(\frac{2}{y}\frac{\mathrm{d}^2 h_c}{\mathrm{d}y^2} - \frac{2}{y^2}\frac{\mathrm{d}h_c}{\mathrm{d}y} - \frac{\mathrm{d}h_c}{\mathrm{d}y}\right)v \tag{9}$$

$$\frac{\partial r}{\partial t} - \frac{\partial r}{\partial x} + \frac{\partial v}{\partial y} + \frac{1}{2}yv = \frac{2}{y}\frac{\mathrm{d}h_c}{\mathrm{d}y}\frac{\partial r}{\partial x} - h_c\left[\frac{1}{2}\left(\frac{\partial q}{\partial x} - \frac{\partial r}{\partial x}\right) + \frac{\partial v}{\partial y}\right] - \left(\frac{2}{y}\frac{\mathrm{d}^2 h_c}{\mathrm{d}y^2} - \frac{2}{y^2}\frac{\mathrm{d}h_c}{\mathrm{d}y} + \frac{\mathrm{d}h_c}{\mathrm{d}y}\right)v \tag{10}$$

$$\frac{\partial v}{\partial t} - \frac{2}{y}\frac{\mathrm{d}h_c}{\mathrm{d}y}\frac{\partial v}{\partial x} + \frac{1}{2}\left[\left(\frac{\partial q}{\partial y} + \frac{1}{2}yq\right) + \left(\frac{\partial r}{\partial y} - \frac{1}{2}yr\right)\right] = 0 \tag{11}$$

将 q,r 和 v 展成抛物圆柱函数即 Weber 函数，即：

$$(q,r,v) = \sum (q_n, r_n, v_n) D_n(y) \tag{12}$$

方程式(9)至式(11)给出：

$$\frac{\partial q_n}{\partial t} + \frac{\partial q_n}{\partial x} - v_{n-1} = A_n \tag{13}$$

$$\frac{\partial r_n}{\partial t} - \frac{\partial r_n}{\partial x} + (n+1)v_{n+1} = B_n \tag{14}$$

$$\varepsilon\frac{\partial v_{n+1}}{\partial t} + \frac{1}{2}[(n+2)q_{n+2} - r_n] = \delta C_{n+1} \tag{15}$$

式中 ε 和 δ 为标识号，而

$$A_n = \frac{1}{n!}$$

$$\sum_{m=0}^{\infty}\left\langle (I^{(1)}_{m,n} - I^{(2)}_{m,n})\frac{\partial q_m}{\partial x} + I^{(2)}_{m,n}\frac{\partial r_m}{\partial x} - [mI^{(2)}_{m-1,n} - I^{(2)}_{m+1,n} - J^{(1)}_{m,n}]v_m\right\rangle \tag{16}$$

$$B_n = \frac{1}{n!}$$

$$\sum_{m=0}^{\infty}\left\langle (I^{(1)}_{m,n} - I^{(2)}_{m,n})\frac{\partial r_m}{\partial x} + I^{(2)}_{m,n}\frac{\partial q_m}{\partial x} - [mI^{(2)}_{m-1,n} - I^{(2)}_{m+1,n} + J^{(2)}_{m,n}]v_m\right\rangle \tag{17}$$

$$C_n = \frac{1}{(n+1)!}\sum_{m=0}^{\infty} I^{(1)}_{m,n+1}\frac{\partial v_m}{\partial x} \tag{18}$$

其中

$$I^{(1)}_{m,n} = \frac{1}{\sqrt{2p}}\int_{-\infty}^{\infty}\frac{2}{y}\frac{\mathrm{d}h_c}{\mathrm{d}y}D_m D_n \mathrm{d}y \tag{19a}$$

$$I^{(2)}_{m,n} = \frac{1}{\sqrt{2p}}\int_{-\infty}^{\infty}\frac{h_c}{2}D_m D_n \mathrm{d}y \tag{19b}$$

$$J^{(1),(2)}_{m,n} = \frac{1}{\sqrt{2p}}\int_{-\infty}^{\infty}\left(\frac{2}{y}\frac{\mathrm{d}^2 h_c}{\mathrm{d}y^2} - \frac{2}{y^2}\frac{\mathrm{d}h_c}{\mathrm{d}y} \mp \frac{2}{y}\frac{\mathrm{d}h_c}{\mathrm{d}y}\right)D_m D_n \mathrm{d}y = K^{(1)}_{m,n} \mp K^{(2)}_{m,n} \tag{19c}$$

方程式(13)至式(15)是要讨论的基本方程。

2 广义 Matsuno 方程

首先,由方程(13)给出:

$$\frac{\partial q_0}{\partial t}+\frac{\partial q_0}{\partial x}=A_0 \tag{20}$$

上式左端是控制 Kelvin 波的算子,由于 A_0 中包含其他 $n\neq0$ 的模,因此 Kelvin 模现在不再独立,它将与其他模耦合。

当 $n>0$ 时,由式(13)至式(15)消去 v_{n+1},给出:

$$\begin{aligned}&\left[\varepsilon\left(\frac{\partial^3}{\partial t^3}-\frac{\partial^3}{\partial t\partial x^2}\right)+\frac{1}{2}(2n+3)\frac{\partial}{\partial t}-\frac{1}{2}\frac{\partial}{\partial x}\right]q_{n+2}\\&=\left[\varepsilon\left(\frac{\partial^2}{\partial t^2}-\frac{\partial^2}{\partial t\partial x}\right)+\frac{1}{2}(n+1)\right]A_{n+2}B_n+\delta\left(\frac{\partial}{\partial t}-\frac{\partial}{\partial x}\right)C_{n+1}\end{aligned} \tag{21}$$

$$\begin{aligned}&\left[\varepsilon\left(\frac{\partial^3}{\partial t^3}-\frac{\partial^3}{\partial t\partial x^2}\right)+\frac{1}{2}(2n+3)\frac{\partial}{\partial t}-\frac{1}{2}\frac{\partial}{\partial x}\right]r_n\\&=\left[\varepsilon\left(\frac{\partial^2}{\partial t^2}-\frac{\partial^2}{\partial t\partial x}\right)+\frac{1}{2}(n+2)\right]B_n+\frac{1}{2}(n+1)(n+2)A_{n+2}-\delta\left(\frac{\partial}{\partial t}+\frac{\partial}{\partial x}\right)(n+1)C_{n+1}\end{aligned} \tag{22}$$

这两个方程连同方程(20)和方程(15)如不考虑背景场的不均匀性是 Matsuno 方程[17],因此现在可称广义 Matsuno 方程。注意到在没有不均匀背景场时,方程(21)和方程(22)左端的算子是一样的,它们描写的是高频重力惯性波和低频 Rossby 波,而现在这些波都不独立,通过不均匀背景场耦合起来了。

3 准长波近似

略去式(21)和式(22)中左端时间变化的3次方项(低频近似[18])和右端时间的2次方变化项,进而在式(15)中令 $\varepsilon=0$,式(15)简化成:

$$(n+2)q_{n+2}-r_n=2\delta C_{n+1} \tag{23}$$

当 $\delta=0$ 时,退化为长波近似。式(23)称准长波近似,它与长波近似的差别在于考虑了经圈速度的平流作用。

在上述简化条件下,式(21)和式(22)取 $\varepsilon=0$,分别简化成:

$$\frac{\partial q_{n+2}}{\partial t}-\frac{1}{2n+3}\frac{\partial q_{n+2}}{\partial x}=\frac{1}{2n+3}((n+1)A_{n+2}+B_n)+\delta\frac{2}{2n+3}\left(\frac{\partial}{\partial t}-\frac{\partial}{\partial x}\right)C_{n+1} \tag{24}$$

$$\frac{\partial r_n}{\partial t}-\frac{1}{2n+3}\frac{\partial r_n}{\partial x}=\frac{n+2}{2n+3}((n+1)A_{n+2}+B_n)-\delta\frac{2}{2n+3}\left(\frac{\partial}{\partial t}+\frac{\partial}{\partial x}\right)(n+1)C_{n+1} \tag{25}$$

式(23),式(20),式(24)和式(25)构成准长波近似下的基本方程。

在下面的分析中,将 Weber 函数的模取 $n=0,1,2$。注意到,关于赤道对称和反对称模的方程是独立的,反对称模方程的解在此不讨论,只分析对称模方程的色散关系。

4 长波近似的分析

为了说明温跃层或高度场的不均匀性以及相应的赤道基本流的不均匀性对热带低频长波的影

响，先给出长波近似结果供参考。

背景场的起伏结构取：

$$h_c = \bar{h}(e^{-\alpha y^2} - 1) + h_c \tag{26}$$

式中 h_0 为赤道值，$\bar{h}$ 是一个参量，总厚度为

$$H = 1 + \bar{h}(e^{-\alpha y^2} - 1) + h_0 \tag{27}$$

相应的地转基本流为

$$U = 4\bar{h}\alpha e^{-\alpha y^2} \tag{28}$$

注意到，当 $\bar{h}$ 取正值时，基本流在赤道是向东的。对海洋，可以把这一基本流看成是赤道潜流，一般在 El Niño 发展前期时它的强度较强；对大气，可看成 El Niño 已接近发展盛期时，Walker 环流已逆转，这时在中、西太平洋海域低空吹西风或距平西风。当无量纲参数分别取 $\alpha = 0.04$ 和 $\alpha = 0.9$ 以及 $\bar{h} = 0.7$，对应着在赤道的基本洋流速度分别为 1 cm/s 和 25 cm/s；相应的大气风速为 0.1 m/s 和 2.5 m/s。

图 1a 是温跃层均匀的情况。这时只有两支波，一支是向东传播的 Kelvin 波；另一支是向西传播的 Rossby 波。随着温跃层坡度，即 α 值加大，Kelvin 波的频率加大，而 Rossby 波的频率减小（图 1(b)）；当 $\alpha = 0.9$ 时，Rossby 波在“地形”影响下转向东传（图 1(c)）。

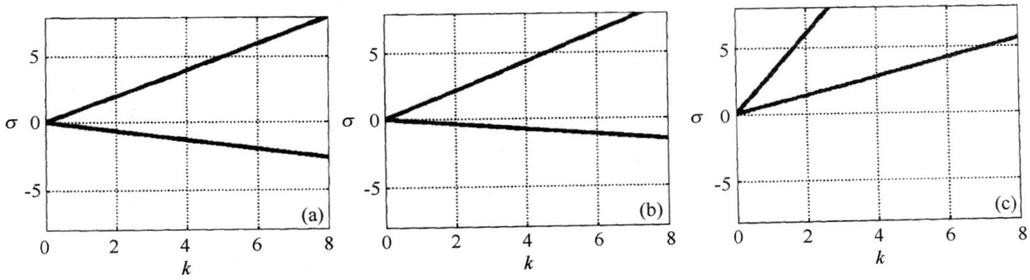

图 1　长波近似的波动色散关系

(a)经典的长波近似；(b)，(c)考虑温跃层径向变化后的准长波近似的结果。(b)和(c)之中的参数分别为 $\alpha = 0.04$ 和 $\alpha = 0.9$，$\bar{h}$ 都为 0.7。对应着赤道处的海洋基本流速分别为 1 cm/s 和 25 cm/s，大气基本流速分别为 0.1 m/s和 2.5 m/s。图中横坐标为无量纲波数，纵坐标为无量纲频率

5　准长波近似的分析

5.1　东向流情况

温跃层或高度场的不均匀分布仍取式(27)，相应的基本流仍为式(28)，当 $\bar{h}$ 值同上而 $\alpha = 0.04$ 时，计算结果见图 2(a)。可见，本质上为 Rossby 波的那一支已转向东传，这时又出现主要由“地形”激发出的新波，它是向西传播的，且在长波波段频率很大。当 α 增大到 0.9 时，本质上为 Rossby 的那支波与本质上为地形 Rossby 波的那支波在波数大于某一临界值后相重合，即变为不稳定的共轭波（图 2(b)）。不稳定增长率当波数 $k = 3$ 时为 $\sigma_i = 0.2$，因此波振幅增长 e 倍所需的时间对海洋约为 30 天，对大气约为 10 天。

注意到只有在经圈速度控制方程中，保留基本流的平流项时才能激发出独立于 Rossby 波的地形

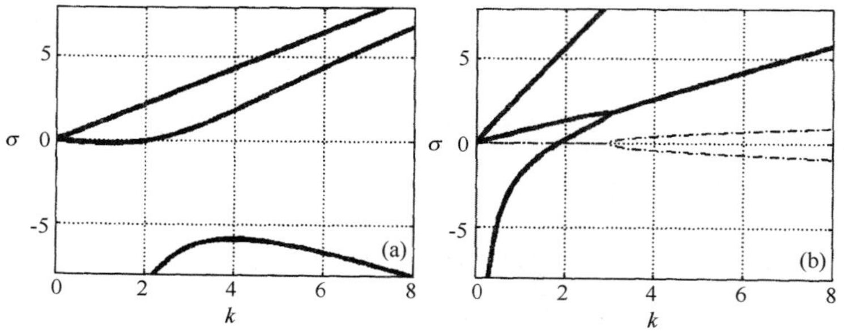

图 2 存在东向基本流时的波动色散关系

表示存在向东基本流情况下的准长波近似的结果。(a)和(b)中各参数同图 1b 和图 1c。(b)的虚线表示不稳定增长率。坐标同图 1

Rossby 波,这是准长波近似与长波近似的一个重要差别。

图 3 是不稳定波发生的参数域,在固定的参数范围内,存在临界波数,当波数超过该临界波数时会产生不稳定;反之,波数小于临界波数时不稳定现象消失。

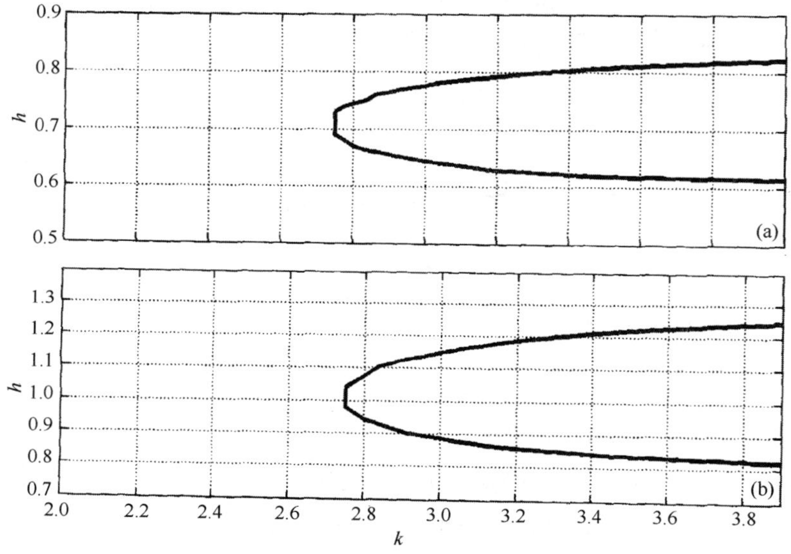

图 3 不稳定波出现的参数域

表示不稳定波发生的参数域。(a)中 $\alpha=1$,(b)中 $\bar{h}=0.7$,横坐标均表示无量纲波数;图中线条的内部是不稳定区

一个有意思的现象是,这支共轭波在未共轭的长波波段,向西的相速很大,但群速却是向东的,也很大,相速和群速分别给在图 4 中。

5.2 西向流情况

如果将式(28)中的 $\bar{h}$ 取成负号,这时基本流在赤道变成向西流动。当参数取成和以上向东流相同时,其色散关系见图 5,这时本质上为 Kelvin 波向东的相速减慢,而本质上为 Rossby 波的向西相速变大,而新生的地形 Rossby 波是一支向东传播的波。计算表明,不论参数如何调整也不会出现不稳定性。

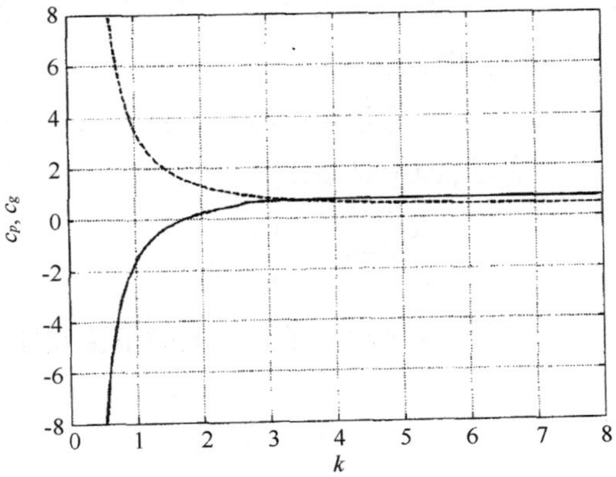

图 4　不稳定波的群速和相速

参数为 $\alpha=0.9, \bar{h}=0.7$。对应着赤道处的基本流速海洋为 25 cm/s，大气为 2.5 m/s。虚线表示地形波的群速度 c_g，实线表示相速度 c_p，纵坐标为波数，图中各量均是无量纲量

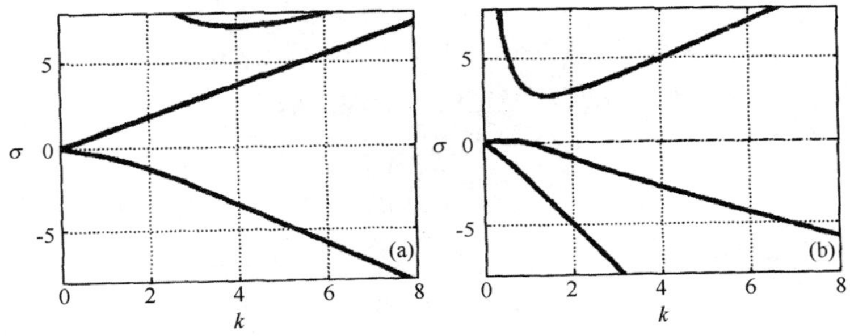

图 5　存在向西流时的波动色散关系

(a)，(b) 存在向西基本流情况下的准长波近似的结果；参数分别为 $\alpha=0.04$ 和 $\alpha=0.9, \bar{h}$ 都为 -0.7。坐标同图 1

6　对解的性质的分析

6.1　波的传播

对于给定的基本流，波的传播速度和方向依赖于基本流在赤道的流向及纬向不均匀程度。当为向东流时，Kelvin 波的相速增大，而 Rossby 波的相速减小。另一方面，由于纬向流分布的不均匀性使经典 Rossby 长波将发生如下的改变，它的传播速度与绝对涡度梯度

$$\frac{d\zeta_a}{dy} = \beta - \frac{d^2U}{dy^2} \tag{29}$$

有关。如由基本流相对涡度梯度减小了行星涡度梯度即 β 的作用，则 Rossby 波向西传播速度将减小，甚至反向传播。

6.2　不稳定波的必要条件

从以上的结果可以看到，由于温跃层或高度场分布不均匀性所产生的地形 Rossby 波在某些参数域中和 Rossby 波在某些波段可以耦合，从而出现不稳定性，因此可以称为地形 Rossby 波和 Rossby 波

的耦合不稳定。但我们注意到,这种不稳定是发生在有 β 作用和基本流作用的正压流体(海洋或大气)中,由于 Rossby 波也参与了不稳定,因此这样的不稳定性应与 Kuo[19] 提出的正压不稳定有一定的联系,或可称为广义正压不稳定。关于正压不稳定的必要条件,Kuo 在 20 世纪 40 年代末指出,当绝对涡度梯度在研究区域中的某一纬度为零,即满足条件

$$\frac{\mathrm{d}\zeta_a}{\mathrm{d}y}=\beta-\frac{\mathrm{d}^2U}{\mathrm{d}y^2}=0 \tag{30}$$

时,扰动运动将有可能出现不稳定。我们来分析上述向东流的不稳定能否满足这一必要条件。

取基本流的形式为

$$U=U_0\mathrm{e}^{-\alpha y^2} \tag{31}$$

式中各量均为有量纲。容易看出,在无量纲时如取 $U_0=4\bar{h}\alpha$ 即为式(28)的基本流。

将式(31)代入式(30),得到:

$$\alpha_c=2\alpha(2\alpha y^2-1)\mathrm{e}^{-\alpha y^2} \tag{32}$$

式中,$\alpha_c=\dfrac{\beta}{U_0}$。若写成无量纲形式,即取 $U_0=CU_0^*$,$y=L_cy^*$,$\alpha=\left(\dfrac{1}{L_c}\right)\alpha^*$,上式为(略去上标的 * 号)

$$1=4U_0\alpha(2\alpha y^2-1)\mathrm{e}^{-\alpha y^2}\equiv F(U_0,\alpha,y) \tag{33}$$

只要在定义域中存在这一条件,则有出现不稳定的可能性。取

$$y=\left(\frac{3}{\alpha}\right)^{\frac{1}{2}} \tag{34}$$

在这个纬度上式(33)右端给出:

$$F=20U_0\alpha\mathrm{e}^{-3} \tag{35}$$

因此,只要流速和背景场的不均匀坡度满足

$$\alpha=\frac{\mathrm{e}^3}{20U_0} \tag{36}$$

不稳定的必要条件成立。若取 $U_0=2$ 则 $\alpha\approx0.5$,这个坡度值与以上的分析接近,表明这一必要条件在定义域中是可以达到的。同时看到,由于 $\alpha>0$,因此必须 $U_0>0$,即向西流不满足不稳定的必要条件。

6.3 不稳定波的充分条件

在一般条件下,要得到赤道低频不稳定波出现的充分条件是困难的。在本文所给的基本流及截断模解的情况下,由于频率方程是 3 阶的,出现共轭复根的条件容易算得。问题的色散关系经过运算后为(系数的具体表达式见附录)

$$\sigma^3+p(k)\sigma+q(k)=0 \tag{37}$$

根据 3 次方程根的判别式,当

$$\left(\frac{p}{2}\right)^2+\left(\frac{q}{3}\right)^3>0 \tag{38}$$

时,方程(37)可以出现共轭复根。容易算得,当参数为 $\alpha=0.9$,$\bar{h}=0.7$ 时,且当 k 大于一定值后,判别式(38)成立。

7 结论

通过浅水模式的低阶截断模分析可以看到,当温跃层或高度场具有径向不均匀分布及相应的赤

道急流时,这样的背景场对赤道低频长波的发展有重要影响。当赤道附近基本流是向东时,可以使经典的 Rossby 波的波速发生改变甚至变成向东传播,同时与由背景场的不均匀性产生的地形 Rossby 波相互作用后,可产生向东传播的不稳定波。文中指出这种不稳定的性质,在本质上是发生在赤道海域或大气中的正压不稳定,但它不完全与中纬度的正压不稳定性相同,它是通过地形 Rossby 波和 Rossby 波两支波相互作用后产生的耦合不稳定。

这种耦合波在不稳定波段是向东传播的,其相速度要比经典的 Kelvin 波小很多,因此很有可能在 El Niño 发展的初始位相海温距平的向东传播是藉这类不稳定耦合波来实现的。在大气中,当 El Niño 发展到盛期低空转为西风后,这种不稳定向东传播的波可能成为低频振荡扰动向东发展的载体。当然,这一看法尚需在观测和理论上进一步研究。

同时指出,不稳定波在稳定的长波波段,向西传播的相速度虽然很大,但群速度是向东的。这些结果表明,单从海洋波动的角度来看,西太平洋暖池在 El Niño 发生初期是十分重要的,因为不仅信号以不稳定增长的形式向东传播,而在长波波段的波能量也是向东传播的。

这些结果由于是从正压模式的最低阶截断模解求得到的,因此是初步的,还有待进一步用更接近实际的模式来研究。

参考文献

[1] Rasmusson E M, Carpenter T H. Variations in tropical sea surface temperature and surface wind field associated with the Southern Oscillation/El Niño. Monthly Weather Review, 1982, 110:354-384.

[2] Rasmusson E M, Wallace J M. Meteorological aspect of the El Niño/Southern Oscillation. Science, 1983, 222:1195-1202.

[3] McPhaden M J. Genesis and evolution of the 1997-1998 El Niño. Science, 1999, 283:950-953.

[4] 李崇银,穆明权. 厄尔尼诺的发生与赤道西太平洋暖池次表层海温异常. 大气科学,1999,23:513-521.

[5] 巢清尘,巢纪平. 热带西太平洋和东印度洋对 El Niño 发展的影响. 自然科学进展,2001,59(5):515-523.

[6] 巢纪平,袁绍宇,巢清尘. 热带西太平洋暖池次表层暖水的起源——对 1997/1998 年 El Niño 事件的分析. 大气科学,2003,27(2):145-151.

[7] Chao Jiping, Yuan Shaoyu, Chao Qingchen, et al. A data analysis study on the evolution of the El Niño/La Niña cycle. Advances of Atmospheric Science, 2002, 19:837-844.

[8] 巢纪平,袁绍宇,蔡怡. 热带印度洋的大尺度海气相互作用事件. 气象学报,2003,61(2): 251-262.

[9] 巢纪平,袁绍宇. 热带印度洋和太平洋海气相互作用事件的相互联系. 自然科学进展,2003,13(12):1280-1285.

[10] 李崇银. 大气低频振荡. 北京: 气象出版社,1993.

[11] Ji Z, Chao J. Instability of the oceanic wave in the tropical region. Acta Meteologica Sinica, 1990, 4:343-348.

[12] 巢纪平,吴钦岳. 地转气流中的重力惯性波. 气象学报,34(4):523-530.

[13] Pedlosky J. Geophysical Fluid Dynamics. New York: Springer-Verlag Press, 1987.

[14] Marco De La Cruz-Heredia, Moore G W K. Barotropic instability due to Kelvin wave-Rossby wave coupling. Journal Atmosphere Science, 1999, 56:2376-2383.

[15] Sakai S. Rossby-Kelvin instability: A new type of ageostrophic instability caused by a resonance between Rossby and gravity waves. Journal of Fluid Mechanics, 1989, 202:149-176.

[16] Philander S G H. Equatorial waves in the presence of equatorial undercurrent. Journal of Physical Oceanography, 1979, 9(2):254-262.

[17] Matsuno T. Quasi-geostrophic motion in the equatorial area. Journal of Meteorology Society of Japan, 1966, 44:25-43.

[18] 巢纪平. 热带斜压大气的适应运动和发展运动. 中国科学,D 辑,1999,29(3): 279-288.

[19] Kuo H L. Dynamic instability of two-dimensional nondivergent flow in a barotropic atmosphere. Journal of Meteorology, 1949, 6: 105-122.

附录

解系数行列式矩阵可以得到关于频率 σ 的 3 次方程：

$$f_3\sigma^3 + f_2\sigma^2 + f_1\sigma + f_0 = 0 \tag{a1}$$

其中

$$
\begin{aligned}
f_3 &= 4I_{1,1}^{(1)}k \\
f_2 &= -6 - 2I_{0,0}^{(2)} + 4I_{0,2}^{(2)} - 2I_{2,2}^{(2)} - 2J_{1,2}^{(1)} - 2J_{0,1}^{(2)} - 4I_{1,1}^{(1)}k^2 \\
&\quad + 8I_{0,0}^{(1)}I_{1,1}^{(1)}k^2 + 2I_{1,1}^{(1)}I_{2,2}^{(1)}k^2 - 2I_{1,1}^{(1)}I_{2,2}^{(2)}k^2 \\
f_1 &= 4k - 10I_{0,0}^{(1)}k - I_{2,2}^{(1)} + 6I_{0,0}^{(2)}k - 2I_{0,0}^{(1)}I_{0,0}^{(2)}k - 2I_{0,2}^{(1)}I_{0,0}^{(2)}k \\
&\quad - I_{2,2}^{(1)}I_{0,0}^{(2)}k - 4I_{0,2}^{(2)}k + 6I_{0,0}^{(1)}I_{0,2}^{(2)}k + 2I_{0,2}^{(1)}I_{0,2}^{(2)}k + I_{2,2}^{(1)}I_{0,2}^{(2)}k \\
&\quad - (I_{0,2}^{(2)})^2k + I_{2,2}^{(2)}k - 4I_{0,0}^{(1)}I_{2,2}^{(2)}k + 6I_{0,0}^{(2)}I_{2,2}^{(2)}k + 2I_{0,2}^{(1)}J_{0,1}^{(1)}k \\
&\quad + 2I_{0,0}^{(2)}J_{0,1}^{(1)}k - 2I_{0,2}^{(2)}J_{0,1}^{(1)}k - 4I_{0,0}^{(1)}J_{1,2}^{(1)}k + I_{0,2}^{(2)}J_{1,2}^{(1)}k + 4J_{0,1}^{(2)}k \\
&\quad - 2I_{0,0}^{(1)}J_{0,1}^{(2)}k - I_{2,2}^{(1)}J_{0,1}^{(2)}k + 2I_{0,0}^{(2)}J_{0,1}^{(2)}k - 2I_{0,2}^{(2)}J_{0,1}^{(2)}k \\
&\quad + I_{2,2}^{(2)}J_{0,1}^{(2)}k - 4I_{1,1}^{(1)}k^3 - 8I_{0,0}^{(1)}I_{1,1}^{(1)}k^3 + 4I_{1,1}^{(1)}(I_{0,0}^{(1)})^2k^3 - 2(I_{0,2}^{(1)})^2I_{1,1}^{(1)}k^3 \\
f_0 &= 2k^2 + 2I_{0,0}^{(1)}k^2 - 4(I_{0,0}^{(1)})^2 + (I_{0,2}^{(1)})^2 + I_{2,2}^{(1)} - I_{0,0}^{(1)}I_{2,2}^{(1)} + 4I_{0,0}^{(2)} \\
&\quad + 2I_{0,0}^{(1)}I_{0,0}^{(2)} - 2I_{0,2}^{(1)}I_{0,0}^{(2)} - 2I_{0,0}^{(1)}I_{0,2}^{(2)} + (I_{0,2}^{(1)})^2I_{0,0}^{(2)} + 2I_{2,2}^{(1)}I_{0,0}^{(2)} \\
&\quad - I_{0,0}^{(1)}I_{0,0}^{(1)}I_{0,0}^{(2)} - 2I_{0,0}^{(1)}I_{0.2}^{(2)} + 2(I_{0,0}^{(1)})^2I_{0,2}^{(2)} + 2I_{0,0}^{(1)}I_{0,2}^{(1)}I_{0,2}^{(2)} \\
&\quad - (I_{0,2}^{(1)})^2I_{0,2}^{(2)} - I_{2,2}^{(1)}I_{0,2}^{(2)} + I_{0,0}^{(1)}I_{2,2}^{(1)}I_{0,2}^{(2)} - 2(I_{0,2}^{(2)})^2 \\
&\quad - I_{0,0}^{(1)}(I_{0,2}^{(2)})^2 + I_{0,2}^{(1)}(I_{0,2}^{(2)})^2 + I_{2,2}^{(2)} + I_{0,0}^{(1)}I_{2,2}^{(2)} - 2(I_{0,0}^{(1)})^2I_{2,2}^{(2)} \\
&\quad + 2I_{0,0}^{(2)}I_{2,2}^{(2)} + 2I_{0,0}^{(2)}I_{2,2}^{(2)} + I_{0,0}^{(1)}I_{2,2}^{(2)} - I_{0,2}^{(1)}I_{0,0}^{(2)}I_{2,2}^{(2)} + 2I_{0,0}^{(1)}J_{0,1}^{(1)} \\
&\quad + 2I_{0,0}^{(1)}I_{0,2}^{(1)}J_{0,1}^{(1)} - 2I_{0,0}^{(2)}J_{0,1}^{(1)} + 2I_{0,2}^{(1)}I_{0,0}^{(2)}J_{0,1}^{(1)} + I_{2,2}^{(1)}I_{0,0}^{(2)}J_{0,1}^{(1)} \\
&\quad - 2I_{0,2}^{(2)}J_{0,1}^{(1)} - 2I_{0,0}^{(1)}I_{0,2}^{(2)}J_{0,1}^{(1)} - I_{0,2}^{(1)}I_{0,2}^{(2)}J_{0,1}^{(1)} + (I_{0,2}^{(2)})^2J_{0,1}^{(1)} \\
&\quad - I_{0,0}^{(2)}I_{2,2}^{(2)}J_{0,1}^{(1)} + 2J_{1,2}^{(1)} - 2(I_{0,0}^{(1)})^2J_{1,2}^{(1)} + 4I_{0,0}^{(2)}J_{1,2}^{(1)} \\
&\quad - I_{0,2}^{(1)}I_{0,0}^{(2)}J_{1,2}^{(1)} - I_{0,2}^{(2)}J_{1,2}^{(1)} + I_{0,0}^{(1)}I_{0,2}^{(2)}J_{1,2}^{(1)} - 2J_{0,1}^{(2)} + 2I_{0,0}^{(1)}J_{0,1}^{(2)} \\
&\quad + (I_{0,2}^{(1)})^2J_{0,1}^{(2)} + I_{2,2}^{(1)}J_{0,1}^{(2)} - I_{0,0}^{(1)}I_{2,2}^{(1)}J_{0,1}^{(2)} - 2I_{0,0}^{(2)}J_{0,1}^{(2)} \\
&\quad + 2I_{0,2}^{(1)}I_{0,0}^{(2)}J_{0,1}^{(2)} + I_{2,2}^{(1)}I_{0,0}^{(2)}J_{0,1}^{(2)} + 2I_{0,2}^{(2)}J_{0,1}^{(2)} - 2I_{0,0}^{(1)}I_{0,2}^{(2)}J_{0,1}^{(2)} \\
&\quad - 2I_{0,2}^{(1)}I_{0,2}^{(2)}J_{0,1}^{(2)} + (I_{0,2}^{(2)})^2J_{0,1}^{(2)} - I_{2,2}^{(2)}J_{0,1}^{(2)} + I_{0,0}^{(1)}I_{2,2}^{(2)}J_{0,1}^{(2)} \\
&\quad - I_{0,0}^{(2)}I_{2,2}^{(2)}J_{0,1}^{(2)} + 4I_{1,1}^{(1)}K^4 - 4(I_{0,0}^{(1)})^2I_{1,1}^{(1)} - 2(I_{0,2}^{(1)})^2I_{1,1}^{(1)} \\
&\quad - 2I_{0,0}^{(1)}(I_{0,2}^{(1)})^2I_{1,1}^{(1)} - 2I_{1,1}^{(1)}I_{2,2}^{(1)} + 2I(I_{0,0}^{(1)})^2I_{1,1}^{(1)}I_{2,2}^{(1)} \\
&\quad - 8I_{1,1}^{(1)}I_{0,0}^{(2)} - 2(I_{0,2}^{(1)})^2I_{1,1}^{(1)}I_{0,0}^{(2)} - 4II_{1,1}^{(1)}I_{2,2}^{(2)}I_{0,0}^{(2)} \\
&\quad + 4I_{0,2}^{(1)}I_{1,1}^{(1)}I_{0,2}^{(2)} + 4I_{0,0}^{(1)}I_{0,2}^{(1)}I_{1,1}^{(1)}I_{0,2}^{(2)} - 4I_{1,1}^{(1)}(I_{0,2}^{(2)})^2 \\
&\quad + 2I_{1,1}^{(1)}I_{2,2}^{(2)} - 2(I_{0,0}^{(1)})^2I_{1,1}^{(1)}I_{2,2}^{(2)} + 4I_{1,1}^{(1)}I_{0,0}^{(2)}I_{2,2}^{(2)}
\end{aligned}
$$

对(a1)进行变换可以得到形如正文的方程(37)

$$\sigma^3 + p(k)\sigma + q(k) = 0$$

其中

$$p = b - \frac{a^2}{3},$$

$$q = \frac{2a^3}{27} - \frac{ab}{3} + c,$$

$$a = \frac{f_2}{f_3},$$

$$b = \frac{f_1}{f_3},$$

$$c = \frac{f_0}{f_3}$$

赤道两层海洋模式中基本流的切变不稳定性*

巢纪平[1,2]　高新全[3,4]　冯立成[5,6]

（1. 国家海洋环境预报中心，北京 100081；2. 国家海洋局第一海洋研究所，青岛 266061；3. 国家气候中心，北京 100081；4. 兰州大学大气科学学院，兰州 730000；5. 中国科学院大气物理研究所大气科学和地球流体力学数值模拟国家重点实验室，北京 100029；6. 中国科学院研究生院，北京 100049）

摘要：设计了一个热带赤道β-平面的两层海洋模式，在准长波近似下，应用最大截断模分析赤道波的基本形态，指出无论是正压模或斜压模 Kelvin 波、Rossby 波及基本流所对应的"地形 Rossby 波"是最基本的波系，在基本流的一定切变条件下，它们之间可以耦合出一类不稳定波。在浅混合层近似和"快波近似"下，正压模和斜压模是可以分离的，因此可以分别分析它们的色散特征，由于它们的特征量不同，在同样波长（扰动的纬向尺度）下，扰动的增长率也不同，通过分析得出在一定参数下，斜压模扰动增长率为正压模的 2 倍。近似分析表明，混合层中流场的增长要快于温跃层，但温跃层的温度增长要比混合层明显。

关键词：两层海洋模式；Kelvin 波；Rossby 波；地形 Rossby 波；耦合不稳定

1　引言

巢纪平等[1]在分析 ENSO 循环中海温距平的传播、发展特征时指出，沿温跃层曲面在赤道附近向东传播的海温距平其强度要比海表温度距平强很多，在参考文献[2,3]中用两层海洋模式分析了混合层和温跃层海温对风应力的响应特征，表明温跃层海温距平的响应要强于混合层，这在一定程度上支持了参考文献[1]的资料分析。注意到在前面两个两层海洋模式的计算中，除研究的是海洋对风应力的强迫响应外，在模式中背景场是静止的，不存在像赤道潜流那样具有切变的基本流。但我们知道，当存在像赤道潜流那样的切变流时，将改变 Matsuno[4]的赤道内波的特征，如 Philander[5]指出，改变了特征的赤道内波可以使潜流不稳定。如果把海洋和大气统一起来看，则 Boyd[6]、Zhang 和 Webster[7]、Wang 和 Xie[8]等都研究过切变流对赤道内波的影响。另一方面，参照参考文献[9]提出的 Kelvin 波-Rossby 波耦合的概念，巢纪平、刘琳等[10]研究了等值正压赤道模式在有基本切变流存在时波的耦合不稳定。本文将发展这一工作，应用两层赤道β-平面模式来研究波的耦合不稳定。

2　模式

模式基本上同参考文献[2,3]，如图 1 所示，参考坐标在垂直方向向上为正，静止海洋在垂直方向

* 气象学报，2007，65(1)：1-17.

取海表的坐标为 $z_1=H_1$，上层海洋底部的坐标为 $z_2=H_2$，静止海洋的总厚度为 $\overline{D}=H_1-H_2$，即将海洋分为上、下两层，厚度、密度和温度各不相同。设 $z_1-z_\eta=\eta$ 为上层混合层的厚度，$z_\eta-z_2=h$ 为温跃层的厚度，总厚度为 $\eta+h=D(x,y,t)$，其中 z_1 为海-气界面，z_η 为混合层-温跃层界面，z_2 为温跃层与准静止层（或称下均匀层）的界面。气候状态为 $\overline{\eta}+\overline{h}=\overline{D}(y)$，如设 $\overline{\eta}=\gamma\overline{D}$，则有 $\overline{h}=(1-\gamma)\overline{D}$，由 γ 的大小来控制两层的相对厚度。

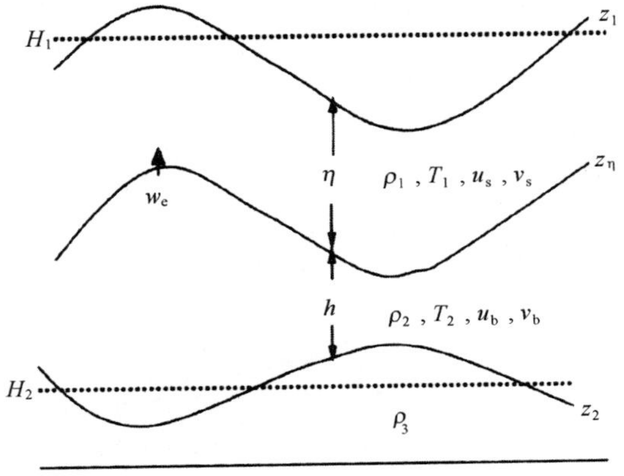

图 1　两层斜压海洋模式结构

基本场为纬向地转流

$$\beta y U_s=-g'\frac{dH_2}{dy}-g\frac{\rho_2}{\rho_3}\frac{dH_1}{dy}+\frac{1}{2}g\alpha\eta\frac{\partial\overline{T_s}}{\partial y}$$

式中，$g'=(\rho_3-\rho_2)/\rho_3$ 为约化重力；α 为海水的热膨胀系数。这个背景流可以简化，设海平面在大尺度上是比较均匀的，但温跃层底有明显的不均匀，有

$$\beta y U_s=-g'\frac{dH_2}{dy}+\frac{1}{2}g\alpha\eta\frac{\partial\overline{T_s}}{\partial y}=\beta y\hat{U}+\beta y\widetilde{U}_s \tag{1}$$

在类似简化条件下，有

$$\beta y U_b=-g'\frac{dH_2}{dy}+\frac{1}{2}g\alpha h\frac{\partial\overline{T}_b}{\partial y}+g\alpha\eta\frac{\rho_1}{\rho_2}\frac{\partial\overline{T_s}}{\partial y}$$

考虑到混合层的深度不大，温度分布水平比较均匀，因此其近似式为

$$\beta y U_b\approx-g'\frac{dH_2}{dy}+\frac{1}{2}g\alpha h\frac{\partial\overline{T}_b}{\partial y}=\beta y\hat{U}+\beta y\widetilde{U}_b \tag{2}$$

由此可见，基本流在上、下层的共同部分，由所研究的模式海洋厚度的不均匀性引起，可称为正压分量，标以 $\hat{A}$，由温度场引起的部分可称斜压分量，标以 $\hat{A}$。下标 s 为混合层，b 为温跃层。

2.1　运动方程

除背景流外，参照参考文献[2,3]将各物理量分成扰动量和平均量两部分，然后略去非线性项，得到线性化运动方程为

$$\frac{\partial u_s}{\partial t}+U_s\frac{\partial u_s}{\partial x}-\beta y v_s+\frac{dU_s}{dy}v_s=-g'\frac{\partial}{\partial x}(\eta+h)+\frac{1}{2}g\alpha\eta\frac{\partial T'_s}{\partial x} \tag{3}$$

$$\frac{\partial v_s}{\partial t} + U_s \frac{\partial v_s}{\partial x} + \beta y u_s = - g' \frac{\partial}{\partial y}(\eta + h) + \frac{1}{2} g\alpha\eta \frac{\partial T'_s}{\partial y} \tag{4}$$

$$\frac{\partial u_b}{\partial t} + U_b \frac{\partial u_b}{\partial x} - \beta y v_b + \frac{\mathrm{d}U_b}{\mathrm{d}y} v_b = - g' \frac{\partial}{\partial x}(\eta + h) + \frac{1}{2} g\alpha h \frac{\partial T'_b}{\partial x} + g\alpha\eta \frac{\rho_1}{\rho_2} \frac{\partial T'_s}{\partial x} \tag{5}$$

$$\frac{\partial v_b}{\partial t} + U_b \frac{\partial v_b}{\partial x} + \beta y u_b = - g' \frac{\partial}{\partial y}(\eta + h) + \frac{1}{2} g\alpha h \frac{\partial T'_b}{\partial y} + g\alpha\eta \frac{\rho_1}{\rho_2} \frac{\partial T'_s}{\partial y} \tag{6}$$

定义垂直平均

$$q_s = \eta^{-1} \int_{z_\eta}^{z_1} q \mathrm{d}z, \qquad q_b = h^{-1} \int_{z_2}^{z_\eta} q \mathrm{d}z \tag{7}$$

引进正压模和斜压模，分别为

$$\hat{u} = \frac{\eta u_s + h u_b}{\eta + h}, \qquad \tilde{u} = u_b - u_s \tag{8}$$

于是有

$$u_s = \hat{u} - \frac{h}{\eta + h} \tilde{u}, \quad u_b = \hat{u} + \frac{\eta}{\eta + h} \tilde{u} \tag{9}$$

变量 v 类似。

新的方程分别为

$$\begin{aligned}
&\frac{\partial \hat{u}}{\partial t} + \frac{\tilde{u}}{(\eta + h)^2}\left(h \frac{\partial \eta}{\partial t} - \eta \frac{\partial h}{\partial t}\right) + \hat{U} \frac{\partial \hat{u}}{\partial x} - \beta y \hat{v} + \left(\widetilde{U}_s \frac{\eta}{\eta + h} + \widetilde{U}_b \frac{h}{\eta + h}\right) \frac{\partial \hat{u}}{\partial x} + \frac{\hat{U}\tilde{u}}{(\eta + h)^2} \\
&\times \left(h \frac{\partial \eta}{\partial t} - \eta \frac{\partial h}{\partial t}\right) + \widetilde{U}_b \frac{h}{\eta + h} \frac{\partial}{\partial x}\left(\frac{\eta}{\eta + h} \tilde{u}\right) - \widetilde{U}_s \frac{\eta}{\eta + h} \frac{\partial}{\partial x}\left(\frac{h}{\eta + h} \tilde{u}\right) + \frac{\mathrm{d}\hat{U}}{\mathrm{d}y} \hat{v} + \left(\frac{\eta}{\eta + h} \frac{\mathrm{d}\widetilde{U}_s}{\mathrm{d}y} + \frac{h}{\eta + h} \frac{\mathrm{d}\widetilde{U}_b}{\mathrm{d}y}\right) \hat{v} \\
&+ \frac{\eta h}{(\eta + h)^2}\left[\frac{\mathrm{d}(\widetilde{U}_b - \widetilde{U}_s)}{\mathrm{d}y}\right] \tilde{v} = - g' \frac{\partial}{\partial x}(\eta + h) + g\alpha \frac{1}{\eta + h}\left(\frac{1}{2} h^2 \frac{\partial T'_b}{\partial x} + \eta h \frac{\rho_1}{\rho_2} \frac{\partial T'_s}{\partial x} + \frac{1}{2} \eta^2 \frac{\partial T_s'}{\partial x}\right)
\end{aligned} \tag{10}$$

$$\begin{aligned}
&\frac{\partial \tilde{u}}{\partial t} + (\hat{U} + \widetilde{U}_b) \frac{\partial}{\partial x}\left(\hat{u} + \frac{\eta}{\eta + h} \tilde{u}\right) - (\hat{U} + \widetilde{U}_s) \frac{\partial}{\partial x}\left(\hat{u} - \frac{h}{\eta + h} \tilde{u}\right) - \beta y \hat{v} + \frac{\mathrm{d}(\hat{U} + \widetilde{U}_b)}{\mathrm{d}y}\left(\hat{v} + \frac{\eta}{\eta + h} \tilde{v}\right) \\
&- \frac{\mathrm{d}(\hat{U} + \widetilde{U}_s)}{\mathrm{d}y}\left(\hat{v} - \frac{h}{\eta + h} \tilde{v}\right) = g\alpha\left(\frac{1}{2} h \frac{\partial T_b'}{\partial x} + \eta \frac{\rho_1}{\rho_2} \frac{\partial T_s'}{\partial x} - \frac{1}{2} \eta \frac{\partial T_s'}{\partial x}\right)
\end{aligned} \tag{11}$$

$$\begin{aligned}
&\frac{\partial \hat{v}}{\partial t} + \frac{\hat{v}}{(\eta + h)^2}\left(h \frac{\partial \eta}{\partial t} - \eta \frac{\partial h}{\partial t}\right) + \hat{U} \frac{\partial \hat{v}}{\partial x} + \beta y \hat{u} + \left(\widetilde{U}_s \frac{\eta}{\eta + h} + \widetilde{U}_b \frac{h}{\eta + h}\right) \frac{\partial \hat{v}}{\partial x} \\
&+ \frac{\hat{U}\tilde{v}}{(\eta + h)^2}\left(h \frac{\partial \eta}{\partial x} - \eta \frac{\partial h}{\partial x}\right) + \widetilde{U}_b \frac{h}{\eta + h} \frac{\partial}{\partial x}\left(\frac{\eta}{\eta + h} \tilde{v}\right) - \widetilde{U}_s \frac{\eta}{\eta + h} \frac{\partial}{\partial x}\left(\frac{h}{h + \eta} \tilde{v}\right) \\
&= - g' \frac{\partial}{\partial y}(\eta + h) + g\alpha \frac{1}{\eta + h}\left(\frac{1}{2} h^2 \frac{\partial T_b'}{\partial y} + \eta h \frac{\rho_1}{\rho_2} \frac{\partial T_s'}{\partial y} + \frac{1}{2} \eta^2 \frac{\partial T_s'}{\partial y}\right)
\end{aligned} \tag{12}$$

$$\frac{\partial \tilde{v}}{\partial t} + (\hat{U} + \widetilde{U}_b) \frac{\partial}{\partial x}\left(\hat{v} + \frac{\eta}{\eta + h} \tilde{v}\right) - (\hat{U} + \widetilde{U}_s) \frac{\partial}{\partial x}\left(\hat{v} - \frac{h}{\eta + h} \tilde{v}\right) + \beta y \tilde{u}$$

$$= g\alpha\left(\frac{1}{2}h\frac{\partial T_b{}'}{\partial y} + \eta\frac{\rho_1}{\rho_2}\frac{\partial T_s{}'}{\partial y} - \frac{1}{2}\eta\frac{\partial T_s{}'}{\partial y}\right) \tag{13}$$

注意到在以上方程中仍然存在非线性项，进一步的简化约定为，将除了压力梯度外各项中的厚度近似地取成：$\eta\to\gamma\overline{D}$，$h\to(1-\gamma)\overline{D}$，$(\eta+h)\to\overline{D}$，并设是纬度的平均值。在这些条件下，以上方程进一步简化成

$$\left\{\frac{\partial\hat{u}}{\partial t} + (\hat{U} + \gamma\tilde{U}_s + (1-\gamma)\tilde{U}_b)\frac{\partial\hat{u}}{\partial x} - \left[\beta y - \frac{\mathrm{d}\hat{U}}{\mathrm{d}y} - \gamma\frac{\mathrm{d}\tilde{U}_s}{\mathrm{d}y} - (1-\gamma)\frac{\mathrm{d}\tilde{U}_b}{\mathrm{d}y}\right]\tilde{v}\right\}$$

$$+ \gamma(1-\gamma)\left\{(\tilde{U}_b - \tilde{U}_s)\frac{\partial\tilde{u}}{\partial x} + \frac{\mathrm{d}(\tilde{U}_b - \tilde{U}_s)}{\mathrm{d}y}\tilde{v}\right\} =$$

$$- g'\frac{\partial}{\partial x}(\eta + h) + g\alpha\overline{D}\left[\frac{1}{2}(1-\gamma)^2\frac{\partial T_b{}'}{\partial x} + \gamma(1-\gamma)\frac{\partial T_s{}'}{\partial x} + \frac{1}{2}\gamma^2\frac{\partial T_s{}'}{\partial x}\right] \tag{14}$$

$$\left\{\frac{\partial\hat{v}}{\partial t} + (\hat{U} + \gamma\tilde{U}_s + (1-\gamma)\tilde{U}_b)\frac{\partial\hat{v}}{\partial x} + \beta y\hat{u}\right\} + \gamma(1-\gamma)(\tilde{U}_b - \tilde{U}_s)\frac{\partial\tilde{v}}{\partial x}$$

$$= - g'\frac{\partial}{\partial y}(\eta + h) + g\alpha\overline{D}\left[\frac{1}{2}(1-\gamma)^2\frac{\partial T_b{}'}{\partial y} + \gamma(1-\gamma)\frac{\partial T_s{}'}{\partial y} + \frac{1}{2}\gamma^2\frac{\partial T_s{}'}{\partial y}\right] \tag{15}$$

$$\left\{\frac{\partial\tilde{u}}{\partial t} + [\hat{U} + \gamma\tilde{U}_b + (1-\gamma)\tilde{U}_s]\frac{\partial\tilde{u}}{\partial x} + \left[\frac{\mathrm{d}[\hat{U} + \gamma\tilde{U}_b + (1-\gamma)\tilde{U}_s]}{\mathrm{d}y} - \beta y\right]\tilde{v}\right\} + (\tilde{U}_b - \tilde{U}_s)\frac{\partial\hat{u}}{\partial x}$$

$$+ \frac{\mathrm{d}(\tilde{U}_b - \tilde{U}_s)}{\mathrm{d}y}\hat{v} = g\alpha\overline{D}\left[\frac{1}{2}(1-\gamma)\frac{\partial T_b{}'}{\partial x} + \gamma\frac{\rho_1}{\rho_2}\frac{\partial T_s{}'}{\partial x} - \frac{1}{2}\gamma\frac{\partial T_s{}'}{\partial x}\right] \tag{16}$$

$$\left\{\frac{\partial\tilde{v}}{\partial t} + [\hat{U} + \gamma\tilde{U}_b + (1-\gamma)\tilde{U}_s]\frac{\partial\tilde{v}}{\partial x} + \beta y\tilde{u}\right\} + (\tilde{U}_b - \tilde{U}_s)\frac{\partial\hat{v}}{\partial x}$$

$$= g\alpha\overline{D}\left[\frac{1}{2}(1-\gamma)\frac{\partial T_b{}'}{\partial y} + \gamma\frac{\rho_1}{\rho_2}\frac{\partial T_s{}'}{\partial y} - \frac{1}{2}\gamma\frac{\partial T_s{}'}{\partial y}\right] \tag{17}$$

可见如果上下层气候温度场的分布不一样，斜压模分量和正压模分量将会彼此影响对方的变化。

2.2　线性化连续性方程

参照参考文献[2,3]关于垂直速度的定义，连续性方程的线性化形式给出为

$$\frac{\partial\eta}{\partial t} + U_s\frac{\partial\eta}{\partial x} + \gamma\overline{D}\left[\frac{\partial u_s}{\partial x} + \frac{\partial v_s}{\partial y}\right] + \frac{\beta\gamma}{g'}\hat{U}yv_s = 0 \tag{18}$$

$$\frac{\partial h}{\partial t} + U_b\frac{\partial h}{\partial x} + (1-\gamma)\overline{D}\left(\frac{\partial u_b}{\partial x} + \frac{\partial v_b}{\partial y}\right) + \frac{\beta(1-\gamma)}{g'}\hat{U}yv_b = 0 \tag{19}$$

将速度的正、斜压模代入，分别得到

$$\left\{\frac{\partial\eta}{\partial t} + (\hat{U} + \tilde{U}_s)\frac{\partial\eta}{\partial x} + \gamma\overline{D}\left(\frac{\partial\hat{u}}{\partial x} + \frac{\partial\hat{v}}{\partial y}\right) + \frac{\beta\gamma}{g'}\hat{U}y\hat{v}\right\} - \left\{\gamma(1-\gamma)\overline{D}\left(\frac{\partial\tilde{u}}{\partial x} + \frac{\partial\tilde{v}}{\partial y}\right) + \frac{\beta\gamma(1-\gamma)}{g'}\hat{U}y\tilde{v}\right\} = 0 \tag{20}$$

$$\left\{\frac{\partial h}{\partial t} + (\hat{U} + \tilde{U}_b)\frac{\partial h}{\partial x} + (1-\gamma)\overline{D}\left(\frac{\partial\hat{u}}{\partial x} + \frac{\partial\hat{v}}{\partial y}\right) + \frac{\beta(1-\gamma)}{g'}\hat{U}y\hat{v}\right\}$$

$$+\left\{\gamma(1-\gamma)\overline{D}\left(\frac{\partial\widetilde{u}}{\partial x}+\frac{\partial\widetilde{v}}{\partial y}\right)+\frac{\beta\gamma(1-\gamma)}{g'}\hat{U}y\,\widetilde{v}\right\}=0 \tag{21}$$

2.3 线性化温度方程

在不考虑水平平流项及非线性垂直输送项后,按约定的简化方案,分别给出

$$\left\{\frac{\partial T_s'}{\partial t}+(\hat{U}+\widetilde{U}_s)\frac{\partial T_s'}{\partial x}-\gamma\overline{D}\frac{\partial\overline{T}_s}{\partial z}\left(\frac{\partial\hat{u}}{\partial x}+\frac{\partial\hat{v}}{\partial y}\right)+\frac{\partial\overline{T}_s}{\partial y}\hat{v}\right\}$$

$$+\left\{\gamma(1-\gamma)\overline{D}\frac{\partial\overline{T}_s}{\partial z}\left(\frac{\partial\widetilde{u}}{\partial x}+\frac{\partial\widetilde{v}}{\partial y}\right)-(1-\gamma)\frac{\partial\overline{T}_s}{\partial y}\widetilde{v}\right\}=0 \tag{22}$$

$$\left\{\frac{\partial T_b'}{\partial t}+(\hat{U}+\widetilde{U}_b)\frac{\partial T_b'}{\partial x}-(1-\gamma)\overline{D}\frac{\partial\overline{T}_b}{\partial z}\left(\frac{\partial\hat{u}}{\partial x}+\frac{\partial\hat{v}}{\partial y}\right)+\frac{\partial\overline{T}_b}{\partial y}\hat{v}\right\}$$

$$-\left\{\gamma(1-\gamma)\overline{D}\frac{\partial\overline{T}_b}{\partial z}\left(\frac{\partial\widetilde{u}}{\partial x}+\frac{\partial\widetilde{v}}{\partial y}\right)-\gamma\frac{\partial\overline{T}_b}{\partial y}\widetilde{v}\right\}=0 \tag{23}$$

3 无量纲方程

3.1 特征量

引进正压模的重力波波速,为

$$C=\sqrt{g'\overline{D}} \tag{24}$$

对正压模的物理量的特征值取成:时间 $t\sim(2\beta C)^{-1/2}$,长度 $(x,y)\sim\left(\frac{C}{2\beta}\right)^{1/2}$,速度 $(\hat{u},\hat{v})\sim C$,$(\eta,h)\sim\overline{D}$,$(H_1,H_2)\sim\overline{D}$,$z\sim\overline{D}$。

在另一方面,取温度变化的特征值为 ΔT_s、ΔT_b,可看成是混合层上、下的温度差和温跃层上、下的温度差。引进斜压模的重力波波速

$$C_s=\sqrt{g\alpha\Delta T_s\overline{D}},\quad C_b=\sqrt{g\alpha\Delta T_b\overline{D}},\quad C=\sqrt{g\alpha\Delta\overline{TD}},\quad C_s\sim C_b\sim\widetilde{C} \tag{25}$$

相关物理量的特征值取成:$(T_s',T_b')\sim(\Delta T_s,\Delta T_b)$,$(\widetilde{u},\widetilde{v})\sim\widetilde{C}$,$(U_s,U_b,\hat{U})\sim C$,$(\widetilde{U}_s,\widetilde{U}_b)\sim(C_s^2,C_b^2)/C\sim\widetilde{C}^2/C$。

3.2 基本场

由上面约定的特征量,基本场(1)和基本场(2)的无量纲形式分别为

$$\frac{1}{2}yU_s=-\frac{\mathrm{d}H_2}{\mathrm{d}y}+\frac{1}{2}\left(\frac{C_s}{C}\right)^2\eta\frac{\partial\overline{T}_s}{\partial y}=\frac{1}{2}y\hat{U}+\frac{1}{2}\left(\frac{C_s}{C}\right)^2y\widetilde{U}_s \tag{26}$$

$$\frac{1}{2}yU_b=-\frac{\mathrm{d}H_2}{\mathrm{d}y}+\frac{1}{2}\left(\frac{C_b}{C}\right)^2h\frac{\partial\overline{T}_b}{\partial y}=\frac{1}{2}y\hat{U}+\frac{1}{2}\left(\frac{C_b}{C}\right)^2y\widetilde{U}_b \tag{27}$$

3.3 运动方程

为了使方程简洁,舍去不重要的差异,设 C_s,C_b 均取 $\widetilde{C}$,运动方程右端温度梯度项中的 H_1,H_2 均取

成 $\overline{D}$,这些近似不会使问题有实质性改变。

正压模方程(14)和方程(15)的无量纲形式为

$$\left\{\frac{\partial \hat{u}}{\partial t}+\left[\hat{U}+\left(\frac{\widetilde{C}}{C}\right)^2(\gamma\widetilde{U}_s+(1-\gamma)\widetilde{U}_b)\right]\frac{\partial \hat{u}}{\partial x}-\left[\frac{1}{2}y-\frac{\mathrm{d}\hat{U}}{\mathrm{d}y}-\left(\frac{\widetilde{C}}{C}\right)^2\frac{\mathrm{d}}{\mathrm{d}y}(\gamma\widetilde{U}_s+(1-\gamma)\widetilde{U}_b)\right]\hat{v}\right\}$$
$$+\gamma(1-\gamma)\left(\frac{\widetilde{C}}{C}\right)^3\left\{(\widetilde{U}_b-\widetilde{U}_s)\frac{\partial \tilde{u}}{\partial x}+\frac{\mathrm{d}(\widetilde{U}_b-\widetilde{U}_s)}{\mathrm{d}y}\cdot\tilde{v}\right\}=\frac{\partial}{\partial x}(\eta+h)$$
$$+\left(\frac{\widetilde{C}}{C}\right)^2\left[\frac{1}{2}(1-\gamma)^2\frac{\partial T_b'}{\partial x}+\gamma(1-\gamma)\frac{\partial T_s'}{\partial x}+\frac{1}{2}\gamma^2\frac{\partial T_s'}{\partial x}\right] \tag{28}$$

$$\left\{\frac{\partial \hat{v}}{\partial t}+\left[\hat{U}+\left(\frac{\widetilde{C}}{C}\right)^2(\gamma\widetilde{U}_s+(1-\gamma)\widetilde{U}_b)\right]\frac{\partial \hat{v}}{\partial x}+\frac{1}{2}y\hat{u}\right\}+\gamma(1-\gamma)\left(\frac{\widetilde{C}}{C}\right)^3(\widetilde{U}_b-\widetilde{U}_s)\frac{\partial \tilde{v}}{\partial x}$$
$$=-\frac{\partial}{\partial y}(\eta+h)+\left(\frac{\widetilde{C}}{C}\right)^2\left[\frac{1}{2}(1-\gamma)^2\frac{\partial T_b'}{\partial y}+\gamma(1-\gamma)\frac{\partial T_s'}{\partial y}+\frac{1}{2}\gamma^2\frac{\partial T_s'}{\partial y}\right] \tag{29}$$

斜压模方程,由方程(16)和方程(17)给出

$$\frac{\partial \tilde{u}}{\partial t}+\left[\hat{U}+\left(\frac{\widetilde{C}}{C}\right)^2(\gamma\widetilde{U}_b+(1-\gamma)\widetilde{U}_s)\right]\frac{\partial \tilde{u}}{\partial x}+\left[\frac{\mathrm{d}\hat{U}}{\mathrm{d}y}+\left(\frac{\widetilde{C}}{C}\right)^2\frac{\mathrm{d}}{\mathrm{d}y}(\gamma\widetilde{U}_b+(1-\gamma)\widetilde{U}_s)-\frac{1}{2}y\right]\tilde{v}$$
$$+\left(\frac{\widetilde{C}}{C}\right)\left[(\widetilde{U}_b-\widetilde{U}_s)\frac{\partial \hat{u}}{\partial x}+\frac{\mathrm{d}(\widetilde{U}_b-\widetilde{U}_s)}{\mathrm{d}y}\hat{v}\right]=\left(\frac{\widetilde{C}}{C}\right)\left[\frac{1}{2}(1-\gamma)\frac{\partial T_b'}{\partial x}+\gamma\frac{\rho_1}{\rho_2}\frac{\partial T_s'}{\partial x}-\frac{1}{2}\gamma\frac{\partial T_s'}{\partial x}\right] \tag{30}$$

$$\frac{\partial \tilde{v}}{\partial t}+\left[\hat{U}+\left(\frac{\widetilde{C}}{C}\right)^2(\gamma\widetilde{U}_b+(1-\gamma)\widetilde{U}_s)\right]\frac{\partial \tilde{v}}{\partial x}+\frac{1}{2}y\tilde{u}+\left(\frac{\widetilde{C}}{C}\right)(\widetilde{U}_b-\widetilde{U}_s)\frac{\partial \hat{v}}{\partial x}$$
$$=\left(\frac{\widetilde{C}}{C}\right)\left[\frac{1}{2}(1-\gamma)\frac{\partial T_b'}{\partial y}+\gamma\frac{\rho_1}{\rho_2}\frac{\partial T_s'}{\partial y}-\frac{1}{2}\gamma\frac{\partial T_s'}{\partial y}\right] \tag{31}$$

3.4 浅混合层近似和快波近似

考虑到混合层厚度相对温跃层厚度来讲很浅,即 $\gamma\ll1$,称为“浅混合层近似”,另外,正压模的重力波速相对斜压模来讲要快很多,即 $(\widetilde{C}/C)^2\ll1$,称为“快波近似”。

在浅混合层近似和快波近似下正压模和斜压模方程分别为

$$\frac{\partial \hat{u}}{\partial t}+\hat{U}\frac{\partial \hat{u}}{\partial x}-\left(\frac{1}{2}y-\frac{\mathrm{d}\hat{U}}{\mathrm{d}y}\right)\hat{v}=-\frac{\partial}{\partial x}(\eta+h) \tag{32}$$

$$\frac{\partial \hat{v}}{\partial t}+\hat{U}\frac{\partial \hat{v}}{\partial x}+\frac{1}{2}y\hat{u}=-\frac{\partial}{\partial y}(\eta+h) \tag{33}$$

$$\frac{\partial \tilde{u}}{\partial t}+\hat{U}\frac{\partial \tilde{u}}{\partial x}+\left(\frac{\mathrm{d}\hat{U}}{\mathrm{d}y}-\frac{1}{2}y\right)\tilde{v}+\left(\frac{\widetilde{C}}{C}\right)\left[(\widetilde{U}_b-\widetilde{U}_s)\frac{\partial \hat{u}}{\partial x}+\frac{\mathrm{d}(\widetilde{U}_b-\widetilde{U}_s)}{\mathrm{d}y}\hat{v}\right]=\frac{1}{2}\left(\frac{\widetilde{C}}{C}\right)\frac{\partial T_b'}{\partial x} \tag{34}$$

$$\frac{\partial \tilde{v}}{\partial t}+\hat{U}\frac{\partial \tilde{v}}{\partial x}+\frac{1}{2}y\tilde{u}+\left(\frac{\widetilde{C}}{C}\right)(\widetilde{U}_b-\widetilde{U}_s)\frac{\partial \hat{v}}{\partial x}=\frac{1}{2}\left(\frac{\widetilde{C}}{C}\right)\frac{\partial T_b'}{\partial y} \tag{35}$$

可以看到正压模分量的运动分量是独立的,但在 $(\widetilde{C}/C)$ 保留的近似下,正压模对斜压模的运动分

量有作用。

连续性方程(20)与方程(21)相加后为

$$\frac{\partial}{\partial t}(\eta + h) + \hat{U}\frac{\partial}{\partial x}(\eta + h) + \left(\frac{\partial \hat{u}}{\partial x} + \frac{\partial \hat{v}}{\partial y}\right) + \frac{1}{2}\hat{U}y\hat{v} = 0 \tag{36}$$

温度变化方程为

$$\frac{\partial T_s'}{\partial t} + \hat{U}\frac{\partial T_s'}{\partial x} + \frac{\partial \overline{T}_s}{\partial y}\left(\hat{v} - \frac{\widetilde{C}}{C}\tilde{v}\right) = 0 \tag{37}$$

$$\frac{\partial T_b'}{\partial t} + \hat{U}\frac{\partial T_b'}{\partial x} - \frac{\partial \overline{T}_b}{\partial z}\left(\frac{\partial \tilde{u}}{\partial x} + \frac{\partial \tilde{v}}{\partial y}\right) + \frac{\partial \overline{T}_b}{\partial y}\hat{v} = 0 \tag{38}$$

3.5 正压模控制方程

重新写出,为

$$\frac{\partial \hat{u}}{\partial t} + \hat{U}\frac{\partial \hat{u}}{\partial x} - \left(\frac{1}{2}y - \frac{\mathrm{d}\hat{U}}{\mathrm{d}y}\right)\hat{v} = -\frac{\partial}{\partial x}(\eta + h) \tag{32'}$$

$$\frac{\partial \hat{v}}{\partial t} + \hat{U}\frac{\partial \hat{v}}{\partial x} + \frac{1}{2}y\hat{u} = -\frac{\partial}{\partial y}(\eta + h) \tag{33'}$$

$$\frac{\partial}{\partial t}(\eta + h) + \hat{U}\frac{\partial}{\partial x}(\eta + h) + \left(\frac{\partial \hat{u}}{\partial x} + \frac{\partial \hat{v}}{\partial y}\right) + \frac{1}{2}\hat{U}y\hat{v} = 0 \tag{36'}$$

可见在浅混合层和快波近似下,正压模方程相等于等值浅水模式的方程,混合层温度变化即方程(37)不参与运动。

3.6 斜压模方程

注意到式(34)、式(35)和式(38)中斜压分量并不独立,正压分量将参与进来,但是考虑到在同一个方程中各物理量应用同样的标尺来衡量,在式(34)、式(35)中的($\hat{u}$,$\hat{v}$)若不用 C 而用 $\widetilde{C}$ 作为特征量,并记以($\hat{u}'$,$\hat{v}'$),则有($\hat{u}$,$\hat{v}$)=($\widetilde{C}/C$)($\hat{u}'$,$\hat{v}'$),因此式(34)、式(35)中与正压模相联系的项正比于$(\widetilde{C}/C)^2$,相对于斜压分量来是小项可以略去,而简化成

$$\frac{\partial \tilde{u}}{\partial t} + \hat{U}\frac{\partial \tilde{u}}{\partial x} + \left(\frac{\mathrm{d}\hat{U}}{\mathrm{d}y} - \frac{1}{2}y\right)\tilde{v} = \frac{1}{2}\left(\frac{\widetilde{C}}{C}\right)\frac{\partial T_b'}{\partial x} \tag{39}$$

$$\frac{\partial \tilde{v}}{\partial t} + \hat{U}\frac{\partial \tilde{v}}{\partial x} + \frac{1}{2}y\tilde{u} = \frac{1}{2}\left(\frac{\widetilde{C}}{C}\right)\frac{\partial T_b'}{\partial y} \tag{40}$$

同时式(38)左边最后一项为 $\frac{\partial \overline{T}_b}{\partial y}\left(\frac{\widetilde{C}}{C}\right)\hat{v}'$,在热带温跃层气候温度的经向变化较均匀,因此这一项的作用不大,如略去则为

$$\frac{\partial T_b'}{\partial t} + \hat{U}\frac{\partial T_b'}{\partial x} - \frac{\partial \overline{T}_b}{\partial z}\left(\frac{\partial \tilde{u}}{\partial x} + \frac{\partial \tilde{v}}{\partial y}\right) = 0 \tag{41}$$

这样由式(39)、式(40)和式(41)构成斜压模的闭合方程组。

由正压模和斜压模的运动分量都算得后,可由式(37)算出混合层的温度距平。

4　正压模方程的色散性

现在来分析正压模方程(32′),方程(33′)和方程(36′)的色散关系。

设 $H_2(y)=H_0(1+h_c(y))$,这里 $H_2(y)$ 是温跃层底部的深度,是温跃层深浅的一个量度;$h_c(y)$ 为基于平均温跃层深度上的起伏结构。由无量纲的基本地转流公式(26)或公式(27)

$$\frac{1}{2}y\hat{U}=-\frac{\mathrm{d}H_2(y)}{\mathrm{d}y}$$

有

$$\frac{1}{2}y\hat{U}=-\frac{\mathrm{d}h_c}{\mathrm{d}y},\quad \hat{U}=-\frac{2}{y}\frac{\mathrm{d}h_c}{\mathrm{d}y},\quad \frac{\mathrm{d}\hat{U}}{\mathrm{d}y}=-\frac{2}{y}\frac{\mathrm{d}^2h_c}{\mathrm{d}y^2}+\frac{2}{y^2}\frac{\mathrm{d}h_c}{\mathrm{d}y}$$

代入式(32′),式(33′)和式(36′),有

$$\frac{\partial\hat{u}}{\partial t}-\frac{2}{y}\frac{\mathrm{d}h_c}{\mathrm{d}y}\frac{\partial\hat{u}}{\partial x}-\frac{1}{2}y\hat{v}+\frac{\partial}{\partial x}(\eta+h)=\left(\frac{2}{y}\frac{\mathrm{d}^2h_c}{\mathrm{d}y^2}-\frac{2}{y^2}\frac{\mathrm{d}h_c}{\mathrm{d}_y}\right)\hat{v}\tag{42}$$

$$\frac{\partial\hat{v}}{\partial t}-\frac{2}{y}\frac{\mathrm{d}h_c}{\mathrm{d}y}\frac{\partial\hat{v}}{\partial x}+\frac{1}{2}y\hat{u}+\frac{\partial}{\partial y}(\eta+h)=0\tag{43}$$

$$\frac{\partial}{\partial t}(\eta+h)-\frac{2}{y}\frac{\mathrm{d}h_c}{\mathrm{d}y}\frac{\partial}{\partial x}(\eta+h)+\left(\frac{\partial\hat{u}}{\partial x}+\frac{\partial\hat{v}}{\partial y}\right)=\frac{\mathrm{d}h_c}{\mathrm{d}y}\hat{v}\tag{44}$$

引进变量 $\hat{q}$,$\hat{r}$,定义成

$$\hat{q}=(\eta+h)+\hat{u},\quad \hat{r}=(\eta+h)-\hat{u}$$

于是有

$$\eta+h=\frac{1}{2}(\hat{q}+\hat{r}),\quad \hat{u}=\frac{1}{2}(\hat{q}-\hat{r})$$

这样,方程式(42)至式(44)变成

$$\frac{\partial\hat{q}}{\partial t}+\frac{\partial\hat{q}}{\partial x}+\frac{\partial\hat{v}}{\partial y}-\frac{1}{2}y\hat{v}=\frac{2}{y}\frac{\mathrm{d}h_c}{\mathrm{d}y}\frac{\partial\hat{q}}{\partial x}+\left(\frac{2}{y}\frac{\mathrm{d}^2h_c}{\mathrm{d}y^2}-\frac{2}{y^2}\frac{\mathrm{d}h_c}{\mathrm{d}y}+\frac{\mathrm{d}h_c}{\mathrm{d}y}\right)\hat{v}\tag{45}$$

$$\frac{\partial\hat{r}}{\partial t}-\frac{\partial\hat{r}}{\partial x}+\frac{\partial\hat{v}}{\partial y}+\frac{1}{2}y\hat{v}=\frac{2}{y}\frac{\mathrm{d}h_c}{\mathrm{d}y}\frac{\partial\hat{r}}{\partial x}-\left(\frac{2}{y}\frac{\mathrm{d}^2h_c}{\mathrm{d}y^2}-\frac{2}{y^2}\frac{\mathrm{d}h_c}{\mathrm{d}y}-\frac{\mathrm{d}h_c}{\mathrm{d}y}\right)\hat{v}\tag{46}$$

$$\frac{\partial\hat{v}}{\partial t}+\frac{1}{2}\left[\left(\frac{\partial\hat{q}}{\partial y}+\frac{1}{2}y\hat{q}\right)+\left(\frac{\partial\hat{r}}{\partial y}-\frac{1}{2}y\hat{r}\right)\right]=\frac{2}{y}\frac{\mathrm{d}h_c}{\mathrm{d}y}\frac{\partial\hat{v}}{\partial x}\tag{47}$$

将变量 $\hat{q}$,$\hat{r}$,$\hat{v}$ 展成抛物圆柱函数(Weber 函数),即

$$(\hat{q},\hat{r},\hat{v})=\sum_{m=0}^{\infty}(\hat{q}_m(x,t),\hat{r}_m(x,t),\hat{v}_m(x,t))D_m(y)\tag{48}$$

同时,考虑 Weber 函数的递推公式(这里 $m\geqslant1$)

$$\frac{\mathrm{d}D_m}{\mathrm{d}y}+\frac{1}{2}yD_m=mD_{m-1},\quad \frac{\mathrm{d}D_{m-1}}{\mathrm{d}y}-\frac{1}{2}yD_{m-1}=-D_m$$

并利用 Weber 函数的正交性质

$$\int_{-\infty}^{\infty}D_m(y)\cdot D_n(y)\mathrm{d}y=\begin{cases}0 & m\neq n\\ n!\ \sqrt{2\pi} & m=n\end{cases}=n!\ \sqrt{2\pi}\delta_{mn}$$

方程式(45)至式(47)给出

$$\frac{\partial \hat{q}_n}{\partial t} + \frac{\partial \hat{q}_n}{\partial x} - \hat{v}_{n-1} = \hat{A}_n \tag{49}$$

$$\frac{\partial \hat{r}_n}{\partial t} - \frac{\partial \hat{r}_n}{\partial x} + (n+1)\hat{v}_{n+1} = \hat{B}_n \tag{50}$$

$$\varepsilon \frac{\partial \hat{v}_{n+1}}{\partial t} + \frac{1}{2}[(n+2)\hat{q}_{n+2} - \hat{r}_n] = \delta \hat{C}_{n+1} \tag{51}$$

式中

$$\hat{A}_n = \frac{1}{n!}\sum_{m=0}^{\infty}\left(I_{m,n}\frac{\partial \hat{q}_m}{\partial x} + J_{m,n}^{(1)}\hat{v}_m\right) \tag{52}$$

$$\hat{B}_n = \frac{1}{n!}\sum_{m=0}^{\infty}\left(I_{m,n}\frac{\partial \hat{r}_m}{\partial x} - J_{m,n}^{(2)}\hat{v}_m\right) \tag{53}$$

$$\hat{C}_{n+1} = \frac{1}{(n+1)!}\sum_{m=0}^{\infty} I_{m,n+1}\frac{\partial \hat{v}_m}{\partial x} \tag{54}$$

其中

$$I_{m,n} = \frac{1}{\sqrt{2\pi}}\int_{-\infty}^{\infty}\frac{2}{y}\frac{\mathrm{d}h_c}{\mathrm{d}y}D_m D_n \mathrm{d}y \tag{55a}$$

$$J_{m,n}^{(1),(2)} = \frac{1}{\sqrt{2\pi}}\int_{-\infty}^{\infty}\left(\frac{2}{y}\frac{\mathrm{d}^2 h_c}{\mathrm{d}y^2} - \frac{2}{y^2}\frac{\mathrm{d}h_c}{\mathrm{d}y} \pm \frac{\mathrm{d}h_c}{\mathrm{d}y}\right)D_m D_n \mathrm{d}y = K_{m,n}^{(1)} \pm K_{m,n}^{(2)} \tag{55b}$$

方程(51)中 ε 和 δ 为标识号,其值取0或1,当 $\varepsilon=\delta=0$ 时,称长波近似(纬向半地转近似),当 $\varepsilon=0$ 而 $\delta=1$ 时称准长波近似,在准长波近似中考虑了基本流对经向扰动运动的平流作用。

取温跃层的起伏结构为 $h_c = \bar{h}(\mathrm{e}^{-\alpha y^2} - 1) + h_0$。式中,$h_0$ 为赤道值,$\bar{h}$ 是一个参量,在此 α 是表征温跃层经向不均匀的参量(不是前面海水的热膨胀系数),因此温跃层的总深度为

$$H_2 = 1 + \bar{h}(\mathrm{e}^{-\alpha y^2} - 1) + h_0$$

于是

$$\frac{\mathrm{d}h_c}{\mathrm{d}y} = -2\alpha\bar{h}y\mathrm{e}^{-\alpha y^2}, \quad \frac{\mathrm{d}^2 h_c}{\mathrm{d}y^2} = -2\alpha\bar{h}(1-2\alpha y^2)\mathrm{e}^{-\alpha y^2}$$

这一温跃层的不均匀结构同参考文献[10]。注意到,温跃层极小值所在的纬度为 $y_m = \sqrt{\frac{1}{2\alpha}}$,当 α 取值越大时它越靠近赤道,如与 H_2 对应的地转流为赤道潜流,则当 α 越大时赤道潜流越窄。当 h_c 取现在的形式时,式(55)的积分结果见附录1。

为过滤掉高频的重力惯性波,现在分析准长波近似下最大截断模 $\hat{q}_0, \hat{q}_2, \hat{r}_0, \hat{v}_1$ 时运动的色散关系,方程为

$$\frac{\partial \hat{q}_0}{\partial t} + (1 - I_{0,0})\frac{\partial \hat{q}_0}{\partial x} = I_{2,0}\frac{\partial \hat{q}_2}{\partial x} + J_{1,0}^{(1)}\hat{v}_1 \tag{56}$$

$$\frac{\partial \hat{q}_2}{\partial t} - \frac{1}{3}\left(1 + \frac{1}{2}I_{2,2}\right)\frac{\partial \hat{q}_2}{\partial x} = \frac{1}{6}I_{0,2}\frac{\partial \hat{q}_0}{\partial x} + \frac{1}{3}I_{0,0}\frac{\partial \hat{r}_0}{\partial x} + \frac{2}{3}\delta I_{1,1}\left(\frac{\partial}{\partial t} - \frac{\partial}{\partial x}\right)\frac{\partial \hat{v}_1}{\partial x} - \frac{1}{3}\left(J_{1,0}^{(2)} - \frac{1}{2}J_{1,2}^{(1)}\right)\hat{v}_1 \tag{57}$$

$$\frac{\partial \hat{q}_2}{\partial t} - \left(3 + 2I_{0,0} - \frac{1}{2}I_{2,2}\right)\frac{\partial \hat{q}_2}{\partial x} = -\frac{1}{2}I_{0,2}\frac{\partial \hat{q}_0}{\partial x} + 2\delta I_{1,1}\frac{\partial^2 \hat{v}_1}{\partial t \partial x} - 2\delta I_{1,1}(1 + I_{0,0})\frac{\partial^2 \hat{v}_1}{\partial x^2} -$$

$$\left(2+J_{1,0}^{(2)}+\frac{1}{2}J_{1,2}^{(1)}\right)\hat{v}_1 \tag{58}$$

$$\delta I_{1,1}\frac{\partial\hat{v}_1}{\partial x}=\frac{1}{2}(2\hat{q}_2-\hat{r}_0) \tag{59}$$

由这 4 个方程构成对变量 $\hat{q}_0,\hat{q}_2,\hat{r}_0,\hat{v}_1$ 的闭合方程组。

或者消去 $\hat{v}_1$ 得到

$$\frac{\partial^2\hat{q}_0}{\partial t\partial x}+(1-I_{0,0})\frac{\partial^2 q_0}{\partial x^2}-I_{2,0}\frac{\partial^2\hat{q}_2}{\partial x^2}-\frac{J_{1,0}^{(1)}}{\delta I_{1,1}}\hat{q}_2+\frac{J_{1,0}^{(1)}}{2\delta I_{1,1}}\hat{r}_0=0 \tag{60}$$

$$\frac{\partial^2\hat{q}_2}{\partial t\partial x}+\left(1-\frac{1}{2}I_{2,2}\right)\frac{\partial^2\hat{q}_2}{\partial x^2}+\frac{J_{1,0}^{(2)}-\frac{1}{2}J_{1,2}^{(1)}}{\delta I_{1,1}}\hat{q}_2+\frac{\partial^2\hat{r}_0}{\partial t\partial x}-(1+I_{0,0})\frac{\partial^2\hat{r}_0}{\partial x^2}-$$
$$\frac{J_{1,0}^{(2)}-\frac{1}{2}J_{1,2}^{(1)}}{2\delta I_{1,1}}\hat{r}_0-\frac{1}{2}I_{0,2}\frac{\partial^2\hat{q}_0}{\partial x^2}=0 \tag{61}$$

$$\frac{\partial^2\hat{q}_2}{\partial t\partial x}+\left(1-\frac{1}{2}I_{2,2}\right)\frac{\partial^2\hat{q}_2}{\partial x^2}-\frac{2+J_{1,0}^{(2)}+\frac{1}{2}J_{1,2}^{(1)}}{\delta I_{1,1}}\hat{q}_2-\frac{\partial^2\hat{r}_0}{\partial t\partial x}+(1+I_{0,0})\frac{\partial^2\hat{r}_0}{\partial x^2}+$$
$$\frac{2+J_{1,0}^{(2)}+\frac{1}{2}J_{1,2}^{(1)}}{2\delta I_{1,1}}\hat{r}_0-\frac{1}{2}I_{0,2}\frac{\partial^2\hat{q}_0}{\partial x^2}=0 \tag{62}$$

构成对变量 $\hat{q}_0,\hat{q}_2,\hat{r}_0$ 的准长波近似下的基本方程组。

注意到,一般来讲,在无背景流时,$\hat{q}_0$ 代表的是 Kelvin 波的模态,而 $\hat{q}_2,\hat{r}_0$ 代表的是最大尺度的 Rossby 波模态,它们是相互独立的,而现在有了背景流,这两种波的模态相互耦合起来了。

由式(60),式(61),式(62)消去 $\hat{q}_0,\hat{q}_2$ 写出关于 $\hat{r}_0$ 的控制方程为

$$\left\{\left[\left(\frac{\partial^2}{\partial t\partial x}+a_1\frac{\partial^2}{\partial x^2}\right)\left(\frac{\partial^2}{\partial t\partial x}+b_1\frac{\partial^2}{\partial x^2}+b_2\right)-\frac{a_2}{2}\frac{\partial^2}{\partial x^2}\left(a_2\frac{\partial^2}{\partial x^2}+a_3\right)\right]\left(2\frac{\partial^2}{\partial t\partial x}-2b_3\frac{\partial^2}{\partial x^2}-\frac{b_2+c_1}{2}\right)\right.$$
$$\left.-(b_2+c_1)\left[\left(\frac{\partial^2}{\partial t\partial x}+a_1\frac{\partial^2}{\partial x^2}\right)\left(\frac{\partial^2}{\partial t\partial x}-b_3\frac{\partial^2}{\partial x^2}-\frac{b_2}{2}\right)+\frac{a_2a_3}{4}\frac{\partial^2}{\partial x^2}\right]\right\}\hat{r}_0=0 \tag{63}$$

或者写成

$$L\hat{r}_0=0 \tag{64}$$

其中

$$L=\frac{\partial^6}{\partial t^3\partial x^3}+C_1\frac{\partial^6}{\partial t^2\partial x^4}+C_2\frac{\partial^4}{\partial t^2\partial x^2}+C_3\frac{\partial^4}{\partial t\partial x^3}+C_4\frac{\partial^6}{\partial t\partial x^5}+C_5\frac{\partial^4}{\partial x^4}+C_6\frac{\partial^6}{\partial x^6}$$

$$C_1=a_1+b_1-b_3,\quad C_2=\frac{1}{4}(b_2-3c_1)$$

$$C_3=\frac{1}{4}[a_1(b_2-3c_1)-2a_2a_3-b_1(b_2+c_1)-2b_3(b_2-c_1)],\quad C_4=a_1b_1-\frac{1}{2}a_2^2-a_1b_3-b_1b_3$$

$$C_5=\frac{1}{2}\left[b_3(a_1c_1-a_1b_2+a_2a_3)+\frac{1}{2}(b_2+c_1)\left(\frac{1}{2}a_2^2-a^1b_1\right)\right],\quad C_6=b_3\left(\frac{1}{2}a_2^2-a_1b_1\right)$$

而

$$a_1 = 1 - I_{0,0},\quad a_2 = I_{2,0},\quad a_3 = J_{1,0}^{(1)}/\delta I_{1,1},\quad b_1 = 1 - \frac{1}{2}I_{2,2},$$

$$b_2 = \left(J_{1,0}^{(2)} - \frac{1}{2}J_{1,2}^{(1)}\right)/\delta I_{1,1},\quad b_3 = 1 + I_{0,0},\quad c_1 = \left(2 + J_{1,0}^{(2)} + \frac{1}{2}J_{1,2}^{(1)}\right)/\delta I_{1,1}$$

由于准长波近似基本上是对波长长的波合适,因此可略去含有$\frac{\partial^6}{\partial x^6}\hat{r}_0$的高阶项,并将式(64)改写成

$$L\hat{r}_0 = \frac{\partial^2}{\partial x^2}L'\hat{r}_0 \tag{65}$$

这样方程对 t,x 的导数都是三阶的,即得

$$L'\hat{r}_0 = 0 \tag{66}$$

其中

$$L' = \frac{\partial^4}{\partial t^3 \partial x} + C_1\frac{\partial^4}{\partial t^2 \partial x^2} + C_2\frac{\partial^2}{\partial t^2} + C_3\frac{\partial^2}{\partial t \partial x} + C_4\frac{\partial^4}{\partial t \partial x^3} + C_5\frac{\partial^2}{\partial x^2}$$

算子方程中各系数的表达式仍同前。

设变量 $\hat{r}_0$ 的标准解为

$$\hat{r}_0 = |\hat{r}_0| \exp\{i(kx - \omega t)\}$$

代入式(66),得到该方程的频散关系为

$$\omega^3 + a\omega^2 + b\omega + c = 0 \tag{67}$$

其中

$$a = C_2\frac{1}{k} - C_1 k,\ b = C_4 k^2 - C_3,\ c = C_5 k$$

温跃层经向不均匀分布当 $\bar{h}$ 为正值时,表示基本流是向东的(图 2)。由图可见,随着参数 α 的增大,Kelvin 波的频率逐渐增加,而本质上为 Rossby 波的那一支波频率逐渐减小,并转向东传(图 2b),这时又出现主要由"地形"激发出的新波——地形 Rossby 波,它是向西传播的,而且在长波波段频率很大。当 α 超过某一临界值以后,本质上为 Kelvin 波的那支波和本质上为 Rossby 波的那支波在波数大于临界波数后相重合,即变成不稳定的共轭波(Kelvin 波-Rossby 波耦合不稳定),共轭波段是向东传播的,不稳定增长率随波数也改变。随着参数 $\bar{h}$ 的增加,正压模在准长波近似下的频散关系特征与参数 α 的增大类似(图略)。当参数 $\alpha=0.75$,$\bar{h}=0.7$ 时,发生不稳定现象的临界波数为 3.97(图 2c),取定波数区间为[3.97,8],平均增长率为 1.24。其中当波数 $k=4$ 时,不稳定增长率为 $\omega_i=0.15$,因此波振幅增长到 e 倍所需要的时间约为 8.6 天。

注意到只有在经圈速度控制方程中,保留基本流的平流项即准长波近似时,才能激发出独立于传统的 Kelvin 波和 Rossby 波的地形 Rossby 波,而长波近似在同样的情况下只有 Kelvin 波和 Rossby 波,这两支波不会产生耦合不稳定,这是准长波近似和长波近似的一个重要差别。

Kelvin 波和 Rossby 波虽然在物理性质上存在很大差异,即前者本质上是属于一类重力波,后者则主要由于参数 β 引起,而且传播方向也相反,但考虑到在长波部分两者的频率都很低,因此基本流将改变它们的传播方向而相互耦合是可能的[11]。注意到这种不稳定是由波与波的相互作用而产生的,可称为波的耦合不稳定[10]。

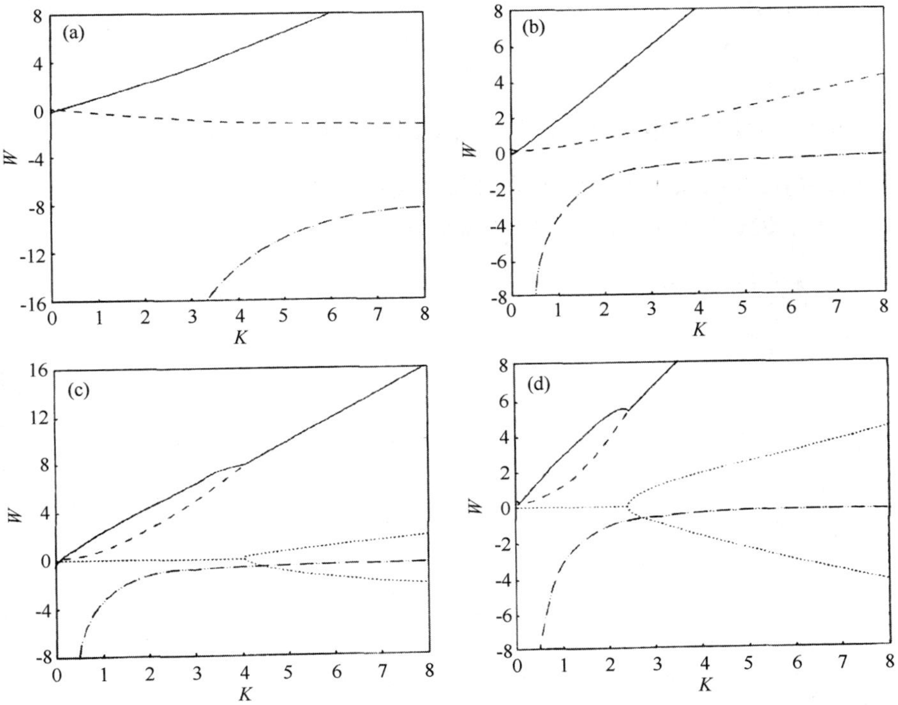

图2　正压模准长波近似下东向基本流波的色散关系($\bar{h}=0.7$)

(a)$\alpha=0.01$;(b)$\alpha=0.3$;(c)$\alpha=0.75$;(d)$\alpha=1.0$;实线代表 Kelvin 波,虚线代表原来的那支 Rossby 波,点划线代表新产生的地形 Rossby 波,点线表示不稳定增长率

比较上图可以看到,要出现耦合不稳定,α 需超过某一临界值,即赤道潜流不能太宽,但当潜流变窄时,不稳定区虽向低波数移动,但高波数(短波)的增加率将变得很大。由于分析是在准长波近似下进行的,短波波段出现的行为只能作为近似的参考,它将由模式中未考虑的物理过程(例如湍流过程)来减弱它们的不稳定增长。

如基本流是向西的则不出现共轭不稳定(图略)。

5　斜压模方程的色散性

方程(39)、方程(40)、方程(41)在形式上和正压模相似,直接对 $\tilde{v}_1$ 给出最后的控制方程为

$$L_{\text{qlw}}\tilde{v}_1 = 0 \tag{68}$$

其中

$$L_{\text{qlw}} = C_{02}\frac{\partial^4}{\partial t^3\partial x} + C_{03}\frac{\partial^4}{\partial t^2\partial x^2} + C_{04}\frac{\partial^4}{\partial t\partial x^3} + C_{05}\frac{\partial^4}{\partial x^4} + C_{08}\frac{\partial^2}{\partial t^2} + C_{09}\frac{\partial^2}{\partial t\partial x} + C_{10}\frac{\partial^2}{\partial x^2}$$

进一步简化可略去含有 $\dfrac{\partial^4}{\partial x^4}\tilde{v}_1$ 的高波数项,即将式(68)改写成

$$L'_{\text{qlw}}\tilde{v}_1 = 0 \tag{69}$$

这样方程对 t,x 的导数都是三阶的,其中

$$L'_{\mathrm{qlw}} = C_{02}\frac{\partial^4}{\partial t^3 \partial x} + C_{03}\frac{\partial^4}{\partial t^2 \partial x^2} + C_{04}\frac{\partial^4}{\partial t \partial x^3} + C_{08}\frac{\partial^2}{\partial t^2} + C_{09}\frac{\partial^2}{\partial t \partial x} + C_{10}\frac{\partial^2}{\partial x^2}$$

算子方程中各系数见附录 2。

对向东的基本流,算得的色散关系见图 3。由于图的结构同正压模的色散关系,分析从略。对于参数 $\alpha=0.9$,$\bar{h}=0.7$(图 3c),发生不稳定现象的临界波数为 2.36,取定波数区间为[2.36,8],平均增长率为 2.06。其中当波数 $k=3$ 时,不稳定增长率为 $\omega_i=0.82$。

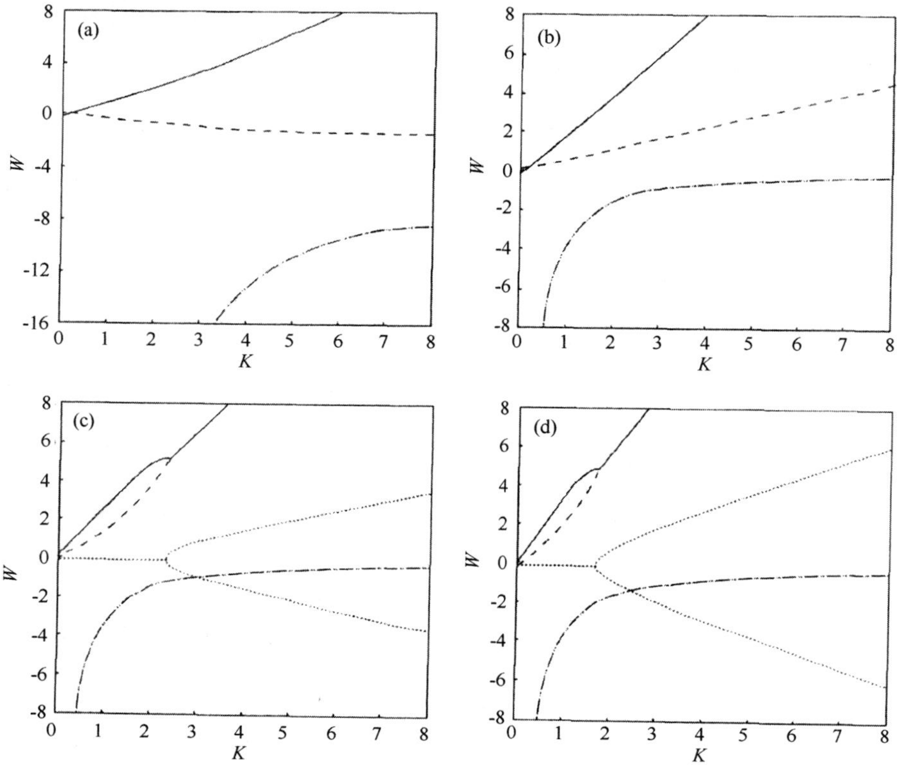

图 3 斜压模准长波近似下东向基本流波的色散关系($\bar{h}=0.7$)

(a)$\alpha=0.01$;(b)$\alpha=0.3$;(c)$\alpha=0.9$;(d) $\alpha=1.5$;其中实线代表 Kelvin 波,虚线代表原来的那支 Rossby 波,点划线代表新产生的地形 Rossby 波;点线表示不稳定增长率

如果基本流是向西的则同样不出现共轭不稳定(图略)。

6 正压不稳定与斜压不稳定的比较

注意到,虽然在上述有关近似下,正压模和斜压模在色散关系的性质或图的结构上是相似的,但由于两个模态的特征量不同,即使对同样的波长两个模态的相速度和增长率也是不同的。考虑到

$$\frac{\hat{k}}{\tilde{k}} \sim \left[\frac{\rho_3 \alpha \Delta \bar{T}}{\rho_3 - \rho_2}\right]^{\frac{1}{4}}, \quad \frac{\hat{\omega}}{\tilde{\omega}} \sim \left[\frac{\rho_3 - \rho_2}{\rho_3 \alpha \Delta \bar{T}}\right]^{\frac{1}{4}}$$

式中符号 $\hat{a}$,$\tilde{a}$ 分别表示正压模和斜压模,可见两个模态的差别依赖于温跃层的强度 $\Delta\bar{T}$。当取 $(\rho_3-\rho_2)/\rho_3=2.04\times10^{-3}$,海水的热膨胀系数取 $\alpha=1.0\times10^{-4}/℃$,$\Delta\bar{T}=5℃$时,$\hat{k}/\tilde{k}\approx0.7$,$\hat{\omega}/\tilde{\omega}\approx1.4$。可

以将图 3 用正压模的波数和频率为坐标重新画出(图 4)。

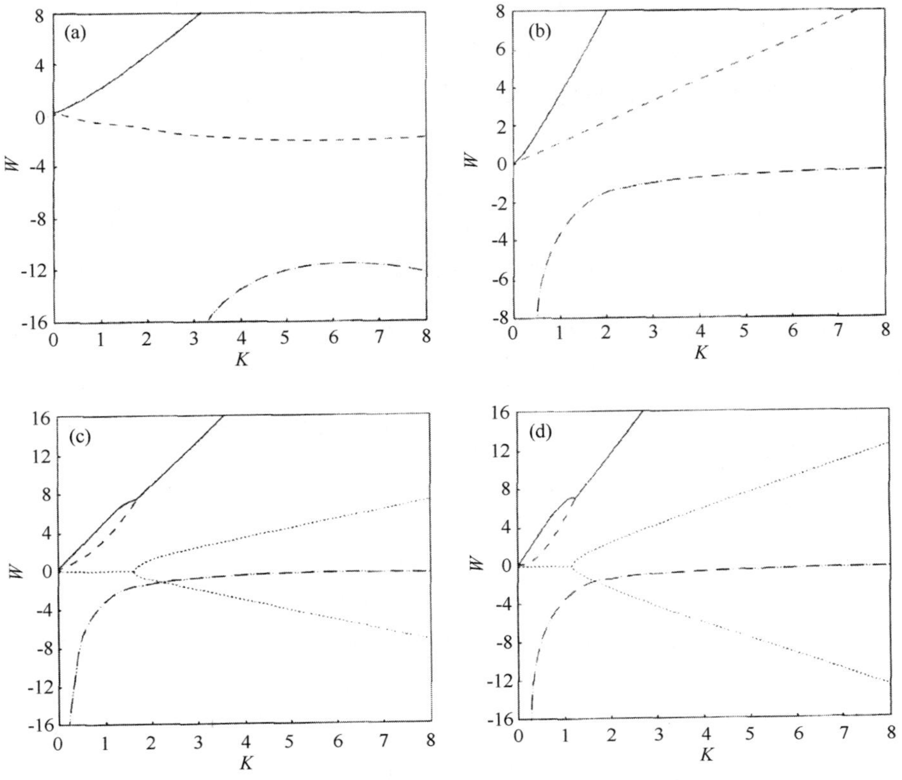

图 4　用正压模波数和频率来表示的图 3 的色散关系($\bar{h}$=0.7)

(a)α=0.01;(b)α=0.3;(c)α=0.9;(d)α=1.5 ;其中实线代表 Kelvin 波,虚线代表原来的那支 Rossby 波,点划线代表新产生的地形 Rossby 波;点线表示不稳定增长率

在相同的 α,$\bar{h}$ 取值下,图 4 与图 3 的结果大体相似。下面对两者进行详细的比较,图 4(a)与图 3(a)相比在同样的波数下图 4(a)中 Kelvin 波的频率大致为图 3(a)中频率的 2 倍;图 4(a)中 Rossby 波的频率绝对值也大于图 3(a);图 4(a)中地形 Rossby 波频率在所给波数范围内出现先增大后减小的变化趋势有别于图 3(a)中一致增大的变化趋势。由图 3(b)和图 4(b)可见,相同的波数下图 4(b)中 Kelvin 波和 Rossby 波的频率大致为图 3(b)中频率的 2 倍,而地形 Rossby 波的差异则不明显。类似地,图 4(c)中 Kelvin 波与 Rossby 波的频率近似为同样波数下图 3(c)中频率的 2 倍。图 4(c)中 Kelvin 波与 Rossby 波出现耦合不稳定的临界波数要小于图 3(c),即向长波方向移动,且其不稳定增长率大致为图 3(c)的 2 倍。而地形 Rossby 波的变化不大。图 4(d)与图 3(d)的差异与图 4(c)和图 3(c)的差异类似。

7　讨论

本文设计了一个热带赤道 β-平面的两层海洋模式,研究了正压模和斜压模下 Kelvin 波、Rossby 波及“地形 Rossby 波”的色散关系特征。指出在准长波近似下并当基本流为东向流时,Kelvin 波和 Rossby 波可以耦合出一类不稳定波,这类不稳定波可以用来解释 ENSO 发展中的某些现象。在浅混合层近似和“快波近似”下,正压模和斜压模是可以分离的,因此可以分别分析它们的色散特征,由于它们的特征量不同,在同样的波长(扰动的纬向尺度)下,扰动的增长率也不同。这里用正压模的波数和频率为坐标(图 4)重画了斜压模的色散关系(图 3),从而能够在相同的特征尺度下比较两者的相对大小,结果表明在一定的参数条件下,斜压模的不稳定增长率大约为正压模的 2 倍。

注意到,在浅混合层近似下,式(9)可近似地写成

$$u_s \approx \hat{u} - \tilde{u}, \quad u_b \approx \hat{u}$$

由于斜压模的增长快于正压模,当时间充分长后,$u_s \sim \tilde{u}$,这表明混合层中的流场要比温跃层中的流场强。在另一方面,由混合层温度距平变化方程(37)可以看到,斜压流的前面多了一个小量因子($\tilde{C}/C$),因此混合层中的温度距平的变化基本上由正压模决定,相比由斜压过程决定的温跃层温度距平的变化来要弱,这些特点与参考文献[3,4]强迫运动的分析是一致的。本文的结果连同对强迫运动的分析,为近年来为什么我们要用次表层温度距平的变化来分析 ENSO 事件中海洋物理场的变化(见参考文献[1])提供了进一步的理论根据。

参考文献

[1] 巢纪平,巢清尘,刘琳.热带太平洋 ENSO 事件和印度洋的 DIPOLE 事件.气象学报,2005,63(5):594-602.

[2] 巢纪平,陈鲜艳,何金海.热带西太平洋对风应力的斜压响应.地球物理学报,2002,45(2):176-187.

[3] 巢纪平,陈鲜艳,何金海.风应力对热带斜压海洋的强迫.大气科学,2002,26(5):577-594.

[4] Matsuno T. Quasi-geost rophic motions in the equatorial area. J Met eor Soc Japan,1966,44:25-43.

[5] Philander S G H. Equatorial waves in the presence of the equat orial undercu rrent. J Phys Oceanogr, 1979, 9(2):254-262.

[6] Boyd J P. The effects of latitudinal shear on equat orial waves. Parts Ⅰ and Ⅱ. J Atmos Sci,1978,35(12):2236-2267.

[7] Zhang C, Webster P J. Effects of zonal flows on equatorially trapped waves. J Atmos Sci,1989,46(24):3632-3652.

[8] Wang B, Xie X. Low-frequency equatorial waves in vertically sheared zonal flow. Part I: Stable waves. J Atmos Sci,1996,53(3):449-467.

[9] Cruz-Heredia M D L, Moore G W K. Barotropic inst ability due to Kelvin wave-Rossby wave coupling. J Atmos Sci,1999,56(14):2376-2383.

[10] 巢纪平,刘琳,于卫东.热带海洋和大气中地形 Rossby 波和 Rossby 波的耦合不稳定.中国科学 D 辑,2005,35(1):79-87.

[11] 巢纪平.厄尔尼诺和南方涛动动力学.北京:气象出版社,1993:309.

附录 1

基本积分：

$$\int_0^\infty \mathrm{e}^{-\beta x^2}\mathrm{d}x = \frac{\sqrt{\pi}}{2\sqrt{\beta}},\quad \int_{-\infty}^\infty \mathrm{e}^{-\beta x^2}\mathrm{d}x = 2\int_0^\infty \mathrm{e}^{-\beta x^2}\mathrm{d}x = \frac{\sqrt{\pi}}{\sqrt{\beta}}\qquad (\beta > 0)$$

因此对于积分公式 $E_n = \frac{1}{\sqrt{2\pi}}\int_{-\infty}^\infty y^n \mathrm{e}^{-\alpha y^2}\mathrm{d}y$，可知

$$E_0 = \frac{1}{\sqrt{2\pi}}\int_{-\infty}^\infty \mathrm{e}^{-\alpha y^2}\mathrm{d}y = 2^{-1/2}\alpha^{-1/2}$$

$$E_2 = \frac{1}{\sqrt{2\pi}}\int_{-\infty}^\infty y^2\mathrm{e}^{-\alpha y^2}\mathrm{d}y = -\frac{\mathrm{d}E_0}{\mathrm{d}\alpha} = 2^{-3/2}\alpha^{-3/2}$$

$$E_4 = \frac{1}{\sqrt{2\pi}}\int_{-\infty}^\infty y^4\mathrm{e}^{-\alpha y^2}\mathrm{d}y = -\frac{\mathrm{d}E_2}{\mathrm{d}\alpha} = 3\times 2^{-5/2}\alpha^{-5/2}$$

$$E_6 = \frac{1}{\sqrt{2\pi}}\int_{-\infty}^\infty y^6\mathrm{e}^{-\alpha y^2}\mathrm{d}y = -\frac{\mathrm{d}E_4}{\mathrm{d}\alpha} = 15\times 2^{-7/2}\alpha^{-7/2}$$

$$E_8 = \frac{1}{\sqrt{2\pi}}\int_{-\infty}^\infty y^8\mathrm{e}^{-\alpha y^2}\mathrm{d}y = -\frac{\mathrm{d}E_6}{\mathrm{d}\alpha} = 105\times 2^{-9/2}\alpha^{-9/2}$$

上述积分可统一写成下面的形式，即

$$E_0 = 2^{-1/2}\alpha^{-1/2}$$

$$E_{2n} = (2n-1)!!\times 2^{-\frac{2n+1}{2}}\alpha^{-\frac{2n+1}{2}}\qquad (n = 1,2,3\cdots)$$

$$E_{2n+1} = 0\qquad (n = 0,1,2,3\cdots)$$

这里

$$0! = 1$$

$$n! = n\times(n-1)\times(n-2)\cdots 3\times 2\times 1$$

$$0!! = 0$$

$$(2n+1)!! = \frac{(2n+1)!}{2^n n!} = 1\times 3\times 5\cdots(2n+1)$$

$$(2n)!! = 2^n n! = 2\times 4\times 6\cdots(2n)$$

由此

$$I_{m,n} = \frac{1}{\sqrt{2\pi}}\int_{-\infty}^\infty \frac{2}{y}\frac{\mathrm{d}h_c}{\mathrm{d}y}D_m D_n\,\mathrm{d}y = -\frac{4\alpha\bar{h}}{\sqrt{2\pi}}\int_{-\infty}^\infty \mathrm{e}^{-\alpha y^2}D_m D_n\,\mathrm{d}y$$

根据 Weber 函数 $D_n(y)$ 的前 n 个（$n=0,1,2,3$）具体表达式算得

$$I_{0,0} = -\frac{4\alpha\bar{h}}{\sqrt{2\pi}}\int_{-\infty}^\infty \mathrm{e}^{-\alpha y^2}D_0 D_0\mathrm{d}y = -\frac{4\alpha\bar{h}}{\sqrt{2\pi}}\int_{-\infty}^\infty \mathrm{e}^{-\left(\alpha+\frac{1}{2}\right)y^2}\mathrm{d}y = -4\alpha\bar{h}2^{-1/2}\left(\alpha+\frac{1}{2}\right)^{-1/2}$$

$$I_{1,0} = I_{0,1} = -\frac{4\alpha\bar{h}}{\sqrt{2\pi}}\int_{-\infty}^\infty \mathrm{e}^{-\alpha y^2}D_1 D_0\mathrm{d}y = -\frac{4\alpha\bar{h}}{\sqrt{2\pi}}\int_{-\infty}^\infty y\,\mathrm{e}^{-\left(\alpha+\frac{1}{2}\right)y^2}\mathrm{d}y = 0$$

$$I_{1,1}=-\frac{4\alpha\bar{h}}{\sqrt{2\pi}}\int_{-\infty}^{\infty}\mathrm{e}^{-\alpha y^2}D_1D_1\mathrm{d}y=-\frac{4\alpha\bar{h}}{\sqrt{2\pi}}\int_{-\infty}^{\infty}y^2\mathrm{e}^{-\left(\alpha-\frac{1}{2}\right)y^2}\mathrm{d}y=-4\alpha\bar{h}2^{-3/2}\left(\alpha+\frac{1}{2}\right)^{-3/2}$$

$$I_{2,0}=-\frac{4\alpha\bar{h}}{\sqrt{2\pi}}\int_{-\infty}^{\infty}\mathrm{e}^{-\alpha y^2}D_2D_0\mathrm{d}y=-4\alpha\bar{h}\left[2^{-3/2}\left(\alpha+\frac{1}{2}\right)^{-3/2}-2^{-1/2}\left(\alpha+\frac{1}{2}\right)^{-1/2}\right]=I_{0,2}$$

$$I_{2,1}=I_{1,2}=-\frac{4\alpha\bar{h}}{\sqrt{2\pi}}\int_{-\infty}^{\infty}\mathrm{e}^{-\alpha y^2}D_2D_1\mathrm{d}y=-\frac{4\alpha\bar{h}}{\sqrt{2\pi}}\int_{-\infty}^{\infty}(y^3-y)\mathrm{e}^{-\left(\alpha-\frac{1}{2}\right)y^2}\mathrm{d}y=0$$

$$I_{2,2}=-\frac{4\alpha\bar{h}}{\sqrt{2\pi}}\int_{-\infty}^{\infty}\mathrm{e}^{-\alpha y^2}D_2D_2\mathrm{d}y=-4\alpha\bar{h}\left[3\times2^{-5/2}\left(\alpha+\frac{1}{2}\right)^{-5/2}-2\times2^{-3/2}\left(\alpha+\frac{1}{2}\right)^{-3/2}+2^{-1/2}\left(\alpha+\frac{1}{2}\right)^{-1/2}\right]$$

$$I_{3,0}=I_{0,3}=-\frac{4\alpha\bar{h}}{\sqrt{2\pi}}\int_{-\infty}^{\infty}\mathrm{e}^{-\alpha y^2}D_3D_0\mathrm{d}y=-\frac{4\alpha\bar{h}}{\sqrt{2\pi}}\int_{-\infty}^{\infty}(y^3-3y)\mathrm{e}^{-\left(\alpha+\frac{1}{2}\right)y^2}\mathrm{d}y=0$$

$$I_{3,1}=-\frac{4\alpha\bar{h}}{\sqrt{2\pi}}\int_{-\infty}^{\infty}\mathrm{e}^{-\alpha y^2}D_3D_1\mathrm{d}y=-12\alpha\bar{h}\left[2^{-5/2}\left(\alpha+\frac{1}{2}\right)^{-5/2}-2^{-3/2}\left(\alpha+\frac{1}{2}\right)^{-3/2}\right]=I_{1,3}$$

$$I_{3,2}=I_{2,3}=-\frac{4\alpha\bar{h}}{\sqrt{2\pi}}\int_{-\infty}^{\infty}\mathrm{e}^{-\alpha y^2}D_3D_2\mathrm{d}y=-\frac{4\alpha\bar{h}}{\sqrt{2\pi}}\int_{-\infty}^{\infty}(y^5-4y^3+3y)\mathrm{e}^{-\left(\alpha+\frac{1}{2}\right)y^2}\mathrm{d}y=0$$

$$I_{3,3}=-\frac{4\alpha\bar{h}}{\sqrt{2\pi}}\int_{-\infty}^{\infty}\mathrm{e}^{-\alpha y^2}D_3D_3\mathrm{d}y$$

$$=-12\alpha\bar{h}\left[5\times2^{-7/2}\left(\alpha+\frac{1}{2}\right)^{-7/2}-6\times2^{-5/2}\left(\alpha+\frac{1}{2}\right)^{-5/2}+3\times2^{-3/2}\left(\alpha+\frac{1}{2}\right)^{-3/2}\right]$$

另一积分式

$$J_{m,n}^{(1),(2)}=\frac{1}{\sqrt{2\pi}}\int_{-\infty}^{\infty}\left(\frac{2}{y}\frac{\mathrm{d}^2h_c}{\mathrm{d}y^2}-\frac{2}{y^2}\frac{\mathrm{d}h_c}{\mathrm{d}y}\pm\frac{\mathrm{d}h_c}{\mathrm{d}y}\right)D_mD_n\mathrm{d}y=K_{m,n}^{(1)}\pm K_{m,n}^{(2)}=\frac{(8\alpha\mp2)\alpha\bar{h}}{\sqrt{2\pi}}\int_{-\infty}^{\infty}y\mathrm{e}^{-\alpha y^2}D_mD_n\mathrm{d}y$$

算得

$$J_{0,0}^{(1),(2)}=\frac{(8\alpha\mp2)\alpha\bar{h}}{\sqrt{2\pi}}\int_{-\infty}^{\infty}y\mathrm{e}^{-\alpha y^2}D_0D_0\mathrm{d}y=\frac{(8\alpha\mp2)\alpha\bar{h}}{\sqrt{2\pi}}\int_{-\infty}^{\infty}y\mathrm{e}^{-\left(\alpha+\frac{1}{2}\right)y^2}\mathrm{d}y=0$$

$$J_{1,0}^{(1),(2)}=J_{0,1}^{(1),(2)}=\frac{(8\alpha\mp2)\alpha\bar{h}}{\sqrt{2\pi}}\int_{-\infty}^{\infty}y\mathrm{e}^{-\alpha y^2}D_1D_0\mathrm{d}y=(8\alpha\mp2)\alpha\bar{h}\cdot2^{-\frac{3}{2}}\left(\alpha+\frac{1}{2}\right)^{-\frac{3}{2}}$$

$$J_{1,1}^{(1),(2)}=\frac{(8\alpha\mp2)\alpha\bar{h}}{\sqrt{2\pi}}\int_{-\infty}^{\infty}y\mathrm{e}^{-\alpha y^2}D_1D_1\mathrm{d}y=\frac{(8\alpha\mp2)\alpha\bar{h}}{\sqrt{2\pi}}\int_{-\infty}^{\infty}y^3\mathrm{e}^{-\left(\alpha+\frac{1}{2}\right)y2}\mathrm{d}y=0$$

$$J_{2,0}^{(1),(2)}=J_{0,2}^{(1),(2)}=\frac{(8\alpha\mp2)\alpha\bar{h}}{\sqrt{2\pi}}\int_{-\infty}^{\infty}y\mathrm{e}^{-\alpha y^2}D_2D_0\mathrm{d}y=\frac{(8\alpha\mp2)\alpha\bar{h}}{\sqrt{2\pi}}\int_{-\infty}^{\infty}(y^3-y)\mathrm{e}^{-\left(\alpha+\frac{1}{2}\right)y^2}\mathrm{d}y=0$$

$$J_{2,1}^{(1),(2)}=J_{1,2}^{(1),(2)}=\frac{(8\alpha\mp2)\alpha\bar{h}}{\sqrt{2\pi}}\int_{-\infty}^{\infty}y\mathrm{e}^{-\alpha y^2}D_2D_1\mathrm{d}y$$

$$=(8\alpha\mp2)\alpha\bar{h}\left[3\times2^{-\frac{5}{2}}\left(\alpha+\frac{1}{2}\right)^{-\frac{5}{2}}-2^{-\frac{3}{2}}\left(\alpha+\frac{1}{2}\right)^{-\frac{3}{2}}\right]$$

$$J_{2,2}^{(1),(2)}=\frac{(8\alpha\mp2)\alpha\bar{h}}{\sqrt{2\pi}}\int_{-\infty}^{\infty}y\mathrm{e}^{-\alpha y^2}D_2D_2\mathrm{d}y=0$$

$$J_{3,0}^{(1),(2)} = J_{0,3}^{(1),(2)} = \frac{(8\alpha \mp 2)\alpha\bar{h}}{\sqrt{2\pi}}\int_{-\infty}^{\infty} y e^{-\alpha y^2} D_3 D_0 dy = 3(8\alpha \mp 2)\alpha\bar{h}\left[2^{-\frac{5}{2}}\left(\alpha + \frac{1}{2}\right)^{-\frac{5}{2}} - 2^{-\frac{3}{2}}\left(\alpha + \frac{1}{2}\right)^{-\frac{3}{2}}\right]$$

$$J_{3,1}^{(1),(2)} = J_{1,3}^{(1),(2)} = \frac{(8\alpha \mp 2)\alpha\bar{h}}{\sqrt{2\pi}}\int_{-\infty}^{\infty} y e^{-\alpha y^2} D_3 D_1 dy = 0$$

$$J_{3,2}^{(1),(2)} = J_{2,3}^{(1),(2)} = \frac{(8\alpha \mp 2)\alpha\bar{h}}{\sqrt{2\pi}}\int_{-\infty}^{\infty} y e^{-\alpha y^2} D_3 D_2 dy = 3(8\alpha \mp 2)\alpha\bar{h}$$

$$\left[5 \times 2^{-\frac{7}{2}}\left(\alpha + \frac{1}{2}\right)^{-\frac{7}{2}} - 4 \times 2^{-\frac{5}{2}}\left(\alpha + \frac{1}{2}\right)^{-\frac{5}{2}} + 2^{-\frac{3}{2}}\left(\alpha + \frac{1}{2}\right)^{-\frac{3}{2}}\right]$$

$$J_{3,3}^{(1),(2)} = \frac{(8\alpha \mp 2)\alpha\bar{h}}{\sqrt{2\pi}}\int_{-\infty}^{\infty} y e^{-\alpha y^2} D_3 D_3 dy = \frac{(8\alpha \mp 2)\alpha\bar{h}}{\sqrt{2\pi}}\int_{-\infty}^{\infty} y^3 (y^2 - 3)^2 e^{-\left(\alpha + \frac{1}{2}\right) y^2} dy = 0$$

附录 2

$$C_{02} = \varepsilon a_2 a_3 (a_1 d_1 - b_1 d_1 + d_2) + \delta a_2 a_3 c_1 d_1 = a_2 a_3 [\varepsilon (a_1 d_1 - b_1 d_1 + d_2) + \delta c_1 d_1]$$

$$\begin{aligned}C_{03} &= \varepsilon a_2 a_3 (a_1 b_1 d_1 - a_1 d_2 + a_2 d_3 - a_3^2 d_1 + b_1 d_2) - \delta a_2 a_3 c_1 (a_1 d_1 - b_1 d_1 + d_2) \\ &= a_2 a_3 [\varepsilon (a_1 b_1 d_1 - a_1 d_2 + a_2 d_3 - a_3^2 d_1 + b_1 d_2) - \delta c_1 (a_1 d_1 - b_1 d_1 + d_2)]\end{aligned}$$

$$\begin{aligned}C_{04} &= -\varepsilon a_2 a_3 (a_1 b_1 d_2 - a_2 a_3 d_4 - a_2 b_1 d_3 - a_3^2 d_2) - \delta a_2 a_3 c_1 (a_1 b_1 d_1 - a_1 d_2 + a_2 d_3 - a_3^2 d_1 + b_1 d_2) \\ &= -a_2 a_3 [\varepsilon (a_1 b_1 d_2 - a_2 a_3 d_4 - a_2 b_1 d_3 - a_3^2 d_2) + \delta c_1 (a_1 b_1 d_1 - a_1 d_2 + a_2 d_3 - a_3^2 d_1 + b_1 d_2)]\end{aligned}$$

$$C_{05} = \delta a_2 a_3 c_1 (a_1 b_1 d_2 - a_2 a_3 d_4 - a_2 b_1 d_3 - a_3^2 d_2)$$

$$\begin{aligned}C_{08} &= -2b_2 c_3 (a_1 b_1 d_1 - a_1 d_2 + a_2 d_3 - a_3^2 d_1 + b_1 d_2) + (a_1 d_1 - b_1 d_1 + d_2)(2a_1 b_2 c_3 + a_2 b_2 c_2 + \\ &\quad 2a_3 a_4 c_3) - 2c_3 (a_1 b_2 d_2 - a_2 a_3 d_5 - a_2 b_2 d_3 + a_3 a_4 d_2) - (2a_1 c_3 + a_2 c_2 - 2b_1 c_3)(a_1 b_2 d_1 + \\ &\quad a_3 a_4 d_1 + b_2 d_2) + b_2 d_1 (2a_1 b_1 c_3 - a_2 a_3 c_3 + a_2 b_1 c_2 - 2a_3^2 c_3) \\ &= -a_2 a_3 a_4 c_2 d_1 - a_2 a_3 b_2 c_3 d_1 + 2a_2 a_3 c_3 d_5 = -a_2 a_3 (a_4 c_2 d_1 + b_2 c_3 d_1 - 2c_3 d_5)\end{aligned}$$

$$\begin{aligned}C_{09} &= 2b_3 c_3 (a_1 b_1 d_2 - a_2 a_3 d_4 - a_2 b_1 d_3 - a_3^2 d_2) + (2a_1 b_2 c_3 + a_2 b_2 c_2 + 2a_3 a_4 c_3)(a_1 b_1 d_1 - a_1 d_2 + \\ &\quad a_2 d_3 - a_3^2 d_1 + b_1 d_2) + (2a_1 c_3 + a_2 c_2 - 2b_1 c_3)(a_1 b_2 d_2 - a_2 a_3 d_5 - a_2 b_2 d_3 + a_3 a_4 d_2) - \\ &\quad (a_1 b_2 d_1 + a_3 a_4 d_1 + b_2 d_2)(2a_1 b_1 c_3 - a_2 a_3 c_3 + a_2 b_1 c_2 - 2a_3^2 c_3) \\ &= a_1 a_2 a_3 b_2 c_3 d_1 - 2a_1 a_2 a_3 c_3 d_5 - a_2^2 a_3 c_2 d_5 + a_2 a_3^2 a_4 c_3 d_1 - a_2 a_3^2 b_2 c_2 d_1 - a_2 a_3 a_4 b_1 c_2 d_1 + \\ &\quad a_2 a_3 a_4 c_2 d_2 + 2a_2 a_3 a_4 c_3 d_3 + 2a_2 a_3 b_1 c_3 d_5 + a_2 a_3 b_2 c_3 d_2 - 2a_2 a_3 b_2 c_3 d_4 \\ &= a_2 a_3 (a_1 b_2 c_3 d_1 - 2a_1 c_3 d_5 - a_2 c_2 d_5 + a_3 a_4 c_3 d_1 - a_3 b_2 c_2 d_1 - a_4 b_1 c_2 d_1 + a_4 c_2 d_2 + \\ &\quad 2a_4 c_3 d_3 + 2b_1 c_3 d_5 + b_2 c_3 d_2 - 2b_2 c_3 d_4)\end{aligned}$$

$$\begin{aligned}C_{10} &= -(2a_1 b_2 c_3 + a_2 b_2 c_2 + 2a_3 a_4 c_3)(a_1 b_1 d_2 - a_2 a_3 d_4 - a_2 b_1 d_3 - a_3^2 d_2) + (2a_1 b_1 c_3 - \\ &\quad a_2 a_3 c_3 + a_2 b_1 c_2 - 2a_3^2 c_3)(a_1 b_2 d_2 - a_2 a_3 d_5 - a_2 b_2 d_3 + a_3 a_4 d_2) \\ &= -2a_1 a_2 a_3 b_1 c_3 d_5 - a_1 a_2 a_3 b_2 c_3 d_2 + 2a_1 a_2 a_3 b_2 c_3 d_4 + a_2^2 a_3^2 c_3 d_5 - a_2^2 a_3 b_1 c_2 d_5 + \\ &\quad a_2^2 a_3 b_2 c_2 d_4 + a_2^2 a_3 b_2 c_3 d_3 + 2a_2 a_3^3 c_3 d_5 - a_2 a_3^2 a_4 c_3 d_2 + 2a_2 a_3^2 a_4 c_3 d_4 + \\ &\quad a_2 a_3^2 b_2 c_2 d_2 + 2a_2 a_3^2 b_2 c_3 d_3 + a_2 a_3 a_4 b_1 c_2 d_2 + 2a_2 a_3 a_4 b_1 c_3 d_3) \\ &= a_2 a_3 (-2a_1 b_1 c_3 d_5 - a_1 b_2 c_3 d_2 + 2a_1 b_2 c_3 d_4 + a_2 a_3 c_3 d_5 - a_2 b_1 c_2 d_5 + a_2 b_2 c_2 d_4 + a_2 b_2 c_3 d_3 + \\ &\quad 2a_3^2 c_3 d_5 - a_3 a_4 c_3 d_2 + 2a_3 a_4 c_3 d_4 + a_3 b_2 c_2 d_2 + 2a_3 b_2 c_3 d_3 + a_4 b_1 c_2 d_2 + 2a_4 b_1 c_3 d_3)\end{aligned}$$

热带扰动在大尺度经圈中的行为*

巢纪平[1,2] 徐昭[3]

(1. 国家海洋局海洋环境科学和数值模拟国家海洋局重点实验室,青岛 266061;2. 国家海洋局海洋环境预报中心,北京 100081;3. 中国海洋大学物理海洋实验室,青岛 266100)

摘要:在赤道β-平面上经圈流的背景流场内,利用等值浅水模式来分析波动的不稳定性。结果显示,在经圈半地转假设下,扰动信号通过变性的 Rossby 波来传递。对于大洋西部,由于向极地方向经圈流的引入,赤道对称的扰动模态对所有的波数 k 都是不稳定的。对于大洋东部的向赤道流,对赤道对称的扰动却是稳定的。由于一般来讲,扰动倾向于赤道对称,因此西边界的向极流,如黑潮,比东边界的向赤道流,如加利福尼亚洋流,更易因扰动的不稳定而产生涡旋。

关键词:经圈流;变性 Rossby 波;不稳定

1 引言

在热带海洋与大气中的大尺度基本运动基本上是沿纬圈的,因此扰动在纬圈背景流中的行为研究较多,如可以由沿赤道纬圈流的经圈方向(南北方向)的水平切变引起惯性不稳定(Dunkerton[1], Stevens[2]),还可以类似中纬度由基本流二阶导数存在而以绝对涡度梯度为零为必要条件的正压不稳定,对于在热带地区的这类不稳定,Prohel[3]参照 Lindzen[4,5]提出的 Rossby 波的超反射概念给予了必要和充分条件的物理解释。海洋中的赤道潜流也存在这类性质的不稳定(Philander[6,7])。最近,巢纪平等[8,9]提出了当纬圈背景流存在南北向的不均匀时,可以出现波与波之间的耦合不稳定现象。

相比之下研究经圈流中扰动行为的工作较少,然而大尺度的经圈流也是存在的,如大气中副热带西侧的气流是向极的,南海季风也基本上是南北向的。在热带海洋中,南、北赤道洋流在临近大洋的西边界时也转向两极流动,形成副热带环流西侧的向极流,甚至可以出现像黑潮和湾流那样的向极强洋流,另外如跨赤道的索马里洋流。在大洋海盆东侧则是向赤道流,如太平洋东部的加利福尼亚流。Walker 和 Pedlosky[10]在β-平面上利用中纬度的准地转位涡方程,通过理论解和模式结果研究了经圈斜压流的不稳定性,两层流体的分析表明(经圈流只在上层流体中存在),与纬圈流的不稳定性不同,扰动出现不稳定时似乎并不需要基本流存在最小的切变值,同时不稳定扰动具有边界层的结构。

本文将以海洋为例把经圈流的不稳定性拓展到热带地区,用一个简单的等值浅水[11]模式来分析这一问题,其结果稍加改动不难用于大气研究。

* 地球物理学报,2008,51(6):1657-1662.

2　模式和运动方程

设热带海洋中有一个由风应力或其他原因驱动的经圈流 $V(x,y)$，经圈流及相应的温跃层厚度是大尺度的，它在纬圈方向是缓变的，对这类问题一般可以用 WKB 方法解决，但如研究的扰动在 x 方向的尺度远比其经圈流的变化尺度小，例如尺度为一个 Rossby 变形半径，则可以不考虑经圈流及温跃层在 x 方向的不均匀效应。

由于考虑的是热带问题，运动在赤道 β 平面上进行。流体可以是斜压的，但对垂直方向任一斜压模，其水平运动的结构方程和浅水运动方程相似，因此称为等值浅水运动，最简单的等值浅水运动相当于两层密度不同的运动。这时重力 g 可改用视重力 g'，相应的重力波波速为 $C=(g'H_0)^{\frac{1}{2}}$，H_0 是流体的垂直特征厚度。由此可得等值浅水运动方程为

$$\frac{\partial u}{\partial t}+u\frac{\partial u}{\partial x}+v\frac{\partial u}{\partial y}-\beta yv=-g'\frac{\partial h}{\partial x}+\overline{F}_x \tag{1}$$

$$\frac{\partial v}{\partial t}+u\frac{\partial v}{\partial x}+v\frac{\partial v}{\partial y}+\beta yu=-g'\frac{\partial h}{\partial y}+\overline{F}_y \tag{2}$$

$$\frac{\partial h}{\partial t}+\frac{\partial(hu)}{\partial x}+\frac{\partial(hv)}{\partial y}=\overline{F}_z \tag{3}$$

$\overline{F}_x,\overline{F}_y,\overline{F}_z$ 分别代表 x,y,z 方向的外力。令

$$u=u',\quad v=V(y)+v',\quad h=H+h' \tag{4}$$

由于某种原因的外力（比如大洋内区的 Sverdrup 输送），维持着某种形式的经向背景流，当背景流被背景高度场及外力场平衡后，扰动运动的线性化方程为

$$\frac{\partial u'}{\partial t}+V\frac{\partial u'}{\partial y}-\beta yv'=-g'\frac{\partial h'}{\partial x} \tag{5}$$

$$\frac{\partial v'}{\partial t}+V\frac{\partial v'}{\partial y}+v'\frac{\mathrm{d}V}{\mathrm{d}y}+\beta yu'=-g'\frac{\partial h'}{\partial y} \tag{6}$$

$$\frac{\partial h'}{\partial t}+h'\frac{\mathrm{d}V}{\mathrm{d}y}+H_0\left(\frac{\partial u'}{\partial x}+\frac{\partial v'}{\partial y}\right)+V\frac{\partial h'}{\partial y}=0 \tag{7}$$

在这里已假定 H 在 x 方向是缓变的，它在 x 方向的导数小到可忽略，取 $H\approx H_0$，为气候的平均值。特征量取 $(u',v')\sim C,h'\sim H_0,t\sim(2\beta C)^{-\frac{1}{2}},(x,y)\sim(C/2\beta)^{\frac{1}{2}}$，对方程(5)，方程(6)，方程(7)进行无量纲化后，得到

$$\frac{\partial u'}{\partial t}+\gamma\widetilde{V}\frac{\partial u'}{\partial y}-\frac{1}{2}yv'=-\frac{\partial h'}{\partial x} \tag{8}$$

$$\frac{\partial v'}{\partial t}+\gamma\widetilde{V}(y)\frac{\partial v'}{\partial y}+\frac{1}{2}yu'+\frac{\partial h'}{\partial y}+\gamma\frac{\mathrm{d}\widetilde{V}}{\mathrm{d}y}v'=0 \tag{9}$$

$$\frac{\partial h'}{\partial t}+\gamma\widetilde{V}(y)\frac{\partial h'}{\partial y}+\left(\frac{\partial u'}{\partial x}+\frac{\partial v'}{\partial y}\right)+\gamma\frac{\mathrm{d}\widetilde{V}}{\mathrm{d}y}h'=0 \tag{10}$$

式中 $\gamma=V_0/C$。巢纪平[12]指出在海洋经圈边界附近，经圈流速很大，因此在这样特殊的地区，建立经圈地转平衡也是可能的。由此方程(8)简化成

$$\frac{1}{2}yv' = \frac{\partial h'}{\partial x} \tag{11}$$

推导出对变量 v' 的单一方程为

$$\left(\frac{\partial}{\partial t} + \gamma\widetilde{V}\frac{\partial}{\partial y} + \gamma\frac{\mathrm{d}\widetilde{V}}{\mathrm{d}y}\right)\frac{\partial^2 v'}{\partial x^2} - \frac{y^2}{4}\left(\frac{\partial}{\partial t} + \gamma\widetilde{V}\frac{\partial}{\partial y} + \gamma\frac{\mathrm{d}\widetilde{V}}{\mathrm{d}y}\right)v' + \frac{1}{2}\frac{\partial v'}{\partial x} - \gamma\widetilde{V}\frac{y}{4}v' = 0 \tag{12}$$

3 Weber 函数展开和截断模

对于热带地区,在 y 方向可用抛物圆柱函数即 Weber 函数 $D_n(y)$ 展开成:

$$v' = \sum_n v'_n(x,t)D_n(y) \tag{13}$$

根据 Weber 函数的正交性,方程(12)给出

$$\frac{\partial^3 v'_n}{\partial t\partial x^2} + \frac{1}{2}\frac{\partial v'_n}{\partial x} - \frac{1}{4}\frac{1}{n!\sqrt{2\pi}}\int_{-\infty}^{\infty} y^2\frac{\partial v'_m}{\partial t}D_mD_n\mathrm{d}y = -\frac{\gamma}{2n!\sqrt{2\pi}}\int_{-\infty}^{\infty}\widetilde{V}(y)\left[(m+1)\frac{\partial^2 v'_{m+1}}{\partial x^2} - \frac{\partial^2 v'_{m-1}}{\partial x^2}\right]\times$$
$$D_mD_n\mathrm{d}y + \frac{\gamma}{8n!\sqrt{2\pi}}\int_{-\infty}^{\infty}\widetilde{V}(y)y^2[(m+1)v'_{m+1} - v'_{m-1}]D_mD_n\mathrm{d}y + \frac{\gamma}{4n!\sqrt{2\pi}}$$
$$\int_{-\infty}^{\infty}\widetilde{V}(y)yv'D_mD_n\mathrm{d}y - \gamma\frac{1}{n!\sqrt{2\pi}}\int_{-\infty}^{\infty}\frac{\mathrm{d}\widetilde{V}}{\mathrm{d}y}\left(\frac{\partial^2}{\partial x^2} - \frac{1}{4}y^2\right)v'_mD_mD_n\mathrm{d}y \tag{14}$$

如果假设经圈流是大尺度副热带环流的边界流,对赤道是对称的,其函数设为

$$\widetilde{V} = Ay\exp(-\alpha y^2) \tag{15}$$

A 为表征边界流强度的特征量,对于西边界的向极流,$A>0$,对于东边界的向赤道流 $A<0$,α 为表征经圈流在 y 方向结构的特征量,$\alpha>0$。将方程(15)代入方程(14),因为背景流场关于赤道对称,因此波解中关于赤道对称和反对称的模态是独立的,可以分开处理。将(13)式的展开式截断到 3 阶,令 $n=1$,$n=3$,$n=5$,即只考虑 v'_1,v'_3,v'_5,为关于赤道对称的模态。本文只考虑对称模态的行为,下文中如不说明均指关于赤道对称的模态。关于非对称模态即 $n=0$,$n=2$,$n=4$ 将在另文中分析。

4 本征值分析

设波解

$$v'_n = |v'_n|\,\mathrm{e}^{-\mathrm{i}(kx-\sigma t)} \tag{16}$$

代入方程(14)即可求得频率 σ 与波数 k 的函数关系,即色散关系。

4.1 不考虑经圈流

如果不考虑经圈流,方程(12)退化成

$$\frac{\partial}{\partial t}\left(\frac{\partial^2}{\partial x^2} - \frac{y^2}{4}\right)v' + \frac{1}{2}\frac{\partial v'}{\partial x} = 0 \tag{17}$$

注意到对于短波近似(k 很大),式(17)可近似的写为

$$\frac{\partial^3 v'}{\partial t\partial x^2} + \frac{1}{2}\frac{\partial v'}{\partial x} = 0 \tag{18}$$

这是 Rossby 短波的控制方程,令 $\zeta' = \dfrac{\partial v'}{\partial x}$ 代入式(18),其特解为 $\zeta' = \mathrm{J}_0(\sqrt{2xt})$,$\mathrm{J}_0$ 为贝塞尔函数。可见

在一个有限的时间内,波能量密集在边界附近一窄的区域内,这一区域可称为惯性边界层,过去的计算表明这一边界层的存在和特征宽度[13]。

在不取短波近似的情况下利用 Weber 函数,求得方程(17)前 3 个模态的解。如图 1 所示,可以看到 3 个模态都是 Rossby 波。即当经圈流不存在时,扰动信号通过 Rossby 波来传递。并且这 3 个模态的虚部为 0,表明波均是稳定的。

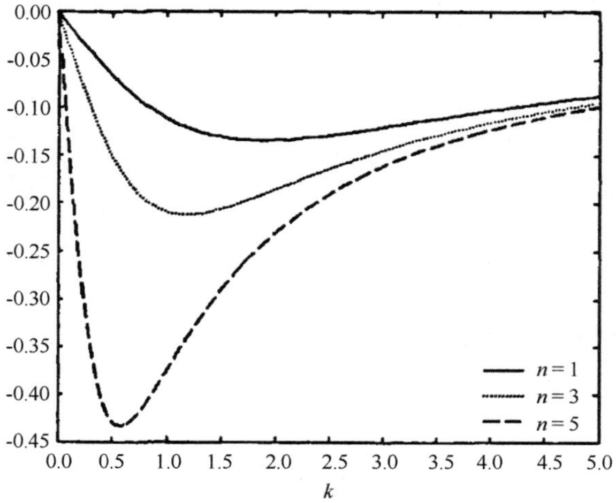

图 1　无经圈流时关于赤道对称的前 3 个模态的频散关系

横,纵坐标为波数和频率。图中均为无量纲量

4.2　考虑经圈流

由于经圈流的存在,使得方程(14)中的有关 σ 的 3 次方程变成复系数方程,3 个模态的解变得复杂。

考虑海盆西边界向极流,取 $A=2,\alpha=1$。图 2 分别为 3 个模态的虚部和实部。从实部的部分可以看出,当 $\gamma\rightarrow 0$ 时,即$\frac{V_0}{C}\rightarrow 0$ 时,σ 的实部趋向于 Rossby 波的频散关系。σ 的虚部则趋向于 0,即趋向稳定。这与不考虑经圈流的结果是一致的。而经圈流的引入,随着 γ 的增大,σ 的实部与虚部都变得更为复杂。σ 的实部为负,相速仍然是向西的,可认为这是变性的 Rossby 波。此外,容易看出 3 个模态 σ 的虚部均大于 0,也就是说任何模态的波动都是不稳定的,不需要其他条件来限制。这与 Walker 和 Pedlosky[10] 在中纬度 β 平面的结论是一致的。

3 个波解显示出,随着 γ 的增大,σ 虚部逐渐增大,表明随着经圈背景流的增强,波动的不稳定性也加强。对 3 个模态分别分析,第一个模态随着 k 的增大,σ 虚部逐渐增大,表明在短波部分更不稳定。第二个模态当 k 趋近于 0 时,σ 虚部逐渐增大,即长波部分出现较大的不稳定。第三个模态 σ 虚部的最大值约出现在 $k=3$。

在实际海洋中利用高度计观测资料,观测到在黑潮过吕宋海峡时有中尺度涡产生,其空间尺度大约为 1°到 3°[14,15]。袁耀初等[16]对东海黑潮的观测资料分析也发现尺度在 1°到 2°的冷涡和暖涡。考虑第一斜压模,其特征尺度约为 1 个纬度。由此计算当 $k=3$ 时的有量纲的不稳定波波长为 3 个纬度,这与实际情况接近。

对于东边界流,例如加利福尼亚海流,其特征速度要比西边界流黑潮小一个量级,故可取

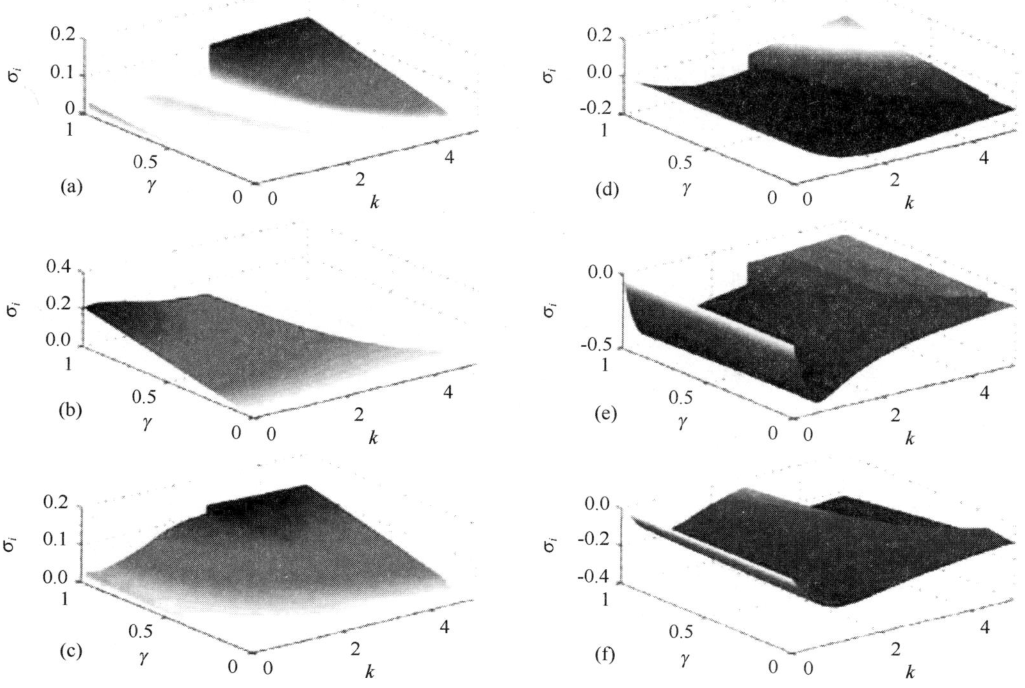

图 2　向极流存在时关于赤道对称的模态频散关系(见书后彩插)

(a),(b),(c)分别为前 3 个模态 σ 的虚部,(d),(e),(f)分别为前 3 个模态的 σ 的实部。x 轴为波数 k,y 轴为背景流速和重力波波速之比 γ,z 轴为 σ。$A=2,\alpha=1$。图中均为无量纲量

$A=-0.2$。图 3 显示的结果与图 2 显示西边界流的结果有很大不同。σ 的实部仍然是负的,可认为这仍然是变性的 Rossby 波。类似的随 γ 的增大,第一模态在短波范围内出现 σ 的实部为正,表明 Rossby 波动出现向东传播的性质。注意到 3 个模态的虚部几乎没有大于 0 的波段或参数域,即 3 个模态都是稳定的,甚至是阻尼的。图 4 表明这一性质与流速无关,增大东边界流的流速使其和西边界流流速有相同的量级。取 $A=2,\alpha=1$,波动的稳定性并没有发生改变。

5　结论

本文以海洋为例把经圈流的小稳定性拓展到热带地区,用一个简单的等值浅水方程组来分析这一问题,其结果稍加改动就不难用于大气研究。

在由某一原因驱动的经圈流范围内,研究的扰动在东—西方向的尺度远比其经圈流的变化尺度小,不考虑经圈流及温跃层在东—西方向的不均匀效应。考虑在赤道 β 平面上,运动是经圈半地转的,利用 Weber 函数对关于 v 的单变量方程进行展开,选取关于赤道对称的模态,截断到 3 阶。

结果显示,当经圈流不存在时,扰动信号通过 Rossby 波来传递,并且这 3 个模态的虚部为 0,即波是稳定的。由于经圈流的存在,使得 3 个模态的解变得复杂。

如果考虑海盆西边界流,由于经圈流的引入,关于赤道对称的模态,观察到 σ 的实部体现为变性的 Rossby 波,而虚部则在任何 k 的情况下都大于 0,表现出不稳定。通过计算不稳定波有量纲的波长,它与在黑潮区观测到的涡旋尺度接近,即 300 km 左右。

如果考虑东边界向赤道流,显示的结果与两边界流的结果有很大不同。关于赤道对称的模态,

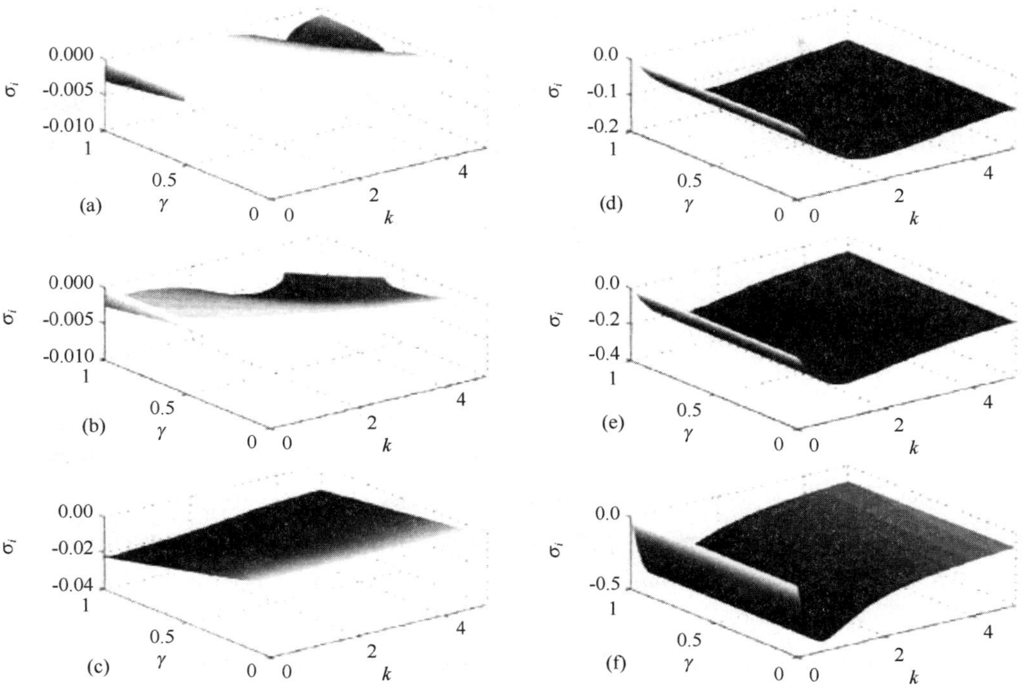

图3 弱向赤道流存在时关于赤道对称的模态频散关系(见书后彩插)

(a),(b),(c)分别为前3个模态σ的虚部;(d),(e),(f)分别为前3个模态的σ的实部。x轴为波数k,y轴为背景流速和重力波波速之比γ,z轴为σ。$A=-0.2,\alpha=1$。图中均为无量纲量

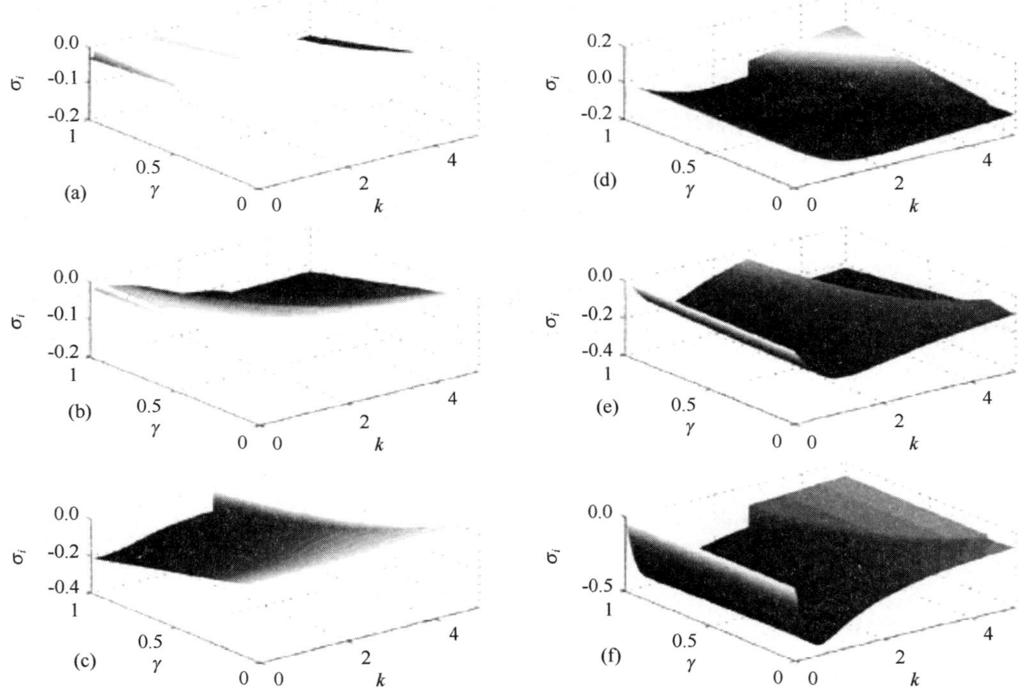

图4 强向赤道流存在时关于赤道对称的模态频散关系(见书后彩插)

(a),(b),(c)分别为前3个模态σ的虚部;(d),(e),(f)分别为前3个模态的σ的实部。x轴为波数k,y轴为背景流速和重力波波速之比γ,z轴为σ。图中均为无量纲量$A=-2,\alpha=1$

σ 的实部虽仍表现为变性的 Rossby 波，但虚部几乎没有大于 0 的波段，即波是稳定的，甚至是阻尼的。

这一简单的分析表明，西边界经圈流和东边界经圈在物理特性上有很大不同，这说明为什么在太平洋黑潮南端的流系会呈现出多种不同的形态，而加利福尼亚接近赤道的海域，流系较为简单。

这些结果由于是从正压模式的最低阶截断模解求得到的，因此是初步的，还有待进一步用更接近实际的模式来研究。

参考文献

[1] Dunkerton T J. On the inertial stability of the equatorial middle atmosphere. *J. Atmos. Sci.*, 1981,38:2354-2364.

[2] Stevens D E. On symmetric instability of zonal mean flows near the equator. *J. Atmos. Sci.*, 1983,40:882-893.

[3] Prohel J A. Linear stability of equatorial zonal flows. *J. Phy Oceanogr*, 1996,26:601-621.

[4] Lindzen R S, Tung K K. Wave overrefection and shear instability. *J Atmos Sci*, 1978,35:1626-1632.

[5] Lindzen R S, et al. The concept of wave overreflection and its application to baroclinic instability. *J. Atmos. Sci.*, 1980, 37:44-63.

[6] Philander S G H. Instabilities of zonal equatorial current. *J. Geophy. Res.*, 1976,81:3725-3735.

[7] Philander S G H. Instabilities of zonal equatorial current, 2. *J. Geophy. Res.*, 1978,83:3679-3682.

[8] 巢纪平，刘琳，于卫东．热带海洋和大气中地形 Rossby 波和 Rossby 波的耦合不稳定．中国科学(D 辑)：地球科学，2005，35：79-87.

[9] 巢纪平，高新全，冯立成．赤道两层海洋模式中基本流的切变不稳定．气象学报，2007，65：1-17.

[10] Walker A, Pedlosky J. Instability of meridional baroclinic currents. *J. Phys. Ocean.*, 2002,32:1075-1093.

[11] Moore D W, Philander S G H. Modelling of the tropical ocean circulation. Goldberg E D, McCave I N, O'Brien J J and Stee I N Eds. The Sea. 6, Wiley (Interscience), New York,1977:319-362.

[12] 巢纪平．热带大气和海洋的半地转适应和发展运动．气象学报，2000，58：257-267.

[13] 巢纪平，陈峰．热带大洋东、西部对风应力经圈不对称的响应．大气科学，2000，24：723-738.

[14] 李燕初，李立，林明森，等．用 TOPEX/POSEIDON 高度计识别台湾西南海域中尺度强涡．海洋学报，2002，24(增刊 1)：163-169.

[15] 林鹏飞，王凡，陈永利，等．南海中尺度涡的时空变化规律 I：统计特征分析. 海洋学报，2007，29(13)：14-22.

[16] 袁耀初，杨成浩，王彰贵．2000 年东海黑潮和琉球群岛以东海流的变异 I．东海黑潮及其附近中尺度涡的变异. 海洋学报，2006，28(2)：1-13.

第6部分　荒漠化、城市化的理论

The Effects of Climate on Development of Ecosystem in Oasis *①

Pan Xiaoling[1](潘晓玲),Chao Jiping[2](巢纪平)

(1. *Institute of Desert Ecology and Environment*, *Xinjiang University*, *Urumqi*, 830046; 2. *On leave National Research Center for Marine Environment Forecasts*, *Beijing* 100081)

ABSTRACT: When vegetation and bare soil coexist, in consideration of some ecological conditions of plant, the total evapotranspiration rate of the oasis and the temperature of vegetation and soil in different climatic and ecological conditions are calculated by using the thermal energy balance equations of vegetation and soil. The evapotranspiration rate depends on climatic and ecological conditions. In some conditions, quasi-bifurcation and multi-equilibrium state appear in the solutions of evapotranspiration rate in the are as covered by small part of vegetation.

Key words: Energy balance equations; Bifurcation; Bi-equilibrium state

1 Introduction

The development of ecological system in oasis, to a great extent, depends on the response of ecological system to environmental conditions, such as climate and water conditions, be-sides the physiological and biochemical conditions of vegetation. The theory concerning the development of ecological system is generally called ecological dynamics. It is substantively the interaction between the processes of ecosystem and climate.

In recent years, many numerical models have been developed on the interaction between biosphere and atmosphere, such as BAT of Dickinson et al.(1986) and SiB of Seller et al. (1986). But after Charney's deserts theory (Charney, 1975), there are few analytical theories about ecology-climate interaction.

This paper attempts in the way of theoretical analysis to study the effects of climate on the development of ecological system in oasis, in which the dynamic process of general atmospheric circulation is temporarily neglected, but the important process of the energy balance in a simple system that includes vegetation, soil and atmosphere, is considered. Basically the response of the development of different kind of vegetation in oasis to climatic conditions is just considered. The whole oasis is assumed to be a close system in the horizontal direction, and the energy budget is balanced among vegetation, soil and atmosphere.

---

* Advances in Atmospheric Sciences, 2001, 18(1): 42-52.

① Sponsored by the National Key Project of Fundamental Research "The ecological environment evolution and control in western China arid area" (01999043500).

2 Energy budget

For the vegetation in oasis, the group energy budget is calculated by averaging all of the single plant's energy budget, for example, taking

$$\bar{A} = \sum_{i}^{N} A_i / N \tag{1}$$

where N is the number of the plants. The another way is to average the connective areas of vegetation. The oasis coverage is assumed to be S, in which a is the coverage of vegetation and $b=S-a$ is the coverage of bare soil. The average value is defined by the connective area, that is

$$\bar{A}^{a,b} = (a,b)^{-1} \sum_{i} A_{l,S}^{(i)}(\mathrm{d}a, \mathrm{d}b) \tag{2}$$

where d represents one small cell.

So the average energy budget equation of vegetation in oasis is

$$\overline{F_l^{rn}}^{a} = \overline{F^h}^{a} + L_v \overline{E_l}^{a} + \overline{F_{l,s}^{a}} \tag{3}$$

where l,s represent vegetation and bare soil respectively, L_v is the evaporating latent heat, E_l is the evaporating rate of vegetation. The net radiation flux in Celsius scale ($T=T^*+T', T^*=273$ K) is approximately

$$\overline{F_l^{rn}}^{a} = (1 - \overline{\alpha_l}^{a}) Q_a - 4\varepsilon\sigma T^{*3}(\overline{T'_l}^{a} - T'_a) \tag{4}$$

where α is albedo, σ is the Stefan-Botzmann constant, ε is the grey body coefficient. The sensible heat flux is

$$\overline{F^h}^{a} = \rho_a C_p C_d V(\overline{T'_l}^{a} - T'_a) \tag{5}$$

where T_a is air temperature (because the scale of climate variation is larger than the scale of oasis, the contant air temperature is taken over oasis), ρ_a is air density, C_p is air specific heat on constant pressure, V is wind speed, C_d is the aerodynamics drag coefficient for heat.

According to Dickinson(1984), the latent heat flux is taken as

$$L_v \bar{E}_l = \rho_a L_v L_{AE}(r_E + \bar{r}_s^{a})^{-1}[B_e^{-1} L_v^{-1} C_p(\overline{T'_l}^{a} - T'_a) + (1-r) q^{sat}(T_a)] \tag{6}$$

where B_e is the Bowen ratio, q^{sat} is the air saturation specific humidity, r is the air relative humidity, $r_E^{-1} = C_d V$, L_{AE} is the leaf surface size coefficient, r_s is the resistance coefficient of stomata.

The last term on the right-hand side of Eq.(3) indicates that there are some vertical mixed processes over the land, so the temperature is different in the air just above vegetation and bare soil. There is also some horizontal heat exchange in the air. Now we assume that the temperature in the air near the ground is proportional to the temperature of vegetation and bare soil. The distribution of vegetation and bare soil is assumed to be uneven, so the horizontal heat exchange in the air near the ground can also be assumed to take the type of turbulence: This process can be parameterized as

$$\overline{F_{l,s}}^{a} \propto \rho_a C_p C_h V(\overline{T'_l}^{a} - \overline{T'_S}^{b}) \tag{7}$$

and

$$C_h \propto \overline{u'l}/VL \tag{8}$$

where u' is the fluctuating value of horizontal velocity in the turbulent process, l is the horizontal mixing length, the denominator is the characteristic value of larger flow motion, V is the characteristic velocity, L is the characteristic scale, i.e. it shows the intensity ratio of tur-bulent and laminar flow, normally its order of magnitude is $10^{-4}-10^{-3}$.

For the bare soil, the energy budget equation is

$$\overline{F_s^{rn}}^{b} = \overline{F_s^{h}}^{b} + L_v \overline{E_s}^{b} + \overline{F_{s,l}}^{b} \tag{9}$$

where

$$\overline{F_s^{rn}}^{b} = (1 - \overline{\alpha_s}^{b}) Q_a - 4\varepsilon\sigma T^{*3}(\overline{T'_s}^{b} - T'_a) \tag{10}$$

$$\overline{F_s^{h}}^{b} = \rho_a C_p C_d V(\overline{T_s}^{b} - T'_a) \tag{11}$$

$$L_v \overline{E_s}^{b} = \rho_a L_v C_d V(\overline{q_s}^{b} - q_a) \approx \rho_a C_p C_d V B_a^{-1}(\overline{T'_s}^{b} - T'_a) \tag{12}$$

$$\overline{F_{s,l}}^{b} = \rho_a C_p C_h V(\overline{T'_s}^{b} - \overline{T'_l}^{a}) \tag{13}$$

3　Temperature of vegetation and bare soil

Substituting Eq.(4)–Eq.(7) into Eq.(3), we have

$$A_l \overline{T'_l}^{a} - B \overline{T'_l}^{a} = C_l \tag{14}$$

where

$$A_l = \rho_a C_p V(C_d + C_h) + 4\varepsilon\sigma T^{*3} + \rho_a C_p L_{AE} B_e^{-1}(r_E + \overline{r_s}^{a})^{-1} \tag{15}$$

$$B = \rho_a C_p C_h V, \tag{16}$$

$$C_l = \{(1 - \overline{\alpha_l}^{a}) Q_a + [\rho_a C_p (r_E^{-1} + L_{AE} B_e^{-1}(r_E + \overline{r_s}^{a})^{-1})] T'_a\} - \rho_a L_v L_{AE}(r_E + \overline{r_s}^{a})^{-1}(1 - r) q^{sat}(T_a) \tag{17}$$

Similarly, substituting Eq.(10)–Eq.(13) into Eq.(9), we have

$$A_s \overline{T'_s}^{b} - B \overline{T'_l}^{a} = C_s \tag{18}$$

where

$$A_s = \rho_a C_p V(C_d + C_h + B_e^{-1} C) + 4\varepsilon\sigma T^{*3} \tag{19}$$

$$B = \rho_a C_p C_h V \tag{20}$$

$$C_s = (1 - \overline{a_s}^{b}) Q_a - \rho_a C_p C_d V(1 + B_e^{-1}) T'_a \tag{21}$$

Then the average temperatures of vegetation and bare soil are respectively represented by

$$\overline{T'_l}^{a} = \frac{(A_s C_l + B C_s)}{(A_l A_s - B^2)} \tag{22}$$

$$\overline{T'_s}^{b} = \frac{(A_l C_s + B C_l)}{(A_l A_s - B^2)} \tag{23}$$

The temperatures of vegetation and bare soil are calculated. The parameter values used here are: $Q_a = 500$ W/m^2, $V = 2.0$ m/s、$C_d = C_h = 2.75\times10^{-3}$, $\sigma = 5.673\times10^{-8}$ J/(s · m^2 · K^4), $\rho_a = 1.293$ kg/m^3, $C_p = 1004$ J/(kg · K), $L_v = 2.5\times10^{16}$ J/kg, other parameters will be given in figures.

Figures 1a, b show the relationship between the vegetation temperature and the two parameters $\overline{r_s}^{a}$, $\overline{\alpha_l}^{a}$ ($\overline{\alpha_s}^{b} = 0.12$, $r = 0.50$). In Fig.1(a), the vegetation temperature is generally lower than the air temperature

20℃, except when the resistance of stomata is very large and the albedo of vegetation is very small. In Fig.1 (b), the vegetation temperature is generally higher than the air temperature 5℃, except when the resistance of stomata is very small and the albedo of vegetation is very large. If there exist some vertical mixing processes between the underlying surface and the above air, vegetation can modify air temperature and reduce the variable range of air temperature.

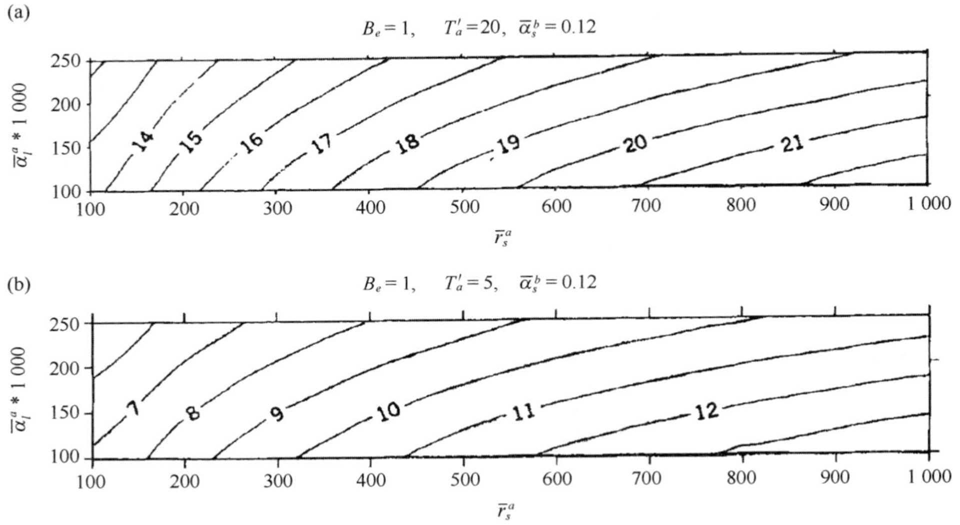

Fig.1 The relationship of vegetation temperature and vegetation albedo and resistance of stomata
(a) air temperature 20℃; (b) air temperature is 5℃

In the above analysis B_e takes 1.0. For the case of $B_e=0.5$ and 1.5, the results are shown in Figs.2(a)-(b) and Figs. 2(c)-(d), and the other parameters are the same as those used before. We can see that there is no great change in the results, so in the following calculations B_e takes 1.0

When the albedo of vegetation and that of soil are 0.15 respectively, and resistance of stomata is 400 s/m, the result shows that the effect of air temperature on vegetation temperature is more important than that of air humidity on the vegetation temperature. Only when the air temperature is very high, the relative humidity is important to some extent. When the air temperature is in the range of 5~20℃, the vegetation temperature is lower than the air temperature; when the air temperature is lower than 5℃, the vegetation temperature is higher than the air temperature (figures omitted).

On the other hand, only when the air temperature is high, the resistance of stomata can increase the temperature of vegetation; when the air temperature is high, the increase of vegetation albedo reduces the vegetation temperature (figures omitted).

Figures 3(a)-(b) indicate the relationship among temperature of bare soil, soil albedo, and resistance of stomata (the vegetation albedo is 0.22, the air temperatures are 20℃ and 5℃).It can be seen from the figure that the temperature of soil is generally higher than the air temperature. except when the soil albedo is large.

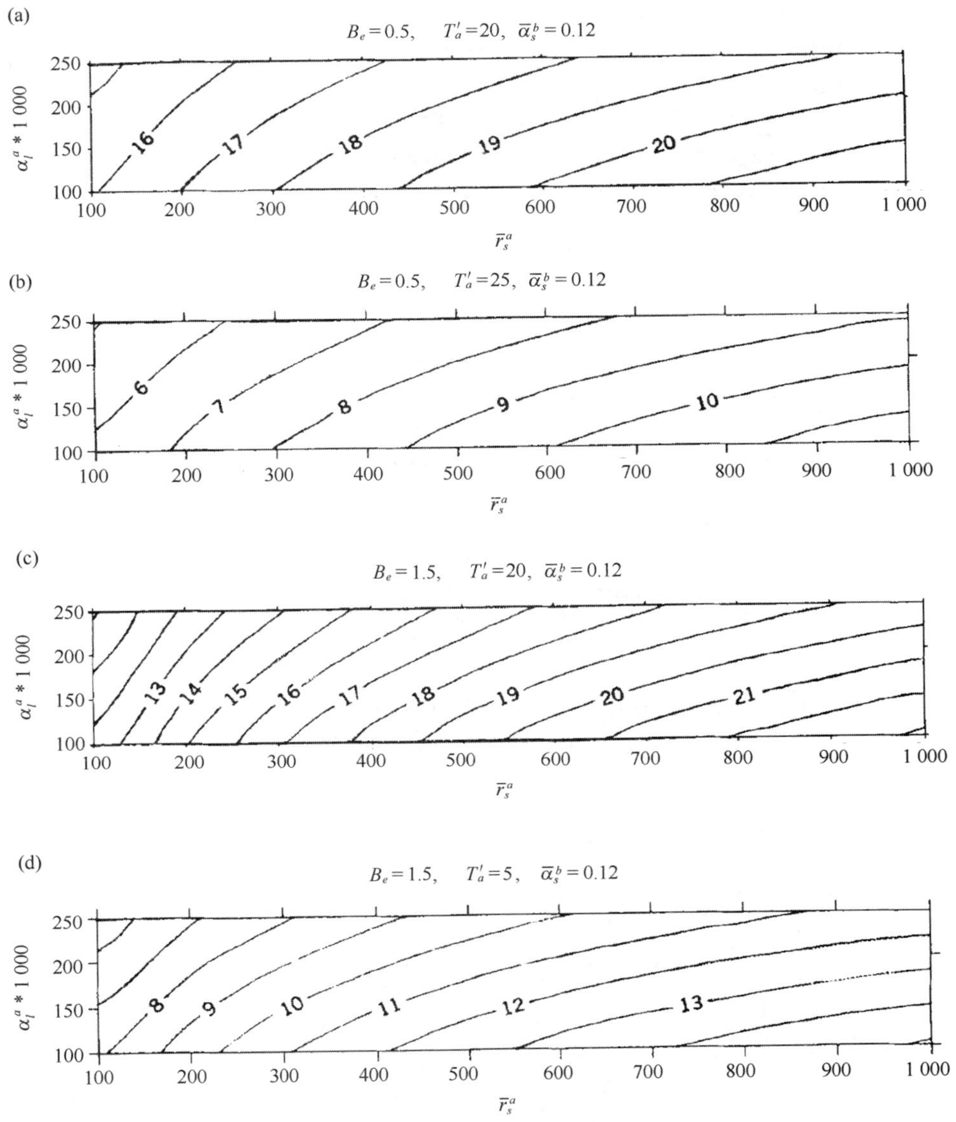

Fig.2　With different B_e, the relationship of vegetion temperature and vegetation albedo and resistance of stomata. (a) $B_e=0.5, T'_a=20$; (b) $B_e=0.5, T'_a=5$; (c) $B_e=1.5, T'_a=20$; (d) $B_e=1.5, T'_a=5$

4　Total energy balance and total evapotranspiration of oasis

The total energy balance equation of oasis is according to this equation, the total evapotranspiration rate of oasis is calculated, for the average temperatures of vegetation and bare soil have been known.

$$\left[1-\left(\frac{a}{S}\right)\overline{\alpha_l}^{a}-\left(1-\left(\frac{a}{S}\right)\overline{\alpha_s}^{b}\right)\right]Q_a-4\varepsilon\sigma T^{*3}\left[\left(\frac{a}{S}\right)\overline{T'_l}^{a}+\left(1-\frac{a}{S}\right)\overline{T'_s}^{b}-T'_a\right]$$
$$=(\rho_a C_p C_d V)\left[\left(\frac{a}{S}\right)\overline{T'_l}^{a}+\left(1-\frac{a}{S}\right)\overline{T'_s}^{b}-T'_a\right]+L_v\overline{E}^{s} \qquad (24)$$

Now we are attempting to get the relationship between oasis evapotranspiration rate and climate variables

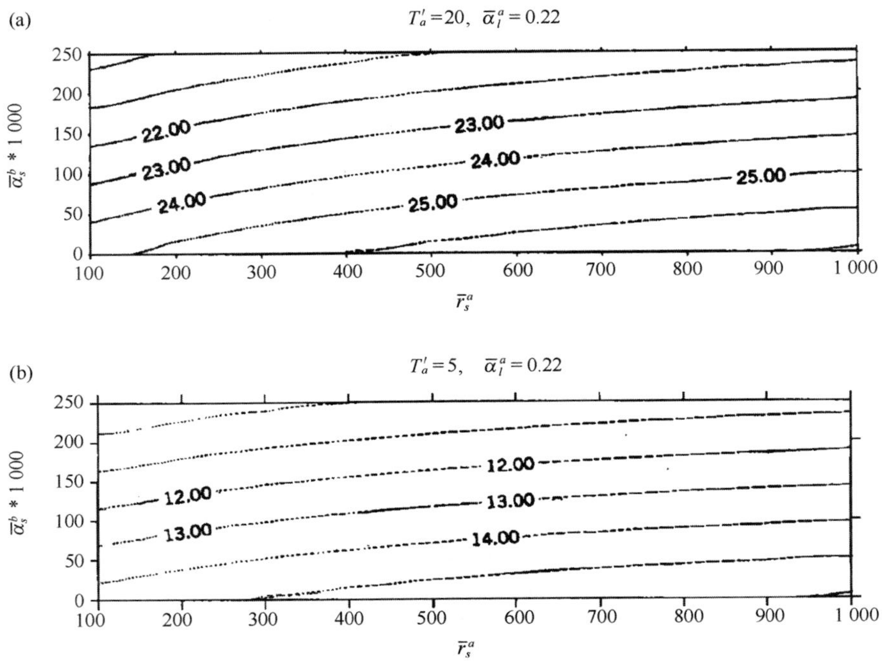

Fig.3 The relationship of soil temperature and resistance of stomata and soil albedo

(a) air temperature is 20℃; (b) air temperature is 5℃

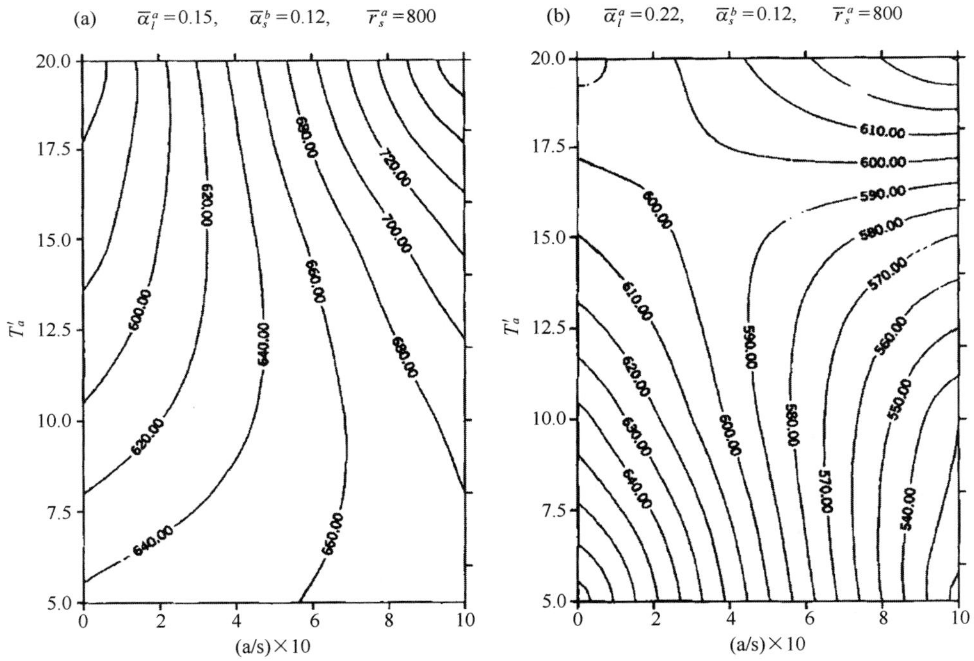

Fig.4 The relationship between evapotranspiration (mm/yr) and air temperature.

(a) vegetation albedo is 0. 15; (b) vegetation albedo is 0. 22

as well as ecological parameters.

Firstly, the effects of air temperature are analyzed when $r = 0.50$, $\overline{r_s}^a = 800$ s/m, $\overline{\alpha_s}^b = 0.12$. Figure 4a shows that the effect of the air temperature on the evapotranspiration rate depends on the vegetation albedo. When $\overline{\alpha_l}^a = 0.15$, the dependence is not very obvious. But when the vegetation albedo is up to $\overline{\alpha_l}^a = 0.22$, the oasis evapotranspiration rate distribution changes obviously (Fig.4(b)). The evapotranspiration rate is 590 mm/a at the air temperature of 10℃, and the corresponding vegetation covering rate is about 55%. But when the air temperature increases to 17℃, vegetation has almost covered the whole oasis. We notice that the same value lines of evapotranspiration rate higher than 590 mm/a appear double. In one of the constant evapotranspiration rate lines, the vegetation covering area reduces as the air temperature increases, for example, it is 600 mm/a at the air temperature of 10℃, and the corresponding covering area is 45%.It reduces rapidly as air temperature increases, and at about 17℃ the vegetation area is almost zero, i.e. desert appears. But there is another 600 mm/a constant line of evapotranspiration rate that appears at high air temperature. When the air temperature is about 20℃, the corresponding area covered by vegetation is 25% and increases as the air temperature does. At about 17℃, the area covered by vegetation expands rapidly, even up to the whole oasis. The changing rule is the same as that when the evapotranspiration rate is larger than 600 mm/a.

The above results indicate that under some environmental humidity condition (such as relative humidity is 50%, semiarid) and some ecological condition (the vegetation albedo is 0.22, the resistance of stomata is 800 s/m), the evapotranspiration rate has bi-equilibrium state. One is in low air temperature region and another is in high air temperature region. In the low temperature region, when the air temperature increases, the area covered by vegetation reduces, oasis develops into desert; in the high temperature region, when the air temperature reduces, the area covered by vegetation increases, desert develops into oasis. The appearance of bi-equilibrium state depends sensitively on the physical and ecological conditions.

5　Multi-equilibrium state and bifurcation of evapotranspiration rate

The distribution of evapotranspiration rate in Fig.4(b) can be divided into three regions. In region I and region II, the constant evapotranspiration rate lines separately lead to the direction of small and large areas covered by vegetation. The 590 mm/a and 600 mm/a constant lines are very close in the starting state (the air temperature is 5℃), but they go to the opposite direction, which is called quasi-bifurcation of solutions. It is understandable in physics. As to 590 mm/a line, the area covered by vegetation is a bit more than the bare soil area. When the air temperature is low, the vegetation temperature is higher than the air temperature, the air temperature increases more than the vegetation temperature does and the evapotranspiration rate increases. To maintain the total value of evapotranspiration rate in the same value, the area covered by vegetation should reduce correspondingly. But when the air temperature is up to the some critical value(for example 17℃), as the air temperature increases, the vegetation temperature will be lower than the air temperature and the evapotranspiration rate will reduce. To maintain the same evapotranspiration rate, the area covered by vegetation should increase. This is the physical explanation to the region II. Similarly, in region Ⅲ if the initial air temperature is 20℃, the vegetation temperature is generally lower than 20℃, but it still has large evapotranspiration rate. When the air temperature reduces, the vegetation temperature reduces too, and the evapotranspirature rate reduces correspondingly. To maintain the same evapotranspiration rate, the area covered by vegetation should increase. The areas covered by vegetation in region II and region III are large, even to the whole oasis, so they can be called oasis solution domain, warm domain and cold domain.

In the region I, when the air temperature increases, the vegetation temperature increases more, then the evapotranspiration rate will increase. To maintain the same evapotranspiration value, the vegetation area should reduce, even changes into desert. This region is called cold desert domain. Since the soil temperature is higher than the vegetation temperature, the evapotranspiration in this region is larger than that in region II.

At the same time, we notice that region I and region III have common evapotranspiration rate (600 mm/yr), which indicates that with the given parameters there are two solutions, one is warm oasis solution and another is cold desert solution.

The appearances of quasi-bifurcation solutions and multi−equilibrium state solutions depend on climatic conditions of environment and ecological condition of vegetation. In Figs 5(a)−(c), the air temperature is 15℃, the soil albedo is 0. 12, other parameters are the same as before, but the resistance of stomata is 400 s/m, 600 s/m, 800 s/m respectively. Fig.5a shows that there is a semi-bifurcation of oasis solutions and desert solution in a small zone covered by vegetation, but the bi-eqilibrium state solution does not exist. Fig.5(b) indicates that evapotranspiration rate has bi-equilibrium state solutions due to increasing of resistance of stomata, one is desert solution with small vegetation albedo, the other is oasis solution with large vegetation albe-

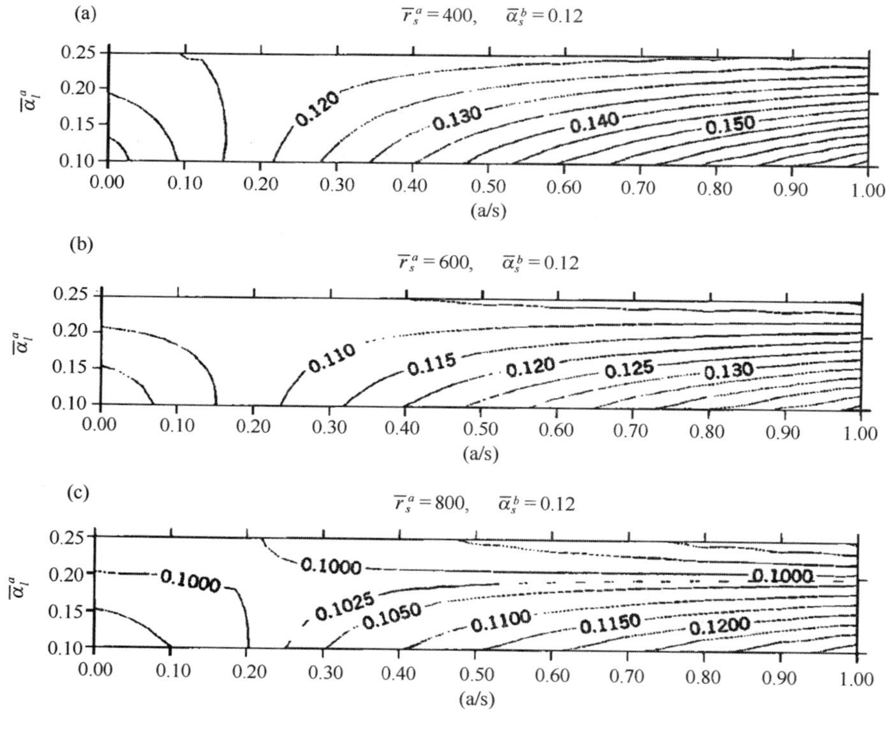

Fig. 5 The relationship of evapotranspiration (mm/h) and vegetation albedo (a) resistance of stomata is 400 s/m; (b) resistance of stomata is 600 s/m; (c) resistance of stomata is 800 s/m

do. Fig.5c shows that in the same covering area of vegetation, one quasi-bifurcation solution and two equilibrium state solutions coexist in high and low value part of vegetation albedo. One of the bi-equilibrium state solutions is 0. 1000 mm/h, the other is 0. 1025 mm/h. They are respectively corresponding to one desert solution and one oasis solution.

Figs. 6(a)-(c) indicate the effect of soil albedo. The air temperature is 15℃, the resistance of stomata is 800 s/m, the soil albedo is 0.15, 0. 20, 0. 25 respectively. In these parameters, the desert zone has only bifurcation and does not have bi-eqilibrium state. But if the soil albedo reduces to 0. 10 (Fig. 6 (d)), two pairs of bi-equilibrium state solutions appear, similar to Fig.5(c).

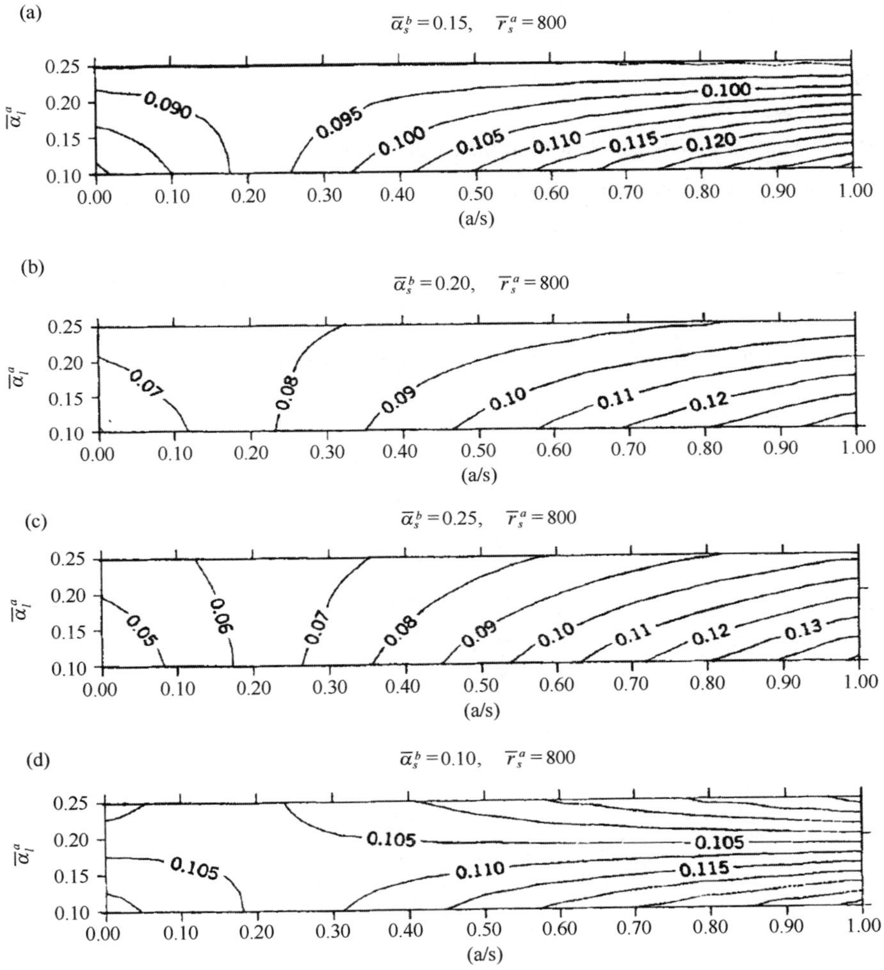

Fig.6　The relationship of evapotranspiration (mm/h) and vegetation albedo

(a) soil albedo is 0. 15; (b) soil albedo is 0. 20; (c) soil albedo is 0. 25 (d) soil albedo is 0. 10

Under what kind of environmental and ecological conditions, the bi-equilibrium state will appear, which is a very important problem and will provide us with some references and guidance to choose different kind of plants to control the desert and should be further studied. In fact, it can be developed into an artificial control theory to modify oasis after thorough researches carried out.

6 Conclusion

Biosphere-atmosphere interaction is one of the main projects in the earth environmental science. Strengthening observation and developing numerical simulation are no doubt important, but some theoretic analysis in simple models is helpful to develop observation and numerical simulation and is practical for people to control oasis.

By using a simple thermal energy balance model, in this paper it is shown that there significantly exist bifurcation solutions and bi-equilibrium state or multi-equilibrium state solutions. It is worth to pay much attention to and put more effort into this phenomenon.

REFERENCES

Charney, J G. 1975. Dynamics of deserts and drought in the Sahel, *Quart. J. Roy. Meteor. Soc.*, 101:193-202.

Dickinson. , R E, et al. 1986. Biosphere-atmosphere transfer scheme (BATS) for the community climate model, National Center for Atmosphere, Boulder, Co., Tech. Note/TN-275 STR.

Dickinson, R E, 1984. Modelling evapotranspiration for three-dimensional global climate models. *Geophys. Monograph*, 29: 58-72.

Seller, P J. et al. 1986. A simple biosphere model (SiB) for use within GCM. *J. Almos. Sci.*, 43:505-531.

大气边界层动力学和植被生态过程耦合的一个简单解析理论*

巢纪平[1]　周德刚[2,3]

(1. 国家海洋环境预报研究中心,北京 100081;2. 中国科学院大气物理研究所,北京 100029;3. 中国科学院研究生院,北京 100039)

摘要:发展了一个大气边界层动力学和植被某些生态过程相互作用的简单模式,求得了这个耦合模式的解析解,分析了植被反照率和冠层阻抗(气孔阻力)对大气运动及植被温度的影响,这一相互作用的方式可为进一步发展大气运动和生态过程相互作用的、更复杂的数值模拟模式提供参考。

关键词:大气边界层运动;植被生态过程;耦合的解析理论

1　引言

大气圈和生物圈相互作用的研究是近年来颇受关注的全球变化及全球变化区域响应的一个重要组成部分,而大气运动的动力学过程和植被的生态过程之间的耦合研究又是地-气相互作用研究的核心问题之一。

自 Charney[1] 发展了一个简单的陆-气相互作用模式并用来研究沙漠化的理论后,下垫面过程特别是植被的生态过程对发展全球或区域气候模式的作用已受到广泛的注意,在这方面,Dickinson 等[2] 和 Sillers 等[3] 把下垫面的生态过程和大气物理过程耦合起来的工作无疑是创造性的,对发展气候模式是重要的。在国内这方面的工作也随之开展,如季劲钧等[4] 发展了用于气候研究的简单陆面过程模式,赵鸣等[5] 引入近地层的土壤-植被-大气相互作用的模式,中国科学院大气物理研究所用于气候研究的陆面过程模式[6] 等。从气候变化的角度来看,陆气相互作用可能比海气相互作用更复杂。研究复杂的陆面过程对气候变化的作用,模式的数值模拟虽然是一个强有力的方法,但适度地发展一些有反馈物理过程的较为简单的陆-气相互作用模式,在解析数学的范围内讨论各种物理过程的相对重要性,这仍然是有意义的研究。事实上,Charney 的沙漠化理论就属于这类工作。

本文的目的在于发展一个能在数学上解析处理的简单大气边界层动力学和植被生态过程相互作用的模式,用以讨论影响气候态变化的植被属性的相对重要性。

2　大气动力学-植被生态过程耦合模式

模式结构是这样考虑的,有一个很薄的贴地层,其厚度相当于株冠高度,植被的生态过程主要发

* 大气科学,2005,29(1):37-46.

生在这一层中,由于层中的能量(例如感热)会与其上的大气温度关联,这意味着大气运动会对这一层中的能量平衡起作用。在株冠高度以上是大气边界层,边界层中除有运动的动力过程外,还有辐射传输过程,即边界层中的运动不是正压的,温度将参与进来,而温度垂直变化方程的底边界的定解条件,又会用到株冠层中的温度。通过这种方式(过程)把大气层和植被层中的物理过程耦合起来了,构成了一个简单的大气边界层运动和植被生态过程相互作用模式。

2.1 植被冠层的能量平衡

考虑在植被冠层主要能量过程为:对太阳辐射的吸收、对来自其上大气层的长波辐射的吸收和本层中长波辐射的发射,由此,辐射平衡表达式为

$$R_n = (1-\alpha_c)Q_a + \varepsilon\sigma T_a^4 - \varepsilon\sigma T_c^4 \tag{1}$$

其中,Q_a 为到达冠层的太阳辐射通量;T_c 为植被的温度;T_a 为植被冠层顶处大气的温度;ε 为灰体系数;σ 为斯蒂芬-玻耳兹曼常数;α_c 为植被的反照率。注意到在长波辐射通量公式中,温度是用绝对温度写出的,如扣去 $T=273$ K 后用摄氏温度写出,并认为在所研究的问题中,摄氏温度的变化区间要比 T 小一个量级,则上面的平衡式可写成

$$R_n = (1-\alpha_c)Q_a + 4\varepsilon\sigma T^3(T_a - T_c) \tag{2}$$

式中温度已改用摄氏温度。

感热通量为

$$H_c = \frac{\rho c_p(T_c - T_a)}{r_a} \tag{3}$$

式中,ρ 为空气密度;c_p 为空气的定压比热;$r_a=(C_d\overline{V})^{-1}$ 为空气阻抗;C_d 为空气的拖曳系数;V 为参考风速值。

潜热通量为

$$\lambda E_c = \frac{\rho L_v}{r_a + r_c}\left[\frac{c_p}{B_e L_v}(T_c - T_a) + (1-r)q_s(T_c)\right] \tag{4}$$

式中,L_v 为相变潜热;B_e 为 Bowen 比;r_c 为植被冠层阻抗(气孔阻力);r 为相对湿度;q_s 为饱和比湿,取植被温度作为参考温度。

假定植被层的能量是平衡的,那么有

$$\begin{aligned}(1-\alpha_c)Q_a - 4\varepsilon\sigma T^3(T_c - T_a) = &\frac{\rho c_p(T_c - T_a)}{r_a}\\ &+ \frac{\rho L_v}{r_a + r_c}\left[\frac{c_p(T_c - T_a)}{B_e L_v} + (1-r)q_s(T_c)\right]\end{aligned} \tag{5}$$

取 $q_s(T_c) \approx q_s(T^*) + \partial q_s/\partial T(T_c - T^*) = \hat{q} + bT_c$,在此 $\hat{q} = q_s(T^*) - \partial q_s/\partial T \times T^*$,而 T^* 为一参考温度,在这个温度值附近,饱和比湿和温度的关系接近线性,b 是由统计定出的一个常数。于是上式可改写成,

$$\begin{aligned}(1-\alpha_c)Q_a - 4\varepsilon\sigma T^3(T_c - T_a) = &\frac{\rho c_p}{r_a}(T_c - T_a)\\ &+ \frac{\rho L_v}{r_a + r_c}\left[\frac{c_p(T_c - T_a)}{B_e L_v} + (1-r)(\hat{q} + bT_c)\right]\end{aligned} \tag{5'}$$

在这一贴地层或植被覆盖层的能量平衡中,对植被的属性引进两个宏观参数,一是植被冠层阻抗

r_c，另一是植被的反照率 α_c。对多数植物来讲，温度高有利于它们生长、变得茂盛，因此有理由认为，当温度高时，茂盛的植被反照率会变小，即设 $\alpha_c=\hat{\alpha}_c-aT_c$，$T_c>0$，如温度不为正，仍有 $\alpha_c=\hat{\alpha}_c$，a 由观测资料统计定出。与这一假定相似的例子是，在研究冰期气候形成时就假定了冰面的反照率与温度有反比关系[7]。

当上面各种能量或热通量处于平衡状态时，得到

$$T_c = a_1 T_a + a_2 \tag{6}$$

式中，

$$a_1 = \frac{4\varepsilon\sigma T^3 + \dfrac{\rho c_p}{r_a}\dfrac{\rho c_p}{B_e(r_a + r_c)}}{4\varepsilon\sigma T^3 + \dfrac{\rho c_p}{r_a} + \dfrac{\rho c_p + \rho L_v(1-r)bB_e}{B_e(r_a + r_c)} - aQ_a},$$

$$a_2 = \frac{(1-\hat{\alpha}_c)Q_a - \dfrac{\rho L_v(1-r)\hat{q}}{r_a + r_c}}{4\varepsilon\sigma T^3 + \dfrac{\rho_{c_p}}{r_a} + \dfrac{\rho_{c_p} + \rho L_v(1-r)bB_e}{B_e(r_a + r_c)} - aQ_a}$$

当在植被覆盖层中存在上面所述的能量平衡时，就把植被温度和植被冠层顶处的大气温度联系起来了，这也表示了大气运动通过边界层底部的温度影响着植被覆盖层的生态温度。

2.2　边界层大气运动方程

考虑到植被的生态过程首先影响的是大气的行星边界层，现给出边界层的动力学方程。在(x,z)剖面上边界层大气运动方程为

$$fu = \mu\frac{\partial^2 v}{\partial z^2} \tag{7}$$

$$f\frac{\partial v}{\partial z} = \frac{g}{T}\frac{\partial T}{\partial x} \tag{8}$$

$$\frac{\partial u}{\partial x} + \frac{\partial w}{\partial z} = 0 \tag{9}$$

由式(9)，引入流函数 ψ，

$$u = -\frac{\partial\psi}{\partial z},\quad w = \frac{\partial\psi}{\partial x} \tag{10}$$

代入式(7)，得到

$$f\frac{\partial\psi}{\partial z} = -\mu\frac{\partial^2 v}{\partial z^2} \tag{11}$$

取条件

$$z\to\infty,\quad \frac{\partial v}{\partial z}\to 0,\quad \psi\to 0 \tag{12}$$

由此，有

$$\psi = -\frac{\mu}{f}\frac{\partial v}{\partial z} \tag{13}$$

由式(8)、式(10)、式(13)给出

$$w = -\frac{\mu g}{f^2 T}\frac{\partial^2 T}{\partial x^2} \tag{14}$$

$$u = \frac{\mu g}{f^2 T}\frac{\partial^2 T}{\partial x \partial z} \tag{15}$$

以及

$$v = -\frac{g}{fT}\frac{\partial}{\partial x}\int_z^{\infty} T\mathrm{d}z \tag{16}$$

在此已假定 $z\to\infty$，$v\to 0$。至此，当大气温度已知后可算出运动场。

2.3 热量传输方程

在辐射能传递、感热垂直输送和垂直运动相平衡下，热量平衡方程为

$$\left(\rho c_p N^2 \frac{T}{g}\right) w = K_T \frac{\partial^2 T}{\partial z^2} + \sum_j \alpha'_j \rho_c (A_j + B_j - 2E_j) + \alpha'' \rho_c Q, \tag{17}$$

式中，$N = \left(\frac{g}{\theta}\frac{\partial\theta}{\partial z}\right)^{\frac{1}{2}}$ 为 Brunt-Vaisala 频率；而 α'_j 和 α'' 分别为对波长为 λ_j 的长波辐射的吸收系数和对太阳辐射的吸收系数；A_j 和 B_j 分别为在波长 λ_j 区间内向下和向上的长波辐射通量；E_j 为在该波长区间内的黑体辐射能量；ρ_c 为吸收介质密度；Q 为太阳辐射通量；K_T 为热量的垂直湍流交换系数。

考虑到式(14)有

$$-\left(\frac{N}{f}\right)^2 K_v \frac{\partial^2 T}{\partial x^2} = K_T \frac{\partial^2 T}{\partial z^2} + \sum_j \alpha'_j \rho_c (A_j + B_j - 2E_j) + \alpha'' \rho_c Q \tag{18}$$

按郭晓岚[8]和巢纪平、陈英仪[7]处理长波辐射通量传输过程的方案，即将长波辐射谱按某一准则分成波长短的谱区和波长长的谱区，经过这样的分割处理后，式(18)的最终形式为

$$-\left(\frac{N}{f}\right)^2 K_v \frac{\partial^2 T}{\partial x^2} = K_T \frac{\partial^2 T}{\partial z^2} + \frac{8r^{\circ}\varepsilon\sigma T^3}{\alpha'_s \rho_c}\frac{\partial^2 T}{\partial z^2} - 2(1-r^{\circ})\alpha'_w \varepsilon\sigma\rho_c T^4 + \alpha''\rho_c Q + C_0 + C_1 z \tag{19}$$

式中，α'_s，α'_w 分别为波长短的辐射区和波长长的辐射区中介质的平均吸收系数；r°表示物质的辐射能量在波长短的辐射区中的部分与总辐射能量之比。

注意到，方程(19)中有两个积分常数，其中之一，考虑到 $z\to\infty$ 时，物理量有限，因此，$C_1=0$，而 C_0 可用下面的方法求得，设在充分大的侧向边界及大气顶部没有能量输入，能量只能在底部边界输入，因此有

$$C_0 = 2(1-r^{\circ})\alpha'_w \varepsilon\sigma\rho_c T^4 - \alpha''\rho_c \overline{Q} - \frac{1}{H}\left(K_T + \frac{8r^{\circ}\varepsilon\sigma T^3}{\alpha'_s \rho_c}\right)\left(\frac{\partial T}{\partial z}\right)_{z\approx 0} \tag{20}$$

式中，H 为区域的垂直厚度。如果区域取得相当大，上式最后一项相当小，可以略去，而在这一区域中，区域平均温度由辐射平衡决定，即 $\varepsilon\sigma T^4 = (1-\overline{\alpha})\overline{Q}$。由此给出

$$C_0 = [2\alpha_w \rho_c (1-r^{\circ})(1-\overline{\alpha}) - \alpha''\rho_c]\overline{Q} \tag{21}$$

于是方程(19)为

$$\left(\frac{N}{f}\right)^2 K_v \frac{\partial^2 T}{\partial x^2} + \left(K_T + \frac{8r^{\circ}\varepsilon\sigma T^3}{\alpha'_s \rho_s}\right)\frac{\partial^2 T}{\partial z^2} - 8(1-r^{\circ})\alpha'_w \rho_c \varepsilon\sigma T^3 T = 2(1-r^{\circ})\alpha'_w \rho_c [\varepsilon\sigma T^4 - (1-\overline{\alpha})\overline{Q}] - \alpha''\rho_c (Q - \overline{Q}), \tag{22}$$

这是大气边界层中动力-辐射耦合模式的基本方程。

方程(22)的定解条件之一为

$$z \to \infty, \quad \frac{\partial T}{\partial z} = 0 \tag{23}$$

考虑到植被覆盖层很薄,可将下边界条件置放在 $z \approx 0$ 处,在这个界面上感热、潜热和辐射通量是平衡的,即有

$$z \approx 0,$$

$$-\rho c_p K_l \frac{\partial T}{\partial z} = (1-\alpha)Q_a + 4\varepsilon\sigma T^3(T - T_e) - \frac{\rho L_v}{r_a + r_c}\left[\frac{c_p}{B_e L_v}(T_e - T) + (1-r)(\hat{q} + bT)\right] \tag{24}$$

式中,T_a 的下标已去掉,T_e 可用式(6)替换成 T。从这个边界条件可以看到,植被的生态过程通过长波辐射通量和潜热通量又影响了边界层中的大气运动,由此可见,这是一个简单的植被生态和行星边界层大气运动相互作用的模式。

2.4　模式的数学表达

现把上面各种物理过程汇总成一个数学问题。首先对方程进行无量纲处理,引进 $z=(\rho c_p K_l)z'$,$x=(N/f)\sqrt{K_v}x'$,方程(22)可写成(略去 x,z 的上标撇号)

$$\frac{\partial^2 T}{\partial x^2} + \hat{a}^2 \frac{\partial^2 T}{\partial z^2} - \hat{d}^2 T = \Omega(x,z) \tag{25}$$

式中

$$\hat{a}^2 = \left(K_T + \frac{8r^{\cdot}\varepsilon\sigma T^3}{\alpha'_s\rho_s}\right)/(\rho c_p K_l)^2,$$

$$\hat{d}^2 = 8(1-r^{\cdot})\alpha'_w\rho_c\varepsilon\sigma T^3,$$

$$\Omega(x,z) = 2(1-r^{\cdot})\alpha'_w\rho_c[\varepsilon\sigma\hat{T}^4 - (1-\bar{\alpha})\bar{Q}] - \alpha''\rho_c(Q - \bar{Q})$$

垂直边界条件为

$$z = 0, \quad -\frac{\partial T}{\partial z} = AT + S(x) \tag{26}$$

$$z = \infty, \quad \frac{\partial T}{\partial z} = 0 \tag{27}$$

式中,

$$A = 4\varepsilon\sigma T^3 + \frac{\rho c_p}{B_e(r_a + r_c)} - \frac{\rho L_v(1-r)b}{(r_a + r_c)} - \left[4\varepsilon\sigma T^3 + \frac{\rho c_p}{B_e(r_a + r_c)}\right]a_1 \tag{28a}$$

$$S = (1-\alpha)Q_a - \frac{\rho L_v(1-r)\hat{q}}{r_a + r_c} - \left[4\varepsilon\sigma T^3 + \frac{\rho c_p}{B_e(r_a + r_c)}\right]a_2 \tag{28b}$$

方程(25)和条件式(26)、条件式(27)构成了一个二阶椭圆型方程的混合边界条件问题。问题的侧向边界条件视问题的提法而定。

3　热力学耦合模式

由方程(25)注意到,如果强迫源在区域中是水平均匀的,对一个无侧向边界强迫的线性问题来讲,响应出的温度分布在 x 方向也是均匀的,由式(14)、式(15)和式(16)可以看到,此时在大气边界

层中垂直运动及水平运动均为零，也即边界层的动力过程将不在这一相互作用模式中起作用。这样，方程(25)中只有辐射能传输与热量垂直湍流交换之间的平衡，即动力学模式退化成单纯的热力学模式。

在这种情况下，方程(25)变成

$$\frac{d^2T}{dz^2}-C^2T=\Omega^{\cdot}(z) \tag{29}$$

式中，$C^2=\hat{d}^2/\hat{a}^2$，$\Omega^{\cdot}=\Omega/\hat{a}^2$。边界条件仍为式(26)、式(27)，但式(26)中的 A 和 S 不再是 x 的函数，分别取常数 $A^{\cdot}$ 和 $S^{\cdot}$，边界条件式(26)改写为

$$z=0,\qquad -\frac{dT}{dz}=A^{\cdot}T+S^{\cdot} \tag{30}$$

方程(29)满足条件式(30)、式(27)的解为

$$T(z)=\frac{S^{\cdot}\,e^{-Cz}}{C-A^{\cdot}}-\frac{1}{2C}\left[\frac{C+A^{\cdot}}{C-A^{\cdot}}\int_0^{\infty}\Omega^{\cdot}(\xi)e^{-C(\xi+z)}d\xi+\int_0^{z}\Omega^{\cdot}(\xi)e^{-C(z-\xi)}d\xi+\int_z^{\infty}\Omega^{\cdot}(\xi)e^{-C(\xi-z)}d\xi\right] \tag{31}$$

解(31)是考虑了大气边界层中的热力过程(即辐射过程)和贴地植被层中生态过程后温度在垂直方向的分布。我们称这一温度为热力学平衡温度，标以 T_t。

假定，太阳辐射垂直方向上的分布为 $Q=S_0-(S_0-Q_a)e^{-z/H}$，取 $S_0=340\ W/m^2$，$Q_a=200\ W/m^2$，$H=3\times10^3\ m$，上述各方程中的参数除反照率 α_c 和气孔阻力 r_c 待定外，其他各参数分别取值如下：$\hat{q}=3.46\times10^{-4}$，$b=6.86\times10^{-4}/℃$，$C_d=2.75\times10^{-3}$，$\overline{V}=2.5\ m/s$，$r=0.5$，$K_v=K_T=3.0\times10^4\ W/(m\cdot ℃)$，$N^2=1.2\times10^{-4}/s^2$，$f=1\times10^{-4}/s$，$\rho=1.29\ kg/m^3$，$\rho_c=6\times10^{-3}\ kg/m^3$，$\alpha'_s=10\ m^2/kg$，$\alpha'_w=0.125\ m^2/kg$，$\alpha''=2.5\times10^{-2}\ m^2/kg$。

分别讨论不同反照率和冠层阻抗对热力学温度的影响。图 1(a)，图 1(b)是 $r_c=200\ s/m$ 和分别取反照率 $\alpha_c=0.16$ 和 $a_c=0.26$ 时的结果，前者算得近地面空气温度为 13.2℃，植被温度为 18.7℃，后者近地面空气温度为 10.6℃，植被温度为 15.8℃，可见当反照率增加后植被和空气的温度均降低。

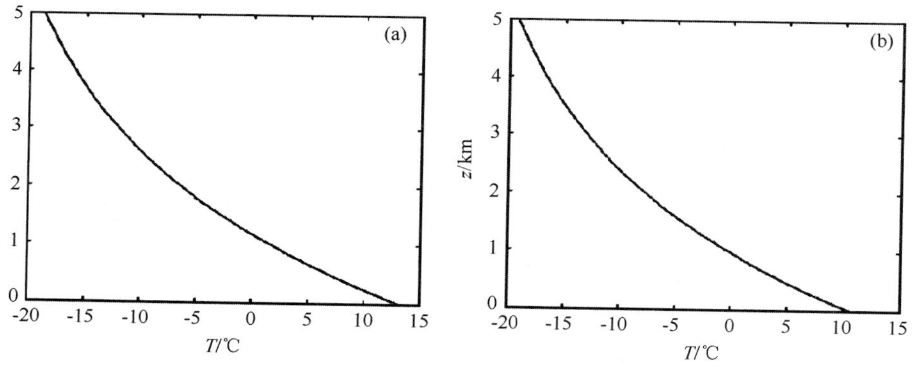

图 1　不同反照率 $\alpha_c=0.16$ (a) 和 $\alpha_c=0.26$ (b) 对热力学平衡温度的影响

先不考虑反照率随温度的反馈，来看冠层阻抗对热力学平衡温度的影响。取 $\alpha_c=0.16$，冠层阻抗分别取 $r_c=150\ s/m$ 和 $250\ s/m$，结果如图 2(a)，(b)中的实线所示，这时近地面温度分别为 11.7℃，14.5℃，相应植被温度分别为 16.8℃，20.5℃。由结果可见，当冠层阻抗增加时，植被和空气的温度相应增加。

当考虑植被温度对反照率有反馈时，比如 $\alpha_c=\hat{\alpha}_c-aT_c$，$a$ 取 0.01，相当植被温度增加 1℃，反照率减小 0.01。如果以图 1(a)中的热力学平衡温度解作为参照，即取 $r_c=200$ s/m 和 $\alpha_c=0.16$，此时植被温度 $T_{c0}=18.7$℃，把反照率的表达式修改为：$\alpha_c=\hat{\alpha}_c-a(T_c-T_{c0})$，则如果取 $\hat{\alpha}_c=0.16$，当 $r_c=200$ s/m，显然热力学温度解就回到了图 1(a)，此时植被温度 $T_c=T_{c0}$，反照率 $\alpha_c=\alpha_c=0.16$，也就是说这时不存在反馈。当取 $r_c=150$ s/m 和 250 s/m，反照率的反馈机制起作用，其结果分别见图 2(a)，图 2(b)中的点线，此时计算出近地面温度和植被温度分别为 10.8℃、15.5℃和 15.9℃、21.6℃。与没有反照率反馈的结果相比较，前者当冠层阻抗从 150 s/m 增加到 250 s/m 时，地面气温增加了 2.8℃，当反照率存在随温度的变化有反馈时，同样的冠层阻抗变化使地面气温增加了 4.7℃。这一对比表明反照率随温度变化的反馈是重要的，如何提出一个更合理的反馈方案是值得进一步研究的。

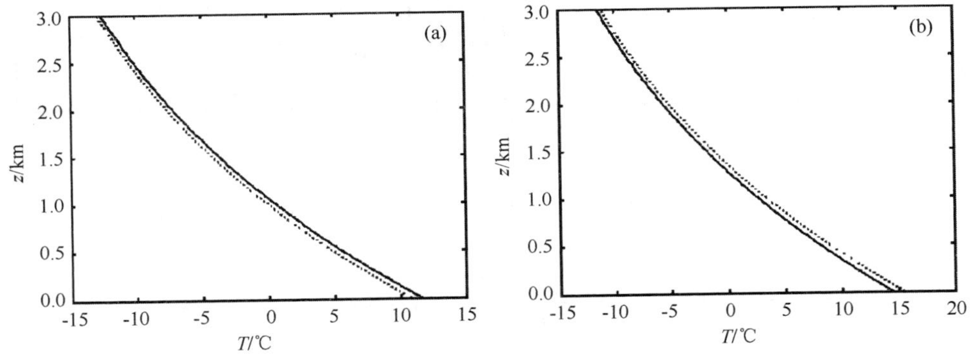

图 2　不同冠层阻抗 $r_c=150$ s/m（a）和 $r_c=250$ s/m（b）对热力学平衡温度的影响

4　动力学耦合模式的解析解

现以热力学耦合模式得到的热力学平衡温度作为背景温度，将动力学耦合模式得出的温度扣去背景温度，即可分析动力学过程在气候中的作用。令 $T'=T-T_t$，方程(25)可写成

$$\frac{\partial^2 T'}{\partial x^2}+\hat{a}^2\frac{\partial^2 T'}{\partial z^2}-\hat{d}^2 T'=\Omega'(x,z) \tag{32}$$

式中，$\Omega'=\Omega-\hat{a}^2\mathrm{d}^2T_t/\mathrm{d}z^2+\hat{d}^2T_t$。同时，认为边界条件式(26)右边的 AT 项中 A 变化不大，近似取作常数 $A°$，式(26)改写为

$$z=0,\quad \frac{\partial T'}{\partial z}+A^{\cdot}\ T'=-S'(x) \tag{33}$$

式中，$S'=S+\mathrm{d}T_t/\mathrm{d}z+AT_t$。式(27)改为

$$z=\infty,\quad \frac{\partial T'}{\partial z}=0 \tag{34}$$

问题的侧向条件为

$$|x|\rightarrow\infty,\quad T'=0 \tag{35}$$

这个数学物理问题的解法见附录 1。

当考虑不同的植被分布时，有相应的 $S'(x)$ 和 $\Omega'(x,z)$ 表达式，进而可以求出不同的动力过程解。这里只简单考虑 $S'(x)$ 的影响而不考虑 $\Omega'(x,z)$，即令 $\Omega'(x,z)=0$，于是附录中式(A23)可简化为

$$T'(x,z)=\frac{\hat{a}}{\pi}\int_{-\infty}^{+\infty}S'(x'')K_0\left(\frac{\hat{d}}{\hat{a}}\sqrt{\hat{a}^2(x''-x)^2+z^2}\right)\mathrm{d}x''$$
$$+\frac{A^{\cdot}\hat{a}}{\pi}\int_{-\infty}^{+\infty}S'(x'')\mathrm{d}x''\int_{z}^{\infty}\mathrm{e}^{A^{\cdot}(z''-z)}K_0\left[\frac{\hat{d}}{\hat{a}}\sqrt{\hat{a}^2(x''-x)^2+z''^2}\right]\mathrm{d}z'' \tag{36}$$

式中，K_0 为第二类修正贝赛尔函数。

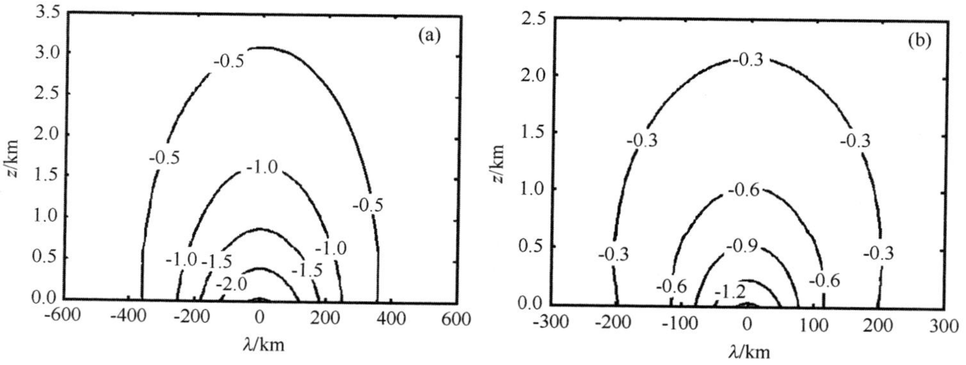

图 3　植被阻抗在不同水平分布范围下的动力学解：(a) $L=150$ km；(b) $L=50$ km

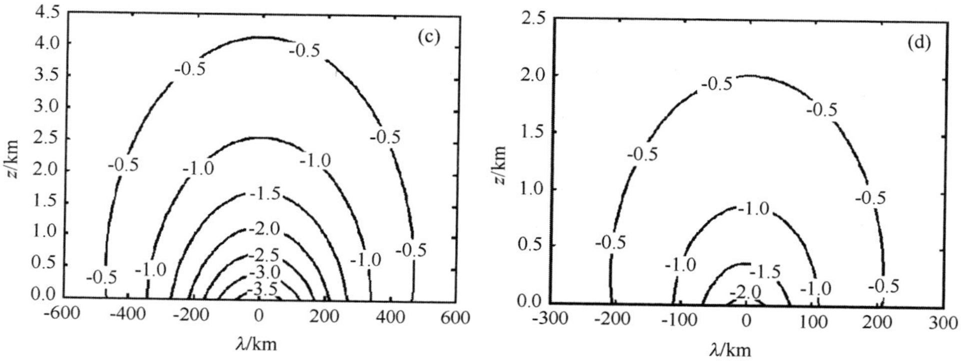

图 4　反照率存在随温度反馈的动力学解：(a) $L=150$ km；(b) $L=50$ km

4.1　植被不同水平分布格局的影响

考虑植被的属性在水平方向上的分布是不均匀的，例如取植被阻抗在水平方向上分布为：$r_c=200-100\mathrm{e}^{-x^2/(2L^2)}$ s/m，$L=150$ km，其他参数与图 1(a) 中所取的参数一样，即以图 1(a) 的热力学解为基础，由此计算出的动力学解如图 3(a) 所示，在 $x=0$ 的中心处温度距平可达到 −2.5℃。如果改变植被阻抗差异的水平分布格局，令 $L=50$ km，其他参数与图 3(a) 相同，其动力学解见图 3(b)。两图相比较，可以发现当植被阻抗差异的水平范围变窄时，其动力学温度解由 −2.5℃ 变为 −1.5℃，幅度变小，解在垂直和水平方向上的分布范围变窄。

4.2　反照率存在随温度反馈的影响

当考虑反照率存在随温度的反馈时，采用上面反照率随温度反馈的表达形式 $\alpha_c=\hat{\alpha}_c-a(T_c-T_{c0})$，$\hat{\alpha}_c=0.16$，取植被的热力学解温度 $T_{c0}=18.7$℃，分别考虑植被不同水平分布范围（与图 3 相同的分布）的影响，如果 $a=0$，则动力学解回到图 3 所示的解，现取 $a=0.01$，动力学解分别如图 4(a)，图 4(b) 所示。与图 3 相比,可以发现考虑反照率随温度反馈后,植被阻抗差异造成的动力学解幅度增加,同时它们在垂直方向和水平方向的影响范围也增加。

5　大气动力学场的计算

注意到,虽然现在温度已解出,但从式(14)不能直接算出垂直速度,这是因为式(14)只在流体内部表明垂直运动和温度场的关系,而垂直运动尚需受边界条件的约束,为此,可对式(10)求旋度,将式(14)、式(15)代入,得到

$$\frac{\partial^2 \psi}{\partial x^2}+\frac{\partial^2 \psi}{\partial z^2}=-\frac{\mu g}{f^2 T}\left(\frac{\partial^2}{\partial x^2}+\frac{\partial^2}{\partial z^2}\right)\frac{\partial T}{\partial x} \tag{37}$$

改写成

$$\frac{\partial^2 \Phi}{\partial x^2}+\frac{\partial^2 \Phi}{\partial z^2}=0 \tag{38}$$

式中,

$$\Phi=\psi+\frac{\mu g}{f^2 T}\frac{\partial T}{\partial x}$$

边界条件

$$z=0,\qquad \psi=0 \tag{39}$$

地面是一条流线,确保垂直运动为零,因此有

$$z=0,\qquad \Phi=\frac{\mu g}{f^2 T}\frac{\partial T(x,0)}{\partial x} \tag{40}$$

右端已知。考虑到 $z\to\infty$,$\psi\to 0$,$\partial T/\partial x\to 0$,于是有

$$z\to\infty,\quad \Phi\to 0 \tag{41}$$

类似地考虑,给出

$$|x|\to\infty,\quad \Phi\to 0 \tag{42}$$

当 Φ 算出后,

$$\psi=\Phi-\frac{\mu g}{f^2 T}\frac{\partial T}{\partial x}$$

方程的解为

$$\begin{aligned}\psi(x,z)&=\int_{-\infty}^{+\infty}\frac{\mu g}{f^2 T}\frac{\partial T(x',0)}{\partial x}\frac{z}{(x'-x)^2+z^2}\mathrm{d}x'-\frac{\mu g}{f^2 T}\frac{\partial T}{\partial x}\\&=\int_{-\frac{\pi}{2}}^{+\frac{\pi}{2}}\frac{\mu g}{f^2 T}\frac{\partial T(x+\tan\alpha z,0)}{\partial x}\mathrm{d}\alpha-\frac{\mu g}{f^2 T}\frac{\partial T}{\partial x}\end{aligned} \tag{43}$$

当流函数求出后,可以计算出 u,w,由式(16)可计算出 v。这里给出与图 3 所示温度动力学解相对应的大气运动的动力学场(见图 5),其中实线表示速度 v 的空间分布,由流函数算得的 u,w 由矢量线给出。由图 5 可以看出,在低层水平辐合,导致在中间有上升气流;植被差异水平分布范围越宽,u,w 的值越小,差异的水平范围越窄,u,w 的值越大。另外,v 和 u 的分布符合科里奥利力场应起的作用,表明 u,v,w 这 3 个场在动力学上是协调的。

值得注意的是,在轴中心的温度虽然低于两侧,但空气是上升的,这显然是动力学的结果。可以预料,如果空气是湿的,水汽可以在上升过程中相变,则潜热的释放,将使温度增加,那么中心附近的

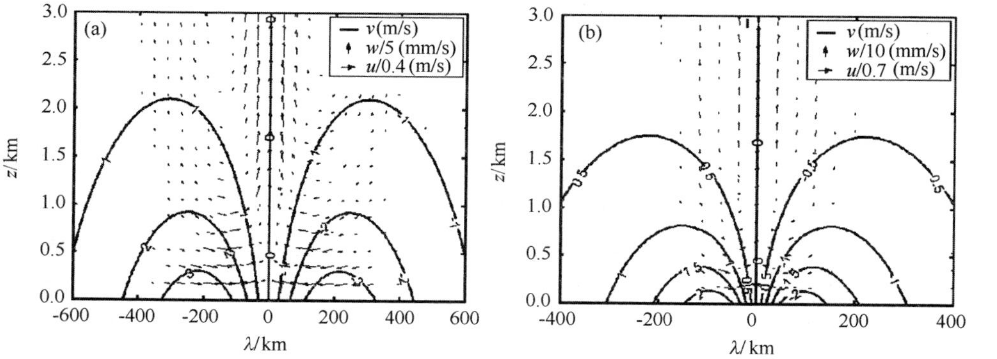

图 5 植被阻抗不同水平分布下的大气动力学场：(a) L= 150 km；(b) L= 50 km

低温将向高温转化，从而改变植被上空温度分布的格局，这是绿洲生态动力学研究中一个值得重视的问题。

6 结论

本文在考虑了行星边界层大气中的动力学过程和辐射传输过程后，将边界层的物理场与下垫面植被冠层中的热量平衡方程相耦合，从而建立一个简单的边界层大气运动和植被生态过程相耦合的理论模式，并对该耦合模式进行了解析求解。用此解析理论分析了不同的植被反照率和阻抗对热力学解和动力学解的影响，进而考虑了反照率随温度的反馈的作用。这一相互作用的方式可为进一步发展大气运动和生态过程相互作用的、更复杂的数值模拟模式提供参考。

由于陆面过程对气候的影响有两种显然的过程是重要的：一是改变地表反照率从而改变局地辐射通量和温度；二是通过蒸发和降水改变系统中热量收支。对于前者本文简单探讨了植被反照率和温度的反馈对大气运动及植被温度的影响；对于后者，通过解析求解出的温度场和垂直运动场的格局初步定性地提出潜热在绿洲生态动力学中的重要作用，它将改变绿洲上空温度分布格局，从而影响绿洲的生消。这两种过程有待于进一步的研究。

参考文献

[1] Charney J G.Dynamics of deserts and drought in the Sahel.*Quart. J. Roy. Meteor.Soc.*, 1975, 101: 193-202.

[2] Dickinson R E.Modeling evapotranspiration for three-dimensional global climate models.*Geophys.Monograph*, 1984, 29: 58-72.

[3] Sillers P J, Mintz Y.The design of a Simple Biosphere model (SIB) for use within general circulation models.*J.Atmos. Sci.*, 1986, 43: 505-531.

[4] Ji J J, Hu Y C.A simple land surface process model for use in climate study.*Acta Meteorologica Sinica*, 1989, 3: 342-351.

[5] 赵鸣，江静，苏炳凯，等.一个引入近地层的土壤-植被-大气相互作用模式.大气科学，1995, 19 (4): 405-414.

[6] Dai Y J, Zeng Q C.A land surface model (IAP94) for climate studies, Part I: Formulation and validation in off-line experiments.*Adv.Atmos.Sci.*, 1997, 14 (4): 433-460.

[7] 巢纪平，陈英仪.二维能量平衡模式中极冰-反照率的反馈对气候的影响，中国科学，1979, 12: 1198-1207.

[8] Kuo H L.On a simplified radiative-conductive heat transfer equation. *Pure & Appl.Geophys.*, 1973, 109: 1870-1876.

附录　动力学耦合模式方程的解析求解

解分成两个部分：A：一部分为齐次方程满足非齐次边界条件的解，即

$$\frac{\partial^2 T'}{\partial x^2}+\hat{a}^2\frac{\partial^2 T'}{\partial z^2}-d^2T'=0 \tag{A1}$$

及条件式(33)至式(35)；B：另一部分为非齐次方程(32)满足齐次边界条件，

$$\frac{\partial T'}{\partial z}=-\mathring{A}T' \tag{A2}$$

及式(34)、式(35)的解。

引进傅立叶变换

$$T'(x,z)=\frac{1}{\sqrt{2\pi}}\int_{-\infty}^{\infty}T'(k,z)\mathrm{e}^{ikx}\mathrm{d}k \tag{A3}$$

而

$$T'(k,z)=\frac{1}{\sqrt{2\pi}}\int_{-\infty}^{\infty}T'(x'',z)\mathrm{e}^{-ikx''}\mathrm{d}x'' \tag{A4}$$

对问题A，有

$$\frac{\mathrm{d}^2T'}{\mathrm{d}z^2}-K^2T'=0 \tag{A5}$$

式中，$K^2=(k^2+\hat{d}^2)/\hat{a}^2$。它满足条件

$$z=0,\quad \frac{\mathrm{d}T'}{\mathrm{d}z}=-\mathring{A}T'-S' \tag{A6}$$

式中

$$S'(k)=\frac{1}{\sqrt{2\pi}}\int_{-\infty}^{\infty}S'(x'')\mathrm{e}^{-ikx''}\mathrm{d}x'',$$

$$z\to\infty,\quad \frac{\mathrm{d}T'}{\mathrm{d}z}=0 \tag{A7}$$

因而解为(标以下标1)

$$T'_1(x,z)=\frac{1}{2\pi}\int_{-\infty}^{\infty}\int_{-\infty}^{\infty}S'(x'')\mathrm{e}^{-A^{\cdot}z}\frac{\mathrm{e}^{-(K-A^{\cdot})z}}{K-A^{\cdot}}\mathrm{e}^{ik(x-x'')}\mathrm{d}k\mathrm{d}x'' \tag{A8}$$

令

$$I(z'',t)=\frac{1}{\sqrt{2\pi}}\int_{-\infty}^{\infty}\mathrm{e}^{-K_z''}\mathrm{e}^{ikt}\mathrm{d}k \tag{A9}$$

$$J(z,t)=\frac{1}{\sqrt{2\pi}}\int_{-\infty}^{\infty}\int_{-\infty}^{\infty}\frac{\mathrm{e}^{-(K-A^{\cdot})z}}{K-A^{\cdot}}\mathrm{e}^{ikt}\mathrm{d}k=\int_{z}^{\infty}\mathrm{e}^{A^{\cdot}z''}I\mathrm{d}z'' \tag{A10}$$

在式(A10)中，对 $K>A^{\circ}$，z''在$[z,+\infty)$范围内积分是收敛的。注意到

$$I=\frac{1}{\sqrt{2\pi}}\int_{-\infty}^{\infty}\mathrm{e}^{-\frac{z''}{\hat{a}}\sqrt{k^2+\hat{d}^2}}\mathrm{e}^{ikt}\mathrm{d}k=\sqrt{\frac{2}{\pi}}\frac{\hat{d}z''}{\sqrt{(\hat{a}t)^2+z''^2}}K_1\left(\frac{\hat{d}}{\hat{a}}\sqrt{(\hat{a}t)^2+z''^2}\right) \tag{A11}$$

式中，$K_1(r)$为第二类修正贝赛尔函数。由此得

$$J=\sqrt{\frac{2}{\pi}}\int_{z}^{\infty}\frac{de^{A^{\cdot}z''}z''}{\sqrt{\hat{a}^{2}(x-x'')^{2}+z''^{2}}}K_{1}\left(\frac{\hat{d}}{\hat{a}}\sqrt{\hat{a}^{2}(x-x'')^{2}+z''^{2}}\right)\mathrm{d}z'' \tag{A12}$$

最后积分为

$$T'_{1}(x,z)=\frac{\hat{d}}{\pi}\int_{-\infty}^{\infty}\int_{z}^{\infty}\frac{S'(x'')z''e^{-A^{\circ}(z-z'')}}{\sqrt{\hat{a}^{2}(x-x'')^{2}+z''^{2}}}K_{1}\left(\frac{\hat{d}}{\hat{a}}\sqrt{\hat{a}^{2}(x-x'')^{2}+z''^{2}}\right)\mathrm{d}z''\mathrm{d}x'' \tag{A13}$$

问题 B,应用傅立叶变换后,方程(31)为

$$\frac{\partial^{2}T'}{\partial z^{2}}-K^{2}T'=\widetilde{\Omega}'(k,z) \tag{A14}$$

其中,

$$\widetilde{\Omega}'(k,z)=\frac{1}{\hat{a}^{2}\sqrt{2\pi}}\int\Omega'(x'',z)\mathrm{e}^{-\mathrm{i}kx}\mathrm{d}x''$$

条件为

$$z=0,\quad \frac{\partial T'}{\partial z}+A^{\cdot}\ T'=0 \tag{A15}$$

$$z\to\infty,\quad \frac{\partial T'}{\partial z}=0 \tag{A16}$$

应用格林函数求解,格林函数为

$$0\leqslant z\leqslant\xi,\quad G(z,\xi)=-\frac{1}{2K}\left[\mathrm{e}^{-K(\xi-z)}+\left(\frac{K+A^{\cdot}}{K-A^{\cdot}}\right)\mathrm{e}^{-K(z+\xi)}\right] \tag{A17a}$$

$$\xi<z\leqslant\infty,\quad G(z,\xi)=-\frac{1}{2K}\left[\mathrm{e}^{-K(z-\xi)}+\left(\frac{K+A^{\cdot}}{K-A^{\cdot}}\right)\mathrm{e}^{-K(z+\xi)}\right] \tag{A17b}$$

由此,得到这一部分的解为(标以下标 2)

$$\begin{aligned}T'_{2}(x,z)&=\frac{1}{2\pi\hat{a}^{2}}\int_{-\infty}^{\infty}\int_{-\infty}^{\infty}\int_{0}^{\infty}\Omega'(x'',\xi)G(z,\xi)\mathrm{e}^{\mathrm{i}k(x-x'')}\mathrm{d}\xi\mathrm{d}k\mathrm{d}x''\\&=-\frac{1}{4\pi\hat{a}^{2}}\int_{-\infty}^{\infty}\int_{-\infty}^{\infty}\left[\int_{0}^{z}\Omega'(x'',\xi)\mathrm{e}^{-K(z-\xi)}\mathrm{d}\xi\right]\frac{1}{K}\mathrm{e}^{\mathrm{i}k(x-x'')}\mathrm{d}k\mathrm{d}x''\\&\quad-\frac{1}{4\pi\hat{a}^{2}}\int_{-\infty}^{\infty}\int_{-\infty}^{\infty}\left[\int_{z}^{\infty}\Omega'(x'',\xi)\mathrm{e}^{-K(\xi-z)}\mathrm{d}\xi\right]\frac{1}{K}\mathrm{e}^{\mathrm{i}k(x-x'')}\mathrm{d}k\mathrm{d}x''\\&\quad-\frac{1}{4\pi\hat{a}^{2}}\int_{-\infty}^{\infty}\int_{-\infty}^{\infty}\left[\int_{0}^{\infty}\Omega'(x'',\xi)\mathrm{e}^{-K(z+\xi)}\mathrm{d}\xi\right]\left(\frac{2}{K-A^{\cdot}}-\frac{1}{K}\right)\mathrm{e}^{\mathrm{i}k(x-x'')}\mathrm{d}k\mathrm{d}x''\end{aligned} \tag{A18}$$

应用与问题 A 相同的方法,令 $t=x-x''$,容易求得:

$$\begin{aligned}T'_{2}(x,z)&=-\frac{1}{2\hat{a}^{2}\sqrt{2\pi}}\int_{-\infty}^{\infty}\int_{0}^{z}\int_{z-\xi}^{\infty}I\mathrm{d}z''\Omega'(x'',\xi)\mathrm{d}\xi\mathrm{d}x''-\frac{1}{2\hat{a}^{2}\sqrt{2\pi}}\int_{-\infty}^{\infty}\int_{z}^{\infty}\int_{\xi-z}^{\infty}I\mathrm{d}z''\Omega'(x'',\xi)\mathrm{d}\xi\mathrm{d}x''\\&\quad-\frac{1}{\hat{a}^{2}\sqrt{2\pi}}\int_{-\infty}^{\infty}\int_{0}^{\infty}\int_{\xi+z}^{\infty}\mathrm{e}^{Az''}I\mathrm{d}z''\Omega'(x'',\xi)\mathrm{e}^{-A^{\cdot}(\xi+z)}\mathrm{d}\xi\mathrm{d}x''+\frac{1}{2\hat{a}^{2}\sqrt{2\pi}}\int_{-\infty}^{\infty}\int_{0}^{\infty}\int_{\xi+z}^{\infty}I\mathrm{d}z''\Omega'(x'',\xi)\mathrm{d}\xi\mathrm{d}x''\\&=-\frac{1}{2\hat{a}^{2}\sqrt{2\pi}}\int_{-\infty}^{\infty}\int_{0}^{z}\int_{z-\xi}^{z+\xi}I\mathrm{d}z''\Omega'(x'',\xi)\mathrm{d}\xi\mathrm{d}x''-\frac{1}{2\hat{a}^{2}\sqrt{2\pi}}\int_{-\infty}^{\infty}\int_{z}^{\infty}\int_{\xi-z}^{\xi+z}I\mathrm{d}z''\Omega'(x'',\xi)\mathrm{d}\xi\mathrm{d}x''\\&\quad-\frac{1}{\hat{a}^{2}\sqrt{2\pi}}\int_{-\infty}^{\infty}\int_{0}^{\infty}J(\xi+z,t)\Omega'(x'',\xi)\mathrm{e}^{-A^{\cdot}(\xi+z)}\mathrm{d}\xi\mathrm{d}x''\end{aligned} \tag{A19}$$

把式(A11)、式(A12)中 I,J 的值带入，可求得总的表达式为

$$T'(x,z)=\frac{\hat{d}^2}{\hat{a}\pi}\int_{-\infty}^{\infty}S'(x'')\mathrm{e}^{-A\cdot z}\mathrm{d}x''\int_{z}^{\infty}\frac{\hat{a}z''\mathrm{e}^{A\cdot z''}}{\hat{d}\sqrt{\hat{a}^2t^2+z''^2}}K_1\left(\frac{\hat{d}^2}{\hat{a}}\sqrt{\hat{a}^2t^2+z''^2}\right)\mathrm{d}z''$$

$$-\frac{\hat{d}^2}{2\hat{a}^3\pi}\int_{-\infty}^{\infty}\mathrm{d}x''\int_{0}^{z}\Omega'(x'',\xi)\mathrm{d}\xi\int_{z-\xi}^{\xi+z}\frac{\hat{a}z''}{\hat{d}\sqrt{\hat{a}^2t^2+z''^2}}K_1\left(\frac{\hat{d}}{\hat{a}}\sqrt{\hat{a}^2t^2+z''^2}\right)\mathrm{d}z''$$

$$-\frac{\hat{d}^2}{2\hat{a}^3\pi}\int_{-\infty}^{\infty}\mathrm{d}x''\int_{z}^{\infty}\Omega'(x'',\xi)\mathrm{d}\xi\int_{\xi-z}^{\xi+z}\frac{\hat{a}z''}{\hat{d}\sqrt{\hat{a}^2t^2+z''^2}}K_1\left(\frac{\hat{d}}{\hat{a}}\sqrt{\hat{a}^2t^2+z''^2}\right)\mathrm{d}z''$$

$$-\frac{\hat{d}^2}{\hat{a}^3\pi}\int_{-\infty}^{\infty}\mathrm{d}x''\int_{0}^{\infty}\Omega'(x'',\xi)\mathrm{e}^{A\cdot(\xi+z)}\mathrm{d}\xi\int_{\xi+z}^{\infty}\frac{\hat{a}\mathrm{e}^{A\cdot z''}z''}{\hat{d}\sqrt{\hat{a}^2t^2+z''^2}}K_1\left(\frac{\hat{d}}{\hat{a}}\sqrt{\hat{a}^2t^2+z''^2}\right)\mathrm{d}z'' \tag{A20}$$

考虑到

$$\int_{z}^{\infty}\frac{\hat{a}z''}{\hat{d}\sqrt{\hat{a}^2t^2+z''^2}}K_1\left(\frac{\hat{d}}{\hat{a}}\sqrt{\hat{a}^2t^2+z''^2}\right)\mathrm{d}z''=\frac{\hat{a}^2}{\hat{d}^2}K_0\left(\frac{\hat{d}}{\hat{a}}\sqrt{\hat{a}^2t^2+z''^2}\right) \tag{A21}$$

$$\int_{z}^{\infty}\frac{\hat{a}\mathrm{e}^{A\cdot z''}z''}{\hat{d}\sqrt{\hat{a}^2t^2+z''^2}}K_1\left(\frac{\hat{d}}{\hat{a}}\sqrt{\hat{a}^2t^2+z''^2}\right)\mathrm{d}z''$$

$$=\frac{\hat{a}^2\mathrm{e}^{A\cdot z}}{\hat{d}^2}K_0\left(\frac{\hat{d}}{\hat{a}}\sqrt{\hat{a}^2t^2+z^2}\right)+\frac{\hat{a}^2A\cdot}{\hat{d}^2}\int_{z}^{\infty}\mathrm{e}^{A\cdot z''}K_0\left(\frac{\hat{d}}{\hat{a}}\sqrt{\hat{a}^2t^2+z''^2}\right)\mathrm{d}z'' \tag{A22}$$

式中 $K_0(r)$ 为第二类修正贝赛尔函数。

式(A20)可进一步简化为

$$T'(x,z)=\frac{\hat{a}}{\pi}\int_{-\infty}^{+\infty}S'(x'')K_0\left(\frac{\hat{d}}{\hat{a}}\sqrt{\hat{a}^2(x''-x)^2+z^2}\right)\mathrm{d}x''$$

$$+\frac{A\cdot\hat{a}}{\pi}\int_{-\infty}^{+\infty}\mathrm{d}x''\int_{z}^{\infty}\left[S'(x'')-\int_{0}^{z''-z}\Omega'(x'',\xi)\mathrm{e}^{-A\circ\xi}\mathrm{d}\xi\right]\mathrm{e}^{A\circ(z''-z)}K_0\left(\frac{\hat{d}}{\hat{a}}\sqrt{\hat{a}(x''-x)^2+z''^2}\right)\mathrm{d}z''$$

$$-\frac{1}{2\pi\hat{a}^2}\int_{-\infty}^{+\infty}\mathrm{d}x''\int_{-z}^{\infty}\Omega'(x'',z''+z)K_0\left(\frac{\hat{d}}{\hat{a}}\sqrt{\hat{a}^2(x''-x)^2+z''^2}\right)\mathrm{d}z''$$

$$-\frac{1}{2\pi\hat{a}^2}\int_{-\infty}^{+\infty}\mathrm{d}x''\int_{z}^{\infty}\Omega'(x'',z''-z)K_0\left(\frac{\hat{d}}{\hat{a}}\sqrt{\hat{a}^2(x''-x)^2+z''^2}\right)\mathrm{d}z'' \tag{A23}$$

热力学和动力学耦合的二维能量平衡模式中荒漠化气候的演变*

巢纪平[1,2] 李耀锟[3,4]

(1. 国家海洋环境预报中心,北京 100081;2. 国家海洋局第一海洋研究所, 青岛 266061;3. 中国科学院大气物理研究所大气科学和地球流体力学数值模拟国家重点实验室, 北京 100029;4. 中国科学院研究生院,北京 100049)

摘要:利用考虑了辐射能量传播的热力学过程和边界层运动的动力学过程相耦合的全球纬向平均的二维能量平衡模式, 半解析地研究了反照率的变化和对于荒漠演化的关系。先根据目前反照率的纬向分布, 模拟出包括温度、纬向风、经向风和垂直运动在内的气候状态, 然后分析以反照率的大小和分布表征的荒漠带的分布, 研究气候对不同反照率的响应, 以此来分析荒漠气候的演变。有意思的是, 这一简单的气候模式可模拟出三圈经圈环流; 另一个有意思的结果是, 当荒漠和植被交界处的温度值高到某一临界值后, 气候会对应着两个平衡状态, 一个是反照率增加荒漠带南移(北半球), 对应着现在的气候状态变化趋势, 另一个则是当反照率增加, 植被带北扩, 荒漠带变窄。这种双平衡气候态是否存在需要用具备更多物理过程的模式来验证。

关键词:二维能量平衡模式;荒漠化;双平衡态

陆面植被覆盖区和荒漠区的反照率差异很大,它们覆盖面积的相对大小直接影响地-气之间局地的或总的辐射能交换和平衡,因此反照率的大小和分布是影响区域和全球气候的一个重要物理因子[1]。近几十年来,我国北方地区的干旱有加剧的趋势[2],干旱和半干旱带正在向东和向南扩展[3]。因此荒漠化的成因和演变趋势已成为气候变化中一个重要的研究问题。

最早用解析方法研究沙漠化形成的是 Charney[4],他指出下垫面反照率的增加会导致沙漠边缘降水减少,促使沙漠扩张。在这之后,有不少学者探讨过地表属性的变化对气候变化的影响,例如 Dickinson[5]考虑了陆面生态过程对气候的影响。同时,许多数值模式也被用以研究这一复杂的过程,国内外已有许多代表性的陆面模式,诸如 Sellers 等[6]提出的 SiB 方案和 Dai 等[7]发展的 IAP94 模式。近 10 年来,陆面模式有了新的发展,如 Sellers 等[8]在 SiB 基础上改进得到的 SiB2 模式,Dai 等[9]发展的新一代通用的陆面模式 CLM(Common Land Model)以及 Oleson 等[10]发展的 NCAR CLM 模式。虽然如今用于模拟陆面过程影响气候变化的模式已有很大进展,对陆面过程的模拟能力也在不断提高,但是陆面模式也仍存在一定的不足[11]。其中困难之一是由于气候模式包括了多种物理过程,且模式又是高度非线性的,因此往往难以识别陆面过程是如何影响气候的。即使在气候和陆面模式高

* 中国科学:地球科学,2010,40(8):1060-1067.

度发展的今天，通过发展一些有反馈过程的简单模式，在理论上分析若干下垫面因子的作用，会对陆-气之间能量平衡的改变如何影响气候变化有更为深刻的认识。因此，这类研究仍有它的必要性和意义。

巢纪平和陈英仪[12]曾发展了一个纬圈平均后的二维（纬度和高度）能量平衡模式。它考虑了垂直方向辐射能的传输过程，通过引入极冰反照率的影响，用解析解讨论了小冰期气候出现的可能性。本文试图在这个能量平衡模式的基础上，通过引入与垂直运动成正比的位能，将它和行星边界层运动联接起来，构成一个新的能量平衡模式，并应用这一发展后的理论模型来讨论荒漠化气候演变的问题。

需要指出的是，Wu 等[13]、巢纪平和周德刚[14]曾在 f 平面上分析过绿洲和荒漠区尺度对比对局地气候的影响。但本文是在纬向平均后的全球尺度上分析反照率分布的改变对气候演变的影响，目的和处理的方法均不与前两文相同。

1　模型的建立

巢纪平和陈英仪[12]在考虑了辐射能量和湍流热量输送过程的基础上，考虑垂直运动对位能的影响，热力学方程及辐射能传输方程分别为

$$\bar{\rho}c_p\widetilde{N}^2\frac{\bar{T}}{g}w=\frac{K}{a^2}\frac{\partial}{\partial x}(1-x^2)\frac{\partial T}{\partial x}+\frac{\partial}{\partial z}\left(k_t\frac{\partial T}{\partial z}\right)+\sum_j\alpha'_j\quad\rho_c(A_j+B_j-2E_j)+\alpha''\rho_cQ \tag{1}$$

$$\frac{\partial A_j}{\partial z}=\alpha'_j\quad\rho_c(A_j-E_j) \tag{2}$$

$$\frac{\partial B_j}{\partial z}=\alpha'_j\quad\rho_c(E_j-B_j) \tag{3}$$

$$\frac{\partial Q}{\partial z}=\alpha''_j\quad\rho_cQ \tag{4}$$

式中，T 为空气温度；a 为地球半径；K 和 k_t 分别为水平和垂直湍流交换系数；ρ_c 是吸收介质的密度；A_j 和 B_j 分别是在波长为 $\Delta\lambda_j$ 区间内向下和向上的长波辐射通量，E_j 为在这个波段内的黑体辐射；Q 为太阳辐射通量；$\widetilde{N}=\left(\frac{g}{\theta}\frac{\mathrm{d}\bar{\theta}}{\mathrm{d}z}\right)^{\frac{1}{2}}$ 为 Brunt-Väsälä 频率；w 为垂直速度；α'_j 和 α''_j 分别为对波长为 λ_j 的长波辐射吸收系数以及对太阳辐射的吸收系数；z 是垂直坐标；$x=\sin\theta$，θ 是纬度；其他符号的定义同参考文献[12]。

在球面上大气行星边界层的运动方程可以这样写出，取柯利奥利力与垂直湍流交换相平衡，即

$$-2\Omega\sin\theta(\bar{\rho}v)=\frac{\partial}{\partial z}\left(\mu\frac{\partial u}{\partial z}\right) \tag{5}$$

纬向风是地转的，即有

$$2\Omega\sin\theta(\bar{\rho}u)=-\frac{\partial p}{a\partial\theta} \tag{6}$$

运动是静力平衡的

$$0=-\frac{\partial p}{\partial z}-\rho g \tag{7}$$

连续方程为

$$\frac{1}{a\cos\theta}\frac{\partial(\cos\theta\bar{\rho}v)}{\partial\theta}+\frac{\partial\bar{\rho}w}{\partial z}=0 \tag{8}$$

由连续方程引入流函数

$$\bar{\rho}v=-\frac{1}{\cos\theta}\frac{\partial\psi}{\partial z},\quad \bar{\rho}w=\frac{1}{a\cos\theta}\frac{\partial\psi}{\partial\theta} \tag{9}$$

假定在推导中略去与柯氏参数变化的项,由方程式(5)至式(9)可以得到

$$\bar{\rho}w=-\frac{\mu g}{a^2\bar{f}^2\bar{T}}\frac{\partial}{\partial x}(1-x^2)\frac{\partial T}{\partial x} \tag{10}$$

将式(10)代入到方程(1)中,并利用 Kuo[15] 提出的长波辐射传输的简化方案来计算长波辐射通量(参见文献[12]),最后得到

$$D\frac{\partial}{\partial x}(1-x^2)\frac{\partial E}{\partial x}+\frac{\partial}{\partial\xi}(k_t+k_r)\frac{\partial E}{\partial\xi}-N^2E=-\widetilde{S}\xi_0Q+\widetilde{S}\overline{Q}_0\left[r\left(1+\frac{\alpha''}{\alpha_w}\right)-\mathrm{e}^{-\xi_0}\int_0^1\Gamma S(x)\,\mathrm{d}x\right] \tag{11}$$

其中

$$\frac{\partial T}{\partial z}\approx\frac{1}{4\sigma\bar{T}^3}\frac{\partial E}{\partial z},\quad \xi=\frac{\alpha''}{\alpha_s\xi_0}\int_z^\infty\alpha_s\rho_c\,\mathrm{d}z$$

$$\xi_0=\frac{\alpha''}{\alpha_s}\int_0^\infty\alpha_s\rho_c\,\mathrm{d}z$$

$$D=\frac{\xi_0^2K}{(\alpha''\rho_c)^2a^2}+\frac{\xi_0^2K_w}{(\alpha''\rho_c)^2a^2},\quad k_r=\frac{8r\sigma\bar{T}^3}{\alpha_s\rho_c}$$

$$\widetilde{S}=\frac{4\xi_0\sigma\bar{T}^3}{\alpha''\rho_c},\quad N^2=\frac{8(1-r)\alpha_w\rho_c\xi_0^2\sigma\bar{T}^3}{(\alpha''\rho_c)}$$

式中,ξ 为光学厚度;ξ_0 为整层大气的光学厚度;Γ 为地表反照率;与 k_t 相比,k_r 可以称为辐射交换系数,而 N^2 则可称为牛顿辐射冷却系数;r 表示在强吸收区中物质的辐射能量占总辐射能量的部分;σ 是史蒂芬-波尔兹曼常数;$\bar{T}$ 为地球的平均温度,α_s 和 α_w 分别为强吸收区和弱吸收区的吸收系数;Q_0 为大气上界的太阳辐射通量,定义:

$$Q_0(x)=\overline{Q}_0S(x),\quad \int_0^1S(x)\,\mathrm{d}x=1 \tag{12}$$

其中,$\overline{Q}_0$ 为大气上界太阳辐射通量的平均值。

由方程(11)可以看出,加入和垂直运动有关的动力过程后,相当于加大了水平湍流的作用。

在另一方面,由行星边界层方程容易求得流函数和温度场的关系为

$$\frac{(1-x^2)}{a^2}\frac{\partial^2\psi}{\partial x^2}+\left(\frac{\alpha''\rho_c}{\xi_0}\right)^2\frac{\partial^2\psi}{\partial\xi^2}=-\left(\frac{\mu g}{\bar{f}^2\bar{T}}\right)\frac{1-x^2}{a}\left[\frac{1}{a^2}\frac{\partial}{\partial x}\left(\frac{\partial}{\partial x}(1-x^2)\frac{\partial T}{\partial x}\right)+\left(\frac{\alpha''\rho_c}{\xi_0}\right)^2\frac{\partial}{\partial x}\left(\frac{\partial^2T}{\partial\xi^2}\right)\right] \tag{13}$$

2 解法

方程(11)是写在球面上的,可以利用勒让德多项式求得它的半解析解,设解的形式为

$$E(x,\xi)=E^{(0)}(\xi)P_0(x)+E^{(2)}(\xi)P_2(x)+\cdots \tag{14}$$

代入方程(11)中,不难求得各项系数随高度 ξ 的变化,在计算温度时选取的是整层大气,即 ξ 从 1 取

到 0。由此可以计算得到温度分布。计算得到温度分布后，再利用差分迭代方法来求解方程(13)，即能得到流函数。边界条件取为在边界处流函数的值均为零。由于主要考虑对流层中的流场分布，因此在计算中，上边界取在对流层顶大约 10 km 高度处。计算得到流函数后，就可以根据式(9)求得流场分布。

为了研究地表反照率分布对气候的影响，取

$$\Gamma(x,x_x,x_v)=\begin{cases}\Gamma_1, & x>x_s,\\ \Gamma_2, & x_s>x>x_v,\\ \Gamma_3, & x<x_v\end{cases} \tag{15}$$

即将北半球划分为 3 个带：北面是冰雪带；中间是荒漠带；南边是植被带。式(15)中，Γ_1 是冰雪区的反照率；Γ_2 是荒漠区的反照率；Γ_3 是植被区的反照率，令 $x_s=\sin\theta_s$，$x_v=\sin\theta_v$。其中，θ_s 和 θ_v 分别是雪线和荒漠交界的纬度、植被带和荒漠交界的纬度。在下面的讨论中，使用 x_s 和 x_v 作为雪线纬度和植被纬度。

参照 Budyko[16] 选取 $\Gamma_1=0.62$；植被的反照率较低，不同的植被反照率也不尽相同，根据刘飞和巢纪平[17]，一般情况下植被反照率取 $\Gamma_3=0.1$，并对不同的植被反照率进行敏感性试验。刘辉志等[18]对半干旱区不同下垫面地表反照率变化特征进行了研究，半干旱区的反照率大约在 0.25。在本文的计算中，Γ_2 的大小可以调整，用以表征荒漠化的程度。

在求解时用了展开式(14)的低阶模。为此先对不同截断模的展开精度进行比较，图 1 是不同截断模时算得的荒漠带的南界纬度，可以看到多项式从 P_4 到 P_{16}，在 $\Gamma_2<0.4$ 时解的变化很小。现将解展开到 P_{16} 用以作本文的分析和讨论。

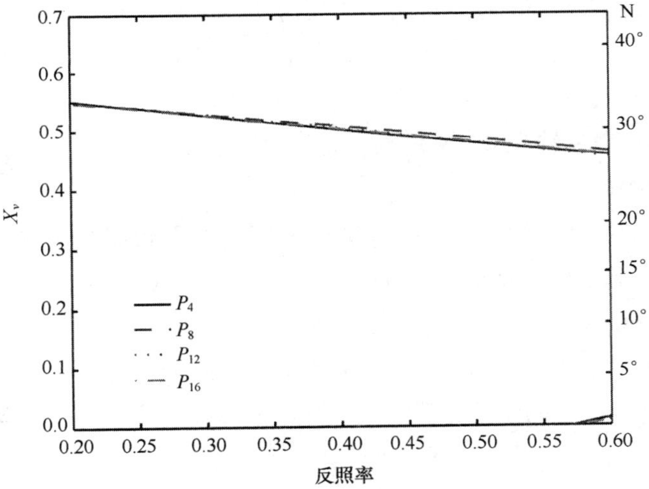

图 1　不同截断模时荒漠带的南界纬度

3　模式计算结果及分析

考虑一种简单的情况，假定雪线的纬度是固定的，维持在现在的状态上，即 $x_s=0.951\ 1$。利用上述确定的参数，先计算求得了温度分布，然后通过式(13)可以计算得到流函数分布。这样得到了一个气候态，在这个气候态下，全球平均温度为 15℃，植被和荒漠的交界纬度在 30°N 附近，这个模拟的气候与目前观测的气候场在结构上是较为一致的。

图 2 给出温度场的分布,图 3 给出了观测的温度场。为了便于比较,已将观测资料中气压层次用光学厚度来表达。由图 2 和图 3 可以看出,计算比较好地模拟出了实际的温度场的结构。但是在对流层高层 10 km 以上,模式与实际的差别比较大。因此,计算流函数时将上边界取在对流层顶约 10 km高度处,只利用到 10 km 以下的温度场来计算,即主要研究对流层中的流场分布。

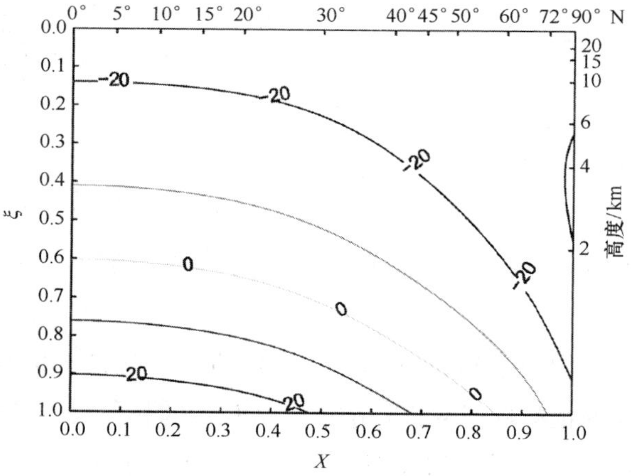

图 2　计算所得的气候态的温度分布

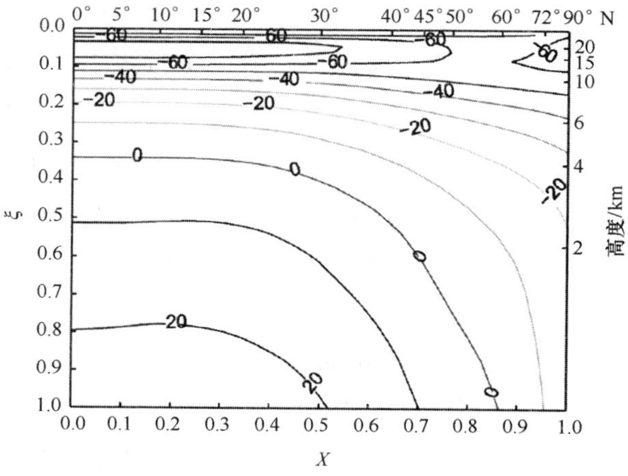

图 3　观测的平均温度分布

图 4 给出了由图 3 计算得到的气候态的流函数场、水平速度场和垂直速度场。从图中可以看出,平均经圈环流在 10°N 和 35°N 附近为上升气流,25°N 和 65°N 附近为下沉气流,上升及下沉速度大约为每秒毫米量级。在上升下沉气流的中间地面依次为北风、南风和北风,高空依次为南风、北风和南风,速度大约为 0.2 m/s。由模拟的流函数的分布,可以看到模拟得到了三圈环流,但是中间的费雷尔环流比较弱,垂直位置明显偏低,由于太弱,且标度也不一样,在垂直运动图上无法标出。

以纬圈风表征的西风急流约在 45°N 的对流层顶,强度约为 15 m/s,比实际观测值小,这可能与计算得到的温度梯度较小有关。

以上的计算表明模式虽然在对流层高层的误差较大,并且模拟得到的费雷尔环流强度偏弱,位置偏低,但是考虑到这是一个简单的模式,不可能再现出真实的气候状态。而且,能够通过这一简单的模式模拟得到和实际一致的温度分布,能够计算得到三圈环流,这说明模式中包括了实际气候形成的重要因子。因此,下面就以上面得到的结论作为基准的气候状态,通过改变反照率的大小来计算荒漠带的变迁,以及由此给气候带来的影响。

图 5 给出了植被纬度随着不同的植被荒漠交界温度 T_v 的变化曲线。

若取 $T_v = 297$ K,从图 5 中可以看到这是一个双平衡态解。北支解随着 Γ_2 的增加,植被纬度会向南缓慢退却,即荒漠带变宽;而南支解随着 Γ_2 的增加,植被纬度北移,即荒漠带变窄,植被带变宽。当 Γ_2 达到一定的数值后,两支解合一,Γ_2 超过这一临界值后无解。无解,在这里应理解成在能量平衡下,不存在定常的平衡态气候,即并不排斥可以出现随时间变化的变动型气候。

从图 5 可以看到,当植被和荒漠交界处的温度不断升高时,即在全球温度升高的过程中,特别是低纬温度增暖,将有利于植被带北移,荒漠带变窄,这会使全球平均气温向好的方向变迁。当然这只是从反照率的改变分析,若在模式中考虑其他过程,如水分循环过程,气候态将如何变迁,尚需进一步研究。

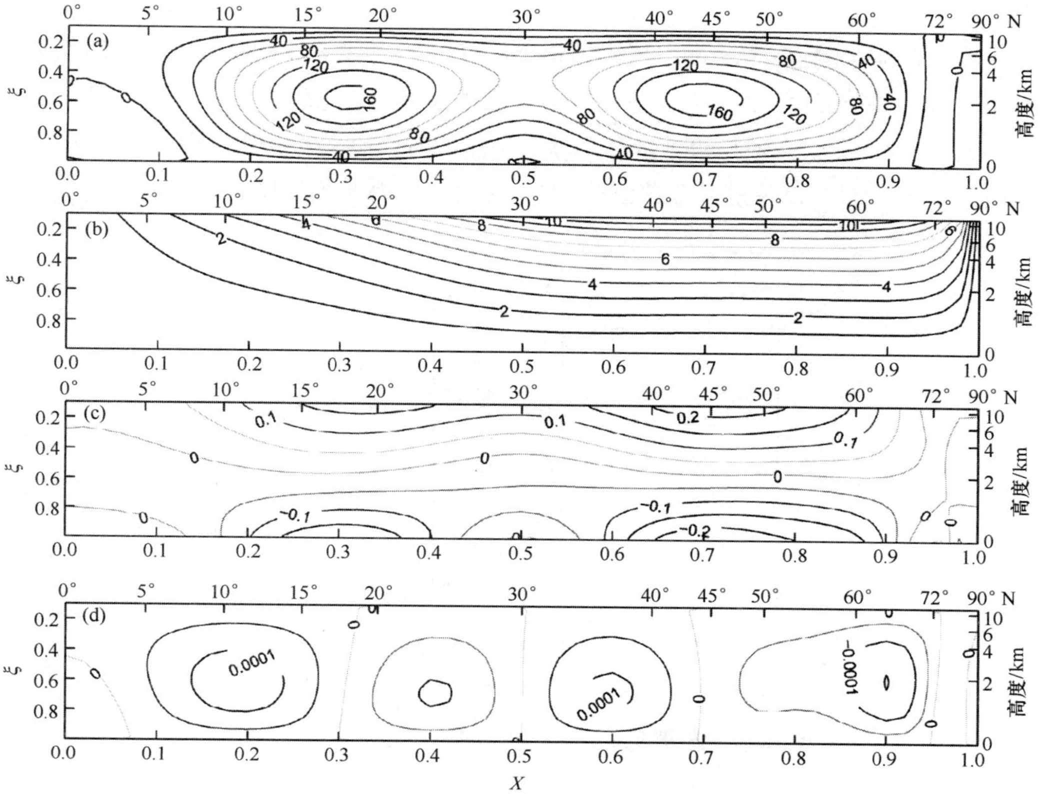

图 4　给定的气候态

(a) 流函数(kg/(m·s));(b) 纬向风速(m/s);(c) 经向风速(m/s);(d) 垂直速度(m/s)

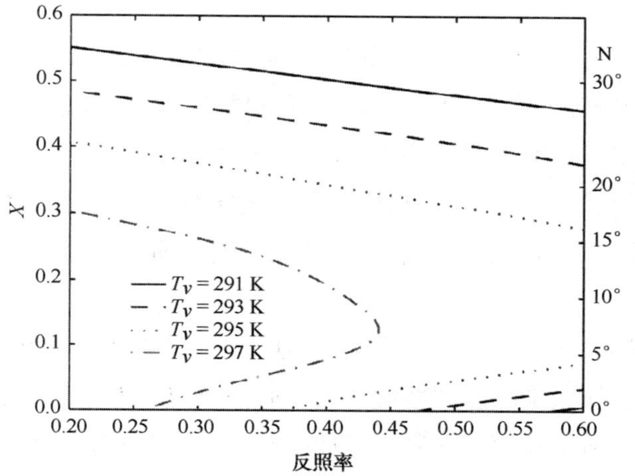

图 5　不同荒漠植被交界温度下植被纬度的变化

由图 5 也可以看到,T_v 的取值变小时,出现分岔解的临界 Γ_2 值增大。例如当取 $T_v = 297$ K 时,两支解合一的反照率将达到 0.47 左右;当取 $T_v = 293$ K 时,两解合一的临界 Γ_2 值超过 0.6,即达到甚至超过冰面的反照率,这显然是难以接受的。因此在一般条件下只存在北支解,这是目前共识的气候变化趋势,即反照率增大,荒漠带变宽。而南支解似乎更适合人类的生活,但它的存在目前难以确定,需进一步研究。

为了验证其他参数对于解的影响，针对不同的 Γ_3 和 Γ_1 进行了敏感性试验。计算结果表明，Γ_3 和 Γ_1 对于解的影响是不大的，即解对于这两个参数的选取是不敏感的。

图 6 给出了在 $T_v=291$ K 时植被纬度和全球平均温度随反照率的变化。图中横坐标是荒漠的反照率，左图纵坐标是植被带纬度及其正弦值；右图纵坐标是全球平均温度。

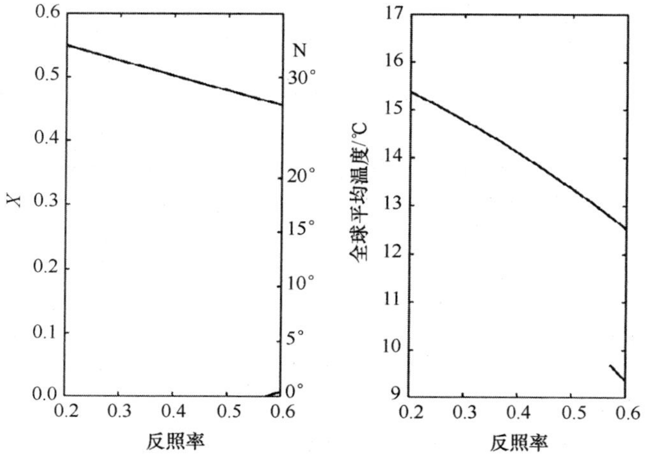

图 6 荒漠植被交界纬度 x_v 和全球平均温度随 Γ_2 的变化

由图 6 可以看出，当反照率增加时，全球平均温度会下降，这是因为反照率的增加会反射更多的太阳辐射，使得到达地面的太阳辐射减少，从而导致全球平均温度的降低。从图 6 还可以看出，随着荒漠反照率从 0.2 增加到 0.6，荒漠的南边界大约向南推进了 5 个纬度，全球平均温度大约会下降 2.5℃。

图 7 给出不同反照率分布下地面温度的距平分布。在荒漠的反照率增加后，中高纬度温度将会降低，低纬度温度将有所升高，大约以 20°N 分界。可见尽管全球平均温度在降低，中高纬度和低纬度的温度对比却在增加，即斜压性加大，有利于中纬度扰动的发展。

下面给出了在不同的荒漠反照率下气候和给定的气候态的距平场，依次来分析反照率改变之后气候的变迁。下面的讨论均是相对于图 4 的变化情况。图 8 和图 9 分别是 $\Gamma_2=0.2$ 和 $\Gamma_2=0.5$ 时的气候相对于图 4 的变化。

从图 8 和图 9 可以看到，随着反照率的增加，平均经圈环流的强度整体上都有所加强。这与图 7 得到的结论是一致的，充分说明了高纬度和低纬度之间温度对比的增加将有利于大气的斜压性的增强，这是有利于平均经圈环流增强的。反照率增加的越大，平均经圈环流强度增加越大，这在运动场上也可以得到明显体现。西风环流的强度也在增加，这也是容易理解的结论。

不同反照率分布下地表经向风的分布如图 10 所示。在荒漠的反照率增加后，地表的经向风则表现为随着反照率的增加将导致原有的北风和南风加强，反照率增加越多，风速增加越大。北风和南风整体的位置也略有南移。在南北风的幅合区，风速增加将会导致垂直速度的增加，这在图 8 和图 9 中垂直速度场的变化上也可以看出来，上升运动的增加会为降雨提供条件，而在南北风的辐散区，将会不利于降水。从整体来说，幅合区和辐散区的位置都是随着反照率的增加向南移动的，因此在其他条件不变的情况下，降水带的位置也应该会随着反照率的增加逐步向南移动。

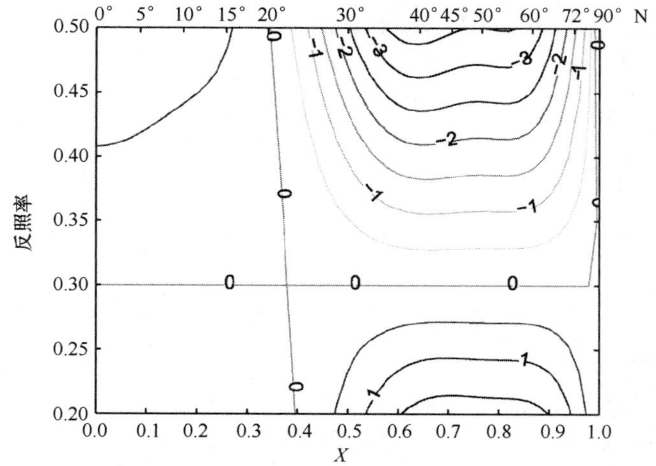

图 7　不同反照率下地面温度相对于 $\Gamma_2=0.3$ 时地面温度的距平

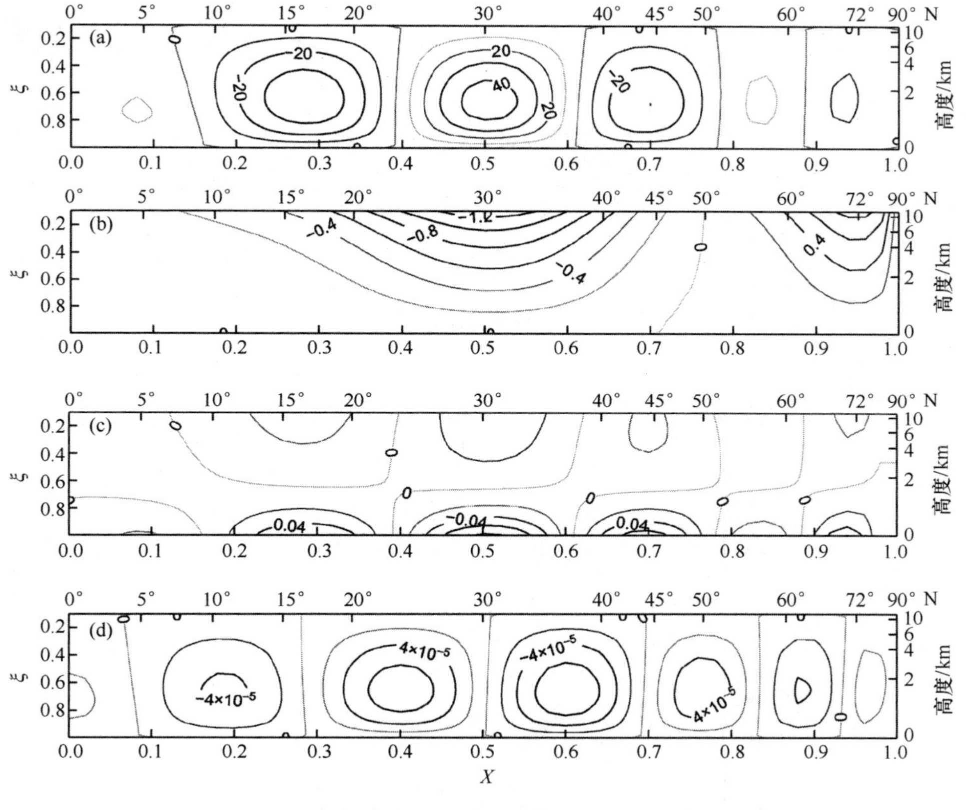

图 8　$\Gamma_2=0.2$ 时与给定气候场的距平场

说明同图 4

从图 10 也可以看到，随着反照率的增加，气候系统的变化并不是非常显著的，环流的位置和强度变化均不大。但是当 $\Gamma_2>0.45$ 以后，气候变化的幅度将变大。

计算中也考虑了 x_s 和 x_v 均可以变化的情况，参数的选择与前面类似。通过计算可以看出（图略），结果与固定 x_s 的情况下相比较，并没有太大的差异。气候同样会出现与前面相似的两个平衡

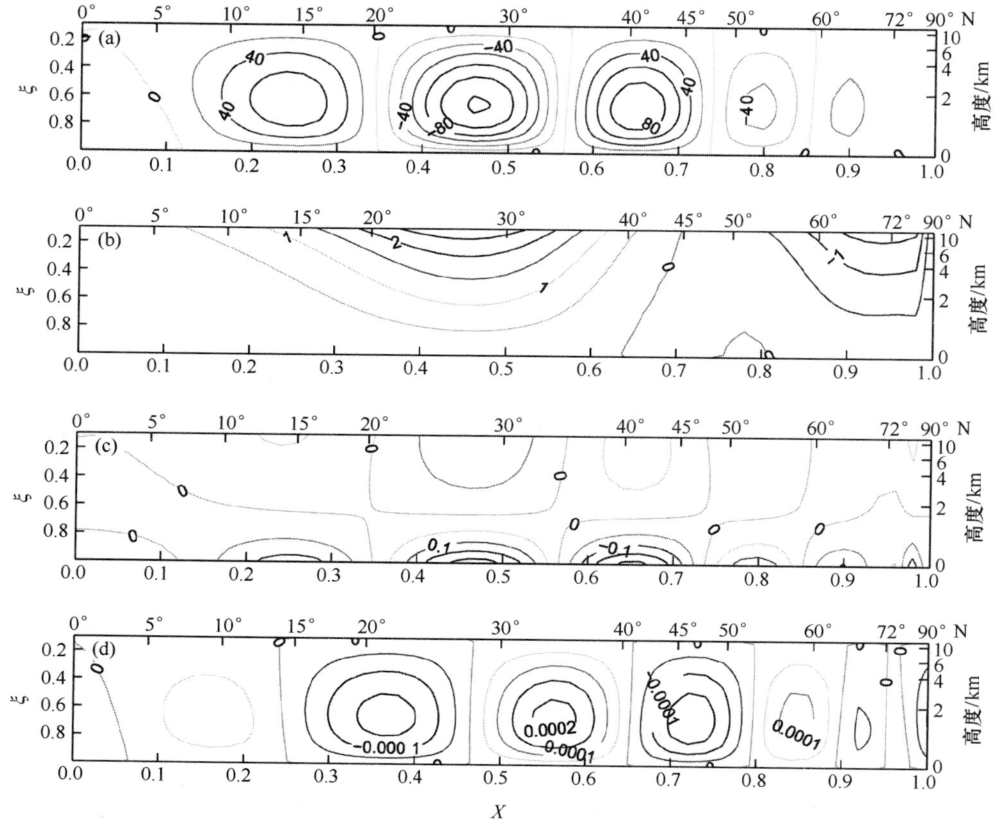

图 9　$\Gamma_2=0.5$ 时与给定气候场的距平场

说明同图 4

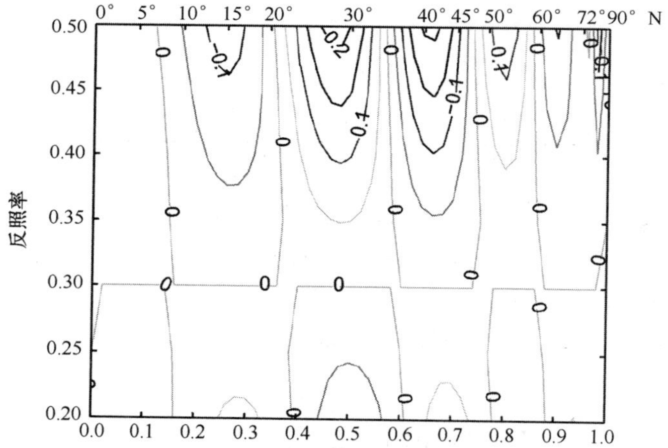

图 10　不同反照率下地面 V 风场相对于 $\Gamma_2=0.3$ 时地面 V 风场的距平

态。另外随着荒漠反照率的增加，雪线纬度和植被荒漠交界处纬度都会向南移动，全球平均温度也会降低，中纬度将会降温，低纬度将会增温。

4　结论

本文设计了一个考虑热力学能量传输和边界层动力学过程的全球纬向平均的二维能量平衡模式,简化了反照率的纬向分布,模拟出的气候状态与实际情形相比,在环流型上是相似的;并且通过改变荒漠地区反照率的大小,计算得到了荒漠带演化和反照率变化之间的联系,据此分析了气候对不同反照率的响应。本文设计的模式虽然较为简单,但是根据计算,模式能够模拟得到实际中的三圈环流,只是费雷尔环流的强度偏弱,这似乎是由于没有引入平流项造成的;当荒漠和植被交界处的温度值大于某一临界值后,存在着两个不同的气候状态,一个对应着荒漠将随着反照率的增加向低纬度移动;另一个则对应着植被向高纬度扩张,荒漠带相应变窄。现实中是否存在这样两个平衡状态,需要通过多种途径来验证。

由于本文重在考察反照率的影响,并没有考虑水汽和其他物理过程的影响,因此所得的结果虽然是定量的,但严格地讲只能为更复杂的气候模式做数值模拟时定性参考。就能量平衡模式来讲,也需要不断改进。

致谢　审稿人提出修改建议,特此致谢。

参考文献

[1]　李巧萍,丁一汇. 植被覆盖变化对区域气候影响的研究进展. 南京气象学院学报,2004,27:131-140.

[2]　符淙斌,马柱国. 全球变化与区域干旱化. 大气科学,2008,32:752-760.

[3]　马柱国,符淙斌. 中国干旱和半干旱带的 10 年际演变特征. 地球物理学报,2005,48:519-525.

[4]　Charney J G. Dynamics of deserts and drought in the Sahel. Q J R Meteorol Soc,1975,101:193-202.

[5]　Dickinson R E. Modeling evapotranspiration for three-dimensional global climate models. In:Hansen J E,Takahashi T, eds. Climate Processes and Climate Sensitivity. Geophys Monogr Ser,1984,29:58-72.

[6]　Sellers P J,Mintz Y,Sud Y C,et al. A simple biosphere model (SiB) for use within GCMs. J Atmos Sci,1986,43: 505-531.

[7]　Dai Y J,Zeng Q C. A land surface model (IAP94) for climate studies. Part I:Formulation and validation in off-line experiments. Adv Atmos Sci,1997,14:433-460.

[8]　Sellers P J,Collatz G J,Randall D A,et al. A revised land surface parameterization (SiB2) for atmospheric GCMs. Part I:Model formulation. J Clim,1996,9:676-705.

[9]　Dai Y J,Zeng X B,Dickinson R E,et al. The common land model. Bull Am Meteorol Soc,2003,84:1013-1023.

[10]　Oleson K W, Dai Y J, Bonan G. Technical Description of the Community Land Model (CLM). Boulder,Colorado: National Center for Atmospheric Research,NCAR/TN-459STR,2004.

[11]　林朝晖,刘辉志,谢正辉,等. 陆面水文过程研究进展. 大气科学,2008,32:935-949.

[12]　巢纪平,陈英仪. 二维能量平衡模式中极冰-反照率的反馈对气候的影响. 中国科学,1979,12:1198-1207.

[13]　Wu L Y,Chao J P,Fu C B,et al. On a simple dynamics model of interaction between oasis and climate. Adv Atmos Sci,2003,20:775-780.

[14]　巢纪平,周德刚. 大气边界层动力学和植被生态过程耦合的一个简单解析理论. 大气科学,2005,29:37-46.

[15]　Kuo H L. On a simplified radiative-conductive heat transfer equation. Pure Appl Geophys,1973,109:1870-1876.

[16]　Budyko M I. The effect of solar radiation variation on the climate of the earth. Tellus,1969,21:611-619.

[17]　刘飞,巢纪平. 全球植被分布对陆面气温影响的半解析分析. 科学通报,2009,54:1761-1766.

[18]　刘辉志,涂钢,董文杰. 半干旱区不同下垫面地表反照率变化特征. 科学通报,2008,53:1220-1227.

一个简单的绿洲和荒漠共存时距平气候形成的动力理论*

巢纪平[1,2] 井宇[3]

(1. 国家海洋环境预报中心,北京 100081;2. 国家海洋局第一海洋研究所,青岛 266061;3. 兰州大学大气科学学院,兰州 730000)

摘要:发展了一个由辐射传输过程和大气边界层动力过程耦合的能量平衡模式,分析了当绿洲和荒漠尺度不同时的区域气候态及气候态的季节变化。结果表明,在背景条件相同时,只有大于一定尺度的绿洲得以维持;在另一方面,当绿洲尺度给定时,在绿洲相对湿度较小时,绿洲得以维持。这些简单的理论结果可为更复杂的数值模拟模式提供参考,为防治荒漠化提供一定的科学依据。

关键词:绿洲;荒漠;定常态;年际变化

全球变化已成为国际关注的焦点,干旱化问题也是其中之一。《联合国气候变化框架公约(UNFCCC)》、《联合国防治荒漠化公约(UNCCD)》和《联合国生物多样性公约(CBD)》等都对干旱区给予了很大的关注[1]。IPCC 第四次报告中专门提及全球变暖背景下干旱化问题的重要性和迫切性,受干旱的影响将使得积雪减少、粮食减产、死亡和疾病的发生率增加等,预计在未来干旱化面积会继续增大并提出了相应的适应策略。

干旱地区占有地球陆地面积的 1/4,而中国荒漠化和沙漠有 128 万多平方千米,西北干旱地区占有我国陆地面积的 1/3。我国西北干旱地区绿洲的面积虽不到 4%~5%,但却集中了该地区 95%以上的人口和财富。绿洲是干旱和半干旱地区工农业生产和人们生活的基地,也是该地区人们生存的生命线。干旱区水资源消耗最大,也是水利用效率最高的地域,绿洲环境建设是干旱区生态环境建设的核心。随着我国西部大开发的进程,加强绿洲的研究和建设,对该地区的经济可持续发展有着特殊的战略意义[2]。

绿洲的发展对我们的环境起着至关重要的作用,而荒漠化与绿洲的消亡密不可分,进而使得我们的生存环境急剧恶化,但是如果人类有序的发展绿洲,是有可能使局地气候向有利于好的一面发展的。因此这方面的研究既有理论意义又有实际价值。很多学者对这方面的研究已早有注意。如关于在副热带地区荒漠化形成的原因,很早 Charney[3] 曾从辐射平衡和地表反照率之间的耦合,原创性的提出了沙漠化理论。而对绿洲中植被的蒸腾、辐射等与大气有交互作用过程的研究有助于我们对绿洲的维持和衰亡有更进一步的认识。在理论上研究植被的蒸腾和辐射过程的是 Dickinson[4] 和 Seller 等[5],并已用到气候模式的陆面过程中。

---

* 中国科学:地球科学,2012,42(3):424-433.

一般情况下,在荒漠带中也有植被即荒漠和绿洲是并存的,由于它们对辐射的响应不同,特别是它们的潜热通量有很大差别,由此造成的区域气候会有多样性。Pan 和 Chao[6] 的研究表明,在热力学的能量守恒下,荒漠和绿洲并存时,在给定的物理参数下当绿洲和荒漠区域比例达到一定时,蒸发率的大小可出现分岔现象,一个解向全区域荒漠化发展,另一个解向区域绿洲化发展。薛具奎和胡隐樵[7] 等的工作从动力学和非平衡态热力学角度分析了近地面绿洲与沙漠的相互作用特征,揭示了绿洲蒸发率对绿洲湿度和绿洲尺度的依赖性,指出绿洲的维持与发展存在一个最小临界尺度。注意到这些研究都是在绿洲和沙漠间一些物理量的交换中进行的。

考虑到前面的学者或是在大尺度的背景场下分别研究荒漠化或植被与大气交互过程,或是在小尺度背景场下分析的只是热力学问题,或是在近地面从动力学和非平衡热力学角度来研究,这些工作多数都未在一个大气动力学和热力学特别是辐射过程并存的模式中来分析绿洲和沙漠的格局对气候的影响,在国内巢纪平和周德刚[8] 曾开始用一个简单的大气边界层动力学和植被生态过程耦合模式求出中尺度环境中绿洲上空的运动形式,刘飞和巢纪平[9] 又用一个三维的动力学惯性模式分析了全球植被分布对陆面气温的影响。为了研究这一有意思的结果,巢纪平和李耀锟[10] 应用更复杂的热力学模式,在绿洲和荒漠并存区域中总能量守恒的约束下,也得到物理量的多平衡态解,并研究了沙漠的经向格局对全局气候的动力学和热力学影响。Wu 等[11] 引入的绿洲和荒漠并存时的热力学模式,在垂直平均条件下分析了绿洲和荒漠区占不同比例时对局地气温的影响。但这些工作都是研究的定常态。

本文的目的在于发展一个在绿洲与荒漠共存的环境下,构成一个简单大气边界层动力学和植被生态过程相互作用的非定常模式,用以讨论绿洲和荒漠不同面积时对区域气候的影响和年际变化过程。

1　绿洲和荒漠共存时的动力系统

设在副热带的半干旱区有两条带,一条带(-a,0)中下垫面基本上被植被覆盖,称为绿洲,在其侧有另一条带(0,b)植被很少,称为荒漠。这两条带中能量的平衡和传输过程有差异,但两条带是连通的,即能量和热量可交换,现从动力系统的观点来分析它们上空气候的演化。

1.1　贴地层能量平衡

从地面到株冠高度这一层称为贴地层,在贴地层中的能量平衡如下。

(1) 绿洲

能量平衡方程为

$$(1-\alpha_0)Q-\varepsilon\sigma T_0^4=\frac{\rho_a c_p(T'_0-T')}{r_E}+\frac{\rho_a L_v[q_{sat}(T_0)-r_0 q_{sat}(T)]}{(r_E+r_C)} \tag{1}$$

式中,左端为向下的短波辐射和向上的长波辐射;α_0 为植被的平均反照率;T_0 为植被或绿洲的平均绝对温度;右端第一项为感热通量;T'_0,T'分别是植被温度和空气温度对平均温度的距平值;r_E 为空气动力学阻力系数;ρ_a 为空气密度;c_p 为空气的定压比热;右端第二项为潜热通量,其中,r_C 为叶面的气孔阻力(stomata);L_v 为相变潜热系数;r_0 为绿洲空气的相对湿度;q_{sat}为饱和比湿。

假定$(T_0,T)=(\overline{T}_0,\overline{T})+(T'_0,T')$,$T^4\approx\overline{T}^4+4\overline{T}^3T'$,$T_0^4$ 同此。另,$Q=\widetilde{Q}+Q'$,$\widetilde{Q}$ 为年平均值,Q'为季节偏差,并设

$$(1-\alpha_0)\widetilde{Q}-\varepsilon\sigma\overline{T}_0^4\approx 0,$$

$$q_{sat}(T_0)\approx q_{sat}(\overline{T}_0)+\left(\frac{\partial q}{\partial T}\right)_{sat}T'_0,$$

$$q_{sat}(T)\approx q_{sat}(\overline{T})+\left(\frac{\partial q}{\partial T}\right)_{sat}T'$$

而 $q_{sat}(\overline{T}_0)\approx q_{sat}(\overline{T})$将这些关系式代入方程(1),得到

$$T'_0=\frac{C_0^{(1)}}{A_0}T'+\frac{C_0^{(2)}}{A_0} \tag{2}$$

其中

$$A_0=4\varepsilon\sigma\overline{T}_3+\frac{\rho_a c_p}{r_E}+\frac{\rho_a L_v}{(r_E+r_C)}\left(\frac{\partial q}{\partial T}\right)_{sat} \tag{3a}$$

$$C_0^{(1)}=\frac{\rho_a c_p}{r_E}+\frac{\rho_a L_v}{(r_E+r_C)}\left(\frac{\partial q}{\partial T}\right)_{sat}r_0 \tag{3b}$$

$$C_0^{(2)}=(1-\alpha_0)Q'-\frac{\rho_a L_v(q_{sat}(\overline{T}_0)-r_0q_{sat}(\overline{T}))}{(r_E+r_C)} \tag{3c}$$

(2) 荒漠

由于在荒漠带没有或只有极小量的植被覆盖,因此叶面气孔的蒸腾效应对潜热通量已不重要,为此采用公式:

$$\frac{\rho_a c_p(T'_s-T')}{r_E B_E}$$

式中,B_E 为 Bowen 比。由此在荒漠带的能量平衡方程在距平形式下为

$$(1-\alpha_s)Q'-4\varepsilon\sigma\overline{T}^3T'_s=\frac{\rho_a c_p(T'_s-T')}{r_E}+\frac{\rho_a c_p(T'_s-T')}{r_E B_E} \tag{4}$$

式中,α_s 为荒漠的平均反照率;T'_s 是荒漠温度距平值,其他符号同式(1)。由此得到:

$$T'_s=\frac{C_s^{(1)}}{A_s}T'+\frac{C_s^{(2)}}{A_s} \tag{5}$$

式中

$$A_s=4\varepsilon\sigma\overline{T}^3+\rho_a c_p(1+B_E) \tag{6a}$$

$$C_s^{(1)}=\rho_a c_p(2+B_E) \tag{6b}$$

$$C_s^{(2)}=r_E B_E(1-\alpha_s)Q' \tag{6c}$$

注意到式(2)和式(5)通过贴地层的能量平衡将贴地层的绿洲温度和荒漠温度和邻近的大气温度联系起来了。

1.2 贴地层顶的热量平衡条件

在贴地层顶被植被覆盖的绿洲地区,由感热通量、潜热通量和长、短波辐射通量相平衡的条件为

$$z=z_b\approx 0,$$

$$\rho_a c_p K_0\frac{\partial T'}{\partial z}=(1-\alpha_0)Q'-4\varepsilon\sigma\overline{T}^3T'_0-\frac{\rho_a L_v}{(r_E+r_C)}\times\left[(1-r_0)+\left(\frac{\partial q}{\partial T}\right)_{sat}(T'_0-r_0T')\right] \tag{7}$$

将式(2)代入消去 T'_0，整理后给出

$$z = z_b \approx 0,$$

$$\rho_a c_p K_0 \frac{\partial T'}{\partial z} = -\widetilde{A}_0 T' + \widetilde{B}_0 \tag{8}$$

式中

$$\widetilde{A}_0 = \frac{\rho_a L_v}{(r_E + r_C)}\left(\left(\frac{\partial q}{\partial T}\right)_{sat} \frac{C_0^{(1)}}{A_0} - r_0\right),$$

$$\widetilde{B}_0 = (1 - \alpha_0) Q' - 4\varepsilon\sigma \overline{T}^3 \frac{C_0^{(2)}}{A_0} - \frac{\rho_a L_v}{(r_E + r_C)}\left((1 - \alpha_0) + \left(\frac{\partial q}{\partial T}\right)_{sat} \frac{C_0^{(2)}}{A_0}\right)$$

在贴地层高度的荒漠带区，其热量平衡条件为

$$\rho_a c_p K_s \frac{\partial T'}{\partial z} = (1 - \alpha_s) Q' - 4\varepsilon\sigma \overline{T}^3 T'_s - \frac{\rho_a c_p}{r_E B_e}(T'_s - T') \tag{9}$$

应用式(5)消去上式中 T'_s，整理后得到

$$z = z_b \approx 0,$$

$$\rho_a c_p K_s \frac{\partial T'}{\partial z} = -\widetilde{A}_s T' + \widetilde{B}_s \tag{10}$$

式中

$$\widetilde{A}_s = -\left(4\varepsilon\sigma \overline{T}^3 + \frac{\rho_a c_p}{r_E B_e}\right)\frac{C_s^{(1)}}{A_s} + \frac{\rho_a c_p}{r_E B_e},$$

$$\widetilde{B}_s = (1 - \alpha_s) Q' - \left(4\varepsilon\sigma \overline{T}^3 + \frac{\rho_a c_p}{r_E B_e}\right)\frac{C_s^{(2)}}{A_s}$$

现把研究区域的下界放在贴地层顶，因此式(7)和式(9)可作为问题求解时的下界条件。

1.3　辐射传输方程

根据 Kuo[12] 及巢纪平和陈英仪[13]，加上内能在辐射能传输中的平衡作用，并设这个过程是随时间缓慢变化的，有

$$\tau \frac{\partial T}{\partial t} = k_l \frac{\partial^2 T}{\partial x^2} + k_h \frac{\partial^2 T}{\partial z^2} + \left(\frac{8\sigma r^{\cdot} \overline{T}^3}{\alpha'_s \rho_c}\right)\frac{\partial^2 T}{\partial z^2} - 2(1 - r^{\cdot})\alpha'_w \sigma \rho_c T^4 + \alpha'' \rho_c Q - \left(N^2 \frac{\overline{T}}{g}\right) w + C_0 + C_1 z = 0 \tag{11}$$

式中，τ 是温度变化的张弛时间；$r^{\cdot}$ 表示在强吸收区中介质的辐射能量占总辐射能量的部分；N 为 Brunt-Väsälä 频率；k_l 为水平交换系数；k_h 为垂直交换系数；α'_s，α'_w 分别为波长短的辐射区和波长长的辐射区中介质的平均吸收系数。α''为对太阳辐射的吸收系数，w 为垂直速度。

考虑到在无穷高处解有限，因此 $C_1 = 0$，可以这样定，设大气的 C_0 平均温度为 $\widetilde{T}$，平均太阳辐射为 $\widetilde{Q}$，由此有

$$C_0 = 2(1 - r^{\cdot})\alpha'_w \sigma \rho_c \widetilde{T}^4 - \alpha'' \rho_c \widetilde{Q}$$

由于 C_0 与定常态相平均，平衡时的这一太阳辐射可称平衡状态下的太阳辐射。于是方程(11)对扰动态写出为

$$\tau\frac{\partial T'}{\partial t}=k_l\frac{\partial^2 T'}{\partial x^2}+\left(k_h+\frac{8\sigma r^{\cdot}\,\overline{T}^3}{\alpha'_s\rho_c}\right)\frac{\partial^2 T'}{\partial z^2}-8(1-r^{\cdot})\alpha'_w\sigma\rho_c\overline{T}^3T'+\alpha''\rho_cQ'-\left(N^2\frac{\overline{T}}{g}\right)w \tag{12}$$

1.4 边界层大气运动

这类问题的着眼点在对流层下部,因此只需用到边界层方程,另外根据巢纪平等[14]的观点对长期过程而言,是大气对温度场适应的,而适应过程是很快的,因此可认为大气运动方程是定常的。由此有

$$fu=\mu\frac{\partial^2 v}{\partial z^2} \tag{13}$$

$$f\frac{\partial v}{\partial z}=\frac{g}{\overline{T}}\frac{\partial T'}{\partial x} \tag{14}$$

$$\frac{\partial u}{\partial x}+\frac{\partial w}{\partial z}=0 \tag{15}$$

其中,f 为科氏力;μ 为湍流扩散系数。由式(15)引进流函数 ψ,有

$$w=\frac{\partial\psi}{\partial x},\quad u=-\frac{\partial\psi}{\partial z} \tag{16}$$

将此式代入式(13),给出

$$f\frac{\partial\psi}{\partial z}=-\mu\frac{\partial^2 v}{\partial z^2} \tag{17}$$

假定

$$z\to\infty,\quad \psi\to 0,\quad \mu\frac{\partial v}{\partial z}\to 0 \tag{18}$$

于是有

$$f\psi=-\mu\frac{\partial v}{\partial z} \tag{19}$$

将此式代入式(14),得到

$$f^2\psi=-\mu\frac{g}{\overline{T}}\frac{\partial T'}{\partial x} \tag{20}$$

最后有

$$u=\mu\frac{g}{f^2\overline{T}}\frac{\partial^2 T'}{\partial x\partial z} \tag{21}$$

$$w=-\mu\frac{g}{f^2\overline{T}}\frac{\partial^2 T'}{\partial x^2} \tag{22}$$

1.5 温度变化的控制方程

将式(22)代入式(12),得到

$$\tau\frac{\partial T'}{\partial t}=\left(k_l+\frac{N^2}{f^2}\mu\right)\frac{\partial^2 T'}{\partial x^2}+\left(k_h+\frac{8\sigma r^{\cdot}\,\overline{T}^3}{\alpha'_s\rho_c}\right)\frac{\partial^2 T'}{\partial z^2}-8(1-r^{\cdot})\alpha'_w\sigma\rho_c\overline{T}^3T'+\alpha''\rho_cQ' \tag{23}$$

这是辐射传输的热力学过程和运动的动力学过程耦合后的扰动温度变化的控制方程,它是本工作研究的基础方程,这即为巢纪平等[14]提出气候态的距平模式。

在这里时间变化项的引进是出自物理考虑，温度场一般来讲总要向外源（在这是太阳辐射）适应[15]，其张弛系数的大小可通过气候模式的数值试验做出估值。

1.6　大气动力学场的计算

注意到，虽然现在温度已解出，但从式（22）不能直接算出垂直速度，这是因为式（22）只在流体内部表明垂直运动和温度场的关系，而垂直运动尚需受边界条件的约束，为此可对式（16）求旋度，将式（21），式（22）代入，得到

$$\frac{\partial^2\psi}{\partial x^2}+\frac{\partial^2\psi}{\partial z^2}=-\frac{\mu g}{f^2\hat{T}}\left(\frac{\partial^2}{\partial x^2}+\frac{\partial^2}{\partial z^2}\right)\frac{\partial T}{\partial x} \tag{24}$$

改写成

$$\frac{\partial^2\Phi}{\partial x^2}+\frac{\partial^2\Phi}{\partial z^2}=0 \tag{25}$$

式中，$\Phi=\psi+\frac{\mu g}{f^2\hat{T}}\frac{\partial T}{\partial x}$

边界条件为

$$z=0,\quad \psi=0 \tag{26}$$

地面是一条流线，确保垂直运动为零，由此有

$$z=0,\quad \Phi=\frac{\mu g}{f^2\hat{T}}\frac{\partial T(x,0)}{\partial x} \tag{27}$$

右端已知。考虑到 $z\to\infty$，$\psi\to0$，$\partial T/\partial x\to0$，于是有

$$z\to\infty,\quad \Phi\to0 \tag{28}$$

类似地考虑，给出

$$x=-a,\quad \Phi=\frac{\mu g}{f^2\hat{T}}\frac{\partial T(-a,z)}{\partial x}$$

$$x=b,\quad \Phi=\frac{\mu g}{f^2\hat{T}}\frac{\partial T(b,z)}{\partial x} \tag{29}$$

当 Φ 算出后，$\psi=\Phi-\frac{\mu g}{f^2\hat{T}}\frac{\partial T}{\partial x}$；当流函数求出后，可以计算出 u 和 w。

1.7　模式的数学表达式

现把上面各种物理过程汇总成一个数学问题。

$$\tau\frac{\partial T'}{\partial t}=\left(k_l+\frac{N^2}{f^2}\mu\right)\frac{\partial^2 T'}{\partial x^2}+\left(k_h+\frac{8\sigma r^{\cdot}\,\overline{T}^3}{\alpha'_s\rho_c}\right)\frac{\partial^2 T'}{\partial z^2}-8(1-r^{\cdot})\alpha'_w\sigma\rho_c\overline{T}^3T'+\alpha''\rho_cQ' \tag{30}$$

当下垫面为绿洲时，下边界条件为式（8）；当下垫面为荒漠时，下边界条件为式（10）。

侧边界条件为

$$z=-a\ 和\ z=b,\quad T'=0 \tag{31}$$

上边界条件为

$$z=h,\quad T'=0 \tag{32}$$

2 定常距平气候态

当不考虑时间变化项时,方程变为

$$0=\left(k_l+\frac{N^2}{f^2}\mu\right)\frac{\partial^2 T'}{\partial x^2}+\left(k_h+\frac{8\sigma r^{\cdot}\ \overline{T}^3}{\alpha'_s\rho_c}\right)\frac{\partial^2 T'}{\partial z^2}-8(1-r^{\cdot})\alpha'_w\sigma\rho_c\overline{T}^3T'+\alpha''\rho_cQ'. \tag{33}$$

边界条件同上。

太阳辐射的平衡值取334.59 W/m^2,计算见参考文献[16~18],$H=3\times10^3$ m,$v=2.4$ m/s,$C_d=2.75$ e^{-3},C_d 为空气的拖曳系数,$\sigma=$ 5.673 e^{-8} J/(s · m^2 · K^4),$\rho_a=1.293$ kg/m^3,$c_p=1004$ J/(kg · K),$L_v=2.5$ e^6 J/kg,$r_c=150$ s/m,$\alpha_l=0.3-0.1\times\sin\left(\frac{x}{a}\pi\right)$,$\alpha_s=0.3+0.1\times\sin\left(\frac{x-a}{b}\pi\right)$,$\varepsilon=0.95$,$r=0.5$,$K_0=9.6$,$K_s=4.1$,$B_E=5$,$R=0.5$,$k_l=1.01$ e^7 J/m,$k_h=2.3$ e^4 J/m,$\alpha'_s=10$ m^2/kg,$\alpha'_w=0.125$ m^2/kg,$\alpha''=0.025$ m^2/kg,$r_E=1/C_dv$,$\xi_0=0.4$,$f=1\times10^{-4}$/s,$g=9.8$ m/s^2,$q_{sat}=0.622\times6.11\mathrm{e}^{-3}\times10^{7.45\,t/(237.3+t)}$,$T=(273.15+t)$K,$N=1.16$ e^2/s,$\rho_c=6$ e^3 kg/m^3,$r^{\cdot}=0.5$。

2.1 太阳辐射距平取平衡值的1%

由于是气候态的距平模式,若取太阳辐射的距平值为上述平衡值的1%,即 $Q'=3.3459$ W/m^2,则计算结果如下。

2.1.1 绿洲荒漠尺度不同的影响

考虑绿洲荒漠在水平方向上分布尺度不同,例如水平范围共200 km,当绿洲尺度分别为65 km,100 km,135 km和165 km时温度场和动力学场的变化可参见图1。

从图1可以看出,在本模式中,当绿洲尺度大于100 km时,正温度距平区在大气低层,并跨越绿洲和荒漠交界区,中心在交界处附近,使它们耦合成一个相互并存的整体。从环流上看,也是在交界处附近上升的一个整体环流,且强度较弱,两区之间的热量和水分(通过潜热)交换不强,这是两区能并存的物理条件。随着绿洲尺度的缩小,荒漠区的负温度区扩大,当绿洲面积小于65 km^{2}时,正的温度距平只在绿洲上空,环流发生明显的变化,绿洲上空是上升运动,荒漠区上空是下沉运动,这样从绿洲尺度大时的两圈环流,过渡到三圈环流。特别注意到,在低层从荒漠区向绿洲区的辐合环流加强,这样必然有大量干冷空气从荒漠区流入绿洲区,使绿洲消亡。

由此可见,在一定的气候条件和物理条件(如植被不同、蒸腾率不同等)下,绿洲若要维持,其面积需达到一个临界尺度。

2.1.2 绿洲区相对湿度不同的影响

当绿洲尺度取100 km,其他参数不变,绿洲区的空气相对湿度取值不同时,对整个区域的影响。

从图2可以看出,在绿洲尺度为100 km时,当绿洲湿度小于50%时,整个区域的低层为正温度距平,低层气流辐合高层气流辐散,但气流较弱,温度高有利于植被的蒸发和蒸腾,使低层大气处于暖湿状态,而由于气流弱,热量和水分不易流失,如果有充足的土壤水分来支持蒸发和蒸腾,绿洲可处在自我维持状态。

由此可见,绿洲上空的空气太湿润并不是绿洲发展的好条件,相反,较为干燥的大气环境,更有利绿洲的生存发展。这是对我国西北干旱区中绿洲得以维持的一个理论说明。

当绿洲相对湿度大于50%时,荒漠区出现负温度距平区并随着绿洲相对湿度的增大负温度距平区迅速扩大，绿洲和荒漠两区分别呈现正、负温度距平。在荒漠区的低层流向绿洲区的气流加强，大量干冷空气流向绿洲，将不利绿洲的生存发展。在环流结构上，也从两圈环流过渡到三圈环流。

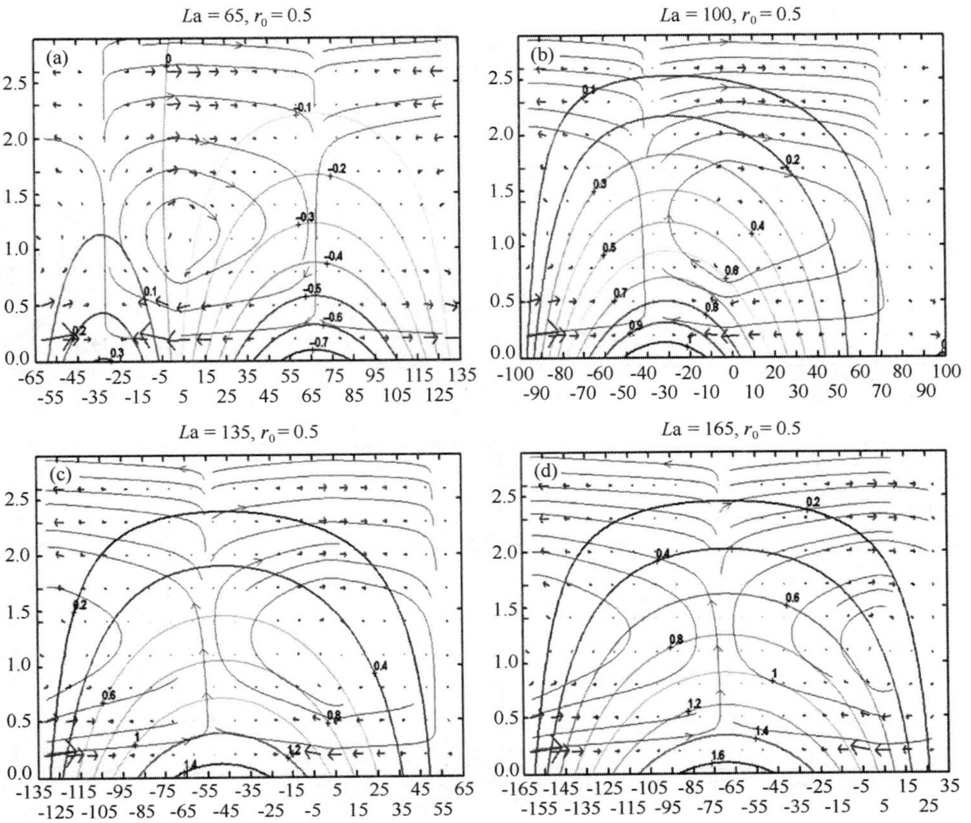

图 1　绿洲相对湿度不变尺度不同时的温度距平和动力学场（见书后彩插）

（a）至（d）绿洲尺度分别为 65 km，100 km，135 km，165 km。图中 La 为绿洲尺度，r_0 为绿洲空气相对湿度，水平方向和垂直方向单位为 km（下图同），温度距平单位为℃，图中垂直速度最大值为 0.17 m/s，水平速度最大值为 0.64 m/s，温度距平最大值为 1.68℃

2.2　达到准定常态的时间

当考虑时间项的变化时，即方程（30），时间张弛系数取 $\tau = 7.776\ e^6$，边界条件、其他参数与不考虑时间项时所取一样。

当方程中加入时间项后，达到准定常态的距平气候与上面定常态（两个时次迭代的温度差达 10^{-6}）的结果相比，当绿洲相对湿度不变，绿洲尺度变化时，除了绿洲和荒漠交界处环流场稍弱外，其他均接近定常态。按上面取的张弛系数，当绿洲尺度为 65 km，100 km，135 km，165 km 时，达到准定常态的时间分别约为 8.58 年，10.1 年，11.17 年，11.58 年。可见，绿洲尺度越大，达到准定常态的时间越长。

当绿洲尺度不变，相对湿度变化时，除了在绿洲和荒漠交界处环流场减弱外，其他结果也接近定常态。当绿洲湿度为 10%，30%，50%，80%时，达到准定常态的时间分别约为 13.25 年，11.92 年，10.1 年，6.75 年。可知绿洲相对湿度越大，达到准定常态的时间越短。

从图 3（a）至图 3（d）可以看出，绿洲尺度为 100 km，相对湿度为 50%时，随着时间的增长，整个区域的环流逐渐减弱,有利于绿洲水汽的维持,荒漠的干冷气流也更不易入侵到绿洲,绿洲有维持的趋势。当绿洲湿度不变,绿洲尺度变化时其他尺度结果类同不加时间项时的分析。

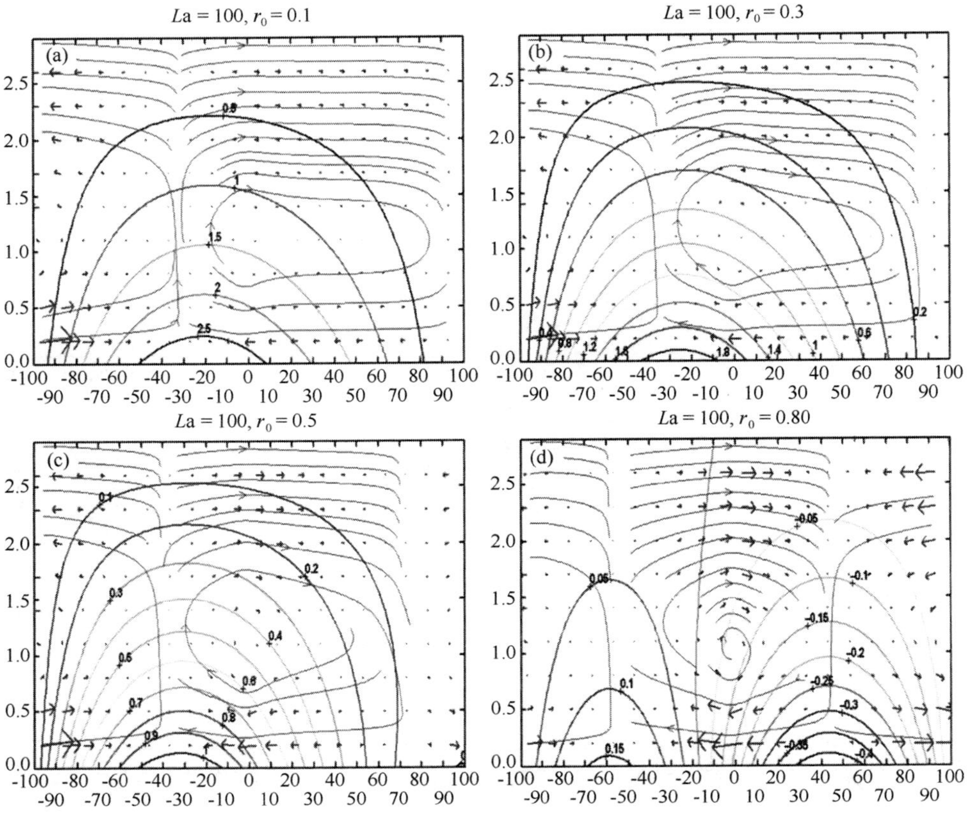

图 2 绿洲尺度不变相对湿度不同时的温度距平和动力学场(见书后彩插)

(a)至(d) 绿洲相对湿度分别为 10%,30%,50%,80%。图中垂直速度最大值为 0.38 m/s,水平速度最大值为 1.38 m/s,温度距平最大值为 2.9℃

从图 3(e)至图 3(h)可以看出,绿洲尺度为 100 km,相对湿度为 80%时,随着时间的增长,只有绿洲区域的正温度距平区环流减弱,而荒漠侧的负温度距平区环流并没有太大变化;并在高空冷温度距平区逐渐由荒漠向交界处扩张,不利于绿洲的维持发展。当绿洲尺度不变,相对湿度变化时,在其他相对湿度条件下的结果也类同不加时间项时的分析。

3 结合实际

图 4 为 2006—2007 年全国的平均相对湿度分布图,从这张图可以看出,我国南部平均相对湿度要大于北部,东部的平均相对湿度要大于西部。

由上面理论上相对湿度对绿洲的影响分析来看,在黄河流域以南,大部分地区相对湿度大于 50%,由上面的分析可知湿度过大并不是很有利于绿洲发展,但我们可以通过扩大绿洲面积超过临界尺度,使得绿洲得以较好地维持;而在黄河流域以北,大部分地区相对湿度小于 50%,在这样的湿度条件如果有充足的土壤水分来支持蒸发和蒸腾,有利于绿洲维持。所以在这些地区,我们依然应该扩大绿洲面积保持水土,另一方面可以对绿洲地区进行一定的灌溉,使得地面土壤有较充足的水分使得

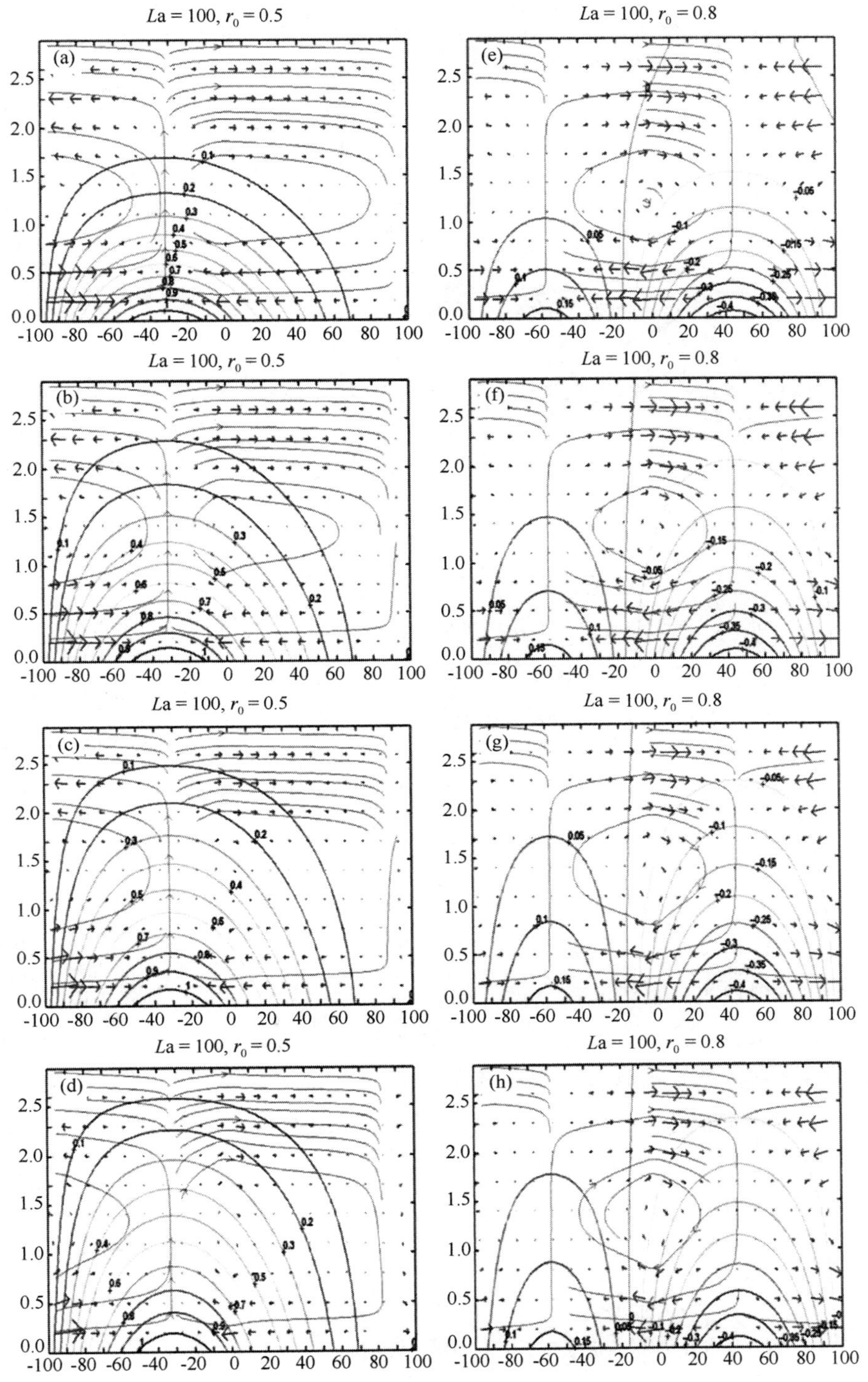

图 3　绿洲的温度距平和动力学场随时间演变(见书后彩插)

(a)至(d) 分别为绿洲尺度为 100 km,相对湿度为 50%时随时间迭代到两年和准定常态时的图;

(e)至(h) 分别为绿洲尺度为 100 km,相对湿度为 80%时随时间迭代到两年和准定常态时的图。

图中垂直速度最大值为 0.13 m/s,水平速度最大值为 0.40 m/s,温度距平最大值为 1.1℃

绿洲维持;但是由上面的分析可知在黄河流域以北地区相对湿度较小,一片绿洲达到准定常态需要更

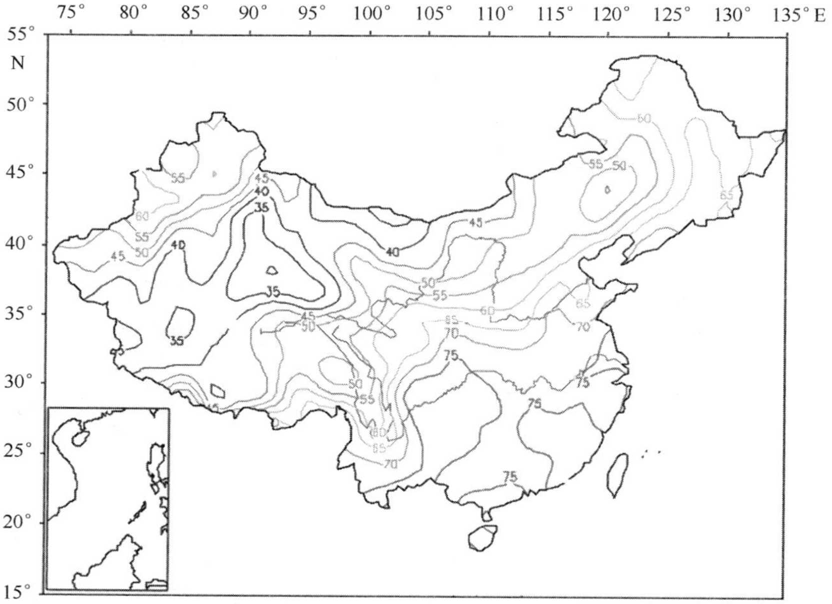

图 4 2006—2007 年全国的平均相对湿度(%)分布(见书后彩插)

长的时间,所以我们更应该加强对这些地区绿洲的防护,否则一旦破坏,将要我们付出很多年的努力来使它重新恢复平衡。

4 结论

本文在考虑了行星边界层大气中动力学过程和辐射传输过程后,在绿洲和荒漠共存的环境里,将边界层物理场与下垫面植被冠层中的热量平衡方程相耦合,从而建立了一个简单的荒漠与绿洲共存时边界层大气运动和植被生态过程相耦合的理论模式。用此分析了绿洲荒漠共存时热力学场和动力学场的变化。分析结果表明:一是绿洲尺度越大,越有利于绿洲维持;二是绿洲相对湿度不太高,如果有充足的土壤水分来支持蒸发和蒸腾,绿洲可处在自我维持状态,而绿洲相对湿度太大反而不是绿洲维持的有利条件;三是绿洲尺度越大,达到准定常态需要的时间越长,绿洲相对湿度越大,达到准定常态需要的时间越短。

致谢:感谢审稿专家提出的宝贵意见。

参考文献

[1] 王涛. 干旱区绿洲化、荒漠化研究的进展与趋势. 中国沙漠,2009,29:1-9.

[2] 胡隐樵,左洪超. 绿洲环境形成机制和干旱区生态环境建设对策. 高原气象,2003,22:537-544.

[3] Charney J G. Dynamics of deserts an drought in the Sahel. Q J R Meteorol Soc,1975,101:193-202.

[4] Dickinson R E. Modeling evapotranspiration for three-dimensional global climate models. In:Hansen J E,Takahashi T,eds. Climate Processes and Climate Sensitivity. Geophys Monogr Ser,1984,29:58-72.

[5] Sellers P J,Mintz Y,Sud Y C,et al. A revised land surface parameterization(SiB) for use within GCMs. J Atmos Sci,1986,43:505-531.

[6] Pan X L,Chao J P. The effects of climate on development of ecosystem in Oasis. Adv Atmos Sci,2001,18:42-52.

[7]　薛具奎,胡隐樵. 绿洲与沙漠相互作用的数值试验研究. 自然科学进展,2001,11：514-517.
[8]　巢纪平,周德刚. 大气边界层动力学和植被生态过程耦合的一个简单解析理论. 大气科学,2005,1：37-46.
[9]　刘飞,巢纪平. 全球植被分布对陆面气温影响的半解析分析. 科学通报,2009,54：1761-1766.
[10]　巢纪平,李耀锟. 热力学和动力学耦合的二维能量平衡模式中荒漠化气候的演变. 中国科学 D 辑：地球科学,2010,40：1060-1067.
[11]　Wu L Y,Chao J P,Fu C B,et al. On a simple dynamics model of interaction between sasis and climate. Adv Atmos Sci,2003,5：775-780.
[12]　Kuo H L. On a simplified radiative-conductive heat transfer equation. Pure Appl Geophys,1973,109：1870-1876.
[13]　巢纪平,陈英仪. 二维能量平衡模式中极冰-反照率的反馈对气候的影响. 中国科学 A 辑：数学,1979,12：1198-1207.
[14]　巢纪平,季劲钧,何家骅,等. 一种长期数值预报方法的物理基础. 中国科学 A 辑：数学,1977,20：162-172.
[15]　许有丰. 二层线性模式长期气象过程的一些计算. 见：顾震潮,主编. 动力气象学论文集. 北京：科学出版社,1961:90-95.
[16]　王炳忠. 太阳辐射计算讲座第一讲太阳能中天文参数的计算. 太阳能,1999,2：8-10.
[17]　王炳忠. 太阳辐射计算讲座第二讲相对于斜面的太阳位置计算. 太阳能,1999,3：8-9.
[18]　王炳忠. 太阳辐射计算讲座第三讲地外水平面辐射量的计算. 太阳能,1999,4：12-13.

二维能量平衡模式对若干气候问题的研究*②

李耀锟[1]　巢纪平[2]

(1.北京师范大学全球变化与地球系统科学研究院,北京 100875;2.国家海洋环境预报中心,北京 100081)

摘要:回顾并分析讨论了二维能量平衡模式在冰界纬度和太阳常数的关系、荒漠化的演变及二氧化碳温室效应等一系列气候问题中的应用。同一个太阳常数最多可以对应3个平衡态,一个对应着目前的气候状态,一个对应着冰期气候,另一个对应着几乎全球冰封的气候。平衡态的数目及其稳定性与反照率的分布关系密切,反照率不连续性越大,平衡态的数目越多。荒漠带会随着荒漠反照率的增大分别向南、向北扩张。恶化荒漠植被交界地区的生态环境,危及人类的生产生活,荒漠化对中低纬度(高纬度)的温度变化影响较弱(显著)。二氧化碳浓度升高并不会使极冰迅速消融,全球平均温度缓慢升高,冰界纬度缓慢向北退却,时常会出现"停滞"现象。

关键词:二维能量平衡模式;多平衡态;冰期;荒漠化;二氧化碳温室效应

1　引言

全球变化是当今最活跃、发展最快的科学领域之一,而气候变化又是全球变化主要的研究领域之一(叶笃正等, 2002)。大气中温室气体和气溶胶含量的变化,及其导致的地-气辐射平衡和地表特性的变化,都会改变气候系统的能量平衡,引起全球气候变化(秦大河等,2007)。气候变化对自然生态系统、国民经济系统和国家安全系统的负面影响远比其正面影响更受关注(秦大河,2004)。与气候变化相伴随的一系列严重的环境问题已经威胁到了人类社会的发展和进步,为了避免可能出现的环境灾难,通过科学合理的方式(有序人类活动)实现经济社会的可持续发展成为人类现实生存发展的迫切需求(叶笃正等,2001)。

尽管温室气体增加对全球变暖的贡献已经得到了公认,然而人类目前尚未真正了解不同时间尺度气候变化的驱动力以及气候系统内部各种复杂的响应和反馈过程(刘玉芝,2006),如冰芯资料分析表明大气中温室气体的浓度跟大气温度并没有直接的因果关系,甚至温度的变化可能会超前温室气体浓度的变化(崔伟宏等,2012)。又如1999—2008年全球变暖的趋势接近于0,显著低于此前的预估(Knight et al., 2009),而大气中温室气体的浓度则在逐步升高。因此,深入分析和了解气候变化的各种驱动因子及反馈过程,不仅有着重要的理论意义,也会为指导有序人类活动提供一定的实际意义。

由于气候系统极为复杂,气候模式甚至是地球系统模式成为研究的有利工具,模式可以定量分析

* 气象学报,2014,72(5):880-891.

② 资助课题:国家重点基础研究计划(973项目)项目(2014CB953903)、中央高校基本科研业务费专项资金(2013YB45)。

某种驱动因子或某种反馈过程对全球气候甚至是局地气候的影响,加深入们对气候变化的认识水平,并对未来做出预估,为决策者提供决策参考。此外,通过一些简单的,但是却具有坚实物理基础的理论模式研究气候变化,有利于突出引起气候变化的因子,深入分析其影响气候变化的物理过程;有利于从理论和物理上对数值模式的结果进行补充,完善和丰富气候变化这一研究领域。最简单的做法是将陆-气系统看作一个质点,利用陆-气系统的能量平衡来求解平均温度的分布特征,即 0 维能量平衡模式,对若干参数进行订正,很容易将温室效应和冰雪反照率反馈机制引入到 0 维能量平衡模式中(Budyko,1969; North et al. , 1981)。在 0 维能量平衡模式的基础上,考虑温度沿纬度的分布,即得到了一维能量平衡模式,Budyko(1969)、Sellers(1969)及 North(1975a)的一维气候模式正是这类研究中最具有代表性的工作,对以后的研究产生了深远的影响。若进一步在一维能量平衡模式中加入经向或垂直方向的物理过程,即可得到二维能量平衡模型(Sellers,1976;巢纪平等,1979),还可以进一步考虑海洋的作用(石广玉等,1996)。若从垂直方向上扩展 0 维模式,就得到了一维辐射-对流模式(Manabe et al. ,1964,1967),一维辐射-对流模式已经不能用解析的方法来求解。若要求得垂直方向的解析解,需要对长波吸收谱进行简化,巢纪平等(1979)在郭晓岚 1973 年简化长波吸收谱工作基础上发展了一个二维能量平衡模式,并利用该模式研究了冰界和太阳常数的关系(巢纪平等,1979)、荒漠化(巢纪平等,2010a)以及二氧化碳的温室效应(巢纪平等, 2010b)这些有着重要价值的问题。本研究将在他们工作的基础上,进一步完善和发展二维能量平衡模式在上述 3 个领域内的应用。

2　二维能量平衡模式

平均温度场的形成受到许多因子的控制,如太阳辐射的纬度分布、辐射能传播及转换过程、湍流热交换、水汽凝结和蒸发等过程造成的潜热、冷暖平流引起的热量交换。完整的温度方程可写为(叶笃正等,1958)

$$\rho_a c_p \frac{dT}{dt} - \frac{dp}{dt} = \varepsilon_1 + \varepsilon_2 + \varepsilon_3 \tag{1}$$

式中,ρ_a 为大气密度;c_p 为空气的定压比热;T 为大气温度;p 为大气压强;ε_1、ε_2、ε_3 分别表示由辐射引起的热通量、由湍流引起的热通量及由凝结或蒸发引起的热通量。

令 k_j 为对波长为 λ_j 的长波辐射吸收系数,k' 为对太阳辐射的平均吸收系数,$A_j(B_j)$ 表示在波长 $\Delta\lambda_i$ 区间内向下(向上)的长波辐射,E_i 为波长 $\Delta\lambda_i$ 区间内的黑体辐射,Q 表示向下的太阳辐射,则由于辐射引起的热通量可写为

$$\varepsilon_1 = \sum_j k_j(A_j + B_j - 2E_j) + k'Q \tag{2}$$

根据布格-朗伯定律

$$\frac{\partial A_j}{\partial z} = k_j(A_j - E_j) \tag{3}$$

$$\frac{\partial B_j}{\partial z} = k_j(E_j - B_j) \tag{4}$$

$$\frac{\partial Q}{\partial z} = k'Q \tag{5}$$

由湍流引起的热通量可以表述为

$$\varepsilon_2 = \rho_a c_p K_h \nabla_h^2 T + \rho_a c_p K \frac{\partial^2 T}{\partial z^2} \tag{6}$$

式中，K_h 为水平端流导温系数（$\kappa_h = \rho_a c_p K_h$ 称为水平湍流导热系数）；K 为垂直湍流导温系数（$\kappa_z = \rho_a c_p K$ 称为垂直湍流导热系数）；∇_h^2 为水平拉普拉斯算子。注意式中已令 K_h、K 均为常数。考虑辐射能传输过程与湍流热输送相平衡，并对式(1)求纬向平均，则在球坐标下式(1)可写为

$$\frac{\kappa_h}{a^2}\frac{\partial}{\partial x}(1-x^2)\frac{\partial T}{\partial x} + \frac{\partial}{\partial z}\left(\kappa_z \frac{\partial T}{\partial z}\right) + \sum_j k_j(A_j + B_j - 2E_j) + k'Q = 0 \tag{7}$$

式中，$x=\sin\phi$，ϕ 为纬度；a 为地球半径。式(7)要对整个吸收谱求积分，较为复杂，且不利于理论分析，如何用解析或半解析的方法处理辐射能传输是一个非常重要的问题，Kibel(1943)建立了严格的温度分布理论，Blinova (1947)将该理论进一步发展并精细化，他们将整个吸收谱作为一个整体考虑，并假定吸收系数为常数，这样就避免了繁杂的积分计算，最终可得到一个温度的4阶微分方程，详细介绍可参见叶笃正等(1958)。郭晓岚1973年将长波吸收谱分为强、弱吸收区两个部分(分别用下标s和w表示)，两个吸收区域的平均吸收系数定义为

$$k_{s,w} = \frac{\int_{s,w} k_j E_j \mathrm{d}\lambda}{E_{s,w}} \tag{8}$$

并定义 $E_s = rE$，$E_w(1-r)E$，$E=\sigma T^4$。

引入光学厚度

$$\xi = \frac{1}{\xi_0}\int_z^\infty k'\mathrm{d}z, \quad \xi_0 = \int_0^\infty k'\mathrm{d}z \tag{9}$$

根据简化规则，式(7)经推导可得

$$D\frac{\partial}{\partial x}(1-x^2)\frac{\partial E}{\partial x} + (\kappa_2 + \kappa_\tau)\frac{\partial^2 E}{\partial \xi^2} - N^2 E = -\widetilde{S}\overline{Q}_0 \xi_0 \mathrm{e}^{-\xi_0 \xi} S(x) + \widetilde{S}\overline{Q}_0 C \tag{10}$$

式中

$$D = \frac{\xi_0^2 \kappa_h}{k'^2 a^2}, \quad \widetilde{S} = \frac{4\xi_0 \sigma \overline{T}^3}{k'},$$

$$\kappa_r = \frac{8r\sigma\overline{T}^3}{k_s}, \quad N^2 = \frac{8(1-r)k_w \xi_0^2 \sigma \overline{T}^3}{k'^2}$$

式中，$\overline{Q}_0$ 为大气上界的平均太阳辐射。与 κ_z 相比，κ_r 可称辐射等效湍流交换系数，它代表了强辐射区的辐射效应，而 N^2 可称为牛顿辐射冷却系数，它代表了弱辐射区的辐射效应。Brunt (1934)指出，强吸收区的作用很像热传导过程，而弱吸收区的作用则可以取为牛顿冷却的形式。1973年郭晓岚从理论上将这两种效应统一到一个方程中，但他并没有给出积分常数 C 的具体形式，巢纪平等(1979)利用全球积分平均给出积分常数 C 为

$$C = r\left(1 + \frac{k'}{k_w}\right)\overline{\prod}_0 + (\overline{\prod}_1 - \overline{\prod}_0 + 1 - \mathrm{e}^{-\xi_0}) \tag{11}$$

模式的上、下边界条件分别取为

$$\xi = 0, \int_0^1 E_0 \mathrm{d}x = \frac{1}{2}\left(1 + \frac{k'}{k_w}\right)\overline{\prod}_0 \overline{Q}_0 \tag{12}$$

$$\xi = 1, (\kappa_2 + \kappa_r)\frac{\partial E}{\partial \xi} = N^2\int_0^1 E\mathrm{d}\xi - 2(1-r)\widetilde{S}E_0 + \widetilde{S}\overline{Q_0}\left(\frac{k'}{k_w}\prod_0 + \prod_1\right)S(x) \tag{13}$$

式中，$\prod_0 = 1 - \Gamma_p$，$\prod_1 = 1 - \Gamma_p - (1 - e^{-\xi_0})$ 分别表示抵达大气上界的净太阳辐射及到达地面的太阳辐射。$\overline{\Pi}_0 = \int_0^1 \prod_0 S(x)\mathrm{d}x$，$\overline{\prod_1} = \int_0^1 \prod_1 S(x)\mathrm{d}x$ 分别为其全球积分平均值。巢纪平等(1979)没有考虑上边界对太阳辐射的反射，即$\prod_0 = 1$。

利用勒让德级数给出式(10)的解，设

$$E(x,\xi) = \sum_n E^{(n)}(\xi)P_n(x) \tag{14}$$

将相关的变量均展开为勒让德级数，最终即可求解式(10)的解析解。该模式被称为郭晓岚－巢纪平－陈英仪模式。陈英仪(1982a,1982b)进一步对解的稳定性及参数敏感性进行了分析。上述二维能量平衡模式在气候变化方面有着较为广泛的应用，如研究冰期对太阳常数的敏感性(巢纪平等, 1979)、行星大气温度分布(吕越华等,1981)、二氧化碳的温室效应(巢纪平等,2010b)及荒漠化效应(巢纪平等,2010a)等。

3　冰界纬度与太阳常数的关系

20 世纪 60 年代气候处于冷期，众多科学家对冷期产生的条件进行了研究，其中最有代表性的是 Budyko(1969)、Sellers(1969)及 North(1975a)的工作(图 1)。一维模式中冰界纬度对太阳常数较为敏感，太阳常数只需减小 2%左右即可使冰界向南推进到 50°N，甚至出现全球冰封的情形；另外，一维模式在太阳常数取某些值(如取目前的观测值)时，会存在多个平衡态的冰界纬度，对应着不同的气候状态。一维模式中出现的这两个现象可归结为两个问题：一个是冰界纬度对太阳常数的敏感性如何；另一个是不同平衡态的稳定性及是否具有物理意义。

Lindzen 等(1977)指出，只要在模式中加入一些更为实际的热量输送过程，就可以使冰界对太阳常数的依赖性降低；Lian 等(1977)发现通过修正温度和反照率的关系可以降低温度变化对太阳常数变化的敏感性；Oerlemans 等(1978)用卫星资料对长波辐射和温度的参数化关系进行了合理的修正，结果表明太阳常数要减小 9%左右才能出现冰期气候。考虑到他们的结果是用一维能量平衡模式得到的，而一般来说，一维模式由于缺少垂直方向的能量调节，对太阳常数的变化较为敏感。巢纪平等(1979)应用郭晓岚－巢纪平－陈英仪模式重新研究这一问题，指出太阳常数要减小 15%左右才能出现冰期气候，即在二维能量平衡模式下冰界纬度对太阳常数的敏感性会大大减低。

Budyko(1972)最早发现了简单一维模型中解的多平衡态特征，指出当平衡态所在位置满足 $\mathrm{d}Q/\mathrm{d}x_s>0$，则平衡态是稳定的，而当 $\mathrm{d}Q/\mathrm{d}x_s<0$ 时，平衡态是不稳定的。Chylek 等(1975)随后也分析解的分岔这一现象。North(1975a,1975b)在其所用的一维模型中发现极地附近存在两个位置很接近的平衡态，其中更靠近极地的平衡态不稳定。他认为这一平衡态不是很合理，并通过调整太阳辐射的分布去掉了这一平衡态。Held 等(1974)分析了简单模型中的反照率反馈机制，他的结论支持 Budyko(1972)的结果，同时他认为极地附近的两个平衡态的出现是由于引入与温度梯度有关的扩散项造成的。当扩散项足够小时，此时极地附近的两个平衡态之间不再存在显著的物理差别。Ghil(1976)发现存在间冰期、冰期和几乎全球冰封 3 个平衡态。Drazin 等(1977)进一步分析了模式中多平衡态解产生的原因，指出如果模式中方程的特征值小于 0，则其对应的平衡态是不稳定的。Lin(1978)认为给定非线性湍流交换系数可以去掉极地附近的不稳定的平衡态。Cahalan 等(1979)则认为极地附近的

分岔点是由于反照率取阶梯函数造成的，在冰界附近对反照率做任何形式的平滑，均可以去掉这一分岔点。Coakley(1979)也提到了采用光滑的反照率可以去掉分岔点。

由于不稳定的平衡点意味着冰界会随着太阳常数的增大而向南推进，这似乎有些不符合物理常识，因而 North 等(1981)在对一维模式的总结回顾中更倾向于认为分岔点可能是为了数学上处理方便引入分段阶梯反照率函数而人为导致的，实际中很有可能并不存在。然而 Held 等(1974)和 Ghil(1976)似乎并没有这样的倾向性。实际中反照率的分布可以存在较大的不连续性，如南极极冰和其周边海水的反照率差异很大，此时取反照率为阶梯函数似乎也更合理一些，因此，分岔点的出现可能表征某些情况下地球气候的演变特征，不宜将其当作没有物理意义的计算解而简单去掉。

巢纪平等(1979)认为一维模式中没有考虑热量在垂直方向的输送，因而对太阳常数较为敏感，他们建立了一个考虑了能量垂直输送的二维模式。研究结果表明，温度对太阳常数并不是十分敏感，太阳常数要减少 15%~20%才能出现冰期气候，同时，解对其他参数也并不是十分敏感，各种参数变化 20%都不会对解产生显著的影响(陈英仪，1982b)。

二维模式中反照率取与一维模式相似的分布，即

$$\Gamma(x,x_s)=\begin{cases}\alpha_1(x) & x>x_s\\ \alpha_2(x) & x<x_s\end{cases} \tag{15}$$

式中 $x_s=\sin\phi$，ϕ_s 为冰界所在纬度，由温度 $T=-10$℃决定，而 α_1、α_2 分别为冰面和非冰面的反照率分布，可以取不同的形式。

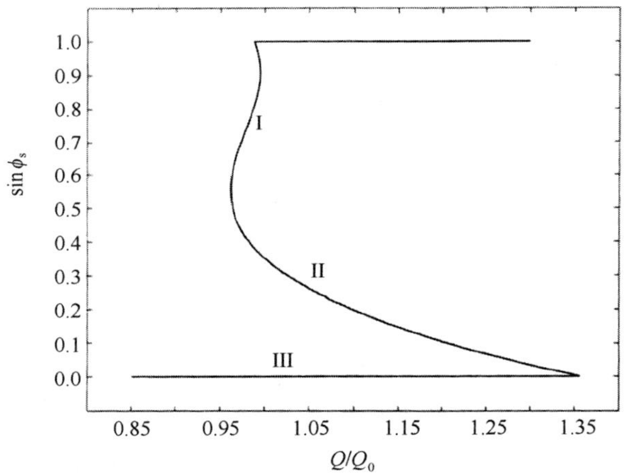

图 1 一维模式中冰界纬度与太阳常数的关系(North，1975a)

(图中横坐标为无量纲的太阳辐射，Q_0 为当前的太阳常数，ϕ_s 为冰界纬度)

图 2 给出了 3 种反照率分布情形下冰界纬度与太阳常数的关系。图中实线中反照率分布取为 $\alpha_1(x)=\alpha_2(x)=a_0+a_2P_2(x)$，即反照率分布为连续函数(North, et al.，1979)，此时冰界与太阳常数的关系较为稳定，表现为随着太阳常数的增大，冰界会向北退却，直到全球无冰。若要出现全球无冰的情形，太阳常数要增大 10%左右；若要出现极冰向南扩张到 50°N 附近，太阳常数大约要减小 20%。

图 2 中虚线表示反照率取为 $\alpha_1(x)=0.62$，$\alpha_2(x)=a_0+a_2P_2(x)$，即将冰面的反照率固定为 0.62，非冰面的反照率仍然按照观测资料给定(Coakley，1979)。在约 30°N 以北，冰界纬度会随着太阳常数的增大而向北退却，直到全球无冰。若要出现全球无冰的情形，太阳常数大约要增加 10%；若要冰界向南发展到 50°N，太阳常数要减小 16%。在 30°N 以南，还存在一支冰界纬度随着太阳常数的增大而

向南扩张的解。两支解在 30°N 附近汇合成一支。

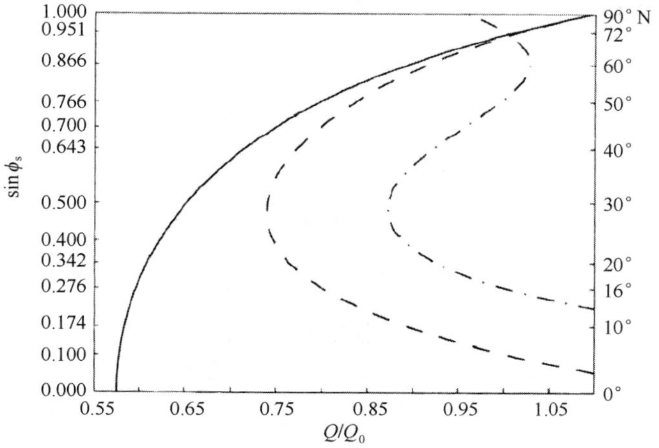

图 2　冰界纬度与太阳常数的关系

（实线为反照率取 $\alpha_1(x)=\alpha_2(x)=a_0+a_2P_2(x)$；虚线为反照率取，$\alpha_1(x)=0.62$，$\alpha_2(x)=a_0+a_2P_2(x)$；点划线为反照率取 $\alpha_1(x)=0.62$，$\alpha_2(x)=0.132$，横坐标为无量纲化的太阳常数，ϕ_s 为冰界纬度，右纵坐标为相应的纬度）

图 2 中点划线表示反照率取为 $\alpha_1(x)=0.62$，$\alpha_2(x)=0.132$，此时冰面和非冰面的反照率均取为常数(Budyko, 1969)。由图可知，存在两个分岔点，分别位于约 30°N、60°N 位置。从赤道到 30°N 及从 60°N 到北极，极冰范围会随着太阳常数的增大而增大，而 30°—60°N，极冰范围会随着太阳常数的增大而减小。在目前的太阳常数下存在 3 个冰界纬度，分别在 72°N、50°N 和 16°N 附近。这 3 个冰界纬度似乎在地球演变历史中均出现过，目前冰界大约在 72°N；在石炭—二叠纪冰期及第四纪冰期，冰界纬度能抵达中纬度地区；而在新元古代冰期，有证据表明冰界能够抵达低纬度地区，甚至可能出现全球冰封的情形(Sumner, et al., 1987; Schmidt, et al., 1991, 1995; Sohl, et al., 1999)。

综上所述，当反照率分布形式变化时，冰界纬度和太阳常数的关系也会发生变化：反照率分布连续时，冰界纬度和太阳常数存在一一对应的关系，即冰界纬度随着太阳常数的增大(减小)而向北退缩(向南推进)；反照率分布不连续时，冰界纬度和太阳常数一一对应的关系被破坏，太阳常数对应 2 个或 3 个冰界纬度，此时会出现某个冰界纬度随太阳常数的增大(减小)而向南推进(向北退缩)的情形，不过此种情况对应的平衡态并不稳定。冰界纬度和太阳常数的关系强烈地依赖反照率的分布，充分应用卫星资料对全球反照率分布给出更为细致的刻画将有助于加深对这一问题的认识。

由二维模式计算的地表温度分布(图 3(a))可知，理论模式较好地表征了地面温度的分布。在热带地区模式模拟的结果较高，这可能是由于模式中没有考虑哈得来环流对热输送的影响造成的，如果引入更接近实际的热输送(Lindzen, et al., 1977)，会改进对热带地区温度分布的模拟情况。直接利用光学厚度定义式(9)计算得到的温度垂直分布与观测相比有较大差别，通过观测资料订正光学厚度，可算得温度的垂直分布如图 3(b)所示，可见二维模型能较好地表征对流层内温度分布的主要特征，但是与观测资料相比仍然存在一定的差别，尤其是在对流层下部的热带区域及对流层上部的极地地区，由于模式没有考虑大气功力过程对温度分布的调节作用，不能苛求其计算结果与观测资料完全一致。

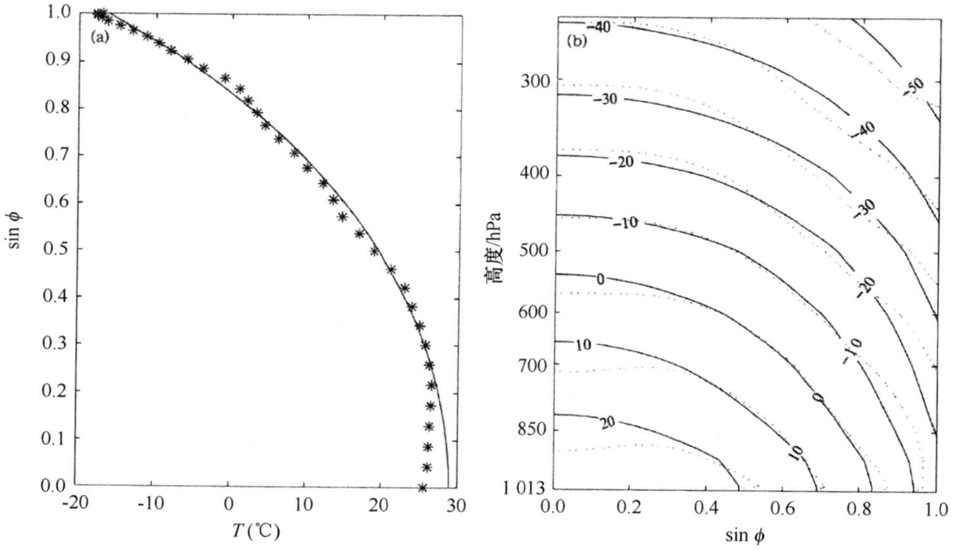

图 3 二维模式计算的温度分布

(a)地表温度的分布,图中实线为理论计算值,星号线为观测资料结果,图中纵坐标为纬度的正弦值(sinφ);(b)温度随高度的分布,图中实线为理论计算结果,点线为观测资料结果

4 荒漠化

联合国防治荒漠化公约(UNCCD)将荒漠化定义为包括气候变异和人类活动在内的种种因素造成的干旱、半干旱和亚湿润干旱地区的土地退化。全球干旱陆地中(不包括极度干旱的沙漠),大约有 3.6×10^9 hm^2 或 70%发生了土地退化(UNCCD,2000)。中国严重缺水的华北和西北是受气候变化影响较大的气候脆弱地区(叶笃正,1986),截至 2009 年底,中国荒漠化土地总面积达 2.62×10^6 km^2,占国土总面积的 27.33%,主要分布在西北、华北地区(国家林业局, 2011)。近几十年来,中国北方地区的干旱有加剧的趋势,干旱和半干旱带有向东和向南扩展的趋势,持续的干旱导致了当地荒漠化等一系列环境问题(马柱国等,2005,2006;符淙斌等, 2008),已经成为经济发展的严重障碍。北方干旱化如何形成、其发展趋势、其对经济社会的影响以及应对对策已经成为国家需求,围绕这些问题,中国科学家取得了一系列研究成果,如从 3 个方面分析研究北方干旱化形成和发展的驱动因子:季风系统干湿变化的自然规律;全球变化(主要是全球变暖)影响下,季风系统的异常响应;人类活动对生存环境的影响(符淙斌等,2002a,2002b)。

Charney(1975)最早从理论上分析了沙漠边缘植被变化对气候的影响,他指出植被退化会导致下垫面的反照率增大,改变地表的能量平衡,引起异常的下沉气流,从而抑制降水,而降水减少则不利于植被生长,这一正反馈过程加速了荒漠化发展。Dickinson(1984)将植被蒸腾作用引入到大气模式中。这一开创性的工作促进了陆面模式的发展和成熟,目前大多数研究都是从利用陆面模式或气候模式来研究荒漠化这一问题,而理论研究则相对匮乏。巢纪平等(2010a)根据极冰、荒漠带及植被带反照率的不同,给定分段的反照率分布,并假定植被和荒漠带的分界线可以通过温度来确定,由此探讨了荒漠化气候的演变。

由地表反照率分布(图 4)可知,虽然在不同地区反照率的分布差异较大,但是一般而言,北半球从极地开始，反照率先减小后增大而后又减小，分别对应着极冰、高纬度植被、中纬度荒漠及低纬度植被，这一点在亚洲和北美洲大陆西侧体现得最为明显。南半球陆面地面较小，整个南极大陆为冰雪所覆盖，往北依次是海洋（反照率可取为0.07）、较小的荒漠分布及热带植被。根据观测的反照率分布，将二维模型所用到的地表反照率分布给定为

$$\Gamma(x,x_s)=\begin{cases}\alpha_1(x) & 1>x>x_s\\ \alpha_2(x) & x_s>x>x_d\\ \alpha_3(x) & x_d>x>x_v\\ \alpha_4(x) & x_v>x>0\end{cases} \tag{16}$$

即根据陆表覆盖将反照率分为4段，α_i，$i=1,2,3,4$ 分别表示极冰、高纬度植被、中纬度荒漠及低纬度植被的反照率，而 x_s、x_d、x_v 分别表示极冰、荒漠和植被分界纬度的正弦值。若令 $\alpha_1=0.75$，$\alpha_2=\alpha_4=0.1$，$\alpha_3=0.25$，并假定在当前气候状态下，$x_s\approx0.95$，$x_d\approx0.766$，$x_v\approx0.5$，可算得地表平均的反照率为0.15左右，与实际较为接近。冰界纬度所在温度仍然按照前文，取为 $T_s=-10$℃，根据与观测度比较，给定 $T_d=5$℃，$T_v=19$℃分别作为自纬度植被与中纬度荒漠及中纬度荒漠和低纬度植被的分界温度。用荒漠带的地在反照率 α_3 表征荒漠化的演化和发展（图5）。随着荒漠带地表反照率 α_3 的增大，x_s（图5(a)实线）和 x_d（图5(a)虚线）均会向北退却，而 x_v（图5(a)点划线）则呈现出先向北退却，而后又向南推进的情形，不过变化幅度并不大。交界纬度的分布会影响到极冰、植被及荒漠的相对大小，从而改变地球的行星反照率（图5(b)）。行星反照率 Γ_p 随荒漠带地表反照率 α_3 的增大先减小后增大。在 α_3 较小时，其反照率增大引起的行星反照率增大要小于由于冰界位置北移造成的行旱反照率减小。因此，此时行星反照率会呈现逐步减小的趋势，当 α_3 增大到0.26左右时，此时行星反照率达到了极小值；当 α_3 继续增大时，极冰反照率的作用对行星反照率的贡献已经相对较小，此时荒漠反照率的增大及其面积增大会导致行星反照率从极小值开始增大。行星反照率的变化最终又会影响到温度的变化（图5(c)），全球积分的地表温度表现出先增大后减小的趋势。

图6给出了不同 α_3 时地表温度相对于当前气候状态（$\alpha_3=0.25$）的偏差。可见在 α_3 较小时，由于极冰范围较大，负的温度距平可达中低纬度，只有热带地区存在一定的升温。随着 α_3 的增大，冰界范围迅速减小，高纬度温度由负距平转为正距平，而低纬度地区则相反，为正的温度距平。高纬度的温度距平要大于中低纬度，这表明极冰-反照率反馈机制对温度的调节作用在高纬度更为明显。

x_d 会随着 α_3 的增大一直向北退却，这表明荒漠化的演变有利于荒漠带向北发展侵入高纬度植被带，但同时 x_s 的位置也会北移，这表明冰界会退缩而高纬度植被则会向北推进，由此可知，荒漠化主要会使高纬度植被带整体向北推移，但是对于中纬度而言，植被带和荒漠带的北移会恶化当地的生存环境，危害当地人的生产生活。对于中低纬度而言，x_v 向北移动表示低纬度植被会向北迁移，这对改善当地的环境是有利的；当荒漠化发展加剧时，低纬度植被带会向南迁移，这又不利于改善当地环境。不过相比 x_d，x_v 的变化范围并不大，其大约在30°N附近几个纬度内变化。将 x_d、x_v 分别看作是荒漠带的北边界和南边界，荒漠会随着其反照率的增大而向北、向南扩张，恶化荒漠和植被交界带的生态环境，而当地的生态环境本来就是较为脆弱的，如此将更有利于荒漠化的发展。由于下垫面物理和生态状态的分布影响了反照率的分布。因此，反照率地理位置变化影响气候的敏感性研究是值得加强的。因为这将告诉人们，有序人类活动（如灌溉、造林等）能在多大程度上改变区域性气候，使其向有利于人类社会生产、生存环境的方向可持续性发展。

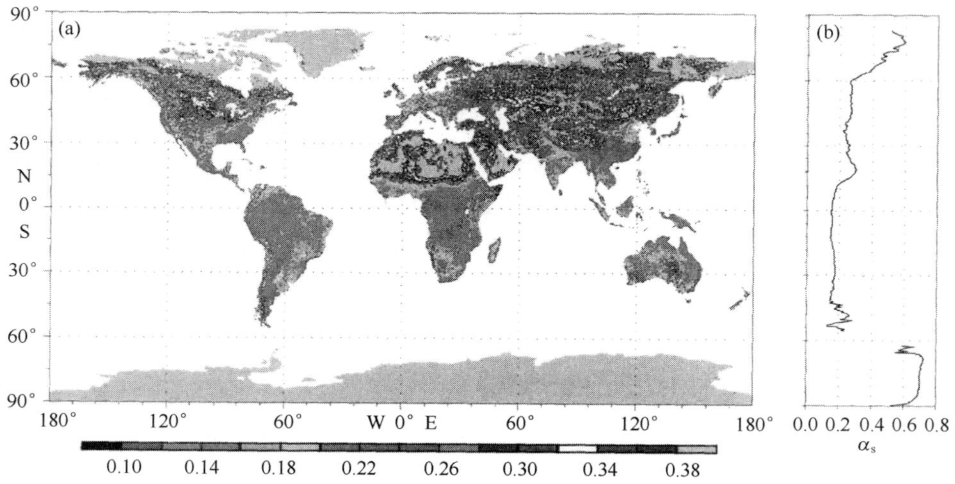

图 4　地表反照率(a)及其纬向分布(b)(见书后彩插)

(反照率资料下载自 ESA GlobAlbedo 项目网站 http://www.GlobAlbedo.org)

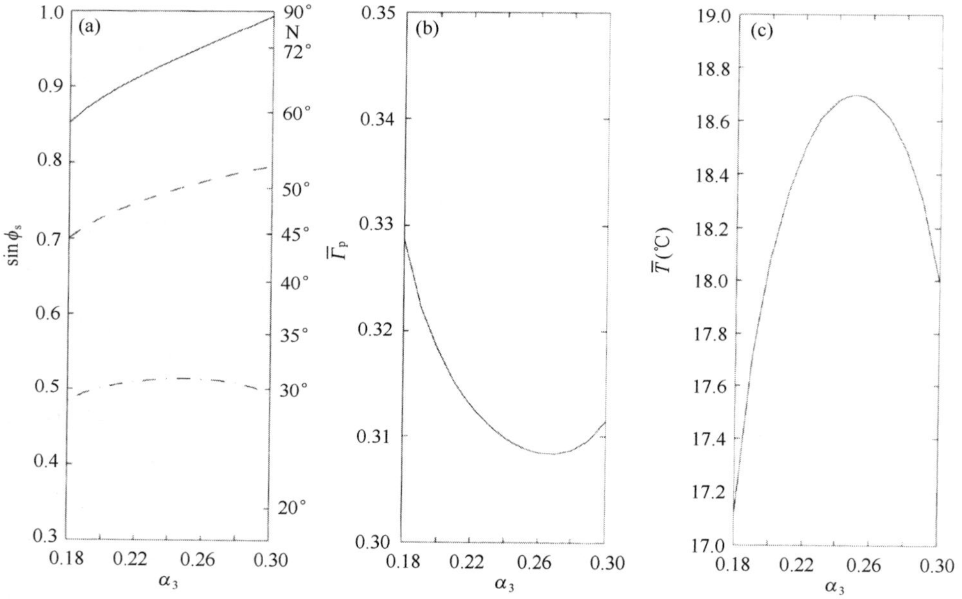

图 5　(a)x_s(实线)、x_d(虚线)、x_v(点划线)随 α_3 的变化(纵坐标为纬度及纬度的正弦值);(b)全球积分的行星反照率随 α_3 的变化;(c)全球积分平均的地表温度随 α_3 的变化

5　二氧化碳的温室效应

通过吸收向上的长波辐射,温室气体可以使地球保持在适当的温度区间内,过多的温室气体则会使地球温度升高。人类生产、生活活动排放的二氧化碳被认为是全球气候变暖的一个非常重要的原因。工业革命以来,大气中二氧化碳的浓度一直在稳步升高,根据世界气象组织(WMO)的报告,2013年几个测站的二氧化碳日均浓度已经突破 400×10^{-6},达到过去 300 万年来的最高值。按照目前的速率,二氧化碳的年均浓度很有可能在 2015 年到 2016 年突破 400×10^{-6}。人们普遍认为,随着二氧化碳

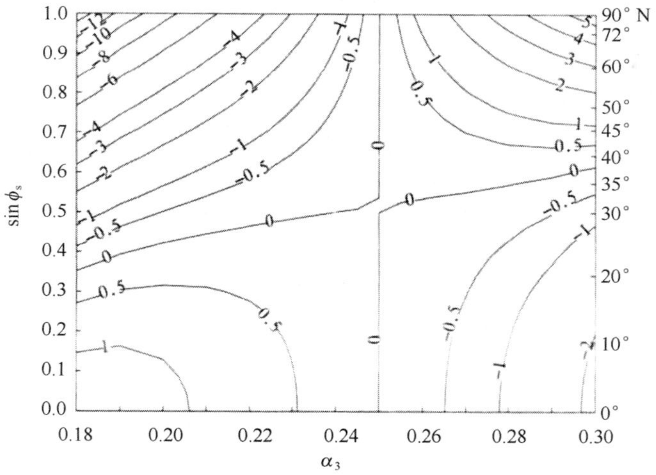

图 6　不同 α_3 条件下的地表温度相对于当前气候状态($\alpha_3=0.25$)的偏差

(纵坐标为纬度和纬度的正弦值)

浓度的不断升高,全球平均温度会继续升高,但升高的速率会有多大,目前似乎尚无定论,特别是最近十多年来,虽然大气二氧化碳浓度一直在升高,全球平均温度却表现出升温停滞的现象(Kerr,2009;Knight,et al. ,2009)。可见大气中二氧化碳等温室气体浓度变化导致的温度变化,尚不能完全定论,仍然有必要通过多种途径来研究大气对二氧化碳温室效应的响应这一问题。

刘玉芝等(2002)讨论了二氧化碳温室效应的饱和度问题,指出虽然在 15 m 这一强吸收带中心确已达到饱和,但是二氧化碳浓度变化对辐射强迫的贡献主要来自于 15 m 带的两翼,其温室效应远没有达到饱和,这表明二氧化碳的增温潜力仍然存在。那么该如何表征二氧化碳浓度继续升高和温度变化的关系呢? 通过气候模式来对这一问题进行分析和探讨不失为一个很好的方法,同时,采用一些简单的、具有坚实物理基础的简单模式作为补充,无疑会进一步加深对这一问题的认识和理解。如刘玉芝(2006)利用 Budyko(1969)一维模型分析了近几十万年来控制冰期—间冰期气候旋回的物理机制。

巢纪平等(2010b)利用郭晓岚-巢纪平-陈英仪模式讨论了大气对二氧化碳温室效应的响应。本节在其工作的基础上进一步利用更贴合物理实际的方法来更新这一工作。大气在某一波段对辐射的吸收系数 k_j 可写为

$$k_j=\eta(\alpha_{H_2O,j}\rho_{H_2O}+\alpha_{CO_2,j}\rho_{CO_2})\tag{17}$$

式中,$\eta=5/3$ 为散射因子;α 表示质量吸收系数;ρ 表示吸收介质的密度。利用 HITRAN(http://hitran.iao.ru/)根据 HITRAN 2004 高分辨率光谱资料提供的吸收系数分布,可计算长波波段(4~100 m)的吸收系数。不过大气对长被辐射的吸收极其复杂,不同波段的吸收系数可相差好几个数量级,为了便于理论上解析处理,采用郭晓岚 1973 年提出的长波辐射能吸收的简化方案将吸收波段分解为强、弱吸收区两个区间。

图 7 给出了温度为 290 K,压强为一个标准大气压时二氧化碳浓度在 0~1 000×10^{-6}内强、弱吸收区吸收系数及强吸收区占总能量的比例分布。可见随着二氧化碳浓度的升高,强吸收区的吸收系数会逐步减小,到二氧化碳浓度超过 400×10^{-6}后,变化的幅度大大减小,基本稳定在 1.36/m 附近;而弱吸收区的吸收系数变化则较为复杂,出现较为明显的波动现象,不过总的趋势也是在逐步减小,最终

大约维持在 1.32/mm 附近；强吸收区占总吸收的比例则会稳步上升。巢纪平等(2010b)在研究时假定强吸收区所占比例不变(0.5)，即强、弱吸收区各占一半，并对弱辐射区吸收系数给了较为简单直观的假定，虽然这么做具有一定的物理考量，但是仍显得较为粗糙，利用更为精确的 HITRAN 2004 分辨率光谱资料，按照式(8)来计算强、弱吸收区的吸收系数，进一步发展了巢纪平等(2010b)的工作。在不同温度、压强下计算结果虽有所不同，但是不会产生明显的变化，通过调整湍流交换系数，仍能保证在当前情况下温度分布符合观测。

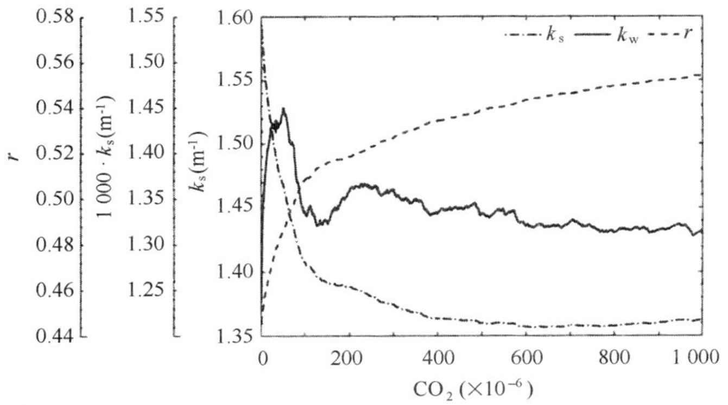

图 7 强、弱吸收区吸收系数及强吸收区占总吸收的比例随二氧化碳浓度的变化关系

(T=290 K，压强为一个标准大气压)

参照前文，将地表反照率分为两段，分别取 $\alpha_1=0.62$，$\alpha_2=0.132$，计算不同二氧化碳浓度下冰界温度的分布，此时的 3 个冰界纬度分别如图 8(a)至图 8(c)所示。可见随着二氧化碳浓度升高，3 个平衡态的演化出现不同特征，目前的气候状态下，冰界纬度会随着二氧化碳浓度的升高而缓慢向北退却，二氧化碳大约要升高到 1 000×10^{-6}，才会出现全球几乎无冰的情形(图 8(a))。极冰并不单调地随着二氧化碳浓度的升高而向北退却，在某些浓度区间内(如(400~500)×10^{-6})，极冰变化很小，在某些区间内极冰甚至会增加。在冰期气候下(图 8(b))，极冰并不会随着二氧化碳浓度的升高而消融，反而会向南推进。注意到，在(0~120)×10^{-6}和(180~270)×10^{-6}两个区间段内，并没有冰界与之对应，这是由于这两支解在区间端点附近发生了融合，此时没有实数解对应着冰界纬度。在全球几乎冰封的气候状态下(图 8(c))，冰界纬度会随着二氧化碳浓度的升高向北退却。图 9 给出了 3 种气候状态下对应的全球平均的地表温度，当前气候下(图 9(a))全球平均的地表温度会随着二氧化碳浓度的升高而升高，但是升高的幅度越来越小，当二氧化碳浓度超过 600×10^{-6}后，全球平均温度几乎不再随二氧化碳的浓度而变化，达到"饱和"的状态。在冰期气候中(图 9(b))，全球平均温度会随着冰界的向南推移而逐步降低；而在全球几乎冰封的气候下(图 9(c))，虽然冰界向北移动，但是全球平均温度仍然会降低，不过幅度不大。

6 展望

系统回顾了二维能量平衡模式的发展历史及其在若干气候问题上的应用，二维能量平衡模式虽然较为简单，但具有坚实的物理基础，有助于加深对气候变化的认识，未来需要进一步通过古气候资料及气候模式来验证理论分析结果。所述气候问题都是目前人们极为关注的，同时也是还没有完全认识清楚的重大问题，未来对这些问题的回答很可能会取得一定的突破和进展，这也在一定程度上代

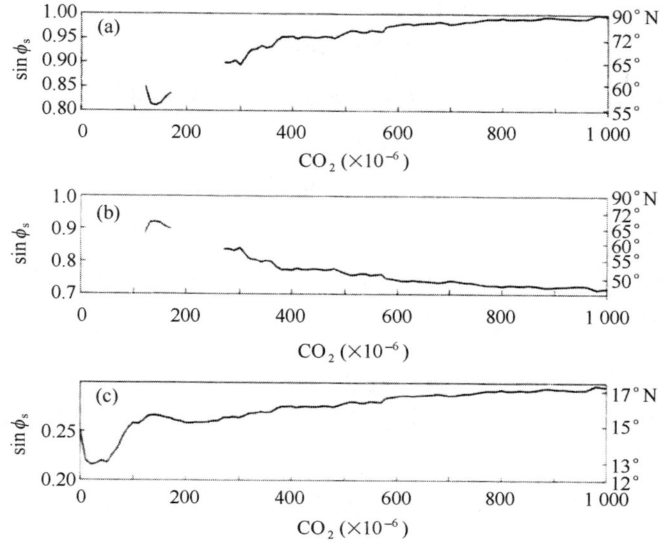

图 8　冰界纬度随二氧化碳浓度的变化

(a)当前气候；(b)冰期气候；(c)近冰封气候

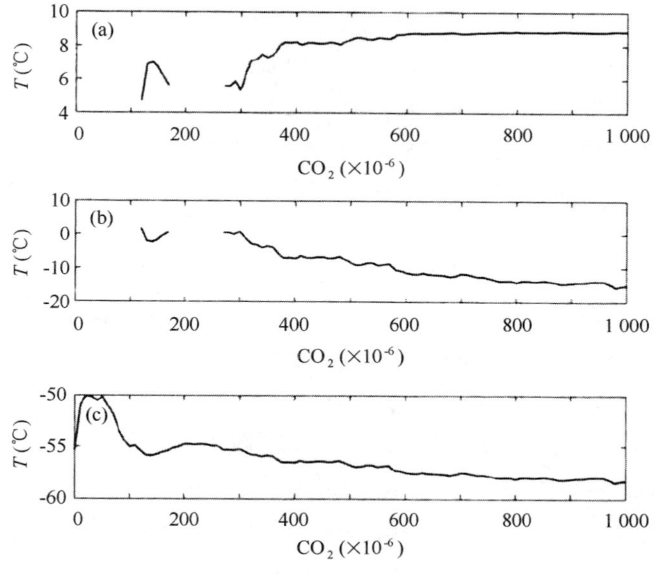

图 9　全球平均的地表温度随二氧化碳浓度的变化

(a)当前气候；(b)冰期气候；(c)近冰封气候

表了理论气候变化的研究方向。

(1)反照率分布和冰界纬度多平衡态及平衡态稳定性的关系，平滑的反照率分布会减小平衡态的数目，不连续的反照率则会增加平衡态的数目，虽然实际情况很可能介乎二者之间，但是哪一种情况更容易出现仍然是值得研究的，此外，平衡态的稳定性如何，不稳定的平衡态是否具有物理意义，其演变趋势如何，这些都是值得进一步研究的问题。对这些问题的回答将有助于理解目前气候所处的状态及其发展的趋势。

(2)荒漠化的发展及其对全球温度分布、大气运动的影响。目前研究主要集中在下垫面反照率分布变化对荒漠化及边界层运动的影响，需要进一步将其与自由大气的运动耦合在一起，讨论荒漠化对大气环流的影响。同时，现有的反照率分布及荒漠和植被交界纬度的确定仍存在一定的简单化和人为性，荒漠化对这些参数的敏感性如何，如何设计既能体现物理本质又能便于理论计算的不同下垫面属性分布及其差异，都是需要进一步研究的问题。

（3）二氧化碳浓度升高对全球平均温度的影响到底该如何确定。现有研究发现，对于不同的平衡态，二氧化碳浓度升高可以对应着全球温度升高或降低的情形，在目前的气候态下，二氧化碳造成的升温似乎存在逐步“饱和”的情形，这些结论的合理性及其中涉及到的物理过程如何，这些问题仍然没有定论，需要进一步分析研究。在更符合实际的下垫面属性分布下，是否仍然会存在此种现象，这些也都是需要进一步研究的问题。

（4）除了文中详述的 3 个问题外，还可以将二维能量平衡模式运用到其他重要的气候问题中，如气溶胶的气候效应等。

致谢：地表反照率资料下载自 ESA GlobAlbedo 项目 http://www.GlobAlbedo.org。

参考文献

巢纪平，陈英仪 . 1979.二维能量平衡模式中极冰-反照率的反馈对气候的影响.中国科学 A 辑，22(2)：1198-1207.

巢纪平，李耀錕 . 2010a.热力学和动力学耦合的二维能量平衡模式中荒漠化气候的演变.中国科学：地球科学，40(8)：1060-1067.

巢纪平，李耀錕 . 2010b.二维能量平衡模式中大气温度对二氧化碳增温效应的响应.气象学报，68(2)：147-152.

陈英仪. 1982a.二维能量平衡模式中极冰对气候的影响——解的稳定性分析(一). 气象学报，40(1)：1-12.

陈英仪. 1982b.二维能量平衡模式中极冰对气候的影响——参数敏感性分析(二). 气象学报，40(2)：175-184.

陈英仪，巢纪平 . 1988.对“对二维气候能量平衡模式中辐射和动力学参数化的一些看法”的回答.大气科学，12(1)：106-112.

崔伟宏，Singer S F，Courtillot V，等 . 2012.自然是气候变化的主要驱动因素.北京：中国科学技术出版社，20-21.

符淙斌，安芷生 . 2002a.我国北方干旱化研究——面向国家需求的全球变化科学问题 . 地学前缘，9(2)：271-275.

符淙斌，温刚. 2002b.中国北方干旱化的几个问题.气候与环境研究，7(1)：22-29.

符淙斌，马柱国 . 2008.全球变化与区域干旱化.大气科学，32(4)：752-760.

国家林业局. 2011.第四次中国荒漠化和沙化状况公报.北京：国家林业局 .

刘玉芝，肖稳安，石广玉 . 2002.论大气二氧化碳温室效应的饱和度.地球科学进展，17(5)：653-658.

刘玉芝 . 2006 地球气候变化的自然强迫——冰芯资料分析与物理模式研究[D].北京：中国科学院大气物理研究所，98.

吕越华，巢纪平 . 1981.行星大气温度分布的气候理论.大气科学，5(2)：145-156.

马柱国，符淙斌 . 2005.中国干旱和半干旱带的 10 年际演变特征.地球物理学报，48(3)：519-525.

马柱固，符淙斌. 2006. 1951—2004 年中国北方干旱化的基本事实.科学通报，51(20)：2429-2439.

秦大河 . 2004.进入 21 世纪的气候变化科学——气候变化的事实、影响与对策.科技导报，22(7)：4-7.

秦大河，陈振林，罗勇，等 . 2007.气候变化科学的最新认知.气候变化研究进展，3(2)：63-73.

石广玉，郭建东，樊小标，等. 1996.近百年全球平均气温变化的物理模式研究.科学通报，41(18)：1681-1684.

叶笃正，朱抱真 . 1958. 大气环流的若干基本问题.北京：科学出版社，73-75.

叶笃正 . 1986.人类活动引起的全球性气候变化及其对我国自然、生态、经济和社会发展的可能影响.中国科学院院刊，1(2)：112-120.

叶笃正，符淙斌，季劲钧，等. 2001.有序人类活动与生存环境.地球科学进展，16(4)：453-460.

叶笃正,符淙斌,董文杰 . 2002.全球变化科学进展与未来趋势.地球科学进展,17(4): 467-469.

Blinova E N. 1947. On the mean annual temperature distribution in the earth's atmosphere with consideration of continents and oceans (in Russian). Izv AN USSR, Ser Geogr Geofiz, 11(1): 3-13.

Brunt D. 1934. Physical and Dynamical Meteorology. New York: Cambridge University Press, 411.

Budyko M I. 1969. The effect of solar radiation variations on the climate of the earth. Tellus, 21(5): 611-619.

Budyko M I. 1972. The future climate. Eos, Transactions American Geophysical Union, 53(10): 868-874.

Cahalan R F. North G R. 1979. A stability theorem for energy-balanee climate models. J Atmos Sci, 36(7): 1178-1188.

Charney J G. 1975. Dynamics of deserts and drought in the Sahel. Quart 1 Roy Meteor Soc, 101(428): 193-202.

Chylek P, Coakley J A Jr. 1975. Analytical analysis of a Budykotype climate model. J Attnos Sci, 32(4): 675-679.

Coakley J A. 1979. A study of climate sensitivity using a simple energy balance model. J Atmos Sci, 36(2): 260-269.

Dickinson R E. 1984. Modeling evapotranspiration for three-dimensional global climate models. Geophys Monogr Ser, 29: 58-72.

Drazin P G, Griffel D H. 1977. On the branching structure of diffusive climatological models. J Atmos Sci, 34(11): 1696-1706.

Ghil M. 1976. Climate stability for a Sellers-type model. J Atmos Sci. 33(1): 3-20.

Held I M. Suarez M J. 1974. Simple albedo feedback models of the icecaps. Tellus, 26(6): 613-629.

Kerr R A. 2009. What happened to global warming? Scientists say just wait a bit. Science, 326(5949): 28-29.

Kibei A. 1943. Distribution of the temperature in the Earth's atmosphere(in Russian). DAN USSR, 39(1): 18-22.

Knight J. Kennedy J J. Folland C, et al. 2009. Do global tempera-ture trends over the last decade falsify climate predictions. Bull Amer Meteor Soc, 90(8):S22-S23.

Lian M S. Cess R D. 1977. Energy balance climate models: a reappraisal of ice-albedo feedback. J Atmos Sci, 34(7): 1058-1062.

Lin C A. 1978. The effect of nonlinear diffusive heat transport in a simple climate model. J Atmos Sci, 35(2): 337-339.

Lindzen R S. Farrell B. 1977. Some realistic modifications of simple climate models. J Atmos Sci, 34(10): 1487-1501.

Manabe S, Strickler R F. 1964. Thermal equilibrium of the atmospherc with a convective adjustment. J Atmos Sci, 21 (4): 361-385.

Manabe S. Wetherald R T. 1967. Thermal equilibrium of the atmosphere with a given distribution of relative humidity. J Atmos Sci, 24(3): 241-259.

North G R. 1975a. Analytical solution to a simple climate model with diffusive heat transport. J Atmos Sci, 32(7): 1301-1307.

North G R. 1975b. Theory of energy-balance climate models. J Atmos Sci, 32(11): 2033-2043.

North G R. Coakley J A. 1979. Differences between seasonal and mean annual energy balance model calculations of climate and climate sensitivity. J Atmos Sci, 36(7): 1189-1204.

North G R, Cahalan R F,Coakley J A. 1981. Energy balance climate models. Rev Geophys, 19(1): 91-121

Oerlemans J, Van Den Dool H M. 1978. Energy balance climate models: stability experiments with a refined albedo and updated coefficients for infrared emission. J Atmos Sci, 35(3): 371-381.

Schmidt P W, Williams G E, Embleton B. 1991. Low palaeolatitude of Late Proterozoic glaciation: early timing of remanence in haematite of the Elatina Formation. South Australia. Earth Planet Sci Lett, 105(4): 355-367.

Schmidt P W. Williams G E. 1995. The Neoproterozoic climatic paradox: Equatorial palaeolatitude for Marinoan glaciation near sea level in South Australia. Earth Planet Sci Lett, 134(1): 107-124.

Sellers W D. 1969. A global climatic model based on the energy balance of the Earth-atmosphere system. J Appl Meteor, 8 (3): 392-400.

Sellers W D. 1976. A two dimensional global climatic model. Mon Wea Rev, 104(3): 233-248.

Sohl L E. Christie-Blick N, Kent D V. 1999. Paleomagnetic polarity reversals in Marinoan (ca. 600 Ma) glacial deposits of Australia: Implications for the duration of low-latitude glaciation in Neoproterozoic time. GeolSoc Am Bull, 111(8): 1120-1139.

Sumner D D, Kirschvink J L. Runnegar B N. 1987. Soft-sediment paleomagnetic fold tests of late Precambrian glaciogenic sediments. EOS, 68: 1251.

UNCCD. 2000. Fact Sheet 2: The Causes of Desertification. United Nations Secretariat of the Convention to Combat Desertification [2006-03-21J. http://www. unccd. int/publicinfo/factsheets/ showFS. php? Number=2[Geo-2-171].

孤立绿洲系统演化的动力学理论研究*②

李耀锟[1]*　巢纪平[2]

(1. 北京师范大学全球变化与地球系统科学研究院,北京 100875;2. 国家海洋环境预报中心,北京 100081)

摘要:假定绿洲和荒漠组成一个与周围环境无物质和能量交换的孤立系统,在能量守恒的条件下分析绿洲的分布和演变特征,结果表明,绿洲的演变会出现多平衡态分布的特征。在初始面积较小时,第一个平衡态表征了绿洲面积增加的解,此时荒漠地表-冠层温差较大,绿洲和荒漠之间存在较强的能量交换,绿洲通过降温增加面积而荒漠升温导致面积减小;第二个平衡态表征了绿洲面积减小的解,荒漠地表-冠层温差较为合适。若迁移后绿洲面积增加,则平衡态不稳定,能量迁移趋向于零,绿洲和荒漠能量各自达到平衡,绿洲面积最终等于初始面积。进一步分析表明,两个平衡态的性质完全不同,在绿洲面积有较大增加的平衡态中,绿洲和荒漠的反照率差异越小、水分差异越大,绿洲和荒漠间的能量交换越强,能量迁移后绿洲面积增加越多,而绿洲面积减小的平衡态则相反,绿洲和荒漠的反照率差异越大、水分差异越小,能量迁移后绿洲越容易维持在初始面积上。初值向平衡点的演化分析表明,荒漠地表温度较高的初值点更容易向绿洲面积增加的平衡态收敛。

关键词:能量守恒;绿洲;荒漠;双平衡态

绿洲是干旱区典型的自然景观。绿洲生态系统的维持和稳定为干旱区居民提供了赖以生存和发展的条件(张强和胡隐樵,2002)。我国西北地处干旱区,占国土面积4%的绿洲聚集着95%以上的人口和90%以上的社会财富(韩德林,1999;张强、胡隐樵,2002),而近50年来,西北地区绿洲化与沙漠化并存,植被覆盖总体有所退化(黄荣辉等,2011),这无疑对西北地区社会经济的可持续发展构成了严峻的挑战。绿洲的稳定与调控成为应对荒漠化的重要手段,绿洲演变的研究有着重要的理论和应用价值。

Charney(1975)最早从理论上将生物通过反照率变化的反馈机制引入到Sahel地区的干旱研究中,他指出由于植被破坏导致反照率增加,会进一步导致干旱的加剧和植被的减少。Dickinson(1984)开创性地将植被的蒸腾作用引入到大气模式中。此后陆面模式得到了较大发展,绿洲研究也逐步呈兴旺之势,然而现有研究主要集中在绿洲灌溉农业、绿洲土壤与土地利用、绿洲与荒漠过渡带的生物、气候等方面,有关绿洲基本理论研究则十分薄弱(王亚俊和曾凡江,2010)。为数不多的工作有,Pan等(2001)考虑绿洲地区的能量平衡分析了蒸发率的分岔解和绿洲面积突变的条件。在此基础上,吴凌云等(2003,2004)将绿洲的能量平衡与边界层大气运动相耦合,从理论上分析了绿洲的气候效应,巢纪平等(2005)进一步利用该理论模型分析了反照率和冠层阻抗对大气运动的影响。张强等

* 中国科学:地球科学,2015,45(3):305-317.

② 国家重点基础研究发展计划项目(编号:2014CB953903)和中央高校基本科研业务费专项资金项目(编号:2013YB45)资助

(2003)将绿洲系统近似为一个定态、开放的热力学系统,分析了绿洲系统各种内外因素对熵流的影响。这些工作提高了对绿洲的理论认识水平,对生产实践活动也具有一定的科学指导意义。因此,进一步加强绿洲的理论研究,有助于进一步完善绿洲研究,使绿洲研究更加系统化;有助于进一步总结各种经验和规律,为西部大开发深入开展提供更为科学合理的指导意见。

影响绿洲演变的因素非常多,从其他角度对绿洲演变的研究还有:樊自立(1993)按照塔里木盆地绿洲的形成历史将其演变分为古绿洲、旧绿洲和新绿洲,分析了自然因素与人为因素对古代绿洲衰亡的影响。穆桂金等(2000)将绿洲演变分为3种类型:人为干扰型、气候波动型及构造活动型。

本文在Pan等(2001)工作的基础上,将绿洲与其荒漠背景看作是一个孤立系统,利用下垫面的能量平衡,在系统能量守恒的条件下,引入绿洲和荒漠的能量交换分析绿洲的演变,得到了一系列有意义的结论。

1 绿洲的能量平衡

绿洲一词有多种定义,经过学术界的讨论,目前取得一些共识,如绿洲存在于干旱区、半干旱区的荒漠背景条件下;有水源保证或者稳定的水资源;适于植被繁茂生长等(高华君,1987;沈玉凌,1994;韩德林,1999)。为了简化分析,本文的"绿洲"在满足以上的特征的基础上,定义成完全为繁茂的植被所覆盖的区域。

绿洲植被冠层的净辐射能量为

$$R_c = (1 - \alpha_c)Q_a + \varepsilon_a\sigma T_a^4 - \varepsilon_c\sigma T_c^4 \tag{1}$$

式中,下标c表示冠层;α_c为植被的反照率;Q_a为到达冠层顶部的太阳辐射;$\varepsilon_a\sigma T_a^4$为大气向下的长波辐射;$\varepsilon_c\sigma T_c^4$为冠层向上的长波辐射。$\varepsilon_a$,$\varepsilon_c$为相应的比辐射率,$\sigma$为斯蒂芬-波尔兹曼常数,$T_a$为冠层顶部的大气温度,$T_c$为植被冠层温度。$T_a$,$T_c$为绝对温度。利用

$$T_c^4 = (T_a + T_c - T_a)^4 \approx T_a^4 + 4T_a^3(T_c - T_a) \tag{2}$$

可将冠层净辐射能式(1)近似写成

$$R_c = (1 - \alpha_c)Q_a + (\varepsilon_a - \varepsilon_c)\sigma T_a^4 + 4\varepsilon_c\sigma T_a^3(T_a - T_c) \tag{3}$$

式中,$(\varepsilon_a-\varepsilon_c)\sigma T_a^4$可称为等温净长波辐射(Monteith,1981)。

植被冠层和周围空气间的感热通量,本质上是湍流扩散过程,用空气动力学的参数化方案来计算感热通量

$$H_c = \frac{\rho_a C_p(T_c - T_a)}{r_a} \tag{4}$$

式中,ρ_a为空气密度;C_p为空气的定压比热;$r_a=(C_d\bar{V})^{-1}$为空气阻抗;C_d为空气的拖曳系数;$\bar{V}$为冠层风速的参考值。

潜热通量来自空气中水分的蒸发和叶面的蒸腾,可近似写成

$$E_a = \frac{\rho_a L_v}{r_a + r_c}[q^s(T_c) - rq^s(T_a)] \tag{5}$$

式中,L_v为相变潜热;r_c为冠层阻抗;$q^s(T)$为温度T时的饱和比湿;r为空气的相对湿度。由于蒸腾,叶面附近空气中水分充分,已达到饱和,故取饱和比湿$q^s(T_c)$,而大气未达到饱和,利用相对湿度将其写为$rq^s(T_a)$。将饱和比湿在$T=T_a$处展开

$$q^{s}(T)=q^{s}(T_{a})+\left.\frac{\partial q^{s}}{\partial T}\right|_{T=T_{a}}(T-T_{a}) \tag{6}$$

于是潜热通量写成

$$E_{c}=\frac{\rho_{a}L_{v}}{r_{a}+r_{c}}[(1-r)q^{s}(T_{a})+\Delta(T_{c}-T_{a})] \tag{7}$$

式中，已令 $\Delta=\left.\frac{\partial q^{s}}{\partial T}\right|_{T=T_{a}}$。潜热通量式(5)实际上是把单片叶子的气孔阻抗(Penman, 1948)推广到了植被冠层表面。单层模型很难区分植物蒸腾和土壤蒸发，因此在计算地面完全为低矮植被覆盖时效果较好(陈曦，2012)。

根据式(3)，式(4)和式(7)，绿洲植被冠层的能量平衡为

$$T_{c}=T_{a}+n_{c}\equiv f_{c}(T_{a}) \tag{8}$$

式中

$$n_{c}=\frac{(1-\alpha_{c})Q_{a}-(\varepsilon_{c}-\varepsilon_{a})\sigma T_{a}^{4}-\frac{\rho_{a}L_{v}}{r_{a}+r_{c}}(1-r)q^{s}(T_{a})}{4\varepsilon_{c}\sigma T_{a}^{3}+\frac{\rho_{a}C_{p}}{r_{a}}+\Delta\frac{\rho_{a}L_{v}}{r_{a}+r_{c}}}$$

上面式(8)给出了能量平衡条件下植被冠层温度与大气温度的制约关系。根据相关研究(张强、胡隐樵，2002；黄荣辉等，2011；陈曦，2012)，取 $T_{a}=10℃$，$Q_{a}=180\ \mathrm{W/m^{2}}$，$\alpha_{c}=0.15$，$C_{d}=2.5\times10^{-3}$，$\overline{V}=2\ \mathrm{m/s}$，$r_{c}=150\ \mathrm{s/m}$，$r=0.3$，$\varepsilon_{a}=0.83$，$\varepsilon_{c}=0.96$ 给出冠层温度与其上空气温度之差(冠-气温差)随各个参数的变化(图 1)。图中一个参数变化时，其他参数均固定在给定的参考值上。

阻抗增大不利于冠层与大气之间的能量交换，因此，冠-气温差随空气动力学阻抗和冠层阻抗的增加而增加(图 1(a)和图 1(b))。植被的反照率的增加导致更多的能量被反射，冠层温度会下降(图 1(c))。相对湿度表征了冠层上方空气的湿润程度，空气相对湿度增大时，冠层蒸发减小温度升高，导致冠-气温差随相对湿度的增大而增大(图 1(d))。

应当指出，以上分析的各个参数均存在复杂的变化特征，且不同参数间还存在着相互关联。陆面模式物理过程更为细致，各个参量的取值较为复杂，依赖的参数非常之多(Dickinson 等，1993；Dai 等，2001)，本文中能量平衡与陆面模式是基本一致的，不同之处在于为了便于理论分析，只是给出了各参量的参考气候态分布。这么做虽然比较粗糙，但是却也能抓住陆面过程的物理本质，便于从理论上分析其整体效应。

2　荒漠的能量平衡

本文中荒漠是指绿洲周围地表完全无植被覆盖的区域，与绿洲类似，荒漠的净辐射能量可以写为

$$R_{s}=(1-\alpha_{s})Q_{a}+\varepsilon_{a}\sigma T_{a}^{4}-\varepsilon_{s}\sigma T_{s}^{4} \tag{9}$$

式中，下标 s 表示荒漠属性；α_{s} 为荒漠的地表反照率；T_{s} 为地表温度；ε_{s} 为地表的比辐射率。

感热、潜热通量分别写为

$$H_{s}=\frac{\rho_{a}C_{p}(T_{s}-T_{a})}{r_{a}} \tag{10}$$

$$E_{s}=\frac{\rho_{a}L_{v}}{r_{a}}[\alpha q^{s}(T_{s})-rq^{s}(T_{a})] \tag{11}$$

潜热通量的计算采用了 α 法，α 表示在土壤表层空气的相对湿度，可用土壤近表层的含水量 w_s（单位体积土壤中含有水的体积）来表示（Barton，1979）

$$\alpha = \begin{cases} \dfrac{1.8w_s}{w_s + 0.3}, & w_s < 0.375; \\ 1, & w_s \geq 0.375 \end{cases} \tag{12}$$

式中，0.375 可视为土壤的饱和持水量，α 还有其他类似的参数化方法（Yasuda 和 Toya，1981；Noilhan 和 Planton，1989）。采用与绿洲类似的处理方法，最终可将荒漠地表的能量平衡归纳为

$$T_s = T_a + n_s \equiv f_s(T_a), \tag{13}$$

式中

$$n_s(T_a) = \frac{(1-\alpha_s)Q_a - (\varepsilon_s - \varepsilon_a)\sigma T_a^4 - (\alpha - r)\dfrac{\rho_a L_v}{r_a}q^s(T_a)}{4\varepsilon_s \sigma T_a^3 + \dfrac{\rho_a C_{pa}}{r_a} + \alpha\Delta\dfrac{\rho_a L_v}{r_a}}$$

取地表反照率 $\alpha_s = 0.25$，土壤含水量 $w_s = 0.1$，分别计算荒漠地表土壤与大气的温差（地-气温差）随各参数的变化（图 2）。土壤湿润会增加土壤中水分的蒸发量，降低土壤温度，减小地-气温差（图 2（b））。其他参数不再具体分析，可参见图 1。

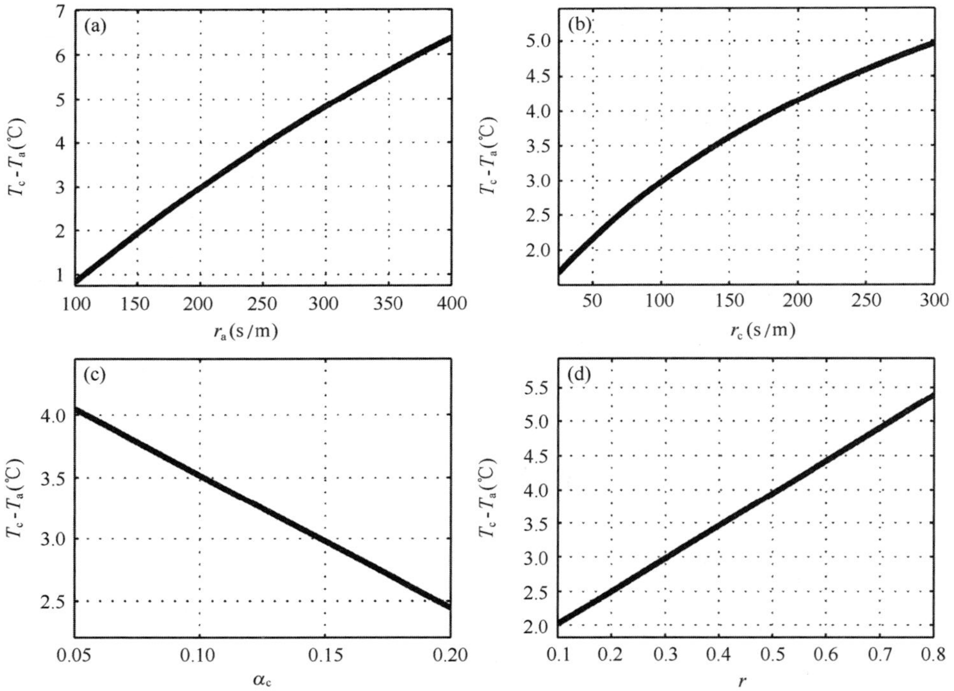

图 1 冠-气温差随空气动力学阻抗（a），冠层阻抗（b），植被反照率（c）及相对湿度（d）的变化

3 能量守恒

以上分析中绿洲和荒漠各自达到了能量平衡，彼此间没有能量交换，本节进一步讨论能量可以在绿洲与荒漠间互相迁移时绿洲的演变特征。令绿洲与其周围的荒漠与外界没有能量交换，为一孤立

系统，则绿洲与荒漠的总能量守恒。

设初始时绿洲的面积为 a_0，则系统的总能量可以写为

$$E_0 = a_0 C_c T_{c,0} + (1 - a_0) C_s T_{s,0} \tag{14}$$

式中，C_c 为植被冠层的热容量；C_s 为土壤的热容量；$T_{c,0}$，$T_{s,0}$ 分别为植被和荒漠各自能量平衡关系式(8)和式(13)对应的温度，写为

$$T_{c,0} = f_c(T_a) \tag{15}$$

$$T_{s,0} = f_s(T_a) \tag{16}$$

则在给定初始面积给定的条件下，式(14)为一已知量。

能量迁移时绿洲的面积相应发生改变，令迁移时绿洲的面积为 a，植被温度为 T_c，地表温度为 T_s，于是迁移时系统的总能量为

$$E = aC_c T_c + (1 - a) C_s T_s \tag{17}$$

由于假定系统与外界没有能量交换，系统的能量守恒，即可得到

$$aC_c T_c + (1 - a) C_s T_s = E_0 \tag{18}$$

式(18)给出了在能量守恒条件下绿洲面积与绿洲植被温度、荒漠地表温度所满足的约束关系。

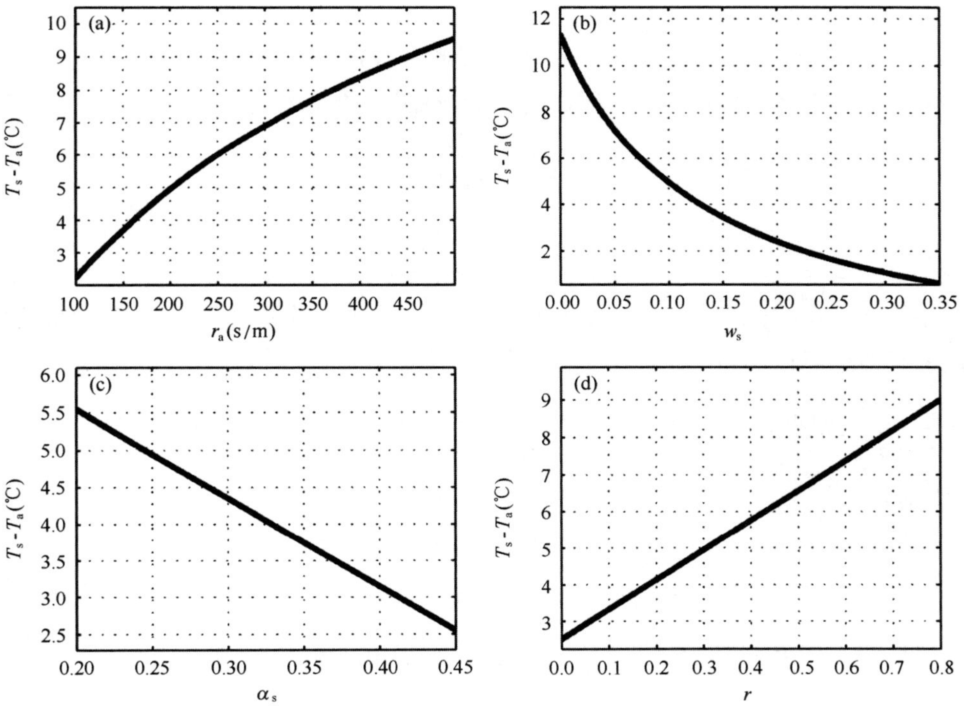

图 2　地-气温差随空气动力学阻抗(a)，土壤含水量(b)，土壤反照率(c)及空气相对湿度(d)的变化

4　动力系统及其平衡态

能量迁移时绿洲植被和荒漠各自的能量平衡关系式(15)和式(16)两式均不再满足，其温度方程可写为

$$\tau_c \frac{dT_c}{dt} = (1 - q_c)\varphi_c + \frac{1 - a}{a} q_s \varphi_s \equiv F(T_c, T_s) \tag{19}$$

$$\tau_s \frac{dT_s}{dt} = (1 - q_s)\varphi_s + \frac{a}{1-a} q_c \varphi_c \equiv G(T_c, T_s) \tag{20}$$

式中

$$\varphi_c = f_c(T_a) - T_c \tag{21}$$

$$\varphi_s = f_s(T_a) - T_s \tag{22}$$

τ_c, τ_s 表示能量迁移的时间尺度。q_c, q_s 分别为植被向荒漠、荒漠向绿洲的能量迁移,设其可写为

$$q_c = \eta a(T_c - T_s) \tag{23}$$

$$q_s = \eta(1 - a)(T_s - T_c) \tag{24}$$

式中,η 为比例系数,称为能量迁移系数。

动力系统式(19)和式(20)的平衡点满足

$$F(T_c^*, T_s^*) = 0 \tag{25}$$

$$G(T_c^*, T_s^*) = 0 \tag{26}$$

式(25)和式(26)两式相加可得

$$a^* \varphi_c^* + (1 - a^*)\varphi_s^* = 0 \tag{27}$$

将式(27)式代入式(25)

$$\eta(T_c^* - T_s^*)(2a^* - 1) = 1 \tag{28}$$

将式(28)代入(23)和(24)两式,可得达到平衡态时的能量迁移为

$$q_c^* = \frac{a^*}{2a^* - 1} \tag{29}$$

$$q_s^* = -\frac{1 - a^*}{2a^* - 1} \tag{30}$$

由式(27),式(29)和式(30)可知,当系统达到平衡态时,系统总的剩余能量为零,能量迁移分别只与绿洲和荒漠的平衡态面积有关。

令 $X = \frac{a^*}{1 - a^*}$,根据式(18),式(27)和式(28)三式消去 T_c, T_s 可得到变量 X 的一个一元三次方程(推导见附录)

$$\alpha X^3 + \beta X^2 + \chi X + \delta = 0 \tag{31}$$

$$\begin{aligned} \alpha &= \pi_1 \\ \beta &= (C_s - C_c)\left[\frac{1}{\eta} + (T_{s,0} - T_{c,0})\right](1 + X_0) - \pi_1 X_0 \\ \chi &= (C_s - C_c)\left[\frac{1}{\eta} + (T_{s,0} - T_{c,0})\right](1 + X_0) - \pi_1 \\ \delta &= \pi_1 X_0 \end{aligned} \tag{32}$$

式中,$X_0 = \frac{a_0}{1 - a_0}$,$\pi_1 = C_s T_{s,0} - C_c T_{c,0}$。当 $\pi_1 = 0$ 时,有 $E_0 = C_c T_{c,0} = C_s T_{s,0}$,此时式(31)退化为二次方程,其两个根为 $X = 0, X = -\chi/\beta$,即 $a = 0$,$a = \frac{1}{2} - \frac{1}{2\eta(T_{s,0} - T_{c,0})}$,即能量迁移后绿洲植被面积要么变为0,要么在0.5附近,这表明荒漠地表-冠层温差和迁移系数的加大均有利于绿洲和荒漠之间的能量交换,较强的能量交换将使绿洲和荒漠各自占据50%的面积,而无论绿洲的初始面积有多大或者多

小。一般而言，$\pi_1>0$，因此式(31)为三次方程。利用三次方程的判别式(A6)，即可讨论三次方程的根在不同参数下的变化特征。

平衡点(T_c^*，T_s^*)处的 Jacobi 矩阵为

$$\boldsymbol{J}=\begin{bmatrix}\dfrac{\partial F}{\partial T_c} & \dfrac{\partial F}{\partial T_s}\\ \dfrac{\partial G}{\partial T_c} & \dfrac{\partial G}{\partial T_s}\end{bmatrix}_{(T_c^*,T_s^*)} \tag{33}$$

其特征值 λ 满足

$$\begin{vmatrix}\dfrac{\partial F}{\partial T_c}-\lambda & \dfrac{\partial F}{\partial T_s}\\ \dfrac{\partial G}{\partial T_c} & \dfrac{\partial G}{\partial T_s}-\lambda\end{vmatrix}_{(T_c^*,T_s^*)}=0 \tag{34}$$

若对某平衡点而言，若特征值的实部全部是非零的，即 $\mathrm{Re}\lambda\neq0$，则系统称为双曲的。若所有特征值 $\mathrm{Re}\lambda<0$，则该平衡点是稳定的；若至少有一个特征值 $\mathrm{Re}\lambda>0$，则该平衡点是不稳定的(刘式适、刘式达，2011)。

5　平衡态随参数的变化

根据戴永久等(1997)、黄荣辉等(2011)，植被冠层、土壤的体积热容量分别取为 $C_c=4.23\times10^3\ \mathrm{J/(m^3\cdot K)}$，$C_s=1.12\times10^6\ \mathrm{J/(m^3\cdot K)}$。推导过程引入了初始面积 a_0，迁移系数 η 两个参数，分析解随这两个参数的分布特征(图 3 和图 4)。图中斜线阴影围住的区域表示三次方程式(31)判别式大于 0，此时方程只有一个实数解，该实数解绿洲面积不在 0 到 1 之间，是无物理意义的计算解，已在图中舍去。斜线阴影区域以外，式(31)有 3 个实数解，其中一个无物理意义，也已舍去，因此方程式(31)存在两个有物理意义的实数解，即动力系统存在着两个平衡态。利用式(34)计算平衡点的 Jacobi 矩阵的特征值 λ，计算结果表明，在竖阴影线围住的区域，所有特征值均满足 $\mathrm{Re}\lambda<0$，即平衡态是稳定的，在无阴影区域，至少有一个 $\mathrm{Re}\lambda>0$，平衡点是不稳定的。

斜阴影区将解分成为两个部分：左边的一段绿洲初始面积较小(小于 0.5)；而右边的一段绿洲初始面积较大(大于 0.5)。根据第一个平衡态(图 3)，当绿洲初始面积小于 0.5 时，能量迁移后绿洲面积会增加到接近 0.5，相比初始面积有了较大的增加，但是荒漠地表-冠层温差较大，可达 10℃以上，较大的温差有利于绿洲和荒漠之间维持较强的能量交换。稳定性分析表明，此时平衡态是稳定的。而当绿洲初始面积大于 0.5 时，能量迁移后绿洲面积会减小，荒漠地表-冠层温差为负数，即冠层温度要高于荒漠地表温度，此时平衡态不稳定，即稍有扰动，绿洲面积将演化到初始面积上，绿洲和荒漠各自达到能量平衡，能量交换最终为 0。

第二个平衡态(图 4)表明，在绿洲初始面积较小(小于 0.5)时，能量迁移后绿洲面积略有减小，荒漠地表-冠层温差不大，绿洲和荒漠之间的能量交换较弱，当迁移系数较小时，绿洲面积甚至可以增加。稳定性分析表明，以绿洲初始面积为界，若迁移后绿洲面积小于初始面积，则平衡态是稳定的(图 4 中左端竖线阴影区)，若迁移后绿洲面积增加，则平衡态是不稳定的。当绿洲初始面积大于 0.5 时，迁移后绿洲面积减小到接近 0.5，同时，荒漠地表-冠层温差为负数，且较大。在该段内，平衡态是稳定的。

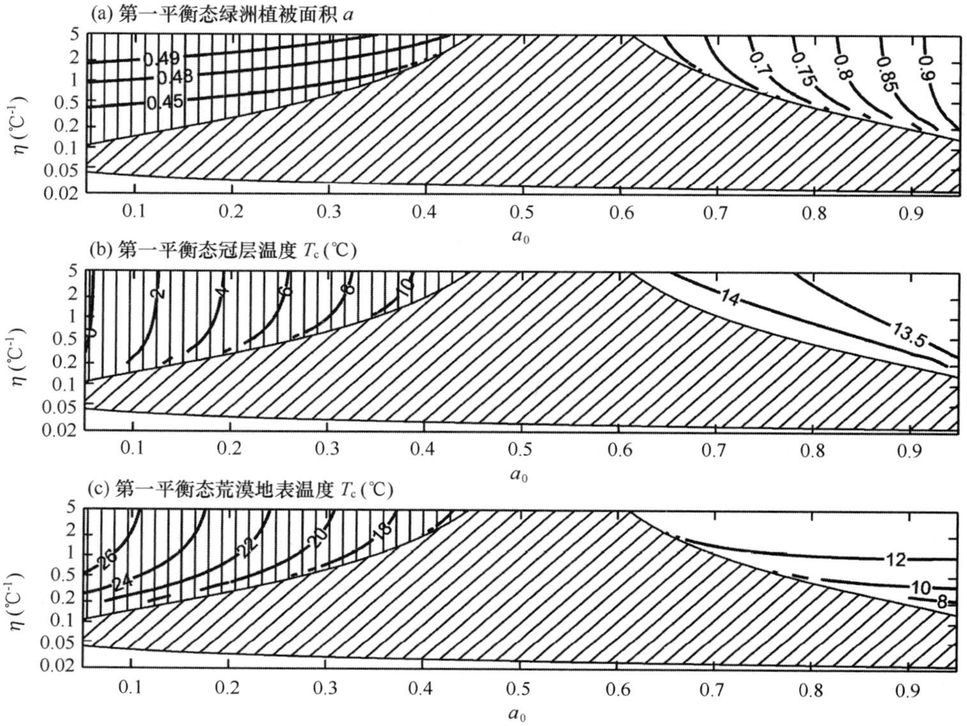

图 3　第一平衡态在参数(η, a_0)空间内的分布

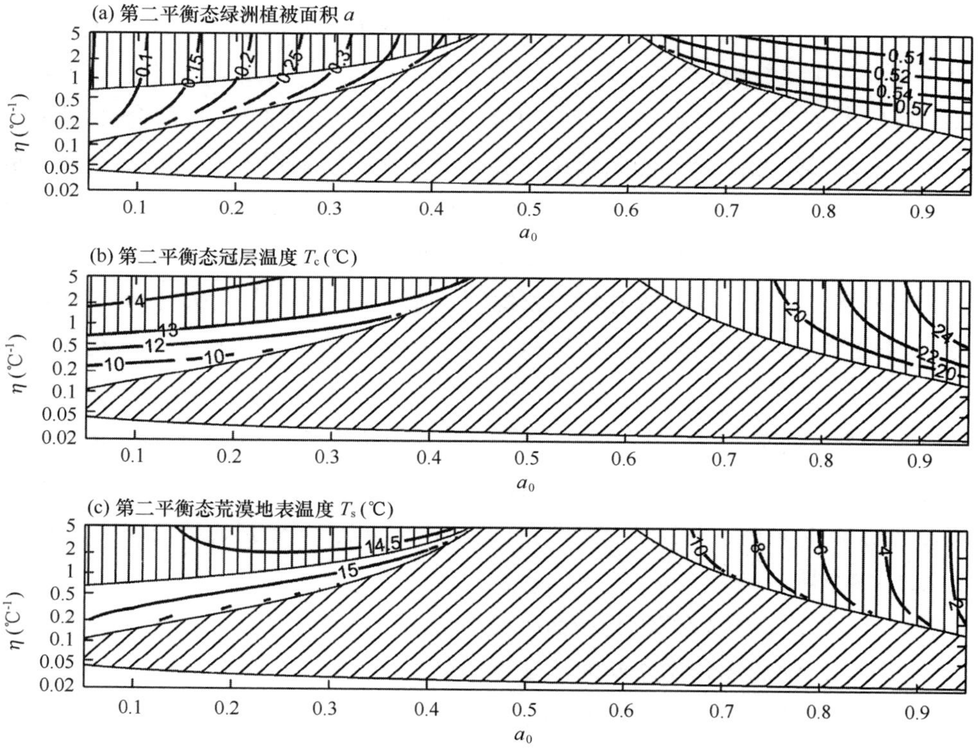

图 4　第二平衡态在参数(η, a_0)空间内的分布

以上分析也可以从式(28)中得到，当 $a^*<0.5$ 时，有 $T_c^*<T_s^*$，即荒漠地表-冠层温差为正数，在 η 固定时，a^* 越接近于 0.5，则 $T_c^*-T_s^*$ 越大。当 $a^*>0.5$ 时，有 $T_c^*>T_s^*$，即荒漠地表-冠层温差为负数，a^* 越接近于 0.5，则 $T_c^*-T_s^*$ 越大。稳定性分析表明，与初始面积相比，第一个平衡态绿洲面积增加是稳定的，而第二个平衡态绿洲面积减小是稳定的。一般而言，绿洲面积要远远小于荒漠面积，且荒漠地表温度要高于植被冠层的温度，因此，下文中只分析绿洲初始面积小于 0.5 时的情形。

两个平衡态表明绿洲演化存在着两个完全不同的轨迹。第一个平衡态是绿洲面积增加的解，即能量迁移将使绿洲面积扩张，对西北干旱区绿洲的保持和发展无疑是有利的，但是该平衡态荒漠地表-冠层温差较大，此时绿洲和荒漠之间存在着较为强烈的能量交换。第二平衡态是绿洲面积减小或维持的解，即能量迁移不利于绿洲面积的增加，这似乎与实际情况更为接近一些。

根据以上分析，取 $\eta=1℃^{-1}$，$a_0=0.2$，分析平衡态在不同植被反照率 α_c，叶面阻抗 r_c 下的分布特征，如图 5 和图 6 所示，图中竖线阴影围住的区域表示平衡态是稳定的。

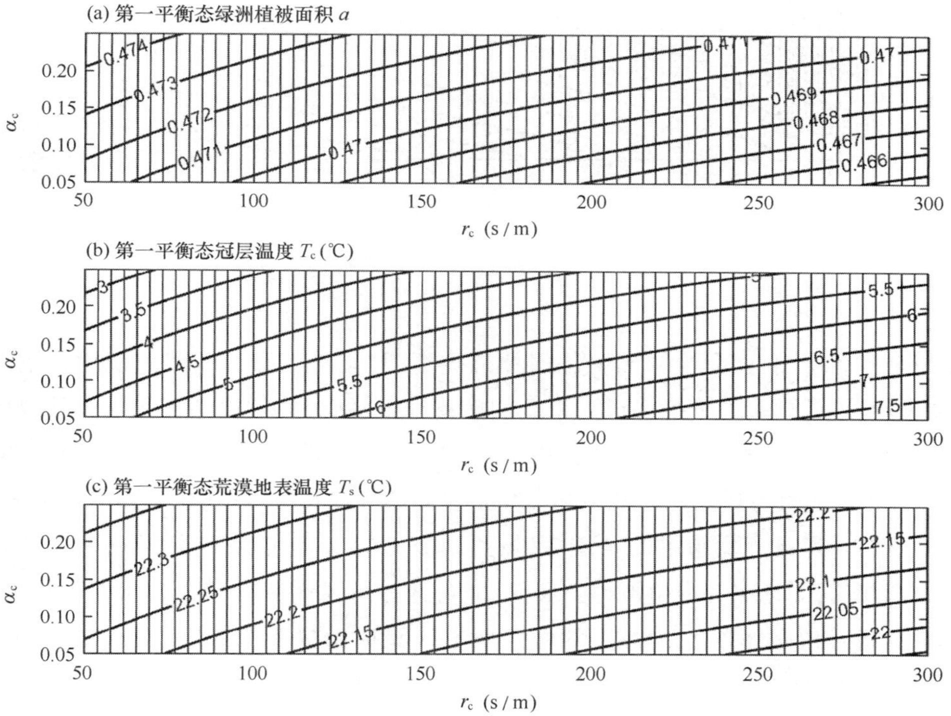

图 5　第一平衡态在参数(r_c，α_c)空间内的分布

第一个平衡态(图 5)表明，迁移后绿洲面积由初始值 0.2 增加到 0.4 以上，增加了 1 倍以上。α_c 越大，r_c 越小，冠层温度越低而荒漠地表温度越高，绿洲与荒漠间的能量交换就越强，绿洲面积增加越多，这一平衡态中，能量迁移后荒漠地表-冠层温差可达 10℃以上，冠层温度甚至会低于大气温度。第一个平衡态在所取的参数空间内均是稳定的。第二个平衡态(图 6)表明，与初始面积相比，迁移后绿洲面积变化幅度不大，均在 0.2 附近浮动。与第一个平衡态相反，α_c 越大，r_c 越小，迁移后绿洲面积越小，甚至会低于初始面积，造成绿洲面积缩减。冠层温度和荒漠地表温度均随 α_c 增加，r_c 减小而降低，荒漠地表-冠层温差也在逐步减小，不过变化幅度不大，在 3～4℃。迁移后绿洲面积减小，平衡态是稳定的，绿洲面积增加对应的平衡态是不稳定的。

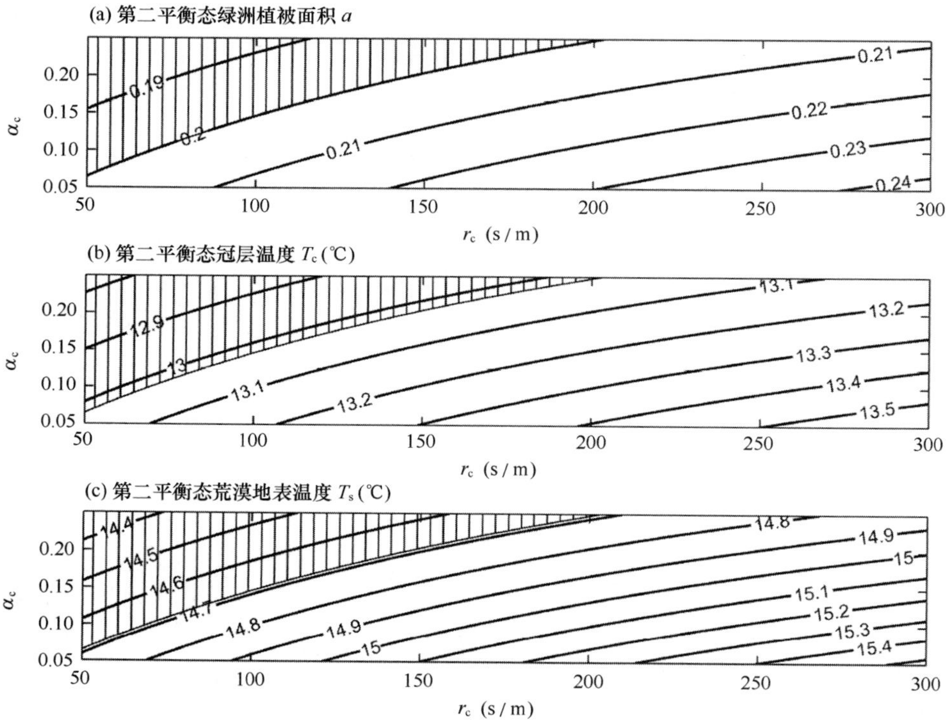

图 6 第二平衡态在参数(r_c,α_c)空间内的分布

对于第一个平衡态而言,无论绿洲中是何种植被类型,能量迁移后绿洲面积均有较大幅度的增加,根据前文理论分析可知,为了维持孤立系统的能量守恒,绿洲通过增加面积来弥补温度降低造成的能量损失。绿洲中植被对水分需求越大(叶面阻抗越小)、反照率越大,绿洲面积增加的幅度越大。Zuo 等(2011)分析观测资料后指出植被增加会减小表面的加热作用,其作用类似一个异常的冷源,这从观测的角度说明了理论分析结果的合理性,虽然观测可能未必完全符合理论分析所需要的条件。第二个平衡态则相反,无论绿洲中植被种类如何,能量迁移后绿洲面积会比初始面积有所减小(增加时不稳定,最终回到初始面积),但是减小的程度非常小,此时植被蒸腾消耗的水分越少越有利于绿洲面积维持或略有减小,即在缺水地区种植耐旱的植被更为合理一些。

图 7 和图 8 给出了平衡态随荒漠反照率 α_s、土壤含水量 w_s 的分布特征。与前类似,斜线阴影围住的区域表明在该参数范围内三次方程只有一个实数解,为无物理意义的计算解。剩余区域内三次方程有 3 个实数解,其中一个无物理意义。由图可知当地表反照率 α_s、土壤含水量 w_s 太大时无平衡态解对应。竖线阴影围住的区域表示平衡态是稳定的。

第一个平衡态(图 7)表明,与初始面积相比,能量迁移后绿洲面积增加了 1 倍左右。地表反照率越小,土壤含水量越小,冠层温度越低而荒漠地表温度则越高(荒漠地表-冠层温差最大可达 30℃以上),绿洲与荒漠之间的能量交换越强,绿洲面积越容易增加。在给定参数域中,第一个平衡态是稳定的。一般而言,荒漠地表反照率要更大一些,这一平衡态意味着荒漠与绿洲在反照率上的差异越小,能量迁移后绿洲面积越大,同时,荒漠与绿洲在湿度上的差异越显著,也越有利于绿洲面积增加。第二个平衡态(图 8)表明,迁移后绿洲面积变化不大,基本能维持在初始面积附近。与第一个平衡态相反,地表反照率越大,土壤越湿润,绿洲面积越容易增加,而冠层温度和荒漠地表温度也越低。荒漠地表-冠层温差的变化范围不大，在 2℃附近。第二个平衡态当绿洲面积减小时是稳定的。这一平衡态中荒漠和绿洲反照率差异较大、荒漠土壤较为湿润均有利于绿洲维持在初始面积上。虽然实际中很难找到完全符合理论分析的例子，但是根据观测资料分析找到一些可以类比的例子，如春季长江中下游到华北地区异常偏湿的土壤会增加水分的蒸发，最终导致长江中下游地区夏季降水增加（Zhang and Zuo，2011），降水增加有利于植被维持在其原有的面积上。

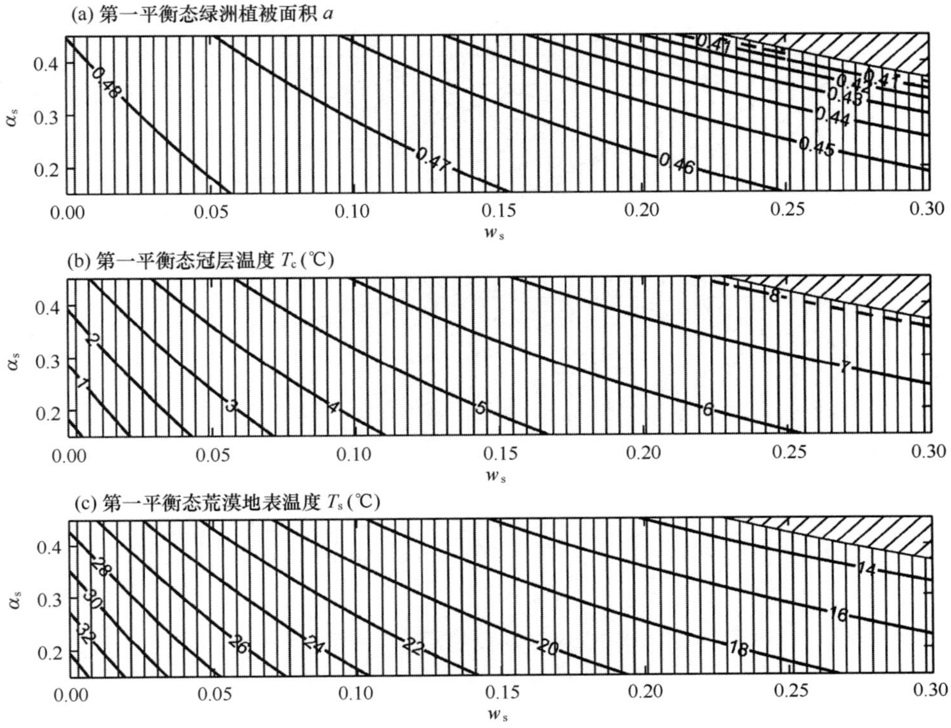

图 7　第一平衡态在参数（w_s，α_s）空间内的分布

综上所述，考虑能量迁移后绿洲演变会出现三个平衡态，舍去一个没有物理意义的计算解，剩余两个平衡态显示了绿洲演变截然不同的两个特征。在给定初始面积和迁移系数的条件下，第一个平衡态表明能量迁移后绿洲面积有较大的增加，面积甚至可以增加 1 倍以上，但对应的荒漠地表-冠层温差则较大，可达 10℃以上。绿洲和荒漠反照率差异越小（植被反照率增加而荒漠地表反照率减小）、水汽差异越大（较小的叶面阻抗意味着植被蒸腾作用较为旺盛，需要较多的水分供应，较低的土壤湿度意味着土壤含水量少，较为干燥）二者之间的能量交换就越强烈，绿洲面积增加的也越多。第二个平衡态表明迁移后绿洲面积不会有明显变化，这一平衡态要求的荒漠地表-冠层温差在较为合适的范围。绿洲和荒漠反照率差异越大（植被反照率较小而荒漠地表反照率较大）、水汽差异越小（叶面阻抗较大的植被蒸腾作用较弱，更为耐旱，而荒漠土壤水分含量增加使二者水分差异减小）绿洲越容易维持在初始面积上。阻抗稳定性分析则表明，第一个平衡态中绿洲面积增加是稳定的，而第二个平衡态中绿洲面积减小是稳定的。

从能量守恒的观点来看，绿洲较为稳定，不容易发生绿洲面积大幅减小的情况，这对绿洲的发展无疑是有利的，但是这并不意味着人类活动不应该受到任何限制。对于人类生产生活而言，绿洲面积增加更为有利，但第一个平衡态出现的可能性似乎较低，实际中较难找到合适的例子，如何让绿洲向

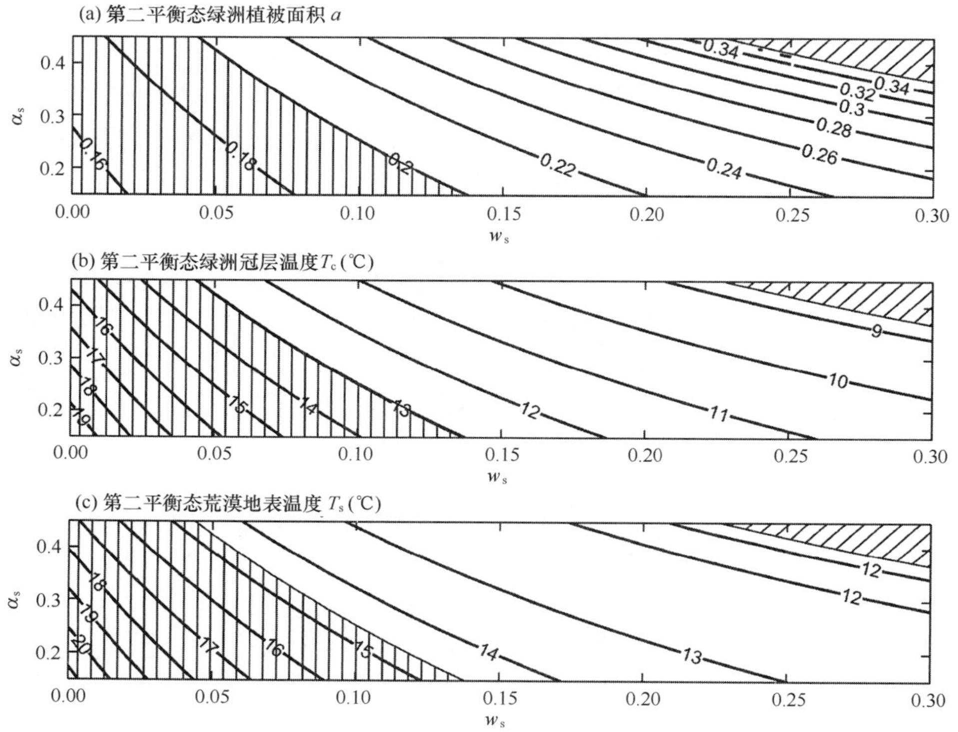

图 8 第二平衡态在参数(w_s, α_s)空间内的分布

着人类有利的方向演变无疑是一个非常有意义的研究课题。

6 不同初值在相空间内的轨迹

考察动力系统的时间积分在相空间中的轨迹。注意到对张弛时间 τ_c,τ_s 的大小没有给出确切的定义,计算表明它们的大小并不会对平衡点的位置,但却影响达到平衡点所需的时间及不同初值点在相空间内的轨迹。如果令 $\tau_c=\tau_s=r_a \cdot K$,其中 K 为热量的湍流交换系数,r_a 为空气的动力学阻抗,若取 $K=10^6$ W/(m · K),可计算得到特征时间 $t=K \cdot r_a^2/(\rho \cdot C_p)$ 约为 1 年。

选取$(\eta=1, a_0=0.2)$计算平衡态解在相空间内的轨迹(图 9)。图中 A,B 两点分别表示迁移后的两个平衡点。通过 A,B 点的虚线分别为各自的分型线。实线表示不同的初值随时间的演变过程。

由图 9 可知,迁移后平衡点 A 对应着绿洲面积有较大增加的解,而平衡点 B 则对应着绿洲面积略有减小的解,此时面积减小的程度非常小,基本仍维持在初始面积上。时间积分表明,两个平衡点都是稳定的,这与前文的分析是一致的。当平衡点 B 的对应的绿洲面积大于初始面积时,平衡点 B 是不稳定的,其附近的初值点会向$(T_{c,0}, T_{s,0})$点收敛,该点表示绿洲和荒漠之间没有能量交换(图略)。由此可见,对于第二个平衡点而言,若绿洲面积增加,则绿洲和荒漠间的能量迁移最终将趋于 0,能量各自达到平衡,绿洲面积最终将回归到初始面积上。

初值收敛到哪一个平衡点是取决于其与分型线在相平面上的相对位置。平衡点 A 附近的初值,最终沿其分型线达到 A 点。平衡点 B 附近的初值,或沿着 A 点的分型线向 A 点收敛,或沿着 B 点分型线向 B 点收敛。通过计算可以发现,B 点斜率(绝对值)较小的分型线决定的不同初值的收敛方向,当初值点在该分型线之上时,即使此时初值更靠近平衡点 B,初值仍将向平衡点 A 收敛;同理,当初值

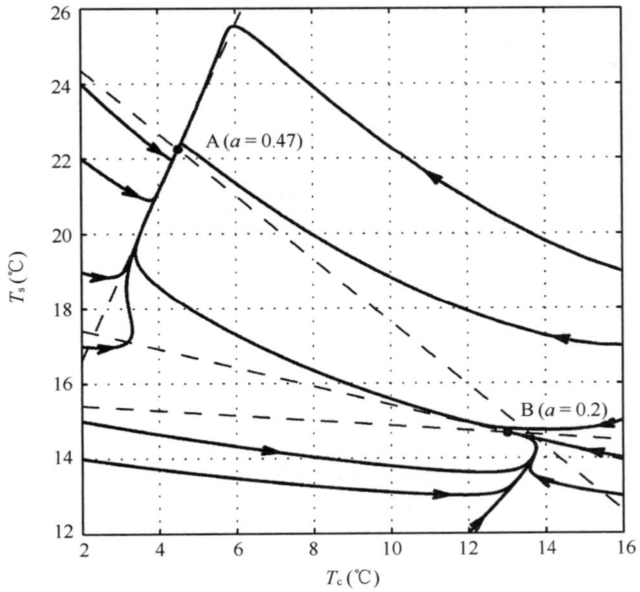

图 9　($\eta=1, a_0=0.2$)时不同初值在相空间内的轨迹

点在该分型线下方时,即使此时初值更靠近平衡点 A,初值仍将向平衡点 B 收敛；当初值恰好在该分型线上时,初值将向平衡点 B 收敛。以上分析表明,初值对荒漠地表温度更为敏感,分型线附近很小的荒漠地表温度差别都会让初值收敛到不同的平衡点上,温度增加,特别是荒漠地表温度增加,对初值向绿洲面积增加的平衡点 A 演化十分有利,这似乎是使绿洲面积有较大增加的一个有效途径。当然,其中涉及到具体的物理过程仍有待于进一步分析讨论。

7　结论

本文首先分别考虑绿洲和荒漠地区的能量平衡,给出了冠-气温差,地-气温差随各参数的变化,进一步引入绿洲和荒漠能量交换的概念,将绿洲演变归结为一个动力系统,分析了动力系统平衡态随参数的分布及其稳定性。

绿洲演变存在三个平衡态,其中一个无物理意义,已舍去。在给定参数的条件下,第一个平衡态是绿洲面积增加的解。植被反照率增加、叶面阻抗减小、土壤湿度减小、地表反照率减小均有利于绿洲面积的增加。这一平衡态要求荒漠地表-冠层温差较大,甚至可达 10℃以上,较大的温差有利于绿洲和荒漠之间维持较强的能量交换。绿洲和荒漠间能量交换达到动态的平衡,平衡态也是稳定的。实际中很难找到天然绿洲面积大幅增加的例子,这表明绿洲面积增加的平衡态需要满足较多的条件,在自然状态下不容易发生。

第二个平衡态是绿洲面积减小的解。以初始面积为界,当能量迁移后绿洲面积增加时,平衡态是不稳定的,此时绿洲和荒漠能量交换将趋向于零,最终各自达到能量平衡,而绿洲面积回到初始面积上。当迁移后绿洲面积减小时,平衡态是稳定的,此时绿洲面积虽不会增加,但是也不会有太大的减小,基本可以维持在初始面积附近。与第一个平衡态不同,植被反照率减小、叶面阻抗增加、土壤湿度增加、地表反照率增加均有利于绿洲面积的维持甚至增加(增加时平衡态不稳定)。这一平衡态要求的荒漠地表-冠层温差在较为合适的范围之内,似乎更符合惯常理解下的气候状态。

不同初值在相空间内的演变表明,平衡点 B 斜率(绝对值)较小的分型线起着至关重要的作用,当初值在该分型线上方时,不同的初值最终都沿着平衡点 A 的分型线向其收敛,绿洲面积有较大增加;而当初值在该分型线下方时,初值最终沿着平衡点 B 的分型线向其收敛,绿洲面积则略有减小。

由于绿洲的演变极为复杂,受到多种因素的影响,本文假定绿洲和荒漠为一孤立系统,且没有考虑水源等影响绿洲演变的重要因子,所得的结果只适用于绿洲和荒漠能量存在交换的情景,即使在此种情况下,绿洲演变仍然会出现两个平衡态,其演变性质完全不同,仍有必要进一步对其分析研究。

附录

由式(27)可得

$$\varphi_s = -X\varphi_c \tag{A1}$$

将式(A1)代入式(28)可得

$$T_s = T_c + \frac{1}{\eta} \cdot \frac{1+X}{1-X} \tag{A2}$$

式(A2)代入式(18)中,消去 T_s 后得到

$$(C_s + XC_c)T_c = (1+X)E_0 - C_s\frac{1}{\eta} \cdot \frac{1+X}{1-X} \tag{A3}$$

式(A2)代入式(A1)中消去 T_s

$$(1+X)T_c = T_{s,0} + XT_{c,0} - \frac{1}{\eta} \cdot \frac{1+X}{1-X} \tag{A4}$$

式(A3),式(A4)两式消去 T_c,即可得到关于 X 的单变量方程

$$\alpha X^3 + \beta X^2 + \chi X + \delta = 0 \tag{A5}$$

而

$$\alpha = \pi_1,$$
$$\beta = (C_s - C_c)\left[\frac{1}{\eta} + (T_{s,0} - T_{c,0})\right](1+X_0) - \pi_1 X_0,$$
$$\chi = (C_s - C_c)\left[\frac{1}{\eta} - (T_{s,0} - T_{c,0})\right](1+X_0) - \pi_1,$$
$$\delta = \pi_1 X_0$$

式中,$X_0 = \frac{a_0}{1-a_0}$,$\pi_1 = C_s T_{s,0} - C_c T_{c,0}$。

方程(A5)是变量 X 的一元三次方程,求解方程即可得到面积 a 的分布。方程(A5)的判别式(范盛金,1989)写为

$$\Delta = B^2 - 4AC \tag{A6}$$

式中

$$A = \beta^2 - 3\alpha\chi,$$
$$B = \beta\chi - 9\alpha\delta,$$
$$C = \chi^2 - 3\beta\delta$$

当 $\Delta<0$ 时,方程(A5)有 3 个实数根,当 $\Delta>0$ 时,方程有 1 个实数解和 1 对共轭虚根。

致谢：本文写作得到了董文杰教授的大力支持以及 3 位评审专家富有建设性的意见，特此致谢。

参考文献

巢纪平，周德刚. 2005. 大气边界层动力学和植被生态过程耦合的一个简单解析理论. 大气科学，29：37-46.

陈曦. 2012. 亚洲中部干旱区蒸散发研究. 北京：气象出版社：7-9，21-24.

戴永久，曾庆存，王斌. 1997. 一个简单的陆面过程模式. 大气科学，21：705-716.

樊自立. 1993. 塔里木盆地绿洲形成与演变. 地理学报，48：421-427.

范盛金. 1989. 一元三次方程的新求根公式与新判别法. 海南师范学院学报：自然科学版，2：91-98.

高华君. 1987. 我国绿洲的分布和类型. 干旱区地理，10：23-29.

韩德林. 1999. 中国绿洲研究之进展. 地理科学，19：313-319.

黄荣辉，陈文，马耀明，等. 2011. 中国西北干旱区陆-气相互作用及其对东亚气候变化的影响. 北京：气象出版社：7-27.

刘式适，刘式达. 2011. 大气动力学. 第 2 版. 北京：北京大学出版社：548.

穆桂金，刘嘉麒. 2000. 绿洲演变及其调控因素初析. 第四纪研究，20：539-547.

沈玉凌. 1994. "绿洲"概念小议. 干旱区地理，17：70-74.

王亚俊，曾凡江. 2010. 中国绿洲研究文献分析及研究进展. 干旱区研究，27：501-506.

吴凌云，巢纪平. 2004. 一个简单陆气耦合模式中的绿洲荒漠化效应. 气候与环境研究，2：350-360.

张强，胡隐樵. 2002. 绿洲地理特征及其气候效应. 地球科学进展，17：477-486.

张强，胡隐樵，侯平. 2003. 绿洲系统维持机制的非线性热力学分析. 中国沙漠，23：174-181.

Barton I J. 1979. A parameterization of the evaporation from nonsaturated surfaces. J Appl Meteorol，18：43-47.

Charney J G. 1975. Dynamics of deserts and drought in the Sahel. Q J R Meteorol Soc，101：193-202.

Dai Y，Zeng X，Dickinson R E. 2001. Common Land Model（CLM）. CLM technical documentation and user's guide.

Dickinson R E. 1984. Modeling evapotranspiration for three-dimensional global climate models. Geophys Monogr Ser，29：58-72.

Dickinson R E，Henderson A H，Kennedy P J，et al. 1993. Biosphere-atmosphere transfer scheme（BATS）version 1e as coupled to community climate model. NCAR Technical Note. Boulder，CO：National Center For Atmospheric Research.

Monteith J L. 1981. Evaporation and surface temperature. Q J R Meteorol Soc，107：1-27.

Noilhan J，Planton S. 1989. A simple parameterization of land surface processes for meteorological models. Mon Weather Rev，117：536-549.

Pan X L，Chao J P. 2001. The effects of climate on development of ecosystem in oasis. Adv Atmos Sci，18：42-52.

Penman H L. 1948. Natural evaporation from open water，bare soil and grass. Proc R Soc A-Math Phys Eng Sci，193：120-145.

Wu L，Chao J，Fu C，et al. 2003. On a simple dynamics model of interaction between oasis and climate. Adv Atmos Sci，20：775-780.

Yasuda N，Toya T. 1981. Evaporation from non-saturated surface and surface moisture availability. Pap Meteor Geophys，32：89-98.

Zhang R，Zuo Z. 2011. Impact of spring soil moisture on surface energy balance and summer monsoon circulation over East Asia and precipitation in East China. J Clim，24：3309-3322.

Zuo Z，Zhang R，Zhao P. 2011. The relation of vegetation over the Tibetan Plateau to rainfall in China during the boreal summer. Clim Dyn，36：1207-1219.

城市热岛效应和气溶胶浓度的动力、热力学分析*②

李耀锟[1,2] 巢纪平[3] 匡贡献[4]

(1. 北京师范大学全球变化与地球系统科学研究院,北京 100875;2. 全球变化研究协同创新中心,北京 100875;3. 国家海洋环境预报中心,北京 100081;4. 北京市 2433 信箱,北京 100081)

摘要:在能量平衡方程中引入气溶胶的吸收和散射作用,并与三维行星边界层运动方程组相耦合,根据温度分布显式求解运动场,探讨三维行星边界层内温度、运动、气溶胶浓度分布特征。结果表明,城市人为热释放直接决定了城市热岛效应的强度,城市面积越大,城市热岛效应的强度也越强,城市面积固定时,城市越分散,城市热岛效应的强度越弱,这为城市建设多采取卫星城的方式提供了一定的理论支撑。气溶胶的散射作用要大于吸收作用,其对城市热岛效应的强度主要起削弱作用,当气溶胶浓度较大时,吸收作用更显著一些,此时城市热岛效应的强度会有一定的增强,但是幅度不大。当城市热岛效应的强度增强时,其所驱动的环流也会增强,造成城区中心气溶胶浓度略有下降。

关键词:行星边界层;能量平衡方程;城市热岛效应;空气污染物;城市群

1 引言

城市热岛效应是指城市温度高于郊野温度的现象。城市地区水泥、沥青等所构成的下垫面反照率小、热容大,能够有效地吸收太阳辐射,加之人类活动释放的热量,使城市区域形成高温中心,并由此向外围递减。而气溶胶是悬浮在大气中的固态粒子或液态小滴物质的统称。人类活动已向大气中排放了大量的人为气溶胶,其中含有的有毒有害的细颗粒物会对人体造成危害,有研究表明由于空气污染增加了肺癌、心脏病和中风的发病率,使我国北方居民的预期寿命减少了 5.5 年(Chen et al., 2013)。随着城市化进程的加快及城市环境日益恶化,城市热岛效应与大气污染物的相互影响成为研究的热点之一。城市热岛效应诱发的环流会改变污染物在边界层内的分布,而空气污染物也会通过散射和吸收太阳辐射对城市热岛效应产生影响(寿亦萱和张大林,2012)。

数值模式是研究城市热岛效应及大气污染物的有利工具,如今国内外已经开发出多种模式,如城市能量平衡模式(TEB)(Lemonsu et al.,2004)、MM5/WRF 城市冠层模式(Kusaka et al.,2000)、WRF 建筑环境参数化模式(Martilli,2002;Dupont et al.,2004)、三维非静力区域边界层模式(Jiang et al., 2002;徐敏等,2002;Fang et al.,2004)、根据污染物扩散方程的气溶胶输送模式(毛节泰,1992;黄美元

* 地球物理学报,2015,58(3):729-740.

② 国家重点基础研究发展计划(973 计划)项目(2014CB953903)、中央高校基本科研业务费专项资金(2013YB45)资助。

等,1996;王自发等,1997;罗淦和王自发,2006)。边界层模式更重视人为热及城市冠层的热力、动力作用,而污染物扩散模式则更强调在已给环流条件下污染物的输送和扩散,如二者相结合即可讨论城市热岛和污染物的相互影响。此外,采用一些简单的物理模型进行分析也是非常必要的,如 Oke(1988)分析了城市冠层的能量平衡过程,直接促进了各种模式的建立和发展。桑建国等(桑建国,1986;桑建国等,2000)从大气热力、动力方程组出发分析城市加热引起的热岛环流特征。Agarwal 和 Tandon(2010)利用一个稳定态的二维数学模型(污染物浓度扩散方程)研究了城市热岛条件下污染物的扩散现象。

然而这些研究主要是从城市热岛效应或污染物浓度某一个方面分析的,尚没有对二者的相互作用进行分析。考虑到作为空气污染物的气溶胶粒子能够吸收和散射太阳辐射,对边界层内的辐射过程产生影响,若在温度方程中引入气溶胶的吸收和散射作用,则可以考虑气溶胶对辐射平衡(温度分布)的影响,将温度方程与边界层运动相耦合,便可讨论温度对边界层运动的影响,边界层运动进一步会对污染物浓度分布产生影响。基于以上分析,本文在巢纪平和陈英仪(1979)建立的考虑辐射能传输过程的能量平衡模式中引入气溶胶粒子的吸收和散射作用,并将其与三维行星边界层运动方程组相结合,显式表达温度对环流的驱动作用,并根据环流分布计算气溶胶粒子的浓度分布特征,由此实现城市热岛效应和气溶胶浓度的物理联系,从动力和热力的角度对其进行分析。

2　模式方程组

根据观测,气溶胶主要分布在大气边界层内(陈鹏飞等,2012),把静止大气作为大气的背景状态,Boussinesq 近似下的边界层运动方程组可写为

$$-fv = -\frac{1}{\bar{\rho}}\frac{\partial p}{\partial x} + \mu\frac{\partial^2 u}{\partial z^2} \tag{1}$$

$$fu = -\frac{1}{\bar{\rho}}\frac{\partial p}{\partial y} + \mu\frac{\partial^2 v}{\partial z^2} \tag{2}$$

$$\frac{1}{\rho}\frac{\partial p}{\partial z} = g\alpha T \tag{3}$$

$$\frac{\partial u}{\partial x} + \frac{\partial v}{\partial y} + \frac{\partial w}{\partial z} = 0 \tag{4}$$

$$\bar{\rho}C_p\frac{\bar{T}}{g}N^2 w = \kappa_h\nabla_h^2 T + (\kappa_z + \kappa_r)\frac{\partial^2 T}{\partial z^2} - \lambda^2 T - k'Q_0 e^{-\int_z^\infty k_v dz} e^{-\int_z^\infty k_{sc} dz}\int_z^\infty k_a dz \tag{5}$$

$$u\frac{\partial C}{\partial x} + v\frac{\partial C}{\partial y} + w\frac{\partial C}{\partial z} = K_h\left(\frac{\partial^2 C}{\partial x^2} + \frac{\partial^2 C}{\partial y^2}\right) + K_z\frac{\partial^2 C}{\partial z^2} - \gamma C \tag{6}$$

式中,$\bar{\rho}$ 为静止大气背景的空气密度;$\bar{T}$ 为静止大气背景的空气温度,它们均只是高度的函数,T, p, u, v, w 分别为空气温度、压强、速度相对于静止大气背景的偏差。静止大气的背景状态中没有基本气流,可去掉基本气流对城市热岛效应和气溶胶浓度的影响,更加清楚地显示出二者间相互影响的过程。另一方面,基本气流较大时,城市热岛效应会被破坏,局地的气溶胶浓度也会大幅度降低,此时二者受基本气流的影响更大一些,而其相互影响则相对较弱。因此,以静止大气为背景来讨论城市热岛效应和气溶胶的相互影响更为合适。f 为科氏参数,α 为空气的热膨胀系数,μ 为垂直方向的湍流黏性系数,C_p 为空气的定压比热,g 为重力加速度,N 为浮力频率,κ_h、κ_z 分别为水平和垂直湍流导热系数,

∇_h^2 为水平拉普拉斯算子，κ_r 为辐射交换系数，λ^2 为牛顿辐射冷却系数，Q_0 为抵达大气上界的太阳辐射，C 为气溶胶的浓度，γ 为气溶胶的消耗参数，K_h、K_z 分别为水平和垂直湍流导温系数。太阳辐射的吸收系数 k' 由两部分构成：一部分是水汽对太阳辐射的吸收 k_v；另一部分是气溶胶粒子对太阳辐射的吸收 k_a，即 $k'=k_v+k_a$，k_{sc} 是气溶胶对太阳辐射的散射系数，气溶胶的吸收系数与散射系数之和为气溶胶的消光系数。温度方程的具体推导过程可参见附录 A。

温度方程的上边界条件取为 $z\to\infty$，$T\to 0$。下边界条件取在城市冠层顶部，仅考虑感热与净辐射、人为热相平衡

$$-\kappa_z\frac{\partial T}{\partial z}=-(1-\Gamma_{uc})Q_0\mathrm{e}^{-\int_0^\infty k_v\mathrm{d}z}\mathrm{e}^{-\int_0^\infty k_{sc}\mathrm{d}z}\int_0^\infty k_a\mathrm{d}z+4\varepsilon_{uc}\sigma\overline{T}^3(T-T_{uc})+Q_a \tag{7}$$

式中，Γ_{uc} 为城市冠层的反照率；ε_{uc} 为城市冠层的灰体系数；T_{uc} 为城市冠层温度相对于静止大气的偏差；Q_a 为城市人为热释放量。

气溶胶浓度方程(6)的上边界可取为 $z\to\infty$，$C\to 0$，下边界取为

$$-K_z\frac{\partial C}{\partial z}=S-V_sC \tag{8}$$

式中，S 为源排放率；V_s 为干沉降速度。气溶胶粒子的沉降速度 V_s 与气溶胶粒子大小等有关，可测量或用参数化的方法求得，本文中为了方便起见，令其为常数，并对不同的沉降速度进行比较。

根据推导可求得速度分布为(参见附录 B)

$$u=-\frac{gh_R\alpha}{\mu}\frac{1}{\left(\frac{1}{h_R^4}+\frac{4}{h_E^4}\right)}\left[A(z)\frac{\partial T_0}{\partial x}-B(z)\frac{\partial T_0}{\partial y}\right] \tag{9}$$

$$v=-\frac{gh_R\alpha}{\mu}\frac{1}{\left(\frac{1}{h_R^4}+\frac{4}{h_E^4}\right)}\left[B(z)\frac{\partial T_0}{\partial x}+A(z)\frac{\partial T_0}{\partial y}\right] \tag{10}$$

$$w=\frac{gh_R\alpha}{\mu}\frac{1}{\left(\frac{1}{h_R^4}+\frac{4}{h_E^4}\right)}\int_0^z A(z)\mathrm{d}z\cdot\nabla_h^2T_0 \tag{11}$$

将垂直速度式(11)代入温度方程(5)中可得

$$(\kappa_h-\kappa_m)\nabla_h^2T+(\kappa_z+\kappa_r)\frac{\partial^2T}{\partial z^2}-\lambda^2T=k'Q_0\mathrm{e}^{-\int_z^\infty k_v\mathrm{d}z}\mathrm{e}^{-\int_z^\infty k_{sc}\mathrm{d}z}\int_z^\infty k_a\mathrm{d}z \tag{12}$$

式中

$$\kappa_m=\bar{\rho}C_p\frac{N^2h_R}{\mu}\frac{1}{\left(\frac{1}{h_R^4}+\frac{4}{h_E^4}\right)}\mathrm{e}^{z/h_R}\int_0^z A(z)\mathrm{d}z \tag{13}$$

温度方程(12)中气溶胶的吸收和散射作用均与气溶胶浓度有关，要与气溶胶浓度方程(6)联立才能求解出温度和气溶胶温度分布，求得温度分布后，即可根据式(9)，式(10)，式(11)三式计算速度分布。

3 结果分析

计算中参数取值为 $\overline{T}=273$ K，$\bar{\rho}=1.293$ kg/m³，$f=1\times10^{-4}$/s，$\varepsilon=1$，$\mu=5$ m²/s，$C_p=1004$ J/(kg · K)，

$N=1.16\times10^{-2}/\mathrm{s}$，$g=9.8\ \mathrm{m/s^2}$，$\kappa_h=6.5\times10^4\ \mathrm{W/(m\cdot K)}$，$\kappa_z=6.5\times10^3\ \mathrm{W/(m\cdot K)}$，$K_h=50\ \mathrm{m^2/s}$，$K_z=5\ \mathrm{m^2/s}$，$\sigma=5.6696\times10^{-8}\ \mathrm{W/(m^2\cdot K^4)}$，$k_s=1.364/\mathrm{m}$，$k_w=1.3327\times10^{-3}/\mathrm{m}$，根据以上参数取值可得，$\kappa_r=3.6076\ \mathrm{W/(m\cdot K)}$，$r=0.53331$，$h_E=316.23\ \mathrm{m}$，$\lambda^2=5.7403\times10^{-3}\ \mathrm{W/(m^3\cdot K)}$，参照附录 B 的讨论，取 $h_R=h_E$。

气溶胶的特征厚度可取为 $H_a=1\ 500\ \mathrm{m}$，一般而言，气溶胶的散射系数要远大于其吸收系数，参照对几次雾霾天气中气溶胶辐射性质的观测（表 1），单次散射反射率取为 $\omega_0=0.85$，气溶胶对可见光的吸收主要由黑碳气溶胶引起的，北京郊区夏季的观测结果表明黑碳浓度占 PM2.5 浓度的比例基本都在 5%以下（荆俊山等，2011），据此气溶胶的吸收系数 $k_a=0.05\alpha_{bc}C$，α_{bc} 为黑碳气溶胶的质量吸收系数，其值可取为 $8.28\ \mathrm{m^2/g}$（吴兑等，2009）。求得气溶胶的吸收系数后，即可根据单次散射反射率求得其消光系数 k_{ex}。城市冠层的反照率取为 $\Gamma_{uc}=0.15$。

表 1　气溶胶浓度、消光系数及单散射反射率的观测值

地点	观测时间段	平均浓度（μg/m³）	消光系数（M/m）	单散射反射率	来源
北京	2011-09-01—2011-12-07	137.1	1001.0	0.94	赵秀娟等（2013）
济南	2009-10-11—2009-11-18	137.2	662.7	0.88	徐政等（2011）
广州	2004-10-04—2004-11-05	103	509	0.83	Andreae 等（2008）

人为热 Q_a 取决于人均能源消耗量及城市人口密度（Oke，1988），参照桑建国等（2000），将其写为

$$Q_a(x,y)=Q_I\exp(-x^2/a_1^2-y^2/b_1^2),\tag{14}$$

取 $Q_I=100\ \mathrm{W/m^2}$，距离参数 $a_1=b_1=16.5\ \mathrm{km}$，表示一个直径为 50 km 的圆形城市。参照胡非等（1999），取拖曳系数 $C_D=0.0625$，参考风速 $V=1\ \mathrm{m/s}$，可算得城市冠层的阻抗 $r_c=16\ \mathrm{s/m}$。为了比较，另给一种城市人为热分布型

$$Q_a(x,y)=Q_I\exp\left[-\left(x\pm\frac{1}{2}L\right)^2/a_2^2-\left(y\pm\frac{1}{2}L\right)^2/b_2^2\right],\tag{15}$$

距离参数 $a_2=b_2=8.3\ \mathrm{km}$，表示 4 个中心分别位于 $\left(\pm\frac{1}{2}L,\pm\frac{1}{2}L\right)$，直径为 25 km 的圆形城市。两种情况下城市的总面积是相同的，人为热排放强度也是相同的，不同之处在于前者是单个城市，后者是 4 个稍小的城市组成的一个城市群（见图 1）。气溶胶排放分布取与人为热相同的形式，参照 Venkatachalappa 等（2003）及 Agarwal 和 Tandon（2010），气溶胶排放强度取为 $S=1\ \mu\mathrm{g/(m^2\cdot s)}$，沉降速度取为 $V_s=1\ \mathrm{mm/s}$，消耗参数取为 $\gamma=2\times10^{-5}/\mathrm{s}$。

根据式（14）确定的单个城市人为热及气溶胶分布算得气溶胶浓度、温度、速度场分布如图 2 所示。气溶胶浓度、温度分布由下边界强迫的分布决定，表现为在排放、加热中心区最大，向四周依次减弱。排放中心区气溶胶浓度可达 87 μg/m³，温度超过 3.5℃。城市热岛驱动了一个气旋性环流，气流向加热中心辐合上升，在 150 m 高度处达到最大，可达 3 cm/s，上升到高空后向四周辐散，在远离加热中心的区域下沉。水平速度在温度梯度最大处达到最大，可达 2 m/s。人为热对大气的加热作用随高度增加而降低，到 500 m 高度已经较为微弱。

由式（15）确定的城市群的计算结果（图 3）与单个城市（图 2）类似，气溶胶浓度分别在每个排放中心达到最大，向四周递减，每个加热中心都驱动产生了相应的气旋性环流。但是也有一定的不同之

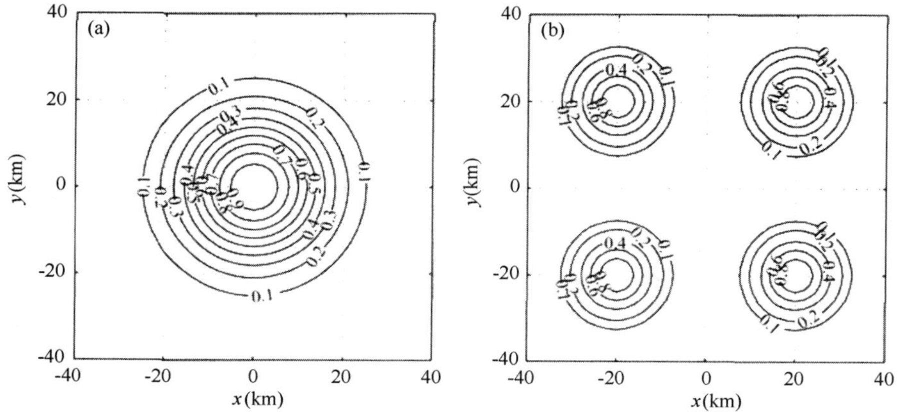

图 1　人为热释放的两种分布型:单一大城市(a)和由 4 个小城市组成的城市群(b)

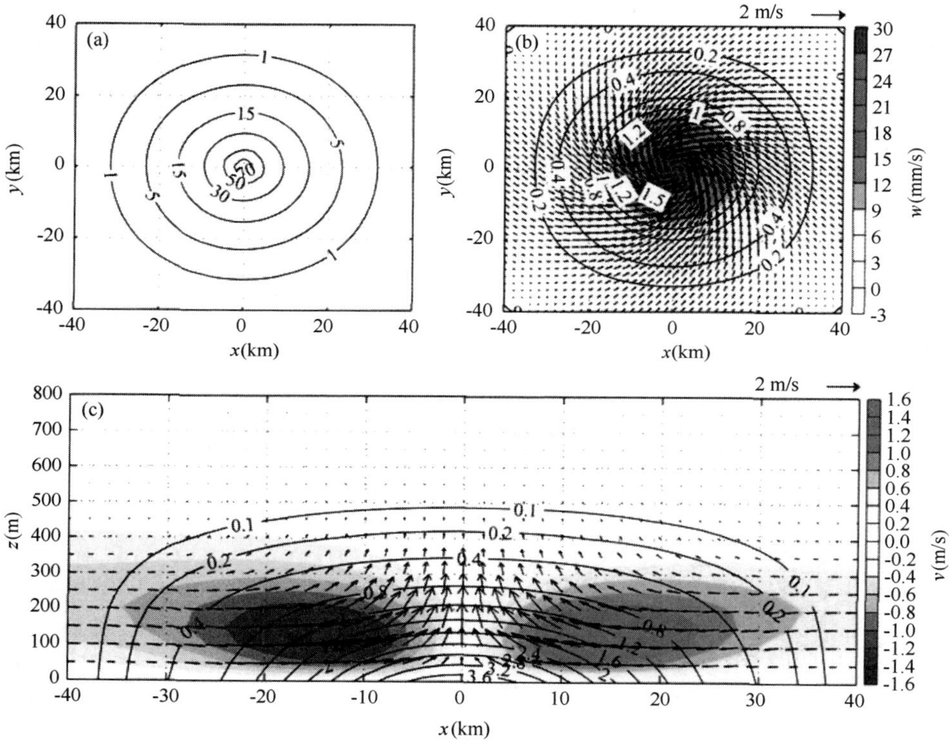

图 2　气溶胶浓度分布(a);150 m 高度处扰动温度(等值线)、水平速度(矢量)和垂直速度(填色)的分布(b);$y=0$ 剖面温度(等值线)、速度(矢量,填色)分布(c),(c)中垂直速度已扩大 100 倍(见书后彩插)

处,城市群中,气溶胶浓度有了一定程度的下降,排放中心最大浓度为 57 μg/m³,比单个城市的情形下降了约 30 μg/m³,整个计算区域内,气溶胶平均浓度从 5.4 μg/m³下降到 4.0 μg/m³;城市群中,城市热岛的强度也受到了一定程度的削弱,从单个城市的 3.9℃下降到 3.0℃,下降幅度接近 1℃。城市热岛的高度也出现了一定的下降,以 0.1℃等温线为例,单个城市在约 500 m 高度处,而多个城市组成的城市群下降到 400 m 高度处。城市热岛强度降低,其驱动的环流强度也有一定程度的减弱。

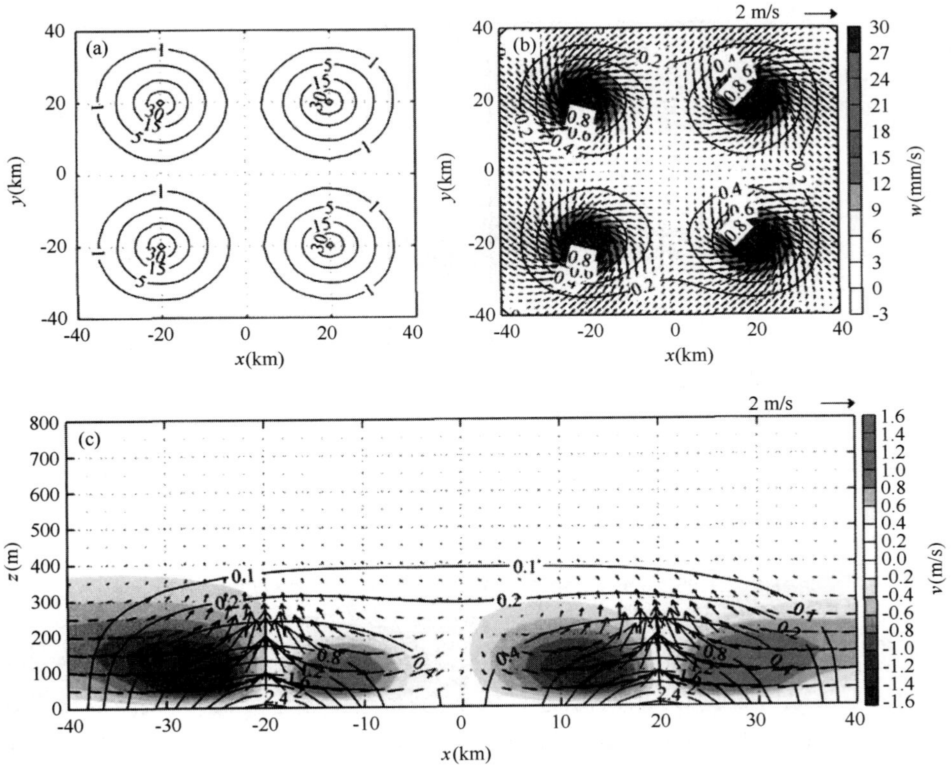

图3　气溶胶浓度分布(a);150 m高度处扰动温度(等值线)、水平速度(矢量)和垂直速度(填色)的分布(b);$y=\frac{1}{2}L$剖面温度(等值线)、速度(矢量,填色)分布(c),(c)中垂直速度已扩大100倍(见书后彩插)

图4(a)进一步给出了城市热岛效应和城市半径的关系,图中实线(虚线)表示单个城市(城市群)引起的城市热岛效应强度。由图可知,单个城市引起的城市热岛强度要高于城市群,并且城市面积越大,二者差距越大,最大可接近1℃;图4(b)为城市总面积固定在1963.5 km^2时城市数目对城市热岛效应强度的影响,由图可见,随着城市数目的增加,城市热岛效应的强度在逐步减弱,从单个城市到6个小城市构成的城市群,城市热岛效应的强度可降低约1℃。与之类似,虽然排放强度固定,随着城市面积不断扩大,中心城区的气溶胶浓度也在增加,当城市面积固定时,中心城区气溶胶浓度在单个城市要比小城市群更高(图略)。综上可知,相同的城市面积下,一个单一大城市中空气污染程度、城市热岛效应均要超过几个城市组成的城市群。而我国许多城市的发展一般都是由中心城区不断向外扩张,这种发展不仅造成了严重的交通问题,而且还加重了城市热岛效应和城市污染问题,因此,采取建设卫星城的方式扩张城市将减小城市发展给环境带来的压力。

两种城市分布情况下气溶胶排放源强度与中心城区气溶胶浓度及城市热岛效应强度的关系如图5所示。气溶胶浓度会随着排放强度的增强而增加,由于没有侧边界的输入和输出,二者之间表现为较为简单的线性关系(图5(a))。同时可以发现小城市群城区中心气溶胶浓度要低于单个大城市,二者的差距随着排放强度的增加而加大,这表明排放越强,小城市群对气溶胶浓度的降低作用越显著。城市热岛效应的强度则随着排放增强表现为逐步减弱而后又有所增加的特征(图5(b)),这是由于在人为热释放固定的条件下,随着排放增强,空气中气溶胶浓度逐步增加,气溶胶粒子散射的太阳辐射

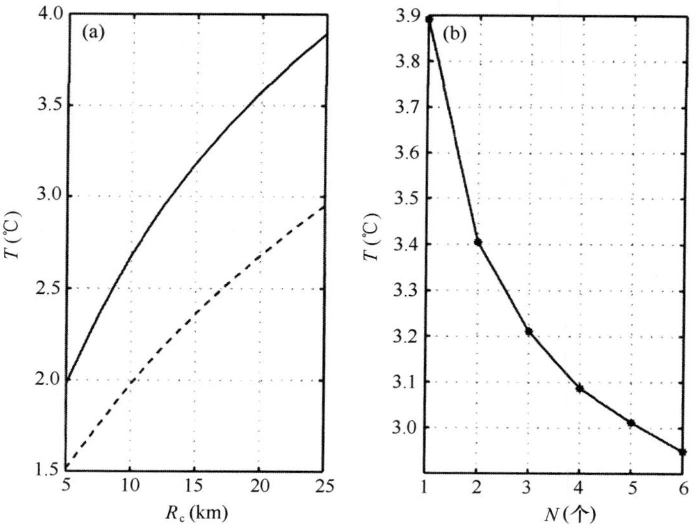

图 4　城市热岛效应强度和城市半径(a)与城市数目(b)的关系

(a)中实线表示单个大城市,虚线表示小城市群

逐步增加,从而使城市冠层吸收的太阳辐射逐步减小,城市冠层温度降低,从而减弱城市热岛的强度。随着气溶胶浓度的增加,当散射作用使得抵达城市冠层的太阳辐射可以忽略时,散射就不再起作用,而气溶胶吸收的太阳辐射会加热空气,此时空气温度又会有所增加。一般而言,气溶胶对太阳辐射的散射要大于其吸收作用,因此其降温作用一般而言要大于增温作用。由于环流是由温度驱动的,当城市热岛效应减弱时,环流的强度也会减弱。

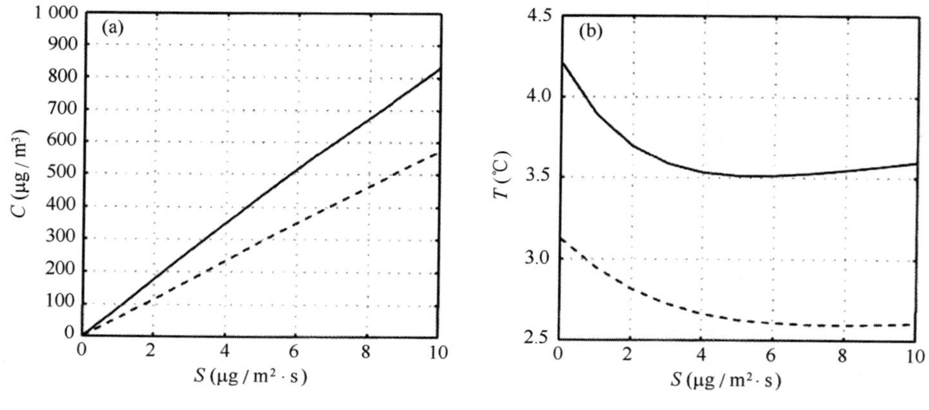

图 5　排放源 S 强度与气溶胶浓度(a)及城市热岛强度(b)的关系

实线表示单个大城市,虚线表示小城市群

在相同的排放强度下,城市热岛效应的强度在小城市群中大约要比单个城市低 0.8℃左右,进一步体现出城市分散对城市热岛效应的抑制作用。另一方面,500 $\mu g/m^3$的气溶胶浓度引起的城市热岛效应强度变化在 0.5℃左右,若与气溶胶排放为零时单个城市热岛效应的强度(约 4℃)相比,500 $\mu g/m^3$的气溶胶浓度对城市热岛效应的强度能造成约 13%的减弱;若与没有气溶胶排放时小城市群热岛效应的强度(约 3℃)相比,500 $\mu g/m^3$的气溶胶浓度对城市热岛效应强度能产生约 17%的减弱。

在排放强度固定的情况下，沉降速度越大，空气中气溶胶浓度也就越低（图 6(a)），若不考虑沉降作用，同样的排放强度下单个中心城区气溶胶浓度可达 90 μg/m³以上（实线），而小城市群中心城区气溶胶浓度可达 60 μg/m³（虚线），均比前文计算值要高；当沉降速度为 10 mm/s 时，单个城市中心城区气溶胶浓度下降到约 50 μg/m³，而小城市群中心城区气溶胶浓度下降到 40 μg/m³，均比前文计算值小了不少。根据前述分析可知，气溶胶浓度降低有利于增强城市热岛效应的强度（图 6(b)），沉降速度从 0 增加到 10 mm/s，单个城市热岛效应强度可增强 0.2℃，而小城市群则增温不到 0.1℃，同时小城市群的城市热岛效应强度要比单个城市低将近 1℃，进一步说明小城市群对城市热岛效应的抑制作用。

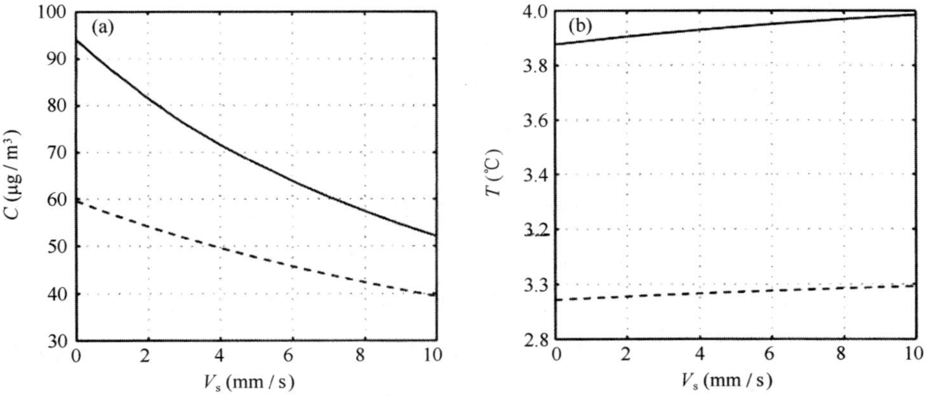

图 6　沉降速度 V_s 与气溶胶浓度（a）及城市热岛强度（b）的关系

实线表示单个大城市，虚线表示小城市群

人为热释放对中心城区气溶胶浓度的影响如图 7（a）所示。随着城市热岛效应的增强，城市热岛诱导的环流强度也在逐步增强，环流对气溶胶的输送和再分配作用也增强，从而有助于降低中心城区的气溶胶浓度。对单个城市而言（图 7a 实线），人为热释放从 50 W/m²增加到 150 W/m²，气溶胶浓度减小了 20 μg/m³左右，而对小城市群而言（图 7a 虚线），气溶胶浓度减小了约 25 μg/m³，可见小城市群有助于降低气溶胶浓度。显然城市热岛效应的强度随着人为热释放的增加而增加（图 7b），但人为热对单个大城市的加热作用要超过小城市群，且二者间的差值会随着人为热的增加而不断增大，说明小城市群会对城市热岛效应的强度起到一定的抑制作用。图 7 进一步说明城市分散对减弱城市热岛效应、降低污染物浓度的有益贡献。

综上可知，排放源的强弱决定了气溶胶浓度的高低，排放源越强气溶胶浓度越高，而沉降速度越大，气溶胶浓度则越低，人为热释放的强度直接决定了城市热岛的强度，人为热释放越大，城市热岛效应越强。另一方面，气溶胶浓度增加对城市热岛效应起到了抑制作用，气溶胶浓度增加会加大气溶胶粒子对太阳辐射的散射作用，减少到达城市冠层的太阳辐射，最终降低城市上方的空气温度。在相同的排放源强度、沉降速度和人为热释放量下，小城市群对城市热岛效应的强度均能起到一定的抑制作用，同时也有利于降低中心城区的污染物浓度。

4　结论

本文在能量平衡方程中引入气溶胶粒子的吸收和散射作用，并将其与三维行星边界层运动方程

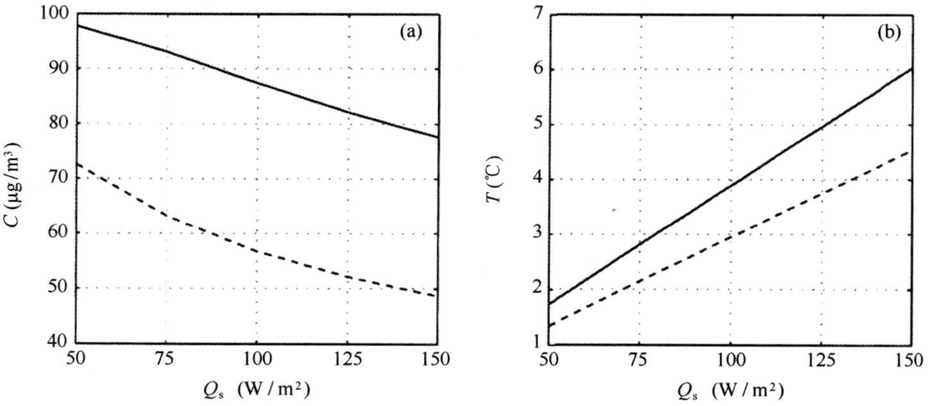

图 7　人为热 Q_a 与气溶胶浓度(a)与城市热岛强度(b)的关系

实线表示单个大城市,虚线表示小城市群

组相耦合,将运动场显式表达为温度的函数,根据边界强迫求解气溶胶浓度、温度及速度场分布特征,从理论上分析了不同城市分布下城市热岛效应和气溶胶浓度的相互影响,主要结论如下。

气溶胶排放源的强度和分布形式决定了空气中气溶胶浓度及其分布特征。气溶胶对城市热岛效应的作用主要体现在气溶胶粒子对太阳辐射的吸收和散射作用上。气溶胶粒子对太阳辐射的散射作用要大于其吸收作用,因此,气溶胶主要表现降温作用,随着气溶胶浓度增加,城市热岛效应会逐步减弱,直到吸收作用超过散射作用,城市热岛效应的强度又会有所增加,但是增加的幅度不大。

人为热释放直接加热了城市上方的空气,并驱动一个相应的气旋性环流,人为释放量越大,城市热岛效应也就越强,驱动的环流强度也随之加强。城市热岛强度对气溶胶分布的影响主要体现在热岛环流对气溶胶浓度的再分配作用上,环流加强会降低中心城区的气溶胶浓度,不过城市热岛效应对气溶胶浓度的影响有限。

在人为热释放强度固定的情况下,城市面积越大,城市热岛效应的强度也越强,而在城市总面积固定时,城市内建筑物越分散,城市热岛效应的强度也就越低。在污染物排放强度固定的情况下,城市面积越大,中心城区的气溶胶浓度也越高,污染越严重,而在城市总面积固定时,城市建设越分散,中心城区的气溶胶浓度也越低。因此对城市建设发展而言,多中心的卫星城模式更有利。

致谢:感谢两位审稿专家富有建设性的意见。

附录 A

考虑辐射能传输和湍流耗散后,温度方程可以写为

$$\bar{\rho}C_p\frac{\mathrm{d}T}{\mathrm{d}t}-\frac{\mathrm{d}p}{\mathrm{d}t}=\kappa_h\nabla_h^2T+\frac{\partial}{\partial z}\left(\kappa_z\frac{\partial T}{\partial z}\right)+\sum_j k_j(A_j+B_j-2E_j)+k'Q \tag{A1}$$

式中,$\bar{\rho}$ 为参考态空气密度;C_p 为其定压比热;T,p 分别为空气的温度和压强;κ_h,κ_z 分别为水平和垂直湍流导热系数;∇_h^2 为水平拉普拉斯算子;k 为对长波辐射的吸收系数;k'为对太阳辐射的吸收系数;A,B 分别表示向下和向上的长波辐射;E 为黑体辐射;下标 j 表示波长 $\Delta\lambda_j$ 范围内。

求解式(A1)需要在整个吸收谱范围内求和,直接计算较为繁琐,参照巢纪平和陈英仪(1979),采

用简化方案求解辐射能传输过程(Kuo,1973),可得

$$\bar{\rho}C_p\frac{\mathrm{d}T}{\mathrm{d}t}-\frac{\mathrm{d}p}{\mathrm{d}t}=\kappa_h\nabla_h^2T+(\kappa_z+\kappa_r)\frac{\partial^2T}{\partial z^2}-\lambda^2T+k'Q+C_0 \tag{A2}$$

式中,$\kappa_r=\dfrac{8r\varepsilon\sigma\bar{T}^3}{k_s}$ 为辐射交换系数,$\lambda^2=8(1-r)k_w\varepsilon\sigma\bar{T}^3$ 为牛顿辐射冷却系数,其中 k_s,k_w 分别为强弱吸收区的吸收系数,r 为强吸收区占总吸收的比例,ε 为灰体系数,σ 为史蒂芬-玻耳兹曼常数,$\bar{T}$ 为参考态空气温度,C_0 为一积分常数。

若令参考态空气为静止大气,即令 $T=\bar{T}(z)+T'$,$Q=\bar{Q}(z)+Q'$,$(u,v,w)=(u',v',w')$,同时忽略高阶项,可将式(A2)写为

$$\bar{\rho}C_p\frac{\bar{T}}{g}N^2w=\kappa_h\nabla_h^2T'+(\kappa_z+\kappa_r)\frac{\partial^2T'}{\partial z^2}-\lambda^2T'+k'Q' \tag{A3}$$

因为

$$Q=Q_0\mathrm{e}^{-\int_z^\infty k_v\mathrm{d}z}\mathrm{e}^{-\int_z^\infty k_{sc}\mathrm{d}z}\mathrm{e}^{-\int_z^\infty k_a\mathrm{d}z}\approx Q_0\mathrm{e}^{-\int_z^\infty k_v\mathrm{d}z}\mathrm{e}^{-\int_z^\infty k_{sc}\mathrm{d}z}\left(1-\int_z^\infty k_a\mathrm{d}z\right) \tag{A4}$$

Q_0 为抵达大气上界的太阳辐射。据此可认为辐射偏差

$$Q'=-Q_0\mathrm{e}^{-\int_z^\infty k_v\mathrm{d}z}\mathrm{e}^{-\int_z^\infty k_{sc}\mathrm{d}z}\int_z^\infty k_a\mathrm{d}z \tag{A5}$$

将其代入方程(A3)中,即可得到(已略去温度偏差之撇号)

$$\bar{\rho}C_p\frac{\bar{T}}{g}N^2w=\kappa_h\nabla_h^2T+(\kappa_z+\kappa_r)\frac{\partial^2T}{\partial z^2}-\lambda^2T-k'Q_0\mathrm{e}^{-\int_z^\infty k_v\mathrm{d}z}\mathrm{e}^{-\int_z^\infty k_{sc}\mathrm{d}z}\int_z^\infty k_a\mathrm{d}z \tag{A6}$$

附录 B

令 $U=u+v\mathrm{i}$,将式(3)代入式(1)、式(2)中消去 p,将式(1)和式(2)两式合写为

$$\frac{\partial^3U}{\partial z^3}-\frac{f}{\mu}\mathrm{i}\frac{\partial U}{\partial z}=\frac{g\alpha}{\mu}\left(\frac{\partial T}{\partial x}+\frac{\partial T}{\partial y}\mathrm{i}\right) \tag{B1}$$

假定在 $z\to\infty$ 时

$$\left(\frac{\partial^2U}{\partial z^2}-\frac{f}{\mu}\mathrm{i}U\right)\to 0 \tag{B2}$$

于是式(B1)可写为

$$\frac{\partial^2U}{\partial z^2}-\frac{f}{\mu}\mathrm{i}U=-\frac{g\alpha}{\mu}\int_z^\infty\left(\frac{\partial T}{\partial x}+\frac{\partial T}{\partial y}\mathrm{i}\right)\mathrm{d}z\equiv F(x,y,z) \tag{B3}$$

将其改写为

$$LU=F(x,y,z) \tag{B4}$$

式中算子 $L=\dfrac{\partial}{\partial z}\left[p(z)\dfrac{\partial}{\partial z}\right]+q(z)$,$p(z)=1$,$q(z)=-\left(\dfrac{1+\mathrm{i}}{h_E}\right)^2$,而 $h_E=\sqrt{\dfrac{2\mu}{f}}$ 为 Ekman 标高。边界条件取为

$$z\to 0,\infty,\quad DU=0 \tag{B5}$$

其中 $D=1$ 可称为边界条件算子。

方程(B4)的解为

$$U=\int_0^{\infty}G(z,\xi)F(x,y,\xi)\mathrm{d}\xi \tag{B6}$$

式中,$G(z,\xi)$ 为 Green 函数,根据 Green 函数的性质(数学手册编写组,1979),可求得

$$G(z,\xi)=\begin{cases}-\dfrac{1}{2\beta}[\mathrm{e}^{\beta(z-\xi)}-\mathrm{e}^{-\beta(z+\xi)}], 0\leqslant z<\xi\\ -\dfrac{1}{2\beta}[\mathrm{e}^{\beta(\xi-z)}-\mathrm{e}^{-\beta(z+\xi)}], \xi<z\leqslant\infty\end{cases} \tag{B7}$$

式中已令 $\beta=(1+\mathrm{i})/h_{\mathrm{E}}$。为了求解式(B6),还需求得 $F(x,y,z)$ 的表达式,若令

$$T(x,y,z)=T_0(x,y)\mathrm{e}^{-z/h_R} \tag{B8}$$

$T_0(x,y)$ 为城市冠层顶部的温度分布,h_{R} 可称为温度垂直分布的特征高度。式(B8)可称为求解方程的辅助温度函数。借助辅助函数,可将 F 写为

$$F(x,y,z)=-\frac{g\alpha}{\mu}\left(\frac{\partial T_0}{\partial x}+\frac{\partial T_0}{\partial y}\mathrm{i}\right)\int_z^{\infty}\mathrm{e}^{-z/h_{\mathrm{R}}}\mathrm{d}z=-\frac{gh_{\mathrm{R}}\alpha}{\mu}\left(\frac{\partial T_0}{\partial x}+\frac{\partial T_0}{\partial y}\mathrm{i}\right)\mathrm{e}^{-z/h_{\mathrm{R}}} \tag{B9}$$

为保证积分值有限,需满足 $h_{\mathrm{R}}>0$,即辅助函数需满足温度分布向上递减的要求。需要指出,式(B8)只是为求解积分表达式引入的一个近似辅助函数,并不意味着以上各式中 T 均按此高度分布。将式(B9)及 Green 函数式(B7)代入解式(B6)中

$$U=-\frac{1}{2\beta}\frac{gh_{\mathrm{R}}\alpha}{\mu}\left(\frac{\partial T_0}{\partial x}+\frac{\partial T_0}{\partial y}\mathrm{i}\right)\int_0^{\infty}\mathrm{e}^{-\beta(z+\xi)}\mathrm{e}^{-\frac{\xi}{h_{\mathrm{R}}}}\mathrm{d}\xi+\frac{1}{2\beta}\frac{gh_{\mathrm{R}}\alpha}{\mu}\left(\frac{\partial T_0}{\partial x}+\frac{\partial T_0}{\partial y}\mathrm{i}\right)\int_0^{z}\mathrm{e}^{-\beta(z-\xi)}\mathrm{e}^{-\frac{\xi}{h_{\mathrm{R}}}}\mathrm{d}\xi$$
$$+\frac{1}{2\beta}\frac{gh_{\mathrm{R}}\alpha}{\mu}\left(\frac{\partial T_0}{\partial x}+\frac{\partial T_0}{\partial y}\mathrm{i}\right)\int_z^{\infty}\mathrm{e}^{-\beta(\xi-z)}\mathrm{e}^{-\frac{\xi}{h_{\mathrm{R}}}}\mathrm{d}\xi \tag{B10}$$

分别计算积分项,并分开实部和虚部即可得到

$$u=-\frac{gh_R\alpha}{\mu}\frac{1}{(1/h_{\mathrm{R}}^4+4/h_{\mathrm{E}}^4)}\mathrm{e}^{z/h_{\mathrm{R}}}\left[A(z)\frac{\partial T_0}{\partial x}-B(z)\frac{\partial T_0}{\partial y}\right] \tag{B11}$$

$$v=-\frac{gh_R\alpha}{\mu}\frac{1}{(1/h_{\mathrm{R}}^4+4/h_{\mathrm{E}}^4)}\mathrm{e}^{z/h_{\mathrm{R}}}\left[B(z)\frac{\partial T_0}{\partial x}+A(z)\frac{\partial T_0}{\partial y}\right] \tag{B12}$$

式中

$$A(z)=\frac{1}{h_{\mathrm{R}}^2}\left[\mathrm{e}^{-z/h_{\mathrm{R}}}-\mathrm{e}^{-z/h_{\mathrm{E}}}\cos\frac{z}{h_{\mathrm{E}}}\right]-\frac{2}{h_{\mathrm{E}}^2}\mathrm{e}^{-z/h_{\mathrm{E}}}\sin\frac{z}{h_{\mathrm{E}}} \tag{B13}$$

$$B(z)=\frac{2}{h_{\mathrm{E}}^2}\left[\mathrm{e}^{-z/h_{\mathrm{R}}}-\mathrm{e}^{-z/h_{\mathrm{E}}}\cos\frac{z}{h_{\mathrm{E}}}\right]+\frac{1}{h_{\mathrm{R}}^2}\mathrm{e}^{-z/h_{\mathrm{E}}}\sin\frac{z}{h_{\mathrm{E}}} \tag{B14}$$

若研究范围为城市区域,可不考虑科氏参数 f 随 y 的变化,在 $z=0, w=0$ 的条件下可求得垂直速度

$$w=\frac{gh_{\mathrm{R}}\alpha}{\mu}\frac{1}{(1/h_{\mathrm{R}}^4+4/h_{\mathrm{E}}^4)}\int_0^z A(z)\mathrm{d}z\cdot\nabla_{\mathrm{h}}^2T_0 \tag{B15}$$

为了求解式(B6)的积分表达式,引入了温度垂直分布的特征高度 h_{R} 这一参量,其对速度垂直分布函数 $A(z)$ 和 $B(z)$ 的影响分别如附图(a)、附图(b)所示。在垂直方向上,$A(z)$ 存在一个值等于零的高度,在这一高度以下,$A(z)$ 均小于零,并存在一个极小值,而在这一高度以上,$A(z)$ 大于零,但是其值已经较小。随着 h_{R} 的增加,零点高度逐步增加,极值的大小也有所变化,但是分布型没有实质性的改

变。在垂直方向上，$B(z)$也存在一个值等于零的高度，该高度以下，值为正数，存在一个极大值，这一高度以上值较小，随着 h_R 的增加，极值的强度发生变化，不过分布型也没有太大的改变。附图(c)给出了 $h_R = h_E$ 时 $A(z)$ 和 $B(z)$ 的分布特征，$A(z)$ 在垂直方向上存在一个极小值，且其值小于零，而 $B(z)$ 在垂直方向上存在一个极大值，其值大于零，这一分布特征基本能够反映 h_R 取其他值时的分布特征，因此计算中可令 $h_R = h_E$。

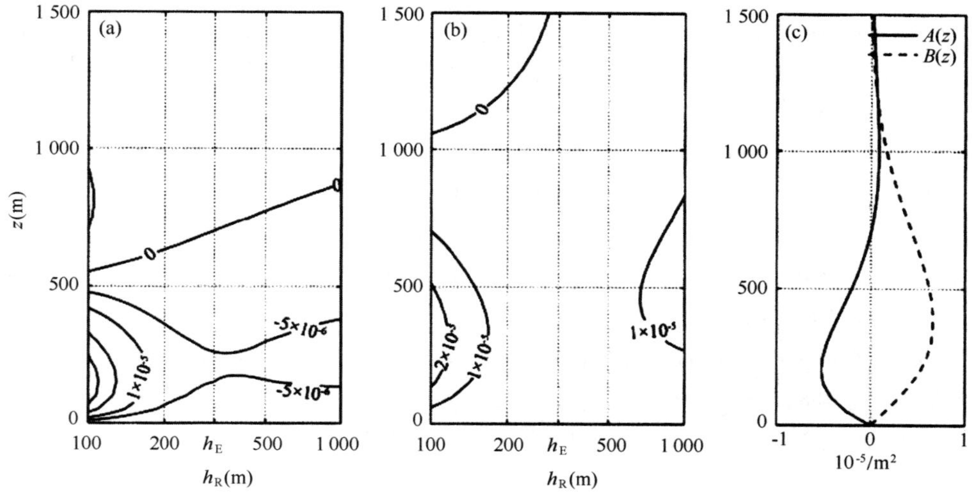

附图 $A(z)$，$B(z)$随 h_R 及高度 z 的分布(a,b)及在 $h_R = h_E$ 时的分布(c)

参考文献

Agarwal M, Tandon A.2010.Modeling of the urban heat island in the form of mesoscale wind and of its effect on air pollution dispersal.*Appl.Math.Model.*, 34(9):2520–2530.

Andreae M O, Schmid O, Yang H, et al.2008.Optical properties and chemical composition of the atmospheric aerosol in urban Guangzhou, China.*Atmos.Environ.*, 42(25):6335–6350.

Chao J P, Chen Y Y.1979.The effect of climate on pole ice and the surface albedo feedback in two dimensions energy balance model.*Scientia Sinica Mathematica* (in Chinese), 22(12):1198–1207.

Chen P F, Zhang Q, Quan J N, et al.2012.Vertical profiles of aerosol concentration in Beijing.*Research of Environmental Sciences* (in Chinese), 25(11):1215–1221.

Chen Y Y, Ebenstein A, Greenstone M, et al.2013.Evidence on the impact of sustained exposure to air pollution on life expectancy from China's Huai River policy.*Proceedings of the National Academy of Sciences of the United States of America*, 110(32):12936–12941.

Dupont S, Otte T L, Ching J K.2004.Simulation of meteorological fields within and above urban and rural canopies with a mesoscale model.*Bound Layer Meteor.*, 113(1):111–158.

Fang X Y, Jiang W M, Miao S G, et al.2004.The multi-scale numerical modeling system for research on the relationship between urban planning and meteorological environment.*Adv. Atmos.Sci.*, 21(1):103–112.

Hu F, Li X, Chen H Y, et al.1999.Turbulence characteristics in the rough urban canopy layer.*Climatic and Environmental Research* (in Chinese), 4(3):252–258.

Huang M Y, Wang Z F, He D Y, et al.1996.Modeling studies on sulfur deposition and transport among different areas in China in summer and winter.*Chinese Science Bulletin* (in Chinese), 41(11):1013–1016.

Jiang W M, Zhou M, Xu M, et al. 2002. Study on development and application of a regional PBL numerical model. *Bound.-Layer Meteor.*, 104(3): 491-503.

Jing J S, Zhang R J, Tao J. 2011. Continuous observation of PM2.5 and black carbon aerosol during summer in Beijing suburb. *Journal of the Meteorological Sciences* (in Chinese), 31(4): 510-515.

Kuo H L. 1973. On a simplified radiative-conductive heat transfer equation. *Pure Appl. Geophys.*, 109(1): 1870-1876.

Kusaka H, Kimura F, Hirakuchi H, et al. 2000. The effects of land-use alteration on the sea breeze and daytime heat island in the Tokyo metropolitan area. *J. Meteor. Soc. Jpn.*, 78(4): 405-420.

Lemonsu A, Grimmond C S B, Masson V. 2004. Modeling the surface energy balance of the core of an old Mediterranean city: Marseille. *J. Appl. Meteor.*, 43(2): 312-327.

Luo G, Wang Z F. 2006. A Global Environmental Atmospheric Transport Model (GEATM): model description and validation. *Chinese Journal of Atmospheric Sciences* (in Chinese), 30(3): 504-518.

Mao J T. 1992. A statistic model on the range of acid deposition. *Acta Scientiae Circumstantiae* (in Chinese), 12(1): 28-36.

Martilli A. 2002. Numerical study of urban impact on boundary layer structure: Sensitivity to wind speed, urban morphology, and rural soil moisture. *J. Appl. Meteor.*, 41(12): 1247-1266.

Mathematics Handbook Team Writing Team. 1979. Mathematics handbook (in Chinese). Beijing: Higher Education Press.

Oke T R. 1988. The urban energy balance. *Prog. Phys. Geog.*, 12(4): 471-508.

Sang J G, Zhang Z K, Zhang B Y. 2000. Dynamical analyses on heat island circulation. *Acta Meteorologica Sinica* (in Chinese), 58(3): 321-327.

Sang J G. 1986. An analytical solution for the effects of heat island. *Acta Meteorologica Sinica* (in Chinese), 44(2): 251-255.

Shou Y X, Zhang D L. 2012. Recent advances in understanding urban heat island effects with some future prospects. *Acta Meteorologica Sinica* (in Chinese), 70(3): 338-353.

Venkatachalappa M, Khan S K, Kakamari K A G. 2003. Time dependent mathematical model of air pollution due to area source with variable wind velocity and eddy diffusivity and chemical reaction. *Proceedings of Indian National Science Academy*, 69A(6): 745-758.

Wang Z F, Huang M Y, He D Y, et al. 1997. Studies on transport of acid substance in China and East Asia. Part I: 3-D Eulerian transport model for pollutants. *Scientia Atmospherica Sinica* (in Chinese), 21(3): 367-375.

Wu D, Mao J T, Deng X J, et al. 2009. Black carbon aerosols and their radiative properties in the Pearl River Delta region. *Science in China Series D: Earth Sciences*, 52(8): 1152-1163.

Xu M, Jiang W M, Ji C P, et al. 2002. Numerical modeling and verification of structures of the boundary layer over Beijing Area. *Journal of Applied Meteorological Science* (in Chinese), 13(S1): 61-68.

Xu Z, Li W J, Yu Y C, et al. 2011. Characteristics of aerosol optical properties at haze and non-haze weather during autumn at Jinan city. *China Environmental Science* (in Chinese), 31(4): 546-552.

Zhao X J, Pu W W, Meng W, et al. 2013. PM2.5 pollution and aerosol optical properties in fog and haze days during autumn and winter in Beijing Area. *Environmental Science* (in Chinese), 34(2): 416-423.

巢纪平,陈英仪.1979.二维能量平衡模式中极冰-反照率的反馈对气候的影响.中国科学数学,22(12):1198-1207.

陈鹏飞,张蔷,权建农,等.2012.北京上空气溶胶浓度垂直廓线特征.环境科学研究,25(11):1215-1221.

胡非,李昕,陈红岩,等.1999.城市冠层中湍流运动的统计特征.气候与环境研究,4(3):252-258.

黄美元,王自发,何东阳,等.1996.我国冬夏季硫污染物沉降与跨地区输送模拟研究.科学通报,41(11):1013-1016.

荆俊山,张仁健,陶俊.2011.北京郊区夏季 PM2.5 和黑碳气溶胶的观测资料分析.气象科学,31(4):510-515.

罗淦,王自发.2006.全球环境大气输送模式(GEATM)的建立及其验证.大气科学,30(3):504-518.

毛节泰.1992.广东、广西地区酸沉降统计模式的研究.环境科学学报,12(1):28-36.

桑建国,张治坤,张伯寅.2000.热岛环流的动力学分析.气象学报,58(3):321-327.

桑建国.1986.城市热岛效应的分析解.气象学报,44(2):251-255.

寿亦萱,张大林.2012.城市热岛效应的研究进展与展望.气象学报,70(3):338-353.

数学手册编写组.1979.数学手册.北京:高等教育出版社.

王自发,黄美元,何东阳,等.1997.关于我国和东亚酸性物质的输送研究Ⅰ.三维欧拉污染物输送实用模式.大气科学,21(3):367-375.

吴兑,毛节泰,邓雪娇,等.2009.珠江三角洲黑碳气溶胶及其辐射特性的观测研究.中国科学:地球科学,39(11):1542-1553.

徐敏,蒋维楣,季崇萍,等.2002.北京地区气象环境数值模拟试验.应用气象学报,13(S1):61-68.

徐政,李卫军,于阳春,等.2011.济南秋季霾与非霾天气下气溶胶光学性质的观测.中国环境科学,31(4):546-552.

赵秀娟,蒲维维,孟伟,等.2013.北京地区秋季雾霾天PM2.5污染与气溶胶光学特征分析.环境科学,34(2):416-423.

An Analytical Solution for Three-Dimensional Sea-Land Breeze*

YaoKun Li[1,2②], JiPing Chao[3]

(1. *College of Global Change and Earth System Science, Beijing Normal University, Beijing* 100875, *China*; 2. *Joint Center for Global Change Studies, Beijing* 100875, *China*; 3. *National Marine Environmental Forecasting Center, Beijing* 100081, *China*)

Abstract: Based on the hydrostatic, incompressible Boussinesq equations in the planetary boundary layer (PBL), the three-dimensional sea-land breeze (SLB) circulation has been elegantly expressed as functions of the surface temperature distribution. The horizontal distribution of the horizontal or vertical motion is determined by the first or second derivative of the surface temperature distribution. For symmetric land-sea and temperature distribution, the full strength of the sea breeze occurs at the inland but not at the coastline and the maximum updraft associates with the heating center. Setting the temperature difference between land and sea (TDLS) varies with the island size, there would exist an optimal island size corresponding to the strongest SLB circulation which weakens with both larger and smaller island size. Each velocity component approaches to a peak at a certain vertical level. Both the peak value and the corresponding vertical level link with the vertical scale of the surface temperature. The more significant the surface temperature influences vertically, the stronger SLB circulation at a higher vertical level it could induce. We apply the Weather Research and Forecasting (WRF) model ideal simulation for two-dimensional sea breeze to verify the theory. Two cases denoting land breeze and sea breeze respectively further support the theory results despite there is certain slight discrepancy due to the highly simplified theoretical equations.

Key words: sea-land breeze, surface temperature distribution, Boussinesq equations, planetary boundary layer

1 Introduction

Sea-land breeze (SLB), a common phenomenon mainly observed in the planetary boundary layer (PBL), is induced by the diurnal rhythms of the thermal contrast between land and sea. It is one of the ol-

* Journal of the Atmospheric Sciences, 2016, 73(1): 41–54.

② Corresponding author: Dr. YaoKun Li. College of Global Change and Earth System Science, Beijing Normal University, Beijing, China. E-mail: liyaokun@bnu.edu.cn dest topics that meteorologists are interested in and has been studied extensively by numerous observational and numerical studies [for example, see review papers by Abbs and Physick (1992), Miller et al. (2003), and Crosman and Horel (2010)]. Moreover analytical studies also make significant contribution for advanced understanding of the SLB. Early analytical modeling supposed that the SLB is mainly the balance between pressure gradient caused by the unequal heating and friction (Jeffreys, 1922). The effect of the earth's rotation was then taken into account in 1947. The deflecting force of the earth's rotation (Coriolis force) was used to explain the influence of the geostrophic wind on the diurnal variations of the SLB (Haurwitz, 1947) and to explain the no perpendicular to the coast when the SLB blowing at full strength (Schmidt, 1947). The force effect of a prescribed surface temperature function on the SLB circulation was then emphasized (Walsh, 1974). The observed difference in the intensities of sea and land breezes (sea breeze is generally stronger than land breeze) was explained analytically by the time-varying diurnal variation of the stratification and the related variation of the eddy diffusion coefficients which were expected to give rise to a temporally asymmetrical circulation even when the surface forcing is symmetrical (Mak and Walsh, 1976). The rotation of the direction of the SLB was also analytically investigated (Neumann, 1977; Simpson, 1996). Furthermore, the atmospheric response was discussed by the relative size of the diurnal heating and cooling frequency ω and the Coriolis parameter f (Rotunno, 1983). When $f > \omega$ the atmospheric response is confined to within a distance of the coastline and when $f < \omega$ the atmospheric response is in the form of internal-inertial waves. However, giving nonperiodic forcing and realistic values of friction, only an elliptic solution (the $f > \omega$ case) could result (Dalu and Pielke, 1989). The horizontal dimension of the SLB circulation was then investigated by Niino (1987). More recently, the linear theory of Rotunno (1983) had been extended and explored analytically by including the effect of background wind on the sea breeze wave response (Qian et al., 2009) and the effect of a base-state thermal wind on the linear dynamics of the sea breeze (Drobinski et al., 2011). Besides, there are also many theoretical studies which solve the SLB circulation numerically (Estoque, 1961; Fisher, 1961; Neumann and Mahrer, 1971; Mahrer and Pielke, 1977; Yan and Anthes, 1987; Arritt, 1993; Feliks, 2004; Antonelli and Rotunno, 2007; Qian et al., 2012). We mainly focus on the analytical solution and do not introduce them explicitly.

For mathematic convenience, the SLB is mainly discussed in two-dimensional space, for example, in x-z plane. The stream function then could be introduced to simplify equations to an equation containing only one variable. The analytical solution then could be obtained to conduct theoretical discussion. It would be much more complex when dealing with the three-dimensional situation. Zhang et al. (1999) had made a good attempt. They concluded the three-dimensional equations to a complex partial differential equation which is so complicated that they had to solve it by numerical relaxation method. Jiang (2012a) extended the linear solutions in Rotunno (1983) and Qian et al. (2009) to the three-dimensional SLB perturbation induced by complex coastlines. He further discussed the complexity introduced by inversion and vertical stratification variation and wind shear (Jiang, 2012b). Above studies applied Fourier transform to combine the system equations into a single wave equation which could be solved analytically under some simple circumstances. Then characteristics of the three-dimensional SLB could be discussed. However when applying the three-dimensional models to complex coastlines it is difficult to obtain analytical solutions and numerical methods have to be used.

SLB is a mesoscale response of the atmosphere to horizontal variations in surface heating (Walsh, 1974) which associates with the temperature difference between land and sea (TDLS). Therefore, it should be written as a function of the temperature distribution even though the form might be complex. For example, Haurwitz (1947) got the expressions for u and v winds but it is difficult to discuss in general terms due to the complexity of the auxiliary constants. Based on the above idea, this paper tries to analytically identify the driving effect of the temperature distribution on the SLB circulation. Consistent with former researchers (for example, see Rotunno (1983)), the horizontal acceleration in the PBL is induced by the imbalance of pressure gradient, friction, and Coriolis force and the basic flow is neglected. Differently, to highlight the forcing effect of the surface temperature and to solve the equation analytically, the temperature equation is not solved and the surface temperature is seen as an independent prescribed variable. The three-dimensional SLB circulation could eventually be derived as an elegant expression of the surface temperature distribution by just hypothesizing the temperature exponentially decreases with height. The circulation characteristics are then discussed. The theoretical analysis could also be used in the studies of urban heat island and the oasis cold island effect.

2 Model framework

SLB circulation mainly prevails in the lower atmosphere therefore Boussinesq equations in the PBL are used to conduct the theoretical analysis. The basic current is set zero because the SLB circulation is mainly driven by the thermal contrast between land and sea. It is more significant in the weak basic current situation. Consider a cylindrical coordinate system in which the ground-plane is at $z = 0$ and z increases upward; r increases outward and θ increases counterclockwise. Incompressible and hydrostatic Boussinesq equations in the cylindrical coordinate system are

$$\frac{\partial u}{\partial t} - fv = -\frac{1}{\rho_0}\frac{\partial p}{\partial r} + \nu\frac{\partial^2 u}{\partial z^2} \tag{1}$$

$$\frac{\partial v}{\partial t} + fu = -\frac{1}{\rho_0 r}\frac{\partial p}{\partial \theta} + \nu\frac{\partial^2 v}{\partial z^2} \tag{2}$$

$$\frac{1}{\rho_0}\frac{\partial p}{\partial z} = \frac{g}{T_0}T \tag{3}$$

$$\frac{1}{r}\frac{\partial ru}{\partial r} + \frac{1}{r}\frac{\partial v}{\partial \theta} + \frac{\partial w}{\partial z} = 0 \tag{4}$$

$$\frac{\partial T}{\partial t} + N^2 w = Q + \kappa\frac{\partial^2 T}{\partial z^2} \tag{5}$$

where f is the Coriolis parameter; g is the acceleration of gravity; ν and κ are the eddy coefficients of viscosity and conduction; ρ_0 and T_0 are the reference density and temperature of the stationary atmospheric background; $T, p, (u, v, w)$ represent the deviations of temperature, pressure and velocity components in (r, θ, z) directions from the stationary atmospheric background, respectively; N^2 is the buoyancy frequency; Q is the heating function. Constant values are assigned to ρ_0, T_0, f, N^2, ν, and κ ($\nu = \kappa$). Equation (3) employs hydrostatic approximation which could describe the SLB circulation well in most cases (Walsh, 1974; Niino, 1987; Jiang, 2012a). Eq. (4) expresses the air mass conservation under the incompressible approximation. Applying the variable separation method, that is, $\eta(r,\theta,z;t) = \hat{\eta}(r,\theta,z)\, e^{i\omega t}$ (for example, see Rotunno (1983)), where η denotes one of the above unknowns and $\omega = 2\pi/\text{day}$ represents the diurnal variation, Eqs. (1), (2), and (5) could be written as

$$\mathrm{i}\omega u - fv = -\frac{1}{\rho_0}\frac{\partial p}{\partial r} + \nu\frac{\partial^2 u}{\partial z^2} \tag{6}$$

$$\mathrm{i}\omega v + fu = -\frac{1}{\rho_0 r}\frac{\partial p}{\partial \theta} + \nu\frac{\partial^2 v}{\partial z^2} \tag{7}$$

$$\mathrm{i}\omega T + N^2 w = Q + \kappa\frac{\partial^2 T}{\partial z^2} \tag{8}$$

The forms of Eqs. (3) and (4) have no variation because they does not explicitly contain the time partial derivative terms. It should be noted that $\hat{\eta}$ has been replaced by η in Eqs. (3), (4), (6), (7) and (8) for simplification.

Set $U = u + v\mathrm{i}$ to substitute Eq. (3) into Eqs. (6) and (7) to eliminate pressure p

$$\frac{\partial^3 U}{\partial z^3} - \frac{f+\omega}{\nu}\mathrm{i}\frac{\partial U}{\partial z} = \frac{g}{\nu T_0}\left(\frac{\partial T}{\partial r} + \frac{1}{r}\frac{\partial T}{\partial \theta}\mathrm{i}\right) \tag{9}$$

Suppose that when $z \to \infty$

$$\left(\frac{\partial^2 U}{\partial z^2} - \frac{f+\omega}{\nu}\mathrm{i}U\right) \to 0 \tag{10}$$

According to the above boundary condition, the integral of Eq. (9) is

$$\frac{\partial^2 U}{\partial z^2} - \frac{f+\omega}{\nu}\mathrm{i}U = -\frac{g\alpha}{\nu}\int_z^\infty \left(\frac{\partial T}{\partial r} + \frac{1}{r}\frac{\partial T}{\partial \theta}\mathrm{i}\right)\mathrm{d}z \equiv F(r,\theta,z) \tag{11}$$

It could be written as

$$LU = F(r,\theta,z) \tag{12}$$

where the operator $L = \frac{\partial}{\partial z}\left[p(z)\frac{\partial}{\partial z}\right] + q(z)$ and $p(z) = 1$, $q(z) = -\left(\frac{1+\mathrm{i}}{h_E}\right)^2$. The term $h_E \equiv \sqrt{\frac{2\nu}{f+\omega}}$ could be called the revised Ekman elevation. Note that h_E is not defined when $f + \omega = 0$ in which case Eq. (11) could be directly solved by its twice-integration with respect to z. The boundary conditions of Eq. (12) are taken as

$$z \to 0, \infty, \quad DU = 0 \tag{13}$$

where $D = 1$ could be called the boundary condition operator.

The solution of Eq. (12) is

$$U = \int_0^\infty G(z,\xi)\, F(r,\theta,\xi)\, \mathrm{d}\xi \tag{14}$$

where $G(z,\xi)$ is the Green function. According to the nature of the Green function, it has the following form

$$G(z,\xi) = \begin{cases} -\dfrac{1}{2\beta}\left[e^{\beta(z-\xi)} - e^{-\beta(z+\xi)}\right], & 0 \leqslant z < \xi \\ -\dfrac{1}{2\beta}\left[e^{\beta(\xi-z)} - e^{-\beta(z+\xi)}\right], & \xi < z \leqslant \infty \end{cases} \tag{15}$$

where $\beta = (1 + \mathrm{i}) / h_E$. To obtain the specific expression of Eq. , it is necessary to solve the specific expression of $F(r,\theta,z)$.

If we set the surface temperature exponentially decays aloft with a vertical scale of h_R , that is

$$T(r,\theta,z) = T_s(r,\theta)\,\mathrm{e}^{-z/h_R} \tag{16}$$

where $T_s(r,\theta)$ represents the surface temperature distribution. It is common practice to assume heating function exponentially declines (for example, see Rotunno (1983) and Jiang (2012a)). Here we apply the same assumption on the temperature distribution to analytically solve Eq. (14). Based on the assumption, the expression of F could be written as

$$F(r,\theta,z) = -\frac{g}{\nu T_0}\left(\frac{\partial T_s}{\partial r} + \frac{1}{r}\frac{\partial T_s}{\partial \theta}\mathrm{i}\right)\int_z^{\infty} \mathrm{e}^{-z/h_R}\mathrm{d}z = -\frac{g h_R}{\nu T_0}\left(\frac{\partial T_s}{\partial r} + \frac{1}{r}\frac{\partial T_s}{\partial \theta}\mathrm{i}\right)\mathrm{e}^{-z/h_R} \tag{17}$$

To make sure the finite integral value in Eq. (17), the vertical scale h_R must be larger than zero, which means the temperature would decrease and tend to zero with increasing height. Eq. (16) might be a little rough for just setting the exponential decline temperature profile but neglecting the possibly existing fluctuations. However, the temperature would eventually tend to be zero when height is high enough, especially for the SLB circulation which is mainly driven by the surface heating and only prevails in the lower 1−2 kilometers. Therefore, it is reasonable to make such an assumption.

Substituting Eqs. (15), (17) into Eq. (14), we get

$$\begin{aligned} U = &-\frac{1}{2\beta}\frac{g h_R}{\nu T_0}\left(\frac{\partial T_s}{\partial r} + \frac{1}{r}\frac{\partial T_s}{\partial \theta}\mathrm{i}\right)\int_0^{\infty} \mathrm{e}^{-\beta(z+\xi)}\,\mathrm{e}^{-\xi/h_R}\mathrm{d}\xi \\ &+ \frac{1}{2\beta}\frac{g h_R}{\nu T_0}\left(\frac{\partial T_s}{\partial r} + \frac{1}{r}\frac{\partial T_s}{\partial \theta}\mathrm{i}\right)\int_0^{z} \mathrm{e}^{-\beta(z-\xi)}\,\mathrm{e}^{-\xi/h_R}\mathrm{d}\xi \\ &+ \frac{1}{2\beta}\frac{g h_R}{\nu T_0}\left(\frac{\partial T_s}{\partial r} + \frac{1}{r}\frac{\partial T_s}{\partial \theta}\mathrm{i}\right)\int_z^{\infty} \mathrm{e}^{-\beta(\xi-z)}\,\mathrm{e}^{-\xi/h_R}\mathrm{d}\xi \end{aligned} \tag{18}$$

The horizontal velocity components then could be derived by separating Eq. (18) into real and imaginary parts, namely

$$u = A(z)\frac{\partial T_s}{\partial r} + B(z)\frac{1}{r}\frac{\partial T_s}{\partial \theta} \tag{19}$$

$$v = A(z)\frac{1}{r}\frac{\partial T_s}{\partial \theta} - B(z)\frac{\partial T_s}{\partial r} \tag{20}$$

where $A(z)$ and $B(z)$ are functions governing the vertical distribution of the horizontal velocity components. Their forms are

$$A(z) = \frac{g h_R}{\nu T_0}\frac{1}{(1/h_R^4 + 4/h_E^4)}\left\{\frac{2}{h_E^2}e^{-z/h_E}\sin\frac{z}{h_E} - \frac{1}{h_R^2}\left[\mathrm{e}^{-z/h_R} - \mathrm{e}^{-z/h_E}\cos\frac{z}{h_E}\right]\right\} \tag{21}$$

$$B(z) = \frac{g h_R}{\nu T_0}\frac{1}{(1/h_R^4 + 4/h_E^4)}\left\{\frac{1}{h_R^2}\mathrm{e}^{-z/h_E}\sin\frac{z}{h_E} + \frac{2}{h_E^2}\left[\mathrm{e}^{-z/h_R} - \mathrm{e}^{-z/h_E}\cos\frac{z}{h_E}\right]\right\} \tag{22}$$

For the SLB circulation, the latitudinal variation of Coriolis parameter could be neglected. Therefore, the vertical velocity could be solved by integrating Eq. under the boundary condition $z = 0, w = 0$, namely

$$w - w|_{z=0} = \int_0^z \frac{\partial w}{\partial z}\mathrm{d}z = -\int_0^z\left(\frac{\partial r u}{\partial r} + \frac{1}{r}\frac{\partial v}{\partial \theta}\right)\mathrm{d}z = -\int_0^z A(z)\,\mathrm{d}z \cdot \nabla_h^2 T_s \tag{23}$$

where ∇_h^2 is the horizontal Laplace operator in cylindrical coordinate system. By introducing the topographic lifting effect, that is, $z = h_s(r,\theta)$, $w = \vec{v}_s \cdot \nabla h_s$ where $h_s(r,\theta)$, $\vec{v}_s = (u_s, v_s)$, and ∇denote the topography distribution, the horizontal velocity components on the lower boundary, and the Hamilton operator in the cylindrical coordinate system, the vertical motion could be written as

$$w = w\big|_{z=h_s} + \int_{h_s}^{z} \frac{\partial w}{\partial z} \mathrm{d}z = \vec{v}_s \cdot \nabla h_s - \int_{h_s}^{z} A(z)\,\mathrm{d}z \cdot \nabla_h^2 T_s \tag{24}$$

Equation (24) could be easily solved for both $h_s(r,\theta)$ and $\vec{v}_s$ are known variables.

Seen from Eqs. (19), (20), and (23), the three-dimensional SLB circulation is completely expressed as functions of the surface temperature distribution. They directly show the driving effect of the surface heating on the SLB circulation. Also they offer a method to theoretically diagnose the SLB circulation through the surface temperature which is very easy to observe. Giving the time-varying temperature distribution, the diurnal variation of the SLB circulation also could be analyzed. Then how should the temperature distribution be determined? Of course it could be obtained by solving the thermal equation Eq. (8). For example, we could assume that the vertical profile of the heating function satisfying the same principle as the vertical temperature profile, that is, $Q = q(r,\theta)\,\mathrm{e}^{-z/h_R}$, where $q(r,\theta)$ is the horizontal structure of the heating function. Then an equation only containing one variable (the surface temperature) could be obtained by just substituting Eq. (23) into Eq. (8) to eliminate the vertical velocity, namely

$$\left(\mathrm{i}\omega - \frac{\kappa}{h_R^2}\right) T_s - N^2 \int_0^z A(z)\,\mathrm{d}z \cdot \nabla_h^2 T_s = q(r,\theta) \tag{25}$$

Eq. (25) is a Helmholtz equation of surface temperature T_s. Giving known $q(r,\theta)$, T_s could be obtained by solving Eq. (25) either analytically or numerically. However, in most cases especially for complex coastlines, it is hard to solve Eq. (25) analytically. On the other hand, once the heating function $q(r,\theta)$ is prescribed (no matter what pattern it is), the surface temperature distribution would be only determined by Eq. (25). Therefore, it is reasonable to directly specify the surface temperature distribution (equivalent that Eq. (25) has already been solved out) to solve the horizontal and vertical velocity components. Although this would nevertheless leave the buoyancy frequency undiscussed, it simplifies the derivation processes and benefits the elegant form of the three-dimensional SLB circulation.

3 Model results

Equations. (19), (20) and (23) could be applied to any given temperature distribution representing different land and sea distributions. A typical land and sea distribution for the SLB circulation is an island surrounded by sea. Therefore, we specify a circular island whose center is located at the coordinate origin and coastline is at $r = r_0$ where r_0 is the island radius. For land r is smaller than r_0 and for sea r is larger than r_0. The three-dimensional SLB circulation then could be reduced to the two-dimensional situation, that is, in (r,z) space due to symmetry. According to the fixed land and sea distribution, we set similar symmetric surface temperature distribution

$$T_s(r,\theta) = T_m \exp(-r/r_0)^2 \tag{26}$$

where $T_m = T_{m0}\,[1 - \exp(-r_0/R)^4]$ is the TDLS intensity, which is revised by a biharmonic term in brackets to make sure T_m is zero when $r_0 = 0$. Here T_{m0} denotes the reference TDLS intensity and is given a constant value. When T_{m0} is greater than zero, it corresponds to the sea breeze and when T_{m0} is smaller than zero, it corresponds to the land breeze. R represents the reference island radius which could be specified by the actual land and sea distribution. We consider TDLS as a function of r_0 because when there is no island in the ocean ($r_0 = 0$), TDLS should be zero and when the island is large enough, TDLS should be a finite value. We would explicitly explain why it takes such a form below. The term $\exp(-r/r_0)^2$ indicates the non-dimensional horizontal surface temperature distribution. It shows that temperature value would be unity when $r = 0$ and $1/e$ when $r = r_0$ and close to zero when r is large enough.

Then substituting Eq.(26) into Eq.(19), (20) and (23) gets

$$u = A(z)\frac{\mathrm{d}T_s}{\mathrm{d}r} = -A(z)\cdot M(r) \tag{27}$$

$$v = -B(z)\frac{\mathrm{d}T_s}{\mathrm{d}r} = B(z)\cdot M(r) \tag{28}$$

$$w = -\int_0^z A(z)\,\mathrm{d}z\,\frac{1}{r}\frac{\partial}{\partial}\left(r\frac{\partial T_s}{\partial r}\right) = \int_0^z A(z)\,\mathrm{d}z\cdot N(r) \tag{29}$$

where $M(r) = -\frac{\mathrm{d}T_s}{\mathrm{d}r} = \left(\frac{2r}{r_0^2}\right)T_s$ and $N(r) = -\frac{1}{r}\frac{\mathrm{d}}{\mathrm{d}r}\left(r\frac{\mathrm{d}T_s}{\mathrm{d}r}\right) = \frac{4}{r_0^2}\left[1-\left(\frac{r}{r_0}\right)^2\right]T_s$ are the negative first and second derivatives of the horizontal surface temperature distribution. They characterize the horizontal distribution of the horizontal and vertical velocity components respectively. Setting $f = 1.0284\times10^{-4}\,\mathrm{s}^{-1}$, $\nu = 5\,\mathrm{m}^2\cdot\mathrm{s}^{-1}$, $g = 9.8\ \mathrm{m}\cdot\mathrm{s}^{-2}$, $T_0 = 283$ K, $r_0 = R = 50$ km, $h_R = 400$ m, and $T_{m0} = 8$℃, the theoretical SLB circulation could be calculated in (r,z) plane (Fig. 1). The r-direction wind component u (maximum speed is about 2.5 $\mathrm{m}\cdot\mathrm{s}^{-1}$) blows from sea to land at the lower layer and reverses its direction with slow speed at the upper

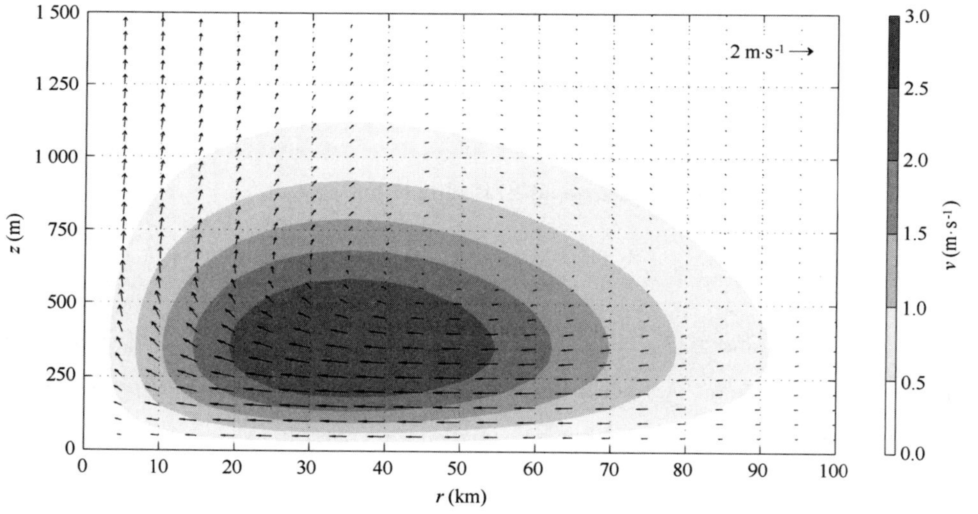

Fig.1 Velocity components of sea-land breeze circulation in (r,z) direction (vector) and in θ direction (shaded)

layer, accompanying strong updrafts near the island center and weak downdrafts over the sea. The (u,w)

wind vector consists of a clockwise circulation centered at about 600 m height above the coastline (near $r = r_0$). *The* θ -direction wind component v (shaded) blows counterclockwise (positive) with the comparable intensity to u. The horizontal wind vector (u, v) does not blow perpendicular to the coast but deviated toward the right. The velocity components have significant horizontal variations which are determined by the first and second derivatives of the surface temperature distribution. Both u and v possess the same horizontal variation characteristic governed by $M(r)$ (see in Fig. 2a) which denotes the negative first derivative of the surface temperature distribution. $M(r)$ increases and then decreases with increasing r. Such a pattern means there

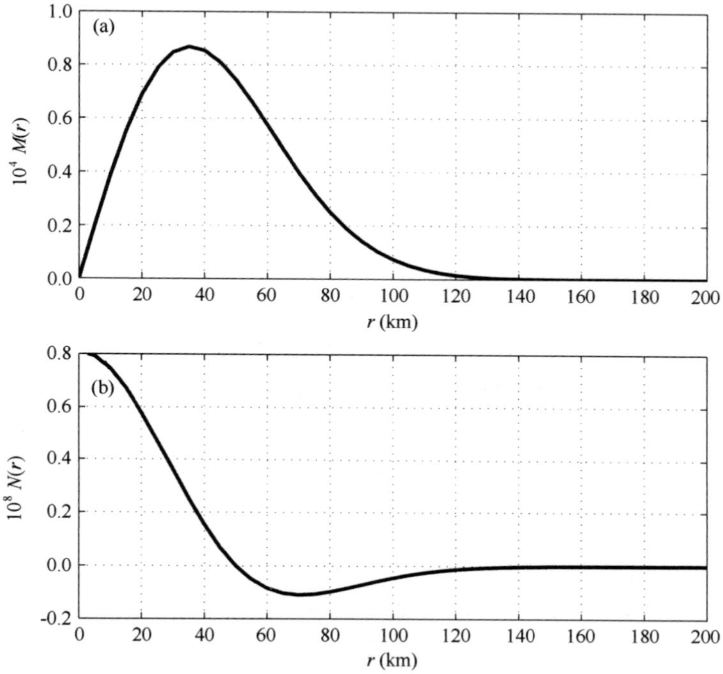

Fig.2　The distributions of $M(r)$ (a) and $N(r)$ (b)

exists a maximum value at a certain r which could be specified by solving the zero point of the first derivative of $M(r)$, namely, $\frac{dM}{dr} = \frac{1}{r_0^2}[2 - (2r/r_0)^2]\ T_s = 0$. After simple derivation, the solution of the equation is $r = \sqrt{2}/2r_0$. Therefore, $M(r)$ would eventually decrease after firstly approaching to its maximum value at $r = \sqrt{2}/2r_0$ which is about 35km for r_0 is 50km. This implies that the sea breeze at full strength would not just blow at the coastline but at the inside land. $M(r)$ mainly dominates in the region where the surface temperature gradient is large enough and it would be very small when r is larger than 120km. The horizontal variation of the vertical velocity w is regulated by $N(r)$ (see in Fig. 2b) which denotes the negative second derivative of the surface temperature distribution. Differently, $N(r)$ gradually decreases to zero and to a minimum value and to infinitely close to zero when r becomes larger and larger. The r at which $N(r)$ approaches to its extreme values could be specified by solving the zero points of its first derivative, that is, $\frac{dN}{dr} = -\frac{8r}{r_0^4}[2 - (r/r_0)^2]\ T_s = 0$. The solutions of the equation are $r = 0$ and $r = \sqrt{2}r_0$. Further analysis indicates $N(r)$ approaches to its maximum and minimum values at the two zero points respectively. The structure demonstrates that the updrafts would be strongest at the heating center and would be gradually weakening to zero when r increasing to the coastline ($r = r_0$), larger than which, it would change sign to become downdrafts which also could approach to its maximum value at another critical value ($r = \sqrt{2}r_0$). This suggests that the updrafts occur on the land while downdrafts happen on the sea and the coastline is a natural boundary of them. Also, it could be easily seen that the updrafts strength is much greater than the downdrafts although the total vertical transports are the same. Therefore the sea breeze correlates with strong but narrow updrafts while weak and wide downdrafts.

As demonstrated before, $M(r)$ and $N(r)$ characterize the first and second derivatives of the surface temperature distribution. Once the surface temperature is specified, the horizontal structures of the horizontal and vertical velocity components are determined. $M(r)$ and $N(r)$ approach to their maximum values when $r = \sqrt{2}/2r_0$ and $r = 0$ respectively. Then substituting the two values into $M(r)$ and $N(r)$, we could get $M_{max}(r_0) \sim T_m/r_0$ and $N_{max}(r_0) \sim T_m/r_0^2$. M_{max} and N_{max} could characterize the intensity of the SLB circulation. However, if T_m is a constant value, the inverse proportions between M_{max}, N_{max} and r_0 imply that the smaller the island size is, the stronger the SLB circulation is. This is obviously unreasonable because when there is no island in the ocean, there should be no SLB circulation. The irrationality could be overcome by setting the TDLS value decreases with the decreasing island size. Therefore, this is also the reason why Eq. (26) revises T_m to zero for $r_0 = 0$. To make sure the limits of $M_{max}(r_0) \sim T_m/r_0$ and $N_{max}(r_0) \sim T_m/r_0^2$ are zero when r_0 tends to zero, T_m should be an infinitesimal of the variable r_0, whose order must be larger than two, the highest order of the infinitesimal variable r_0 on the denominator. Therefore, T_m is taken as the bi-harmonic form. We also could appoint other forms such as $T_m(r_0) = T_{m0}(r_0/R)^3$. For example, Jiang (2012a) took similar heating function. However it would make the surface temperature tend to infinity when r_0 tends to infinity if without any constraint. With the revised TDLS intensity, the SLB circulation would disappear for the no land situation. Fig. 3 further embodies such relationships. M_{max} exhibits a peak at about $r_0 = 60$ km, which decreases with the increase or decrease of r_0 and becomes to zero for enough small or large r_0 (Fig. 3a). Similarly, N_{max} also has a peak at about $r_0 = 50$ km, which decreases for larger or smaller r_0 (Fig. 3b). This suggests that the horizontal and vertical motions would be strongest at certain r_0 which could be called the optimal island sizes. The optimal island size for the horizontal motion is larger than that for the vertical motion. The dependence of the SLB strength on the island size illustrated above is in qualitative agreement with the previous numerical and theoretical studies (Neumann and Mahrer, 1974; Abe and Yoshida, 1982; Xian and Pielke, 1991; Jiang, 2012a). This further demonstrates the reasonability of the revised T_m and the correctness of the theory. The optimal island scale is also determined by the island scale R and a smaller R corresponds to a smaller optimal scale and vice versa (figure omitted).

On the other hand, the vertical variation of the SLB circulation is controlled by $A(z)$ and $B(z)$. The expressions of $A(z)$ and $B(z)$ are a combination of the exponential and trigonometric functions. $A(z)$ has two extreme values (Fig. 4a) which is located at about 200 m and 900 m respectively. The lower maximum value (larger than zero) is much larger than the upper minimum value (smaller than zero), suggesting a strong u wind toward land at the lower layer and its weak return flow toward sea at the upper layer. It also could be

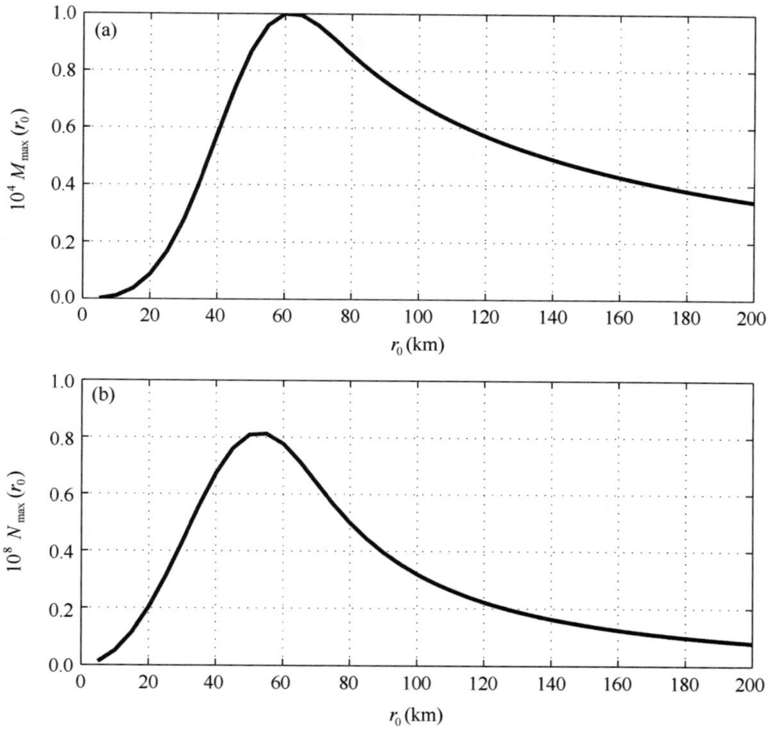

Fig.3　The variation of $M_{max}(r_0)$ (a) and $N_{max}(r_0)$ (b) with r_0

seen intuitively from Fig. 1. Differently, $B(z)$ is larger than zero and has only one maximum value at about 400 m level (Fig. 4b), indicating the counterclockwise v wind in the whole layer. Therefore, the horizontal wind constitutes a cyclonic circulation cell linking with the low pressure driven by the heating at the lower layer while an anti-cyclonic circulation cell at the upper layer. It is the integral of $A(z)$ contributing to the variation of the vertical motion rather than $A(z)$ itself. The integral of $A(z)$ is larger than zero and has a maximum value near 600 m level (Fig. 4c), implying the maximum updrafts occur near the 600 m level. In addition, the heights that $A(z)$ and $B(z)$ approach to their maximum values are different. There is only one extra introduced parameter h_R which as mentioned before indicates the vertical scale that the surface temperature exponentially decays aloft with. Physically, a large h_R means more significant influence of the surface temperature on the SLB circulation. According to Eqs. (21) and (22), h_R could affects the maximum values of $A(z)$, $B(z)$, and the integral of $A(z)$ (A_{max}, B_{max}, and iA_{max} in abbreviated form), and the vertical level they approach to their maximum values. A_{max}, B_{max}, and iA_{max} could be specified by solving the zero points of their first derivatives with respect to z. However, no analytical solutions could be derived due to the complex expressions. Figure 5 exhibits such relationships numerically. The arrows in Fig. 5 denote the direction along which h_R increases. Both A_{max}, B_{max}, iA_{max} and the corresponding vertical levels increase with increasing h_R. This implies that stronger vertical influence of the surface temperature would benefit a stronger SLB circulation occurring at a higher level. Also seen Fig. 5, same increments of A_{max}, B_{max}, and iA_{max} associate with different height increments. For example, a same increment of A_{max} associates with decreasing increment of the vertical level for increasing h_R. This suggests that comparing with the upward moving of the SLB circulation, the strong surface temperature influence would strengthen its intensity more. A reasonable SLB circulation strength needs a moderate h_R which should be comparable to the revised Ekman elevation according to the derivation.

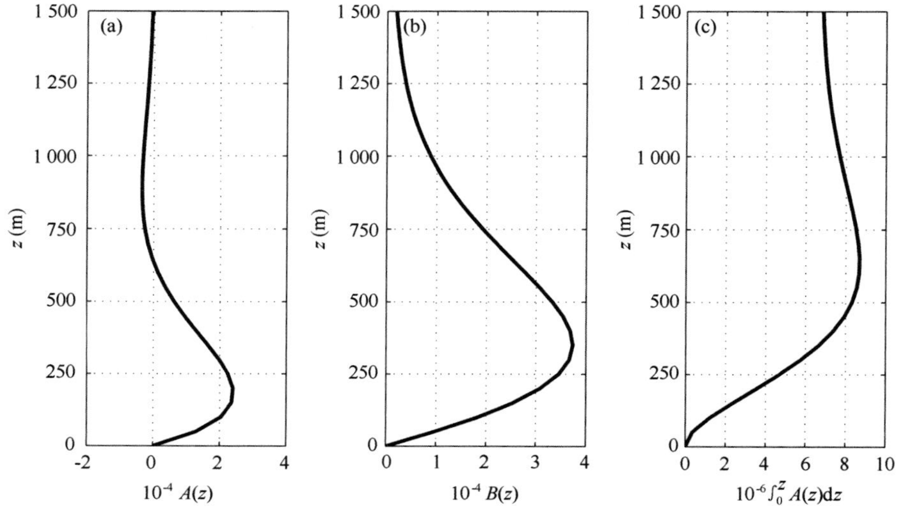

Fig.4 The distribution of $A(z)$ (a) and $B(z)$ (b) and the integral of $A(z)$ (c)

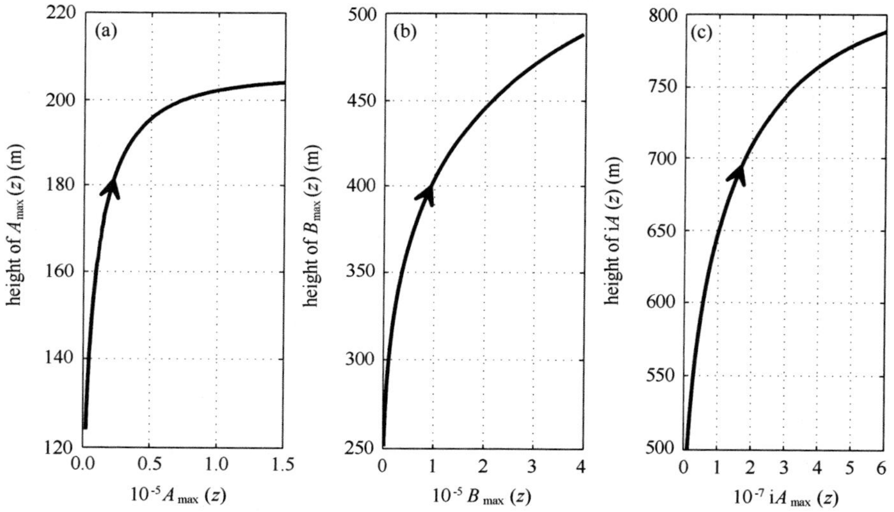

Fig.5 The relationship between maximum value of $A(z)$ and the height $A(z)$ approaches to maximum value (a) and the case for $B(z)$ (b) and for the integral of $A(z)$ (c). The arrows denote the direction along which the vertical temperature decay scale increases

4 WRF model verification

According to the above discussion, we elegantly express the three-dimensional SLB circulation as functions of the surface temperature distribution and then discuss the basic structure of the SLB circulation. In this section, we further apply the National Center for Atmospheric Research (NCAR) Advanced Weather Research and Forecasting (WRF) model version 3. 6. 1 to verify the theoretical calculation. The equation set for the WRF model is fully compressible, Eulerian and non-hydrostatic with a run-time hydrostatic option. The model uses terrain-following, hydrostatic-pressure vertical coordinate with the top of the model being a constant pressure surface. The horizontal grid is the Arakawa-C grid. The time integration scheme in the model uses the third-order Runge-Kutta scheme, and the spatial discretization employs 2nd- to 6th- order schemes. The model supports both idealized and real-data applications with various lateral boundary condition options (Skamarock et al., 2008). It has been used extensively in the idealized sea breeze simulations (Antonelli and Rotunno, 2007; Gibbs, 2008; Crosman and Horel, 2012; Steele et al., 2013).

The WRF model ideal sea breeze case allows the users to gain a physical understanding of the sea breeze itself by only considering the most basic and necessary parameters required to reproduce the appropriate feature (Gibbs, 2008). Details on the WRF model configured as a two dimensional sea breeze simulation are given in Table 1. The model setup is for a 2D case with 202 grid points and 35 vertical levels ($x-z$ plane) with a horizontal grid spacing of 2000 m and a stretched vertical grid spacing of 85–130 m. The land occupies 50 grid points in the middle of the domain. The central longitude is taken as the prime meridian which means start hour is the local time as well as the universal time coordinated (UTC). There is a diurnal cycle and the latitude and longitude are set for radiation to work. Full physics are employed, using the WRF double moment (WDM) 5-class microphysics scheme, the Dudhia and rapid radiative transfer model (RRTM) radiation schemes, the revised Monin–Obukhov surface layer scheme, and the thermal diffusion land surface scheme. The ocean is initially set 7 K warmer than the land surface temperature. The simulation covers a 24 h timeframe at 00 UTC with no wind at the beginning.

Table 1　WRF model two dimensional sea breeze details (WRF namelist selection in italics)

Model parameters	Two dimensional sea breeze configuration
Parameterizations	WDM 5-class microphysics scheme (*mp_physics* = 14); RRTM longwave radiation scheme (*ra_lw_physics* = 1); Dudhia shortwave radiation scheme (*ra_sw_physics* = 1); Monin–Obuhkov surface layer (*sf_sfclay_physics* = 1); thermal diffusion land surface scheme (*sf_surface_physics* = 1); YSU PBL scheme (*bl_pbl_physics* = 1); constant eddy coefficient scheme (*km_opt* = 1)
Domain	404km(x)×6km(y)×10km(z)
x-grid spacing	2000m (202 grid points)
y-grid spacing	2000m (3 grid points)
z-grid spacing	85–130m stretched (35 gird points)
Boundary conditions	Periodic
Time step	30 s
Simulation length	24 h

续表

Model parameters	Two dimensional sea breeze configuration
Damping	W-Rayleigh damping (*damp_opt* = 3) with damping coefficient 0. 1 (*dampcoef* = 0. 1) damping depth from model top (*zdamp* = 5000)
Heat flux	heat and moisture fluxes from the surface (*isfflx* = 1)
Fixed initialization parameter	central latitude 45°N central longitude 0° Coriolis parameter (f) = 1. 0284^{-4} s^{-1} initial temperature difference 7K

The simulation depicts a clear picture of the diurnal evolution of the SLB circulation. A shallow weak land breeze circulation is established over the coastline from approximately 01 UTC. It breaks down and a sea breeze circulation emerges simultaneously from 09 UTC. The sea breeze strengthens and then weakens until 21 UTC when it is replaced by the land breeze. Two times (03 UTC and 12 UTC) are selected to characterize the land breeze and sea breeze circulation respectively. Figure 6 shows the near surface temperature associated with the two cases. Due to the symmetry of land and sea distribution, coordinate origin of the abscissa (represented by variable r) in Fig. 6 starts at the middle of the land and increase toward the sea and the boundary between land and sea is at r=50 km. In the case for the land breeze, the weak TDLS intensity (about −1. 8℃, see from Fig. 6a) corresponds to a weak land breeze circulation (maximum horizontal speed is about 0. 3 m · s^{-1}, see in Fig. 7b). The land breeze circulation mainly dominates in the range of 20−80 km. The wind blows from land to sea at the lower level, then ascends to the higher level on the sea, then returns to the land at the higher level, and then descends to the lower level on the land, and finally forms a complete anticlockwise land breeze circulation cell which could vertically extend up to about 1000 m level. The land breeze at the lower level is stronger than the return flow at the higher level. The center of the circulation cell sits at about 500m level above the boundary between land and sea.

We calculate the theoretical land breeze (see in Fig. 7a) driven by the temperature distribution in the WRF ideal simulation. It could be seen that the analytical solution is consistent with the WRF model results qualitatively. For example, an analogical anticlockwise circulation cell horizontally ranges from about 40 to 60km where the temperature gradient is obvious and vertically extends up to about 1000 m. The downdraft and updraft branches dominate on land and sea respectively. The land breeze occurs at the lower level with the intensity about 0. 3 m · s^{-1} and the return flow at the higher level is weak. Of course, there exists certain discrepancy between the theoretical and the WRF model results. For example, the analytical land breeze is slightly stronger than its WRF model counterpart but with a narrower spatial extent.

The situation is similar in the case for the sea breeze circulation. TDLS decreases from 8 to 0℃ with increasing r (Fig. 6b). It varies gently at the inland and far off the sea but sharply near the boundary between land and sea. The circulation pattern associated with the temperature distribution in the WRF ideal simulation indicates a typical sea breeze (Fig. 8b). Updrafts induced by the warmer land are compensated by the downdrafts over the cooler sea and the wind blows from sea to land at the lower level and reverse its direction at

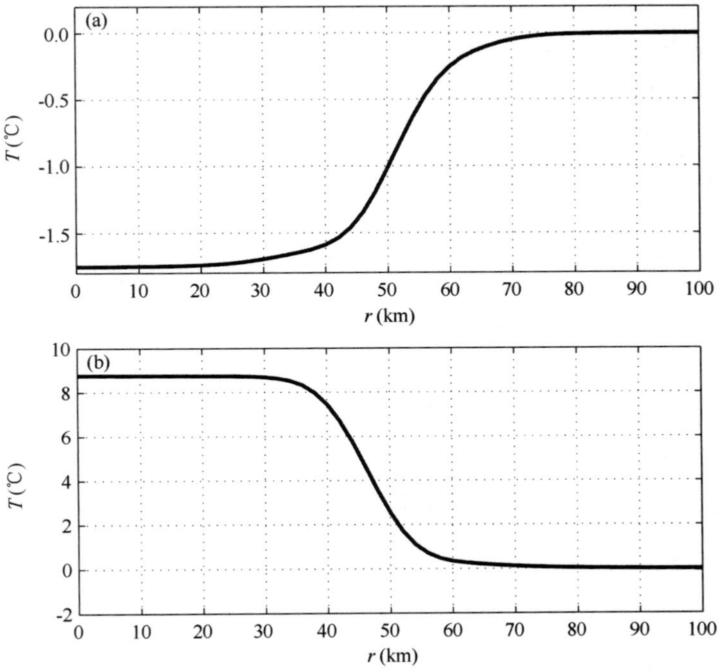

Fig. 6　The near surface temperature in the WRF ideal simulation for the land breeze at 03 UTC (a) and for the sea breeze at 12 UTC (b)

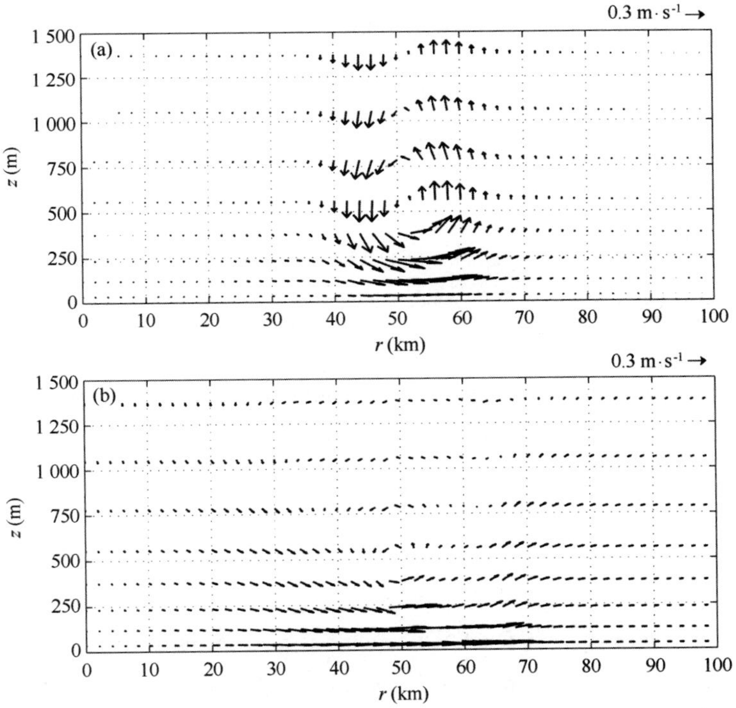

Fig. 7　The land breeze circulation at 03 UTC predicted by the theory calculation (a) and by the WRF model simulation (b)

the upper level. A clockwise sea breeze circulation centered near at 750 m level above $r=40$ km then forms. The WRF model also predicts a sea breeze front (SBF) denoting the boundary between the cooler and warmer air masses. It is located at about $r=35$ km where the sea breeze weakens sharply toward the inland. The analytical sea breeze circulation (Fig.8a) calculated by the WRF model temperature distribution shows good consistence with its WRF model counterpart. For example, an analogical circulation cell with almost equivalent intensity and spatial scope dominates near the boundary of land and sea. The circulation cell center also occurs at the higher level above the land. However, theoretical solution predicts neither the existence of the SBF nor the asymmetry of the motion about the circulation cell center.

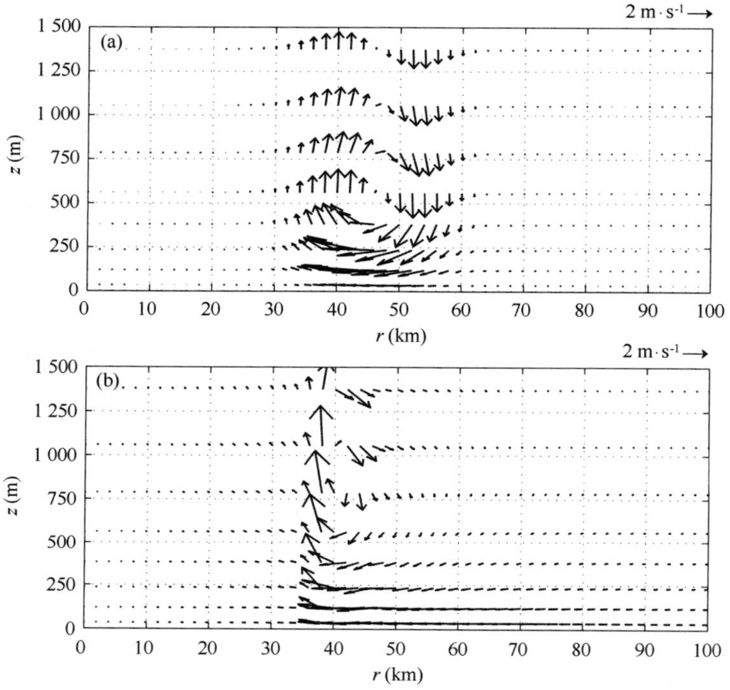

Fig. 8 The sea breeze circulation at 12 UTC predicted by the theory calculation (a) and by the WRF model simulation (b)

Based on the comparison with the WRF model, it could be seen that the theory is correct and it does have the ability to characterize the SLB circulation although it does have certain imperfectness due to the highly simplified equations for obtaining the analytical solution.

5 Conclusion

SLB circulation is an interesting phenomenon which has been extensively investigated either analytically or numerically. Recent theoretical progresses have extended to the complete three-dimensional circulation in which the analytical solution could be only derived for some simple coastline cases. On the other hand, how to directly express the driving effect of the surface temperature on the SLB circulation is still an unsolved question which is worth further exploring. Therefore, we try to theoretically answer the question by applying the hydrostatic and incompressible Boussinesq equations under the stationary atmospheric background in the

PBL. The time partial derivatives could be eliminated by introducing separation of variables. Then the horizontal motion could be reduced into a second-order differential equation which could be solved analytically by just introducing a reasonable hypothesis which predicts the surface temperature exponentially decays with height. Then the vertical motion could be solved analytically by integrating the incompressible continuity equation. Eventually, the theory elegantly exhibits the driving effect of the surface temperature on the three-dimensional SLB circulation.

According to the theory, the horizontal variations of the horizontal and vertical motions are determined by the first and second derivatives of the surface temperature distribution rather than the temperature itself. We calculate the SLB circulation in a simple circular island case (the island radius r_0 is set to 50 km) by prescribing the horizontal surface temperature decays exponentially. The maximum horizontal speed occurs at $\sqrt{2}/2r_0$ which is not located at the coastline but the inland and decreases both toward land or sea. The maximum updrafts is at the center of the island (associated with the heating center) while the maximum downdrafts happens at $\sqrt{2}r_0$; the strong, narrow updrafts and the weak, wide downdrafts are bounded to the coastline. Because TDLS should be zero when there is no land in sea and be finite when the island is large enough, we choose TDLS varying with the island size to satisfy such common sense. This could lead an optimal island radius associated with the strongest SLB circulation. The optimal island size for the SLB circulation is consistent with previous numerical and theoretical studies.

The vertical distributions of the three-dimensional velocity components are determined by the vertical scale of the surface temperature. The u-wind blows toward sea at the lower level and toward land at the upper level. However, the v-wind blows clockwise at the entire layer. This indicates a cyclonic circulation at the lower level but an anti-cyclonic circulation at the upper level. The vertical distribution of the vertical motion is larger than zero which means updrafts and downdrafts occupy the entire layer on land and sea respectively. Also, each velocity component approaches to its maximum value at a certain level. Both the vertical level and the maximum value increase with the increasing vertical scale of the surface temperature, which means that a stronger surface temperature would induce a stronger SLB circulation locating at a higher level. However, the variation is uneven. A same increment of the maximum value corresponds to a large height increment when the vertical scale is small but a small height increment when the vertical scale is large. This suggests that with increasing influence of the surface temperature, the increment of the maximum value would gradually increase while the increment of the height corresponding to the maximum value would gradually decrease. Therefore a stronger influence of the surface temperature would strengthen the circulation intensity more.

To further verify the theory, we apply the widely used WRF model two-dimensional sea breeze simulation to conduct an ideal experiment. An island with a size of 100 km is placed in the middle of the domain. We integrate the model for total 24 hours from 00 UTC and choose two times (03 UTC and 12 UTC) to characterize the land breeze and sea breeze respectively. At 03 UTC, a weak land breeze circulation is established near the boundary between land and sea where the horizontal temperature gradient is significant. The downdrafts and updrafts occur on land and sea respectively with corresponding wind blowing toward sea at the lower level and toward land at the upper level. At 12 UTC, a strong sea breeze with a reversed direction has replaced the weak land breeze. The colder sea air mass accompanying with the sea breeze encounters the warmer land air mass to form an obvious SBF locating at the inland. Using the near surface temperature distribution in the WRF ideal simulation, the theory could give analogical circulation patterns corresponding to the land breeze and sea breeze respectively. Both the horizontal scale and the intensity are consistent with the WRF model results. Despite the theory does not predict the asymmetry of the circulation pattern and the SBF, it features the nature of the SLB circulation.

Acknowledgement

This paper is supported by 973 Program (No. 2014CB953903) and the Fundamental Research Funds for the Central Universities (No. 2013YB45). The authors greatly thank Dr. Zhaoming Liang for his warm help on the WRF model compilation and simulation. The authors appreciate the useful comments from Editor Dr. Grabowski and from two reviewers.

References

Abbs, D. J. and W. L. Physick. 1992. Sea-breeze observations and modelling: a review. Australian Meteorological Magazine, 41:7-19.

Abe, S. and T. Yoshida. 1982. The Effect of the Width of a Peninsula to the Sea-breeze. J. Meteorol. Soc. Jpn., 60:1074-1084.

Antonelli, M. and R. Rotunno. 2007. Large-eddy simulation of the onset of the sea breeze. J. Atmos. Sci., 64:4445-4457.

Arritt, R. W. 1993. Effects of the large-scale flow on characteristic features of the sea breeze. Journal of Applied Meteorology, 32:116-125.

Crosman, E. T. and J. D. Horel. 2010. Sea and lake breezes: a review of numerical studies. Bound.-Lay. Meteorol., 137:1-29.

Crosman, E. T. and J. D. Horel. 2012. Idealized large-eddy simulations of sea and lake breezes: sensitivity to lake diameter, heat flux and stability. Bound.-Lay. Meteorol., 144:309-328.

Dalu, G. A. and R. A. Pielke. 1989. An analytical study of the sea breeze. J. Atmos. Sci., 46:1815-1825.

Drobinski, P. and R. Rotunno and T. Dubos. 2011. Linear theory of the sea breeze in a thermal wind. Q. J. Roy. Meteor. Soc., 137:1602-1609.

Estoque, M. A., 1961. A theoretical investigation of the sea breeze. Q. J. Roy. Meteor. Soc., 87:136-146.

Feliks, Y. 2004. Nonlinear Dynamics and Chaos in the Sea and Land Breeze. J. Atmos. Sci., 61:2169-2187.

Fisher, E. L. 1961. A theoretical study of the sea breeze. Journal of Meteorology, 18:216-233.

Gibbs, A. J. 2008. Idealized numerical modeling of a land/sea breeze. http://twister.ou.edu/MM2015docschapter3/gibbs_seabreeze.pdf.

Haurwitz, B. 1947. Comments on the sea-breeze circulation. Journal of Meteorology, 4:1-8.

Jeffreys, H. 1922. On the dynamics of wind. Q. J. Roy. Meteor. Soc., 48:29-48.

Jiang, Q. 2012a. A linear theory of three-dimensional land-sea breezes. J. Atmos. Sci., 69:1890-1909.

Jiang, Q. 2012b. On offshore propagating diurnal waves. J. Atmos. Sci., 69:1562-1581.

Mahrer, Y. and R. A. Pielke. 1977. The Effects of Topography on Sea and Land Breezes in a two-dimensional numerical model. Mon. Weather Rev., 105:1151-1162.

Mak, M. K. and J. E. Walsh. 1976. On the relative intensities of sea and land breezes. J. Atmos. Sci., 33:242-251.

Miller, S. T. K., B. D. Keim, R. W. Talbot, and H. Mao. 2003. Sea breeze: Structure, forecasting, and impacts. Rev. Geophys., 41:1011.

Neumann, J. and Y. Mahrer. 1971. A Theoretical Study of the Land and Sea Breeze Circulation. J. Atmos. Sci., 28:532-542.

Neumann, J. and Y. Mahrer. 1974. A Theoretical Study of the Sea and Land Breezes of Circular Islands. J. Atmos. Sci., 31: 2027–2039.

Neumann, J. 1977. On the Rotation Rate of the Direction of Sea and Land Breezes. J. Atmos. Sci., 34:1913–1917.

Niino, H. 1987.The linear theory of land and sea breeze circulation. J. Meteorol. Soc. Jpn., 65:901–921.

Qian, T.and C. C. Epifanio and F. Zhang. 2009. Linear Theory Calculations for the Sea Breeze in a Background Wind: The Equatorial Case. J. Atmos. Sci., 66:1749–1763.

Qian, T.and C. C. Epifanio and F. Zhang. 2012. Topographic effects on the tropical land and sea breeze. J. Atmos. Sci., 69: 130–149.

Rotunno, R. 1983. On the linear theory of the land and sea breeze. J. Atmos. Sci., 40:1999–2009.

Schmidt, F. H. 1947. An elementary of the land–and sea–breeze circulation. Journal of Meteorology, 4:9–20.

Simpson, J. E. 1996. Diurnal changes in sea–breeze direction. Journal of Applied Meteorology, 35:1166–1169.

Skamarock, W. C., J. B. Klemp, J. Dudhia, D. O. Gill, D. M. Barker, M. G. Duda, X. Y. Huang, W. Wang, and J. G. Powers. 2008. A description of the advanced research WRF version 3NCAR/TN–475+STR.

Steele, C. J., S. R. Dorling, R. V. Glasow, and J. Bacon. 2013. Idealized WRF model sensitivity simulations of sea breeze types and their effects on offshore windfields. Atmos. Chem. Phys., 13:443–461.

Walsh, J. E. 1974. Sea Breeze Theory and Applications. J. Atmos. Sci., 31:2012–2026.

Xian, Z. and R. A. Pielke. 1991. The effects of width of landmasses on the development of sea breezes. Journal of Applied Meteorology, 30:1280–1304.

Yan, H. and R. A. Anthes. 1987. The effect of latitude on the sea breeze. Mon. Weather Rev., 115:936–956.

Zhang, M., L. Zhang, S. F. Ngan, and Y. Kenneth. 1999. A calculation method on the sea breeze circulation. Chinese Journal of Atmospheric Sciences (in Chinese), 23:693–702.

附　录

"天气变化万千，海洋汹涌澎湃，十分壮观。一个人有成就也好，有挫折也好，比起这些壮观的自然界来又多么渺小，所以有成就不必沾沾自喜，受挫折也不必灰心丧气。一个人的心胸虽然比不上天空、海洋，但还是宽广一些好，要容得下喜、怒、哀、乐，也许这样才能不断前进。"

——巢纪平

中科院院士巢纪平——平凡中铸就辉煌

特约记者 潘俊杰 高 琳

初次印象

一开始，巢纪平不肯接受我们的采访，后来在我们的坚持下，通了几封 E-mail 之后，他终于答应了。

在他的办公室里见面的时候，他首先向我们解释说："一开始之所以不愿意接受你们采访，并不是因为我架子大，而是我没有什么可以讲的。但是看到你们很年轻，太年轻了，接受采访也算尽一点微薄的支持吧。年轻人的成长是需要有人支持的，我年轻时也不例外。"说完，他不由得笑了。

当我们问到他的学术研究的时候，他说："我可以告诉你们我干了些什么，或者说我是怎么干出来的，如果说能够对年轻人有所帮助的话，我都可以谈。但是我不会对我的学术研究进行评价，那不是我的事情。"

专科毕业进入中科院

巢纪平说："作为年轻人，你会遇到很多挫折，但也会遇到很多机遇。我之所以能够到中国科学院，是因为当时南京大学系主任朱炳海老师比较欣赏我，就推荐我到科学院，而不是完全按照成绩来定的。"

1954 年，在朱炳海的极力推荐下，22 岁的巢纪平被分配到中科院地球物理所。当时巢纪平是气象专科生，加之选修课"无线电"不及格，这种情况一般是进不了中科院的。但朱炳海看到了巢纪平在气象专业上的潜力与天赋，不拘一格推荐人才。巢纪平说："所以一个人有时要靠机遇，我就是机遇好，其实比我强的同学有的是。"

挤出时间"种好自留地"

从 1959 年起，巢纪平开始研究积云动力学。他回忆说："我曾在 1961 年写过一篇关于中小尺度的一个基本理论问题的文章，就送给赵九章先生看。他看了三遍。第一遍看了以后，他说巢纪平在胡说八道，可是又觉得巢纪平不是一个胡说八道的人，所以过了一阵又看了第二遍。看了以后觉得这个问题很重要，又看了第三遍，看完以后就把我叫过去了。他说：‘我看懂了你的文章，也觉出了它的重要性’。”这篇论文后来列入《中国大百科全书·大气科学卷》的大事年表，1962 年巢纪平建立中小尺度运动方程组。

让巢纪平感动的是，老科学家赵九章如此重视一个年轻人的一点学识。赵九章在学术上从不固执己见，总是鼓励学生发表不同的见解，尤其赞赏在学术观点上有独立见解的学生。他非常赏识巢纪平的才华，一直支持巢纪平的学术研究。

还有一位就是叶笃正先生。巢纪平说：“我应当感谢他。我独立写的第一篇理论文章能够坚持下来，没有被否决，多亏了他。”因为当时在有些人看来，巢纪平搞这个问题应该算是“不务正业”，是在经营自己的“自留地”。巢纪平《斜压西风带中大地形有限扰动的动力学》这篇论文的理论手稿，叶笃正先生看过后支持他做，并把为自己工作的两个统计员调去帮巢纪平的忙，那时没有现在的计算条件，当时用手摇计算器算这个题目至少要一年（现在的计算机几秒钟就算完了），没有叶先生支持，这篇文章一年的时间是算不完的。巢纪平说，这篇论文发表后，基本上使自己在中科院站住了脚，没有被淘汰。

巢纪平很强调思想方法和学习方法，方法对了，可以花比较少的时间学比较多的知识。“学习并不是为了把知识装满自己的脑袋，而是必须要探索新的问题。”他说，思想方法是可以不断地改进的，“钻牛角尖”并不都是科学上的执着精神，某种意义上有时是思想方法不对，不够辩证，“我自己还是比较用功的，但很少‘钻牛角尖’”。

无论逆境顺境，都必须要坚持

巢纪平说，你做一件事情，做得比较好的话，就会有机遇；如果你什么都不干，那就不会有机遇。如果你确定了你要做什么，那么不管所处的环境是多么不顺，你都必须要坚持。他说：“我们当年的政治运动太多，经常受到批判。遇到这样的情况，如果你轻易放弃了科学研究，随大流，那么你也就完了。”

1958 年“大跃进”的时候，26 岁的巢纪平奉命到兰州搞人工降水，一干就是 3 年。飞机在 7000 米的高空作业，下了飞机之后，他们只能吃窝窝头，当时也没有肉吃。在这样的环境下，他并没有泄气，白天下了飞机后，晚上还接着读书，做研究。当时，他不仅仅满足于实际作业，而十分注意对一些现象的分析，探讨其发生的机理。大跃进结束的时候，有些学术刊物也恢复了，别人苦于没有论文可发，而他就把自己攒下的一组文章发了出去。积云动力学就是在人工降雨工作中得到启示后在那种艰苦环境下写出来的。

1964 年，巢纪平和周晓平合著了《积云动力学》一书。据他的学生、中国气科院研究员林永辉说，改革开放以后，巢老师到美国进行学术交流，才发现这本书早已被美国空军组织人力翻译成了英文。在当时中国人写的专著引起了美国人的关注，还是非常少见的。

在“文革”中，巢纪平坚信这段时期会过去：“我并不知道会是几年能够过去，后来历史证明是 10 年。如果你不抓住这个时间，你所要做的事情也随着时间一去不复返了。”

点滴都是真性情

巢纪平说,我现在天天上网,因为很多工作都在网上完成。记者说怪不得每次发 E-mail 您都回得那么快。记者注意到,他的电脑的音箱配置不错,音乐效果很好。他的学生林永辉说,巢老师很喜欢《英雄》交响曲,这部曲子对他影响比较大,经常在做研究的时候听。

“那您退休后主要读些什么书?”记者问。“我现在没有退休,对院士来讲,没有退休,得干一辈子。”他马上纠正记者的说法。年轻的时候,巴尔扎克、莫泊桑大部分的作品他都读过。“但对现在一些小说不喜欢,深度不够。”巢纪平说:“说到小说,最使我感动的是罗曼·罗兰的《约翰·克利斯朵夫》,里面充满了人生的坎坷和奋斗。不记得还有印象更深的小说了”。

他考记者:“你知道最近名列美国畅销书排行榜第一名的是哪本书吗?”记者说:“《达芬奇的密码》。”“答对了。”他得意地说,“我买到了”。

巢纪平很喜欢读武侠小说。他说,古龙、金庸的小说全看完了,金庸的书有不少哲学思想,古龙的书也有,只是表现手法不同而已,温瑞安的书太粗放了,年轻人可能喜欢。梁羽生的作品不是很喜欢,虽然书中的诗词写得很好,我脾气急,看不了那种过程太慢、太细的书,不是他书写得不好,是与我性格不合。“现在的武侠小说不看了,写得再好的也不多。一个新派武侠小说的时代可能已经过去了。”他练过书法,尤擅隶书,还得过奖。他说,好几年没有写了,“觉得时间不够用啊,得抓紧时间。时间是过得很快的。我现在和你们相比,时间就少得很啊,不说是日薄西山,也可以说是离地平线不远了”。说完,他笑了起来。

在 2003 年中科院大气所成立 75 周年庆祝大会上,巢纪平坚决不戴红花,说:“大家随便些好不好?”在接下来的即兴发言中,他说:“我是回娘家,这支贵宾戴的花是绝对不戴的,能邀请我参加这个会就很感谢了。”

巢纪平曾说过,在美好的感觉中,幸福是瞬间的,愉悦是片刻的,欢乐是短暂的,唯有能持久的是平静。一个人没有永远的欢乐,也没有永远的愉悦,更没有永远的幸福,但可以拥有永远的平静,宁静而致远。

热情的扶植 亲切的教导

——缅怀一代数学宗师华罗庚*

巢纪平

从《人民日报》看到华先生不幸逝世的消息和他的遗照后,心情久久不能平静,我一次又一次播放贝多芬的英雄交响曲来寄托我的哀思。

我虽不是华先生的学生也不是学数学的,但他对我这样一个晚辈却倾注了满腔的热情。华先生曾对我有过三次亲切的教导,这在相当程度上决定了我的科学生涯。1954 年我从南京大学气象专修科毕业,分配到科学院工作。1962 年赵九章所长和原地球物理所党组织要把我越级晋升为副研究员,当时华先生不仅是论文的主要审查者,而且是答辩考试的“主考官”。考试完毕,华先生亲自打电话把我叫到了他的办公室。我刚走进他办公室,不免有些拘谨,但是,当我看到这位著名的科学家是那样的平易近人时,精神很快地就放松了。在谈话中他一方面肯定了我的努力和成绩,同时又指出了我论文中在数学上不严谨之处,并表示要亲自指导我。他这种爱护青年提携晚辈的精神,使我激动不已,至今铭记心中。

1980 年,在美国的普林斯顿,他在高等研究所讲学,我在普林斯顿大学作为高级访问学者从事研究工作。我准备去看他,我想,我们第一次见面距今已事过 18 年了,大概他早已经不认识我了。见面之后,我说:“华先生还认识我吗?”他慈祥地笑了:“巢纪平,我是你的主考官怎么能不认识呢!”异乡重逢,倍感亲切,我们谈了很久,虽然天已很晚了,但他的兴致依然很高。我问他这十几年来为什么要搞与国民经济直接联系的应用数学时,他态度严肃地说:“这一次你能应邀到普林斯顿大学来当访问教授,我是高兴的,说明当年我没有看错你。但你一定要记住,一个科学家如果不把自己的科学研究工作与人民的需要,与国民经济的效益紧密地结合在一起,是不足为训的!你一定要记住这一点。”

我第三次见到华先生,是今年 3 月在讨论“中共中央关于科技体制改革的决定”草案的会议上,他一见到我,就把我叫到一边,说:“我知道你已经离开了科学院,不要后悔,不要考虑自己少写几篇论文,更不要去考虑自己的学术地位会不会受到影响,你的路没有走错,这条路是符合中央精神的,当然这不是说科学院所有的同志都要这样,我对你很关心,希望你能听我的话。”这是到中南海开会前仅有的 10 分钟,在场虽然有很多老科学家,他都没有去交谈,而把这宝贵的 10 分钟用来教导我这个“年轻人”了。我万万没有想到,这竟是华先生对我最后的一次关怀和教导!

华先生,您逝去得太早了,国民经济体制、科技体制、教育体制正面临着一场新的革命,人民需要您这样的科学家,我们这些后一代的科技工作者更需要您再为我们引路,您却离开我们了,怎不使人悲痛万分!

华先生,我再也听不到您对我的教诲了,但您对我的热情的扶持、亲切的教导,我会永远铭记在心的。安息吧,华先生!

(本文作者现为国家海洋局预报研究中心研究员)

* 科学报,1985-6-30(4).

怀念在地理所工作的那3年*

巢纪平

我从1954年跨进科学院,1984年离开,在科学院整整工作了30个年头,度过了青年和壮年。我是1976年到地理所,1979年回大气物理所的,在那里工作了3年。3年只是30年的十分之一。但这十分之一的3年,我深切地怀念它,从未敢忘怀,也不会忘怀。

经历了10年"文革"的灾难后,我认为还是离开科学院的好。我那时已经是副研究员了,按当时政策规定,高研离院需经过院领导批准。调离报告到了当时主管工作的郁文同志手里(借此机会我对郁文同志的逝世表示深切的怀念和哀悼)。郁文同志请他的秘书转告我,可以到院内有关所工作,但不会批准我出院。

我比较可去工作的有关研究所后,选择去地理所。首先,那时的所长是我尊敬的、德高望重的黄秉维先生。我去他家拜见了他,黄先生说话带有广东口音,并不善辞令,他表示欢迎我去,并说,地理科学需从定性描述向定量计算发展,特别是气候学,希望我去后在这方面做些工作。同时主持日常工作的左大康副所长,为人厚道,不仅有过领导学生运动的革命历史,而且在水、热平衡方面也做了不少好的工作,是个双肩挑干部。我相信在他们领导下,会有一个好的工作环境。再有,我知道当时的党委书记李子川同志是位正派的党的领导干部。我还要提一下当时地理所气候室副主任兼支部书记丘宝剑同志,他跟我私交甚深。在那个年代,做科研工作不太重视物质条件,不像现在要考虑计算机、数据库等支持平台,更不会去考虑科研经费,更多的是考虑人际环境是否和谐。地理所是一个人际关系和谐的所,而气候室是一个和谐的、上进的研究室。那个研究室的同志在季风气候、树木年轮古气候和小气候方面都有好的研究基础。所以,我选择去地理所。

但我没有想到的是,去后不久党委和所领导就任命我为气候室主任。主任这个"官"我并不看重,使我心里震撼和感动的是,我这个在"文革"中受到重点冲击,关过"专政队"连党籍都尚未恢复的"黑旗",所领导和党委居然有这样的魄力,并给予这么高度的信任,让我来主持这个室。

1978年职称恢复,我这个研究员是在地理所提的。实际上,所领导知道,由于专业我不会在所里长留,但是地理所的党政领导是公正的,他们没有排外思想,一切按政策和条件办事,提了我。我记得那年被提为研究员的人并不多,我这个刚来的"外来户"却占了一个名额。

我要感谢地理所的另一件事,是工作方面的。那时在国际上还刚开始研究长期数值预报,我提过一个有创新思想的方法,称为距平-滤波方法,这个工作和地理所的研究方向完全不一致,但所领导仍然支持我,在室里同事们的帮助下,1977年和1979年分别在《中国科学》以地理研究所长期数值预报组的名义发表了"一种长期数值预报方法的物理基础"和"长期数值天气预报的滤波方法"两文,在1979年全国科技大会,经地理所申报后,被评为是国际水平的工作。这个在地理所完成的研究,在国际同行中有一定的影响,一度被称为长期数值预报的"北京模式"。

对地理所,我是心有愧疚的,没有为它做更多的事情。我只做了两件事。一是在工作上调整了气

---

* 地理学发展之路,科学出版社,2016,60-61.

候室的研究方向,秉承黄秉维先生的思想,加强了这个室小气候的观测、分析和理论研究,从国外引进一批小气候观测仪器。特别需要指出的是,第一次青藏高原研讨会,1979 年在兰州召开时,我是参加该会的地理所代表之一,地理所在第一次青藏高原试验中是有贡献的,这是历史。第二件事,按政策和条件为地理所气候室提名并经批准提了多个有高级职称的学科带头人。

30 年过去了,地理所所名虽然改了,但学科水平已今非昔比,是一个先进并具有一定国际声誉的所。一批年轻的科研人员已经或正在成长,多个不同子学科的研究团队已经并正在形成。地理科学已成为引领国民经济可持续发展的科技支撑。

30 年过去了,每当我经过 917 大楼旧址时,会情不自禁地怀念起那座大楼,以及大楼外那片宁静的原野,原野上朴实农民的劳动和儿童的嘻耍,那是一个多么和谐的科研"小社会"。楼空了,但人未走,科研在发展。

30 年过去了,我依然深深地怀念地理所。

彩 插

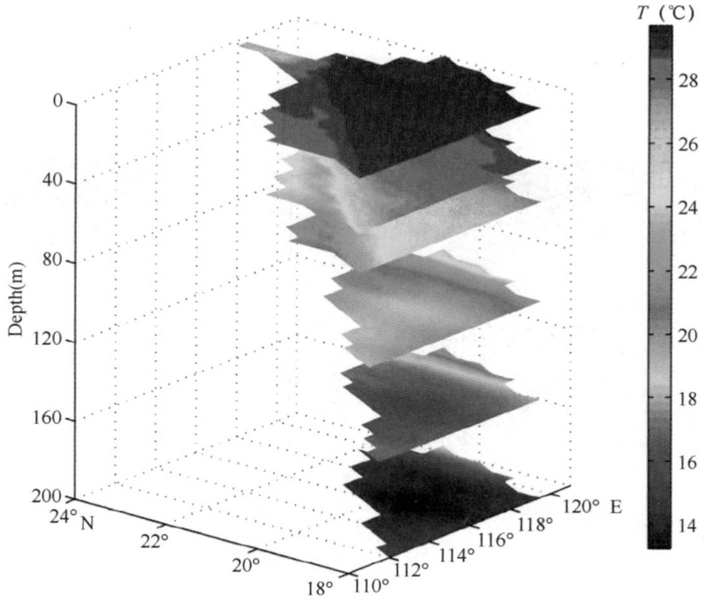

Fig. 2 Temperature slice (from 0 to 200 m below the sea surface) of the NSCS in July p.127

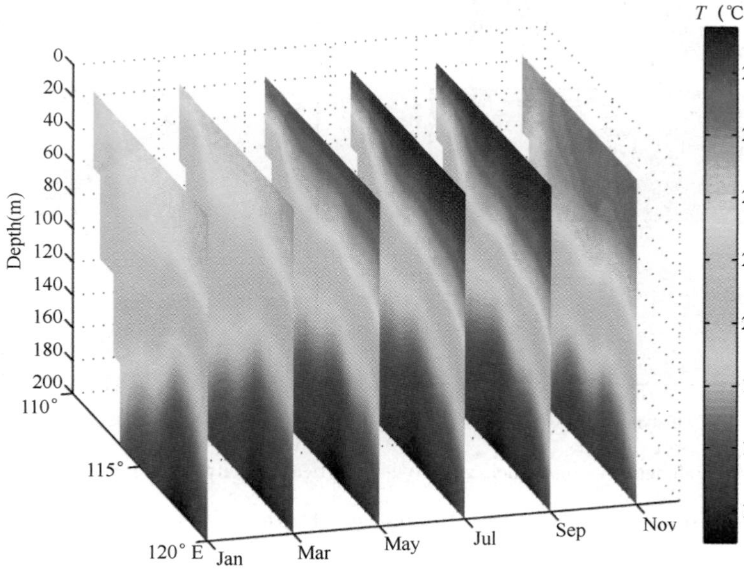

Fig. 3 Longitude-depth cross-section of temperature from January to December at 19.5°N p.127

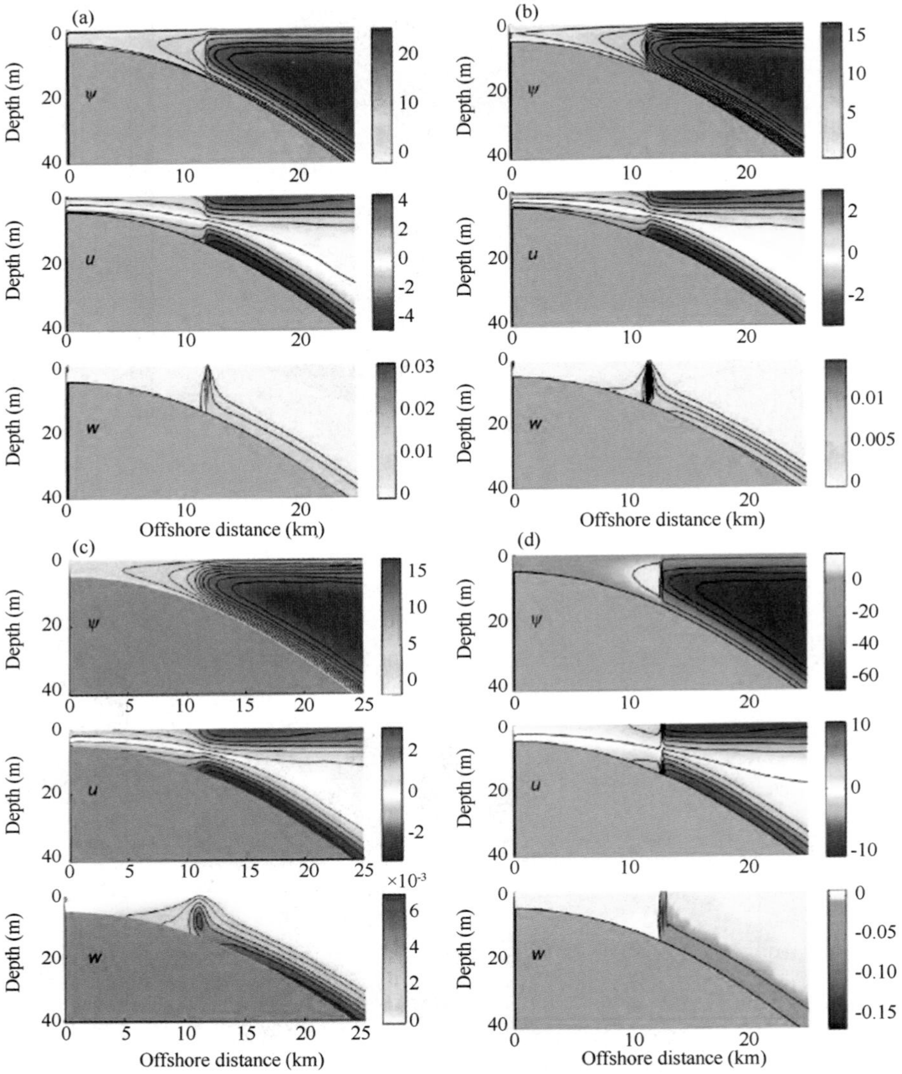

Fig. 5 Streamfunction, cross-shelf velocity and vertical velocity fields for $u_{\infty}=-0.06$ m s^{-1}. (a)-(d) correspond to stratifications 1-4 respectively. The units of u and w are m s^{-1} p.129

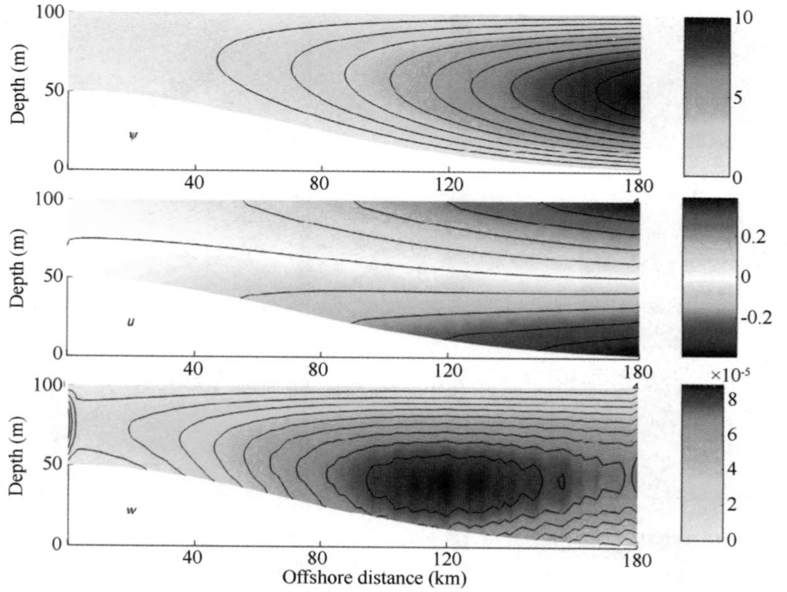

Fig. 6 Streamfunction, cross-shelf velocity and vertical velocity fields been obtained with the FEM. The units of u and w are $m \cdot s^{-1}$ p.130

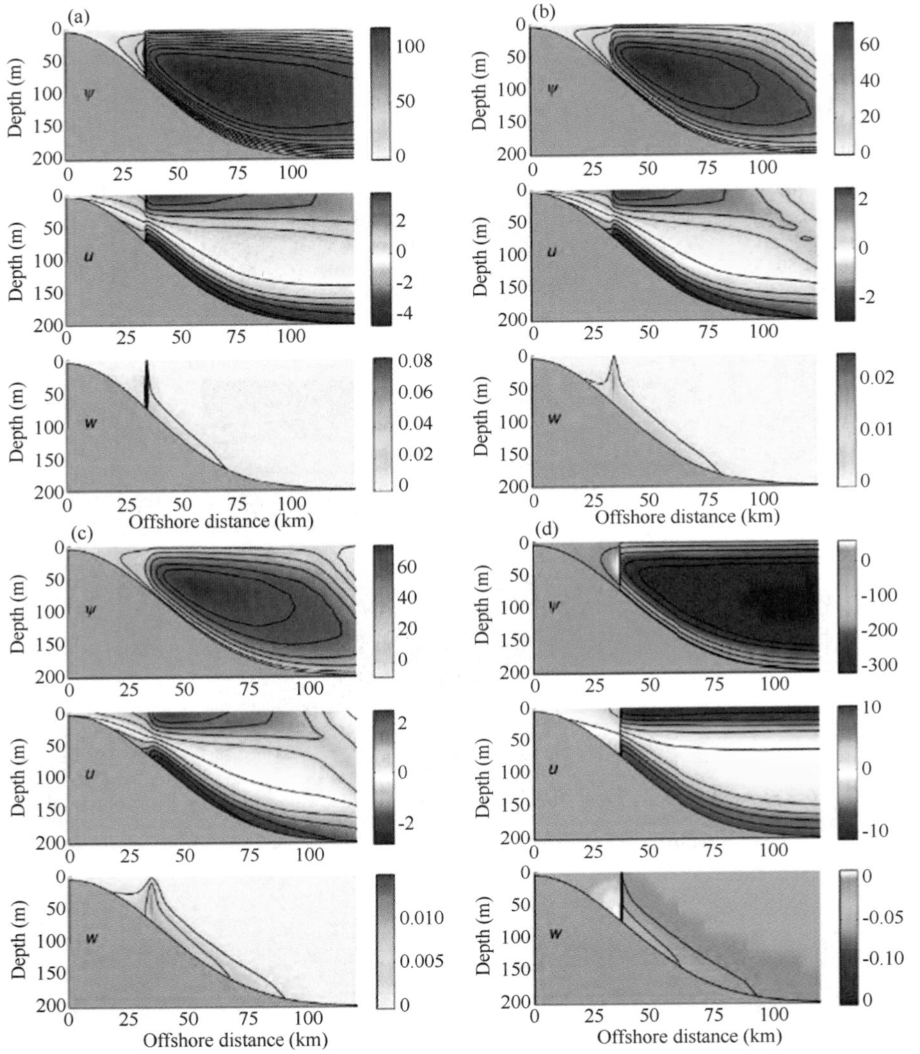

Fig. 7 Streamfunction, cross-shelf velocity and vertical velocity fields for $u_\infty = -0.3\ \mathrm{m \cdot s^{-1}}$. (a)–(d) correspond to stratifications 1–4 respectively. The units of u and w are $\mathrm{m \cdot s^{-1}}$ p.131

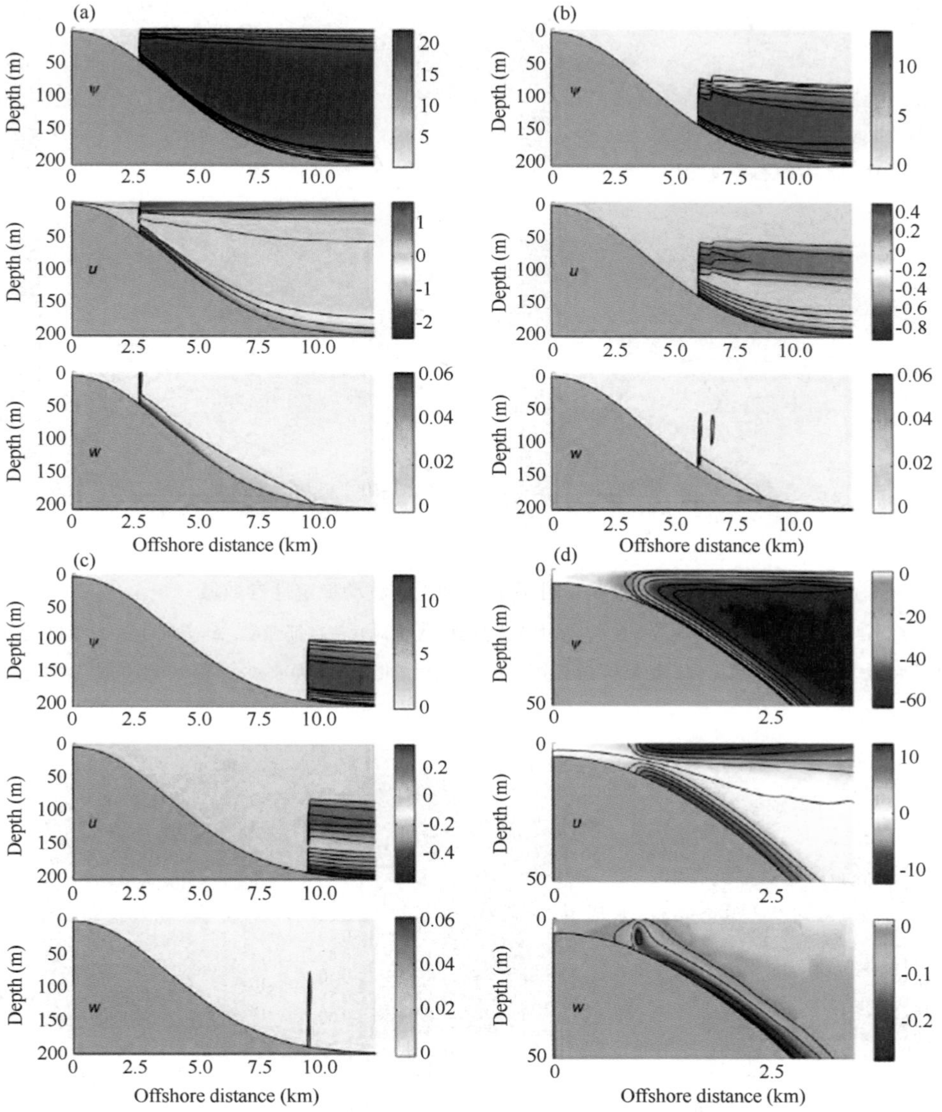

Fig. 9 Streamfunction, cross-shelf velocity and vertical velocity fields over the deeper continental shelf (depth of 200 m, offshore distance of 13 km) for $u_{\infty}=-0.06\ \mathrm{m\cdot s^{-1}}$ (a)-(d) correspond to stratifications 1-4 respectively. The units of u and w are $\mathrm{m\cdot s^{-1}}$ p.133

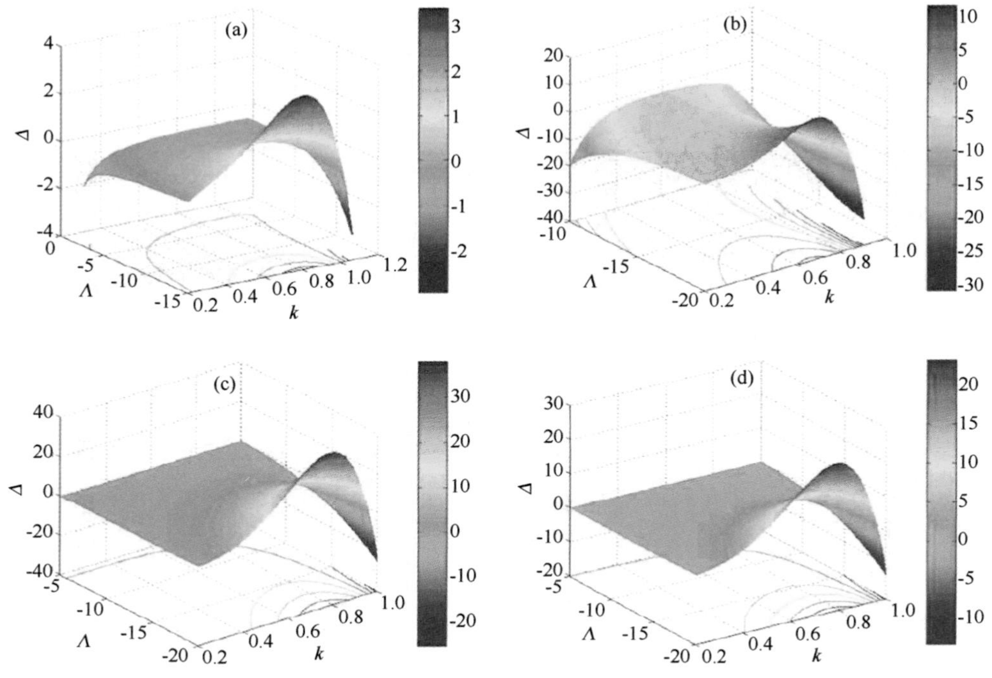

图 1　固定参数 α 时关于 Λ 和 k 的波动稳定性参数域

(a) $\alpha=0.1$;(b) $\alpha=0.3$;(c) $\alpha=0.5$;(d) $\alpha=0.7$。图中 k 为波数, Λ 为风吹流特征强度, $\Delta=(p/2)^2+(Q/3)^3$ 为方程(54)根性质的判别式,$\Delta>0$ 的区域出现共轭复根, 这时有一个根是指数增长的, 即波动不稳定, $\Delta\leqslant 0$ 时波动稳定 p.152

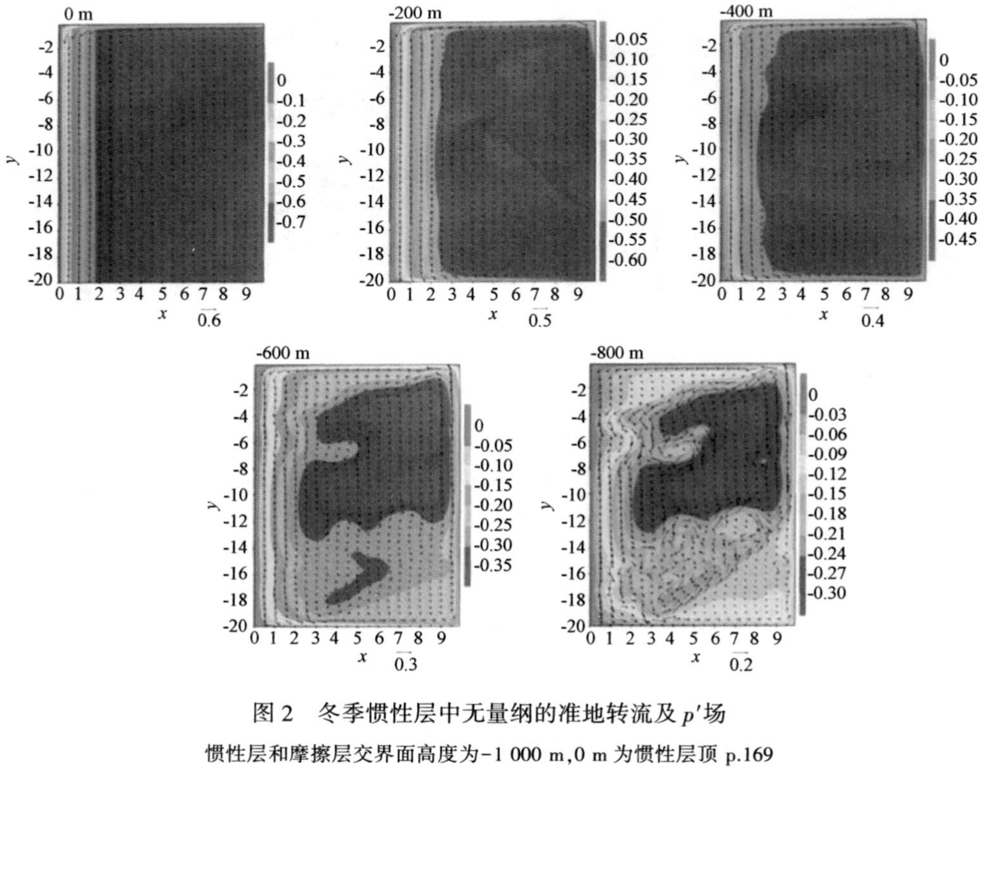

图 2　冬季惯性层中无量纲的准地转流及 p' 场

惯性层和摩擦层交界面高度为-1 000 m,0 m 为惯性层顶 p.169

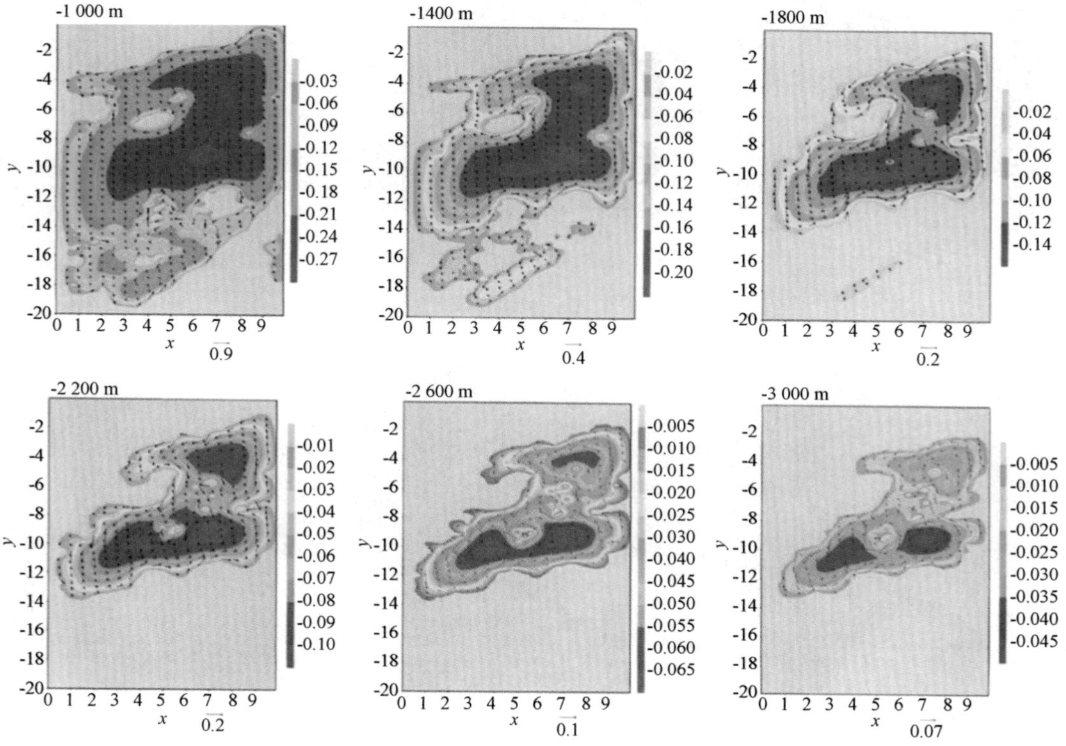

图 3　冬季摩擦层中无量纲的准地转流及 p' 场 p.169

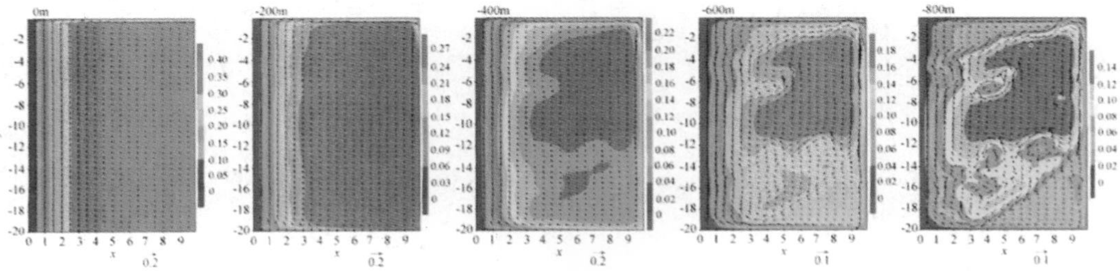

图 5　夏季惯性层中无量纲的准地转流及 p' 场

惯性层和摩擦层交界面高度为-1 000 m p.171

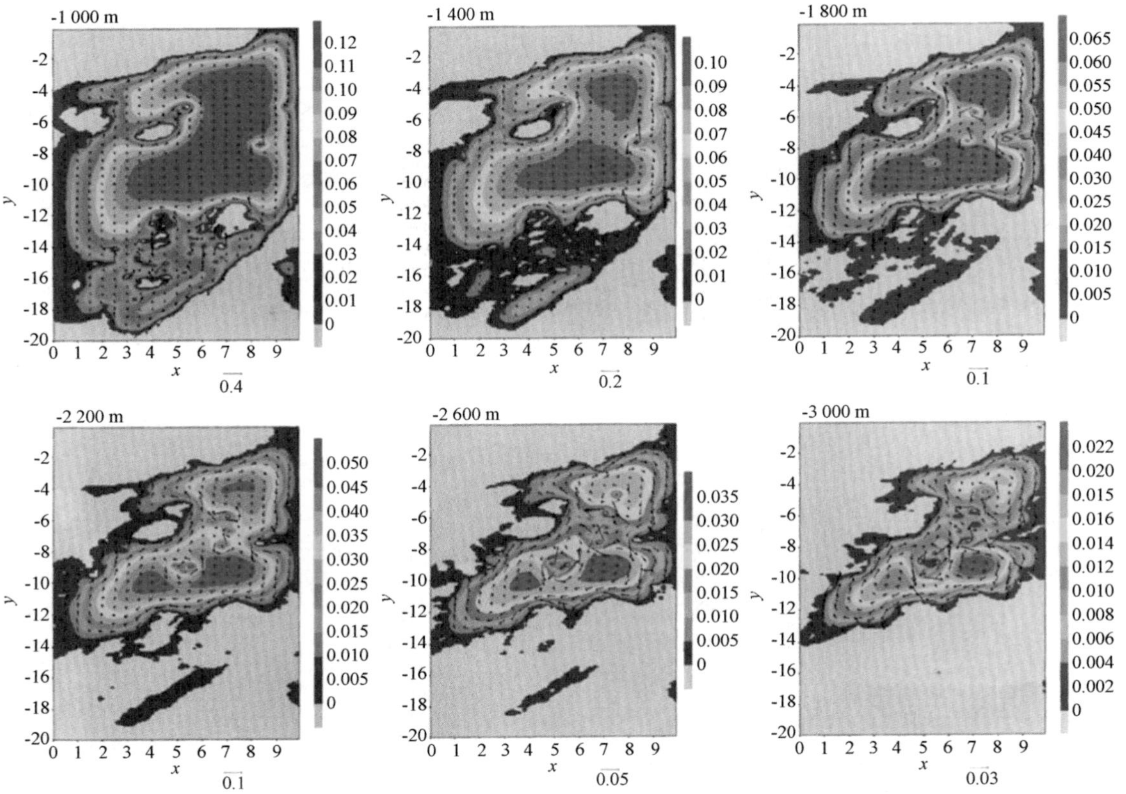

图 6　夏季摩擦层中无量纲的准地转流及 p' 场 p.171

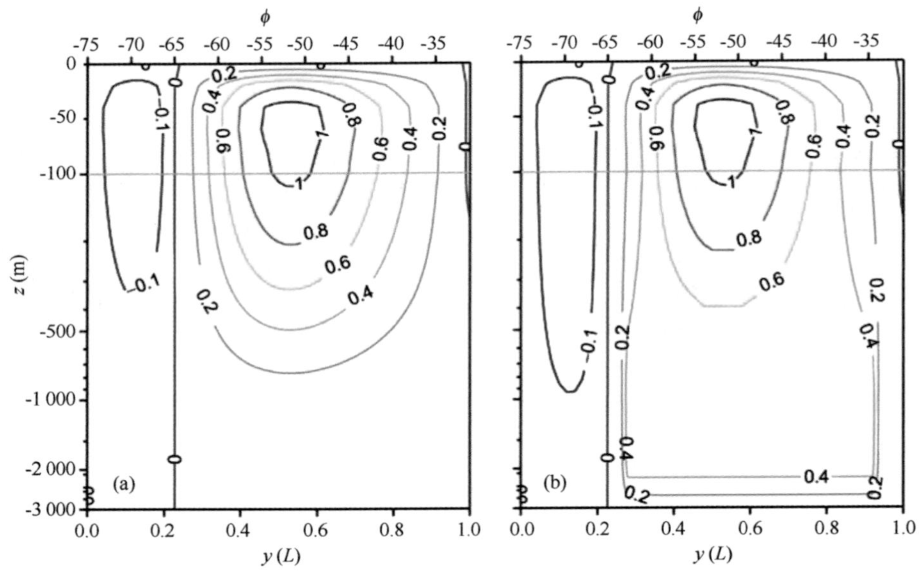

图 4　流函数的两个平衡态的分布

（a）平衡态一；（b）平衡态二 p.368

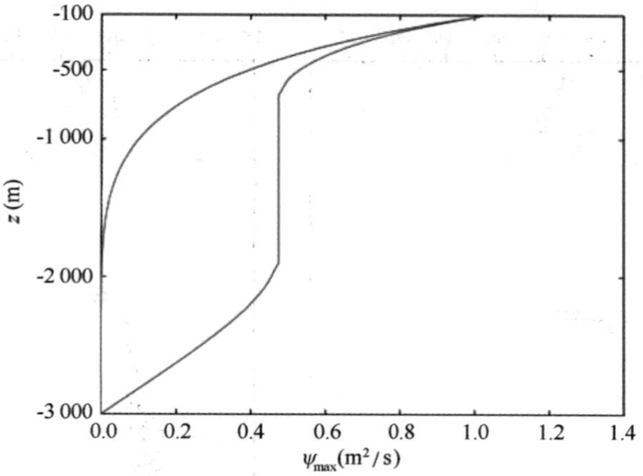

图 5　流函数分岔 p.369

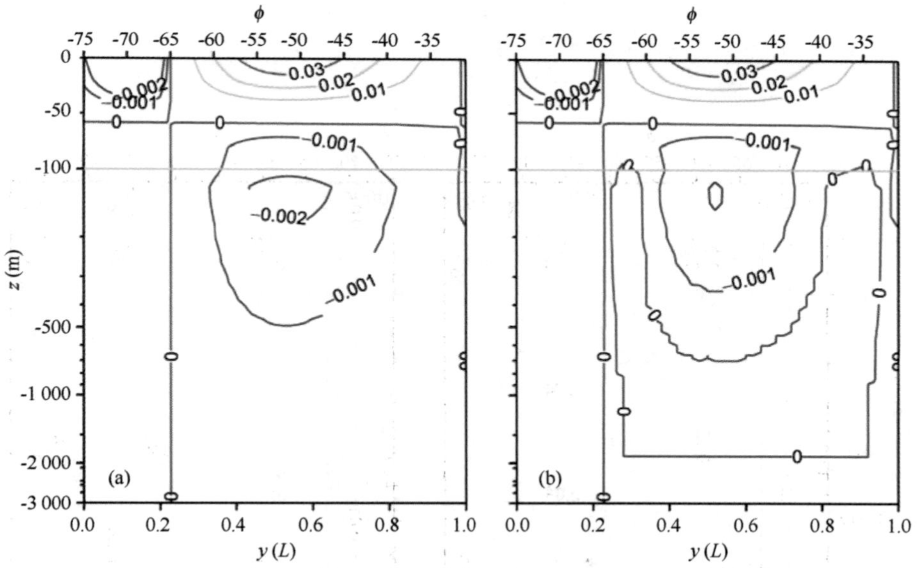

图 6　经圈流的两个平衡态的分布

（a）平衡态一；（b）平衡态二 p.369

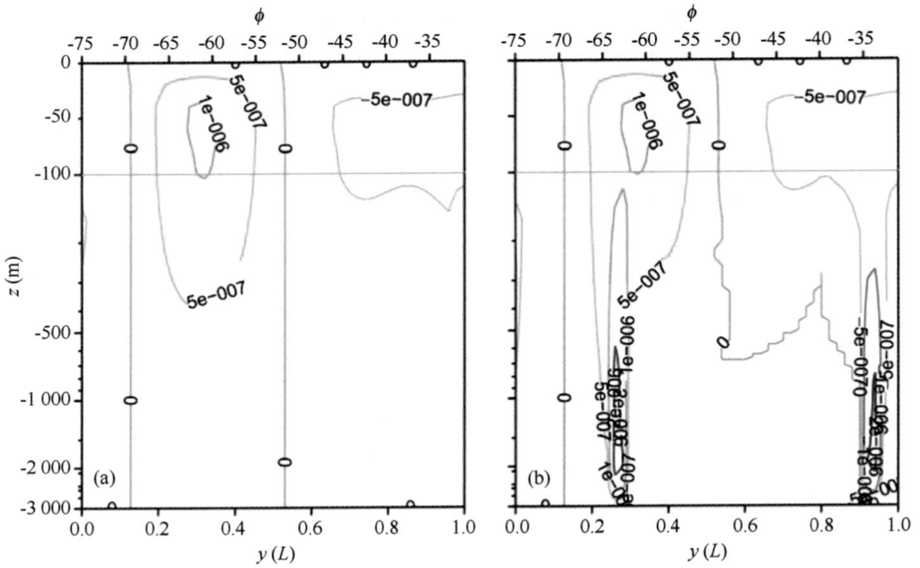

图 7　垂直速度的两个平衡态的分布

（a）平衡态一；（b）平衡态二 p.370

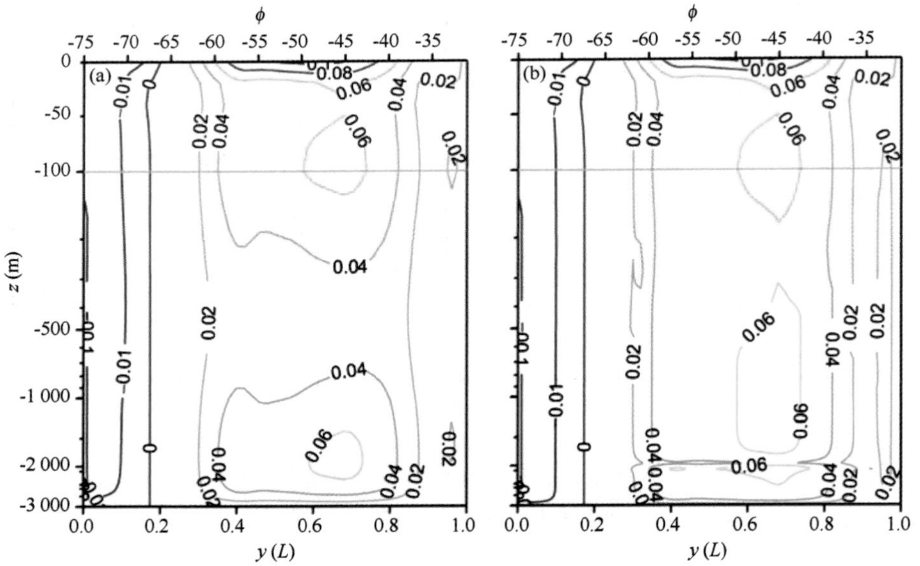

图 8　纬圈流的两个平衡态的分布

（a）平衡态一；（b）平衡态二 p.370

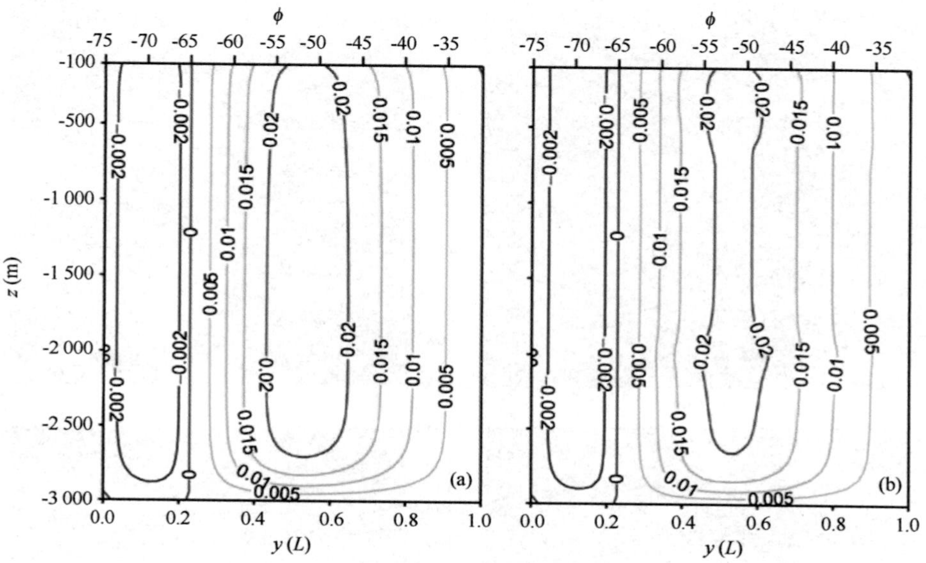

图 9 扰动温度的两个平衡态的分布

（a）平衡态一；（b）平衡态二 p.371

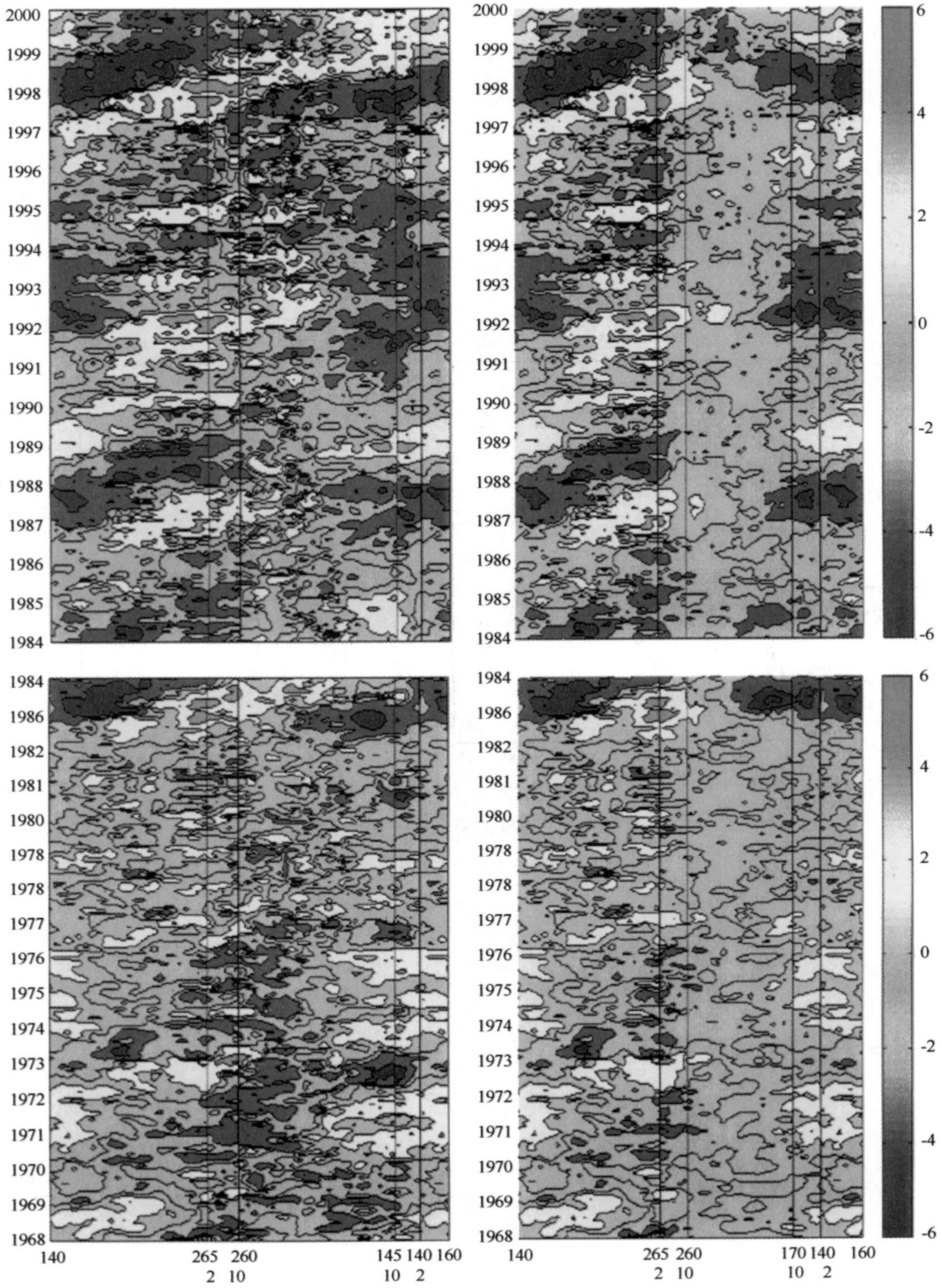

Fig.2 The evolution of the El Niño (La Niña) events p.533

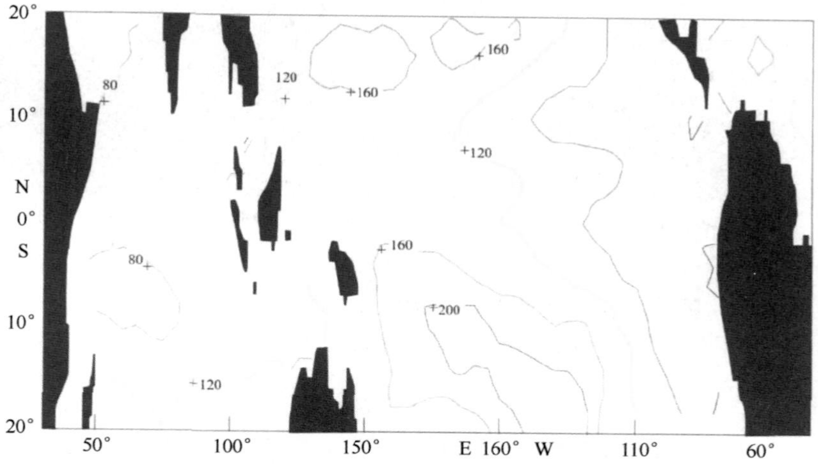

图 1 年平均 MSTA 的深度分布(单位:m) p.538

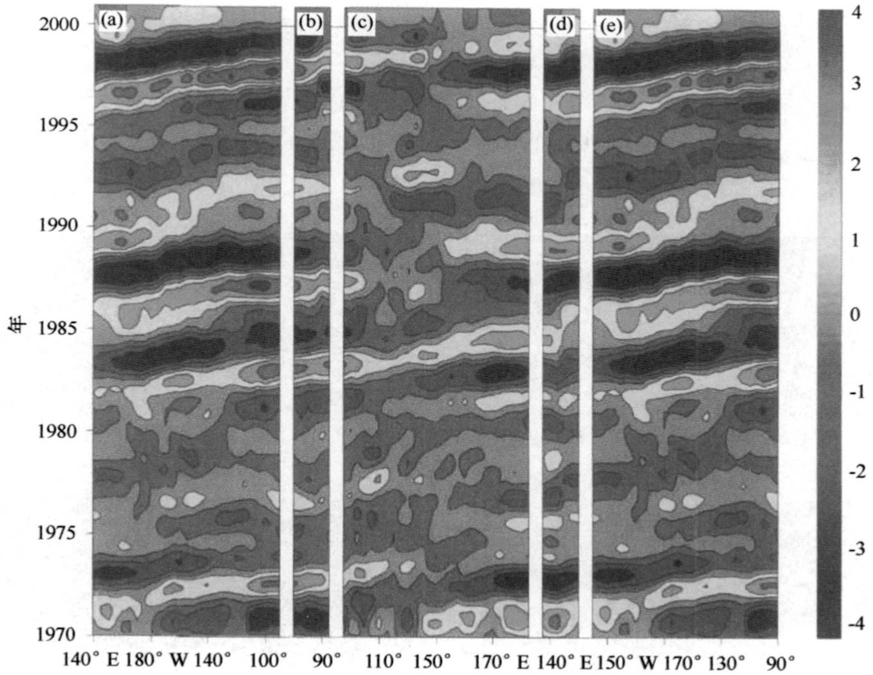

图 2 1970—2000 年海温距平在 MSTA 面上的演变

（a-e 分别为第 1 分区至第 5 分区） p.538

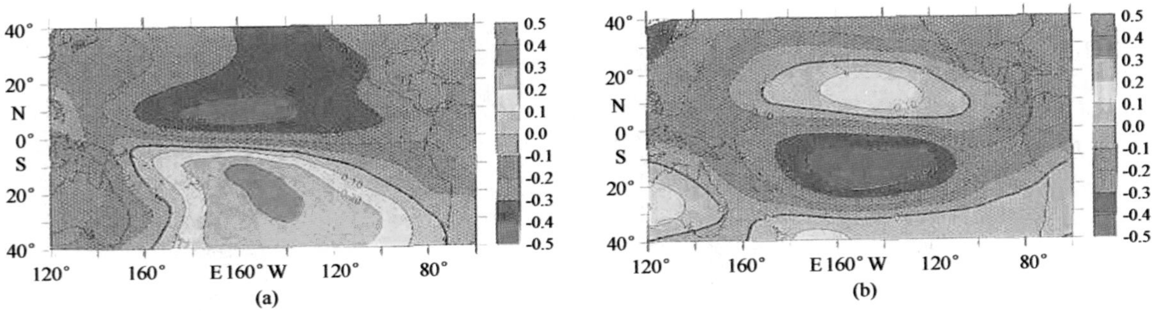

版图 A　流函数和 Niño-3 指数的同期相关系数空间分布

(a) 850 hPa 高度；(b) 200 hPa 高度 p.574

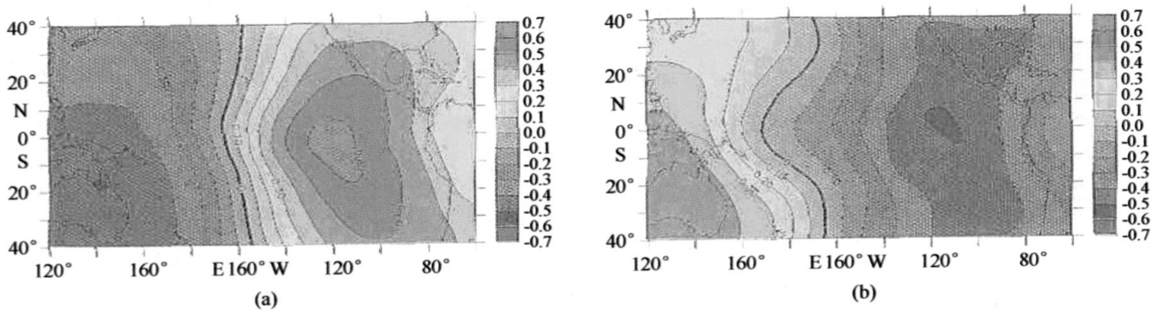

版图 B　速度势和 Niño-3 指数的同期相关系数空间分布

(a) 850 hPa 高度；(b) 200 hPa 高度 p.574

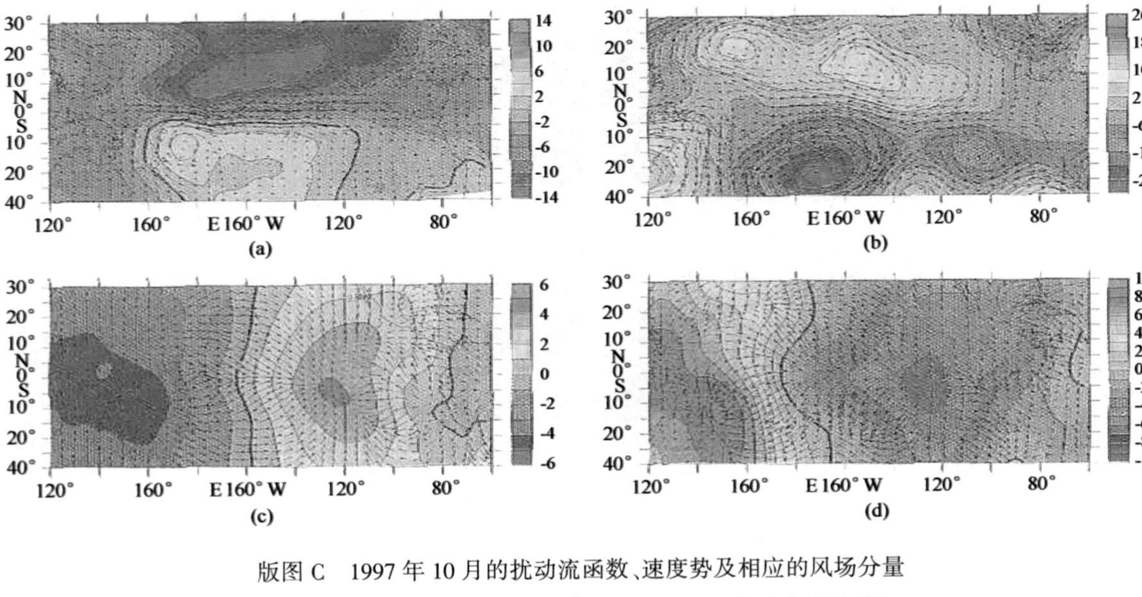

版图 C　1997 年 10 月的扰动流函数、速度势及相应的风场分量

(a) 850 hPa 流函数及无辐散风场；(b) 200 hPa 流函数及无辐散风场；

(c) 850 hPa 速度势及无旋风场；(d) 200 hPa 速度势及无旋风场 p.574

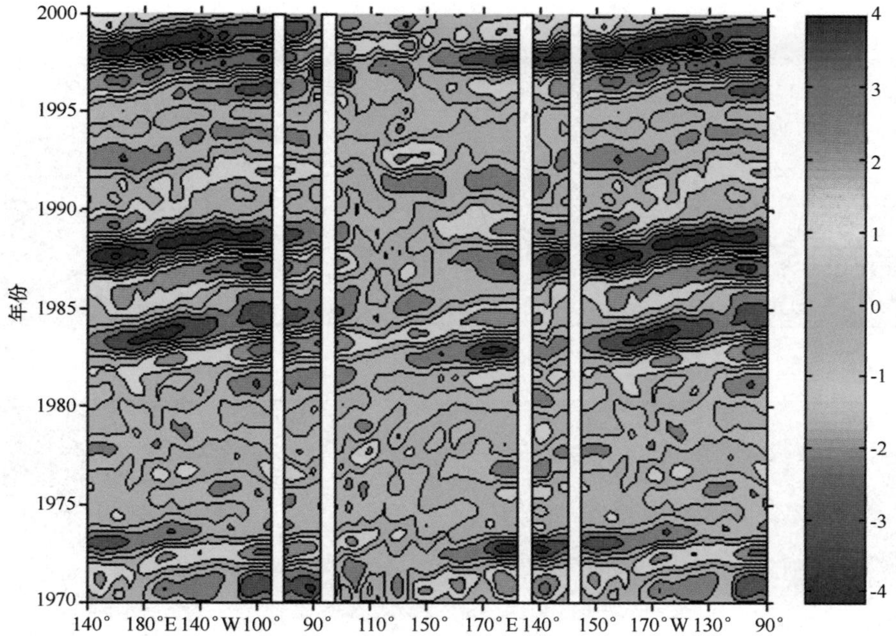

图 1　在 MSTA 上海温距平的演变

虚线表示变化趋势；横坐标第一分格是沿赤道从西太平洋 140°E 到东太平洋 115°W；第二分格是在 115°W 上从赤道到 10°N(由于空间较小纬度标度未给出)；第三分格是 10°N 从 115°W 回到 140°E；第四分格在 140°E 上从 10°N 回到赤道(由于空间较小纬度未标出)；第五分格同第一分格 p.577

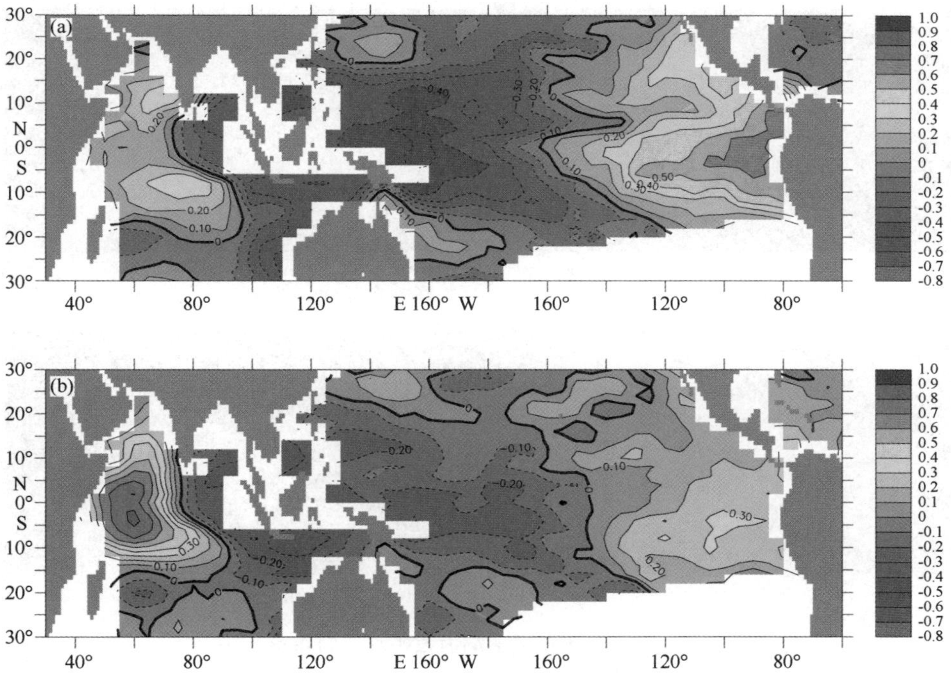

图 6　东太平洋(a)和西印度洋(b)次表层海温距平分别与两大洋 MSTA 面上海温距平的相关系数 p.580

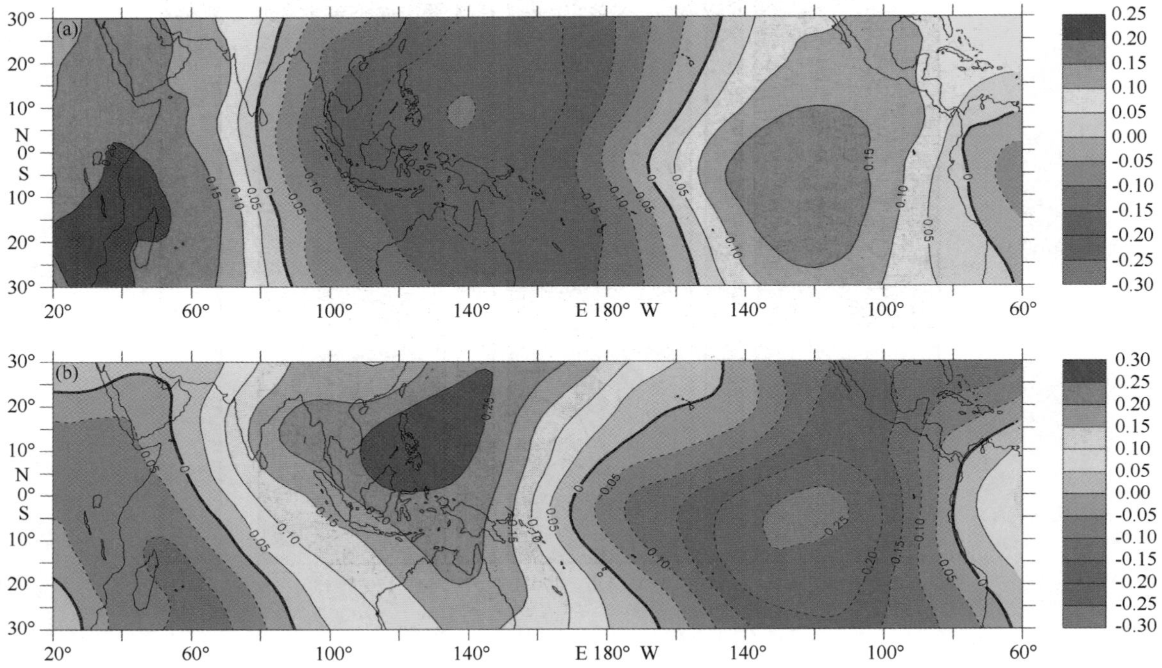

图 7　西印度洋 MSTA 曲面上的海温距平分别和 200 hPa(a),
850 hPa(b)速度势的相关系数 p.581

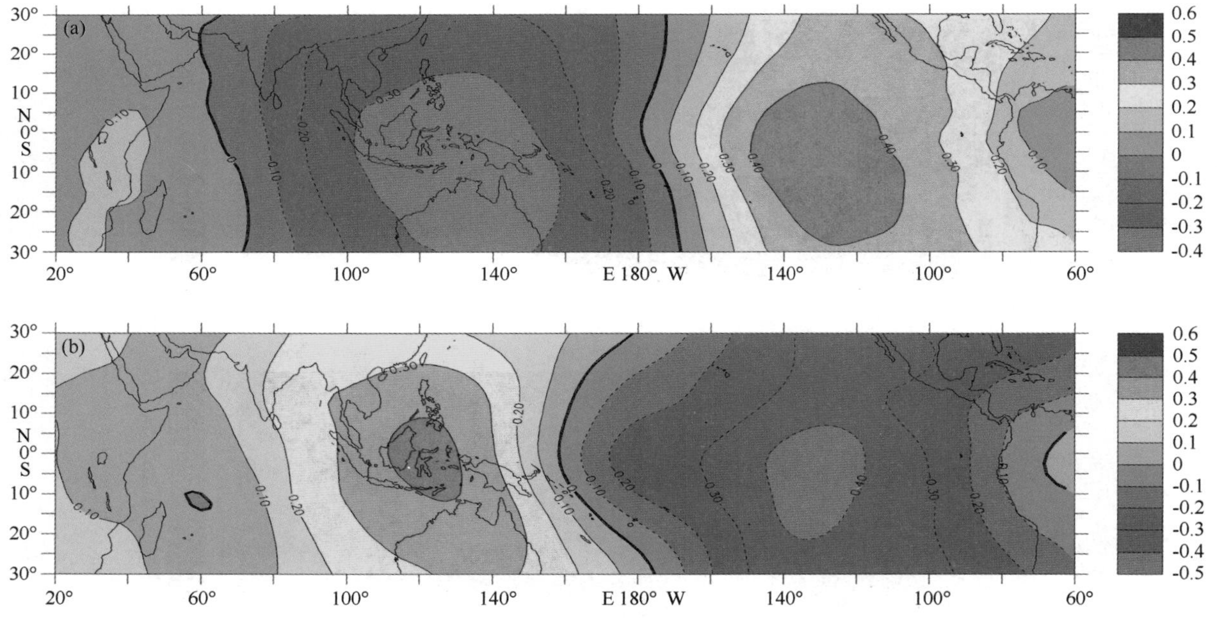

图 8　东太平洋 MSTA 曲面上的海温距平分别和 200 hPa(a),
850 hPa(b)速度势的相关系数 p.581

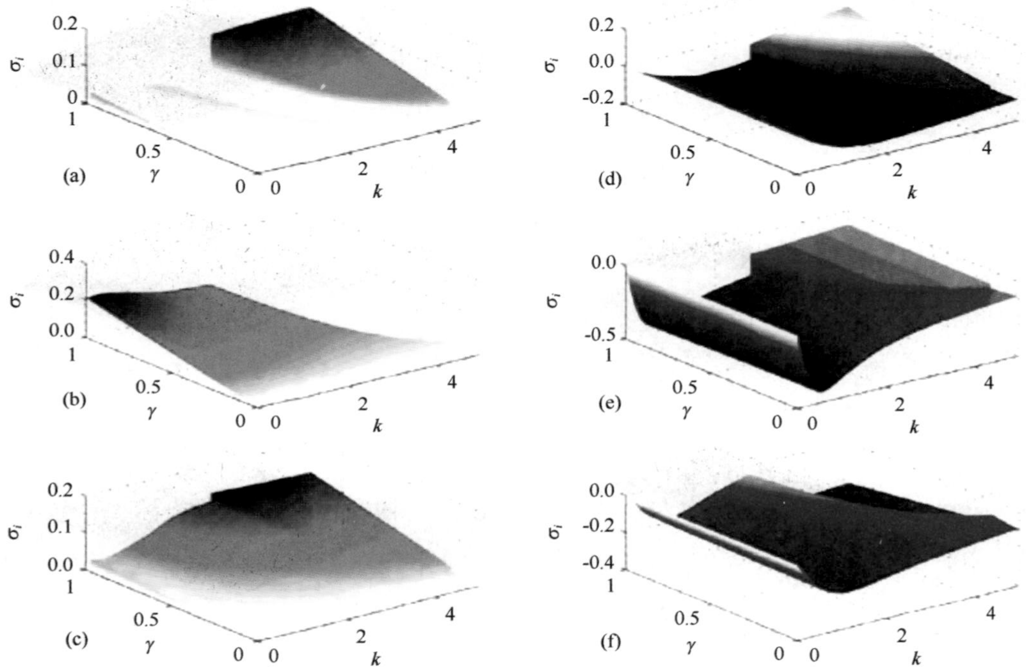

图 2　向极流存在时关于赤道对称的模态频散关系

(a),(b),(c)分别为前 3 个模态 σ 的虚部,(d),(e),(f)分别为前 3 个模态的 σ 的实部。x 轴为波数 k,y 轴为背景流速和重力波波速之比 γ,z 轴为 σ。$A=2,\alpha=1$。图中均为无量纲量 p.660

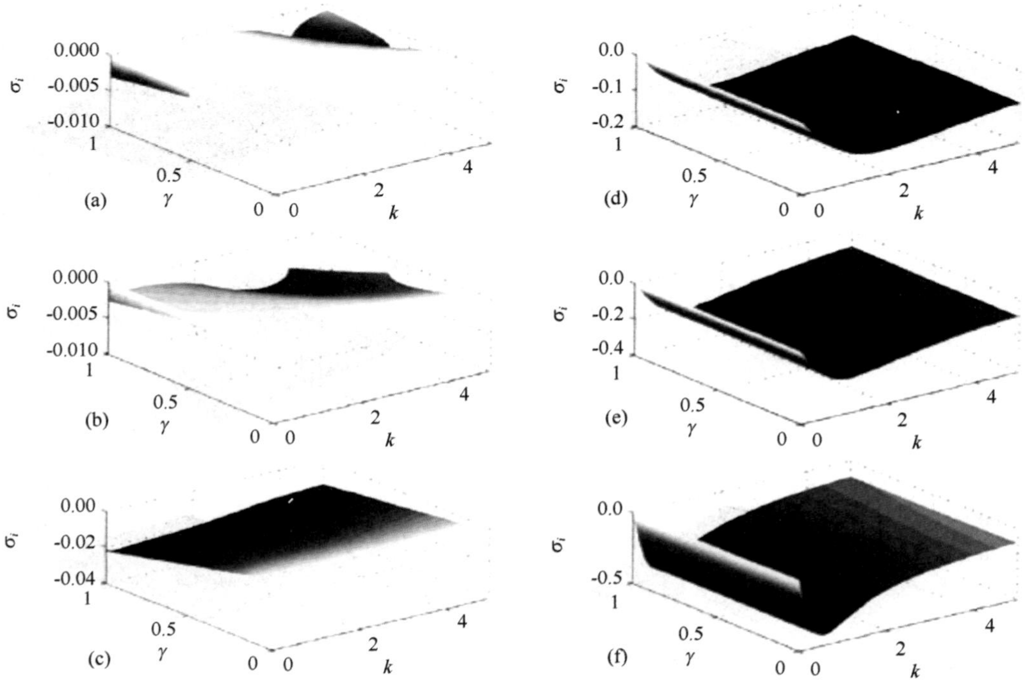

图 3　弱向赤道流存在时关于赤道对称的模态频散关系

(a),(b),(c)分别为前 3 个模态 σ 的虚部;(d),(e),(f)分别为前 3 个模态的 σ 的实部。x 轴为波数 k,y 轴为背景流速和重力波波速之比 γ,z 轴为 σ。$A=-0.2$,$\alpha=1$。图中均为无量纲量 p.661

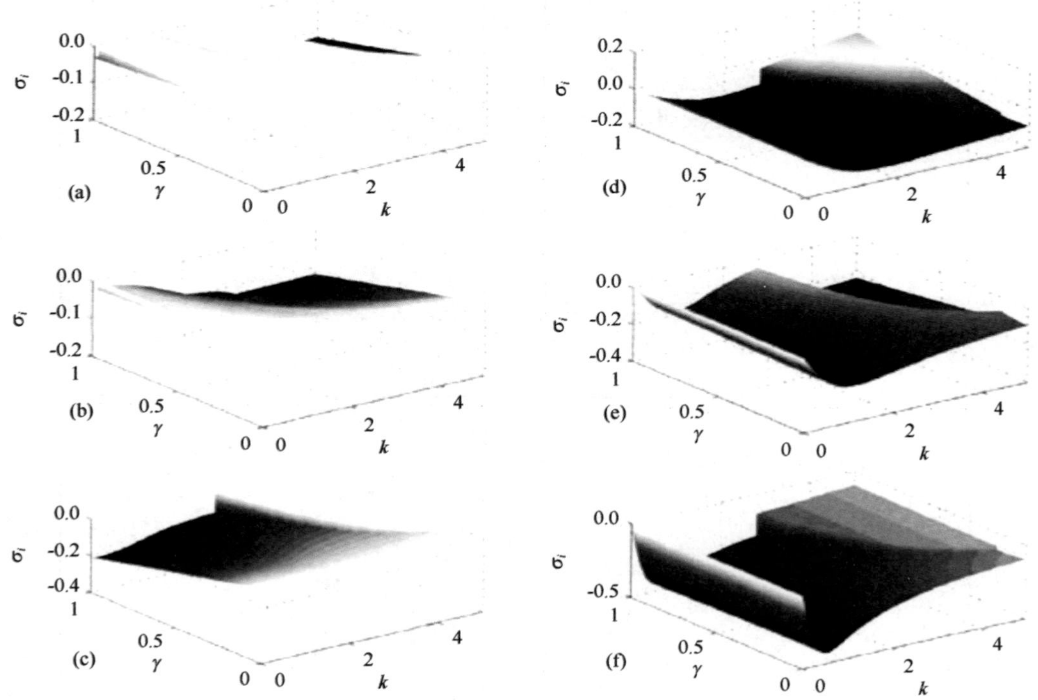

图 4　强向赤道流存在时关于赤道对称的模态频散关系

(a),(b),(c)分别为前 3 个模态 σ 的虚部;(d),(e),(f)分别为前 3 个模态的 σ 的实部。x 轴为波数 k, y 轴为背景流速和重力波波速之比 γ,z 轴为 σ。图中均为无量纲量 $A=-2$,$\alpha=1$ p.661

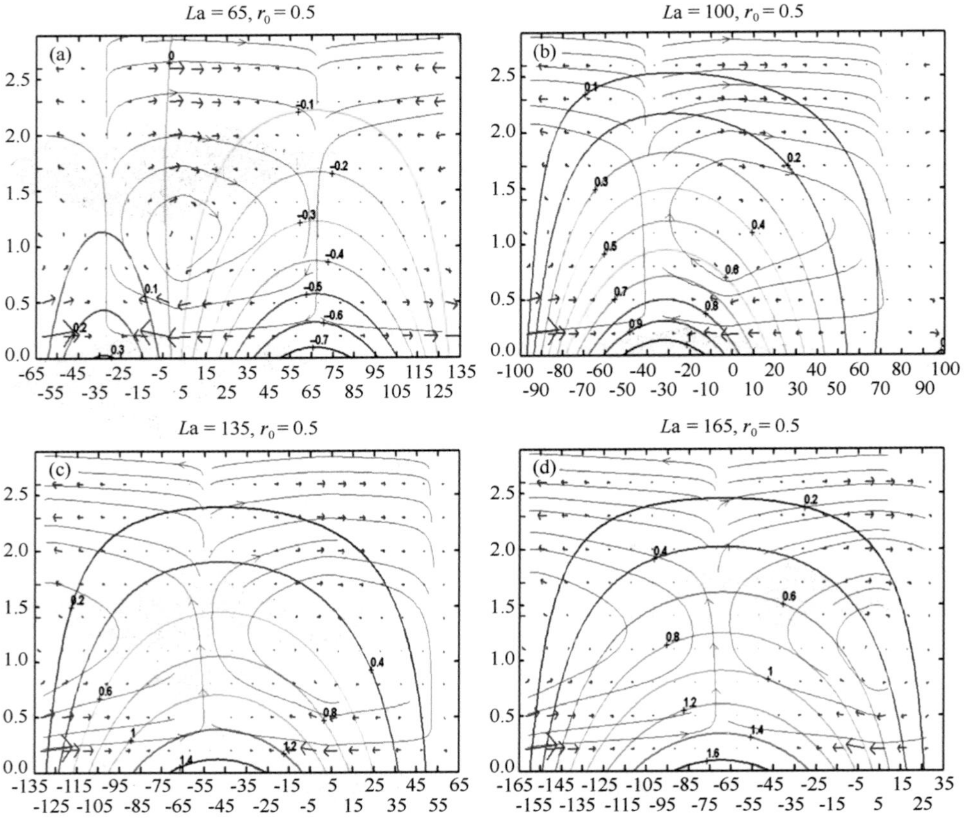

图 1　绿洲相对湿度不变尺度不同时的温度距平和动力学场

(a)至(d) 绿洲尺度分别为 65 km, 100 km, 135 km, 165 km。图中 *La* 为绿洲尺度，r_0 为绿洲空气相对湿度，水平方向和垂直方向单位为 km（下图同），温度距平单位为℃，图中垂直速度最大值为 0.17 m/s，水平速度最大值为 0.64 m/s，温度距平最大值为 1.68℃ p.705

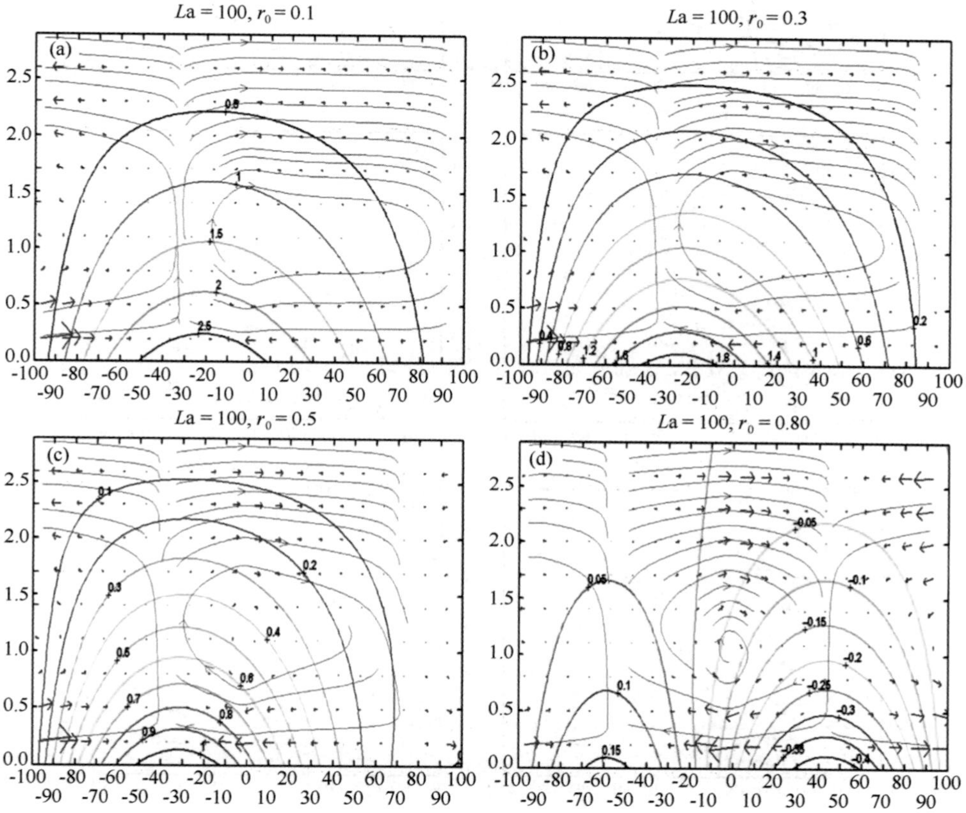

图 2 绿洲尺度不变相对湿度不同时的温度距平和动力学场

(a)至(d) 绿洲相对湿度分别为 10%,30%,50%,80%。图中垂直速度最大值为 0.38 m/s,水平速度最大值为 1.38 m/s,温度距平最大值为 2.9℃ p.706

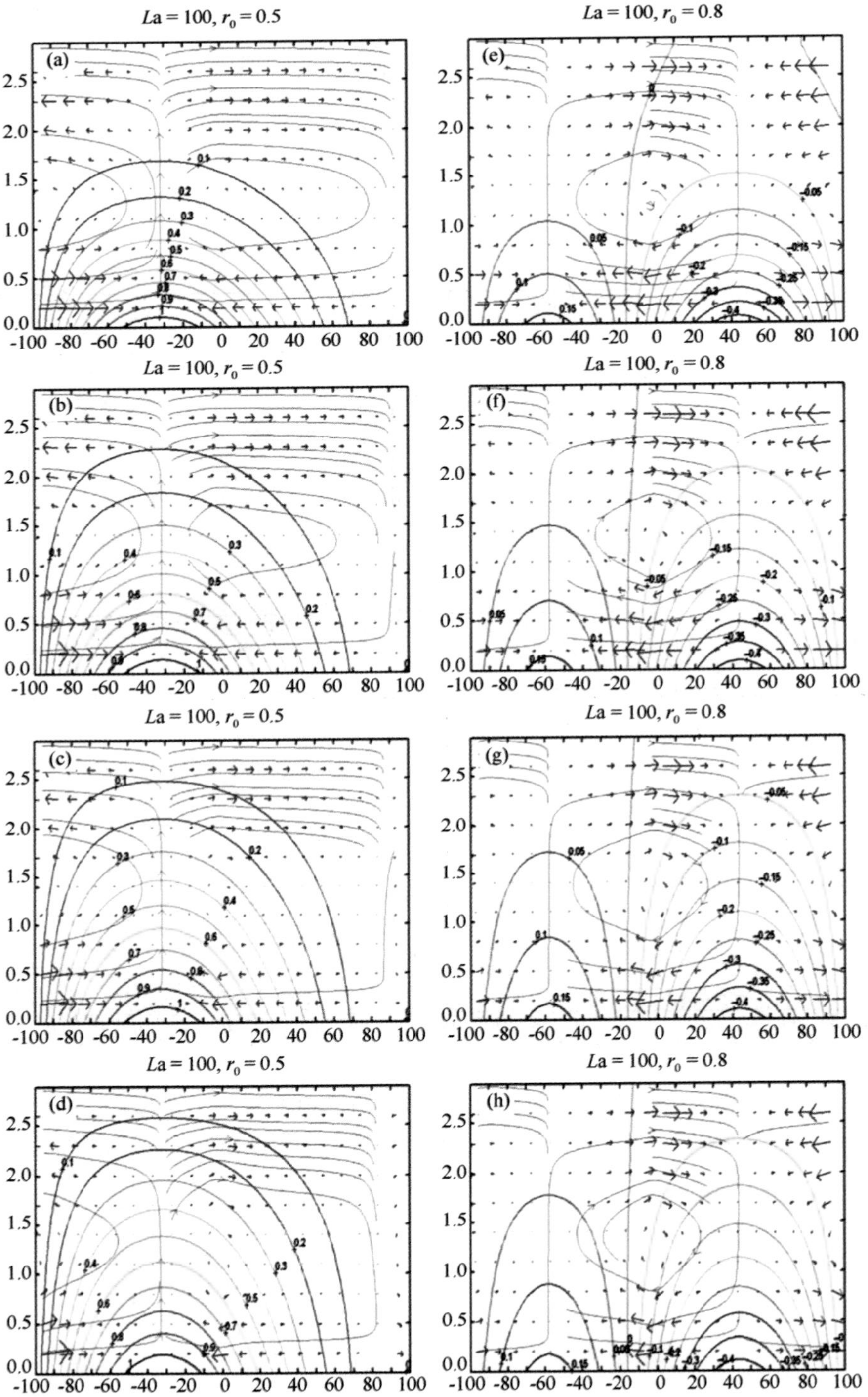

图 3　绿洲的温度距平和动力学场随时间演变

(a)至(d) 分别为绿洲尺度为 100 km，相对湿度为 50%时随时间迭代到两年和准定常态时的图；(e)至(h) 分别为绿洲尺度为 100 km，相对湿度为 80%时随时间迭代到两年和准定常态时的图。图中垂直速度最大值为 0.13 m/s，水平速度最大值为 0.40 m/s，温度距平最大值为 1.1℃ p.707

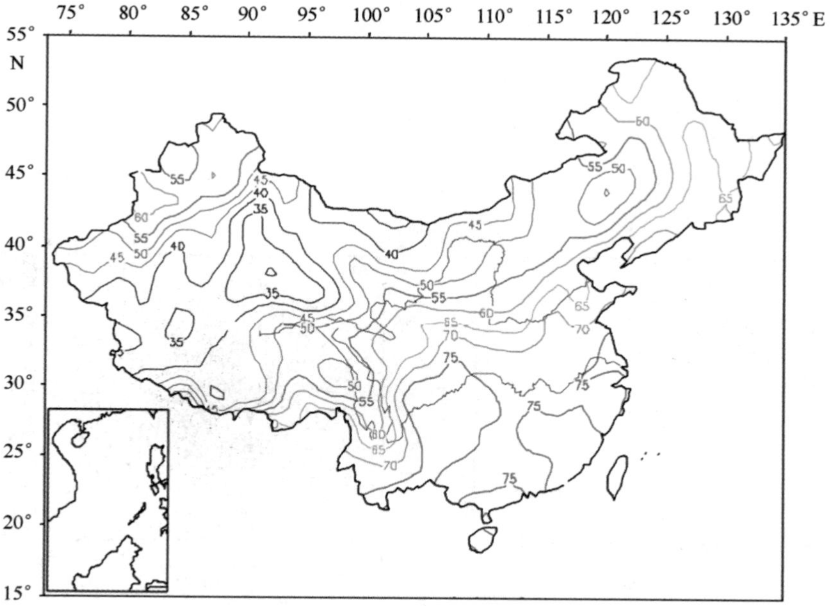

图 4　2006—2007 年全国的平均相对湿度(%)分布 p.708

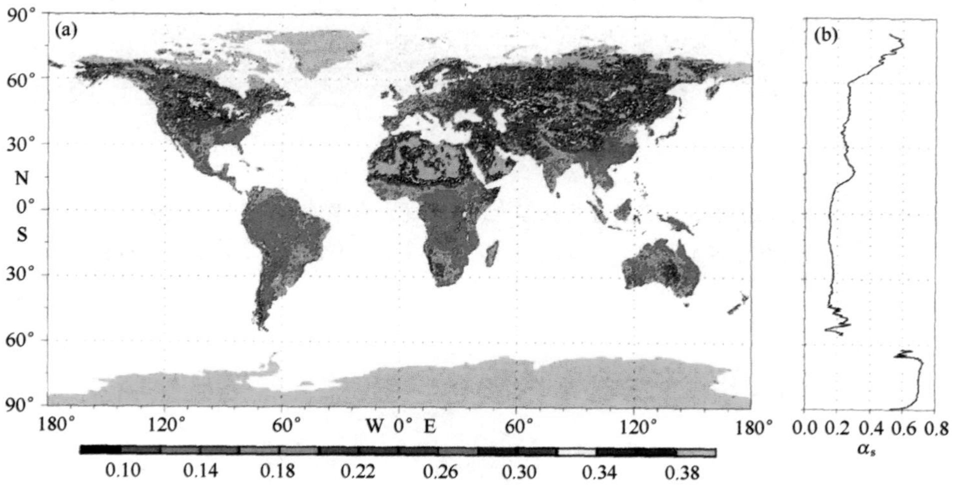

图 4　地表反照率(a)及其纬向分布(b)

(反照率资料下载自 ESA GlobAlbedo 项目网站 http:// www. GlobAlbedo. org) p.718

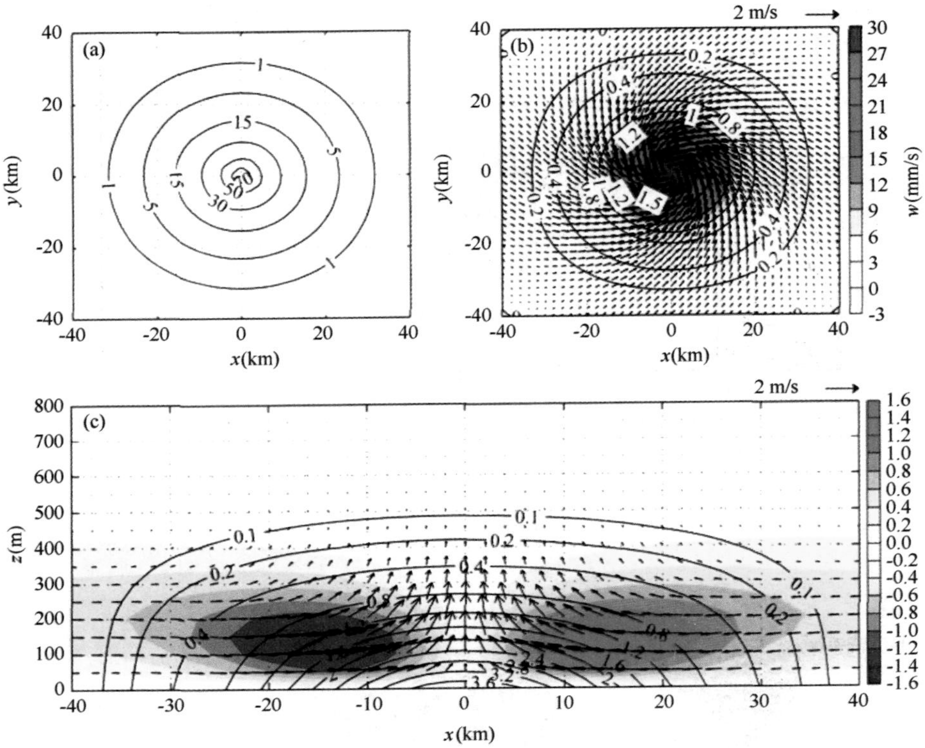

图 2　气溶胶浓度分布(a)；150 m 高度处扰动温度(等值线)、水平速度(矢量)和垂直速度(填色)的分布(b)；$y=0$ 剖面温度(等值线)、速度(矢量，填色)分布(c)，(c)中垂直速度已扩大 100 倍 p.744

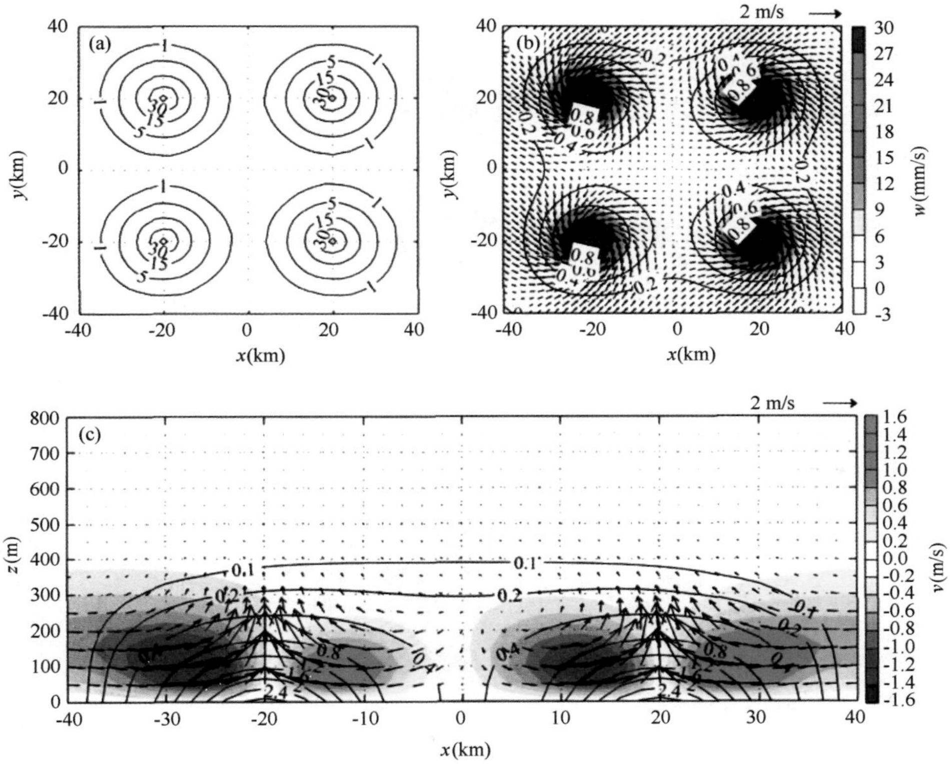

图 3　气溶胶浓度分布(a);150 m 高度处扰动温度(等值线)、水平速度(矢量)和垂直速度(填色)的分布(b);$y=\frac{1}{2}L$ 剖面温度(等值线)、速度(矢量,填色)分布(c),(c)中垂直速度已扩大 100 倍 p.745